Secondary Metabolites from Fungi—in Honour of Prof. Dr. Ji-Kai Liu's 60th Birthday

Secondary Metabolites from Fungi—in Honour of Prof. Dr. Ji-Kai Liu's 60th Birthday

Editors

Tao Feng
Frank Surup

MDPI • Basel • Beijing • Wuhan • Barcelona • Belgrade • Manchester • Tokyo • Cluj • Tianjin

Editors
Tao Feng
School of Pharmaceutical Sciences
South-Central Minzu University
Wuhan
China

Frank Surup
Department of Microbial Drugs
Helmholtz Centre for Infection Research GmbH
Braunschweig
Germany

Editorial Office
MDPI
St. Alban-Anlage 66
4052 Basel, Switzerland

This is a reprint of articles from the Special Issue published online in the open access journal *Journal of Fungi* (ISSN 2309-608X) (available at: www.mdpi.com/journal/jof/special_issues/Metabolites_Fungi).

For citation purposes, cite each article independently as indicated on the article page online and as indicated below:

LastName, A.A.; LastName, B.B.; LastName, C.C. Article Title. *Journal Name* **Year**, *Volume Number*, Page Range.

ISBN 978-3-0365-6144-8 (Hbk)
ISBN 978-3-0365-6143-1 (PDF)

Cover image courtesy of Tao Feng

Contents

About the Editors

Tao Feng

Tao Feng holds an appointment as Professor at the School of Pharmaceutical Sciences, South-Central Minzu University (China). He obtained his Ph.D. at Kunming Institute of Botany (CAS) in medicinal chemistry in 2010. Since then, he has been working on the chemistry of natural products of fungi, with a particular focus on specific groups of natural products that occur only in fungi. He is the author of over 100 PubMed-listed publications.

Frank Surup

Frank Surup obtained his Ph.D. at the University of Göttingen, Chair of Organic and Biomolecular Chemistry in 2007. He serves as a research associate at the research groups of Prof. Müller and Stadler Helmholtz Centre for Infection Research, Braunschweig, Germany. His research interest is secondary metabolites including structure elucidation, organic chemistry, and bioactivity.

as the original template and more than 200 derivatives have been synthesized. Moreover, a new derivative with significant pancreatic lipase inhibitory activity (IC_{50} = 14 nM) has been considered to be a drug candidate. A unique biosynthetic pathway of vibralactone, including several very interesting reactions that may involve unusual enzymes, was elucidated [11–13].

The fungal natural products with structural and biological activity diversity discovered by Liu's team are too numerous to list. More than twenty molecules have been rated "hot off the press", playing important roles in synthetic chemistry, pharmacology, and drug development [14–38].

Liu's workload and achievements are so broad that describing them in a brief editorial is hard. It can be said that Liu has left an indelible mark on the field, with over 450 publications and over 100 trainees from academia and industry who have carried on Liu's passion for fungal natural products. We hope that the above description provides at least an overview of Liu's main discoveries and his enormous impact on the field of higher fungal natural products. Nonetheless, at 60, Liu is still devoted to the work of fungal natural products chemistry with full enthusiasm. His professionalism and positive attitude toward life have deeply affected his students, colleagues, and fellow researchers. Therefore, it is with the deepest respect and affection for Liu that we, on Liu's 60th birthday, publish this Special Issue of the *Journal of Fungi* in Liu's name. We look forward to the exciting new discoveries reported in the future by Liu's team and the labs of Liu's scientific family.

Liu's comprehensive research interests are reflected by the broad scope of the contributions in this SI, covered by six reviews and sixteen original research articles.

Terpenoids are among the most important class of natural products and are thus the subject of two review articles. Whereas Dai et al. review the isolation, structural determination, bioactivities, and synthesis of sesquiterpenoids produced by fungi over the past five years [39], Fa-Lei Zhang et al. provide an overview of the structures, biological activities, evolution, organic synthesis, and biosynthesis of fungal diterpenoids reported in the period from 2010 to 2020 [40].

The review of Wen et al. underlines the importance of endophytic fungi as an alternative source of secondary metabolites, some of which have potential in the development of new pharmacologicals [41].

Nearer in its scope are the sulfur-containing compounds from plant endophytic fungi. Fan et al. reported 143 new sulfur-containing compounds that were reported from 1985 to 2022 and their fungal producers, plant sources, chemical structures, and bioactivities. These natural products mainly belong to the classes of polyketides, nonribosomal peptides, terpenoids, and hybrids [42].

The review of Hou et al. covers sorbicillinoids, a family of hexaketide metabolites with a characteristic sorbyl side-chain residue. Sixty-nine sorbicillinoids from fungi, newly identified from 2016 to 2021, are summarized in this review, including their structures and bioactivities [43].

Last but not least, Dai et al. report on the biosynthesis of fungal natural products, which involves two separate pathway crosstalk. This stresses that fungal natural product biosynthetic genes are not always arranged simply within one single biosynthetic gene cluster, thus increasing the structural complexity and chemical diversity of fungal NPs and expanding the scope of bioactivities [44].

For the original research articles, new natural products from Ascomycota are the first focus of this SI. Both Garcia et al. and Yan et al. report new cytochalasans from *Sparticola triseptata* and the endophytic fungus *Phomopsis* sp. [45,46]. These showed antiproliferative, cytotoxic, and anti-migratory activities, respectively, as expected for these PKS-NRPS hybrids.

During screening for new natural products with inhibitory activities on acetylcholinesterase, the endophytic fungus *Phaeosphaeria* sp., isolated from *Huperzia serrata*, was investigated, and new small polyketides were discovered by Xiao et al. [47].

Verrucosidin is a toxic pyrone-type mycotoxin of polyketide origin. New derivatives were discovered by Han et al. from *Penicillium cellarum* using a state-of-the-art MS/MS-based molecular networking approach [48].

New antibacterial chloro-containing polyketides from the alga-derived fungus *Asteromyces cruciatus* were elucidated by Zhuravleva et al. The new isoprenylated cyclohexanol structures contain a characteristic conjugated alkyne–alkene moiety, presumably responsible for the observed activities [49].

From another marine-derived fungus, *Acremonium chrysogenum*, Duan et al. obtained sorbicillinoid derivatives with radical scavenging activities. The analysis of the biosynthetic gene cluster, the proposed biosynthetic pathway, and the radical scavenging activity complete the study [50].

Further studies on endophytic fungi by Ai et al. and Yu et al. report new eremophilane sesquiterpenoids and new sativene, as well as longifolene sesquiterpenoids plus two new xanthones from *Boeremia exigua* isolated from *Fritillaria hupehensis* and the Kiwifruit-associated fungus *Bipolaris* sp., highlighting the vast chemical diversity to be explored by these plant-associated fungi [51,52].

Three manuscripts report on the isolation of secondary metabolites from Basidiomycota. First, Dai et al. reported eighteen previously undescribed bergamotane sesquiterpenes and one new victoxinine derivative from the edible mushroom *Craterellus odoratus*. Some of the compounds possess immunosuppressive activity [53].

Peng et al. reported on anti-adipogenic lanostane-type triterpenoids from the edible mushroom *Ganoderma applanatum*, which is utilized for medical purpose [54]. Other lanostane-type metabolites are presented from the species *Ganoderma australe* by Lin Zhou [55].

The phenalenone skeleton, found in several plant or fungal compounds, presumably plays a role in protecting these organisms against various external threats. Ibrahim et al. explored the activity of fungal phenalenone derivatives as potential CK2 inhibitors using computational methods [56].

The genetic engineering of fungal natural product biosynthesis genes and gene clusters constitutes the last focus of our SI. Feng et al. engineered *Aspergillus oryzae* for the heterologous expression of bacterial polyketide synthase genes, having a modular architecture [57].

The biosynthetic gene cluster encoding ilicicolin biosynthesis from the well-studied organism *Trichoderma reesei* was studied by heterologous expression in the fungal host *Aspergillus oryzae* by Shenouda et al. from the Cox group [58]. Furthermore, they developed *Trichoderma reesei* as a microbial cell factory for the heterologous expression of secondary metabolites for waste valorization [59].

After revealing the biosynthetic pathway of the potential protein kinase C inhibitor balanol through overexpression in previous studies, Li et al. now report the regulation of balanol biosynthesis by BlnR. This further improved the titers of balanol in the herb fungus *Tolypocladium ophioglossoides* [60].

Finally, we thank all of the contributors to this Special Issue and warmheartedly wish Prof. Ji-Kai Liu all the best for the future.

Conflicts of Interest: The authors declare no conflict of interest.

References

1. Liu, J.K. Natural Terphenyls: Developments since 1877. *Chem. Rev.* **2006**, *106*, 2209–2223. [CrossRef] [PubMed]
2. Liu, J.K. N-containing compounds of Macromycetes. *Chem. Rev.* **2005**, *105*, 2723–2744. [CrossRef] [PubMed]
3. Jiang, M.Y.; Feng, T.; Liu, J.K. N-Containing compounds of macromycetes. *Nat. Prod. Rep.* **2011**, *28*, 783–808. [CrossRef] [PubMed]
4. Zhou, Z.Y.; Liu, J.K. Pigments of fungi (Macromycetes). *Nat. Prod. Rep.* **2010**, *27*, 1531–1570. [CrossRef] [PubMed]
5. Zhou, Z.Y.; Shi, G.Q.; Fontaine, R.; Wei, K.; Feng, T.; Wang, F.; Wang, G.Q.; Qu, Y.; Li, Z.H.; Dong, Z.J.; et al. Evidence for the natural toxins from the mushroom *Trogia venenata* as a cause of sudden unexpected death in Yunnan Province, China. *Angew. Chem. Int. Ed. Engl.* **2012**, *51*, 2368–2370. [CrossRef]
6. Stone, R. Heart-stopping revelation about how Chinese mushroom kills. *Science* **2012**, *335*, 1293. [CrossRef]
7. Stone, R. Will a midsummer's nightmare return? *Science* **2010**, *329*, 132–134. [CrossRef]

8. Bedry, R.; Baudrimont, I.; Deffieux, G.; Creppy, E. Pomies Wild-mushroom intoxication as a cause of rhabdomyolysis. *N. Engl. J. Med.* **2001**, *345*, 798–802. [CrossRef]
9. Yin, X.; Feng, T.; Shang, J.H.; Zhao, Y.L.; Wang, F.; Li, Z.H.; Dong, Z.J.; Luo, X.D.; Liu, J.K. Chemical and toxicological investigations of a previously unknown poisonous European mushroom *Tricholoma terreum*. *Chem. Eur. J.* **2014**, *20*, 7001–7009. [CrossRef]
10. Liu, D.Z.; Wang, F.; Liao, T.G.; Tang, J.G.; Steglich, W.; Zhu, H.J.; Liu, J.K. Vibralactone: A lipase inhibitor with an unusual fused β-lactone produced by cultures of the basidiomycete *Boreostereum vibrans*. *Org. Lett.* **2006**, *8*, 5749–5752. [CrossRef]
11. Zhao, P.J.; Yang, Y.L.; Du, L.; Liu, J.K.; Zeng, Y. Elucidating the biosynthetic pathway for vibralactone: A pancreatic lipase inhibitor with a fused bicyclic β-lactone. *Angew. Chem. Int. Ed.* **2013**, *52*, 2298–2302. [CrossRef] [PubMed]
12. Yang, Y.L.; Zhou, H.; Du, G.; Feng, K.N.; Feng, T.; Fu, X.L.; Liu, J.K.; Zeng, Y. A monooxygenase from *Boreostereum vibrans* catalyzes oxidative decarboxylation in a divergent vibralactone biosynthesis pathway. *Angew. Chem. Int. Ed.* **2016**, *55*, 5463–5466. [CrossRef] [PubMed]
13. Feng, K.N.; Yang, Y.L.; Xu, Y.X.; Zhang, Y.; Feng, T.; Huang, S.X.; Liu, J.K.; Zeng, Y. A hydrolase-catalyzed cyclization forms the fused bicyclic beta-lactone in vibralactone. *Angew. Chem. Int. Ed.* **2020**, *59*, 7209–7213. [CrossRef] [PubMed]
14. Wang, F.; Yang, X.Y.; Lu, Y.T.; Li, Z.H.; Xu, Y.H.; Hu, J.; Liu, J.K.; Xiong, W.Y. The natural product antroalbol H promotes phosphorylation of liver kinase B1 (LKB1) at threonine 189 and thereby enhances cellular glucose uptake. *J. Biol. Chem.* **2019**, *294*, 10415–10427. [CrossRef] [PubMed]
15. Liu, Q.; Shu, X.; Wang, L.; Sun, A.; Liu, J.; Cao, X. Albaconol, a plant-derived small molecule, inhibits macrophage function by suppressing NF-κB activation and enhancing SOCS1 expression. *Cell. Death. Dis.* **2008**, *5*, 271–278. [CrossRef]
16. Luo, X.J.; Li, L.L.; Deng, Q.P.; Yu, X.F.; Yang, L.F.; Luo, F.J.; Xiao, L.B.; Chen, X.Y.; Ye, M.; Liu, J.K.; et al. Grifolin, a potent antitumour natural product upregulates death-associated protein kinase 1 DAPK1 via p53 in nasopharyngeal carcinoma cells. *Eur. J. Cancer* **2011**, *47*, 316–325. [CrossRef]
17. Ye, M.; Luo, X.J.; Li, L.L.; Shi, Y.; Tan, M.; Weng, X.X.; Li, W.; Liu, J.K.; Cao, Y. Grifolin, a potential antitumor natural product from the mushroom *Albatrellus confluens*, induces cell-cycle arrest in G1 phase via the ERK1/2 pathway. *Cancer Lett.* **2007**, *258*, 199–207. [CrossRef]
18. Wang, Y.H.; Tang, J.G.; Wang, R.R.; Yang, L.M.; Dong, Z.J.; Du, L.; Shen, X.; Liu, J.K.; Zheng, Y.T. Flazinamide, a novel β-carboline compound with anti-HIV actions. *Biochem. Biophys. Res. Commun.* **2007**, *355*, 1091–1095. [CrossRef]
19. Feng, T.; Gan, X.Q.; Zhao, Y.L.; Zhang, S.B.; Chen, H.P.; He, J.; Zheng, Y.S.; Sun, H.; Huang, R.; Li, Z.H.; et al. Tricholopardins A and B, anti-inflammatory terpenoids from the fruiting bodies of *Tricholoma pardinum*. *J. Nat. Prod.* **2019**, *82*, 45–50. [CrossRef]
20. Ding, J.H.; Feng, T.; Li, Z.H.; Yang, X.Y.; Guo, H.; Yin, X.; Wang, G.Q.; Liu, J.K. Trefolane A, a sesquiterpenoid with a new skeleton from cultures of the basidiomycete *Tremella foliacea*. *Org. Lett.* **2012**, *14*, 4976–4978. [CrossRef]
21. Yang, X.Y.; Feng, T.; Li, Z.H.; Sheng, Y.; Yin, X.; Leng, Y.; Liu, J.K. Conosilane A, an unprecedented sesquiterpene from the cultures of the basidiomycete *Conocybe siliginea*. *Org. Lett.* **2012**, *14*, 5382–5384. [CrossRef] [PubMed]
22. Yin, X.; Feng, T.; Li, Z.H.; Dong, Z.J.; Li, Y.; Liu, J.K. Highly oxygenated meroterpenoids from fruiting bodies of the mushroom *Tricholoma terreum*. *J. Nat. Prod.* **2013**, *76*, 1365–1368. [CrossRef] [PubMed]
23. Chen, H.P.; Zhao, Z.Z.; Zhang, Y.; Bai, X.; Zhang, L.; Liu, J.K. (+)- and (-)-ganodilactone, a pair of meroterpenoid dimers with pancreatic lipase inhibitory activities from the macromycete *Ganoderma leucocontextum*. *RSC Adv.* **2016**, *6*, 64469–64473. [CrossRef]
24. Chen, H.P.; Zhao, Z.Z.; Li, Z.H.; Dong, Z.J.; Wei, K.; Bai, X.; Zhang, L.; Wen, C.N.; Feng, T.; Liu, J.K. Novel natural oximes and oxime esters with a vibralactone backbone from the basidiomycete *Boreostereum vibrans*. *ChemistryOpen* **2016**, *5*, 142–149. [CrossRef] [PubMed]
25. Feng, T.; Cai, J.L.; Li, X.M.; Zhou, Z.Y.; Huang, R.; Zheng, Y.S.; Li, Z.H.; Liu, J.K. Phellibarin D with an unprecedented triterpenoid skeleton isolated from the mushroom *Phellinus rhabarbarinus*. *Tetrahedron Lett.* **2016**, *57*, 3544–3546. [CrossRef]
26. Zhao, Z.Z.; Chen, H.P.; Wu, B.; Zhang, L.; Li, Z.H.; Feng, T.; Liu, J.K. Matsutakone and matsutoic acid, two (nor)steroids with unusual skeletons from the edible mushroom *Tricholoma matsutake*. *J. Org. Chem.* **2017**, *82*, 7974–7979. [CrossRef]
27. Li, W.; He, J.; Feng, T.; Yang, H.X.; Ai, H.L.; Li, Z.H.; Liu, J.K. Antroalbocin A, an antibacterial sesquiterpenoid from higher fungus *Antrodiella albocinnamomea*. *Org. Lett.* **2018**, *20*, 8019–8021. [CrossRef]
28. Wang, S.; Li, Z.H.; Ai, H.L.; He, J.; Feng, T.; Liu, J.K. Polyellisin, a novel polyketide from cultures of the basidiomycete *Polyporus ellisii*. *RSC Adv.* **2018**, *8*, 31538–31541. [CrossRef]
29. Wang, W.X.; Li, Z.H.; Feng, T.; Li, J.; Sun, H.; Huang, R.; Yuan, Q.X.; Ai, H.L.; Liu, J.K. Curtachalasins A and B, Two cytochalasans with a tetracyclic skeleton from the endophytic fungus *Xylaria curta* E10. *Org. Lett.* **2018**, *20*, 7758–7761. [CrossRef]
30. Yang, H.X.; Ai, H.L.; Feng, T.; Wang, W.X.; Wu, B.; Zheng, Y.S.; Sun, H.; He, J.; Li, Z.H.; Liu, J.K. Trichothecrotocins A-C, antiphytopathogenic agents from potato endophytic fungus *Trichothecium crotocinigenum*. *Org. Lett.* **2018**, *20*, 8069–8072. [CrossRef]
31. Zhang, S.B.; Huang, Y.; He, S.J.; Chen, H.P.; Wu, B.; Li, S.Y.; Zhao, Z.Z.; Li, Z.H.; Wang, X.; Zuo, J.P.; et al. Heterocyclic compounds from the mushroom *Albatrellus confluens* and their inhibitions against lipopolysaccharides-induced B lymphocyte cell proliferation. *J. Org. Chem.* **2018**, *83*, 10158–10165. [CrossRef] [PubMed]
32. Wang, W.X.; Lei, X.X.; Ai, H.L.; Bai, X.; Li, J.; He, J.; Li, Z.H.; Zheng, Y.S.; Feng, T.; Liu, J.K. Cytochalasans from the endophytic fungus *Xylaria cf. curta* with resistance reversal activity against fluconazole-resistant *Candida albicans*. *Org. Lett.* **2019**, *21*, 1108–1111. [CrossRef] [PubMed]

33. Wang, W.X.; Lei, X.X.; Yang, Y.L.; Li, Z.H.; Ai, H.L.; Li, J.; Feng, T.; Liu, J.K. Xylarichalasin A, a halogenated hexacyclic cytochalasan from the fungus *Xylaria cf. curta*. *Org. Lett.* **2019**, *21*, 6957–6960. [CrossRef] [PubMed]
34. Li, J.; Wang, W.X.; Chen, H.P.; Li, Z.H.; He, J.; Zheng, Y.S.; Sun, H.; Huang, R.; Yuan, Q.X.; Wang, X.; et al. (+/-)-Xylaridines A and B, highly conjugated alkaloids from the fungus *Xylaria longipes*. *Org. Lett.* **2019**, *21*, 1511–1514. [CrossRef]
35. Li, J.; Wang, W.X.; Li, Z.H.; He, J.; Huang, R.; Zheng, Y.S.; Li, L.Q.; Wang, X.; Feng, T.; Liu, J.K. Xylaridines C and D, unusual thiopyranodipyridine alkaloids from the fungus *Xylaria longipes*. *Org. Lett.* **2019**, *21*, 6145–6148. [CrossRef]
36. Liu, Y.P.; Dai, Q.; Wang, W.X.; He, J.; Li, Z.H.; Feng, T.; Liu, J.K. Psathyrins: Antibacterial diterpenoids from *Psathyrella candolleana*. *J. Nat. Prod.* **2020**, *83*, 1725–1729. [CrossRef]
37. Yang, H.X.; Wu, X.; Chi, M.J.; Li, Z.H.; Feng, T.; Ai, H.L.; Liu, J.K. Structure and cytotoxicity of trichothecenes produced by the potato-associated fungus *Trichothecium crotocinigenum*. *Bioorg. Chem.* **2021**, *111*, 104874. [CrossRef]
38. Ai, H.L.; Shi, B.B.; Li, W.; He, J.; Li, Z.H.; Feng, T.; Liu, J.K. Bipolarithizole A, an antifungal phenylthiazole-sativene merosesquiterpenoid from the potato endophytic fungus *Bipolaris eleusines*. *Org. Chem. Front.* **2022**, *9*, 1814–1819. [CrossRef]
39. Dai, Q.; Zhang, F.L.; Feng, T. Sesquiterpenoids specially produced by fungi: Structures, bological activities, chemical and biosynthesis (2015–2020). *J. Fungi* **2021**, *7*, 1026. [CrossRef]
40. Zhang, F.L.; Feng, T. Diterpenes specially produced by fungi: Structures, biological activities, and biosynthesis (2010–2020). *J. Fungi* **2022**, *8*, 244. [CrossRef]
41. Wen, J.; Okyere, S.K.; Wang, S.; Wang, J.C.; Xie, L.; Ran, Y.N.; Hu, Y.C. Endophytic fungi: An effective alternative source of plant-derived bioactive compounds for pharmacological studies. *J. Fungi* **2022**, *8*, 205. [CrossRef] [PubMed]
42. Fan, Y.; Ma, Z.; Zhang, Y.; Wang, Y.; Ding, Y.; Wang, C.; Cao, S. Sulfur-containing compounds from endophytic fungi: Sources, structures and bioactivities. *J. Fungi* **2022**, *8*, 628. [CrossRef] [PubMed]
43. Hou, X.; Zhang, X.; Xue, M.; Zhao, Z.; Zhang, H.; Xu, D.; Lai, D.; Zhou, L. Recent advances in sorbicillinoids from fungi and their bioactivities (covering 2016–2021). *J. Fungi* **2022**, *8*, 62. [CrossRef] [PubMed]
44. Dai, G.Z.; Shen, Q.Y.; Zhang, Y.M.; Bian, X.Y. Biosynthesis of fungal natural products involving two separate pathway crosstalk. *J. Fungi* **2022**, *8*, 320. [CrossRef]
45. Yan, B.C.; Wang, W.G.; Kong, L.M.; Tang, J.W.; Du, X.; Li, Y.; Puno, P.T. Cytochalasans from the endophytic fungus *Phomopsis* sp. shj2 and their antimigratory activities. *J. Fungi* **2022**, *8*, 543. [CrossRef]
46. Garcia, K.Y.M.; Quimque, M.T.J.; Lambert, C.; Schmidt, K.; Primahana, G.; Stradal, T.E.B.; Ratzenböck, A.; Dahse, H.M.; Phukhamsakda, C.; Stadler, M.; et al. Antiproliferative and cytotoxic cytochalasins from *Sparticola triseptata* inhibit actin polymerization and aggregation. *J. Fungi* **2022**, *8*, 560. [CrossRef]
47. Xiao, Y.W.; Liang, W.Z.; Zhang, Z.B.; Wang, Y.; Zhang, S.S.; Liu, J.T.; Chang, J.; Ji, C.J.; Zhu, D. Polyketide derivatives from the endophytic fungus *Phaeosphaeria* sp. LF5 isolated from *Huperzia serrata* and their acetylcholinesterase inhibitory activities. *J. Fungi* **2022**, *8*, 232.
48. Han, J.; Chen, B.; Zhang, R.; Zhang, J.; Dai, H.; Wang, T.; Sun, J.; Zhu, G.; Li, W.; Li, E.; et al. Exploring verrucosidin derivatives with glucose-uptake-stimulatory activity from *Penicillium cellarum* using MS/MS-based molecular networking. *J. Fungi* **2022**, *8*, 143.
49. Zhuravleva, O.I.; Oleinikova, G.K.; Antonov, A.S.; Kirichuk, N.N.; Pelageev, D.N.; Rasin, A.B.; Menshov, A.S.; Popov, R.S.; Kim, N.Y.; Chingizova, E.A.; et al. New antibacterial chloro-containing polyketides from the alga-derived fungus *Asteromyces cruciatus* KMM 4696. *J. Fungi* **2022**, *8*, 454. [CrossRef]
50. Duan, C.; Wang, S.; Huo, R.; Li, E.; Wang, M.; Ren, J.; Pan, Y.; Liu, L.; Liu, G. Sorbicillinoid derivatives with the radical scavenging activities from the marine-derived fungus *Acremonium chrysogenum* C10. *J. Fungi* **2022**, *8*, 530. [CrossRef]
51. Yu, J.J.; Jin, Y.X.; Huang, S.S.; He, J. Sesquiterpenoids and xanthones from the kiwifruit-associated fungus *Bipolaris* sp. and their anti-pathogenic microorganism activity. *J. Fungi* **2022**, *8*, 9. [CrossRef] [PubMed]
52. Ai, H.L.; Lv, X.; Ye, K.; Wang, M.X.; Huang, R.; Shi, B.B.; Li, Z.H. Four new highly oxygenated eremophilane sesquiterpenes from an endophytic fungus *Boeremia exigua* isolated from *Fritillaria hupehensis*. *J. Fungi* **2022**, *8*, 492. [CrossRef] [PubMed]
53. Dai, Q.; Zhang, F.L.; Li, Z.H.; He, J.; Feng, T. Immunosuppressive sesquiterpenoids from the edible mushroom *Craterellus odoratus*. *J. Fungi* **2021**, *7*, 1052. [CrossRef] [PubMed]
54. Peng, X.R.; Wang, Q.; Su, H.G.; Zhou, L.; Xiong, W.Y.; Qiu, M.H. Anti-adipogenic lanostane-type triterpenoids from the edible and medicinal mushroom *Ganoderma applanatum*. *J. Fungi* **2022**, *8*, 331. [CrossRef]
55. Zhou, L.; Guo, L.L.; Isaka, M.; Li, Z.H.; Chen, H.P. [20(22)*E*]-Lanostane triterpenes from the fungus *Ganoderma australe*. *J. Fungi* **2022**, *8*, 503. [CrossRef]
56. Ibrahim, S.R.M.; Bagalagel, A.A.; Diri, R.M.; Noor, A.O.; Bakhsh, H.T.; Muhammad, Y.A.; Mohamed, G.A.; Omar, A.M. Exploring the activity of fungal phenalenone derivatives as potential CK2 inhibitors using computational methods. *J. Fungi* **2022**, *8*, 443. [CrossRef]
57. Feng, J.; Hauser, M.; Cox, R.J.; Skellam, E. Engineering *Aspergillus oryzae* for the heterologous expression of a bacterial modular polyketide synthase. *J. Fungi* **2021**, *7*, 1085. [CrossRef]
58. Shenouda, M.L.; Ambilika, M.; Cox, R.J. *Trichoderma reesei* contains a biosynthetic gene cluster that encodes the antifungal agent ilicicolin H. *J. Fungi* **2021**, *7*, 1034. [CrossRef]

59. Shenouda, M.L.; Ambilika, M.; Skellam, E.; Cox, R.J. Heterologous expression of secondary metabolite genes in *Trichoderma reesei* for waste valorization. *J. Fungi* **2022**, *8*, 355. [CrossRef]
60. Li, R.Q.; Liu, X.; Zhang, M.; Xu, W.Q.; Li, Y.Q.; Chen, X.A. Gram-level production of balanol through reulatory pathway and medium optimization in herb fungus *Tolypocladium ophioglossoides*. *J. Fungi* **2022**, *8*, 510. [CrossRef]

Journal of *Fungi*

MDPI

Review

Diterpenes Specially Produced by Fungi: Structures, Biological Activities, and Biosynthesis (2010–2020)

Fa-Lei Zhang and Tao Feng *

School of Pharmaceutical Sciences, South-Central University for Nationalities, Wuhan 430074, China; flzhang@mail.scuec.edu.cn
* Correspondence: tfeng@mail.scuec.edu.cn

Abstract: Fungi have traditionally been a very rewarding source of biologically active natural products, while diterpenoids from fungi, such as the cyathane-type diterpenoids from *Cyathus* and *Hericium* sp., the fusicoccane-type diterpenoids from *Fusicoccum* and *Alternaria* sp., the guanacastane-type diterpenoids from *Coprinus* and *Cercospora* sp., and the harziene-type diterpenoids from *Trichoderma* sp., often represent unique carbon skeletons as well as diverse biological functions. The abundances of novel skeletons, biological activities, and biosynthetic pathways present new opportunities for drug discovery, genome mining, and enzymology. In addition, diterpenoids peculiar to fungi also reveal the possibility of differing biological evolution, although they have similar biosynthetic pathways. In this review, we provide an overview about the structures, biological activities, evolution, organic synthesis, and biosynthesis of diterpenoids that have been specially produced by fungi from 2010 to 2020. We hope this review provides timely illumination and beneficial guidance for future research works of scholars who are interested in this area.

Keywords: diterpenes; isolation; structure; biological activity; biosynthesis

Citation: Zhang, F.-L.; Feng, T. Diterpenes Specially Produced by Fungi: Structures, Biological Activities, and Biosynthesis (2010–2020). *J. Fungi* **2022**, *8*, 244. https://doi.org/10.3390/jof8030244

Academic Editor: Gary A. Strobel

Received: 25 January 2022
Accepted: 27 February 2022
Published: 28 February 2022

1. Introduction

Fungi are widely distributed in terrestrial environments, freshwater, and marine habitats; more than one million distinctive fungal species exist, but only approximately 100,000 of these have been classified [1]. These eukaryotic microbes produce specialized metabolites that participate in a variety of ecological functions, such as quorum sensing, chemical defense, allelopathy, and maintenance of symbiotic interactions [2]. There are more than 40,000 terpenoid compounds in nature, which compose the largest family of natural products [3]. Terpenoids exist in all domains of life, but are particularly prevalent in plants, fungi, and marine invertebrates, and are essential constituents of secondary metabolism [3,4].

Diterpenoids are a class of C20 compounds derived from isoprenoid precursor geranylgeranyl diphosphate (GGPP) under the catalysis of diterpene synthases (DTSs) [5–11]. Prenyltransferase (PT) and terpene synthase (TPS) are key enzymes in the formation of the basic carbon skeletons of terpenoids [8,12]. The PT enzymes determine the prenyl carbon chain length, whereas the TPS enzymes generate the structural complexity of the molecular scaffolds, forming various ring structures [8]. Fungi are among the most important microbial resources for drug discovery, owing to their capability to produce structurally diverse and biologically important secondary metabolites [13,14]. It is also well known that fungi possess extraordinary biosynthetic gene clusters that may encode highly diverse biosynthetic pathways of natural products [15–18].

Between 2010 and 2020, about 400 fungal-specific diterpenes have been reported. In addition to 172 cyathane diterpenes reviewed by Bailly et al. [19] and Gao et al. [20], a total of 232 diterpenes were collected in this review (Chart 1). These diterpenoids are mainly tricyclic or tetracyclic skeletal structures such as cyathane-type, fusicoccane-type, guanacastane-type, and harziene-type diterpenoids (Chart 1). Judging from the distribution of fungal diterpenoid resources, the diterpenes from the genera *Trichoderma*, *Penicillium*, *Cyathus*, *Hericium*, and *Crinipellis* account for 60% of the total (Chart 2). In addition, systematic studies on the chemical constituents of fungi have shown that a large number of fungal diterpenoids exhibited significant biological functions such as anti-inflammatory, cytotoxic, antimicrobial, and antiviral activities (Chart 3). For instance, the semi-synthetic pleuromutilin analogues tiamulin **193** and valnemulin **194** have been used for over three decades as antibiotics to treat economically important infections in swine and poultry [21–25].

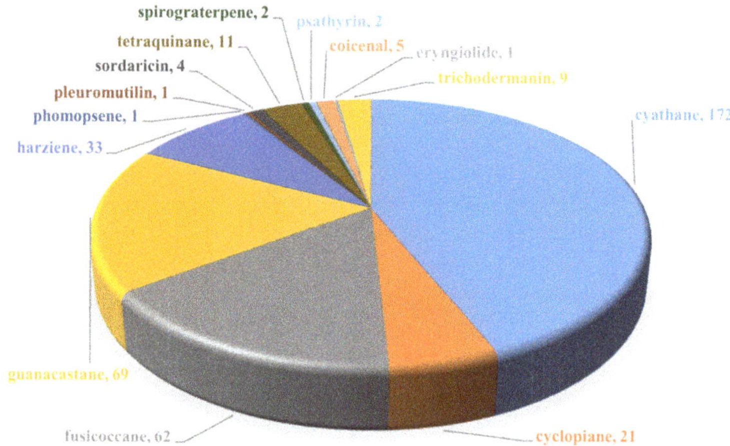

Chart 1. Fungal diterpenoids (2010–2020) classified by skeleton.

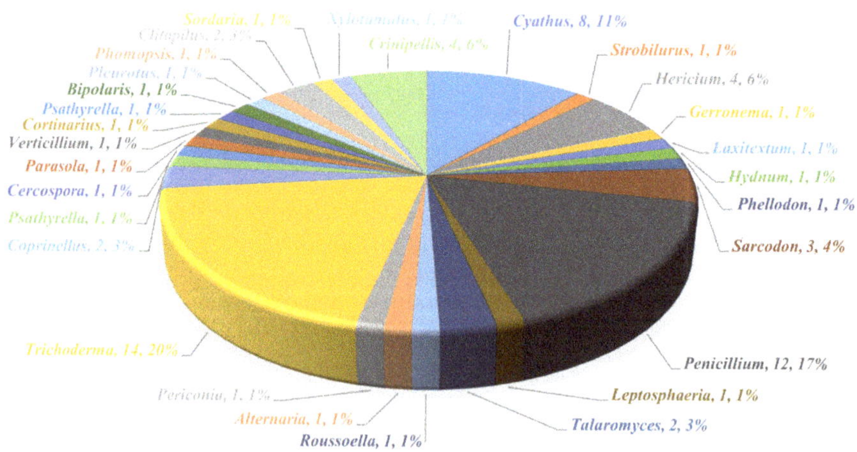

Chart 2. Source genera of fungal diterpenoids (2010–2020).

Consequently, a wealth of novel skeletons, biosynthetic pathways, and bioactivities have provided new opportunities for drug discovery, genome mining, enzymology, and chemical synthesis. During the period covered in this review, there have been several more specialized reviews of fungal metabolites [26–28], including benzene carbaldehydes [29], trichothecenes [30,31], protoilludane sesquiterpenoids [32], meroterpenoids [33–35], meroterpenoid cyclases [36], terpenoids [37], and natural product biosynthetic genes and enzymes of fungi [17,18,38,39]. In addition, the isolation and chemistry of diterpenoids from terrestrial sources have been summarized [40]. In this review, we provide an overview of diterpenoids that were specially produced by fungi during the period from 2010 to 2020, and focus on their structures, biological activities, and biosynthesis, and we also conduct an evolutionary analysis.

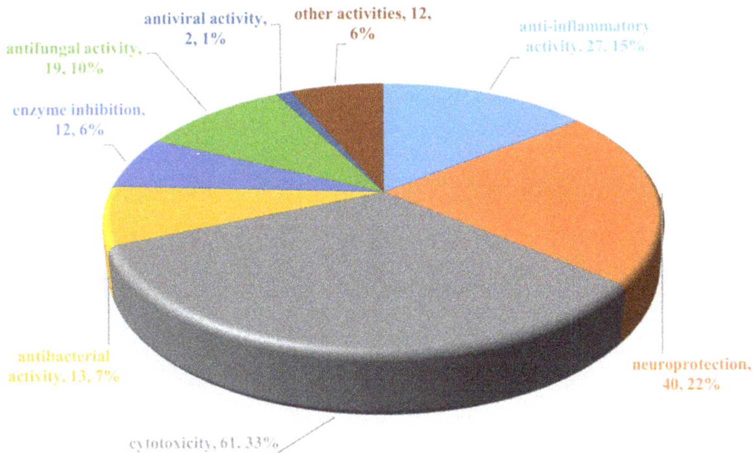

Chart 3. The proportion of one activity as compared with the whole occurrence of activities of bioactive fungal diterpenoids (2010–2020).

In particular, literature investigation of known databases such as PubMed and Web of Science was conducted from 2010 to July 2020 using the keywords "diterpenes/diterpenoids" paired with "fungi", "fungal diterpenoids" paired with "structure elucidation", or "fungal diterpenoids" paired with "biosynthesis". There were no language restrictions imposed. The references were further scrutinized and, finally, 210 references were selected. The data inclusion criteria included: (1) diterpenes/diterpenoids isolated from fungi, (2) carbon skeleton obtained only from fungi or rarely from other sources, (3) studies on the biological activities of diterpenes/diterpenoids and their derivatives that had been carried out in vitro or in vivo, (4) studies on the biosynthesis of diterpenes/diterpenoids and their derivatives. The data exclusion criteria included: (1) carbon skeleton of diterpenes/diterpenoids obtained in abundance from other sources, such as plants, bacteria and so on, (2) duplication of data and titles and/or abstracts not meeting the inclusion criteria.

2. Cyathane

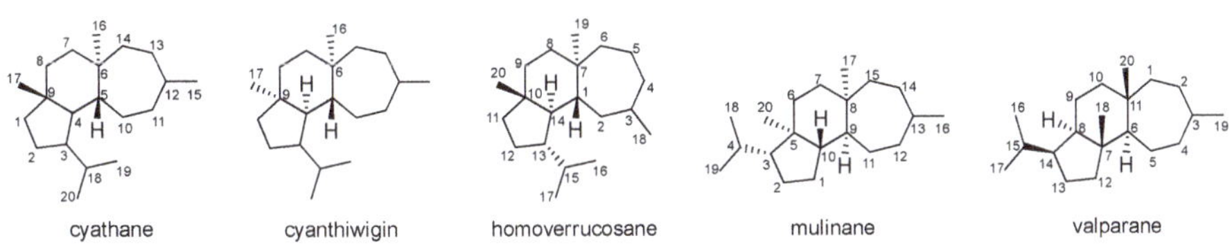

Cyathane diterpenes are a group of natural products that possess unusual, angularly fused 5/6/7 tricyclic cores, and they are characteristic of certain basidiomycete species including *Cyathus*, *Hericium*, and *Sarcodon* (Figure 1). For example, there have been more than 170 compounds isolated from fungi such as *Cyathus africanus* and *Hericium erinaceus* [19,20,41]. These compounds have a common biosynthetic precursor and can be produced via biosynthesis, hemi-synthesis, or total synthesis [42–47]. The cyathane diterpenoids include the classes of cyathins, striatins, sarcodonins, scabronines, and erinancines, according to their origins. Among them, the striatals, striatins, and erinacines, called cyathane-xylosides, which represent an unusual group of cyathane diterpenoids attached to a modified pentose

(*D*-xylose) moiety, have been isolated from cultures of *Cyathus* and *Hericium* species [20]. The cyathane structure is different from the homoverrucosane, mulinane, and valparane diterpenoids which also possess a 5/6/7 tricarbocyclic system [48,49]. The cyathanes are most similar to cyanthiwigins and can be differentiated by the orientation of the angular methyl groups, mainly present in some sponges [50–59]. These compounds display a diverse range of biological activities, including anticancer, antimicrobial, anti-MRSA (methicillin-resistant *Staphylococcus aureus*), anti-inflammatory, anti-proliferative, and nerve growth factor (NGF)-like properties [19,20,60,61]. An overview of cyathane-type diterpenes including isolation, structure diversity, synthesis, and bioactivity has been reviewed by Bailly et al. [19] and Gao et al. [20]. Therefore, in this review, we no longer summarize the details of cyathane diterpenoids.

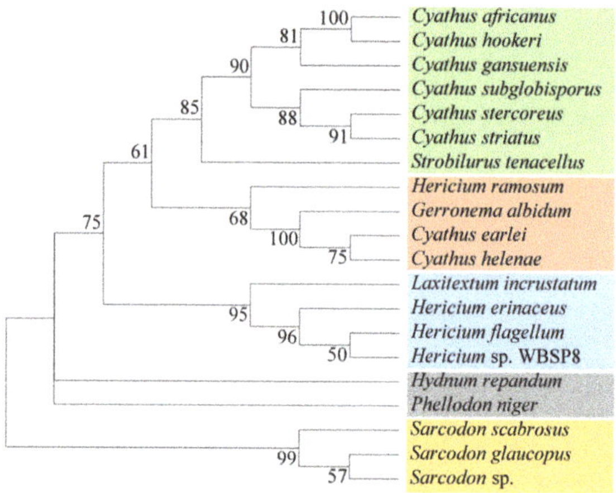

Figure 1. The evolutionary analysis tree constructed with selected fungi producing cyathane diterpenoids. The evolutionary analysis was reconstructed by the maximum likelihood method from the internal transcribed spacer (ITS) sequences as follows: *Cyathus africanus* (JX103204.1), *C. earlei* (KY964272.1), *C. gansuensis* (KC869661.1), *C. helenae* (DQ463334.1), *C. hookeri* (KC005989.1), *C. stercoreus* (MH543350.1), *C. striatus* (KU865513.1), *C. subglobisporus* (MH156046.1), *Gerronema albidum* (MF318924.1), *Hericium erinaceus* (KU855351.1), *H. flagellum* (MG649451.1), *H. ramosum* (U27043.1), *H.* sp. WBSP8 (MN243091.1), *Hydnum repandum* (LC377888.1), *Laxitextum incrustatum* (KT722621.1), *Phellodon niger* (MH310794.1), *Sarcodon glaucopus* (MT955152.1), *S. scabrosus* (MN992643.1), *Strobilurus tenacellus* (MF063128.1). Since the ITS sequence of *Sarcodon cyrneus* was not available, *Sarcodon* sp. (MK049936.1) was selected, since it is in the same family with *S. cyrneus*. The evolutionary history was inferred by using the maximum likelihood method and the general time reversible model [62]. The bootstrap consensus tree inferred from 1000 replicates is taken to represent the evolutionary history of the taxa analyzed [63]. Branches corresponding to partitions reproduced in less than 50% bootstrap replicates are collapsed. The percentage of replicate trees in which the associated taxa clustered together in the bootstrap test (1000 replicates) are shown next to the branches [63]. Initial tree(s) for the heuristic search were obtained automatically by applying the Neighbor-Join and BioNJ algorithms to a matrix of pairwise distances estimated using the maximum composite likelihood (MCL) approach, and then selecting the topology with superior log likelihood value. A discrete Gamma distribution was used to model evolutionary rate differences among sites (5 categories (+*G*, parameter = 1.2219)). The rate variation model allowed for some sites to be evolutionarily invariable ([+*I*], 0.00% sites). This analysis involved 20 nucleotide sequences. Codon positions included were 1st + 2nd + 3rd + noncoding. There were 924 positions in the final dataset. The evolutionary analysis was conducted in MEGA X (version 10.2.2) [64].

To understand the source genera of cyathane diterpenoids, we performed a phylogenetic analysis by using the maximum likelihood method and the general time reversible model [62–64] for all the species involved in the reviews by Bailly et al. [19] and Gao et al. [20]. The results show that source genera are grouped based on their regiospecificity, i.e., genera *Cyathus*, *Hericium*, and *Sarcodon* were clustered into different clades (Figure 1). Taxonomically, *Cyathus africanus*, *C. hookeri*, *C. gansuensis*, *C. subglobisporus*, *C. stercoreus*, and *C. striatus* all belonged to the genus *Cyathus*. They were close to each other, and first, they gathered into one branch, then, they gathered into one branch with *Strobilurus tenacellus* of the genus *Strobilurus*, and finally gathered into one branch with other genera (Figure 1). *C. earlei* and *C. helenae* also belonged to the genus *Cyathus*, they were close to each other, and first, they gathered into one branch, then, they gathered into one branch with *Gerronema albidum* of the genus *Gerronema*. Similarly, *Hericium erinaceus*, *H. flagellum*, and *Hericium* sp. WBSP8, *Sarcodon scabrosus*, *S. glaucopus*, and other species were close to each other. Existing studies have shown that most fungal metabolites are encoded by biosynthetic gene clusters (BGCs) [17]. The natural product BGCs of species in the same genus tend to be highly homologous, and BGC functional divergence gives rise to the evolution of new secondary metabolites, indicating that species-level sampling in these three genera for natural products mining will yield significant returns for cyathane diterpenoids discovery.

3. Cyclopiane

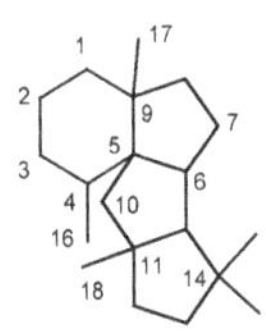

cyclopiane skeleton

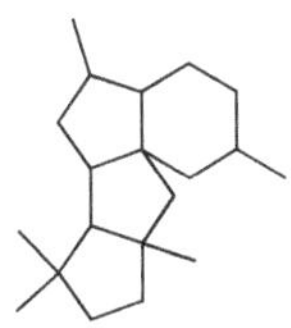

aberrarane skeleton

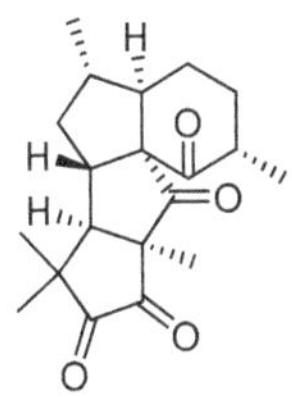

aberrarone

Cyclopiane diterpenoids comprise a class of tetracyclic diterpenes with unique scaffolds. They are characterized by a highly fused 6/5/5/5 ring system. The structural variations of cyclopiane diterpenoids are mainly owing to oxidation occurring at various sites to generate hydroxy groups [65]. In general, cyclopiane diterpenoids have mainly been isolated from different species of the genus *Penicillium* (Figure 2) and have been classified into two groups according to the functionality at C-1, i.e., conidiogenols and conidiogenones. The former featured with a hydroxy group at C-1, while the later possessed a carbonyl group at C-1 [66]. Specifically, *Penicillium commune* MCCC 3A00940, *P.* sp. F23-2, *P.* sp. YPGA11, *P. cyclopium*, *P. roqueforti* IFM 48062, *P.* sp. TJ403-2, *P. chrysogenum* QEN-24S, and *Leptosphaeria* sp. XL026 have been reported to produce conidiogenol-type diterpenoids, while *P. commune* MCCC 3A00940, *P. chrysogenum* MT-12, *P.* sp. YPGA11, and *P. cyclopium* have been reported to produce conidiogenone-type diterpenoids (Figure 2). Structurally, cyclopiane diterpenoids differ from the aberrarane-type diterpenoid aberrarone, which has shown in vitro antimalarial activity against a chloroquine-resistant strain of the protozoan parasite *Plasmodium falciparum* isolated from the Caribbean sea whip *Pseudopterogorgia elisabethae* [67]. The molecular structure of aberrarone was established by spectral analysis and subsequently confirmed by X-ray crystallographic analysis. Some cyclopiane compounds exhibited pronounced biological activities, such as conidiation induction, cytotoxic, anti-inflammatory, antimicrobial, and antiallergic effects.

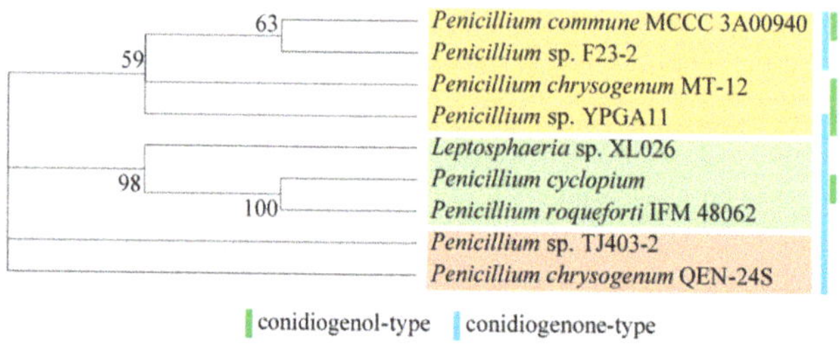

Figure 2. The evolutionary analysis tree constructed with selected fungi producing cyclopiane diterpenoids. The evolutionary analysis was reconstructed by the maximum likelihood method from the ITS sequences as follows: *Penicillium commune* MCCC 3A00940 (KY978585.1), *P.* sp. F23-2 (EU770318.1), *P.* sp. YPGA11 (MG835908.1), *P.* sp. TJ403-2 (MK613138.1), *P. chrysogenum* MT-12 (MF765611.1), *P. chrysogenum* QEN-24S (GU985086.1), *P. roqueforti* IFM 48062 (AB041202.1), and *Leptosphaeria* sp. XL026 (MK603060.1). Since the ITS sequence of strain *P. cyclopium* IMI 229034 was not available, *P. cyclopium* IFM 41611 (AB041169.1) was selected, since it was in the same family as *P. cyclopium*. The evolutionary analysis was conducted in MEGA X (version 10.2.2) [64].

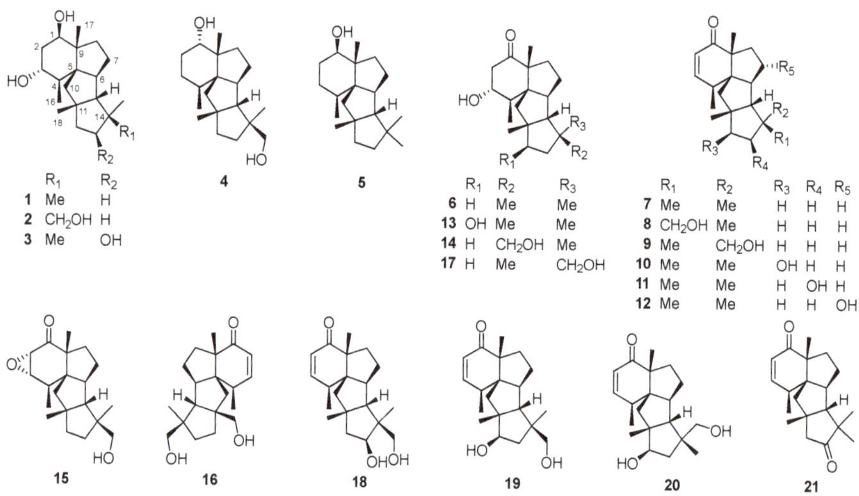

3.1. Conidiogenol Type

Conidiogenol **1** is a potent and selective inducer of conidiogenesis in the liquid culture of *Penicillium cyclopium* under non-nutrient limiting conditions [66]. Conidiogenol B **2** has been obtained from the deep-sea derived fungus *P. commune* MCCC 3A00940 [68]. Conidiogenols C **3** and D **4** have been isolated from a deep-sea derived fungus *P.* sp. YPGA11 [65].

The absolute structure of cyclopiane diterpenoids was first confirmed by Abe and co-workers, in 2018, with the aid of the crystal sponge method [69]. Using the genome-mining approach, a chimeric enzyme of prenyltransferase-diterpene synthase (PT-TS) discovered from *P. chrysogenum* MT-12 was designated as *P. chrysogenum* cyclopiane-type diterpene synthase (PcCS). The new diterpene alcohol metabolite **5** was produced after the gene heterologously expressed in *Aspergillus oryzae*, and the crystalline sponge method also revealed the absolute configuration of **5** [69]. The PT domain of PcCS first generated geranylgeranyl diphosphate (GGPP) from dimethylallyl pyrophosphate (DMAPP) and isopentenyl pyrophosphate (IPP) (Scheme 1A). Then, GGPP was converted into **5** by a cyclization reaction catalyzed by the TS domain of PcCS (Scheme 1B).

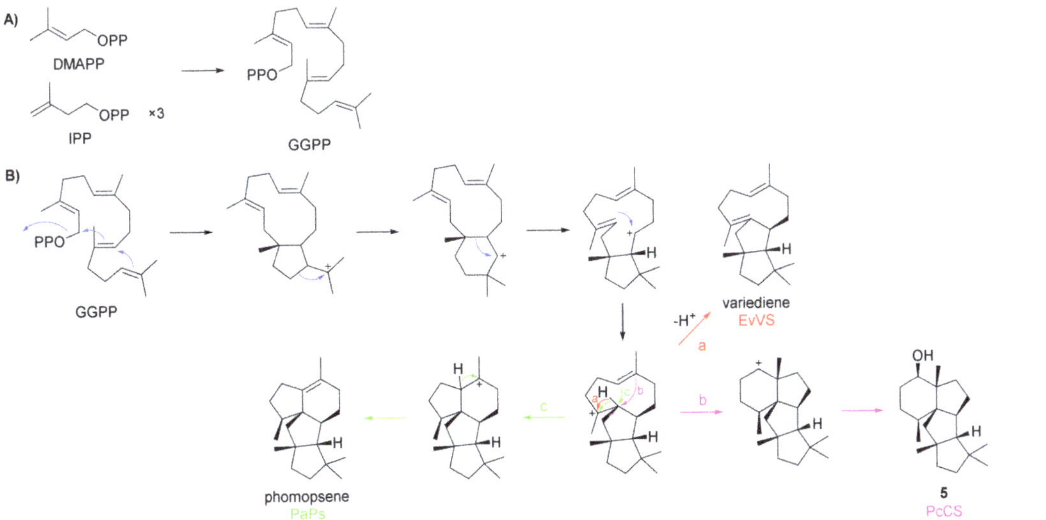

Scheme 1. (**A**) Reaction catalyzed by the prenyltransferase domain of PcCS; (**B**) reaction catalyzed by the terpene synthase domains of PcCS [69].

3.2. Conidiogenone Type

Conidiogenone **6**, first isolated from *Penicillium cyclopium*, was also an inducer of conidiation [66,70]. The biosynthetic pathway of (–)-conidiogenone **6** has been fully elucidated by the heterologous expression of biosynthetic genes in *Aspergillus oryzae* and by in vitro enzyme assay with ^{13}C-labeled substrates [71]. After construction of deoxyconidiogenol by the action of bifunctional terpene synthases (PchDS gene obtained from *Penicillium chrysogenum*, and PrDS gene identified from *Penicillium roqueforti* showed significant homology to PchDS), one cytochrome P450 catalyzed two rounds of oxidation to furnish conidiogenone **6**. The cyclization mechanism catalyzed by terpene synthase, involving successive 1,2-alkyl shifts, was fully elucidated using ^{13}C-labeled geranylgeranyl pyrophosphate (GGPP) as a substrate (Scheme 2).

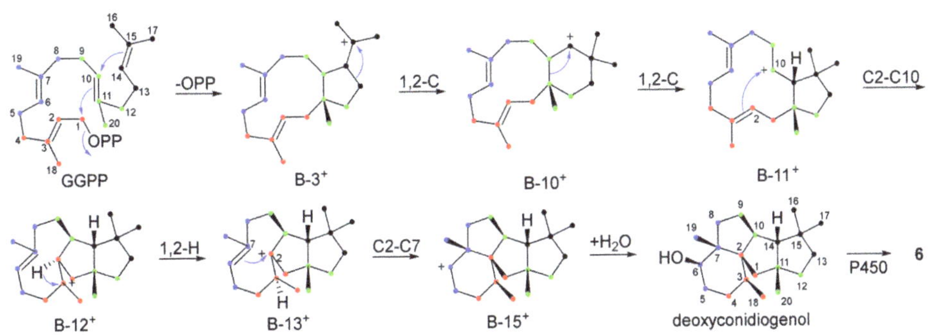

Scheme 2. Proposed cyclization mechanism catalyzed by PchDS/PrDS [71].

A series of new conidiogenone-type diterpenoids have been obtained from several *Penicillium* species including conidiogenones B–G **7**–**12** from the fungus *P.* sp. F23-2 [72], conidiogenones H **13** and I **14** from the endophytic fungus *P. chrysogenum* QEN-24S [73], conidiogenones J **16** and K **15** from the fungus *P. commune* [68], and conidiogenone L **17** from *P.* sp. YPGA11 [65]. Conidiogenone B **7** showed potent activity against methicillin-resistant *Staphylococcus aureus* (MRSA), *Pseudomonas fluorescens*, *P. aeruginosa*, and

Staphylococcus epidermidis (each with a MIC value of 8 μg/mL) [73]. Conidiogenone C **8** showed potent cytotoxicity against HL-60 and BEL-7402 cell lines with IC_{50} values of 0.038 and 0.97 μM, and conidiogenone G **12** showed potent cytotoxicity against HL-60 cell line with an IC_{50} value of 1.1 μM [72]. Provoked by the novelty of structures and potent bioactivities, total syntheses of **1**, **6**, and **7** were achieved, which led to further determination of their absolute configurations [74].

Three new cyclopiane diterpenes 13β-hydroxy conidiogenone C **18** and 12β-hydroxy conidiogenones C **19** and D **20** have been isolated and identified from a sea sediment-derived fungus *Penicillium* sp. TJ403-2 [75]. Their absolute configurations were further established by X-ray crystallography experiment. Compounds **18–20** were evaluated for their anti-inflammatory activity against LPS-induced NO production, and compound **18** showed notable inhibitory potency with an IC_{50} value of 2.19 μM, which was three-fold lower than the positive control indomethacin (IC_{50} 8.76 μM). Further Western blot and immunofluorescence experiments demonstrated that **18** inhibited the NF-κB-activated pathway.

Leptosphin C **21** has been isolated from the solid cultures of an endophytic fungus *Leptosphaeria* sp. XL026 [76]. Its structure was elucidated by extensive spectroscopic methods and single-crystal X-ray diffraction.

4. Fusicoccane

4.1. Structural and Biological Diversity

Fusicoccane diterpenoids, characterized by 5/8/5, 5/8/6, 5/9/4, and 5/9/5 fused carbocyclic ring systems, include the fusicoccins, cotylenins, brassicicenes, heterodimers, and homodimers [77–80]. They were first isolated as glycosides from the phytopathogenic fungus *Fusicoccum amygdali*, in 1964 [81]. Substances exhibiting this structural motif have been isolated from a variety of sources including fungi such as *Talaromyces purpureogenus*, *Alternaria brassicicola* XXC, and *Trichoderma citrinoviride* cf-27 (Figure 3), and rarely from liverworts, algae, ferns, streptomycetes, and higher plants, some of which showed remarkable biological effects relevant for drug discovery, such as antibacterial, antitumor, anti-inflammatory, and antifungal activities [82–89].

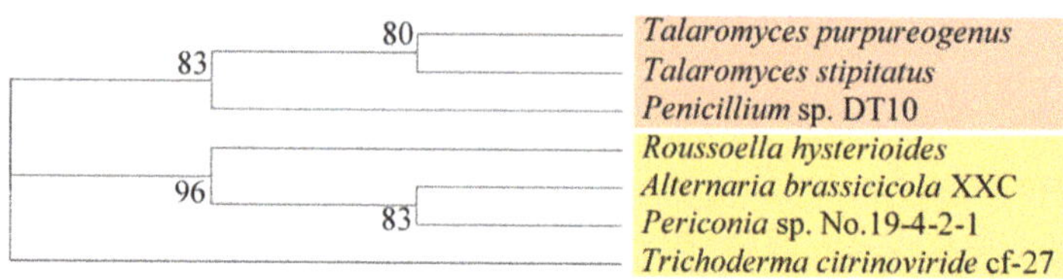

Figure 3. The evolutionary analysis tree constructed with selected fungi producing fusicoccane diterpenoids. The evolutionary analysis was reconstructed by the maximum likelihood method from the ITS sequences as follows: *Alternaria brassicicola* XXC (KR779774.1), *Penicillium* sp. DT10 (MH458525.1), *Periconia* sp. No. 19-4-2-1 (KP873157.1), *Roussoella hysterioides* KT1651 (KJ474829.1), *Talaromyces stipitatus* (MH857968.1), *Talaromyces purpurogenus* (MH120320.1), and *Trichoderma citrinoviride* cf-27 (KT259441.1). The evolutionary analysis was conducted in MEGA X (version 10.2.2) [64].

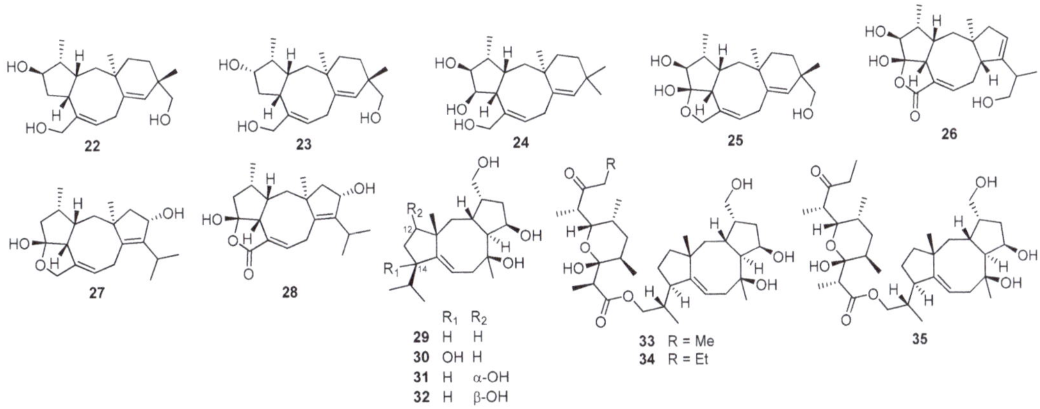

Talaronoids A–D **22–25**, four diterpenoids with an unexpected tricyclic 5/8/6 carbon skeleton isolated from *Talaromyces stipitatus*, represent a new class of fusicoccane diterpenoids with a benzo[*a*]cyclopenta[*d*]cyclooctane skeleton [80]. Plausible biosynthetic pathways of talaronoids A–D **22–25** have been proposed starting from geranylgeranyl diphosphate with a Wagner–Meerwein rearrangement as the key step (Scheme 3). Talaronoids A–D **22–25** have shown moderate butyrylcholinesterase (BCHe) inhibitory activity with IC_{50} values of 14.71, 26.47, 31.51, and 11.37 μM, respectively. A new diterpenoid roussoellol C **26** that exhibited moderate antiproliferative activities against human breast adenocarcinoma (MCF-7) cell line with an IC_{50} value of 6.5 μM has been isolated from an extract of laboratory cultures of the marine-derived fungus *Talaromyces purpurogenus* [90]. Roussoellols A **27** and B **28** have been isolated from the plant-inhabiting ascomycetous fungus *Roussoella hysterioides* KT1651 [91]. Compound **28** inhibited the hyphal growth of the phytopathogen *Cochliobolus miyabeanus* at 10 μg/mL.

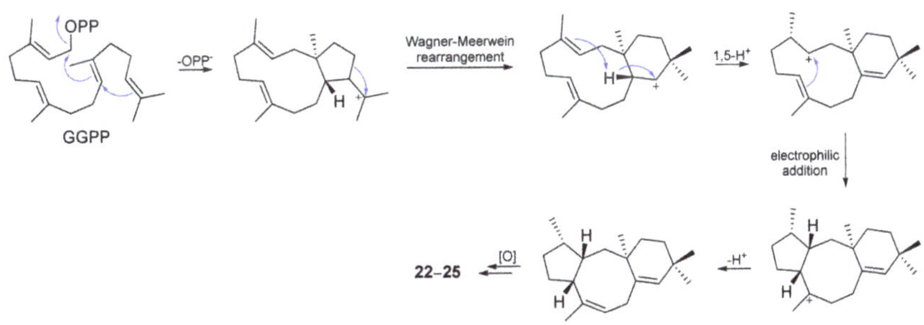

Scheme 3. Proposed biosynthetic pathways of talaronoids A–D **22–25** [80].

Six new fusicoccane-type diterpenoids, 14-hydroxycyclooctatin **30**, 12α-hydroxycyclooctatin **31**, 12β-hydroxycyclooctatin **32**, fusicomycin A **33**, fusicomycin B **34**, and isofusicomycin A **35**, along with a known compound, cyclooctatin **29** [82,92], have been isolated from the fermentation broth of *Streptomyces violascens* [93]. Compounds **33–35** have demonstrated cytotoxicity against five human cancer cell lines (BGC823, H460, HCT116, HeLa, and SMMC7721), with IC_{50} values ranging from 3.5 to 14.1 μM. Cell adhesion, migration, and invasion assays have shown that fusicomycin B **34** inhibited the migration and invasion of human hepatocellular carcinoma SMMC7721 cells in a dose-dependent manner. Through further investigation, it was revealed that **34** inhibited the enzymatic activity of matrix metalloproteinase-2 (MMP-2) and matrix metalloproteinase-9 (MMP-9), in addition to downregulating the expressions of MMP-2 and MMP-9 at both the protein and mRNA levels to influence the migration and invasion of cancer cells.

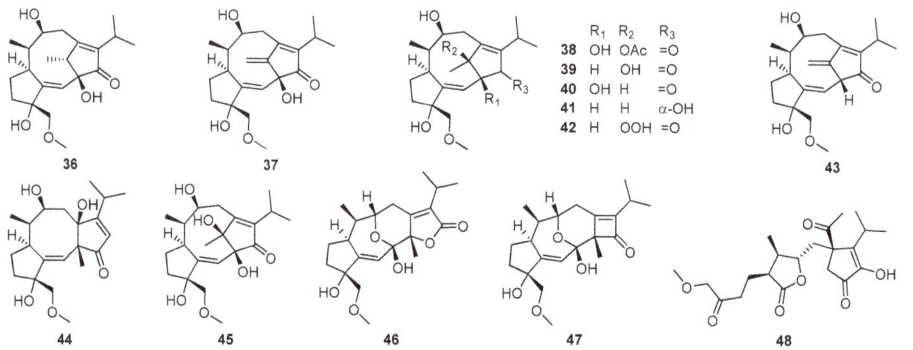

Between 1999 and 2014, eleven new fusicoccane-like diterpenoids were isolated from the phytopathogenic fungus *Alternaria brassicicola* [94–96]. With the aid of computational predictions, experimental validation, and biosynthetic logic-based strategies, Zhang and co-workers first rectified the conclusion that all brassicicenes were originally proposed to have a 5/8/5 fused skeleton and, thus, reassigned brassicicenes C–H **36–41**, J **42**, and K **43** to have a unique bridgehead double-bond-containing 5/9/5 fused skeleton [97]. Meanwhile, brassicicenes L–N **44–46** were three highly modified fusicoccanes also isolated from the fungus *Alternaria brassicicola* [97]. Afterward, alterbrassicene A **47** [78] and alterbrassicicene A **48** [98], two unprecedented fusicoccane-derived diterpenoids featuring a 5/9/4-fused carbocyclic skeleton and a newly transformed monocyclic carbon skeleton (Scheme 4), respectively, were obtained from the same fungal strain and found to function on different targets in the NF-κB signaling pathway of anti-inflammatory activity. Later, the biogenetically related intermediates, brassicicenes O **49** and P **50**, were also discovered [78].

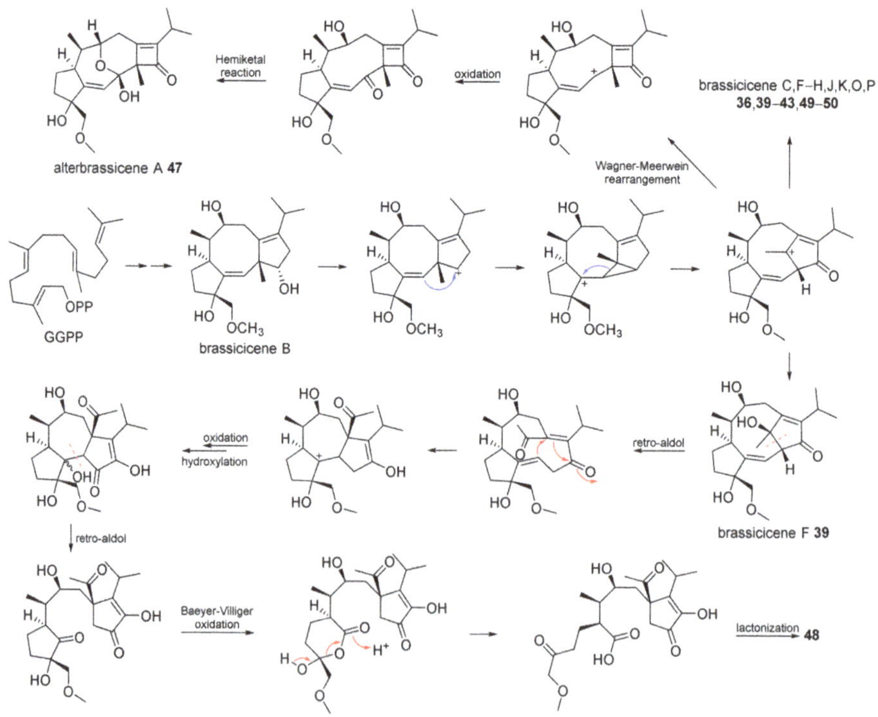

Scheme 4. Hypothetical biosynthetic pathways for alterbrassicene A **47** and alterbrassicicene A **48** [78,98].

Brassicicenes Q–X **51–58** have been isolated from the phytopathogenic fungus *Alternaria brassicicola* [99]. Brassicicene S **53** was found to show significant anti-inflammatory activity against the production of NO, TNF-α, and IL-1β at 10 μM. Further Western blot and immunofluorescence experiments found the mechanism of **53** inhibiting the NF-κB-activated pathway.

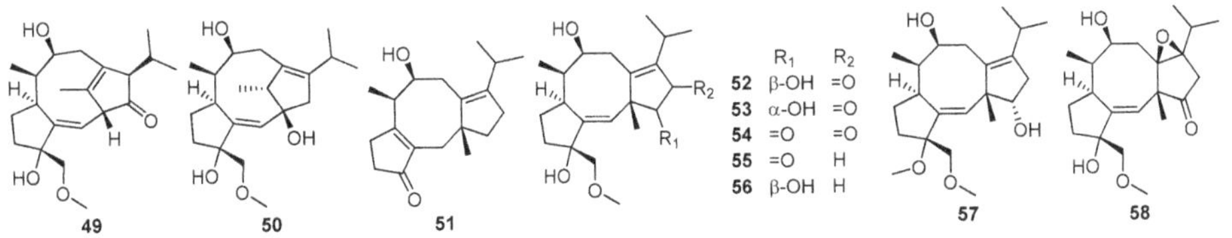

49 **50** **51** **57** **58**

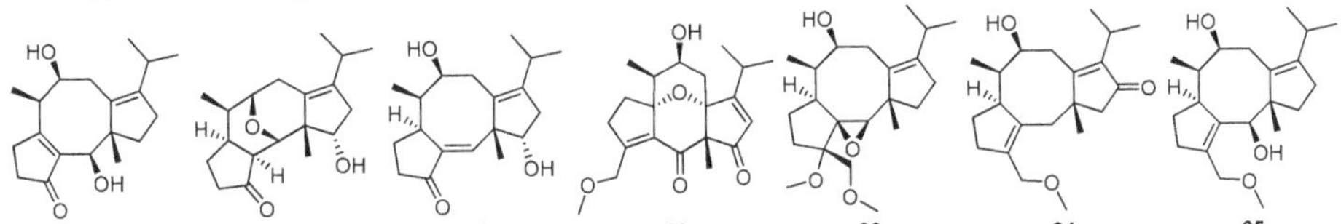

59 **60** **61** **62** **63** **64** **65**

Seven new modified fusicoccane-type diterpenoids **59–65**, together with two known congeners, have been obtained from *A. brassicicola* [100]. Alterbrassicicenes B **60** and C **62** represented the first examples of fusicoccane-type diterpenoids featuring two previously undescribed tetracyclic 5/6/6/5 ring systems, while 1β,2β-epoxybrassicicene I **63** featured a previously undescribed tetracyclic 5/8/5/3 ring system. Alterbrassicicene E **65** showed moderate anti-inflammatory activity against NO production in lipopolysaccharide (LPS)-induced RAW264.7 cell with an IC_{50} value of 24.3 μM. In addition, alterbrassicicene B **60**, 3-ketobrassicicene W **61**, 1β,2β-epoxybrassicicene I **63**, and alterbrassicicene E **65** exerted weak cytotoxicity against certain human tumor cell lines (OCVAR, MDA-MB-231, HeLa, HT-29, and Hep3B cells) with IC_{50} values ranging from 25.0 to 38.2 μM.

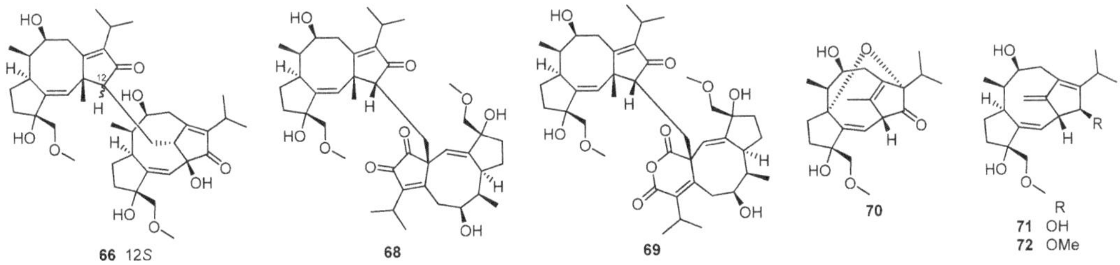

66 12*S*
67 12*R*
68 **69** **70**
R
71 OH
72 OMe

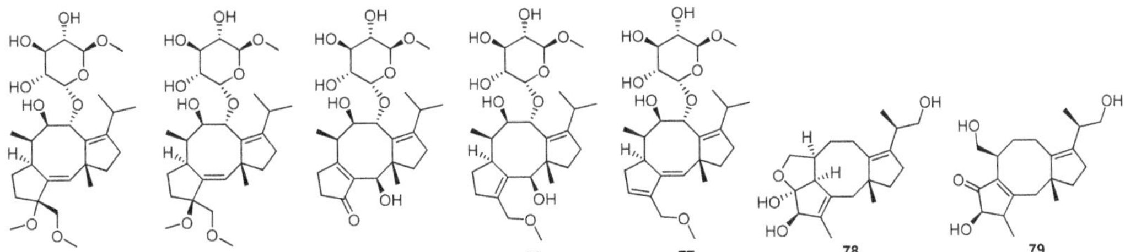

73 **74** **75** **76** **77** **78** **79**

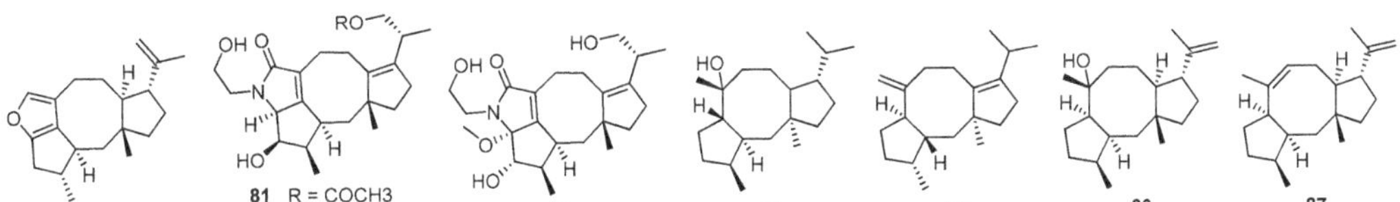

80 **81** R = COCH3 **82** R = H **83** **84** **85** **86** **87**

Alterbrassinoids A–D **66–69**, the first examples of fusicoccane-derived diterpene dimers furnished by forming an undescribed C-12–C-18′ linkage, have been isolated from modified cultures of *Alternaria brassicicola* [79]. Alterbrassinoids A **66** and B **67** represented unprecedented heterodimers, whereas alterbrassinoids C **68** and D **69** represented unprecedented homodimers, and alterbrassinoid D **69** also featured an undescribed anhydride motif. Alterbrassinoids A–D **66–69** showed moderate activities against five cancer cells (including OCVAR, MDAMB-231, HeLa, HT-29, and Hep3B). Afterward, three rearranged fusicoccane diterpenoids bearing a rare bridgehead double-bond-containing tricyclo[9.2.1.$0^{3,7}$]tetradecane (5/9/5 ring system) core skeleton, namely alterbrassicenes B–D **70–72**, were obtained from the same fungus *A. brassicicola* [101]. Their structures were assigned via spectroscopic methods, ECD calculations, and single-crystal X-ray diffraction. Compounds **70–72** showed moderate cytotoxicity against several human tumor cell lines with IC_{50} values ranging from 15.87 to 36.85 μM.

Five new diterpenoid glycosides, dongtingnoids A–E **73–77**, two new diterpenoid aglycones, dongtingnoids F **78** and G **79**, and two known analogues, cotylenins E and J, belonging to the fusicoccane family, have been isolated from the fungus *Penicillium* sp. DT10 [102]. Dongtingnoids A **73**, D **76**, and E **77** showed comparable seed-germination-promoting activities to the growth regulator cotylenin E [103,104]. Such diterpene glucosides have been used for the production of an intermediate compound suitable for semi-synthesis by a mutant constructed by disruption of a specific gene by homologous recombination [105,106].

Trichocitrin **80**, representing the first *Trichoderma*-derived and furan-bearing fusicoccane diterpene, has been isolated from the culture of marine brown alga-endophytic *Trichoderma citrinoviride* [107]. A new class of fusicoccane-type diterpenoid alkaloids with an unusual 5/5/8/5 tetracyclic system, i.e., pericolactines A–C **81–83**, have been isolated from *Periconia* sp. [108].

4.2. Biosynthesis of Fusicoccane Diterpenes

A unique chimeric enzyme PaFS, possessing both a geranylgeranyl diphosphate (GGDP) synthase domain and a diterpene cyclase domain, has been identified from *Phomopsis amygdali* [109]. A biosynthetic gene cluster of brassicicene C **36**, a fusicoccadiene synthase (AbFS) containing 11 genes (*orf1* to *orf11*, Scheme 5A), has been identified in *Alternaria brassicicola* ATCC 96836 from genome database search [110,111]. In vivo and in vitro studies have clearly revealed the function of *Orf8* and *Orf6* as a fusicoccadiene synthase similar to PaFS and methyltransferase, respectively. In this gene cluster, five genes (*orf1*, *orf2*, *orf5*, *orf7*, and *orf11*) encoded cytochrome P450s. *Orf9* was a key dioxygenase to determine the aglycon structures of fusicoccin and brassicicene [112].

Other fusicoccane-type diterpene synthases have been identified from bacteria or fungus, such as CotB2 from bacteria responsible for the biosynthesis of cyclooctat-9-en-7-ol **84** [113], and SdnA from fungus responsible for the biosynthesis of cycloaraneosene **85** [114]. The same 5/8/5 tricyclic skeleton occurred in the sesterterpene ophiobolin F for which the terpene synthase AcOS has been reported from *Aspergillus clavatus* [115]. Oikawa and co-workers applied the *Aspergillus oryzae* heterologous expression system to functionally characterize cryptic bifunctional terpene synthase genes found in fungal genomes and identified the sesterfisherol (contains a characteristic 5/6/8/5 tetracyclic system) synthase gene (*NfSS*) from *Neosartorya fischeri* [116].

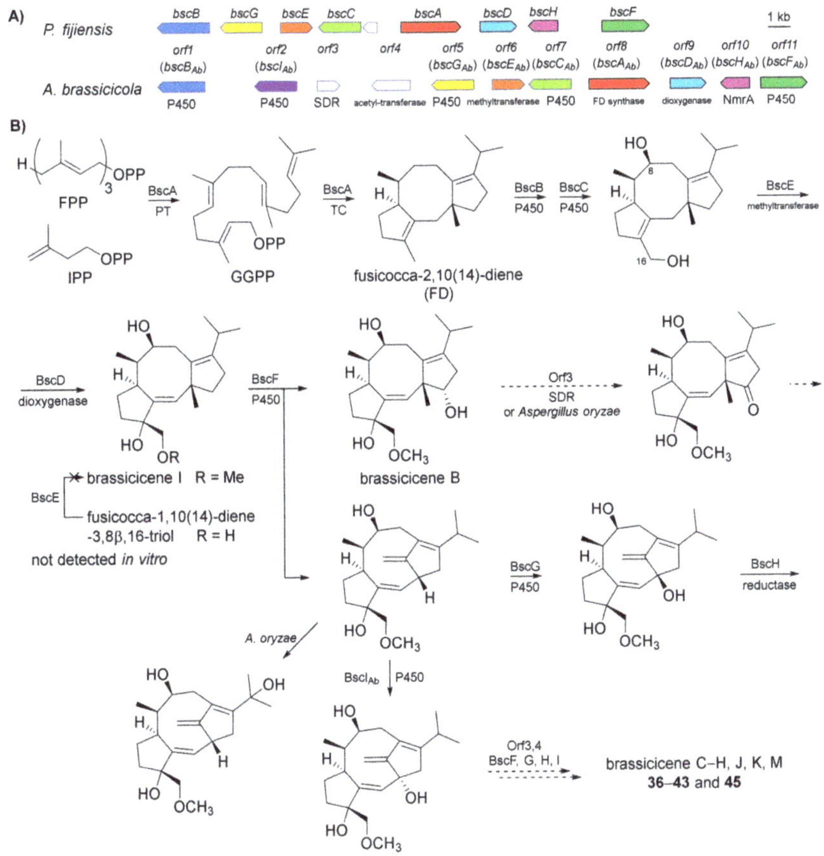

Scheme 5. (**A**) Biosynthetic gene clusters of the brassicicenes in *P. fijiensis* and *A. brassicicola*; (**B**) proposed biosynthetic pathway for brassicicenes (dashed arrows are those deduced from expected protein function) [117].

A unique P450 enzyme *bscF* has been identified in the phytopathogen *Pseudocercospora fijiensis* that generated two structurally different products from the single substrate. In addition to the heterologous expression of the eight genes, *bscA-bscH* elucidated the biosynthetic pathway for brassicicenes (Scheme 5B) [117].

A new fusicoccane-type diterpene synthase MgMS has been identified from the fungus *Myrothecium graminearum* by the genome mining method, which catalyzed the formation of the new diterpene alcohol myrothec-15(17)-en-7-ol **86** with all the seven stereocenters being introduced in the cyclization steps and conserved in the structure of the product. Based on this, its novel cyclization mode was unambiguously assigned (Scheme 6) [118].

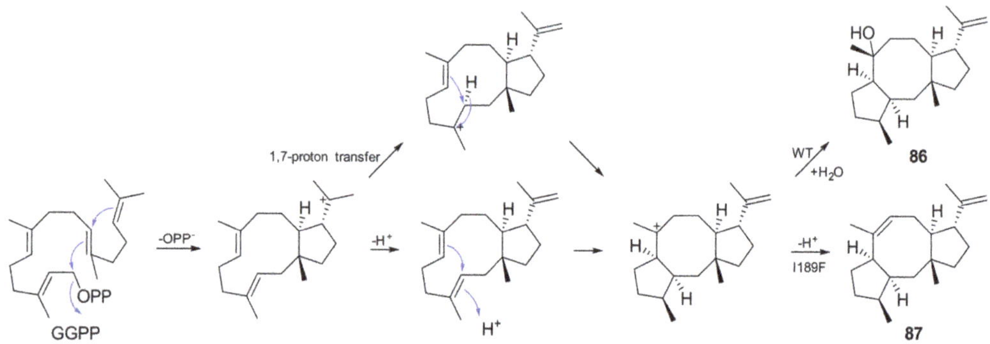

Scheme 6. Mechanistic hypothesis for the cyclization of GGPP to myrothec-15(17)-en-7-ol **86** and myrotheca-7,15(17)-diene **87** [118].

5. Guanacastane

The discovery of 5/7/6 ring-fused guanacastane diterpenoids has been limited to several fungal species in the different genera *Cercospora*, *Cortinarius*, *Coprinellus*, *Coprinus*, *Psathyrella*, and *Verticillium* (Figure 4). *Coprinellus heptemerus* and *C. radians* M65 belong to the same genus, and first, they gather into one branch. *Psathyrella candolleana* and *Cercospora* sp. gather into one branch although they come from different genera. They are all able to produce guanacastane diterpenoids, indicating that highly homologous BGCs may also exist in fungi of different genera. The first member guanacastepene A **88**, a new diterpene antibiotic against methicillin-resistant *Staphylococcus aureus* (MRSA) and vancomycin-resistant *Enterococcus faecalis* (VREF), has been isolated from an unidentified endophytic fungus [119]. Meanwhile, fourteen new analogues guanacastepenes B–O **89–102** have been isolated from the same resource [120]. The novel skeleton has attracted great interests for organic synthesis [121–136]. The biological activities of guanacastanes have mainly been reported to possess cytotoxicity and antimicrobial effects.

Figure 4. The evolutionary analysis tree constructed with selected fungi producing guanacastane diterpenoids. The evolutionary analysis was reconstructed by the maximum likelihood method from the ITS sequences as follows: *Coprinus heptemerus* D99052 (JN159553.1), *Coprinus radians* M65 (HM045514.1), *Coprinus plicatilis* 82 (*Parasola plicatilis*) (FM163216.1), *Psathyrella candolleana* (MF401519.1), *Cercospora* sp. (KF577929.1), and *Verticillium dahlia* (HQ839784.1). Since the ITS sequence of *Cortinarius pyromyxa* was not available, *Cortinarius misermontii* (NR_130230.1) was selected since their ITS sequences were the most similarly. The evolutionary analysis was conducted in MEGA X (version 10.2.2) [64].

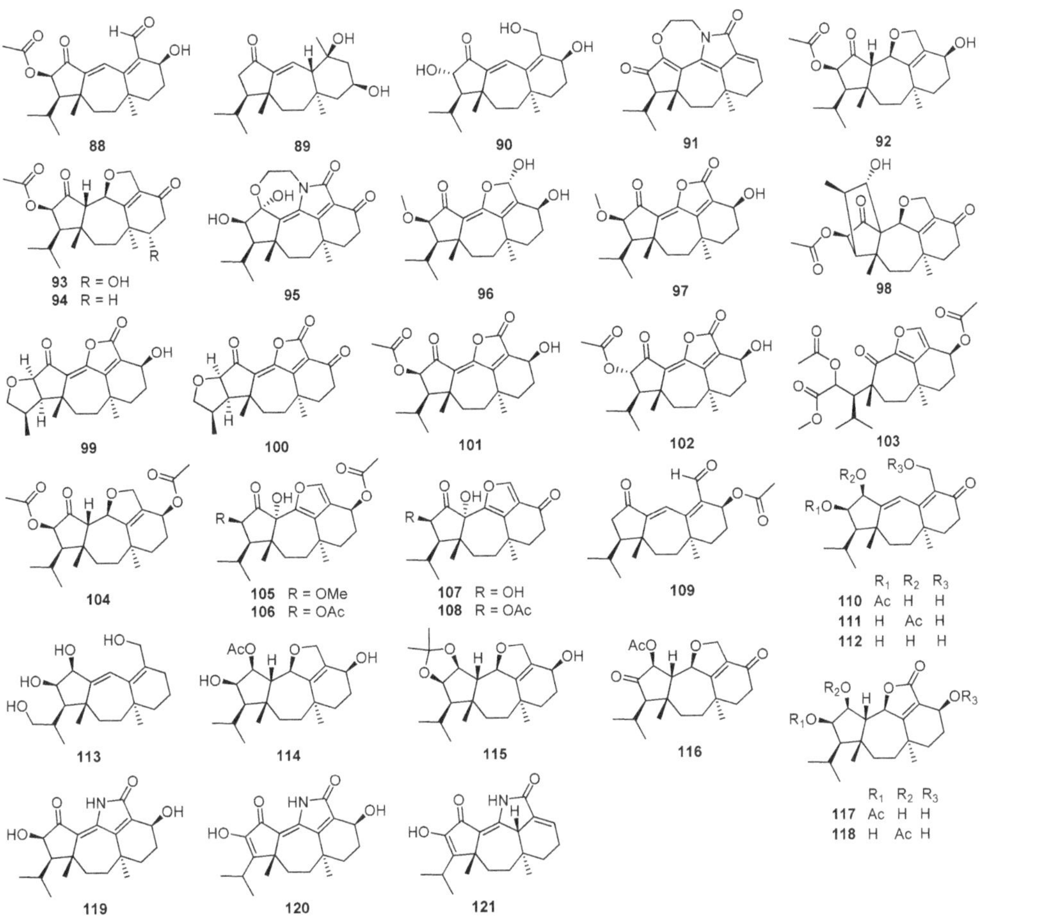

Heptemerones A–G **103–109** have been isolated from cultures of *Coprinus heptemerus* [137,138]. Radianspenes A–M **110–122** have been obtained from *Coprinus radians* [139]. Among the biological activities of these isolates, the inhibition of fungal germination was the most potent, and depended highly on the composition of the assay medium [137]. Radianspene C **112** showed inhibitory activity against human breast carcinoma (MDA-MB-435) cell with an IC_{50} value of 0.91 μM [139]. Investigation of secondary metabolites from the fungal *Coprinus plicatilis* led to the discovery of several new guanacastane-type diterpenoids, named plicatilisins A–D **123–126** [140] and E–H **127–130** [141]. In vitro cytotoxic activities against the human cancer cell lines (HepG2, HeLa, MDA-MB-231, BGC-823, HCT 116, and U2OS) showed that plicatilisin A **123** exhibited significant cytotoxicity with IC_{50} values ranging from 1.2 to 6.0 μM [140].

Guanacastepenes P–T **131–135** have been isolated from cultures of the fungus *Psathyrella candolleana* [142]. Guanacastepene R **133** exhibited inhibitory activity against both human and mouse isozymes of 11β-hydroxysteroid dehydrogenase (11β-HSD1) with IC_{50} values of 6.2 and 13.9 μM, respectively. Cercosporenes A–F **136–141**, including two homodimers **140** and **141**, have been isolated from the fungus *Cercospora* sp. [143]. Cercosporene F **141** was cytotoxic to five human tumor cell lines (HeLa, A549, MCF-7, HCT116, and T24) with IC_{50} values of 8.16–46.1 μM, and induced autophagy in HCT116 cell.

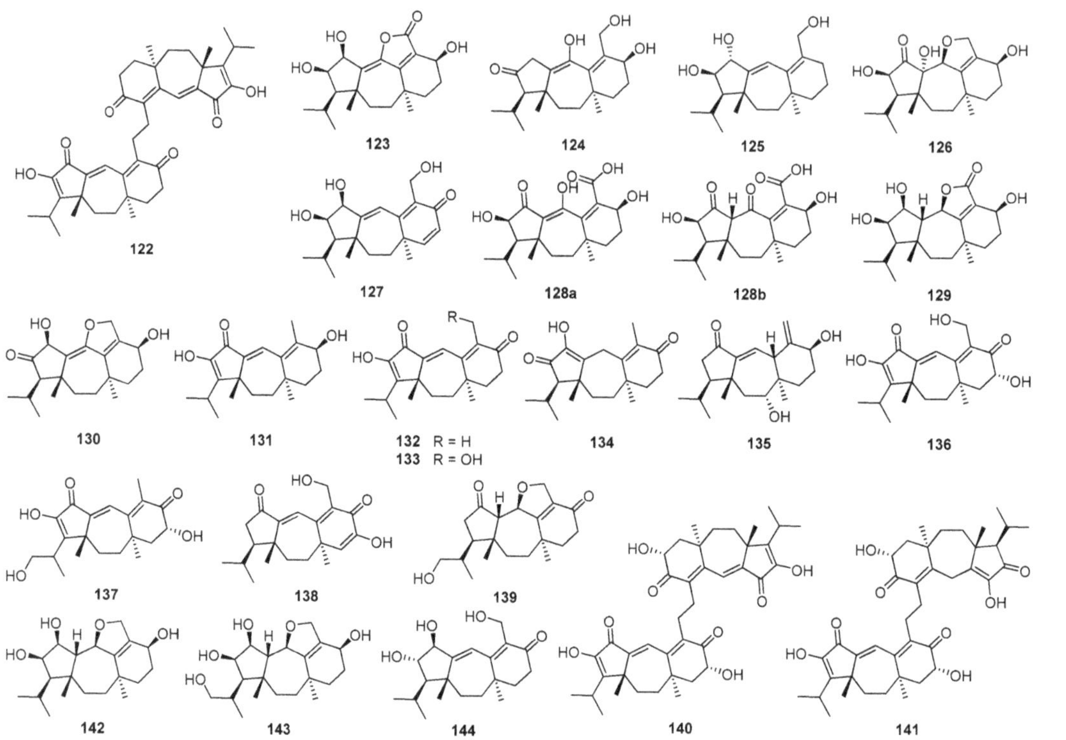

122, 123, 124, 125, 126, 127, 128a, 128b, 129, 130, 131, 132 R = H, 133 R = OH, 134, 135, 136, 137, 138, 139, 140, 141, 142, 143, 144

Eleven new guanacastane-type diterpenoids dahlianes A–K **142–152** have been obtained from the fungus *Verticillium dahlia* that was isolated from the gut of insect *Coridius chinensis* [144,145]. In the cytotoxicity evaluation against human tumor cell lines, dahlianes B **143** and C **144** exhibited significant cytotoxicity against human breast cancer cell MCF-7 with IC_{50} values of 3.35 and 4.72 μM, respectively [144]. In addition, the isolates were evaluated for their cytotoxicity toward drug-sensitive and DOX resistant MCF-7 cells by MTT assay. As a result, dahliane G **148** showed an 80-fold potentiation effect on the sensitization of doxorubicin at the concentration of 15 μM when screening the reversal activity on doxorubicin-resistant human breast cancer cell (MCF-7/DOX) [145].

Pyromyxones A–D **153–156** have been isolated from fruiting bodies of *Cortinarius pyromyxa*, which possessed an undescribed *nor*-guanacastane skeleton of a 5/7/6 tricyclic system [146]. Pyromyxones A **153**, B **154**, and D **156** exhibited weak activity against Gram-positive *Bacillus subtilis* and Gram-negative *Aliivibrio fischeri*, as well as the phytopathogenic fungi *Botrytis cinerea*, *Septoria tritici*, and *Phytophthora infestans* [146].

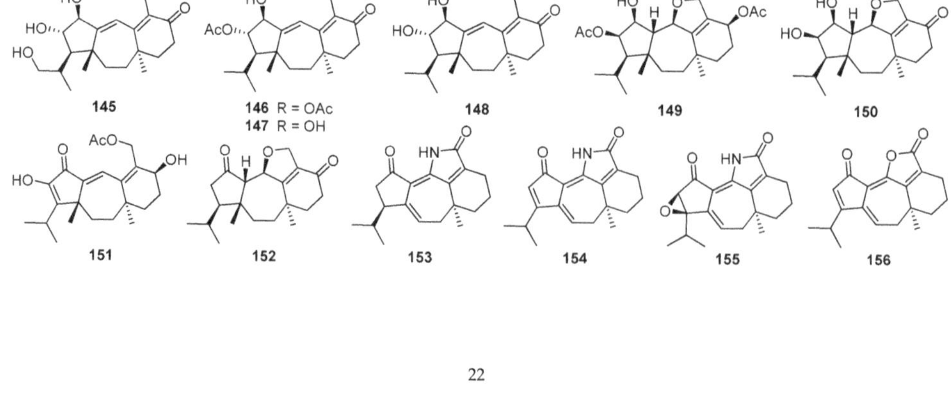

145, 146 R = OAc, 147 R = OH, 148, 149, 150, 151, 152, 153, 154, 155, 156

6. Harziene

Harziene is a small group of diterpenoids that have a unique 4/7/5/6 tetracyclic scaffold. They have mainly been obtained from different *Trichoderma* species and rarely from liverworts [147]. Harziandione **157** was the first harziene diterpenoid isolated from the liquid culture of *T. harzianum,* in 1992 [148]. Harzianone **158**, a new harziene diterpene, has been isolated from an alga-endophytic isolate of *T. longibrachiatum* [149]. The structure with absolute configuration of **158** was unambiguously identified by NMR and mass spectrometric methods as well as quantum chemical calculations. In addition, the absolute configuration of harziandione **157** was supported by optical rotation calculation, and the structure of isoharziandione isolated from culture filtrate of a strain of *Trichoderma viride* [150] was revised to harziandione **157** on the basis of ^{13}C NMR data comparison and calculation.

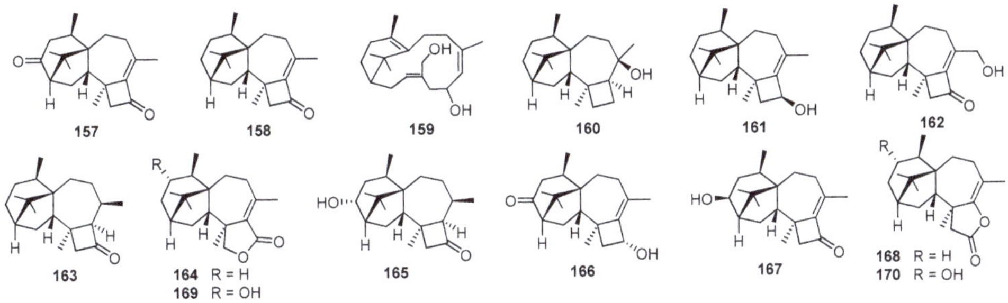

The terpene cyclization mechanism of harzianone **158** has been studied by feeding experiments using selectively ^{13}C- and ^{2}H-labeled synthetic mevalonolactone isotopologues, followed by the analysis of the incorporation patterns of ^{13}C NMR spectroscopy and GC/MS, and the structure of harzianone **158** was further supported from a ^{13}C-^{13}C COSY experiment of the in vivo generated fully ^{13}C-labeled diterpenoid (Scheme 7) [151].

Four new harziene-related compounds **159–162** have been isolated from an endophytic fungus *Trichoderma atroviridae* UB-LMA [152]. Among them, **159** is a potential derivative of geranylgeranyl diphosphate and may represent the biosynthetic precursor of this scarce family of compounds (Scheme 7). Recently, the first total synthesis of nominal harziene diterpenoid **160** has been achieved; stereochemical analysis and subsequent synthesis of the epimeric tertiary alcohol led to the reassignment of configuration for compound **160** as shown for harzianol I **180** [153].

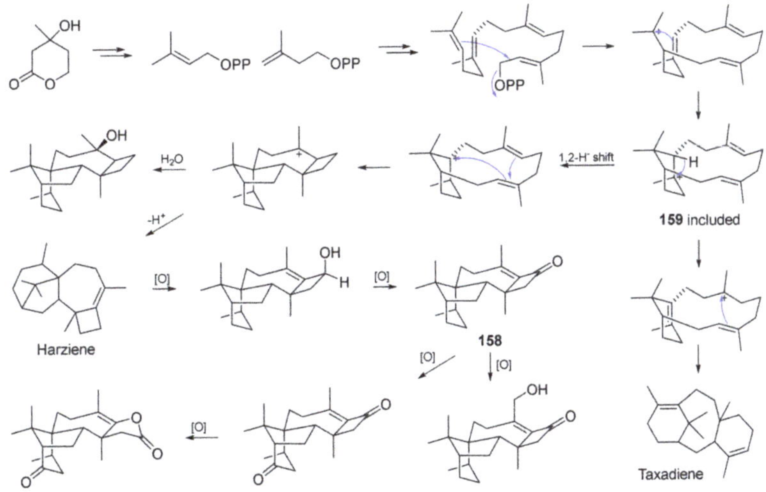

Scheme 7. Biosynthetic mechanism to harziene and taxadiene scaffolds [151,152].

(9*R*,10*R*)-Dihydro-harzianone **163** and harzianelactone **164** have been isolated from the endophytic fungus *Trichoderma* sp. Xy24 [154]. Compound **163** was the reductive product of harzianone **158** while **164** possessed a 6/5/7/5-fused ring core containing a lactone. The latter was the Baeyer–Villiger monooxygenase catalyzed oxidation product of harzianone **158**. Compound **163** exhibited cytotoxicity against HeLa and MCF-7 cell lines with IC_{50} values of 30.1 and 30.7 μM, respectively.

3*R*-Hydroxy-9*R*,10*R*-dihydroharzianone **165** has been isolated from an endophytic fungus *Trichoderma harzianum* X-5 [155]. 11-Hydroxy-9-harzien-3-one **166**, isolated from *T. asperellum* cf44-2, showed inhibitory activity against pathogenic bacteria *Vibrio parahaemolyticus* with a 6.2 mm zone [156]. 3*S*-Hydroxyharzianone **167**, isolated from *T. asperellum* A-YMD-9-2, could highly inhibit four marine phytoplankton species (*Chattonella marina*, *Heterosigma akashiwo*, *Karlodinium veneficum*, and *Prorocentrum donghaiense*) with the IC_{50} values ranging from 3.1 to 7.7 μg/mL [157]. Deoxytrichodermaerin **168**, a harziene lactone possessing potent inhibition against the four phytoplankton species (*C. marina*, *H. akashiwo*, *K. veneficum*, and *P. donghaiense*), has been obtained from an endophyte *Trichoderma longibrachiatum* A-WH-20-2 [158].

Two new harziene diterpene lactones, i.e., harzianelactones A **169** and B **170**, and five new ones, i.e., harzianones A–D **171–174** and harziane **175**, have been isolated from the soft coral-derived fungus *Trichoderma harzianum* XS-20090075 [159]. These compounds exhibited potent phytotoxicity against seedling growth of amaranth and lettuce. Harzianone E **176**, which exhibited weak antibacterial activity against *Photobacterium angustum*, has been obtained from the culture of coral-derived fungus *T. harzianum* treated with 10 μM sodium butyrate [160]. Harzianols F–J **177–181** and three known derivatives have been obtained from the liquid fermentation of an endophytic fungus *T. atroviride* B7 [161]. Among them, compound **180** exhibited significant antibacterial effect against *Staphylococcus aureus*, *Bacillus subtilis*, and *Micrococcus luteus* with EC_{50} values of 7.7, 7.7, and 9.9 μg/mL, respectively. Meanwhile, cytotoxic activity of **180** against three cancer cell lines was also observed [161].

Furanharzianones A **182** and B **183** are two new harziene-type diterpenoids with an unusual 4/7/5/6/5 ring system, while harzianols A–E **184–188** and harziane acid **189** are six new oxidized derivatives of harzianone [162,163]. These compounds have all been obtained from microbial transformation by the bacterial strain *Bacillus* sp. IMM-006.

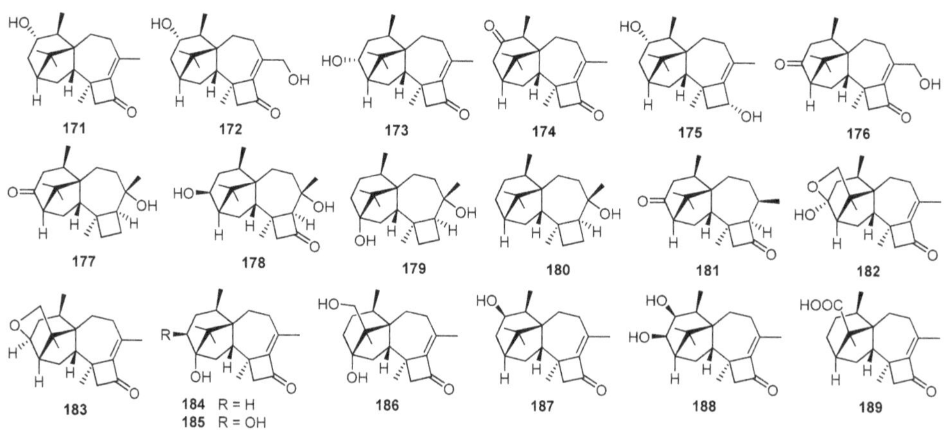

7. Phomopsene

Structure determination of the novel diterpene hydrocarbon phomopsene **190** has been provided by enzymatic synthesis with the recombinant terpene synthase PaPS from *Phomopsis amygdali*, and screening of fungal broth extracts regarding characteristic NMR signals of phomopsene **190** resulted in the isolation of a new diterpene, methyl phomopsenonate **191** (Scheme 8) [164].

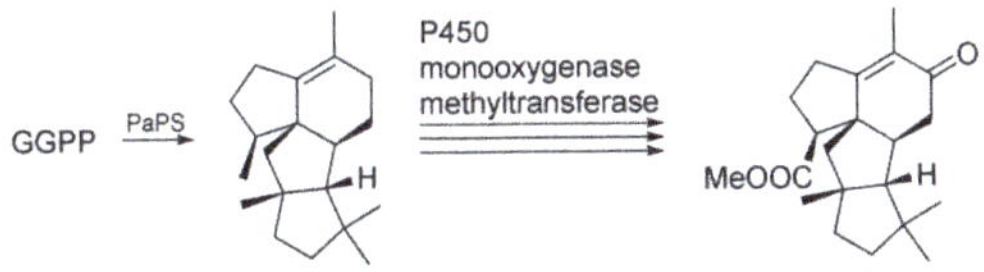

Scheme 8. Proposed biosynthetic pathway of phomopsene **190** and methyl phomopsenonate **191** [164].

The cyclization mechanism of tetracyclic diterpene phomopsene **190** with phomopsene synthase (PaPS) has been examined through systematically deuterium-labeled geranylgeranyl diphosphate (GGPP), starting from site-specific deuterium-labeled isopentenyl diphosphates (IPPs) using IPP isomerase and three prenyltransferases (Scheme 9) [165].

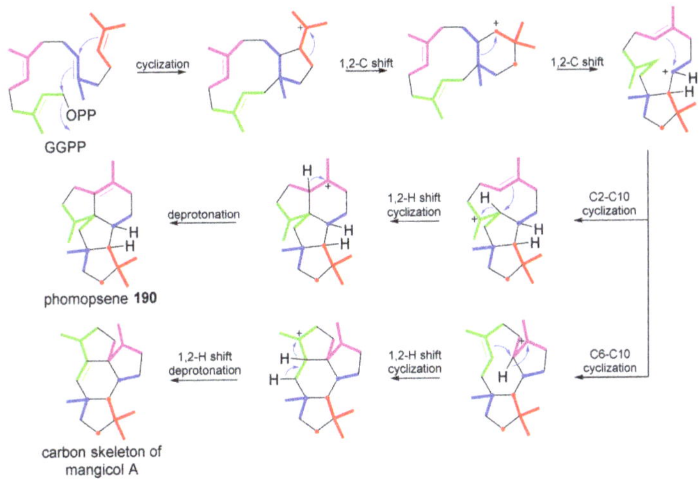

Scheme 9. Proposed cyclization mechanism catalyzed by PaPS [165].

Otherwise, other phomopsene synthases have been identified from actinomycetes such as *Allokutzneria albata* (PmS), *Nocardia testacea* (NtPS), and *Nocardia rhamnosiphila* (NrPS) [166,167]. All enzymes were subjected to in-depth mechanistic studies involving isotopic labeling experiments, metal-cofactor variation, and site-directed mutagenesis.

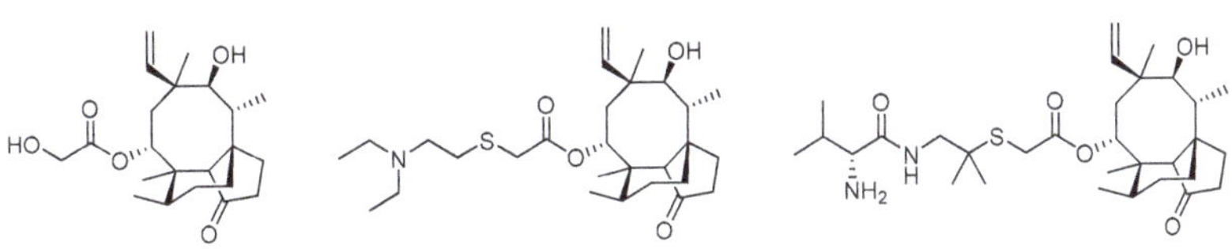

8. Pleuromutilin

Pleuromutilin **192** is a diterpene with a tricyclic skeleton possessing antimicrobial properties. It was first discovered from two basidiomycete fungal species including *Pleurotus mutilis* (synonymous to *Clitopilus scyphoides* f. *mutilus*) and *Pleurotus passeckerianus* (synonymous to *Clitopilus passeckerianus*) [168], and then produced by a number of other related species [169]. Its chemical structure and cyclisation mechanism has been elucidated by independent works [170–172], while total synthesis has been achieved by [173,174]. The semi-synthetic pleuromutilin analogues tiamulin **193** and valnemulin **194** have been used for over three decades to treat economically important infections in swine and poultry without showing any significant development of resistance in their target bacteria [21–25]. In recent years, extensive research including structure–activity relationship studies have been conducted to generate new orally available pleuromutilin derivatives having been used systemically in human medicine to treat acute bacterial skin and skin structure infections, as well as multidrug-resistant tuberculosis [175–178].

The gene cluster for the antibiotic pleuromutilin **192** has been isolated in *Clitopilus passeckerianus* [179]. Total de novo biosynthesis of pleuromutilin **192** was achieved through the expression of the entire gene cluster in the secondary host *Aspergillus oryzae*, proving that the seven genes isolated were sufficient for biosynthesis of the diterpene antibiotic. Heterologous expression of genes from the pleuromutilin gene cluster in *A. oryzae* revealed the biosynthesis of the antibiotic pleuromutilin **192** (Scheme 10) and generated bioactive semi-synthetic derivatives [180].

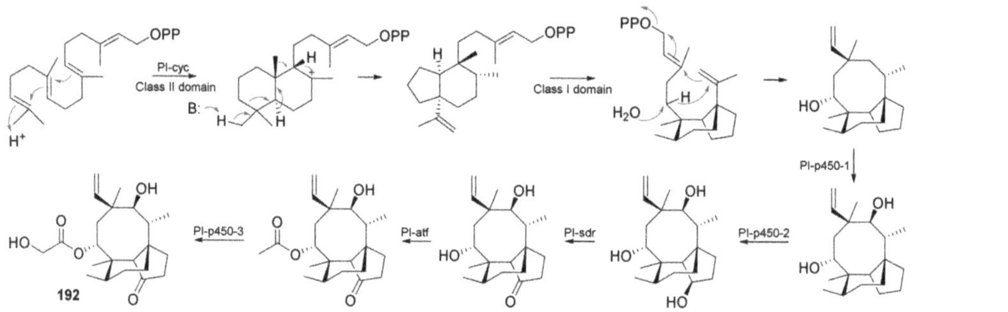

Scheme 10. Proposed biosynthetic pathway to pleuromutilin **192** in *Clitopilus passeckerianus* [180].

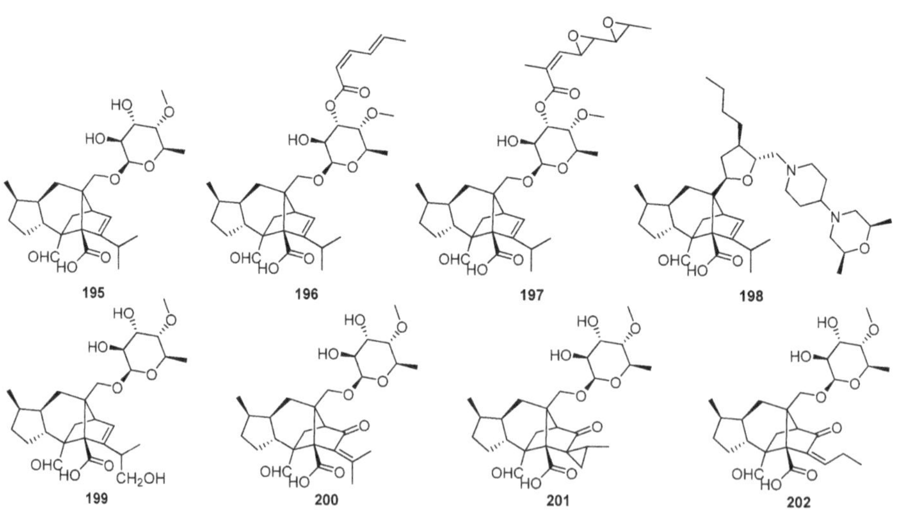

9. Sordaricin

Sordarin **195**, an antifungal antibiotic possessing a unique 5/6/5/5-fused ring system, was discovered in 1971 as a metabolite of *Sordaria araneosa* [181]. A number of related semisynthetic sordarin derivatives have also been reported and some have been developed as antifungal agents such as zofimarin **196**, hypoxysordarin (FR231956) **197**, and FR290581 **198** [182–186]. Sordarin **195** and related compounds have been shown to inhibit protein synthesis by a mechanism involving selective binding to the elongation factor 2 (EF-2) and ribosome complex in fungi [187–189].

Sordarins C–F **199–202**, possessing a unique 5/6/5/5 or 5/6/5/5/3 ring system varied at the C-11 position and the branch attached to C-12 of the sordaricin-type diterpene skeleton, have been isolated from the fungus *Xylotumulus gibbisporus* [190]. Genome mining of the sordarin biosynthetic gene cluster from *Sordaria araneosa* has been carried out, and the results suggest that the identified *sdn* gene cluster is responsible for the biosynthesis of sordarin **195** and hypoxysordarin **197** (Scheme 11) [114].

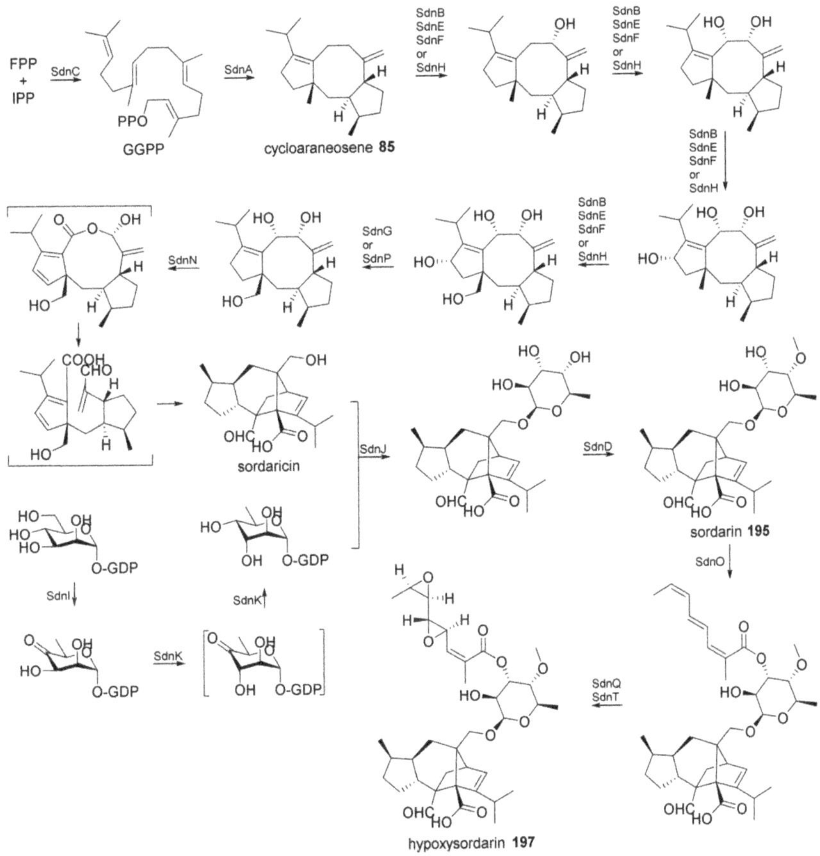

Scheme 11. The biosynthetic pathway for sordarin **195** and hypoxysordarin **197** [114].

10. Tetraquinane

Several antibiotic crinipellin-related diterpenoids containing a 5/5/5/5 tetraquinane skeleton have been obtained from the basidiomycetous fungus *Crinipellis stipitaria* [191,192]. Up to now, the total synthesis of (±)-crinipellin B **203** and (−)-crinipellin A **204** have been reported [186,193–195].

Four novel diterpenoids, namely (4*β*)-4,4-*O*-dihydrocrinipellin A **205**, (4*β*,8*α*)-4,4-*O*,8,8-*O*-tetrahydrocrinipellin B **206**, crinipellins C **207** and D **208**, along with three known diterpenoids have been isolated from the fungus *Crinipellis* sp. 113 [196]. Antitumor assays demonstrated that the compounds possess moderate activities against HeLa cell.

Four new tetraquinane diterpenoids crinipellins E–H **209–212** have been isolated from fermentations of a *Crinipellis* species [197]. Crinipellins E–G **209–211** inhibited the LPS/IFN-γ induced CXCL10 promoter activity in transiently transfected human MonoMac6 cell in a dose-dependent manner with IC_{50} values of 15, 1.5, and 3.15 μM, respectively. Moreover, crinipellins E–G **209–211** reduced mRNA level and synthesis of proinflammatory mediators such as cytokines and chemokines in LPS/IFN-γ stimulated MonoMac6 cell.

A new crinipellin derivative crinipellin I **213** together with the known crinipellin A **204** have been obtained from the fungus *Crinipellis rhizomaticola* [198]. Crinipellin A **204** exhibited a wide range of antifungal activity in vitro against *Colletotrichum coccodes, Magnaporthe oryzae, Botrytis cinerea*, and *Phytophthora infestans* (MICs of 1, 8, 31, and 31 μg/mL, respectively).

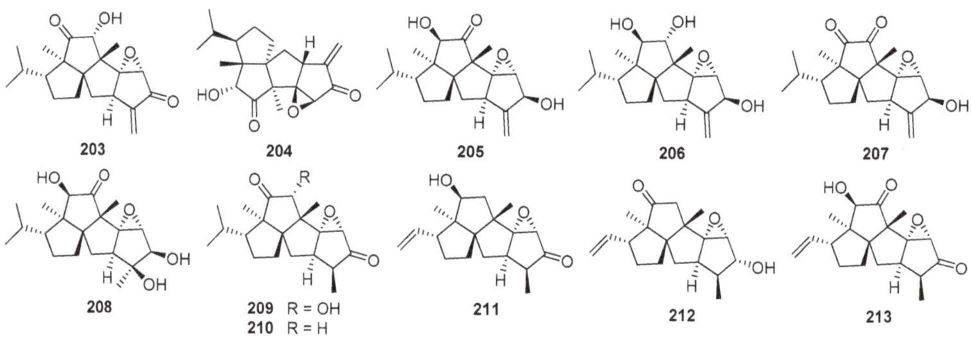

11. Others

11.1. Spirograterpene

A novel spiro-tetracyclic diterpene featuring a 5/5/5/5 spirocyclic carbon skeleton, i.e., spirograterpene A **214**, has been isolated from the deep-sea-derived fungus *Penicillium granulatum* [199]. Spiroviolene **215**, bearing the same carbon skeleton to that of **214**, has been obtained from a bacterial terpene synthase [200]. Spirograterpene A **214** showed an antiallergic effect on immunoglobulin E (IgE)-mediated rat mast RBL-2H3 cell with 18% inhibition as comparedwitho 35% inhibition for the positive control (loratadine) at the same concentration of 20 μg/mL [199].

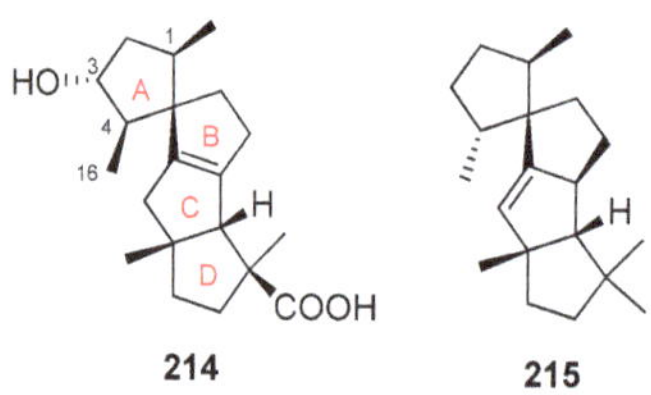

11.2. Psathyrin

Two skeletally novel tetracyclic diterpenoids that possess a novel 5/5/4/6 tetracyclic system, psathyrins A **216** and B **217**, have been characterized from cultures of the basidiomycete *Psathyrella candolleana*. They displayed weak antibacterial activities against *Staphylococcus aureus* and *Salmonella enterica*. The biosynthetic pathway of **216** and **217** was proposed to start from GGPP and the final products were obtained through a series of reactions (Scheme 12) [201].

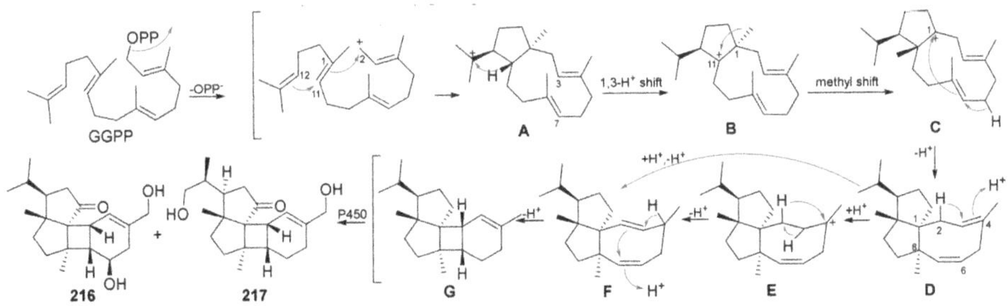

Scheme 12. Proposed biosynthetic pathway for psathyrins A **216** and B **217** [201].

11.3. Coicenal

Coicenals A–D **218–221**, possessing a previously undescribed 6/6 fused carbon skeleton, have been isolated from the solid culture of the plant pathogenic fungus *Bipolaris coicis* [202]. Coicenals A **218** and B **219** could be transformed into **221** and compound **222** by treatment with acetyl chloride, respectively. Coicenals A–D **218–221** showed moderate inhibitory activity against NO release with IC_{50} values of 16.34, 23.55, 10.82, and 54.20 μM, respectively.

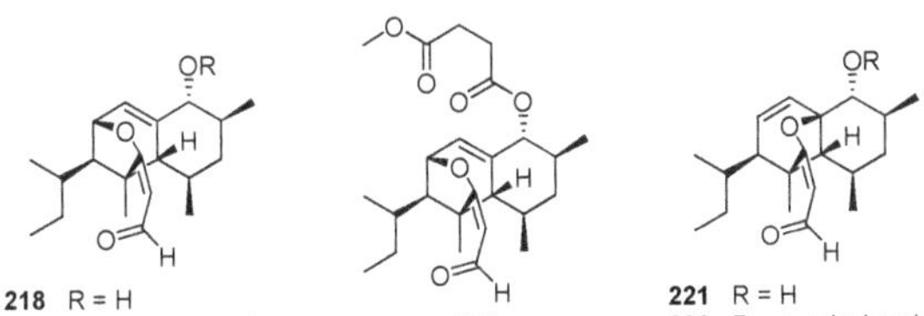

218 R = H
219 R = succinyl moiety
220
221 R = H
222 R = succinyl moiety

11.4. Eryngiolide

Eryngiolide A **223** has been isolated from the solid culture of the edible mushroom *Pleurotus eryngii* [203]. It is the first member of diterpenoids with a novel skeleton deriving from a cyclododecane core fused with two γ-lactone units [203]. It has exhibited moderate cytotoxicity against two human cancer cell lines (Hela and HepG2) in vitro. Biogenetically, eryngiolide A **223** could be the first diterpene not synthesized from GGPP unit, which indicated a completely new route for diterpene biosynthesis in nature (Scheme 13).

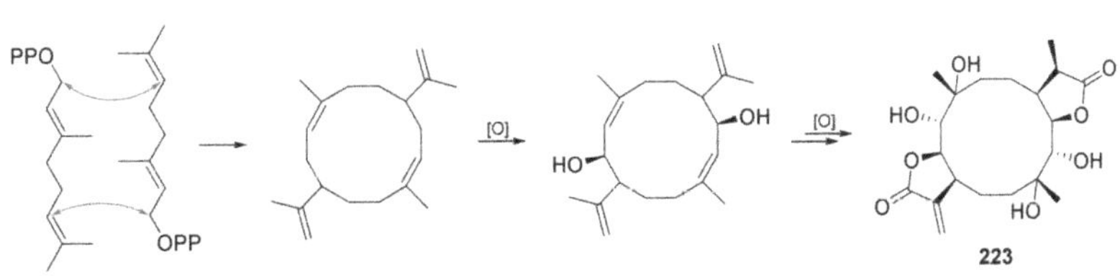

Scheme 13. Plausible biogenetic origin of eryngiolide A **223** [203].

11.5. Trichodermanin

Trichodermanin A **224**, a structurally unique diterpenoid with skeletal carbons arranged compactly in a 6/5/6/6 ring system, has been isolated from cultures of *Trichoderma atroviride* [204]. Its absolute configuration was elucidated by single crystal X-ray diffraction. Wickerols A **225** and B **226** were two novel diterpenoids produced by *Trichoderma atroviride* and the absolute configuration of **226** was confirmed by X-ray crystallographic analysis [205,206]. Wickerol A **225** showed potent antiviral activity against the A/H1N1 flu virus (A/PR/8/34 and A/WSN/33 strains) with an IC_{50} value of 0.07 μg/mL, but not active against the A/H3N2 virus. Wickerol B **226** also showed anti-influenza virus activity against A/PR/8/34 virus with an IC_{50} value of 5.0 μg/mL [206].

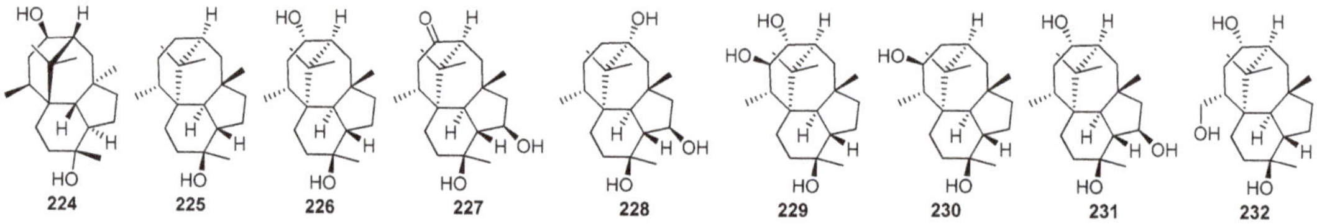

224 225 226 227 228 229 230 231 232

The new skeleton of wickerols A **225** and B **226** was revealed by the feeding experiments of [1-^{13}C]-, [2-^{13}C]-, and [1,2-$^{13}C_2$]-acetates, respectively [206]. The cyclization mechanism of wickerol B **226** was predicted, as shown in Scheme 14. First, pyrophosphate was ejected from the terminus of the boat-like transition state of GGPP, forming a verticillen-12-yl cation intermediate, the same as the first step of phomactatriene and taxadiene biosynthesis [207]. 1,2-Rearrangements of *β*-methyl and *α*-hydride occurred at the six-membered ring part, then, the ring inversion and cyclization progressed to form the 6/5/9 ring intermediate. A rearrangement proceeded to expand the ring from five to six membered, and the last step resulted in the formation of the 6/5/6/6 ring skeleton. The C-8 position of wickerol A **225** was oxidized by a cytochrome P450 to give wickerol B **226**.

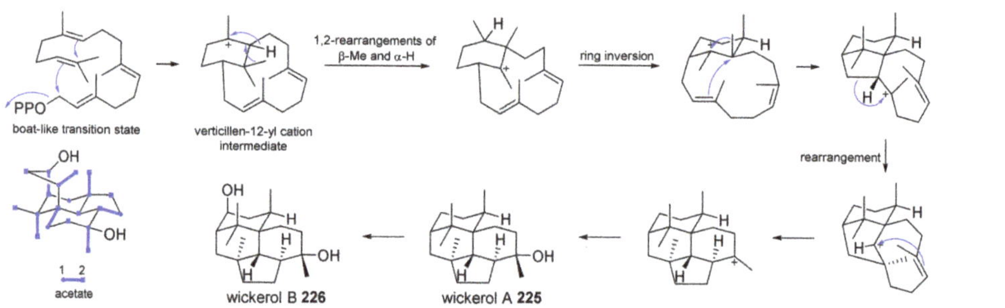

Scheme 14. Incorporation patterns of [1-^{13}C]-, [2-^{13}C]-, and [1,2-$^{13}C_2$]-acetates enriched wickerol B **226**, and proposed mechanism of cyclization from GGPP to wickerols [206].

Trichodermanins C–H **227**–**232** are new diterpenes with a 6/5/6/6 tetracyclic system that have been isolated from the marine sponge-derived fungus *Trichoderma harzianum* [208,209]. Trichodermanin C **227** potently inhibited the growth of murine P388 leukemia, human HL-60 leukemia, and murine L1210 leukemia cell lines with IC_{50} values of 7.9, 6.8, and 7.6 μM, respectively [208].

12. Conclusions and Future Prospects

Diterpenoids show huge potential for drug discovery and development due to their extensive biological functions and structural diversity. Fungal diterpenoids are a diverse family of hybrid natural products with potent bioactivities and intriguing structural architectures. A large number of fungal diterpenoids have exhibited significant anti-inflammatory, cytotoxic, anti-MRSA, antimicrobial, antiviral, antihypertensive, and platelet aggregation-inhibitory activities. Consequently, these bioactive diterpenoids are always hot trending topics for the synthesis community [173,174,186]. Nevertheless, the structural complexity and limited availability of natural products remain obstacles to synthesizing a large collection of natural products and their structural analogues in sufficient amounts. Thus, a synthetic biology method based on the combination of heterologous biosynthesis and genome mining is a promising approach to translate enormous amounts of biosynthetic gene information to richly diverse natural products. Interestingly, while fungi have evolved their systems to create terpenoid diversity, they have also biosynthesized some of the same classes of terpenoids found in plants, bacteria, and other organisms. These relationships provide accessible and renewable prokaryotic systems for eukaryotic natural product biosynthesis and enzymology. In conclusion, we hope it is evident from this review that most of the fungal diterpenoids are biologically active with a few key scaffolds paving a path towards potential drug discovery and development.

Author Contributions: Conceptualization, T.F.; discussion of the contents, F.-L.Z., and T.F.; writing—original draft preparation, F.-L.Z.; writing—review and editing, and editing, T.F. All authors have read and agreed to the published version of the manuscript.

Funding: This research was funded by the Natural Science Foundation of China (grant number 81872762).

Institutional Review Board Statement: Not applicable.

Informed Consent Statement: Not applicable.

Data Availability Statement: Not applicable.

Conflicts of Interest: The authors declare no conflict of interest.

References

1. Blackwell, M. The fungi: 1, 2, 3 ... 5.1 million species? *Am. J. Bot.* **2011**, *98*, 426–438. [CrossRef]
2. Spiteller, P. Chemical ecology of fungi. *Nat. Prod. Rep.* **2015**, *32*, 971–993. [CrossRef] [PubMed]
3. Dictionary of Natural Products. 2020. Available online: http://dnp.chemnetbase.com (accessed on 26 August 2021).
4. Buckingham, J.; Cooper, C.M.; Purchase, R. *Natural Products Desk Reference*; CRC Press: Boca Raton, FL, USA, 2016; pp. 1–219.
5. Christianson, D.W. Structural and chemical biology of terpenoid cyclases. *Chem. Rev.* **2017**, *117*, 11570–11648. [CrossRef]
6. Quin, M.B.; Flynn, C.M.; Schmidt-Dannert, C. Traversing the fungal terpenome. *Nat. Prod. Rep.* **2014**, *31*, 1449–1473. [CrossRef]
7. Minami, A.; Ozaki, T.; Liu, C.; Oikawa, H. Cyclopentane-forming di/sesterterpene synthases: Widely distributed enzymes in bacteria, fungi, and plants. *Nat. Prod. Rep.* **2018**, *35*, 1330–1346. [CrossRef] [PubMed]
8. Mitsuhashi, T.; Abe, I. Chimeric terpene synthases possessing both terpene cyclization and prenyltransfer activities. *ChemBioChem* **2018**, *19*, 1106–1114. [CrossRef] [PubMed]
9. Dickschat, J.S. Bacterial diterpene biosynthesis. *Angew. Chem. Int. Ed.* **2019**, *58*, 15964–15976. [CrossRef] [PubMed]
10. Mafu, S.; Zerbe, P. Plant diterpenoid metabolism for manufacturing the biopharmaceuticals of tomorrow: Prospects and challenges. *Phytochem. Rev.* **2018**, *17*, 113–130. [CrossRef]
11. Rudolf, J.D.; Alsup, T.A.; Xu, B.; Li, Z. Bacterial terpenome. *Nat. Prod. Rep.* **2020**, *38*, 905–980. [CrossRef]
12. Pemberton, T.A.; Chen, M.; Harris, G.G.; Chou, W.K.; Duan, L.; Koksal, M.; Genshaft, A.S.; Cane, D.E.; Christianson, D.W. Exploring the influence of domain architecture on the catalytic function of diterpene synthases. *Biochemistry* **2017**, *56*, 2010–2023. [CrossRef] [PubMed]
13. Keller, N.P.; Turner, G.; Bennett, J.W. Fungal secondary metabolism—From biochemistry to genomics. *Nat. Rev. Microbiol.* **2005**, *3*, 937–947. [CrossRef] [PubMed]
14. Bills, G.F.; Gloer, J.B. Biologically active secondary metabolites from the fungi. *Microbiol. Spectr.* **2016**, *4*, 1087–1119. [CrossRef] [PubMed]
15. Sanchez, J.F.; Somoza, A.D.; Keller, N.P.; Wang, C.C.C. Advances in *Aspergillus* secondary metabolite research in the post-genomic era. *Nat. Prod. Rep.* **2012**, *29*, 351–371. [CrossRef]
16. Fischer, M.J.; Rustenhloz, C.; Leh-Louis, V.; Perriere, G. Molecular and functional evolution of the fungal diterpene synthase genes. *BMC Microbiol.* **2015**, *15*, 221. [CrossRef]
17. Rokas, A.; Mead, M.E.; Steenwyk, J.L.; Raja, H.A.; Oberlies, N.H. Biosynthetic gene clusters and the evolution of fungal chemodiversity. *Nat. Prod. Rep.* **2020**, *37*, 868–878. [CrossRef]
18. Gressler, M.; Löhr, N.A.; Schäfer, T.; Lawrinowitz, S.; Seibold, P.S.; Hoffmeister, D. Mind the mushroom: Natural product biosynthetic genes and enzymes of Basidiomycota. *Nat. Prod. Rep.* **2021**, *38*, 702–722. [CrossRef]
19. Bailly, C.; Gao, J.M. Erinacine A and related cyathane diterpenoids: Molecular diversity and mechanisms underlying their neuroprotection and anticancer activities. *Pharmacol. Res.* **2020**, *159*, 104953. [CrossRef] [PubMed]
20. Tang, H.Y.; Yin, X.; Zhang, C.C.; Jia, Q.; Gao, J.M. Structure diversity, synthesis, and biological activity of cyathane diterpenoids in higher fungi. *Curr. Med. Chem.* **2015**, *22*, 2375–2391. [CrossRef]
21. Laber, G.; Schütze, E. *In Vivo* efficacy of 81.723 hfu, a new pleuromutilin derivative against experimentally induced airsacculitis in chicks and turkey poults. *Antimicrob. Agents Chemother.* **1975**, *7*, 517–521. [CrossRef]
22. Burch, D.G.; Jones, G.T.; Heard, T.W.; Tuck, R.E. The synergistic activity of tiamulin and chlortetracycline: In-feed treatment of bacterially complicated enzootic pneumonia in fattening pigs. *Vet. Rec.* **1986**, *119*, 108–112. [CrossRef]
23. Stipkovits, L.; Ripley, P.H.; Tenk, M.; Glávits, R.; Molnár, T.; Fodor, L. The efficacy of valnemulin (Econor®) in the control of disease caused by experimental infection of calves with *Mycoplasma bovis*. *Res. Vet. Sci.* **2005**, *78*, 207–215. [CrossRef] [PubMed]

24. Rittenhouse, S.; Biswas, S.; Broskey, J.; McCloskey, L.; Moore, T.; Vasey, S.; West, J.; Zalacain, M.; Zonis, R.; Payne, D. Selection of retapamulin, a novel pleuromutilin for topical use. *Antimicrob. Agents Chemother.* **2006**, *50*, 3882–3885. [CrossRef] [PubMed]
25. Zhao, D.H.; Zhang, Z.; Zhang, C.Y.; Liu, Z.C.; Deng, H.; Yu, J.J.; Guo, J.P.; Liu, Y.H. Population pharmacokinetics of valnemulin in swine. *J. Vet. Pharmacol. Ther.* **2014**, *37*, 59–65. [CrossRef]
26. Zhang, L.; Fasoyin, O.E.; Molnár, I.; Xu, Y. Secondary metabolites from hypocrealean entomopathogenic fungi: Novel bioactive compounds. *Nat. Prod. Rep.* **2020**, *37*, 1181–1206. [CrossRef] [PubMed]
27. Kuang, Y.; Li, B.; Wang, Z.; Qiao, X.; Ye, M. Terpenoids from the medicinal mushroom *Antrodia camphorata*: Chemistry and medicinal potential. *Nat. Prod. Rep.* **2021**, *38*, 83–102. [CrossRef] [PubMed]
28. Jiang, M.H.; Wu, Z.E.; Guo, H.; Liu, L.; Chen, S.H. A review of terpenes from marine-derived fungi: 2015–2019. *Mar. Drugs* **2020**, *18*, 321. [CrossRef]
29. Ran, H.; Li, S.M. Fungal benzene carbaldehydes: Occurrence, structural diversity, activities and biosynthesis. *Nat. Prod. Rep.* **2021**, *38*, 240–263. [CrossRef]
30. Zhu, M.Z.; Cen, Y.F.; Ye, W.; Li, S.N.; Zhang, W.M. Recent advances on macrocyclic trichothecenes, their bioactivities and biosynthetic pathway. *Toxins* **2020**, *12*, 417. [CrossRef]
31. Proctor, R.H.; McCormick, S.P.; Gutierrez, S. Genetic bases for variation in structure and biological activity of trichothecene toxins produced by diverse fungi. *Appl. Microbiol. Biotechnol.* **2020**, *104*, 5185–5199. [CrossRef]
32. Cadelis, M.M.; Copp, B.R.; Wiles, S. A review of fungal protoilludane sesquiterpenoid natural products. *Antibiotics* **2020**, *9*, 928. [CrossRef]
33. El-Demerdash, A.; Kumla, D.; Kijjoa, A. Chemical diversity and biological activities of meroterpenoids from marine derived-fungi: A comprehensive update. *Mar. Drugs* **2020**, *18*, 317. [CrossRef]
34. Zhao, M.; Tang, Y.; Xie, J.; Zhao, Z.; Cui, H. Meroterpenoids produced by fungi: Occurrence, structural diversity, biological activities, and their molecular targets. *Eur. J. Med. Chem.* **2021**, *209*, 112860. [CrossRef] [PubMed]
35. Jiang, M.; Wu, Z.; Liu, L.; Chen, S. The chemistry and biology of fungal meroterpenoids (2009–2019). *Org. Biomol. Chem.* **2020**, *19*, 1644–1704. [CrossRef] [PubMed]
36. Barra, L.; Abe, I. Chemistry of fungal meroterpenoid cyclases. *Nat. Prod. Rep.* **2020**, *38*, 566–585. [CrossRef] [PubMed]
37. Zhao, Y.; Cui, J.; Liu, M.Y.J.; Zhao, L. Progress on terpenoids with biological activities produced by plant endophytic fungi in China between 2017 and 2019. *Nat. Prod. Commun.* **2020**, *15*, 1934578X20937204. [CrossRef]
38. Zhang, L.; Yue, Q.; Wang, C.; Xu, Y.; Molnár, I. Secondary metabolites from hypocrealean entomopathogenic fungi: Genomics as a tool to elucidate the encoded parvome. *Nat. Prod. Rep.* **2020**, *37*, 1164–1180. [CrossRef]
39. Lyu, H.N.; Liu, H.W.; Keller, N.P.; Yin, W.B. Harnessing diverse transcriptional regulators for natural product discovery in fungi. *Nat. Prod. Rep.* **2020**, *37*, 6–16. [CrossRef]
40. Hanson, J.R.; Nichols, T.; Mukhrish, Y.; Bagley, M.C. Diterpenoids of terrestrial origin. *Nat. Prod. Rep.* **2019**, *36*, 1499–1512. [CrossRef]
41. Friedman, M. Chemistry, nutrition, and health-promoting properties of *Hericium erinaceus* (Lion's Mane) mushroom fruiting bodies and mycelia and their bioactive compounds. *J. Agric. Food Chem.* **2015**, *63*, 7108–7123. [CrossRef]
42. Enquist, J.A.; Stoltz, B.M. Synthetic efforts toward cyathane diterpenoid natural products. *Nat. Prod. Rep.* **2009**, *26*, 661–680. [CrossRef]
43. Kim, K.; Cha, J.K. Total synthesis of cyathin A_3 and cyathin B_2. *Angew. Chem. Int. Ed.* **2009**, *48*, 5334–5336. [CrossRef] [PubMed]
44. Kanoh, N.; Sakanishi, K.; Iimori, E.; Nishimura, K.; Iwabuchi, Y. Asymmetric total synthesis of (−)-scabronine G via intramolecular double Michael reaction and Prins cyclization. *Org. Lett.* **2011**, *13*, 2864–2867. [CrossRef]
45. Kobayakawa, Y.; Nakada, M. Enantioselective total synthesis of (−)-cyathin B_2. *J. Antibiot.* **2014**, *67*, 483–485. [CrossRef]
46. Nakada, M. Enantioselective total syntheses of cyathane diterpenoids. *Chem. Rec.* **2014**, *14*, 641–662. [CrossRef]
47. Wu, G.J.; Zhang, Y.H.; Tan, D.X.; He, L.; Cao, B.C.; He, Y.P.; Han, F.S. Synthetic studies on enantioselective total synthesis of cyathane diterpenoids: Cyrneines A and B, glaucopine C, and (+)-allocyathin B_2. *J. Org. Chem.* **2019**, *84*, 3223–3238. [CrossRef] [PubMed]
48. Marcos, I.S.; Moro, R.F.; Gil-Mesón, A.; Díez, D. Chapter 5. 7-6-5 Tricarbocyclic diterpenes: Valparanes, mulinanes, cyathanes, homoverrucosanes, and related ones. In *Studies in Natural Products Chemistry*; ur Rahman, A., Ed.; Elsevier Inc.: Amsterdam, The Netherlands, 2016; Volume 48, pp. 137–207.
49. De Jesus Dzul-Beh, A.; Uc-Cachon, A.H.; Borquez, J.; Loyola, L.A.; Pena-Rodriguez, L.M.; Molina-Salinas, G.M. Mulinane- and azorellane-type diterpenoids: A systematic review of their biosynthesis, chemistry, and pharmacology. *Biomolecules* **2020**, *10*, 1333. [CrossRef] [PubMed]
50. Sennett, S.H.; Pompeni, S.A.; Wright, A.E. Diterpene metabolites from two chemotypes of the marine sponge *Myrmekioderma styx*. *J. Nat. Prod.* **1992**, *55*, 1421–1429. [CrossRef] [PubMed]
51. Green, D.; Goldberg, I.; Stein, Z.; Ilan, M.; Kashman, Y. Cyanthiwigin A–D, novel cytotoxic diterpenes from the sponge *Epipolasis reiswigi*. *Nat. Prod. Lett.* **1992**, *1*, 193–199. [CrossRef]
52. Peng, J.; Walsh, K.; Weedman, V.; Bergthold, J.D.; Lynch, J.; Lieu, K.L.; Braude, I.A.; Kelly, M.; Hamann, M.T. The new bioactive diterpenes cyanthiwigins E–AA from the Jamaican sponge *Myrmekioderma styx*. *Tetrahedron* **2002**, *58*, 7809–7819. [CrossRef]
53. Pfeiffer, M.W.B.; Phillips, A.J. Conversion of cyanthiwigin U to related cyanthiwigins: Total syntheses of cyanthiwigin W and cyanthiwigin Z. *Tetrahedron Lett.* **2008**, *49*, 6860–6861. [CrossRef]
54. Miller, L.C.; Ndungu, J.M.; Sarpong, R. Parallel kinetic resolution approach to the cyathane and cyanthiwigin diterpenes using a cyclopropanation/cope rearrangement. *Angew. Chem. Int. Ed.* **2009**, *48*, 2398–2402. [CrossRef] [PubMed]

55. Enquist, J.A., Jr.; Virgil, S.C.; Stoltz, B.M. Total syntheses of cyanthiwigins B, F, and G. *Chem. Eur. J.* **2011**, *17*, 9957–9969. [CrossRef] [PubMed]
56. Wang, C.; Wang, D.; Gao, S. Total synthesis of cyanthiwigins A, C, G, and H. *Org. Lett.* **2013**, *15*, 4402–4405. [CrossRef]
57. Kim, K.E.; Stoltz, B.M. A second-generation synthesis of the cyanthiwigin natural product core. *Org. Lett.* **2016**, *18*, 5720–5723. [CrossRef]
58. Chang, Y.; Shi, L.; Huang, J.; Shi, L.; Zhang, Z.; Hao, H.D.; Gong, J.; Yang, Z. Stereoselective total synthesis of (±)-5-*epi*-cyanthiwigin I via an intramolecular Pauson–Khand reaction as the key step. *Org. Lett.* **2018**, *20*, 2876–2879. [CrossRef]
59. Kim, K.E.; Adams, A.M.; Chiappini, N.D.; Du Bois, J.; Stoltz, B.M. Cyanthiwigin natural product core as a complex molecular scaffold for comparative late-stage C–H functionalization studies. *J. Org. Chem.* **2018**, *83*, 3023–3033. [CrossRef] [PubMed]
60. Cao, C.Y.; Zhang, C.C.; Shi, X.W.; Li, D.; Cao, W.; Yin, X.; Gao, J.M. Sarcodonin G derivatives exhibit distinctive effects on neurite outgrowth by modulating NGF signaling in PC12 cells. *ACS Chem. Neurosci.* **2018**, *9*, 1607–1615. [CrossRef]
61. Dixon, E.; Schweibenz, T.; Hight, A.; Kang, B.; Dailey, A.; Kim, S.; Chen, M.Y.; Kim, Y.; Neale, S.; Groth, A.; et al. Bacteria-induced static batch fungal fermentation of the diterpenoid cyathin A_3, a small-molecule inducer of nerve growth factor. *J. Ind. Microbiol. Biotechnol.* **2011**, *38*, 607–615. [CrossRef]
62. Nei, M.; Kumar, S. *Molecular Evolution and Phylogenetics*; Oxford University Press: New York, NY, USA, 2000.
63. Felsenstein, J. Confidence limits on phylogenies: An approach using the bootstrap. *Evolution* **1985**, *39*, 783–791. [CrossRef]
64. Kumar, S.; Stecher, G.; Li, M.; Knyaz, C.; Tamura, K. MEGA X: Molecular evolutionary genetics analysis across computing platforms. *Mol. Biol. Evol.* **2018**, *35*, 1547–1549. [CrossRef]
65. Cheng, Z.; Li, Y.; Xu, W.; Liu, W.; Liu, L.; Zhu, D.; Kang, Y.; Luo, Z.; Li, Q. Three new cyclopiane-type diterpenes from a deep-sea derived fungus *Penicillium* sp. YPGA11 and their effects against human esophageal carcinoma cells. *Bioorg. Chem.* **2019**, *91*, 103129. [CrossRef] [PubMed]
66. Roncal, T.; Cordobés, S.; Ugalde, U.; He, Y.; Sterner, O. Novel diterpenes with potent conidiation inducing activity. *Tetrahedron Lett.* **2002**, *43*, 6799–6802. [CrossRef]
67. Rodriíguez, I.I.; Rodriíguez, A.D.; Zhao, H. Aberrarone: A gorgonian-derived diterpene from *Pseudopterogorgia elisabethae*. *J. Org. Chem.* **2009**, *74*, 7581–7584. [CrossRef]
68. Niu, S.; Fan, Z.; Tang, X.; Liu, Q.; Shao, Z.; Liu, G.; Yang, X.W. Cyclopiane-type diterpenes from the deep-sea-derived fungus *Penicillium commune* MCCC 3A00940. *Tetrahedron Lett.* **2018**, *59*, 375–378. [CrossRef]
69. Mitsuhashi, T.; Kikuchi, T.; Hoshino, S.; Ozeki, M.; Awakawa, T.; Shi, S.P.; Fujita, M.; Abe, I. Crystalline sponge method enabled the investigation of a prenyltransferase-terpene synthase chimeric enzyme, whose product exhibits broadened NMR signals. *Org. Lett.* **2018**, *20*, 5606–5609. [CrossRef] [PubMed]
70. Roncal, T.; Cordobes, S.; Sterner, O.; Ugalde, U. Conidiation in *Penicillium cyclopium* is induced by conidiogenone, an endogenous diterpene. *Eukaryot. Cell* **2002**, *1*, 823–829. [CrossRef]
71. Shiina, T.; Nakagawa, K.; Fujisaki, Y.; Ozaki, T.; Liu, C.; Toyomasu, T.; Hashimoto, M.; Koshino, H.; Minami, A.; Kawaide, H.; et al. Biosynthetic study of conidiation-inducing factor conidiogenone: Heterologous production and cyclization mechanism of a key bifunctional diterpene synthase. *Biosci. Biotechnol. Biochem.* **2019**, *83*, 192–201. [CrossRef]
72. Du, L.; Li, D.; Zhu, T.; Cai, S.; Wang, F.; Xiao, X.; Gu, Q. New alkaloids and diterpenes from a deep ocean sediment derived fungus *Penicillium* sp. *Tetrahedron* **2009**, *65*, 1033–1039. [CrossRef]
73. Gao, S.S.; Li, X.M.; Zhang, Y.; Li, C.S.; Wang, B.G. Conidiogenones H and I, two new diterpenes of cyclopiane class from a marine-derived endophytic fungus *Penicillium chrysogenum* QEN-24S. *Chem. Biodivers.* **2011**, *8*, 1748–1753. [CrossRef]
74. Hou, S.H.; Tu, Y.Q.; Wang, S.H.; Xi, C.C.; Zhang, F.M.; Wang, S.H.; Li, Y.T.; Liu, L. Total syntheses of the tetracyclic cyclopiane diterpenes conidiogenone, conidiogenol, and conidiogenone B. *Angew. Chem. Int. Ed.* **2016**, *128*, 4532–4536. [CrossRef]
75. Li, F.; Sun, W.; Zhang, S.; Gao, W.; Lin, S.; Yang, B.; Chai, C.; Li, H.; Wang, J.; Hu, Z.; et al. New cyclopiane diterpenes with anti-inflammatory activity from the sea sediment-derived fungus *Penicillium* sp. TJ403-2. *Chin. Chem. Lett.* **2020**, *31*, 197–201. [CrossRef]
76. Chen, H.Y.; Liu, T.K.; Shi, Q.; Yang, X.L. Sesquiterpenoids and diterpenes with antimicrobial activity from *Leptosphaeria* sp. XL026, an endophytic fungus in *Panax notoginseng*. *Fitoterapia* **2019**, *137*, 104243. [CrossRef]
77. De Boer, A.H.; De Vries-Van Leeuwen, I.J. Fusicoccanes: Diterpenes with surprising biological functions. *Trends Plant Sci.* **2012**, *17*, 360–368. [CrossRef] [PubMed]
78. Hu, Z.; Sun, W.; Li, F.; Guan, J.; Lu, Y.; Liu, J.; Tang, Y.; Du, G.; Xue, Y.; Luo, Z.; et al. Fusicoccane-derived diterpenoids from *Alternaria brassicicola*: Investigation of the structure-stability relationship and discovery of an IKKβ inhibitor. *Org. Lett.* **2018**, *20*, 5198–5202. [CrossRef]
79. Li, F.L.; Lin, S.; Zhang, S.T.; Hao, X.C.; Li, X.N.; Yang, B.Y.; Liu, J.J.; Wang, J.P.; Hu, Z.X.; Zhang, Y.H. Alterbrassinoids A–D: Fusicoccane-derived diterpenoid dimers featuring different carbon skeletons from *Alternaria brassicicola*. *Org. Lett.* **2019**, *21*, 8353–8357. [CrossRef]
80. Zhang, M.; Yan, S.; Liang, Y.; Zheng, M.; Wu, Z.; Zang, Y.; Yu, M.; Sun, W.; Liu, J.; Ye, Y.; et al. Talaronoids A–D: Four fusicoccane diterpenoids with an unprecedented tricyclic 5/8/6 ring system from the fungus *Talaromyces stipitatus*. *Org. Chem. Front.* **2020**, *7*, 3486–3492. [CrossRef]
81. Ballio, A.; Chain, E.B.; De Leo, P.; Erlanger, B.F.; Mauri, M.; Tonolo, A. Fusicoccin: A new wilting toxin produced by *Fusicoccum amygdali* Del. *Nature* **1964**, *203*, 297. [CrossRef]

82. Aoyagi, T.; Aoyama, T.; Kojima, F.; Hattori, S.; Honma, Y.; Hamada, M.; Takeuch, T. Cyclooctatin, a new inhibitor of lysophospholipase, produced by *Streptomyces melanosporofaciens* MI614-43F2. Taxonomy, production, isolation, physico-chemical properties and biological activities. *J. Antibiot.* **1992**, *45*, 1587–1591. [CrossRef]
83. Muromtsev, G.S.; Voblikova, V.D.; Kobrina, N.S.; Koreneva, V.M.; Krasnopolskaya, L.M.; Sadovskaya, V.L. Occurrence of fusicoccanes in plants and fungi. *J. Plant Growth Regul.* **1994**, *13*, 39–49. [CrossRef]
84. Rasoamiaranjanahary, L.; Marston, A.; Guilet, D.; Schenk, K.; Randimbivololona, F.; Hostettmann, K. Antifungal diterpenes from *Hypoestes serpens* (Acanthaceae). *Phytochemistry* **2003**, *62*, 333–337. [CrossRef]
85. Komala, I.; Ito, T.; Nagashima, F.; Yagi, Y.; Kawahata, M.; Yamaguchi, K.; Asakawa, Y. Zierane sesquiterpene lactone, cembrane and fusicoccane diterpenoids, from the Tahitian liverwort *Chandonanthus hirtellus*. *Phytochemistry* **2010**, *71*, 1387–1394. [CrossRef] [PubMed]
86. Gilabert, M.; Ramos, A.N.; Schiavone, M.A.M.; Arena, M.E.; Bardoón, A. Bioactive sesqui- and diterpenoids from the Argentine liverwort *Porella chilensis*. *J. Nat. Prod.* **2011**, *74*, 574–579. [CrossRef] [PubMed]
87. Kawamura, A.; Iacovidou, M.; Hirokawa, E.; Soll, C.E.; Trujillo, M. 17-Hydroxycyclooctatin, a fused 5–8–5 ring diterpene, from *Streptomyces* sp. MTE4a. *J. Nat. Prod.* **2011**, *74*, 492–495. [CrossRef] [PubMed]
88. Kenmoku, H.; Tada, H.; Oogushi, M.; Esumi, T.; Takahashi, H.; Noji, M.; Sassa, T.; Toyota, M.; Asakawa, Y. Seed dormancy breaking diterpenoids from the liverwort *Plagiochila sciophila* and their differentiation inducing activity in human promyelocytic leukemia HL-60 cells. *Nat. Prod. Commun.* **2014**, *9*, 915–920. [CrossRef]
89. Takahashi, M.; Kawamura, A.; Kato, N.; Nishi, T.; Hamachi, I.; Ohkanda, J. Phosphopeptide-dependent labeling of 14–3–3ζ proteins by fusicoccin-based fluorescent probes. *Angew. Chem. Int. Ed.* **2012**, *51*, 509–512. [CrossRef]
90. Wang, W.; Wan, X.; Liu, J.; Wang, J.; Zhu, H.; Chen, C.; Zhang, Y. Two new terpenoids from *Talaromyces purpurogenus*. *Mar. Drugs* **2018**, *16*, 150. [CrossRef]
91. Takekawa, H.; Tanaka, K.; Fukushi, E.; Matsuo, K.; Nehira, T.; Hashimoto, M. Roussoellols A and B, tetracyclic fusicoccanes from *Roussoella hysterioides*. *J. Nat. Prod.* **2013**, *76*, 1047–1051. [CrossRef]
92. Aoyama, T.; Naganawa, H.; Muraoka, Y.; Aoyagi, T.; Takeuchi, T. The structure of cyclooctatin, a new inhibitor of lysophospholipase. *J. Antibiot.* **1992**, *45*, 1703–1704. [CrossRef]
93. Zheng, D.; Han, L.; Qu, X.D.; Chen, X.; Zhong, J.L.; Bi, X.X.; Liu, J.; Jiang, Y.; Jiang, C.L.; Huang, X.S. Cytotoxic fusicoccane-type diterpenoids from *Streptomyces violascens* isolated from *Ailuropoda melanoleuca* feces. *J. Nat. Prod.* **2017**, *80*, 837–844. [CrossRef]
94. Mackinnon, S. Components from the phytotoxic extract of *Alternaria brassicicola*, a black spot pathogen of canola. *Phytochemistry* **1999**, *51*, 215–221. [CrossRef]
95. Pedras, M.S.; Chumala, P.B.; Jin, W.; Islam, M.S.; Hauck, D.W. The phytopathogenic fungus *Alternaria brassicicola*: Phytotoxin production and phytoalexin elicitation. *Phytochemistry* **2009**, *70*, 394–402. [CrossRef] [PubMed]
96. Kenmoku, H.; Takeue, S.; Oogushi, M.; Yagi, Y.; Sassa, T.; Toyota, M.; Asakawa, Y. Seed dormancy breaking diterpenoids, including novel brassicicenes J and K, from fungus *Alternaria brassicicola*, and their necrotic/apoptotic activities in HL-60 cells. *Nat. Prod. Commun.* **2014**, *9*, 351–354. [CrossRef] [PubMed]
97. Tang, Y.; Xue, Y.; Du, G.; Wang, J.; Liu, J.; Sun, B.; Li, X.N.; Yao, G.; Luo, Z.; Zhang, Y. Structural revisions of a class of natural products: Scaffolds of aglycon analogues of fusicoccins and cotylenins isolated from fungi. *Angew. Chem. Int. Ed.* **2016**, *55*, 4069–4073. [CrossRef] [PubMed]
98. Li, F.; Sun, W.; Guan, J.; Lu, Y.; Zhang, S.; Lin, S.; Liu, J.; Gao, W.; Wang, J.; Hu, Z.; et al. Alterbrassicicene A, a highly transformed fusicoccane-derived diterpenoid with potent PPAR-γ agonistic activity from *Alternaria brassicicola*. *Org. Lett.* **2018**, *20*, 7982–7986. [CrossRef] [PubMed]
99. Li, F.; Sun, W.; Guan, J.; Lu, Y.; Lin, S.; Zhang, S.; Gao, W.; Liu, J.; Du, G.; Wang, J.; et al. Anti-inflammatory fusicoccane-type diterpenoids from the phytopathogenic fungus *Alternaria brassicicola*. *Org. Biomol. Chem.* **2018**, *16*, 8751–8760. [CrossRef]
100. Li, F.; Lin, S.; Zhang, S.; Pan, L.; Chai, C.; Su, J.C.; Yang, B.; Liu, J.; Wang, J.; Hu, Z.; et al. Modified fusicoccane-type diterpenoids from *Alternaria brassicicola*. *J. Nat. Prod.* **2020**, *83*, 1931–1938. [CrossRef]
101. Li, F.L.; Pan, L.F.; Lin, S.; Zhang, S.T.; Li, H.Q.; Yang, B.Y.; Liu, J.J.; Wang, J.P.; Hu, Z.X.; Zhang, Y.H. Fusicoccane-derived diterpenoids with bridgehead double-bond-containing tricyclo[9.2.1.0^{3,7}]tetradecane ring systems from *Alternaria brassicicola*. *Bioorg. Chem.* **2020**, *100*, 103887. [CrossRef]
102. Bie, Q.; Chen, C.M.; Yu, M.Y.; Guo, J.R.; Wang, J.P.; Liu, J.J.; Zhou, Y.; Zhu, H.C.; Zhang, Y.H. Dongtingnoids A–G: Fusicoccane diterpenoids from a *Penicillium* species. *J. Nat. Prod.* **2019**, *82*, 80–86. [CrossRef]
103. Harada, J.; Tanaka, T.; Sassa, T. Sprouting of dormant tubers of *Sagittaria trifolia*, a perennial paddy weed, caused by cotylenin E, a new plant growth regulator. *J. Weed Sci. Technol.* **1981**, *26*, 37–39. [CrossRef]
104. Takeuchi, Y.; Sassa, T.; Kawaguchi, S.; Ogasawara, M.; Yoneyama, K.; Konnai, M. Stimulation of germination of *Monochoria vaginalis* seeds by seed coat puncture and cotylenins. *J. Weed Sci. Technol.* **1995**, *40*, 221–224. [CrossRef]
105. Noike, M.; Ono, Y.; Araki, Y.; Tanio, R.; Higuchi, Y.; Nitta, H.; Hamano, Y.; Toyomasu, T.; Sassa, T.; Kato, N.; et al. Molecular breeding of a fungus producing a precursor diterpene suitable for semi-synthesis by dissection of the biosynthetic machinery. *PLoS ONE* **2012**, *7*, e42090. [CrossRef]
106. Arens, J.; Engels, B.; Klopries, S.; Jennewein, S.; Ottmann, C.; Schulz, F. Exploration of biosynthetic access to the shared precursor of the fusicoccane diterpenoid family. *Chem. Commun.* **2013**, *49*, 4337–4339. [CrossRef] [PubMed]

107. Liang, X.R.; Miao, F.P.; Song, Y.P.; Guo, Z.Y.; Ji, N.Y. Trichocitrin, a new fusicoccane diterpene from the marine brown alga-endophytic fungus *Trichoderma citrinoviride* cf-27. *Nat. Prod. Res.* **2016**, *30*, 1605–1610. [CrossRef]
108. Wu, Y.H.; Chen, G.D.; He, R.R.; Wang, C.X.; Hu, D.; Wang, G.Q.; Guo, L.D.; Yao, X.S.; Gao, H. Pericolactines A–C, a new class of diterpenoid alkaloids with unusual tetracyclic skeleton. *Sci. Rep.* **2015**, *5*, 17082. [CrossRef] [PubMed]
109. Toyomasu, T.; Tsukahara, M.; Kaneko, A.; Niida, R.; Mitsuhashi, W.; Dairi, T.; Kato, N.; Sassa, T. Fusicoccins are biosynthesized by an unusual chimera diterpene synthase in fungi. *Proc. Natl. Acad. Sci. USA* **2007**, *104*, 3084–3088. [CrossRef]
110. Minami, A.; Tajima, N.; Higuchi, Y.; Toyomasu, T.; Sassa, T.; Kato, N.; Dairi, T. Identification and functional analysis of brassicicene C biosynthetic gene cluster in *Alternaria brassicicola*. *Bioorg. Med. Chem. Lett.* **2009**, *19*, 870–874. [CrossRef]
111. Hashimoto, M.; Higuchi, Y.; Takahashi, S.; Osada, H.; Sakaki, T.; Toyomasu, T.; Sassa, T.; Kato, N.; Dairi, T. Functional analyses of cytochrome P450 genes responsible for the early steps of brassicicene C biosynthesis. *Bioorg. Med. Chem. Lett.* **2009**, *19*, 5640–5643. [CrossRef] [PubMed]
112. Ono, Y.; Minami, A.; Noike, M.; Higuchi, Y.; Toyomasu, T.; Sassa, T.; Kato, N.; Dairi, T. Dioxygenases, key enzymes to determine the aglycon structures of fusicoccin and brassicicene, diterpene compounds produced by fungi. *J. Am. Chem. Soc.* **2011**, *133*, 2548–2555. [CrossRef]
113. Kim, S.Y.; Zhao, P.; Igarashi, M.; Sawa, R.; Tomita, T.; Nishiyama, M.; Kuzuyama, T. Cloning and heterologous expression of the cyclooctatin biosynthetic gene cluster afford a diterpene cyclase and two P450 hydroxylases. *Chem. Biol.* **2009**, *16*, 736–743. [CrossRef]
114. Kudo, F.; Matsuura, Y.; Hayashi, T.; Fukushima, M.; Eguchi, T. Genome mining of the sordarin biosynthetic gene cluster from *Sordaria araneosa* Cain ATCC 36386: Characterization of cycloaraneosene synthase and GDP-6-deoxyaltrose transferase. *J. Antibiot.* **2016**, *69*, 541–548. [CrossRef]
115. Chiba, R.; Minami, A.; Gomi, K.; Oikawa, H. Identification of ophiobolin F synthase by a genome mining approach: A sesterterpene synthase from *Aspergillus clavatus*. *Org. Lett.* **2013**, *15*, 594–597. [CrossRef] [PubMed]
116. Ye, Y.; Minami, A.; Mandi, A.; Liu, C.; Taniguchi, T.; Kuzuyama, T.; Monde, K.; Gomi, K.; Oikawa, H. Genome mining for sesterterpenes using bifunctional terpene synthases reveals a unified intermediate of di/sesterterpenes. *J. Am. Chem. Soc.* **2015**, *137*, 11846–11853. [CrossRef] [PubMed]
117. Tazawa, A.; Ye, Y.; Ozaki, T.; Liu, C.; Ogasawara, Y.; Dairi, T.; Higuchi, Y.; Kato, N.; Gomi, K.; Minami, A.; et al. Total biosynthesis of brassicicenes: Identification of a key enzyme for skeletal diversification. *Org. Lett.* **2018**, *20*, 6178–6182. [CrossRef]
118. Lin, F.L.; Lauterbach, L.; Zou, J.; Wang, Y.H.; Lv, J.M.; Chen, G.D.; Hu, D.; Gao, H.; Yao, X.S.; Dickschat, J.S. Mechanistic characterization of the fusicoccane-type diterpene synthase for myrothec-15(17)-en-7-ol. *ACS Catal.* **2020**, *10*, 4306–4312. [CrossRef]
119. Brady, S.F.; Singh, M.P.; Janso, J.E.; Clardy, J. Guanacastepene, a fungal-derived diterpene antibiotic with a new carbon skeleton. *J. Am. Chem. Soc.* **2000**, *122*, 2116–2117. [CrossRef]
120. Brady, S.F.; Bondi, S.M.; Clardy, J. The guanacastepenes: A highly diverse family of secondary metabolites produced by an endophytic fungus. *J. Am. Chem. Soc.* **2001**, *123*, 9900–9901. [CrossRef]
121. Dudley, G.B.; Danishefsky, S.J. A four-step synthesis of the hydroazulene core of guanacastepene. *Org. Lett.* **2001**, *3*, 2399–2402. [CrossRef]
122. Dudley, G.B.; Tan, D.S.; Kim, G.; Tanski, J.M.; Danishefsky, S.J. Remarkable stereoselectivity in the alkylation of a hydroazulenone: Progress towards the total synthesis of guanacastepene. *Tetrahedron Lett.* **2001**, *42*, 6789–6791. [CrossRef]
123. Mehta, G.; Umarye, J.D. Studies toward the total synthesis of diterpene antibiotic guanacastepene A: construction of the hydroazulenic core. *Org. Lett.* **2002**, *4*, 1063–1066. [CrossRef]
124. Lin, S.; Dudley, G.B.; Tan, D.S.; Danishefsky, S.J. A stereoselective route to guanacastepene A through a surprising epoxidation. *Angew. Chem. Int. Ed.* **2002**, *41*, 2188–2191. [CrossRef]
125. Brummond, K.M.; Gao, D. Unique strategy for the assembly of the carbon skeleton of guanacastepene A using an allenic Pauson–Khand-type reaction. *Org. Lett.* **2003**, *5*, 3491–3494. [CrossRef] [PubMed]
126. Du, X.; Chu, H.V.; Kwon, O. Synthesis of the [5–7–6] tricyclic core of guanacastepene A via an intramolecular Mukaiyama aldol reaction. *Org. Lett.* **2003**, *5*, 1923–1926. [CrossRef] [PubMed]
127. Hughes, C.C.; Kennedy-Smith, J.J.; Trauner, D. Synthetic studies toward the guanacastepenes. *Org. Lett.* **2003**, *5*, 4113–4115. [CrossRef] [PubMed]
128. Hughes, C.C.; Miller, A.K.; Trauner, D. An electrochemical approach to the guanacastepenes. *Org. Lett.* **2005**, *7*, 3425–3428. [CrossRef]
129. Battiste, M.A.; Pelphrey, P.M.; Wright, D.L. The cycloaddition strategy for the synthesis of natural products containing carbocyclic seven-membered rings. *Chem. Eur. J.* **2006**, *12*, 3438–3447. [CrossRef]
130. Iimura, S.; Overman, L.E.; Paulini, R.; Zakarian, A. Enantioselective total synthesis of guanacastepene N using an uncommon 7-endo Heck cyclization as a pivotal step. *J. Am. Chem. Soc.* **2006**, *128*, 13095–13101. [CrossRef]
131. Li, C.C.; Wang, C.H.; Liang, B.; Zhang, X.H.; Deng, L.J.; Liang, S.; Chen, J.H.; Wu, Y.D.; Yang, Z. Synthetic study of 1,3-butadiene-based IMDA approach to construct a [5–7–6] tricyclic core and its application to the total synthesis of C8-*epi*-guanacastepene O. *J. Org. Chem.* **2006**, *71*, 6892–6897. [CrossRef]
132. McGowan, C.A.; Schmieder, A.K.; Roberts, L.; Greaney, M.F. Synthesis of the guanacastepene A–B hydrazulene ring system through photochemical ring transposition. *Org. Biomol. Chem.* **2007**, *5*, 1522–1524. [CrossRef]

133. Michalak, K.; Michalak, M.; Wicha, J. Construction of the tricyclic 5-7-6 scaffold of fungi-derived diterpenoids. Total synthesis of (±)-heptemerone G and an approach to Danishefsky's intermediate for guanacastepene A synthesis. *J. Org. Chem.* **2010**, *75*, 8337–8350. [CrossRef]
134. Oonishi, Y.; Taniuchi, A.; Sato, Y. Rhodium(I)-catalyzed hydroacylation/cycloisomerization cascade reaction: Application to the construction of the tricyclic core of guanacastepenes. *Synthesis* **2010**, *2010*, 2884–2892. [CrossRef]
135. Gampe, C.M.; Carreira, E.M. Total syntheses of guanacastepenes N and O. *Angew. Chem. Int. Ed.* **2011**, *50*, 2962–2965. [CrossRef]
136. Michalak, K.; Michalak, M.; Wicha, J. A facile construction of the tricyclic 5-7-6 scaffold of fungi-derived diterpenoids. The first total synthesis of (±)-heptemerone G and a new approach to Danishefsky's intermediate for a guanacastepene A synthesis. *Tetrahedron Lett.* **2010**, *51*, 4344–4346. [CrossRef]
137. Kettering, M.; Valdivia, C.; Sterner, O.; Anke, H.; Thines, E. Heptemerones A–G, seven novel diterpenoids from *Coprinus heptemerus*: Producing organism, fermentation, isolation and biological activities. *J. Antibiot.* **2005**, *58*, 390–396. [CrossRef] [PubMed]
138. Valdivia, C.; Kettering, M.; Anke, H.; Thines, E.; Sterner, O. Diterpenoids from *Coprinus heptemerus*. *Tetrahedron* **2005**, *61*, 9527–9532. [CrossRef]
139. Ou, Y.X.; Li, Y.Y.; Qian, X.M.; Shen, Y.M. Guanacastane-type diterpenoids from *Coprinus radians*. *Phytochemistry* **2012**, *78*, 190–196. [CrossRef]
140. Liu, Y.Z.; Li, Y.Y.; Ou, Y.X.; Xiao, S.Y.; Lu, C.H.; Zheng, Z.H.; Shen, Y.M. Guanacastane-type diterpenoids with cytotoxic activity from *Coprinus plicatilis*. *Bioorg. Med. Chem. Lett.* **2012**, *22*, 5059–5062. [CrossRef]
141. Liu, Y.Z.; Lu, C.H.; Shen, Y.M. Guanacastane-type diterpenoids from *Coprinus plicatilis*. *Phytochem. Lett.* **2014**, *7*, 161–164. [CrossRef]
142. Yin, X.; Feng, T.; Li, Z.H.; Leng, Y.; Liu, J.K. Five new guanacastane-type diterpenes from cultures of the fungus *Psathyrella candolleana*. *Nat. Prod. Bioprospect.* **2014**, *4*, 149–155. [CrossRef]
143. Feng, Y.; Ren, F.X.; Niu, S.B.; Wang, L.; Li, L.; Liu, X.Z.; Che, Y.S. Guanacastane diterpenoids from the plant endophytic fungus *Cercospora* sp. *J. Nat. Prod.* **2014**, *77*, 873–881. [CrossRef]
144. Wu, F.B.; Li, T.X.; Yang, M.H.; Kong, L.Y. Guanacastane-type diterpenoids from the insect-associated fungus *Verticillium dahliae*. *J. Asian Nat. Prod. Res.* **2016**, *18*, 117–124. [CrossRef]
145. Zhang, H.; Yang, M.H.; Li, Y.; Cheng, X.B.; Pei, Y.H.; Kong, L.Y. Seven new guanacastane-type diterpenoids from the fungus *Verticillium dahliae*. *Fitoterapia* **2019**, *133*, 219–224. [CrossRef]
146. Lam, Y.T.H.; Palfner, G.; Lima, C.; Porzel, A.; Brandt, W.; Frolov, A.; Sultani, H.; Franke, K.; Wagner, C.; Merzweiler, K.; et al. *Nor*-guanacastepene pigments from the Chilean mushroom *Cortinarius pyromyxa*. *Phytochemistry* **2019**, *165*, 112048. [CrossRef] [PubMed]
147. Wang, X.; Jin, X.Y.; Zhou, J.C.; Zhu, R.X.; Qiao, Y.N.; Zhang, J.Z.; Li, Y.; Zhang, C.Y.; Chen, W.; Chang, W.Q.; et al. Terpenoids from the Chinese liverwort *Heteroscyphus coalitus* and their anti-virulence activity against *Candida albicans*. *Phytochemistry* **2020**, *174*, 112324. [CrossRef]
148. Ghisalberti, E.L.; Hockless, D.C.R.; Rowland, C.; White, A.H. Harziandione, a new class of diterpene from *Trichoderma harzianum*. *J. Nat. Prod.* **1992**, *55*, 1690–1694. [CrossRef]
149. Miao, F.P.; Liang, X.R.; Yin, X.L.; Wang, G.; Ji, N.Y. Absolute configurations of unique harziane diterpenes from *Trichoderma* species. *Org. Lett.* **2012**, *14*, 3815–3817. [CrossRef]
150. Mannina, L.; Segre, A.L.; Ritieni, A.; Fogliano, V.; Vinale, F.; Randazzo, G.; Maddau, L.; Bottalico, A. A new fungal growth inhibitor from *Trichoderma viride*. *Tetrahedron* **1997**, *53*, 3135–3144. [CrossRef]
151. Barra, L.; Dickschat, J.S. Harzianone biosynthesis by the biocontrol fungus *Trichoderma*. *ChemBioChem* **2017**, *18*, 2358–2365. [CrossRef]
152. Adelin, E.; Servy, C.; Martin, M.T.; Arcile, G.; Iorga, B.I.; Retailleau, P.; Bonfill, M.; Ouazzani, J. Bicyclic and tetracyclic diterpenes from a *Trichoderma* symbiont of *Taxus baccata*. *Phytochemistry* **2014**, *97*, 55–61. [CrossRef]
153. Hönig, M.; Carreira, E.M. Total synthesis and structural revision of a harziane diterpenoid. *Angew. Chem. Int. Ed.* **2020**, *59*, 1192–1196. [CrossRef]
154. Zhang, M.; Liu, J.M.; Zhao, J.L.; Li, N.; Chen, R.D.; Xie, K.B.; Zhang, W.J.; Feng, K.P.; Yan, Z.; Wang, N.; et al. Two new diterpenoids from the endophytic fungus *Trichoderma* sp. Xy24 isolated from mangrove plant *Xylocarpus granatum*. *Chin. Chem. Lett.* **2016**, *27*, 957–960. [CrossRef]
155. Song, Y.P.; Fang, S.T.; Miao, F.P.; Yin, X.L.; Ji, N.Y. Diterpenes and sesquiterpenes from the marine algicolous fungus *Trichoderma harzianum* X-5. *J. Nat. Prod.* **2018**, *81*, 2553–2559. [CrossRef]
156. Song, Y.P.; Liu, X.H.; Shi, Z.Z.; Miao, F.P.; Fang, S.T.; Ji, N.Y. Bisabolane, cyclonerane, and harziane derivatives from the marine-alga-endophytic fungus *Trichoderma asperellum* cf44-2. *Phytochemistry* **2018**, *152*, 45–52. [CrossRef] [PubMed]
157. Song, Y.P.; Miao, F.P.; Liang, X.R.; Yin, X.L.; Ji, N.Y. Harziane and cadinane terpenoids from the alga-endophytic fungus *Trichoderma asperellum* A-YMD-9-2. *Phytochem. Lett.* **2019**, *32*, 38–41. [CrossRef]
158. Zou, J.X.; Song, Y.P.; Ji, N.Y. Deoxytrichodermaerin, a harziane lactone from the marine algicolous fungus *Trichoderma longibrachiatum* A-WH-20-2. *Nat. Prod. Res.* **2021**, *35*, 216–221. [CrossRef]
159. Zhao, D.L.; Yang, L.J.; Shi, T.; Wang, C.Y.; Shao, C.L.; Wang, C.Y. Potent phytotoxic harziane diterpenes from a soft coral-derived strain of the fungus *Trichoderma harzianum* XS-20090075. *Sci. Rep.* **2019**, *9*, 13345. [CrossRef]
160. Shi, T.; Shao, C.L.; Liu, Y.; Zhao, D.L.; Cao, F.; Fu, X.M.; Yu, J.Y.; Wu, J.S.; Zhang, Z.K.; Wang, C.Y. Terpenoids From the coral-derived fungus *Trichoderma harzianum* (XS-20090075) induced by chemical epigenetic manipulation. *Front. Microbiol.* **2020**, *11*, 572. [CrossRef]

161. Li, W.Y.; Liu, Y.; Lin, Y.T.; Liu, Y.C.; Guo, K.; Li, X.N.; Luo, S.H.; Li, S.H. Antibacterial harziane diterpenoids from a fungal symbiont *Trichoderma atroviride* isolated from *Colquhounia coccinea* var. *mollis*. *Phytochemistry* **2020**, *170*, 112198. [CrossRef] [PubMed]
162. Zhang, M.; Liu, J.; Chen, R.; Zhao, J.; Xie, K.; Chen, D.; Feng, K.; Dai, J. Two furanharzianones with 4/7/5/6/5 ring system from microbial transformation of harzianone. *Org. Lett.* **2017**, *19*, 1168–1171. [CrossRef] [PubMed]
163. Zhang, M.; Liu, J.; Chen, R.; Zhao, J.; Xie, K.; Chen, D.; Feng, K.; Dai, J. Microbial oxidation of harzianone by *Bacillus* sp. IMM-006. *Tetrahedron* **2017**, *73*, 7195–7199. [CrossRef]
164. Toyomasu, T.; Kaneko, A.; Tokiwano, T.; Kanno, Y.; Kanno, Y.; Niida, R.; Miura, S.; Nishioka, T.; Ikeda, C.; Mitsuhashi, W.; et al. Biosynthetic gene-based secondary metabolite screening: A new diterpene, methyl phomopsenonate, from the fungus *Phomopsis amygdali*. *J. Org. Chem.* **2009**, *74*, 1541–1548. [CrossRef]
165. Shinde, S.S.; Minami, A.; Chen, Z.; Tokiwano, T.; Toyomasu, T.; Kato, N.; Sassa, T.; Oikawa, H. Cyclization mechanism of phomopsene synthase: Mass spectrometry based analysis of various site-specifically labeled terpenes. *J. Antibiot.* **2017**, *70*, 632–638. [CrossRef] [PubMed]
166. Lauterbach, L.; Rinkel, J.; Dickschat, J.S. Two bacterial diterpene synthases from *Allokutzneria albata* produce bonnadiene, phomopsene, and allokutznerene. *Angew. Chem. Int. Ed.* **2018**, *57*, 8280–8283. [CrossRef] [PubMed]
167. Rinkel, J.; Steiner, S.T.; Dickschat, J.S. Diterpene biosynthesis in actinomycetes: Studies on cattleyene synthase and phomopsene synthase. *Angew. Chem. Int. Ed.* **2019**, *58*, 9230–9233. [CrossRef]
168. Kavanagh, F.; Hervey, A.; Robbins, W.J. Robbins, W.J. Antibiotic substances from Basidiomycetes VIII. *Pleurotus Multilus* (Fr.) Sacc. and *Pleurotus passeckerianus* Pilat. *Proc. Natl. Acad. Sci. USA* **1951**, *37*, 570. [CrossRef]
169. Hartley, A.J.; De Mattos-Shipley, K.; Collins, C.M.; Kilaru, S.; Foster, G.D.; Bailey, A.M. Investigating pleuromutilin-producing *Clitopilus* species and related basidiomycetes. *FEMS Microbiol. Lett.* **2009**, *297*, 24–30. [CrossRef]
170. Arigoni, D. La struttura di un terpene di nuovo genere. *Gazz. Chim. Ital.* **1962**, *92*, 884–901.
171. Birch, A.J.; Holzapfel, C.W.; Rickards, R.W. The structure and some aspects of the biosynthesis of pleuromutilin. *Tetrahedron* **1966**, *22*, 359–387. [CrossRef]
172. Arigoni, D. Some studies in the biosynthesis of terpenes and related compounds. *Pure Appl. Chem.* **1968**, *17*, 331–348. [CrossRef]
173. Hu, Y.J.; Li, L.X.; Han, J.C.; Min, L.; Li, C.C. Recent advances in the total synthesis of natural products containing eight-membered carbocycles (2009–2019). *Chem. Rev.* **2020**, *120*, 5910–5953. [CrossRef]
174. Min, L.; Liu, X.; Li, C.C. Total synthesis of natural products with bridged bicyclo[m.n.1] ring systems via type II [5 + 2] cycloaddition. *Acc. Chem. Res.* **2020**, *53*, 703–718. [CrossRef] [PubMed]
175. Ling, C.; Fu, L.; Gao, S.; Chu, W.; Wang, H.; Huang, Y.; Chen, X.; Yang, Y. Design, synthesis, and structure–activity relationship studies of novel thioether pleuromutilin derivatives as potent antibacterial agents. *J. Med. Chem.* **2014**, *57*, 4772–4795. [CrossRef] [PubMed]
176. Wang, X.; Ling, Y.; Wang, H.; Yu, J.; Tang, J.; Zheng, H.; Zhao, X.; Wang, D.; Chen, G.; Qiu, W.; et al. Novel pleuromutilin derivatives as antibacterial agents: Synthesis, biological evaluation and molecular docking studies. *Bioorg. Med. Chem. Lett.* **2012**, *22*, 6166–6172. [CrossRef]
177. Dong, Y.J.; Meng, Z.H.; Mi, Y.Q.; Zhang, C.; Cui, Z.H.; Wang, P.; Xu, Z.B. Synthesis of novel pleuromutilin derivatives. Part 1: Preliminary studies of antituberculosis activity. *Bioorg. Med. Chem. Lett.* **2015**, *25*, 1799–1803. [CrossRef] [PubMed]
178. Novak, R. Are pleuromutilin antibiotics finally fit for human use? *Ann. N. Y. Acad. Sci.* **2011**, *1241*, 71–81. [CrossRef] [PubMed]
179. Bailey, A.M.; Alberti, F.; Kilaru, S.; Collins, C.M.; de Mattos-Shipley, K.; Hartley, A.J.; Hayes, P.; Griffin, A.; Lazarus, C.M.; Cox, R.J.; et al. Identification and manipulation of the pleuromutilin gene cluster from *Clitopilus passeckerianus* for increased rapid antibiotic production. *Sci. Rep.* **2016**, *6*, 25202. [CrossRef] [PubMed]
180. Alberti, F.; Khairudin, K.; Venegas, E.R.; Davies, J.A.; Hayes, P.M.; Willis, C.L.; Bailey, A.M.; Foster, G.D. Heterologous expression reveals the biosynthesis of the antibiotic pleuromutilin and generates bioactive semi-synthetic derivatives. *Nat. Commun.* **2017**, *8*, 1831. [CrossRef]
181. Hauser, D.; Sigg, H.P. Isolation and decomposition of sordarin. *Helv. Chim. Acta* **1971**, *54*, 1178–1190. [CrossRef]
182. Ogita, T.; Hayashi, T.; Sato, A.; Furutani, W. Antibiotic Substance Zofimarin. JPN Patent No. JPS 6240292A, 21 February 1987.
183. Michael, D.; Sarah, M.; Timm, A.; Olov, S. Hypoxysordarin, a new sordarin derivative from *Hypoxylon croceum*. *Z. Naturforsch. C* **1999**, *54*, 474–480. [CrossRef]
184. Hori, Y.; Nitta, K.; Kobayashi, M.; Takase, S.; Hino, M. Novel Sordarin Derivative as a Therapeutic Antimicrobial Agent. JPN Patent No. WO/2001/000639, 4 January 2001.
185. Hanadate, T.; Tomishima, M.; Shiraishi, N.; Tanabe, D.; Morikawa, H.; Barrett, D.; Matsumoto, S.; Ohtomo, K.; Maki, K. FR290581, a novel sordarin derivative: Synthesis and antifungal activity. *Bioorg. Med. Chem. Lett.* **2009**, *19*, 1465–1468. [CrossRef]
186. Büschleb, M.; Dorich, S.; Hanessian, S.; Tao, D.; Schenthal, K.B.; Overman, L.E. Synthetic strategies toward natural products containing contiguous stereogenic quaternary carbon atoms. *Angew. Chem. Int. Ed.* **2016**, *55*, 4156–4186. [CrossRef]
187. Justice, M.C.; Hsu, M.J.; Tse, B.; Ku, T.; Balkovec, J.; Schmatz, D.; Nielsen, J. Elongation factor 2 as a novel target for selective inhibition of fungal protein synthesis. *J. Biol. Chem.* **1998**, *273*, 3148–3151. [CrossRef] [PubMed]
188. Domínguez, J.M.; Martín, J.J. Identification of elongation factor 2 as the essential protein targeted by sordarins in *Candida albicans*. *Antimicrob. Agents Chemother.* **1998**, *42*, 2279–2283. [CrossRef]

189. Domínguez, J.M.; Kelly, V.A.; Kinsman, O.S.; Marriott, M.S.; Gómez de las Heras, F.; Martín, J.J. Sordarins: A new class of antifungals with selective inhibition of the protein synthesis elongation cycle in yeasts. *Antimicrob. Agents Chemother.* **1998**, *42*, 2274–2278. [CrossRef] [PubMed]
190. Chang, Y.C.; Lu, C.K.; Chiang, Y.R.; Wang, G.J.; Ju, Y.M.; Kuo, Y.H.; Lee, T.H. Diterpene glycosides and polyketides from *Xylotumulus gibbisporus*. *J. Nat. Prod.* **2014**, *77*, 751–757. [CrossRef]
191. Kupka, J.; Anke, T.; Oberwinkler, F.; Schramm, G.; Steglich, W. Antibiotics from basidiomycetes. VII. Crinipellin, a new antibiotic from the basidiomycetous fungus *Crinipellis stipitaria* (Fr.) Pat. *J. Antibiot.* **1979**, *32*, 130–135. [CrossRef]
192. Anke, T.; Heim, J.; Knoch, F.; Mocek, U.; Steffan, B.; Steglich, W. Crinipellins, the first natural products with a tetraquinane skeleton. *Angew. Chem. Int. Ed.* **1985**, *24*, 709–711. [CrossRef]
193. Piers, E.; Renaud, J. Total synthesis of the tetraquinane diterpenoid (±)-crinipellin B. *J. Org. Chem.* **1993**, *58*, 11–13. [CrossRef]
194. Piers, E. Tetraquinane diterpenoids: Total synthesis of (±)-crinipellin B. *Synthesis* **1998**, *1998*, 590–602. [CrossRef]
195. Kang, T.; Song, S.B.; Kim, W.Y.; Kim, B.G.; Lee, H.Y. Total synthesis of (−)-crinipellin A. *J. Am. Chem. Soc.* **2014**, *136*, 10274–10276. [CrossRef] [PubMed]
196. Li, Y.Y.; Shen, Y.M. Four novel diterpenoids from *Crinipellis* sp. 113. *Helv. Chim. Acta* **2010**, *93*, 2151–2157. [CrossRef]
197. Rohr, M.; Oleinikov, K.; Jung, M.; Sandjo, L.P.; Opatz, T.; Erkel, G. Anti-inflammatory tetraquinane diterpenoids from a *Crinipellis* species. *Biorg. Med. Chem.* **2017**, *25*, 514–522. [CrossRef] [PubMed]
198. Han, J.W.; Oh, M.; Lee, Y.J.; Choi, J.; Choi, G.J.; Kim, H. Crinipellins A and I, two diterpenoids from the basidiomycete fungus *Crinipellis rhizomaticola*, as potential natural fungicides. *Molecules* **2018**, *23*, 2377. [CrossRef] [PubMed]
199. Niu, S.; Fan, Z.W.; Xie, C.L.; Liu, Q.; Luo, Z.H.; Liu, G.; Yang, X.W. Spirograterpene A, a tetracyclic spiro-diterpene with a fused 5/5/5/5 ring system from the deep-sea-derived fungus *Penicillium granulatum* MCCC 3A00475. *J. Nat. Prod.* **2017**, *80*, 2174–2177. [CrossRef] [PubMed]
200. Rabe, P.; Rinkel, J.; Dolja, E.; Schmitz, T.; Nubbemeyer, B.; Luu, T.H.; Dickschat, J.S. Mechanistic investigations of two bacterial diterpene cyclases: Spiroviolene synthase and tsukubadiene synthase. *Angew. Chem. Int. Ed.* **2017**, *56*, 2776–2779. [CrossRef]
201. Liu, Y.P.; Dai, Q.; Wang, W.X.; He, J.; Li, Z.H.; Feng, T.; Liu, J.K. Psathyrins: Antibacterial diterpenoids from *Psathyrella candolleana*. *J. Nat. Prod.* **2020**, *83*, 1725–1729. [CrossRef]
202. Wang, Q.X.; Qi, Q.Y.; Wang, K.; Li, L.; Bao, L.; Han, J.J.; Liu, M.M.; Zhang, L.X.; Cai, L.; Liu, H.W. Coicenals A–D, four new diterpenoids with new chemical skeletons from the plant pathogenic fungus *Bipolaris coicis*. *Org. Lett.* **2013**, *15*, 3982–3985. [CrossRef]
203. Wang, S.J.; Li, Y.X.; Bao, L.; Han, J.J.; Yang, X.L.; Li, H.R.; Wang, Y.Q.; Li, S.J.; Liu, H.W. Eryngiolide A, a cytotoxic macrocyclic diterpenoid with an unusual cyclododecane core skeleton produced by the edible mushroom *Pleurotus eryngii*. *Org. Lett.* **2012**, *14*, 3672–3675. [CrossRef]
204. Sun, P.X.; Zheng, C.J.; Li, W.C.; Jin, G.L.; Huang, F.; Qin, L.P. Trichodermanin A, a novel diterpenoid from endophytic fungus culture. *J. Nat. Med.* **2011**, *65*, 381–384. [CrossRef]
205. Omura, S.; Shiomi, K.; Masuma, R.; Ui, H.; Nagai, T.; Yamada, H. Wickerol and Process for Production Thereof. JPN Patent No. WO 2009116604, 24 September 2009.
206. Yamamoto, T.; Izumi, N.; Ui, H.; Sueki, A.; Masuma, R.; Nonaka, K.; Hirose, T.; Sunazuka, T.; Nagai, T.; Yamada, H.; et al. Wickerols A and B: Novel anti-influenza virus diterpenes produced by *Trichoderma atroviride* FKI-3849. *Tetrahedron* **2012**, *68*, 9267–9271. [CrossRef]
207. Tokiwano, T.; Fukushi, E.; Endo, T.; Oikawa, H. Biosynthesis of phomactins: Common intermediate phomactatriene and taxadiene. *Chem. Commun.* **2004**, *40*, 1324–1325. [CrossRef]
208. Yamada, T.; Suzue, M.; Arai, T.; Kikuchi, T.; Tanaka, R. Trichodermanins C–E, new diterpenes with a fused 6-5-6-6 ring system produced by a marine sponge-derived fungus. *Mar. Drugs* **2017**, *15*, 169. [CrossRef] [PubMed]
209. Yamada, T.; Fujii, A.; Kikuchi, T. New diterpenes with a fused 6-5-6-6 ring system isolated from the marine sponge-derived fungus *Trichoderma harzianum*. *Mar. Drugs* **2019**, *17*, 480. [CrossRef] [PubMed]

Journal of *Fungi*

MDPI

Review

Sesquiterpenoids Specially Produced by Fungi: Structures, Biological Activities, Chemical and Biosynthesis (2015–2020)

Quan Dai †, Fa-Lei Zhang † and Tao Feng *

School of Pharmaceutical Sciences, South-Central University for Nationalities, Wuhan 430074, China; quandai@mail.scuec.edu.cn (Q.D.); flzhang@mail.scuec.edu.cn (F.-L.Z.)
* Correspondence: tfeng@mail.scuec.edu.cn
† These authors contributed equally to this paper.

Abstract: Fungi are widely distributed in the terrestrial environment, freshwater, and marine habitat. Only approximately 100,000 of these have been classified although there are about 5.1 million characteristic fungi all over the world. These eukaryotic microbes produce specialized metabolites and participate in a variety of ecological functions, such as quorum detection, chemical defense, allelopathy, and maintenance of symbiosis. Fungi therefore remain an important resource for the screening and discovery of biologically active natural products. Sesquiterpenoids are arguably the richest natural products from plants and micro-organisms. The rearrangement of the 15 high-ductility carbons gave rise to a large number of different skeletons. At the same time, abundant structural variations lead to a diversification of biological activity. This review examines the isolation, structural determination, bioactivities, and synthesis of sesquiterpenoids that were specially produced by fungi over the past five years (2015–2020).

Keywords: sesquiterpenoids; fungus; structures; structural diversity; biological activity; synthesis

Citation: Dai, Q.; Zhang, F.-L.; Feng, T. Sesquiterpenoids Specially Produced by Fungi: Structures, Biological Activities, Chemical and Biosynthesis (2015–2020). *J. Fungi* **2021**, *7*, 1026. https://doi.org/10.3390/jof7121026

Academic Editor: Laurent Dufossé

Received: 14 November 2021
Accepted: 28 November 2021
Published: 30 November 2021

1. Introduction

Fungi are undoubtedly important resources for natural products discovery. With the advancement of natural product research, the importance of its biological resources has been infinitely enlarged. In the giant natural product system of fungi, sesquiterpenes, due to their carbon skeletons and amounts, are the largest of all types. The C-15-hydrocarbon skeletal system of various sesquiterpenoids isolated from fungi, bacteria, and plants are synthesized from farnesyl pyrophosphate (FPP) under the catalysis of sesquiterpene synthases [1,2]. Sesquiterpene synthases catalyze different initial cyclization reactions to produce secondary or tertiary cyclic carbocation intermediates, which can then be further cyclized and reassembled until carbocation quenching at the active center, followed by the enzymatic release of the final sesquiterpenoid scaffold (Figure 1) [3]. A huge number of sesquiterpenoids were, consequently, produced [4–6]. Among various other resources, fungal species have an enormous contribution owing to their potential to carry out the bio-transformations and drug synthesis under environmentally acceptable conditions. For instance, hydroxymethylacylfulvene (HMAF) is a semisynthetic antitumor agent based on the naturally occurring illudin S occurring in the mushroom *Omphalotus olearius* [7]. It has been advanced into human clinical trials for the treatment of cancers [8,9]. Trichothecenes, a class of tricyclic sesquiterpenes produced by a wide variety of fungi, are toxic to animals and humans and frequently present in cereal crops. They have attracted much attention in the areas such as agriculture, food contamination, and health care [10–13].

Our research group has been engaged in the study of the chemical composition of fungi for decades [14,15], while a large number of sesquiterpenoids have been reported [6]. It has been found that the vast majority of skeletons, such as alliacane, bergamotane, hirsutane, tremulane, etc., are specially produced by fungi. Many compounds displayed significant biological activities, and it is obvious that cytotoxic activity accounts for the largest proportion (Figure 2). In addition, with the development of synthetic biology, the biosynthesis of many fungal sesquiterpenoids has been figured out. This review gives an overview about the structures, biological activities, chemical synthesis and biosynthesis of sesquiterpenoids specially produced by fungi presented from 2015 to 2020.

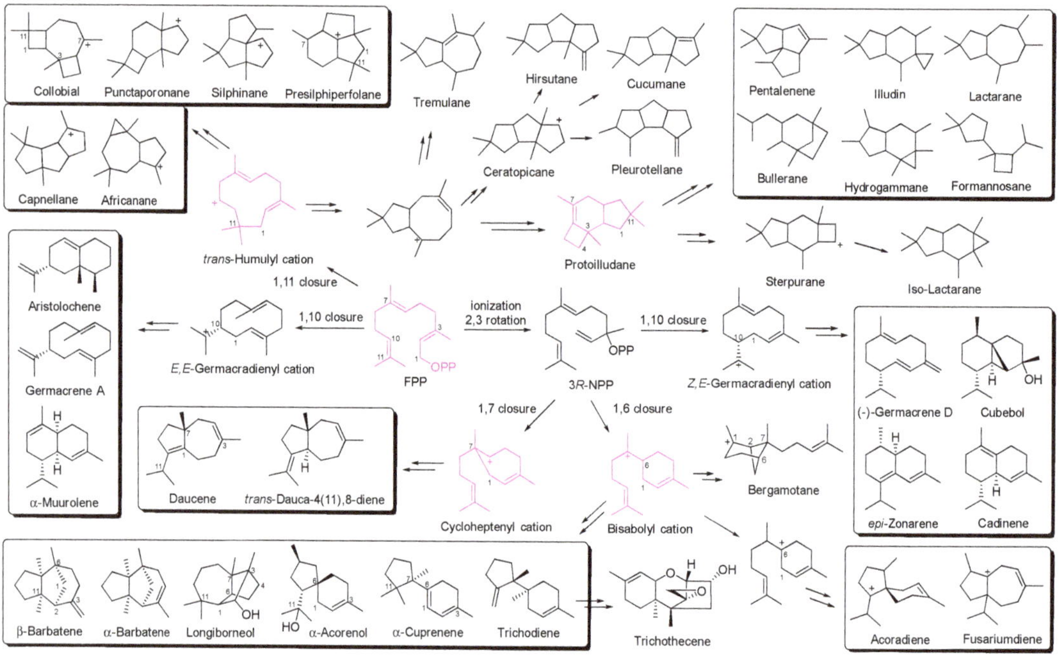

Figure 1. Cyclization of FPP by characterized fungal sesquiterpene synthases (Reference [3]).

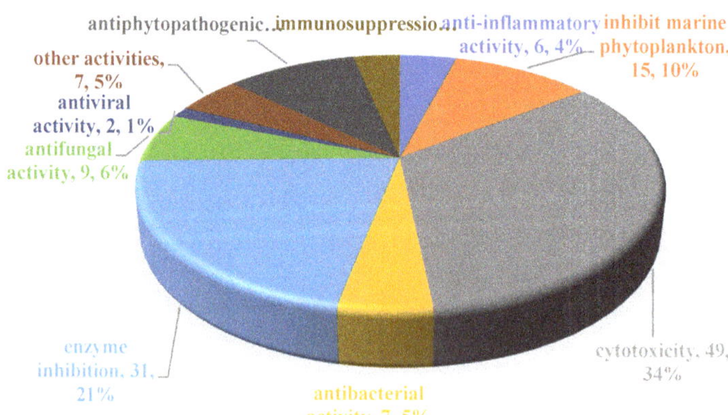

Figure 2. The proportion of one activity compared to the whole occurrence of activities of bioactive fungal sesquiterpenoids.

2. Composition and Bioactivities

2.1. *Alliacane, Cadinene, Azulene, and Zierane*

Nine alliacane sesquiterpenoids inonoalliacanes A–I **1a/1b–6a/6b–7–9** were isolated from the culture broth of the basidiomycete *Inonotus* sp. BCC 22670 [16]. Inonoalliacane A **1** exhibited moderate antibacterial activity against *Bacillus cereus* with a minimum inhibitory concentration (MIC) value of 25 μg/mL. Inonoalliacane B **2** showed antiviral activity against herpes simplex virus type 1 (HSV-1) with IC_{50} of 17 μg/mL.

Clitocybulols G–O **10–18**, highly oxidized alliacane sesquiterpenoids, were isolated from the solid culture of the edible fungus *Pleurotus cystidiosus* [17]. Clitocybulols G **10** and L **15** showed weak inhibitory activity against protein tyrosine phosphatase-1B (PTP1B) with IC_{50} values of 49.5, 38.1 μM, respectively.

In the ^{1}H NMR-guided fractionation of extracts from the edible mushroom *Lactarius deliciosus*, two new azulene-type sesquiterpenoids **19** and **20** were characterized [18]. Pestabacillin A **21** bearing a zierane-type sesquiterpene skeleton was isolated from the co-culture of the endophytic fungus *Pestalotiopsis* sp. with *Bacillus subtilis* [19]. Furthermore, the absolute configuration of **21** was confirmed by single-crystal X-ray diffraction analysis.

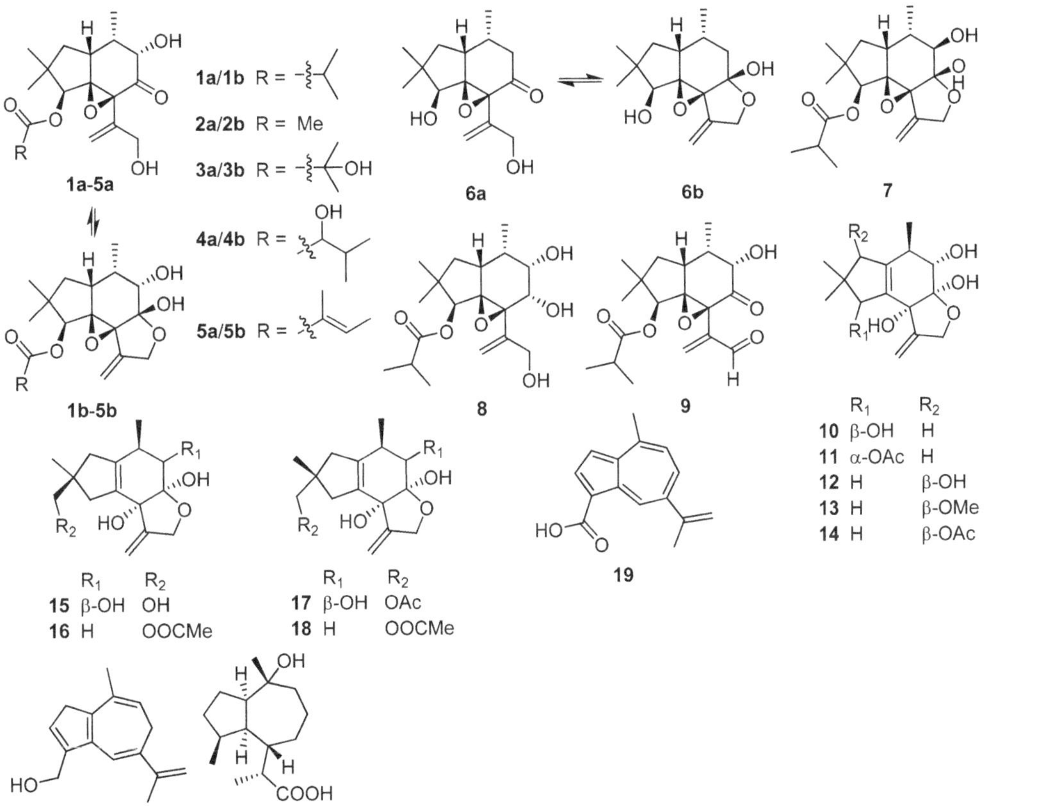

2.2. Bergamotane, Spiroaminal, and Spiroaxane

Bergamotane sesquiterpenes bearing a bridged 6/4 bicyclic ring incorporated with an isopentyl unit, are naturally occurring in plants and fungi [20,21]. A new class of polyoxygenated bergamotanes with notable features inspired by a 6/4/5/5 tetracyclic ring system was very rare in nature and all examples of the polycyclic bergamotanes only derived from fungi [22–25].

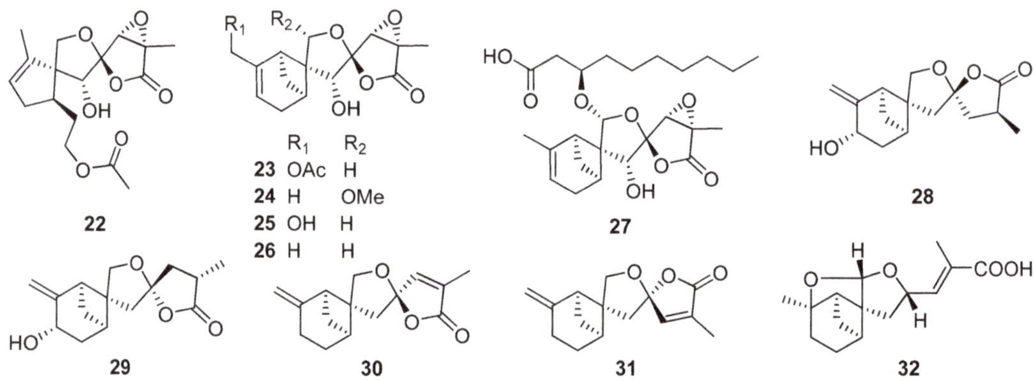

Purpurolide A **22**, an unprecedented sesquiterpene lactone with a rarely encountered 5/5/5 spirocyclic skeleton, along with five new 6/4/5/5 tetracyclic sesquiterpene lactones (purpurolides B–F **23–27**), was isolated from the cultures of the endophytic fungus *Penicillium purpurogenum* [26,27]. The structures and absolute configurations of **22–27** were established by spectroscopic analysis, a single-crystal X-ray diffraction, and calculations of the ^{13}C NMR and ECD data. The plausible biosynthetic pathway of **22–27** is shown in Scheme 1. Compounds **22–27** showed significant inhibitory activity against pancreatic lipase with IC_{50} values of 1.22–7.88 μM.

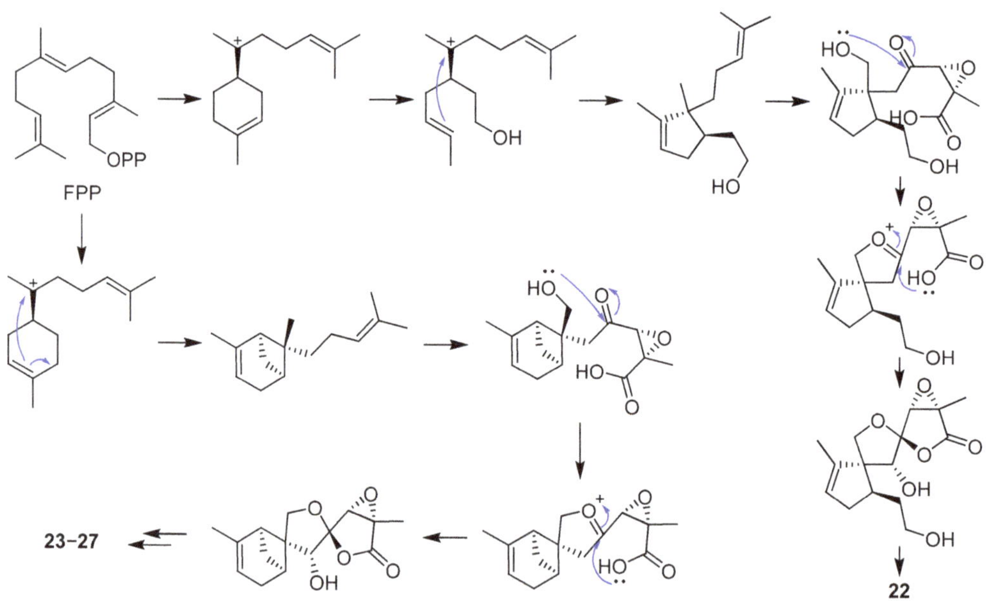

Scheme 1. Plausible biogenetic pathways for **22–27** (Reference [26]).

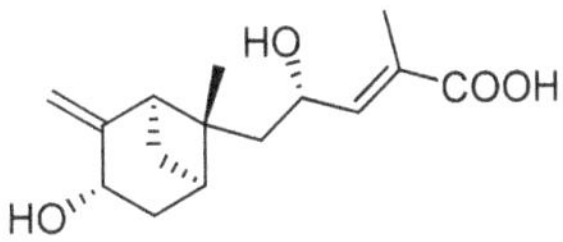

33

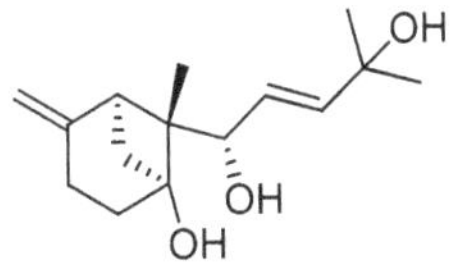

34

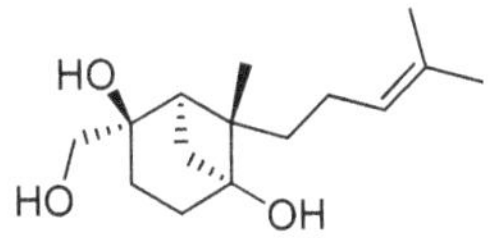

35

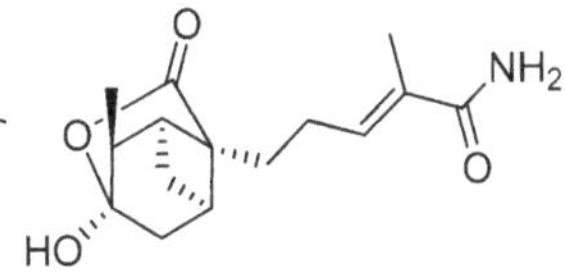

36

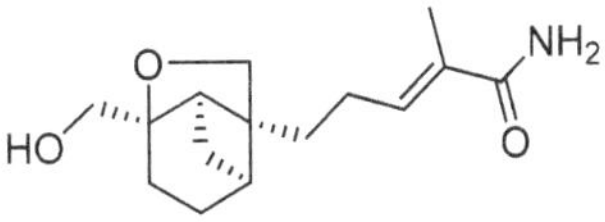

37

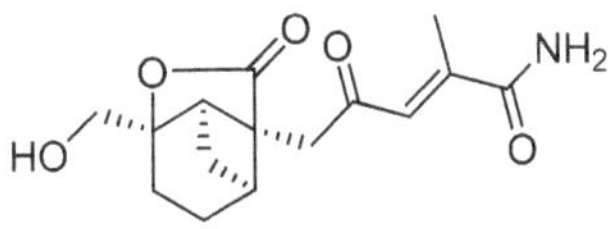

38

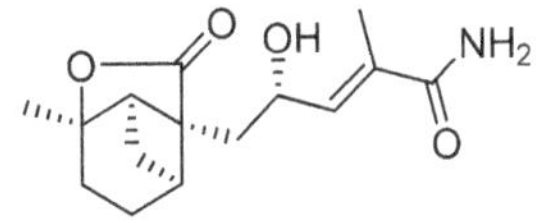

39

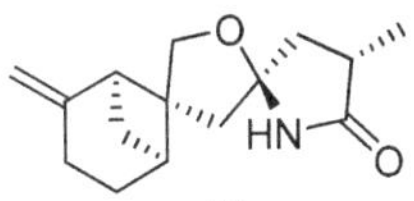

40

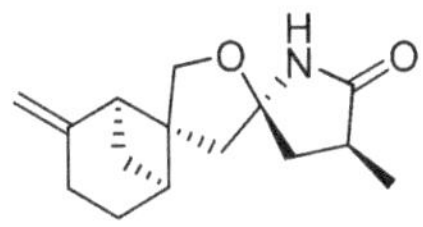

41

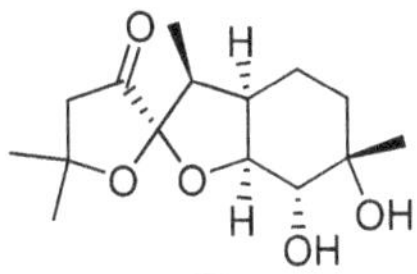

42

Expansolides C **28** and D **29** were two new bergamotane sesquiterpene lactones isolated from the plant pathogenic fungus *Penicillium expansum* [28]. The epimeric mixture of expansolides C **28** and D **29** (in a ratio of 2:1 at the temperature of the bioassay) exhibited more potent α-glucosidase inhibitory activity (IC_{50} 0.50 mM) as compared with the positive control acarbose (IC_{50} 1.90 mM) in an in vitro bioassay.

Donacinolides A **30** and B **31** and donacinoic acids A **32** and B **33**, four new rare tetracyclic bergamotane-type sesquiterpenoids, were isolated from the mushroom-associated fungus *Montagnula donacina* [29]. Two new β-bergamotane sesquiterpenoids **34** and **35** were isolated from the marine-derived fungus *Aspergillus fumigatus* [30]. Brasilamides K–N **36–39** were isolated from the plant endophytic fungus *Paraconiothynium Brasiliense* [31].

Sporulaminals A **40** and B **41**, a pair of unusual epimeric spiroaminal derivatives bearing a 6/4/5/5 tetracyclic ring system derived from bergamotane sesquiterpenoid (Scheme 2), were isolated from a marine-derived fungus *Paraconiothyrium sporulosum* [32]. Pleurospiroketal F **42**, a new perhydrobenzannulated 5,5-spiroketal sesquiterpene was isolated from solid-state fermentation of *Pleurotus citrinopileatus*, and the absolute configuration of **42** was determined by single-crystal X-ray diffraction analysis [33].

Flammuspirones A–J **43–52**, ten spiroaxane sesquiterpenoids, were obtained from the edible mushroom *Flammulina velutipes* [34]. Flammuspirones A **43** and C **45** showed inhibition on HMG-CoA reductase with IC_{50} of 114.7 and 77.6 μM, respectively. Flammuspirones C–E **45–47** and H **50** showed inhibitory activity on DPP-4 with IC_{50} values in the range from 70.9 to 83.7 μM.

Talaminoid A **53** was obtained from the fungus *Talaromyces minioluteus* [35]. Talaminoid A **53** showed a significant suppressive effect on the production of nitric oxide (NO) on lipopolysaccharide (LPS) induced BV-2 cell, with IC_{50} of 5.79 μM. In addition, talaminoid A **53** exhibited significant anti-inflammatory activities against the production of TNF-α and IL-6. Further immunofluorescence experiments revealed the mechanism of action to be inhibitory the NF-κB-activated pathway. A new sesquiterpenoid **54** was isolated from the fungus *Pholiota nameko* [36]. Tramspiroins A–D **55–58** have been isolated from the cultures of Basidiomycete *Trametes versicolor* [37].

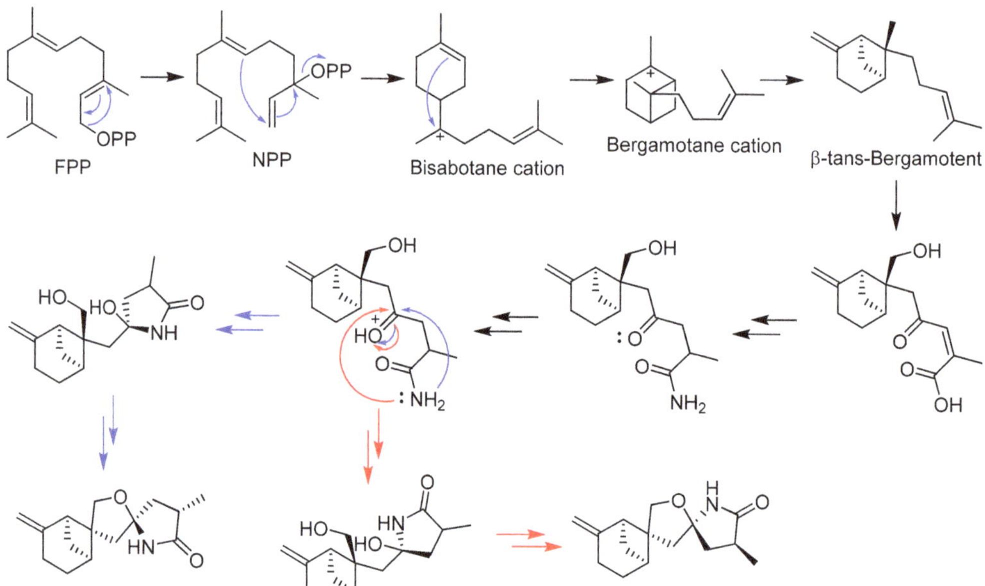

Scheme 2. Plausible biosynthetic pathways of sporulaminals A **40** and B **41** (Reference [32]).

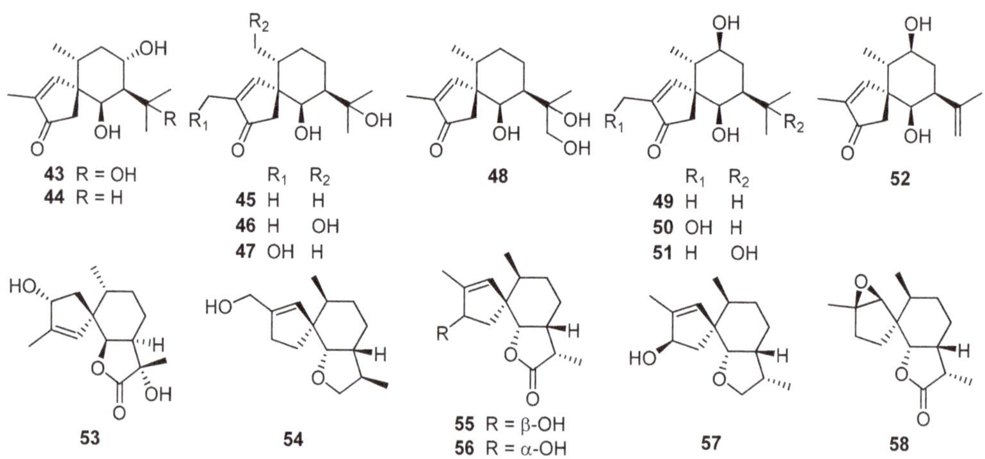

2.3. Carotane, Cyclonerane, Cyclofarnesane, and Longifolene

A new dimeric sesquiterpene divirensol H **59** and two exceptionally novel trimeric sesquiterpenes trivirensols A **60** and B **61** were purified from an endophytic fungus *Trichoderma virens* [38]. Divirensol H **59** showed significant activities against fungi *Penicillium italicum*, *Fusarium oxysporum*, *Fusarium graminearum*, *Colletotrichum musae*, and *Colletotrictum gloeosporioides* with MIC values of 6.25 to 25 μg/mL. Rhinomilisin A **62** and four new heptelidic acid derivatives, rhinomilisin B–E **63–66**, were isolated from the endophytic fungus *Rhinocladiella similis* [39]. Rhinomilisins A **62** showed moderate cytotoxicity activity against the mouse lymphoma cell line L5178Y with an IC_{50} value of 5.0 μM.

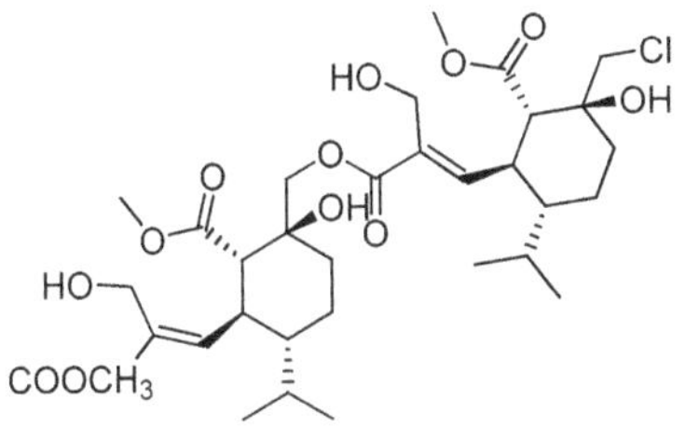

59

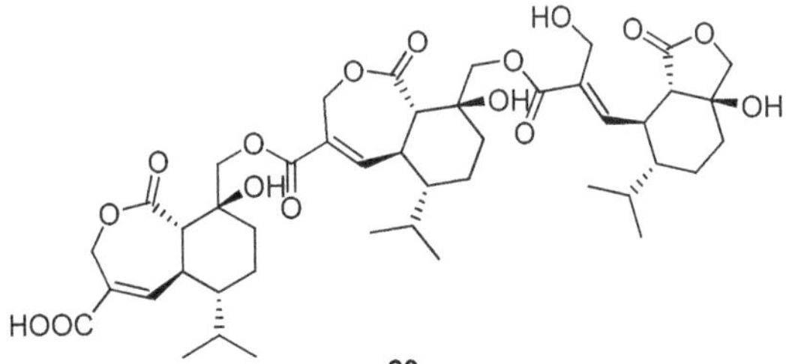

60

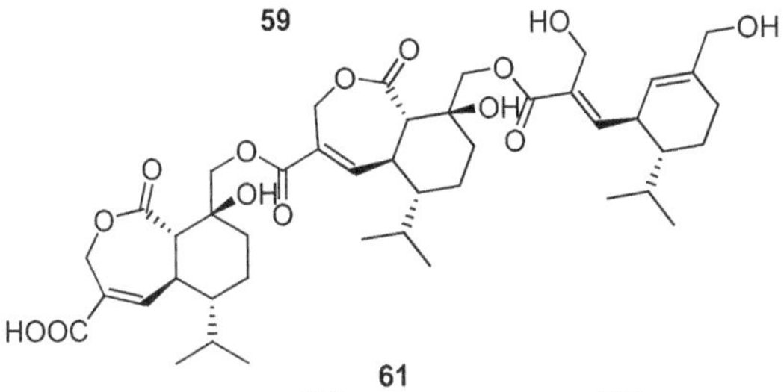

61

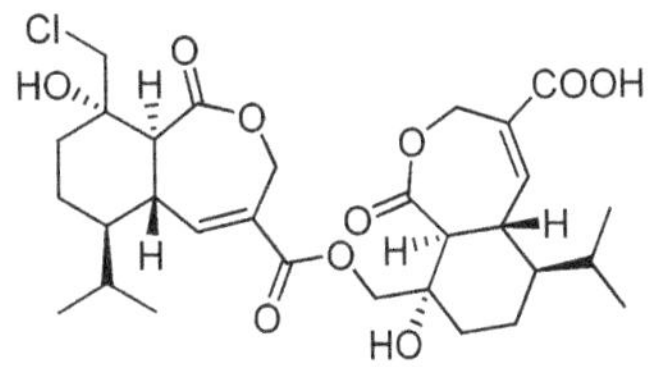

62

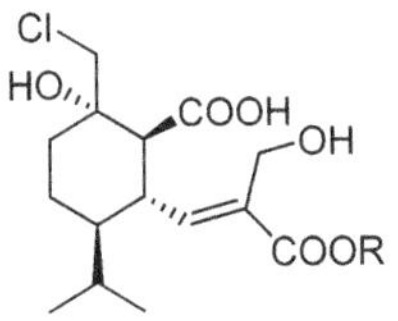

63 R = H
64 R = Me

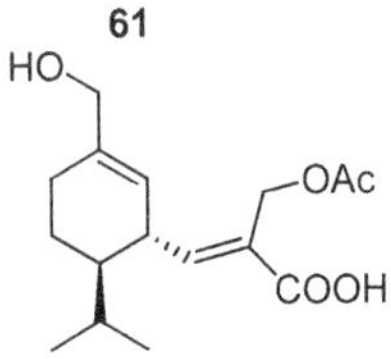

65

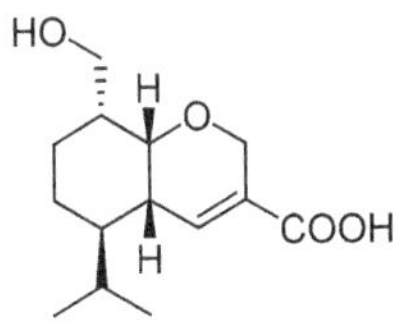

66

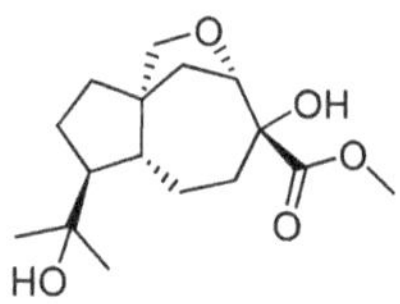
67
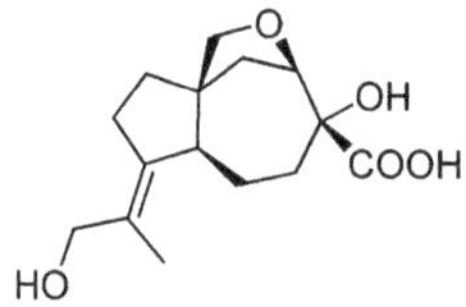
68
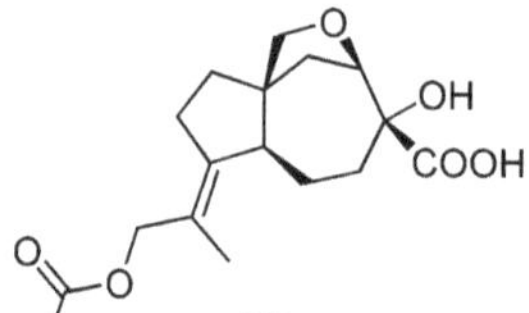
69
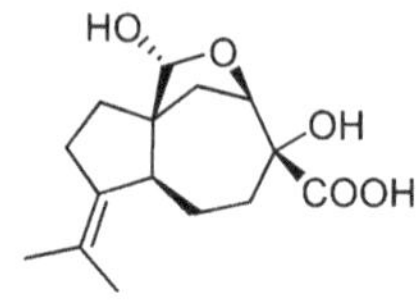
70
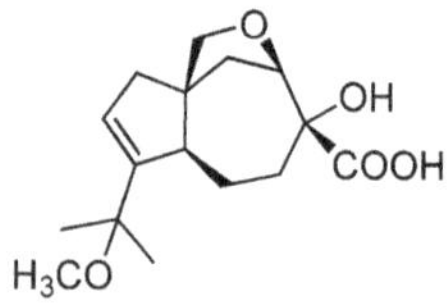
71
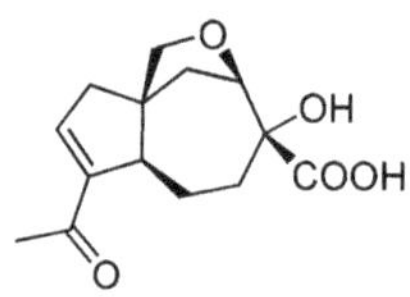
72
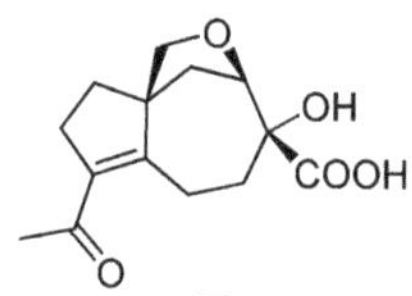
73
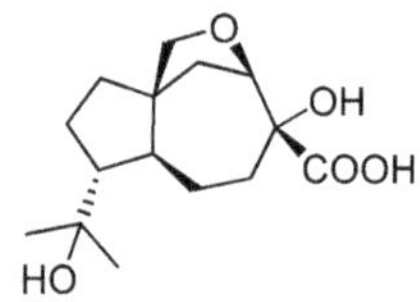
74
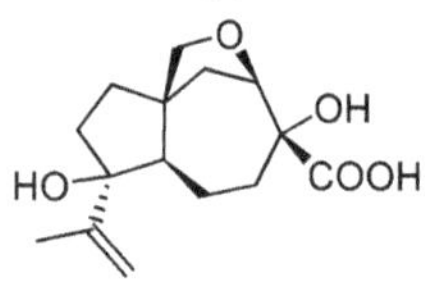
75
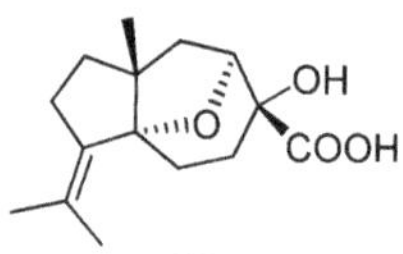
76
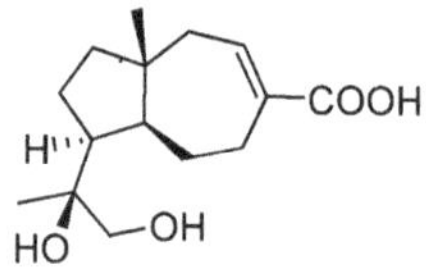
77
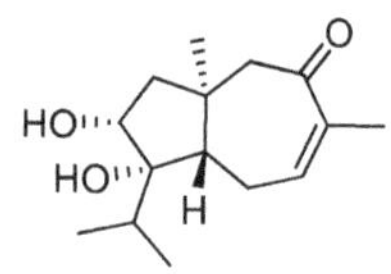
78
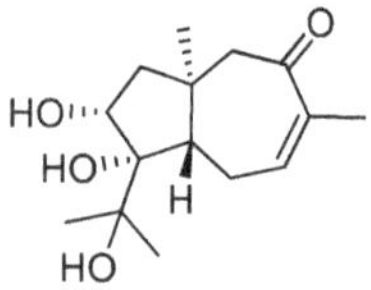
79
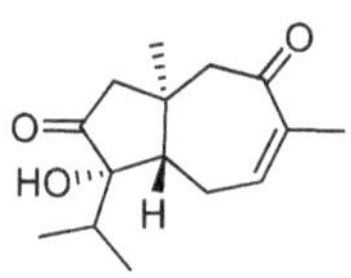
80
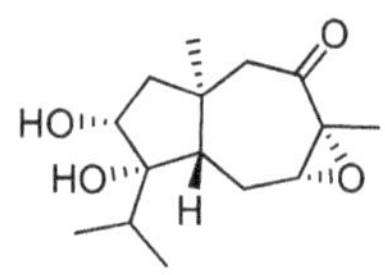
81
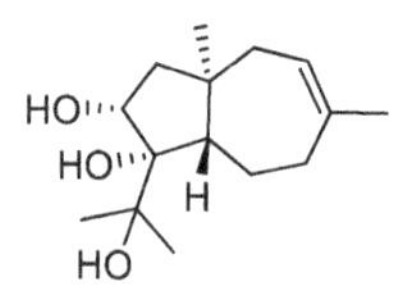
82
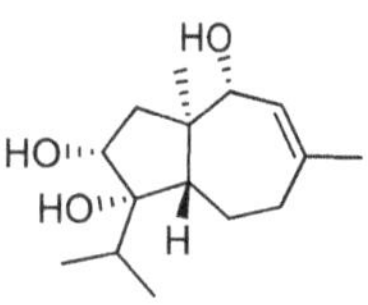
83
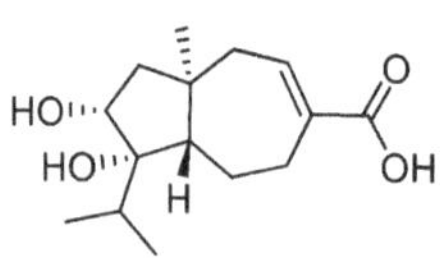
84
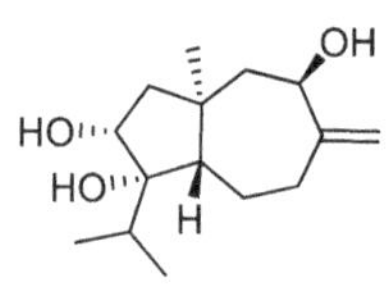
85
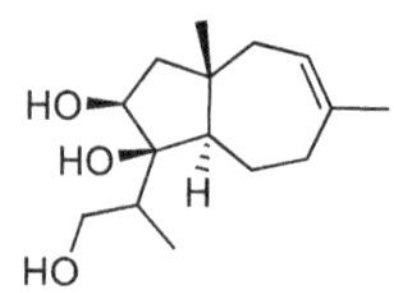
86
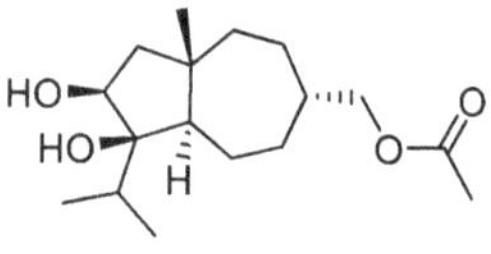
87
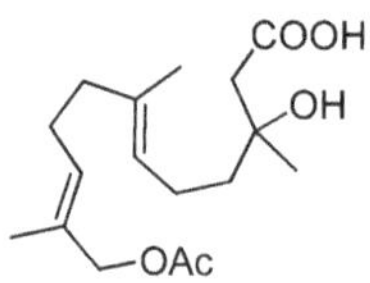
88
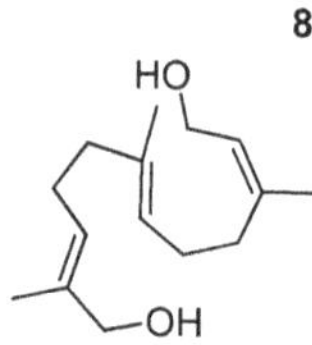
89

Peniterester **67**, a new tricyclic sesquiterpene was isolated from the secondary metabolites of an artificial mutant *Penicillium* sp. T2-M20 [40]. Peniterester **67** showed significant activities against *Bacillus subtilis*, *Escherichia coli*, and *Staphylococcus aureus* in vitro with MICs of 8.0, 8.0, and 4.0 μg/mL, respectively.

Piltunines A–F **68–73** and penigrisacids A–D **74–77**, ten new carotane sesquiterpenoids, were isolated from the marine-derived fungus *Penicillium griseofulvum* and *Penicillium piltunense*, respectively [41,42]. Penigrisacid D **75** showed a weak effect on ECA-109 tumor cells with an IC_{50} value of 28.7 μM [41]. Trichocarotins A–H **78–85**, eight new carotane sesquiterpenes, were isolated from the culture of the fungus *Trichoderma virens* [43]. Trichocarotins C–E **80–82** and H **85** displayed potent inhibition against the four marine phytoplankton species (*Chattonella marina*, *Heterosigma akashiwo*, *Karlodinium veneficum*, and *Prorocentrum donghaiense*) tested, especially against *C. marina* with IC_{50} values ranging from 0.24 to 1.2 μg/mL.

Trichocaranes E **86** and F **87** were isolated from cultures of the insect pathogenic fungus *Isaria fumosorosea* [44]. Trichocaranes E **86** and F **87** showed potent cytotoxic activities against six tumor cell lines MDA, MCF-7, SKOV-3, Hela, A549, and HepG2 with IC_{50} values in a concentration range of 0.13–4.57 μg/mL. Two new carotane-type biogenetically related sesquiterpenes, aspterrics A **88** and B **89**, were isolated from the deep-sea-derived fungus *Aspergillus terreus* [45].

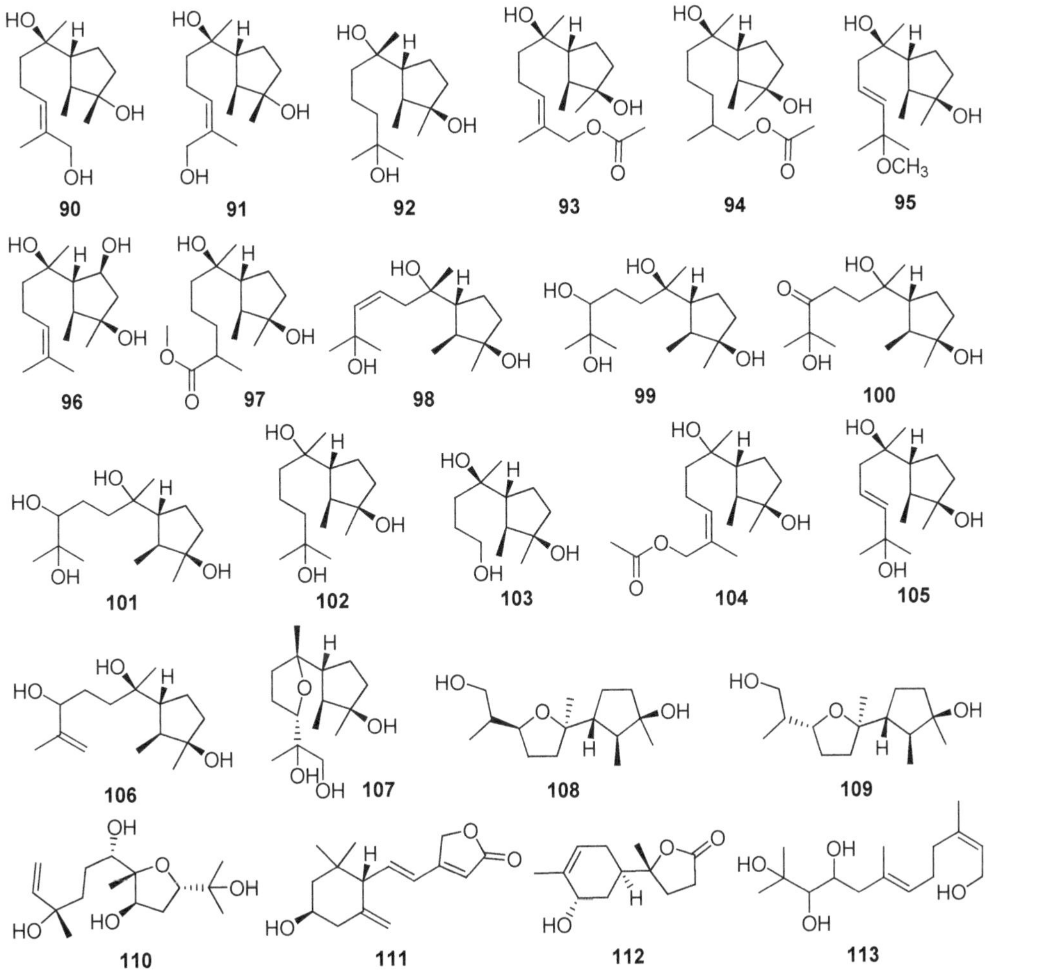

Two new cycloneranes **90** and **91** were isolated from the marine alga endophytic fungus *Trichoderma citrinoviride* [46]. The compound **90** had an inhibition to the marine phytoplankton species *Karlodinium veneficum* with an IC_{50} value of 8.1 μg/mL. Six new cycloneranes **92–97** were isolated from the fungus *Trichoderma harzianum* [47–49]. The three new ones **95–97** all exhibited growth inhibition of the four phytoplankton species (*Chattonella marina, Heterosigma akashiwo, Karlodinium veneficum,* and *Prorocentrum donghaiense*) with IC_{50} values ranging from 0.66 to 75 μg/mL [49].

Cyclonerotriol B **98** was isolated from the soil fungus *Fusarium avenaceum* [50]. Cyclonerodiol B **99** was isolated from the mangrove plant endophytic fungus *Trichoderma* sp. Xy24 [51]. Cyclonerodiol B **99** exhibited significant neural anti-inflammatory activity by inhibiting LPS-induced NO production in BV2 cells with the inhibitory rates of 75.0% at

0.1 μM, which are more potent than curcumin, positive control with the inhibitory rate of 21.1% at 0.1 μM.

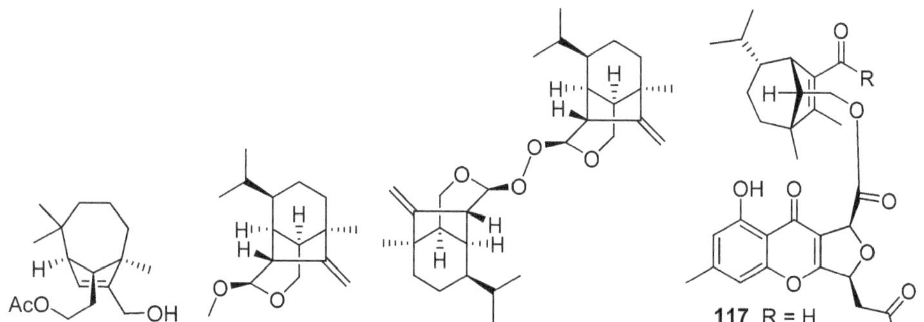

114 **115** **116** **117** R = H **118** R = OH

Ten new cycloneranes **100–109** were isolated from the algicolous endophytic fungus *Trichoderma asperellum* [52,53]. The seven new ones, **100–104**, **108**, and **109**, all exhibited growth inhibition of the four phytoplankton species (*Chattonella marina*, *Heterosigma akashiwo*, *Karlodinium veneficum*, and *Prorocentrum donghaiense*) with IC_{50} values ranging from 2.4 to 76 μg/mL [52].

A new sesquiterpenoid **110** was isolated and identified from an endophytic fungus *Umbelopsis dimorpha* grown on host-plant *Kadsura angustifolia* and wheat bran [54]. Inonofarnesane **111**, a new cyclofarnesane sesquiterpenoid, was isolated from cultures of the wood-rotting basidiomycete *Inonotus* sp. BCC 23706 [55].

One new norbisabolane sesquiterpenoid degradation, isopolisin B **112**, was isolated from the fungus *Pestalotiopsis heterocornis* [56]. Koninginol D **113** as a new farnesane sesquiterpenoid was isolated from the endophytic fungus *Trichoderma koningiopsis* [57].

Bipolenin F **114**, a new *seco*-longifolene sesquiterpenoid, and two new *seco*-sativene sesquiterpenoids, bipolenins D **115** and E **116**, and two novel sesquiterpenoid-xanthone adducts, bipolenins I **117** and J **118**, were obtained from cultures of potato endophytic fungus *Bipolaris eleusines* [58,59]. Bipolenins I **117** and J **118** exhibited potent inhibitory activity against the plant pathogens *Alternaria solani* with MIC values of 8 and 16 μg/mL, respectively [59].

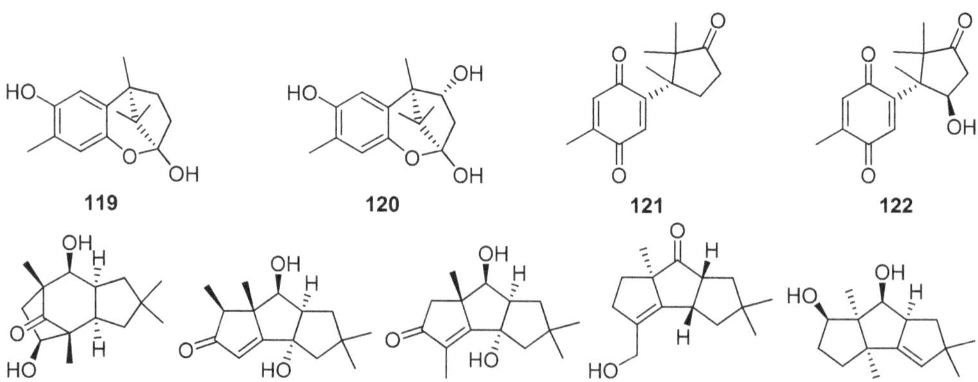

119 **120** **121** **122**

123 **124** **125** **126** **127**

2.4. Cerapicane, Cucumane, Cuparene, Hirsutane, Isohirsutane, and Triquinane

Cuparane-type sesquiterpenoids of fungal origin possess a skeleton with a six-membered ring connected to a five-membered ring, of which the six-membered ring is always aromatic. Linear triquinane sesquiterpenoids have a basic skeleton 1*H*-cyclopenta[α]pentalene [60]. Many compounds displayed a wide range of biological activities, such as cytotoxic, antimicrobial, and anti-inflammatory activities. A review gives an overview about the isolation, structure, biological activities, and chemical synthesis of linear triquinane sesquiterpenoids [61].

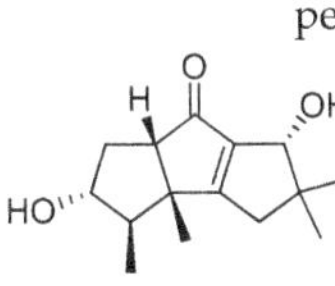

128

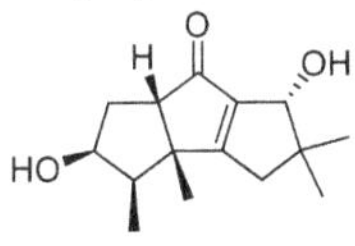

129

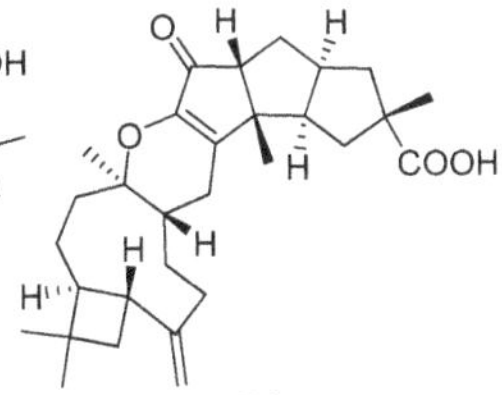

130

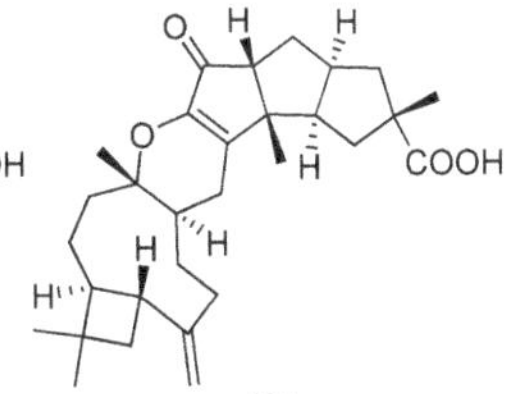

131

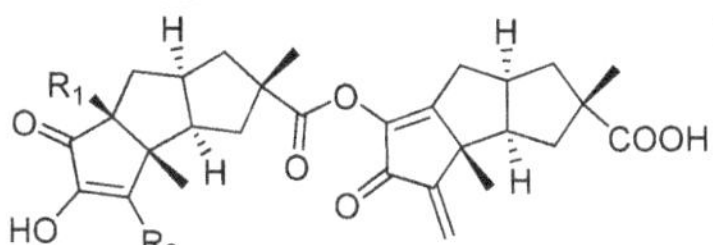

	R_1	R_2
132	OH	Me
133	H	CH_2OMe

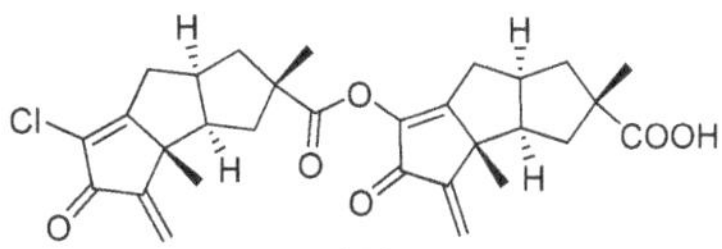

134

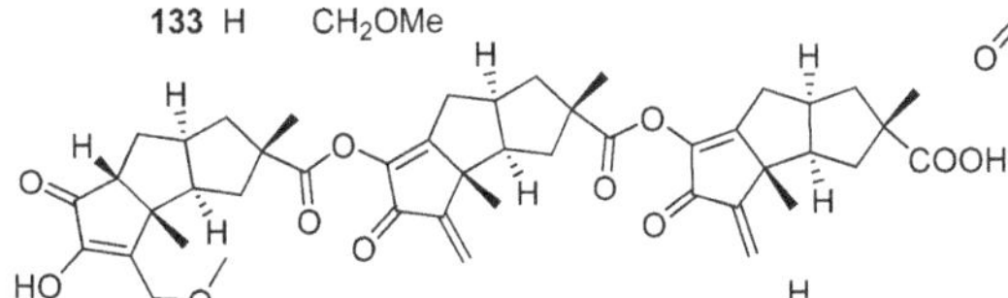

135

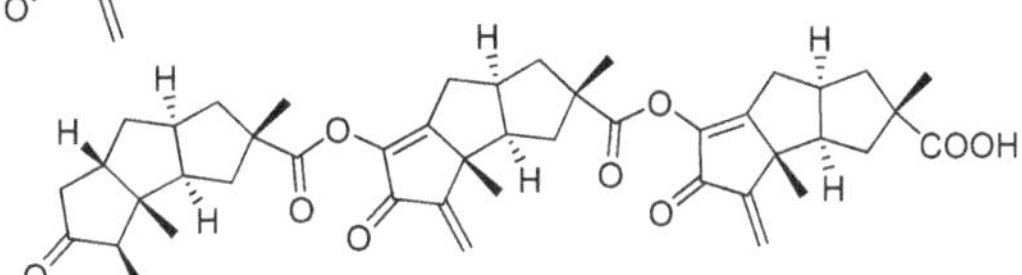

136

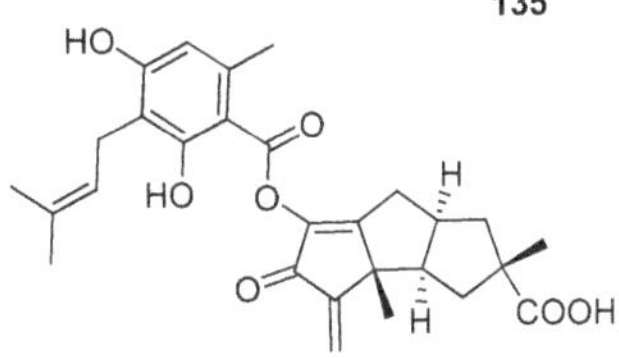

137

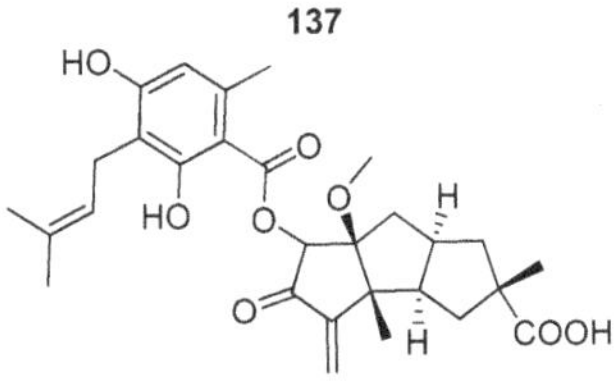

138

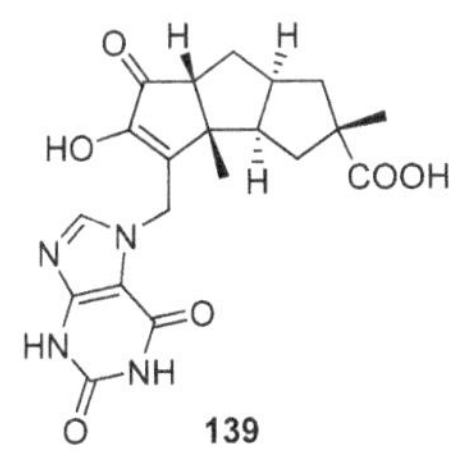

139

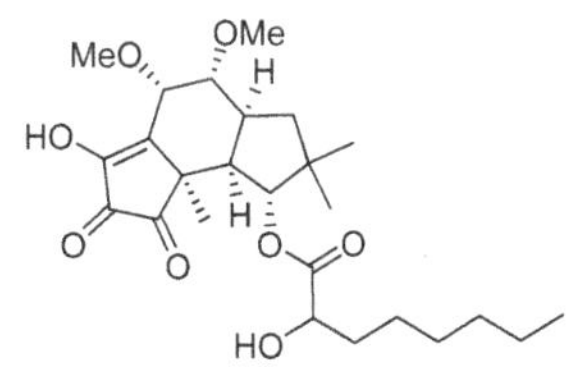

140

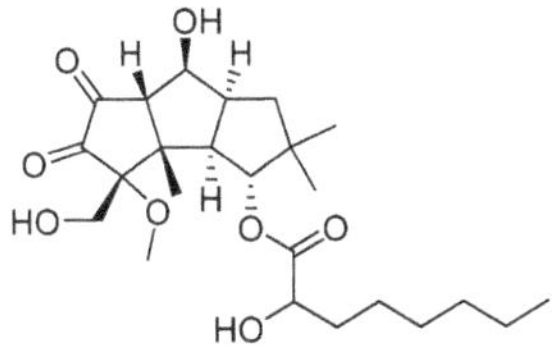

141

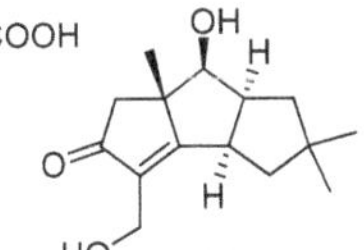

142

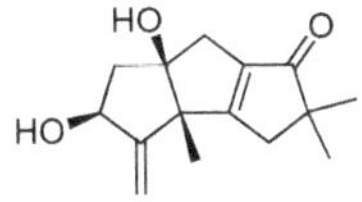

143

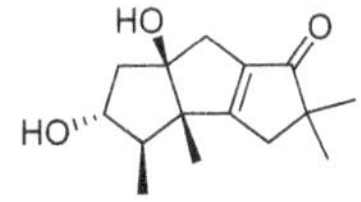

144

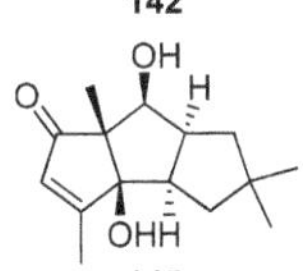

145

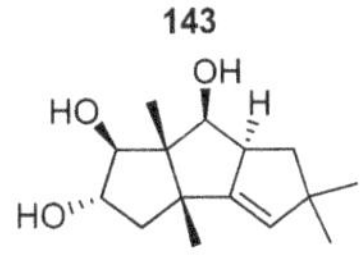

146

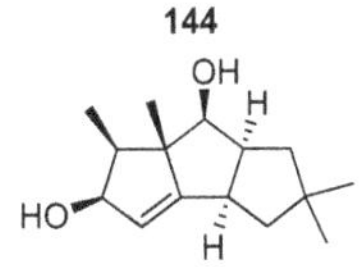

147

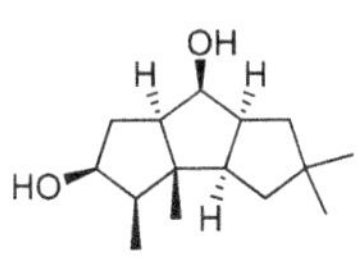

148

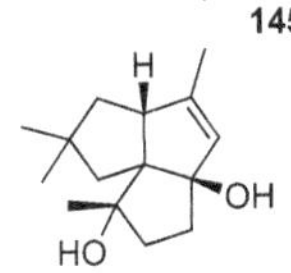

149

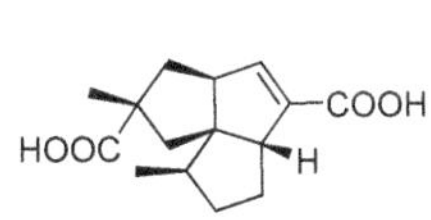

150

Enokipodins A–D **119–122**, highly oxygenated cuparene-type sesquiterpenes were obtained from the fungi *Flammulina rossica* and *Flammulina velutipes*. In addition, enokipodins B **121** and D **122** are oxidized compounds of enokipodins A **119** and C **120**, respectively [62].

One new cerapicane cerrenin A **123**, and two new isohirsutane sesquiterpenoids cerrenins B **124** and C **125**, were isolated from the broth extract of *Cerrena* sp. which was isolated from *Pogostemon cablin* [63]. Trefoliol C **126**, one new cucumane sesquiterpenoid, was isolated from cultures of the basidiomycetes *Tremella foliacea* [64]. A new sesquiterpenoid **127** was isolated from the crude extract of *Antrodiella albocinnamomea* [65]. Two new hirsutane-type sesquiterpenoids, chondrosterins N **128** and O **129**, were isolated from the marine fungus *Chondrostereum* sp. [66].

Ten new hirsutane-type sesquiterpenoids, sterhirsutins C–L **130–139**, were isolated from the culture of *Stereum hirsutum* [67]. Sterhirsutins C **130** and D **131** possessed an unprecedented chemical skeleton with a 5/5/5/6/9/4 fused ring system, and the absolute configuration of sterhirsutin C **130** was assigned by single-crystal X-ray diffraction experiment. Sterhirsutin L **139** was the first sesquiterpene coupled with a xanthine moiety. Sterhirsutins C–L **130–139** showed cytotoxicity against K562 and HCT116 cell lines, and sterhirsutin K **138** induced autophagy in HeLa cells. Sterhirsutin G **133** inhibited the activation of the IFNβ promoter in Sendai virus-infected cells.

Cerrenins D **140** and E **141**, two new triquinane-type sesquiterpenoids, were obtained from the endophytic fungus *Cerrena* sp. A593 [68]. Chondrosterins K–M **142–144** were isolated from the marine fungus *Chondrostereum* sp. [69]. Chondrosterins K–M **142–144** showed different degrees of cytotoxicities against various cancer cell lines (CNE1, CNE2, HONE1, SUNE1, A549, GLC82, and HL7702) in vitro, with IC_{50} values ranging from 12.03 to 58.83 μM.

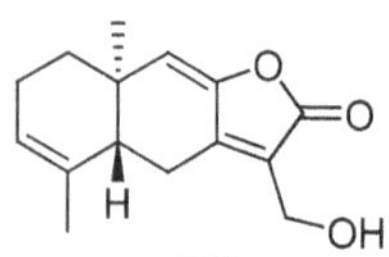

151

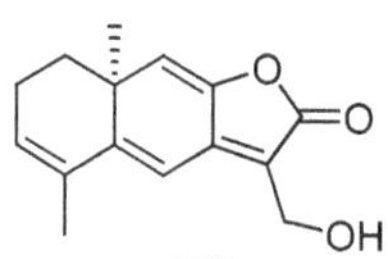

152

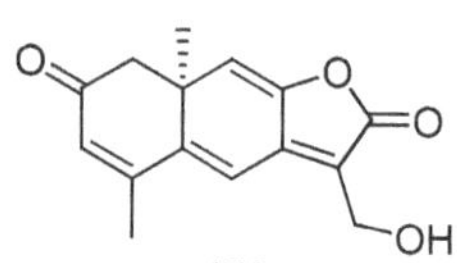

153

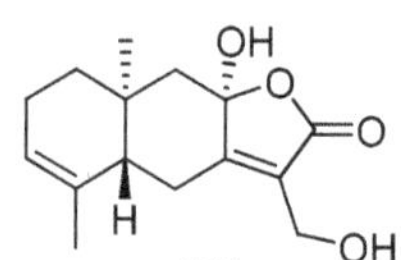

154

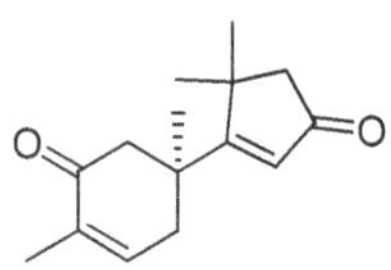

156

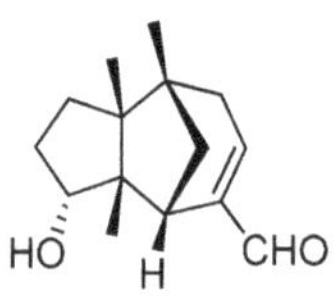

157

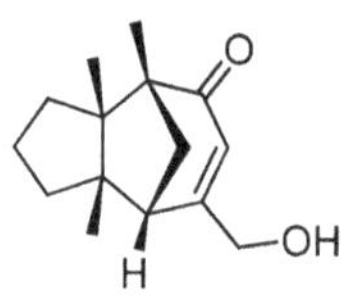

158

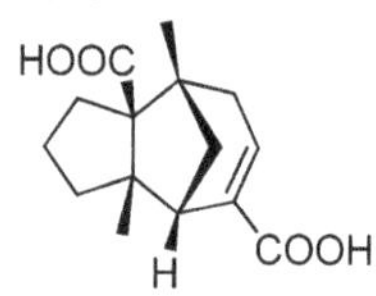

159

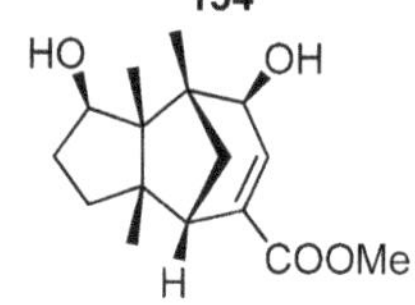

160

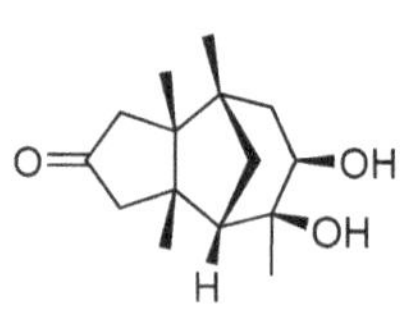

161

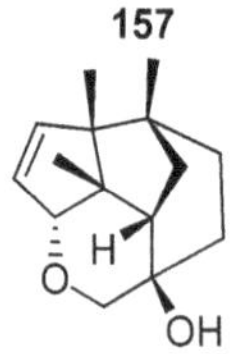

162

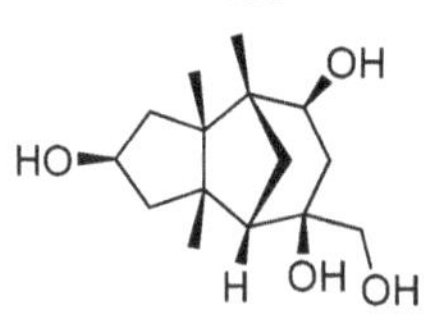

163

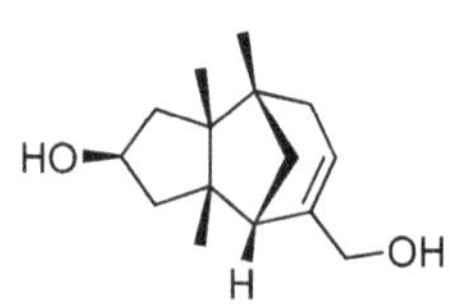

164

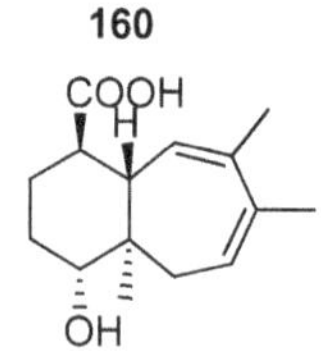

165

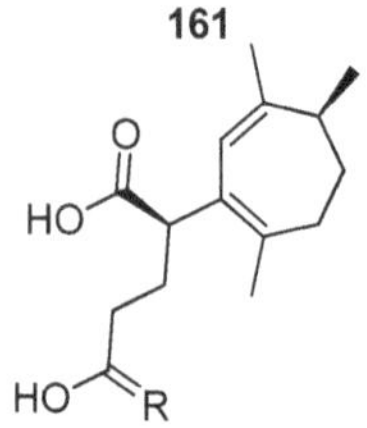

166 R = H_2
167 R = O

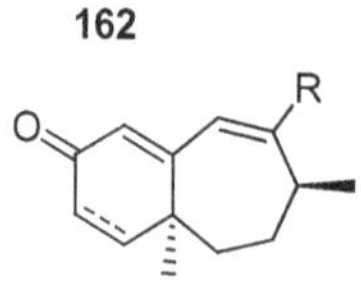

168 R = CHO, $\Delta^{7(8)}$
169 R = $CH(OMe)_2$, $\Delta^{7(8)}$
170 R = CH_2OH, $\Delta^{7(8)}$
171 R = COOH

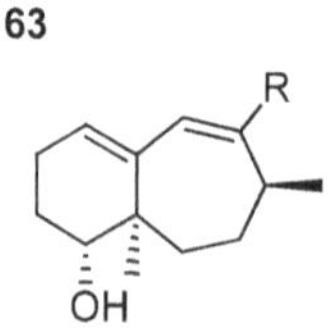

172 R = COOH
173 R = CH_2OH

Antrodins A–E **145–149** were isolated from the fermentation of *Antrodiella albocinnamomea* [70]. Tremutin H **150** was isolated from cultures of the basidiomycetes *Irpex lacteus* [71]. The absolute configuration of **150** was determined by single-crystal X-ray diffraction analysis, and **150** shows a weak inhibitory effect on NO production with an IC_{50} value of 22.7 μM.

2.5. Eudesmanolide, Gymnomitrane, and Humulane

Humulane-type sesquiterpenoids are found rarely in nature. They have been recognized as being biogenetic precursors of many types of sesquiterpenoids [6]. The macrocyclic nature of members of the humulane group has proved to be troublesome for the determination of their absolute configurations.

Four new 12,8-eudesmanolides **151–154** were isolated from a mangrove rhizosphere-derived fungus *Eutypella* sp. 1–15 [72]. Periconianone A **155**, a polyoxygenated sesquiterpenoid with a new 6/6/6 tricarbocyclic skeleton, was isolated from the endophytic fungus *Periconia* sp., and the biosynthesis of the unusual six-membered carbonic ring of **155** was postulated to be formed through intramolecular aldol condensation (Scheme 3) [73]. The first enantioselective total synthesis of the periconianone A **155** based on a postulated biogenesis has been reported (Scheme 4) [74].

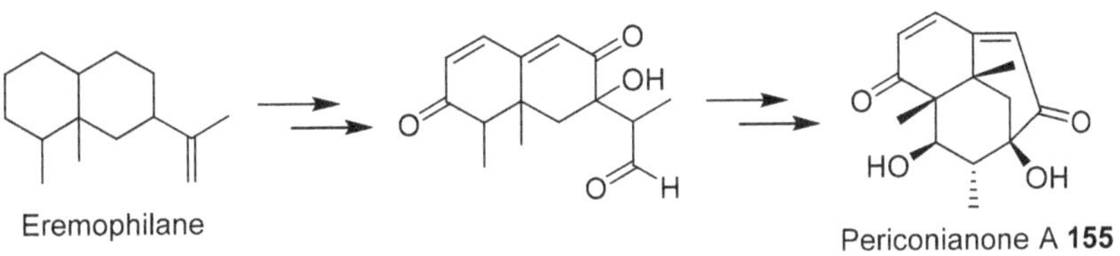

Scheme 3. Hypothetical biosynthetic pathway of periconianone A **155** (Reference [73]).

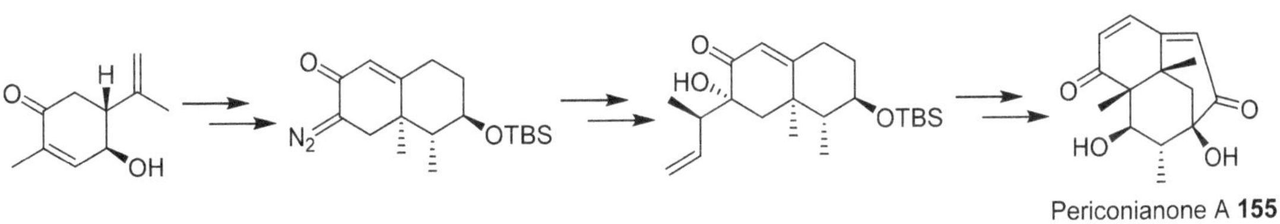

Scheme 4. Total synthesis of periconianone A **155** (Reference [74]).

An unusual type sesquiterpene **156** possessed an unusual 14(7-6)-cuparane scaffold (Scheme 5), and six rarely-encountered gymnomitrane-type sesquiterpenoids **157–162**, were isolated from the medicinal mushroom *Ganoderma lingzhi* [75]. A new gymnomitrane-type sesquiterpenoid **163** was isolated from the fruiting body of *Ganoderma lucidum* [76]. This compound **163** significantly inhibited the growth of epidermal growth factor receptor-tyrosine kinase inhibitor EGFR-TKI-resistant human lung cancer A549 and human prostate cancer PC3 cell lines. Antrodin F **164** was isolated from the fermentation of *Antrodiella albocinnamomea* [70].

Nine new humulane-derived sesquiterpenoids, ochracenes A–I **165–173**, were isolated from the Antarctic fungus *Aspergillus ochraceopetaliformis* [77]. A biogenetic pathway for them was given in Scheme 6. The two unprecedented 8,9-secocyclic sesquiterpenoids, ochracenes B **166** and C **167**, exhibited inhibitory effects on LPS-induced NO release in RAW 264.7 mouse macrophage cell with IC_{50} values of 14.6 and 18.3 μM, respectively.

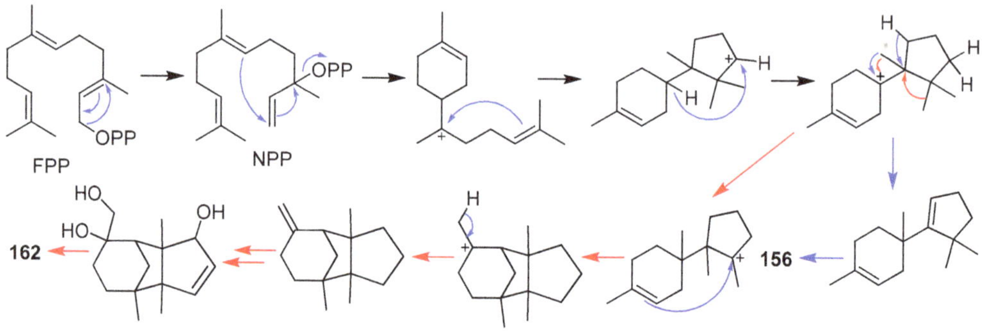

Scheme 5. Proposed biosynthetic pathway of **156** and **162** (Reference [75]).

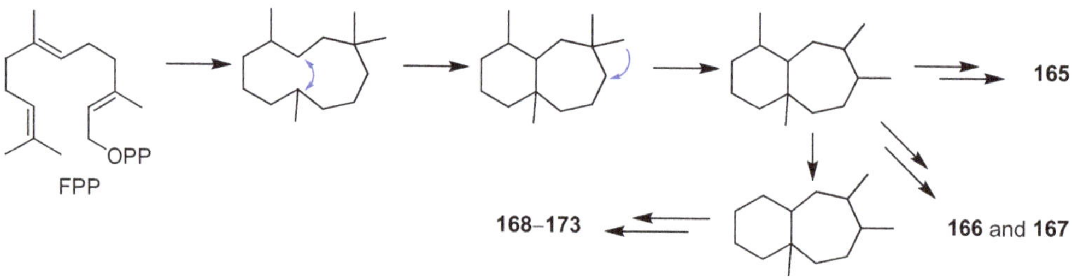

Scheme 6. Postulated biogenetic pathway for ochracenes A–I **165–173** (Reference [77]).

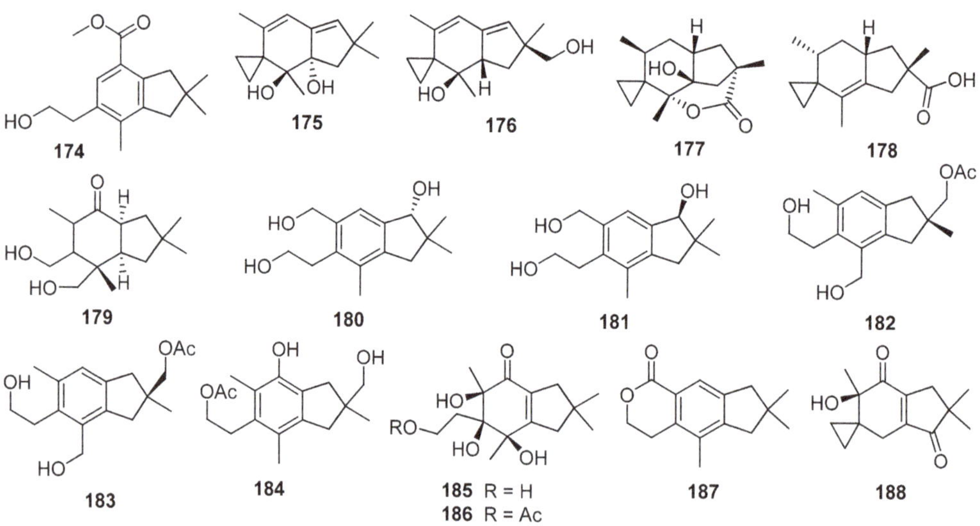

2.6. Illudane, Illudalane, Protoilludane, Marasmane, and Norilludane

A review offers a comprehensive description of the investigations that started with the discovery of illudins in 1950, led to HMAF clinical trials against various tumors as a single agent and in combination therapy beginning in 2002, and culminated in the past decade of advances in chemical synthesis and mechanisms of toxicity of AFs, including biotransformation processes, DNA alkylation products, unique influences of DNA repair capacities, and enzyme inhibition properties [9]. The 4/6/5 ring-fused protoilludane-type sesquiterpenoids are the precursors of many other sesquiterpenoids, representing the largest group of sesquiterpene metabolites of fungal origin.

Phellinignin D **174** was isolated from the fungus *Phellinus igniarius*, which possessed a new carbon skeleton that might derive from an illudane framework by methyl shift and aromatization [78]. Phellinignin D **174** showed moderate cytotoxicities to three human cancer cell lines (HL-60, SMMC-7721, and SW480) with the IC_{50} values of 21.1, 12.3, and 13.9 μM, respectively.

Illudadienes A **175** and B **176** were obtained from the wood-decomposing fungus *Granulobasidium vellereum* [79]. Phellinuin J **177** and sulphureuine A **178** were isolated from cultures of *Phellinus tuberculosus* and *Laetiporus sulphureus* [80]. Agrocybins H–K **179–184** were obtained from the edible mushroom *Agrocybe salicacola* [81]. Craterellins D **185** and E **186** were isolated from cultures of *Craterellus cornucopioides* [82]. Illudalane derivative, granulolactone **187**, and a 15-norilludane, granulodione **188**, were isolated from an agar plate culture of *Granulobasidium vellereum* [83].

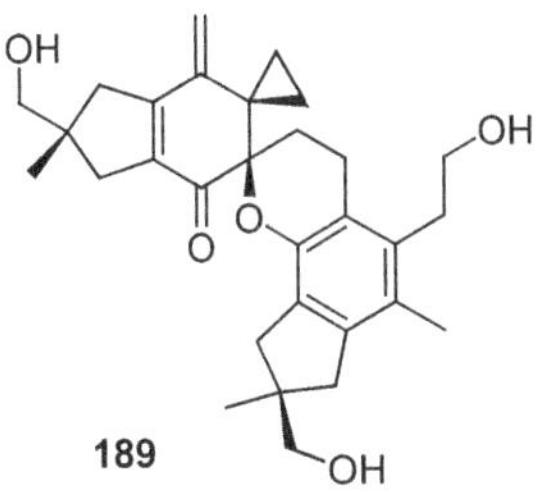

189

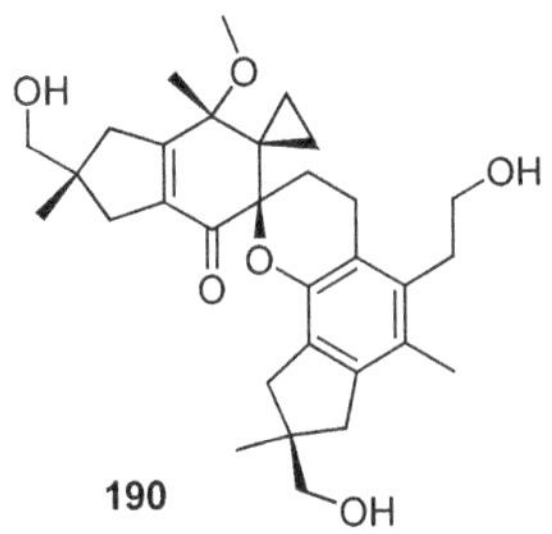

190

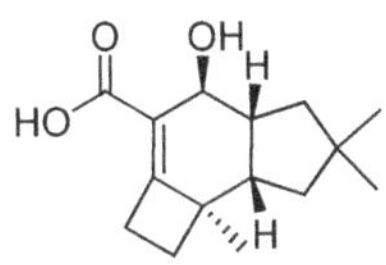

191

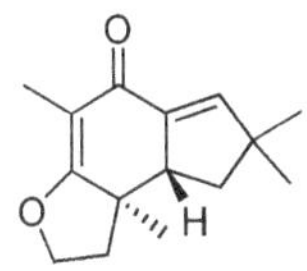

192

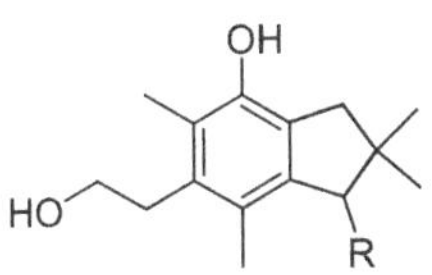

193 R = α-OMe
194 R = β-OMe

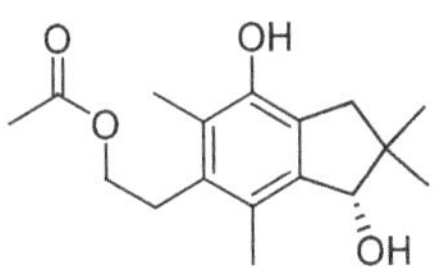

195

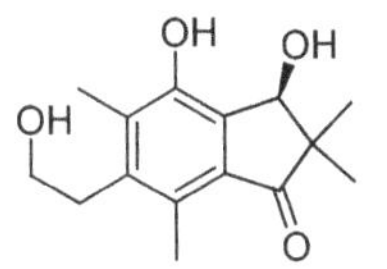

196

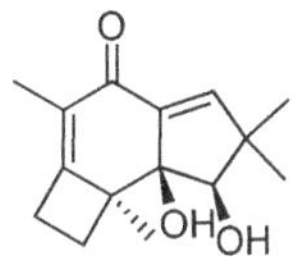

197

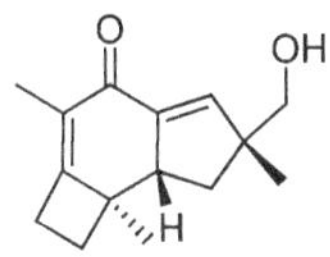

198

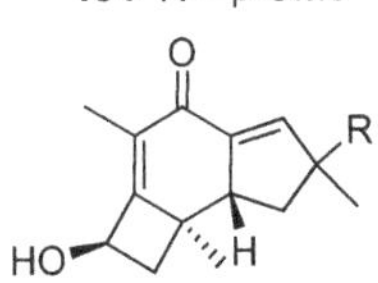

199 R = β-CH_2OH
200 R = α-CH_2OH

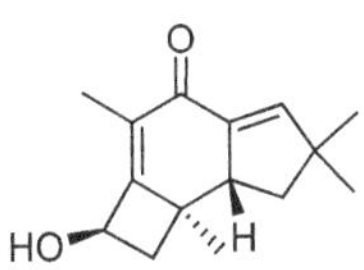

201

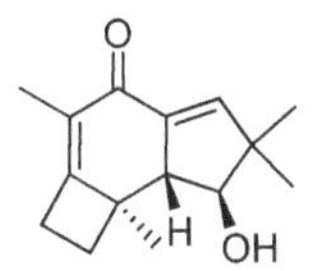

202

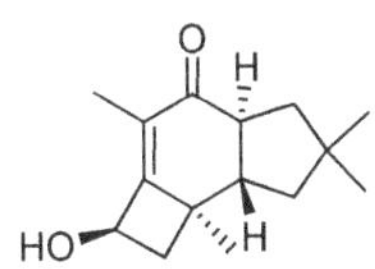

203

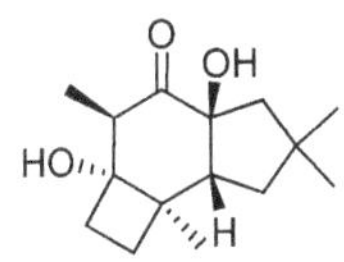

204

Two new disesquiterpenoid derivatives, bovistol B **189** and C **190**, and a new protoilludane derivative, pasteurestin C **191**, were isolated from the fermentation broth of the edible mushroom *Cyclocybe aegerita* [84]. Four illudalanes **192–195**, an unusual 2,3-*seco*-protoilludane **196**, and eight protoilludanes **197–204** were identified from the liquid culture of the endophytic fungus *Phomopsis* sp. TJ507A [85]. Phomophyllins A–G **196–202**, and phomophyllin I **204** displayed β-site amyloid precursor protein cleaving enzyme 1 (BACE1) inhibitory activities ranging from 19.4% to 43.8% at the concentration of 40 μM.

Epicoterpenes A–E **205–209**, and armilliphatic A **210** were isolated from *Armillaria* sp. by co-culture with the endophytic fungus *Epicoccum* sp. associated with *Gastrodia elata* [86]. Epicoterpene D **208** was the first example of an *ent*-protoilludane sesquiterpenoid scaffold bearing a five-membered lactone. Two new protoilludane sesquiterpene aryl esters **211** and **212** were isolated from the mycelium of *Armillaria mellea* [87]. Compound **212** showed cytotoxic activity for HepG2 cells with an IC_{50} value of 18.03 μg/mL. Three new sesquiterpene aryl esters, melleolide N **213**, Q **214**, and R **215**, were isolated from the EtOH extract of the mycelium of *Armillaria mellea* [88]. And **213–215** showed cytotoxicity to several human cancer cell lines.

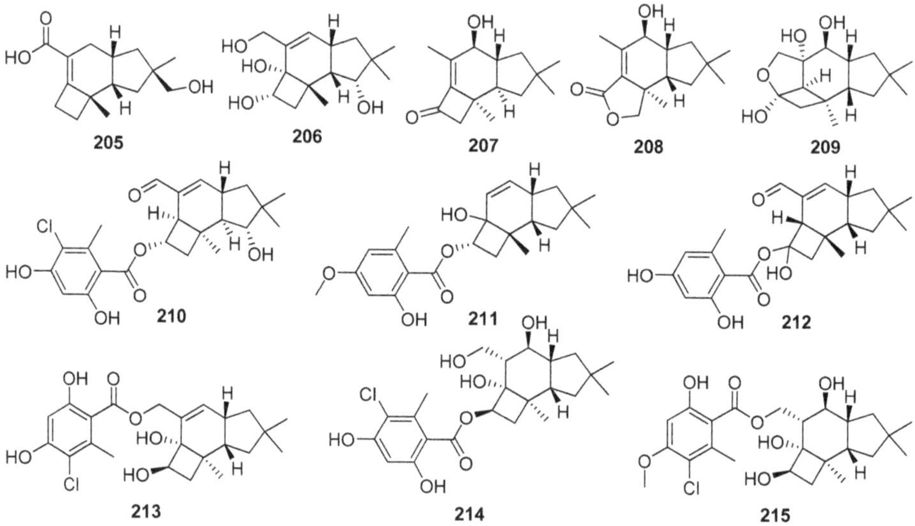

Unified total syntheses of marasmane, mellolide, and protoilludane sesquiterpenoids have been achieved through a key organocatalytic enantioselective annulation (Scheme 7) [89]. The elaboration of key bicyclic lactone **216** was the molecular springboard from which the first enantioselective total syntheses of protoilludanes echinocidin B **220** and echinocidin D **221**, and the mellolide armillaridin **219**, as well as the synthesis of the marasmane isovelleral **222**, were accomplished. The vanadium(II)/zinc(II) reductive coupling yielded the final ring of the densely functionalized *cis*-fused carbocyclic core. Finally, the unexpected semi-Pinacol-type ring contraction to establish cyclopropyl aldehyde **218** from cyclobutanediol **217** was potentially biomimetic in origin.

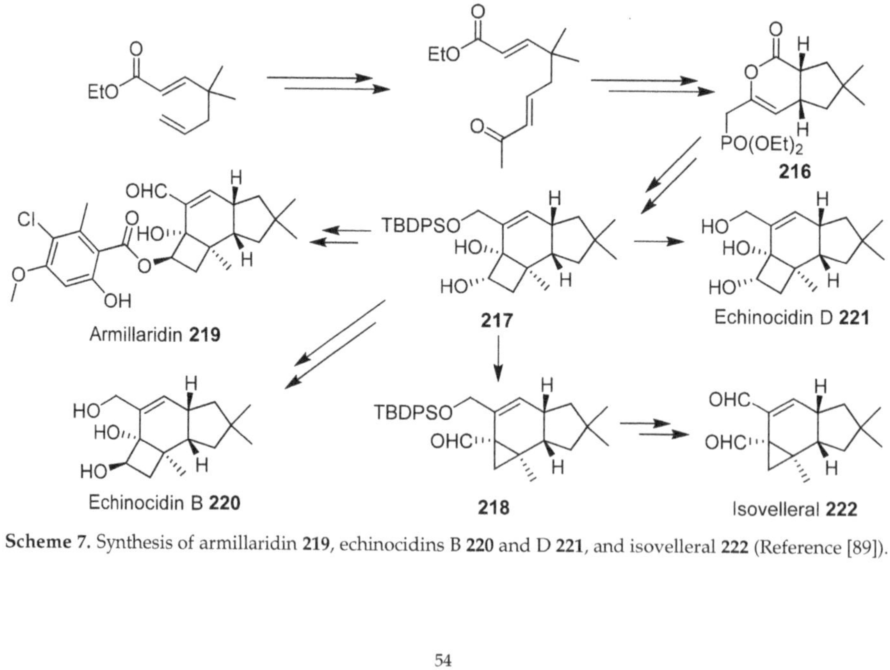

Scheme 7. Synthesis of armillaridin **219**, echinocidins B **220** and D **221**, and isovelleral **222** (Reference [89]).

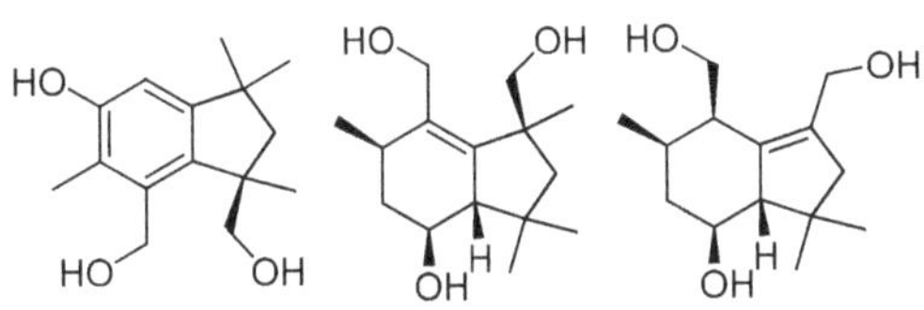

223 224 225 226 227 228

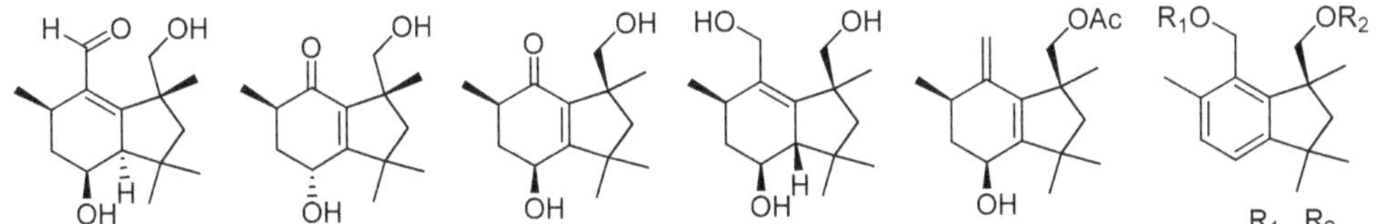

229 230 231 232 233

	R_1	R_2
234	H	Ac
235	Ac	Ac

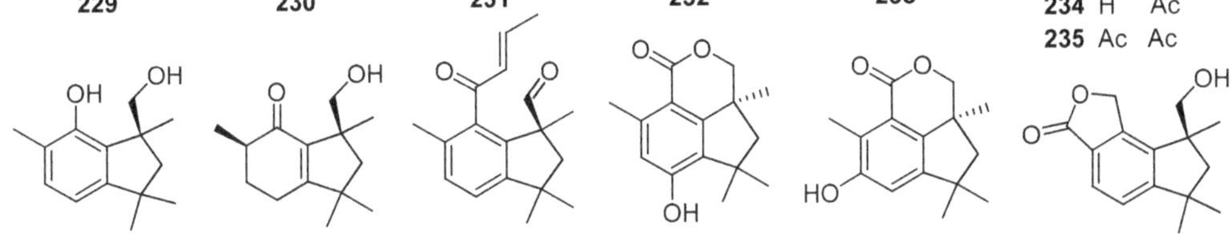

236 237 238 239 240 241

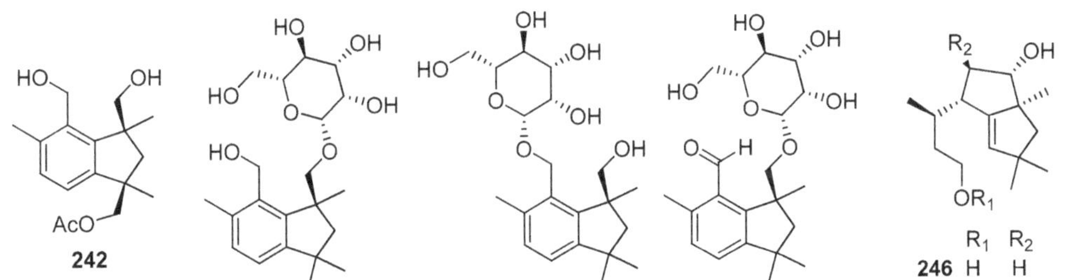

242 243 244 245

	R_1	R_2
246	H	H
247	Ac	H
248	H	OH

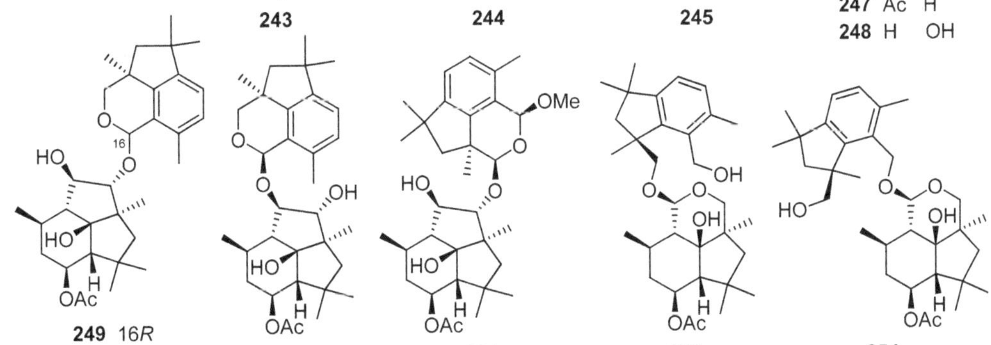

249 16*R*
250 16*S*
251 252 253 254

2.7. Botryane and Seco-Probotryane

A botryane-type sesquiterpenoid **223** was identified from the liquid culture of the endophytic fungus *Phomopsis* sp. TJ507A [85]. Arthrinins E–G **224–226**, three new sesquiterpenoids possessing non-isoprenoid botryane skeleton, were isolated from the endophytic fungus *Arthrinium* sp. HS66 [90]. Five new botryanes **227–231** were obtained from an endophytic fungus *Nemania bipapillata* [91]. Five new botryanes **232–236** were isolated from *Trichoderma oligosporum* [92]. Compounds **236** showed moderate cytotoxicity activity against K562 cells with an inhibitory rate of 45–60% at 6.25 μM (Taxol was used as a positive control with 60.3% inhibition at 2.0 μM).

A new 10-norbotryane derivative **237** and three new botryanes **238–240** were isolated from the ascomycete *Hypoxylon rickii* [93,94]. Five new botryanes **241–245**, along with 4,5-*seco*-Probotryenols A–C **246–248** derived from cleavage of the probotryane skeleton at C-4/C-5, were isolated from *Stachybotrys bisbyi* [95]. Six new heterodimeric botryane ethers, hypocriols A–F **249–254**, were isolated from the insect-associated *Hypocrea* sp. EC1-35 [96]. A plausible biosynthetic pathway for **249–254** was given (Scheme 8). Hypocriols A–D **249–252** and F **254** showed significant activity against the HeLa cell, with IC_{50} values of 7.7, 3.1, 11.8, 3.8, and 4.6 μM, respectively. Hypocriol F **254** inhibited the proliferation of the HCT116 cell, showing an IC_{50} value of 2.7 μM.

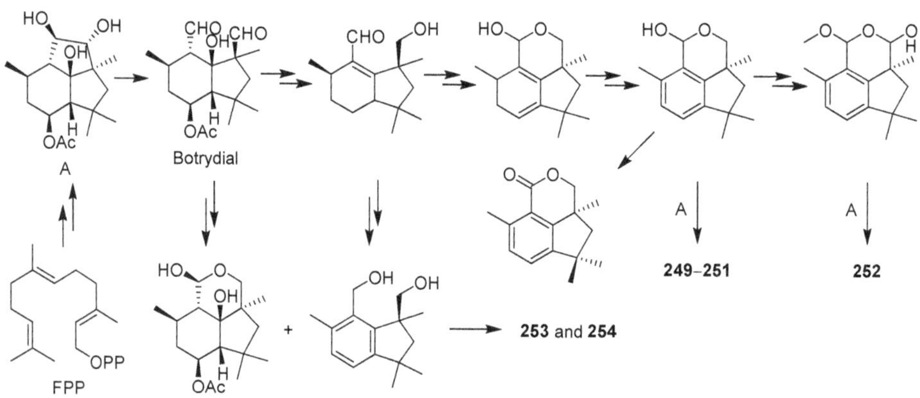

Scheme 8. Plausible biosynthetic pathways for hypocriols A–F **249–254** (Reference [96]).

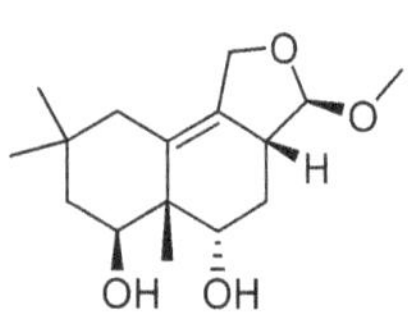

255

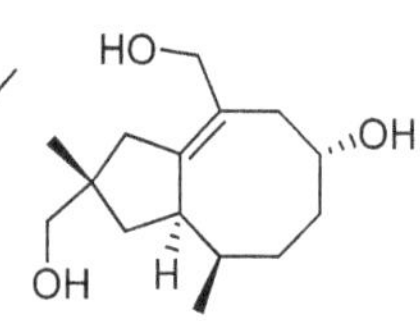

256

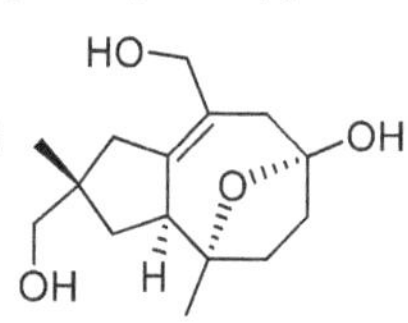

257

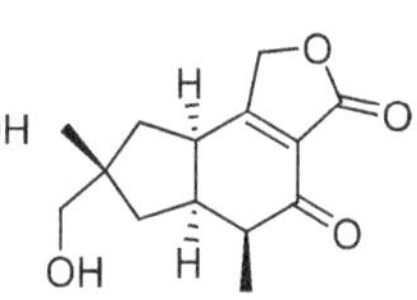

258

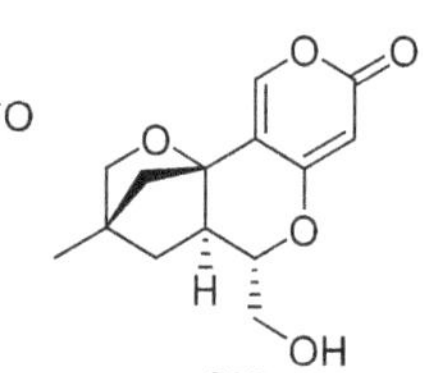

259

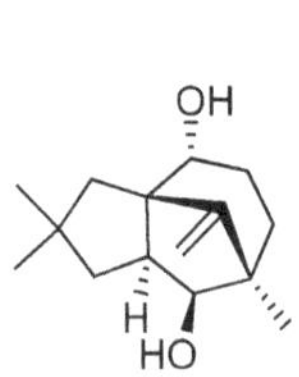

260

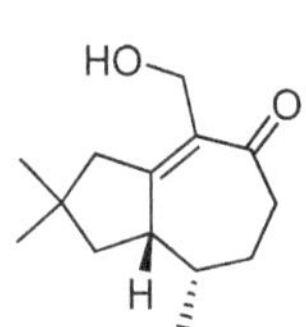

261

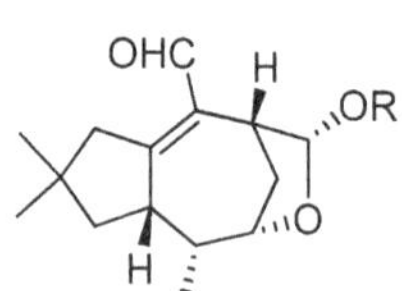

262 R = Me
263 R = Ac

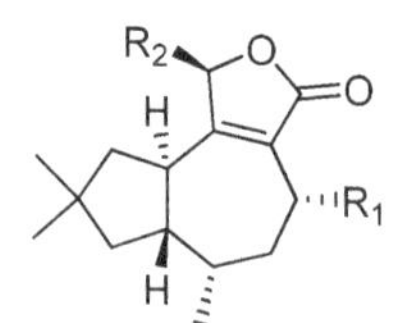

264 R_1 = OH, R_2 = H
265 R_1 = H, R_2 = OH

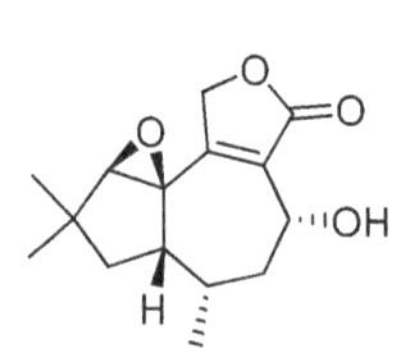

266

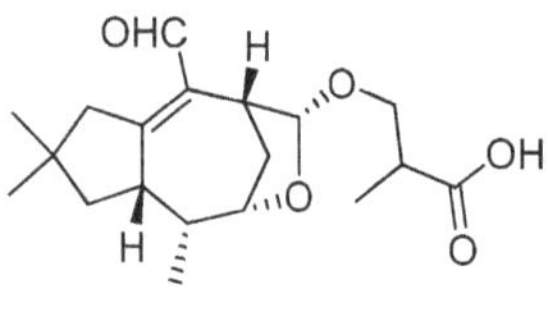

267

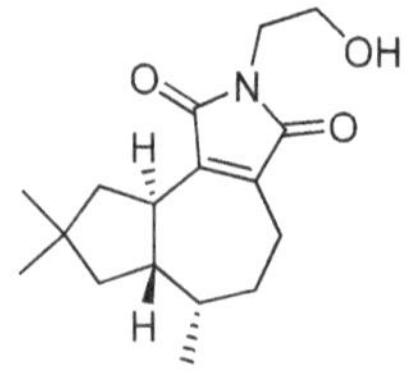

268

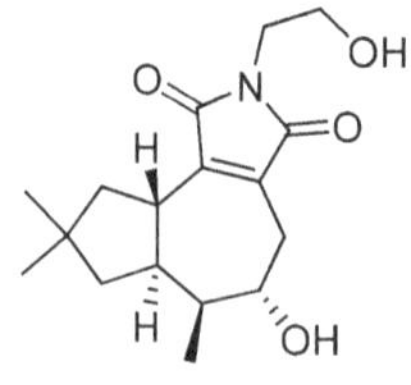

269

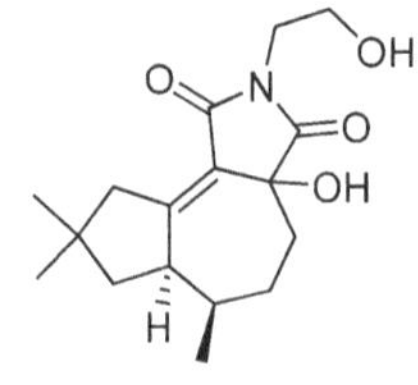

270

2.8. Tremulane, Sterpurane, Phlebiane, Merulane, and Irlactane

Tremulane-type sesquiterpenoids are a class of sesquiterpenoids with a 5/7-ringfused perhydroazulene carbon skeleton. The first example was isolated from the wood-decaying fungus *Phellinus tremulae* in 1993 [97]. The biosynthesis pathway was elucidated through a ^{13}C-labeled feeding experiment revealed that tremulanes are derived from *trans,trans*-farnesyl pyrophosphate via humulene and a key step of methyl migration [98].

A new irlactane-type, irlactin K **255**, was isolated from the fermentation broth of the medicinal fungus *Irpex lacteus* [99]. The absolute configuration of **255** was established by single-crystal X-ray diffraction analysis. Irlactin K **255** could be derived from the tremulane type sesquiterpene irlactin E via a ring rearrangement [100]. Conosiligins A–D **256–259**, four ring-rearranged sesquiterpenoids, were isolated from cultures of the basidiomycete *Conocybe siliginea* [101]. Conosiligins A **256** and B **257** possessed a 5/8-fused ring system, while conosiligin C **258** has a 5/6-fused backbone conjugated with a γ-lactone. Conosiligin D **259** was a 5,6-seco tremulane derivative with the loss of a skeletal carbon, featuring a tetracyclic system involving a pyranone moiety (Scheme 9). Conosiligins C **258** and D **259** inhibited Con A-induced T cell proliferation with IC_{50} values of 12.3 and 6.6 μM, respectively.

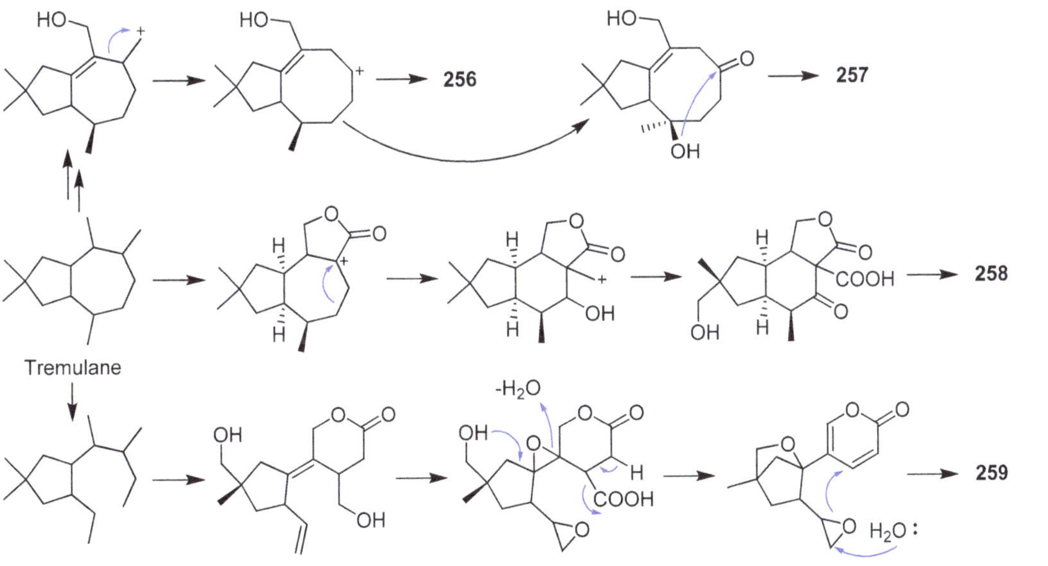

Scheme 9. Proposed biosynthetic pathway for conosiligins A–D **256–259** (Reference [101]).

Antroalbocin A **260** possessing a bridged tricyclic system was isolated from cultures of the higher fungus *Antrodiella albocinnamomea* [102]. The structure with the absolute configuration was determined by extensive spectroscopic methods and single-crystal X-ray diffraction analysis and a plausible biosynthetic pathway for **260** was proposed (Scheme 10).

Twenty-two tremulanes, irlactins F–J **261–265**, L–M **266–267**, irlactam A **268**, and irpexolactins A–N **269–282**, were isolated from cultures of the medicinal fungus *Irpex lacteus* [99,103–105]. Irlactin I **264** exhibited moderate cytotoxicities on HL-60, SMMC-7721, A-549, MCF-7, and SW480 cells with IC_{50} values of 16.23, 20.40, 25.55, 19.05, and 18.58 μM, respectively [104].

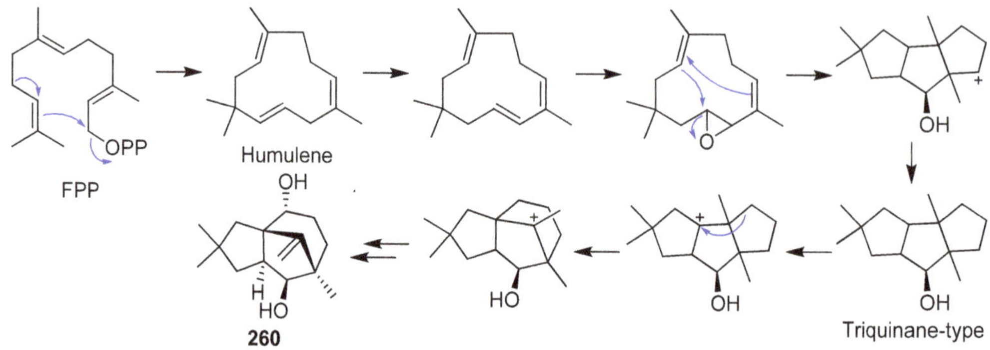

Scheme 10. Proposed biosynthetic pathway for antroalbocin A **260** (Reference [102]).

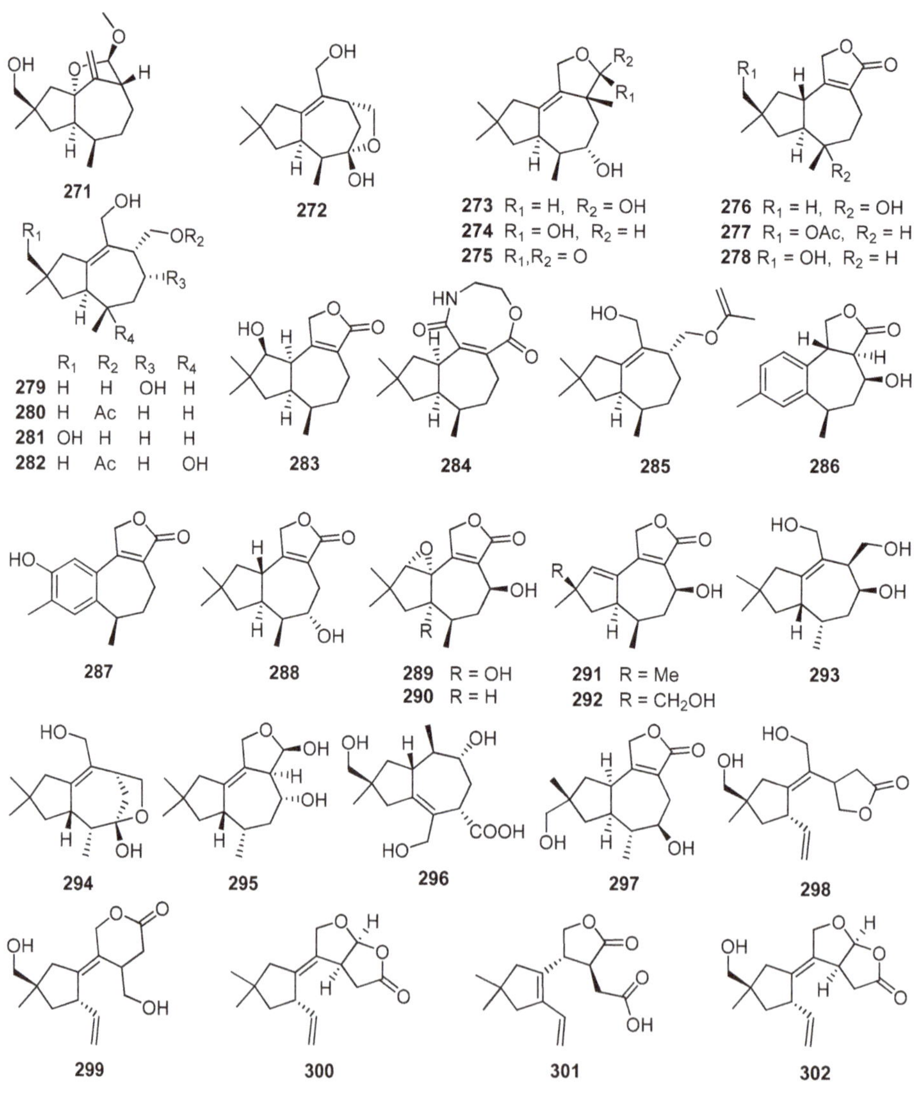

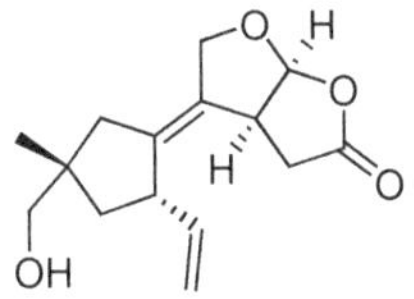
303

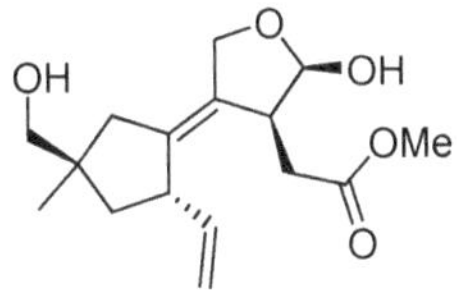
304

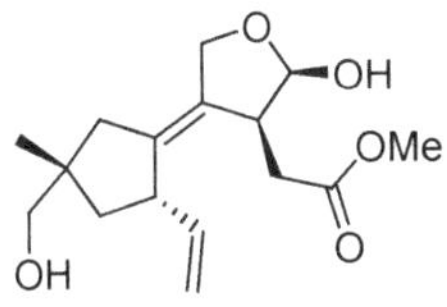
305

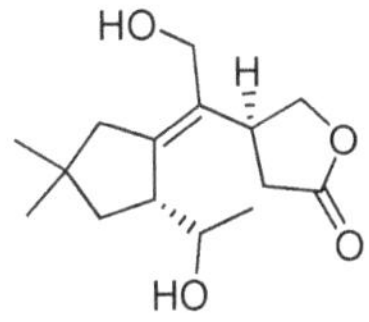
306

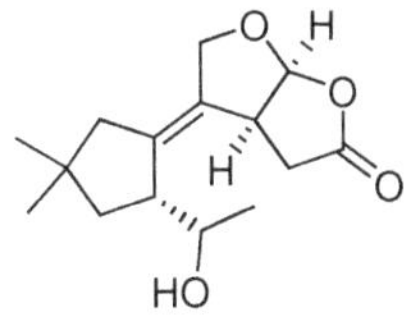
307

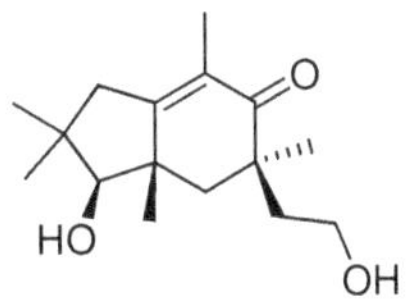
308

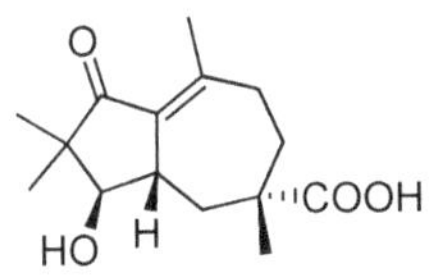
309

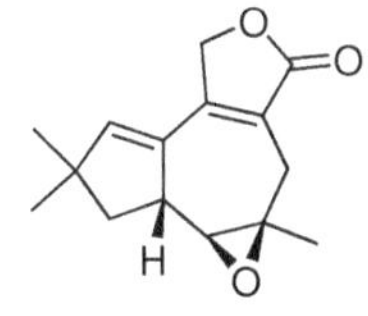
310

Phellinignins A–C **283–285** were new tremulane sesquiterpenoids that have been isolated from *Phellinus igniarius* [78]. Phellinignins A **283** and B **284** showed certain cytotoxicities to three human cancer cell lines (HL-60, SMMC-7721, and SW480) with the IC_{50} values of 0.7–17.4 μM, respectively. Tremutins A–G **286–292** were isolated from cultures of the basidiomycetes *Irpex lacteus* [71]. Tremutins A **286** and B **287** possessed an unusual 6/7-fused ring system that might be derived from a tremulane framework (Scheme 11), **289** and **290** were the first tremulane examples with a 1,2-epoxy moiety to be reported. Tremutin A **286** inhibited the lipopolysaccharide (LPS)-induced proliferation of B lymphocyte cells with an IC_{50} value of 22.4 μM. Tremutin B **287** inhibited concanavalin A (Con A)-induced T cell proliferation and LPS-induced B lymphocyte cell proliferation with IC_{50} values of 16.7 and 13.6 μM, respectively.

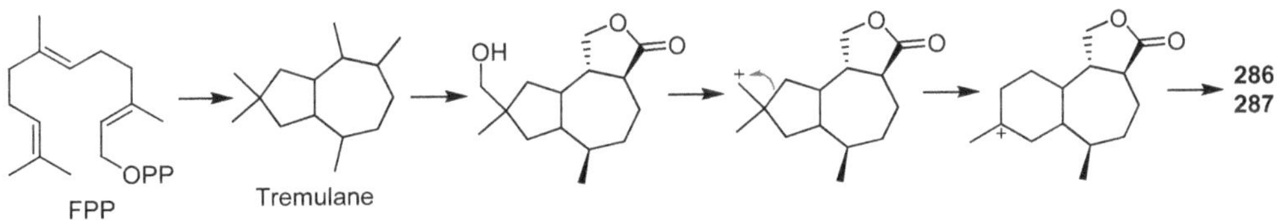

Scheme 11. Proposed biosynthetic pathway for tremutins A **286** and B **287** (Reference [71]).

Nigrosirpexin A **293** was produced by *Nigrospora oryzae* co-cultured with *Irpex lacteus* [106]. Two new tremulanes **294** and **295** were obtained from different cocultures of *Nigrospora oryzae* and *Irpex lacteus* in a solid medium [107]. 5-Demethyl conocenol C **294** showed antifungal activities against *Didymella glomerate* and *Colletotrichum gloeosporioides* with MICs of 1 and 8 μg/mL, respectively.

Davotremulanes A–D **296–299** were isolated from a plant-associated fungus X1-2 [108]. Davotremulanes A **296** and B **297** displayed selectively moderate activities to the A549 cell line with IC_{50} at 15.3, 25.2 μg/mL. A new tremulane sesquiterpenoid analogue **300** was isolated from the cultures of endophytic fungus *Colletotrichum capsica* [109]. Leptosphin B **301** was isolated from the endophytic fungus *Leptosphaeria* sp. XL026 [110]. Leptosphin B **301** showed moderate antibacterial activity against *Bacillus cereus* with a MIC value of 12.5 μg/mL.

Six 5,6-*seco*-tremulane analogues **302–307** were isolated from the culture broth of the medicinal fungus *Irpex lacteus* [111]. Two sesquiterpenes with new carbon skeletons, *seco*-sterpurane **308** and phlebiane **309**, and a novel merulane sesquiterpene **310** were isolated from cultures of the basidiomycete *Phlebia tremellosa* [112]. The plausible biogenetic pathways of **309** and **310** is shown in Scheme 12.

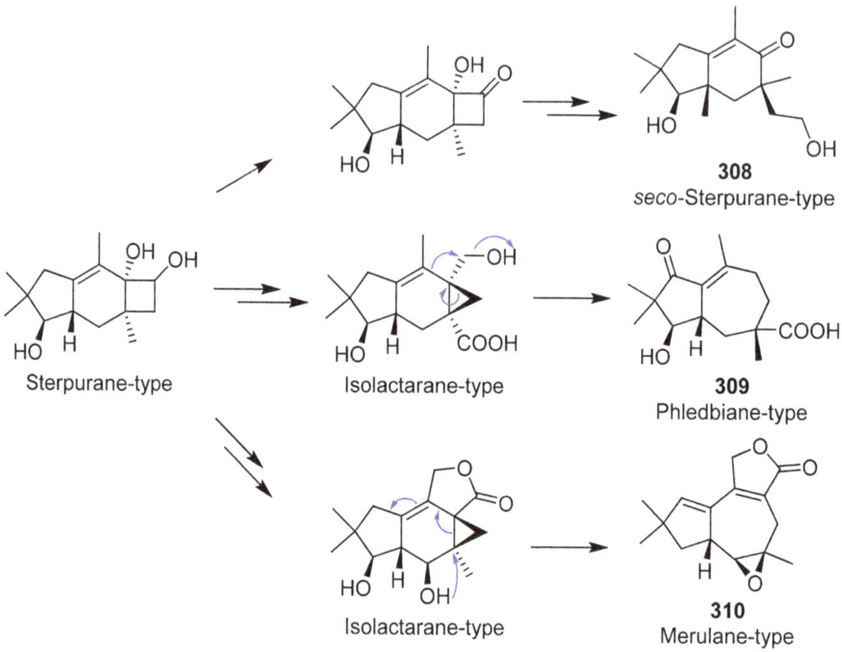

Scheme 12. Plausible biogenetic pathways of **308–310** (Reference [112]).

2.9. Trichothecene, Merosesquiterpenoid, Norsesquiterpenoid, and Pyrone

Trichothecenes are a family of sesquiterpenoid mycotoxins produced by multiple genera of fungi, including plant and insect pathogens, and they are toxic to animals and humans and frequently detected in cereal crops [113]. Because of their diversity in structure and biological activity, trichothecenes are of concern in agriculture, food contamination, health care, and building protection.

Trichoderminol **311** was isolated from the filamentous fungus *Trichoderma albolutescens* [114]. Trichobreols A–E **312–316** were isolated from the marine-derived fungus *Trichoderma* cf. *brevicompactum* [115,116]. Trichobreols A–E **312–316** inhibited the growth of two yeast-like fungi, *Candida albicans*, and *Cryptococcus neoformans*, with a range of MIC values of 1.6 to 50 μg/mL [115,116]. Three new macrocyclic trichothecenes, miophytocen D **317**, roridin F **318**, and satratoxin I **319**, were isolated from a deadly poisonous mushroom *Podostroma cornu-damae* [117]. Satratoxin I **319** showed cytotoxic potency to etoposide against four human breast cancer cell lines (Bt549, HCC70, and MDA-MB-231), with IC_{50} values of 1.8, 7.7, and 3.6 μM, respectively.

Epiroridin acid **320**, verrucarins Y **321** and Z **322**, and dihydromyrothecine C **323**, four new macrocyclic trichothecenes, were isolated from the endophytic fungus *Myrothecium roridum* [118–121]. The cytotoxic mechanisms result showed that the epiroridin acid **320** induced the apoptosis of cancer cell HepG-2 via activation of caspase-9 and caspase-3, up-regulation of *bax* gene expression, down-regulation of *bcl*-2 gene expression, and disruption of the mitochondrial membrane potential of the HepG-2 cell [118].

Chartarenes A–D **324–327** were isolated from the sponge-derived fungus *Stachybotrys chartarum* [122]. Chartarenes A–D **324–327** exerted potent or selective inhibition against a panel of tumor cell lines including HCT-116, HepG2, BGC-823, NCI-H1650, and A2780, with IC_{50} values ranging from 0.68 to 10 μM. In addition, chartarenes B **326**, C **327**, and D **324** showed potent inhibition against tumor-related kinases FGFR3, IGF1R, PDGFRb, and TRKB, with IC_{50} values ranging from 0.1 to 12.9 μM.

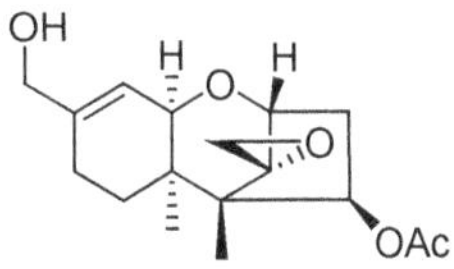

311

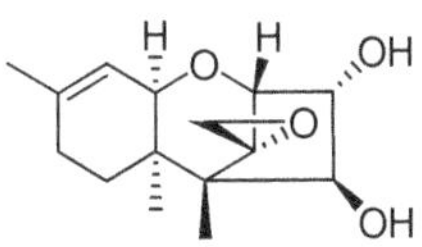

312

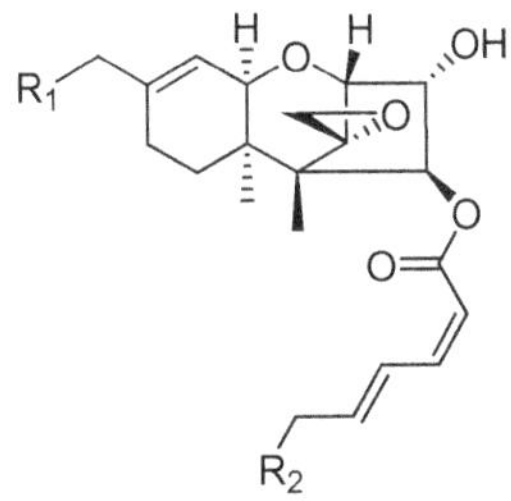

313 R_1,R_2 = H
314 R_1 = OH, R_2 = H
315 R_1 = H, R_2 = OH

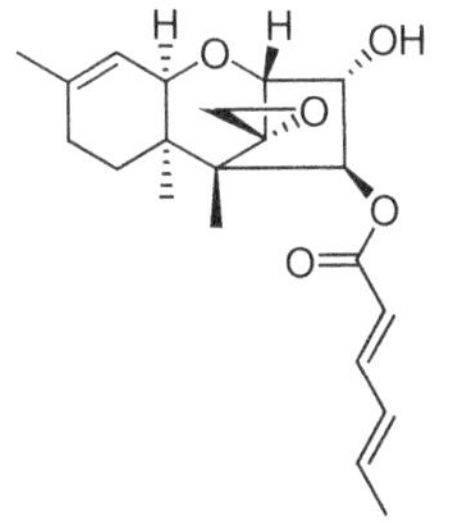

316

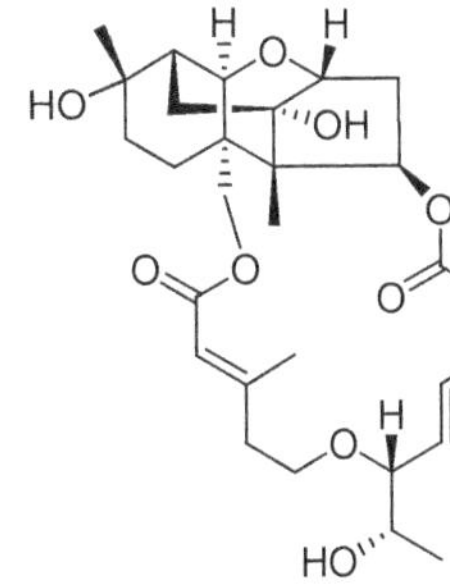

317

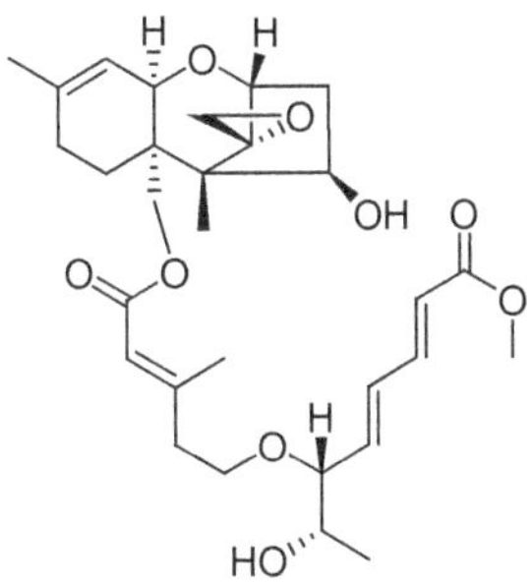

318

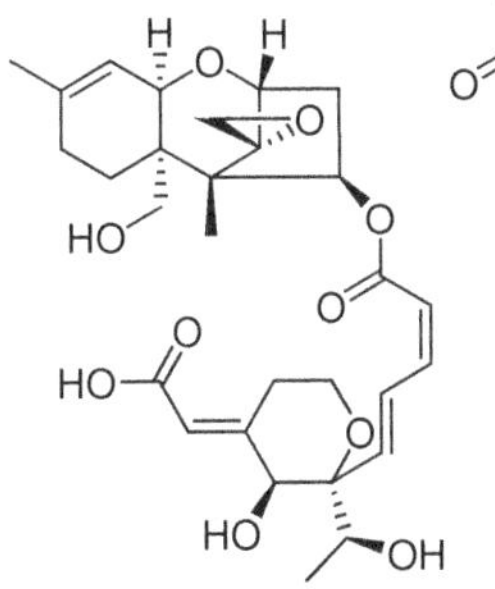

319

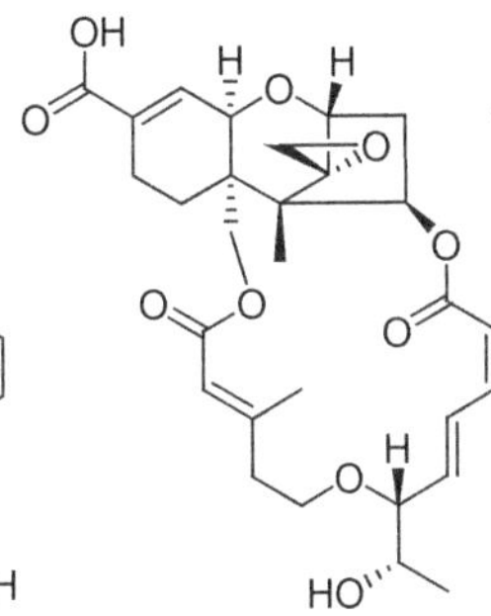

320

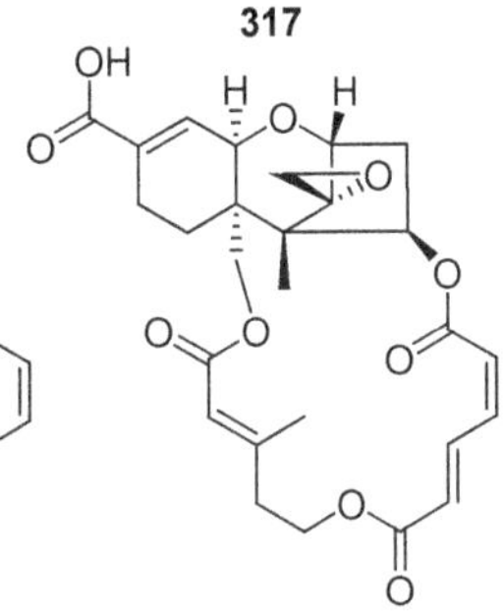

321

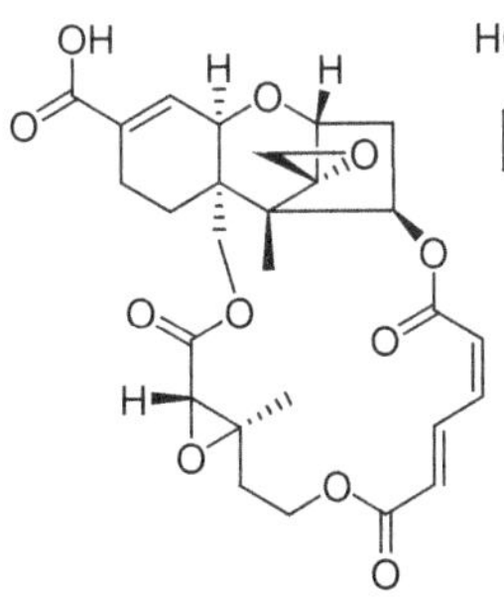

322

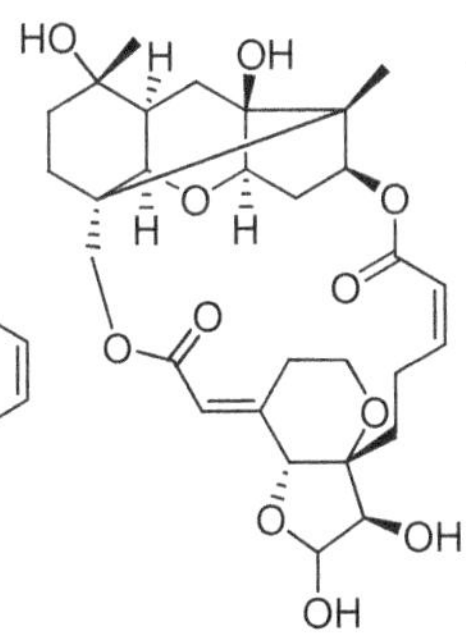

323

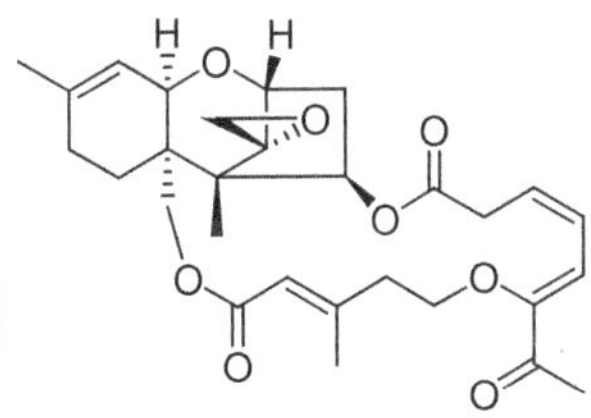

324

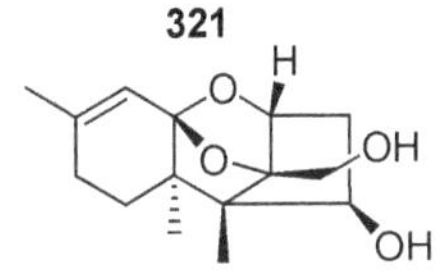

327

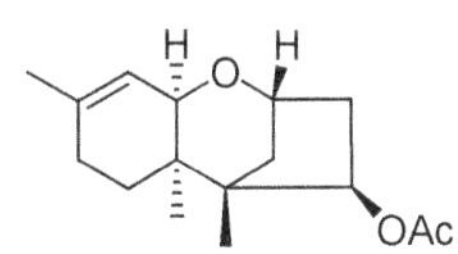

328

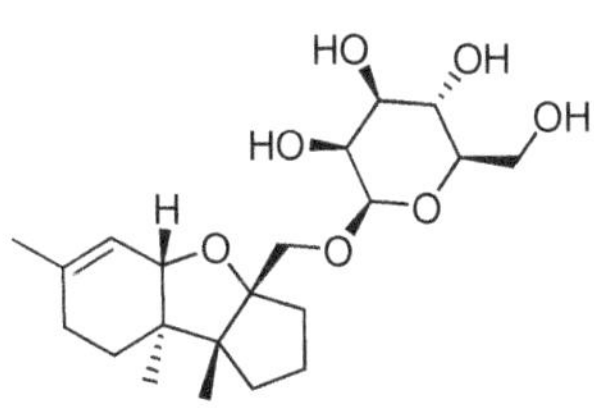

325

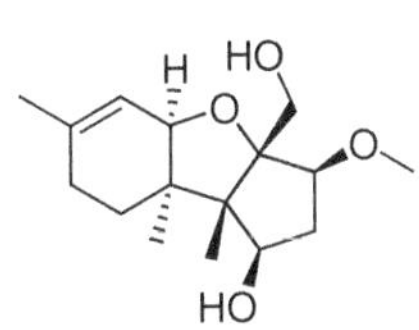

326

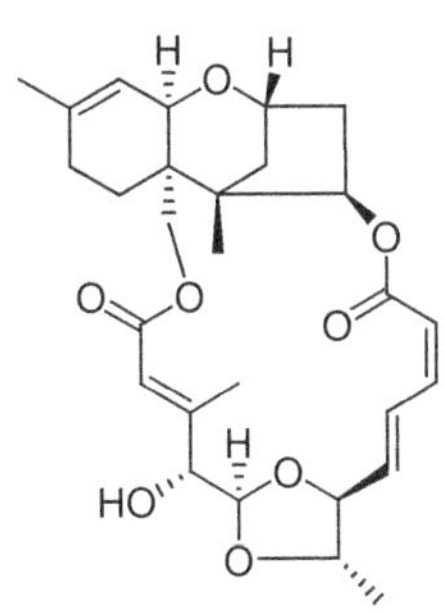

329

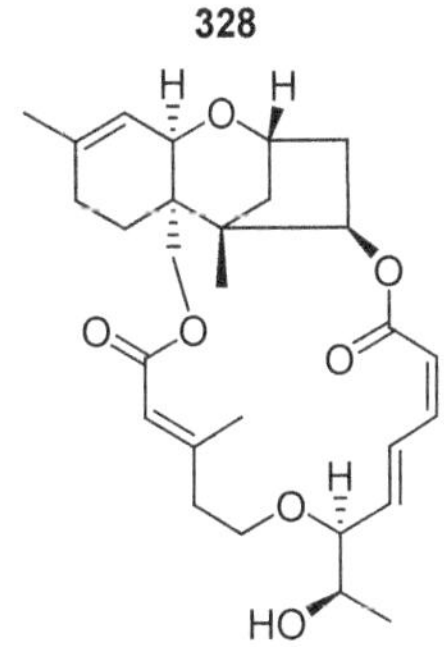

330

12-Deoxytrichodermin **328**, 12-deoxyroridin J **329**, and 12-deoxyepiisororidin E **330** were isolated from the fungus *Calcarisporium arbuscular*, and *Trichoderma* sp., respectively [123,124]. The structure-activity relationship investigation of **328–330** with other known natural trichothecenes against a human colon cancer cell line (COLO201) and filamentous fungus *Cochliobolus miyabeanus* revealed that the 12-epoxide functionality is essential for the antifungal activity [124].

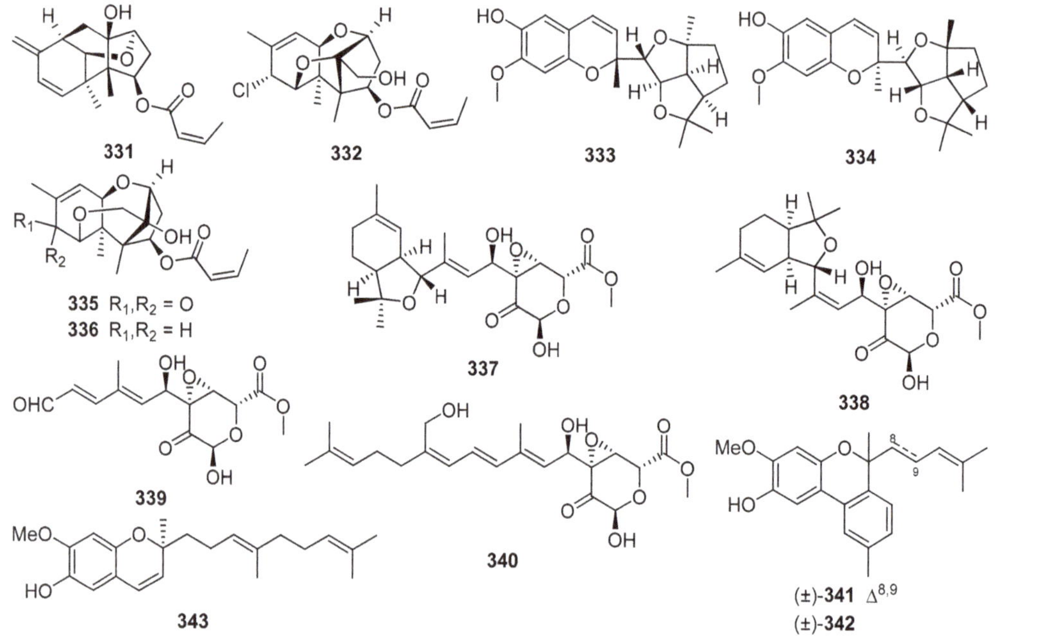

Four novel trichothecene sesquiterpenoids possessing new ring systems, trichothecrotocins A **331** and B **332**, trichothecrotocins K **335** and L **336**, and a merosesquiterpenoid racemate possesses a novel 6/6–5/5/5 fused ring system, (±)-trichothecrotocin C (**333** and **334**), and seven new merosesquiterpenoids, trichothecrotocins D–J **337–343**, were obtained from potato endophytic fungus *Trichothecium crotocinigenum* by bioguided isolation (Schemes 13 and 14) [125,126]. Compounds **337–340** were rare meroterpenoids featuring a *seco*-phenyl group, while **337** and **338** possessed a novel 6–6/5 fused ring system. Compounds **331–335**, and **337–340** showed antiphytopathogenic activities with MIC values of 8–128 μg/mL [125,126].

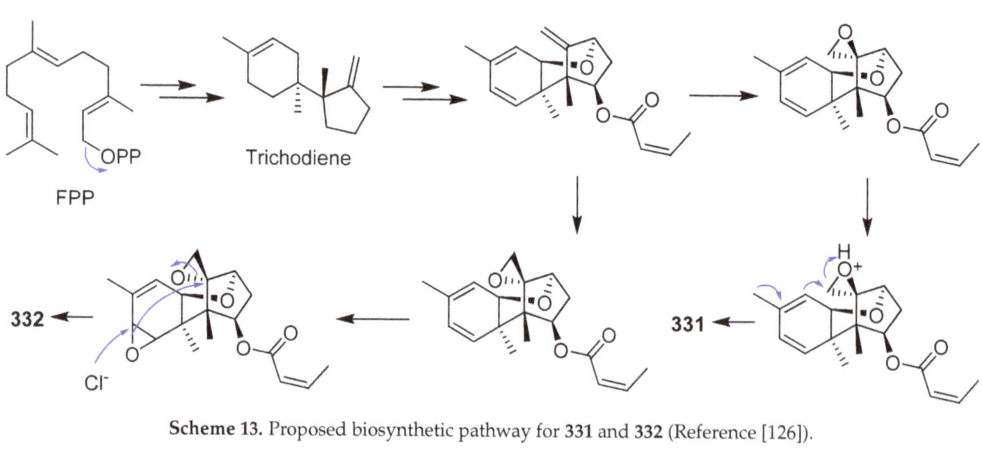

Scheme 13. Proposed biosynthetic pathway for **331** and **332** (Reference [126]).

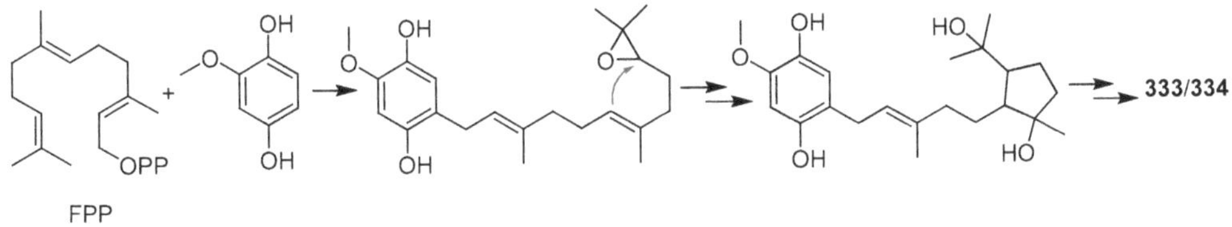

Scheme 14. Proposed biosynthetic pathway for **333** and **334** (Reference [125]).

The semisynthesis of several trichodermin and trichodermol derivatives has been developed (Scheme 15) [127]. Some derivatives with a short chain at the C-4 position displayed selective antimicrobial activity against *Candida albicans* and they showed MIC values similar to those displayed by trichodermin. It was important to highlight the cytotoxic selectivity observed for compounds **350**, **354**, and **356**, which presented average IC_{50} values of 2 μg/mL and were cytotoxic against tumorigenic cell line MCF-7 (breast carcinoma) and not against Fa2N4 (non-tumoral immortalized human hepatocytes).

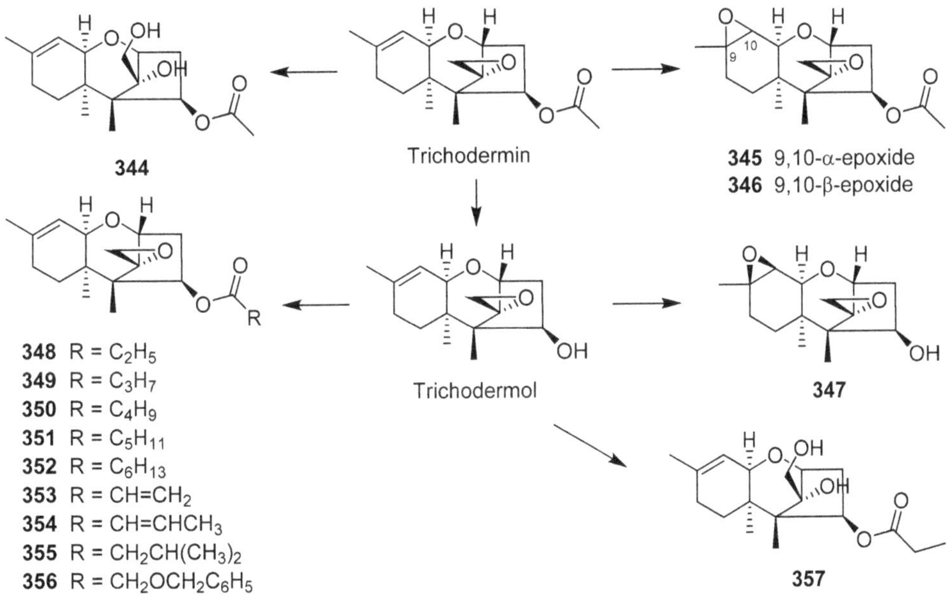

Scheme 15. Chemical transformations were carried out on trichodermin and trichodermol for the preparation of compounds **344–357** (Reference [127]).

Three novel highly oxygenated α-pyrone merosesquiterpenoids, emerones A–C **358–360**, have been obtained from the fungus *Emericella* sp. XL029 [128]. Structurally, emerone A **358** possessed an unprecedented 5/7 bicyclic ring architecture, emerone B **359** had an unusual substituted 10-membered ring, and emerone C **360** had an undescribed norsesquiterpene skeleton. Ochraceopone F **361**, a new α-pyrone merosesquiterpenoid possessing an angular tetracyclic carbon skeleton, was isolated from the marine fungus *Aspergillus flocculosus* [129].

Five new highly oxygenated α-pyrone merosesquiterpenoids, ochraceopones A–E **362–366**, were isolated from an Antarctic soil-derived fungus *Aspergillus ochraceopetaliformis* [130]. Ochraceopones A–D **363–366** were the first examples of α-pyrone merosesquiterpenoids possessing a linear tetracyclic carbon skeleton. Ochraceopone A **363** exhibited antiviral activities against the H3N2 influenza virus with IC_{50} values of 12.2 μM. Yaminterritrem C **367** was isolated from a deep-sea-derived fungus *Penicillium chryso-*

genum [131]. Verruculides A **368** and B **369** were isolated from a culture broth of the Indonesian ascidian-derived *Penicillium verruculosum* [132]. Verruculide A **368** inhibited the activity of PTP1B with an IC_{50} value of 8.4 μM.

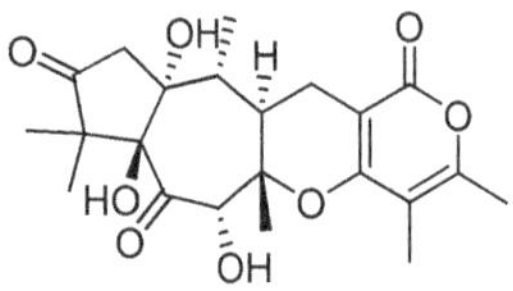

358

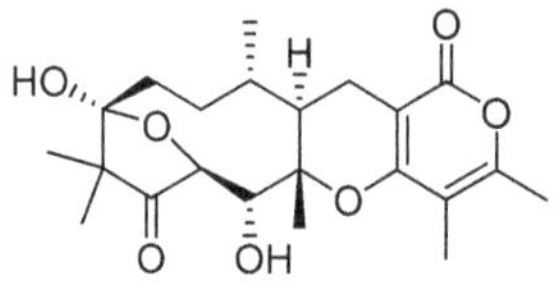

359

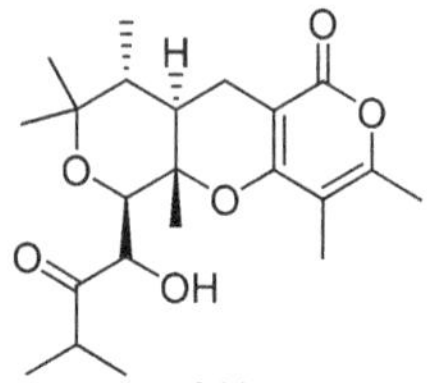

360

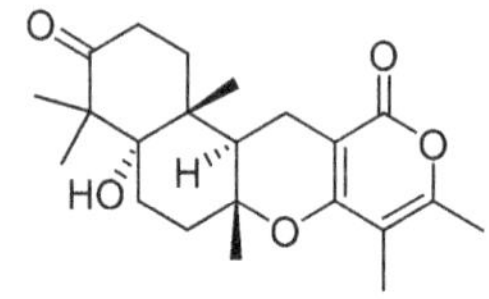

361

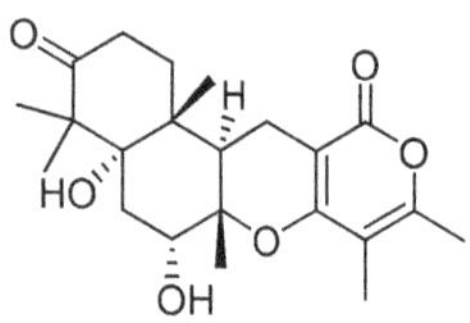

362

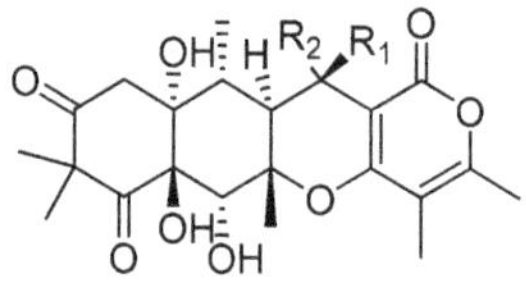

363 R_1 = OMe, R_2 = H
364 R_1 = OH, R_2 = H
365 R_1,R_2 = H

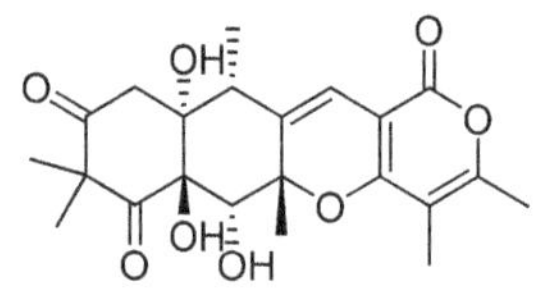

366

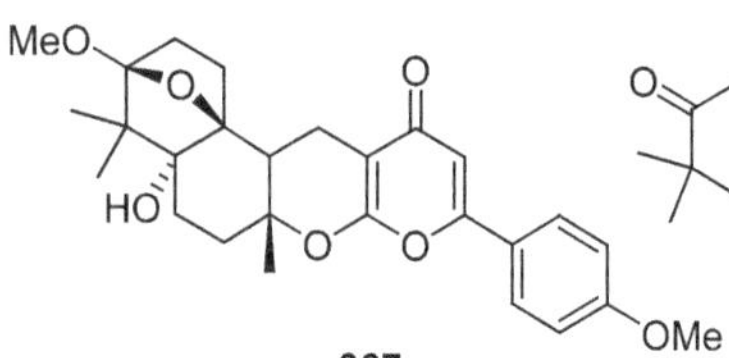

367

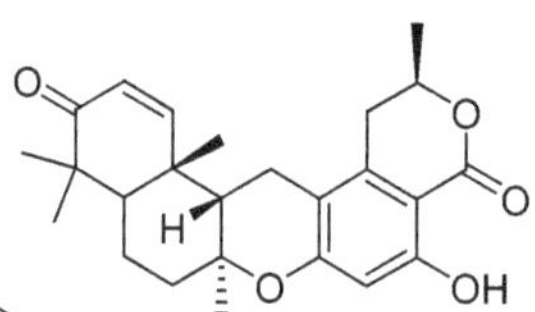

368

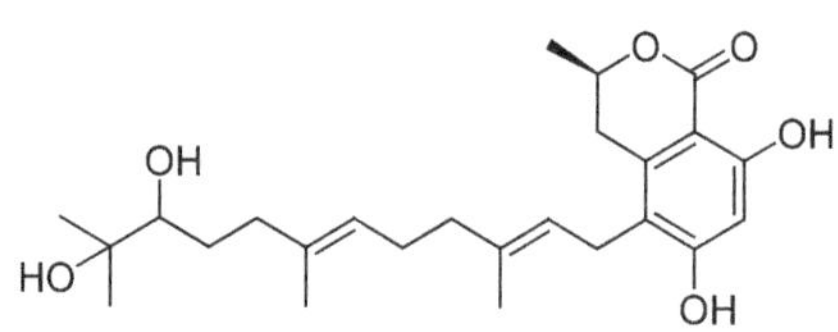

369

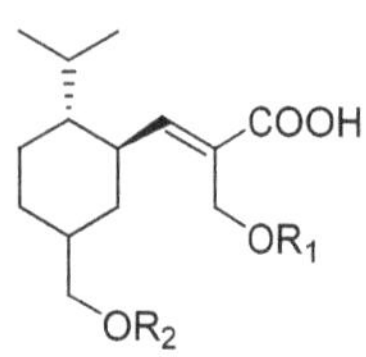

370 R_1,R_2 = H
371 R_1 = Ac, R_2 = H

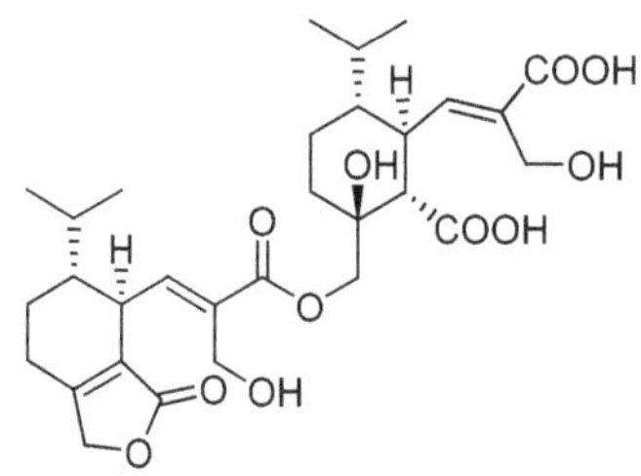

372

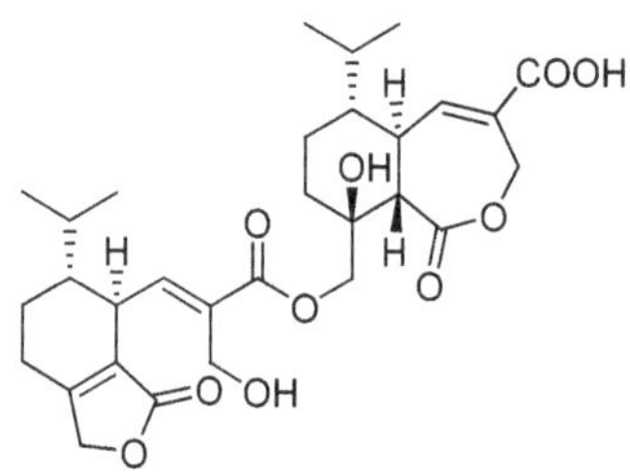

373

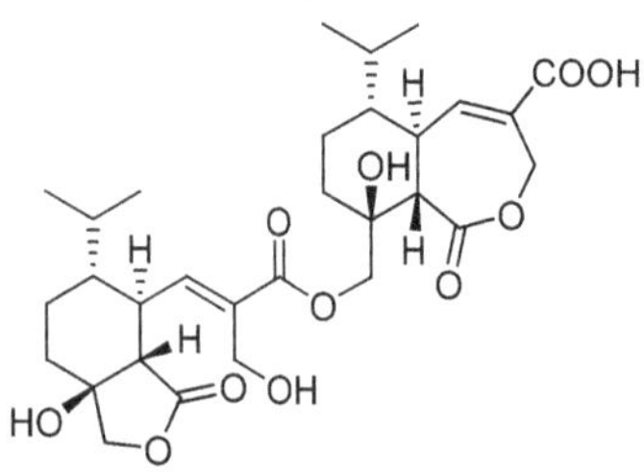

374

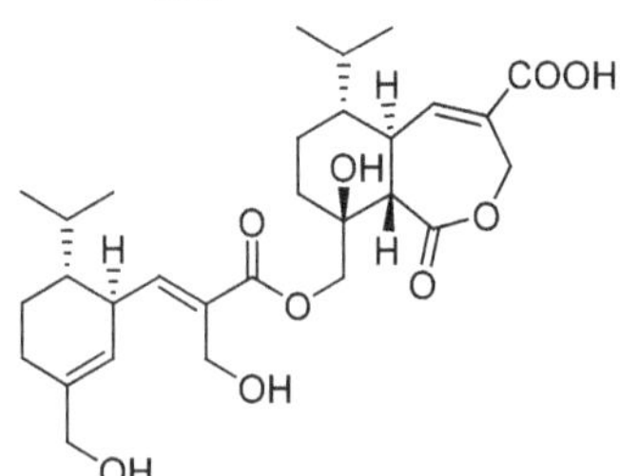

375

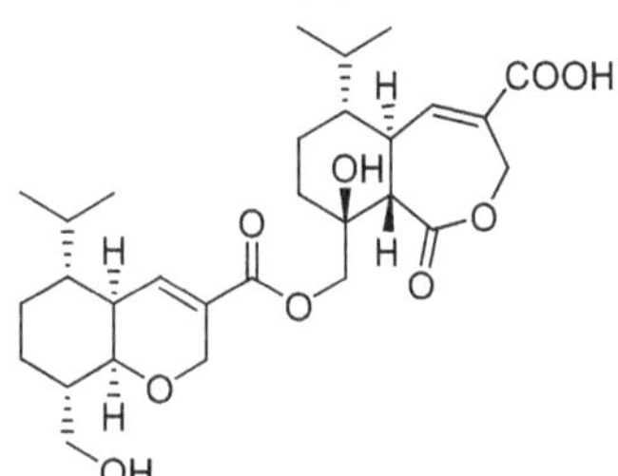

376

Two new sesquiterpenes **370** and **371** and seven new dimeric norsesquiterpene congeners, divirensols A–G **372–378**, along with seven new firstin-class trimeric sesquiterpenes, trivirensols A–G **379–385**, were obtained from the Australian termite nest-derived fungus *Trichoderma virens* [133,134]. A pair of rare naturally enantiomeric norsesquiterpenoids, (±)-preuisolactone A (**386** and **387**) featuring an unprecedented tricyclo[4.4.0^{1,6}.0^{2,8}]decane carbon scaffold were isolated from *Preussia isomera* (plausible biosynthetic pathway shown in Scheme 16) [135]. (±)-Preuisolactone A (**386** and **387**) exhibited remarkable antibacterial activity against *Micrococcus luteus* with a MIC value of 10.2 μM.

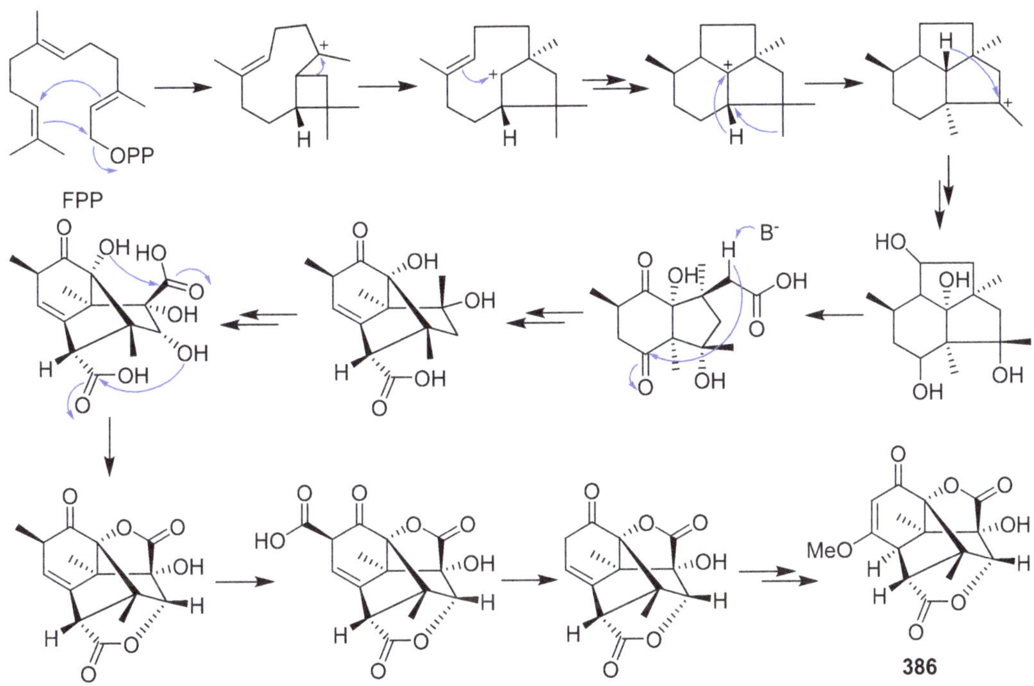

Scheme 16. Proposed biosynthetic pathway for (+)-preuisolactone A **386** (Reference [135]).

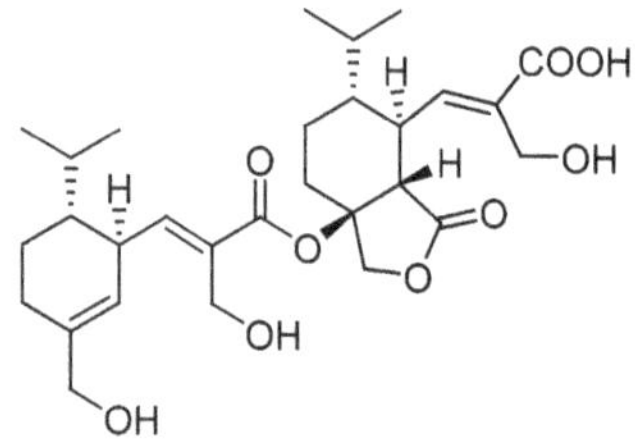

377

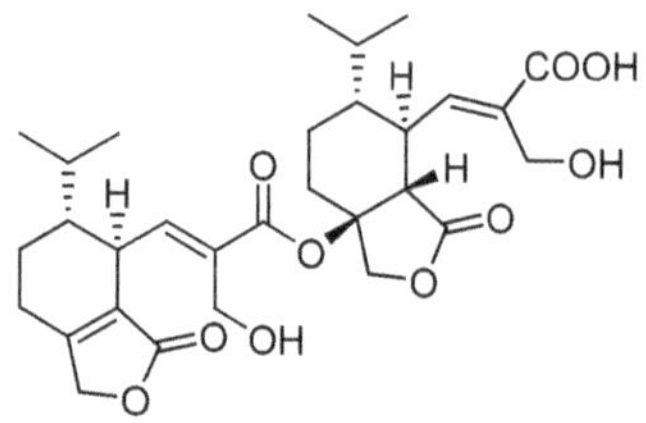

378

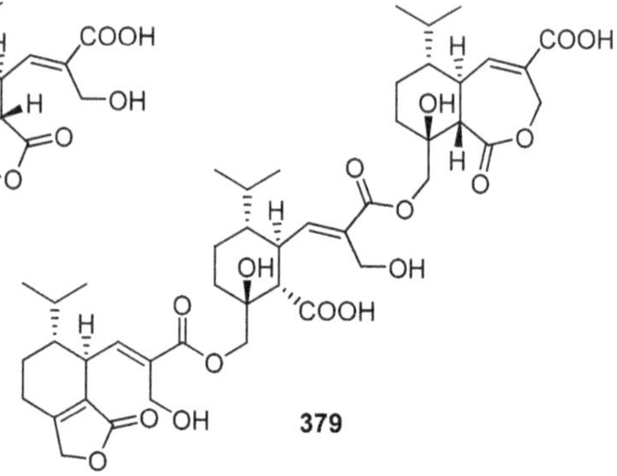

379

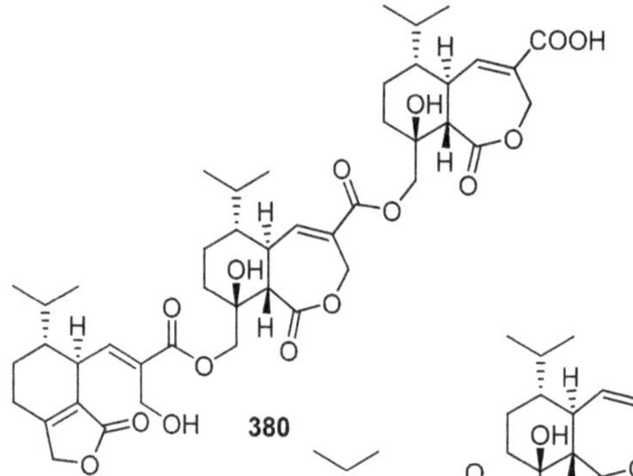

380

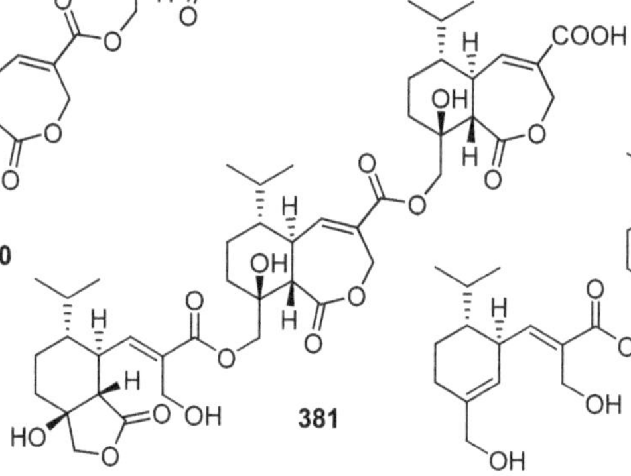

381

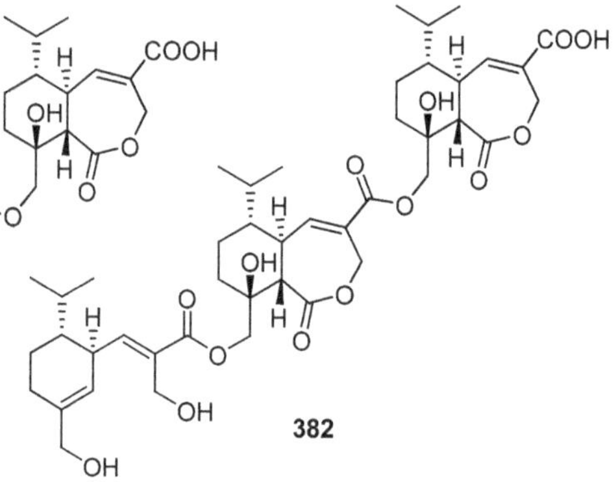

382

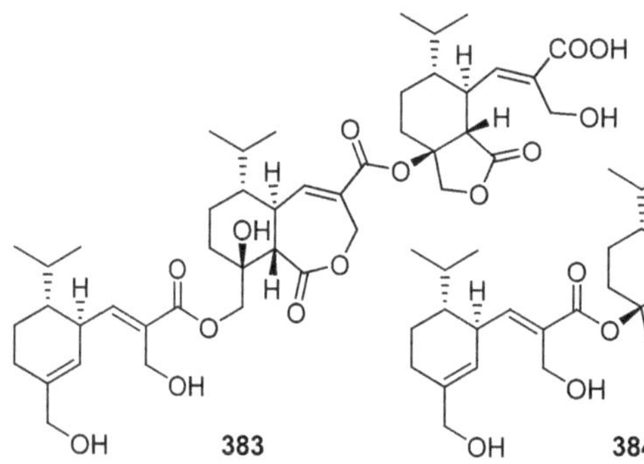

383

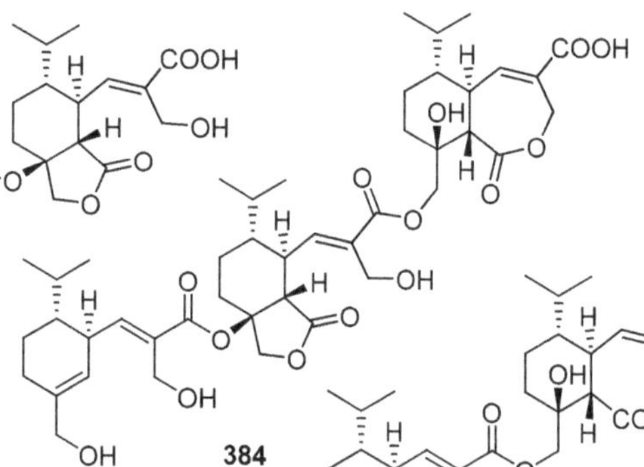

384

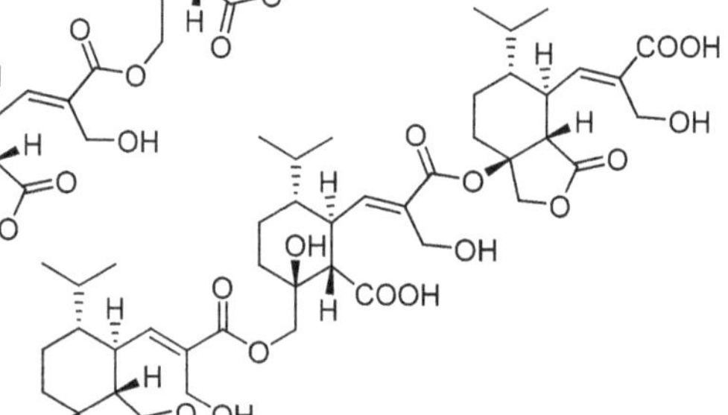

385

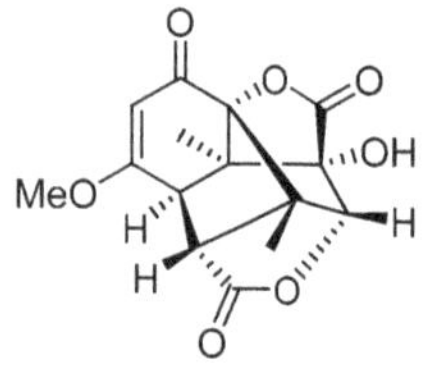

386

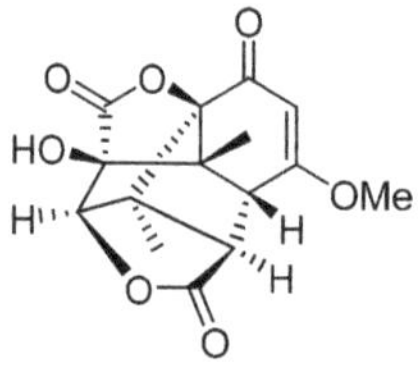

387

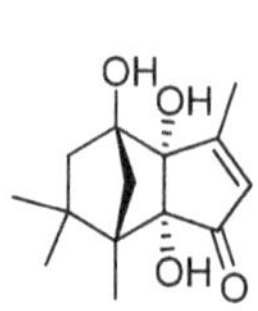

388

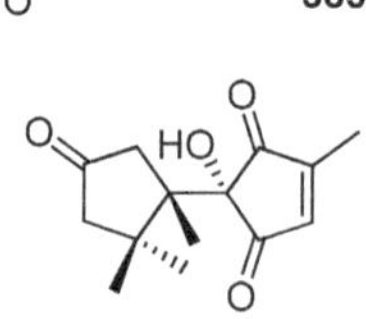

389

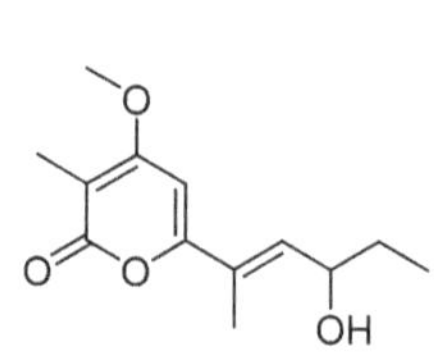

390

Hitoyol A **388**, an unprecedented norsesquiterpenoid with an exo-tricyclo[5.2.1.0^{2,6}]decane skeleton, along with a novel skeletal hitoyol B **389** containing 4-cyclopentene-1,3dione, was isolated from the fungus *Coprinopsis cinerea* [136]. Hitoyol A **388** was possibly biosynthesized through decarboxylation-induced cyclization of lagopodin B, a known cuparene-type sesquiterpenoid (Scheme 17). Hitoyol B **389** showed weak antimalarial activity against *Plasmodium falciparum* with an IC_{50} of 59 μM.

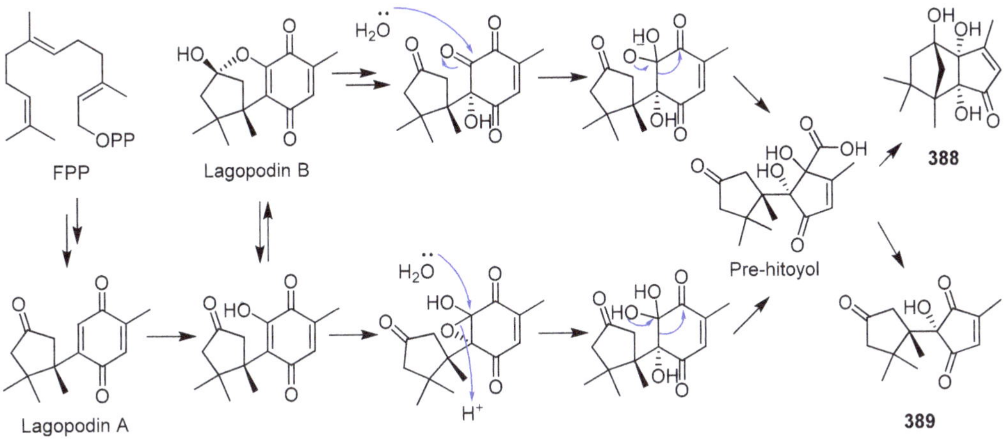

Scheme 17. Plausible biosynthetic pathway for hitoyol A **388** and hitoyol B **389** (Reference [136]).

An α-pyrone 9-hydroxyxylarone **390** was isolated from a culture broth of endophytic fungus *Xylaria* sp. NC1214 [137]. Four new polyenic α-pyrone mycotoxins, avertoxins A–D **391–394**, were obtained from an endophytic fungus *Aspergillus versicolor* [138]. Avertoxins B **392** and C **393** showed activity against human tumor HCT116 and HeLa cell lines with an IC_{50} value of 10 μM. And avertoxin B **392** was an active inhibitor against human acetylcholinesterase with the IC_{50} value of 14.9 μM.

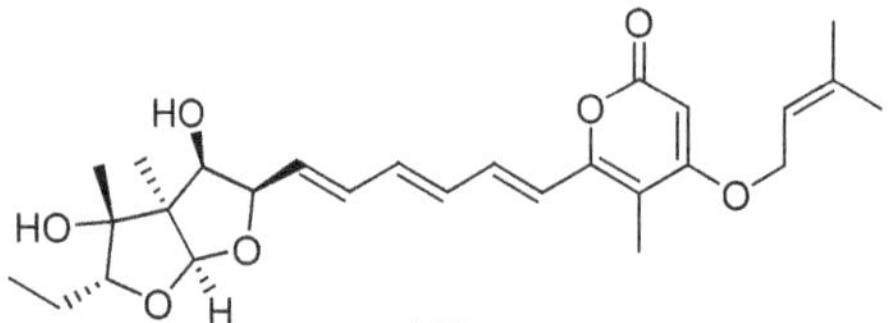

391

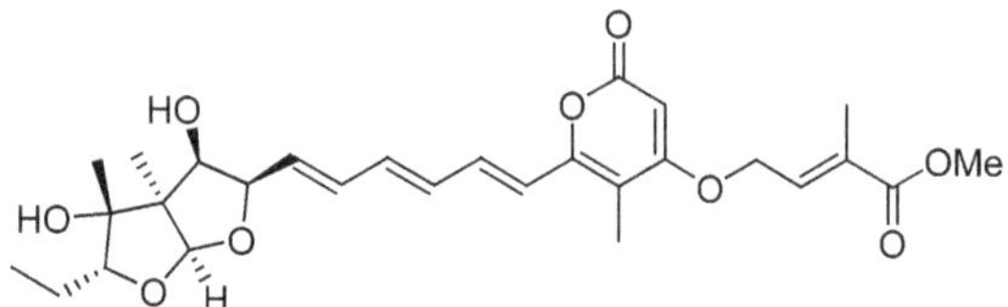

392

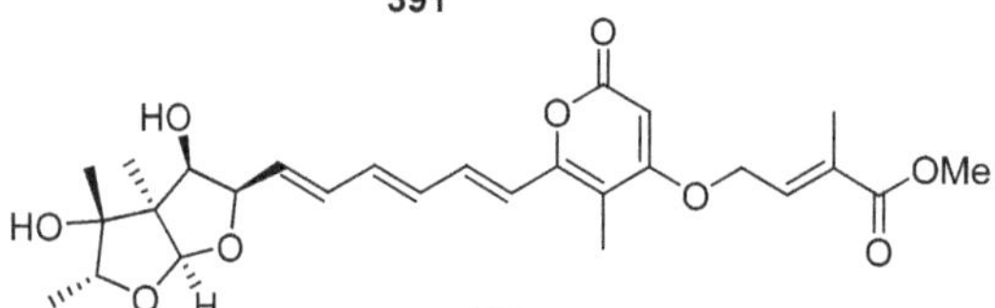

393

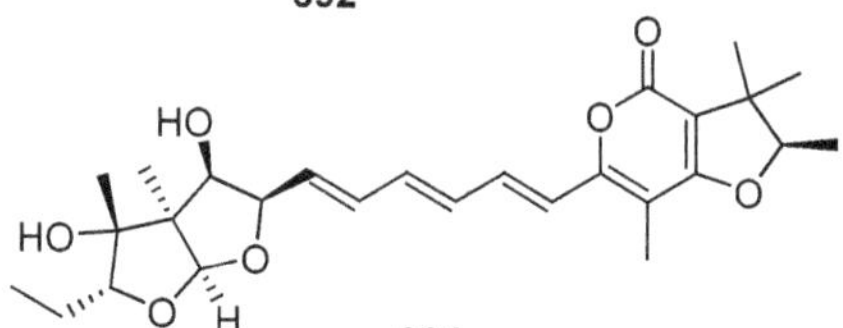

394

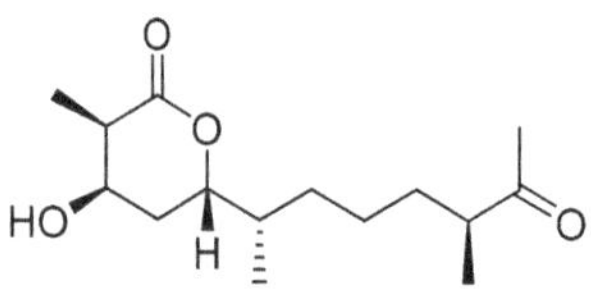

395

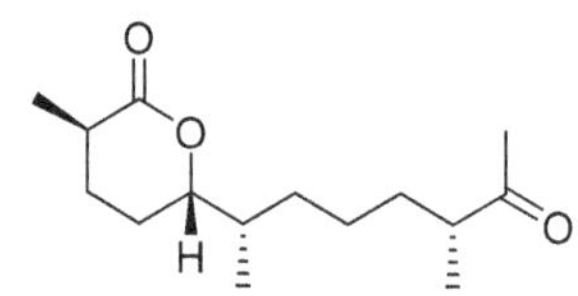

396

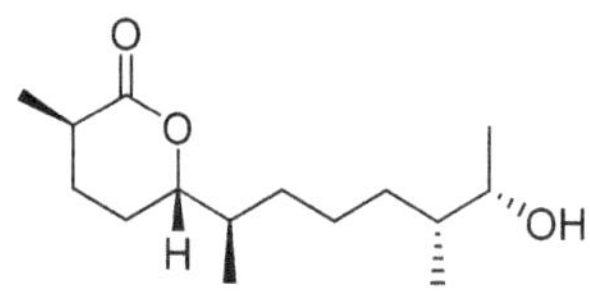

397

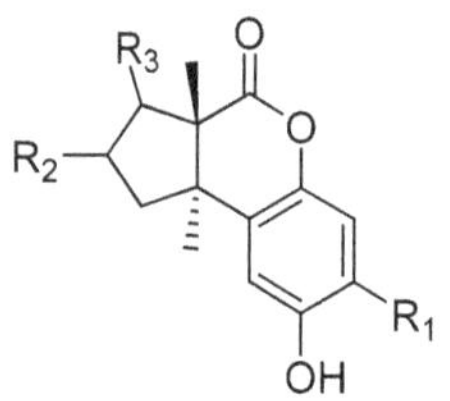

	R_1	R_2	R_3
398	CH_2OH	H	H
399	CH_2OCH_2Me	H	H
400	CH_2OCH_2Me	α-OH	H
401	CH_2OMe	α-OH	H
402	CH_2OH	α-OH	H
403	CH_2OH	β-OH	H
404	CH_2OH	H	=O
405	OH	H	H
406	CHO	H	H

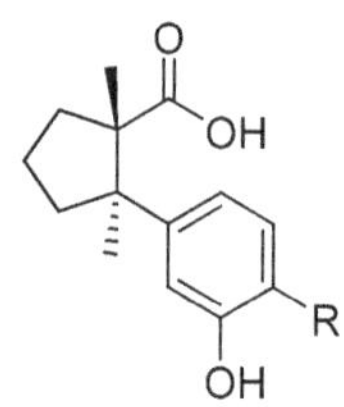

407 R = CHO
408 R = CH_2OH

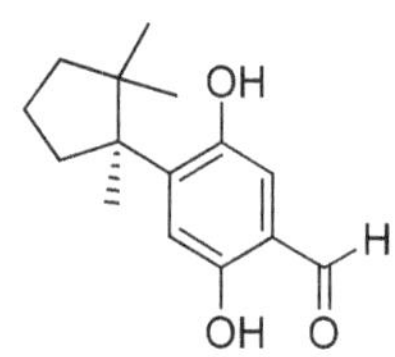

409

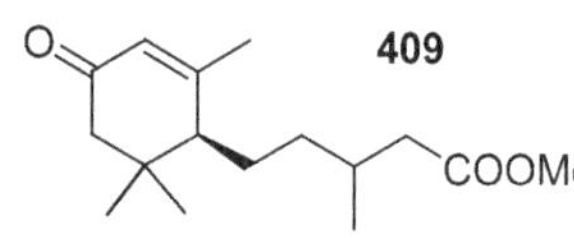

410

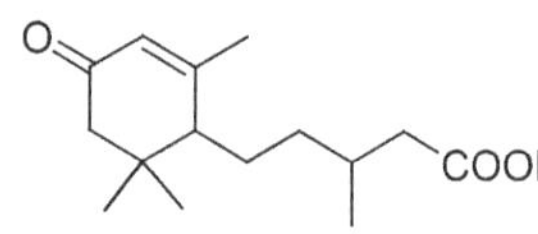

411

2.10. Other Types

Three new sesquiterpenoids, chermesiterpenoids A–C **395–397**, were isolated and identified from the marine red algal-derived fungus *Penicillium chermesinum* [139]. Chermesiterpenoid B **396** showed antimicrobial activities against the aquatic pathogens *Vibrio anguillarum*, *Vibrio parahaemolyticus*, *Micrococcus luteus*, and human pathogen *Escherichia coli* with minimum inhibitory concentration (MIC) values of 0.5, 16, 64, and 64 μg/mL, respectively. Similarly, chermesiterpenoid C **397** showed activities against the aquatic pathogens *V. anguillarum*, *V. parahaemolyticus*, and *M. luteus* with MIC values of 1, 32, and 64 μg/mL, respectively. Chermesiterpenoids A–C **395–397** exhibited activity against the plant pathogenic fungus *Colletottichum gloeosporioides* with MIC values of 64, 32, and 16 μg/mL, respectively.

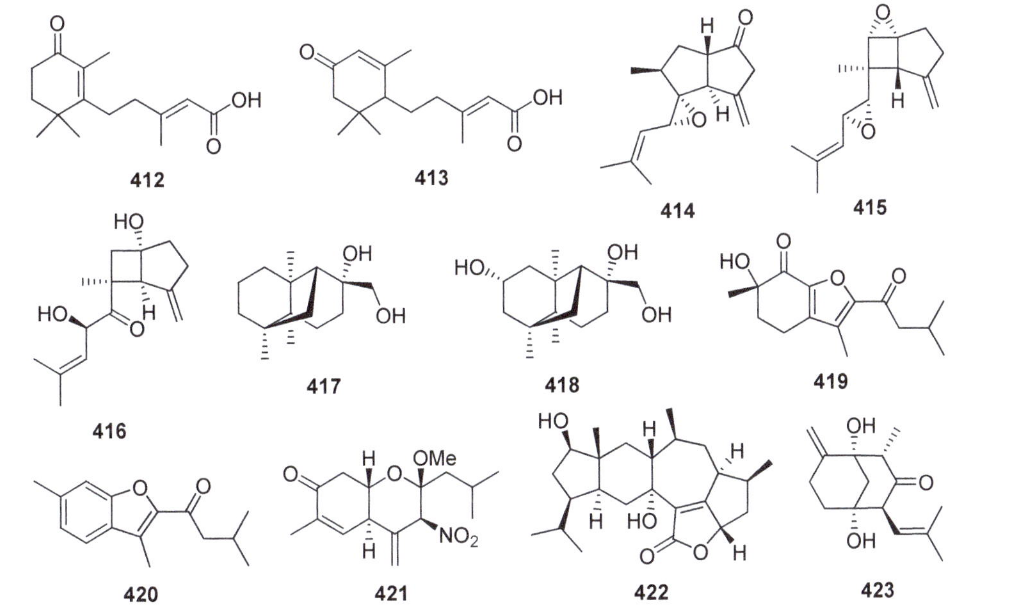

Fomitopins A–L **398–409** were isolated via bioassay-guided purification from the bracket fungus *Fomitopsis pinicola* [140]. Fomitopin K **408** exhibited the most potent anti-inflammatory activity with IC_{50} of 0.81 μM for inhibition of superoxide anion generation and IC_{50} of 0.74 μM for inhibition of elastase release. Fomitopins J **407** and L **409** also exhibited moderate inhibition of superoxide anion generation with IC_{50} of 1.66 and 1.72 μM, respectively.

1-Methoxypestabacillin B **410** was obtained from the solid cultures of a mangrove endophytic fungus *Diaporthe* sp. SCSIO 41011 [141]. Pestabacillin B **411** was isolated from the co-culture of the endophytic fungus *Pestalotiopsis* sp. with *Bacillus subtilis* [19]. Two new abscisic acid-type sesquiterpenes **412** and **413** were isolated from the fermentation extract of *Amycolatopsis alba* [142]. Pseudapenes A–C **414–416** possessing unique carbon skeletons were isolated from the marine-derived fungus *Pseudallescheria apiosperma* [143].

Emericellins A **417** and B **418**, representing a new type of sesquiterpenoid with an unprecedented tricyclo[1,2,4,4]hendecane scaffold (Scheme 18), were isolated from the liquid cultures of an endophytic fungus *Emericella* sp. associated with the leaves of *Panax notoginseng* [144]. Emericellins A **417** and B **418** displayed moderate activities against three fungal strains (*Verticillium dahliae* Kleb, *Helminthosporium maydis*, and *Botryosphaeria dothidea*) and three bacterial strains (*Bacillus subtilis*, *Bacillus cereus*, and *Escherichia coli*) with MIC values of 25–50 μg/mL.

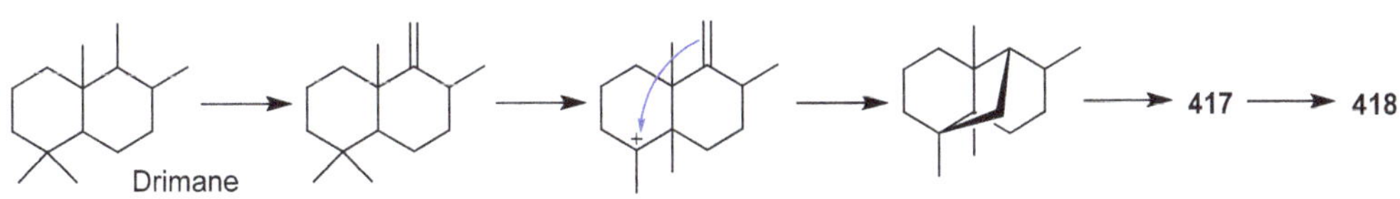

Scheme 18. The proposed formation of **417** and **418** from the drimane-type sesquiterpenoid skeleton (Reference [144]).

Stereumenes A–C **419–421** were isolated and identified from the fungus *Stereum* sp. [145]. Stereumene B **420** showed weak nematicidal activity against *Caenorhabditis elegans*, which killed 41.1% of *C. elegans* at 200 mg/L in 24 h. Sesteralterin **422** was obtained from the culture extract of an *Alternaria alternata* strain isolated from the surface of the marine red alga *Lomentaria hakodatensis* [146]. Colletotrichine A **423** was obtained from the endophyte fungus *Colletotrichum gloeosporioides* [147].

Four novel mixed terpenes, stereumamides A–D **424**–**427**, which were sesquiterpenes combined with α-amino acids to form quaternary ammonium hybrids, were isolated from the mycelium of mushroom *Stereum hirsutum* [148]. Stereumamides A **424** and D **427** showed antibacterial activity against *Escherichia coli*, *Staphylococcus aureus*, and *Salmonella typhimurium*, with MIC values of 12.5–25.0 μg/mL.

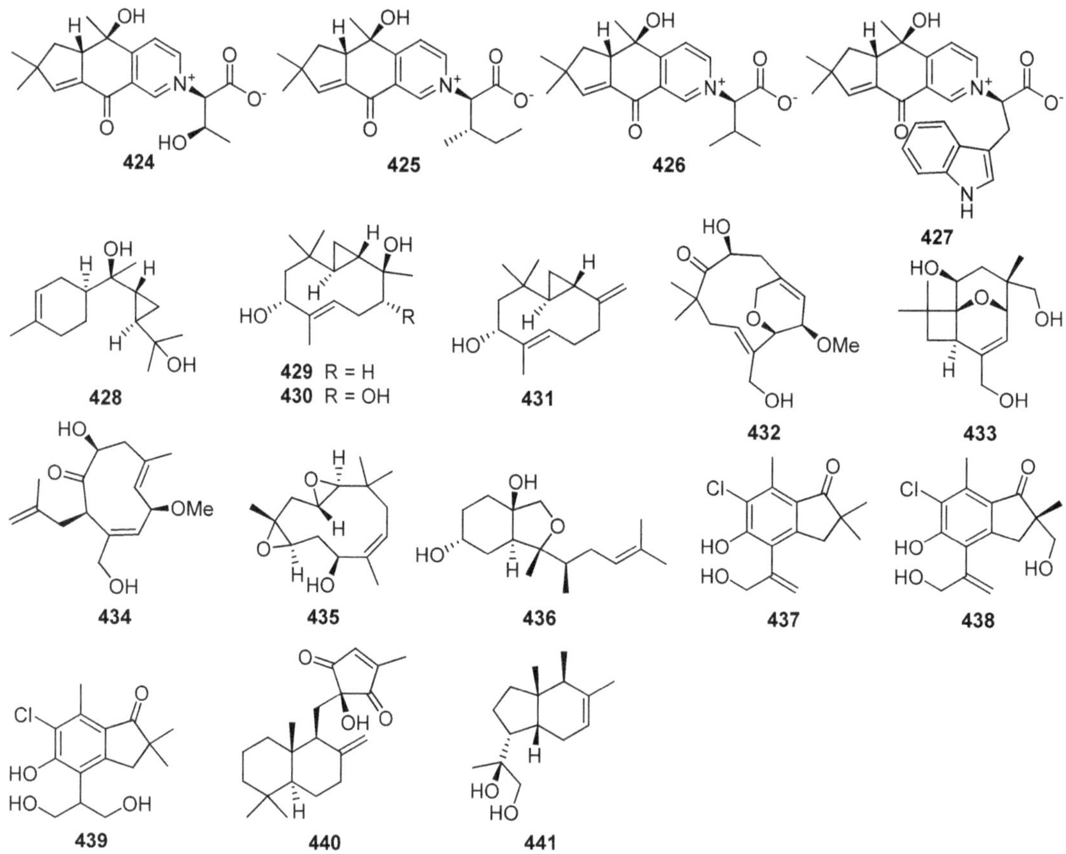

Phellilane L **428**, a new cyclopropane-containing sesquiterpenoid, was isolated from the medicinal mushroom *Phellinus linteus* [149]. The first asymmetric, protecting group-free total synthesis of the sesquiterpenoid phellilane L **428**, featuring a highly stereoselective one-pot synthesis involving intermolecular alkylation/cyclization/lactonization on epoxyiodide **428a** to construct the key cyclopropane-γ-lactone intermediate **428b** has been reported (Scheme 19) [149].

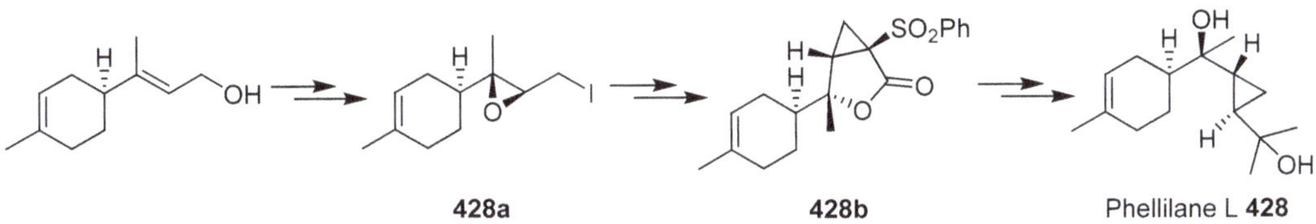

Scheme 19. Total synthesis of phellilane L **428** (Reference [149]).

Hypocoprins A–C **429–431** have a distinctive ring system consisting of fused cyclopropane and cyclodecene units were isolated from the Coprophilous fungus *Hypocopra rostrate* [150]. Pestaloporonins A–C **432–434**, three new sesquiterpenoids related to the caryophyllene-derived punctaporonins, were isolated from cultures of a fungicolous isolate of *Pestalotiopsis* sp. MYC-709 [151]. Among them, pestaloporonins A **432** and B **433** contained new bicyclic and tricyclic ring systems, respectively, and the absolute configuration of **432** was confirmed by single-crystal X-ray crystallographic analysis.

Phomanoxide **435** was isolated from the solid substrate fermentation cultures of the fungus *Phoma* sp. [152]. Colletotrichine B **436** was produced by the fungal *Colletotrichum gloeosporioides* [153]. Three new chlorinated sesquiterpenes, lepistatins A–C **437–439**, were isolated from the culture broth of Basidiomycete *Lepista sordida* [154]. The structures of lepistatins A–C **437–439** feature the indanone core structure but differ from other indanone-containing sesquiterpenes of fungal origin by the alkyl substitution pattern. This indicates that lepistatins A–C **437–439** probably possessed a new sesquiterpene scaffold derived from the common precursor, trans-humulyl cation, by an alternative cyclization (Scheme 20).

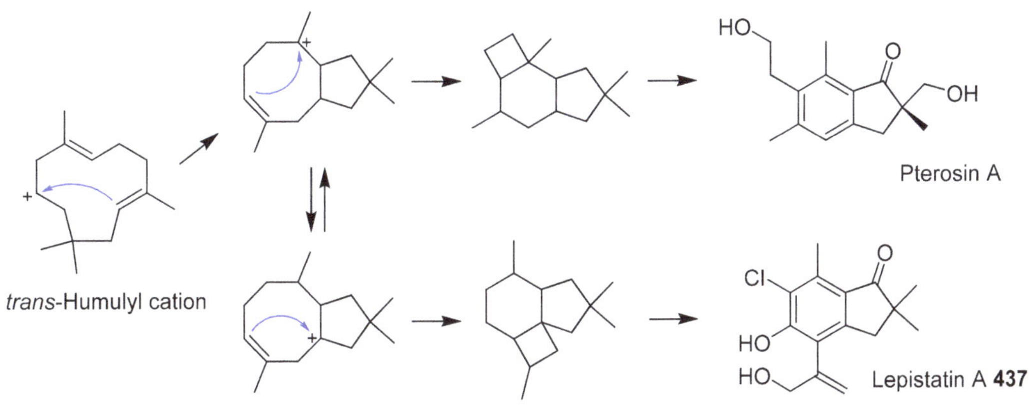

Scheme 20. Plausible biosynthetic pathway for lepistatin A **437** and pterosin A (Reference [154]).

A novel sesquiterpene methylcyclopentenedione, penicilliumin B **440**, was obtained from a deep sea-derived fungus *Penicillium* sp. F00120 [155]. Penicilliumin B **440**, presenting the first example with the sesquiterpene cyclopentenedione skeleton as natural products (Scheme 21), was structurally determined by analysis of the NMR and MS spectroscopic data, while the absolute configurations were assigned by single-crystal X-ray experiments. Penicilliumin B **440** with low toxicity showed significant potential to inhibit the kidney fibrogenic action in vitro by a mechanism dependent on disruption of oxidative stress. Seiricardine D **441** was a new bicyclic sesquiterpene obtained from the endophytic fungus *Cytospora* sp. [156]. Twenty new sesquiterpenes (**442–461**) were isolated from the endophytic fungus *Pseudolagarobasidium acaciico* [157]. Among them, compounds **459** and **460** displayed cytotoxicity against several cancer and normal cell lines.

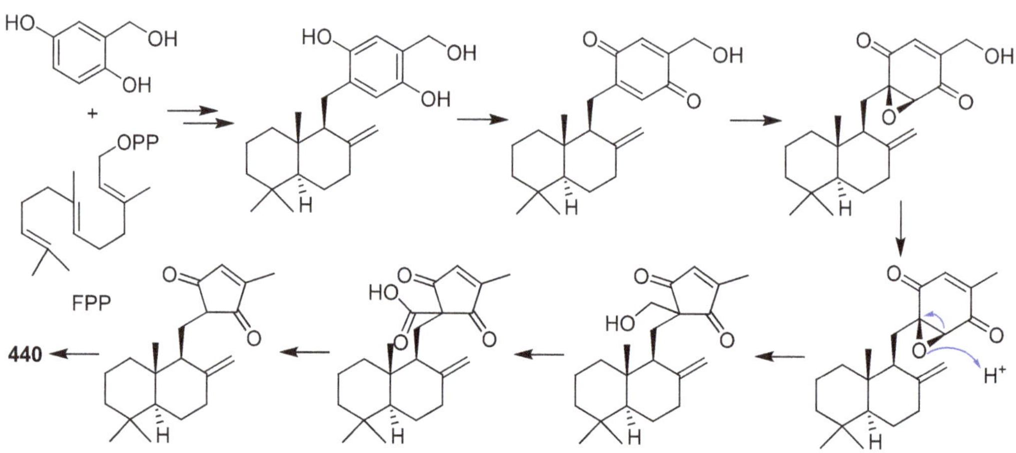

Scheme 21. Plausible biosynthetic pathway of penicilliumin B **440** (Reference [155]).

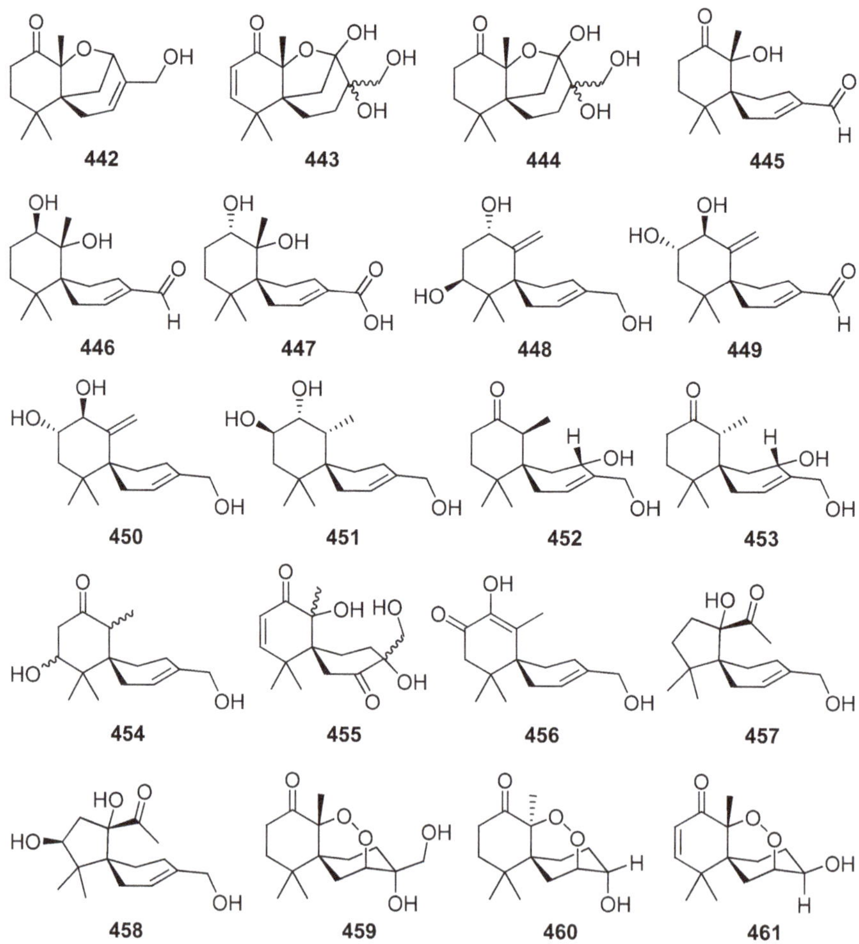

3. Biosynthesis

3.1. Asperterpenoid A

A putative three-gene cluster for asperterpenoid A was identified [158]. Stepwise reconstitution of this gene cluster in *Aspergillus oryzae* reveals that *astC* encodes a sesterterpene cyclase to synthesize preasperterpenoid A **462**, which was dually oxidized by a P450 enzyme AstB to give asperterpenoid A **463** along with a minor product asperterpenoid B **464**, and asperterpenoid A **463** was further oxidized by another P450 enzyme AstA to afford a new sesterterpenoid asperterpenoid C **465** (Scheme 22). Asperterpenoids A **463** and B **464** exhibit potent inhibitory activity against *Mycobacterium tuberculosis* protein tyrosine phosphatase B with IC_{50} values of 3–6 μM.

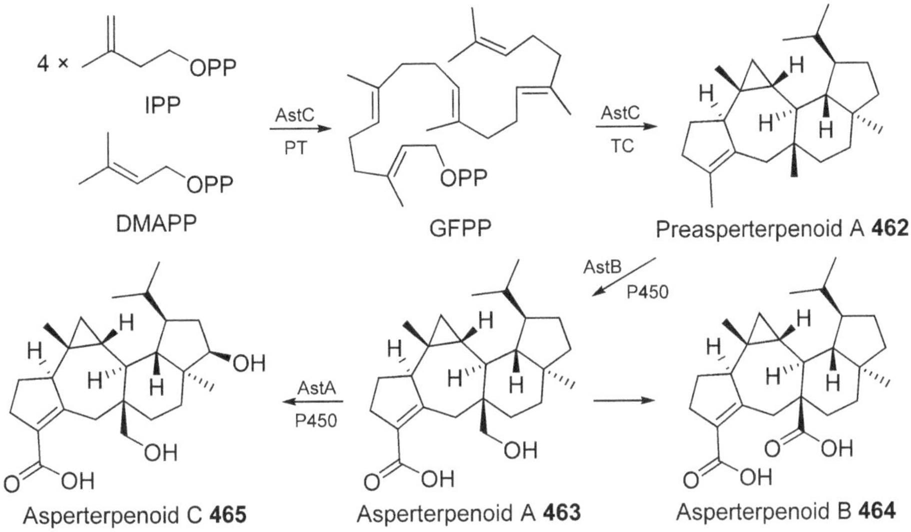

Scheme 22. Complete biosynthetic pathway of asperterpenoids A–C **463–465** (Reference [158]).

3.2. Cuparene

Use of the *ku*70-deficient strain of *Coprinopsis cinerea* enabled confirmation within the native context of the central role the sesquiterpene synthase Cop6 plays in lagopodin biosynthesis [159]. Furthermore, yeast in vivo bioconversion and in vitro assays of two cytochrome P450 monooxygenases Cox1 and Cox2 allowed elucidation of the network of oxidation steps that build structural complexity onto the α-cuprenene framework during the biosynthesis of lagopodins (Scheme 23). Three new compounds **466–468** were identified as intermediates formed by the redox enzymes.

3.3. Fusariumdiene and Fusagramineol

The novel sesquiterpenes fusariumdiene **469**, *epi*-fusagramineol **470**, and fusagramineol **471** with 5/7 bicyclic and 5/6/3 tricyclic ring systems, respectively, as well as five known sesquiterpenes **472–476** have been produced by exploiting the potential power of sesquiterpene synthase FgJ03939 from *Fusarium graminearum* in a farnesyl diphosphate-overexpressing *Saccharomyces cerevisiae* chassis (Scheme 24) [160].

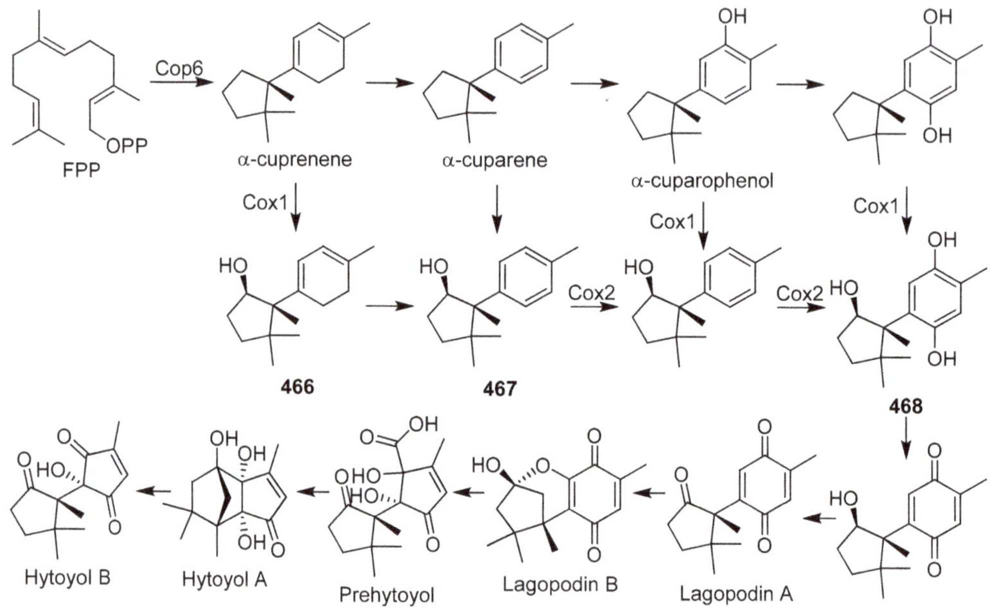

Scheme 23. The proposed biosynthetic pathway of lagopodins and hytoyols (Reference [159]).

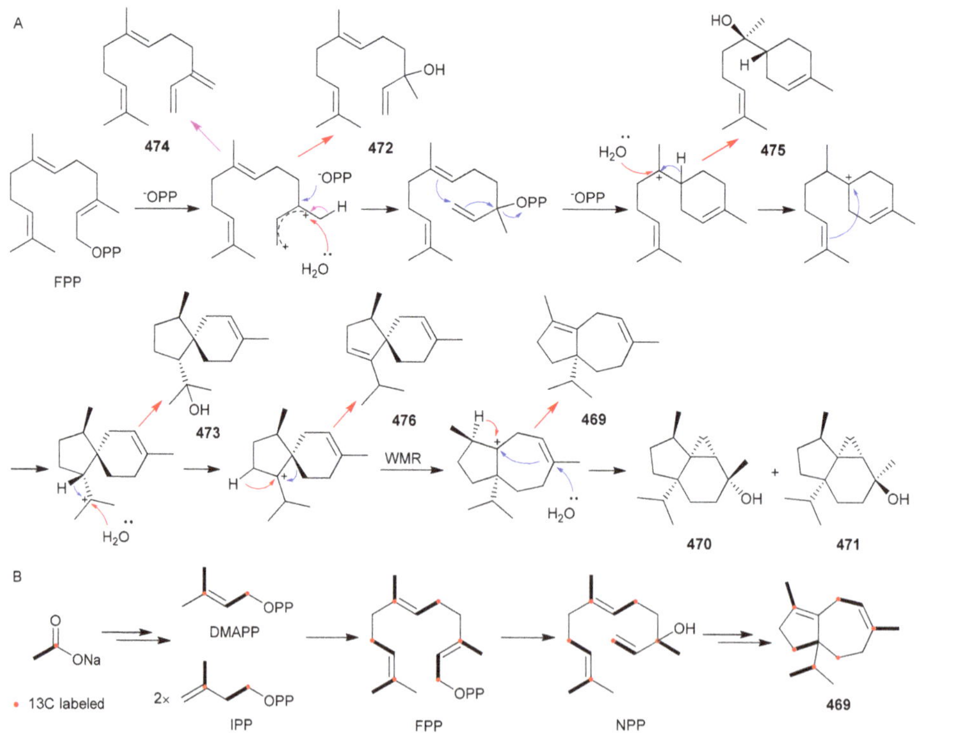

Scheme 24. (**A**) Proposed mechanisms for the enzymatic cyclization of FPP to compounds **469**–**476**; (**B**) Summary of feeding experiments with [1-^{13}C,^{2}H$_3$] sodium acetate (Reference [160]).

3.4. Hirsutenoid

The identification and successful cloning of the previously elusive hirsutene synthase from the wood-rotting mushroom *Stereum hirsutum* provide the biosynthetic pathways of hirsutane-type sesquiterpenoids (Scheme 25) [161]. The hirsutene synthase, as an unexpected fusion protein of a sesquiterpene synthase (STS) with a C-terminal 3-hydroxy-3-methylglutaryl-coenzyme A (3-hydroxy-3-methylglutaryl-CoA) synthase (HMGS) domain, was part of a biosynthetic gene cluster that includes P450s and oxidases that were expressed and could be cloned from cDNA.

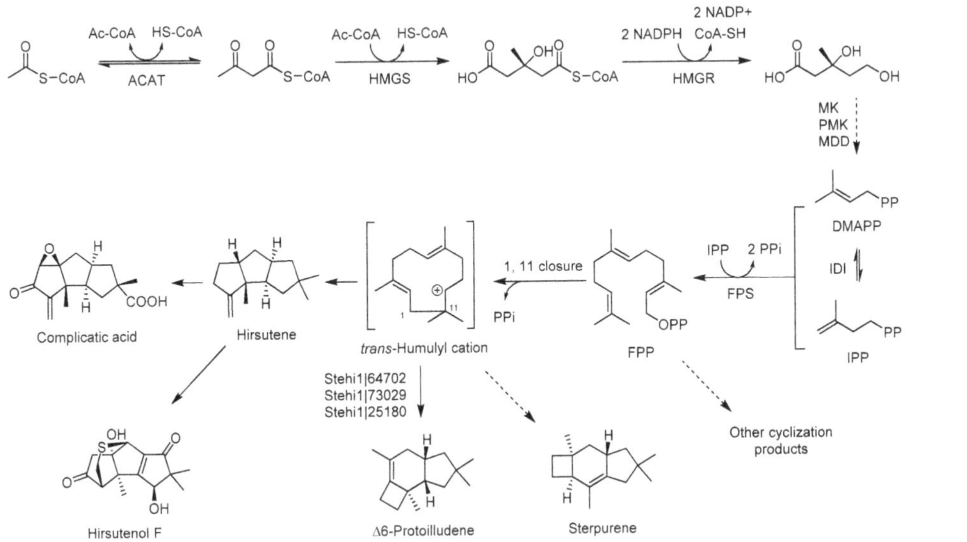

Scheme 25. Overview of pathways to *trans*-humulyl cation-derived sesquiterpenoids in *S. hirsutum* (Reference [161]).

3.5. Koraidiol

Two known oxygenated sesquiterpenoid products, culmorin **477** and culmorone **478**, and a new compound, koraidiol **479**, were successfully generated and characterized by a combinatorial biosynthesis approach which was utilized by the combination of a promiscuous myxobacterial P450 (CYP260B1) with two sesquiterpene cyclases (FgJ01056, FgJ09920) of filamentous fungi *Fusarium graminearum* (Scheme 26) [162].

3.6. Protoilludenes

Sixteen sesquiterpene synthases genes as full-length cDNAs have been isolated by using RT-PCR, and heterologous expression revealed that the sesquiterpene synthases could produce a series of sesquiterpene scaffolds with distinct metabolic profiles (Scheme 27) [163].

3.7. Trichothecenes

The acyltransferase-encoding gene *tri18*-encoded acyltransferase (TRI18) and a previously characterized acyltransferase (TRI3) were required in the saprotroph *Trichoderma arundinaceum* for conversion of the trichothecene biosynthetic intermediate trichodermol **480** to harzianum A **482**, an antifungal trichothecene analog with an octa-2,4,6-trienedioyl acyl group [164]. Previous studies indicate that *tri18* may not be necessary for the biosynthesis of harzianum A **482** because all catalytic activities required for its formation can be accounted for by activities of enzymes (TRI5, TRI4, TRI22, TRI17, and TRI3) encoded by other *tri* genes [165,166]. Further analysis proposed that TRI3 catalyzes trichothecene 4-O-acetylation, and subsequently, TRI18 catalyzes replacement of the resulting acetyl group with octa-2,4,6-trienedioyl to form harzianum A **482** (Scheme 28) [164].

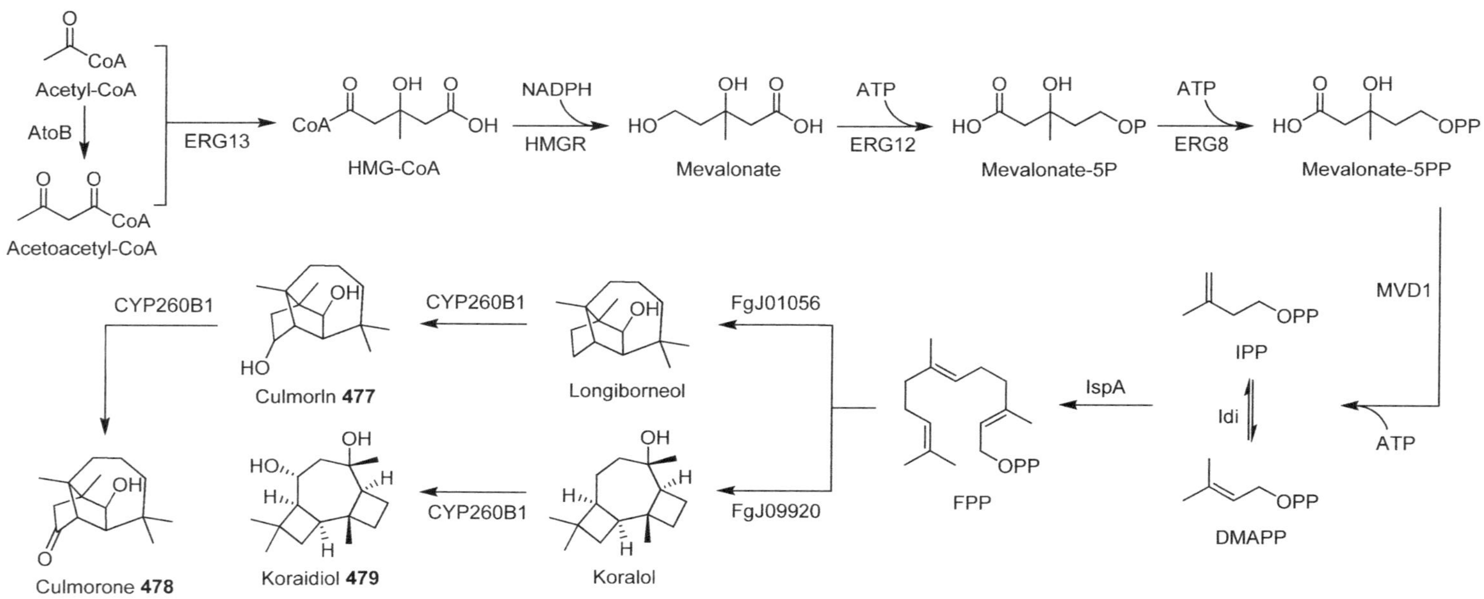

Scheme 26. Biosynthesis pathways of culmorin **477**, culmorone **478**, and koraidiol **479** in *E. coli* (Reference [162]).

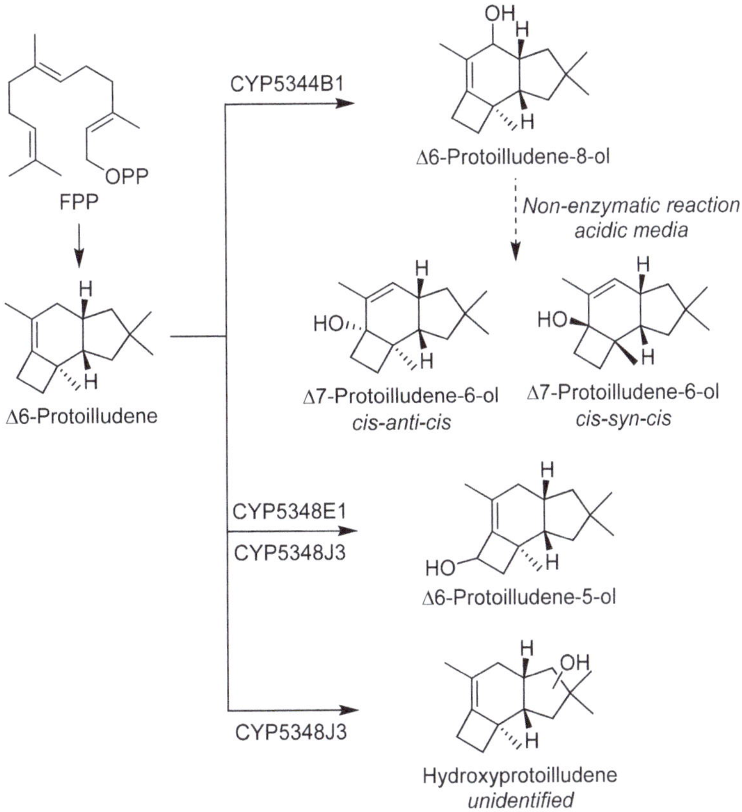

Scheme 27. Reaction pathways of protoilludene metabolism by PpSTS-08 and PpCYPs (Reference [163]).

An artificial metabolic route to an unnatural trichothecene was designed by taking advantage of the broad substrate specificities of the T-2 toxin biosynthetic enzymes of *Fusarium sporotrichioides* [167]. By feeding 7-hydroxyisotrichodermin, a shunt pathway metabolite of *F. graminearum*, to a trichodiene synthase-deficient mutant of *F. sporotrichioides*, 7-hydroxy T-2 toxin **483** was obtained as the final metabolite (Scheme 29). The toxicity of 7-hydroxy T-2 toxin **483** was 10 times lower than that of T-2 toxin in HL-60 cells.

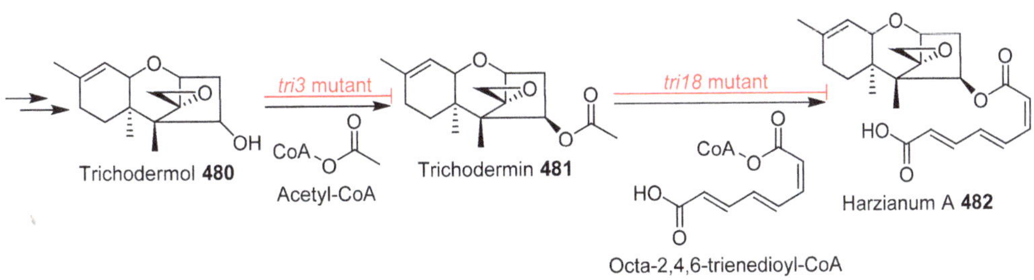

Scheme 28. Biosynthesis pathways of trichodermol **480** to harzianum A **482**. The symbol ⊥ indicates that the pathway is partially or completely blocked at the step indicated in the *tri3* and *tri18* mutants (Reference [164]).

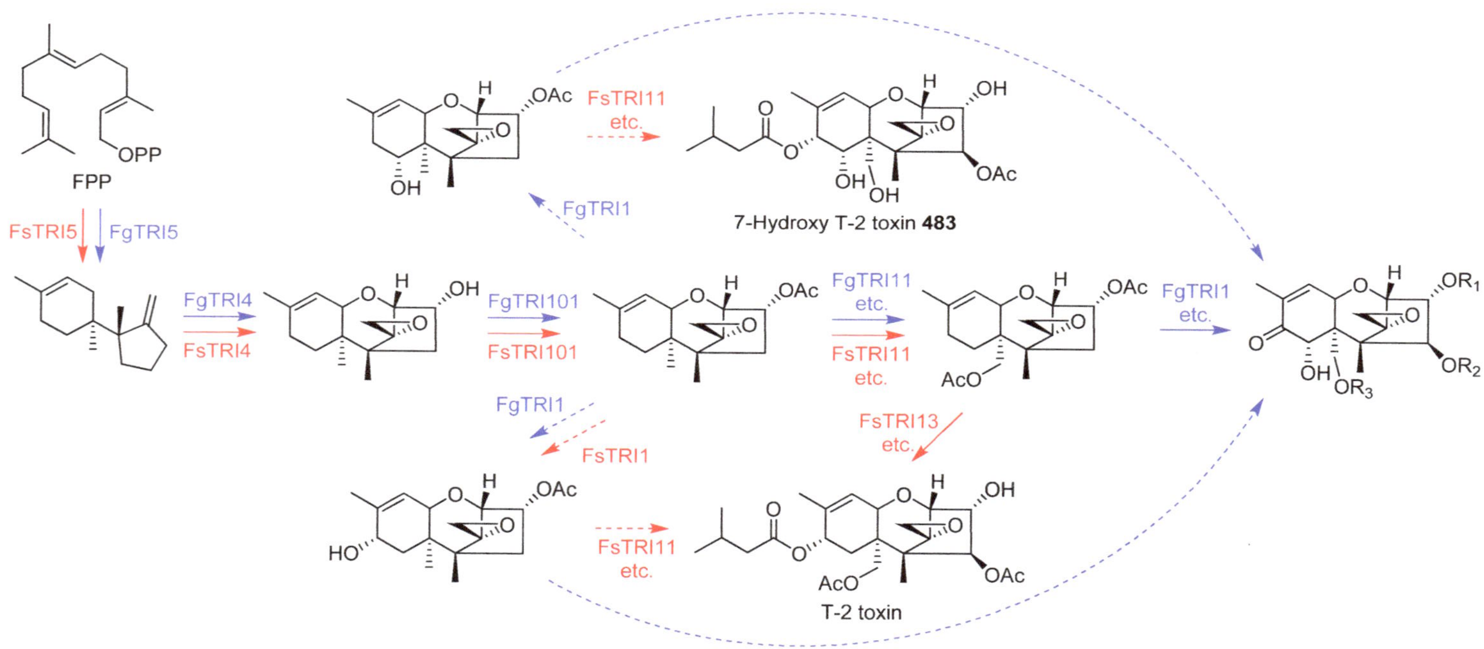

Scheme 29. Biosynthetic approach to generate 7-hydroxy T-2 toxin **483.** Red arrows indicate the metabolic pathway of *F. sporotrichioides*, whereas blue arrows indicate that of *F. graminearum*. Solid arrows indicate the main pathway, and dotted arrows indicate the shunt pathway (Reference [167]).

The candidate gene, *Clm2*, a second structural gene required for culmorin biosynthesis in the plant pathogen *Fusarium graminearum*, encodes a regio- and stereoselective cytochrome P450 monooxygenase for C-11 of longiborneol **484** (Scheme 30) [168]. *Clm2* gene disruptants were grown in liquid culture and assessed for culmorin production via HPLC-evaporative light scattering detection. The analysis indicated a complete loss of culmorin **485** from the liquid culture of the Δ*Clm2* mutants. Culmorin production resumed in a Δ*Clm2* complementation experiment. A detailed analysis of the secondary metabolites extracted from the largescale liquid culture of disruptant Δ*Clm2*D20 revealed five new natural products: **486–490**. The structures of the new compounds were elucidated by a combination of HRMS, 1D and 2D NMR, and single-crystal X-ray crystallography analysis.

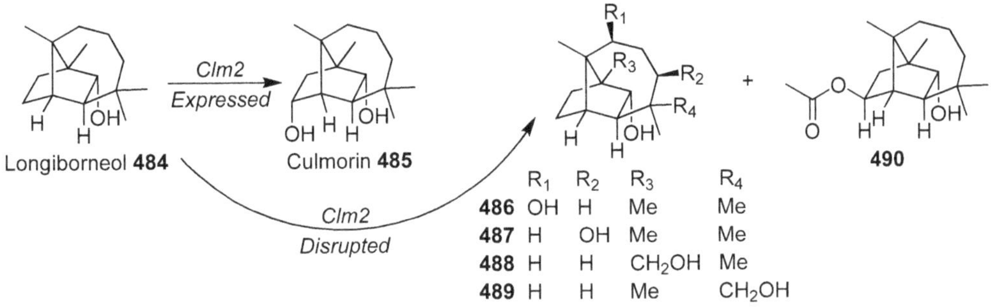

Scheme 30. Biosynthetic pathways of **484–490** (Reference [168]).

4. Conclusions and Future Prospects

Natural products, in particular bioactive molecules as precursor pharmaceutical compounds, have attracted particular attention in the field of health promotion and drug discovery and development. Compared with other sources, fungal species play a decisive role in bio-transformations and drug synthesis owing to their wide varieties, easy cultivation, diverse chemical compositions, and distinct biological activities. This process has been accelerated by considerable advances in microbial genome research and in understanding the structure of genes and their corresponding products. Genome mining-based natural products discovery programs mainly use the most identifiable terpene synthases and prenyltransferases to locate and quickly identify new terpenoids. In the last five years, nearly 500 new sesquiterpenes, including about 20 new skeletons were identified from fungi. These sesquiterpenoids exhibit various biological activities, such as anti-tumor, anti-viral, anti-microbial, anti-inflammatory, etc. These efforts have clearly led to a global promotion of discovery and characterization of fungal terpenoids and offer optimism for the future of fungal terpenoid discovery.

This review summarized the isolation, chemical structures, plausible biosynthetic pathways, bioactivity, chemical synthesis, and biosynthesis of 490 recent sesquiterpenoids. This could be a useful reference for modern researchers studying this category of compounds.

Author Contributions: Q.D. and F.-L.Z. contributed equally to this paper. Conceptualization, T.F.; resources, Q.D. and F.-L.Z.; discussion of the contents, Q.D., F.-L.Z. and T.F.; writing—original draft preparation, Q.D. and F.-L.Z.; writing—review and editing, and editing, T.F. All authors have read and agreed to the published version of the manuscript.

Funding: This research was funded by the Natural Science Foundation of China (grant number 81872762) and the Research Fund of Central University, South-Central University for Nationalities (grant number CZP21001).

Institutional Review Board Statement: Not applicable.

Informed Consent Statement: Not applicable.

Data Availability Statement: Not applicable.

Conflicts of Interest: The authors declare no conflict of interest.

References

1. Christianson, D.W. Unearthing the roots of the terpenome. *Curr. Opin. Chem. Biol.* **2008**, *12*, 141–150. [CrossRef] [PubMed]
2. Minami, A.; Ozaki, T.; Liu, C.; Oikawa, H. Cyclopentane-forming di/sesterterpene synthases: Widely distributed enzymes in bacteria, fungi, and plants. *Nat. Prod. Rep.* **2018**, *35*, 1330–1346. [CrossRef]
3. Schmidt-Dannert, C. *Biosynthesis of Terpenoid Natural Products in Fungi. Advances in Biochemical Engineering-Biotechnology*; Schrader, J., Bohlmann, J., Eds.; Springer: New York, NY, USA, 2015; Volume 148, pp. 19–61.
4. Li, D.; Wang, K.W. Natural new sesquiterpenes: Structural diversity and bioactivity. *Curr. Org. Chem.* **2016**, *20*, 994–1042. [CrossRef]
5. Fraga, B.M. Natural sesquiterpenoids. *Nat. Prod. Rep.* **2012**, *29*, 1334–1366. [CrossRef] [PubMed]
6. Chen, H.P.; Liu, J.K. Secondary metabolites from higher fungi. *Prog. Chem. Org. Nat. Prod.* **2017**, *106*, 1–201. [CrossRef]
7. Gonzalez Del Val, A.; Platas, G.; Arenal, F.; Orihuela, J.C.; Garcia, M.; Hernandez, P.; Royo, I.; De Pedro, N.; Silver, L.L.; Young, K.; et al. Novel illudins from *Coprinopsis episcopalis* (syn. Coprinus episcopalis), and the distribution of illudin-like compounds among filamentous fungi. *Mycol. Res.* **2003**, *107*, 1201–1209. [CrossRef]
8. Alexandre, J.; Raymond, E.; Kaci, M.O.; Brain, E.C.; Lokiec, F.; Kahatt, C.; Faivre, S.; Yovine, A.; Goldwasser, F.; Smith, S.L.; et al. Phase I and pharmacokinetic study of irofulven administered weekly or biweekly in advanced solid tumor patients. *Clin. Cancer Res.* **2004**, *10*, 3377–3385. [CrossRef]
9. Tanasova, M.; Sturla, S.J. Chemistry and biology of acylfulvenes: Sesquiterpene-derived antitumor agents. *Chem. Rev.* **2012**, *112*, 3578–3610. [CrossRef]
10. McMullen, M.; Jones, R.; Gallenberg, D. Scab of wheat and barley: A re-emerging disease of devastating impact. *Plant Dis.* **1997**, *81*, 1340–1348. [CrossRef]
11. Eriksen, G.S.; Pettersson, H. Toxicological evaluation of trichothecenes in animal feed. *Anim. Feed Sci. Technol.* **2004**, *114*, 205–239. [CrossRef]
12. Qinghua, W.; Vlastimil, D.; Kami, K.; Zonghui, Y. Trichothecenes: Structure-toxic activity relationships. *Curr. Drug Metab.* **2013**, *14*, 641–660. [CrossRef]
13. Pascari, X.; Maul, R.; Kemmlein, S.; Marin, S.; Sanchis, V. The fate of several trichothecenes and zearalenone during roasting and enzymatic treatment of cereal flour applied in cereal-based infant food production. *Food Control* **2020**, *114*, 107245. [CrossRef]
14. Zhou, Z.Y.; Liu, J.K. Pigments of fungi (macromycetes). *Nat. Prod. Rep.* **2010**, *27*, 1531–1570. [CrossRef] [PubMed]
15. Jiang, M.Y.; Feng, T.; Liu, J.K. N-Containing compounds of macromycetes. *Nat. Prod. Rep.* **2011**, *28*, 783–808. [CrossRef] [PubMed]
16. Isaka, M.; Sappan, M.; Supothina, S.; Srichomthong, K.; Komwijit, S.; Boonpratuang, T. Alliacane sesquiterpenoids from submerged cultures of the basidiomycete *Inonotus* sp. BCC 22670. *Phytochemistry* **2017**, *136*, 175–181. [CrossRef]
17. Tao, Q.Q.; Ma, K.; Bao, L.; Wang, K.; Han, J.J.; Zhang, J.X.; Huang, C.Y.; Liu, H.W. New sesquiterpenoids from the edible mushroom *Pleurotus cystidiosus* and their inhibitory activity against alpha-glucosidase and PTP1B. *Fitoterapia* **2016**, *111*, 29–35. [CrossRef]
18. Tala, M.F.; Qin, J.C.; Ndongo, J.T.; Laatsch, H. New azulene-type sesquiterpenoids from the fruiting bodies of *Lactarius deliciosus*. *Nat. Prod. Bioprospect.* **2017**, *7*, 269–273. [CrossRef] [PubMed]
19. Liu, S.; Dai, H.F.; Heering, C.; Janiak, C.; Lin, W.H.; Liu, Z.; Proksch, P. Inducing new secondary metabolites through co-cultivation of the fungus *Pestalotiopsis* sp. with the bacterium *Bacillus subtilis*. *Tetrahedron Lett.* **2017**, *58*, 257–261. [CrossRef]
20. Cane, D.E. Enzymic formation of sesquiterpenes. *Chem. Rev.* **1990**, *90*, 1089–1103. [CrossRef]
21. Fraga, B.M. Natural sesquiterpenoids. *Nat. Prod. Rep.* **2013**, *30*, 1226–1264. [CrossRef]
22. Massias, M.; Rebuffat, S.; Molho, L.; Chiaroni, A.; Riche, C.; Bodo, B. Expansolides A and B: Tetracyclic sesquiterpene lactones from *Penicillium expansum*. *J. Am. Chem. Soc.* **1990**, *112*, 8112–8115. [CrossRef]
23. Oh, H.; Gloer, J.B.; Shearer, C.A. Massarinolins A-C: New bioactive sesquiterpenoids from the aquatic fungus *Massarina tunicata*. *J. Nat. Prod.* **1999**, *62*, 497–501. [CrossRef] [PubMed]
24. Che, Y.; Gloer, J.B.; Koster, B.; Malloch, D. Decipinin A and decipienolides A and B: New bioactive metabolites from the coprophilous fungus *Podospora decipiens*. *J. Nat. Prod.* **2002**, *65*, 916–919. [CrossRef] [PubMed]
25. MacíAs, F.A.; Varela, R.M.; Simonet, A.M.; Cutler, H.G.; Cutler, S.J.; Hill, R.A. Absolute configuration of bioactive expansolides A and B from *Aspergillus fumigatus* Fresenius. *Tetrahedron Lett.* **2003**, *44*, 941–943. [CrossRef]
26. Wang, Y.N.; Xia, G.Y.; Wang, L.Y.; Ge, G.B.; Zhang, H.W.; Zhang, J.F.; Wu, Y.Z.; Lin, S. Purpurolide A, 5/5/5 spirocyclic sesquiterpene lactone in nature from the endophytic fungus *Penicillium purpurogenum*. *Org. Lett.* **2018**, *20*, 7341–7344. [CrossRef]
27. Xia, G.Y.; Wang, L.Y.; Zhang, J.F.; Wu, Y.Z.; Ge, G.B.; Wang, Y.N.; Lin, P.C.; Lin, S. Three new polyoxygenated bergamotanes from the endophytic fungus *Penicillium purpurogenum* IMM 003 and their inhibitory activity against pancreatic lipase. *Chin. J. Nat. Med.* **2020**, *18*, 75–80. [CrossRef]
28. Ying, Y.M.; Fang, C.A.; Yao, F.Q.; Yu, Y.; Shen, Y.; Hou, Z.N.; Wang, Z.; Zhang, W.; Shan, W.G.; Zhan, Z.J. Bergamotane sesquiterpenes with α-glucosidase inhibitory activity from the plant pathogenic fungus *Penicillium expansum*. *Chem. Biodivers.* **2017**, *14*, e1600184. [CrossRef]
29. Zhao, Z.Z.; Zhao, K.; Chen, H.P.; Bai, X.; Zhang, L.; Liu, J.K. Terpenoids from the mushroom-associated fungus *Montagnula donacina*. *Phytochemistry* **2018**, *147*, 21–29. [CrossRef]

30. Wang, Y.; Li, D.H.; Li, Z.L.; Sun, Y.J.; Hua, H.M.; Liu, T.; Bai, J. Terpenoids from the marine-derived fungus *Aspergillus fumigatus* YK-7. *Molecules* **2016**, *21*, 31. [CrossRef]
31. Guo, Z.; Ren, F.X.; Che, Y.S.; Liu, G.; Liu, L. New bergamotane sesquiterpenoids from the plant endophytic fungus *Paraconiothyrium brasiliense*. *Molecules* **2015**, *20*, 14611–14620. [CrossRef]
32. Zhang, L.H.; Feng, B.M.; Chen, G.; Li, S.G.; Sun, Y.; Wu, H.H.; Bai, J.; Hua, H.M.; Wang, H.F.; Pei, Y.H. Sporulaminals A and B: A pair of unusual epimeric spiroaminal derivatives from a marine-derived fungus *Paraconiothyrium sporulosum* YK-03. *RSC Adv.* **2016**, *6*, 42361–42366. [CrossRef]
33. Tao, Q.Q.; Ma, K.; Bao, L.; Wang, K.; Han, J.J.; Wang, W.Z.; Zhang, J.X.; Huang, C.Y.; Liu, H.W. Sesquiterpenoids with PTP1B inhibitory activity and cytotoxicity from the edible mushroom *Pleurotus citrinopileatus*. *Planta Med.* **2016**, *82*, 639–644. [CrossRef]
34. Tao, Q.Q.; Ma, K.; Yang, Y.L.; Wang, K.; Chen, B.S.; Huang, Y.; Han, J.J.; Bao, L.; Liu, X.B.; Yang, Z.L.; et al. Bioactive sesquiterpenes from the edible mushroom *Flammulina velutipes* and their biosynthetic pathway confirmed by genome analysis and chemical evidence. *J. Org. Chem.* **2016**, *81*, 9867–9877. [CrossRef]
35. Chen, C.M.; Sun, W.G.; Liu, X.R.; Wei, M.S.; Liang, Y.; Wang, J.P.; Zhu, H.C.; Zhang, Y.H. Anti-inflammatory spiroaxane and drimane sesquiterpenoids from *Talaromyces minioluteus* (*Penicillium minioluteum*). *Bioorg. Chem.* **2019**, *91*, 103166. [CrossRef]
36. Yang, X.Y.; Niu, W.R.; Li, R.T.; Cui, X.M.; Liu, J.K. Two new sesquiterpenes from cultures of the higher fungus *Pholiota nameko*. *Nat. Prod. Res.* **2018**, *33*, 1992–1996. [CrossRef]
37. Wang, S.R.; Zhang, L.; Chen, H.P.; Li, Z.H.; Dong, Z.J.; Wei, K.; Liu, J.K. Four new spiroaxane sesquiterpenes and one new rosenonolactone derivative from cultures of Basidiomycete *Trametes versicolor*. *Fitoterapia* **2015**, *105*, 127–131. [CrossRef] [PubMed]
38. Hu, Z.B.; Tao, Y.W.; Tao, X.Y.; Su, Q.H.; Cai, J.C.; Qin, C.; Ding, W.J.; Li, C.Y. Sesquiterpenes with phytopathogenic fungi inhibitory activities from fungus *Trichoderma virens* from *Litchi chinensis* Sonn. *J. Agric. Food Chem.* **2019**, *67*, 10646–10652. [CrossRef] [PubMed]
39. Liu, S.; Zhao, Y.P.; Heering, C.; Janiak, C.; Muller, W.E.G.; Akone, S.H.; Liu, Z.; Proksch, P. Sesquiterpenoids from the endophytic fungus *Rhinocladiella similis*. *J. Nat. Prod.* **2019**, *82*, 1055–1062. [CrossRef]
40. Duan, R.T.; Yang, R.N.; Li, H.T.; Tang, L.H.; Liu, T.; Yang, Y.B.; Zhou, H.; Ding, Z.T. Peniterester, a carotane-type antibacterial sesquiterpene from an artificial mutant *Penicillium* sp. T2-M20. *Fitoterapia* **2020**, *140*, 104422. [CrossRef]
41. Xing, C.P.; Xie, C.L.; Xia, J.M.; Liu, Q.M.; Lin, W.X.; Ye, D.Z.; Liu, G.M.; Yang, X.W. Penigrisacids A-D, four new sesquiterpenes from the deep-sea-derived *Penicillium griseofulvum*. *Mar. Drugs* **2019**, *17*, 507. [CrossRef] [PubMed]
42. Afiyatullov, S.S.; Zhuravleva, O.I.; Antonov, A.S.; Leshchenko, E.V.; Pivkin, M.V.; Khudyakova, Y.V.; Denisenko, V.A.; Pislyagin, E.A.; Kim, N.Y.; Berdyshev, D.V.; et al. Piltunines A-F from the marine-derived fungus *Penicillium piltunense* KMM 4668. *Mar. Drugs* **2019**, *17*, 647. [CrossRef]
43. Shi, Z.Z.; Fang, S.T.; Miao, F.P.; Yin, X.L.; Ji, N.Y. Trichocarotins A–H and trichocadinin A, nine sesquiterpenes from the marine-alga-epiphytic fungus *Trichoderma virens*. *Bioorg. Chem.* **2018**, *81*, 319–325. [CrossRef]
44. Zhang, J.; Liu, S.S.; Yuan, W.Y.; Wei, J.J.; Zhao, Y.X.; Luo, D.Q. Carotane-type sesquiterpenes from cultures of the insect pathogenic fungus *Isaria fumosorosea*. *J. Asian Nat. Prod. Res.* **2017**, *21*, 234–240. [CrossRef] [PubMed]
45. Li, Y.L.; Liu, W.; Xu, W.; Zeng, X.; Cheng, Z.B.; Li, Q. Aspterrics A and B, new sesquiterpenes from deep sea-derived fungus *Aspergillus terreus* YPGA10. *Rec. Nat. Prod.* **2020**, *14*, 18–22. [CrossRef]
46. Liu, X.H.; Hou, X.L.; Song, Y.P.; Wang, B.G.; Ji, N.Y. Cyclonerane sesquiterpenes and an isocoumarin derivative from the marine-alga-endophytic fungus *Trichoderma citrinoviride* A-WH-20-3. *Fitoterapia* **2020**, *141*, 104469. [CrossRef] [PubMed]
47. Shi, T.; Shao, C.L.; Liu, Y.; Zhao, D.L.; Cao, F.; Fu, X.M.; Yu, J.Y.; Wu, J.S.; Zhang, Z.K.; Wang, C.Y. Terpenoids from the coral-derived fungus *Trichoderma harzianum* (XS-20090075) induced by chemical epigenetic manipulation. *Front. Microbiol.* **2020**, *11*, 572. [CrossRef]
48. Fang, S.T.; Wang, Y.J.; Ma, X.Y.; Yin, X.L.; Ji, N.Y. Two new sesquiterpenoids from the marine-sediment-derived fungus *Trichoderma harzianum* P1-4. *Nat. Prod. Res.* **2019**, *33*, 3127–3133. [CrossRef]
49. Song, Y.P.; Fang, S.T.; Miao, F.P.; Yin, X.L.; Ji, N.Y. Diterpenes and sesquiterpenes from the marine algicolous fungus *Trichoderma harzianum* X-5. *J. Nat. Prod.* **2018**, *81*, 2553–2559. [CrossRef]
50. Jiang, C.X.; Li, J.; Zhang, J.M.; Jin, X.J.; Yu, B.; Fang, J.G.; Wu, Q.X. Isolation, identification, and activity evaluation of chemical constituents from soil fungus *Fusarium avenaceum* SF-1502 and endophytic fungus *Fusarium proliferatum* AF-04. *J. Agric. Food Chem.* **2019**, *67*, 1839–1846. [CrossRef]
51. Zhang, M.; Zhao, J.L.; Liu, J.M.; Chen, R.D.; Xie, K.B.; Chen, D.W.; Feng, K.P.; Zhang, D.; Dai, J.G. Neural anti-inflammatory sesquiterpenoids from the endophytic fungus *Trichoderma* sp. Xy24. *J. Asian Nat. Prod. Res.* **2017**, *19*, 651–658. [CrossRef] [PubMed]
52. Song, Y.P.; Miao, F.P.; Liu, X.H.; Yin, X.L.; Ji, N.Y. Cyclonerane derivatives from the algicolous endophytic fungus *Trichoderma asperellum* A-YMD-9-2. *Mar. Drugs* **2019**, *17*, 252. [CrossRef] [PubMed]
53. Song, Y.P.; Liu, X.H.; Shi, Z.Z.; Miao, F.P.; Fang, S.T.; Ji, N.Y. Bisabolane, cyclonerane, and harziane derivatives from the marine-alga-endophytic fungus *Trichoderma asperellum* cf44-2. *Phytochemistry* **2018**, *152*, 45–52. [CrossRef]
54. Qin, D.; Wang, L.; Han, M.J.; Wang, J.Q.; Song, H.C.; Yen, X.; Duan, X.X.; Dong, J.Y. Effects of an endophytic fungus *Umbelopsis dimorpha* on the secondary metabolites of host-plant *Kadsura angustifolia*. *Front. Microbiol.* **2018**, *9*, 2845. [CrossRef]

55. Isaka, M.; Yangchum, A.; Supothina, S.; Boonpratuang, T.; Choeyklin, R.; Kongsaeree, P.; Prabpai, S. Aromadendrane and cyclofarnesane sesquiterpenoids from cultures of the basidiomycete *Inonotus* sp. BCC 23706. *Phytochemistry* **2015**, *118*, 94–101. [CrossRef]
56. Lei, H.; Lin, X.P.; Han, L.; Ma, J.; Ma, Q.J.; Zhong, J.L.; Liu, Y.H.; Sun, T.M.; Wang, J.H.; Huang, X.S. New metabolites and bioactive chlorinated benzophenone derivatives produced by a marine-derived fungus *Pestalotiopsis heterocornis*. *Mar. Drugs* **2017**, *15*, 69. [CrossRef]
57. Chen, S.C.; Li, H.H.; Chen, Y.C.; Li, S.N.; Xu, J.L.; Guo, H.; Liu, Z.M.; Zhu, S.; Liu, H.X.; Zhang, W.M. Three new diterpenes and two new sesquiterpenoids from the endophytic fungus *Trichoderma koningiopsis* A729. *Bioorg. Chem.* **2019**, *86*, 368–374. [CrossRef]
58. Yang, M.S.; Cai, X.Y.; He, Y.Y.; Lu, M.Y.; Liu, S.; Wang, W.X.; Li, Z.H.; Ai, H.L.; Feng, T. Seco-sativene and seco-longifolene sesquiterpenoids from cultures of endophytic fungus *Bipolaris eleusines*. *Nat. Prod. Bioprospect.* **2017**, *7*, 147–150. [CrossRef] [PubMed]
59. He, J.; Yang, M.S.; Wang, W.X.; Li, Z.H.; Elkhateeb, W.A.M.; Wen, T.C.; Ai, H.L.; Feng, T. Anti-phytopathogenic sesquiterpenoid-xanthone adducts from potato endophytic fungus *Bipolaris eleusines*. *RSC Adv.* **2019**, *9*, 128–131. [CrossRef]
60. Le Bideau, F.; Kousara, M.; Chen, L.; Wei, L.; Dumas, F. Tricyclic sesquiterpenes from marine origin. *Chem. Rev.* **2017**, *117*, 6110–6159. [CrossRef]
61. Qiu, Y.; Lan, W.J.; Li, H.J.; Chen, L.P. Linear triquinane sesquiterpenoids: Their isolation, structures, biological activities, and chemical synthesis. *Molecules* **2018**, *23*, 2095. [CrossRef]
62. Tabuchi, A.; Fukushima-Sakuno, E.; Osaki-Oka, K.; Futamura, Y.; Motoyama, T.; Osada, H.; Ishikawa, N.K.; Nagasawa, E.; Tokimoto, K. Productivity and bioactivity of enokipodins A-D of *Flammulina rossica* and *Flammulina velutipes*. *Biosci. Biotechnol. Biochem.* **2020**, *84*, 876–886. [CrossRef] [PubMed]
63. Liu, H.X.; Tan, H.B.; Chen, K.; Chen, Y.C.; Li, S.N.; Li, H.H.; Zhang, W.M. Cerrenins A-C, cerapicane and isohirsutane sesquiterpenoids from the endophytic fungus *Cerrena* sp. *Fitoterapia* **2018**, *129*, 173–178. [CrossRef]
64. Ding, J.H.; Li, Z.H.; Wei, K.; Dong, Z.J.; Ding, Z.H.; Feng, T.; Liu, J.K. Two new sesquiterpenoids from cultures of the basidiomycete *Tremella foliacea*. *J. Asian Nat. Prod. Res.* **2016**, *18*, 46–50. [CrossRef] [PubMed]
65. Chen, Z.M.; Wang, S.L. Two new compounds from cultures of the basidiomycete *Antrodiella albocinnamomea*. *Nat. Prod. Res.* **2015**, *29*, 1985–1989. [CrossRef]
66. Huang, L.; Lan, W.J.; Li, H.J. Two new hirsutane-type sesquiterpenoids chondrosterins N and O from the marine fungus *Chondrostereum* sp. *Nat. Prod. Res.* **2018**, *32*, 1578–1582. [CrossRef]
67. Qi, Q.Y.; Ren, J.W.; Sun, L.W.; He, L.W.; Bao, L.; Yue, W.; Sun, Q.M.; Yao, Y.J.; Yin, W.B.; Liu, H.W. Stucturally diverse sesquiterpenes produced by a Chinese Tibet fungus *Stereum hirsutum* and their cytotoxic and immunosuppressant activities. *Org. Lett.* **2015**, *17*, 3098–3101. [CrossRef]
68. Liu, H.X.; Tan, H.B.; Chen, Y.C.; Li, S.N.; Li, H.H.; Zhang, W.M. Cytotoxic triquinane-type sesquiterpenoids from the endophytic fungus *Cerrena* sp. A593. *Nat. Prod. Res.* **2019**, *34*, 2430–2436. [CrossRef] [PubMed]
69. Huang, L.; Lan, W.J.; Deng, R.; Feng, G.K.; Xu, Q.Y.; Hu, Z.Y.; Zhu, X.F.; Li, H.J. Additional new cytotoxic triquinane-type sesquiterpenoids chondrosterins K-M from the marine fungus *Chondrostereum* sp. *Mar. Drugs* **2016**, *14*, 157. [CrossRef] [PubMed]
70. Chen, Z.M.; Chen, H.P.; Wang, F.; Li, Z.H.; Feng, T.; Liu, J.K. New triquinane and gymnomitrane sesquiterpenes from fermentation of the basidiomycete *Antrodiella albocinnamomea*. *Fitoterapia* **2015**, *102*, 61–66. [CrossRef]
71. Wang, M.; Du, J.X.; Yang, H.X.; Dai, Q.; Liu, Y.P.; He, J.; Wang, Y.; Li, Z.H.; Feng, T.; Liu, J.K. Sesquiterpenoids from cultures of the basidiomycetes *Irpex lacteus*. *J. Nat. Prod.* **2020**, *83*, 1524–1531. [CrossRef]
72. Wang, Y.Z.; Wang, Y.; Wu, A.A.; Zhang, L.; Hu, Z.Y.; Huang, H.Y.; Xu, Q.Y.; Deng, X.M. New 12,8-eudesmanolides from *Eutypella* sp. 1–15. *J. Antibiot.* **2017**, *70*, 1029–1032. [CrossRef]
73. Zhang, D.W.; Ge, H.L.; Zou, J.H.; Tao, X.Y.; Chen, R.D.; Dai, J.G. Periconianone A, a new 6/6/6 carbocyclic sesquiterpenoid from endophytic fungus *Periconia* sp. with neural anti-inflammatory activity. *Org. Lett.* **2014**, *16*, 1410–1413. [CrossRef]
74. Liffert, R.; Linden, A.; Gademann, K. Total synthesis of the sesquiterpenoid periconianone A based on a postulated biogenesis. *J. Am. Chem. Soc.* **2017**, *139*, 16096–16099. [CrossRef]
75. Zhao, Z.Z.; Liang, X.B.; Feng, W.S.; Wu, Y.; Zhi, Y.L.; Xue, G.M.; Chen, H.P.; Liu, J.K. Unusual constituents from the medicinal mushroom *Ganoderma lingzhi*. *RSC Adv.* **2019**, *9*, 36931–36939. [CrossRef]
76. Binh, P.T.; Descoutures, D.; Dang, N.H.; Nguyen, N.P.; Dat, N.T. A new cytotoxic gymnomitrane sesquiterpene from *Ganoderma lucidum* fruiting bodies. *Nat. Prod. Commun.* **2015**, *10*, 1911–1912. [CrossRef]
77. Wang, J.F.; He, W.J.; Kong, F.D.; Tian, X.P.; Wang, P.; Zhou, X.J.; Liu, Y.H. Ochracenes A-I, humulane-derived sesquiterpenoids from the antarctic fungus *Aspergillus ochraceopetaliformis*. *J. Nat. Prod.* **2017**, *80*, 1725–1733. [CrossRef] [PubMed]
78. Wu, P.F.; Ding, R.; Tan, R.; Liu, J.; Hu, E.M.; Li, C.Y.; Liang, G.Y.; Yi, P. Sesquiterpenes from cultures of the fungus *Phellinus igniarius* and their cytotoxicities. *Fitoterapia* **2020**, *140*, 104415. [CrossRef] [PubMed]
79. Nord, C.; Menkis, A.; Broberg, A. Cytotoxic illudane sesquiterpenes from the fungus *Granulobasidium vellereum* (Ellis and Cragin) Jülich. *J. Nat. Prod.* **2015**, *78*, 2559–2564. [CrossRef] [PubMed]
80. He, J.B.; Tao, J.; Miao, X.S.; Feng, Y.P.; Bu, W.; Dong, Z.J.; Li, Z.H.; Feng, T.; Liu, J.K. Two new illudin type sesquiterpenoids from cultures of *Phellinus tuberculosus* and *Laetiporus sulphureus*. *J. Asian Nat. Prod. Res.* **2015**, *17*, 1054–1058. [CrossRef] [PubMed]
81. Dai, Q.; Zhang, F.L.; Du, J.X.; Li, Z.H.; Feng, T.; Liu, J.K. Illudane sesquiterpenoids from edible mushroom *Agrocybe salicacola* and their bioactivities. *ACS Omega* **2020**, *5*, 21961–21967. [CrossRef]

82. Guo, H.; Diao, Q.P.; Zhang, B.; Feng, T. Two new illudane sesquiterpenoids and one new menthane monoterpene from cultures of *Craterellus cornucopioides*. *J. Asian Nat. Prod. Res.* **2019**, *21*, 123–128. [CrossRef]
83. Kokubun, T.; Scott-Brown, A.; Kite, G.C.; Simmonds, M.S.J. Protoilludane, illudane, illudalane, and norilludane sesquiterpenoids from *Granulobasidium vellereum*. *J. Nat. Prod.* **2016**, *79*, 1698–1701. [CrossRef]
84. Surup, F.; Hennicke, F.; Sella, N.; Stroot, M.; Bernecker, S.; Pfutze, S.; Stadler, M.; Ruhl, M. New terpenoids from the fermentation broth of the edible mushroom *Cyclocybe aegerita*. *Beilstein J. Org. Chem.* **2019**, *15*, 1000–1007. [CrossRef]
85. Xie, S.S.; Wu, Y.; Qiao, Y.B.; Guo, Y.; Wang, J.P.; Hu, Z.X.; Zhang, Q.; Li, X.N.; Huang, J.F.; Zhou, Q.; et al. Protoilludane, illudalane, and botryane sesquiterpenoids from the endophytic fungus *Phomopsis* sp. TJ507A. *J. Nat. Prod.* **2018**, *81*, 1311–1320. [CrossRef] [PubMed]
86. Li, H.T.; Tang, L.H.; Liu, T.; Yang, R.N.; Yang, Y.B.; Zhou, H.; Ding, Z.T. Protoilludane-type sesquiterpenoids from *Armillaria* sp. by co-culture with the endophytic fungus *Epicoccum* sp. associated with *Gastrodia elata*. *Bioorg. Chem.* **2020**, *95*, 103503. [CrossRef] [PubMed]
87. Li, Z.J.; Wang, Y.C.; Jiang, B.; Li, W.L.; Zheng, L.H.; Yang, X.G.; Bao, Y.L.; Sun, L.G.; Huang, Y.X.; Li, Y.X. Structure, cytotoxic activity and mechanism of protoilludane sesquiterpene aryl esters from the mycelium of *Armillaria mellea*. *J. Ethnopharmacol.* **2016**, *184*, 119–127. [CrossRef]
88. Chen, C.C.; Kuo, Y.H.; Cheng, J.J.; Sung, P.J.; Ni, C.L.; Chen, C.C.; Shen, C.C. Three new sesquiterpene aryl esters from the mycelium of *Armillaria mellea*. *Molecules* **2015**, *20*, 9994–10003. [CrossRef]
89. Hovey, M.T.; Cohen, D.T.; Walden, D.M.; Cheong, P.H.Y.; Scheidt, K.A. A carbene catalysis strategy for the synthesis of protoilludane natural products. *Angew. Chem. Int. Ed.* **2017**, *56*, 9864–9867. [CrossRef]
90. Su, X.Z.; Tang, J.W.; Hu, K.; Li, X.N.; Sun, H.D.; Puno, P.T. Arthrinins E-G, three botryane sesquiterpenoids from the plant endophytic fungus *Arthrinium* sp. HS66. *Nat. Prod. Bioprospect.* **2020**, *10*, 201–207. [CrossRef] [PubMed]
91. Medina, R.P.; Araujo, A.R.; Batista, J.M., Jr.; Cardoso, C.L.; Seidl, C.; Vilela, A.F.L.; Domingos, H.V.; Costa-Lotufo, L.V.; Andersen, R.J.; Silva, D.H.S. Botryane terpenoids produced by *Nemania bipapillata*, an endophytic fungus isolated from red alga *Asparagopsis taxiformis-Falkenbergia* stage. *Sci. Rep.* **2019**, *9*, 12318. [CrossRef]
92. Chen, B.S.; Li, E.W.; Liu, L.; Liao, M.F.; Zhu, Z.X.; Zhuang, W.Y.; Bao, L.; Liu, H.W. Botryane sesquiterpenoids, cyclopentadepsipeptides, xanthones, and trichothecenes from *Trichoderma oligosporum*. *Planta Med.* **2018**, *84*, 1055–1063. [CrossRef]
93. Kuhnert, E.; Surup, F.; Wiebach, V.; Bernecker, S.; Stadler, M. Botryane, noreudesmane and abietane terpenoids from the ascomycete *Hypoxylon rickii*. *Phytochemistry* **2015**, *117*, 116–122. [CrossRef]
94. Wiebach, V.; Surup, F.; Kuhnert, E.; Stadler, M. Rickicaryophyllane A, a caryophyllane from the ascomyceteous fungus *Hypoxylon rickii* and a 10-norbotryane congener. *Nat. Prod. Commun.* **2016**, *11*, 909–912. [CrossRef] [PubMed]
95. Bao, Y.R.; Chen, G.D.; Gao, H.; He, R.R.; Wu, Y.H.; Li, X.X.; Hu, D.; Wang, C.X.; Liu, X.Z.; Li, Y.; et al. 4,5-*seco*-Probotryenols A-C, a new type of sesquiterpenoids from *Stachybotrys bisbyi*. *RSC Adv.* **2015**, *5*, 46252–46259. [CrossRef]
96. Ren, F.X.; Zhu, S.M.; Wang, B.; Li, L.; Liu, X.Z.; Su, R.B.; Che, Y.S. Hypocriols A-F, heterodimeric botryane ethers from *Hypocrea* sp., an insect-associated fungus. *J. Nat. Prod.* **2016**, *79*, 1848–1856. [CrossRef] [PubMed]
97. Ayer, W.A.; Cruz, E.R. The tremulanes, a new group of sesquiterpenes from the aspen rotting fungus *Phellinus tremulae*. *J. Org. Chem.* **1993**, *58*, 7529–7534. [CrossRef]
98. Ayer, W.A.; Cruz, E.R. 2-Carbomethoxyoxepin: 1-carbomethoxybenzene 1,2-oxide and the biosynthesis of methyl salicylate in *Phellinus tremulae*. *J. Nat. Prod.* **1995**, *58*, 622–624. [CrossRef]
99. Chen, H.P.; Ji, X.; Li, Z.H.; Feng, T.; Liu, J.K. Irlactane and tremulane sesquiterpenes from the cultures of the medicinal fungus *Irpex lacteus* HFG1102. *Nat. Prod. Bioprospect.* **2020**, *10*, 89–100. [CrossRef] [PubMed]
100. Ding, J.H.; Feng, T.; Cui, B.K.; Wei, K.; Li, Z.H.; Liu, J.K. Novel sesquiterpenoids from cultures of the basidiomycete *Irpex lacteus*. *Tetrahedron Lett.* **2013**, *54*, 2651–2654. [CrossRef]
101. He, J.; Pu, C.J.; Wang, M.; Li, Z.H.; Feng, T.; Zhao, D.K.; Liu, J.K. Conosiligins A–D, ring-rearranged tremulane sesquiterpenoids from *Conocybe siliginea*. *J. Nat. Prod.* **2020**, *83*, 2743–2748. [CrossRef]
102. Li, W.; He, J.; Feng, T.; Yang, H.X.; Ai, H.L.; Li, Z.H.; Liu, J.K. Antroalbocin A, an antibacterial sesquiterpenoid from higher fungus *Antrodiella albocinnamomea*. *Org. Lett.* **2018**, *20*, 8019–8021. [CrossRef]
103. Ding, J.H.; Li, Z.H.; Feng, T.; Liu, J.K. Two new sesquiterpenes from cultures of the fungus *Irpex lacteus*. *J. Asian Nat. Prod. Res.* **2020**, *23*, 348–352. [CrossRef]
104. Ding, J.H.; Li, Z.H.; Feng, T.; Liu, J.K. Tremulane sesquiterpenes from cultures of the basidiomycete *Irpex lacteus*. *Fitoterapia* **2018**, *125*, 245–248. [CrossRef] [PubMed]
105. Ding, J.H.; Li, Z.H.; Feng, T.; Liu, J.K. A new tremulane sesquiterpenoid from the fungus *Irpex lacteus*. *Nat. Prod. Res.* **2019**, *33*, 316–320. [CrossRef]
106. Zhou, Q.Y.; Yang, X.Q.; Zhang, Z.X.; Wang, B.Y.; Hu, M.; Yang, Y.B.; Zhou, H.; Ding, Z.T. New azaphilones and tremulane sesquiterpene from endophytic *Nigrospora oryzae* cocultured with *Irpex lacteus*. *Fitoterapia* **2018**, *130*, 26–30. [CrossRef] [PubMed]
107. Wu, Y.M.; Zhou, Q.Y.; Yang, X.Q.; Luo, Y.J.; Qian, J.J.; Liu, S.X.; Yang, Y.B.; Ding, Z.T. Induction of antiphytopathogenic metabolite and squalene production and phytotoxin elimination by adjustment of the mode of fermentation in cocultures of phytopathogenic *Nigrospora oryzae* and *Irpex lacteus*. *J. Agric. Food Chem.* **2019**, *67*, 11877–11882. [CrossRef]
108. Guo, Z.Y.; Li, X.S.; Zhang, L.; Feng, Z.W.; Deng, Z.S.; He, H.B.; Zou, K. Cytotoxic tremulanes and 5,6-secotremulanes, four new sesquiterpenoids from a plant-associated fungus X1-2. *Nat. Prod. Res.* **2016**, *30*, 2582–2589. [CrossRef] [PubMed]

109. Wang, F.Q.; Ma, H.R.; Hu, Z.X.; Jiang, J.C.; Zhu, H.; Cheng, L.; Yang, Q.W.; Zhang, H.; Zhang, G.; Zhang, Y.H. Secondary metabolites from *Colletotrichum capsici*, an endophytic fungus derived from *Siegesbeckia pubescens* Makino. *Nat. Prod. Res.* **2017**, *31*, 1849–1854. [CrossRef]
110. Chen, H.Y.; Liu, T.K.; Shi, Q.; Yang, X.L. Sesquiterpenoids and diterpenes with antimicrobial activity from *Leptosphaeria* sp. XL026, an endophytic fungus in *Panax notoginseng*. *Fitoterapia* **2019**, *137*, 104243. [CrossRef] [PubMed]
111. Chen, H.P.; Zhao, Z.Z.; Li, Z.H.; Feng, T.; Liu, J.K. *Seco*-tremulane sesquiterpenoids from the cultures of the medicinal fungus *Irpex lacteus* HFG1102. *Nat. Prod. Bioprospect.* **2018**, *8*, 113–119. [CrossRef]
112. Nakashima, K.; Tomida, J.; Hirai, T.; Kawamura, Y.; Inoue, M. Sesquiterpenes with new carbon skeletons from the basidiomycete *Phlebia tremellosa*. *J. Nat. Med.* **2019**, *73*, 480–486. [CrossRef]
113. Zhu, M.Z.; Cen, Y.F.; Ye, W.; Li, S.N.; Zhang, W.M. Recent advances on macrocyclic trichothecenes, their bioactivities and biosynthetic pathway. *Toxins* **2020**, *12*, 417. [CrossRef]
114. Ryu, S.M.; Lee, H.M.; Song, E.G.; Seo, Y.H.; Lee, J.; Guo, Y.; Kim, B.S.; Kim, J.J.; Hong, J.S.; Ryu, K.H.; et al. Antiviral activities of trichothecenes isolated from *Trichoderma albolutescens* against *pepper mottle virus*. *J. Agric. Food Chem.* **2017**, *65*, 4273–4279. [CrossRef]
115. Yamazaki, H.; Yagi, A.; Takahashi, O.; Yamaguchi, Y.; Saito, A.; Namikoshi, M.; Uchida, R. Antifungal trichothecene sesquiterpenes obtained from the culture broth of marine-derived *Trichoderma* cf. *brevicompactum* and their structure-activity relationship. *Bioorg. Med. Chem. Lett.* **2020**, *30*, 127375. [CrossRef]
116. Yamazaki, H.; Takahashi, O.; Kirikoshi, R.; Yagi, A.; Ogasawara, T.; Bunya, Y.; Rotinsulu, H.; Uchida, R.; Namikoshi, M. Epipolythiodiketopiperazine and trichothecene derivatives from the NaI-containing fermentation of marine-derived *Trichoderma* cf. *brevicompactum*. *J. Antibiot.* **2020**, *73*, 559–567. [CrossRef]
117. Lee, S.R.; Seok, S.; Ryoo, R.; Choi, S.U.; Kim, K.H. Macrocyclic trichothecene mycotoxins from a deadly poisonous mushroom, *Podostroma cornu-damae*. *J. Nat. Prod.* **2019**, *82*, 122–128. [CrossRef] [PubMed]
118. Ye, W.; Chen, Y.C.; Li, H.H.; Zhang, W.M.; Liu, H.X.; Sun, Z.H.; Liu, T.M.; Li, S.N. Two trichothecene mycotoxins from *Myrothecium roridum* induce apoptosis of HepG-2 cells via caspase activation and disruption of mitochondrial membrane potential. *Molecules* **2016**, *21*, 781. [CrossRef]
119. Liu, H.X.; Liu, W.Z.; Chen, Y.C.; Sun, Z.H.; Tan, Y.Z.; Li, H.H.; Zhang, W.M. Cytotoxic trichothecene macrolides from the endophyte fungus *Myrothecium roridum*. *J. Asian Nat. Prod. Res.* **2016**, *18*, 684–689. [CrossRef]
120. Shen, L.; Zhu, L.; Tan, Q.W.; Wan, D.; Xie, J.; Peng, J.N. New cytotoxic trichothecene macrolide epimers from endophytic *Myrothecium roridum* IFB-E012. *J. Antibiot.* **2016**, *69*, 652–655. [CrossRef] [PubMed]
121. Mondol, M.A.; Surovy, M.Z.; Islam, M.T.; Schuffler, A.; Laatsch, H. Macrocyclic trichothecenes from *Myrothecium roridum* strain M10 with motility inhibitory and zoosporicidal activities against *Phytophthora nicotianae*. *J. Agric. Food Chem.* **2015**, *63*, 8777–8786. [CrossRef] [PubMed]
122. Li, Y.; Liu, D.; Cheng, Z.B.; Proksch, P.; Lin, W.H. Cytotoxic trichothecene-type sesquiterpenes from the sponge-derived fungus *Stachybotrys chartarum* with tyrosine kinase inhibition. *RSC Adv.* **2017**, *7*, 7259–7267. [CrossRef]
123. Matsumoto, M.; Tanaka, S.; Tonouchi, A.; Hashimoto, M. 12-Deoxyroridin J and 12-deoxyepiisororidin E from *Calcarisporium arbuscular*. *Tetrahedron Lett.* **2018**, *59*, 1992–1995. [CrossRef]
124. Matsumoto, M.; Nishiyama, M.; Maeda, H.; Tonouchi, A.; Konno, K.; Hashimoto, M. Structure-activity relationships of trichothecenes against COLO201 cells and *Cochliobolus miyabeanus*: The role of 12-epoxide and macrocyclic moieties. *Bioorg. Med. Chem. Lett.* **2019**, *29*, 982–985. [CrossRef] [PubMed]
125. Yang, H.X.; Ai, H.L.; Feng, T.; Wang, W.X.; Wu, B.; Zheng, Y.S.; Sun, H.; He, J.; Li, Z.H.; Liu, J.K. Trichothecrotocins A-C, antiphytopathogenic agents from potato endophytic fungus *Trichothecium crotocinigenum*. *Org. Lett.* **2018**, *20*, 8069–8072. [CrossRef]
126. Yang, H.X.; He, J.; Zhang, F.L.; Zhang, X.D.; Li, Z.H.; Feng, T.; Ai, H.L.; Liu, J.K. Trichothecrotocins D–L, antifungal agents from a potato-associated *Trichothecium crotocinigenum*. *J. Nat. Prod.* **2020**, *83*, 2756–2763. [CrossRef]
127. Barua, J.E.; de la Cruz, M.; de Pedro, N.; Cautain, B.; Hermosa, R.; Cardoza, R.E.; Gutierrez, S.; Monte, E.; Vicente, F.; Collado, I.G. Synthesis of trichodermin derivatives and their antimicrobial and cytotoxic activities. *Molecules* **2019**, *24*, 3811. [CrossRef] [PubMed]
128. Chen, H.Y.; Liu, T.K.; Yang, J.; Yang, X.L. Emerones A-C: Three novel merosesquiterpenoids with unprecedented skeletons from *Emericella* sp. XL029. *Org. Biomol. Chem.* **2019**, *17*, 8450–8455. [CrossRef]
129. Shin, H.J.; Choi, B.K.; Trinh, P.T.H.; Lee, H.S.; Kang, J.S.; Van, T.T.T.; Lee, H.S.; Lee, J.S.; Lee, Y.J.; Lee, J. Suppression of RANKL-induced osteoclastogenesis by the metabolites from the marine fungus *Aspergillus flocculosus* isolated from a sponge *Stylissa* sp. *Mar. Drugs* **2018**, *16*, 14. [CrossRef]
130. Wang, J.F.; Wei, X.Y.; Qin, X.C.; Tian, X.P.; Liao, L.; Li, K.M.; Zhou, X.F.; Yang, X.W.; Wang, F.Z.; Zhang, T.Y.; et al. Antiviral merosesquiterpenoids produced by the antarctic fungus *Aspergillus ochraceopetaliformis* SCSIO 05702. *J. Nat. Prod.* **2016**, *79*, 59–65. [CrossRef] [PubMed]
131. Chen, S.T.; Wang, J.F.; Wang, Z.; Lin, X.P.; Zhao, B.X.; Kaliaperumal, K.; Liao, X.J.; Tu, Z.C.; Li, J.L.; Xu, S.H.; et al. Structurally diverse secondary metabolites from a deep-sea-derived fungus *Penicillium chrysogenum* SCSIO 41001 and their biological evaluation. *Fitoterapia* **2017**, *117*, 71–78. [CrossRef]

132. Yamazaki, H.; Nakayama, W.; Takahashi, O.; Kirikoshi, R.; Izumikawa, Y.; Iwasaki, K.; Toraiwa, K.; Ukai, K.; Rotinsulu, H.; Wewengkang, D.S.; et al. Verruculides A and B, two new protein tyrosine phosphatase 1B inhibitors from an Indonesian ascidian-derived *Penicillium verruculosum*. *Bioorg. Med. Chem. Lett.* **2015**, *25*, 3087–3090. [CrossRef] [PubMed]
133. Jiao, W.H.; Dewapriya, P.; Mohamed, O.; Khalil, Z.G.; Salim, A.A.; Lin, H.W.; Capon, R.J. Divirensols: Sesquiterpene dimers from the Australian termite nest-derived fungus *Trichoderma virens* CMB-TN16. *J. Nat. Prod.* **2019**, *82*, 87–95. [CrossRef] [PubMed]
134. Jiao, W.H.; Salim, A.A.; Khalil, Z.G.; Dewapriya, P.; Lin, H.W.; Butler, M.S.; Capon, R.J. Trivirensols: Selectively bacteriostatic sesquiterpene trimers from the Australian termite nest-derived fungus *Trichoderma virens* CMB-TN16. *J. Nat. Prod.* **2019**, *82*, 3165–3175. [CrossRef]
135. Xu, L.L.; Chen, H.L.; Hai, P.; Gao, Y.; Xie, C.D.; Yang, X.L.; Abe, I. (+)- and (−)-Preuisolactone a: A pair of caged norsesquiterpenoidal enantiomers with a tricyclo[4.4.$0^{1,6}$.$0^{2,8}$]decane carbon skeleton from the endophytic fungus *Preussia isomera*. *Org. Lett.* **2019**, *21*, 1078–1081. [CrossRef]
136. Otaka, J.; Hashizume, D.; Masumoto, Y.; Muranaka, A.; Uchiyama, M.; Koshino, H.; Futamura, Y.; Osada, H. Hitoyol A and B, two norsesquiterpenoids from the basidiomycete *Coprinopsis cinerea*. *Org. Lett.* **2017**, *19*, 4030–4033. [CrossRef]
137. Wei, H.; Xu, Y.M.; Espinosa-Artiles, P.; Liu, M.X.; Luo, J.G.; U'Ren, J.M.; Arnold, A.E.; Gunatilaka, A.A. Sesquiterpenes and other constituents of *Xylaria* sp. NC1214, a fungal endophyte of the moss *Hypnum* sp. *Phytochemistry* **2015**, *118*, 102–108. [CrossRef]
138. Wang, M.Z.; Sun, M.W.; Hao, H.L.; Lu, C.H. Avertoxins A-D, prenyl asteltoxin derivatives from *Aspergillus versicolor* Y10, an endophytic fungus of *Huperzia serrata*. *J. Nat. Prod.* **2015**, *78*, 3067–3070. [CrossRef]
139. Hu, X.Y.; Li, X.M.; Yang, S.Q.; Liu, H.; Meng, L.H.; Wang, B.G. Three new sesquiterpenoids from the algal-derived fungus *Penicillium chermesinum* EN-480. *Mar. Drugs* **2020**, *18*, 194. [CrossRef] [PubMed]
140. Tai, S.H.; Kuo, P.C.; Hung, C.C.; Lin, Y.H.; Hwang, T.L.; Lam, S.H.; Kuo, D.H.; Wu, J.B.; Hung, H.Y.; Wu, T.S. Bioassay-guided purification of sesquiterpenoids from the fruiting bodies of *Fomitopsis pinicola* and their anti-inflammatory activity. *RSC Adv.* **2019**, *9*, 34184–34195. [CrossRef]
141. Luo, X.W.; Chen, C.M.; Li, K.L.; Lin, X.P.; Gao, C.H.; Zhou, X.F.; Liu, Y.H. Sesquiterpenoids and meroterpenoids from a mangrove derived fungus *Diaporthe* sp. SCSIO 41011. *Nat. Prod. Res.* **2019**, *35*, 282–288. [CrossRef]
142. Li, X.M.; Li, X.M.; Lu, C.H. Abscisic acid-type sesquiterpenes and ansamycins from *Amycolatopsis alba* DSM 44262. *J. Asian Nat. Prod. Res.* **2017**, *19*, 946–953. [CrossRef]
143. Li, H.J.; Jiang, C.W.; Xu, M.Y.; Yan, D.F.; Xu, J.; Lan, W.J. Pseudapenes A-C, sesquiterpenoids from the marine-derived fungus *Pseudallescheria apiosperma* F52-1. *Tetrahedron Lett.* **2019**, *60*, 150953. [CrossRef]
144. Pang, X.J.; Zhang, S.B.; Xian, P.J.; Wu, X.; Yang, D.F.; Fu, H.Y.; Yang, X.L. Emericellins A and B: Two sesquiterpenoids with an unprecedented tricyclo [4,4,2,1] hendecane scaffold from the liquid cultures of endophytic fungus *Emericella* sp. XL 029. *Fitoterapia* **2018**, *131*, 55–58. [CrossRef]
145. Yan, J.M.; Wang, X.; Tian, M.Q.; Liu, C.M.; Zhang, K.Q.; Li, G.H. Chemical constituents from the fungus *Stereum* sp. YMF1.04183. *Phytochem. Lett.* **2017**, *22*, 6–8. [CrossRef]
146. Shi, Z.Z.; Miao, F.P.; Fang, S.T.; Liu, X.H.; Yin, X.L.; Ji, N.Y. Sesteralterin and tricycloalterfurenes A–D: Terpenes with rarely occurring frameworks from the marine-alga-epiphytic fungus *Alternaria alternata* k21-1. *J. Nat. Prod.* **2017**, *80*, 2524–2529. [CrossRef]
147. Chen, X.W.; Yang, Z.D.; Sun, J.H.; Song, T.T.; Zhu, B.Y.; Zhao, J.W. Colletotrichine A, a new sesquiterpenoid from *Colletotrichum gloeosporioides* GT-7, a fungal endophyte of *Uncaria rhynchophylla*. *Nat. Prod. Res.* **2018**, *32*, 880–884. [CrossRef]
148. Duan, Y.C.; Feng, J.; Bai, N.; Li, G.H.; Zhang, K.Q.; Zhao, P.J. Four novel antibacterial sesquiterpene-α-amino acid quaternary ammonium hybrids from the mycelium of mushroom *Stereum hirsutum*. *Fitoterapia* **2018**, *128*, 213–217. [CrossRef] [PubMed]
149. Ota, K.; Yamazaki, I.; Saigoku, T.; Fukui, M.; Miyata, T.; Kamaike, K.; Shirahata, T.; Mizuno, F.; Asada, Y.; Hirotani, M.; et al. Phellilane L, sesquiterpene metabolite of *Phellinus linteus*: Isolation, structure elucidation, and asymmetric total synthesis. *J. Org. Chem.* **2017**, *82*, 12377–12385. [CrossRef]
150. Jayanetti, D.R.; Yue, Q.; Bills, G.F.; Gloer, J.B. Hypocoprins A-C: New sesquiterpenoids from the coprophilous fungus *Hypocopra rostrata*. *J. Nat. Prod.* **2015**, *78*, 396–401. [CrossRef]
151. Hwang, I.H.; Swenson, D.C.; Gloer, J.B.; Wicklow, D.T. Pestaloporonins: Caryophyllene-derived sesquiterpenoids from a fungicolous isolate of *Pestalotiopsis* sp. *Org. Lett.* **2015**, *17*, 4284–4287. [CrossRef] [PubMed]
152. Zhang, J.Y.; Liu, L.; Wang, B.; Zhang, Y.; Wang, L.L.; Liu, X.Z.; Che, Y.S. Phomanolides A and B from the fungus *Phoma* sp.: Meroterpenoids derived from a putative tropolonic sesquiterpene via hetero-diels-alder reactions. *J. Nat. Prod.* **2015**, *78*, 3058–3066. [CrossRef]
153. Chen, X.W.; Yang, Z.D.; Li, X.F.; Sun, J.H.; Yang, L.J.; Zhang, X.G. Colletotrichine B, a new sesquiterpenoid from *Colletotrichum gloeosporioides* GT-7, a fungal endophyte of *Uncaria rhynchophylla*. *Nat. Prod. Res.* **2019**, *33*, 108–112. [CrossRef] [PubMed]
154. Kang, H.S.; Ji, S.A.; Park, S.H.; Kim, J.P. Lepistatins A-C, chlorinated sesquiterpenes from the cultured basidiomycete *Lepista sordida*. *Phytochemistry* **2017**, *143*, 111–114. [CrossRef] [PubMed]
155. Lin, X.P.; Wu, Q.Y.; Yu, Y.Y.; Liang, Z.; Liu, Y.H.; Zhou, L.L.; Tang, L.; Zhou, X.F. Penicilliumin B, a novel sesquiterpene methylcyclopentenedione from a deep sea-derived *Penicillium* strain with renoprotective activities. *Sci. Rep.* **2017**, *7*, 10757. [CrossRef] [PubMed]
156. Deng, Q.; Li, G.; Sun, M.Y.; Yang, X.B.; Xu, J. A new antimicrobial sesquiterpene isolated from endophytic fungus *Cytospora* sp. from the Chinese mangrove plant *Ceriops tagal*. *Nat. Prod. Res.* **2018**, *34*, 1404–1408. [CrossRef]

157. Wibowo, M.; Prachyawarakorn, V.; Aree, T.; Mahidol, C.; Ruchirawat, S.; Kittakoop, P. Cytotoxic sesquiterpenes from the endophytic fungus *Pseudolagarobasidium acaciicola*. *Phytochemistry* **2016**, *122*, 126–138. [CrossRef]
158. Huang, J.H.; Lv, J.M.; Wang, Q.Z.; Zou, J.; Lu, Y.J.; Wang, Q.L.; Chen, D.N.; Yao, X.S.; Gao, H.; Hu, D. Biosynthesis of an anti-tuberculosis sesterterpenoid asperterpenoid A. *Org. Biomol. Chem.* **2019**, *17*, 248–251. [CrossRef]
159. Masuya, T.; Tsunematsu, Y.; Hirayama, Y.; Sato, M.; Noguchi, H.; Nakazawa, T.; Watanabe, K. Biosynthesis of lagopodins in mushroom involves a complex network of oxidation reactions. *Org. Biomol. Chem.* **2019**, *17*, 234–239. [CrossRef]
160. Bian, G.K.; Hou, A.W.; Yuan, Y.J.; Hu, B.; Cheng, S.; Ye, Z.L.; Di, Y.T.; Deng, Z.X.; Liu, T.G. Metabolic engineering-based rapid characterization of a sesquiterpene cyclase and the skeletons of fusariumdiene and fusagramineol from *Fusarium graminearum*. *Org. Lett.* **2018**, *20*, 1626–1629. [CrossRef]
161. Flynn, C.M.; Schmidt-Dannert, C. Sesquiterpene synthase-3-hydroxy-3-methylglutaryl coenzyme A synthase fusion protein responsible for hirsutene biosynthesis in *Stereum hirsutum*. *Appl. Environ. Microbiol.* **2018**, *84*. [CrossRef]
162. Yuan, Y.J.; Litzenburger, M.; Cheng, S.; Bian, G.K.; Hu, B.; Yan, P.; Cai, Y.S.; Deng, Z.X.; Bernhardt, R.; Liu, T.G. Sesquiterpenoids produced by combining two sesquiterpene cyclases with promiscuous myxobacterial CYP260B1. *ChemBioChem* **2019**, *20*, 677–682. [CrossRef] [PubMed]
163. Ichinose, H.; Kitaoka, T. Insight into metabolic diversity of the brown-rot basidiomycete *Postia placenta* responsible for sesquiterpene biosynthesis: Semi-comprehensive screening of cytochrome P450 monooxygenase involved in protoilludene metabolism. *Microb. Biotechnol.* **2018**, *11*, 952–965. [CrossRef]
164. Lindo, L.; McCormick, S.P.; Cardoza, R.E.; Busman, M.; Alexander, N.J.; Proctor, R.H.; Gutierrez, S. Requirement of two acyltransferases for 4-*O*-acylation during biosynthesis of harzianum A, an antifungal trichothecene produced by *Trichoderma arundinaceum*. *J. Agric. Food Chem.* **2019**, *67*, 723–734. [CrossRef]
165. Cardoza, R.E.; Malmierca, M.G.; Hermosa, M.R.; Alexander, N.J.; McCormick, S.P.; Proctor, R.H.; Tijerino, A.M.; Rumbero, A.; Monte, E.; Gutiérrez, S. Identification of loci and functional characterization of trichothecene biosynthesis genes in filamentous fungi of the genus *Trichoderma*. *Appl. Environ. Microbiol.* **2011**, *77*, 4867–4877. [CrossRef] [PubMed]
166. Proctor, R.H.; McCormick, S.P.; Kim, H.-S.; Cardoza, R.E.; Stanley, A.M.; Lindo, L.; Kelly, A.; Brown, D.W.; Lee, T.; Vaughan, M.M.; et al. Evolution of structural diversity of trichothecenes, a family of toxins produced by plant pathogenic and entomopathogenic fungi. *PLoS Pathog.* **2018**, *14*, e1006946. [CrossRef]
167. Kamata, K.; Sato, H.; Maeda, K.; Furihata, K.; Aikawa, S.; Adachi, K.; Tanaka, A.; Tokai, T.; Nakajima, Y.; Yoshida, Y.; et al. Exploring an artificial metabolic route in *Fusarium sporotrichioides*: Production and characterization of 7-hydroxy T-2 toxin. *J. Nat. Prod.* **2018**, *81*, 1041–1044. [CrossRef]
168. Bahadoor, A.; Schneiderman, D.; Gemmill, L.; Bosnich, W.; Blackwell, B.; Melanson, J.E.; McRae, G.; Harris, L.J. Hydroxylation of longiborneol by a *Clm2*-encoded CYP450 monooxygenase to produce culmorin in *Fusarium graminearum*. *J. Nat. Prod.* **2016**, *79*, 81–88. [CrossRef] [PubMed]

Journal of *Fungi*

MDPI

Review

Endophytic Fungi: An Effective Alternative Source of Plant-Derived Bioactive Compounds for Pharmacological Studies

Juan Wen [1,2,†], Samuel Kumi Okyere [1,2,†], Shu Wang [1,2], Jianchen Wang [1,2], Lei Xie [1,2], Yinan Ran [1,2] and Yanchun Hu [1,2,3,*]

1 Key Laboratory of Animal Diseases and Environmental Hazards of Sichuan Province, College of Veterinary Medicine, Sichuan Agricultural University, Chengdu 611130, China; juanwen881010@163.com (J.W.); samuel20okyere@gmail.com (S.K.O.); shuw0326@163.com (S.W.); wangjianchen01@163.com (J.W.); wsxielei@gmail.com (L.X.); ranyinan17@163.com (Y.R.)
2 Key Laboratory of Animal Disease and Human Health of Sichuan Province, Sichuan Agricultural University, Chengdu 611130, China
3 New Ruipeng Pet Healthcare Group Co., Ltd., Shenzhen 518000, China
* Correspondence: hychun114@163.com; Tel.: +86-2886291162
† These authors contributed equally to this work.

Abstract: Plant-associated fungi (endophytic fungi) are a biodiversity-rich group of microorganisms that are normally found asymptomatically within plant tissues or in the intercellular spaces. Endophytic fungi promote the growth of host plants by directly producing secondary metabolites, which enhances the plant's resistance to biotic and abiotic stresses. Additionally, they are capable of biosynthesizing medically important "phytochemicals" that were initially thought to be produced only by the host plant. In this review, we summarized some compounds from endophyte fungi with novel structures and diverse biological activities published between 2011 and 2021, with a focus on the origin of endophytic fungi, the structural and biological activity of the compounds they produce, and special attention paid to the exploration of pharmacological activities and mechanisms of action of certain compounds. This review revealed that endophytic fungi had high potential to be harnessed as an alternative source of secondary metabolites for pharmacological studies.

Keywords: endophytic fungi; secondary metabolites; structural feature; biological activities; drug discovery

Citation: Wen, J.; Okyere, S.K.; Wang, S.; Wang, J.; Xie, L.; Ran, Y.; Hu, Y. Endophytic Fungi: An Effective Alternative Source of Plant-Derived Bioactive Compounds for Pharmacological Studies. *J. Fungi* **2022**, *8*, 205. https://doi.org/10.3390/jof8020205

Academic Editors: Tao Feng and Frank Surup

Received: 21 January 2022
Accepted: 16 February 2022
Published: 20 February 2022

1. Introduction

The term "endophytic fungi" refers to fungi that live in plant tissues throughout the entire or partial life cycle by establishing a mutually beneficial symbiotic relationship with its host plant without causing any adverse effect or disease [1,2]. They are natural components of the plant micro-ecosystem that positively affect the physiological activities of the host plant in several ways, including producing hormones such as indoleacetic acid, biosynthesizing and acquiring nutrients for plant growth and development, secreting stress-adaptor metabolites to protect the host plant from the invasion of herbivores, pathogens, and improving the host's adaptability to abiotic stressors. In return, plants provide habitats and nutrients for endophytic fungi [3,4]. Endophytic fungi are capable of producing a rich variety of bioactive substances and can produce compounds that are identical or similar to pharmacological activities identified from plants [5]. They produce a range of metabolites of different chemical classes, including alkaloids, flavonoids, steroids, terpenoids, and phenolic compounds. Some compounds show pleiotropic and interesting pharmacological activities, including antimicrobial, antioxidant, anti-diabetic, anti-malarial, and antitumor properties. The discovery of these structurally novel and diverse active compounds provides a valuable resource for studying natural medical products from the microbiome [6–8]. In the search for bioactive molecules as pro-drug compounds or in the development of medicines, endophytic fungi can serve as an alternative source for valuable active plant compounds. Endophytic fungi can be harnessed to produce bioactive compounds for human pharmaceutical use when the bioactive secondary metabolites are not commercially available, derived from slow-growing or rare and endangered plants, and difficult to synthesize due to heavy molecular weight or structural complexity. Endophytic fungal secondary metabolites have drawn extensive attention among medicinal plants, mangroves, and marine microorganisms [9,10].

Endophytic fungi are a highly biodiverse and versatile microbial community that seems to be ubiquitous in nature. Studies have shown that almost all plants contain endophytic fungi, including colonized plants in the Arctic and Antarctic regions, deserts, oceans, and tropical rainforests [11,12]. They have been isolated and cultured from the roots and above-ground parts of various plants, including algae, mosses, ferns, gymnosperms, and angiosperms. Evidence from microorganism's records in the fossil plant tissue indicated that the plant-endophytic fungal interactions have existed for approximately 400 million years, and during this time, endophytic fungi co-evolved unique biosynthetic pathways and metabolic mechanisms to synthesize complex secondary metabolites [13]. To date, only 5% of 1.5 million fungal species on Earth have been described in detail, and out of this percentage (69,000 fungal species), only 16% (11,500 species) have been cultured and studied. About 0.035–5.1 million fungal species have been found on Earth according to results from next-generation sequencing technologies [14]. Approximately 300,000 known species of higher plants exist on Earth, and each of which is a host for one or more obligate endophytic fungi. The high number of bioactive secondary metabolites found in endophytic fungi is due to their rich species diversity [15,16]. Endophytic fungi have been studied for more than 100 years, with the first endophytic strain isolated from the seeds of ryegrass (*Lolium temulentum* L.) by Vogl et al. in 1898 [17]. Stierle et al. [18] discovered the paclitaxel-producing endophytic fungus (*Taxomyces andreanae*) from the Pacific yew and then from other plant species successively. This discovery aroused the attention of mycologists and pharmaceutical chemists on endophytic fungi as a new source of bioactive substances and stimulated the interest in endophytic fungi as a sustainable source of plant metabolites. As shown in Table 1, many compounds that were isolated from endophytic fungi were also identified in some plant species as well as exhibited similar biological activity even though there were isolated from different sources, confirming endophytic fungi as an alternative source of bioactive compounds [19–32]. An overview of the recent literature surveys revealed that 51% of the bioactive substances isolated from endophytic fungi were previously unknown, with about 38% being isolated from soil microbiota [19]. Over the past decade, there has been a surge in the number of patents for endophytic fungi with new molecular secondary metabolites, which play a key role in the pharmaceutical industry, phytoremediation, and biomedicine [20,21]. Researchers are now searching for an economical, environmentally safe, and sustainable way to obtain new bioactive secondary metabolites from endophytic fungi.

This article reports 220 new compounds with rare or novel structures or skeleton structures from endophytic fungi from 82 journal articles between 2011 and 2021 and briefly describes the sources of endophytic fungi, chemical structures, and biological activities of these compounds. Among all the new compounds reported in this review, terpenoids (35%) were largest in proportion, followed by alkaloids (26%). The proportion of different types of compounds among all the new compounds are presented in Figure 1. These new compounds were obtained from different species of endophytic fungi, which had diverse chemical skeletons and exhibited diverse and interesting biological activities. Additionally, the most common pharmacological activities these compounds showed were antimicrobial and antitumor activities. However, some of the compounds showed anti-angiogenic, anti-phytotoxic, and α-glucosidase inhibitory effects. Therefore, this review summarized different insights into the prospects and challenges of endophytic fungi as an alternative source of plant-derived bioactive compounds for drug development. In addition, this review will affirm that endophytic fungi produce similar bioactive compounds just as their host plants to give knowledge for the development of drug candidates from endophytic fungi using different strategies, thus making Endophytic fungi a treasure trove of new secondary metabolites.

Table 1. Several endophytic fungi of host plants have been reported to produce compounds with similar activity.

No.	Endophytic Fungus	Host Plant	Regions/Countries	Compound	Biological Activity	Ref.
1	*Lophiostoma* sp.	*Eucalyptus exserta*	Guangzhou, China.	Scorpinone	Antibacterial	[22]
2	*Mycosphaerella* sp.	*Myrciaria floribunda*	Amazon rainforest, Brazil.	Myriocin	Antifungal	[23]
3	*Mucor* sp.	*Centaurea stoebe*	Idaho, USA	Terezine E	Antifungal and cytotoxicity	[24]
4	*Aspergillus calidoustus*	*Acanthospermum australe*	Jalapao State Park, Tocantins, Brazil.	Ophiobolin K 6-epi-ophiobolin K	Antifungal, trypanocidal and cytotoxicity	[25]
5	*Phomopsis* sp.	*Garcinia kola (Heckel) nut*	Yaounde, Cameroon	Cytochalasins H	Antibacterial and cytotoxicity	[26]
6	*Aspergillus nidulans*	*Nyctanthes arbor-tristis Linn*	Karachi, Pakistan	Sterigmatocystin	Antiproliferative activity	[27]
7	*Trichoderma asperellum* and *Trichoderma brevicompactum*	*Vinca herbacea*	Hamedan, Iran	4b-hydroxy-12,13-epoxytrichothec-9-ene	Antimicrobial and antiproliferative activity	[28]
8	*Phyllosticta elongata*	*Cipadessa baccifera*	Western Ghats, India	Camptothecin	Anticancer agent	[29]
9	*Fusarium verticillioides*	*Huperzia serrata*	Gucheng Mountain, Sichuan, China	Huperzine A	Treatment of Alzheimer's disease	[30]
10	*Fusarium solani*	*Cassia alata*	Bangladesh	Napthaquinones Azaanthraquinones	Cytotoxicity, antimicrobial and antioxidant activity	[31]
11	*Fusarium* sp. and *Lasiodiplodia theobromae*	*Avicennia lanata*	Terengganu, Malaysia	Anhydrofusarubin dihydrojavanicin	Antitrypanosomal	[32]
12	*Corynespora cassiicola*	*Gongronema latifolium*	Nigeria	Corynesidone D	Anti-inflammatory/ anticancer agent	[33]
13	*Pestalotiopsis theae*	*Camellia sinensis Theaceae*	Hangzhou, China	punctaporonin H	Antibacterial and cytotoxicity	[34]
14	*Phialocephala fortinii*	*Podophyllum peltatum*	Tamilnadu, India	Podophyllotoxin	Antiviral, antioxidant, and antirheumatic activities	[35]

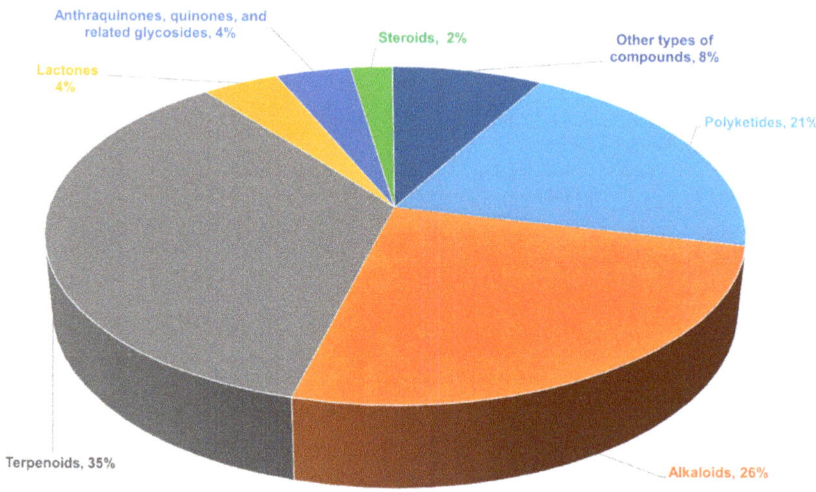

Figure 1. The proportion of different types of compounds among all new compounds.

2. Bioactive New Metabolites Isolated from Endophytic Fungi and Their Biological Activities

2.1. Polyketides

2.1.1. Chromones

The induction of endophyte metabolism by adding Host components was used to add the same phytocomponents (2R, 3R)-3, 5, 7- trihydroxyflavanone 3-acetate in Botryosphaeria ramosa L29 potato dextrose broth culture to induce the production of 5-hydroxy2,3-dihydroxymethyl-7-methoxychromone **1** (Figure 2), 5-hydroxy-3-acetoxymethyl-2-methyl-7- methoxychromone **2** (Figure 2) and 5,7-dihydroxy-3-hydroxymethyl-2-methylchromone **3** (Figure 2), where Compounds **1–3** displayed acceptable antimicrobial activities against *Fusarium oxysporum* with MIC values of 50 μg/mL, 50 μg/mL, and 6.25 μg/mL, respectively. These values were superior compared to those of the positive drug—triadimefon—for the antimicrobial test (with an MIC value of 100 μg/mL) [36]. This indicated that the induction of endophytes metabolism to produce bioactive components of interest might be an ideal strategy for easy identification of drug candidates from these microbes; however, there is the need for long-term studies on how specific components influence endophytes metabolism and the bioactive compounds there are linked with. Phaeosphaonesa A **4** (Figure 2), isolated from *Phaeosphaeria fuckelii*, contains a β-(oxy)thiotryptophan motif structure that is rare in nature. Compound **4** showed stronger inhibition activity of mushroom tyrosinase than the positive control kojic acid (IC_{50} value of 40.4 μM) at 100 μM concentration, with an IC_{50} value of 33.2 μM [37]. Two aromatic chromones, Chaetosemins B–C **5–6** (Figure 2), were isolated from *Chaetomium seminudum* brown rice cultures, and compounds **5–6** contained L-cysteine and D-cysteine units, respectively. Compound **5** showed antifungal activity against *Magnaporthe oryzae* and *Gibberella saubinetti*, with MIC values of 6.25 μM and 12.5 μM, respectively. Compound **6** showed significant antioxidant activity at a concentration of 50 μM with a DPPH radical scavenging rate of 50.7% [38]. Pestaloficiols M–P **7–10** (Figure 2), which are new isoprenylated chromone derivatives, were isolated from brown rice culture extract of the plant endophytic fungus *Pestalotiopsis fici*. The structures of these compounds were elucidated primarily by MS and NMR techniques. Compounds **7–8** displayed inhibitory effects on HIV-1 replication in C8166 cells, with EC_{50} values of

56.5 μM and 10.5 μM, respectively (the EC $_{50}$ value of the positive control Indinavir Sulfate was 8.2 μM), whereas compounds **9–10** showed cytotoxic activity against the human tumor cell line HeLa, with IC_{50} values of 56.2 μM and 74.9 μM, respectively (the positive control 5-fluorouracil has an IC_{50} of 10.0 μM). Compound **10** exhibited a potent antifungal activity against *Aspergillus fumigatus* at IC_{50} = 7.35 μM) [39].

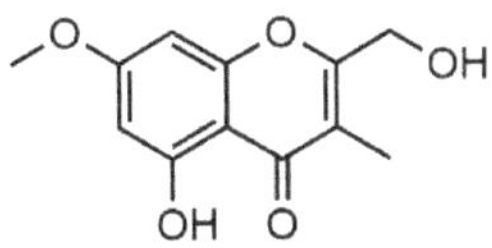

5-hydroxy-2,3-dihydroxymethyl -7-methoxychromone **(1)**

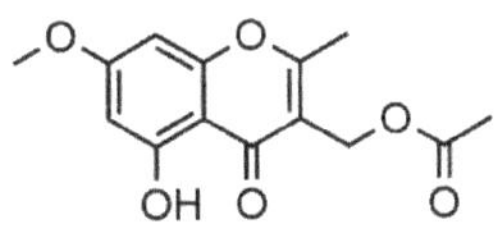

5-hydroxy-3-acetoxymethyl-2methyl -7-methoxychromone **(2)**

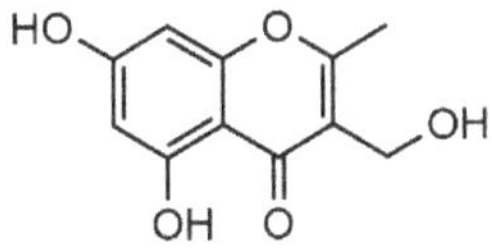

5,7-dihydroxy-3-hydroxymethyl -2-methylchromone **(3)**

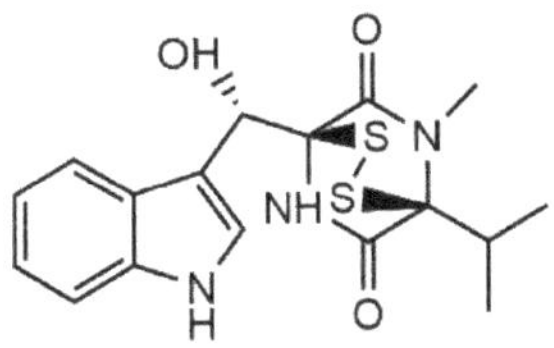

phaeosphaones A **(4)**

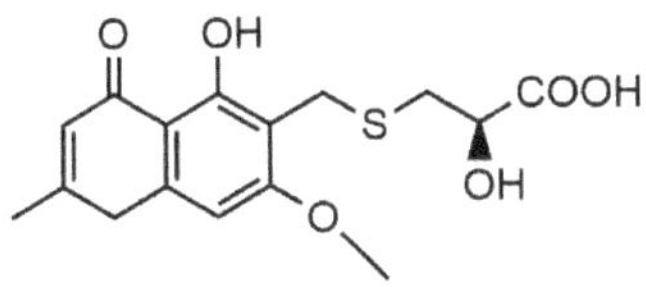

chaetosemins B **(5)**

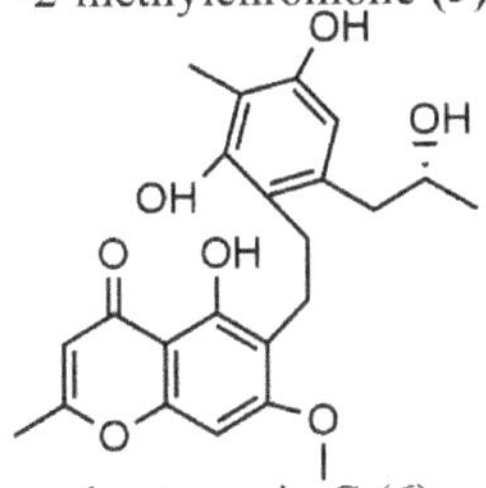

chaetosemin C **(6)**

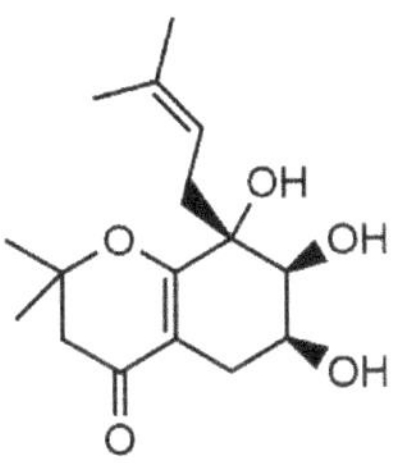

pestaloficiols M **(7)**

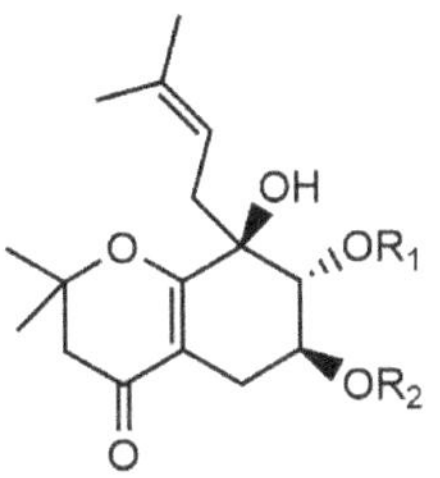

pestaloficiols N:R1=H, R2=H **(8)**
pestaloficiols O:R1=$C_{15}H_{31}CO$,R2=H**(9)**
pestaloficiols P:R1=H, R2=$C_{15}H_{31}CO$**(10)**

Figure 2. Chemical structures of chromones.

2.1.2. α-Pyrones

Two tetrasubstituted α-pyrone derivatives—*Neurospora udagawae* udagawanones A-B **11–12** (Figure 3)—were isolated from oak endophytic fungi, with both containing unique oxidation functional groups at the C-2 position. Compound **11** exhibited potent antifungal activity against *Rhodoturula glutinis* with MIC = 66 μg/mL). Additionally, compounds **11** and **13** showed moderate cytotoxic activity against KB3.1 cells with IC_{50} = 27 μg/mL [40]. The study revealed moderate activity of compounds **11** and **12** against fungi and mammalian cells, and this may be as a result of the method (serial dilution antimicrobial assay) used; therefore, it is suggested that other biological tests be employed to verify these findings. The nigerapyrones A–B **13–14** (Figure 3) were obtained from *Aspergillus niger* MA-132, which was isolated from the mangrove plant *Avicennia marina*. Compounds **13–14** both showed potent antifungal activities against two tumor cell lines (HL60 and A549), with IC_{50} values ranging from 0.3 to 5.41 μM [41]. The ficipyrones A–B **15–16** (Figure 3) were isolated from solid cultures of *Pestalotiopsis fici*. Compound **15** showed significant antifungal activity against *Gibberella zeae* CGMCC 3.2873, with an IC_{50} value of 15.9 μM, but had no activity against *Fusarium culmorum* CGMCC 3.4595 and *Verticillium aiboatrum* CGMCC 3.4306 [42].

The endophytic fungus *Aspergillus oryzae* was isolated from the rhizome of *Paris polyphylla* in Dali, Yunnan, China, and 4-hydroxy-6-[(2S, 3S)-3-hydroxybutan-2-yl]-3-methyl-2H-pyran-2-one **17** (Figure 3) and (R)-4-hydroxy-6-(l-hydroxy-2-methylpropyl)-3-methyl-2H-pyran-2-one **18** (Figure 3) were obtained from this fungi. However, the biological activities of these compounds were not tested in the study; hence, investigating the biological activities of these compounds is needed, as it may yield a very important source of drug activity [43].The pyran-2-one scaffold compounds **19–21** (Figure 3) were isolated by adding 10 mg/L DNA methyltransferase inhibitor 5-aza-2-deoxycytidine to *Penicillium herquei* liquid cultures, whereas the MTT method was used to measure the cytotoxicity of all compounds in MDA-ME-231 and MV-411 cell lines. Compounds **19–21** showed weak cytotoxicity only against the MV4-11 cell line with IC_{50} values of 90.09 μM, 74.16 μM, and 70.00 μM, respectively [44].

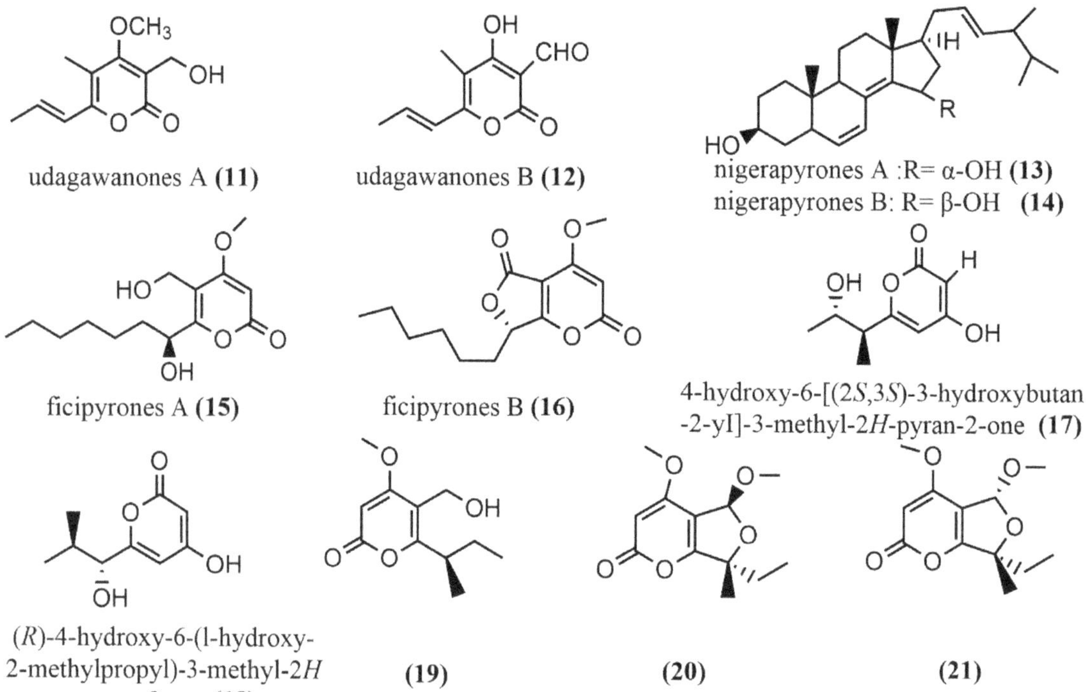

Figure 3. Chemical structures of α-pyrone compounds.

2.1.3. Other Polyketides

The phomaketides A–E **22–26** (Figure 4), pseurotins A_3 **27** (Figure 4), and pseurotins G **28** (Figure 4) were isolated from fermentation broth and mycelial extracts of the marine red algae endophytic fungus *Phoma* sp. NTOU4195. The mouse macrophages RAW264 were induced using the endothelial progenitor cells of human umbilical cord blood, lipopolysaccharide (LPS), to assess the anti-angiogenic and anti-inflammatory activities of all compounds. Compound **22** showed potent anti-angiogenic activity by inhibiting endothelial cell proliferation, with an IC_{50} value of 8.1 μM. Compound **24** at the concentration of 20 μM induced effective nitric oxide (NO) inhibition activity against LPS-induced RAW264.7 cells, with an IC_{50} value of 8.8 μM [45]. There were two tetracyclic polyketide compounds, simplicilone A–B **29–30** (Figure 4), containing helical centers obtained from the broth culture of the endophytic fungus *Simplicillium* sp., which was isolated from the bark of the medicinal plant *Duguetia staudtii* (Engl. and Diels) Chatrou in the Cameroon region. Compounds **29–30** showed weak cytotoxic activities against the KB3.1 cell line, with IC_{50} values of 1.25 μg/mL and 2.29 μg/mL, respectively, but had no antimicrobial activity against the tested bacteria (*Staphylococcus aureus* DSM 346 and *Bacillus subtilis* DSM 10) [46]. 5R-hydroxyrecifeiolide **31** (Figure 4), 5S-hydroxyrecifeiolide **32** (Figure 4), and ent-cladospolide F–H **33–35** (Figure 4) were also isolated from the endophytic fungal strain *Cladosporium cladosporioides* MA-299, which was obtained from the leaves of the mangrove plant *Bruguiera gymnorrhiza* from Hainan Island, China. Compounds **31–35** showed potent antimicrobial activities against *Escherichia coli* and *Staphylococcus aureus*, with MIC values ranging from 1.0 to 64 μg/mL. Compound **33** showed moderate inhibition activity against acetylcholinesterase, with an IC_{50} value of 40.26 μM [47]. The antimicrobial polyketide compound, palitantin **36** (Figure 4), was obtained from *Aspergillus fumigatiaffnis* and isolated from healthy leaves of *Tribulus terrestris L.* In addition, compound **36** showed effective antimicrobial activity against the multi-drug-resistant pathogens *Enterococcus faecalis* UW 2689 and *Streptococcus pneumoniae* 25697, both with an MIC value of 64 μg/mL [48]. The four polyketide derivatives—isotalaroflavone **37** (Figure 4), (+/−)-50-dehydroxytalaroflavone **38–39** (Figure 4), and bialternacin G **40** (Figure 4)—were obtained from the endophytic fungus *Alternaria alternata* ZHJG5 isolated from the leaves of *Cercis chinensis*, which was collected from the Nanjing Botanical Garden, Nanjing, China. They exhibited potent antimicrobial activity against *Xanthomonas oryzae* pv. *oryzicola (Xoc)* and *Ralstonia solanacearum*, with MIC values ranging from 0.5 to 64 μg/mL. Compound **37** at the concentration of 200 μg/mL showed a significant protective effect against the bacterial blight of rice caused by *Xanthomonas oryzae* pv. *oryza*, with a protection rate of 75.1% [49]. Four polyketide derivatives containing the benzoisoquinoline-9-one moiety structure peyronetides A–D **41–44** (Figure 4) were isolated from the mycelial crude acetone extract of *Peyronellaea* sp. FT431. Compounds **41–42** showed moderate to weak cytotoxic activity against human kidney cancer cell line TK10 and human ovarian cancer cell line A2780cisR, with IC_{50} values ranging from 6.7 to 29.2 μM [50]. The aromatic polyketide compound, (−)alternamgin **45** (Figure 4), was obtained from potato dextrose broth cultures of the endophytic fungus *Alternaria* sp. MG1 isolated from *Vitis quinquangularis*. This compound was of particular interest because it had the rare dibenzopyrone functionality of 6/6/6/6/5/6/6/6 heptacyclic backbone. Compound **45** displayed a weak cytotoxic activity against cells from two tested cell lines (Hela and HepG2), both with IC_{50} values exceeding 20 μM [51].

In summary, Polyketides, such as chromones and α-pyrone, and their derivatives identified from plant sources have also been found in endophytic fungi in recent studies. Chromones and their derivatives isolated from both plant and endophytic fungi sources all showed antimicrobial properties against specific pathogens; therefore, chromones from endophytic fungus can be used in the development of antimicrobials in the place of plant chromones to reduce the depletion of plants' resources in the ecosystem.

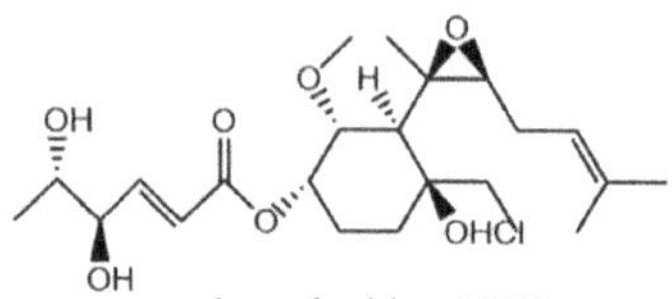

phomaketides A(**22**)

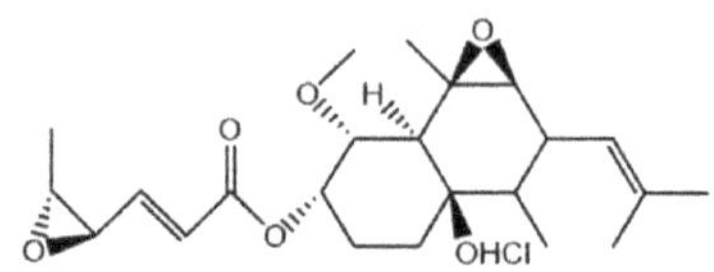

phomaketides B(**23**)

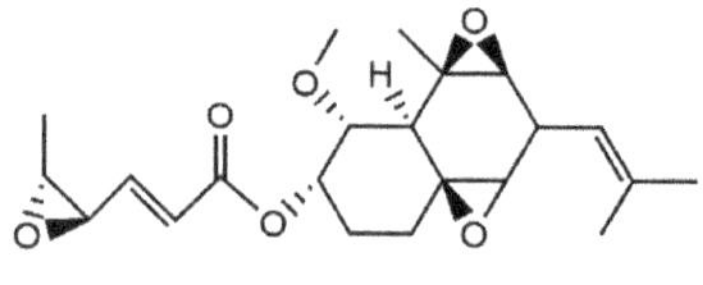

phomaketides C(**24**)

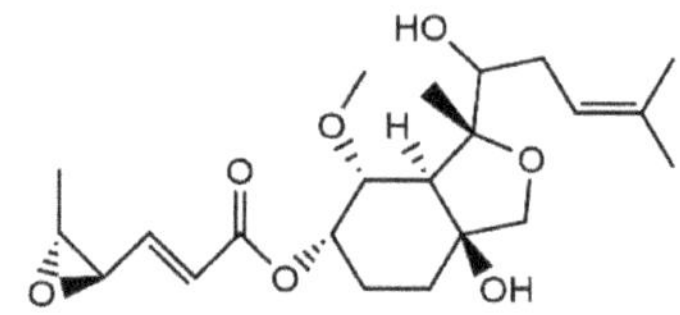

phomaketides D(**25**)

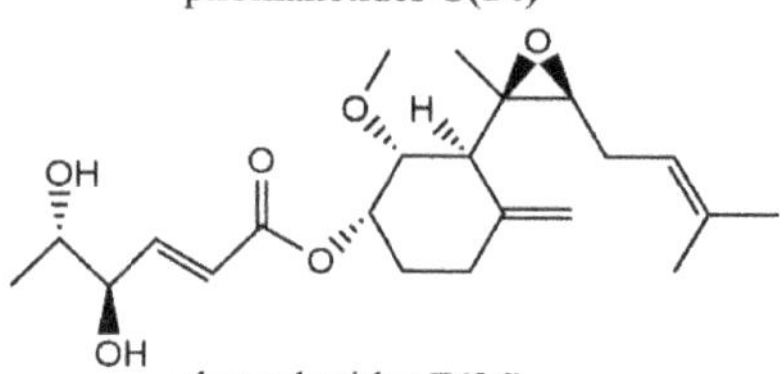

phomaketides E(**26**)

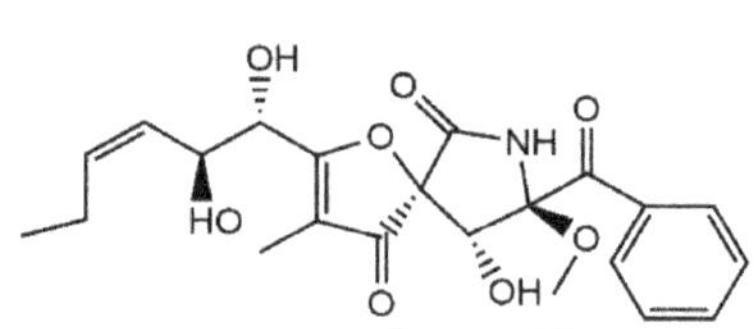

pseurotins A3(**27**)

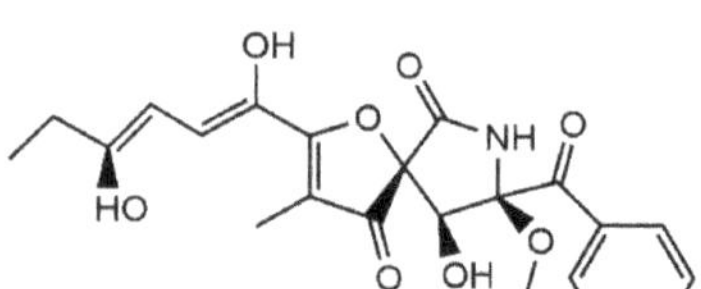

pseurotins G(**28**)

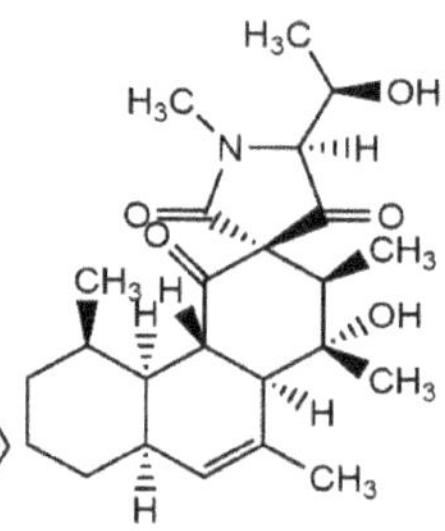

simplicilone A(**29**)

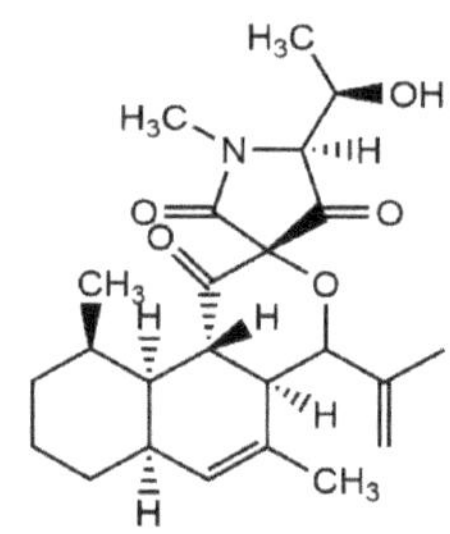

simplicilone B(**30**)

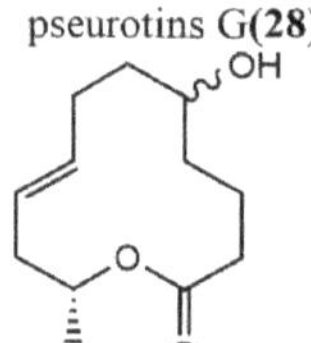

hydroxyrecifeiolide: 5*R*(**31**)
hydroxyrecifeiolide: 5S(**32**)

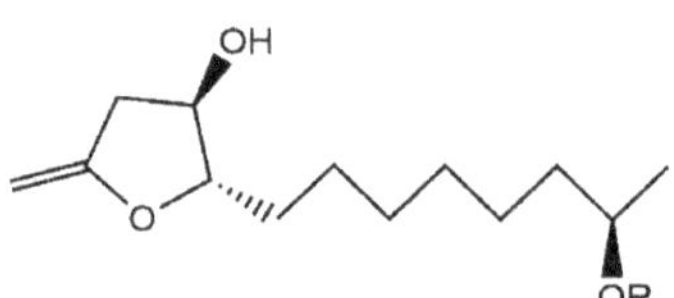

ent-cladospolide F:R=H(**33**)
cladospolide G:R=$COCH_3$ (**34**)

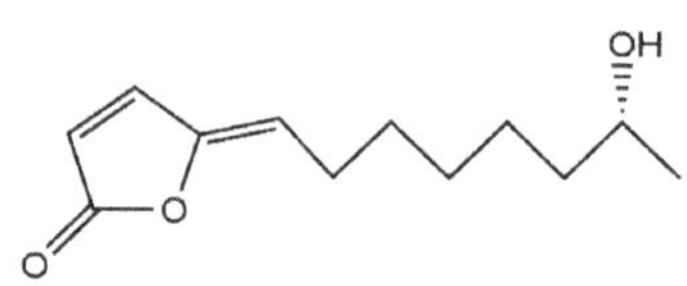

cladospolide H (**35**)

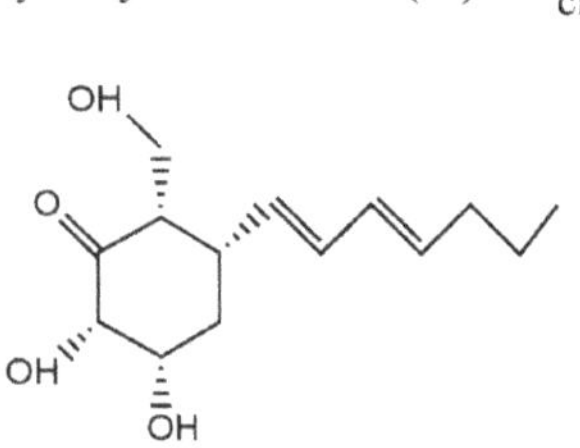

(-) palitantin (**36**)

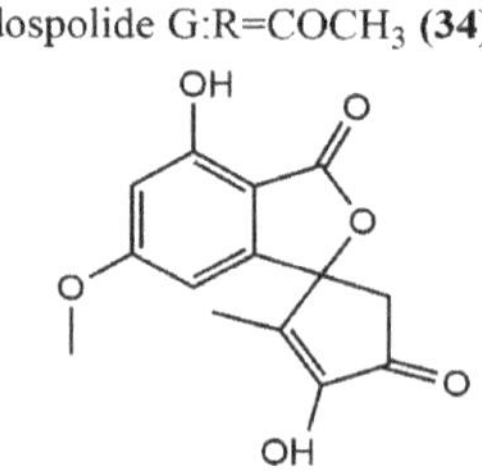

isotalaroflavone (**37**)

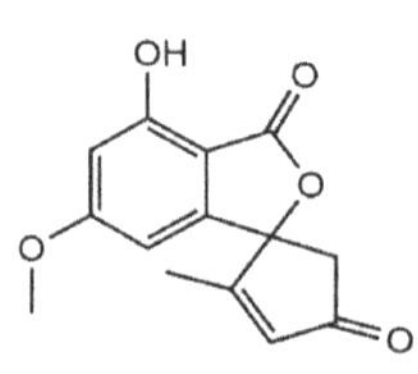

(+)-5/-dehydroxytalaroflavone (**38**)
(-)-5/-dehydroxytalaroflavone (**39**)

Figure 4. *Cont.*

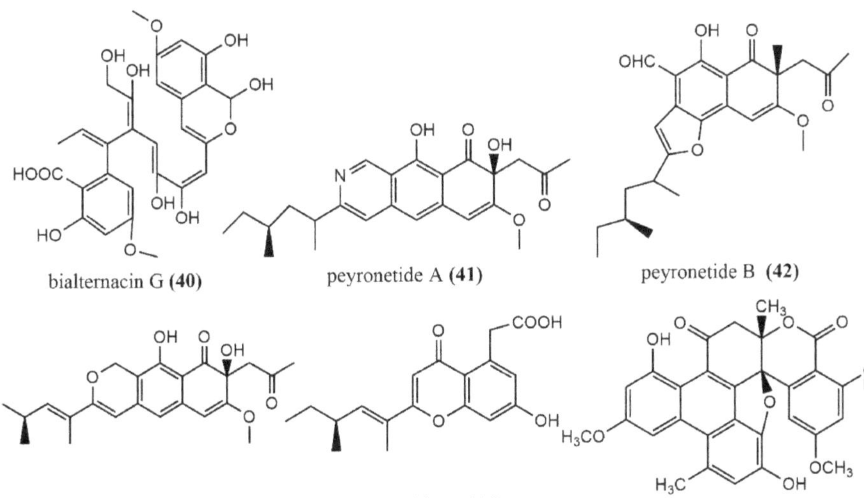

Figure 4. Chemical structures composition of other polyketides.

2.2. *Alkaloids*

2.2.1. Cytochalasin

The methylation-deficient backbone, Phomopsisin A–C **46–48** (Figure 5), was obtained from brown rice cultures of *Phomopsis* sp. sh917, which was isolated from *Isodon eriocalyx var. laxiflora* stems. Compound **46** contained an unusual 5/6/11/5 tetracyclic ring system 2H-isoxazole moiety and showed significant inhibition activity against LPS-induced NO production in RAW264.7 cells, with an IC_{50} value of 32.38 μM, which was more potent than the positive control L-NMMA (IC_{50} value of 42.34 μM) [52]. The highly oxidized cytochalasin alkaloids—armochaetoglobins S–Z **49–57** (Figure 5) and 7-O-acetylarmochaetoglobin S **50** (Figure 5)—were identified and isolated from *Chaetomium globosum* TW1-1. The effects of all compounds on five tested human cancer cell lines (HL-60, A-549, SMMC-7721, MCF-7, and SW-480) were measured using the MTT method. Compounds **56–57** showed potent cytotoxic activities, with IC_{50} values ranging from 10.45 to 30.42 μM [53]. Furthermore, diaporthichalasins D–H **58–62** (Figure 5) were obtained from solid cultures of the endophytic fungus *Diaporthe* sp. SC-J0138 isolated from the leaves of the pteridophyte *Cyclosorus parasiticus*, and the MTS method was used to evaluate the cytotoxic activities of these compounds on four human cancer cell lines (A549, HeLa, HepG2, and MCF-7). Compound **58** exhibited significant cytotoxic activity against all tested human cancer cell lines; compounds **59–62** exhibited selective cytotoxic activities against some cell lines [54]. Cytochrysins A–C **63–65** (Figure 5) were obtained from rice cultures of *Cytospora chrysosperma* HYQZ-931, an endophytic fungus isolated from the desert plant *Hippophae rhamnoides*. Compound **63** showed significant antimicrobial activity to *Enterococcus faecium*, with an MIC value of 25 μg/mL. Compound **65** showed potent antimicrobial activity to *Staphylococcus aureus*, with an MIC value of 25 μg/mL [55].

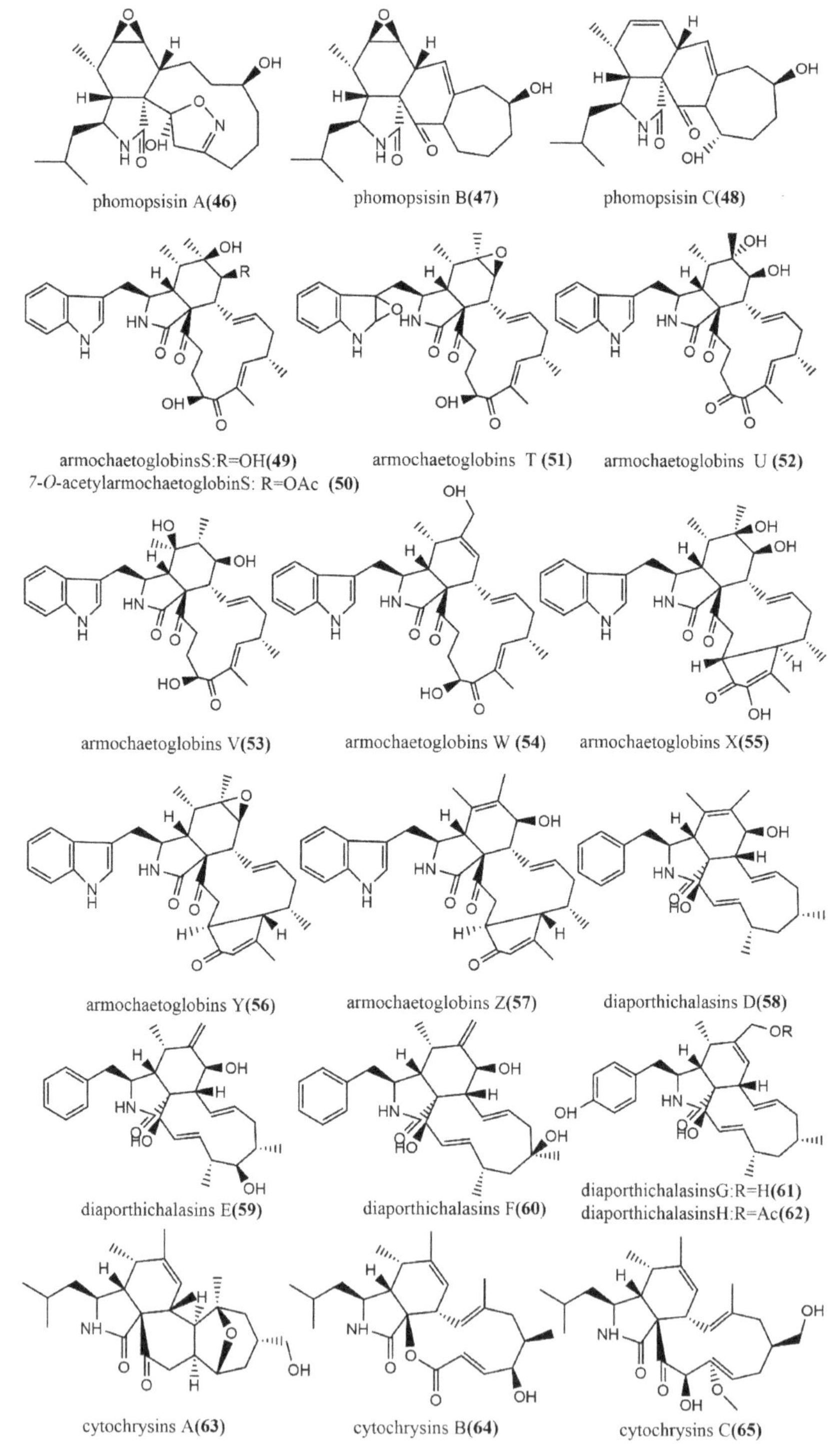

Figure 5. Chemical structures composition of cytochalasins.

2.2.2. Indole Alkaloids

Six prenylated indole alkaloids, asperthrins A–F **66–71** (Figure 6), were derived from the marine endophytic fungus *Aspergillus* sp. YJ191021. Compound **66** showed moderate antimicrobial activity against *Vibrio anguillarum*, with an MIC value of 8 μg/mL. Additionally, the compounds **66** and **69** showed potent–weak anti-inflammatory activities against propionibacterium acnes-induced human mononuclear cell line (THP-1), with IC_{50} values of 1.46 μM and 30.5 μM, respectively, while compound **66** showed higher anti-inflammatory activity than the positive control Tretinoin at an IC_{50} value of 3.38 μM [56]. The α-pyrone meroterpenoid-type alkaloid, oxalicine C **72** (Figure 6), was obtained from *Penicillium chrysogenum* XNM-12, which was isolated from the marine brown algae *Leathesia nana*. Compound **72** showed potent antimicrobial activity against the phytopathogenic fungus *Ralstonia solanacearum*, with an MIC of 8 μg/mL [57]. Scalarane **73** (Figure 6) was isolated from *Hypomontagnella monticulosa* Zg15SU through the potato dextrose liquid culture. Compound **73** showed potent cytotoxic activity against cancer cell lines Panc-1, NBT-T2, and HCT116, with IC_{50} values of 0.05, 0.75, and 0.05 μg/mL, respectively [58]. Asperlenines A–C **74–76** (Figure 6) were isolated from *Aspergillus lentulus* DTO 327G5 cultures, and the antimicrobial activity of all compounds was evaluated using the broth-microdilution method against five tested agricultural pathogens (*Xanthomonas oryzae* pv. *Oryzae*, *Xanthomonas oryzae* pv. *Oryzicola*, *Rhizoctonia solani*, *Fusarium oxysporum*, and *Colletotrichum gloeosporioides)*. Compounds **74–76** showed moderate to weak antimicrobial activities against *Xanthomonas oryzae* pv. *Oryzae* and *Xanthomonas oryzae* pv. *Oryzicola*, with MIC values ranging from 25 to 100 μg/mL [59].

2.2.3. Diketopiperazine Derivatives

The thiodiketopiperazine alkaloid, phaeosphaones D **77** (Figure 7), featuring an unusual β-(oxy) thiotryptophan motif, was obtained from endophytic fungus *Phaeosphaeria fuckelii* isolated from the medicinal plant *Phlomis umbrosa*. Compound **77** showed stronger mushroom tyrosinase inhibition activity than the positive control kojic acid (IC_{50} value of 40.4 μM), with an IC_{50} value of 33.2 μM. [60]. The oxepine-containing diketopiperazine-type alkaloids, varioloids A-B **78–79** (Figure 7), were obtained from *Paecilomyces variotii* EN-291, which was isolated from the marine red alga *Grateloupia turuturu*. Compounds **78–79** showed potent antifungal effects against *Fusarium graminearum*, with MIC values of 8 μg/mL and 4 μg/mL, respectively [61]. Aspergiamides A–F **80–85** (Figure 7) were isolated from the endophytic fungus *Aspergillus* sp. 16-5 of mangroves, and all compounds were evaluated for their inhibition activities against protein-tyrosine phosphatase 1B (PTP1B) and α-glucosidase. Compounds **80** and **81** showed potent to moderate α-glucosidase inhibition activities, with IC_{50} values of 18.2 μM and 40.7 μM, respectively. Compounds **80–85** did not show significant PTP1B inhibition activities (<10% inhibition) at 100 μg/mL [62]. Five sulfide diketopiperazines derivatives, penicibrocazines A–E **86–90** (Figure 7), were obtained from the endophytic fungus *Penicillium brocae* MA-231 isolated from the mangrove plant *Avicennia marina*. The antimicrobial effects of all compounds were evaluated by the agar diffusion method against five tested pathogens (*Aeromonas hydrophilia*, *Escherichia coli*, *Staphylococcus aureus*, *Vibrio arveyi*, and *V. parahaemolyticus*). Compounds **86–90** showed potent antimicrobial activities against *S. aureus*, with MIC values ranging from 0.25 to 32 μg/mL [63]. Spirobrocazines A–C **91–93** (Figure 7) were isolated from the mangrove-derived *Penicillium brocae* MA-231. Compounds **91–93** contained a 6/5/6/5/6 cyclic system with a rare spirocyclic center at C-2. All compounds showed moderate antimicrobial activities against *S. aureus*, *Aeromonas hydrophilia*, and *Vibrio harveyi*, with MIC values ranging from 16 to 64 μg/mL [64].

2.2.4. Other Types of Alkaloids

The quinazoline alkaloid (-)-(1R,4R)-1,4-(2,3)-indolmethane-1-methyl-2,4-dihydro-1H-pyrazino-[2,1-b]-quinazoline-3,6-dione **94** (Figure 8) was obtained from the endophytic fungus *Penicillium vinaceum* X1, which was isolated from corms of *Crocus sativus* (Iridaceae).

The in vitro cytotoxicity of compound **94** was evaluated against three human tumor cell lines (A549, LOVO, and MCF-7), to which compound **94** showed weak cytotoxic activities against all human tumor cell lines, with IC_{50} values of 76.83, 68.08, and 40.55 μg/mL, respectively [65]. The enantiomeric bromotyrosine alkaloids S-Acanthodendrilline **95** (Figure 8) and R-Acanthodendrilline **96** (Figure 8) were isolated from the ethyl acetate extract of the sponge endophytic fungus *Acanthodendrilla* sp. The cytotoxic activities of compounds **95–96** against human non-small cell lung cancer H292 and normal human immortalized fibroblast HaCaT cell lines were evaluated using the MTT method. Compound **95** (IC_{50} value of 58.5 μM) was approximately three times more potent than compound **96** (IC_{50} value of 173.5 μM) against the H292 cell line. Compounds **95–96** exhibited efficient and selective cytotoxic activities against H292 and HaCaT cell lines, with IC_{50} values ranging from 58.5 to 173.5 μM and >400 μM, respectively [66]. Three phenylpyridone derivatives, citridones E–G **97–99** (Figure 8), were obtained from the endophytic fungal strain *Penicillium sumatrense* GZWMJZ-313 9, which was isolated from the leaves of *Garcinia multiflora*. These compounds showed moderate to weak antimicrobial activities against *Staphylococcus aureus* ATCC6538, *Pseudomonas aeruginosa* ATCC10145, and *Escherichia coli* ATCC11775, with MIC values ranging from 32 to 128 μg/mL [67]. Two isoprenylisoindole alkaloids, diaporisoindoles A-B **100–101** (Figure 8), were obtained from the endophytic fungus *Diaporthe* sp. SYSU-HQ3, which was isolated from a fresh branch of the mangrove plant *Excoecaria agallocha*. Compound **100** showed potent inhibition activity against *Mycobacterium tuberculosis* protein-tyrosine phosphatase B, with an IC_{50} value of 4.2 μM [68].

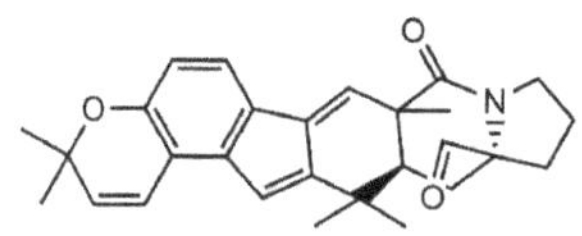

asperthrins A(**66**)

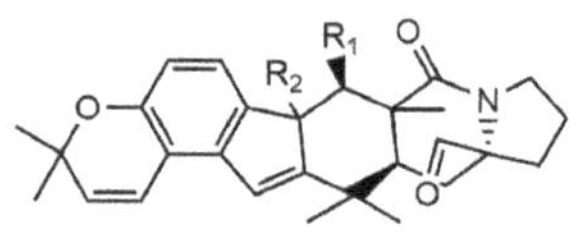

asperthrinsB:R_1=H, R_2=β-OH 3*R*(**67**)
asperthrinsC:R_1=OMe, R_2=α-OH 3*S* 10*S*(**68**)

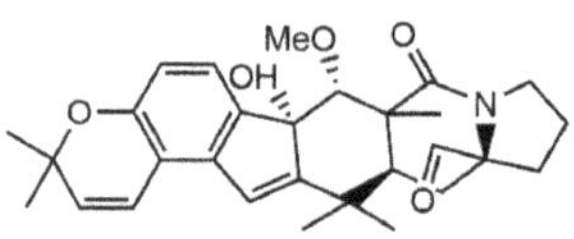

asperthrins D(**69**)

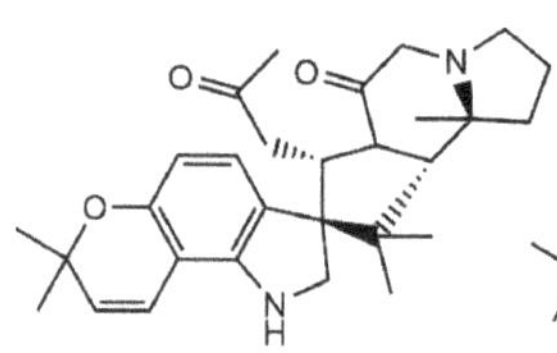

asperthrins E(**70**)

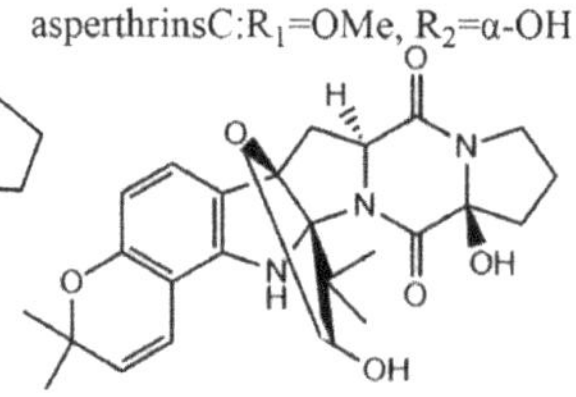

asperthrins F(**71**)

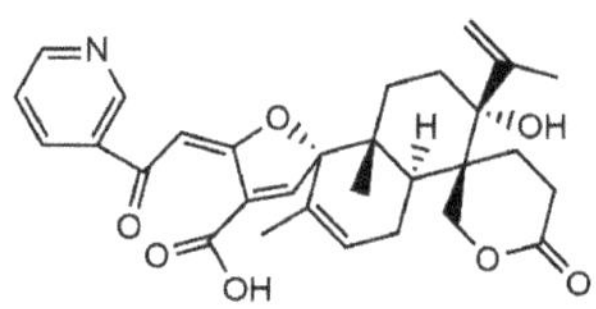

oxalicine C(**72**)

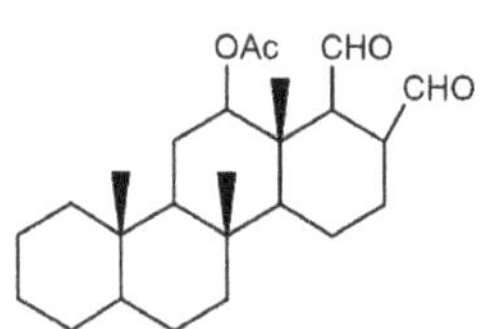

scalaradial (**73**)

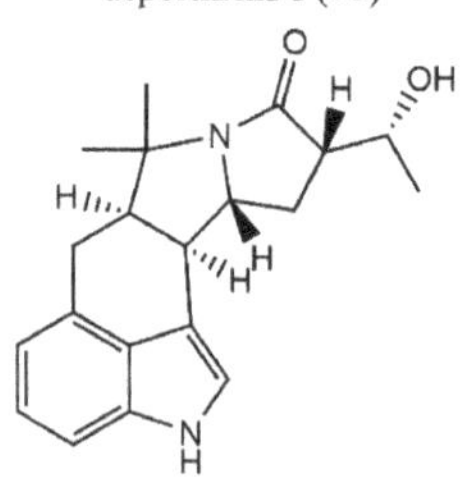

asperleninesA(**74**)

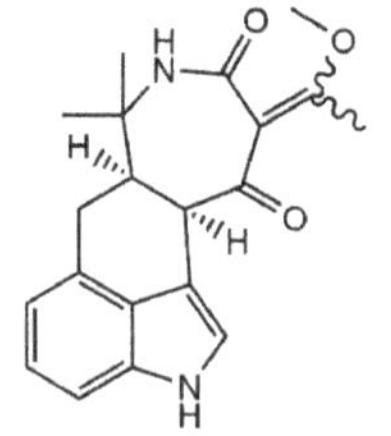

asperleninesB(**75**)

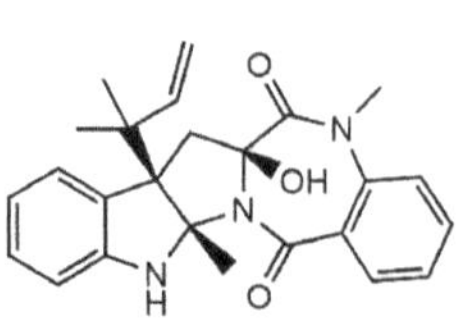

asperleninesC(**76**)

Figure 6. Chemical structures of indole alkaloids.

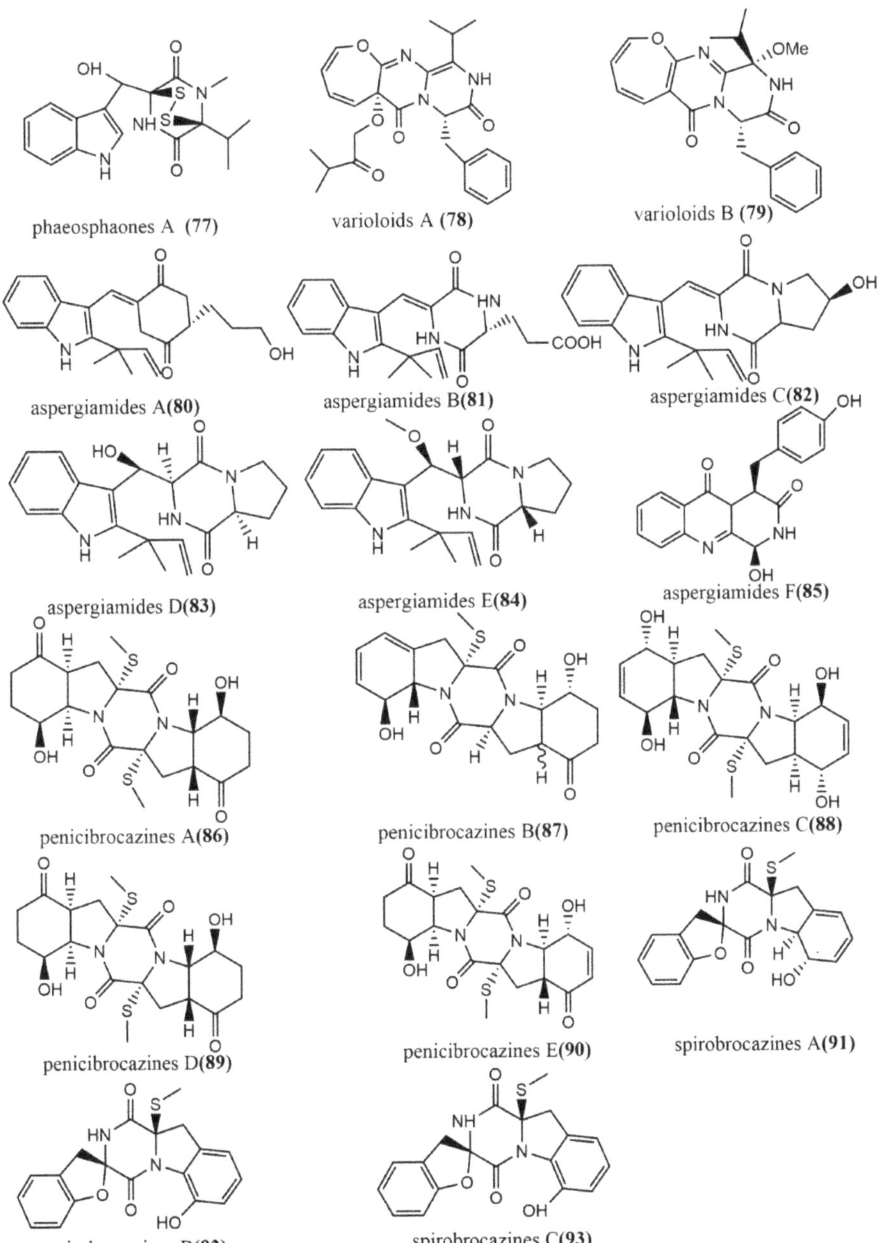

Figure 7. Chemical structures of diketopiperazine derivatives.

In a nutshell, anti-angiogenic and anti-inflammatory activities were the main activities of alkaloids in both plants and endophytic fungi. In addition, phomaketides and their derivatives that were isolated from fungal endophytes possess antimicrobial activity just as those isolated in plants; therefore, alkaloids producing endophytic fungi can be used in the development of anti-angiogenic, anti-inflammatory, and antimicrobial drugs for both human and animal use.

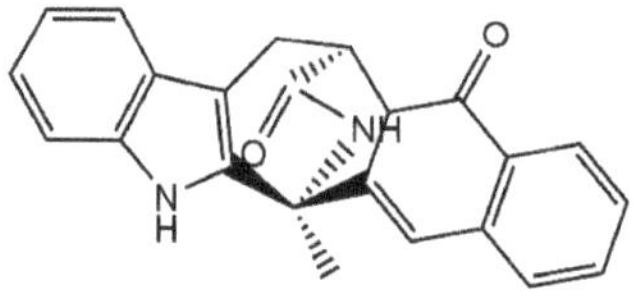

(-)-(1R,4R)-1,4-(2,3)-indolmethane-1-methyl-2,4-dihydro-1H-pyrazino-[2,1-b]-quinazoline-3,6-dione **(94)**

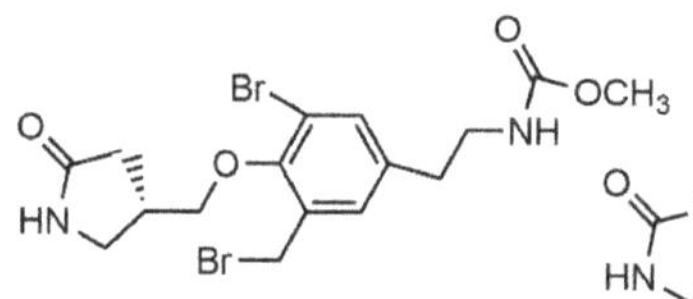

S-Acanthodendrilline **(95)**

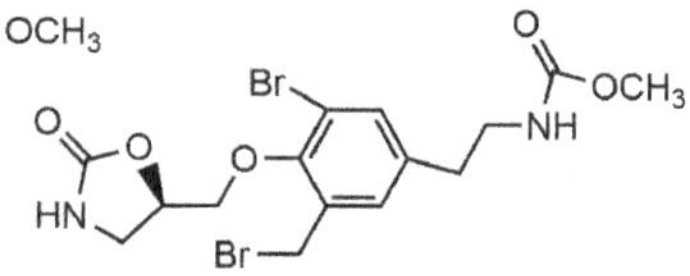

R- Acanthodendrilline **(96)**

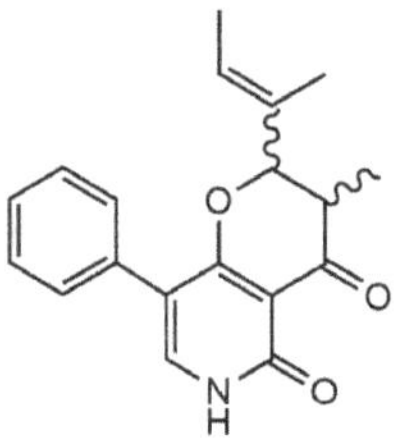

citridones E **(97)**

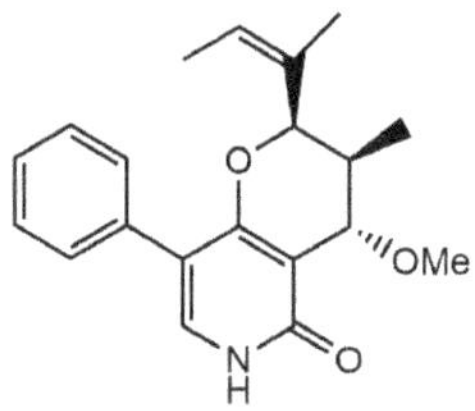

citridones F **(98)**

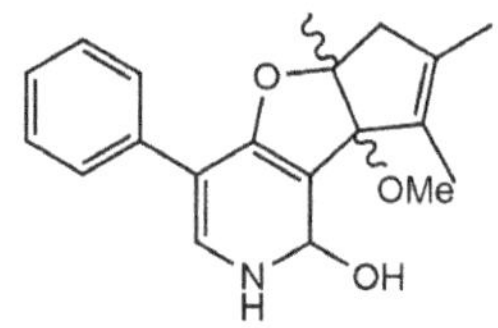

citridones G**(99)**

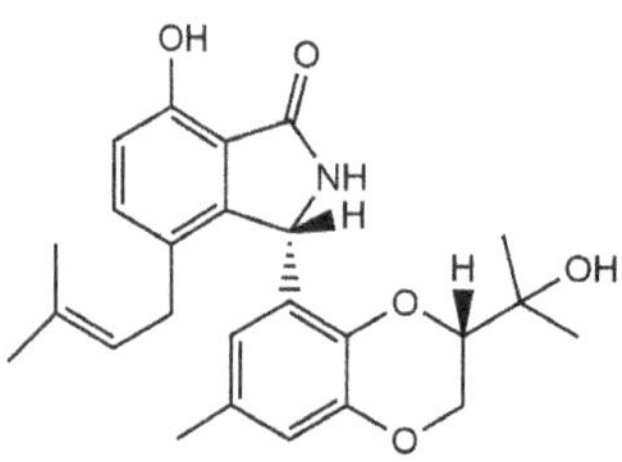

diaporisoindoles A**(100)**

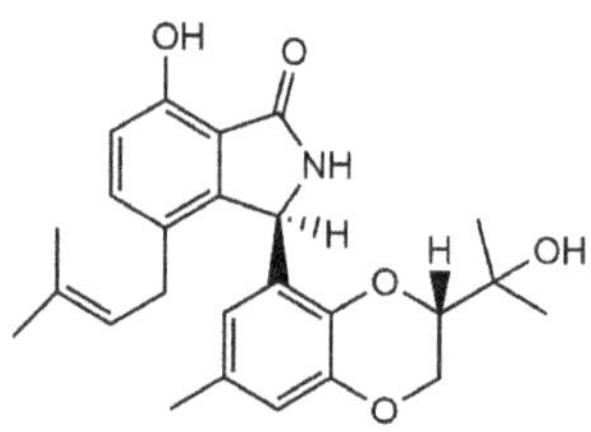

diaporisoindoles B**(101)**

Figure 8. Chemical structure of other types of alkaloids.

2.3. *Terpenoids*

2.3.1. Sesquiterpenoids and Their Derivatives

The 1-methoxypestabacillin B **107** (Figure 9) was obtained from brown rice cultures of endophytic fungus *Diaporthe* sp. SCSIO 41011 isolated from the stem of the mangrove plant *Rhizophora stylosa*. Compound **107** was evaluated for the reversal of HIV incubation period and anti-influenza A virus activities, to which compound **107** did not show antiviral activity. However, its structure could serve as the backbone for the synthesis of more potent antiviral compounds [69]. The eremophilane-type sesquiterpenoids rhizoperemophilanes A-N **102–115** (Figure 9) were isolated from the ethyl acetate extract of *Rhizopycnis vagum* Nitaf22. Compound **111** contained a C-4/C-11 epoxide, and compound **115** had a 3-nor-eremophilane lactone-lactam skeleton. All compounds were evaluated for their cytotoxic activities against five tested human cancer cells (BGC823, Daoy, HCT116, HepG2, and NCI-H1650) and inhibition activities against radicle growth in rice seedlings. Compound **115** showed high selective cytotoxicity against NCI-H1650 and BGC823 cell lines, with IC_{50} values of 15.8 μM and 48.2 μM, respectively, while no significant cytotoxic activity was observed for other compounds at $IC_{50} > 50$ μm. Compounds **106–107** and **113–114** showed strong phytotoxic activities against radicle growth in rice seedlings at a concentration of 200 μg/mL, where the inhibition exceeded 50% [70]. The bisabolane-type sesquiterpene, trichoderic acid **116,** (Figure 9) and acorane-type sesquiterpene, 2β-hydroxytrichoacorenol **117** (Figure 9), were obtained from *Trichoderma* sp. PR-35 culture, an endophytic fungus isolated from stems of *Paeonia delavayi*. Compounds **116–117** were tested for antimicrobial activity against two pathogens (*Escherichia coli*, and *Shigella sonnei*) using an agar diffusion method. Compounds **116–117** showed moderate to weak antimicrobial activities, with MIA values ranging from 50 to 175 μg/mL [69]. The ring flores aurantii alkane-type sesquiterpene, cyclonerotriol B **118** (Figure 9), and the α-pinene skeleton-containing sesquiterpene, 3β-hydroxy-β-acorenol **119** (Figure 9), were obtained from *Fusarium proliferatum* AF-04 isolated from *Chlorophytum comosum* roots via a combination of high-performance liquid chromatography (HPLC) and a bioassay-guided method. Compounds **118–119** showed weak antimicrobial activities (MIC values > 100 μg/mL) against *Bacillus subtilis, Clostridium perfringens, E. coli*, and methicillin-resistant *Staphylococcus aureus* (MRSA) [71]. The aromatic bisabolene-type sesquiterpene (7S, 8S)-8-hydroxysydowic acid **120** (Figure 9) was obtained from the brown rice culture of the endophytic fungus *Aspergillus sydowii* EN-434 isolated from the marine red alga *Symphyocladia latiuscula* from Qingdao, China. Compound **120** showed potent DPPH radical scavenging activity, with an IC_{50} value of 113.5 μmol/L [72]. The ophiobolane sesquiterpenes ophiobolins P–T **121–125** (Figure 9) were isolated from the acetone extract of the endophytic fungus *Ulocladium* sp. using the one-strain many-compound (OSMAC) strategy. Compounds **121–125** were evaluated for their cytotoxicity and antibacterial activities against two tested human cancer cell lines (KB and HepG2 cell lines) and three tested pathogens (*Bacillus subtilis*, MRSA, and *Bacille* Calmette-Guerin). Compounds **121–125** showed moderate antimicrobial activities against *B. subtilis* and multi-drug-resistant *S. aureus*, with MIC values ranging from 15.6 to 62.5 μM. Compound **125** showed moderate antimicrobial activity against *Bacille* Calmette-Guerin, with an MIC value of 31.3 μM. Additionally, compound **125** showed potent cytotoxic activity against the HepG2 cell line, with an IC_{50} value of 0.24 μM, which was stronger than the positive control etoposide (IC_{50} value of 2.02 μM) [73]. The daucane-type sesquiterpenes trichocarotins I-M **126–130** (Figure 9) were obtained from *Trichoderma virens* QA-8 isolated from the roots of *Artemisia argyi* H. Lév. and Vaniot, and these compounds showed significant antimicrobial activities against *E. coli* EMBLC-1, with MIC values ranging from 0.5 to 16 μg/mL [74].

2.3.2. Diterpenoids

The ring diterpene diaporpenoid A **131** (Figure 10), containing a 5/10/5-fused tricyclic ring system, was isolated from the MeOH extract obtained from cultures of the mangrove endophytic fungus *Diaporthe* sp. QYM12. Compound **131** showed significant anti-inflammatory activity by inhibiting LPS-induced NO production in a mouse macrophage cell line RAW264.7, with an IC_{50} value of 21.5 μM [75]. The pimarane-type diterpene Libertellenone M **132** (Figure 10) was isolated from the marine source endophytic fungus *Phomopsis* sp. S12. Compound **132** inhibited pro-inflammatory cytokines IL1β and IL-18 mRNA expression in colon tissue, significantly reduced the cleavage of pro-caspase1, and dose-dependently inhibited the NF-κB nuclear translocation in macrophages. Clinical indications of acute colitis induced by 3% dextran sulphate sodium in mice were attenuated by intravenous administration of different doses of compound **132** (10 or 20 mg/kg), which is a potent inhibitor of NLRP3 inflammatory vesicles and may be a new medicine for treating acute colitis [76]. Three pimarane-type diterpenoids—pedinophyllol K **133** (Figure 10), pedinophyllol L **134** (Figure 10), and libertellenone T **135** (Figure 10)—were isolated from the endophytic fungal *Phomopsis* sp. S12 culture using the OSMAC strategy. The anti-inflammatory activities of all compounds were assessed using an LPS-induced inflammation model of mouse macrophage RAW264.7. Compound **135** dose-dependently inhibited the expression of inflammatory factors IL-1β and IL-6 at the mRNA level. Additionally, the anti-inflammatory activity of compounds **133–134** was similar to that of compound **135** in terms Of IL-6 inhibition [77]. Two tetranorlabdane diterpenoids botryosphaerins G–H **136–137** (Figure 10) were obtained from the ethyl acetate extract of *Botryosphaeria* sp. P483 isolated from the branches of the herb *Huperzia serrata* (Thunb.) Trev. and tested for their antifungal activities against *Gaeumannomyces graminis, Fusarium solani*, and *Pyricularia oryzae* by the disk diffusion method. Compound **137** showed effective antifungal activity at a concentration of 100 μg/disk with an inhibitory zone diameter of 9 mm. (The inhibitory zone diameter of positive control carbendazim was 15–18 mm.) Compounds **136–137** were evaluated for their nematicidal activities against *Panagrellus redivivus* and *Caenorhabditis elegans* and showed weak nematicidal activities, with 30% and 28% fatality rates at a 24h action concentration of 400 mg/L, respectively [78]. The isopimarane diterpene sphaeropsidin A **138** (Figure 10) was isolated from the ethyl acetate extract of the endophytic fungus *Smardaea* sp. AZ0432 of *Ceratodon purpureus*. The in vitro cytotoxic activities of compound **138** against five human cancer cell lines (NCI- H460, MDA-MB-231, MCF-7, PC-3M, and SF-268) and human embryonic lung fibroblast cell line WI-38 were evaluated using the resazurin colorimetric assay. The results showed that compound **138** showed a high cell selectivity when it was applied at a concentration of 10 μM for 72 h and inhibited the migration of MDA-MB-231 cells by 50% at a subcytotoxic concentration of 1.5 μM [79]. (10S)-12,16-epoxy-17(15→16)-abeo-3,5,8,12,15-abietapentaene-2,7,11,14-tetraone **139** (Figure 10) was obtained from the cultures of the endophytic fungus *Pestalotiopsis adusta* isolated from stems of the medicinal plant *Clerodendrum canescens*. The cytotoxicity of compound **139** to the HL-60 tumor cell line was evaluated using the MTT assay, by which compound **139** showed moderate cytotoxic activity, with an IC_{50} value of 12.54 μM [80]. (The IC_{50} value of the positive control cisplatin was 9.20 μM.) The trichodermanin A **140** (Figure 10), a diterpene containing a 6-5-6-6 ring system, was obtained from the endophytic fungus *Trichoderma atroviride* S361 of *Cephalotaxus fortunei* and was not tested for any biological activities [81]. Therefore, further studies are needed to identify the potential biological activity of this compound in the future. The new tetranorlabdane diterpenoids, asperolides A–C **141–143** (Figure 10), were isolated from the ethyl acetate extract of the marine brown alga *Aspergillus wentii* EN-48 and the cytotoxic activities of compounds **141–143** to seven tested human cancer cell lines (NCI-H460, MDA-MB-231, HeLa, MCF-7, SMMC-7721, HepG2, and SW1990) were evaluated using the MTT method. Compounds **141–143** showed moderate cytotoxic activities, with IC_{50} values $\leq$ 10 Mm [82].

2.3.3. Triterpenoids

The 24-homo-30-nor-cycloartane triterpenoid **154** (Figure 11) was isolated from the endophytic fungus *Mycoleptodiscus indicus* FT1137. Compound **154** showed no activity against the human ovarian cancer cell line A2780 at a concentration of 20 μg/mL [83]. Three Lanostane-type triterpenes—sclerodols A–B **144–145** (Figure 11) and lanosta-8,23-dien-3β,25-diol **146** (Figure 11)—were obtained from *Eucalyptus grandis* cultures derived from the endophytic fungus *Scleroderma* UFSMSc1, and the antifungal activities of compounds **144–146** against *Candida albicans* and *Candida parapsolosis* were evaluated by the agar diffusion method. Compounds **144–146** showed moderate to weak antifungal activities, with MIC values ranging from 12.5 to 50 μg/mL. The antifungal effects of these compounds against *C. albicans* were associated with the inhibition of the selenocysteine methyltransferase (SMT) activity [84]. Fusidic acid **147** (Figure 11) was obtained from the cultures of the endophytic fungus *Acremonium pilosum* F47, isolated from the stem of *Mahonia fortunei* using the bioactivity-guided assay, and the antimicrobial activities of compound **147** against four human pathogens were tested (*S. aureus* ATCC 6538, *B. subtilis* ATCC 9372, *P. aeruginosa* ATCC 27853, and *E. coli* ATCC 25922) and evaluated. Compound **147** showed effective antimicrobial activities against *S. aureus* ATCC 6538 and *B. subtilis ATCC* 9372. The acetylation of the C-16 hydroxyl group of compound **147** was essential for antimicrobial action [85]. Two new ring *A*-cleaved lanostane-type triterpenoids, glometenoid A–B **148–149** (Figure 11), were obtained from the ethyl acetate extract of the mason pine endophytic fungus *Glomerella* sp. F00244. The cytotoxic activity of compounds **148–149** against the human ovarian cancer cell line HeLa was tested using the MTT assay. Compound **148** showed weak cytotoxic activity at a concentration of 10 μM with 21% inhibition [83]. Nine highly oxygenated schitriterpenoids—kadhenrischinins A–H **150–157** (Figure 11) and 7β-schinalactone C **158** (Figure 11)—were isolated from *Penicillium* sp. SWUKD4.1850, and compounds **154–157** contained a unique 3-one-2-oxabicyclo [1–3]-octane motif. All compounds were tested for their cytotoxic activities against the HepG2 tumor cell lines using the MTT assay, and these compounds showed weak cytotoxic activities, with IC_{50} values ranging from 14.3 to 40 μM [86]. Two tetracyclic triterpenoids—integracide E **159** (Figure 11) and isointegracide E **160** (Figure 11)—were isolated from the mycelia of *Hypoxylon* sp. 6269. Compound **159** showed weak inhibition activity against the HIV-1 integrase, with an IC_{50} value of 31.63 μM [87]. The tetracyclic triterpenoids, integracides H–J **161–163** (Figure 11), were obtained from the endophytic fungus *Fusarium* sp., which was isolated from the roots of *Mentha longifolia* L. (Labiatae) and were evaluated for antileishmanial activity against *L. donovani* promastigotes. Compound **161** showed significant antileishmanial activity, with an IC_{50} value of 4.75 μM, exceeding the positive control Pentamidine (IC_{50} value of 6.35 μM) [88]. The tetracyclic triterpenoids, integracides F–G **164–165** (Figure 11), were obtained from the endophytic fungus *Fusarium* sp. of *Mentha longifolia* L. (Labiatae). Compounds **164–165** were evaluated for their antileishmanial and cytotoxic activities to BT-549 and SKOV-3 cells and *Leishmania donovani* promastigotes. Compounds **164–165** showed significant cytotoxic activities against SKOV-3 and BT-549 cell lines, with IC_{50} values ranging from 0.16 to 1.97 μg/mL and 0.12 to 1.76 μg/mL, respectively. (The IC_{50} value of the positive control Pentamidine was 2.1 μg/mL.) Compounds **164–165** showed potent antileishmanial activities against *L. donovani* promastigotes, with IC_{50} values of 3.74 μg/mL and 2.53 μg/mL, respectively [89].

2.3.4. Meroterpenoids

Guignardones P–S **166–169** (Figure 12) were obtained from *Guignardia mangiferae* A348 cultures, and the cytotoxic activities of compounds **166–169** against three human cancer cell lines (SF-268, MCF-7, and NCI-H460) were tested using an MTT assay. Compounds **167** and **169** only showed weak cytotoxic activities against MCF-7 cell lines, with IC_{50} values ranging from 83.7 to 92.1 μM [90]. Six 3, 5-demethylorsellinic acid-based meroterpenoids emeridones A–F **170–175** (Figure 12) were isolated from *Emericella* sp. TJ29 cultures. Compound **171** possessed a 2,6 dioxabicyclo [2.2.1] heptane and a spiro [bicycle [3.2.2] nonane-2,1′-cyclohexane] moiety. The cytotoxic activities of all compounds against five human cancer cell lines (HL-60, SMMC7721, A549, MCF-7, and SW-480) were tested using the MTT assay, and compounds **172**, **173**, and **175** showed moderate cytotoxic activities against all tested cell lines, with IC_{50} values ranging from 8.19 to 18.8 μM [91]. Phyllomeroterpenoids A–C **176–178** (Figure 12) were isolated from the crude extract of *Phyllosticta* sp. J13-2-12Y fermentation broth. Compounds **176–178** showed moderate antimicrobial activities against *Staphylococcus aureus* 209P, *Candida aureus* 209P, and *Candida albicans* FIM709, with MIC values ranging from 32 to 128 μg/mL [92]. Austin **179** (Figure 12) was obtained from the ethyl acetate extract of *Talaromyces purpurogenus* H4 and *Phanerochaete* sp. H2 co-cultures, which showed moderate trypanocidal activity against *T. cruzi* at a concentration of 100 μg/mL, with an IC_{50} value of 36.6 μM. Notably, neither of the two endophytic fungi produced compound **179** when cultured separately under similar conditions [93].

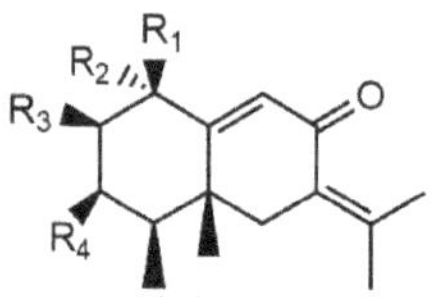

rhizoperemophilanes A :R1=R3=H,R2=R4=OH(**102**)
rhizoperemophilane B:R1=R3=OH,R2=R4=H (**103**)
rhizoperemophilane C: R1=R3=R4=OH,R2=H(**104**)

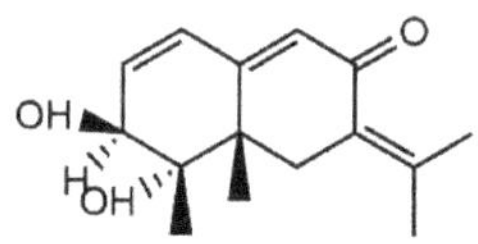

rhizoperemophilanes D:R1=H,R2=R3=OH (**105**)
rhizoperemophilanes E:R1=OH,R2=2H(**106**)
rhizoperemophilanes F:R1=H,R2=O (**107**)

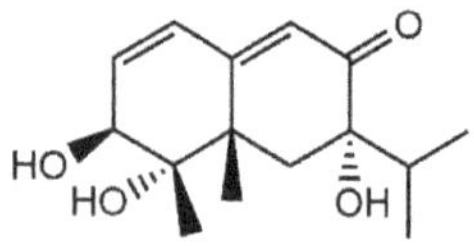

rhizoperemophilanes G (**108**)

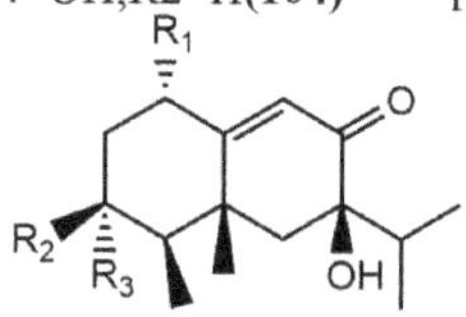

rhizoperemophilanesH:R1=R2=OH,R3=H(**109**)
rhizoperemophilanes I:R1=H,R2+R3=O(**110**)

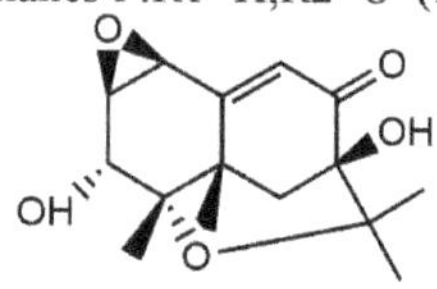

rhizoperemophilanes J (**111**)

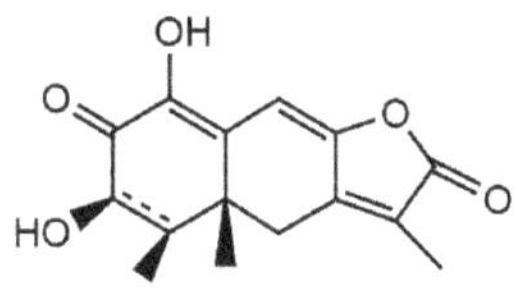

rhizoperemophilanes K (**112**)

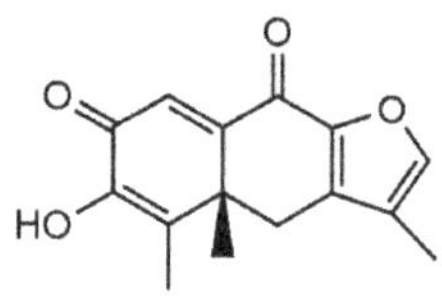

rhizoperemophilanes L(**113**)

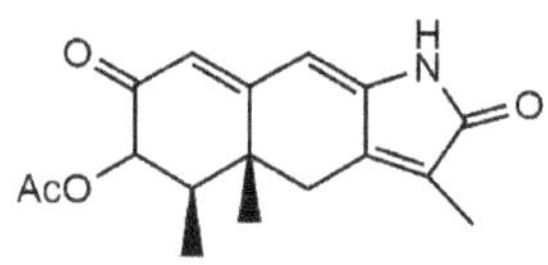

rhizoperemophilanes M (**114**)

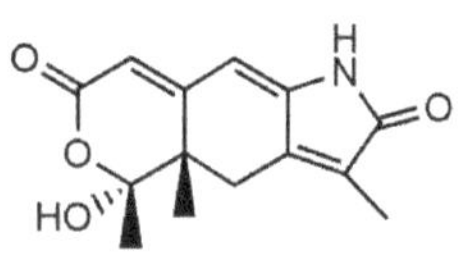

rhizoperemophilanes N (**115**)

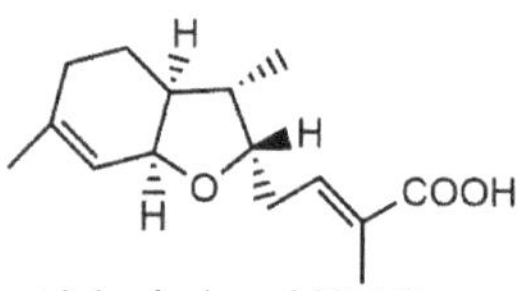

trichoderic acid(**116**)

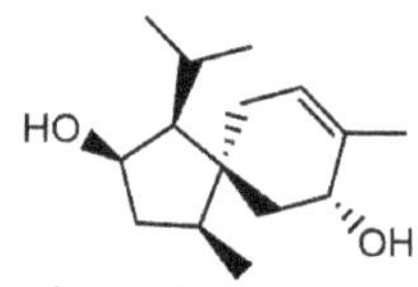

2β-hydroxytrichoacorenol (**117**)

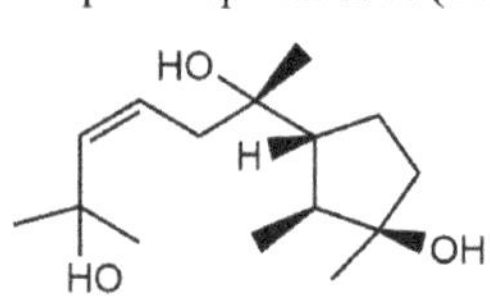

cyclonerotriol B (**118**)

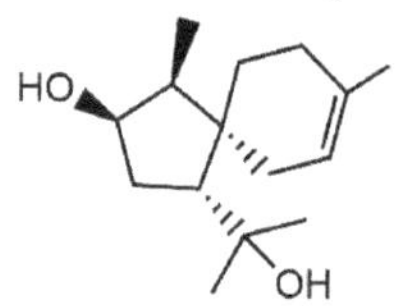

3β-hydroxy-β-acorenol(**119**)

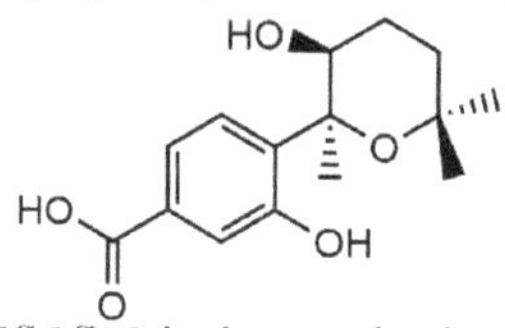

(7*S*,8*S*)-8-hydroxysydowic acid(**120**)

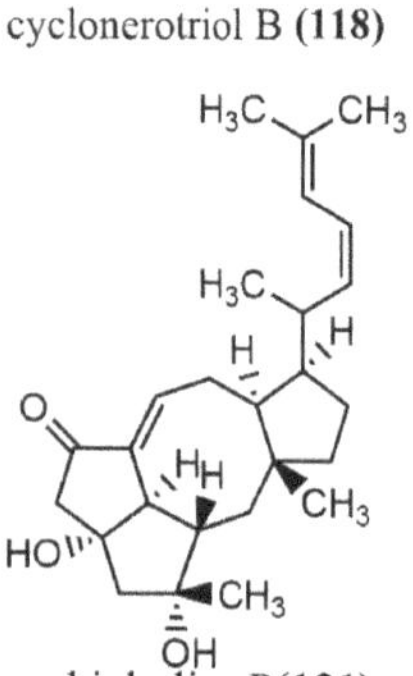

ophiobolins P(**121**)

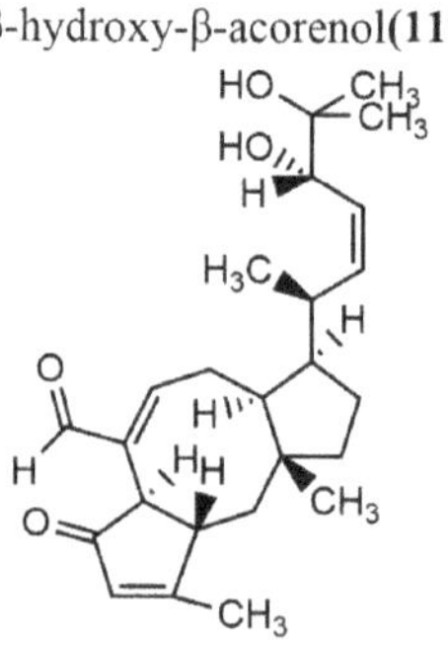

ophiobolins Q(**122**)

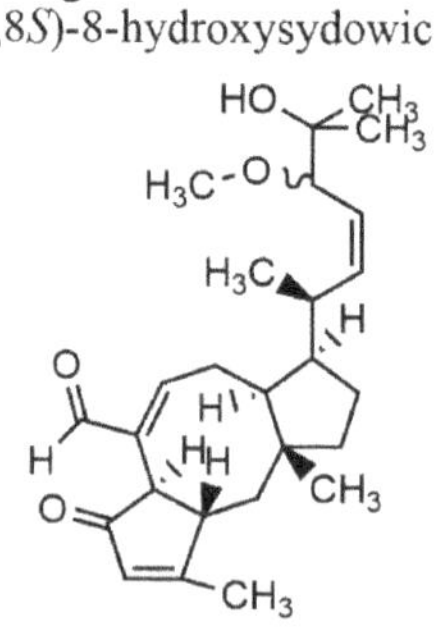

ophiobolins R(**123**)

Figure 9. *Cont.*

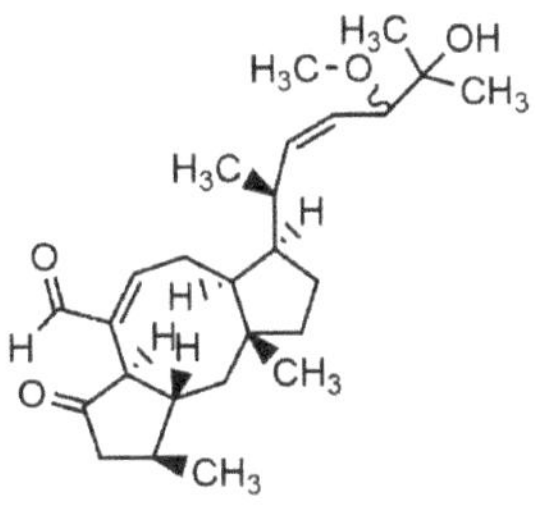

ophiobolins S(**124**)

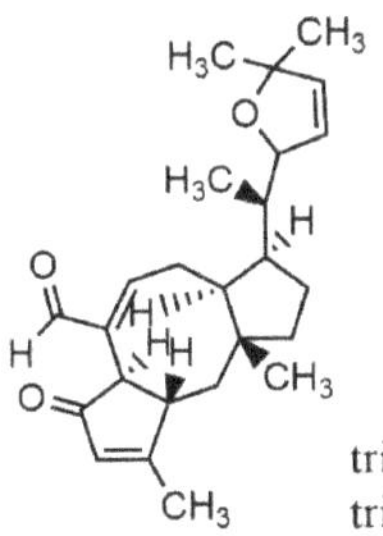

ophiobolins T(**125**)

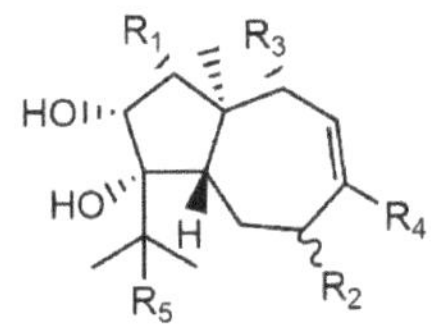

trichocarotins I :R1=OH,R2=R3=R5=H,R4=CH3(**126**)
trichocarotins J:R1=R3=R5=H,R2=αOH,R4=CH3 (**127**)
trichocarotinsM:R1=R2=R3=R5=H,R4=CH2OH (**128**)

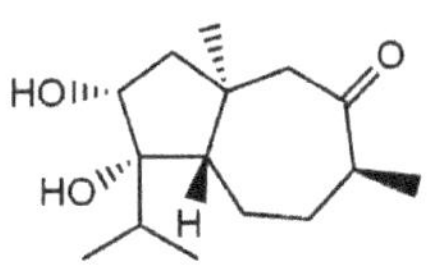

trichocarotins K(**129**)

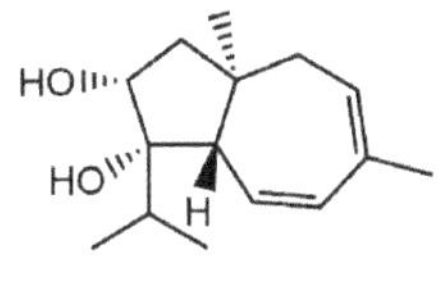

trichocarotins L(**130**)

Figure 9. Chemical structures of sesquiterpenoids and derivatives.

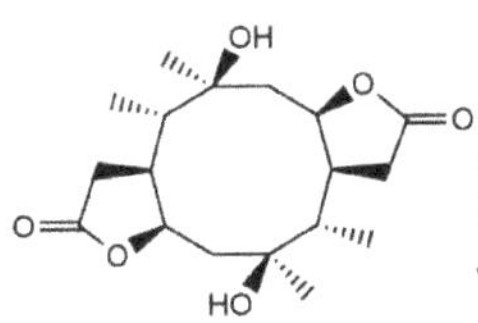

diaporpenoid A(**131**)

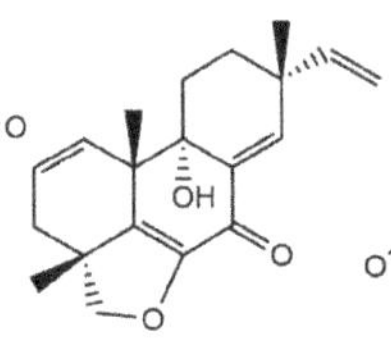

libertellenone M(**132**)

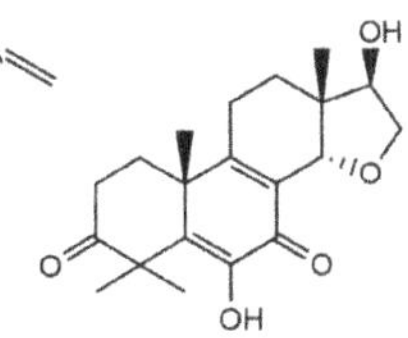

pedinophyllol K(**133**)

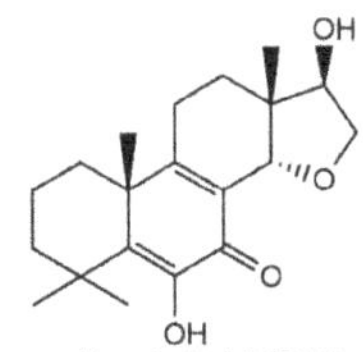

pedinophyllol L(**134**)

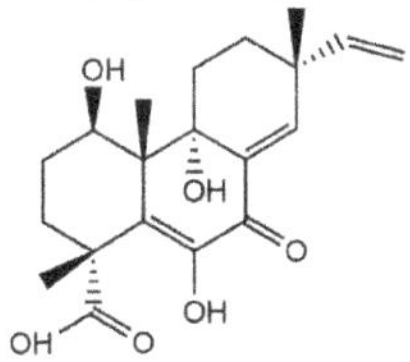

libertellenone T(**135**)

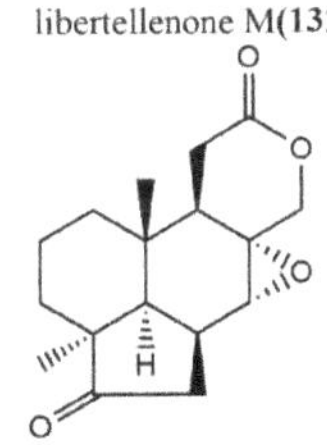

botryosphaerins G(**136**)

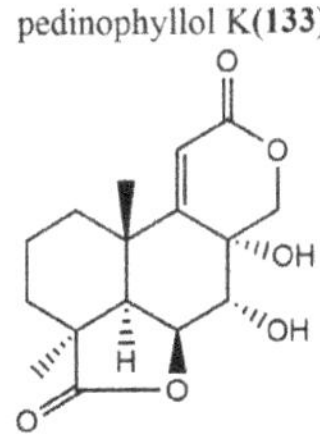

botryosphaerins H(**137**)

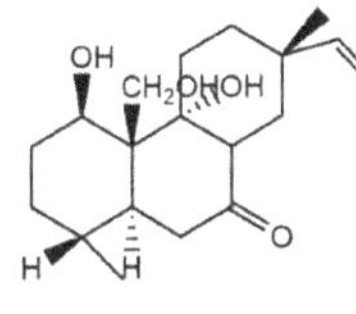

sphaeropsidins A (**138**)

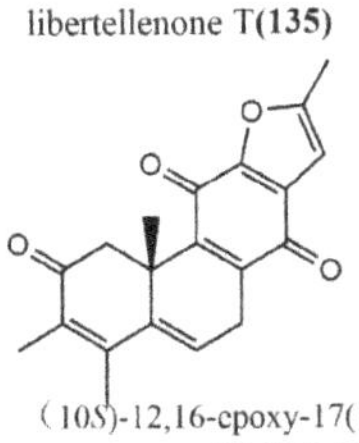

(10S)-12,16-epoxy-17(15→16)-abeo-3,5,8,12,15-abietapentaen-2,7,11,14-tetraone (**139**)

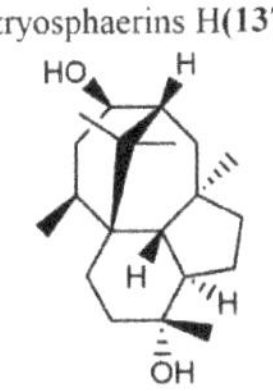

trichodermanin A(**140**)

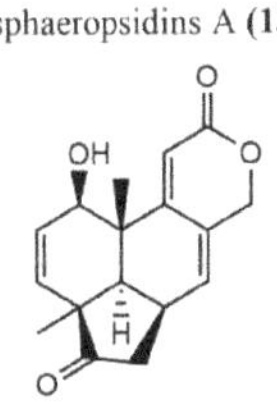

asperolides A(**141**)

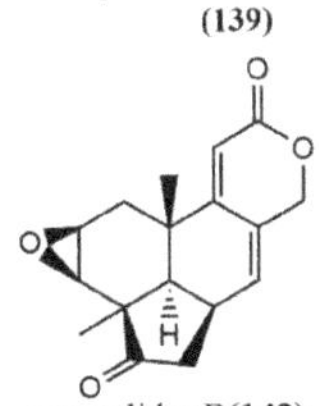

asperolides B(**142**)

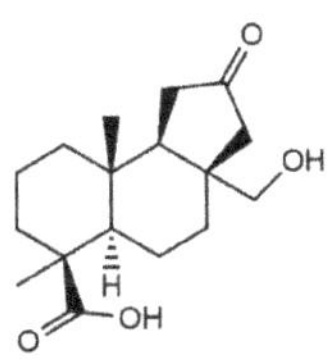

asperolides C(**143**)

Figure 10. Chemical structures of diterpenoids and derivatives.

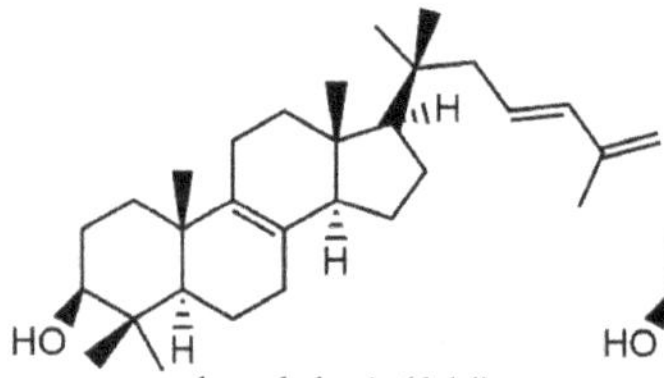

sclerodols A **(144)**

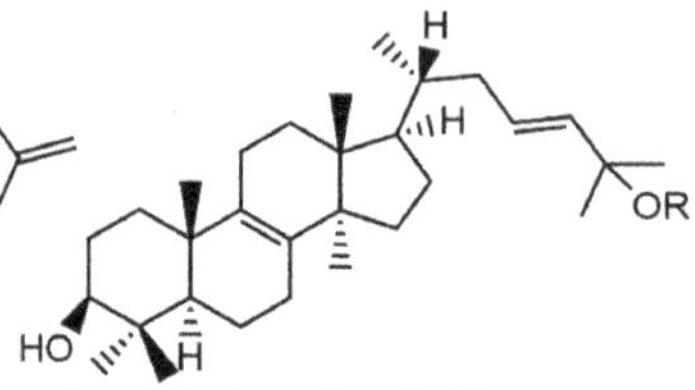

sclerodols B:R=CH_3**(145)**
lanosta-8,23-dien-3β,25-diol:R=H**(146)**

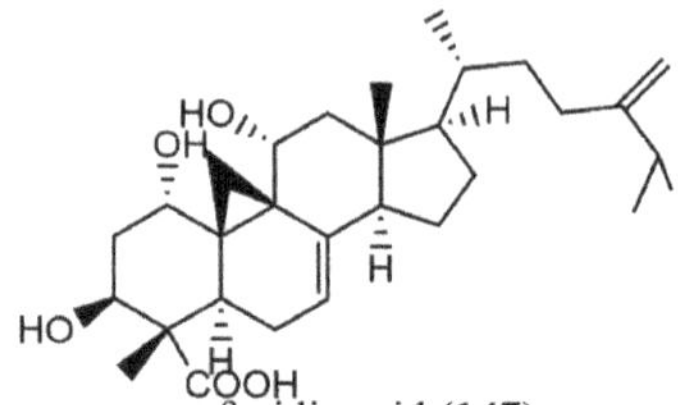

fusidic acid **(147)**

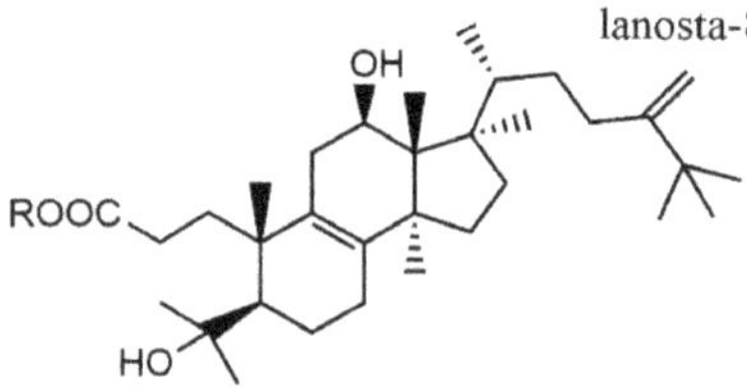

glometenoid A:R=H**(148)**
glometenoid B:R=Me**(149)**

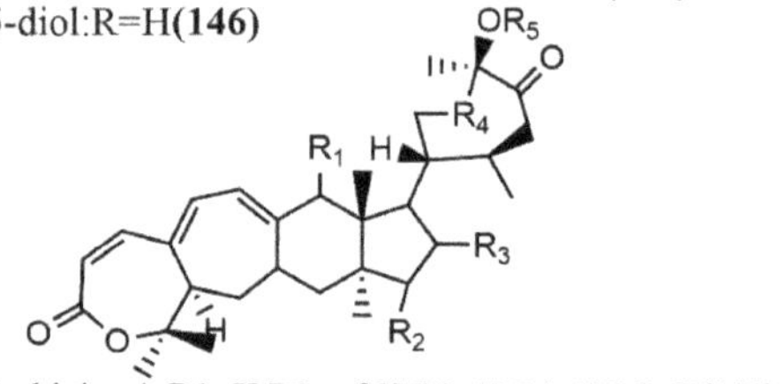

Kadhenrischinins A:R1=H,R2=αOH,R3=H,R4=H,R5=H**(150)**
kadhenrischinins B:R1=H,R2=H,R3=H,R4=αOH,R5=H**(151)**
kadhenrischinins C:R1=H,R2=H,R3=αOH,R4=H,R5=H**(152)**
kadhenrischininsD:R1=αOH ,R2=H,R3=H,R4=H,R5=H**(153)**

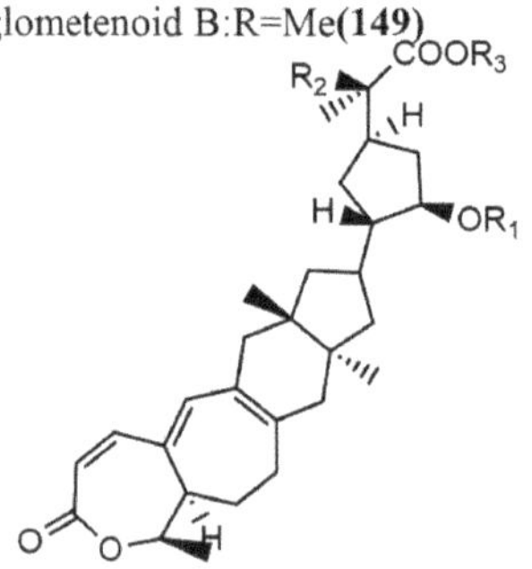

kadhenrischinins E： R1=H,R2=Ac,R3=H**(154)**
kadhenrischinins F:R1=H,R2=H,R3=H**(155)**
kadhenrischinins G:R1=Ac,R2=Ac,R3=H**(156)**
kadhenrischininsH:R1=H,R2=Ac,R3=Me**(157)**

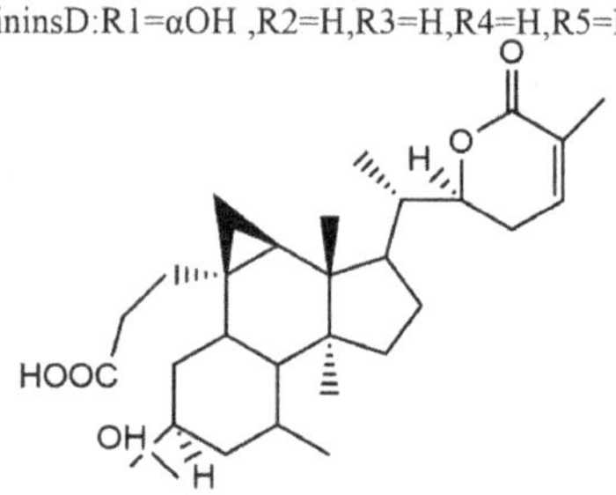

7β-schinalactone C**(158)**

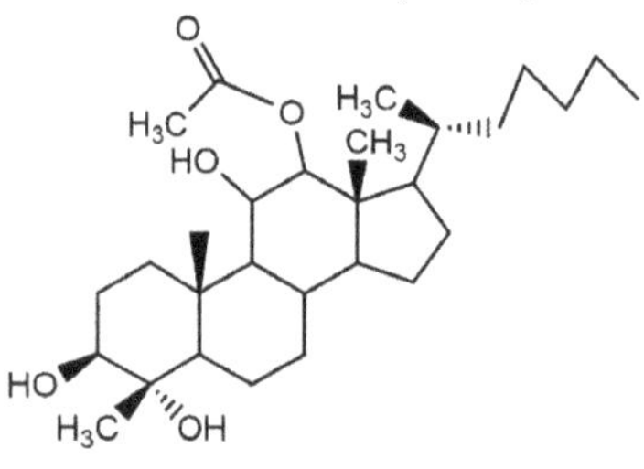

integracide E **(159)**

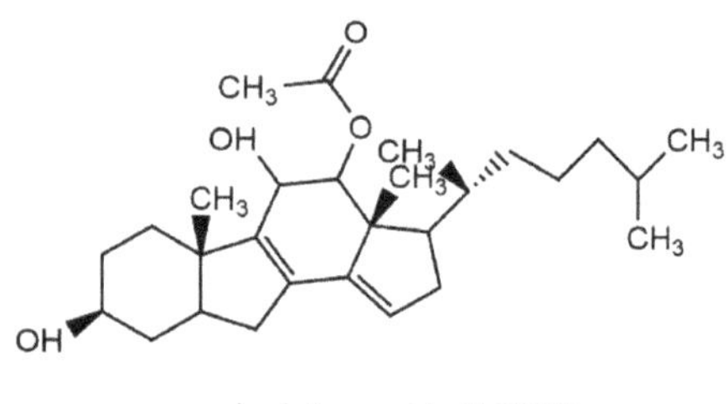

isointegracide E **(160)**

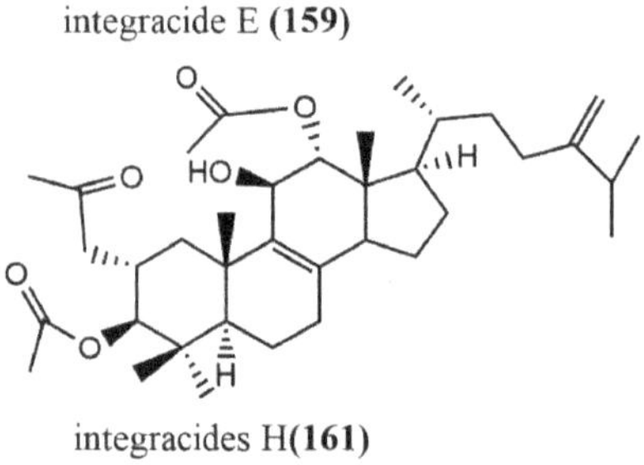

integracides H**(161)**

Figure 11. *Cont.*

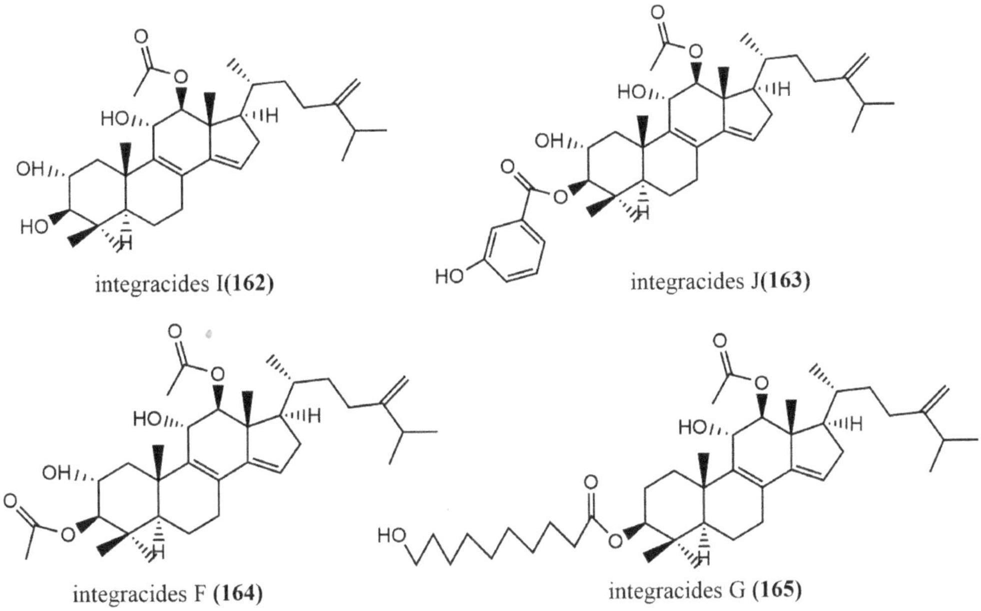

Figure 11. Chemical structures of terpenoids.

To sum up, Meroterpenoids and their derivatives, which are mainly known for their antifungal properties in most plants species, have been found in endophytic fungi. However, recent studies have also reported anti-oxidative, anti-inflammatory, and anti-cancer activities from these compounds. Therefore, these microorganisms can be used in the development of drugs candidates for human, animal, and other agricultural activities.

2.4. Lactones

Helicascolide F **180** (Figure 13) was obtained from *Talaromyces assiutensis* JTY2 isolated from *Ceriops tagal* leaves. The cytotoxic activities of compound **180** against three human cancer cell lines (HeLa, MCF-7, and A549) were tested using an MTT assay, in which compound **180** showed a moderate cytotoxic effect on all tested cell lines, with an IC_{50} value range of 14.1–38.6 μM [94]. Two β-lactones, polonicin A–B **181–182** (Figure 13), were obtained from the brown rice culture of the endophytic fungus *Penicillium polonicum* in the fruit of *Camptotheca acuminata*. Compound **181** showed effective glucose uptake activity at a concentration of 30 μg/mL on rat skeletal myoblast cell line L6, which enhanced 1.8-fold compared to that of the control. Compound **182** was used to assess its effect on GLUT4 translocation by using the fluorescent protein, IRAP-mOrange, which is stably expressed in L6 cells. It showed a 2.1-fold increase in fluorescence intensity on L6 cell membranes compared to the untreated controls [95]. The spirodilactone compound chaetocuprum **183** (Figure 13) was obtained from cultures of the endophytic fungus *Chaetomium cupreum* of wild *Anemopsis californica* from New Mexico, U.S.A. Compound **183** showed a weak antimicrobial activity against *S. aureus*, with an MIC value of 50 μg/mL [96]. A phytotoxic bicyclic lactone, (3aS,6aR)-4,5-dimethyl-3,3a,6,6a-tetrahydro-2H-cyclopenta [b] furan-2-one **184** (Figure 13), was obtained from the fermentation broth of *Xylaria curta* 92092022. Compound **184** contained a rare 5/5 rings-fusion system and was tested for antimicrobial activities against four pathogens (*Pseudomonas aeruginosa* ATCC 15442, *Staphylococcus aureus* NBRC 13276, *Aspergillus clavatus* F318a, and *Candida albicans* ATCC 2019) and the phytotoxicity against lettuce seedlings. Compound **184** showed moderate antimicrobial activities against *Pseudomonas aeruginosa* ATCC 15442 and *Staphylococcus aureus* NBRC

13276 at a concentration of 100 μg/disk, with inhibitory zone diameters of 13 mm and 12 mm, respectively. At the concentration of 25 μg mL^{-1}, compound **184** showed 50% inhibition on lettuce roots with a root length of 1.6 ± 0.3 cm (3.2 ± 0.5 cm for the control). At a concentration of 200 μg mL^{-1}, compound **184** strongly inhibited lettuce seed germination, with 90% inhibition [97]. Lasiodiplactone A **185** (Figure 13) was obtained from the mangrove endophytic fungus *Lasiodiplodia theobromae* ZJ-HQ1 and contained a unique tetracyclic system (12/6/6/5) of RAL 12 (12-membered β-resorcylic acid lactone) with a pyran ring and a furan ring. Compound **185** showed significant anti-inflammatory activity by inhibiting the LPS-induced NO production in RAW 264.7 cells, with an IC_{50} value of 23.5 μM, which was stronger than the positive control indomethacin (IC_{50} = 26.3 μM). Additionally, compound **185** showed potent α-glucosidase inhibition activity, with an IC_{50} value of 29.4 μM, which was superior to the commonly used clinical drug acarbose (IC_{50} = 36.7 μM) [98]. (+)-phomalactone **186** (Figure 13), hydroxypestalopyrone **187** (Figure 13), and pestalopyrone **188** (Figure 13) were isolated from the endophytic fungus *Aspergillus pseudonomiae* J1 cultures and evaluated for in vitro anti-trypanosomal activity against the *Trypanosoma cruzi* Y strain using an anti-epimastigote assay. Compounds **186–188** showed moderate to weak anti-trypanosomal activities, with IC_{50} values of 0.86 μM, 88.33 μM, and 580.19 μM, respectively [99].

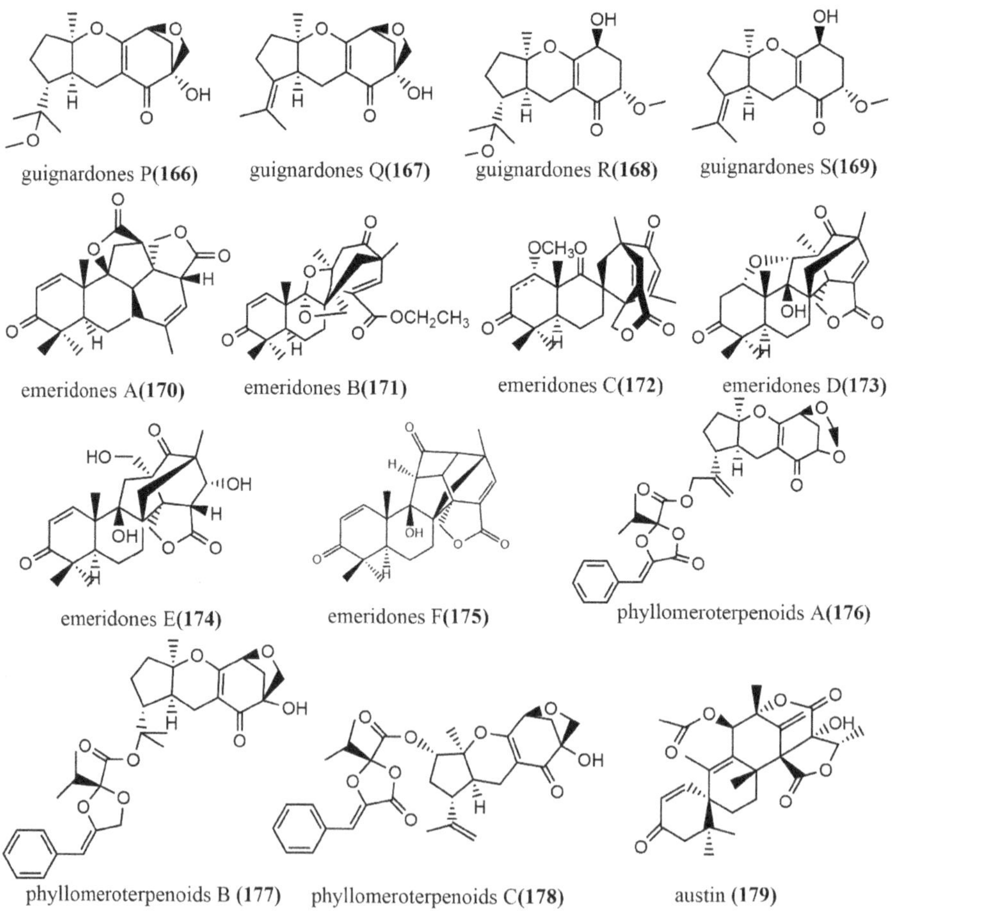

Figure 12. Chemical structures of Meroterpenoids.

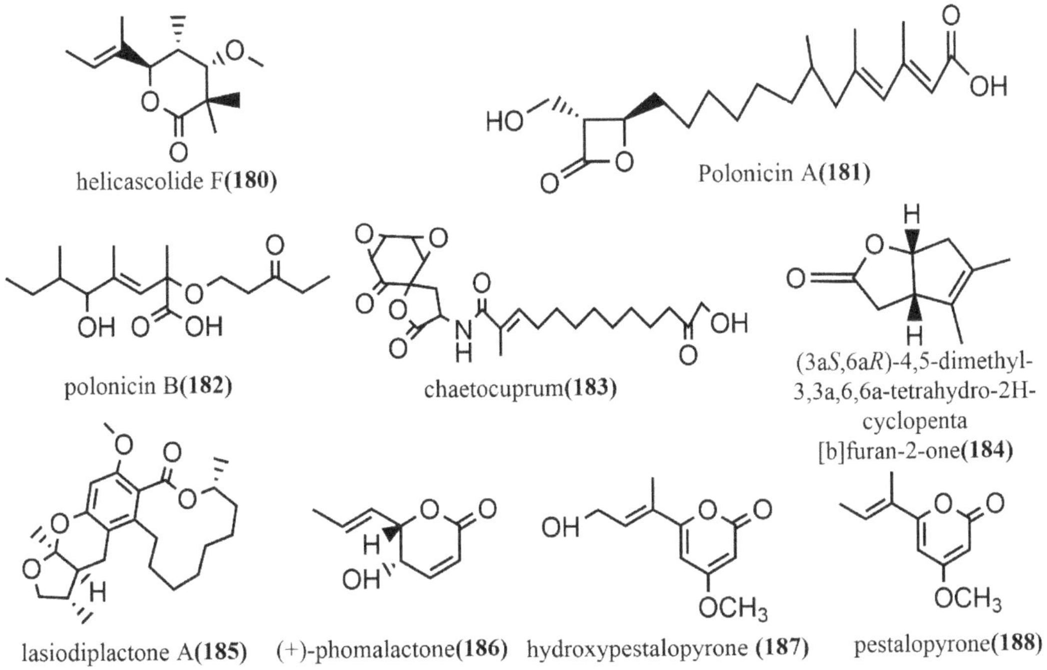

Figure 13. Chemical structures of Lactones.

In summary, this review reported that fungal endophytes could produce Lactones and their derivatives through their metabolic activities. In addition, these compounds possessed biological activities, such as antimicrobial, anti-cancer, allelopathic, and anti-inflammatory; thus, fungal endophytes that produce these compounds may be utilized in the pharmacological setup as alternatives to plant-derived compounds.

2.5. *Anthraquinones, Quinones, and Related Glycosides*

6,8-di-O-methylbipolarin **189** (Figure 14), aversin **190** (Figure 14), and 6,8-di-O-methylaverufin **191** (Figure 14) were obtained from rice cultures of the marine red algae endophytic fungus *Acremonium vitellinum* from Qingdao, China. Compounds **189**–**191** showed moderate insecticidal activities against the third-instar larvae of *Helicoverpa armigera*, with LC_{50} values of 0.72 mg/mL, 0.78 mg/mL, and 0.87 mg/mL, respectively. (The LC_{50} value for the positive control, matrine, was 0.29 mg/mL.) Additionally, the molecular mechanism of the insecticidal activity of compound **191** was investigated based on transcriptome sequencing. The identification of 5,732 differentially expressed genes was performed, of which 2,904 genes were downregulated and 2,828 genes were upregulated. The upregulated genes were primarily involved in cell autophagy, apoptosis, DNA mismatch repair, and replication [100]. A new quinone, identified as 1,3-dihydroxy-4-(1,3,4-trihydroxybutan-2-yl)-8-methoxy-9H-xanthen-9-one **192** (Figure 14), was obtained from *Phomopsis* sp. isolated from the rhizome of *Paris polyphyllavar.* in Yunnan, China. Compound **192** showed significant cytotoxic activities against A549 and PC3 cell lines, with IC_{50} values of 5.8 μM and 3.6 μM, respectively [101]. The anthraquinone derivative eurorubrin **193** (Figure 14) was obtained from the ethyl acetate extract of the endophytic fungus *Eurotium cristatum* EN-220 of the seaweed *Sargassum thunbergii* and tested for its antimicrobial activities against three tested pathogens (*E. coli*, *Physalospora obtuse*, and *Valsa mali*), including its fatal activity against brine shrimp larvae. Compound **193** only showed a weak antimicrobial activity against *E. coli*, with an MIC value of 64 μg/mL. At the concentration of 10 μg/mL, compound **193** showed moderate fatal activity against brine shrimp larvae, with a fatality rate of 41.4% [102]. Isorhodoptilometrin-1-methyl ether **194** (Figure 14), emodin **195** (Figure 14), and 1-methyl emodin **196** (Figure 14) were obtained from cultures of the endophytic fungus *Aspergillus versicolor* of the red seaweed *Halimeda opuntia*. Compounds **194–196** were evaluated for their inhibiting activities against the hepatitis C virus NS3/4A protease, where Compounds **195–196** showed weak inhibition activities, with IC_{50} values ranging from 22.5 to 40.2 μg/mL [103]. The quinone altersolanol A **197** (Figure 14) was isolated from the endophytic fungus *Stemphylium globuliferum* of the medicinal plant *Mentha pulegium* (Lamiaceae). Compound **197** inhibited the proliferation of K562 and A549 cells in a time-dependent, dose-dependent manner and caused apoptosis by cleaving Caspase-3 and Caspase-9 and decreasing anti-apoptotic protein expression [104].

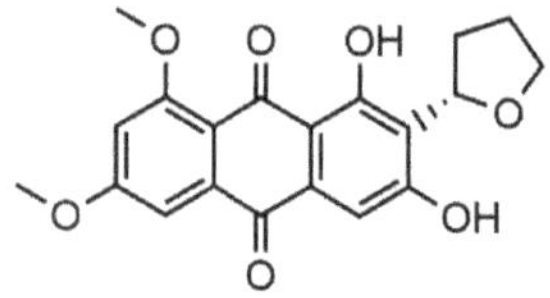

6,8-di-O-methylbipolarin **(189)**

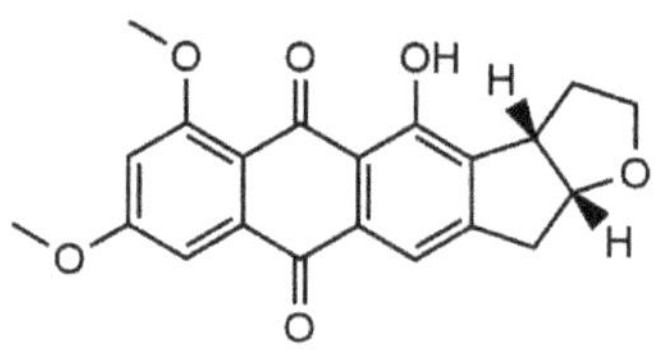

aversin **(190)**

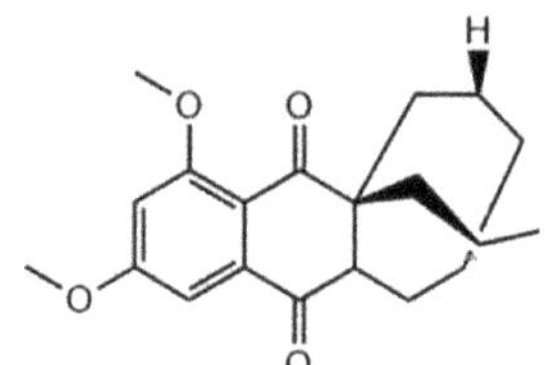

6,8-di-O-methylaverufin **(191)**

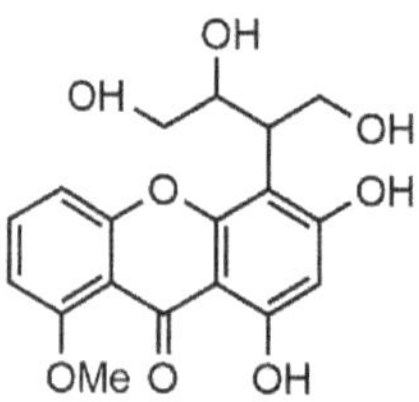

1,3-dihydroxy-4-(1,3,4-trihydroxybutan-2yl)-8-methoxy-9H-xanthen-9-one **(192)**

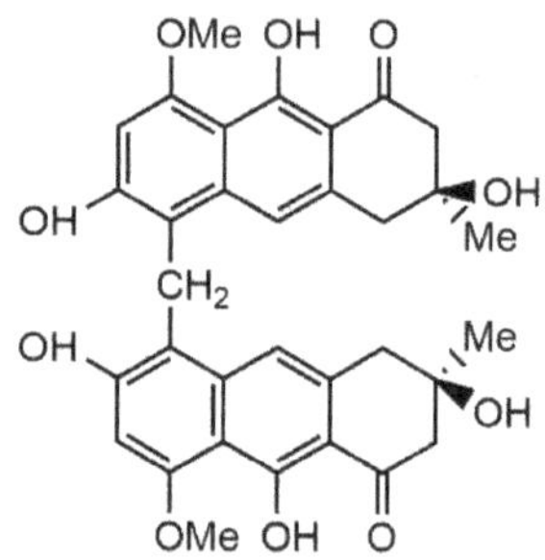

eurorubrin **(193)**

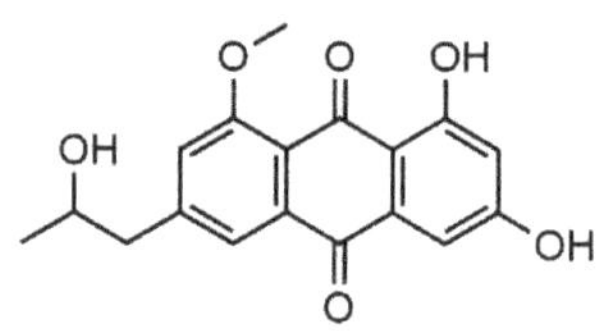

isorhodoptilometrin-1-methyl ether**(194)**

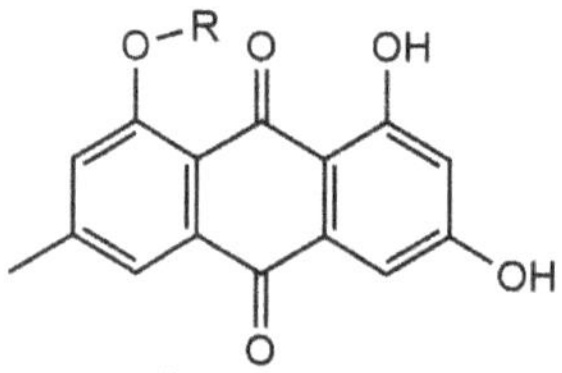

emodin :R=H**(195)**
1-methyl emodin:R=CH_3**(196)**

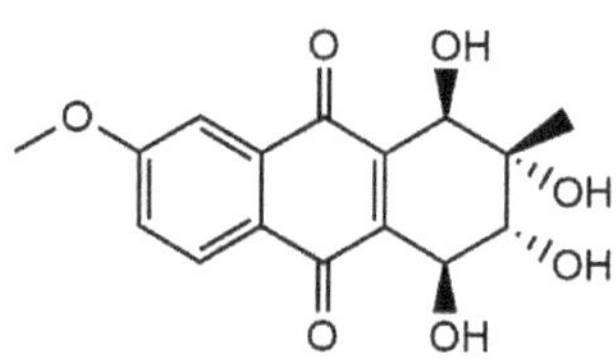

altersolanol A**(197)**

Figure 14. Chemical structure of anthraquinones, quinones, and related glycosides.

Anthraquinones, quinones, and related glycosides are known for their anti-viral and anti-apoptotic activity both in vitro and in vivo. Interestingly, these compounds have been identified and isolated from fungal endophytes by various studies and have similarly shown anti-viral and anti-apoptotic activities. Thus, endophytes that produce these compounds may serve as cheap and environmentally friendly alternative sources for the development of antimicrobial drugs instead to plant sources.

2.6. Steroids

Phomosterols A–B **198–199** (Figure 15) were isolated from the endophytic fungus *Phoma* sp. SYSU-SK-7 of mangrove plants. Compounds **198–199** had an unusual aromatic B ring skeleton and showed significant inhibition activities against LPS-induced NO production in RAW 264.7 cells, with IC_{50} values of 13.5 μM and 25.0 μM, respectively. Additionally, compounds **198–199** showed potent α-glucosidase inhibition activities with IC_{50} values of 51.2 μM and 46.8 μM, respectively, exceeding the positive control 1-deoxynojirimycin (IC_{50} value of 62.8 μM) [105]. The ergosterol derivative fusaristerol A **200** (Figure 15) was obtained from the endophytic fungus *Fusarium* sp., which was isolated from the root of *Mentha longifolia* L. This compound showed significant antimicrobial activity against *Candida albicans*, with an MIC value of 8.3 μg/disc. Additionally, compound **200** showed moderate cytotoxic activity against human colorectal cancer cell line HCT 116, with an IC_{50} value of 0.21 μM, compared to the positive control adriamycin (IC_{50} value of 0.06 μM) [106]. (5,6,15,22E)-6-ethoxy-5,15-dihydroxyergosta-7,22-dien-3-one **201** (Figure 15) and (14,22E)-9,14-dihydroxyergosta-4,7,22-triene-3,6-dione **202** (Figure 15) were isolated from the endophytic fungus *Phomopsis* sp. of *Aconitum carmichaeli* in Yunnan, China. Compounds **201–202** were analyzed against six tested pathogenic fungi (*Candida albicans, Aspergillus niger, Fusarium avenaceum, Pyricularia oryzae, Hormodendrum compactum,* and *Trichophyton gypseum*) using a broth microdilution assay. Compounds **201–202** showed weak antifungal activities against *C. albicans* and *F. avenaceum*, with MIC values ranging from 64 to 128 μg/mL [107].

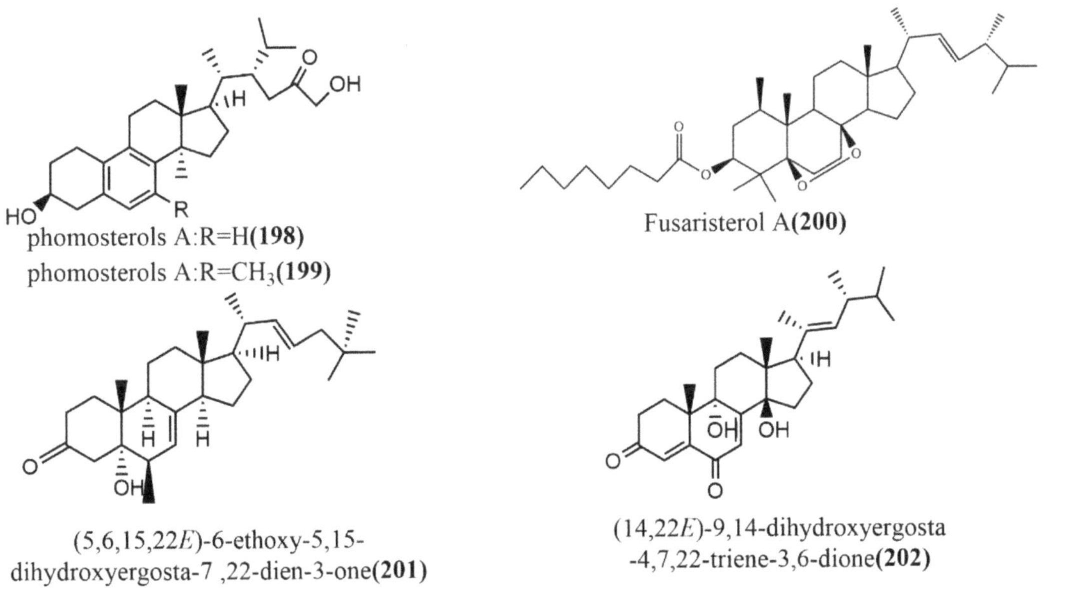

Figure 15. Chemical structures of steroids.

To summarize, endophytic fungi are alternative sources of steroids and their derivatives; thus, they may be harnessed for the production of various drugs since they have shown antimicrobial and anticancer activity in previous studies.

2.7. Other Types of Compounds

Four lignans, terrusnolides A–D **203–206** (Figure 16), were obtained from the endophytic fungus *Aspergillus* sp. isolated from the root of *Tripterygium wilfordii*. Compounds **203–206** showed significant inhibition of LPS-induced IL-1β, TNF-α, and NO production in RAW264.7 cells, with IC_{50} values ranging from 16.21 to 35.23 μM, 19.83 to 42.57 μM, and 16.78 to 38.15 μM, respectively, which were comparable to the positive control indomethacin (IC_{50} value of 15.67–21.34 μM) [108]. The indene derivative methyl 2-(4-hydroxybenzyl)-1,7-dihydroxy-6-(3-methylbut-2-enyl)-1H-indene-carboxylate **207** (Figure 16) obtained from the endophytic fungus *Aspergillus flavipes* Y-62 isolated from *Suaeda glauca* Bunge in Zhoushan, Zhejiang, China, showed weak antimicrobial activities against *Pseudomonas aeruginosa*, *Klebsiella pneumonia*, and *Staphylococcus aureus*, with MIC values ranging from 32 to 128 μg/mL [109]. The polychlorinated triphenyl diether simatorone **208** (Figure 16) was isolated from *Microsphaeropsis* sp. cultures, and its antimicrobial activities against three pathogens (*Escherichia coli*, *Bacillus megaterium*, and *Microbotryum violaceum*) were evaluated using an agar diffusion assay. Compound **208** showed effective antimicrobial activities against *B. megaterium* and *E. coli* with inhibitory zone diameters of 14 mm and 18 mm, respectively [110]. Two alkylated furan derivatives—5-(undeca-3′,5′,7′-trien-1′-yl) furan-2-ol **209** (Figure 16) and 5-(undeca-3′,5′,7′-trien-1′-yl) furan-2-carbonate **210** (Figure 16)—were obtained from the methanol extract of the endophytic fungus *Emericella* sp. XL029 isolated from *Panax notoginseng* leaves in Hebei, China. Compounds **209–210** both showed potent antifungal activities against six tested plant pathogenic fungi (*Rhizoctorzia solani*, *Verticillium dahliae Kleb*, *Helminthosporium maydis*, *Fusarium oxysporum*, *Fusarium tricinctum*, and *Botryosphaeria dothidea*), with MIC values ranging from 25 to 3.1 μg/mL [111]. The new azaphilone, isochromophilone G **211** (Figure 16), was obtained from the endophytic fungus *Diaporthe perseae* sp. isolated from *Pongamia pinnata* (L.) Pierre. Compound **211** showed significant DPPH and ABTS radical scavenging activities, with IC_{50} values of 7.3 μmol/mL and 1.6 μmol/mL, respectively [112]. The furan derivative, 3-(5-oxo-2,5-dihydrofuran-3-yl) propanoic acid **212** (Figure 16), was obtained from the endophytic fungus *Aspergillus tubingensis* DS37 isolated from *Decaisnea insignis* (Griff.) Hook & Thomson, and showed significant inhibition activities against *Fusarium graminearum* and *Streptococcus lactis*, with MIC values of 16 μg/mL and 32 μg/mL, respectively [113]. The pyrrolidinone derivative, nigrosporamide A **213** (Figure 16), was isolated from the endophytic fungus *Nigrospora sphaerica* ZMT05 of *Oxya chinensis* Thunberg and showed a three-fold higher α-glucosidase inhibition activity than the positive control acarbose (IC_{50} value of 446.7 μM) with an IC_{50} value of 120.3 μM. Compound **213** has the potential to be a lead compound for the development of α-glucosidase inhibitors [114]. The production of the terrein derivative asperterrein **214** (Figure 16) was induced by co-culturing endophytic fungi *Aspergillus terreus* EN-539 and *Paecilomyces lilacinus* EN-531 of the marine red alga *Laurencia okamurai*. Compound **214** showed weak antimicrobial activities against *Physalospora piricola* and *Staphylococcus aureus*, with MIC values ranging from 32 to 64 μg/mL. Additionally, compound **214** was not detected in the sterile cultures of the two fungi alone [115]. The endophytic fungus *Lachnum palmae* of *Przewalskia tangutica* was isolated to halogenated dihydroisocoumarins palmaerones A–F **215–220** (Figure 16) under the guidance of UPLC-ESIMS. The antimicrobial activities of all compounds against five tested pathogens (*Cryptococcus neoformans*, *Penicillium sp.*, *Candida albicans*, *Bacillus subtilis*, and *Staphylococcus aureus*) were evaluated using the broth microdilution method. Compounds **215–220** showed potent to weak antimicrobial activities against all tested pathogens, with MIC values ranging from 10 to 55 μg/mL. Additionally, compounds **215** and **219** showed moderate inhibition of LPS-induced NO production in RAW264.7 macrophages, with IC_{50} values of 26.3 μM and 38.7 μM, respectively [116].

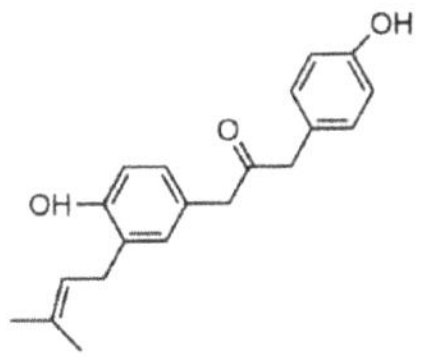
terrusnolides A(**203**)

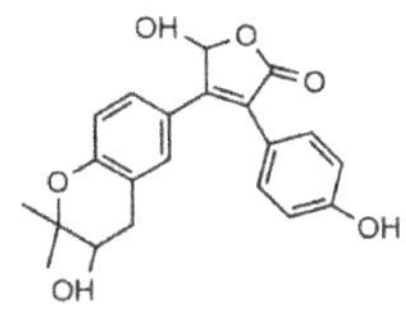
terrusnolides B(**204**)

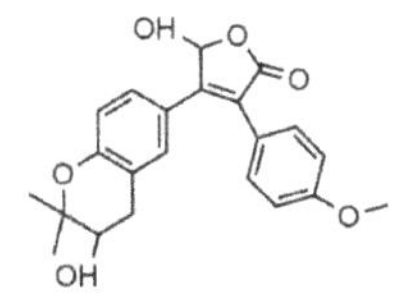
terrusnolides C(**205**)

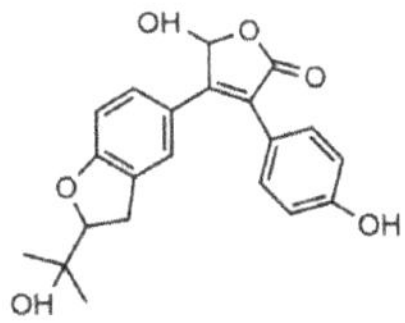
terrusnolides D(**206**)

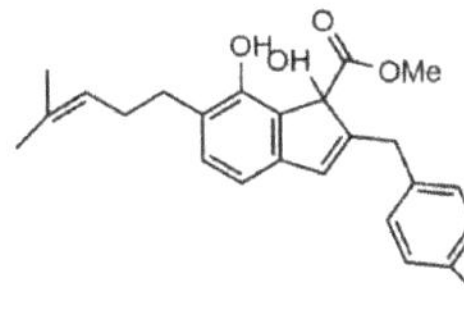
methyl 2-(4-hydroxybenzyl)-1,7-dihydroxy-6-(3-methylbut-2-enyl)-1H-indene-1-carboxylate(**207**)

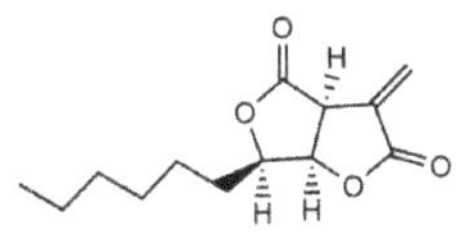
simatorone(**208**)

5-(undeca-3′,5′,7′-trien-1′-yl) furan-2-ol (**209**)

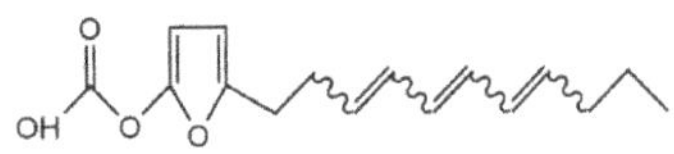
5-(undeca-3′,5′,7′-trien-1′-yl)furan-2-carbonate(**210**)

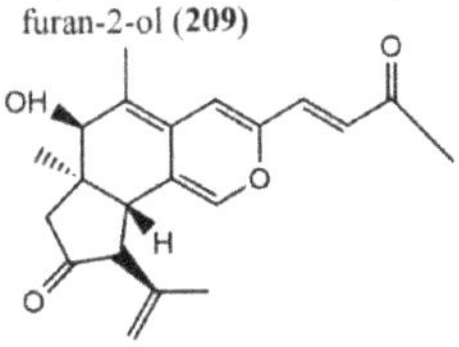
isochromophilone G(**211**)

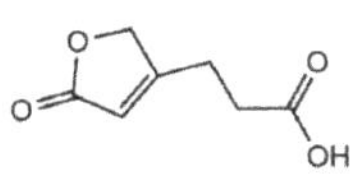
3-(5-oxo-2,5-dihydrofuran-3-yl) propanoic acid(**212**)

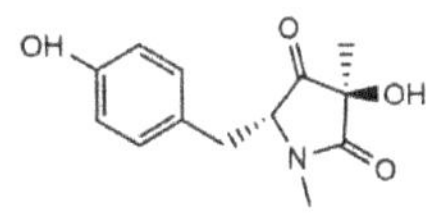
nigrosporamide A(**213**)

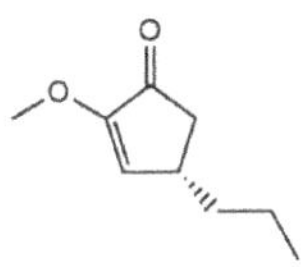
asperterrein (**214**)

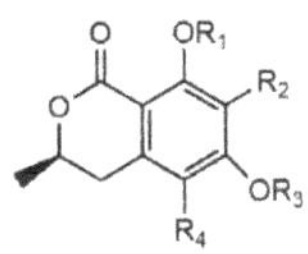
palmaerones A:R1=CH_3,R2=H,R3=H,R4=Br(**215**)
palmaerones B:R1=CH_3,R2=Br,R3=H,R4=H(**216**)
palmaeronesC:R1=CH_3,R2=Br,R3=CH_3,R4=H(**217**)
palmaerones D:R1=H,R2=Br,R3=H,R4=H(**218**)
palmaerones E:R1=CH_3,R2=OH,R3=H,R4=Br(**219**)
palmaerones F:R1=CH_3,R2=H,R3=H,R4=Cl(**220**)

Figure 16. Chemical structures of other new compounds.

Over the past few years, plants have been a major source of numerous compounds that possess biological activities; however, this review revealed that most of these compounds were also produced by various endophytes, especially fungi. Therefore, the isolation and development of these compounds as novel drug candidates would be of great importance to the pharmacological industry since endophytes are easy to manage, keep, and work with compared with plants. Thus, we conclude that endophytic fungi may serve as alternative sources of bioactive compounds of pharmacological interest.

All the information about the new compounds have been summarized below in Table 2.

Table 2. Brief summary of new compounds.

Compound	Molecular Formula	Color and Morphology	Endophytic Fungus	Host Plant	Site and Nation	Pharmacological Activity	Ref.
			Polyketides Chromones				
1	$C_{12}H_{13}O_6$	colorless powder	*Botryosphaeria ramosa* L29	leaf of *Myoporum bontioides*	Leizhou Peninsula, China	Displayed acceptable antimicrobial activities against *Fusarium oxysporum*	[36]
2	$C_{14}H_{15}O_6$	white powder					
3	$C_{11}H_{11}O_5$						
4	$C_{17}H_{19}N_3O_3S_2$	yellow crystals	*Phaeosphaeria fuckelii*	*Phlomis umbrosa*	Mount Hua, China	Mushroom tyrosinase inhibitory activity	[37]
5	$C_{15}H_{16}O_7S$	yellow powder	*Chaetomium seminudum*			Showed antifungal activity (5–6); Exhibited radical scavenging activity against DPPH; Showed significant antioxidant activity ((5)	[38]
6	$C_{16}H_{18}O_7S$						
7	$C_{16}H_{24}O_5$	colorless oil	*Pestalotiopsis fici* W106-1	*Camellia sinensis*	Hangzhou, China	Displayed inhibitory effects on HIV-1 replication in C8166 cells ((7–8); Showed low to moderate cytotoxic activity (9–10); Displayed significant antifungal activity (9)	[39]
8							
9	$C_{32}H_{54}O_6$						
10	$C_{32}H_{54}O_6Na$						
			Polyketides α-pyrones				
11	$C_{11}H1_4O_4$	colorless crystals	*Neurospora udagawae*	shoot of *Quercus macranthera*		Exhibited moderate antifungal (vs. *Rhodoturula glutinis*) activity and cytotoxicity against KB3.1 cells (12)	[40]
12	$C_{10}H_{10}O_4$	colorless oil					
13	$C_{28}H_{42}O_4$	colorless amorphous powder	*Aspergillus niger* MA-132	*Avicennia marina*	Hainan Province, China	Showed potent antifungal and cytotoxic activities	[41]
14							
15	$C_{14}H_{22}O_5$	yellow oil	*Pestalotiopsis fici*	branches of *Camellia sinensis* (Theaceae)	Hangzhou, China	Displayed significant antifungal against *Gibberella zeae*	[42]
17	$C_{10}H_{14}O_4$	yellow oil	*Aspergillus oryzae*	*Paris polyphylla* var. *yunnanensis*	Dali, Yunnan Province, China	The biological activities of compounds 17–18 were not tested	[43]
18							
19	$C_{11}H_{16}O_4$	yellow gum	*Penicillium herque*	*Cordyceps sinensis*	Xiahe, China	Weak cytotoxic activity	[44]
20	$C_{12}H_{16}O_5$						
21							
			Polyketides: Other polyketides				

Table 2. *Cont.*

Polyketides: Other polyketides							
22	$C_{22}H_{35}ClO_7$	white powder	*Phoma* sp. NTOU4195	*Pterocladiella capillacea*	Taiwan, China	Showed potent anti-angiogenic activity (22); Exhibited inhibition of nitric oxide production in lipopolysaccharide (LPS)-stimulated RAW264.7 macrophage cells (24)	[45]
23							
24							
25							
26							
27	$C_{22}H_{34}O_6$						
28	$C_{22}H_{26}NO_8$						
29	$C_{24}H_{35}NO_5$	colorless oil	*Simplicillium subtropicum* SPC3	fresh bark of *Duguetia staudtii*	Cameroon	Weak cytotoxic activity	[46]
30	$C_{24}H_{35}NO_6$						
31	$C_{12}H_{20}O_3$	colorless oil	*Cladosporium cladosporioides* MA-299	leaves of the mangrove plant *Bruguiera gymnorrhiza*	Hainan Island, China	Showed potent antimicrobial((vs. *Escherichia coli* and *Staphylococcus aureus*) activity and moderate inhibition activity against acetylcholinesterase (33)	[47]
32							
33	$C_{12}H_{22}O_4$	pale yellow powder					
34	$C_{14}H_{24}O_5$	pale yellow oil					
35	$C_{12}H_{20}O_4$	colorless crystals					
36	$C_{14}H_{20}O_5$	colorless powder	*Aspergillus fumigatiaffnis*	*Tribulus terestris*		Weak antimicrobial activities	[48]
37	$C_{14}H_{12}O_6Na$	white amorphous solid	*Alternaria alternata* ZHJG5	leaf of *Cercis hinensis*	Nanjing, China	Exhibited potent antimicrobial activity; Showed significant protective effect against the bacterial blight of rice (37)	[49]
38	$C_{14}H_{12}O_5Na$	white powder					
39	$C_{14}H_{12}O_5Na$						
40	$C_{29}H_{22}O_{12}Na$						
41	$C_{24}H_{27}NO_5$	brown solid	*Peyronellaea* sp. FT431	healthy leaf of a Hawaiian indigenous plant, *Verbena* sp.	Lyon, France	Showed weak to moderate cytotoxic activity (41–42)	[50]
42	$C_{24}H_{26}O_7$						
43	$C_{24}H_{26}O_7$						
44	$C_{18}H_{20}O_5$						
45	$C_{29}H_{22}O_9$	red wine colored lump crystal	*Alternaria* sp. MG1	*Vitis quinquangularis*		Showed weak cytotoxicity	[51]
Alkaloids Cytochalasin							
46	$C_{22}H_{32}N_2O_5$	white amorphous solid	*Phomopsis* sp. sh917	Fresh stems of *I. eriocalyx* var. *laxiflora*	Kunming, China	Significant inhibitory activity against NO production in LPS-induced RAW264.7 cells (46)	[52]
47	$C_{22}H_{33}NO_4$						
48							

Table 2. *Cont.*

49	$C_{32}H_{40}N_2O_6$	colorless amorphous powder	*Chaetomium globosum* TW1-1	*Armadillidium vulgare*	Hubei Province, China	Showed potential cytotoxic activities against cancer cell lines (HL-60, A-549, SMMC-7721, MCF-7, and SW-480)	[53]
50	$C_{32}H_{38}N_2O_6$						
51							
52	$C_{32}H_{40}N_2O_6$						
53	$C_{32}H_{38}N_2O_5$	white amorphous powder					
54	$C_{32}H_{83}N_2O_6$						
55	$C_{32}H_{36}N_2O_4$						
56		colorless amorphous powder					
57	$C_{34}H_{42}N_2O_7Na$						
58	$C_{28}H_{37}NO_3$	white amorphous solid	*Diaporthe* sp. SC-J0138	*Cyclosorus parasiticus* (Thelypteridaceae)	Guangdong Province, China	Showed significant cytotoxic activities against four human cancer cell lines (A549, HeLa, HepG2, and MCF-7) (58); Exhibited selective cytotoxic activity (59–62)	[54]
59	$C_{28}H_{37}NO$						
60							
61	$C_{28}H_{37}NO_4$						
62							
63	$C_{25}H_{37}NO_4$	colorless crystal	*Cytospora chrysosperma* HYQZ-931	*Hippophae rhamnoides*		Exhibited significant antibacterial activity (63,65)	[55]
64		white amorphous powder					
65	$C_{26}H_{41}NO_5$						
Alkaloids Indole alkaloids							
66	$C_{26}H_{28}N_3O_4$	brilliant yellowish powder	*Aspergillus* sp. YJ191021		Zhejiang Province, China	Exhibited moderate antibacterial activity (66); Displayed notable anti-inflammatory; Exhibited notable cytotoxicity (66–69)	[56]
67	$C_{26}H_{29}N_3O_5$	white powder					
68	$C_{27}H_{31}N_3O_6Na$						
69	$C_{27}H_{31}N_3O_6$						
70	$C_{28}H_{31}N_3O_6$						
71	$C_{26}H_{31}N_3O_6$						
72	$C_{30}H_{33}NO_7$	white amorphous powder	*Penicillium chrysogenum* XNM-12	*Leathesia nana*	Shandong Province, China	Exhibited moderate antibacterial effects against *Ralstonia solanacearum*	[57]
73	$C_{23}H_{38}N_1NaO_3$	amorphous powder	*Hypomontagnella monticulosa* Zg15SU	fresh rhizome of *Zingiber griffithii* Baker	Indonesia	Showed potent cytotoxic activity	[58]

Table 2. *Cont.*

74	$C_{20}H_{22}N_2NaO_4$	yellowish powder	*Aspergillus lentulus* DTO 327G5	*Caenagrion*	Shanghai, China	Displayed weak to moderate antibacterial activity	[59]
75	$C_{19}H_{21}O_4N_2$	white powder					
76	$C_{24}H_{25}N_3NaO_3$						
			Alkaloids Diketopiperazine derivatives				
77	$C_{20}H_{27}N_3O_3S_2Na$	white solid powder	*Phaeosphaeria fuckelii*	*Phlomis umbrosa*	Mount Hua, China	Showed strong inhibitory effects on mushroom tyrosinase	[60]
78	$C_{26}H_{29}N_3O_5$	colorless oil	*Paecilomyces variotii* EN-291	*Grateloupia turuturu*	Qingdao Province, China	Exhibited potent antifungal effects	[61]
79	$C_{22}H_{23}N_3O_4$						
80	$C_{21}H_{25}O_3N_3$	yellow powder	*Aspergillus* sp. 16-5c	leaf of *S. apetala*	Hainan Island, China	Showed potent to moderate α-glucosidase inhibitory activity (80–81)	[62]
81	$C_{21}H_{23}O_4N_3$	white powder					
82	$C_{21}H_{23}O_3N_3$	yellow powder					
83	$C_{21}H_{25}O_3N_3$						
84	$C_{22}H_{27}O_3N_3$						
85	$C_{18}H_{15}O_4N_3$	white powder					
86	$C_{19}H_{24}N_2O_6S$	colorless crystals	*Penicillium brocae* MA-231	*Avicennia marina*		Displayed moderate to high activities against *Staphylococcus aureus*	[63]
87	$C_{19}H_{22}N_2O_5S$	yellowish solid					
88	$C_{20}H_{26}N_2O_6S_2$	colorless crystals					
89	$C_{20}H_{26}N_2O_6S_2$	colorless solid					
90	$C_{20}H_{24}N_2O_6S_2$						
91	$C_{19}H_{18}N_2O_4S$	colorless crystals	*Penicillium brocae* MA-231			Showed moderate antimicrobial activities against *S. aureus* and *Aeromonas hydrophilia*	[64]
92	$C_{19}H_{16}N_2O_4S$						
93	$C_{18}H_{14}N_2O_4$						
			Alkaloids: Other types of alkaloids				
94	$C_{21}H_{16}N_4O_2$	colorless needles	*Penicillium vinaceum* (X17)	corm of *Crocus sativus*	Shanghai, China	Showed weak cytotoxic activities against three human tumor cell lines (A549, LOVO, and MCF-7)	[65]
95	$C_{14}H_{16}Br_2N_2O_5$	colorless amorphous powder	*Acanthodendrilla* sp.		Thailand	Exhibited efficient and selective cytotoxic activities against two human tumor cell lines (H292 and HaCaT)	[66]
96							

Table 2. *Cont.*

97	$C_{19}H_{20}NO3$	colorless needles crystal	*Penicillium sumatrense* GZWMJZ-313	leaf of *Garcinia multiflora*	Guizhou, China	Showed moderate to weak antimicrobial activities against *Staphylococcus aureus*, *Pseudomonas aeruginosa*, and *Escherichia coli*	[67]
98	$C_{20}H_{24}NO_3$	white powder					
99	$C_{20}H_{21}NO_3$						
100	$C_{25}H_{29}O_5N$	white powder	*Diaporthe* sp. SYSUHQ3	fresh branch of the mangrove plant *Excoecaria agallocha*		Showed potent inhibition activity against *Mycobacterium tuberculosis* protein-tyrosine phosphatase B	[68]
101	$C_{25}H_{29}O_5N$						
			Terpenoids Sesquiterpenoids and their derivatives				
102	$C_{15}H_{22}O_3Na$	colorless oil	*Rhizopycnis vagum Nitaf22*	*Nicotiana tabacum*		Exhibited high selective cytotoxicity against NCI-H1650 and BGC823 cell lines (115); Showed strong phytotoxic activities against radicle growth in rice seedlings (106–107, 113–114)	[70]
103	$C_{15}H_{23}O_3$						
104	$C_{15}H_{22}O_4Na$	colorless amorphous solid					
105	$C_{15}H_{20}O_3Na$	colorless oil					
106	$C_{15}H_{24}O_3Na$						
107	$C_{15}H_{22}NaO_3$						
108	$C_{15}H_{21}O_4$						
109	$C_{15}H_{25}O_4$						
110	$C_{15}H_{22}O_3Na$						
111	$C_{15}H_{19}O_5$	colorless amorphous solid					
112	$C_{15}H_{13}O_5$	brown amorphous solid	*Rhizopycnis vagum Nitaf22* *Rhizopycnis vagum Nitaf22*	*Nicotiana tabacum* *Nicotiana tabacum*		Exhibited high selective cytotoxicity against NCI-H1650 and BGC823 cell lines (115); Showed strong phytotoxic activities against radicle growth in rice seedlings (106–107, 113–114)	[70]
113	$C_{15}H_{14}O_4Na$	yellowish oil					
114	$C_{17}H_{18}NO_4$	greenish-yellow amorphous solid					
115	$C_{14}H_{15}NO_4$	light-yellowish amorphous solid					
116	$C_{15}H_{22}O_3$	colorless oil	*Trichoderma* sp. PR-35	healthy stem of *Paeonia delavayi*	Yunnan Province, China	Showed moderate to weak antimicrobial activities against *Escherichia coli* and *Shigella sonnei*	[69]
117	$C_{15}H_{26}O_2$						
118	$C_{15}H_{28}O_3Na$	colorless oil	*Fusarium proliferatum* AF-04		Lanzhou, China	Displayed weak antimicrobial against *Bacillus subtilis*, *Clostridium perfringens*, *E. coli*, and MRSA	[71]
119	$C_{15}H_{26}O_2Na$						
120	$C_{15} H_{20} O_5$	colorless oil	*Aspergillus sydowii* EN-434	*Symphyocladialatiuscula*	Qingdao Province, China	Exhibited radical scavenging activity against DPPH	[72]

Table 2. *Cont.*

121	$C_{25}H_{37}O_4$	amorphous powder	*Ulocladium* sp.		Yunnan Province, China	Showed moderate antimicrobial activities against *B. subtilis* and multi-drug-resistant *S. aureus* (121–125); Exhibited high selective cytotoxicity against the HepG2 cell line (125)	[73]
122							
123	$C_{26}H_{38}O_4Na$						
124	$C_{26}H_{40}O_5Na$						
125	$C_{25}H_{34}O_3Na$						
126	$C_{15}H_{26}O_3$	colorless crystals	*Trichoderma virens* QA-8	fresh inner root tissue of the grown medicinal herb *Artemisia argyi* H. Lév. and Vaniot	Hubei Province, China	Showed significant antimicrobial activities against *E. coli*	[74]
127		amorphous powder					
128		colorless oil					
129	$C_{15}H_{24}O_2$						
130	$C_{15}H_{26}O_3$	colorless waxy solid					
			Terpenoids Diterpenoids				
131	$C_{20}H_{32}O_6Na$	colorless oil	*Diaporthe* sp. QYM12	healthy leaves of *Kandelia candel*	Hainan Province, China	Showed significant anti-inflammatory effects through the inhibition of NO production	[75]
132	$C_{21}H_{28}O_6$	colorless crystals	*Phomopsis* sp. S12	seed of *Illigera rhodantha*		Showed excellent inhibitory effects on the production of IL-1β and IL-18; Effects on the NF-κB signaling pathway	[76]
133	$C_{20}H_{26}O_5$	colorless needle crystal	*Phomopsis* sp. S12	seed of *Illigera rhodantha*		Exhibited anti-inflammatory activity against the production of IL-1b and IL-6 induced by lipopolysaccharide (LPS) in macrophages	[77]
134	$C_{20}H_{28}O_4$	colorless oil					
135	$C_{20}H_{26}O_6$						
136	$C_{16}H_{20}O_5$	colorless needles	*Botryosphaeria* sp. P483	Chinese Herbal Medicine *Huperzia serrata*	Kunming, China	Showed effective antifungal antifungal activities against *Gaeumannomyces graminis*, *Fusarium solani*, and *Pyricularia oryza* (136); Showed weak nematicidal activities	[78]
137	$C_{16}H_{20}O_6$	colorless solid					
138	$C_{20}H_{28}O_6$	white amorphous solid	*Smardaea* sp. AZ0432	photosynthetic tissue of the moss *Ceratodon purpureus*	Chiricahua Mountains of southeastern Arizona, USA	Exhibited selective cytotoxicity	[79]
139	$C_{20}H_{16}O_5$	yellowish needles	*Pestalotiopsisadusta*	Fresh, healthy stems of *Clerodendrum canescens*	Yandang, Zhejiang Province, China	Demonstrated cytotoxic activities against the HL-60 tumor cell line	[80]
140	$C_{20}H_{34}O_2$	colorless needles	*Trichoderma atroviride* S361	*Cephalotaxus fortunei*	Jiande, Zhejiang, China	Bioactivity tests were not performed	[81]

Table 2. *Cont.*

141	$C_{16}H_{16}O_5$	colorless needles	*Aspergillus wentii* EN-48	unidentified marine brown algal species of the genus *Sargassum*	Qingdao Province, China	Showed moderate cytotoxic activities against seven human tumor cell lines (NCI-H460, MDA-MB-231, HeLa, MCF-7, SMMC-7721, HepG2, and SW1990)	[82]
142	$C_{16}H_{16}O_5$						
143	$C_{16}H_{24}O_5$						
Terpenoids Triterpenoids							
144	$C_{30}H_{48}O$	colorless solid	*Scleroderma* UFSMSc1	*Eucalyptus grandis*		Showed moderate to weak antifungal activities against *Candida albicans* and *Candida parapsolosis*	[84]
145							
146							
147	$C_{29}H_{46}O_5$	white powder	*Acremonium pilosum* F47	pedicel of the Chinese medicinal plant *Mahonia fortunei*	Qingdao Province, China	Displayed effective antimicrobial activities against *S. aureus* and *B. subtili*	[85]
148	$C_{30}H_{50}O_6$	yellow amorphous powder	*Glomerella* sp. F00244	stem of mason pine	Fujian Province, China	Showed weak cytotoxic activity (148)	[83]
149	$C_{31}H_{52}O_6$	white amorphous powder					
150	$C_{30}H_{40}O_6$	yellowish needle crystals	*Penicillium* sp. SWUKD4.1850	healthy branches of *Kadsura angustifolia*	Yunnan Province, China	Exhibited moderate in vitro cytotoxic activities	[86]
151	$C_{30}H_{40}O_6$	white needle crystals					
152	$C_{30}H_{40}O_6$	white amorphous solid					
153	$C_{30}H_{41}O_6$						
154	$C_{32}H_{44}O_7$	white amorphous powder					
155	$C_{30}H_{42}O_6$	white powder					
156	$C_{34}H_{46}O_8$	yellow amorphous solid					
157	$C_{31}H_{44}O_6$						
158	$C_{30}H_{46}O_6$	white amorphous powder					
159	$C_{32}H_{50}O_5$	white amorphous powder	*Hypoxylon* sp. 6269	*Artemisia annua*		Weak inhibition activity against the HIV-1 integrase (159)	[87]
160	$C_{29}H_{44}O_4$						
161	$C_{36}H_{55}O_7$	white amorphous powder	*Fusarium* sp.	roots of *Mentha longifolia*	Saudi Arabia	Showed significant antileishmanial activity (161)	[88]
162	$C_{32}H_{51}O^5$						
163	$C_{39}H_{55}O_7$						
164	$C_{34}H_{53}O_6$	colorless powder	*Fusarium* sp.	roots of *Mentha longifolia*	Saudi Arabia	Displayed potent cytotoxic activity towards BT-549 and SKOV-3; Showed potent antileishmanial activities against *L. donovani* promastigotes	[89]
165	$C_{42}H_{68}O_7$	white amorphous powder					

Table 2. *Cont.*

Terpenoids Meroterpenoids							
166	$C_{18}H_{26}O_5$	colorless crystal	*Guignardia mangiferae* A348	Medicinal Plant *Smilax glabra*	Luofu Mountain Natural Reservation, Guangdong Province, China	Showed weak cytotoxic activities against MCF-7 cell lines(167,169)	[90]
167	$C_{17}H_{22}O_4$						
168	$C_{18}H_{28}O_5$	white powder					
169	$C_{17}H_{24}O_4$						
170	$C_{25}H_{30}O_5$	colorless amorphous powder	*Emericella* sp. TJ29	root of the plant *Hypericum perforatum*	the Shennongjia areas of Hubei Province, China	Showed moderate cytotoxic activities against five human tumor cell lines (HL-60, SMMC7721, A549, MCF-7, and SW-480) (172, 173, 175)	[91]
171	$C_{27}H_{34}O_6$	white powder					
172	$C_{26}H_{32}O_6$	colorless crystals					
173	$C_{25}H_{30}O_6$	colorless crystals					
174	$C_{25}H_{32}O_6$	white powder					
175	$C_{25}H_{28}O_6$	colorless crystals					
176	$C_{31}H_{35}O_9$	yellowish oil	*Phyllosticta* sp. J13-2-12Y	leaf of *Acorus tatarinowii*	Guangxi Province, China	Exhibited moderate antimicrobial activities against *Staphylococcus aureus* 209P, *Candida aureus* 209P, and *Candida albicans* FIM709	[92]
177	$C_{31}H_{37}O_9$						
178	$C_{31}H_{34}O_9$						
179	$C_{27}H_{32}O_9$	white powder	Co-culture *Talaromyces purpurogenus* H4 and *Phanerochaete* sp. H2	*Handroanthus impetiginosus*	Alfenas, Minas Gerais, Brazil.	Showed moderate trypanocidal activity against *T. cruzi*	[93]
Lactones							
180	$C_{13}H_{22}O_3$	colorless gum	*Talaromyces assiutensis* JTY2	leaf of *Ceriops tagal*	South China Sea, China	Showed moderate cytotoxic activities against three human cancer cell lines (HeLa, MCF-7, and A549)	[94]
181	$C_{21}H_{34}O_5$	yellow oil	*Penicillum polonicum*	fruits of *Camptotheca acuminata* Decne	Wuhan, China	Showed effective glucose uptake activity on rat skeletal muscle myoblast L6 (181); Significantly promoted GLUT4 translocation in L6 cells	[95]
182	$C_{16}H_{28}O_5$	light red oil					
183	$C_{24}H_{33}NO_8$	colorless crystal	*Chaetomium cupreum*	*Anemopsis californica*	New Mexico, U.S.A.	Showed weak antimicrobial activity against *S. aureus*	[96]
184			*Xylaria curta* 92092022		Taiwan, China	Showed moderate antimicrobial activities against *Pseudomonas aeruginosa* and *Staphylococcus aureus*; Displayed strongly inhibited lettuce seed germination	[97]

Table 2. *Cont.*

185	$C_{24}H_{34}O_5$	white powder	*Lasiodiplodia theobromae* ZJ-HQ1	healthy leaves of the marine mangrove *Acanthus ilicifolius*	South China Sea, China	Exhibited inhibitory effects on lipopolysaccharide-induced nitric oxide production in RAW 264.7 macrophage cells; Showed moderate inhibitory activity against α-glucosidase	[98]
186	$C_8H_{10}O_3$		*Aspergillus pseudonomiae* J1	*Euphorbia umbellata* (Pax) *Bruyns* (Euphorbiaceae)	Bahia, Brazil	Showed moderate to weak anti-trypanosomal activity	[99]
187	$C_{10}H_{12}O_4$						
188	$C_{10}H_{12}O_3$						
Anthraquinones, quinones, and related glycosides							
189	$C_{20}H_{19}O_7$	Brilliant yellowish oil	*Acremonium vitellinum*	*Acanthus ilicifolius Linn*	Qingdao Province, China	Showed moderate insecticidal activities against the third-instar larvae of Helicoverpa ar-migera	[100]
190	$C_{20}H_{16}O_7$	yellow solid					
191	$C_{22}H_{21}O_7$						
192	$C_{18}H_{18}O_8Na$	yellow amorphous powder	*Phomopsis* sp.	*Paris polyphyllavar*	Yunnan Province, China	Showed significant cytotoxic activities against A549 and PC3 cell lines	[101]
193	$C_{21}H_{20}O_{10}$	red amorphous powder	*Eurotium cristatum* EN-220	*Sargassum thunbergii*	Qingdao Province, China	Showed weak antimicrobial activity against *E. coli* only; Showed moderate fatal activity against brine shrimp larvae	[102]
194	$C_{18}H_{15}O_6$	orange yellow powder	*Aspergillus versicolor*	*Halimeda opuntia*	South Sinai, Egypt	Weak inhibitory activity against hepatitis C virus NS3/4A protease	[103]
195	$C_{12}H_{11}O_4$	red powder					
196	$C_{16}H_{11}O_5$	orange powder					
197	$C_{16}H_{21}O_7$	red powder	*Stemphylium globuliferum*	healthy stems of *Mentha pulegium*	Beni Mellal, Morocco	Showed significant inhibition of proliferation of K562 and A549 cells	[104]
Steroids							
198	$C_{27}H_{40}O_3$	white crystals	*Phoma* sp. SYSU-SK-7		Guangdong Province, China	Exhibited inhibitory effects on lipopolysaccharide-induced nitric oxide production in RAW 264.7 macrophage cells; Showed moderate inhibitory activity against α-glucosidase	[105]
199	$C_{28}H_{41}O_3$	white solid					
200	$C_{38}H_{64}O_4$	white amorphous powder	*Fusarium* sp.	*Mentha longifolia*	Egypt	Showed moderate cytotoxic activity against human colorectal cancer cell line HCT 116	[106]
201	$C_{28}H_{40}O_4$		*Phomopsis* sp.	*Aconitum carmichaeli*	Huize County, Yunnan Province, China	Showed weak antifungal activities against *C. albicans* and *F. avenaceum*	[107]
202	$C_{30}H_{48}O_4$						

Table 2. *Cont.*

Other types of compounds							
203	$C_{20}H_{22}O_3$	yellow oil	*Aspergillus* sp.	root of *Tripterygium wilfordii*	Wuhan, China	Showed significant inhibition of LPS-induced IL-1β, TNF-α, and NO production in RAW264.7 cells	[108]
204	$C_{24}H_{26}O_6$						
205	$C_{24}H_{26}O_6$	colorless oil					
206	$C_{23}H_{24}O_6$						
207	$C_{23}H_{24}O_5$	brown powder	*Aspergillus flavipes* Y-62	stems of plant *Suaeda glauca* Bunge	Zhoushan coast, Zhejiang province, China	Exhibited antimicrobial activities against the Gram-negative pathogens *Pseudomonas aeruginosa* and *Klebsiella pneumoniae*	[109]
208	$C_{23}H_{21}O_5$	white powder	*Microsphaeropsis* sp.			Showed effective antimicrobial activities against *B. megaterium* and *E. coli*	[110]
209	$C_{15}H_{20}O_2$	brown amorphous powder	*Emericella* sp. XL029	leaf of *Panax notoginseng*	Shijiazhuang, Hebei province, China	Showed potent antifungal activities against six tested plant pathogenic fungi (*Rhi-zoctorzia solani*, *Verticillium dahliae* Kleb, *Helminthosporium maydis*, *Fusarium oxysporum*, *Fusarium tricinctum*, and *Botryosphaeria dothidea*)	[111]
210	$C_{16}H_{20}O_4$						
211	$C_{18}H_{18}O_6Cl$	yellow powder	*Diaporthe* perseae sp.	stem of Chinese mangrove *Pongamia pinnata*	Hainan city, China	Showed significant DPPH and ABTS radical scavenging activities	[112]
212	$C_7H_7O_4$	colorless flake crystal	*Aspergillus tubingensis* DS37	*Decaisnea insignis* (Griff.) Hook. f. and Thomson		Showed significant inhibition activities against *Fusarium graminearum* and *Streptococcus lactis*	[113]
213	$C_{13}H_{15}NO_4Na$	amorphous powder	*Nigrospora sphaerica* ZMT05	*Oxya chinensisThunber*	Guangdong Province, China.	Showed significant α-glucosidase inhibitory activity	[114]
214	$C_9H_{14}O_2$	colorless oil	Co-culture *Aspergillus terreus* EN-539 & *Paecilomyces lilacinus* EN-531	*Laurencia okamurai*	Qingdao, China	Showed weak antimicrobial activities against *Physalospora piricola* and *Staphylococcus aureus*	[115]
215	$C_{11}H_{11}BrO_4$	white amorphous powder	*Lachnum palmae*	*Przewalskia tangutica*		Exhibited potent to weak antimicrobial activities against *Cryptococcus neoformans*, *Penicillium sp.*, *Candida albicans*, *Bacillus subtilis, and Staphy-lococcus aureus* (215–220); Showed moderate inhibitory effects on NO production in LPS-induced RAW 264.7 cells (215,219)	[116]
216							
217	$C_{12}H_{13}BrO_4$						
218	$C_{10}H_9BrO_4$						
219	$C_{11}H_{11}BrO_5$						
220	$C_{11}H_{11}ClO_4$						

3. Future Prospects and Challenges of Using Endophytic Fungi as an Alternative Source of Plant Bioactive Compounds

Endophytic fungi are hidden and subtle dwellers in several plant tissues and intercellular spaces and can produce diverse chemical structures and efficient, low-toxic new secondary metabolites that were initially thought to be produced by the host plants. The current reports on the biosynthesis of plant metabolites by endophytic fungi, in conjunction with recent research advances in fermentation culture, extraction, isolation, and structure identification techniques, permit us to rapidly uncover new valuable compounds. Generally, fungi are chemically diverse, easily cultured, and biologically active modalities that have great flexibility to be regulated by adding precursors, elicitors, and specific enzymes to effectively increase the quantity and yield of bioactive compounds. Table 3 represents the culture conditions and specific bioactive secondary metabolites and yields produced by various endophytic fungi. Endophytic fungi can convert active compounds of the host plant into more potent derivatives. This makes endophytic fungi an alternative and sustainable source of plant bioactive compounds [117,118]. The search for new compounds in endophytic fungi requires specific theories and ingenious bioprospecting strategies. Along with the continuously growing literature reports, the most promising host plants can be selected. It includes the selection of (A) plants from special habitats or growing in biodiversity-rich areas, including mangrove plants in tropical marine intertidal zones, and (B) medicinal and indigenous plants with ethnopharmacological uses, including *Camptotheca acuminata* and *Ageratina adenophora*. These selection criteria provide a reference for the current and future screening of host plants for endophytic fungi with new bioactive compounds [119,120]. This review has summarized 220 new compounds obtained between 2011 and 2021 from endophytic fungi using different culture methods, including the common culture, co-culture with bacteria or other fungi, and the addition of metal ions. These new compounds have unique molecular structures, and these rare structures allow these compounds to possess diverse biological activities, including significant antimicrobial and cytotoxic activities and α-glucosidase inhibition. These compounds have the potential to be modified as pro-drug molecules or directly developed as drugs for treating certain diseases. However, most of the current studies on the activity of new compounds with endophytic fungal sources are limited to in vitro studies; therefore, animal experiments and human intervention clinical trials are needed to further investigate the in vivo activities and mechanisms of action of the new compounds.

Unfortunately, endophytic fungi as new sources of bioactive secondary metabolites encounter various limitations, including the attenuated yield of secondary metabolites due to long-term storage and repeated passages under laboratory culture conditions, silencing of biosynthetic gene clusters or low level of expression (activation of gene clusters depends on environmental factors). Thus, the ability of endophytic fungi to produce new compounds of interest has been underestimated [129]. The expression could be upregulated by physicochemical and genetic manipulation techniques to increase the production of specific metabolites in endophytic fungi and to produce analogs of new active secondary metabolites. Methods including the OSMAC strategy (activation of silent biosynthetic gene clusters mediated by changes in medium composition, temperature, and aeration efficiency to produce desired metabolites), co-culture (mimicking natural ecosystems and triggering silent gene clusters to promote metabolite secretion and enhance bioactive metabolite production by microbial interaction-induced stress responses), and chemical epigenetic modification methods have been used to isolate new compounds. It was found that the addition of micromolar or even nanomolar small-molecule chemicals to cultures inhibits or activates relevant enzymes and remodels the fungal epigenome to increase the diversity of its secondary metabolites, including DNA methyltransferases (DNMTs) and histone deacetylase inhibitors (HDACs) [130,131]. The addition of epigenetic modifiers (5 μM SAHA and 10 μM AZA) to the endophytic fungus *Xylaria psidii* isolated from leaves of *Vitis vinifera* showed elevated resveratrol concentrations of 52.32 μg/mL and 48.94 μg/mL, respectively, by HPLC analysis (control concentration was 35.43 μg/mL). The treatments with 5 μM SAHA and 10 μM AZA showed stronger antioxidant activity with 30.92% and 33.82% DPPH radical scavenging, respectively, compared to the wild strain (19.26%) [132]. Unlike the chemical epigenetic modification methods reported, introducing exogenous substances as precursors into the cultures, including methyl jasmonate, causes the production of new compounds containing their structural units [133]. However, the addition of host plant components to the culture to induce the production of new compounds has rarely been reported. Additionally, it is necessary to elucidate the pathways by which endophytic fungi biosynthesize secondary metabolites, including the enzymes and genes involved via "omics" techniques—genomics, transcriptomics, and metabolomics—in regulating and manipulating the biosynthetic process to increase the number of new compounds [134].

Table 3. Culture conditions and yields of bioactive secondary metabolites produced by endophytic fungi.

No.	Endophytic Fungus	Host Plant	Culture Conditions	Secondary Metabolites	Yield	Ref.
1	*Hansfordia biophila*	*Hedychium acuminatum* Roscoe	Inoculated in potato glucose broth (PDB) medium and shaken at 120 rpm at 25 °C for 7 days.	Tannin	41.6 $\mu m \cdot mL^{-1}$	[121]
2	*Aspergillus terreus*	*Ficus elastica*	Inoculated into PDB medium and incubated at 30 °C for 20 days on a rotatory shaker incubator at 140 rpm.	Camptothecin	320 μg/L	[122]
3	*Guignardia mangiferae* HAA11	*Taxus x media*	Inoculated into (PDB) medium and incubated at 200 rpm at 28 °C for 5 days.	Paclitaxel	720 ng/L	[123]
4	*Papulasora* sp.S6	*Phellodendron amurense* Rupr	Mutagenesis by UV, X-ray rays, and $NaNO_2$, inoculated in PDB medium, and shaken at 100 rpm at 28 °C for 7 days.	Berberine	12.28 mg/L	[124]
5	*Actinoplanes teichomyceticus*		Improvement of the output of teicoplanin by genome shuffling; Inoculated teicoplanin medium and cultured at 28 °C for 15–20 days.	Teicoplanin	3016 $\mu m \cdot mL^{-1}$	[125]
6	*Phialocephala fortinii*	*Podophyllum peltatum*	Inoculated in Sabouraud's dextrose agar (SDA) and cultured at 23 °C for 4–6 weeks.	Podophyllotoxin	189 μg/L	[126]
7	*Entrophospora infrequens* RJMEF001	*Nothapodytes foetida*	Inoculated into wheat bran containing Sabouraud's broth, and incubation was carried out at 28 ± 2 °C for 28 days.	Camptothecin	503 ± 25 μg/100 g dry cell mass (in Sabouraud broth)	[127]
8	*Epicoccum nigrum* SZMC 23769	*Hypericum perforatum*	Fungal isolates were grown in potato dextrose broth (PDB) for 7 days at 25 °C.	Hypericin, Emodin	117.1 μg/mL, 87.7 μg/mL	[128]

4. Conclusions

Pharmaceutical chemists are turning their focus on the development of safe, efficient, and low-toxic new drugs from natural sources. Endophytic fungi may serve as renewable sources of novel bioactive compounds with pharmacological activities, as the number of new compounds to be isolated in the future tends to increase exponentially and rapidly. In addition, numerous studies have also reported that these bioactive compounds isolated from the endophytic fungi are also present in plants and have similar biological activities as the compounds from plant sources. Therefore, we conclude that endophytic fungi may be the best alternative for harnessing pharmacological bioactive compounds for the development of drugs for both human and animal use. Hence, there is a need for the identification of more compounds with pharmacological activity from endophytic fungi and elucidate their mechanisms of action through biological, pharmacodynamic, biochemical, bioinformatics, and pre-clinical approaches.

Author Contributions: J.W. (Juan Wen) and S.K.O. drafted the manuscript; S.K.O., S.W., Y.R. and J.W. (Juan Wen) generated the figures and tables; S.K.O., Y.H., J.W. (Jianchen Wang) and L.X. discussed literatures; J.W. (Juan Wen) designed the work, drafted and revised the manuscript, Y.H. supervised and funded the paper. All authors have read and agreed to the published version of the manuscript.

Funding: This research was supported by the Sichuan Province Science and Technology Support Program (Grant No. 2020YFS0337).

Institutional Review Board Statement: Not applicable.

Informed Consent Statement: Not applicable.

Data Availability Statement: Not applicable.

Conflicts of Interest: The authors declare no conflict of interest.

References

1. Hyde, K.D.; Xu, J.C.; Rapior, S.; Jeewon, R.; Lumyong, S.; Niego, A.G.T.; Abeywickrama, P.D.; Aluthmuhandiram, J.V.S.; Brahamanage, R.S.; Brooks, S. The amazing potential of fungi: 50 ways we can exploit fungi industrially. *Fungal Divers.* **2019**, *97*, 1–136. [CrossRef]
2. Patchett, A.; Newman, J.A. Comparison of Plant Metabolites in Root Exudates of *Lolium perenne* Infected with Different Strains of the Fungal Endophyte *Epichlo festucae* var. *lolii*. *J. Fungi* **2021**, *7*, 148. [CrossRef] [PubMed]
3. Odelade, K.A.; Babalola, O.O. Bacteria, fungi and archaea domains in rhizospheric soil and their effects in enhancing agricultural productivity. *Int. J. Environ. Res. Public Health* **2019**, *16*, 3873. [CrossRef]
4. Igiehon, N.O.; Babalola, O.O.; Cheseto, X.; Torto, B. Effects of rhizobia and arbuscular mycorrhizal fungi on yield, size distribution and fatty acid of soybean seeds grown under drought stress. *Microbiol. Res.* **2021**, *242*, 126640. [CrossRef] [PubMed]
5. Jin, J.; Zhao, Q.; Zhang, X.M.; Li, W.J. Research progress on bioactive products from endophytes. *J. Microbiol.* **2018**, *38*, 103–113. [CrossRef]
6. Deshmukh, S.K.; Gupta, M.K.; Prakash, V.; Saxena, S. Endophytic fungi: A Source of Potential Antifungal Compounds. *J. Fungi.* **2018**, *4*, 77. [CrossRef]
7. Kouipou, R.M.T.; Boyom, F.F. Endophytic fungi from *Terminalia* species: A comprehensive review. *J. Fungi* **2019**, *5*, 43. [CrossRef]
8. Popli, D.; Anila, V.; Subramanyama, A.B.; Namratha, M.N.; Ranjitha, V.R.; Raoa, S.N.; Ravishankar, V.; Govindappaa, M. Endophyte fungi, Cladosporium species-mediated synthesis of silver nanoparticles possessing in vitro antioxidant, anti-diabetic and anti-Alzheimer activity. *Artif. Cell Nanomed. Biotechnol.* **2018**, *46*, 676–683. [CrossRef]
9. Zheng, R.H.; Li, S.J.; Zhang, X.; Zhao, C.Q. Biological activities of some new secondary metabolites isolated from endophytic fungi: A review study. *Int. J. Mol. Sci.* **2021**, *22*, 959. [CrossRef]
10. Rustamova, N.; Bozorov, K.; Efferth, T.; Yili, A. Novel secondary metabolites from endophytic fungi: Synthesis and biological properties. *Phytochem. Rev.* **2020**, *19*, 425–448. [CrossRef]
11. Ding, W.J.; Wang, S.S.; Ren, J.Q.; Li, G.; Zhan, J.P. Progress on plant endophyte. *Curr. Biotechnol.* **2015**, *5*, 425–428. [CrossRef]
12. Jin, L.R.; Yang, L.; Li, W.J.; Xu, D.; Yang, N.N.; Li, G.Q.; Wan, P. Diversity and Biocontrol Potential of Culturable Endophytic Fungi in Cotton. *Front. Microbiol.* **2021**, *12*, 698930. [CrossRef] [PubMed]
13. Wang, S.S.; Liu, J.M.; Sun, J.; Huang, Y.T.; Jin, N.; Li, M.M.; Liang, Y.T.; Fan, B.; Wang, F.Z. Analysis of Endophytic Bacterial Diversity From Different Dendrobium Stems and Discovery of an Endophyte Produced Dendrobine-Type Sesquiterpenoid Alkaloids. *Front. Microbiol.* **2020**, *12*, 775665. [CrossRef]
14. Kharwar, R.N.; Mishra, A.; Stierle, A.; Kharwar, R.N.; Gond, S.K.; Stierle, D. Anticancer compounds derived from fungal endophytes: Their importance and future challenges. *Nat. Prod. Rep.* **2011**, *28*, 1208–1228. [CrossRef] [PubMed]
15. Hawksworth, D.L. The fungal dimension of biodiversity: Magnitude, significance, and conservation. *Mycol. Res.* **1991**, *95*, 641–655. [CrossRef]
16. Wu, B.; Hussain, M.; Zhang, W.W.; Stadler, M.; Liu, X.Z.; Xiang, M.C. Current insights into fungal species diversity and perspective on naming the environmental DNA sequences of fungi. *Mycology* **2019**, *10*, 127–140. [CrossRef] [PubMed]
17. Vogl, A. Mehl und die andiron mehlprodukt der cerealien und leguminosen. *Nahrunsm. Unters. Hug. Waren.* **1898**, *12*, 25–29.
18. Stierle, A.; Strobel, G.; Stierle, D. Taxol and taxane production by *Taxomyces andreanae*, an endophytic fungus of Pacific yew. *Science* **1993**, *260*, 214–216. [CrossRef] [PubMed]
19. Cruz, J.S.; Silva, C.A.; Hamerski, L. Natural Products from Endophytic Fungi associated with *Rubiaceae* Species. *J. Fungi* **2020**, *6*, 128. [CrossRef]
20. Gokhale, M.; Gupta, D.; Gupta, U.; Faraz, R.; Sandhu, S.S. Patents on endophytic fungi. *Curr. Med. Chem.* **2017**, *11*, 120–140. [CrossRef]

21. Oberhofer, M.; Wackerlig, J.; Zehl, M.; Büyük, H.; Cao, J.J.; Urban, E.; Zotchev, S.B. Endophytic *Akanthomyces* sp. LN303 from edelweiss produces emestrin and two new 2-hydroxy-4 pyridone alkaloids. *ACS Omega* **2021**, *6*, 2184–2191. [CrossRef]
22. Mao, Z.L.; Zhang, W.H.; Wu, C.Y.; Feng, H.; Peng, Y.H.; Shahid, H.; Cui, Z.N.; Ding, P.; Shan, T.J. Diversity and antibacterial activity of fungal endophytes from *Eucalyptus exserta*. *BMC Microbiol.* **2021**, *21*, 155. [CrossRef]
23. Pereira, C.B.; Oliveira, D.M.; Hughes, A.F.; Kohlhoff, M.; Vieira1, M.L. Endophytic fungal compounds active against *Cryptococcus neoformans* and *C. gattii*. *J. Antibiot.* **2015**, *68*, 436–444. [CrossRef] [PubMed]
24. Abdou, R.; Shabana, S.; Rateb, M.E. Terezine E, bioactive prenylated tryptophan analogue from an endophyte of *Centaurea stoebe*. *Nat. Prod. Res.* **2020**, *34*, 503–510. [CrossRef]
25. Carvalho, C.R.D.; Vieira, M.D.L.A.; Cantrell, C.L.; Wedge, D.E.; Alves, T.M.; Zani, C.L.; Pimenta, R.S.; Rosaa, L.H. Biological activities of ophiobolin k and 6-epi-ophiobolin k produced by the endophytic fungus *Aspergillus calidoustus*. *Nat. Prod. Res.* **2016**, *30*, 478–481. [CrossRef] [PubMed]
26. Jouda, J.B.; Tamokou, J.; Mbazoa, C.D.; Douala-Meli, C.; Sarkar, P.; Bag, P.K.; Wandji, J. Antibacterial and cytotoxic cytochalasins from the endophytic fungus *Phomopsis* sp. harbored in garcinia kola (heckel) nut. *BMC Complement. Altern. Med.* **2016**, *16*, 462. [CrossRef] [PubMed]
27. Sana, T.; Siddiqui, B.S.; Shahzad, S.; Farooq, A.D.; Siddiqui, F.; Sattar, S.; Begumet, S. Antiproliferative activity and characterization of metabolites of *Aspergillus nidulans*: An endophytic fungus from *Nyctanthes arbor-tristis* linn. against three human cancer cell lines. *Med. Chem.* **2019**, *15*, 352–359. [CrossRef] [PubMed]
28. Sahar, L.; Doustmorad, Z. Antiproliferative and antimicrobial activities of secondary metabolites and phylogenetic study of endophytic *Trichoderma* species from *Vinca* plants. *Front. Microbiol.* **2018**, *9*, 1484. [CrossRef]
29. Dhakshinamoorthy, M.; Ponnusamy, S.K.; Kannaian, U.P.; Srinivasan, B.; Shankar, S.N.; Packiam, K.K. Plant-microbe interactions implicated in the production of camptothecin—An anticancer biometabolite from *Phyllosticta elongata* MH458897 a novel endophytic strain isolated from medicinal plant of Western Ghats of India. *Environ. Res.* **2021**, *201*, 111564. [CrossRef]
30. Han, W.X.; Li, W.Z.; Li, X.F.; Zhang, H.; Yang, S.Z.; Du, J.F.; Jing, H.; Cheng, C.B. Isolation and identification of endophytic fungus producing Huperzine A from *Huperzia serrata*. *Microbiol. China* **2017**, *44*, 2153–2160. [CrossRef]
31. Khan, N.; Afroz, F.; Begum, M.N.; Rony, S.R.; Sharmin, S.; Moni, F.; Hasan, C.M.; Shaha, K.; Sohrab, M.H. Endophytic *Fusarium solani*: A rich source of cytotoxic and antimicrobial napthaquinone and aza-anthraquinone derivatives. *Toxicol. Rep.* **2018**, *5*, 970–976. [CrossRef] [PubMed]
32. Mazlan, N.W.; Tate, R.; Yusoff, Y.M.; Clements, C.; Edrada-Ebel, R. Metabolomics-guided isolation of anti-trypanosomal compounds from endophytic fungi of the mangrove plant *Avicennia lanata*. *Curr. Med. Chem.* **2020**, *27*, 1815–1835. [CrossRef]
33. Okoye, F.B.; Lu, S.; Nworu, C.S.; Esimone, C.O.; Proksch, P.; Chadlic, A.; Debbaba, A. Depsidone and diaryl ether derivatives from the fun gus *Corynespora cassiicola*, an endophyte of *Gongronema latifolium*. *Tetrahedron Lett.* **2013**, *54*, 4210–4214. [CrossRef]
34. Liu, G.R.; Huo, R.Y.; Zhai, Y.N.; Liu, L. New bioactive sesquiterpeniods from the plant endophytic fungus *Pestalotiopsis theae*. *Front. Microbiol.* **2021**, *12*, 641504. [CrossRef] [PubMed]
35. Seetharaman, P.; Gnanasekar, S.; Chandrasekaran, R.; Chandrakasan, G.; Kadarkarai, M.; Sivaperumal, S. Isolation and characterization of anticancer flavone chrysin (5,7-dihydroxy flavone)-producing endophytic fungi from *Passiflora incarnata* L. leaves. *Ann. Microbiol.* **2017**, *67*, 321–331. [CrossRef]
36. Hu, Z.B.; Wu, Z.H.; Su, Q.H.; Li, M.Z.; Wu, S.S.; Meng, R.Q.; Ding, W.J.; Li, C.Y. Metabolites with phytopathogenic fungi inhibitory activities from the mangrove endophytic fungus *Botryosphaeria ramose*. *Bioorg. Chem.* **2020**, *104*, 104300. [CrossRef]
37. Hashad, N.; Ibrahim, R.; Mady, M.; Abdel-Aziz, M.S.; Moharram, F.A. Review: Bioactive metabolites and host-specific toxins from endophytic fungi, alternaria alternate. *Vietnam. J. Chem.* **2021**, *59*, 733–759. [CrossRef]
38. Li, H.; Tian, J.M.; Tang, H.Y.; Pan, S.Y.; Zhang, A.L.; Gao, J.M. Chaetosemins A–E, new chromones isolated from an Ascomycete *Chaetomium seminudum* and their biological activities. *RSC Adv.* **2015**, *37*, 29185–29192. [CrossRef]
39. Liu, S.C.; Liu, L. Isoprenylated chromones from the plant endophytic fungus *Pestalotiopsis fici*. *Mycosystema* **2011**, *29*, 585–587. [CrossRef]
40. Apgma, B.; Ajcc, A.; Anb, C.; Ma, C.; Ba, C.; Laep, A. Tetrasubstituted α-pyrone derivatives from the endophytic fungus, *Neurospora udagawae*. *Phytochem. Lett.* **2020**, *35*, 147–151. [CrossRef]
41. Dong, L.; Li, X.M.; Li, C.S.; Wang, B.G. Nigerasterols A and B, Antiproliferative Sterols from the Mangrove-Derived Endophytic Fungus *Aspergillus niger* MA-132. *Helv. Chim. Acta* **2013**, *96*, 1055–1061. [CrossRef]
42. Liu, S.; Liu, X.; Guo, L.; Che, Y.S.; Liu, L. 2H-Pyran-2-one and 2H-Furan-2-one Derivatives from the plant Endophytic Fungus *Pestalotiopsis fici*. *Chem. Biodivers.* **2013**, *10*, 2007–2013. [CrossRef] [PubMed]
43. Yu, H.M.; Hou, Y.T.; Zhu, L.J.; Chen, Y.J.; Dong, M.; Zhou, M.; Shen, D.; Chen, M.H. Two new α-pyronoids from endophytic fungus *Aspergillus oryzae* derived from *Paris polyphylla* var. *yunnanensis*. *Chin. Tradit. Herb. Drugs* **2020**, *51*, 4891–4895. [CrossRef]
44. Guo, D.L.; Qiu, L.; Feng, D.; He, X.; Li, X.H.; Cao, Z.X.; Gu, Y.C.; Mei, L.; Deng, F.; Deng, Y. Three new α-pyrone derivatives induced by chemical epigenetic manipulation of *Penicillium herquei*, an endophytic fungus isolated from *Cordyceps sinensis*. *Nat. Prod. Res.* **2020**, *34*, 958–964. [CrossRef]
45. Lee, M.S.; Wang, S.W.; Wang, G.J.; Pang, K.L.; Lee, C.K.; Kuo, Y.H.; Cha, H.Y.; Lin, R.K.; Lee, T.H. Angiogenesis Inhibitors and Anti-Inflammatory Agents from *Phoma* sp. NTOU4195. *J. Nat. Prod.* **2016**, *79*, 2983–2990. [CrossRef] [PubMed]
46. Anoumedem, G.; Mountessou, B.; Kouam, S.F.; Narmani, A.; Surup, F. Simplicilones A and B Isolated from the Endophytic Fungus *Simplicillium subtropicum* SPC3. *Antibiotics* **2020**, *9*, 753. [CrossRef]

47. Zhang, F.Z.; Li, X.M.; Li, X.; Yang, S.Q.; Meng, L.H.; Wang, B.G. Polyketides from the Mangrove-Derived Endophytic Fungus *Cladosporium cladosporioides*. *Mar. Drugs* **2019**, *17*, 296. [CrossRef]
48. Ola, A.R.; Tawo, B.D.; Belli, H.L.; Proksch, P.; Tommy, D.; Hakim, E.H. A new antibacterial polyketide from the endophytic fungi *Aspergillus fumigatiaffinis*. *Nat. Prod. Commun.* **2018**, *13*, 1573–1574. [CrossRef]
49. Zhao, S.S.; Wang, B.; Tian, K.L.; Ji, W.X.; Zhang, T.Y.; Ping, C.; Yan, W.; Ye, Y.H. Novel metabolites from the *Cercis chinensis* derived endophytic fungus *Alternaria alternata* ZHJG5 and their antibacterial activities. *Pest. Manag. Sci.* **2021**, *77*, 2264–2271. [CrossRef] [PubMed]
50. Li, C.S.; Sarotti, A.M.; Wu, X.H.; Yang, B.J.; Turkson, J.; Chen, Y.F.; Liu, Q.S.; Cao, S.G. An Unusual Benzoisoquinoline-9-one Derivative and Other Related Compounds with Antiproliferative Activity from Hawaiian Endophytic Fungus *Peyronellaea* sp. FT431. *Molecules* **2019**, *24*, 196. [CrossRef]
51. Wu, J.C.; Hou, Y.; Xu, Q.H.; Jin, X.J.; Chen, Y.X.; Fang, J.G.; Hu, B.R.; Wu, Q.X. (±)-Alternamgin, a Pair of Enantiomeric Polyketides, from the Endophytic Fungi *Alternaria* sp. MG1. *Org. Lett.* **2019**, *21*, 1551–1554. [CrossRef] [PubMed]
52. Tang, J.W.; Hu, K.; Su, X.Z.; Li, X.N.; Puno, P.T. Phomopsisins A-C: Three new cytochalasans from the plant endophytic fungus *Phomopsis* sp. sh917. *Tetrahedron* **2020**, *76*, 131475. [CrossRef]
53. Chen, C.M.; Tong, Q.Y.; Zhu, H.C.; Tan, D.D.; Zhang, J.W.; Xue, Y.B.; Yao, G.M.; Luo, Z.W.; Wang, J.P.; Wang, Y.Y. Nine new cytochalasan alkaloids from *Chaetomium globosum* TW1-1 (Ascomycota, Sordariales). *Sci. Rep.* **2016**, *6*, 18711. [CrossRef]
54. Yang, X.L.; Wu, P.; Xue, J.H.; Li, H.X.; Wei, X.Y. Cytochalasans from endophytic fungus *Diaporthe* sp. SC-J0138. *Fitoterapia* **2020**, *145*, 104611. [CrossRef] [PubMed]
55. Mou, Q.L.; Yang, S.X.; Xiang, T.; Liu, W.W.; Yang, J.; Guo, L.P.; Wang, W.J.; Yang, X.L. New cytochalasan alkaloids and cyclobutane dimer from an endophytic fungus *Cytospora chrysosperma* in *Hippophae rhamnoides* and their antimicrobial activities. *Tetrahedron Lett.* **2021**, *87*, 153207. [CrossRef]
56. Yang, J.; Gong, L.Z.; Guo, M.M.; Jiang, Y.; Ding, Y.; Wang, Z.J.; Xin, X.J.; An, F.L. Bioactive Indole Diketopiperazine Alkaloids from the Marine Endophytic Fungus *Aspergillus* sp. YJ191021. *Mar. Drugs* **2021**, *19*, 157. [CrossRef] [PubMed]
57. Xu, K.; Wei, X.L.; Xue, L.; Zhang, Z.F.; Zhang, P. Antimicrobial Meroterpenoids and Erythritol Derivatives Isolated from the Marine-Algal-Derived Endophytic Fungus *Penicillium chrysogenum* XNM-12. *Mar. Drugs* **2020**, *18*, 578. [CrossRef] [PubMed]
58. Lutfia, L.; Munir, E.; Yurnaliza, Y.; Basyuni, M. Chemical analysis and anticancer activity of sesterterpenoid from an endophytic fungus *Hypomontagnella monticulosa* Zg15SU and its host *Zingiber griffithii* Baker. *Heliyon* **2021**, *7*, e06292. [CrossRef] [PubMed]
59. Wang, Z.J.; Jiang, Y.; Xin, X.J.; An, F.L. Bioactive indole alkaloids from insect derived endophytic *Aspergillus lentulus*. *Fitoterapia* **2021**, *153*, 104973. [CrossRef] [PubMed]
60. Zhai, Y.J.; Huo, G.M.; Zhang, Q.; Li, D.; Wang, D.C.; Qi, J.Z.; Gao, J.M. Phaeosphaones: Tyrosinase Inhibitory Thiodiketopiperazines from an Endophytic *Phaeosphaeria fuckelii*. *J. Nat. Prod.* **2020**, *83*, 1592–1597. [CrossRef]
61. Zhang, P.; Li, X.M.; Wang, J.N. Oxepine-Containing Diketopiperazine Alkaloids from the Algal-Derived Endophytic Fungus *Paecilomyces variotii* EN-291. *Helv. Chim. Acta* **2015**, *98*, 800–804. [CrossRef]
62. Ye, G.T.; Huang, C.Y.; Li, J.L.; Chen, T.; Tang, J.; Liu, W.B.; Long, Y.H. Isolation, Structural Characterization and Antidiabetic Activity of New Diketopiperazine Alkaloids from Mangrove Endophytic Fungus *Aspergillus* sp. 16-5c. *Mar. Drugs* **2021**, *19*, 402. [CrossRef] [PubMed]
63. Meng, L.H.; Peng, Z.; Li, X.M.; Wang, B.G. Penicibrocazines A–E, Five New Sulfide Diketopiperazines from the Marine-Derived Endophytic Fungus *Penicillium brocae*. *Mar. Drugs.* **2015**, *13*, 276–287. [CrossRef] [PubMed]
64. Meng, L.H.; Wang, C.Y.; Mándi, A.; Li, X.M.; Hu, X.Y.; Kassack, M.U. Three Diketopiperazine Alkaloids with Spirocyclic Skeletons and One Bisthiodiketopiperazine Derivative from the Mangrove-Derived Endophytic Fungus *Penicillium brocae* MA-231. *Org. Lett.* **2016**, *18*, 5304–5307. [CrossRef]
65. Zheng, C.J.; Li, L.; Zou, J.P.; Han, T.; Qin, L.P. Identification of a quinazoline alkaloid produced by *Penicillium vinaceum*, an endophytic fungus from *Crocus sativus*. *Pharm. Biol.* **2012**, *50*, 129–133. [CrossRef]
66. Sirimangkalakitti, N.; Yokoya, M.; Chamni, S.; Chanvorachote, P.; Plubrukrn, A.; Saito, N.; Suwanborirux, K. Synthesis and Absolute Configuration of Acanthodendrilline, a New Cytotoxic Bromotyrosine Alkaloid from the Thai Marine Sponge *Acanthodendrilla* sp. *Chem. Pharm. Bull.* **2016**, *64*, 258–262. [CrossRef]
67. Xu, Y.C.; Wang, L.P.; Zhu, G.L.; Zuo, M.X.; Gong, Q.Y.; He, W.W.; Li, M.P.; Yuan, C.M.; Hao, X.J.; Zhu, W.M. New phenylpyridone derivatives from the *Penicillium sumatrense* GZWMJZ-313, a fungal endophyte of *Garcinia multiflora*. *Chin. Chem. Lett.* **2019**, *30*, 431–434. [CrossRef]
68. Cui, H.; Lin, Y.; Luo, M.C.; Lu, Y.J.; Huang, X.S.; She, Z.G. Diaporisoindoles A−C: Three Isoprenylisoindole Alkaloid Derivatives from the Mangrove Endophytic Fungus *Diaporthe* sp. SYSU-HQ3. *Org. Lett.* **2017**, *19*, 5621–5624. [CrossRef]
69. Wu, S.H.; Zhao, L.X.; Chen, Y.W.; Huang, R.; Miao, C.P.; Wang, J. Sesquiterpenoids from the Endophytic Fungus *Trichoderma* sp. PR-35 of *Paeonia delavayi*. *Chem. Biodivers.* **2011**, *9*, 1717–1723. [CrossRef]
70. Wang, A.; Yin, R.Y.; Zhou, Z.Y.; Gu, G.; Dai, J.G.; Lai, D.W.; Zhou, L.G. Eremophilane-Type Sesquiterpenoids from the Endophytic Fungus *Rhizopycnis vagum* and Their Antibacterial, Cytotoxic, and Phytotoxic Activities. *Front. Chem.* **2020**, *26*, 596889. [CrossRef]
71. Jiang, C.X.; Li, J.; Zhang, J.M.; Jin, X.J.; Yu, B.; Fang, J.G.; Wu, Q.X. Isolation, Identification, and Activity Evaluation of Chemical Constituents from Soil Fungus *Fusarium avenaceum* SF-1502 and Endophytic Fungus *Fusarium proliferatum* AF-04. *J. Agric. Food Chem.* **2019**, *67*, 839–1846. [CrossRef] [PubMed]

72. Hu, X.Y.; Li, X.M.; Meng, L.H.; Wang, B.G. Antioxidant bisabolane-type sesquiterpenoids from algal-derived fungus *Aspergillus sydowii* EN-434. *J. Oceanol. Limnol.* **2020**, *38*, 1532–1536. [CrossRef]
73. Wang, Q.X.; Bao, L.; Yang, X.L.; Liu, D.L.; Guo, H.; Dai, H.Q.; Song, F.H.; Zhang, L.X.; Guo, L.D.; Li, S.J.; et al. Ophiobolins P–T, five new cytotoxic and antibacterial sesterterpenes from the endolichenic fungus *Ulocladium* sp. *Fitoterapia* **2013**, *90*, 220–227. [CrossRef] [PubMed]
74. Shi, X.S.; Song, Y.P.; Meng, L.H.; Yang, S.Q.; Li, X.M. Isolation and Characterization of Antibacterial Carotane Sesquiterpenes from *Artemisia argyi* Associated Endophytic *Trichoderma virens* QA-8. *Antibiotics* **2021**, *10*, 213. [CrossRef] [PubMed]
75. Chen, Y.; Zou, G.; Yang, W.C.; Zhao, Y.Y.; Tan, Q.; Chen, L.; Wang, J.M.; Ma, C.Y.; Kang, W.Y.; She, Z.G. Metabolites with Anti-Inflammatory Activity from the Mangrove Endophytic Fungus *Diaporthe* sp. QYM12. *Mar. Drugs* **2021**, *19*, 56. [CrossRef]
76. Fan, M.M.; Xiang, G.; Chen, J.W.; Gao, J.; Xue, W.X.; Wang, Y.X.; Li, W.H.; Zhou, L.; Jiao, R.H.; Shen, Y.; et al. Libertellenone M, a diterpene derived from an endophytic fungus *Phomopsis* sp. S12, protects against DSS-induced colitis via inhibiting both nuclear translocation of NF-κB and NLRP3 inflammasome activation. *Int. Immunopharmacol.* **2020**, *80*, 106144. [CrossRef]
77. Xu, K.; Zhang, X.; Chen, J.W.; Shen, Y.; Jiang, N.; Tan, R.X.; Jiao, R.H.; Ge, H.M. Anti-inflammatory diterpenoids from an endophytic fungus *Phomopsis* sp. S12. *Tetrahedron Lett.* **2019**, *60*, 151045. [CrossRef]
78. Chen, Y.M.; Yang, Y.H.; Li, X.N.; Zou, C.; Zhao, P.J. Diterpenoids from the Endophytic Fungus *Botryosphaeria* sp. P483 of the Chinese Herbal Medicine *Huperzia serrata*. *Molecules* **2015**, *20*, 16924–16932. [CrossRef]
79. Wang, X.N.; Bashyal, B.P.; Wijeratne, E.M.; Ren, J.M.; Liu, M.X.; Gunatilaka, M.K. Smardaesidins A-G, Isopimarane and 20-Nor-isopimarane Diterpenoids from *Smardaea* sp., a Fungal Endophyte of the Moss *Ceratodon purpureus*. *J. Nat. Prod.* **2011**, *74*, 2052–2061. [CrossRef]
80. Xu, M.F.; Jia, O.Y.; Wang, S.J.; Zhu, Q. A new bioactive diterpenoid from pestalotiopsis adusta, an endophytic fungus from *Clerodendrum canescens*. *Nat. Prod. Res.* **2016**, *30*, 2642–2647. [CrossRef]
81. Sun, P.X.; Zheng, C.J.; Li, W.C.; Jin, G.L.; Huang, F.; Qin, L.P. Trichodermanin A, a novel diterpenoid from endophytic fungus culture. *J. Nat. Med.* **2011**, *65*, 381–384. [CrossRef] [PubMed]
82. Sun, H.F.; Li, X.M.; Meng, L.; Cui, C.M.; Gao, S.S.; Li, C.S.; Huang, C.G.; Wang, B.G. Asperolides A−C, Tetranorlabdane Diterpenoids from the Marine Alga-Derived Endophytic Fungus *Aspergillus wentii* EN-48. *J. Nat. Med.* **2012**, *75*, 148–152. [CrossRef] [PubMed]
83. Guo, K.; Fang, H.Q.; Gui, F.; Wang, Y.Z.; Xu, Q.Y.; Deng, X.M. Two New Ring A-Cleaved Lanostane-Type Triterpenoids and Four Known Steroids Isolated from Endophytic Fungus *Glomerella* sp. F00244. *Helv. Chim. Acta* **2016**, *99*, 601–607. [CrossRef]
84. Morandini, L.; Neto, A.T.; Pedroso, M.; Antoniolli, Z.I.; Burrow, R.; Morel, A.F. Lanostane-type triterpenes from the fungal endophyte Scleroderma UFSMSc1 (Persoon) Fries. *Bioorg. Med. Chem. Lett.* **2016**, *47*, 1173–1176. [CrossRef]
85. Tian, C.; Gao, H.; Peng, X.P.; Li, G.; Lou, H.X. Fusidic acid derivatives from the endophytic fungus *Acremonium pilosum* F47. *J. Asian Nat. Prod. Res.* **2021**, *23*, 1148–1155. [CrossRef]
86. Qin, D.; Shen, W.Y.; Wang, Y.Q.; Han, M.J.; Chai, F.N.; Duan, X.X.; Yan, X.; Guo, J.L.; Gao, T.C.; Zuo, S.H.; et al. Enhanced production of unusual triterpenoids from *Kadsura angustifolia* fermented by a symbiont endophytic fungus, *Penicillium* sp. SWUKD4.1850. *Phytochemistry* **2019**, *158*, 56–66. [CrossRef]
87. Liang, H.Q.; Zhang, D.W.; Guo, S.X.; Jie, Y. Two new tetracyclic triterpenoids from the endophytic fungus *Hypoxylon* sp. 6269. *J. Asian Nat. Prod. Res.* **2018**, *20*, 1485662. [CrossRef]
88. Ibrahim, R.M.; Abdallah, H.M.; Mohamed, G.A.; Ross, S.A. Integracides H-J: New tetracyclic triterpenoids from the endophytic fungus *Fusarium* sp. *Fitoterapia* **2016**, *112*, 161–167. [CrossRef]
89. Ibrahim, R.M.; Abdallah, H.M.; Mohamed, G.A.; Ross, S.A. Integracides F and G: New tetracyclic triterpenoids from the endophytic fungus *Fusarium* sp. *Phytochem. Lett.* **2016**, *15*, 125–130. [CrossRef]
90. Sun, Z.H.; Liang, F.L.; Wu, W.; Chen, Y.C.; Pan, Q.L.; Li, H.H.; Ye, W.; Liu, H.X.; Li, S.N.; Tan, G.H.; et al. Guignardones P-S, New Meroterpenoids from the Endophytic Fungus *Guignardia mangi ferae* A348 Derived from the Medicinal Plant *Smilax glabra*. *Molecules* **2015**, *20*, 22900–22907. [CrossRef]
91. Li, Q.; Chen, C.M.; Cheng, L.; Wei, M.S.; Dai, C.; He, Y.; Gong, J.J.; Zhu, R.Q.; Li, X.N.; Liu, J.J.; et al. Emeridones A−F, a Series of 3,5-Demethylorsellinic Acid-Based Meroterpenoids with Rearranged Skeletons from an Endophytic Fungus *Emericella* sp. TJ29. *J. Org. Chem.* **2019**, *84*, 1534–1541. [CrossRef]
92. Yang, H.G.; Zhao, H.; Li, J.J.; Chen, S.M.; Mou, L.M.; Zou, J.; Chen, G.D.; Qin, S.Y.; Wang, C.X.; Hu, D.; et al. Phyllomeroterpenoids A-C, Multi-biosynthetic Pathway Derived Meroterpenoids from the TCM Endophytic Fungus *Phyllosticta* sp. and their Antimicrobial Activities. *Sci. Rep.* **2017**, *7*, 12925. [CrossRef] [PubMed]
93. Nascimento, J.; Silva, F.M.; Magallanes-Noguera, C.A.; Kurina-Sanz, M.; Silva, E. Natural trypanocidal product produced by endophytic fungi through co-culturing. *Folia Microbiol.* **2020**, *65*, 323–328. [CrossRef] [PubMed]
94. Li, Y.L.; Yi, J.L.; Cai, J.; Zhou, X.M.; Chen, L.; Zhuo, X.; Lai, X.Y. Two new bioactive secondary metabolites from the endophytic fungus *Talaromyces assiutensis* JTY2. *Nat. Prod. Res.* **2021**, *17*, 1–6. [CrossRef] [PubMed]
95. Wen, Y.Z.; Lv, Y.B.; Hao, J.; Chen, H.; Yang, X.; Huang, Y.; Liu, C.; Huang, H.Q.; Ma, Y.R.; Yang, X.Z. Two new compounds of *Penicillium polonicum*, an endophytic fungus from *Camptotheca acuminata* Decne. *Nat. Prod. Res.* **2019**, *34*, 1879–1883. [CrossRef]
96. Bussey, R.O.; Kaur, A.; Todd, D.A.; Egan, J.M.; El-Elimat, T.; Graf, T.N.; Raja, H.A.; Oberlies, N.H.; Cech, N.J. Comparison of the chemistry and diversity of endophytes isolated from wild-harvested and greenhouse-cultivated yerba mansa (*Anemopsis californica*). *Phytochem. Lett.* **2015**, *11*, 202–208. [CrossRef]

97. Tchoukoua, A.; Ota, T.; Akanuma, R.; Ju, Y.M.; Supratman, U.; Murayama, T.; Koseki, T.; Shiono, Y. A phytotoxic bicyclic lactone and other compounds from endophyte *Xylaria curta*. *Nat. Prod. Res.* **2017**, *31*, 2113–2118. [CrossRef]
98. Chen, S.H.; Liu, Z.M.; Liu, H.J.; Long, Y.H.; Chen, D.N.; Lu, Y.J.; She, Z.G. Lasiodiplactone A, a novel lactone from the mangrove endophytic fungus *Lasiodiplodia theobromae* ZJ-HQ1. *Org. Biomol. Chem.* **2017**, *15*, 6338–6341. [CrossRef]
99. Gusmao, A.S.; Abreu, L.S.; Tavares, J.F.; Freitas, H.F.; Pita, S.; Santos, E.; Caldas, I.; Vieira, A.A.; Silva, E.O. Computer-guided trypanocidal activity of natural lactones produced by endophytic fungus of *Euphorbia umbellata*. *Chem. Biodivers.* **2021**, *18*, e2100493. [CrossRef] [PubMed]
100. Yuan, X.L.; Wang, X.F.; Xu, K.; Li, W.; Chen, D.; Zhang, P. Characterization of a New Insecticidal Anthraquinone Derivative from an Endophyte of *Acremonium vitellinum* against *Helicoverpa armigera*. *J. Agric. Food Chem.* **2020**, *68*, 11480–11487. [CrossRef]
101. Yang, Y.C.; Yang, H.Y.; Li, Y.; Ye, Y.Q.; Hu, Q.F.; Gao, X.M.; Du, G.A. New Xanthone from the Fermentation Products of Endophytic Fungus of *Phomopsis* Species. *Asian J. Chem.* **2014**, *26*, 4591–4593. [CrossRef]
102. Du, F.Y.; Li, X.M.; Song, J.Y.; Li, C.S.; Wang, B.G. Anthraquinone Derivatives and an Orsellinic Acid Ester from the Marine Alga-Derived Endophytic Fungus *Eurotium cristatum* EN-220. *Helv. Chim. Acta* **2014**, *97*, 973–978. [CrossRef]
103. Hawas, U.W.; El-Beih, A.A.; El-Halawany, A.M. Bioactive Anthraquinones from Endophytic Fungus *Aspergillus versicolor* Isolated from Red Sea Algae. *Arch. Pharm. Res.* **2012**, *35*, 1749–1756. [CrossRef] [PubMed]
104. Teiten, M.H.; Mack, F.; Debbab, A.; Aly, A.H.; Dicato, M.; Proksch, P.; Diederich, M. Anticancer effect of altersolanol A, a metabolite produced by the endophytic fungus *Stemphylium globuliferum*, mediated by its proapoptotic and antiinvasive potential via the inhibition of NF-kB activity. *Bioorg. Med. Chem.* **2013**, *21*, 3850–3858. [CrossRef] [PubMed]
105. Chen, Y.; Yang, W.C.; Zou, G.; Yan, Z.Y.; Qiu, P.; Long, Z.G.; She, Z.G. Metabolites with anti-inflammatory and α-glucosidase inhibitory activities from the mangrove endophytic fungus *Phoma* sp. SYSU-SK-7. *Tetrahedron Lett.* **2020**, *61*, 152578. [CrossRef]
106. Ibrahim, S.; Mohamed, G.A.; Haidari, R.; El-Kholy, A.; Zayed, M.F. Fusaristerol A: A New Cytotoxic and Antifungal Ergosterol Fatty Acid Ester from the Endophytic Fungus *Fusarium* sp. Associated with *Mentha longifolia* Roots. *Pharmacogn. Mag.* **2018**, *56*, 308–311. [CrossRef]
107. Wu, S.H.; Huang, R.; Miao, C.P.; Chen, Y.W.; Chen, Y.W. Two New Steroids from an Endophytic Fungus *Phomopsis* sp. *Chem. Biodivers.* **2013**, *10*, 1276–1283. [CrossRef]
108. Qi, C.X.; Gao, W.X.; Wang, J.P.; Liu, M.T.; Zhang, J.W.; Chen, C.M.; Hu, Z.X.; Xue, Y.B.; Li, D.X.; Zhang, Y.H. Errusnolides A-D, new butenolides with anti-inflammatory activities from an endophytic *Aspergillus* from *Tripterygium wilfordii*. *Fitoterapia* **2018**, *130*, 134–139. [CrossRef] [PubMed]
109. Akhter, N.; Pan, C.Q.; Liu, Y.Q.; Shi, Y.T.; Wu, B. Isolation and structure determination of a new indene derivative from endophytic fungus *Aspergillus* flavipes Y-62. *Nat. Prod. Res.* **2019**, *33*, 2939–2944. [CrossRef]
110. Hussain, H.; Root, N.; Jabeen, F.; Al-Harrasi, A.; Ahmad, M.; Mabood, F. Microsphaerol and Seimatorone: Two New Compounds Isolated from the Endophytic Fungi, *Microsphaeropsis* sp. and *Seimatosporium* sp. *Chem. Biodivers.* **2015**, *46*, 289–294. [CrossRef]
111. Wu, X.; Pang, X.J.; Xu, L.L.; Zhao, T.; Long, X.Y.; Zhang, Q.Y.; Qin, H.L.; Yang, D.F.; Yang, X.L. Two new alkylated furan derivatives with antifungal and antibacterial activities from the plant endophytic fungus *Emericella* sp. XL029. *Nat. Prod. Res.* **2017**, *32*, 2625–2631. [CrossRef] [PubMed]
112. Niaz, S.I.; Khan, D.; Naz, R.; Safdar, K.; Zain, S.; Abidin, U.; Khan, I.U.; Gul, R.; Khan, W.U.; Aman, M.; et al. Antimicrobial and antioxidant chlorinated azaphilones from mangrove *Diaporthe perseae* sp. isolated from the stem of Chinese mangrove *Pongamia pinnata*. *J. Asian Nat. Prod. Res.* **2020**, *23*, 1077–1084. [CrossRef] [PubMed]
113. Yang, X.X.; Wang, N.N.; Kang, Y.F.; Ma, Y.M. New furan derivative from an endophytic *Aspergillus tubingensis* of *Decaisnea insignis* (Griff.) Hook.f. & Thomson. *Nat. Prod. Res.* **2019**, *33*, 2777–2783. [CrossRef] [PubMed]
114. Zhu, X.W.; Chen, J.Q.; Zhu, S.R.; He, Y.Y.; Ding, W.J.; Li, C.Y. Two new compounds from *Nigrospora sphaerica* ZMT05, a fungus derivated from *Oxya chinensis* Thunber. *Nat. Prod. Res.* **2018**, *32*, 2375–2381. [CrossRef] [PubMed]
115. Li, H.L.; Li, X.M.; Yang, S.Q.; Cao, J.; Li, Y.H.; Wang, B.G. Induced terreins production from marine red algal-derived endophytic fungus *Aspergillus terreus* EN-539 co-cultured with symbiotic fungus *Paecilomyces lilacinus* EN-531. *J. Antibiot.* **2020**, *73*, 108–111. [CrossRef] [PubMed]
116. Zhao, M.; Yuan, L.Y.; Guo, D.L.; Ye, Y.; Da-Wa, Z.M.; Wang, X.L.; Ma, F.W.; Chen, L. Bioactive halogenated dihydroisocoumarins produced by the endophytic fungus *Lachnum palmae* isolated from Przewalskia tangutica. *Phytochemistry* **2018**, *148*, 97–103. [CrossRef] [PubMed]
117. Singh, A.; Singh, D.K.; Kharwar, R.N.; White, J.F.; Gond, S.K. Fungal Endophytes as Efficient Sources of Plant-Derived Bioactive Compounds and Their Prospective Applications in Natural Product Drug Discovery: Insights, Avenues, and Challenges. *Microorganisms* **2021**, *9*, 197. [CrossRef] [PubMed]
118. Chandra, S. Endophytic fungi: Novel sources of anticancer lead molecules. *Appl. Microbiol. Biotechnol.* **2012**, *95*, 47–59. [CrossRef] [PubMed]
119. Yu, H.S.; Zhang, L.; Li, L.; Zheng, C.J.; Guo, L.; Li, W.C.; Sun, P.X.; Qin, L.P. Recent developments and future prospects of antimicrobial metabolites produced by endophytes. *Microbiol. Res.* **2010**, *165*, 437–449. [CrossRef] [PubMed]
120. Strobel, G. The Emergence of Endophytic Microbes and Their Biological Promise. *J. Fungi* **2018**, *4*, 57. [CrossRef]
121. Hastuti, U.S.; Asna, P.M.A.; Rahmawati, D. Histologic observation, identification, and secondary metabolites analysis of endophytic fungi isolated from a medicinal plant, *hedychium acuminatum* Roscoe. *AIP Conf. Proc.* **2018**, *2002*, 020070. [CrossRef]

122. El-Sayed, A.; Khalaf, S.A.; Azez, H.A.; Hussein, H.A.; El-Baz, A.F. Production, bioprocess optimization and anticancer activity of camptothecin from *Aspergillus terreus* and *Aspergillus flavus*, endophytes of ficus elastica. *Process Biochem.* **2021**, *107*, 59–73. [CrossRef]
123. Xiong, Z.Q.; Yang, Y.Y.; Zhao, N.; Wang, Y. Diversity of endophytic fungi and screening of fungal paclitaxel producer from anglojap yew, *Taxus* x *media*. *BMC Microbiol.* **2013**, *13*, 71. [CrossRef] [PubMed]
124. Gao, Y.; Yin, H.; Sun, Y.; Zhang, Z.; Cui, Y. Mutagenesis of a berberine-producing endophytic fungus. *J. Fungal Res.* **2008**, *6*, 216–225.
125. Xu, B.; Wang, M.R.; Xia, Y.; Yang, K.; Zhang, C.Y. Improvement of the output of teicoplanin by genome shuffling. *Chin. J. Antibiot.* **2006**, *31*, 237–242. [CrossRef]
126. Eyberger, A.L.; Dondapati, R.; Porter, J.R. Endophyte Fungal Isolates from *Podophyllum peltatum* Produce Podophyllotoxin. *J. Nat. Prod.* **2006**, *68*, 1121–1124. [CrossRef] [PubMed]
127. Venugopalan, A.; Srivastava, S. Endophytes as in vitro production platforms of high value plant secondary metabolites. *Biotechnol. Adv.* **2015**, *33*, 873–887. [CrossRef] [PubMed]
128. Vigneshwari, A.; Rakk, D.; Németh, A.; Kocsubé, S. Host metabolite producing endophytic fungi isolated from *Hypericum perforatum*. *PLoS ONE* **2019**, *14*, e0217060. [CrossRef]
129. Gupta, S.; Chaturvedi, P.; Kulkarni, M.G.; Staden, J.V. A critical review on exploiting the pharmaceutical potential of plant endophytic fungi. *Biotechnol. Adv.* **2020**, *39*, 107462. [CrossRef]
130. Wei, Q.; Bai, J.; Yan, D.J.; Bao, X.Q.; Li, W.T.; Hu, Y.C. Genome mining combined metabolic shunting and osmac strategy of an endophytic fungus leads to the production of diverse natural products. *Acta Pharm. Sin. B* **2021**, *11*, 572–587. [CrossRef]
131. Zhang, L.H.; Niaz, S.I.; Khan, D.; Wang, Z.; Zhu, Y.H.; Zhou, H.Y.; Lin, Y.C.; Li, J.; Liu, L. Induction of Diverse Bioactive Secondary Metabolites from the Mangrove Endophytic Fungus *Trichoderma* sp. (Strain 307) by Co-Cultivation with *Acinetobacter johnsonii* (Strain B2). *Mar. Drugs* **2017**, *15*, 35. [CrossRef] [PubMed]
132. Vagish, D.; Shreya, K.; Sanjai, S. Isolation and enhancement of resveratrol production in *Xylaria psidii* by exploring the phenomenon of epigenetics: Using DNA methyltransferases and *Histone deacetylase* as epigenetic modifiers. *Mol. Biol. Rep.* **2019**, *46*, 4123–4137. [CrossRef]
133. Pu, X.; Qu, X.X.; Chen, F.; Bao, J.K.; Zhang, G.L.; Luo, Y.G. Camptothecin-producing endophytic fungus *Trichoderma atroviride* ly357: Isolation, identification, and fermentation conditions optimization for camptothecin production. *Appl. Microbiol. Biotechnol.* **2013**, *97*, 9365–9375. [CrossRef] [PubMed]
134. Pelo, S.P.; Adebo, O.A.; Green, E. Chemotaxonomic profiling of fungal endophytes of *Solanum mauritianum* (alien weed) using gas chromatography high resolution time-of-flight mass spectrometry (GC-HRTOF-MS). *Metabolomics* **2021**, *17*, 43. [CrossRef] [PubMed]

Journal of *Fungi*

MDPI

Review

Sulfur-Containing Compounds from Endophytic Fungi: Sources, Structures and Bioactivities

Yaqin Fan [1,†], Zhiheng Ma [1,†], Yan Zhang [1], Yufei Wang [2], Yousong Ding [3], Cong Wang [2,*] and Shugeng Cao [4,*]

1 Shandong Provincial Key Laboratory of Applied Mycology, School of Life Sciences, Qingdao Agricultural University, Qingdao 266109, China; fanyaqin@qau.edu.cn (Y.F.); 20202206023@stu.qau.edu.cn (Z.M.); 20212206060@stu.qau.edu.cn (Y.Z.)
2 Key Laboratory of Chemistry and Engineering of Forest Products, State Ethnic Affairs Commission, Guangxi Collaborative Innovation Center for Chemistry and Engineering of Forest Products, School of Chemistry and Chemical Engineering, Guangxi Minzu University, Nanning 530006, China; 171416020323@stu.haust.edu.cn
3 Department of Medicinal Chemistry, Center for Natural Products, Drug Discovery and Development, College of Pharmacy, University of Florida, Gainesville, FL 32610, USA; yding@cop.ufl.edu
4 Department of Pharmaceutical Sciences, Daniel K. Inouye College of Pharmacy, University of Hawai'i at Hilo, 200 W. Kawili St., Hilo, HI 96720, USA
* Correspondence: wangcong@gxmzu.edu.cn (C.W.); scao@hawaii.edu (S.C.)
† These authors contributed equally to this work.

Abstract: Endophytic fungi have attracted increasing attention as an under-explored source for the discovery and development of structurally and functionally diverse secondary metabolites. These microorganisms colonize their hosts, primarily plants, and demonstrate diverse ecological distribution. Among endophytic fungal natural products, sulfur-containing compounds feature one or more sulfur atoms and possess a range of bioactivities, e.g., cytotoxicity and antimicrobial activities. These natural products mainly belong to the classes of polyketides, nonribosomal peptides, terpenoids, and hybrids. Here, we reviewed the fungal producers, plant sources, chemical structures, and bioactivities of 143 new sulfur-containing compounds that were reported from 1985 to March 2022.

Keywords: sulfur; plant endophyte; endophytic fungi

Citation: Fan, Y.; Ma, Z.; Zhang, Y.; Wang, Y.; Ding, Y.; Wang, C.; Cao, S. Sulfur-Containing Compounds from Endophytic Fungi: Sources, Structures and Bioactivities. *J. Fungi* **2022**, *8*, 628. https://doi.org/10.3390/jof8060628

Academic Editors: Tao Feng and Frank Surup

Received: 29 May 2022
Accepted: 10 June 2022
Published: 13 June 2022

1. Introduction

Sulfur is one of the prime elements on Earth and the eighth most abundant element in the human body. It is a group 6A (or VIA) member of the periodic table, with a larger atomic size and a weaker electronegativity than oxygen. Sulfur has unique characteristics, such as five different oxidation states, and sulfur-containing molecules often participate in biological redox reactions and electron transfer processes. Notably, two essential amino acids, L-methionine and L-cysteine, both contain a sulfur atom, further highlighting the importance and indispensability of sulfur in biology [1]. Indeed, one fifth (20%) of the FDA-approved drugs contain at least one sulfur atom. These sulfur-containing drugs have different structure skeletons such as sulfonamides, β-lactams, thioethers, thiazoles, thiophenes, phenothiazines, sulfoxides, S=C and S=P structures, thionucleotides, sulfones, sulfates and macrocyclic disulfides. Of note, many sulfur-containing drugs are natural products or their derivatives (i.e., rosuvastatin, ecteinascidin 743 and ixabepilone) [2].

Fungi are a major group of microorganisms that produce a broad array of compounds with novel structures and unique bioactivities. One type of fungi colonizes the intercellular and/or intracellular regions of healthy plant tissues at a particular time and has no interference with and causes no pathogenic symptoms to the host [3]. These endophytic microorganisms are an important but less-explored source for the discovery of structurally novel natural products in drug research. This paper reviews new sulfur-containing compounds isolated from endophytic fungi since 1985 (Table 1). Based on their major chemical features, these compounds will be categorized into peptides, disulfides, polyketides, hybrids and terpenoids. The fungal strains that producing sulfur-containing compounds, host plants, structure uniqueness and biological activities of these compounds will be discussed (Table 1).

Table 1. Sulfur-containing compounds isolated from plant endophyte fungi.

Compound Structures	Producing Strain	Host Plant etc.	Bioactivity	Reference(s)
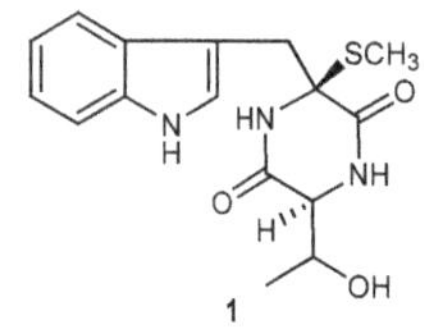 **1**	*Bionectria* sp. Y1085	*Huperzia serrata*	Antibacterial	[4]
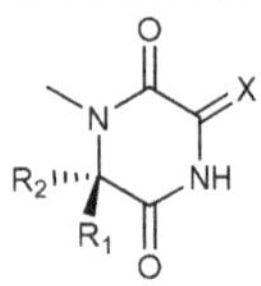 **2** R_1=Me, R_2=SMe, X=H,β-SMe **3** R_1=SMe, R_2=Me, X=H,β-SMe **4** R_1=Me, R_2=SMe, X=O 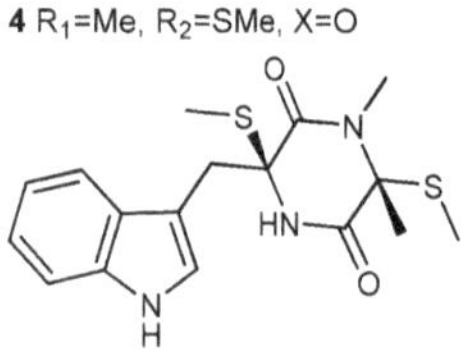 **5** 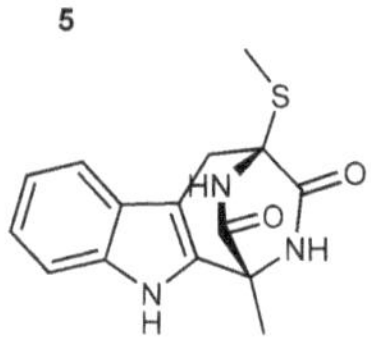 **6**	*Lasiodiplodia pseudotheobromae*	Flower of *Illigera rhodantha* (Hernandiaceae)	Antibacterial (**5**)	[5]
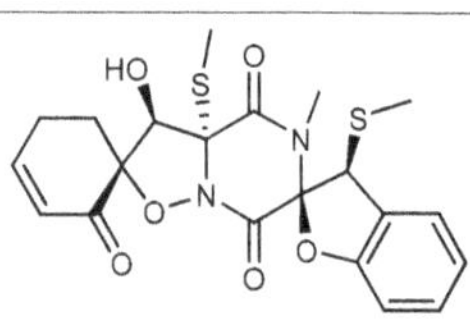 **7** 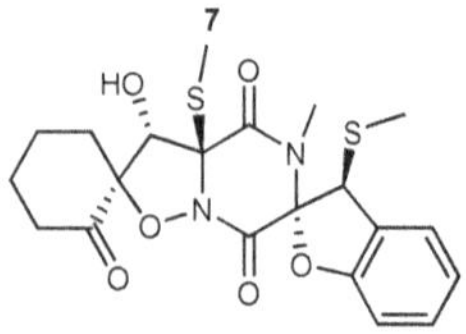 **8**	*Botryosphaeria mamani*	Fresh leaves of *Bixa orellana* L. (Bixaceae)	Anticancer	[6]

Table 1. *Cont.*

Compound Structures	Producing Strain	Host Plant etc.	Bioactivity	Reference(s)
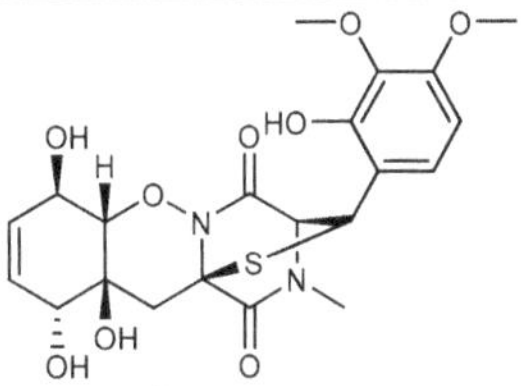 **9**	*Penicillium raciborskii* (TRT59)	*Rhododendron tomentosum*		[7]
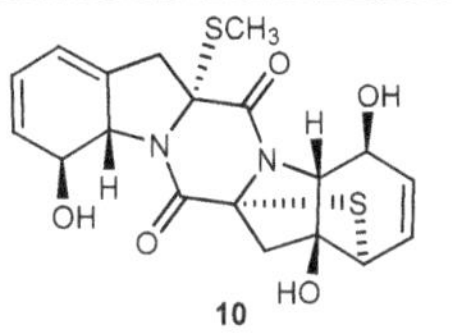 **10** 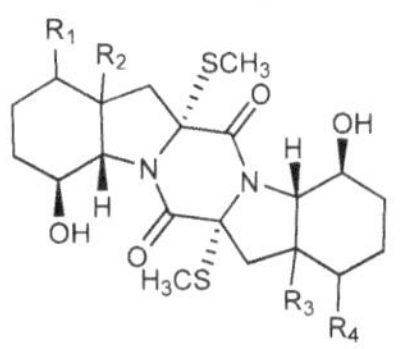 **11** R_1=R_4==O, R_2=R_3=(β)H **12** R_1=R_4=(α)OH, R_2=R_3=(α)H **13** R_1==O, R_2=(β)H, R_3=(α)H, R4=(α)OH 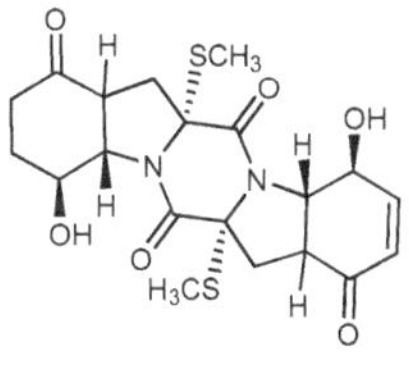 **14** 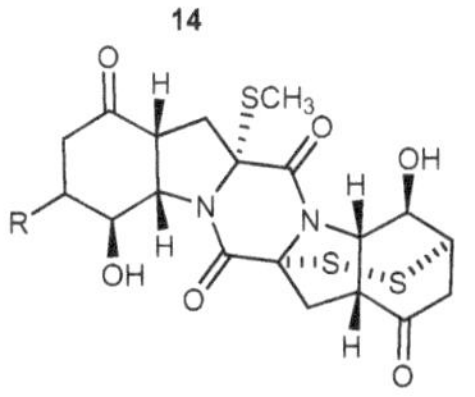 **15** R=H **16** R=(β)OH 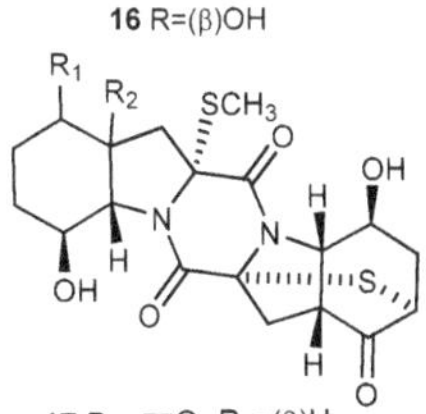 **17** R_1==O, R_2=(β)H **18** R_1=(α)OH, R_2=(α)H	*Epicoccum nigrum*	Leaves of *Lysidice rhodostegia*	Inhibition of β-Glucuronidase release (**11** and **15**)	[8]

Table 1. *Cont.*

Compound Structures	Producing Strain	Host Plant etc.	Bioactivity	Reference(s)
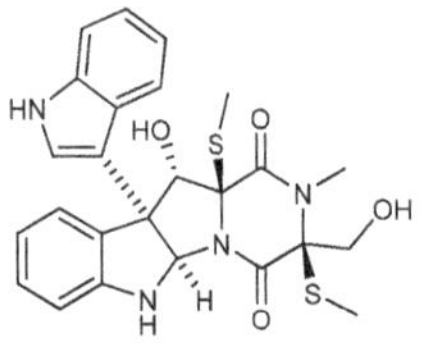 19	*Tilachlidium* sp. (CANU-T988)	Decaying wood sample collected in Christchurch	Cytotoxicity	[9]
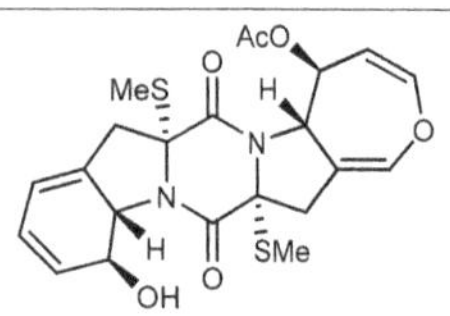 20	*Aspergillus terreus* BCC 4651	Tree hole	Weak antimycobacterial activity	[10]
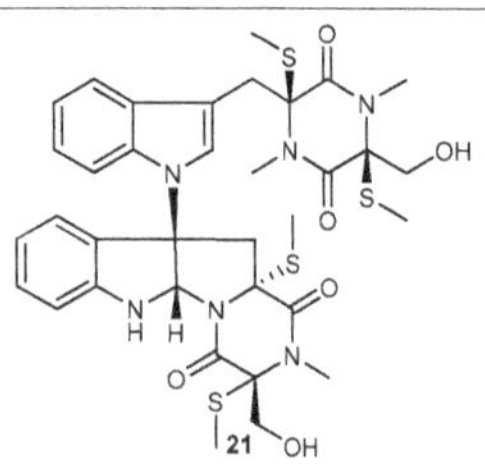 21 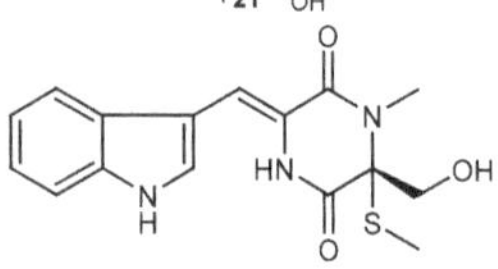 22 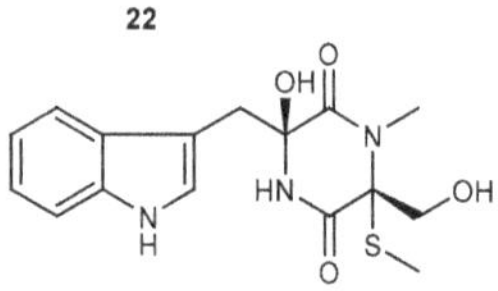 23	*Chaetomium* sp. 88194	*Cymbidium goeringii*	Cytotoxicity (**21**)	[11]
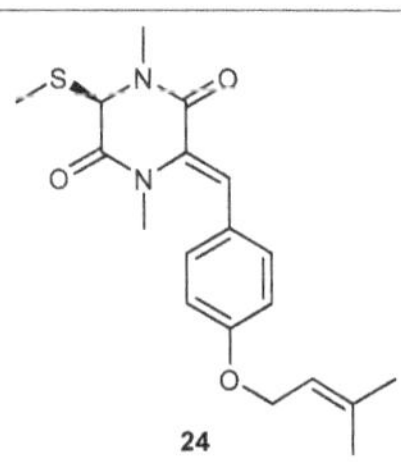 24 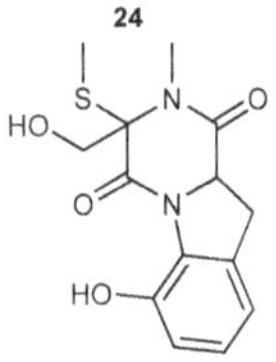 25	*Penicillium crustosum* and *Colletotrichum gloeosporioides*, respectively	*Viguiera robusta*		[12]

Table 1. *Cont.*

Compound Structures	Producing Strain	Host Plant etc.	Bioactivity	Reference(s)
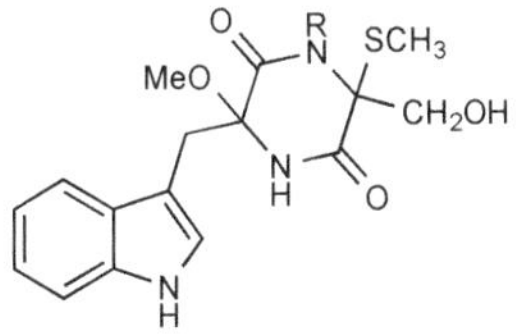 **26** R=CH_3 **27** R=H	*Chaetomium* sp. SYP-F7950	*Panax notoginseng*	Cytotoxic (**26**)	[13]
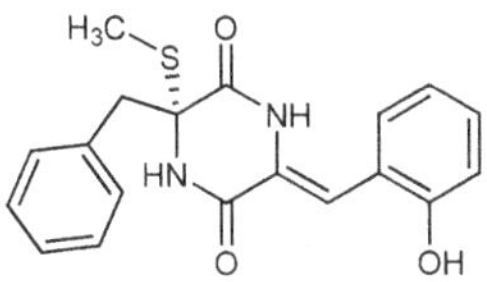 **28** 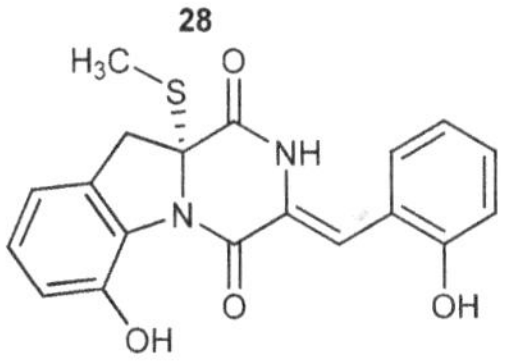 **29** 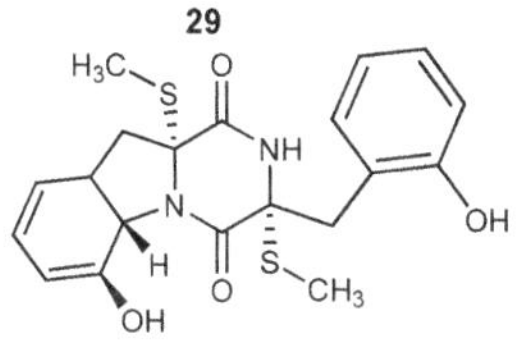 **30** 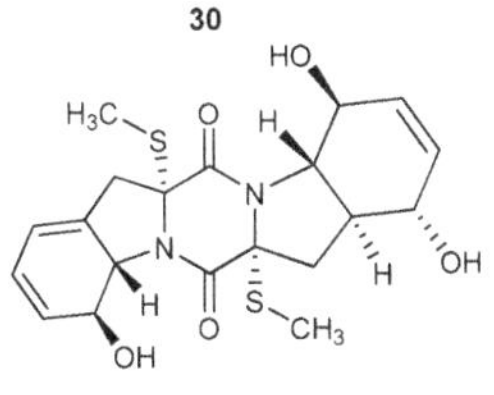 **31**	*Penicillium brocae* MA-231	Fresh tissue of the marine mangrove plant *Avicennia marina*	Antibacterial (**30** and **31**)	[14]
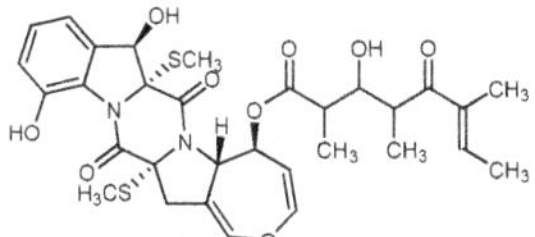 **32** 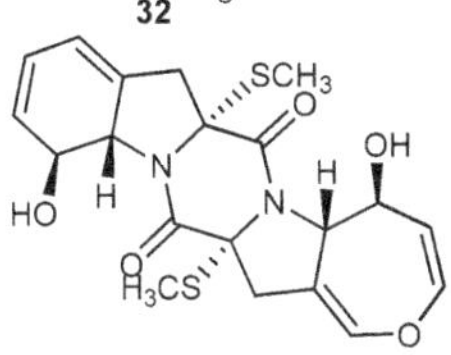 **33**	*Menisporopsis theobromae* BCC 3975	Seed	Antimycobacterial Cytotoxic (**32**)	[15]

Table 1. *Cont.*

Compound Structures	Producing Strain	Host Plant etc.	Bioactivity	Reference(s)
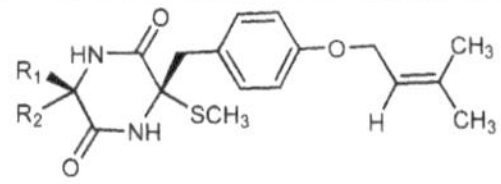 **34** R_1=H, R_2=SCH_3 **35** R_1=SCH_3, R_2=H	*Tolypocladium* sp.	*Quercus virginiana Miller*	PAF inhibition (**35**)	[16]
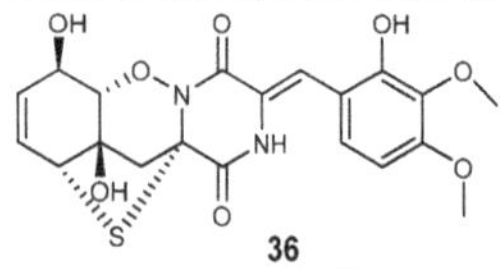 **36** 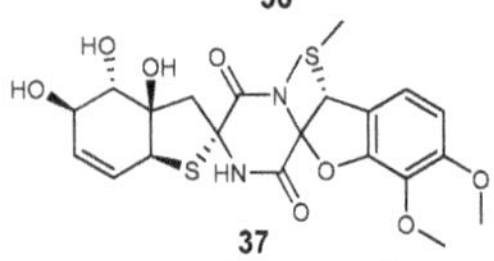 **37** 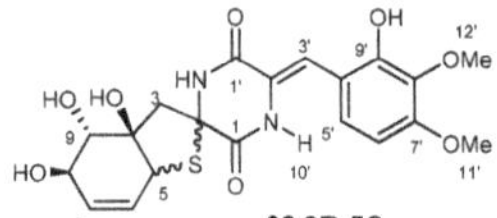 **38** 2R,5S **39** 2S,5R	*Penicillium janthinellum* HDN13-309	Root of *Sonneratia caseolaris*	Cytoprotective (**38** and **39**)	[17]
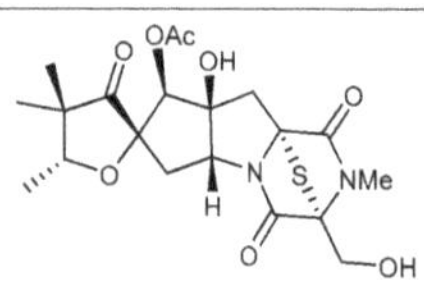 **40**	*Phoma lingam* isolate Leroy	Rapeseed		[18]
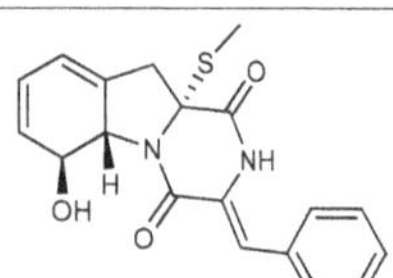 **41** 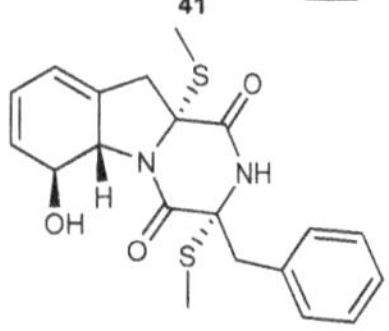 **42**	*Phoma* sp. OUCMDZ-1847	Mangrove plant *Kandelia candel*	Cytotoxic (**42**)	[19]
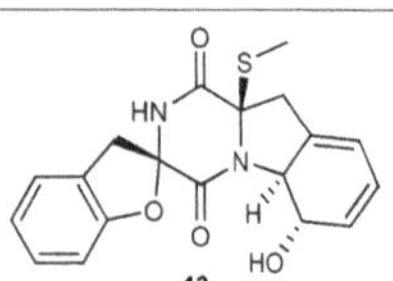 **43** 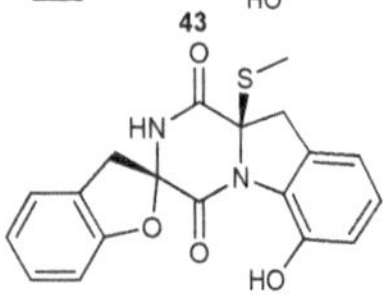 **44**	*Penicillium brocae* MA-231	*Avicennia marina*	Antimicrobial (**43**)	[20]

Table 1. *Cont.*

Compound Structures	Producing Strain	Host Plant etc.	Bioactivity	Reference(s)
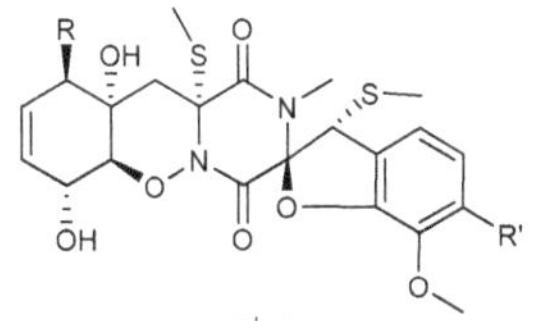 **45** R=Cl, R'=OMe **46** R=OH, R'=OMe **47** R=OH, R'=H	*Penicillium janthinellum* HDN13-309	Root of *Sonneratia caseolaris*		[21]
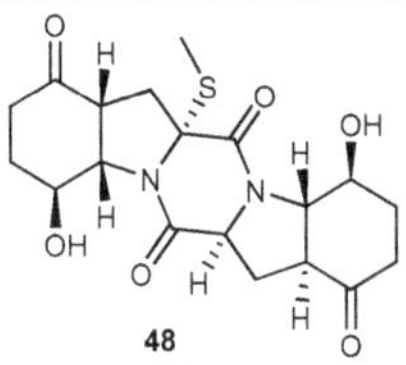 **48** 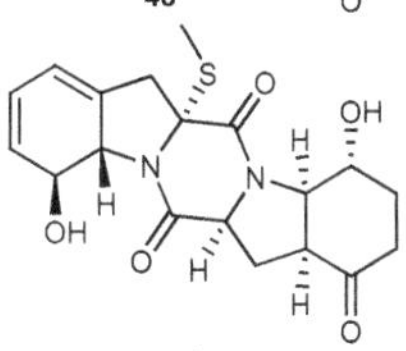 **49** 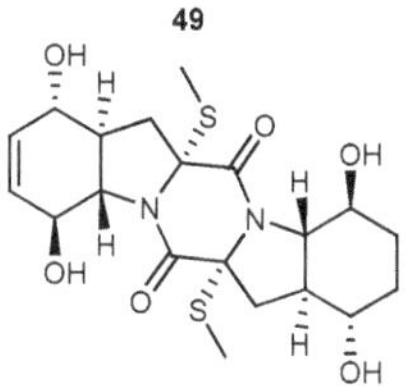 **50** 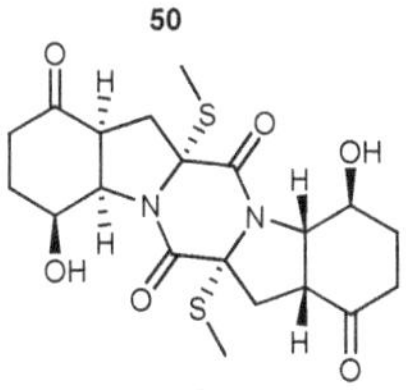 **51** 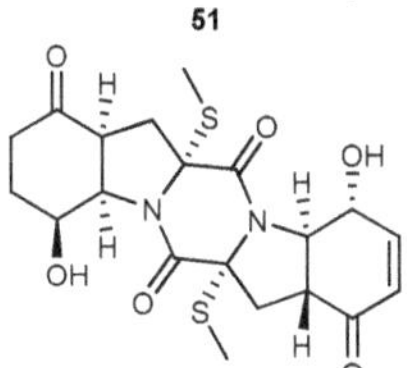 **52**	*Penicillium brocae* MA-231	Fresh tissue of the marine mangrove plant *Avicennia marina*	Antimicrobial	[22]

Table 1. *Cont.*

Compound Structures	Producing Strain	Host Plant etc.	Bioactivity	Reference(s)
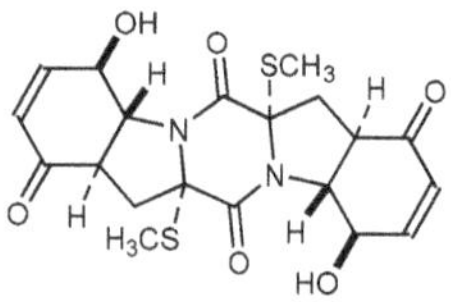 **53** 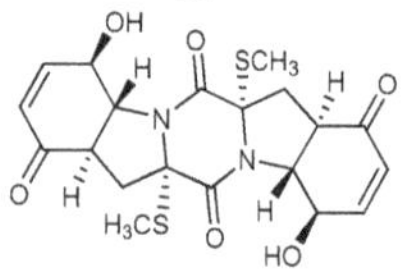 **54**	*Exserohilum holmii*	*Dactyloctenium aegyptium*		[23]
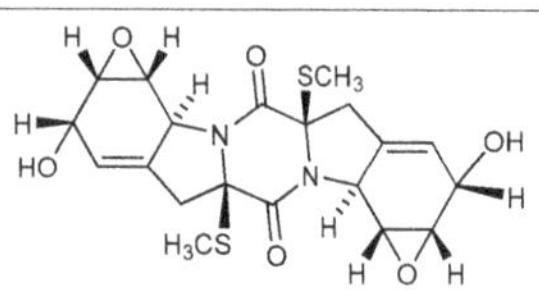 **55**	*Nigrospora sphaerica*	Germinating fescue seed		[24]
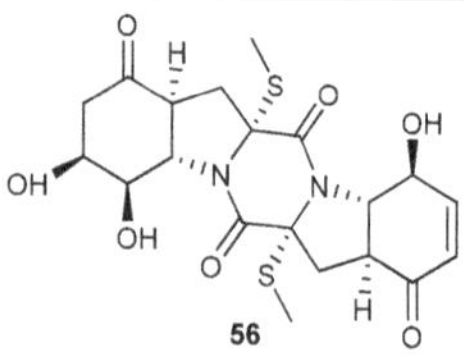 **56** 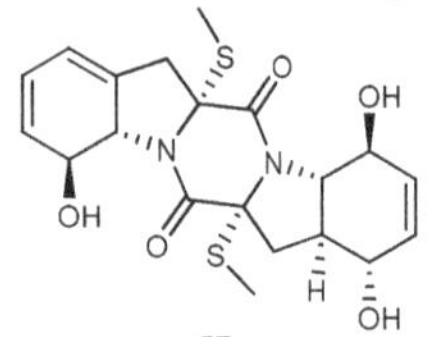 **57** 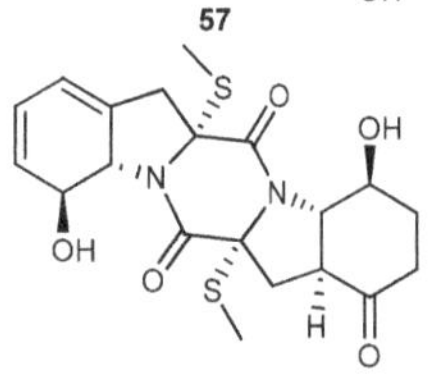 **58**	*Setosphaeria rostrata*	Fresh asymptomatic leaf tissues of the medicinal plant *Costus speciosus*	Inhibiting porcine pancreatic alpha-amylase (**57**)	[25]
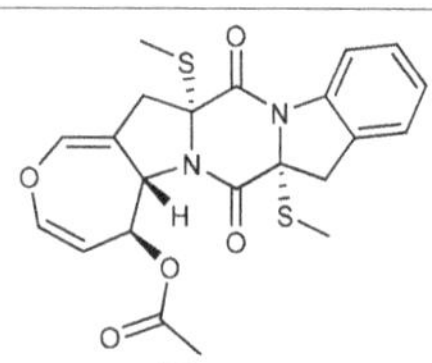 **59**	*Aspergillus versicolor* 0312	Stems of *Paris polyphylla var. yunnanensis*	Cytotoxic	[26]

Table 1. *Cont.*

Compound Structures	Producing Strain	Host Plant etc.	Bioactivity	Reference(s)
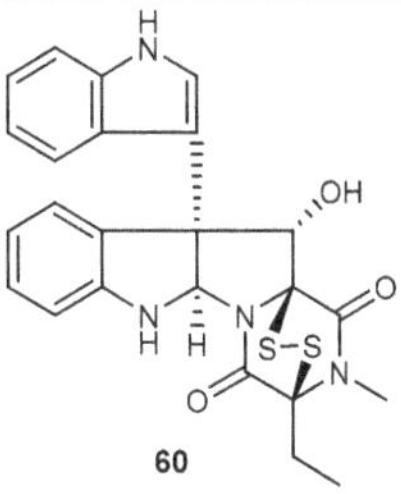 **60**	*Bionectria* sp. Y1085	*Huperzia serrata*	Antibacterial	[9]
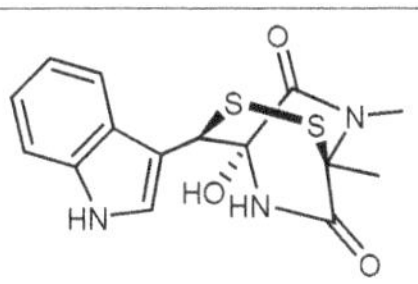 **61**	*Lasiodiplodia pseudotheobromae*	Apparently normal flower of *Illigera rhodantha* (Hernandiaceae)		[5]
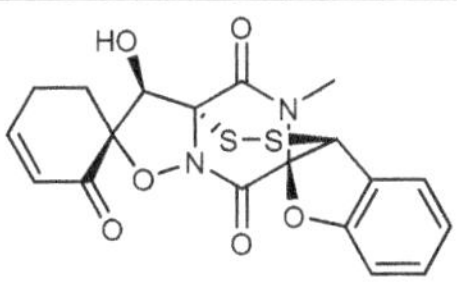 **62**	*Botryosphaeria mamani*	Fresh leaves of *Bixa orellana* L. (Bixaceae)	Cytotoxic	[6]
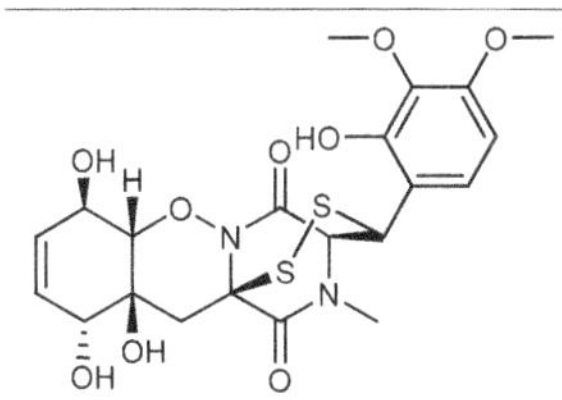 **63** 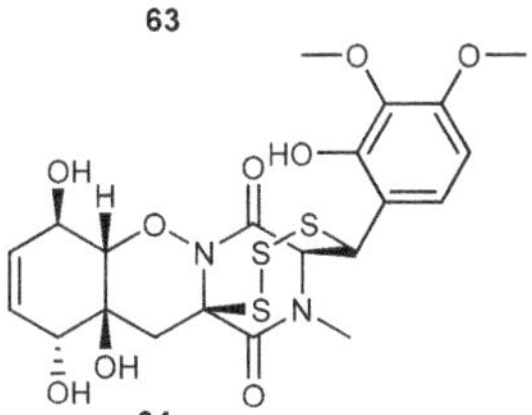 **64**	*Penicillium raciborskii* (TRT59)	*Rhododendron tomentosum*	Cytotoxic (**64**) Antifungal (**64**)	[7]

Table 1. *Cont.*

Compound Structures	Producing Strain	Host Plant etc.	Bioactivity	Reference(s)
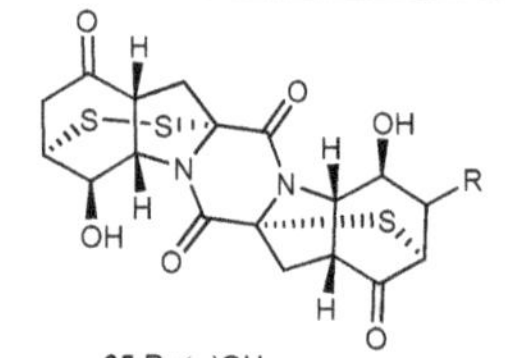 **65** R=(α)OH **66** R=H **67** R=(β)OCH$_2$CH$_3$ 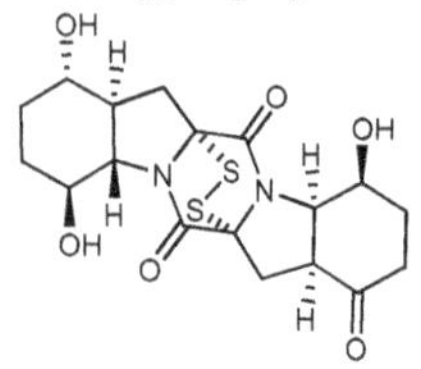 **68**	*Epicoccum nigrum*	Leaves of *Lysidice rhodostegia*	Inhibiting the release of β-glucuronidase (**67**)	[8]
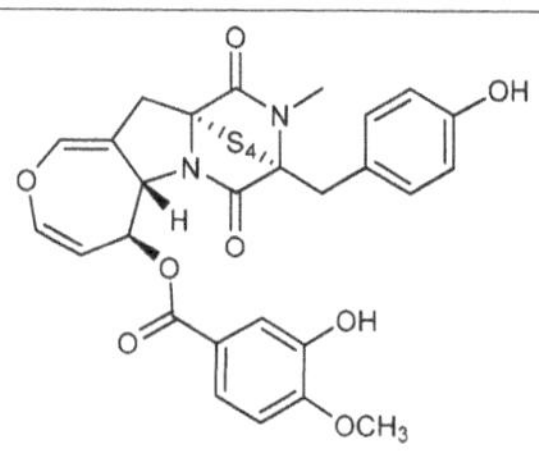 **69**	*Emericella* sp. AST0036	Healthy leaf tissue of *Astragalus lentiginosus*	Cytotoxic	[27]
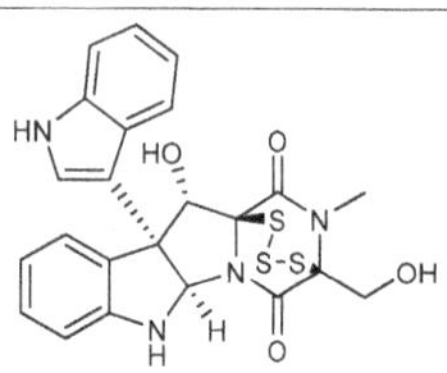 **70** 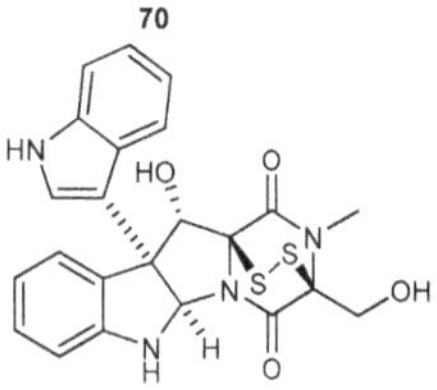 **71**	*Tilachlidium* sp. (CANU-T988)	Decaying wood sample collected in Christchurch	Cytotoxicity	[9]
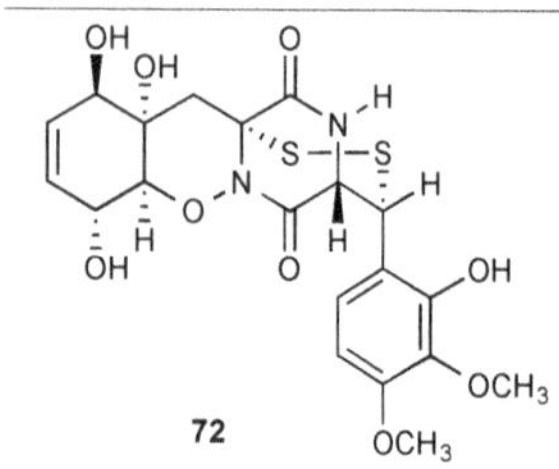 **72**	*Trichoderma* sp. BCC 5926	Bamboo leaf	Antibacterial	[28]

Table 1. *Cont.*

Compound Structures	Producing Strain	Host Plant etc.	Bioactivity	Reference(s)
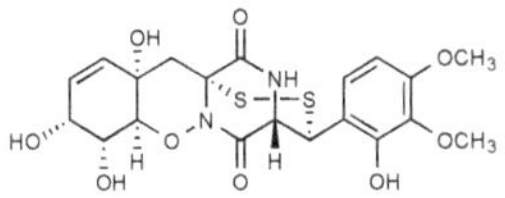 **73**	*Trichoderma harzianum*	*Zingiber officinale*		[29]
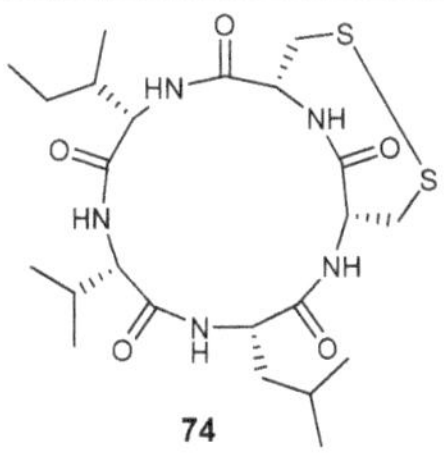 **74**	*Aspergillus tamarii*	*Ficus carica*	Cytotoxic Antimicrobial	[30]
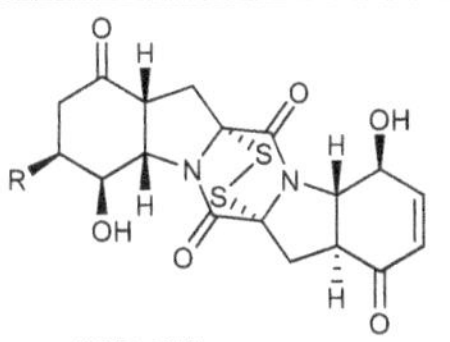 **75** R=OMe **76** R=H 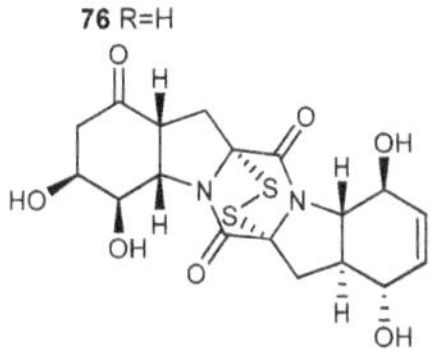 **77** 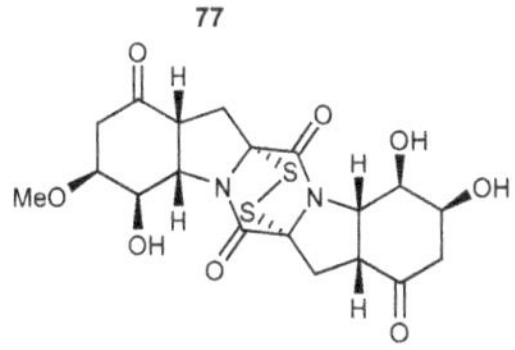 **78** 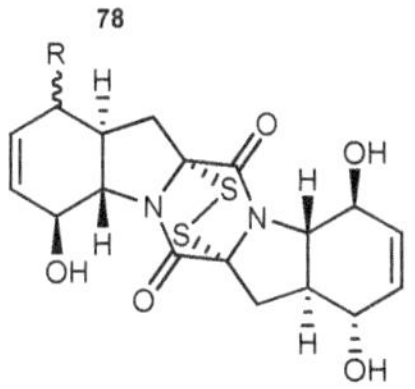 **79** R=(α)OH, (β) H **80** R= =O	*Penicillium brocae* MA-231	Fresh tissue of the marine mangrove plant *Avicennia marina*	Cytotoxic (**75**, **76**, **79** and **80**)	[31]
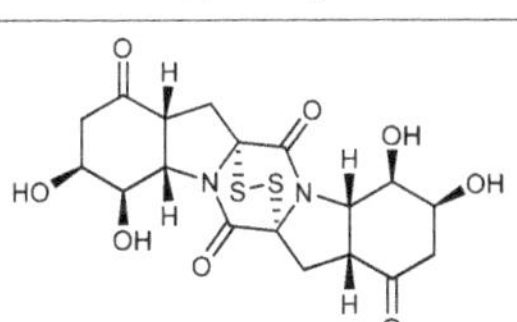 **81**	*Phoma* sp. OUCMDZ-1847	Mangrove plant *Kandelia candel*		[19]

Table 1. *Cont.*

Compound Structures	Producing Strain	Host Plant etc.	Bioactivity	Reference(s)
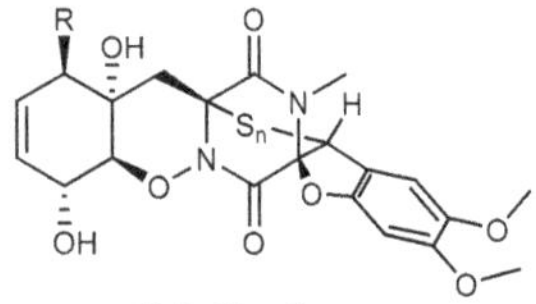 **82** R=Cl, n=2 **83** R=OH, n=2 **84** R=OH, n=3	*Penicillium janthinellum* HDN13-309	Root of *Sonneratia caseolaris*	Cytotoxic	[21]
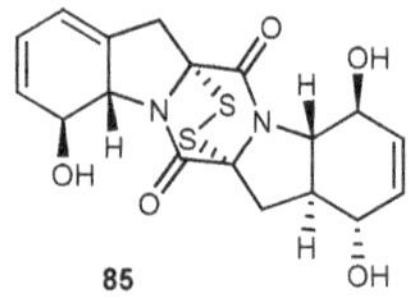 **85**	*Penicillium brocae* MA-231	Fresh tissue of the marine mangrove plant *Avicennia marina*	Cytotoxic Antimicrobial	[20]
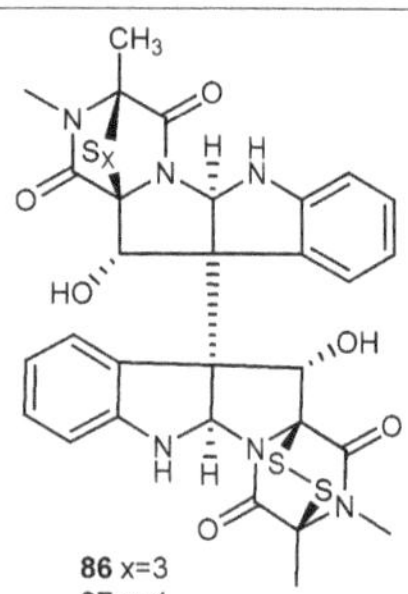 **86** x=3 **87** x=4 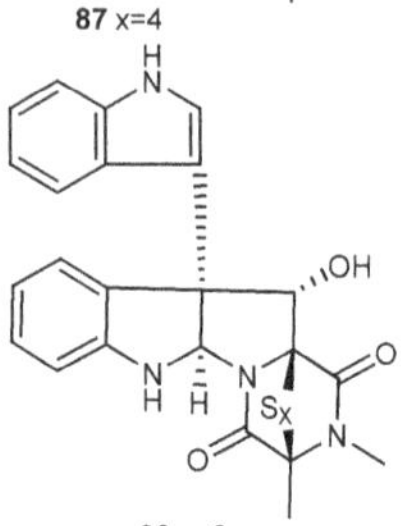 **88** x=2 **89** x=3 **90** x=4	*Gliocladium roseum* 1A	Submerged wood	Nematicidal	[32]
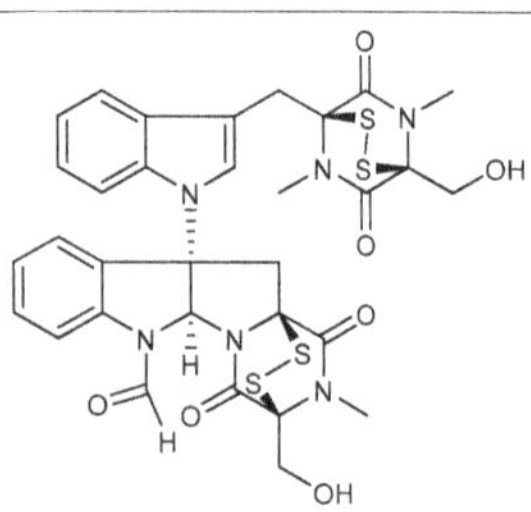 **91**	*Chaetomium* sp. M336	*Huperzia serrata Trev*	Cytotoxic Antibacterial	[33]

Table 1. *Cont.*

Compound Structures	Producing Strain	Host Plant etc.	Bioactivity	Reference(s)
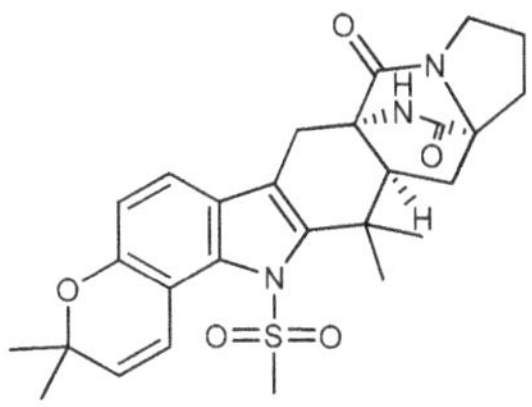 **92**	*Aspergillus versicolor* F210	Bulbs of *Lycoris radiata*	Anticancer	[34]
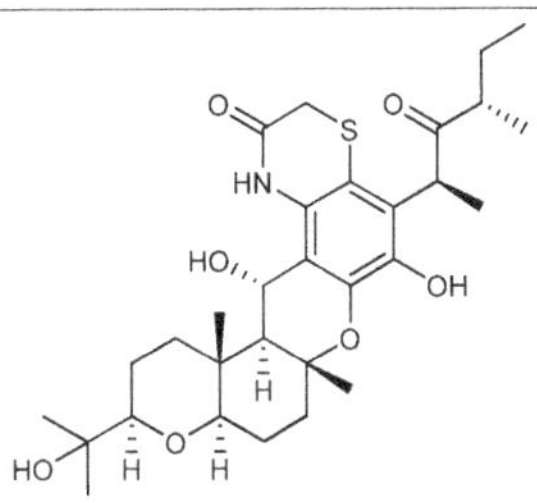 **93**	*Bipolaris sorokiniana* A606	*Pogostemon cablin*	Antiproliferative	[35]
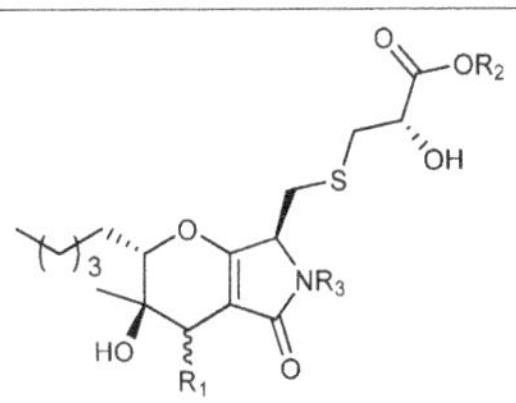 **94** R_1=(α)OH, R_2=R_3=H **95** R_1=(β)OH, R_2=H, R_3=OMe **97** R_1=(β)OH, R_2=Me, R_3=OMe 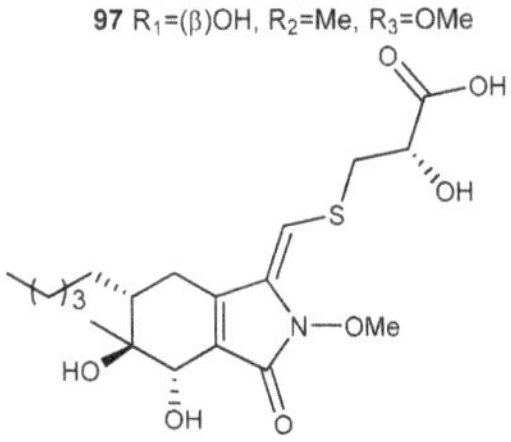 **96**	*Paraphaeosphaeria neglecta* FT462	*Lycopodiella cernua*	Antibacterial (**94**) Inhibiting NF-kB (**94**), iNOS (94 and **95**)	[36]
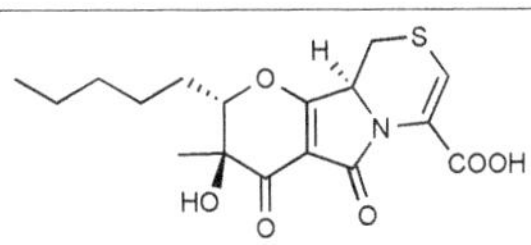 **98**	*Paraphaeosphaeria neglecta* FT462	*Lycopodiella cernua* (L.) Pic		[37]
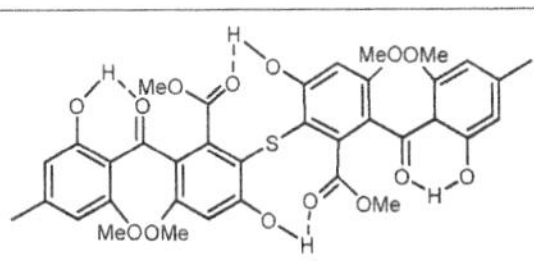 **99**	*Guignardia* sp. IFB-E028	*Hopea hainanensis*	Cytotoxic Antimicrobial	[38]

Table 1. *Cont.*

Compound Structures	Producing Strain	Host Plant etc.	Bioactivity	Reference(s)
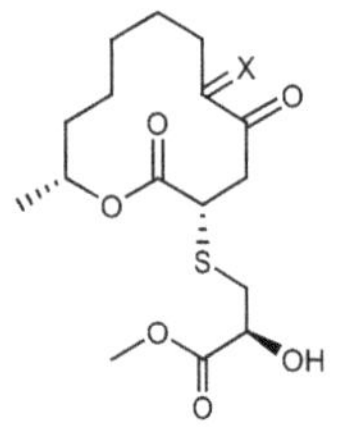 **100** x=H,H; **101** x=O 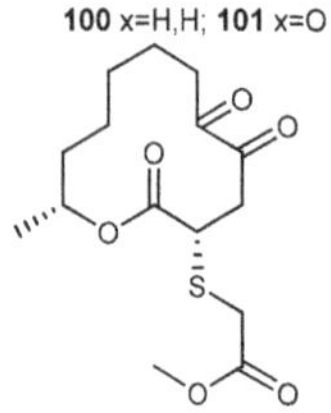 **102** 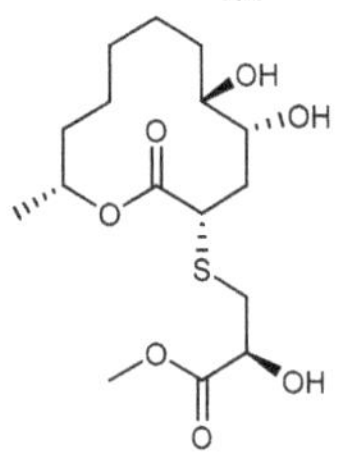 **103**	*Cladosporium cladosporioides* MA-299	*Bruguiera gymnorrhiza*	Antimicrobial	[39]
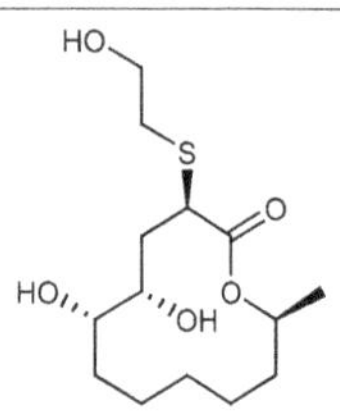 **104**	*Cladosporium* sp. SCNU-F0001	Mangrove plant		[40]
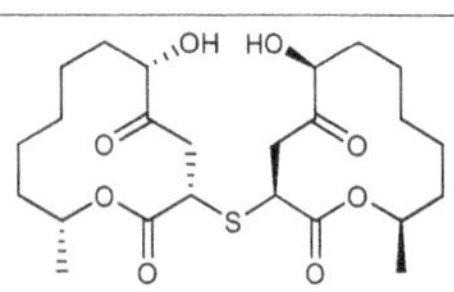 **105** 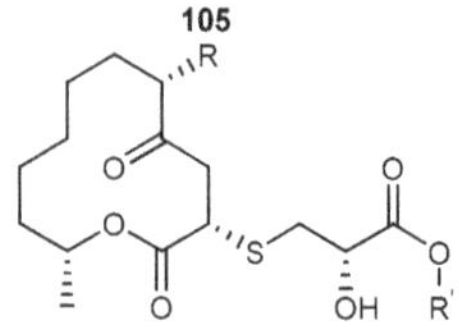 **106** R=OH R'=H; **107** R=R'=H **108** R=OH R'=X; **109** R=OH R'=Y	*Cladosporium oxysporum*	Root of *Avicennia marina (Forssk.) Vierh. (Acanthaceae)*	Antimicrobial	[41]

Table 1. *Cont.*

Compound Structures	Producing Strain	Host Plant etc.	Bioactivity	Reference(s)
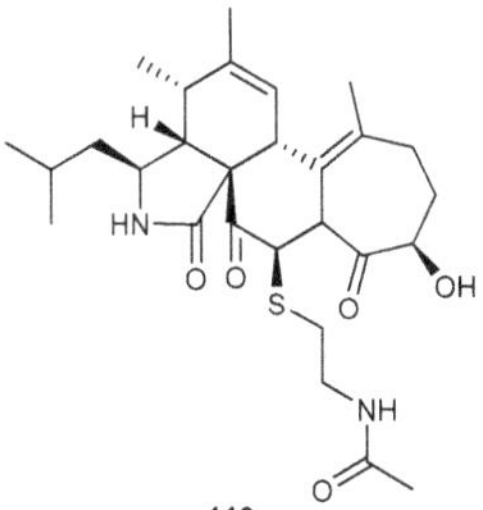 **110** 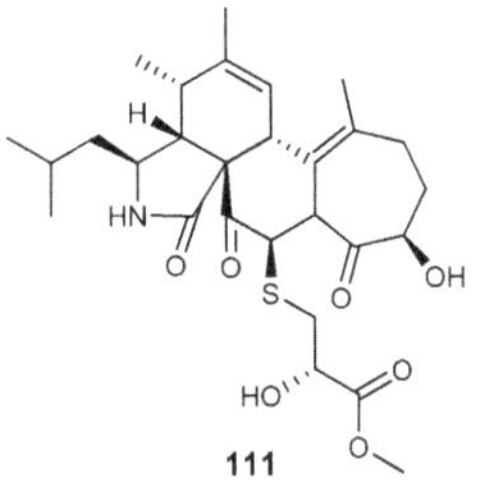 **111**	*Aspergillus micronesiensis*	*Phyllanthus glaucus*	Cytotoxic Antibacteria	[42]
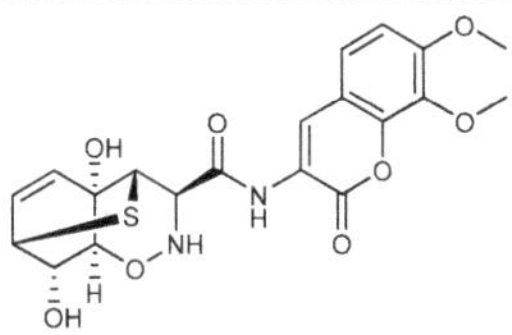 **112**	*Trichoderma harzianum* D13	Root of mangrove plant *Excoecaria agallocha Linn*		[43]
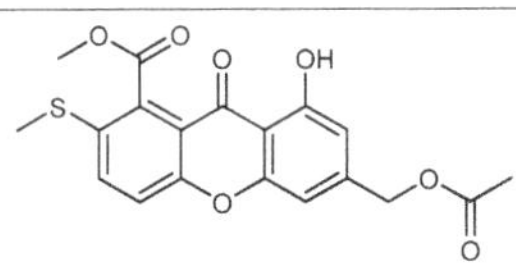 **113** 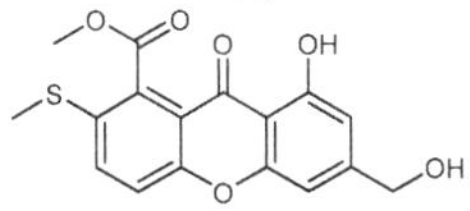 **114**	*Aspergillus sydowii*	Livewort *Scapania ciliata S. Lac*		[44]
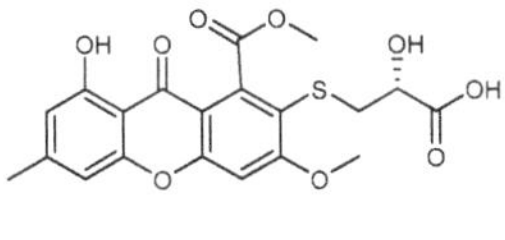 **115** 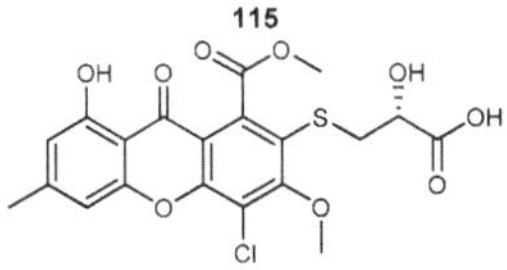 **116**	*Pseudopestalotiopsis theae*	Leaves of *Caloncoba welwitschii*		[45]

Table 1. *Cont.*

Compound Structures	Producing Strain	Host Plant etc.	Bioactivity	Reference(s)
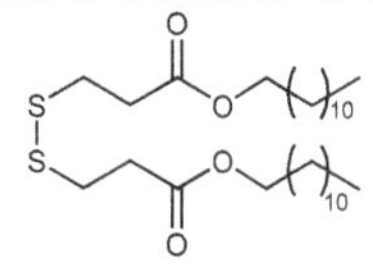 **117**	*Sphaceloma* sp. LN-15	Leaves of *Melia azedarach* L.		[46]
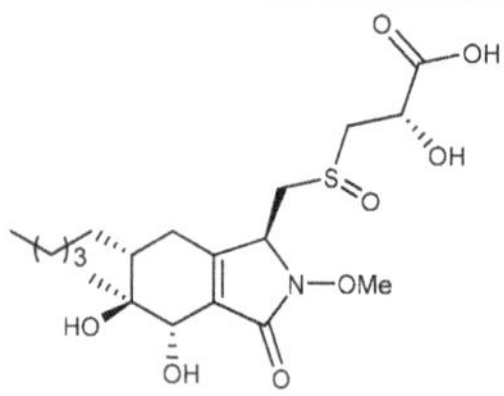 **118**	*Paraphaeosphaeria neglecta* FT462	*Lycopodiella cernua*		[36]
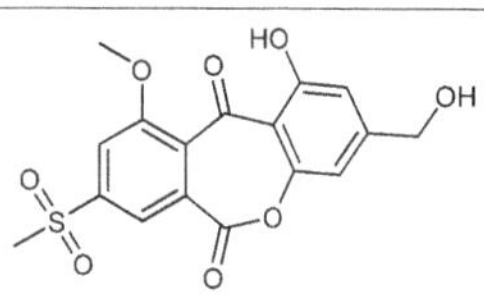 **119**	*Neosartorya udagawae* HDN13-313	Root of the mangrove plant *Aricennia marina*	Decreasing the lipid accumulation elicited by oleic acid	[47]
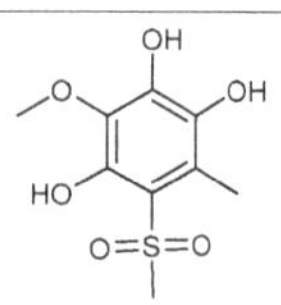 **120**	*Neosartorya udagawae* HDN13-313	Root of the mangrove plant *Avicennia marina*		[47]
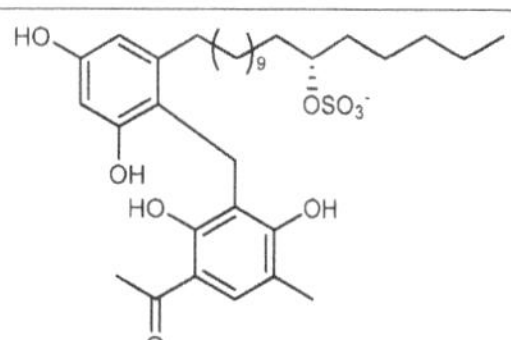 **121** 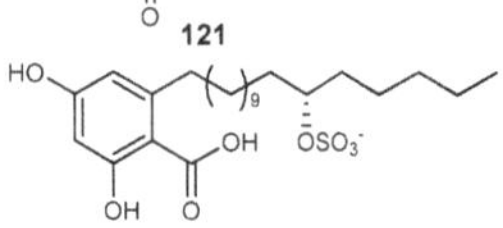 **122**	*Penicillium crustosum* PRB-2 and *Xylaria* sp. HDN13-249.	Root of *Sonneratia caseolaris*	Antibacterial	[48]
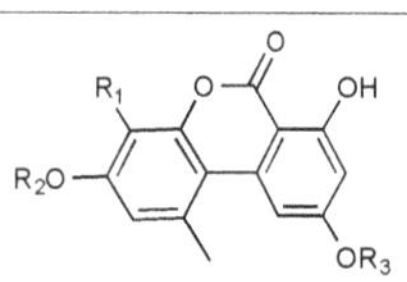 **123** R_1=H, R_2=H, R_3=SO_3H **124** R_1=H, R_2=SO_3H, R_3=CH_3	*Alternaria* sp.	*Polygonum senegalense Meisn.* (Polygonaceae)	Cytotoxic (**123**) Inhibiting protein kinases (**123**)	[49]

Table 1. *Cont.*

Compound Structures	Producing Strain	Host Plant etc.	Bioactivity	Reference(s)
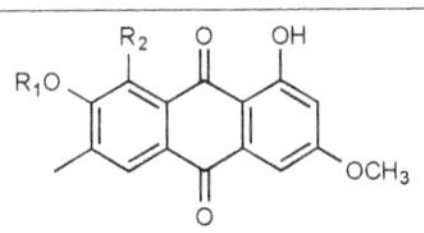 **125** $R_1=SO_3Na^+$, $R_2=H$ **126** $R_1=SO_3Na^+$, $R_2=OH$	*Ampelomyces* sp.	*Urospermum picroides*		[50]
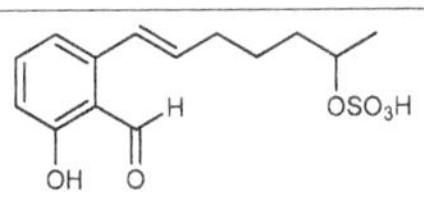 **127**	*Pestalotiopsis* sp. AcBC2	*Aegiceras corniculatum*		[51]
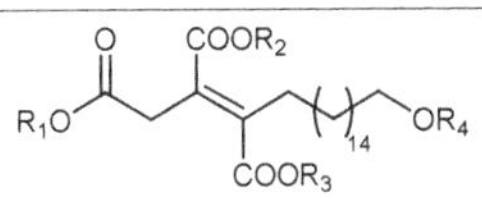 **128** $R_1=R_2=R_3=H$, $R_4=SO_3H$ **129** $R_1=R_2=R_3=CH_3$, $R_4=SO_3H$	MF6046	Surface-sterilized leaves of *Berberis oregana* (Berberidaceae)	Inhibiting FPTase (**128**)	[52]
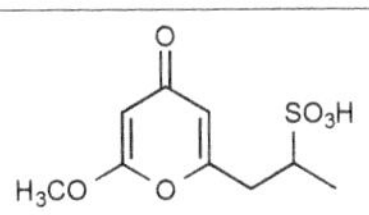 **130**	*Fusarium* sp. (CTGU-ZL-34).	*Davidia involucrata*	Cytotoxic	[53]
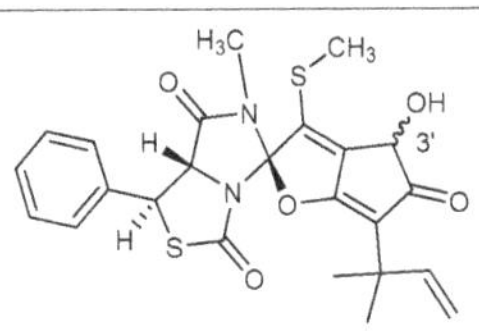 **131** 3'S; **132** 3'R	*Pestalotiopsis* sp. HS30	*Isodon xerophilus*	Antitumor	[54]
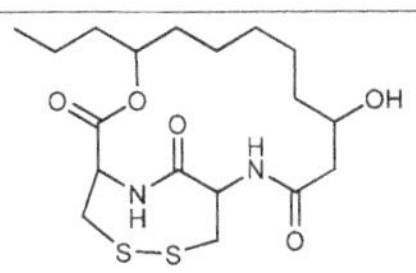 **133**	*Phomopsis glabrae*	Leaves of *Pongamia pinnata* (family Fabaceae)	Anticancer	[55]
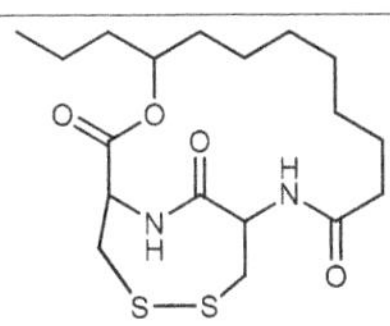 **134**	*Ascochyta* sp. AJ 117309	Raw leaf of *Taxus cuspidata var. nana Rehd*	Cytotoxic	[56]
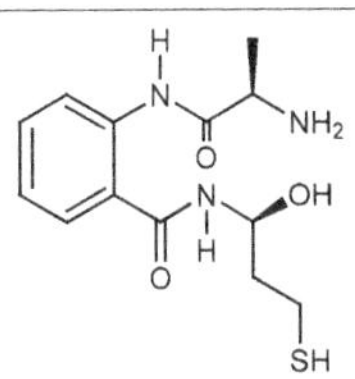 **135**	*Fusarium chlamydosporium*	Leaves of *Anvillea garcinia* (Burm.f.) DC. (Asteraceae)	Cytotoxic Antimicrobial	[57]

Table 1. *Cont.*

Compound Structures	Producing Strain	Host Plant etc.	Bioactivity	Reference(s)
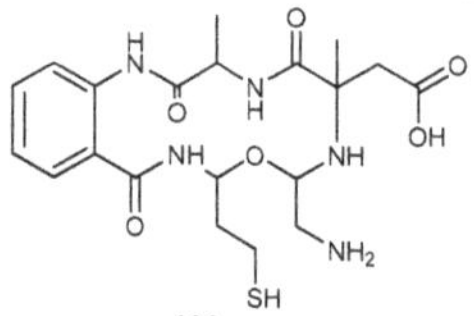 **136**	*Fusarium chlamydosporium*	*Anvillea garcinii* (Burm.f.) DC. leaves	Antibacterial Antifungal Cytotoxic	[58]
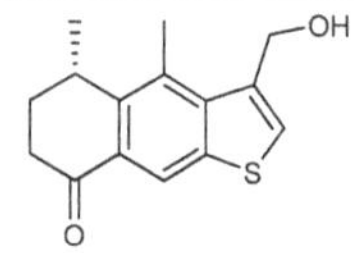 **137**	*Leptosphaeria* sp. XL026	*Panax notoginseng*	Antifungal Antibacterial	[59]
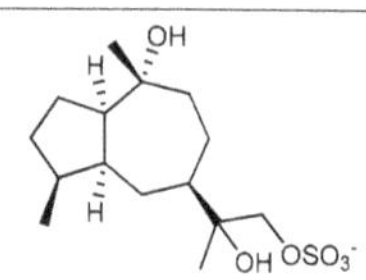 **138** 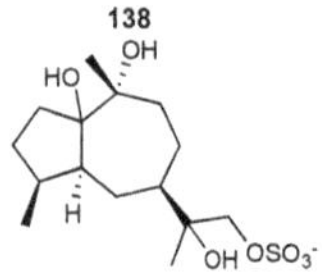 **139**	S49	Bark of *Cephalotaxus hainanensis* tree		[60]
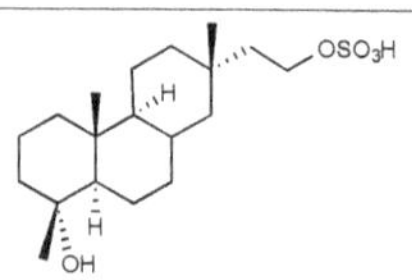 **140** 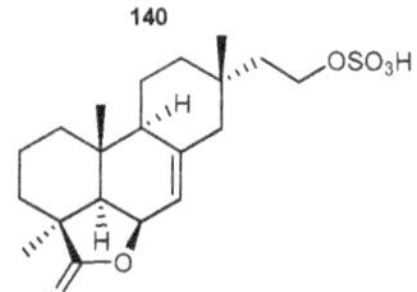 **141**	*Xylaria* sp. YM 311647	*Azadirachta indica*	Antifungal	[61]
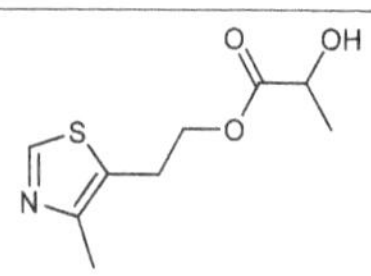 **142**	*Colletotrichum gloeosporioides* A12	*Aquilaria sinensis*		[62]
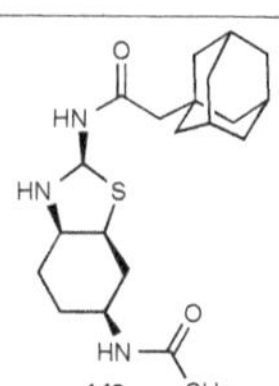 **143**	*Emericella Sp*	*Azadirachta indica*	Anticandidal	[63]

2. Peptides

2.1. Sulfide (R-S-R′)

A rare diketopiperazine bionectin D (**1**) (Figure 1) was obtained from a fungal strain *Bionectria* sp. Y1085 that was isolated from the plant *Huperzia serrata*. Bionectin D (**1**) consists of a tryptophan and a threonine moiety, and the α-carbon of its tryptophan moiety carries a single methylthio substitution. Compound **1** exhibited antibacterial activity against *Staphylococcus aureus*, *Escherichia coli*, and *Salmonella typhimurium* ATCC 6539 with the same minimal inhibitory concentration (MIC) of 25 μg/mL [4]. Lasiodiplines A-C (**2–4**) and E-F (**5–6**) are new sulfureous diketopiperazines that were produced by *Lasiodiplodia pseudotheobromae* F2 isolated from the apparently normal flower of *Illigera rhodantha*. The structure elucidation of these compounds was accomplished using a combination of spectroscopic and computational approaches, and the structure of **2** was further confirmed in conjunction with low-temperature (100 K) single-crystal X-ray diffraction. Lasiodiplines E (**5**) displayed antibacterial activity against *Veillonella parvula*, *Actinmyces israelili*, *Streptococcus* sp., *Bacteroides vulgates* and *Peptostreptococcus* sp. with the MIC values of 0.25, 32.0, 0.12, 0.12 and 0.12 μg/mL, respectively [5].

Botryosulfuranols A and B (**7–8**), two spirocyclic thiodiketopiperazines, were purified from *Botryosphaeria mamani*. The fungal strain was isolated from the fresh leaves of *Bixa orellana* L. (Bixaceae) collected in Peru. These two unique compounds, each of which contains two spiro centers, were derived from two L-phenylalanines with two methylthio substitutions at the α-carbon and β-carbon of the two building blocks, respectively. Botryosulfuranols A (**7**) was active against four cancer cell lines (HT-29, HepG2, Caco-2, HeLa) with IC_{50} values of 8.0, 11.4, 18.2, 23.5 and 9.3 μM, respectively. Botryosulfuranols B (**8**) was active against three cancer cell lines (HT-29, HepG2, HeLa) with the IC_{50} values of 63.2, 56.1, 61.2, 49.9 and 64.7 μM, respectively [6]. Outovirin A (**9**) was a thiodiketopiperazine derived from two molecules of L-phenylalanine. It was produced by *Penicillium raciborskii*, an endophytic fungus isolated from *Rhododendron tomentosum* [7]. Compound **9** contains a nitrogen-oxygen bond in the oxazinane ring between diketopiperazine and conduritol-like rings, and it has a sulfide bridge between the α- and β-carbons rather than the typical α−α bridging. Nine new thiodiketopiperazines, epicoccin I (**10**), ent-epicoccin G (**11**), and epicoccins J-P (**12–18**), have been isolated from the endophytic fungus *Epicoccum nigrum*. Compounds **10**, **17**, and **18** all have a sulfide bridge between the α-carbon and the 2′/3′-position of the reduced benzene ring. Ent-epicoccin G (**11**) and epicoccins M (**15**) showed potent in vitro activities against the release of β-glucuronidase in rat polymorphonuclear leukocytes induced by the platelet-activating factor, with IC_{50} values of 3.07 and 4.16 μM, respectively [8].

Tilachlidium sp. (CANU-T988) isolated from a decaying wood sample was reported to produce T988 B (**19**). Compound **19** has an unusual dimerized indole moiety with a 3-3 linkage, and it displayed potent cytotoxicity against P388 leukemia cells with an IC_{50} of 2.18 μM [9]. Bisdethiobis(methylsulfanyl)apoaranotin (**20**) was produced by *Aspergillus terreus* BCC 4651, which was isolated from a tree hole in Nam Nao National Park, Thailand. Compound **20** was derived from two molecules of L-phenylalanine with one benzene ring being oxidized to a 4,5-dihydrooxepine ring. Compound **20** exhibited weak antimycobacterial activity [10]. Chaetocochin G (**21**), oidioperazine E (**22**), and chetoseminudin E (**23**) were obtained from *Chaetomium* sp 88194, which was isolated from *Cymbidium goeringii*, a plant native to China, Japan and Korea. Chaetocochin G (**21**) is a dimer of serine-tryptophan diketopiperazines. Its structure including the absolute configuration was established by spectroscopic data interpretation and single-crystal X-ray diffraction analysis. Chaetocochin G (**21**) showed cytotoxicity against MCF-7 [11]. Fusaperazine E (**24**) and colletopiperazine (**25**) were obtained from *Penicillium crustosum* and *Colletotrichum gloeosporioides*, respectively. Both strains were isolated from *Viguiera robusta* Gardn. (Asteraceae). [12].

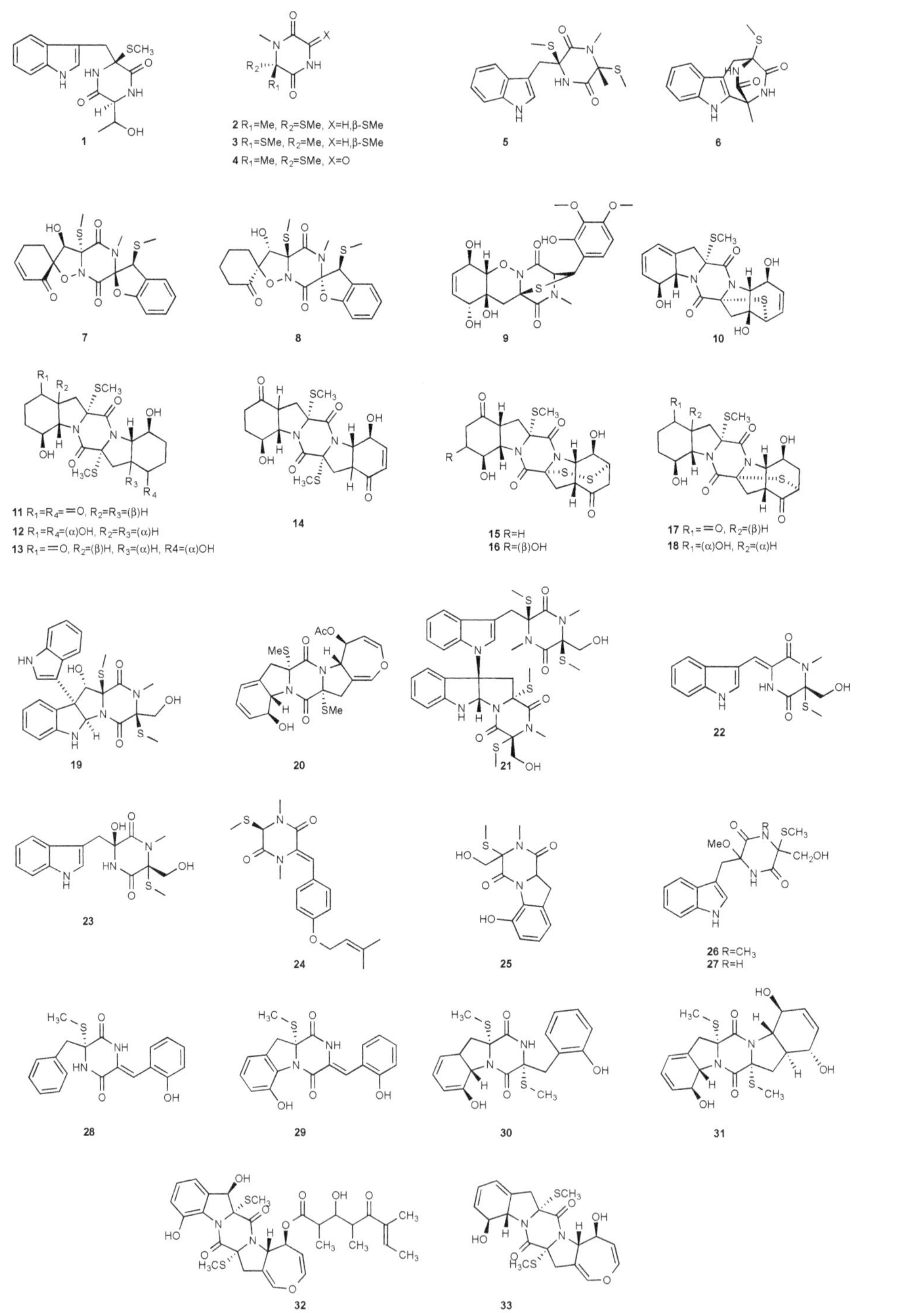

Figure 1. Structures of compounds **1–33**.

Chetoseminudin F and G (**26–27**) were purified from *Chaetomium* sp. SYP-F7950, which was isolated from the root of *Panax notoginseng* collected from Wenshan, Yunnan, P. R. China. Chetoseminudin F (**26**) displayed cytotoxicity against MDA-MB-231 with an IC_{50} of 26.49 μM [13]. Four thiodiketopiperazines penicibrocazines F–I (**28–31**) were purified from *Penicillium brocae* MA-231, which was isolated from the fresh tissue of the marine mangrove plant *Avicennia marina* collected at Hainan Island, P. R. China. Penicibrocazines H (**30**) displayed activity against *V. harveyi, E. coli, A. hydrophilia* and *V. parahaemolyticus* with MICs of 16.0, 16.0, 32.0, and 16.0 μg/mL, respectively. Penicibrocazines I (**31**) displayed activity against *V. harveyi* with an MIC of 32.0 μg/mL [14].

Two new compounds 6-octenoic acid, 3-hydroxy-2,4,6-trimethyl-5-oxo-, (5S,5aS,7aR,8R, 14aR)-5,5a,7a,8,14a,15-hexahydro-8,12-dihydroxy-7a,14a-bis(methylthio)-7,14-dioxo-7H,14H-oxepino[3″,4″:4′,5′]pyrrolo[1′,2′:4,5]pyrazino[1,2-a]indol-5-yl ester (6E) (**32**) and bisdethio-bis(methylthio)deacetylapoaranotin (**33**) were purified from the seed fungus *Menisporopsis theobromae* BCC3975. Compound **32** is a hybrid of diketopiperazine and polyketide. Both compounds showed antimycobacterial activity with MICs of 1.24 and 7.14 μM, respectively. Compound **32** displayed cytotoxicity against NCI-H187 cell line and antimalarial activity with IC_{50} of 20.3 and 2.95 μM, respectively [15].

Two new compounds, Sch 54794 (**34**) and Sch 54796 (**35**) (Figure 2), were separated from the fermentation culture of *ToJypocJadium* sp. The microorganism *ToJypocJadium* sp. was isolated from dead twigs from a *Quercus virginiana* Miller, an old live oak tree in the state of Tamalupas, Mexico. The structures of Sch 54794 (**34**) and Sch 54796 (**35**) were determined as *cis* and *trans* isomers in the spectroscopic analysis. The *trans* isomer, which was similar to other diketopiperazines reported as platelet-activating factor (PAF) inhibitors in the literature, displayed weak inhibitory activity in PAF assay with an IC_{50} of 50 μM. However, the *cis* isomer appeared inactive (IC_{50} > 100 μM) [16].

Four new dioxopiperazine alkaloids, penispirozines A−D (**36–39**), were produced by *Penicillium janthinellum* HDN13-309, which was isolated from the root of the mangrove plant *Sonneratia caseolaris*. Penispirozine A (**36**) contains an unusual pyrazino[1,2]oxazadecaline coupled with a thiophane ring system, and compound **37** possesses a 6/5/6/5/6 pentacyclic ring system with two rare spirocyclic centers. Penispirozines C (**38**) and penispirozines D (**39**) increased the expression of superoxide dismutase 2 (SOD2) and heme oxygenase-1 (HO-1) at 10 μM [17]. A fermentation broth of *Phoma lingam* isolate Leroy obtained from rapeseeds generated a new compound sirodesmin H (**40**) [18]. The octahydrocyclopenta[*b*]pyrrole moiety in **40** might be derived from L-phenylalanine, which reacted with an isoprenyl group (C_5) to form a spiro-furanone system. Two new thiodiketopiperazines phomazines A (**41**) and B (**42**) were purified from *Phoma* sp. OUCMDZ-1847, which was isolated from the mangrove plant *Kandelia candel* at Wenchang, Hainan, P. R. China. Compound **42** displayed inhibitory activity against MGC-803 cells with an IC_{50} of 8.5 μM [19].

Two new pentacyclic diketopiperazines spirobrocazines A (**43**) and B (**44**) were obtained from *Penicillium brocae* MA-231, which was derived from the marine mangrove plant *Avicennia marina* [20]. Compound **43** exhibited moderate antibacterial activities against *Escherichia coli, S. aureus* and *Vibrio harveyi* with MIC values of 32.0, 16.0 and 64.0 μg/mL, respectively. Three new epipolythiodioxopiperazines, penicisulfuranols D–F (**45–47**), were isolated from a marine mangrove plant, *Sonneratia caseolaris*-derived *Penicillium janthinellum* HDN13-309 [21]. The piperazine-2,5-dione core in each of these compounds (**45–47**) was flanked by a 1,2-oxazadecaline moiety and a *spiro*-benzofuran ring. Compounds **45–47** were tested inactive against HeLa and HL-60 cell lines. Five pentacyclic diketopiperazines, penicibrocazines A–E (**48–52**), were obtained from *Penicillium brocae* MA-231, a fungus obtained from the fresh tissue of the marine mangrove plant *Avicennia marina*. In the antimicrobial screening, penicibrocazine B (**49**), penicibrocazine C (**50**) and penicibrocazine D (**51**) showed activity against *Staphylococcus aureus*, with MIC values of 32.0, 0.25, 8.0 μg/mL, respectively, which are comparable with that of the positive control, chloromycetin (MIC = 4.0 μg/mL). Penicibrocazines C (**50**) also showed activity against *Micrococcus luteus* with an MIC of 0.25 μg/mL, which is stronger than that of the positive control, chloromycetin (MIC = 2.0 μg/mL). Moreover, penicibrocazines B (**49**) and D (**51**) exhibited activity against the plant pathogen *Gaeumannomyces graminis* with MIC values of 0.25 and 8.0 μg/mL, respectively, while the positive control amphotericin B has an MIC of 16.0 μg/mL [22].

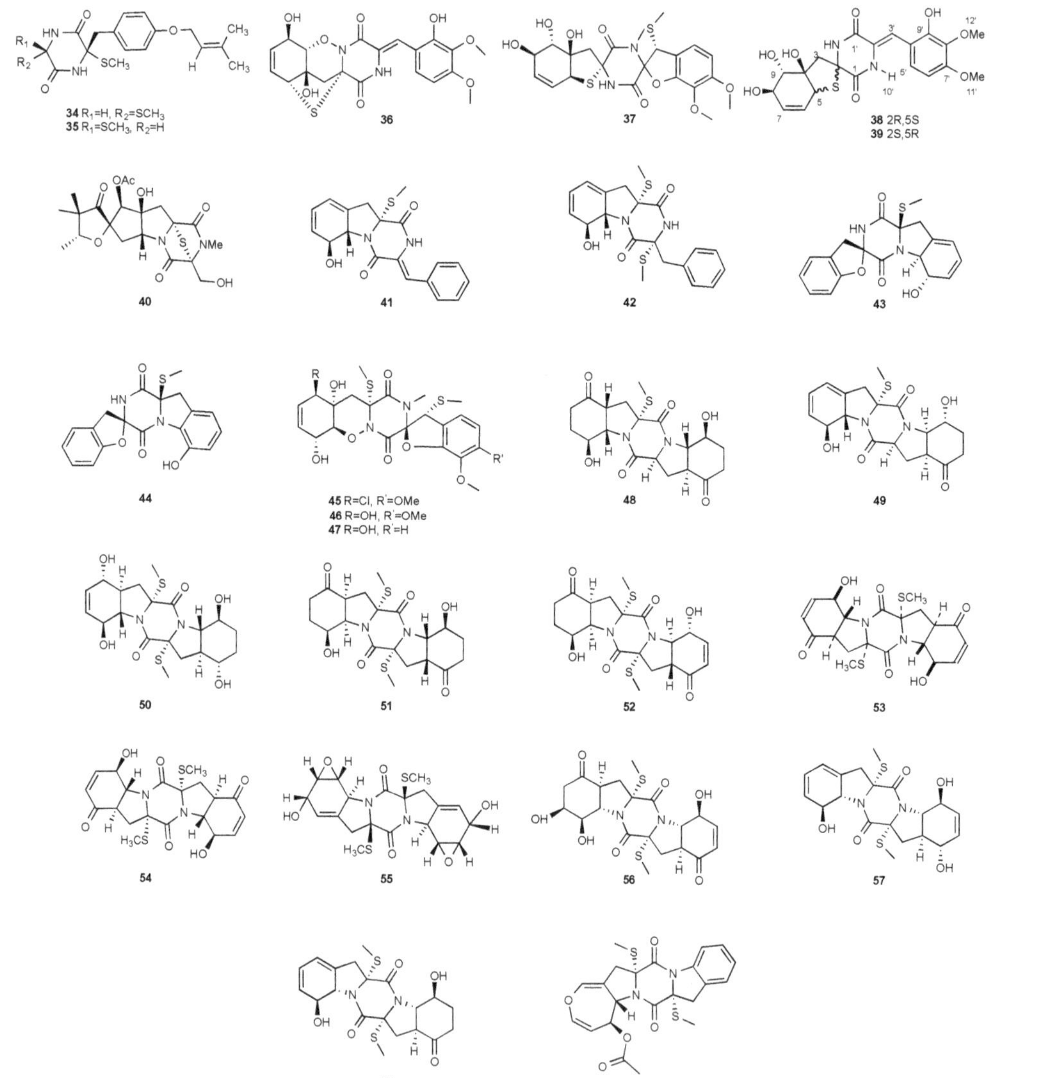

Figure 2. Structures of compounds **34–59**.

The chemical investigation of a culture of *Exserohilum holmii*, a pathogenic fungus of the weedy plant *Dactyloctenium aegyptium*, yielded two linearly fused pentacyclic diketopiperazines exserohilone (**53**) and 9,10-Dihydroexserohilon (**54**) [23]. The fermentation of *Nigrospora sphaerica*, which was isolated from a germinating fescue seed, on shredded wheat medium generated a novel pentacyclic diketopiperazine, epoxyexserohilone (**55**), a congener of the known phytotoxin, exserohilone [24]. The investigation of *Setosphaeria rostrata* led to the discovery of three pentacyclic diketopiperazines, rostratazines A-C (**56–58**).

The fungal strain was isolated from the fresh leaf tissues of the medicinal plant *C. speciosus* collected from Colombo, Sri Lanka. Rostratazine B (**57**) inhibited porcine pancreatic alpha-amylase activity with an IC_{50} of 578 μM [25]. A pentacyclic diketopiperazine with a 4,5-dihydrooxepine moiety versicolor A (**59**) was isolated from *Aspergillus versicolor* 0312. The fungal strain was isolated from the stems of *Paris polyphylla var. yunnanensis* collected in Kunming, Yunnan Province, P. R. China. Compound **59** displayed cytotoxicity against the contraction of the MOLT-4 cell line with an IC_{50} of 29.6 μM [26].

2.2. Disulfide (R-S-S-R′) and Multisulfide (R-S$_n$-S-R′, n = 3 or More)

Bionectin E (**60**) (Figure 3) was obtained from *Bionectria* sp. Y1085, which was isolated from *Huperzia serrata*. Similar to compound **19** (T988 B) [9], compound **60** has an indole moiety attached to the tryptophan-derived 1,2,3,3a,8,8a-hexahydropyrrolo[2,3-*b*]indole. Interestingly, the other amino acid in the α−α′-bridged disulfide diketopiperazine is a dehydroxylated threonine. Compound **60** showed antibacterial activity against *E. coli*, *S. saureus* and *Salmonella typhimurium* with the same MIC value of 12.5 μg/mL [1]. Derived from the apparently normal flower of *Illigera rhodantha*, *Lasiodiplodia pseudotheobromae* F2 produced Lasiodipline D (**61**) [5]. The α position of the alanine moiety in compound **61** was connected to the β position of the tryptophan moiety via a disulfide bond. Botryosulfuranol C (**62**) was obtained from the same fungal strain *Botryosphaeria mamani* as compounds **7** (botryosulfuranols A) and **8** (botryosulfuranols B), but it has an α−β-bridged disulfide bond instead of the sulfide bond in **7** and **8**. Botryosulfuranol C (**62**) showed cytotoxicity against HepG2, HT29, Hela, IEC6 and Vero with IC_{50} values ranging from 15.9 to 115.7 μM [6].

Two new epithiodiketopiperazine natural products, outovirins B (**63**) and C (**64**), resembling the antifungal natural product gliovirin have been identified in an extract of *Penicillium raciborskii*, an endophytic fungus isolated from *Rhododendron tomentosum* [7]. Compounds **63** and **64** were almost identical to compound **9** (outovirins A) except for an α−β-bridged disulfide and a trisulfide bond in compounds **63** and **64**, respectively. Compound **64** inhibited the growth of all tested fungal isolates (*Fusarium oxysporum*, *Botrytis cinerea*, and *Verticillium dahliae*) at a low concentration of 0.38 mM (207 μg/mL), but a more significant growth inhibition was observed at 0.76 mM (413 μg/mL). Compound **64** was the most active against *Botrytis cinerea* (57% inhibition) and slightly less effective against *Verticillium dahliae* (45% inhibition). Four new pentacyclic thiodiketopiperazines, epicoccins Q-T (**65–68**), were discovered from the same fungal strain, *Epicoccum nigrum*, as compounds **10–18**. Epicoccins S (**67**) showed activity against the release of β-glucuronidase with an IC_{50} of 4.95 μM [8].

Secoemestrin D (**69**), a new epitetrathiodioxopiperizine, was obtained from *Emericella* sp. AST0036, a fungal endophyte of *Astragalus lentiginosus*. Compound **69** contains an α−α-bridged tetrasulfide bond. A benzoic acid moiety was attached to the 4,5-dihydrooxepine ring. Secoemestrin D (**69**) exhibited potent cytotoxic activity against a panel of seven cancer cell lines with IC_{50} values ranging from 0.06 to 0.24 μM [27]. *Tilachlidium* sp. (CANU-T988), a fungal strain isolated from a decaying wood sample collected in Christchurch, New Zealand, produced two new thiodiketopiperazine derivatives, T988 A (**70**) and C (**71**), which have an indole ring connected to the 1,2,3,3a,8,8a-hexahydropyrrolo[2,3-*b*]indole, structurally similar to compounds **19** and **60**. Compound **71** has an α−α-bridged disulfide bond, while compound **70** has an α−α-bridged trisulfide bond. Compounds **70** and **71** displayed cytotoxicity against P388 with IC_{50} values of 0.25 and 0.56 μM, respectively [9]. Pretrichodermamide A (**72**) was obtained from *Trichoderma* sp. BCC 5926, which was collected on a bamboo leaf from Khao Yai National Park, Nakhon Ratchasima Province, Thailand. Under alkaline conditions, compound **72** with an α−β-bridged disulfide bond underwent a rapid transformation to a stable amide, which is composed of a 1,2-oxazadecaline moiety and a coumarin derivative. Compound **72** exhibited antibacterial activity against *Mycobacterium tuberculosis* H37Ra with an MIC of 12.5 μg/mL [28]. A new epidithiodiketopiperazine, pretrichodermamide G (**73**), was afforded by *Trichoderma harzianum* associated with the medicinal plant *Zingiber officinale* [29]. Although compound **73** is quite similar to compound **72**, no chemical transformation under alkaline conditions was reported.

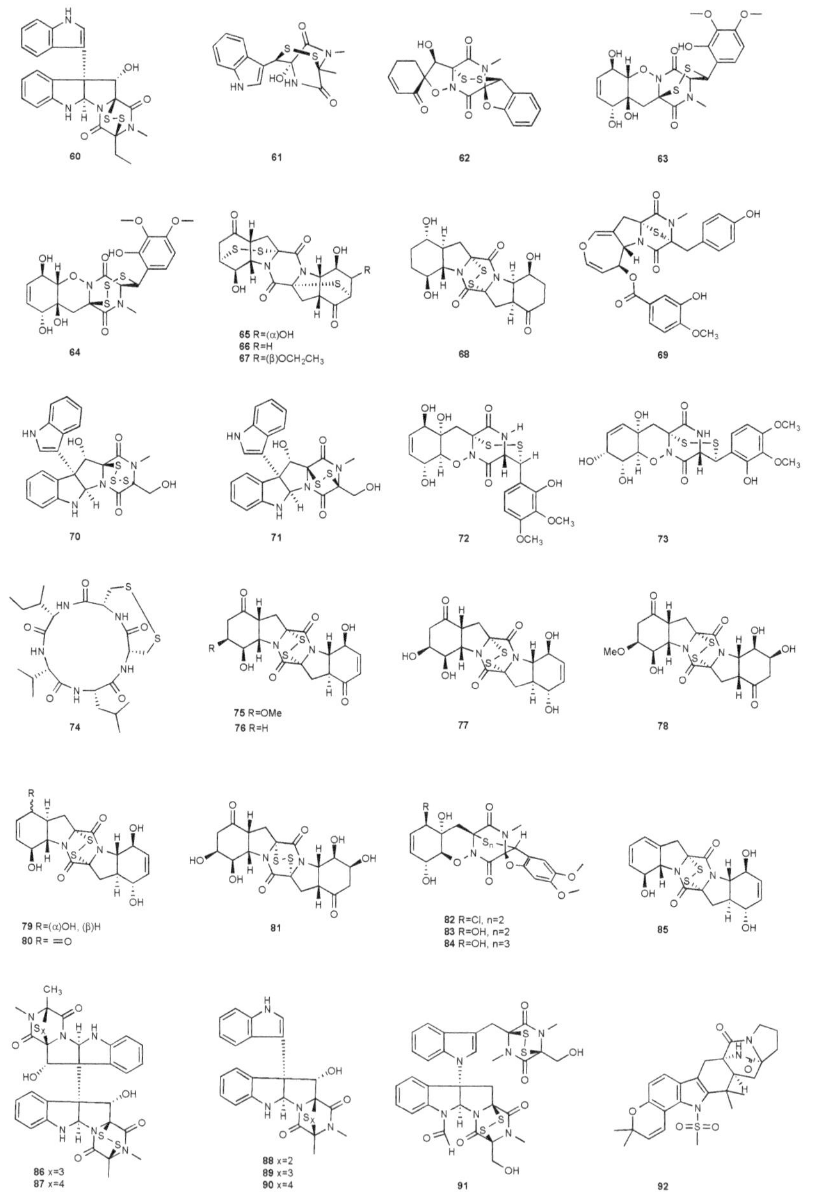

Figure 3. Structures of compounds **60–92**.

The investigation of *Aspergillus tamarii* FR02 led to the isolation of a new cyclic pentapeptide, disulfide cyclo-(Leu-Val-Ile-Cys-Cys), named malformin E (**74**). *A. tamarii* FR02 was isolated from the root of *Ficus carica*. Malformin E (**74**) exhibited cytotoxic activities against MCF-7, A549 and HepG2 with IC_{50} values of 0.65, 2.42 and 36.02 μM, respectively. Malformin E (**74**) also showed antimicrobial and antifungal activities against *Bacillus sub-*

tilis, *Staphylococcus aureus*, *Pseudomonas aeruginosa*, *Escherichia coli*, *Penicillium chrysogenum*, *Candida albicans* and *Fusarium solani* with MIC values ranging from 0.45 to 7.24 μM [30].

Six pentacyclic diketopiperazines, brocazines A-F (**75–80**), were discovered from *Penicillium brocae* MA-231, a fungus obtained from the fresh tissue of the marine mangrove plant *Avicennia marina*. Brocazines A (**75**), B (**76**), E (**79**) and F (**80**) were cytotoxic to a panel of nine tumor cell lines with IC_{50} values ranging from 0.89 to 9.0 μM. [31]. A culture of *Phoma* sp. OUCMDZ-1847 afforded one new phomazine C (**81**), which should be biogenetically generated from the same precursor as compounds **41** and **42** [19]. *Penicillium janthinellum* HDN13-309 produced epipolythiodioxopiperazines, penicisulfuranols A–C (**82–84**), together with compounds **45–47**. Compounds **82–84** exhibited cytotoxicity against HeLa and HL-60 with IC_{50} of 0.1–3.9 μM [21].

Brocazine G (**85**), a new diketopiperazine, along with compounds **43** and **44** was obtained from *Penicillium brocae* MA-231 associated with the fresh tissue of the marine mangrove plant *Avicennia marina*. It showed cytotoxicity against A2780 with an IC_{50} of 59 μM. Brocazine G (**85**) also showed inhibitory activity against *E. coli*, *Aeromonas hydrophilia* and *V. harveyi* with the same MIC of 32.0 μg/mL [20]. Five new epipolysulfanyldioxopiperazines, gliocladines A–E (**86–90**), were isolated from *Gliocladium roseum* 1A, a fungal strain isolated from submerged wood collected from fresh water in Yunnan Province, P. R. China. Both compounds **86** and **87** are dimers with each monomer being derived from L-alanine and L-tryptophan, while each of compounds **88–90** is a diketopiperazine with an indole ring connected to the 1,2,3,3a,8,8a-hexahydropyrrolo[2,3-*b*]indole. These compounds exhibited nematicidal activities toward *C. elegans*, *P. redivivus* and *B. xylophilus* with ED_{50} values ranging from 25 to 250 μg/mL [32].

An analog of compounds **86** and **87**, 6-Formamide-chetomin (**91**), was obtained from a culture of *Chaetomium* sp. M336, isolated from the plant *H. serrata* (Thunb. ex Murray) Trev. Compound **91** was cytotoxic to HeLa, SGC-7901 and A549 cells with IC_{50} values of 21.6–27.1 μM. It exhibited activity against *Escherichia coli*, *Staphylococcus aureus*, *Salmonella typhimurium* ATCC 6539 and *Enterococcus faecalis* with the same MIC of 0.78 μg/mL [33].

2.3. Sulfoxide (R-SO-R′) and Sulfone (R-SO_2-R′)

An indole alkaloid with a rare methylsulfonyl unit, 21-*Epi*-taichunamide D (**92**), was obtained from *Aspergillus versicolor* F210 (*Lycoris radiata*). The strain was isolated from the bulbs of *Lycoris radiata* collected from Yichang City in Hubei Province, P. R. China. Compound **92** inhibited anticancer activity toward HL-60 and A549 cells with IC_{50} values of 26.8 and 32.5 μM, respectively [34].

3. Polyketides

3.1. Sulfide

A new cytotoxic compound, isocochlioquinones D (**93**) (Figure 4), was purified from *Bipolaris sorokiniana* A606. The endophytic fungus was isolated from the medicinal plant *Pogostemon cablin*, also known as patchouli or "Guanghuoxiang" in traditional Chinese medicine (TCM) [35]. Isocochlioquinones D (**93**) is a hybrid of a polyketide and a sesquiterpenoid with a rare benzothiazin-3-one moiety. Compound **93** demonstrated antiproliferative activity toward SF-268, MCF-7, NCI-H460 and HepG-2 with IC_{50} values of 32.8, 28.3, 42.6 and 38.6 μM, respectively.

Paraphaeosphaeria neglecta FT462 yielded paraphaeosphaerides E (**94**), F (**95**), H (**96**) and methyl ester of paraphaeosphaeride F (**97**) [36]. *P. neglecta* FT462 was isolated from the Hawaiian plant *Lycopodiella cernua*, synonym *Palhinhaea cernua* (Lycopodiaceae). Paraphaeosphaeride E (**94**) was active against *E. coli* JW2496 at 20 μg/mL. Paraphaeosphaeride E (**94**) inhibited nuclear factor kappa B (NF-κB) with an IC_{50} of 7.1 μM. Paraphaeosphaerides E (**94**) and F (**95**) also showed inducible nitric oxide synthase (iNOS) with IC_{50} values of 47.9 and 43.2 μM, respectively. Paraphaeosphaeride A (**98**) with the unique 4-pyranone-γ-lactam-1,4-thiazine moiety was obtained from *P. neglecta* FT462 [37].

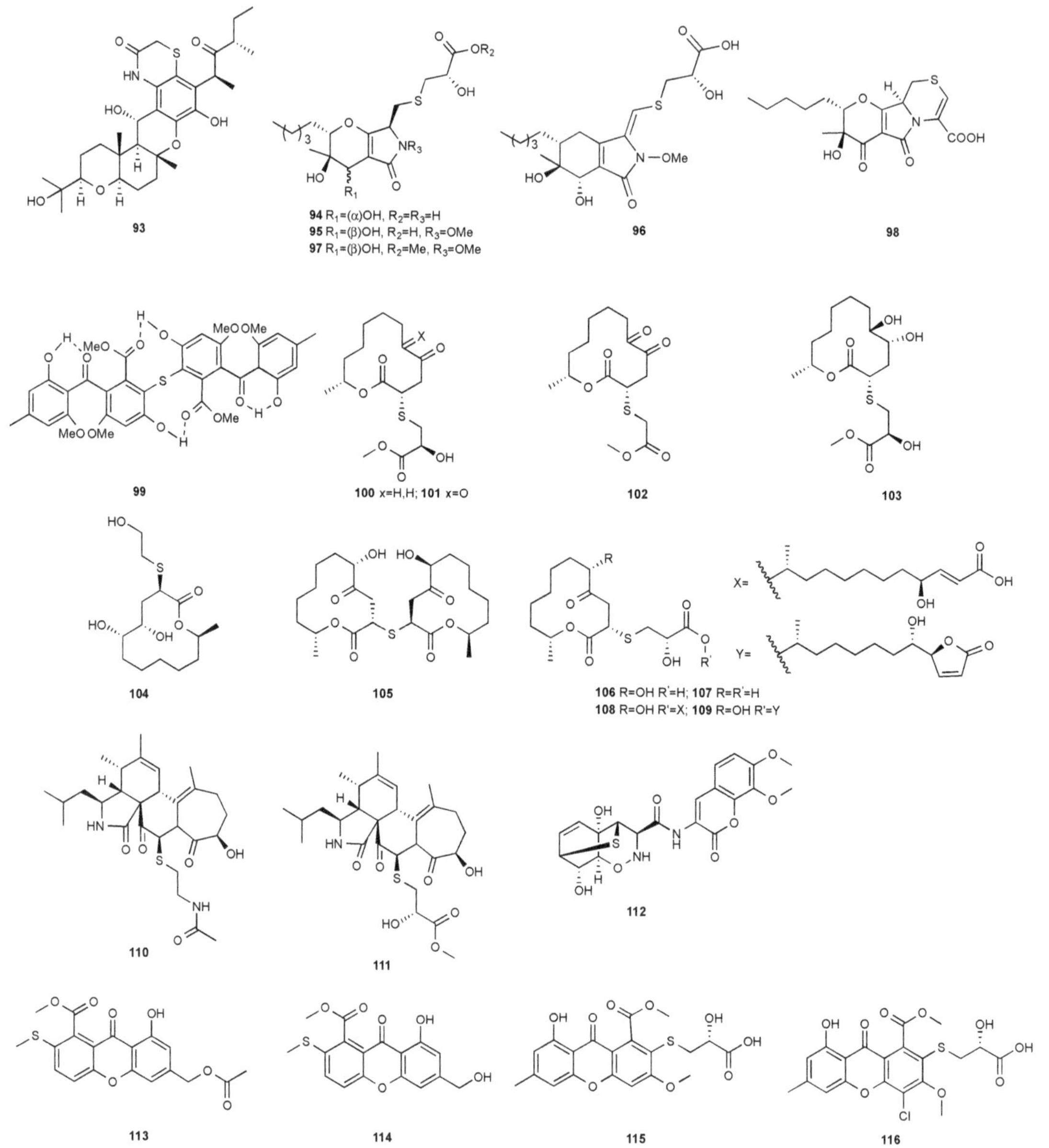

Figure 4. Structures of compounds **93–116**.

The first natural sulfur-containing benzophenone dimer, named guignasulfide (**99**), was isolated from the culture of *Guignardia* sp. IFB-E028, an endophytic fungus residing in the healthy leaves of *Hopea hainanensis*. Guignasulfide (**99**) exhibited cytotoxicity against HepG2 with an IC_{50} of 5.27 μM. It also showed antimicrobial activity against *Helicobacter pylori* with an MIC of 42.9 μM [38].

Cladosporium cladosporioides MA-299 yielded four 12-membered macrolides, thiocladospolides A-D (**100–103**). *C. cladosporioides* MA-299 is an endophytic fungus obtained from the leaves of the mangrove plant *Bruguiera gymnorrhiza*. Thiocladospolide A (**100**) was active against *E. tarda*, *E. ictarda* and *C. glecosporioides* with MIC values of 1, 8 and 2 μg/mL, respectively. Thiocladospolide B (**101**) was active against *C. glecosporioides, P. piricola Nose* and *F. oxysporum f. sp.cucumerinum* with MIC values of 2, 32 and 1 μg/mL, respectively. Thiocladospolide C (**102**) was active against the same three strains as **101** with MIC values of 1, 32 and 32 μg/mL, respectively. Thiocladospolide D (**103**) was active against *E. ictarda, C. glecosporioides, P. piricola* Nose and *F. oxysporum f. sp.cucumerinum* with MIC values of 1, 1, 32, and 1 μg/mL, respectively [39]. The investigation of the mangrove-derived fungus *Cladosporium* sp. SCNU-F0001 afforded a new 12-membered macrolide, thiocladospolide E (**104**) [40]. A mangrove-derived fungus, *Cladosporium oxysporum*, yielded five 12-membered macrolides, thiocladospolides F–J (**105–109**), and they showed a broad spectrum of antimicrobial activity with MIC values ranging from 4 to 32 μg/mL [41].

Two cytochalasan analogs, cyschalasins A (**110**) and B (**111**), were obtained from *Aspergillus micronesiensis*, which was isolated from the root of the traditional Chinese medicinal plant *Phyllanthus glaucus* collected from LuShan Mountain, Jiangxi Province, P. R. China. Cyschalasins A (**110**) and B (**111**) exhibited cytotoxicity against HL60, A549, Hep3B, MCF-7 and SW480 with IC_{50} values in the range of 3.0 to 19.9 μM except for **110**, which was inactive toward A549 at 20 μM. Cyschalasins A (**110**) and B (**111**) also demonstrated antimicrobial activity with MIC_{50} values ranging from 10.6 to 94.7 μg/mL [42].

An amide of a coumarin moiety and L-phenylalanine-derived 1,2-oxazadecaline moiety, trichodermamide G (**112**), was isolated from *Trichoderma harzianum* D13. The fungal strain was isolated from the internal tissues of the root of *Excoecaria agallocha*, distributed in the mangrove regions of various parts of India [43].

Two sulfur-containing xanthones, sydoxanthone A (**113**) and sydoxanthone B (**114**), were purified from *A. sydowii*, occurring in the liverwort *Scapania ciliata* S. Lac. Sydoxanthone B (**114**) was active on the concanavalin A-induced and lipopolysaccharide (LPS)-induced proliferation of mouse splenic lymphocytes with IC_{50} of 22.53 and 15.30 μg/mL, respectively [44]. Sydoxanthones D (**115**) and E (**116**) were discovered from *Pseudopestalotiopsis theae*, which was isolated from the leaves of *Caloncoba welwitschii* [45].

3.2. Disulfide

A new natural compound, a symmetrical disulfide dimer dodecyl 3,3″-dithiodipropionate (**117**) (Figure 5), was isolated from the ethyl acetate extract of fermentation broth of an endophytic fungus, *Sphaceloma* sp. LN-15. The fungal strain was isolated from the leaves of *Melia azedarach* L., commonly known as the chinaberry tree, pride of India, Persian lilac, and some other names [46]. The structure of **117** was determined by NMR and MS and was further confirmed by chemical synthesis.

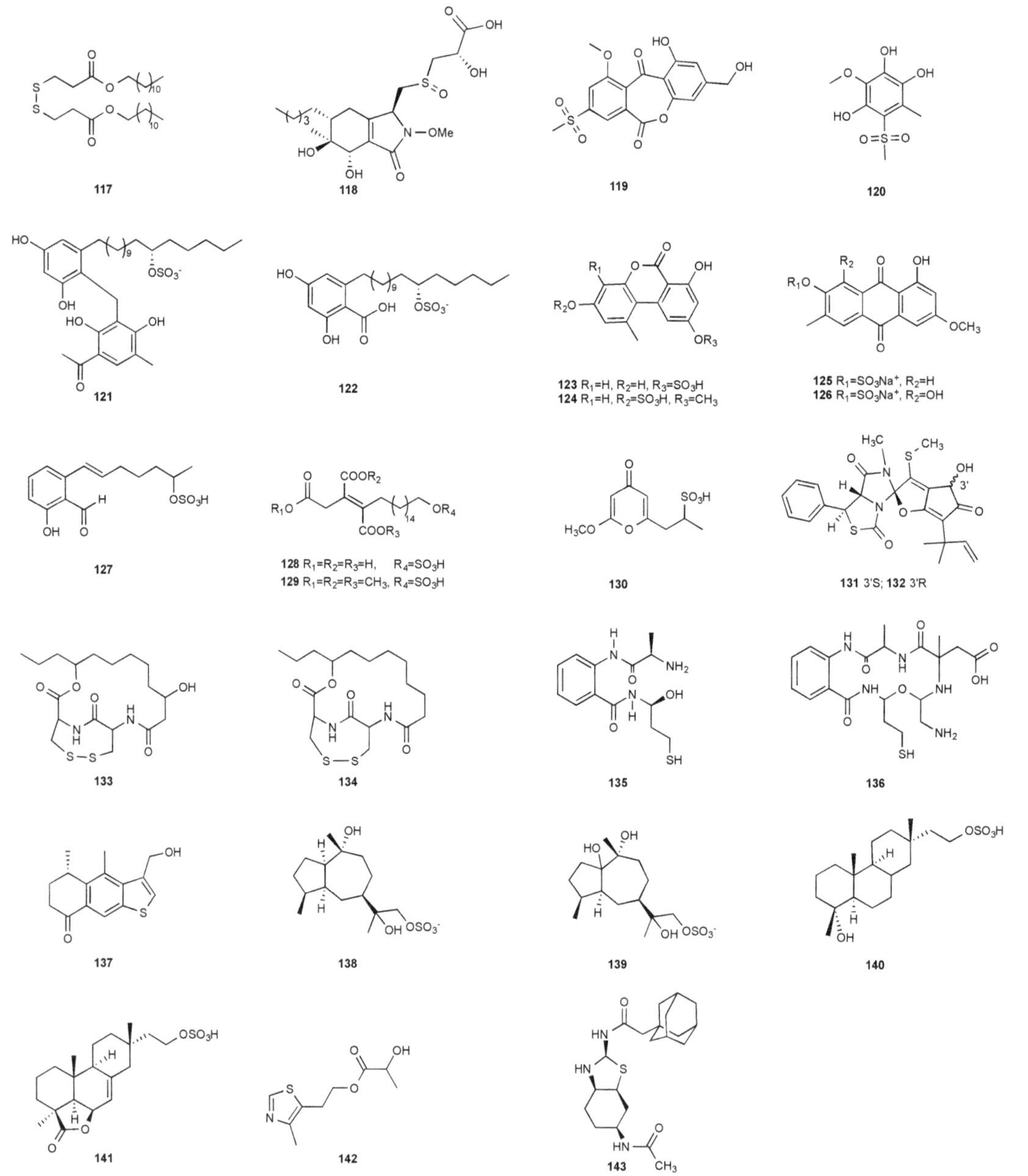

Figure 5. Structures of compounds **117–143**.

3.3. Sulfoxide

LC-UV/MS-based metabolomics analysis of the Hawaiian endophytic fungus *Paraphaeosphaeria neglecta* FT462 led to the identification of unique mercaptolactated γ-pyranol–γ-lactams, paraphaeosphaerides G (**118**). The fungal strain was isolated on potato dextrose agar (PDA) medium from a healthy leaf of the Hawaiian indigenous plant *Lycopodiella cernua* (L.) Pic. Serm, which was collected in the Mokuleia Forest Reserve in 2014 [36].

3.4. Sulfones

Two new polyketides modified with a rare methyl sulfonyl group, neosartoryone A (**119**) and 3-methoxy-6-methyl-5-(methylsulfonyl)benzene-1,2,4-triol (**120**), were isolated from *Neosartorya udagawae* HDN13-313 cultivated with the DNA methyltransferase inhibitor 5-azacytidine. *N. udagawae* HDN13-313 was isolated from the root of the mangrove plant *Aricennia marina* [47]. Compound **119** decreased the lipid accumulation elicited by oleic acid at 10 μM.

3.5. Sulfates and Sulfonates

Two new alkyl sulfate-containing aromatic compounds, penixylarins B (**121**) and D (**122**), were isolated from a mixed culture of the Antarctic deep-sea-derived fungus *Penicillium crustosum* PRB-2 and the fungus *Xylaria* sp. HDN13-249 [48]. *Xylaria* sp. HDN13-249 was isolated from the root of *Sonneratia caseolaris* collected from the mangrove conservation area of Hainan, P. R. China. Penixylarins B (**121**) showed weak antibacterial activity against *Bacillus subtilis* with an MIC_{50} of 100 μM.

Alternariol 5-*O*-sulfate (**123**) and alternariol 5-*O*-methyl ether-4′-*O*-sulfate (**124**) were produced by *Alternaria* sp., which was isolated from fresh healthy leaves of the wild Egyptian medicinal plant *Polygonum senegalense Meisn.* (Polygonaceae) [49]. Alternariol 5-*O*-sulfate (**123**) was cytotoxic against L5178Y with an EC_{50} of 4.5 μg/mL. Compound **123** also showed inhibition toward a panel of protein kinases at the micromolar level.

The extracts of cultures grown in liquid or on solid rice media of the fungal endophyte *Ampelomyces* sp. isolated from the medicinal plant *Urospermum picroides* exhibited considerable cytotoxic activity against L5178Y cells. The extract obtained from liquid cultures afforded two sulfated anthraquinones, macrosporin-7-*O*-sulfate (**125**) and 3-*O*-methylalaternin-7-*O*-sulfate (**126**) [50]. However, neither compound showed any cytotoxic or antimicrobial activities.

A 2-hydroxyl 6-alkylated benzaldehyde derivative, pestalols E (**127**), was isolated from the endophytic fungus *Pestalotiopsis* sp. AcBC2, which was derived from the Chinese mangrove plant *Aegiceras corniculatum*, commonly known as black mangrove or river mangrove [51].

Oreganic acid (**128**) and its trimethyl esters (**129**) were obtained from the extract of an endophytic fungus MF6046 isolated from living leaves of *Berberis oregano* [49]. Oreganic acid (**128**) is a highly potent and specific farnesyl protein transferase (FPTase) inhibitor (IC_{50} = 14 nM) [49].

A novel metabolite containing a sulfonate group, fusaodavinvin (**130**), was isolated from an endophytic fungus *Fusarium* sp. (CTGU-ZL-34). The fungal strain was isolated from a healthy plant *Davidia involucrata*. Compound **130** displayed inhibitory activity against A549, HepG2, Caski and MCF-7 cell lines with IC_{50} values of 11.5, 15.3, 15.2 and 60.5 μg/mL, respectively [53].

4. Hybrids

4.1. Sulfides

A fungal strain *Pestalotiopsis* sp. HS30 was isolated from the fresh stems of *Isodon xerophilus* collected at Kunming Botanical Garden, Yunnan Province, P. R. China [54]. Pestaloamides A (**131**) and B (**132**), two novel alkaloids featuring an unprecedented spiro[imidazothiazoledione-alkylidenecyclopentenone] scaffold, were obtained from the cultures of *Pestalotiopsis* sp. HS30. Compounds **131** and **132** were derived from a polyketide and a Phe-Cys dipeptide together with C_2 and C_5 moieties. Both compounds could enhance the cell surface engagement of NKG2D ligands in HCT116 cells at 40 μM [54].

4.2. Disulfides

PM181110 (**133**) was a new depsipeptide obtained from *Phomopsis glabrae*, which was isolated from the leaves of *Pongamia pinnata* (Fabaceae) [55]. Compound **133** was derived from two molecules of L-cysteine and one C_{12} polyketide. It exhibited potent cytotoxic activity toward 40 human cancer cell lines at the nanomolar level (mean IC_{50} = 89 nM) and 24 human tumor xenografts with the mean IC_{50} of 245 nM [55].

FE399 (**134**), a dehydroxylated **133**, was isolated from *Ascochyta* sp. AJ 117309, an endophytic strain separated from a raw leaf of *Taxus cuspidata var. nana* Rehd. [56]. Compound **134** also demonstrated potent cytotoxic activity against SWS948, K562T, Colon26, CHO-K1 and P388 cells with IC_{50} values ranging from 75 to 400 ng/mL [56].

4.3. Thiols

Fusarium chlamydosporium, an endophytic fungus isolated from the leaves of *Anvillea garcinii* (Burm.f.) DC. (Asteraceae), produced a new benzamide derivative, fusarithioamide A (**135**), which is composed of a 2-aminobenzamide moiety, an L-alanine and a 3-mercaptopropan-1-ol moiety derived from L-cysteine. Compound **135** displayed cytotoxicity against SK-MEL, KB, BT-549 and SKOV-3 cells with IC_{50} values of 9.3, 7.7, 0.4 and 0.8 μM, respectively. It was also active against *S. aureus*, *B. cereus*, *E. coli*, *P. aeruginosa* and *C. albicans* with MIC values of 4.4, 3.1, 6.9, 100 and 2.6 μg/mL, respectively [57].

Fusarithioamide B (**136**), a new aminobenzamide derivative with an unprecedented carbon skeleton, was separated from an EtOAc extract of *Fusarium chlamydosporium* isolated from *Anvillea garcinii* (Burm.f.) DC. leaves (Asteraceae) [58]. Fusarithioamide B (**136**) displayed antifungal activity toward *C. albicans* with an MIC of 1.9 μg/mL. It also showed high antibacterial activity against *E. coli*, *S. aureus* and *B. cereus* with MIC values of 3.4, 2.9 and 3.9 μg/mL, respectively. Compound **136** exhibited cytotoxic activity toward BT-549, MCF-7, HCT-116, SKOV-36, KB and SK-MEL with IC_{50} values of 0.09, 0.21, 0.59, 1.23, 6.9 and 11.2 μM, respectively [58].

5. Terpenoids

5.1. Sulfide/Thiophene

Leptosphin A (**137**), a new sesquiterpenoid with a benzo[b]thiophene moiety, was obtained from a culture of Leptosphaeria sp. XL026 isolated from the leaves of *Panax notoginseng* [59]. Leptosphin A (**137**) displayed antifungal and antibacterial activity with MIC values ranging from 25 to 100 μg/mL [59].

5.2. Sulfates

An endophytic fungus S49 was isolated from the bark of *Cephalotaxus hainanensis*, known as Hainan plum-yew. S49 afforded two new sesquiterpenoids 1,10,11,12-guaianetetrol (**138**) and 1,10,11,12-guaianetetrol (**139**) [60]. Two new isopimarane diterpenoids, 16-*O*-sulfo-18-norisopimar-7-en-4α,16-diol (**140**) and 9-deoxy-hymatoxin A (**141**), were isolated from the culture broth of an endophytic fungus, *Xylaria* sp. YM 311647, obtained from *Azadirachta indica*. Compounds (**140**) and (**141**) were active against *C. albicans* YM 2005, *A. niger* YM 3029, *P. oryzae* YM 3051, *F. avenaceum* YM 3065 and *H. compactum* YM 3077 with MIC values in the range of 32–128 μg/mL, while compound **141** had the same MIC of 16 μg/mL toward *C. albicans* and *P. oryzae* [61].

6. Others

A new thiazole derivative, colletotricole A (**142**), was obtained from *Colletotrichum gloeosporioides* A12, an endophytic fungus derived from *Aquilaria sinensis* [62]. A sulfur-containing anticandidal compound, *N*-[(2*S*,3a*R*,6*S*,7a*S*)-6-acetamido-octahydro-l,3-benzothiazoi-2-yl]2-(adamantan-l-yl) acetamide (**143**), was isolated from *Emericella* sp. from *Azadirachta indica* [63].

7. Discussion and Conclusions

From 1985 to March 2022, 143 new sulfur-containing compounds were obtained from endophytic fungi. This review summarized the fungal producers, host plants, chemical structures and biological activities of these fungal metabolites (Table 1). The majority of these compounds (109 out of 143) were reported in 2010, 2014, 2015, 2017, 2019 and 2020 (Figure 6). There was a trend that more sulfur-containing compounds were reported in recent years except 2021. Only one sulfur-containing compound was reported in 2021, most likely due to the outbreak of COVID-19 in 2020. A total of 24 journals reported these compounds (Figure 7). The *J. Nat. Prod.* has published the highest number of papers (16) that reported sulfur-containing compounds, followed by *Phytochemistry* (8) (Figure 7). This is not unexpected because both *J. Nat. Prod.* and *Phytochemistry* are prominent natural product journals.

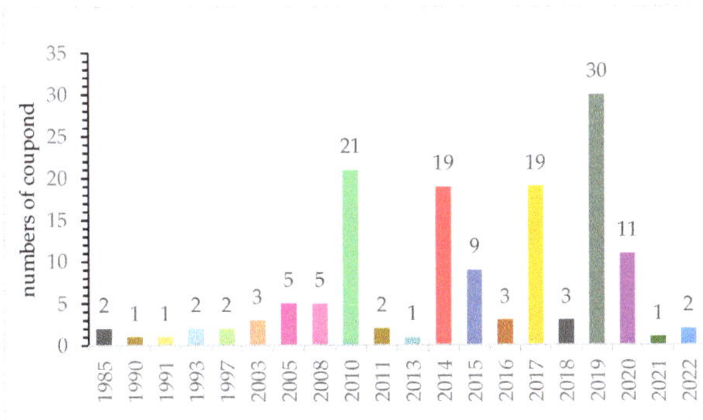

Figure 6. Annual numbers of sulfur-containing compounds identified from 1985 to 2022. (Keywords: sulfur-containing compound, plant endophytic fungi; Databases: SciFinder, PubMed).

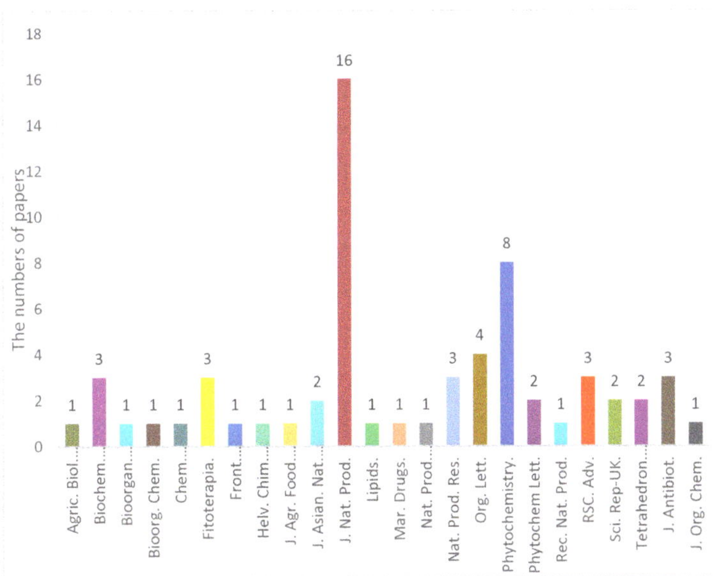

Figure 7. The journal names and numbers for the papers that reported sulfur-containing compounds.

These sulfur-containing compounds demonstrate functional and structural diversity and exhibited many bioactivities. Among the reported biological activities, 42% of these compounds were antimicrobial, while 37% were cytotoxic (Figure 8), which is not surprising because the majority of the FDA-approved antimicrobial and anticancer drugs are either natural products or derived from natural products. For example, Secoemestrin D (**69**), a diketopiperazine, was very active against a panel of seven cancer cell lines with IC_{50} values ranging from 0.06 to 0.24 μM [27], while PM181110 (**133**) [55] and FE399 (**134**) [56], hybrids of polyketides and peptides, exhibited potent anticancer activity with IC_{50} values at the nM level. These compounds also possess other bioactivities. For instance, oreganic acid (**128**), a fatty acid derivative, inhibited FPTase with an IC_{50} of 14 nM [49]. The majority of sulfur-containing compounds (92) were peptides, followed by polyketides (38), hybrids (6), terpenoids (5) and others (2) (Figure 9). All 92 of these peptides are diketopiperazines, and the sulfur atoms in these molecules are mainly derived from L-cysteine that contains a reactive sulph-hydryl group.

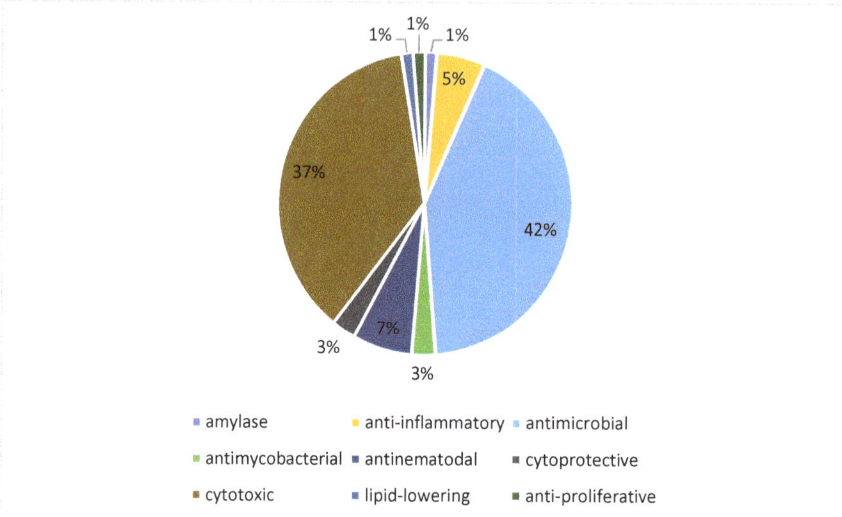

Figure 8. The percentages of the biological activity among sulfur-containing compounds from endophytic fungi.

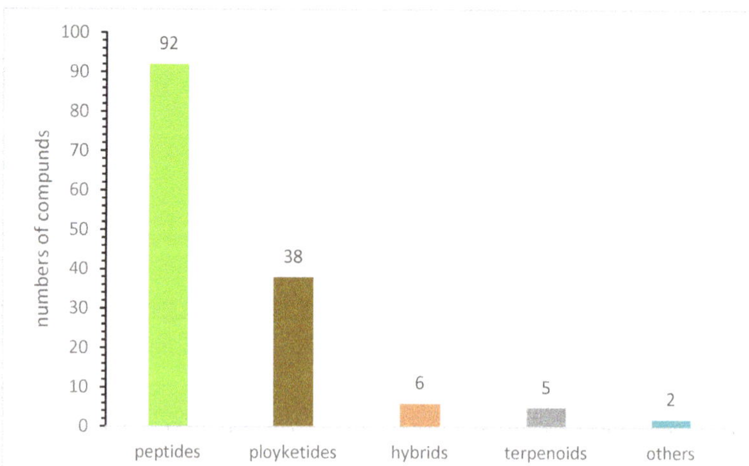

Figure 9. The structural classes of sulfur-containing compounds isolated from endophytic fungi.

8. Prospects

Some plants are rich in sulfur, for example, allium vegetables, legumes and cruciferous plants. These plants should be great sources of endophytic fungi that produce sulfur-containing compounds. Large amounts of sulfur are released during volcanic eruptions. Hence, plants in volcanic areas and hot springs might also be excellent sources for endophytic fungi producing sulfur-containing compounds.

Most of the compounds reviewed in this article were tested for their antimicrobial and antiproliferative or anticancer activities. We believe that other biological properties could be identified if fungal metabolites were evaluated in a broader range of biological settings. For example, sinuxylamides A and B were obtained from *Xylaria* sp. FM1005, an endophytic fungus isolated from *Sinularia densa* (leather coral) collected in the offshore region of the Big Island, Hawaii [64]. Sinuxylamides A and B showed no antibacterial activity or cytotoxicity at 40 μM, but they strongly inhibited the binding of fibrinogen to purified integrin IIIb/IIa in a dose-dependent manner with IC_{50} values of 0.89 and 0.61 μM, respectively.

Diketopiperazines are expected to be biosynthetically assembled from two amino acid building blocks by nonribosomal peptide synthetases [65]. On the other hand, the biogenesis of many sulfur-containing compounds remains incompletely understood. For example, the structures of compounds **20** [10], **40** [18], **98** [37], **136** [58], **142** [62] and **143** [63] are unique. It would be interesting to investigate how these molecules are biogenetically synthesized. Presumably, the 4,5-dihydrooxepine ring in **20** is derived from the benzene ring of L-phenylalanine through ring expansion. On the other hand, the spiro[cyclopenta[*b*]pyrrole-5,2′-furan] moiety in **40** might be formed through the constriction of the benzene ring of L-phenylalanine followed by the merge of the octahydrocyclopenta[*b*]pyrrole ring with an isoprenyl (C_5) group. We previously isolated compound **98** [37]. The precursor of the side chain at the 14-position in compound **98** could be L-cysteine, which is converted to mercaptolactate. The nucleophilic addition of the mercaptolactate thiol to C-14 of paraphaeosphaeride C generates an intermediate that is oxidized to another intermediate. It is also plausible that the second intermediate is generated from mercaptopyruvate and paraphaeosphaeride C. The nitrogen atom in the second intermediate undergoes intramolecular nucleophilic addition to the ketone of the mercaptopyruvate moiety, leading to the formation of the third intermediate. The dehydration of the third intermediate yields the final product **98** [37]. However, the experimental details of the biosynthesis of compound **98** are still not available. Compound **136** is composed of five fragments, including a 2-amino benzoic acid moiety, an L-alanine, a 2-amino-2-methylsuccinic acid fragment that might be derived from an isoprenyl group (C_5), and L-glycine and L-cysteine-derived 3-mercaptopropanoic acid moieties. Compound **142** carries a 2-hydroxyl propanoic acid ester. The thiazole ring in **142** is probably derived from acetate and L-cysteine, while the linker ($-CH_2-CH_2-$) might be derived from another acetate. It would be interesting to investigate how **142** is synthesized biogenetically. Investigating the biosynthesis of diamond-like compound **143** should be very challenging and interesting. Recent advances in genome mining and synthetic biology offer new opportunities to discover new natural products [66]. It becomes routine to sequence the (meta)genomes of fungal isolates, and capable bioinformatics tools (e.g., antiSMASH fungal version) [67] are increasingly available for identifying potential biosynthetic gene clusters (BGCs) of fungal natural products [68]. These predicted BGCs can suggest new chemotypes, enzymology and bioactivities. Subsequently, native and engineered BGCs can be expressed in multiple synthetic biology chasses, such as *Aspergillus nidulans* [69] and Saccharomyces cerevisiae [70]. In this regard, biosynthetic research is critical for laying the basis for the genome mining of BGCs of new fungal sulfur-containing compounds with bioactivities, particularly those whose biogenesis remains unclear.

Author Contributions: Y.D., C.W. and S.C.: Conceptualization. Y.F., Z.M., Y.Z. and Y.W.: discussion of the contents. C.W. and S.C.: writing—original draft preparation. Y.F., Z.M., Y.Z., Y.W., Y.D., C.W. and S.C.: writing—review and editing. All authors have read and agreed to the published version of the manuscript.

Funding: This work was financially supported by the National Natural Science Foundation of China (42006096), the Natural Science Foundation of Shandong Province (ZR2020QD098), the China Postdoctoral Science Foundation (2020M682266) and the Talents of High-Level Scientific Research

Foundation of Qingdao Agricultural University (Grants 6651120031). Y.D. is partially supported by NIH R35 GM128742.

Institutional Review Board Statement: Not applicable.

Informed Consent Statement: Not applicable.

Data Availability Statement: Not applicable.

Conflicts of Interest: The authors declare no conflict of interest.

References

1. Francioso, A.; Conrado, A.B.; Mosca, L.; Fontana, M. Chemistry and Biochemistry of Sulfur Natural Compounds: Key Intermediates of Metabolism and Redox Biology. *Oxid. Med. Cell. Longev.* **2020**, *2020*, 8294158. [CrossRef] [PubMed]
2. Scott, K.A.; Njardarson, J.T. Analysis of US FDA-Approved Drugs Containing Sulfur Atoms. *Top. Curr. Chem.* **2018**, *376*, 5. [CrossRef] [PubMed]
3. Soares, D.A.; Rosa, L.H.; da Silva, J.F.M.; Pimenta, R.S. A review of bioactive compounds produced by endophytic fungi associated with medicinal plants. *Boletim do Museu Paraense Emílio Goeldi Ciências Naturais* **2017**, *12*, 331–352.
4. Yang, Y.H.; Yang, D.S.; Li, G.H.; Pu, X.J.; Mo, M.H.; Zhao, P.J. Antibacterial diketopiperazines from an endophytic fungus *Bionectria* sp. Y1085. *J. Antibiot.* **2019**, *72*, 752–758. [CrossRef]
5. Wei, W.; Jiang, N.; Mei, Y.N.; Chu, Y.L.; Ge, H.M.; Song, Y.C.; Ng, S.W.; Tan, R.X. An antibacterial metabolite from *Lasiodiplodia pseudotheobromae* F2. *Phytochemistry* **2014**, *100*, 103–109. [CrossRef]
6. Barakat, F.; Vansteelandt, M.; Triastuti, A.; Jargeat, P.; Jacquemin, D.; Graton, J.; Mejia, K.; Cabanillas, B.; Vendier, L.; Stigliani, J.-L.; et al. Thiodiketopiperazines with two spirocyclic centers extracted from *Botryosphaeria mamane*, an endophytic fungus isolated from *Bixa orellana* L. *Phytochemistry* **2019**, *158*, 142–148. [CrossRef]
7. Kajula, M.; Ward, J.M.; Turpeinen, A.; Tejesvi, M.V.; Hokkanen, J.; Tolonen, A.; Häkkänen, H.; Picart, P.; Ihalainen, J.; Sahl, H.; et al. Bridged epipolythiodiketopiperazines from *Penicillium raciborskii*, an endophytic fungus of *Rhododendron tomentosum* Harmaja. *J. Nat. Prod.* **2016**, *79*, 685–690. [CrossRef]
8. Wang, J.M.; Ding, G.Z.; Fang, L.; Dai, J.G.; Yu, S.S.; Wang, Y.H.; Chen, X.G.; Ma, S.G.; Qu, J.; Xu, S.; et al. Thiodiketopiperazines produced by the endophytic fungus *Epicoccum nigrum*. *J. Nat. Prod.* **2010**, *73*, 1240–1249. [CrossRef]
9. Feng, Y.; Blunt, J.W.; Cole, A.L.; Munro, M.H. Novel cytotoxic thiodiketopiperazine derivatives from a *Tilachlidium* sp. *J. Nat. Prod.* **2004**, *67*, 2090–2092. [CrossRef]
10. Haritakun, R.; Rachtawee, P.; Komwijit, S.; Nithithanasilp, S.; Isaka, M. Highly conjugated ergostane-type steroids and aranotin-type diketopiperazines from the fungus *Aspergillus terreus* BCC 4651. *Helv. Chim. Acta* **2012**, *95*, 308–313. [CrossRef]
11. Wang, F.Q.; Tong, Q.Y.; Ma, H.R.; Xu, H.F.; Hu, S.; Ma, W.; Xue, Y.B.; Liu, J.J.; Wang, J.P.; Song, H.P.; et al. Indole diketopiperazines from endophytic *Chaetomium* sp. 88194 induce breast cancer cell apoptotic death. *Sci. Rep.* **2015**, *5*, 9294. [CrossRef] [PubMed]
12. Guimarães, D.O.; Borges, W.S.; Vieira, N.J.; De Oliveira, L.F.; Da Silva, C.H.; Lopes, N.P.; Dias, L.G.; Durán-Patrón, R.; Collado, I.G.; Pupo, M.T. Diketopiperazines produced by endophytic fungi found in association with two Asteraceae species. *Phytochemistry* **2010**, *71*, 1423–1429. [CrossRef] [PubMed]
13. Peng, F.; Hou, S.Y.; Zhang, T.Y.; Wu, Y.Y.; Zhang, M.Y.; Yan, X.M.; Xia, M.Y.; Zhang, Y.X. Cytotoxic and antimicrobial indole alkaloids from an endophytic fungus *Chaetomium* sp. SYP-F7950 of *Panax notoginseng*. *RSC Adv.* **2019**, *9*, 28754–28763. [CrossRef] [PubMed]
14. Meng, L.H.; Li, X.M.; Liu, Y.; Xu, G.M.; Wang, B.G. Antimicrobial alkaloids produced by the mangrove endophyte *Penicillium brocae* MA-231 using the OSMAC approach. *RSC Adv.* **2017**, *7*, 55026–55033. [CrossRef]
15. Chinworrungsee, M.; Kittakoop, P.; Saenboonrueng, J.; Kongsaeree, P.; Thebtaranonth, Y. Bioactive compounds from the seed fungus *Menisporopsis theobromae* BCC 3975. *J. Nat. Prod.* **2006**, *69*, 1404–1410. [CrossRef]
16. Chu, M.; Mierzwa, R.; Truumees, I.; Gentile, F.; Patel, M.; Gullo, V.; Chan, T.M.; Puar, M.S. Two novel diketopiperazines isolated from the fungus *Tolypocladium* sp. *Tetrahedron Lett.* **1993**, *34*, 7537–7540. [CrossRef]
17. Zhu, M.; Yang, Z.; Wang, H.; Gan, Q.; Zhang, G.; Che, Q.; Zhu, T.J.; Gu, Q.; Han, B.N.; Li, D. Penispirozines A–H, three classes of dioxopiperazine alkaloids with spirocyclic skeletons isolated from the mangrove-derived *Penicillium janthinellum*. *J. Nat. Prod.* **2020**, *83*, 2647–2654. [CrossRef]
18. Pedras, M.S.C.; Séguin-Swartz, G.; Abrams, S.R. Minor phytotoxins from the blackleg fungus *Phoma lingam*. *Phytochemistry* **1990**, *29*, 777–782. [CrossRef]
19. Kong, F.; Wang, Y.; Liu, P.; Dong, T.; Zhu, W. Thiodiketopiperazines from the marine-derived fungus *Phoma* sp. OUCMDZ-1847. *J. Nat. Prod.* **2014**, *77*, 132–137. [CrossRef]
20. Meng, L.H.; Wang, C.Y.; Mándi, A.; Li, X.M.; Hu, X.Y.; Kassack, M.U.; Kurtán, T.; Wang, B.G. Three diketopiperazine alkaloids with spirocyclic skeletons and one bisthiodiketopiperazine derivative from the mangrove-derived endophytic fungus *Penicillium brocae* MA-231. *Org. Lett.* **2016**, *18*, 5304–5307. [CrossRef]
21. Zhu, M.; Zhang, X.; Feng, H.; Dai, J.; Li, J.; Che, Q.; Gu, Q.; Zhu, T.J.; Li, D. Penicisulfuranols A–F, alkaloids from the mangrove endophytic fungus *Penicillium janthinellum* HDN13-309. *J. Nat. Prod.* **2017**, *80*, 71–75. [CrossRef] [PubMed]

22. Meng, L.H.; Zhang, P.; Li, X.M.; Wang, B.G. Penicibrocazines A–E, five new sulfide diketopiperazines from the marine-derived endophytic fungus *Penicillium brocae*. *Mar. Drugs* **2015**, *13*, 276–287. [CrossRef] [PubMed]
23. Sugawara, K.; Sugawara, F.; Strobel, G.A.; Fu, Y.; He, C.H.; Clardy, J. Exserohilone: A novel phytotoxin produced by *Exserohilum holmii*. *J. Org. Chem.* **1985**, *50*, 5631–5633. [CrossRef]
24. Cutler, H.G.; Hoogsteen, K.; Littrell, R.H.; Arison, B.H. Epoxyexserohilone, a novel metabolite from *Nigrospora sphaerica*. *Agric. Biol. Chem.* **1991**, *55*, 2037–2042. [CrossRef]
25. Centko, R.M.; Ratnaweera, P.B.; Tysoe, C.; Withers, S.G.; de Silva, E.D.; Andersen, R.J. Alpha-glucosidase and alpha-amylase inhibiting thiodiketopiperazines from the endophytic fungus *Setosphaeria rostrata* isolated from the medicinal plant *Costus speciosus* in Sri Lanka. *Phytochem. Lett.* **2017**, *22*, 76–80. [CrossRef]
26. He, T.; Wang, Y.; Du, L.; Li, F.; Hu, Q.; Cheng, G.; Wang, W. Overexpression of global regulator LaeA induced secondary metabolite production in *Aspergillus versicolor* 0312. *Rec. Nat. Prod.* **2020**, *14*, 387–394. [CrossRef]
27. Xu, Y.M.; Espinosa-Artiles, P.; Liu, M.X.; Arnold, A.E.; Gunatilaka, A.L. Secoemestrin D, a cytotoxic epitetrathiodioxopiperizine, and emericellenes A–E, five sesterterpenoids from *Emericella* sp. AST0036, a fungal endophyte of *Astragalus lentiginosus*. *J. Nat. Prod.* **2013**, *76*, 2330–2336. [CrossRef]
28. Seephonkai, P.; Kongsaeree, P.; Prabpai, S.; Isaka, M.; Thebtaranonth, Y. Transformation of an irregularly bridged epidithiodiketopiperazine to trichodermamide A. *Org. Lett.* **2006**, *8*, 3073–3075. [CrossRef]
29. Harwoko, H.; Daletos, G.; Stuhldreier, F.; Lee, J.; Wesselborg, S.; Feldbrügge, M.; Müller, W.E.G.; Kalscheuer, R.; Ancheeva, E.; Proksch, P. Dithiodiketopiperazine derivatives from endophytic fungi *Trichoderma harzianum* and *Epicoccum nigrum*. *Nat. Prod. Res.* **2021**, *35*, 257–265. [CrossRef]
30. Ma, Y.M.; Liang, X.A.; Zhang, H.C.; Liu, R. Cytotoxic and antibiotic cyclic pentapeptide from an endophytic *Aspergillus tamarii* of *Ficus carica*. *J. Agric. Food Chem.* **2016**, *64*, 3789–3793. [CrossRef]
31. Meng, L.H.; Li, X.M.; Lv, C.T.; Huang, C.G.; Wang, B.G. Brocazines A–F, cytotoxic bisthiodiketopiperazine derivatives from *Penicillium brocae* MA-231, an endophytic fungus derived from the marine mangrove plant *Avicennia marina*. *J. Nat. Prod.* **2014**, *77*, 1921–1927. [CrossRef] [PubMed]
32. Dong, J.Y.; He, H.P.; Shen, Y.M.; Zhang, K.Q. Nematicidal Epipolysulfanyldioxopiperazines from *Gliocladium roseum*. *J. Nat. Prod.* **2005**, *68*, 1510–1513. [CrossRef] [PubMed]
33. Yu, F.X.; Chen, Y.; Yang, Y.H.; Li, G.H.; Zhao, P.J. A new epipolythiodioxopiperazine with antibacterial and cytotoxic activities from the endophytic fungus *Chaetomium* sp. M336. *Nat. Prod. Res.* **2018**, *32*, 689–694. [CrossRef] [PubMed]
34. Li, H.; Xu, Q.; Sun, W.; Zhang, R.; Wang, J.; Lai, Y.; Hu, Z.X.; Zhang, Y. 21-Epi-taichunamide D and (±)-versicaline A, three unusual alkaloids from the endophytic *Aspergillus versicolor* F210. *Tetrahedron Lett.* **2020**, *61*, 152219. [CrossRef]
35. Wang, M.; Sun, Z.H.; Chen, Y.C.; Liu, H.X.; Li, H.H.; Tan, G.H.; Li, S.N.; Guo, X.L.; Zhang, W.M. Cytotoxic cochlioquinone derivatives from the endophytic fungus *Bipolaris sorokiniana* derived from *Pogostemon cablin*. *Fitoterapia* **2016**, *110*, 77–82. [CrossRef]
36. Li, C.S.; Sarotti, A.M.; Huang, P.; Dang, U.T.; Hurdle, J.G.; Kondratyuk, T.P.; Pezzuto, J.M.; Turkson, J.; Cao, S. NF-κB inhibitors, unique γ-pyranol-γ-lactams with sulfide and sulfoxide moieties from Hawaiian plant *Lycopodiella cernua* derived fungus *Paraphaeosphaeria neglecta* FT462. *Sci. Rep.* **2017**, *7*, 10424. [CrossRef]
37. Li, C.S.; Ding, Y.; Yang, B.J.; Miklossy, G.; Yin, H.Q.; Walker, L.A.; Turkson, J.; Cao, S. A new metabolite with a unique 4-pyranone–γ-lactam–1, 4-thiazine moiety from a Hawaiian-plant associated fungus. *Org. Lett.* **2015**, *17*, 3556–3559. [CrossRef]
38. Wang, F.W.; Ye, Y.H.; Ding, H.; Chen, Y.X.; Tan, R.X.; Song, Y.C. Benzophenones from *Guignardia* sp. IFB-E028, an Endophyte on *Hopea hainanensis*. *Chem. Biodivers.* **2010**, *7*, 216–220. [CrossRef]
39. Zhang, F.Z.; Li, X.M.; Yang, S.Q.; Meng, L.H.; Wang, B.G. Thiocladospolides A–D, 12-membered macrolides from the mangrove-derived endophytic fungus *Cladosporium cladosporioides* MA-299 and structure revision of pandangolide. *J. Nat. Prod.* **2019**, *82*, 1535–1541. [CrossRef]
40. Huang, C.; Chen, T.; Yan, Z.; Guo, H.; Hou, X.; Jiang, L.; Long, Y. Thiocladospolide E and cladospamide A, novel 12-membered macrolide and macrolide lactam from mangrove endophytic fungus *Cladosporium* sp. SCNU-F0001. *Fitoterapia* **2019**, *137*, 104246. [CrossRef]
41. Wang, W.; Feng, H.; Sun, C.; Che, Q.; Zhang, G.; Zhu, T.; Li, D. Thiocladospolides F-J, antibacterial sulfur containing 12-membered macrolides from the mangrove endophytic fungus *Cladosporium oxysporum* HDN13-314. *Phytochemistry* **2020**, *178*, 112462. [CrossRef] [PubMed]
42. Chen, H.Y.; Liu, T.K.; Shi, Q.; Yang, X.L. Sesquiterpenoids and diterpenes with antimicrobial activity from *Leptosphaeria* sp. XL026, an endophytic fungus in *Panax notoginseng*. *Fitoterapia* **2019**, *137*, 104243. [CrossRef] [PubMed]
43. Zhao, D.L.; Zhang, X.F.; Huang, R.H.; Wang, D.; Wang, X.Q.; Li, Y.Q.; Zheng, C.J.; Zhang, P.; Zhang, C.S. Antifungal nafuredin and epithiodiketopiperazine derivatives from the mangrove-derived fungus *Trichoderma harzianum* D13. *Front. Microbiol.* **2020**, *11*, 1495. [CrossRef] [PubMed]
44. Song, X.-Q.; Zhang, X.; Han, Q.-J.; Li, X.-B.; Li, G.; Li, R.-J.; Jiao, Y.; Zhou, J.-C.; Lou, H.-X. Xanthone derivatives from Aspergillus sydowii, an endophytic fungus from the liverwort Scapania ciliata S. Lac and their immunosuppressive activities. *Phytochemistry Lett.* **2013**, *6*, 318–321. [CrossRef]
45. Akone, S.H.; Wang, H.; Mouelle, E.N.M.; Mándi, A.; Kurtán, T.; Koliye, P.R.; Hartmann, R.; Bhatia, S.; Yang, J.; Müller, W.E.G. Prenylated cyclohexene-type meroterpenoids and sulfur-containing xanthones produced by *Pseudopestalotiopsis theae*. *Phytochemistry* **2022**, *197*, 113124. [CrossRef]

46. Zhang, A.L.; He, L.Y.; Gao, J.M.; Xu, X.; Li, S.Q.; Bai, M.S.; Qin, J.C. Metabolites from an endophytic fungus *sphaceloma* sp. LN-15 isolated from the leaves of *Melia azedarach*. *Lipids* **2009**, *44*, 745–751. [CrossRef]
47. Yu, G.; Wang, Q.; Liu, S.; Zhang, X.; Che, Q.; Zhang, G.; Zhu, T.; Gu, Q.; Li, D. Methylsulfonylated polyketides produced by *Neosartorya udagawae* HDN13-313 via exogenous addition of small molecules. *J. Nat. Prod.* **2019**, *82*, 998–1001. [CrossRef]
48. Yu, G.; Sun, Z.; Peng, J.; Zhu, M.; Che, Q.; Zhang, G.; Zhu, T.; Gu, Q.; Li, D. Secondary metabolites produced by combined culture of *Penicillium crustosum* and a *Xylaria* sp. *J. Nat. Prod.* **2019**, *82*, 2013–2017. [CrossRef]
49. Aly, A.H.; Edrada-Ebel, R.; Indriani, I.D.; Wray, V.; Müller, W.E.; Totzke, F.; Zirrgiebel, U.; Schächtele, C.; Kubbutat, M.H.G.; Lin, W.H. Cytotoxic metabolites from the fungal endophyte *Alternaria* sp. and their subsequent detection in its host plant *Polygonum senegalense*. *J. Nat. Prod.* **2008**, *71*, 972–980. [CrossRef]
50. Aly, A.H.; Edrada-Ebel, R.; Wray, V.; Müller, W.E.; Kozytska, S.; Hentschel, U.; Proksch, P.; Ebel, R. Bioactive metabolites from the endophytic fungus *Ampelomyces* sp. isolated from the medicinal plant *Urospermum picroides*. *Phytochemistry* **2008**, *69*, 1716–1725. [CrossRef]
51. Sun, J.F.; Lin, X.; Zhou, X.F.; Wan, J.; Zhang, T.; Yang, B.; Yang, X.W.; Tu, Z.; Liu, Y. Pestalols A–E, new alkenyl phenol and benzaldehyde derivatives from endophytic fungus *Pestalotiopsis* sp. AcBC2 isolated from the Chinese mangrove plant *Aegiceras corniculatum*. *J. Antibiot.* **2014**, *67*, 451–457. [CrossRef] [PubMed]
52. Jayasuriya, H.; Bills, G.F.; Cascales, C.; Zink, D.L.; Goetz, M.A.; Jenkins, R.G.; Silverman, K.C.; Lingham, R.B.; Singh, S.B. Oreganic acid: A potent novel inhibitor of Ras farnesyl-protein transferase from an endophytic fungus. *Bioorg. Med. Chem. Lett.* **1996**, *6*, 2081–2084. [CrossRef]
53. Zhang, L.; Liu, Y.; Deng, Z.; Guo, Z.; Chen, J.; Tu, X.; Zou, K. Fusaodavinvin, a novel metabolite containing sulfur from the endophytic fungus *Fusarium* sp. (CTGU-ZL-34). *Nat. Prod. Commun.* **2013**, *8*, 83–84. [CrossRef] [PubMed]
54. Wu, Z.; Zhang, X.; Anbari, W.H.A.; Zhou, Q.; Zhou, P.; Zhang, M.; Zeng, F.; Chen, C.; Tong, Q.; Wang, J. Cysteine residue containing merocytochalasans and 17, 18-seco-aspochalasins from *Aspergillus micronesiensis*. *J. Nat. Prod.* **2019**, *82*, 2653–2658. [CrossRef]
55. Su, X.Z.; Zhu, Y.Y.; Tang, J.W.; Hu, K.; Li, X.N.; Sun, H.D.; Li, Y.; Puno, P.T. Pestaloamides A and B, two spiro-heterocyclic alkaloid epimers from the plant endophytic fungus *Pestalotiopsis* sp. HS30. *Sci. China Chem.* **2020**, *63*, 1208–1213. [CrossRef]
56. Verekar, S.A.; Mishra, P.D.; Sreekumar, E.S.; Deshmukh, S.K.; Fiebig, H.H.; Kelter, G.; Maier, A. Anticancer activity of new depsipeptide compound isolated from an endophytic fungus. *J. Antibiot.* **2014**, *67*, 697–701. [CrossRef]
57. Obayashi, Y.; Yoshimura, T.; Ikenoue, Y.; Fudo, R.; Murata, M.; Ando, T. Group of Antitumor Compounds and Method for Producing the Same. U.S. Patent 5,843,755, 1 December 1998.
58. Ibrahim, S.R.M.; Elkhayat, E.S.; Mohamed, G.A.A.; Fat'hi, S.M.; Ross, S.A. Fusarithioamide A, a new antimicrobial and cytotoxic benzamide derivative from the endophytic fungus *Fusarium chlamydosporium*. *Biochem. Bioph. Res. Commun.* **2016**, *479*, 211–216. [CrossRef]
59. Ibrahim, S.R.; Mohamed, G.A.; Al Haidari, R.A.; Zayed, M.F.; El-Kholy, A.A.; Elkhayat, E.S.; Ross, S.A. Fusarithioamide B, a new benzamide derivative from the endophytic fungus *Fusarium chlamydosporium* with potent cytotoxic and antimicrobial activities. *Bioorg. Med. Chem.* **2018**, *26*, 786–790. [CrossRef]
60. Mei, W.L.; Chen, P.; Wang, H.; Huang, J.L.; Dai, H.F. Two new sesquiterpenes from endophytic fungus S49 of *Cephalotaxus hainanensis*. *J. Asian Nat. Prod. Res.* **2010**, *12*, 582–585. [CrossRef]
61. Wu, S.H.; He, J.; Li, X.N.; Huang, R.; Song, F.; Chen, Y.W.; Miao, C.P. Guaiane sesquiterpenes and isopimarane diterpenes from an endophytic fungus *Xylaria* sp. *Phytochemistry* **2014**, *105*, 197–204. [CrossRef]
62. Liu, H.X.; Tan, H.B.; Chen, Y.C.; Li, S.N.; Li, H.H.; Zhang, W.M. Secondary metabolites from the *Colletotrichum gloeosporioides* A12, an endophytic fungus derived from *Aquilaria sinensis*. *Nat. Prod. Res.* **2018**, *32*, 2360–2365. [CrossRef] [PubMed]
63. Bisen, P.S.; Mittal, S.; Shrivastava, D.; Govil, S.; Kumar, S. Isolation of a Novel Anticandidal Sulphur Containing Molecule from Emericella sp. India. Patent IN2012DE03504, 12 December 2014.
64. Uz Zaman, K.H.A.; Park, J.H.; DeVine, L.; Hu, Z.; Wu, X.; Kim, H.S.; Cao, S. Secondary Metabolites from the Leather Coral-Derived Fungal Strain Xylaria sp. FM1005 and Their Glycoprotein IIb/IIIa Inhibitory Activity. *J. Nat. Prod.* **2021**, *84*, 466–473. [CrossRef] [PubMed]
65. Baccile, J.A.; Le, H.H.; Pfannenstiel, B.T.; Bok, J.W.; Gomez, C.; Brandenburger, E.; Hoffmeister, D.; Keller, N.P.; Schroeder, F.C. Diketopiperazine Formation in Fungi Requires Dedicated Cyclization and Thiolation Domains. *Angew. Chem. Int. Ed.* **2019**, *58*, 14589–14593. [CrossRef] [PubMed]
66. Clevenger, K.D.; Bok, J.W.; Ye, R.; Miley, G.P.; Verdan, M.H.; Velk, T.; Chen, C.; Yang, K.; Robey, M.T.; Gao, P.; et al. A scalable platform to identify fungal secondary metabolites and their gene clusters. *Nat. Chem. Biol.* **2017**, *13*, 895–901. [CrossRef]
67. Blin, K.; Shaw, S.; Kloosterman, A.M.; Charlop-Powers, Z.; van Wezel, G.P.; Medema, M.H.; Weber, T. antiSMASH 6.0: Improving cluster detection and comparison capabilities. *Nucleic Acids Res.* **2021**, *49*, W29–W35. [CrossRef]
68. Van der Lee, T.A.J.; Medema, M.H. Computational strategies for genome-based natural product discovery and engineering in fungi. *Fungal Genet. Biol.* **2016**, *89*, 29–36. [CrossRef]
69. Chiang, Y.-M.; Oakley, C.E.; Ahuja, M.; Entwistle, R.; Schultz, A.; Chang, S.-L.; Sung, C.T.; Wang, C.C.C.; Oakley, B.R. An Efficient System for Heterologous Expression of Secondary Metabolite Genes in *Aspergillus nidulans*. *J. Am. Chem. Soc.* **2013**, *135*, 7720–7731. [CrossRef]
70. Harvey, C.J.B.; Tang, M.; Schlecht, U.; Horecka, J.; Fischer, C.R.; Lin, H.-C.; Li, J.; Naughton, B.; Cherry, J.; Miranda, M.; et al. HEx: A heterologous expression platform for the discovery of fungal natural products. *Sci. Adv.* **2018**, *4*, eaar5459. [CrossRef]

Journal of *Fungi*

MDPI

Review

Recent Advances in Sorbicillinoids from Fungi and Their Bioactivities (Covering 2016–2021)

Xuwen Hou, Xuping Zhang, Mengyao Xue, Zhitong Zhao, Huizhen Zhang, Dan Xu, Daowan Lai and Ligang Zhou *

State Key Laboratory of Agrobiotechnology, Department of Plant Pathology, College of Plant Protection, China Agricultural University, Beijing 100193, China; xwhou@cau.edu.cn (X.H.); zhangxuping5@cau.edu.cn (X.Z.); mengyaoxue@cau.edu.cn (M.X.); zhitongzhao@cau.edu.cn (Z.Z.); huizhenzhang@cau.edu.cn (H.Z.); cauxudan@cau.edu.cn (D.X.); dwlai@cau.edu.cn (D.L.)

* Correspondence: lgzhou@cau.edu.cn; Tel.: +86-10-6273-1199

Abstract: Sorbicillinoids are a family of hexaketide metabolites with a characteristic sorbyl side chain residue. Sixty-nine sorbicillinoids from fungi, newly identified from 2016 to 2021, are summarized in this review, including their structures and bioactivities. They are classified into monomeric, dimeric, trimeric, and hybrid sorbicillinoids according to their basic structural features, with the main groups comprising both monomeric and dimeric sorbicillinoids. Some of the identified sorbicillinoids have special structures such as ustilobisorbicillinol A, and sorbicillasins A and B. The majority of sorbicillinoids have been reported from fungi genera such as *Acremonium, Penicillium, Trichoderma*, and *Ustilaginoidea*, with some sorbicillinoids exhibiting cytotoxic, antimicrobial, anti-inflammatory, phytotoxic, and α-glucosidase inhibitory activities. In recent years, marine-derived, extremophilic, plant endophytic, and phytopathogenic fungi have emerged as important resources for diverse sorbicillinoids with unique skeletons. The recently revealed biological activities of sorbicillinoids discovered before 2016 are also described in this review.

Keywords: monomeric sorbicillinoids; bisorbicillinoids; trisorbicillinoids; hybrid sorbicillinoids; fungi; occurrence; biological activities

Citation: Hou, X.; Zhang, X.; Xue, M.; Zhao, Z.; Zhang, H.; Xu, D.; Lai, D.; Zhou, L. Recent Advances in Sorbicillinoids from Fungi and Their Bioactivities (Covering 2016–2021). *J. Fungi* **2022**, *8*, 62. https://doi.org/10.3390/jof8010062

Academic Editor: Célia F. Rodrigues

Received: 20 December 2021
Accepted: 5 January 2022
Published: 7 January 2022

1. Introduction

Sorbicillinoids are a family of fungal metabolites related to the hexaketide sorbicillin, and typically contain a sorbyl side chain in the structures with highly oxygenated frameworks [1,2]. According to their structural characteristics and biosynthesis, sorbicillinoids are divided into four groups: monomeric, dimeric, trimeric and hybrid sorbicillinoids [2]. Since sorbicillin (**1**) was first discovered from *Penicillium notatum* in 1948 [3], about 159 sorbicillinoids have been reported from fungi, especially those from genera *Penicillium* and *Trichoderma*.

Sorbicillinoids have potential pharmaceutical and agrochemical value as antimicrobial, antivirus, and anticancer agents, as well as pigments and food colorants. Sorbicillinoids and their biological activities have been well-reviewed before 2016 [1,2]. In 2011, Harned and Volp reviewed the structures of 62 sorbicillinoids [1]. Successively, 28 additional sorbicillinoids were reviewed by Meng et al. in 2016 [2]. Since then, dozens of new analogues have emerged.

In this mini-review, we focus on the recently identified structures of 69 sorbicillinoids along with their biological activities including newly revealed bioactivities of the sorbicillinoids discovered before 2016, in order to increase the diversity of identified sorbicillinoids as well as to speed up their applications.

2. Occurrence

Sorbicillinoids have a diverse distribution in fungi. In total, 69 sorbicillinoids have been isolated since 2016. They have mainly been found in plant endophytic, marine-derived, extremophilic, phytopathogenic fungi, and soil-derived fungi, mainly from the genera of *Acremonium*, *Aspergillus*, *Clonostachys*, *Penicillium*, *Ustilaginoidea*, and *Verticillium* [4,5]. All these fungi belong to the ascomycetes. The structures of sorbicillinoids are shown in Figures 1–5.

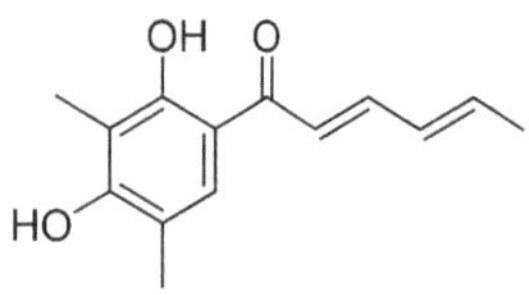
1. Sorbicillin

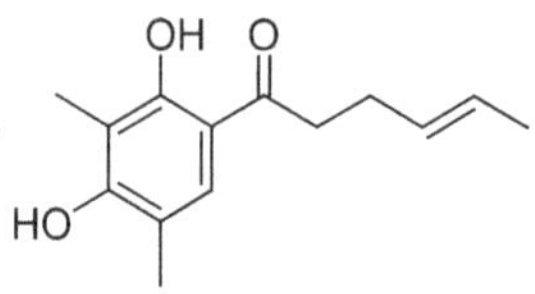
2. Dihydrosorbicillin

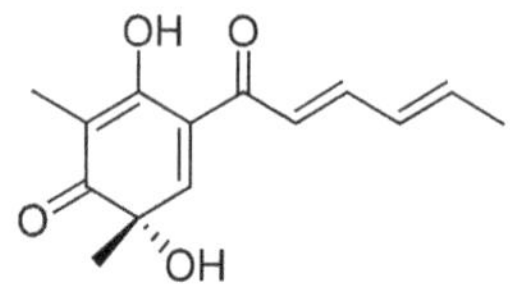
3. Sorbicillinol

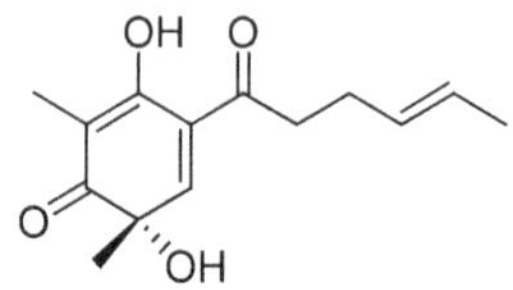
4. Dihydrosorbicillinol

Figure 1. Basic structures of the monomeric sorbicillinoids (**1**–**4**).

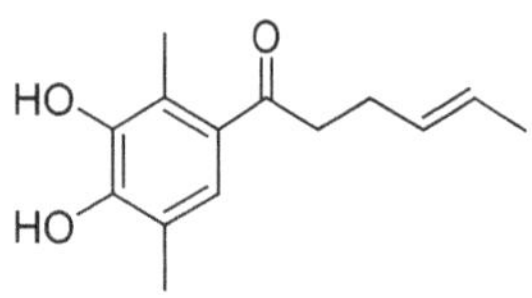
5. 2-Deoxysohirnone C

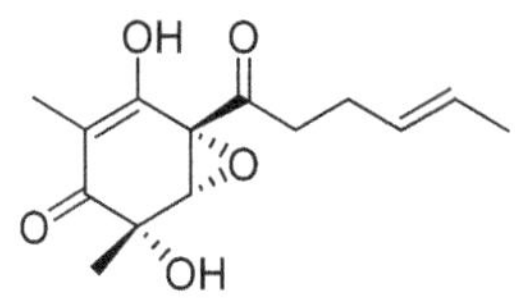
6. 2',3'-Dihydro-epoxysorbicillinol

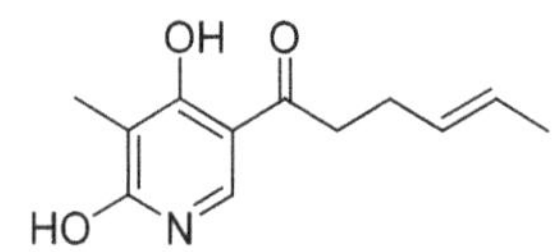
7. (4*E*)-1-(4,6-Dihydroxy-5-methylpyridin-3-yl)hex-4-en-1-one

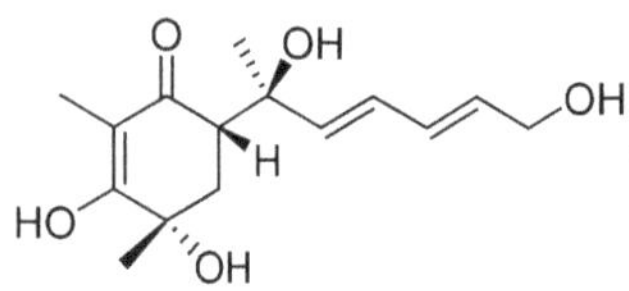
8. Saturnispol E

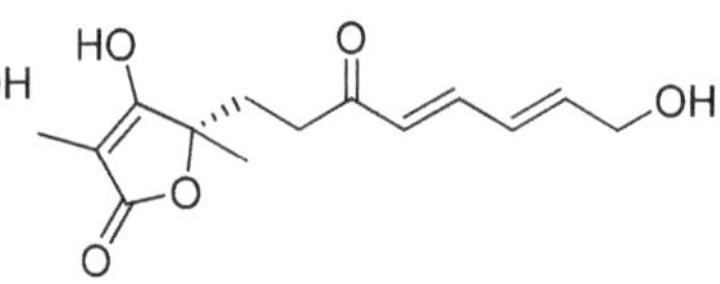
9. Saturnispol F

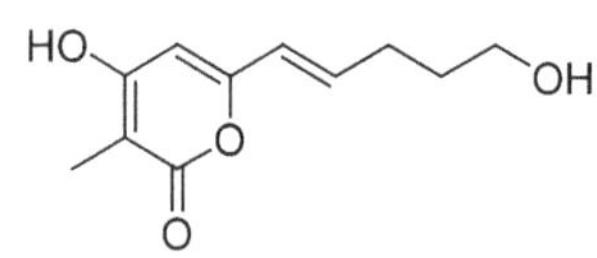
10. Saturnispol G

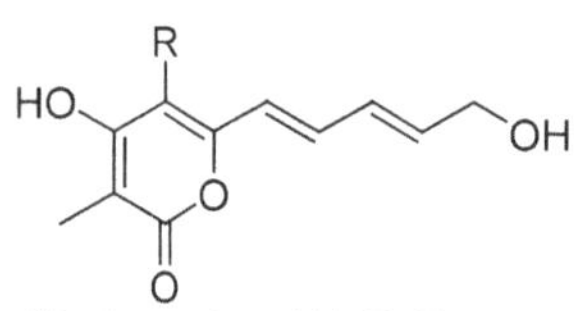
11. Saturnispol H, R=H
12. Ustilopyrone A, R=CH_3

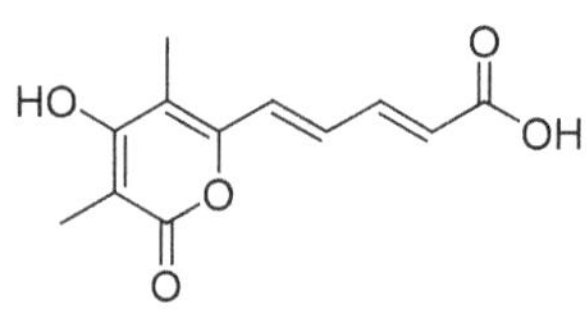
13. Ustilopyrone B

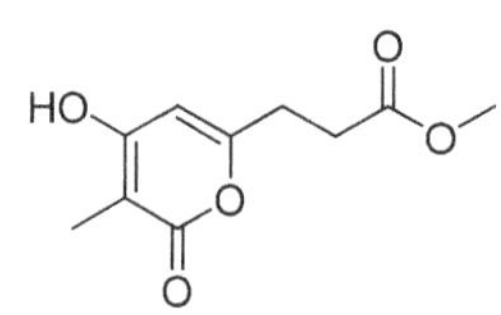
14. Scirpyrone K

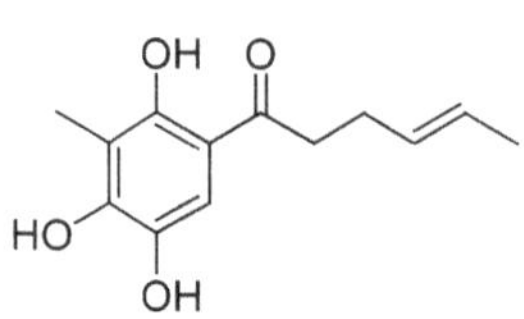
15. 5-Hydroxy-dihydrodemethylsorbicillin

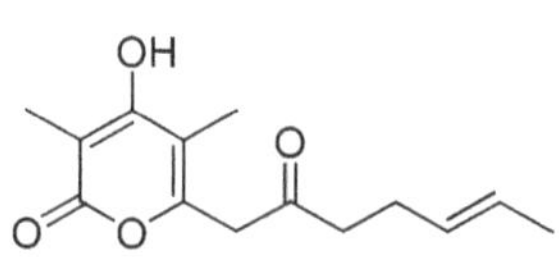
16. Sorbicillpyrone A

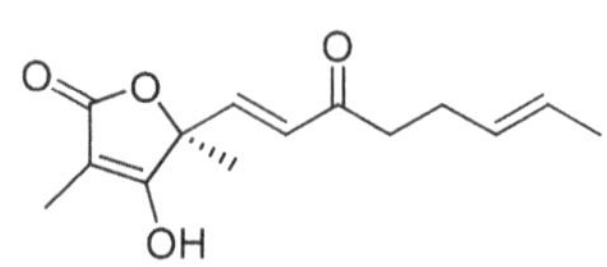
17. 5,6-Dehydrovertinolide

Figure 2. *Cont.*

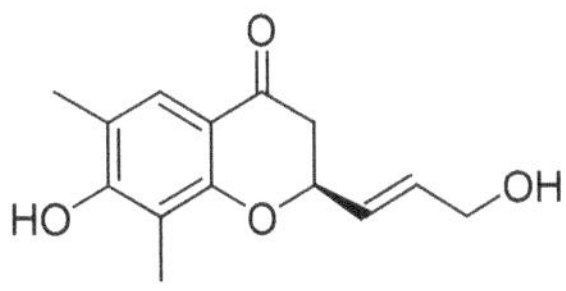

18. Trichosorbicillin B

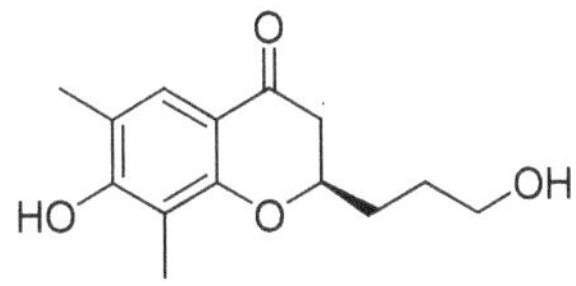

19. Trichosorbicillin C

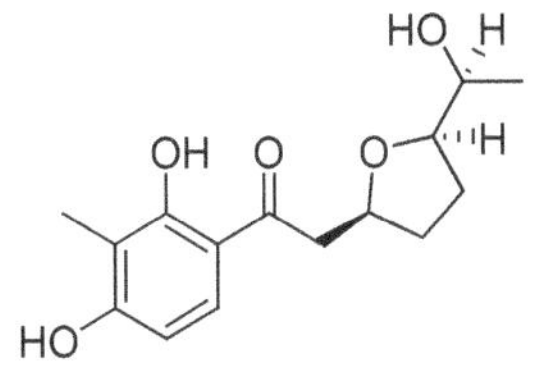

20. Trichosorbicillin D

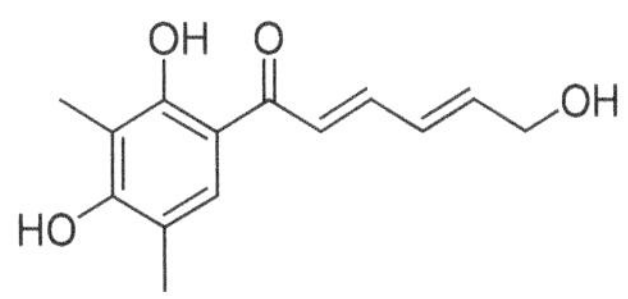

21. 12-Hydroxysorbicillin

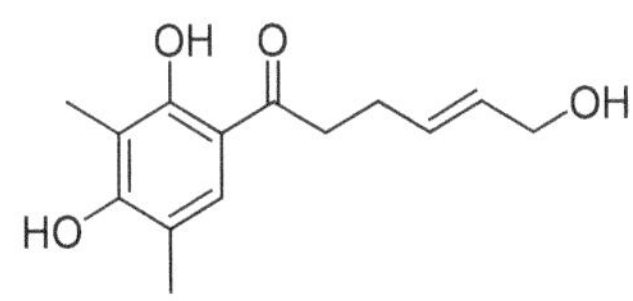

22. 8,9-Dihydro-12-hydroxysorbicillin

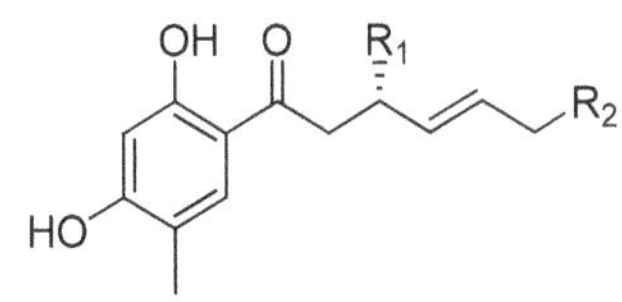

23. Trichosorbicillin E, R_1=H, R_2=OH
24. Isotrichosorbicillin E, R_1=OH, R_2=H

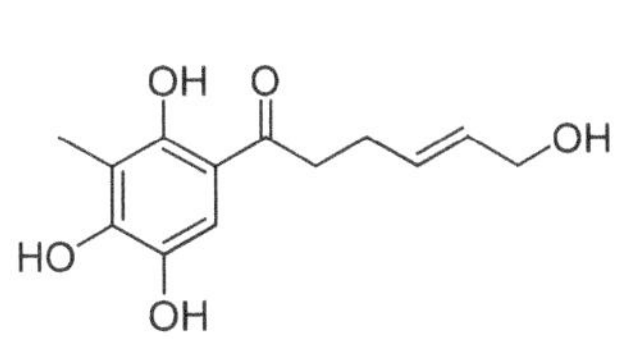

25. Trichosorbicillin F

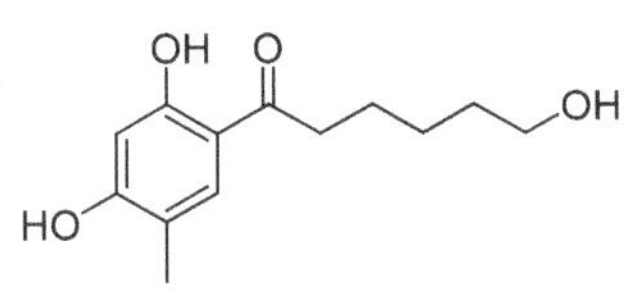

26. Trichosorbicillin G

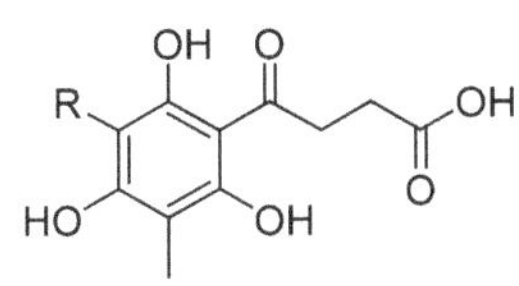

27. Trichosorbicillin H, R=H
28. 3-Methyltrichosorbicillin H, R=CH_3

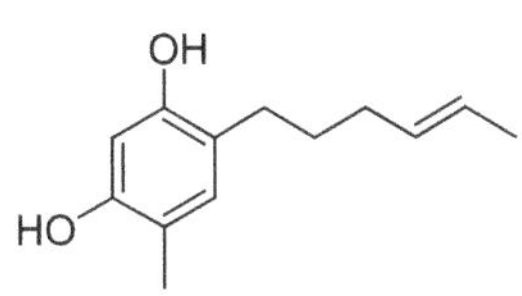

29. Trichosorbicillin I

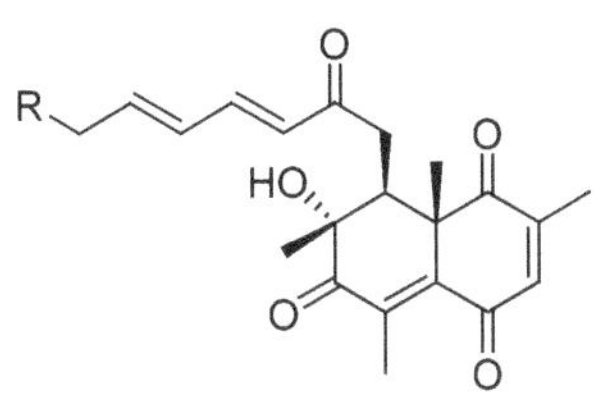

30. Trichoreeseione A, R=H
31. Trichoreeseione B, R=OH

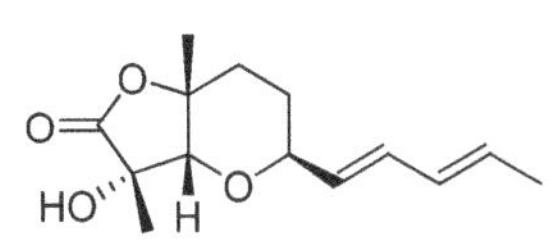

32. Trichoreesin A

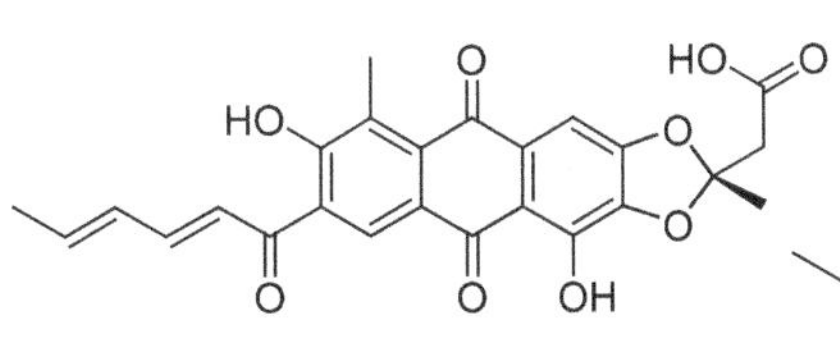

33. Ustilanthracin A

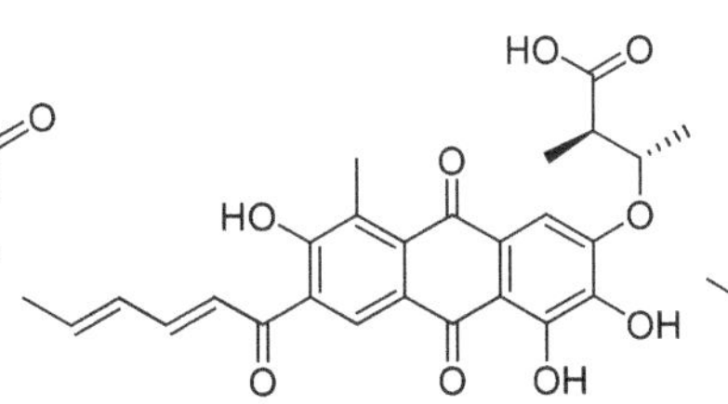

34. Ustilanthracin B

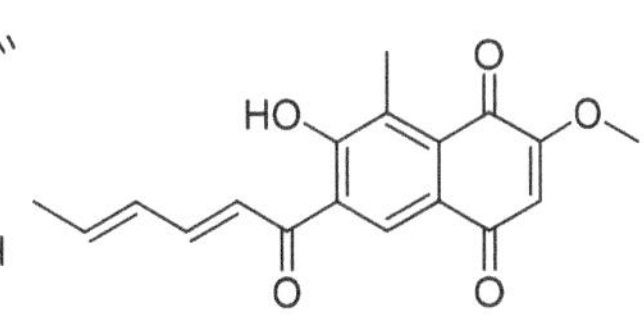

35. Ustinaphalin

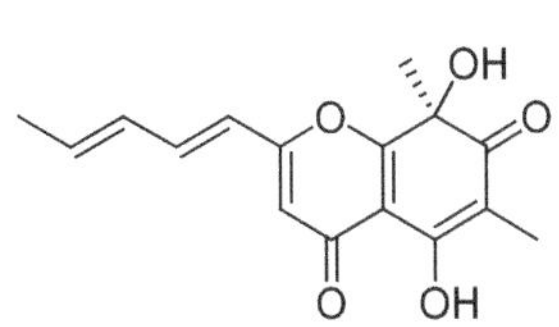

36. Ustisorbicillinol F

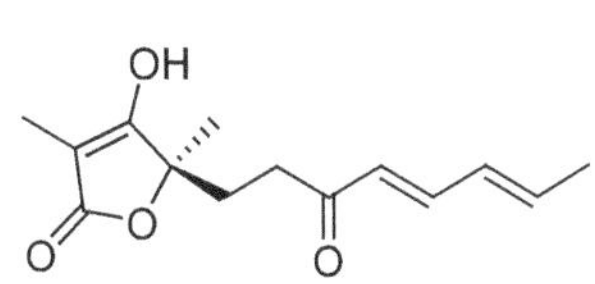

37. (+)-(*R*)-Vertinolide

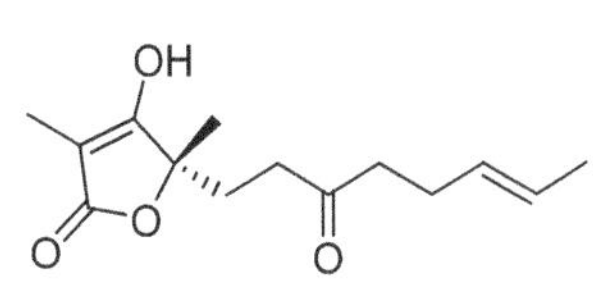

38. (-)-(*S*)-Dihydrovertinolide

Figure 2. Structures of the monomeric sorbicillinoids (**5**–**38**) isolated from fungi.

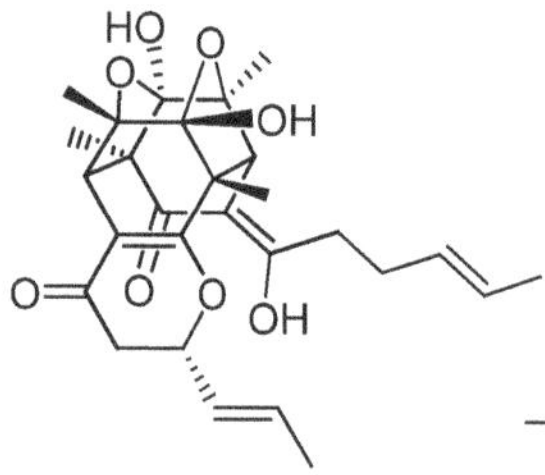

39. Epitetrahydrotrichodimer ether

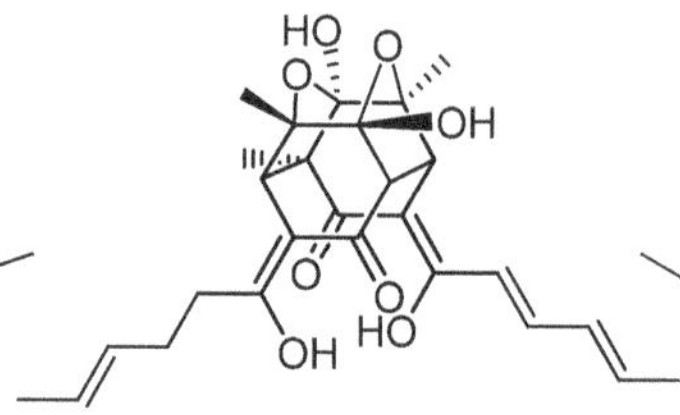

40. Demethyldihydrotrichodimerol

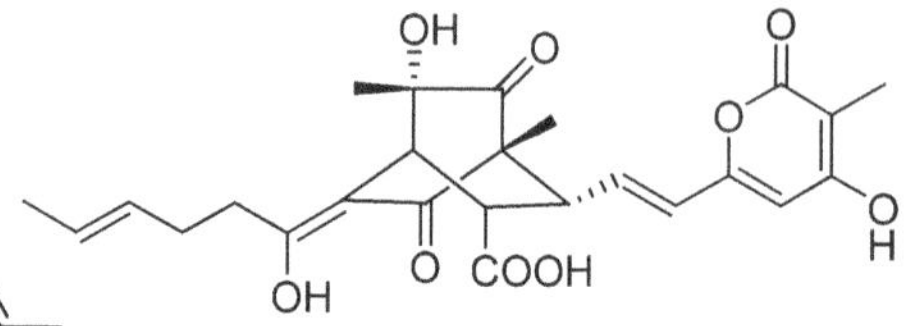

41. Bisorbicillpyrone A

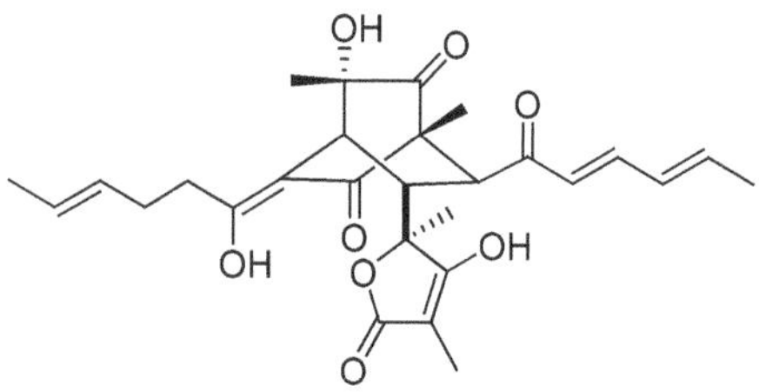

42. 10,11-Dihydrobislongiquinolide

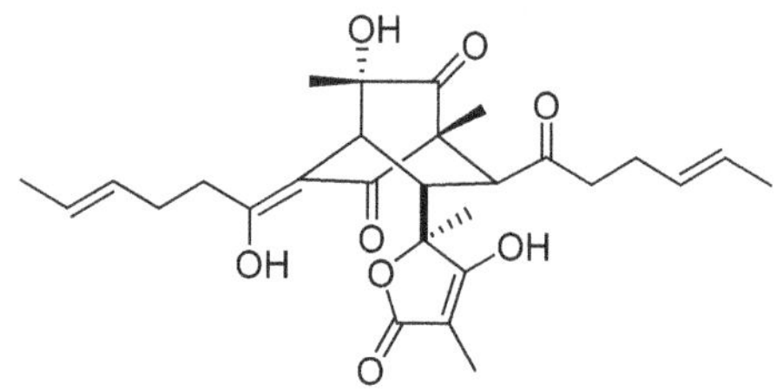

43. 10,11,16,17-Tetrahydrobislongiquinolide

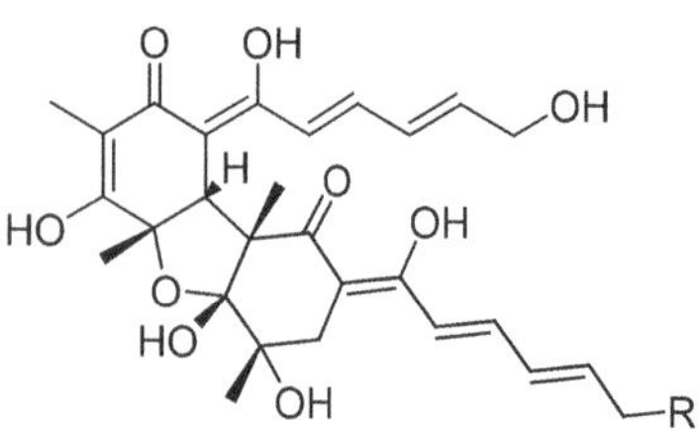

44. Saturnispol A, R=OH
45. Saturnispol B, R=H

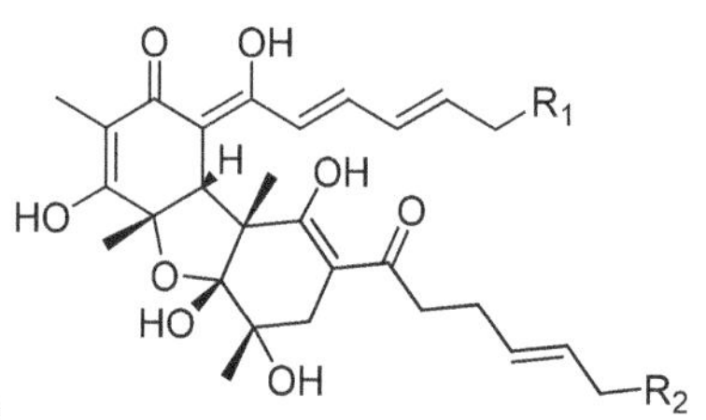

46. Trichobisvertinol A, R_1=H, R_2=H
47. Trichobisvertinol B, R_1=OH, R_2=H
48. Trichobisvertinol C, R_1=OH, R_2=OH

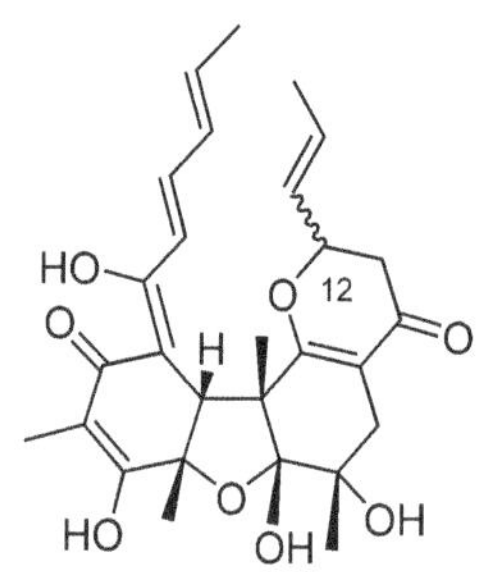

49. Trichobisvertinol D, 12*R*
50. 12-*epi*-Trichobisvertinol D, 12*S*

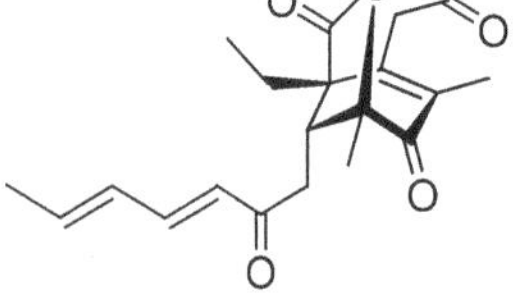

51. Trichodermolide B

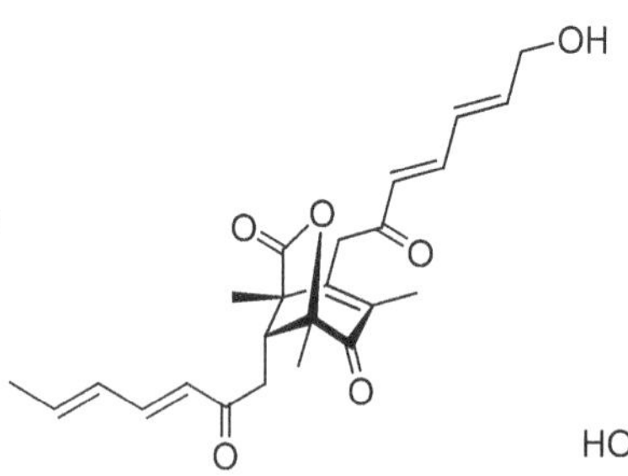

52. 13-Hydroxy-trichodermolide

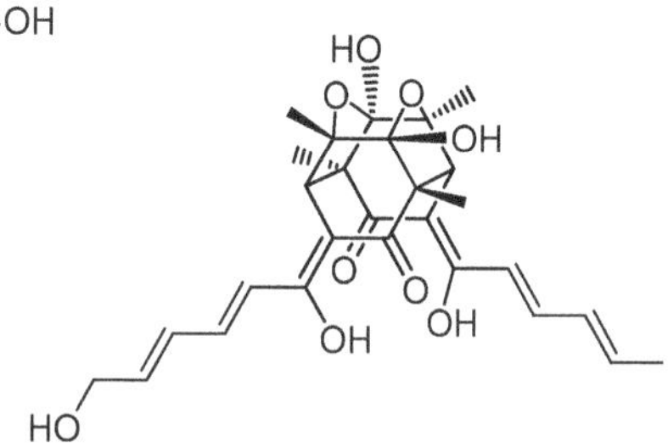

53. 24-Hydroxy-trichodimerol

Figure 3. *Cont.*

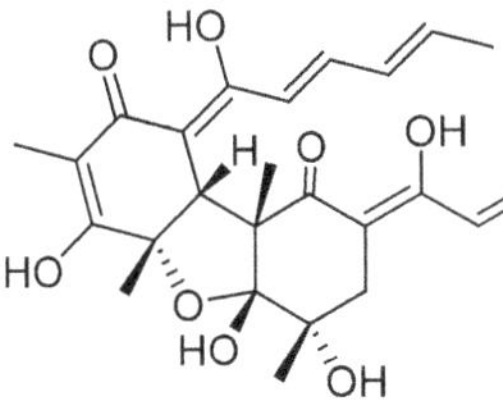

54. 15-Hydroxy-bisvertinol

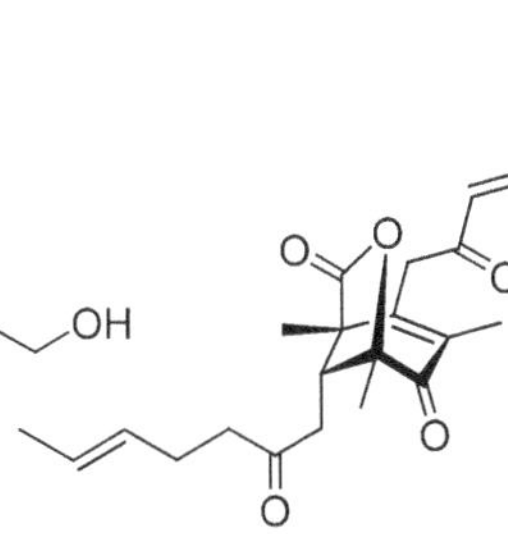

55. 13-Hydroxy-dihydrotrichodermolide

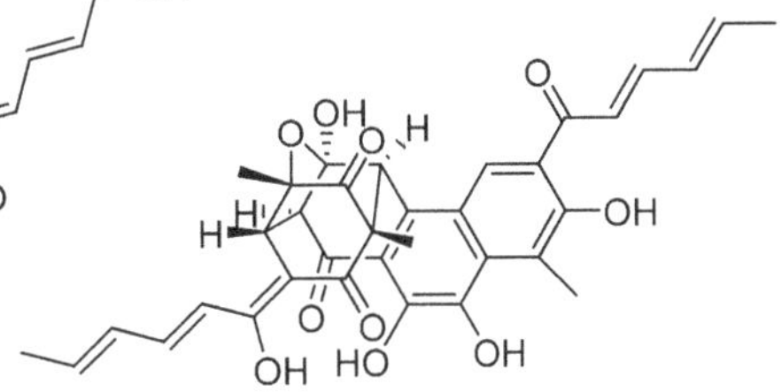

56. Ustilobisorbicillinol A

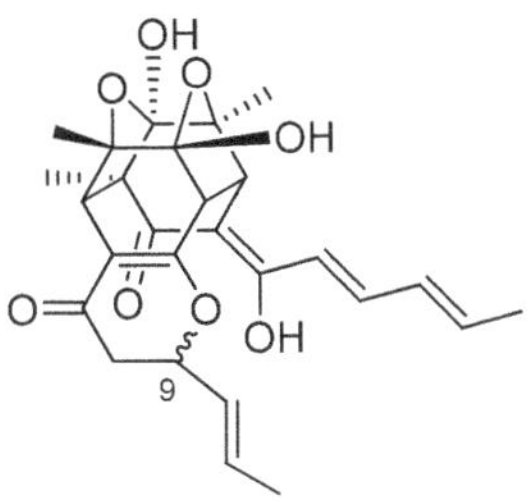

57. Ustisorbicillinol C, 9*S*
58. Ustisorbicillinol D, 9*R*

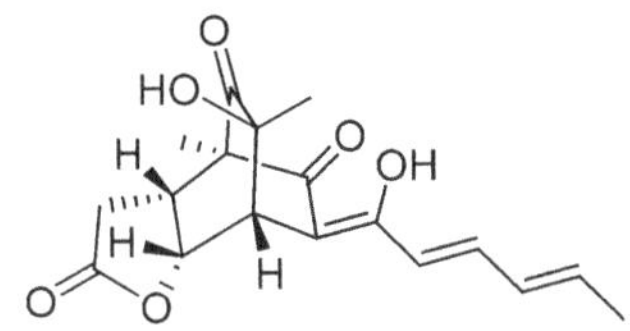

59. Ustisorbicillinol E

Figure 3. Structures of the bisorbicillinoids (**39–59**) isolated from fungi.

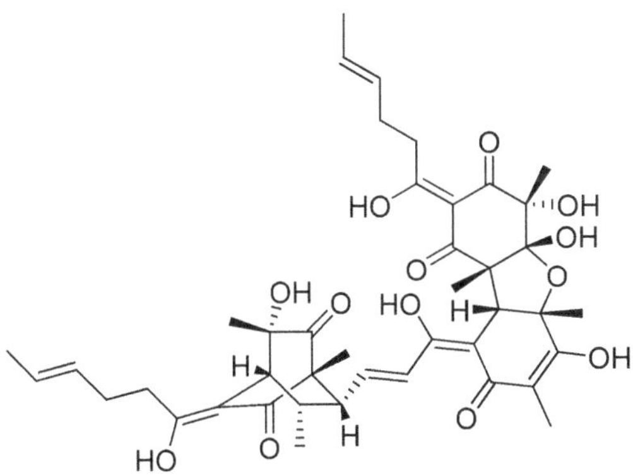

60. 10,11,27,28-Tetrahydrotrisorbicillinone C

Figure 4. Structure of the trisorbicillinoid (**60**) isolated from fungi.

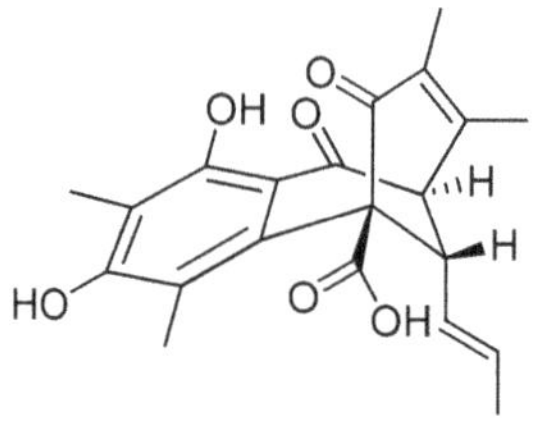

61. 10-Methylsorbiterrin A

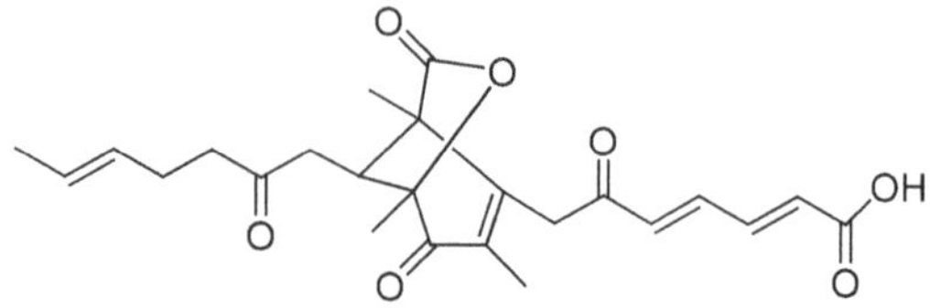

62. Dihydrotrichodermolidic acid

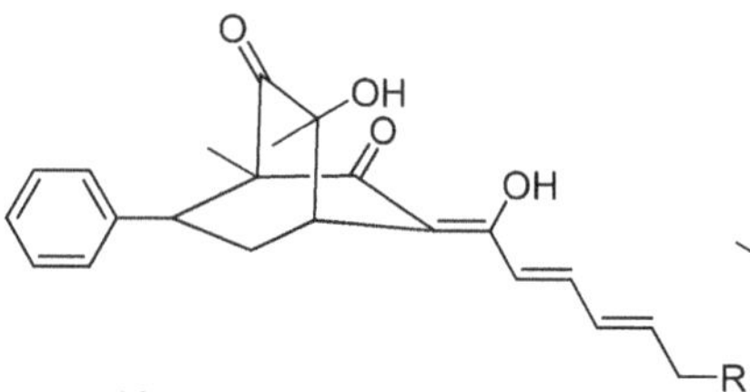

63. Saturnispol C, R=H
64. Saturnispol D, R=OH

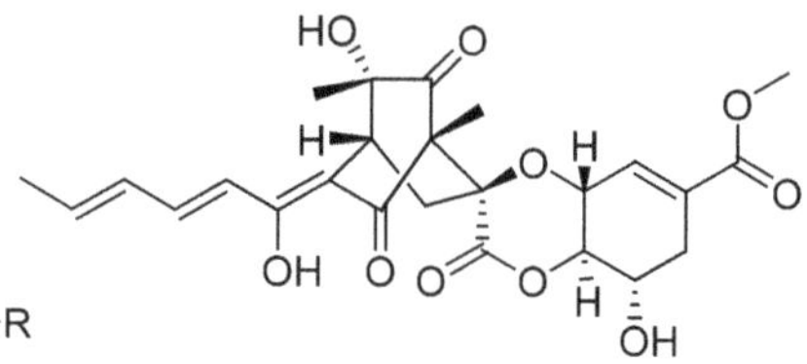

65. Spirosorbicillinol D

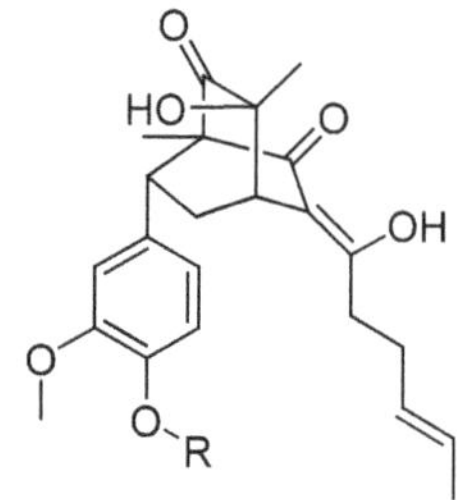

66. Sorbicatechol C, R=CH_3
67. Sorbicatechol D, R=H

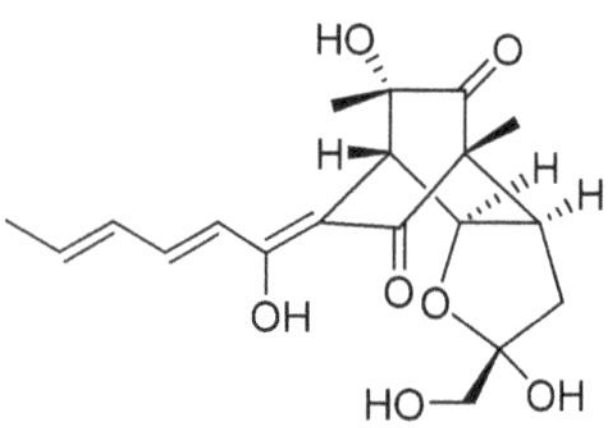

68. Sorbicillfuran A

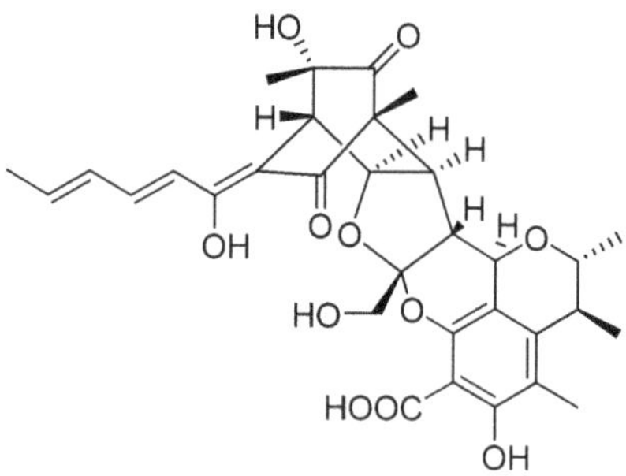

69. Sorbicillfuran B

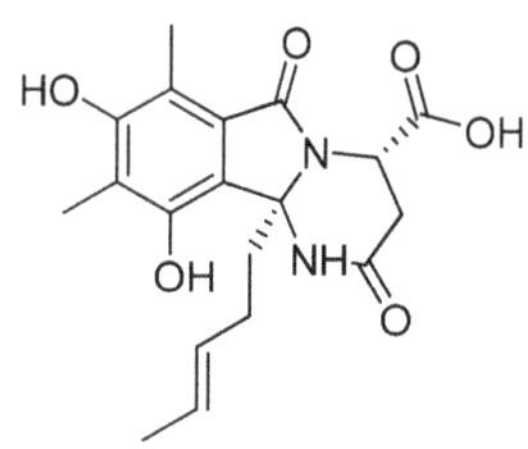

70. Sorbicillasin A

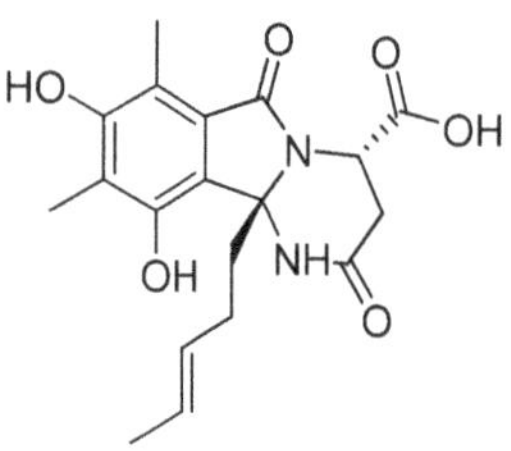

71. Sorbicillasin B

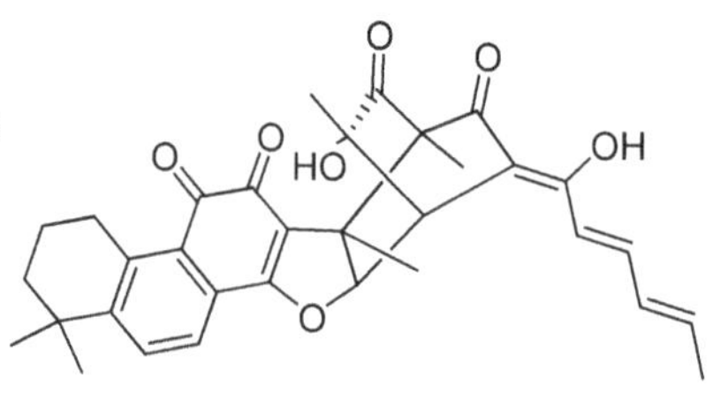

72. Tanshisorbicin

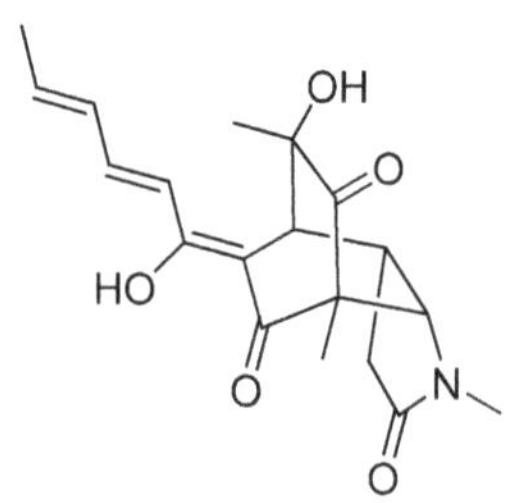

73. Trichosorbicillin A

Figure 5. Structures of the hybrid sorbicillinoids (**61**–**73**) isolated from fungi.

2.1. Monomeric Sorbicillinoids

Sorbicillinoid monomers are the basic units of the sorbyl-containing metabolites catalyzed by polyketide synthases such as SorA and SorB [6]. The initial monomeric sorbicillinoid is sorbicillin (**1**), which is subsequently converted to dihydrosorbicillin (also called 2′,3′-dihydrosorbicillin, (**2**), sorbicillinol (**3**), dihydrosorbicillinol (also called 2′,3′-dihydrosorbicillinol, **4**), and other sorbicillinoids (Figure 1) [7]. The biosynthesis of the monomeric sorbicillinoids was revealed mainly based on genome research. An FAD-dependent monooxygenase encoding gene (*sorbC*) was cloned from *Penicillium chrysogenum* E01-10/3 and expressed as a soluble protein in *Escherichia coli*. The enzyme efficiently performed the oxidative dearomatization of sorbicillin (**1**) and dihydrosorbicillin (**2**) to produce sorbicillinol (**3**) and dihydrosorbicillinol (**4**), respectively [8].

Since 2016, thirty-four monomeric sorbicillinoids (Figure 2 and Table S1) have been isolated from fungi of the genera *Penicillium*, *Trichoderma*, *Ustilaginoidea*, *Phialocephala*, and *Clonostachys*. 2-deoxysohirnone C (**5**) was isolated from *Penicillium* sp. GD6 from the mangrove plant *Bruguiera gymnorrhiza* [9], and later isolated from *Penicillium* sp. SCSIO06871 from deep-sea sediment collected from the Indian Ocean [10].

2′,3′-dihydro-epoxysorbicillinol (**6**) was isolated as a new natural compound from *Trichoderma longibrachiatum* SFC100166, which was isolated from foreshore soil [11].

(4*E*)-1-(4,6-dihydroxy-5-methylpyridin-3-yl)hex-4-en-1-one (**7**) is a nitrogen-containing monomeric sorbicillinoid that was isolated from *Penicillium* sp. DM815 from the rhizosphere soil of a *Hibiscus tiliaceus* mangrove [12].

Four monomeric sorbicillinoids, namely saturnispols E (**8**), F (**9**), G (**10**) and H (**11**), were isolated from *Trichoderma saturnisporum* DI-IA from the marine sponge *Dictyonella incisa* collected at a depth of 10 m in Seferihisar Bay in Turkey [13]. Saturnispol H (**11**) is also named 5-demethylustilopyrone A (**11**), which was later isolated from the rice false smut pathogen *Ustilaginoidea virens* [14].

Both ustilopyrones A (**12**) and B (**13**), with pyrone structures, were isolated from rice false smut pathogen *Ustilaginoidea virens* [14]. Subsequently, ustilopyrone B (**13**) was re-isolated from *Penicillium* sp. SCSIO06871 from deep-sea sediment [10].

Scipyrone K (**14**), with a 3,4,6-trisubstituted α-pyrone structure, was isolated from the fungus *Phialocephala* sp. FL30r obtained from a deep seawater sample [15].

Three sorbicillinoids, namely 5-hydroxy-dihydrodemethylsorbicillin (**15**), sorbicillpyrone A (**16**), and 5,6-dehydrovertinolide (**17**), were isolated from *Penicillium* sp. SCSIO06871 from the deep-sea sediment [10].

Twelve monomeric sorbicillinoids including trichosorbicillins B (**18**), C (**19**), and D (**20**); 12-hydroxysorbicillin (**21**); 8,9-dihydro-12-hydroxysorbicillin (**22**); trichosorbicillin E (**23**); isotrichosorbicillin E (**24**); trichosorbicillins F (**25**), G (**26**), and H (**27**); 3-methyltrichosorbicillin H (**28**); and trichosorbicillin I (**29**) were isolated from marine-derived *Trichoderma reesei* 4670 associated with a sponge [16].

Trichoreeseiones A (**30**) and B (**31**) were isolated from an unidentified sponge-derived fungus *Trichoderma reesei* HN-2016-018. Both sorbicillinoids, with a characteristic naphthalenetrione ring, were first reported in the sorbicillinoid family [17].

Trichoreesin A (**32**) was the first bicyclic vertinolide derivative isolated from *Trichoderma reesei* Z56-8, an epiphytic fungus from the marine brown alga *Sargassum* sp. [18].

Ustilanthracins A (**33**) and B (**34**) were isolated from the rice false smut pathogen *Ustilaginoidea virens*. Both compounds share the same skeleton, but differ in the carboxyl-containing side chain, where dioxygenated butyric acid and 2-methyl-3-oxygenated butyric acid are found in ustilanthracins A (**33**) and B (**34**), respectively [19]. Both ustinaphthalin (**35**) and ustisorbicillinol F (**36**) were successively isolated from rice false smut pathogen *Ustilaginoidea virens* [14,19].

Vertinolides, with the presence of a γ-lactone terminus and a lack of any carbon rings, represent a class of degrading products of monomeric sorbicillinoids [20]. Three vertinolides, namely trichoreesin A (**32**), (+)-(*R*)-vertinolide (**37**), and (−)-(*S*)-dihydrovertinolide (**38**), have been isolated from fungi since 2016 [18,21,22]. (+)-(*R*)-vertinolide (**37**) is a new natural product isolated from *Trichoderma citrinoviride* from indoor air [21]. (*R*)-vertinolide (**37**) differs in stereochemistry from (*S*)-vertinolide isolated from *Verticillium intertextum* [23]. (−)-(*S*)-dihydrovertinolide (**38**) was isolated from the endophytic fungus *Clonostachys rosea* B5-2, which was isolated from the mangrove plant *Bruguiera gymnorrhiza*, collected in the coast of Santolo Garut Beach, West-Java, Indonesia [5,22].

2.2. *Bisorbicillinoids*

Bisorbicillinoids (also called dimeric sorbicillinoids) are formed by either an intermolecular Diels–Alder or Michael reaction of two monomeric sorbicillinoids [24]. Since 2016, twenty-one bisorbicillinoids have been isolated from fungi (Figure 3 and Table S2). These compounds are mainly distributed in the fungi genera *Penicillium, Trichoderma*, and *Ustilaginoidea.*

Three bisorbicillinoids, namely epitetrahydrotrichodimer ether (**39**), demethyldihydrotrichodimerol (**40**), and bisorbicillpyrone A (**41**), were isolated from *Penicillium* sp. SCSIO06871 from the deep-sea sediment. Among them, bisorbicillpyrone (**41**) is the first example of an α-pyrone-containing bisorbicillinoid [10].

Both 10,11-dihydrobislongiquinolide (**42**) and 10,11,16,17-tetrahydrobislongiquinolide (**43**) were produced by overexpression of the global regulator LaeA in the fungus *Penicillium dipodomyis* YJ-11 from a marine sediment sample collected in Jiaozhou Bay in Qingdao, China [25].

Saturnispols A (**44**) and B (**45**) were isolated from *Trichoderma saturnisporum* DI-IA from the marine sponge *Dictyonella incisa* collected in Seferihisar Bay in Turkey [13]. Saturnispols A (**44**) and B (**45**) are also named 15,24-dihydroxybisvertinol (**44**) and 24-hydroxybisvertinol (**45**), respectively. They were successively isolated from the marine-derived *Trichoderma reesei* 4670 from a sponge collected in Shantou, Guangdong, China [16]. Saturnispol B (**45**) was also isolated from an unidentified sponge-derived fungus *Trichoderma reesei* HN-2016-018 [17].

Five dimers, including trichobisvertinols A (**46**), B (**47**), C (**48**), and D (**49**), and 12-*epi*-trichobisvertinol D (**50**), were isolated from the marine-derived *Trichoderma reesei* 4670 from a sponge collected in Shantou, Guangdong, China [16]. Both trichobisvertinol D (**49**) and 12-*epi*-trichobisvertinol D (**50**) are epimeric to each other. Interestingly, they were isolated from *Ustilaginoidea virens* at the same time, and were named ustisorbicillinols A (**49**) and B (**50**), respectively [14].

Four dimeric sorbicillinoids, namely trichodermolide B (**51**), 13-hydroxy-trichodermolide (**52**), 24-hydroxy-trichodimerol (**53**), and 15-hydroxy-bisvertinol (**54**), were isolated from the sponge-derived fungus *Trichoderma reesei* HN-2016-018. Among them, trichodermolide B (**51**) and 13-hydroxy-trichodermolide (**52**) contain a unique bicycle [3.2.1] lactone skeleton. Trichodermolide B (**51**) with a propan-2-one moiety was firstly recorded in sorbicillinoid family [17]. 13-Hydroxy-dihydrotrichodermolide (**55**) is a structurally similar compound isolated from *Penicillium chrysogernum* 581F1 from the marine sponge *Theonella swinhoei* [26].

Ustilobisorbicillinol A (**56**) is a bisorbicillinoid featuring a unique cage structure that incorporates one sorbicillinol and one sorbyl-containing phenanthrenone unit. It was isolated from a culture of *Ustilaginoidea virens*, the rice false smut pathogen [19]. Three other bisorbicillinoids, namely ustisorbicillinols C (**57**), D (**58**), and E (**59**), were also isolated from *Ustilaginoidea virens.* Both ustisorbicillinols C (**57**) and D (**58**) are epimeric to each other [14].

2.3. *Trisorbicillinoids*

Trisorbicillinoids (or called trimeric sorbicillinoids) are formed by either an intermolecular Diels–Alder or Michael reaction of three monomeric sorbicillinoids [24]. Only one trisorbicillinoid, 10,11,27,28-tetrahydrotrisorbicillinone C (**60**), has been isolated from *Penicillium chrysogernum* 581F1 from the marine sponge *Theonella swinhoei* since 2016 (Figure 4) [26].

2.4. *Hybrid Sorbicillinoids*

Hybrid sorbicillinoids are derived from either an asymmetrical Diels–Alder reaction of a monomeric sorbicillinoid diene and a second non-sorbicillinoid dienophile [24]. About 13 hybrid sorbicillinoids have been isolated from fungi since 2016 (Figure 5 and Table S3). Two hybrids, 10-methylsorbiterrin (**61**) and dihydrotrichodermolidic acid (**62**), were isolated from *Penicillium* sp. SCSIO06871 from the deep-sea sediment [10].

Both saturnispols C (**63**) and D (**64**) were isolated from *Trichoderma saturnisporum* DI-IA from the marine sponge *Dictyonella incisa* collected in Seferihisar Bay in Turkey. Biogenetically, it was proposed that the [4+2] Diels–Alder cycloaddition of sorbicillinol with a phenylethylene generated saturnispol C (**63**), followed by hydroxylation, to yield saturnispol D (**64**) [13].

Spirosorbicillinol D (**65**) is a hybrid sorbicillinoid from *Trichoderma longibrachiatum* SFC100166 isolated from foreshore soil [11].

Sorbicatechols C (**66**) and D (**67**) were isolated from *Penicillium allii-sativi* from deep-sea water [27].

Sorbicillfurans A (**68**) and B (**69**) were isolated from the static culture of the fungus *Penicillium citrinum* SCSIO41402, which was isolated from a marine alga *Coelarthrum* sp. collected in Yongxing Island, South China Sea. Both compounds possess a tetrahydrofuran unit. It was suggested that both sorbicillfurans A (**68**) and B (**69**) are derived from the precursor sorbicillinol added with furfuryl alcohol by a Diels–Alder (DA) reaction, followed by the oxidization modification to yield sorbicillfuran A (**68**), and by another DA cycloaddition reaction to generate sorbicillfuran B (**69**) [28].

Two nitrogen-containing sorbicillinoids with hexahydropyrimido [2,1-*a*] isoindole moiety named sorbicillasins A (**70**) and B (**71**) were isolated from the deep-sea fungus *Phialocephala* sp. FL30r obtained from an underwater sample. Sorbicillasins A (**70**) and B (**71**) are probably formed by adding a whole molecule of L-asparagine to 2′,3′-dihydrosorbicillin via sequential intermolecular/intramolecular nucleophilic reactions [15].

When tanshinone IIA was fed to the fermentation cultures of sorbcillinol-producing fungus *Hypocrea* sp., the hybrid sorbicillinoid produced was tanshisorbicin (**72**), which is considered a [4+2] cycloaddition adduct between tanshinone IIA and sorbicillinol (**3**) [29].

Trichosorbicillin A (**73**) is a nitrogen-containing sorbicillinoid isolated from the marine-derived *Trichoderma reesei* 4670 from a sponge collected in Shantou, Guangdong, China. It was hypothesized to arise from a net [4+2] cycloaddition or double Michael reaction between sorbicillinol (**3**) and 1-methyl-1,3-dihydro-2*H*-pyrrol-2-one [16].

3. Biological Activities

The recently isolated sorbicillinoids mainly display cytotoxic, antibacterial, antifungal, anti-inflammatory, phytotoxic, and α-glucosidase inhibitory activities (Tables S4–S10). The structures of some sorbicillinoids (**74**–**91**) discovered before 2016 with newly revealed biological activities are shown in Figure 6.

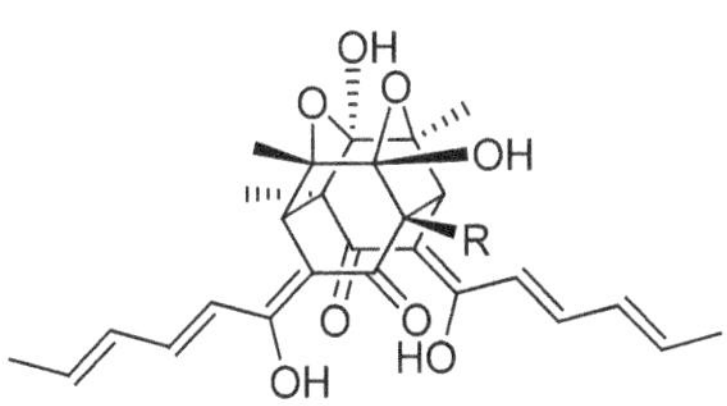

74. Trichodimerol, R=CH_3
75. Demethyltrichodimerol, R=H

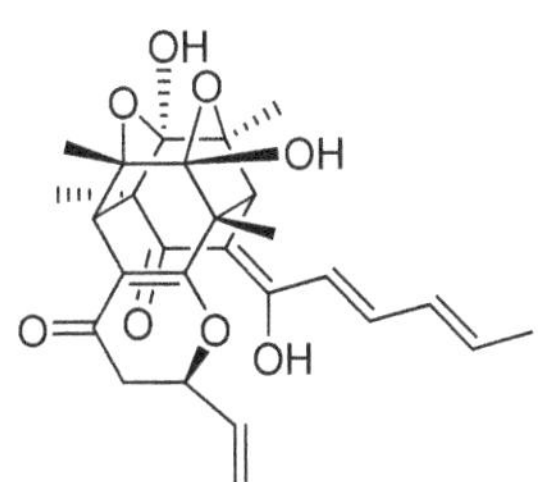

76. Dihydrotrichodimer ether A

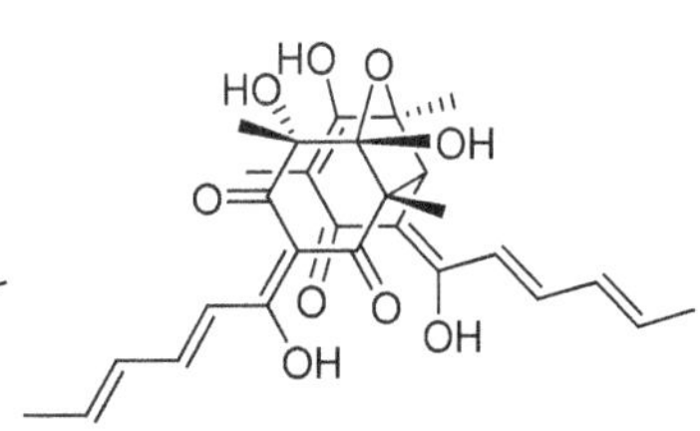

77. Bisvertinolone

Figure 6. *Cont.*

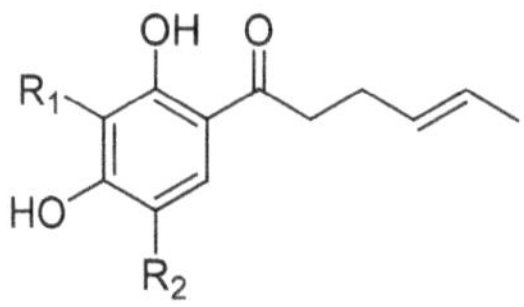

78. Sohirnone A, R_1=H, R_2=CH_3
79. Dihydrodemethylsorbicillin, R_1=CH_3, R_2=H

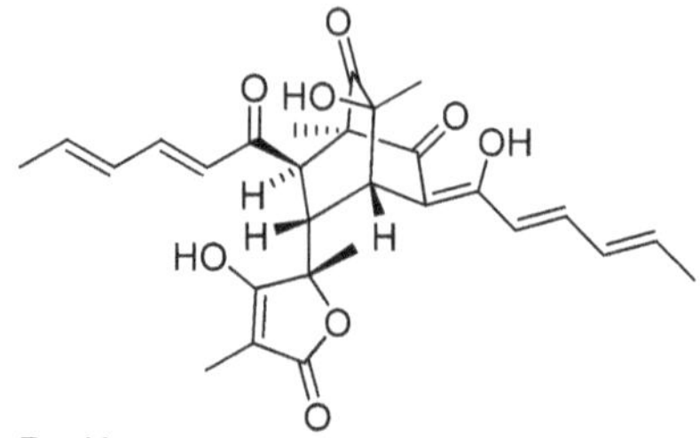

80. Bisbongiquinolide

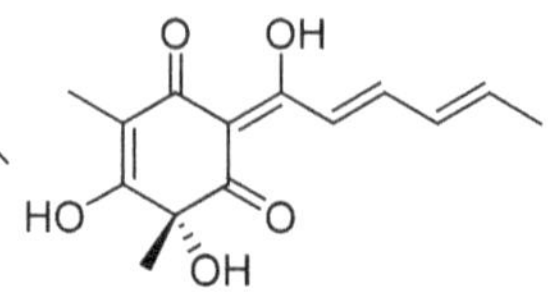

81. Oxosorbicillinol

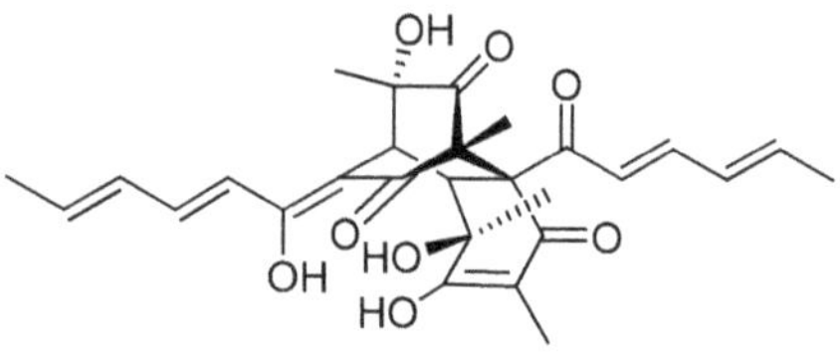

82. Bisorbicillinol

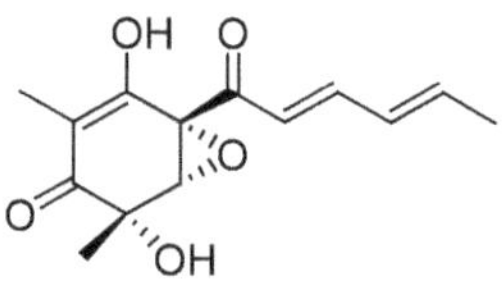

83. Epoxysorbicillinol

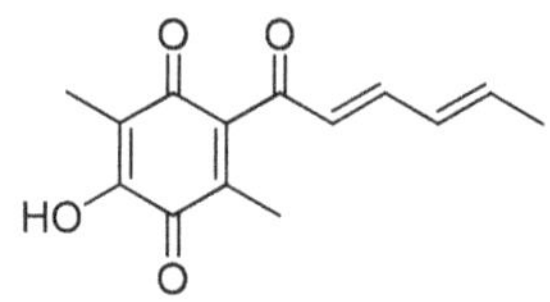

84. Sorrentanone

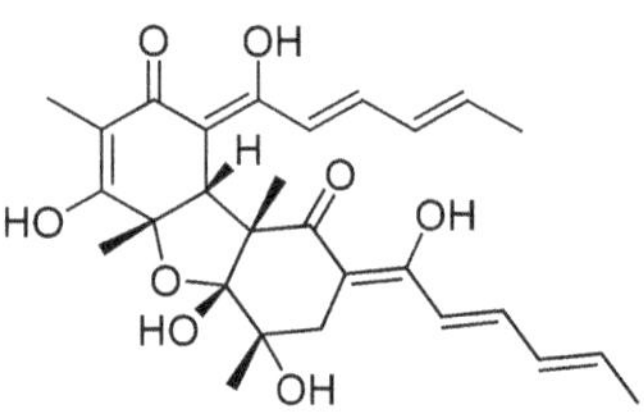

85. Bisvertinol

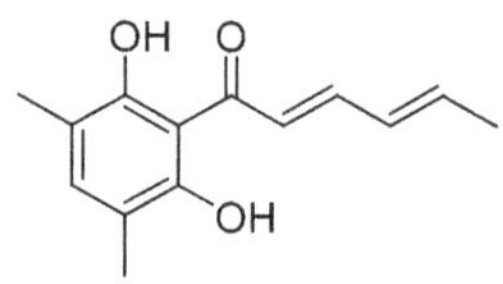

86. (2*E*,4*E*)-1-(2,6-Dihydroxy-3,5-dimethylphenyl)hexa-2,4-dien-1-one

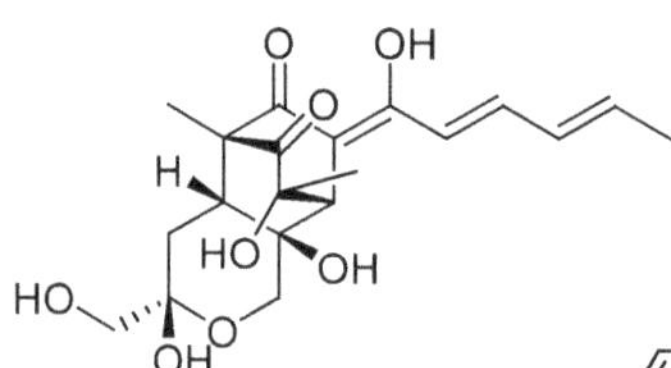

87. Trichodermanone C

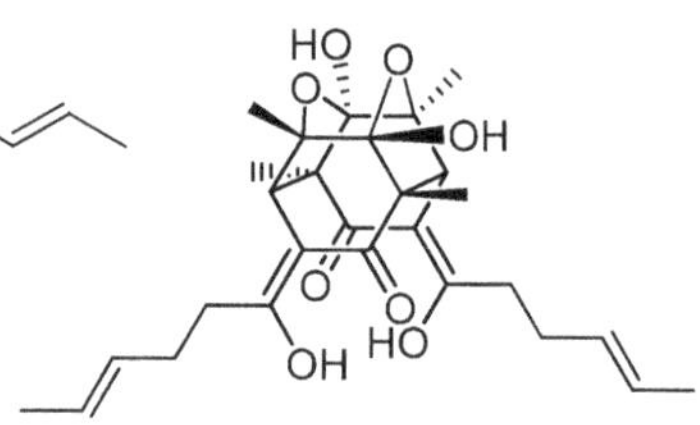

88. Tetrahydrotrichodimerol

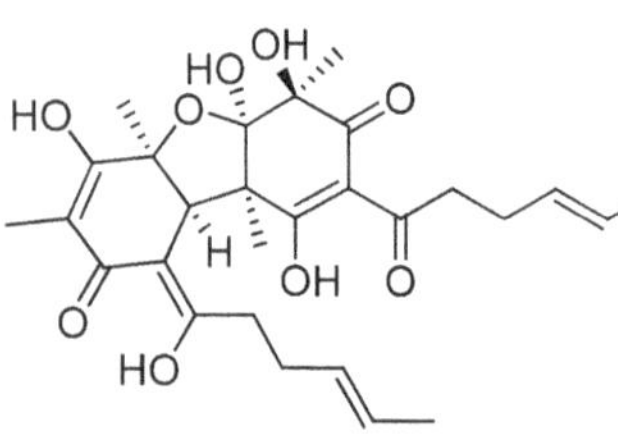

89. Tetrahydrobisvertinolone

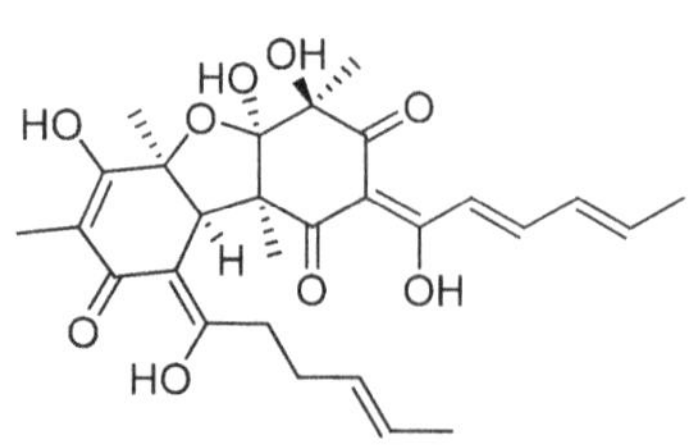

90. 10,11-Dihydrobisvertinolone

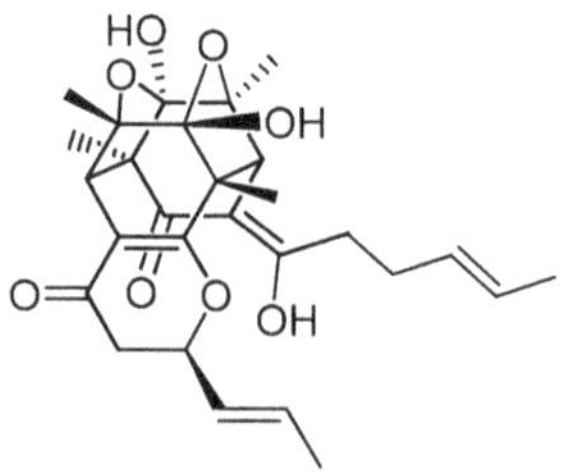

91.Tetrahydrotrichodimer ether

Figure 6. Structures of some sorbicillinoids (**74–91**) discovered before 2016 with newly revealed biological activities.

3.1. Cytotoxic Activity

Some recently revealed sorbicillinoids displayed obviously cytotoxic activities (Table S4). Sorbicatechol D (**67**) and sorbicillin (**1**) were screened to show antiproliferative activity on HT-29 tumor cells in a dose-dependent manner. The mechanism investigation uncovered that they can significantly induce cell cycle G2–M phase arrest by increasing the protein levels of p-H3 and cyclin B1 [27]. Sorbicillin (**1**) was once again isolated from the culture broth of the fungus *Penicillium decumbens* from a limestone soil. It exhibited selective cytotoxic activity against the human hepatocellular carcinoma (QGY-7703) cells with an IC_{50} value of 32.5 μM [30]. Similar cytotoxic activity results of sorbicillin (**1**) have been reported previously [31–33].

Sorbicillfuran B (**69**) showed weak cytotoxic activity against human leukemia cell line HL-60 cells with an IC_{50} value of 9.6 μM [28]. Five cytotoxic bisorbicillinoids, namely ustilobisorbicillinol A (**56**), trichodimerol (**74**), demethyltrichodimerol (**75**), dihydrotrichodimer ether (**76**), and bisvertinolone (**77**), were isolated from the rice false smut pathogen *Ustilaginlidea virens* [14,19]. Among them, trichodimerol (**74**), demethyltrichodimerol (**75**), dihydrotrichodimer ether A (**76**), and bisvertinolone (**77**) showed moderate cytotoxic activities on human carcinoma cells with IC_{50} values of 8.83–74.7 μM [14]. Ustilobisorbicillinol A (**56**) showed notable cytotoxicity against the five tested tumor cell lines, with IC_{50} values in the range of 4.48–18.6 μM. It was further tested for its influence on cell-cycle progression with the gastric cancer cell line BGC823. Interestingly, it markedly induced G0/G1- and G2/M-phase cell-cycle arrest. Ustilobisorbicillinol A (**56**) was also investigated for its effect on apoptosis in BGC823 cells, as cell shrinkage and detached from culture surface was observed after treatment with ustilobisorbicillinol A (**56**). The apoptotic rate of BGC823 cells was examined using flow cytometry. Compared to the control group, treatment with ustilobisorbicillinol A (**56**) at 9 μM for 48 h induced significant apoptosis incidence in BGC823 cells (74.7%). Moreover, treatment with ustilobisorbicillinol A (**56**) altered the expression levels of cleaved caspase-3 and PARP, suggesting the caspase-mediated apoptotic pathway is involved in the induced apoptosis of BGC823 cells [19].

24-hydroxy-trichodimerol (**53**) displayed cytotoxic activities against human tumor cells (A549, MCF-7, and HCT116) with IC_{50} values of 5.1, 9.5, and 13.7 mM, respectively [17].

3.2. Antibacterial Activity

Due to the long-term use of some antibiotics, the bacterial or fungal pathogens easily develop drug resistance, and it is necessary to look for new alternatives. Some sorbicillinoids exhibited obvious antibacterial activities, showing their potential as the antimicrobials (Table S5). Two monomeric sorbicillinoids, saturnispols F (**9**) and H (**11**), showed significant antibacterial activity. Saturnispol F (**9**) displayed inhibition of bacteria with minimum inhibitory concentration (MIC) values of 3.32 μg/mL against *Staphylococcus aureus*, 1.63 μg/mL against vancomycin-resistant *Enterococci faecalis* (VRE), 6.65 μg/mL against *Pseudomonas aeruginosa*, and 6.65 μg/mL against *Klebsiella pneumoniae*. Saturnispol H (**11**) displayed inhibition of bacteria with MIC values of 12.9 μg/mL against vancomycin-resistant *Enterococci faecalis* and 12.9 μg/mL against *Bacillus subtilis* [13].

Both sohirnone A (**78**) and dihydrodemethylsorbicillin (**79**) exhibited significant antibacterial activities against *Staphylococcus aureus* with MIC values of 10.0 μg/mL and 5.0 μg/mL, respectively [10].

Five sorbicillinoids ustisorbicillinol B (or 12-*epi*-trichobisvertinol D (**50**)), demethyltrichodimerol (**75**), dihydrotrichodimer ether A (**76**), bisvertinolone (**77**), and oxosorbicillinol (**81**) from *Ustilaginoidea virens* showed antibacterial activities against six human/plant pathogenic bacteria. Among them, bisvertinolone (**77**) was the most effective [14]. A similar antibacterial activity of oxosorbicillinol (**81**) was reported previously [34]. Bisvertinolone (**77**), isolated from *Aspergillus protuberus* MUT3638, was also previously reported to exhibit significant activity against *Staphylococcus aureus* with an MIC value of 30 μg/mL [35]. Two bisorbicillinoids, bisvertinolone (**77**) and bislongiquinolide *saturnisporum* (**80**), were screened to show antibacterial activities against *Pseudomonas lachrymans* with MIC values of 3.13 and 1.56 μM, respectively, and against *Escherichia coli* with MIC values of 6.25 and 12.5 μM, respectively [36].

Tanshisorbicin (**72**) showed obvious antibacterial activity on *Mycobacterium bovis*, *Staphylococcus aureus* (ATCC 6538), methicillin-resistant *Staphylococcus aureus* (MRSA), and *Bacillus subtilis* (ATCC 6633). The anti-MRSA activity of tanshisorbicin (**72**) was found to be significantly higher than that of tanshinone IIA [29].

Antibacterial mechanisms showed that sorbicillinoids could generate singlet oxygen ($^1O_2$) under UV light irradiation and ultimately displayed photoinactivation activity on Gram-positive bacteria including *Staphylococcus aureus*, *Bacillus subtilis*, and *Micrococcus luteus*, but not Gram-negative ones such as *Escherichia coli* and *Proteus vulgaris*, showing their potential as photosensitizers for antimicrobial photodynamic therapy using a nontoxic dose of UV irradiation [37].

3.3. Antifungal Activity

Some recently discovered sorbicillinoids were screened for antifungal activities (Table S6). Sorbicillin (**1**) displayed antifungal activity toward *Candida albicans* Y0109 with an MIC value of 50 μM [30].

Bisvertinolone (**77**), oxosorbicillinol (**81**), bisorbicillinol (**82**), and epoxysorbicillinol (**83**) from *Trichoderma longibrachiatum* SFC100166 were screened for antifungal activity on phytopathogenic fungi *Cladosporium coccodes, Colletotrichum coccodes, Cylindrocarpon destructans, Magnaporthe oyrzae,* and *Phytopathora infestans*, with MIC values ranging from 6.3 to 100 μg/mL. When tomato plants were treated with the above compounds (**77**,**81**–**83**), bisvertinolone (**77**) strongly reduced the development of tomato late blight disease compared to the untreated control [11].

Demethyltrichodimerol (**75**), bisvertinolone (**77**), and oxosorbicillinol (**81**) displayed moderate antifungal activities by inhibiting the spore germination of rice blast pathogen *Magnaporthe oryzae*. Among them, bisvertinolone (**77**) was the most effective sorbicillinoid [14].

3.4. Anti-Inflammatory Activity

Inflammation is a common response of the human body to injuries caused by microbial pathogens, trauma, or toxic compounds. Bioactive metabolites produced by fungi have received considerable attention as new therapeutic agents [38]. Many sorbicillinoids were screened for anti-inflammatory activities and their potential use in the treatment of inflammatory diseases (Table S7). Trichodimerol (**74**) and sorrentanone (**84**) were isolated from the endophytic fungus *Trichoderma* sp. Xy24 from the mangrove plant *Xylocarpus granatum*. Both compounds displayed anti-inflammatory activity by inhibiting LPS-induced NO production in BV2 microglia cells, with the inhibitory rates of 75.1% and 100.0% at 10 μM, respectively, much more potent than the positive control curcumin [39].

Eighteen mono- and dimeric sorbicillinoids, including trichosorbicillin B (**18**), trichosorbicillin C (**19**), 12-hydroxysorbicillin (**21**), 8,9-dihydro-12-hydroxysorbicillin (**22**), trichosorbicillin E (**23**), isotrichosorbicillin E (**24**), trichosorbicillin F (**25**), trichosorbicilin I (**29**), 24-hydroxybisvertinol (also named saturnispol B, **45**), trichobisvertinol A (**46**), trichobisvertinol B (**47**), trichobisvertinol C (**48**), trichobisvertinol D (**49**), 12-*epi*-trichobisvertinol D (**50**), sohirnone A (**78**), bisvertinol (**85**), 2′,3′-dihydrosorbicillin (also called dihydrosorbicillin, **2**), and (2*E*,4*E*)-1-(2,6-Dihydroxy-3,5-dimethylphenyl)hexa-2,4-dien-1-one (**86**) from the sponge-derived fungus *Trichoderma reesei* 4670, were systematically screened for potent anti-inflammatory activity by inhibiting the production of NO in RAW264.7 cells activated by lipopolysaccharide, with IC_{50} values in the range of 0.94 to 38 μM. The structure−activity relationship analysis indicated that the anti-inflammatory activities of the sorbicillinoids mainly depend on the structural types and the functional groups of the sorbyl side chain [16].

Trichodermanone C (**87**) is a hybrid sorbicillinoid showing an anti-inflammatory activity with inhibition of nitrite levels in lipopolysaccharide (LPS)-stimulated J774A.1 macrophages [40].

Epitetrahydrotrichodimer ether (**39**) and tetrahydrotrichodimerol (**88**) are two dimeric sorbicillinoids isolated from *Penicillium* sp. DM815 from the rhizosphere soil of mangrove *Hibiscus tiliaceus* that significantly reduced the level of NO produced by inducible nitric oxide synthase (iNOS) [12].

3.5. Phytotoxic Activity

Plant pathogenic and endophytic fungi usually produce metabolites poisonous to their host plants. These phytotoxic metabolites from fungi are called phytotoxins [41]. It is considered that the amounts of phytotoxins produced by the endophytic fungi are much lower than those of the phytopathogenic fungi [42].

Four sorbicillinoids (Table S8), namely trichodimerol (**74**), demethyltrichodimerol (**75**), bisvertinolone (**77**), and bislongiquinolide (also named trichotetronine, **80**) from rice false smut pathogen *Ustilaginoidea virens*, showed phytotoxic activity by inhibiting radicle and germ elongation of rice and lettuce seedlings, with bisvertinolone (**77**) displaying the strongest inhibition. These phytotoxic sorbicillinoids might play an important role in the development of rice false smut symptoms [14].

(−)-(*S*)-dihydrovertinolide (**38**) inhibited the shoot growth by 23% and root growth by 65% of lettuce (*Lactuca sativa*) seedlings [22].

3.6. α-Glucosidase Inhibitory Activity

Diabetes is considered as one of the biggest current health crises. Controlling carbohydrate digestibility by inhibiting starch digestive enzyme (i.e., α-amylase and α-glucosidase) activities is an efficient strategy to control postprandial hyperglycemia [43]. Some sorbicillinoids have been screened for their α–glucosidase inhibitory activity (Table S9).

Six sorbicillinoids, including 5-hydroxy-dihydrodemethylsorbicillin (**15**), bisorbicillpyrone A (**41**), dihydrodemethylsorbicillin (**79**), tetrahydrotrichodimerol (**88**), tetrahydrobisvertinolone (**89**), and 10,11-dihydrobisvertinolone (**90**), exhibited α-glucosidase inhibitory activity, with IC_{50} values ranging from 115.8 to 208.5 μM. Among these, 5-hydroxy-dihydrodemethylsorbicillin (**15**) showed the strongest inhibitory activity against α-glucosidase with an IC_{50} value of 36.0 μM, stronger than that of acarbose [10].

2′,3′-dihydrosorbicillin (**2**), which was isolated from the fungus *Aspergillus* sp. HNWSW-20 from Chinese agarwood (*Aquilaria sinensis*), showed α-glucosidase inhibitory activity [44].

3.7. Other Biological Activities

Other biological activities of the sorbicillinoids recently revealed from fungi mainly include antiallergic, antioxidant, neuroprotective and neuritogenic, antihuman-immunodeficiency-virus (HIV), and antimicroalgal activities, as well as inhibitory activities against acetylcholinesterase (AChE) and protein tyrosine phosphatase 1B (Table S10).

Bisorbicillinol (**82**) is a bisorbicillinoid previously isolated from a few fungi such as *Trichoderma* sp. USF-2690 [45], *Trichoderma* sp. f-13 [31], and *Penicillium notatum* [34]. Bisorbicillinol (**82**) from *Trichoderma* sp. USF2690 was found to be an inhibitor of β-hexosaminidase release and tumor necrosis factor (TNF)-α, and 9nterleukin (IL)-4 secretion from rat basophilic leukemia (RBL-2H3) cells, with IC_{50} values of 2.8, 2.9, and 2.8 μM, respectively. The results showed that the inhibitory mechanism of β-hexosaminidase release and TNF-α secretion involve inhibition of Lyn, a tyrosine kinase. This indicated that bisorbicillinol (**82**) should be a candidate antiallergic agent [46].

Scipyrone K (**14**), isolated from the fungus *Phialocephala* sp. FL30r obtained from a deep seawater sample, exhibited weak radical scavenging activity against 2,2-diphenyl-1-picrylhydrazyl (DPPH) with an IC_{50} value of 27.9 μM [15].

Sorbicillin (**1**) was proven to have neuroprotective and neuritogenic activity on PC-12 Adh cells of the 6-hydroxydopamine-induced Parkinson's disease cell model at 1 and 10 μg/mL. The water fraction of halotolerant *Penicillium flavigenum* isolated from Salt Lake in Konya, Turkey, also showed similar activity. The water extract was revealed to contain sorbicillin-like active metabolites by LC-MS compared to a sorbicillin (**1**) standard [47]. Sorbicillin (**1**) and 2′,3′-dihydrosorbicillin (**2**) showed acetylcholinesterase inhibitory activities with inhibition rates of 15.47% and 1.78%, respectively, at a concentration of 50 μg/mL [44].

At a concentration of 40 μM, both 2′,3′-dihydrosorbicillin (**2**) and sohirnone A (**78**) exhibited moderate inhibitory activity of protein tyrosine phosphatase 1B (PTP1B) with inhibitory ratios of 10.58% and 8.47%, respectively, to show their antidiabetic potential [48].

Sorrentanone (**84**) showed a significant inhibitory effect of HIV-1 virus with an IC_{50} value of 4.7 μM, so is worthy of further investigation as a lead anti-HIV compound [38].

Glucagon-like peptide-1 (GLP-1), a gut incretin hormone that stimulates insulin and inhibits glucagon secretion on pancreatic β-cells and α-cells, is considered a target protein related to diabetes. Eukaryotic elongation factor-2 kinase (eEF2K) is a potential therapeutic target for cancer. Both 13-hydroxy-dihydrotrichodermolide (**55**) and 10,11,27,28-tetrahydrotrisorbicillinone C (**60**) displayed high affinities to target proteins GLP-1R and eEF2K with K_d values of 0.0285 and 0.0162 μM for GLP-1R, and 0.118 and 0.0746 μM for eEF2K, respectively. These findings indicate that 13-hydroxy-dihydrotrichodermolide (**55**) and 10,11,27,28-tetrahydrotrisorbicillinone C (**60**) are promising new drug candidates for diabetes and cancer treatment [26].

Both tetrahydrobisvertinolone (**89**) and tetrahydrotrichodimer ether (**91**) exhibited weak acetylcholinesterase (AChE) inhibitory activity with 51.1% and 55.1% inhibitions at a concentration of 50 μg/mL, respectively [10].

Trichoreesin A (**32**) showed antimicroalgal activity against the marine algae *Chattonella marina*, *Heterosigma akashiwo*, and *Prorocentrum donghaiense* with IC_{50} values of 13, 29, and 2.8 μg/mL, respectively [18].

4. Conclusions

From 2016 to 2021, 69 new sorbicillinoids were isolated from fungi. Mainly belonging to the monomeric and dimeric sorbicillinoids, some sorbicillinoids have special structures such as ustilobisorbicillinol A (**56**) [19], and sorbicillasins A (**70**) and B (**71**) [15], increasing their diversity. The majority of sorbicillinoids were reported from the fungi genera of *Acremonium*, *Penicillium*, *Trichoderma*, and *Ustilaginoidea*. This provides a basis for fungal chemotaxonomy, which should be further studied in detail. It is worth mentioning that 21 sorbicillinoids were firstly isolated from the rice false smut pathogen *Ustilaginoidea virens* [14,19], which can produce many types of bioactive secondary metabolites [49–58]. Some sorbicillinoids exhibited cytotoxic (Table S4), antibacterial (Table S5), antifungal (Table S6), anti-inflammatory (Table S7), phytotoxic (Table S8), and α-glucosidase-inhibitory (Table S9) and PTP1B-inhibitory activities (Table S10). They may be utilized as pigments and food colorants as well. Due to the limitation of activity screening models by each research group, many sorbicillinoids need to be further screened for their biological activities. Furthermore, the comparative investigations on the biological activities of sorbicillinoids and other classes of compounds along with their action mechanisms need to be further conducted [59–61]. In recent years, more and more new members of sorbicillinoids have been revealed from plant endophytic, marine-derived, extremophilic, phytopathogenic, and soil-derived fungi. All these sorbicillinoids may be rich resources of biologically active substances with significant pharmaceutical, food colorant, and agricultural value [2].

Fungal sorbicillinoids were studied extensively from 2016 to 2021. Apart from the discovery of new sorbicillinoids and clarification of their biological activities and action mechanisms, other related studies include biosynthetic gene clusters [6], biosynthetic pathways and their related enzymes [5,24,62–65], relevant regulatory mechanisms [7,25,66–68], biochemical engineering to increase the production of sorbicillinoids [59], chemoenzymatic synthesis [69], development of chemical synthesis methods [70], and applications of sorbicillinoids in the agriculture, pharmaceutical, and food industries [37,60,61]. Among them, the most promising is clarification of the Diels–Alder reactions during the biosynthesis of sorbicillinoids. Through co-expression of *sorA*, *sorB*, *sorC*, and *sorD* from *Trichoderma reesei* QM6a, the biosynthetic pathway to epoxysorbicillinol and dimeric sorbicillinoids resembling Diels–Alder-like and Michael-addition-like products was reconstituted in *Aspergillus oryzae* NSAR1 [24].

Supplementary Materials: The following supporting information can be downloaded at: https://www.mdpi.com/article/10.3390/jof8010062/s1, Table S1: Occurrence of the monomeric sorbicillinoids (**5–38**) in fungi; Table S2: Occurrence of the bisorbicillinoids (**39–59**) in fungi; Table S3: Occurrence of the hybrid sorbicillinoids (**61–73**) in fungi; Table S4: Cytotoxic activity of the screened sorbicillinoids in fungi; Table S5: Antibacterial activity of the sorbicillinoids screened from fungi; Table S6: Antifungal activity of the sorbicillinoids screened from fungi; Table S7: Anti-inflammatory activity of the sorbicillinoids screened from fungi; Table S8: Phytotoxic activity of the sorbicillinoids screened from fungi; Table S9: α-Glucosidase inhibitory activity of the sorbicillinoids screened from fungi; Table S10: Other biological activities of the sorbicillinoids screened from fungi.

Author Contributions: Conceptualization, L.Z.; writing—original draft preparation, X.H., X.Z., M.X., Z.Z., H.Z., D.X. and L.Z.; writing—review and editing, D.L. All authors have read and agreed to the published version of the manuscript.

Funding: This research was funded by the National Natural Science Foundation of China (32072373), and the National Key R&D Program of China (2017YFC1600905).

Institutional Review Board Statement: Not applicable.

Informed Consent Statement: Not applicable.

Data Availability Statement: Not applicable.

Conflicts of Interest: The authors declare no conflict of interest.

References

1. Harned, A.M.; Volp, K.A. The sorbicillinoid family of natural products: Isolation, biosynthesis and synthetic studies. *Nat. Prod. Rep.* **2011**, *28*, 1790–1810. [CrossRef]
2. Meng, J.; Wang, X.; Xu, D.; Fu, X.; Zhang, X.; Lai, D.; Zhou, L.; Zhang, G. Sorbicillinoids from fungi and their bioactivities. *Molecules* **2016**, *21*, 715. [CrossRef]
3. Cram, D.J. Mold metabolites. II. The structure of sorbicillin, a pigment produced by the mold *Penicillium notatum*. *J. Am. Chem. Soc.* **1948**, *70*, 4240–4243. [CrossRef] [PubMed]
4. Zhang, X.; Li, S.-J.; Li, J.-J.; Liang, Z.-Z.; Zhao, C.-Q. Novel natural products from extremophilic fungi. *Mar. Drugs* **2018**, *16*, 194. [CrossRef]
5. Han, P.; Zhang, X.; Xu, D.; Zhang, B.; Lai, D.; Zhou, L. Metabolites from *Clonostachys* fungi and their biological activities. *J. Fungi* **2020**, *6*, 229. [CrossRef] [PubMed]
6. Derntl, C.; Guzman-Chavez, F.; Mello-de-Sousa, T.M.; Busse, H.-J.; Driessen, A.J.M.; Mach, R.L.; Mach-Aignr, A.R. In vivo study of the sorbicillinoid gene cluster in *Trichoderma reeesei*. *Front. Mcrobiol.* **2017**, *8*, 2037. [CrossRef]
7. Guzman-Chavez, F.; Salo, O.; Nygard, Y.; Lankhorst, P.P.; Bovenberg, R.A.L.; Driessen, A.J.M. Mechanism and regulation of sorbicillin biosynthesis by *Penicillium chrysorgenum*. *Microb. Biotechnol.* **2017**, *10*, 958–968. [CrossRef] [PubMed]
8. Al Fahad, A.; Abood, A.; Fisch, K.M.; Osipow, A.; Davison, J.; Avramovic, M.; Butts, C.P.; Piel, J.; Simpson, T.J.; Cox, R.J. Oxidative dearomatisation: The key step of sorbicillinoid biosynthesis. *Chem. Sci.* **2014**, *5*, 523–527. [CrossRef] [PubMed]
9. Jiang, C.-S.; Zhou, Z.-F.; Yang, X.-H.; Lan, L.-F.; Gu, Y.-C.; Ye, B.-P.; Guo, Y.-W. Antibacterial sorbicillin and diketopiperazines from the endogenous fungus *Penicillium* sp. GD6 associated Chinese mangrove *Bruguiera gymnorrhiza*. *Chin. J. Nat. Med.* **2018**, *16*, 358–365. [CrossRef]
10. Pang, X.; Zhou, X.; Lin, X.; Yang, B.; Tian, X.; Wang, J.; Xu, S.; Liu, Y. Structurally various sorbicillinoids from the deep-sea sediment derived fungus *Penicillium* sp. SCSIO06871. *Bioorg. Chem.* **2021**, *107*, 104600. [CrossRef]
11. Ngo, M.T.; Nguyen, M.V.; Han, J.W.; Park, M.S.; Kim, H.; Choi, G.J. In vitro and in vivo antifungal activity of sorbicillinoids produced by *Trichoderma longibrachiatum*. *J. Fungi* **2021**, *7*, 428. [CrossRef]
12. Ding, W.; Wang, F.; Li, Q.; Xue, Y.; Dong, Z.; Tian, D.; Chen, M.; Zhang, Y.; Hong, K.; Tang, J. Isolation and characterization of anti-inflammatory sorbicillinoids from the mangrove-derived fungus *Penicillium* sp. DM815. *Chem. Biodivers.* **2021**, *18*, e2100229. [CrossRef]
13. Meng, J.; Cheng, W.; Heydari, H.; Wang, B.; Zhu, K.; Konuklugil, B.; Lin, W. Sorbicillinoid-based metabolites from a sponge-derived fungus *Trichoderma saturnisporum*. *Mar. Drugs* **2018**, *16*, 226. [CrossRef] [PubMed]

14. Meng, J.; Gu, G.; Dang, P.; Zhang, X.; Wang, W.; Dai, J.; Liu, Y.; Lai, D.; Zhou, L. Sorbicillinoids from the fungus *Ustilaginoidea virens* and their phytotoxic, cytotoxic, and antimicrobial activities. *Front. Chem.* **2019**, *7*, 435. [CrossRef] [PubMed]
15. Zhang, Z.; He, Q.; Che, Q.; Zhang, G.; Zhu, T.; Gu, Q.; Li, D. Sorbicillasins A–B and scirpyrone K from a deep-sea-derived fungus, *Phialocephala* sp. FL30r. *Mar. Drugs* **2018**, *16*, 245. [CrossRef]
16. Zhang, P.; Deng, Y.; Lin, X.; Chen, B.; Li, J.; Liu, H.; Chen, S.; Liu, L. Anti-inflammatory mono- and dimeric sorbicillinoids from the marine-derived fungus *Trichoderma reesei* 4670. *J. Nat. Prod.* **2019**, *82*, 947–957. [CrossRef] [PubMed]
17. Rehman, S.U.; Yang, L.-J.; Zhang, Y.-H.; Wu, J.-S.; Shi, T.; Halder, W.; Shao, C.-L.; Wang, C.-Y. Sorbicillinoid derivatives from sponge-derived fungus *Trichoderma reesei* (HN-2016-018). *Front. Microbiol.* **2020**, *11*, 1334. [CrossRef]
18. Ma, X.-Y.; Shi, Z.-Z.; Ji, N.-Y. Sorbicillinoids from the alga-epiphytic fungus *Trichoderma reesei* Z56-8. *Nat. Prod. Res.* **2021**. [CrossRef]
19. Lai, D.; Meng, J.; Zhang, X.; Xu, D.; Dai, J.; Zhou, L. Ustilobisorbicillinol A, a cytotoxic sorbyl-containing aromatic polyketide from *Ustilagninoidea virens*. *Org. Lett.* **2019**, *21*, 1311–1314. [CrossRef]
20. Sugaya, K.; Koshino, H.; Hongo, Y.; Yasunaga, K.; Onose, J.; Yoshikawa, K.; Abe, N. The biosynthesis of sorbicillinoids in *Trichoderma* sp. USF-2690: Prospect for the existence of a common precursor to sorbicillinol and 5-epihydroxyvertinolide, a new sorbicillinoid member. *Tetrahedron Lett.* **2008**, *49*, 654–657. [CrossRef]
21. McMullin, D.R.; Renaud, J.B.; Barasubiye, T.; Sumarah, M.W.; Miller, J.D. Metabolites of *Trichoderma* species isolated from damp building materials. *Can. J. Microbiol.* **2017**, *63*, 621–632. [CrossRef]
22. Supratman, U.; Suzuki, T.; Nakamura, T.; Yokoyama, Y.; Harneti, D.; Maharani, R.; Salam, S.; Abdullah, F.; Koseki, T.; Shiono, Y. New metabolites produced by endophyte *Clonostachys rosea* B5-2. *Nat. Prod. Res.* **2021**, *35*, 1525–1531. [CrossRef]
23. Trifonov, L.S.; Bieri, J.H.; Prewo, R.; Dreiding, A.S.; Rast, D.M.; Hoesch, L. The constitution of vertinolide, a new derivative of tetronic acid, produced by *Verticillium intertextum*. *Tetrahedron* **1982**, *38*, 397–403. [CrossRef]
24. Kahlert, L.; Bassiony, E.F.; Cox, R.J.; Shellam, E.J. Diels-Alder reactions during the biosynthesis of sorbicillinoids. *Angew. Chem. Int. Ed.* **2020**, *59*, 5816–5822. [CrossRef]
25. Yu, J.; Han, H.; Zhang, X.; Ma, C.; Sun, C.; Che, Q.; Gu, Q.; Zhu, T.; Zhang, G.; Li, D. Discovery of two new sorbicillinoids by overexpression of the global regulator LaeA in a marine-derived fungus *Penicillium dipodomyis* YJ-11. *Mar. Drugs* **2019**, *17*, 446. [CrossRef]
26. Cao, M.-J.; Zhu, T.; Liu, J.-T.; Ouyang, L.; Yang, F.; Lin, H.-W. New sorbicillinoid derivatives with GLP-1R and eEF2K affinities from a sponge-derived fungus *Penicillium Chrysogenum* 581F1. *Nat. Prod. Res.* **2020**, *34*, 2880–2886. [CrossRef] [PubMed]
27. Xie, C.-L.; Zhang, D.; Lin, T.; He, Z.-H.; Yan, Q.-X.; Cai, Q.; Zhang, X.-K.; Yang, X.-W.; Chen, H.-F. Antiproliferative sorbicillinoids from the deep-sea-derived *Penicillium allii-sativi*. *Front. Microbiol.* **2021**, *11*, 636948. [CrossRef] [PubMed]
28. Wang, J.; Li, K.; Luo, X.; Wu, Z.; Gu, T.; Liao, S.; Lin, X.; Yang, B.; Liu, Y.; Fang, W.; et al. Sorbicillfurans A and B, two novel sorbicillinoid adducts from the fungus *Penicillium citrinum* SCSIO41402. *Org. Biomol. Chem.* **2019**, *17*, 8721. [CrossRef]
29. He, W.; Liu, M.; Li, X.; Zhang, X.; Abdel-Mageed, W.M.; Li, L.; Wang, W.; Zhang, J.; Han, J.; Dai, H.; et al. Fungal biotransformation of tanshinone results in [4+2] cycloaddition with sorbicillinol: Evidence for enzyme catalysis and increased antibacterial activity. *Appl. Microbiol. Biotechnol.* **2016**, *100*, 8349–8357. [CrossRef]
30. Lin, S.; Wu, Y.-Z.; Chen, K.-Y.; Ye, J.; Yang, X.-W.; Zhang, W.-D. Polyketides from the fungus *Penicillium decumbens*. *J. Asian Nat. Prod. Res.* **2018**, *20*, 445–450. [CrossRef] [PubMed]
31. Du, L.; Zhu, T.; Li, L.Y.; Cai, S.; Zhao, B.; Gu, Q. Cytotoxic sorbicillinoids and bisorbicillinoids from a marine-derived fungus *Trichoderma* sp. *Chem. Pharm. Bull.* **2009**, *57*, 220–223. [CrossRef]
32. Ying, Y.-M.; Zhan, Z.-J.; Ding, Z.-S.; Shan, W.-G. Bioactive metabolites from *Penicillium* sp. P-1, a fungal endophyte in *Huperzia serrata*. *Chem. Nat. Compd.* **2011**, *47*, 541–544. [CrossRef]
33. Yao, Y.; Li, J.; Jiang, C.-S.; Zhao, X.-X.; Miao, Z.-H.; Liu, H.-T.; Zheng, P.; Yao, W.-X.; Li, W.-Q. Trichodimerol and sorbicillin induced apoptosis of HL-60 cells is mediated by reactive oxygen species. *Pharmazie* **2015**, *70*, 394–398.
34. Maskey, R.P.; Grün-Wollny, I.; Grün-Wollny, H. Sorbicillin analogues and related dimeric compounds from *Penicillium notatum*. *J. Nat. Prod.* **2005**, *68*, 865–870. [CrossRef]
35. Corral, P.; Esposito, F.P.; Tedesco, P.; Falco, A.; Tortorella, E.; Tartaglione, L.; Festa, C.; D'Auria, M.V.; Gnavi, G.; Varese, G.C.; et al. Identification of a sorbicillinoid-producing *Aspergillus* strain with antimicrobial activity against *Staphylococcus aureus*: A new polyextremophilic marine fungus from Barents Sea. *Mar. Biotechnol.* **2018**, *20*, 502–511. [CrossRef]
36. Meng, J.; Wang, B.; Cheng, W. Study on the secondary metabolites of *Thichoderma sturnisporum*. *Chin. J. Mar. Drugs* **2017**, *36*, 27–31.
37. Yang, Z.; Qiao, Y.; Li, J.; Wu, F.-G.; Lin, F. Novel type of water-soluble photosensitizer from *Trichoderma reesei* for photodynamic inactivation of Gram-positive bacteria. *Langmuir* **2020**, *36*, 13227–13235. [CrossRef]
38. Venkatesha, S.H.; Achaya, B.; Moudgil, K.D. Natural products as source of anti-inflammatory drugs. *Inflammation* **2017**, *4*, 1661–1690.
39. Zhao, J.-L.; Zhang, M.; Liu, J.-M.; Tan, Z.; Chen, R.-D.; Xie, K.-B.; Dai, J.-G. Bioactive steroids and sorbicillinoids isolated from the endophytic fungus *Trichoderma* sp. Xy24. *J. Asian Nat. Prod. Res.* **2017**, *19*, 1028–1035. [CrossRef] [PubMed]
40. Marra, R.; Nicoletti, R.; Pagano, E.; DellaGreca, M.; Salvatore, M.M.; Borrelli, F.; Lombardi, N.; Vinale, F.; Woo, S.L.; Andolfi, A. Inhibitory effect of trichodermanone C, a sorbicillinoid produced by *Trichoderma citrinoviride* associated to the green alga *Cladophora* sp., on nitrite production in LPS-stimulated macrophages. *Nat. Prod. Res.* **2019**, *33*, 3389–3397. [CrossRef]

41. Xu, D.; Xue, M.; Shen, Z.; Jia, X.; Hou, X.; Lai, D.; Zhou, L. Phytotoxic secondary metabolites from fungi. *Toxins* **2021**, *13*, 261. [CrossRef]
42. Zhao, J.; Shan, T.; Mou, Y.; Zhou, L. Plant-derived bioactive compounds produced by endophytic fungi. *Mini-Rev. Med. Chem.* **2011**, *11*, 159–168. [CrossRef]
43. Sivakumar, P.M.; Prabhawathi, V.; Zarrabi, A.; Akthar, S.; Prabhakar, P.K. Current trends in the therapeutic strategies for diabetes management. *Curr. Med. Chem.* **2021**, *28*, 4616–4637. [CrossRef]
44. Deng, J.; Dai, H.; Wang, Y.; Chen, H.; Tan, Z.; Mei, W. Isolation and identification of the fungus *Aspergillus* sp. HNWSW-20 from Chinese agarwood and its secondary metabolites. *Chin. J. Trop. Crops* **2018**, *39*, 1618–1624.
45. Abe, N.; Murata, T.; Hirota, A. Novel DPPH radical scavengers, bisorbicillinol and demethyltrichodimerol, from a fungus. *Biosci. Biotechnol. Biochem.* **1998**, *62*, 661–666. [CrossRef]
46. Sugaya, K.; Terajima, T.; Takahashi, A.; Onose, J.; Abe, N. Bisorbicillinol inhibits Lyn tyrosine kinase for allergic response on RBL-2H3 cells. *Bioorg. Med. Chem. Lett.* **2019**, *29*, 832–835. [CrossRef] [PubMed]
47. Tilki, E.K.; Ozturk, S.E.; Ozarda, M.G.; Canturk, Z.; Dikmen, M. Investigation of the neuroprotective and neuritogenic effects of halotolerant *Penicillium flavigenum*-derived sorbicillin-like compounds on PC-12 Adh cells. *Cytotechnology* **2021**, *73*, 801–813. [CrossRef] [PubMed]
48. Han, X.; Tang, X.; Qiao, D.; Li, P.; Li, G. Studies on bioactive components of a marine-derived fungus *Penicillium chrysogenum*. *Chin. J. Mar. Drugs* **2019**, *38*, 77–81.
49. Lu, S.; Sun, W.; Meng, J.; Wang, A.; Wang, X.; Tian, J.; Fu, X.; Dai, J.; Liu, Y.; Lai, D.; et al. Bioactive bis-naphtho-γ-pyrones from rice false smut pathogen *Ustilaginoidea virens*. *J. Agric. Food Chem.* **2015**, *63*, 3501–3508. [CrossRef]
50. Meng, J.; Sun, W.; Mao, Z.; Xu, D.; Wang, X.; Lu, S.; Lai, D.; Liu, Y.; Zhou, L.; Zhang, G. Main ustilaginoidins and their distribution in rice false smut balls. *Toxins* **2015**, *7*, 4023–4034. [CrossRef]
51. Sun, W.; Dong, X.; Xu, D.; Meng, J.; Fu, X.; Wang, X.; Lai, D.; Zhou, L.; Liu, Y. Preparative separation of main ustilaginoidins from rice false smut balls by high-speed counter-current chromatography. *Toxins* **2016**, *8*, 20. [CrossRef]
52. Wang, X.; Fu, X.; Lin, F.; Sun, W.; Meng, J.; Wang, A.; Lai, D.; Zhou, L.; Liu, Y. The contents of ustiloxins A and B along with their distribution in rice false smut balls. *Toxins* **2016**, *8*, 262. [CrossRef]
53. Wang, X.; Wang, J.; Lai, D.; Wang, W.; Dai, J.; Zhou, L.; Liu, Y. Ustiloxin G, a new cyclopeptide mycotoxin from rice false smut balls. *Toxins* **2017**, *9*, 54. [CrossRef]
54. Sun, W.; Wang, A.; Xu, D.; Wang, W.; Meng, J.; Dai, J.; Liu, Y.; Lai, D.; Zhou, L. New ustilaginoidins from rice false smut balls caused by *Villosiclava virens* and their phytotoxic and cytotoxic activities. *J. Agric. Food Chem.* **2017**, *65*, 5151–5160. [CrossRef]
55. Lai, D.; Meng, J.; Xu, D.; Zhang, X.; Liang, Y.; Han, Y.; Jiang, C.; Liu, H.; Wang, C.; Zhou, L.; et al. Determination of the absolute configurations of the stereogenic centers of ustilaginoidins by studying the biosynthetic monomers from a gene knockout mutant of *Villosiclava virens*. *Sci. Rep.* **2019**, *9*, 1855. [CrossRef]
56. Wang, A.; Li, P.; Han, P.; Gu, G.; Shan, T.; Lai, D.; Zhou, L. New nitrogen-containing metabolites from cultures of rice false smut pathogen *Villosiclava virens*. *Nat. Prod. Res.* **2021**, *35*, 272–281. [CrossRef] [PubMed]
57. Meng, J.; Zhao, S.; Dang, P.; Zhou, Z.; Lai, D.; Zhou, L. Ustilaginoidin M1, a new bis-naphtho-γ-pyrone from the fungus *Villosiclava virens*. *Nat. Prod. Res.* **2021**, *35*, 1555–1560. [CrossRef] [PubMed]
58. Xu, D.; Yin, R.; Zhou, Z.; Gu, G.; Zhao, S.; Xu, J.-R.; Liu, J.; Peng, Y.; Lai, D.; Zhou, L. Elucidation of ustilaginoidin biosynthesis reveals a previously unrecognised class of ene-reductases. *Chem. Sci.* **2021**, *12*, 14883–14892. [CrossRef] [PubMed]
59. Li, C.; Lin, F.; Sun, W.; Yuan, S.; Zhou, Z.; Wu, F.-G.; Chen, Z. Constitutive hyperproduction of sorbicillinoids in *Trichoderma reesei* ZC121. *Biotechnol. Biofuels* **2018**, *11*, 291. [CrossRef]
60. Ramesh, C.; Vinithkumar, N.V.; Kirubagraran, R.; Venil, C.K.; Dufosse, L. Multifaceted applications of microbial pigments: Current knowledge, challenges and future directions for public health implications. *Microorganisms* **2019**, *7*, 186. [CrossRef]
61. Lin, L.; Xu, J. Fungal pigments and their roles associated with human health. *J. Fungi* **2020**, *6*, 280. [CrossRef]
62. Druzhinina, I.S.; Kubicek, E.M.; Kubicek, C.P. Several steps of lateral gene transfer followed by events of 'birth-and-death' evolution shaped a fungal sorbicillinoid biosynthetic gene cluster. *BMC Evol. Biol.* **2016**, *16*, 269. [CrossRef]
63. Salo, O.; Guzman-Chavez, F.; Ries, M.I.; Lankhorst, P.P.; Bovenberg, R.A.L.; Vreeken, R.J.; Driessen, A.J.M. Identification of a polyketide synthase involved in sorbicillin biosynthesis by *Penicillium chrysogenum*. *Appl. Environ. Microbiol.* **2016**, *82*, 3971–3978. [CrossRef] [PubMed]
64. Chen, G.; Chu, J. Characterization of two polyketide synthases involved in sorbicillinoid biosynthesis by *Acremonium chrysogenum* using the CRISPR/Cas9 system. *Appl. Biochem. Biotechnol.* **2019**, *188*, 1134–1144. [CrossRef] [PubMed]
65. Kahlert, L.; Cox, R.J.; Skellam, E. The same but different: Multiple functions of the fungal flavin dependent monooxygenase SorD from *Penicillium chrysogenum*. *Chem. Commun.* **2020**, *56*, 10934. [CrossRef] [PubMed]
66. Monroy, A.A.; Stappler, E.; Schuster, A.; Sulyok, M.; Schmoll, M. A CRE1-regulated cluster is responsible for light dependent production of dihydrotrichotetronin in *Trichoderma reesei*. *PLoS ONE* **2017**, *12*, e0182530. [CrossRef]
67. Hinterdobler, W.; Schuster, A.; Tisch, D.; Ozkan, E.; Bazafkan, H.; Schinnerl, J.; Brecker, L.; Bohmdorfer, S.; Schmoll, M. The role of PKAc1 in gene regulation and trichodimerol production in *Trichoderma reesei*. *Fungal Biol. Biotechnol.* **2019**, *6*, 12. [CrossRef] [PubMed]
68. Hitzenhammer, E.; Buschl, C.; Sulyok, M.; Schuhmacher, R.; Kluger, B.; Wischnitzki, E.; Schmoll, M. YPR2 is a regulator of light modulated carbon and secondary metabolism in *Trichoderma reesei*. *BMC Genom.* **2019**, *20*, 211. [CrossRef] [PubMed]

69. Sib, A.; Gulder, T.A.M. Stereoselective total synthesis of bisorbicillinoid natural products by enzymatic oxidative dearomatization/dimerization. *Angew. Chem. Int. Ed.* **2017**, *56*, 12888–12891. [CrossRef] [PubMed]
70. Yan, Q.; Banwell, M.G.; Coote, M.L.; Lee, R.; Willis, A.C. Establishing the true structure of the sorbicillinoid-derived isolate rezishanone C by total synthesis. *Chem. Asian J.* **2017**, *12*, 1480–1484. [CrossRef]

Journal of *Fungi*

Review

Biosynthesis of Fungal Natural Products Involving Two Separate Pathway Crosstalk

Guangzhi Dai, Qiyao Shen, Youming Zhang and Xiaoying Bian *

Helmholtz International Lab for Anti-Infectives, Shandong University-Helmholtz Institute of Biotechnology, State Key Laboratory of Microbial Technology, Shandong University, Qingdao 266237, China; daiguangzhi@sdu.edu.cn (G.D.); syndeeyao@foxmail.com (Q.S.); zhangyouming@sdu.edu.cn (Y.Z.)
* Correspondence: bianxiaoying@sdu.edu.cn

Abstract: Fungal natural products (NPs) usually possess complicated structures, exhibit satisfactory bioactivities, and are an outstanding source of drug leads, such as the cholesterol-lowering drug lovastatin and the immunosuppressive drug mycophenolic acid. The fungal NPs biosynthetic genes are always arranged within one single biosynthetic gene cluster (BGC). However, a rare but fascinating phenomenon that a crosstalk between two separate BGCs is indispensable to some fungal dimeric NPs biosynthesis has attracted increasing attention. The hybridization of two separate BGCs not only increases the structural complexity and chemical diversity of fungal NPs, but also expands the scope of bioactivities. More importantly, the underlying mechanism for this hybridization process is poorly understood and needs further exploration, especially the determination of BGCs for each building block construction and the identification of enzyme(s) catalyzing the two biosynthetic precursors coupling processes such as Diels–Alder cycloaddition and Michael addition. In this review, we summarized the fungal NPs produced by functional crosstalk of two discrete BGCs, and highlighted their biosynthetic processes, which might shed new light on genome mining for fungal NPs with unprecedented frameworks, and provide valuable insights into the investigation of mysterious biosynthetic mechanisms of fungal dimeric NPs which are constructed by collaboration of two separate BGCs.

Keywords: biosynthetic pathway crosstalk; natural product; biosynthetic gene cluster; polyketide; fungi; bioactivity

Citation: Dai, G.; Shen, Q.; Zhang, Y.; Bian, X. Biosynthesis of Fungal Natural Products Involving Two Separate Pathway Crosstalk. *J. Fungi* **2022**, *8*, 320. https://doi.org/10.3390/jof8030320

Academic Editors: Tao Feng and Frank Surup

Received: 27 February 2022
Accepted: 17 March 2022
Published: 21 March 2022

1. Introduction

Fungi produce a plethora of chemically diverse natural products (NPs) with various bioactivities and a wide variety of applications in medicine and agriculture [1]. Because of this, fungal NPs have been historically recognized as an invaluable source of inspiration for the development of drug leads for the treatment of infections and cancer, as well as the prevention of crop damage, such as lovastatin (the cholesterol-lowering drug), mycophenolic acid (the immunosuppressive drug), and pyripyropene A (the insecticide) [2–4]. In addition, fungal NPs with novel molecular scaffolds provide excellent templates for the chemical synthesis of new bioactive compounds [5].

In fungi, the natural products are commonly synthesized by the genes arranged in a contiguous fashion as a biosynthetic gene cluster (BGC) [6]. The core biosynthetic enzymes encoded by the BGCs mainly include polyketide synthases (PKSs), non-ribosomal peptide synthetases (NRPSs), terpene synthases, and ribosomally synthesized and post-translationally modified peptide (RiPP) biosynthetic enzymes to enlarge the variety of carbon skeletons of the products. Tailoring enzymes such as the monooxygenase, hydroxylase, and methyltransferase further increase the diversity and complexity of the molecules [7]. With the development of genome sequencing technology, bioinformatic methods, genome mining algorithms, and scalable expression platforms, the genomic-driven approaches have revolutionized NPs discovery, greatly expand the access to the chemical repertoire of fungal-derived NPs, and provide unprecedented opportunities to investigate their biosynthetic mechanisms [8]. Several vital factors that regulate the expression of fungal BGCs, including the environmental signals, transcriptional regulation, and epigenetic regulation, have been summarized by Keller [9], which might be instructive for genome mining and activation of the silent BGCs.

In most reported studies, the microbial NPs biosynthetic pathways are generally completed by one single BGC. However, in some interesting but rare cases, the products have been demonstrated to be constructed via the hybridization of two precursors, which are biosynthesized by two separate BGCs. For instance, the antithrombotic myxadazoles, a family of novel chimeric compounds isolated from *Myxococcus* sp. SDU36, consists of N-ribityl 5,6-dimethylbenzimidazole and a linear fatty acid chain endowed with an isoxazole ring. Interestingly, a non-canonical PKS/NRPS biosynthetic pathway and the vitamin B12 metabolism pathway were proved to interwind together through the construction of isoxazole-benzimidazole hybrids [10]. Tasikamides, recently isolated from *Streptomyces tasikensis* P46, contain a rare hydrazone group that joins the cyclic peptide scaffold to an alkyl 5-hydroxylanthranilate (AHA) moiety. This research also addressed that the biosynthesis of tasikamides required the coupling of two separate gene clusters, with one BGC encoding a NRPS pathway for assembling the cyclic peptide scaffold, and the other BGC encoding the AHA-synthesizing pathway [11]. These works illustrate that functional crosstalk between two different biosynthetic pathways is not only of considerable value in increasing the structural diversity of NPs, but also a more effective way to construct new drug leads of natural origin.

Several key issues need to be considered when we investigate the biosynthetic mechanism of two separate BGC co-participation in NP biosynthesis: (a) identification of each BGC and characterization of the necessary gene function; (b) isolation of sufficient quantity of key intermediates through gene knockout, enzymatic catalyzation, and heterologous expression; and (c) determination of whether the coupling process of the two separate BGCs is spontaneous or enzymatic. Thus, in-depth understanding of the biosynthetic mechanism for the convergence of two distinct biosynthetic pathways will provide an alternative to accelerate the discovery of NPs with novel skeletons, as well as shed new light on which enzyme(s) could catalyze the formation of C-C, C-N, or N-N bonds that link two biosynthetic precursors together. In this mini-review, we summarized some fungal NPs with novel skeletons produced by the crosstalk between two discrete biosynthetic gene clusters, mainly including polyketides, meroterpenoids, and non-ribosomal peptides. We highlighted their biosynthetic processes, which might provide valuable insights into the coupling mechanism of two separate BGCs in fungal NP biosynthesis.

2. Fungal Polyketide Biosynthesis Involving Two Separate Pathway Crosstalk

2.1. The Biosynthesis of Penilactones A and B

The representative examples about non-enzymatic Michael addition mediated the coupling process of polyketide–polyketide hybrids were highly oxygenated fungal polyketides penilactones A (**1**) and B (**2**), which were firstly isolated from an Antarctic deep-sea derived fungus *Penicillium crustosum* PRB-2 by Li and co-workers [12]. The biosynthetic pathway was proposed to be originated from the hybridization of one *o*-quinone methide unit (**5**) and a γ-butyrolactone moiety through 1,4-Michael addition to complete the carbon skeleton construction of **1** and **2**. A biomimetic total synthesis was subsequently achieved to confirm the biosynthetic hypothesis [13]. Considering the enzymes for Michael addition involved in **1** and **2** biosynthesis have not been reported yet, Li and co-workers identified two separate BGCs (termed as *cla* and *tra* BGCs in this review) responsible for the biosynthesis of **1** and **2** through the gene deletion and heterologous expression in *Aspergillus nidulans* [14]. The core non-reducing polyketide synthase (NR-PKS) ClaF in the *cla* BGC is responsible for the biosynthesis of crucial intermediate clavatol (**3**) (Figure 1). The nonheme Fe^{II}/2-oxoglutarate-dependent oxygenase ClaD oxidized **3** into hydroxyclavatol (**4**), which spontaneously underwent dehydration into the crucial intermediate *o*-quinone methide (**5**).

For *tra* BGC, a hybrid polyketide synthase-nonribosomal peptide synthetase (PKS-NRPS) TraA and a *trans*-acting enoyl reductase (ER) TraG collaboratively catalyzed the formation of crustosic acid (**6**) precursor. The nonheme Fe^{II}/2-oxoglutarate-dependent oxygenase TraH subsequently catalyzed the oxidative decarboxylation of **6** into dehydroterrestric acid, the terminal double-bond of which was finally reduced by the flavin-dependent oxidoreductase TraD into an important intermediate terrestric acid (**7**) [15]. The results of precursors feeding experiments in Δ*traA* mutant confirmed that both **6** and **7** were the on-pathway intermediates, and could be transformed into 5-carboxylmethyltetronic acid (**8**) and 5-methyltetronic acid (**9**), respectively. The fascinating issue associated with the biosynthesis of **1** and **2** is which enzyme(s) could catalyze the Michael addition to couple two building blocks together. Surprisingly, incubation of **8** with **4** in water at 25 °C led to the formation of penilactone D (**10**) as the major product, and **2** as the minor product (Figure 1). Incubation of **9** with **4** generated peniphenone D (**11**) as the major product as well as **1** as the minor product. Both **10** and **11** were separately incubated with **4**, the formation of **1** and **2** could be observed (Figure 1). These results unambiguously indicated that the Michael addition involving in the biosynthesis of **1** and **2** is nonenzymatic and could happen spontaneously.

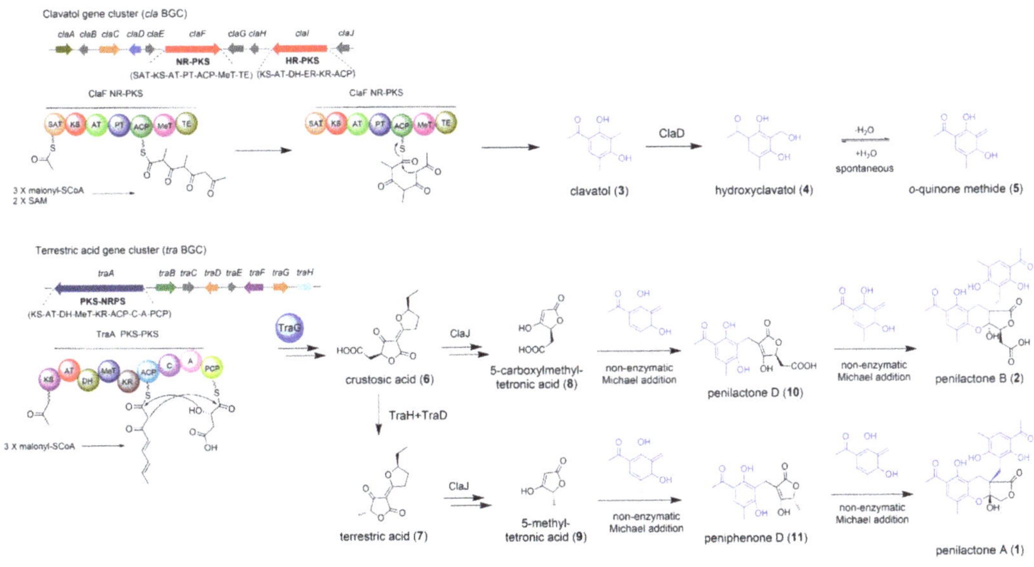

Figure 1. The biosynthetic pathway of fungal polyketides penilactones A (**1**) and B (**2**). The Michael addition that triggers the coupling of *cla* and *tra* BGCs is nonenzymatic.

In general, the combination of enzymatic and nonenzymatic reactions originated from the crosstalk between two separate biosynthetic pathways significantly enriched the structural diversity of fungal NPs. The biosynthesis of fungal polyketides penilactones A and B provides an excellent example to investigate that how two different BGCs interwind at a gene cluster level.

2.2. The Biosynthesis of Dalmanol A and Acetodalmanol A

The mantis-associated fungus *Daldinia eschscholzii* TL01 is known to produce novel polyketides including dalmanol A (**12**) and acetodalmanol A (**13**) with immunosuppressive bioactivity [16–18]. The structural characteristic of **12** and **13** implied that their carbon skeletons construction involved in co-participation of two building blocks naphthalene and chromane. Tan and co-workers conducted the pioneering work to identify two separate BGCs (termed as *chr* and *nap* BGCs in this review) by gene deletion as well as heterologous expression [19]. The *chr* BGC is responsible for chromane biosynthesis and the *nap* BGC biosynthesizes naphthalene. This work addressed the two-gene-cluster crosstalk based biosynthetic pathways of dalmanol A and acetodalmanol A (Figure 2).

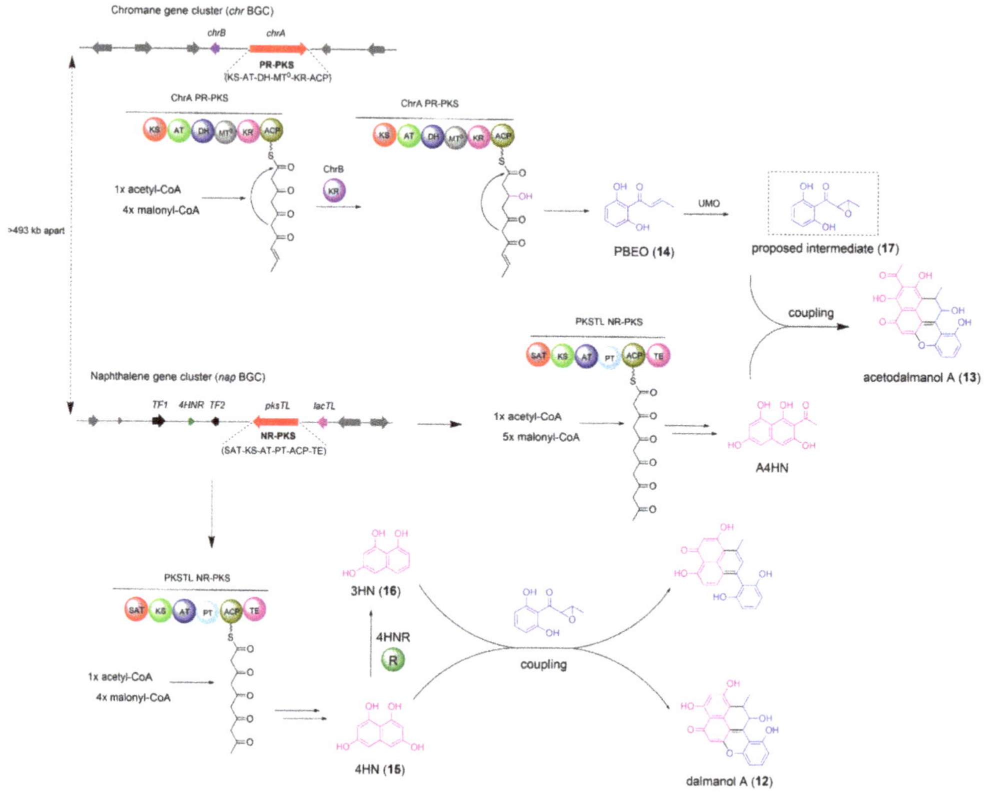

Figure 2. The biosynthetic pathway of fungal polyketides dalmanol A (**12**) and acetodalmanol A (**13**).

In plants and fungi, the assembly of chromane-based aromatic polyketides have been reported to be biosynthesized by both type III PKS and partially reducing type I PKS (PR-PKS) [20,21]. The results of the quantitative reverse transcription PCR analysis, targeted gene deletion, and heterologous expression experiments supported that the PR-PKS ChrA and the keto-reductase ChrB are indeed responsible for the formation of vital intermediate 1-(2,6-dihydroxyphenyl)but-2-en-1-one (PBEO,**14**) (Figure 2). Thus, the genes encoding PR-PKS ChrA, the keto-reductase ChrB, and a transporter constitute the *chr* BGC, which is located on Scaffold_36 in the genome (Figure 2). The NR-PKS gene *pksTL* in *Daldinia eschscholzii* TL01 has been confirmed to be responsible for the biosynthesis of naphthalene-based polyketide 1,3,6,8-tetrahydroxynaphthalene (4HN, **15**) by gene deletion [22]. Co-expression of the NR-PKS pksTL and 4HN reductase 4HNR in *A. oryzae* led to the accumulation of 1,3,6-tetrahydroxynaphthalene (3HN, **16**), a biosynthetic precursor for the assembly of dalesconols, which are polyketides also isolated from *Daldinia eschscholzii* TL01. The genes encoding the NR-PKS pksTL, 4HN reductase 4HNR, two transcription factors, and a laccase constitute the *nap* BGC, which is within Scaffold_20 in the genome and locates at least 493 kb away from the *chr* BGC. Only when *pksTL*, *chrA*, and *chrB* were co-introduced into *A. oryzae*, the polyketides 12 and 13 could be successfully produced. The critical biosynthetic process for the coupling of *chr* and *nap* BGCs was proposed to be the epoxidation of PBEO (14) to the proposed intermediate 17 (Figure 2). The results of enzymatic activity inhibition experiments revealed an unspecific P450 monooxygenase located elsewhere in the genome of *A. oryzae* host might be responsible for the epoxidation of 14 to 17, which triggered the cross-cluster coupling of both *chr* and *nap* BGCs.

Overall, this work further illustrated that two separated BGCs crosstalk is a promising access to improve the structural diversity of fungal NPs. Understanding the regulatory mechanism of the multiple-gene-cluster coupling is of great significance in establishing the synthetic biology approaches to discover NPs with novel skeletons and potential biological activities.

2.3. The Biosynthesis of Azasperpyranone A

Azaphilones, a group of structurally related fungal polyketides, contain a highly oxygenated bicyclic pyrone quinone moiety, and exhibit a broad range of bioactivities including anticancer, antifungal, and antiviral activities [23]. Azasperpyranone A (**18**), recently isolated from *A. terreus*, contains a highly oxygenated pyranoquinone moiety possessing a 6/6/6/6 tetracyclic ring system, and shows potential anticancer activity [24]. Scrutiny of the structural feature of azasperpyranone An (**18**) revealed that two building blocks 5-methyl orsellinic aldehyde (**19**) and preasperpyranone (**20**) are the biosynthetic precursors. Lu and co-workers have identified two separate BGCs by gene deletion, including the BGC A responsible for polyhydric phenol formation and the BGC B involving the azaphilonoid scaffold construction (Figure 3) [24].

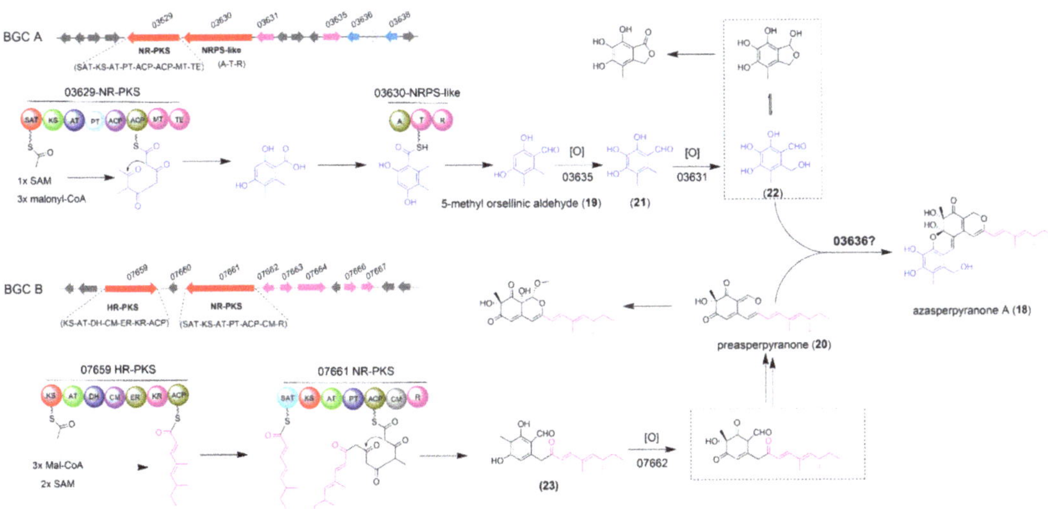

Figure 3. The biosynthetic pathway of fungal polyketide azasperpyranone A (**18**).

In a previous study, the full-length, intron-free open reading frames of two core genes *ATEG_03629* and *ATEG_03630* from BGC A, which encode a NR-PKS and a NRPS-like enzyme, respectively, were co-transformed into *Saccharomyces cerevisiae* to produce the intermediate **19** (Figure 3) [25]. Then the FAD-dependent monooxygenase (FMO) encoded by *ATEG_03635* gene catalyzed the hydroxylation of **19** to afford the intermediate **21**, which was subsequently oxidized into the crucial precursor **22** by the P450 monooxygenase encoded by *ATEG_03631* gene. In BGC B, the two core genes *ATEG_07659* encoding a highly reducing PKS (HR-PKS) and *ATEG_07661* encoding a NR-PKS were heterologously co-expressed in *A. nidulans* to produce **23** [26,27]. Compound **23** was rapidly converted into an important precursor preasperpyranone **20** by the FMO encoded by gene *ATEG_07662*. The remaining gap for the entire biosynthetic pathway of azasperpyranone A is that which enzyme could couple the two vital intermediates **20** and **22** together by catalyzing the formation of C-C and C-O bonds. Deletion of the gene *ATEG_03636* with unknown function abolished the production of **18**, but accumulated two precursors **19** and **20**, which implied the formation of **18** was more likely catalyzed by ATEG_03636 rather than caused by the spontaneous reaction between **20** and **22** (Figure 3). The authors also found that the ATEG_07667 in BGC B could indirectly regulate the cluster-specific regulators ATEG_03638 in BGC A and ATEG_07666 in BGC B to collaboratively synthesize the anti-cancer compound **18**. This interesting collaborative model in the fungal NPs biosynthesis provides new clues for the investigation of regulatory mechanism for other novel natural products which were biosynthesized by two separate BGCs crosstalk.

3. Fungal Meroterpenoids Biosynthesis Involving Two Separate Pathway Crosstalk

The Biosynthesis of Austinol

Meroterpenoids are an important class of fungal NPs [28], some of them have been developed as the drug candidate for treatment of Alzheimer's disease such as territrem [29], and the drug lead for insecticide such as pyripyropene A [30]. Fungal meroterpenoids are usually biosynthesized by a single BGC which encodes a polyketide synthase, a terpene cyclase, a prenyltransferase, and other essential tailoring enzymes to produce polyketide and terpenoid precursors [31]. Deletion of *sumO* gene, encoding the small ubiquitin-like protein SUMO, significantly altered the profiles of secondary metabolites in *A. nidulans*. Austinol (**23**) and dehydroaustinol (**24**) were identified from the Δ*sumO* mutant [32]. Nielsen et al. first reported the NR-PKS AusA responsible for the formation of polyketide precursor 3,5-dimethylorsellinic acid (DMOA) (**25**) in austinol biosynthesis [33]. This conclusion was also verified by Wang et al. [34]. Interestingly, Wang and co-workers found that no prenyltransferase gene was located near NR-PKS gene *ausA*, which suggested that the genes responsible for the biosynthesis of **23** and **24** might be separated in the genome of *A. nidulans* LO2026 [34]. Using the UbiA sequence as a query to blast the prenyltransferase homologs in the genome of *A. nidulans*, Wang et al. found the top two candidate genes *AN9259.4* (designated as *ausN*, on chromosome VIII) and *AN8142.4*. Deletion of *ausN* abolished the production of **23** and **24**, and accumulated the polyketide precursor **25**. By a set of gene deletions around *ausA* and *ausN*, they identified the two separate BGCs: the BGC A containing the necessary genes *ausA-D* and the BGC B consisting of the biosynthetic genes *ausE-N*. The biosynthetic pathway for **23** and **24** was also proposed (Figure 4). Firstly, the polyketide synthase AusA is responsible for the formation of **25**, then the aromatic prenyltransferase AusN catalyzed the C-alkylation of **25** using farnesyl pyrophosphate to form intermediate **26**. The epoxidase AusM catalyzed the epoxidation of the prenylated polyketide intermediate **26**, followed by cyclization catalyzed by a terpene cyclase AusL to form the tetracyclic intermediate **27**. The formation mechanism of the lactone system and spiro-ring in compound **28** has been investigated by Abe group [35]. The co-operation of the non-heme iron-dependent dioxygenase AusE, the hydroxylase AusB, and the Baeyer–Villiger monooxygenase AusC transformed the substrate **27** into **28** [35]; The hypothetical protein AusJ might be responsible for the acid-catalyzed keto-rearrangement and ring contraction of the tetraketide portion in intermediate **28** to generate intermediate **29**. The authors speculated that the AusK is responsible for reducing the C-5′ keto of **29** to hydroxyl group, and the hypothetical protein AusH might function as an accessory enzyme collaboratively working with AusK to alter AusK stereospecificity for its product **30**. The Baeyer−Villiger monooxygenase AusI inserted an oxygen atom between the C-4′ and vicinal carbon at C-3′ of **30** to create a lactone ring in **31**. Finally, the P450 monooxygenase AusG might catalyze the C-11 hydroxylation of **31** to form final product austinol (**23**) (Figure 4).

DMOA-derived fungal meroterpenoids possess complicated structures and attracted researchers' attention to investigate the biosynthetic pathway. Some fungal meroterpenoids such as anditomin and andrastin have been studied in detail [36–38]. All of the necessary genes for anditomin or andrastin biosynthesis are clustered in one single BGC. However, the biosynthesis of austinol and dehydroaustinol in *A. nidulans* LO2026 needs the pathway crosstalk between BGC A and BGC B. This provides an intriguing and valuable insight that fungi could use a variety of strategies to expand the skeletal diversification and subtlety regulate the crosstalk between separate biosynthetic pathways to multiply the number of NPs produced by these BGCs.

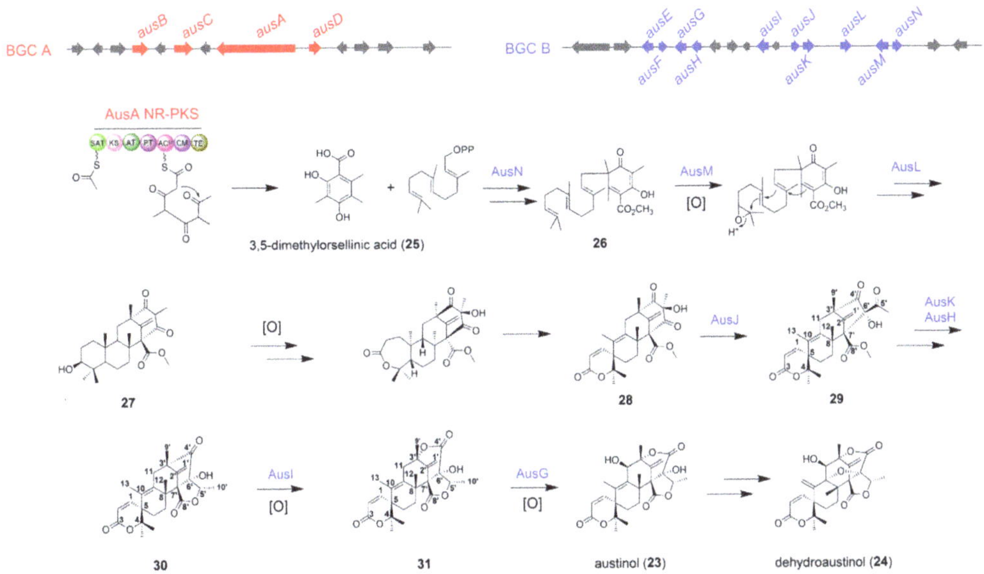

Figure 4. The biosynthetic pathway of fungal meroterpenoids austinol (**23**) and dehydroaustinol (**24**) in *A. nidulans* LO2026.

4. Fungal Non-Ribosomal Peptide Biosynthesis Involving Two Separate Pathway Crosstalk

4.1. The Biosynthesis of Spirotryprostatin A

Many NPs bearing the spiro-carbon system exhibit potential biological activities. Intrigued by the privileged structure and usefulness of spiro carbon system, a great deal of attention has been paid to their catalytic enantioselective synthesis [39,40]. Understanding the mechanism of spiro-carbon biosynthesis and identifying the versatility of enzymes responsible for the spiro-carbon formation are also of great significance. Spirotryprostatins belong to non-ribosomal peptides that isolated from *A. fumigatus*, and known for their pharmaceutical importance and application in cancer treatment [41]. The formation of the spiro-ring moiety in spirotryprostatins remained unknown and aroused Watanabe and co-workers' interest to solve this mystery [42]. The authors utilized *S. cerevisiae* and *A. niger* as the heterologous hosts to efficiently express the whole biosynthetic pathways of spirotryprostatins, and obtained crucial intermediates to identify two pathways for spiro-carbon formation, namely an epoxide route catalyzed by the FMO FqzB and a radical route catalyzed by the cytochrome P450 FtmG (Figure 5) [42]. Spirotryprostatins possess the diketopiperazine frameworks, and show the structural similarity to fumitremorgins and fumiquinazolines, suggesting the peptide backbone of these compounds could be biosynthesized by the NRPS using L-proline and L-tryptophan as the biosynthetic precursors [43,44]. When *ftmA-E* five genes were expressed in *A. niger*, no intermediates featuring the spiro-carbon were isolated, implying other indispensable genes are needed to be introduced. Based on the prior researches about the biosynthetic mechanism of spiro-carbon formation [45–47], the authors creatively introduced the FMO FqzB-encoding gene *fqzB*, which is located within the fumiquinazoline BGC (designed as fqz BGC in this review), into the *A. niger*/ *ftmA-E* mutant, and successfully isolated pirotryprostatin A (**32**) (Figure 5).

In vitro enzymatic assay revealed that FqzB could transform fumitremorgin C (**33**) into **32** by epoxidation mediated semipinacol-type rearrangement (Figure 5). These fascinating results not only emphasize the important function of FMOs as the intersection to trigger the coupling of two separate BGCs, but also highlight that the crosstalk between different biosynthetic pathways allows the structural diversification in NP biosynthesis.

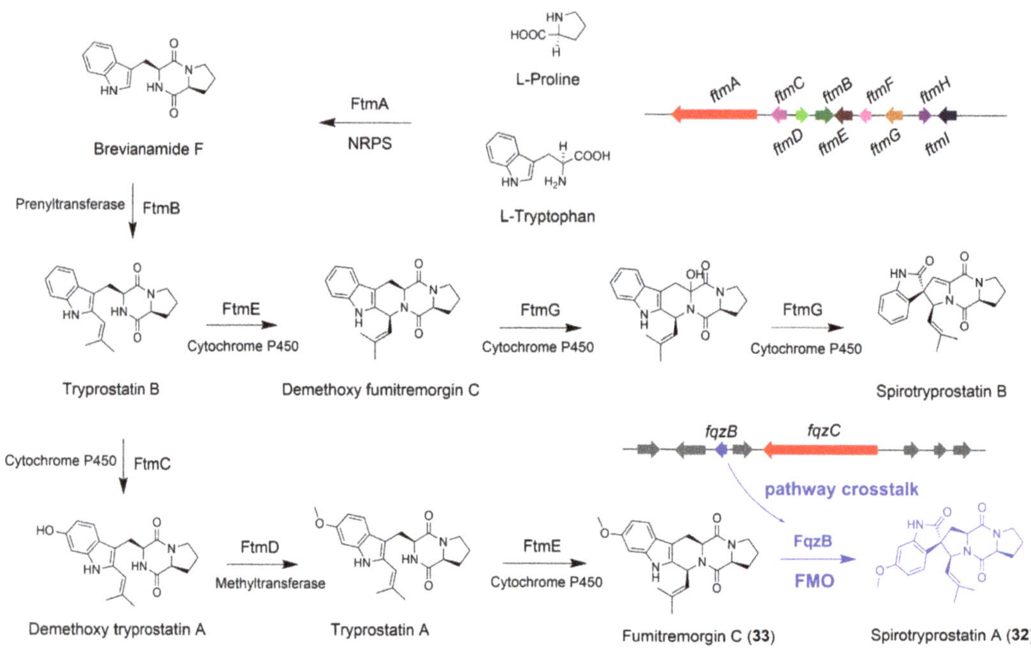

Figure 5. The biosynthetic pathway of fungal non-ribosomal peptide spirotryprostatin A (**32**). The FMO FqzB from the *fqz* BGC catalyzes the formation of spiro-carbon in spirotryprostatin A.

4.2. The Biosynthesis of Echinocandin B

Echinocandins, a family of fungal lipohexapeptides, are firstly isolated from *Emericella rugulosa* NRRL 11440, and exhibit excellent antifungal activities to the opportunistic pathogenic *Candida* strains. Structural modifications of echinocandin B (**34**), especially for the fatty acid moiety, successfully led to the generation of FDA-approved drug anidulafungin, which is a semisynthetic derivative of **34** and contains a substituted terphenyl acyl chain. To better understand how microbes use simple precursors to synthesize complex NPs, Tang and co-workers performed the groundbreaking work to identify and characterize the BGC of echinocandin B [48]. Four nonproteinogenic amino acids including 4*R*,5*R*-dihydroxyl-L-ornithine, 3*S*-hydroxyl-4*S*-methyl-L-proline, 4*R*-hydroxyl-L-proline, and 3*S*,4*S*-dihydroxyl-L-homotyrosine, as well as a long chain fatty acyl amide were contained in echinocandin B (Figure 6). These unusual structural units implied an interesting biosynthetic mechanism of echinocandin B.

By bioinformatics analysis, the NRPS EcdA containing six modules was identified. Gene deletion of *ecdA* confirmed its vital role in the peptide backbone formation of echinocandin B. Other biosynthetic genes, such as *ecdI* encoding a fatty-acyl-AMP ligase (EcdI), *ecdG* and *ecdK* encoding two α-ketoglutarate dependent oxygenases, and *ecdH* encoding a heme-iron-dependent cytochrome P450 oxygenase, were all in proximity to *ecdA* to constitute the *ecd* BGC. However, the genes responsible for the biosynthesis of L-homotyrosine were not present in the vicinity of *ecd* BGC, indicating a separate BGC should reside elsewhere in the genome. The putative 2-(4-hydroxybenzyl)-malic acid (**36**) is proposed to be the biosynthetic intermediate of L-homotyrosine (**39**) (Figure 6). Considering the isopropyl-malate synthase (IPMS) is reported to catalyze the condensation of α-ketovalerate with acetyl-CoA in the leucine biosynthesis, one IPMS homology in *E. rugulosa* genome was proposed to catalyze the condensation of 4-hydroxyphenyl-pyruvate and acetate to form **36**. Using the IPMS gene from *Mycobacterium tuberculosis* as BLAST query, the authors successfully identified the *hty* BGC, which is about 42.5 kb away from the *ecd* BGC, encoding four enzymes for de novo generation of the special building block L-homotyrosine (Figure 6) [48]. The isopropyl malate dehydrogenase HtyA catalyzed aldol-type condensation of 4-hydroxyphenyl-pyruvate (**35**) and acetyl-CoA to form **36**. Then, the aconitase homology HtyD executed the isomerization of 36 to 3-(4-hydroxybenzyl)-malic acid (**37**). Thereafter, **37** underwent decarboxylation and oxidation to form 2-oxo-4-(4-hydroxybenzyl) butanoic acid (**38**) by isopropyl malate dehydrogenase homologue HtyC. Finally, the transaminase HtyB catalyzed the transamination of **38** to form **39**. In general, the biosynthesis of echinocandin B needs the coupling of two sperate BGCs, the *ecd* and *hty* BGCs. The *hty* BGC provides an important biosynthetic precursor L-homotyrosine which was recognized by the fourth A domain of the NRPS EcdA. Understanding the biosynthetic mechanism of echinocandin B will facilitate us to take advantage of synthetic biology techniques to bioengineer NRPSs to generate bioactive compounds [49–51].

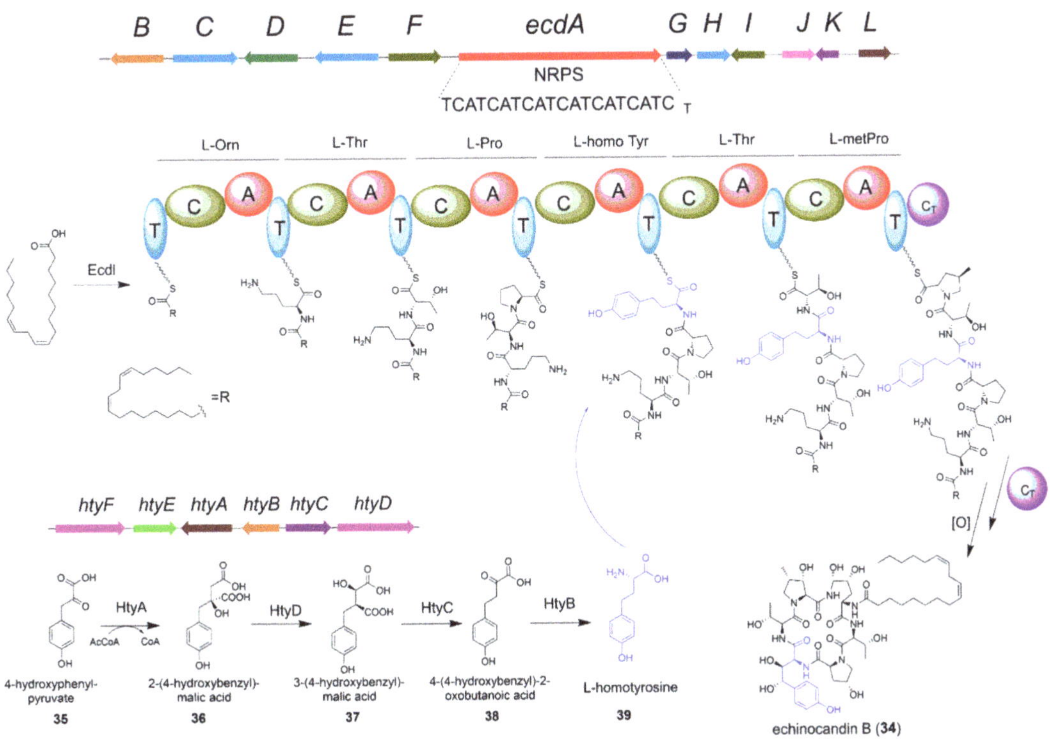

Figure 6. The biosynthetic pathway of fungal non-ribosomal peptide echinocandin B (**34**). The separate *hty* BGC is responsible for the L-homotyrosine moiety formation.

5. Representative Fungal NPs Might Be Biosynthesized by Two Separate Pathways Crosstalk

5.1. Delitschiapyrone A

Delitschiapyrone A (**40**) is a fungal polyketide bearing an unprecedented 6/6/7/5/6-fused pentacyclic ring system (Figure 7), and isolated from a solid culture of the leaf-associated fungus *Delitschia* sp. FL1581 [52]. The absolute configuration of **40** was determined by spectroscopic analysis, X-ray crystallography data, and experimental and calculated ECD. A naphthalenone unit and an α-pyrone moiety were proposed to be linked together via the Diels–Alder addition followed by an α-ketol-type rearrangement to forge the pentacylic ring system of **40** (Figure 7), which suggested the biosynthetic pathway of **40** might be concerned with the crosstalk between two separate BGCs (one for naphthalenone biosynthesis and the other for α-pyrone biosynthesis). A bioinspired total synthesis of delitschiapyrone A has been achieved by simply stirring a heterogeneous mixture of two Diels−Alder substrates, which gave a hint that the intermolecular Diels−Alder reaction might be spontaneous. Recently, Houk and co-workers investigated the mechanisms and dynamics of biosynthetic formation of **40** by density functional theory (DFT) calculations and quasiclassical molecular dynamics simulations with DFT and xTB, and drew a conclusion that **40** is not formed from the proposed Diels−Alder/α-ketol rearrangement cascade but instead formed directly from a single cycloaddition reaction [53], which is of great significance to the subsequent study on the biosynthetic pathway of **40** through gene deletion, heterologous expression, and enzymatic assays.

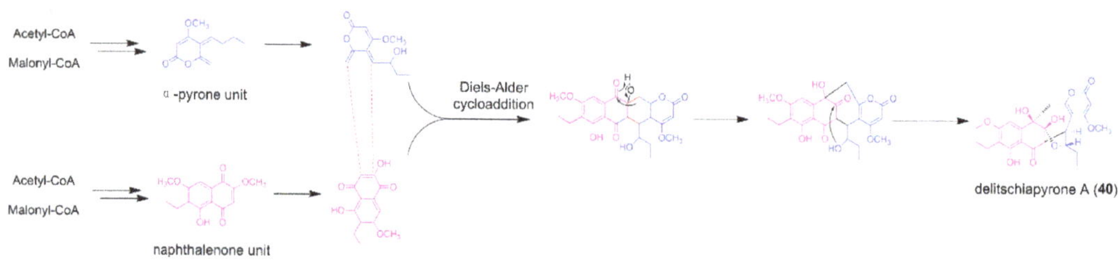

Figure 7. The proposed biosynthetic pathway of delitschiapyrone A (**40**).

5.2. Herpotrichone A

Herpotrichone A (**41**) is a fungal polyketide isolated from the isopod-associated fungus *Herpotrichia* sp. SF09 with an unprecedented pentacyclic 6/6/6/6/3 skeleton, and shows outstanding anti-neuro inflammatory activities in lipopolysaccharide (LPS)-induced BV-2 microglial cells (Figure 8) [54]. Interestingly, compound **41** is also an intermolecular [4 + 2] adduct that involves the coupling of epoxycyclohexenone and α-pyrone two building blocks via the Diels–Alder addition. Recently, the biosynthetic pathways of epoxycyclohexenone-derived fungal polyketides trichoxide, sordarial, and flavoglaucin have been investigated in detail, which provide new insights into the biosynthesis of the naphthalenone unit in **41** [55–57]. The biosynthesis of some fungal α-pyrone-linked NPs such as alternapyrones and citreoviridin have been reported. The α-pyrone moiety in alternapyrones and citreoviridin is indeed formed by the spontaneous intramolecular cyclization of PKS AlpA and CtvA, respectively [58,59]. Thus, herpotrichone A might share the similar strategy to forge the α-pyrone unit by utilization of an unidentified PKS. Overall, fungal NPs with intermolecular Diels–Alder addition features usually have novel carbon skeletons. The unexpected architectures of these compounds may open an interesting topic such as the characterization of two separate BGCs crosstalk, and discovery of more fungal intermolecular Diels–Alderases.

5.3. Citrifuran A

Citrifuran A (**42**) is produced by the centipede intestine-associated *Aspergillus* sp. through solid fermentation (Figure 7), and showed moderate inhibitory activities against LPS-induced NO production in RAW 264.7 macrophages [60]. The novel skeleton of citrifuran A was constructed by coupling of azaphilone and furanone moieties via Michael addition (Figure 9). It is obvious that the two separate BGCs' crosstalk is indispensable to the biosynthesis of **42**. The furanone moiety was also contained in fungal polyketide gregatin A. For gregatin A biosynthesis, a single PKS GrgA with the aid of a *trans*-acting enoylreductase GrgB could biosynthesize the C_4 and C_{11} carbon chains. More interestingly, a predicted hydrolase GrgF is responsible for the fusion two carbon chains to produce the furanone skeleton of gregatin A [61]. This unusual chain-fusing reaction might be also suitable for the biosynthesis of furanone scaffold in **42** (Figure 9). The fungal polyketide-derived mycotoxin citrinin also possesses azaphilone building block. The individual biosynthetic steps of citrinin have been studied by a combination of targeted gene knockout and heterologous gene expression in *A. oryzae* [62], which might provide new clues for investigation of the biosynthetic mechanism of **42**. Though the biosynthetic pathways for azaphilone and furanone have been investigated, the enzyme for catalyzation of Michael addition has not yet been identified, and needs further exploration.

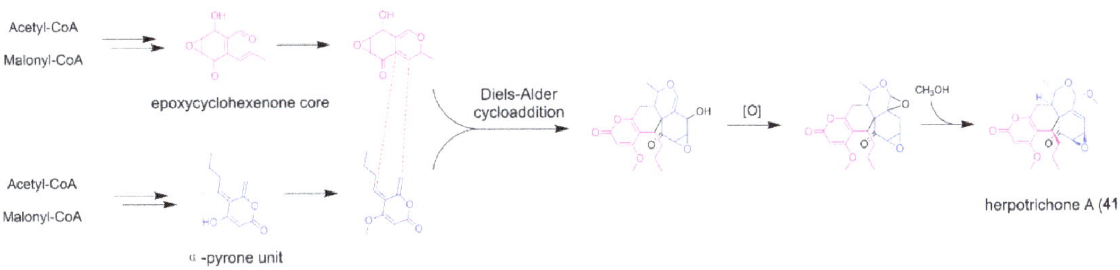

Figure 8. The proposed biosynthetic pathway of herpotrichone A (**41**).

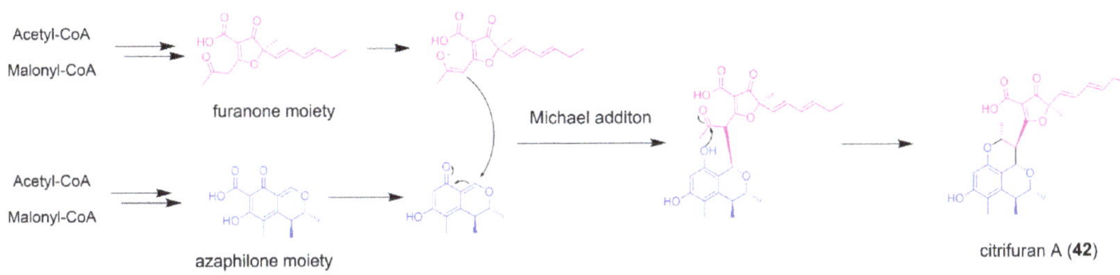

Figure 9. The proposed biosynthetic pathway of citrifuran A (**42**).

5.4. *Acautalide A*

The fungal polyketide acautalide A (**43**) is produced from the solid-state cultivation of isopod *Armadillidium vulgare*-associated *Acaulium* sp. H-JQSF on rice medium, and exhibits neuroprotective bioactivity with antiparkinsonic potential in the 1-methyl-4-phenylpyridinium-challenged nematode model [63]. The architectural features of 43 indicated that two biosynthetic precursors 10-keto-acaudiol and octadeca-9,11,13-trienoic acid intertwined together through intermolecular Diels–Alder cycloaddition (Figure 10). The 10-keto-acaudiol unit is proposed to be an early-stage precursor in the biosynthesis of acaulide and acaulins, two fungal macrodiolides previously isolated from the *Acaulium* sp. H-JQSF [64,65]. However, the biosynthetic mechanism of 10-keto-acaudiol remains unclear. The octadeca-9,11,13-trienoic acid motif might be derived from the biosynthesis of fungal polyunsaturated fatty acids, either biosynthesized from the fungal desaturation of octadecanoic acids in rice. Interestingly, the obtained **43** was in enantiomerically pure form, thus the Diels–Alder cycloaddition might be truly enzymatic. A fungal Diels–Alderase could be expected to catalyze the intermolecular [4 + 2]-cycloaddition in the assembly line of **43** [63].

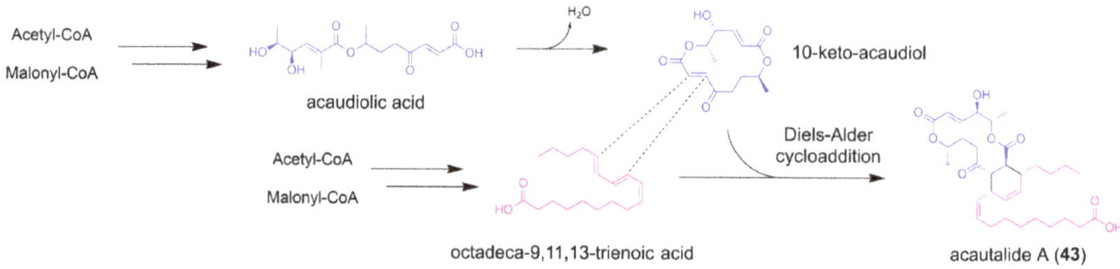

Figure 10. The proposed biosynthetic pathway of acautalide A (**43**).

6. Discussion

One reason that fungi endow great potentials to produce natural products with complex structures and excellent biological activities might be ascribed to the complex metabolic regulatory networks and interaction of different biosynthetic gene clusters. Understanding how simple precursors are synthesized and assembled together to construct complex natural products in organisms may promote the development of new combinational and synthetic biology strategies to create new molecules.

Over the course of evolution, fungi have evolved different strategies to increase the diversity of their NPs to protect themselves and acclimatize to the surrounding ecological environment [66]. For example, a great progress has been made in the discovery and identification of fungal NPs with homo-dimeric or hetero-dimeric skeletons, which effectively expand structural diversity of NPs and accelerates the occurrence of new biological activities. For fungal homodimer NPs, the building blocks are mostly biosynthesized by one single BGC, and catalyzed by the crucial enzymes including cytochrome P450 enzymes, intermolecular Diels–Alderases, and multicopper oxidases to afford the homo-dimeric skeletons, such as the rugulosin A, bisorbicillinol, and viriditoxin biosynthesis [67–70]. On the contrary, a rare but intriguing phenomenon is that two different building blocks, usually produced by two separate BGCs, were coupled together to generate fungal heterodimers NPs. However, the underlying mechanisms including how two structural units are biosynthesized, and whether the two BGCs crosstalk process is enzymatic or spontaneous are still mysterious and need further exploration. The coupling reactions between the two different building blocks are various. The Diels–Alder reaction and Michael addition reaction have been reported to splice the separate biosynthetic precursors together [14,52]. To in-depth understand the biosynthetic process of two separate BGC crosstalk, we should characterize the biosynthetic gene function in two BGCs and acquire crucial biosynthetic precursors through gene knockout and heterologous expression. With the important intermediates in hand, we can further investigate whether this hybridization process is spontaneous or enzymatic. However, the determination of which enzymes responsible for the two separate BGC crosstalk process is sometimes challenging, because the corresponding biosynthetic genes might not be located within the two gene clusters, and distributed elsewhere in the genome of targeted strain. Moreover, these enzymes may be hypothetical proteins, and it is difficult to be identified through bioinformatic analysis.

In general, more endeavors are needed to carry out the research for the discovery of fungal heterodimer NPs constructed by two building blocks, which not only provides an outstanding opportunity for investigation of the currently underestimated hidden biosynthetic crosstalk, but also facilitates the discovery of new BGCs, new regulatory mechanisms, and enzyme catalysts with novel catalytic mechanisms.

Author Contributions: Conceptualization, G.D. and X.B.; writing—original draft preparation, G.D.; writing—review and editing, G.D. and X.B.; discussion of the contents, G.D.; Q.S.; Y.Z., and X.B.; funding acquisition, G.D. and X.B. All authors have read and agreed to the published version of the manuscript.

Funding: This research was funded by National Key R&D Program of China (2019YFA0905700), National Natural Science Foundation of China (32100040, 32070060), the Shandong Provincial Natural Science Foundation, China (ZR2019JQ11), and Qingdao Postdoctoral Application Research Project (62450070311109).

Institutional Review Board Statement: Not applicable.

Informed Consent Statement: Not applicable.

Data Availability Statement: Not applicable.

Conflicts of Interest: The authors declare no conflict of interest.

Abbreviations

NP	natural product
BGC	biosynthetic gene cluster
PKS	polyketide synthase
NR-PKS	non-reducing polyketide synthase
HR-PKS	highly reducing polyketide synthase
NRPS	non-ribosomal peptide synthetases
RiPP	post-translationally modified peptide
FMO	FAD-dependent monooxygenase
ER	enoyl reductase
AHA	alkyl 5-hydroxylanthranilate
IPMS	isopropyl-malate synthase
DMOA	3,5-dimethylorsellinic acid
LPS	lipopolysaccharide
PBEO	1-(2,6-dihydroxyphenyl)but-2-en-1-one
DFT	density functional theory

References

1. González-Medina, M.; Owen, J.R.; El-Elimat, T.; Pearce, C.J.; Oberlies, N.H.; Figueroa, M.; Medina-Franco, J.L. Scaffold diversity of fungal metabolites. *Front. Pharmacol.* **2017**, *8*, 180. [CrossRef] [PubMed]
2. Ma, S.M.; Li, J.W.; Choi, J.W.; Zhou, H.; Lee, K.K.; Moorthie, V.A.; Xie, X.; Kealey, J.T.; Da Silva, N.A.; Vederas, J.C.; et al. Complete reconstitution of a highly reducing iterative polyketide synthase. *Science* **2009**, *326*, 589–592. [CrossRef] [PubMed]
3. Zhang, W.; Du, L.; Qu, Z.; Zhang, X.; Li, F.; Li, Z.; Qi, F.; Wang, X.; Jiang, Y.; Men, P.; et al. Compartmentalized biosynthesis of mycophenolic acid. *Proc. Natl. Acad. Sci. USA* **2019**, *116*, 13305–13310. [CrossRef] [PubMed]
4. Itoh, T.; Tokunaga, K.; Matsuda, Y.; Fujii, I.; Abe, I.; Ebizuka, Y.; Kushiro, T. Reconstitution of a fungal meroterpenoid biosynthesis reveals the involvement of a novel family of terpene cyclases. *Nat. Chem.* **2010**, *2*, 858–864. [CrossRef] [PubMed]
5. Karageorgis, G.; Foley, D.J.; Laraia, L.; Waldmann, H. Principle and design of pseudo-natural products. *Nat. Chem.* **2020**, *12*, 227–235. [CrossRef]
6. Osbourn, A. Secondary metabolic gene clusters: Evolutionary toolkits for chemical innovation. *Trends Genet.* **2010**, *26*, 449–457. [CrossRef]
7. Smanski, M.J.; Zhou, H.; Claesen, J.; Shen, B.; Fischbach, M.A.; Voigt, C.A. Synthetic biology to access and expand nature's chemical diversity. *Nat. Rev. Microbiol.* **2016**, *14*, 135–149. [CrossRef]
8. van Santen, J.A.; Kautsar, S.A.; Medema, M.H.; Linington, R.G. Microbial natural product databases: Moving forward in the multi-omics era. *Nat. Prod. Rep.* **2021**, *38*, 264–278. [CrossRef]
9. Keller, N.P. Fungal secondary metabolism: Regulation, function and drug discovery. *Nat. Rev. Microbiol.* **2019**, *17*, 167–180. [CrossRef]
10. Li, Y.; Zhuo, L.; Li, X.; Zhu, Y.; Wu, S.; Shen, T.; Hu, W.; Li, Y.Z.; Wu, C. Myxadazoles, Myxobacterium-derived isoxazole-benzimidazole hybrids with cardiovascular activities. *Angew. Chem. Int. Ed. Engl.* **2021**, *60*, 21679–21684. [CrossRef]
11. Ma, G.L.; Candra, H.; Pang, L.M.; Xiong, J.; Ding, Y.; Tran, H.T.; Low, Z.J.; Ye, H.; Liu, M.; Zheng, J.; et al. Biosynthesis of tasikamides via pathway coupling and diazonium-mediated hydrazone formation. *J. Am. Chem. Soc.* **2022**, *144*, 1622–1633. [CrossRef] [PubMed]
12. Wu, G.W.; Ma, H.Y.; Zhu, T.J.; Li, J.; Gu, Q.Q.; Li, D.H. Penilactones A and B, two novel polyketides from Antarctic deep-sea derived fungus *Penicillium crustosum* PRB-2. *Tetrahedron* **2012**, *68*, 9745–9749. [CrossRef]
13. Spence, J.T.; George, J.H. Biomimetic total synthesis of ent-penilactone A and penilactone B. *Org. Lett.* **2013**, *15*, 3891–3893. [CrossRef] [PubMed]

14. Fan, J.; Liao, G.; Kindinger, F.; Ludwig-Radtke, L.; Yin, W.B.; Li, S.M. Peniphenone and penilactone formation in *Penicillium crustosum* via 1,4-Michael additions of ortho-quinone methide from hydroxyclavatol to γ-butyrolactones from crustosic acid. *J. Am. Chem. Soc.* **2019**, *141*, 4225–4229. [CrossRef] [PubMed]
15. Fan, J.; Liao, G.; Ludwig-Radtke, L.; Yin, W.B.; Li, S.M. Formation of terrestric acid in *Penicillium crustosum* requires redox-assisted decarboxylation and stereoisomerization. *Org. Lett.* **2020**, *22*, 88–92. [CrossRef] [PubMed]
16. Zhang, Y.L.; Zhang, J.; Jiang, N.; Lu, Y.H.; Wang, L.; Xu, S.H.; Wang, W.; Zhang, G.F.; Xu, Q.; Ge, H.M.; et al. Immunosuppressive polyketides from mantis-associated *Daldinia eschscholzii*. *J. Am. Chem. Soc.* **2011**, *133*, 5931–5940. [CrossRef]
17. Zhang, Y.L.; Ge, H.M.; Zhao, W.; Dong, H.; Xu, Q.; Li, S.H.; Li, J.; Zhang, J.; Song, Y.C.; Tan, R.X. Unprecedented immunosuppressive polyketides from *Daldinia eschscholzii*, a mantis-associated fungus. *Angew. Chem. Int. Ed. Engl.* **2008**, *47*, 5823–5826. [CrossRef]
18. Zhang, A.H.; Tan, R.; Jiang, N.; Yusupu, K.; Wang, G.; Wang, X.L.; Tan, R.X. Selesconol, a fungal polyketide that induces stem cell differentiation. *Org. Lett.* **2016**, *18*, 5488–5491. [CrossRef]
19. Zhou, Z.Z.; Zhu, H.J.; Lin, L.P.; Zhang, X.; Ge, H.M.; Jiao, R.H.; Tan, R.X. Dalmanol biosyntheses require coupling of two separate polyketide gene clusters. *Chem. Sci.* **2018**, *10*, 73–82. [CrossRef]
20. Hertweck, C. The biosynthetic logic of polyketide diversity. *Angew. Chem. Int. Ed. Engl.* **2009**, *48*, 4688–4716. [CrossRef]
21. Abe, I.; Oguro, S.; Utsumi, Y.; Sano, Y.; Noguchi, H. Engineered biosynthesis of plant polyketides: Chain length control in an octaketide-producing plant type III polyketide synthase. *J. Am. Chem. Soc.* **2005**, *127*, 12709–12716. [CrossRef] [PubMed]
22. Fang, W.; Ji, S.; Jiang, N.; Wang, W.; Zhao, G.Y.; Zhang, S.; Ge, H.M.; Xu, Q.; Zhang, A.H.; Zhang, Y.L.; et al. Naphthol radical couplings determine structural features and enantiomeric excess of dalesconols in *Daldinia eschscholzii*. *Nat. Commun.* **2012**, *3*, 1039. [CrossRef] [PubMed]
23. Bowen, J.I.; Wang, L.; Crump, M.P.; Willis, C.L. Core steps to the azaphilone family of fungal natural products. *ChemBioChem* **2021**, *22*, 3027–3036.
24. Huang, X.; Zhang, W.; Tang, S.; Wei, S.; Lu, X. Collaborative biosynthesis of a class of bioactive azaphilones by two separate gene clusters containing four PKS/NRPSs with transcriptional crosstalk in fungi. *Angew. Chem. Int. Ed. Engl.* **2020**, *59*, 4349–4353. [CrossRef]
25. Wang, M.; Beissner, M.; Zhao, H. Aryl-aldehyde formation in fungal polyketides: Discovery and characterization of a distinct biosynthetic mechanism. *Chem. Biol.* **2014**, *21*, 257–263. [CrossRef] [PubMed]
26. Chiang, Y.M.; Szewczyk, E.; Davidson, A.D.; Keller, N.; Oakley, B.R.; Wang, C.C. A gene cluster containing two fungal polyketide synthases encodes the biosynthetic pathway for a polyketide, asperfuranone, in *Aspergillus nidulans*. *J. Am. Chem. Soc.* **2009**, *131*, 2965–2970. [CrossRef] [PubMed]
27. Chiang, Y.M.; Oakley, C.E.; Ahuja, M.; Entwistle, R.; Schultz, A.; Chang, S.L.; Sung, C.T.; Wang, C.C.; Oakley, B.R. An efficient system for heterologous expression of secondary metabolite genes in *Aspergillus nidulans*. *J. Am. Chem. Soc.* **2013**, *135*, 7720–7731. [CrossRef]
28. Zhao, M.; Tang, Y.; Xie, J.; Zhao, Z.; Cui, H. Meroterpenoids produced by fungi: Occurrence, structural diversity, biological activities, and their molecular targets. *Eur. J. Med. Chem.* **2021**, *209*, 112860. [CrossRef]
29. Moodie, L.W.K.; Sepčić, K.; Turk, T.; FrangeŽ, R.; Svenson, J. Natural cholinesterase inhibitors from marine organisms. *Nat. Prod. Rep.* **2019**, *36*, 1053–1092. [CrossRef]
30. Omura, S.; Tomoda, H.; Kim, Y.K.; Nishida, H. Pyripyropenes, highly potent inhibitors of acyl-CoA:cholesterol acyltransferase produced by *Aspergillus fumigatus*. *J. Antibiot.* **1993**, *46*, 1168–1169. [CrossRef]
31. Barra, L.; Abe, I. Chemistry of fungal meroterpenoid cyclases. *Nat. Prod. Rep.* **2021**, *38*, 566–585. [CrossRef] [PubMed]
32. Szewczyk, E.; Chiang, Y.M.; Oakley, C.E.; Davidson, A.D.; Wang, C.C.; Oakley, B.R. Identification and characterization of the asperthecin gene cluster of *Aspergillus nidulans*. *Appl. Environ. Microbiol.* **2008**, *74*, 7607–7612. [CrossRef]
33. Nielsen, M.L.; Nielsen, J.B.; Rank, C.; Klejnstrup, M.L.; Holm, D.K.; Brogaard, K.H.; Hansen, B.G.; Frisvad, J.C.; Larsen, T.O.; Mortensen, U.H. A genome-wide polyketide synthase deletion library uncovers novel genetic links to polyketides and meroterpenoids in *Aspergillus nidulans*. *FEMS Microbiol. Lett.* **2011**, *321*, 157–166. [CrossRef] [PubMed]
34. Lo, H.C.; Entwistle, R.; Guo, C.J.; Ahuja, M.; Szewczyk, E.; Hung, J.H.; Chiang, Y.M.; Oakley, B.R.; Wang, C.C. Two separate gene clusters encode the biosynthetic pathway for the meroterpenoids austinol and dehydroaustinol in *Aspergillus nidulans*. *J. Am. Chem. Soc.* **2012**, *134*, 4709–4720. [CrossRef] [PubMed]
35. Matsuda, Y.; Awakawa, T.; Wakimoto, T.; Abe, I. Spiro-ring formation is catalyzed by a multifunctional dioxygenase in austinol biosynthesis. *J. Am. Chem. Soc.* **2013**, *135*, 10962–10965. [CrossRef]
36. Matsuda, Y.; Wakimoto, T.; Mori, T.; Awakawa, T.; Abe, I. Complete biosynthetic pathway of anditomin: Nature's sophisticated synthetic route to a complex fungal meroterpenoid. *J. Am. Chem. Soc.* **2014**, *136*, 15326–15336. [CrossRef]
37. Zhang, X.; Wang, T.T.; Xu, Q.L.; Xiong, Y.; Zhang, L.; Han, H.; Xu, K.; Guo, W.J.; Xu, Q.; Tan, R.X.; et al. Genome mining and comparative biosynthesis of meroterpenoids from two phylogenetically distinct fungi. *Angew. Chem. Int. Ed. Engl.* **2018**, *57*, 8184–8188. [CrossRef]
38. Matsuda, Y.; Quan, Z.; Mitsuhashi, T.; Li, C.; Abe, I. Cytochrome P450 for citreohybridonol synthesis: Oxidative derivatization of the andrastin scaffold. *Org. Lett.* **2016**, *18*, 296–299. [CrossRef]
39. Xu, P.W.; Yu, J.S.; Chen, C.; Cao, Z.Y.; Zhou, F.; Zhou, J. Catalytic enantioselective construction of spiro quaternary carbon stereocenters. *ACS Catalysis* **2019**, *9*, 1820–1882. [CrossRef]

40. Zi, W.; Zuo, Z.; Ma, D. Intramolecular dearomative oxidative coupling of indoles: A unified strategy for the total synthesis of indoline alkaloids. *Acc. Chem. Res.* **2015**, *48*, 702–711. [CrossRef]
41. Cui, C.B.; Kakeya, H.; Osada, H. Novel mammalian cell cycle inhibitors, spirotryprostatins A and B, produced by *Aspergillus fumigatus*, which inhibit mammalian cell cycle at G2/M phase. *Tetrahedron* **1996**, *52*, 12651–12666. [CrossRef]
42. Tsunematsu, Y.; Ishikawa, N.; Wakana, D.; Goda, Y.; Noguchi, H.; Moriya, H.; Hotta, K.; Watanabe, K. Distinct mechanisms for spiro-carbon formation reveal biosynthetic pathway crosstalk. *Nat. Chem. Biol.* **2013**, *9*, 818–825. [CrossRef] [PubMed]
43. Yin, W.B.; Baccile, J.A.; Bok, J.W.; Chen, Y.; Keller, N.P.; Schroeder, F.C. A nonribosomal peptide synthetase-derived iron(III) complex from the pathogenic fungus *Aspergillus fumigatus*. *J. Am. Chem. Soc.* **2013**, *135*, 2064–2067. [CrossRef] [PubMed]
44. Yan, D.; Chen, Q.; Gao, J.; Bai, J.; Liu, B.; Zhang, Y.; Zhang, L.; Zhang, C.; Zou, Y.; Hu, Y. Complexity and diversity generation in the biosynthesis of fumiquinazoline-related peptidyl alkaloids. *Org. Lett.* **2019**, *21*, 1475–1479. [CrossRef] [PubMed]
45. Gao, X.; Chooi, Y.H.; Ames, B.D.; Wang, P.; Walsh, C.T.; Tang, Y. Fungal indole alkaloid biosynthesis: Genetic and biochemical investigation of the tryptoquialanine pathway in *Penicillium aethiopicum*. *J. Am. Chem. Soc.* **2011**, *133*, 2729–2741. [CrossRef]
46. Haynes, S.W.; Ames, B.D.; Gao, X.; Tang, Y.; Walsh, C.T. Unraveling terminal C-domain-mediated condensation in fungal biosynthesis of imidazoindolone metabolites. *Biochemistry* **2011**, *50*, 5668–5679. [CrossRef] [PubMed]
47. Li, S.; Finefield, J.M.; Sunderhaus, J.D.; McAfoos, T.J.; Williams, R.M.; Sherman, D.H. Biochemical characterization of NotB as an FAD-dependent oxidase in the biosynthesis of notoamide indole alkaloids. *J. Am. Chem. Soc.* **2012**, *134*, 788–791. [CrossRef]
48. Cacho, R.A.; Jiang, W.; Chooi, Y.H.; Walsh, C.T.; Tang, Y. Identification and characterization of the echinocandin B biosynthetic gene cluster from *Emericella rugulosa* NRRL 11440. *J. Am. Chem. Soc.* **2012**, *134*, 16781–16790. [CrossRef] [PubMed]
49. Zhong, L.; Diao, X.; Zhang, N.; Li, F.; Zhou, H.; Chen, H.; Bai, X.; Ren, X.; Zhang, Y.; Wu, D.; et al. Engineering and elucidation of the lipoinitiation process in nonribosomal peptide biosynthesis. *Nat. Commun.* **2021**, *12*, 296. [CrossRef]
50. Kegler, C.; Bode, H.B. Artificial splitting of a non-ribosomal peptide synthetase by inserting natural docking domains. *Angew. Chem. Int. Ed. Engl.* **2020**, *59*, 13463–13467. [CrossRef]
51. Bozhueyuek, K.A.J.; Watzel, J.; Abbood, N.; Bode, H.B. Synthetic zippers as an enabling tool for engineering of non-ribosomal peptide synthetases. *Angew. Chem. Int. Ed. Engl.* **2021**, *60*, 17531–17538. [CrossRef] [PubMed]
52. Luo, J.G.; Wang, X.B.; Xu, Y.M.; U'Ren, J.M.; Arnold, A.E.; Kong, L.Y.; Gunatilaka, A.A. Delitschiapyrone A, a pyrone-naphthalenone adduct bearing a new pentacyclic ring system from the leaf-associated fungus *Delitschia* sp. FL1581. *Org. Lett.* **2014**, *16*, 5944–5947. [CrossRef] [PubMed]
53. Zou, Y.; Houk, K.N. Mechanisms and dynamics of synthetic and biosynthetic formation of delitschiapyrones: Solvent control of ambimodal periselectivity. *J. Am. Chem. Soc.* **2021**, *143*, 11734–11740. [CrossRef] [PubMed]
54. Han, W.B.; Wang, G.Y.; Tang, J.J.; Wang, W.J.; Liu, H.; Gil, R.R.; Navarro-Vázquez, A.; Lei, X.; Gao, J.M. Herpotrichones A and B, two intermolecular [4 + 2] adducts with anti-neuroinflammatory activity from a *Herpotrichia* species. *Org. Lett.* **2020**, *22*, 405–409. [CrossRef] [PubMed]
55. Zhao, Z.; Ying, Y.; Hung, Y.S.; Tang, Y. Genome mining reveals *Neurospora crassa* can produce the salicylaldehyde sordarial. *J. Nat. Prod.* **2019**, *82*, 1029–1033. [CrossRef]
56. Nies, J.; Ran, H.; Wohlgemuth, V.; Yin, W.B.; Li, S.M. Biosynthesis of the prenylated salicylaldehyde flavoglaucin requires temporary reduction to salicyl alcohol for decoration before reoxidation to the final product. *Org. Lett.* **2020**, *22*, 2256–2260. [CrossRef]
57. Liu, L.; Tang, M.C.; Tang, Y. Fungal highly reducing polyketide synthases biosynthesize salicylaldehydes that are precursors to epoxycyclohexenol natural products. *J. Am. Chem. Soc.* **2019**, *141*, 19538–19541. [CrossRef]
58. Lin, T.S.; Chiang, Y.M.; Wang, C.C. Biosynthetic pathway of the reduced polyketide product citreoviridin in *Aspergillus terreus* var. *aureus* revealed by heterologous expression in *Aspergillus nidulans*. *Org. Lett.* **2016**, *18*, 1366–1369. [CrossRef]
59. Li, H.; Hu, J.; Wei, H.; Solomon, P.S.; Vuong, D.; Lacey, E.; Stubbs, K.A.; Piggott, A.M.; Chooi, Y.H. Chemical ecogenomics-guided discovery of phytotoxic α-pyrones from the fungal wheat pathogen *Parastagonospora nodorum*. *Org. Lett.* **2018**, *20*, 6148–6152. [CrossRef]
60. Yin, G.P.; Wu, Y.R.; Yang, M.H.; Li, T.X.; Wang, X.B.; Zhou, M.M.; Lei, J.L.; Kong, L.Y. Citrifurans A-D, four dimeric aromatic polyketides with new carbon skeletons from the fungus *Aspergillus* sp. *Org. Lett.* **2017**, *19*, 4058–4061. [CrossRef]
61. Wang, W.G.; Wang, H.; Du, L.Q.; Li, M.; Chen, L.; Yu, J.; Cheng, G.G.; Zhan, M.T.; Hu, Q.F.; Zhang, L.; et al. Molecular Basis for the Biosynthesis of an Unusual Chain-Fused Polyketide, Gregatin A. *J. Am. Chem. Soc.* **2020**, *142*, 8464–8472. [CrossRef] [PubMed]
62. He, Y.; Cox, R.J. The molecular steps of citrinin biosynthesis in fungi. *Chem. Sci.* **2016**, *7*, 2119–2127. [CrossRef] [PubMed]
63. Tong, Z.W.; Xie, X.H.; Wang, T.T.; Lu, M.; Jiao, R.H.; Ge, H.M.; Hu, G.; Tan, R.X. Acautalides A-C, neuroprotective Diels-Alder adducts from solid-state cultivated *Acaulium* sp. H-JQSF. *Org. Lett.* **2021**, *23*, 5587–5591. [CrossRef] [PubMed]
64. Wang, T.T.; Wei, Y.J.; Ge, H.M.; Jiao, R.H.; Tan, R.X. Acaulide, an osteogenic macrodiolide from *Acaulium* sp. H-JQSF, an isopod-associated fungus. *Org. Lett.* **2018**, *20*, 1007–1010. [CrossRef] [PubMed]
65. Wang, T.T.; Wei, Y.J.; Ge, H.M.; Jiao, R.H.; Tan, R.X. Acaulins A and B, trimeric macrodiolides from *Acaulium* sp. H-JQSF. *Org. Lett.* **2018**, *20*, 2490–2493. [CrossRef]
66. Jensen, P.R. Natural products and the gene cluster revolution. *Trends Microbiol.* **2016**, *24*, 968–977. [CrossRef]
67. Pradeepraj, D.; Li, S.Y. Functional expression and regulation of eukaryotic cytochrome P450 enzymes in surrogate microbial cell factories. *Eng. Microbiol.* **2022**, *2*, 100011. [CrossRef]

68. Kahlert, L.; Bassiony, E.F.; Cox, R.J.; Skellam, E.J. Diels-Alder reactions during the biosynthesis of sorbicillinoids. *Angew. Chem. Int. Ed. Engl.* **2020**, *59*, 5816–5822. [CrossRef]
69. Han, Y.B.; Bai, W.; Ding, C.X.; Liang, J.; Wu, S.H.; Tan, R.X. Intertwined biosynthesis of skyrin and rugulosin A underlies the formation of cage-structured bisanthraquinones. *J. Am. Chem. Soc.* **2021**, *143*, 14218–14226. [CrossRef]
70. Hu, J.; Li, H.; Chooi, Y.H. Fungal dirigent protein controls the stereoselectivity of multicopper oxidase-catalyzed phenol coupling in viriditoxin biosynthesis. *J. Am. Chem. Soc.* **2019**, *141*, 8068–8072. [CrossRef]

Journal of *Fungi*

MDPI

Article

Immunosuppressive Sesquiterpenoids from the Edible Mushroom *Craterellus odoratus*

Quan Dai [1,2], Fa-Lei Zhang [1,2], Zheng-Hui Li [1,2], Juan He [1,2,*] and Tao Feng [1,2,*]

1 School of Pharmaceutical Sciences, South-Central University for Nationalities, Wuhan 430074, China; quandai@mail.scuec.edu.cn (Q.D.); flzhang@mail.scuec.edu.cn (F.-L.Z.); 2015051@mail.scuec.edu.cn (Z.-H.L.)
2 National Demonstration Center for Experimental Ethnopharmacology Education, South-Central University for Nationalities, Wuhan 430074, China
* Correspondence: 2015048@mail.scuec.edu.cn (J.H.); tfeng@mail.scuec.edu.cn (T.F.)

Abstract: The aim of this work was to comprehensively understand the chemical constituents of the edible mushroom *Craterellus ordoratus* and their bioactivity. A chemical investigation on this mushroom led to the isolation of 23 sesquiterpenoids including eighteen previously undescribed bergamotane sesquiterpenes, craterodoratins A–R (**1–18**), and one new victoxinine derivative, craterodoratin S (**19**). The new structures were elucidated by detailed interpretation of spectrometric data, theoretical nuclear magnetic resonance (NMR) and electronic circular dichroism (ECD) calculations, and single-crystal X-ray crystallographic analysis. Compounds **1** and **2** possess a ring-rearranged carbon skeleton. Compounds **3**, **10**, **12–15**, **19**, **20** and **23** exhibit potent inhibitory activity against the lipopolysaccharide (LPS)-induced proliferation of B lymphocyte cells with the IC_{50} values ranging from 0.67 to 22.68 μM. Compounds **17** and **20** inhibit the concanavalin A (ConA)-induced proliferation of T lymphocyte cell with IC_{50} values of 31.50 and 0.98 μM, respectively. It is suggested that *C. ordoratus* is a good source for bergamotane sesquiterpenoids, and their immunosuppressive activity was reported for the first time. This research is conducive to the further development and utilization of *C. ordoratus*.

Keywords: *Craterellus ordoratus*; sesquiterpenoids; isolation and structural elucidation; immunosuppressive activity

Citation: Dai, Q.; Zhang, F.-L.; Li, Z.-H.; He, J.; Feng, T. Immunosuppressive Sesquiterpenoids from the Edible Mushroom *Craterellus odoratus*. *J. Fungi* **2021**, *7*, 1052. https://doi.org/10.3390/jof7121052

Academic Editor: Gary A. Strobel

Received: 13 November 2021
Accepted: 7 December 2021
Published: 8 December 2021

1. Introduction

Edible mushrooms are a large and fascinating group of fungi. Many species of wild edible mushrooms are valued ingredients due to their unique taste and short growth cycle. Yunnan Province is located in southwest China. Its unique climate and geological diversity make this area very favorable for the growth of wild mushrooms. It is estimated that more than 40% of the world's and 90% of Chinese edible mushrooms (about 900 species) grow in Yunnan [1,2]. Many wild edible mushrooms are regarded as local delicacies such as *Tricholoma matsutake*, *Collybia albuminosa*, *Cantharellus cibarius*, and several species of the genus *Boletus*. Studying the chemical constituents of these edible mushrooms, therefore, has become our long-term research project [2–4]. The systematic mining of chemical components and evaluation of their biological activity will be beneficial to the scientific development and utilization of these edible fungi.

Species of the genus *Craterellus* (Cantharellaceae) include well-known edible mushrooms. At present, 142 records of *Craterellus* have been found and about 74 species are currently recognized as members of the genus. Ten species are originally described from Asia and four of these species have been reported in China [5]. Of them, *C. odoratus* is an edible mushroom in the family widespread in mainland China and characterized by possessing a bright orange or yellow cap [5–8]. In China, this wild edible mushroom is especially popular in Yunnan Province, where it is a delicacy on the dinner table from

June to August each year. In northern Thailand, *C. odoratus* is also a wild edible mushroom. Its nutritive value including the content of protein, fat, crude fiber, carbohydrate, and mineral contents was assessed. From the perspective of secondary metabolites, this mushroom has not been systematically studied, only a few polyketides and terpenoids have been reported [9–13]. Therefore, it is necessary to conduct systematic chemical component mining and biological activity evaluation on this mushroom. In this study, a total of 23 sesquiterpenoids (**1–23**, Figure 1) including eighteen new bergamotane-type ones, namely craterodoratins A–R (**1–18**), and one new victoxinine derivative, namely craterodoratin S (**19**), were isolated from *C. odoratus*. Their structures with absolute configurations were elucidated by extensive spectroscopic methods (including 1D and 2D NMR, MS, UV, and IR technologies), single-crystal X-ray diffraction analysis, as well as NMR and ECD calculations.

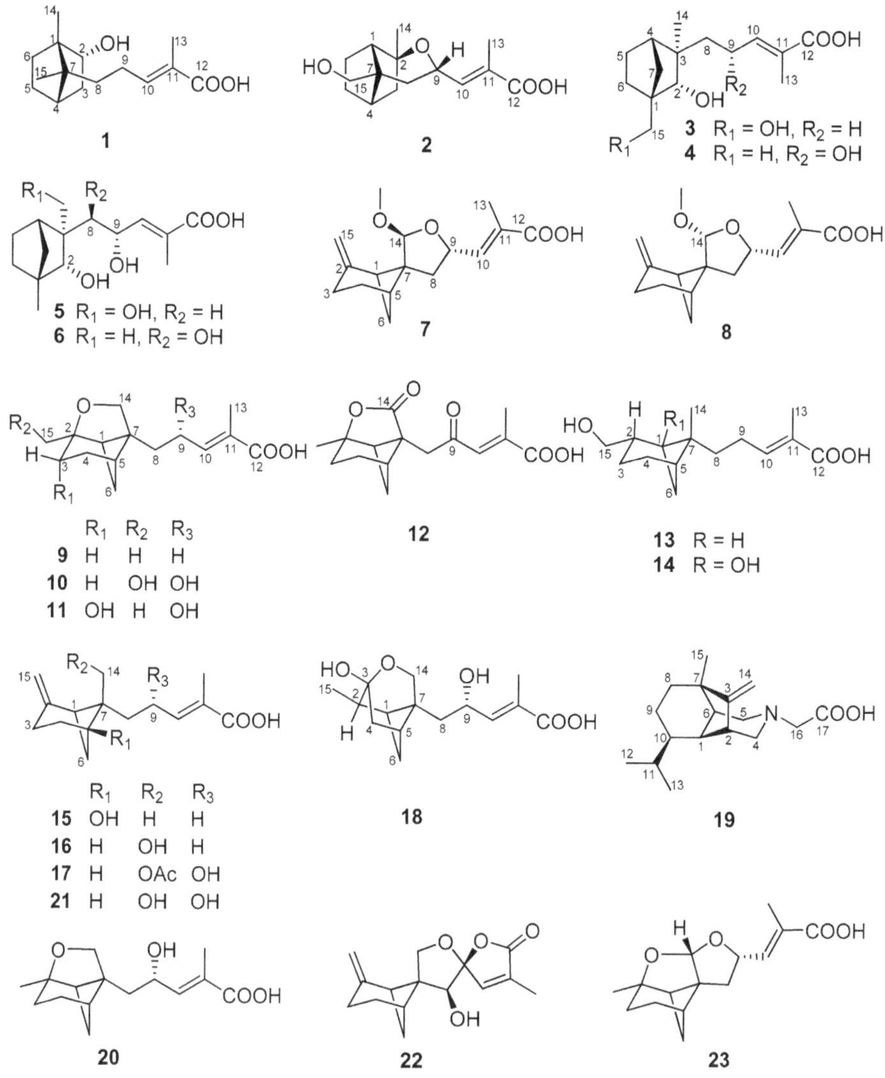

Figure 1. Chemical structures of compounds **1–23**.

Previous pharmacological studies on bergamotane metabolites have demonstrated α-glucosidase inhibitory activity [14], antibacterial activity [9,15,16], inhibition of pancreatic lipase [17], and phytotoxic effects against Johnson grass and Sorghum [18–21]. In this study, all compounds were evaluated for their cytotoxic activities against five human cancer cell lines, and for their immunosuppressive activity on T-cell and B-cell proliferation. Herein the isolation, structural elucidation, and bioactivities of these isolates are reported.

2. Materials and Methods

2.1. General Experimental Procedures

IR spectra were obtained on a Shimadzu Fourier transform infrared spectrometer using KBr pellets. UV spectra were obtained by using a double beam spectrophotometer UH5300 (Hitachi High-Technologies, Tokyo, Japan). Optical rotations were measured on a Rudolph Autopol IV polarimeter (Hackettstown, NJ, USA). High resolution electrospray ionization mass spectra (HRESIMS) were recorded on an Agilent 6200 Q-TOF MS system or a Thermo Scientific Q Exactive Orbitrap MS system. Circular dichroism (CD) spectra were measured with an Applied Photophysics spectrometer (Chirascan, New Haven, CT, USA). NMR spectra were recorded with a Bruker Avance III 600 MHz spectrometer (Bruker, Karlsruhe, Germany). Sephadex LH-20 (GE Healthcare, Pittsburgh, PA, USA), Silica gel (200–300 mesh), and RP-18 gel (20–45 μM, FuJi) were used for column chromatography (CC). HPLC was performed on an Agilent 1260 liquid chromatography system equipped with Zorbax SB-C18 columns (5 μM, 9.4 mm × 150 mm, or 21.2 mm × 150 mm).

2.2. Fungal Material

The fungus *Craterellus odoratus* was collected from the Southern part of the Gaoligong Mountains in Yunnan Province, China, in July 2007. The fungus was identified by Prof. Mu Zang at the Kunming Institute of Botany. A voucher specimen (HFC2007-20180714-DQ1) has been deposited in the School of Pharmaceutical Sciences, South-Central University for Nationalities.

2.3. Fermentation, Extraction and Isolation

The rice culture medium was composed of glucose 5%, yeast 5%, pork peptone 0.15%, KH_2PO_4 0.05%, and $MgSO_4$ 0.05%. The initial pH was adjusted to 6.0. Cultures were grown in an Erlenmeyer flask (220 rpm, 24 °C) for 6 days until the mycelium biomass reached a maximum. This was then transferred to a rice medium and incubated at 24 °C in the dark for 40 days. The rice medium contained 50 g of rice and 50 mL of water, in a 250 mL Erlenmeyer flask, sterilized at 121 °C for 15 min. A total of 400 flasks were used in this work.

The solid rice culture broth of *C. odoratus* (20 kg) was extracted six time with MeOH to give a crude extract. The extract was partitioned between water and ethyl acetate (EtOAc). The EtOAc layer was concentrated under reduced pressure to give an organic extract (167 g). It was subjected to CC over silica gel (200–300 mesh) eluted with a solvent system of $CHCl_3$/MeOH (from 1:0 to 0:1, v/v) to obtain nine fractions A–I. Fraction D (20 g) was fractionated by MPLC over RP-18 silica gel eluted with MeOH/H_2O (from 5:95 to 100:0, v/v) to give 13 subfractions (D_1–D_{13}). Fraction D_6 (5 g) was repeatedly fractionated by CC over silica gel eluted with $CHCl_3$/MeOH (10:1) to give compounds **9** (7 mg), **10** (4 mg), and **11** (18 mg). Fraction D_4 (3 g) was fractionated by CC over Sephadex LH-20 (MeOH) and then purified by prep-HPLC (CH_3CN/H_2O 35:65 in 30 min) to give compounds **1** (3.0 mg, retention time (t_R) = 16.3 min), **8** (3.2 mg, t_R = 20.1 min), and **4** (3.4 mg, t_R = 24.7 min). Fraction D_3 (5 g) was fractionated by CC over Sephadex LH-20 (MeOH) and then purified by prep-HPLC (CH_3CN/H_2O 32:68 in 25 min) to give compounds **17** (2.3 mg, t_R = 13.2 min), **20** (3.0 mg, t_R = 15.4 min), and **18** (2.6 mg, t_R = 22.0 min). Fraction E (9 g) was fractionated by MPLC over RP-18 eluted with MeOH/H_2O (from 5:95 to 100:0, v/v) to give seven subfractions (E_1–E_7). Fraction E_3 (800 mg) was subjected to CC over silica gel (80–100 mesh), Sephadex LH-20 (MeOH), and then purified by prep-HPLC

(CH_3CN/H_2O from 25:75 to 61:39 in 25 min) to give compounds **3** (2.1 mg, t_R = 11.2 min), **6** (3 mg, t_R = 15.3 min), **7** (2.5 mg, t_R = 17.9 min), and **16** (3.8 mg, t_R = 23.1 min). Fraction E_5 (36 mg) was prepared by prep-HPLC (CH_3CN/H_2O 30:70 in 25 min) to give compounds **14** (2.1 mg, t_R = 14.8 min), **13** (2.4 mg, t_R = 18.6 min), and **15** (2.6 mg, t_R = 23.0 min). Fraction F (7 g) was separated by MPLC over RP-18 to give five subfractions F_1–F_5. Fraction F_3 (23 mg) was prepared by prep-HPLC (CH_3CN/H_2O 38:62 in 25 min) to give compounds **2** (2.3 mg, t_R = 15.1 min), **12** (2.0 mg, t_R = 17.6 min), and **5** (3.1 mg, t_R = 20.3 min). Fraction F_4 (30 mg) was prepared by prep-HPLC (CH_3CN/H_2O 40:60 in 25 min) to give compounds **21** (2.3 mg, t_R = 13.0 min), **22** (1.8 mg, t_R = 17.0 min), and **19** (3.1 mg, t_R = 21.3 min). Fraction G (1.2 g) was fractionated by CC over silica gel eluted with $CHCl_3/CH_3OH$ (10:1) to give subfractions G_1-G_6. Compound **23** (3.2 mg) precipitated as acicular crystals from fraction G_3.

Craterodoratin A (**1**). Colorless oil; $[\alpha]_D^{21}$ −1.9 (*c* 0.11, MeOH); UV (MeOH) λ_{max} (log ε) 210 (1.04) nm; IR (KBr) ν_{max} 3365, 2945, 2833, 1660, 1454, 1114, 1031 cm^{-1}; ^1H and ^{13}C NMR spectroscopic data, see Table 1; HRESIMS *m/z* 253.17986 $[M + H]^+$ (calcd for $C_{15}H_{25}O_3^+$, 253.17982).

Table 1. ^1H (600 MHz) and ^{13}C (150 MHz) NMR Spectroscopic Data for **1–4** in Methanol-d_4.

Entry	1		2		3		4	
	δ_C, Type	δ_H (J in Hz)	δ_C, Type	δ_H (J in Hz)	δ_C, Type	δ_H (J in Hz)	δ_C, Type	δ_H (J in Hz)
1	49.9, C	–	43.5, CH	2.25, d (4.3)	55.8, C	–	41.5, C	–
2	76.3, CH	4.02, d (9.1)	81.8, C	–	79.7, CH	3.51, s	83.5, CH	3.44, s
3a 3b	37.6, CH_2	2.16, m 1.00, dd, (13.3, 3.3)	48.2, CH_2	1.98, dd, (14.3, 4.3) 1.28, d, (13.7)	43.2, C	–	48.5, C	–
4	41.9, CH	1.79, m	42.3, CH	1.86, t, (4.3)	47.1, CH	1.83, m	46.5, CH	1.80, s
5a 5b	27.5, CH_2	1.69, m 1.27, m	20.3, CH_2	1.73, m 1.51, m	26.5, CH_2	1.70, m 1.39, m	25.8, CH_2	1.68, m 1.41, m
6a 6b	25.8, CH_2	1.97, m 1.23, m	26.5, CH_2	1.72, m 1.16, m	21.9, CH_2	1.62, m 1.10, m	25.0, CH_2	1.69, m 0.99, m
7a 7b	50.9, C	–	52.5, C	–	37.2, CH_2	1.54, d, (10.1) 1.11, m	40.4, CH_2	1.48, m 1.11, d, (10.3)
8a 8b	30.9, CH_2	1.46, td, (12.8, 12.3, 5.1) 1.17, td, (12.8, 12.3, 5.1)	33.7, CH_2	1.98, dd, (14.3, 4.3) 1.74, m	42.3, CH_2	1.36, m	48.9, CH_2	1.71, m 1.45, m
9a 9b	24.5, CH_2	2.25, m 2.07, m	66.4, CH	4.64, m	24.7, CH_2	2.15, m	65.3, CH	4.58, m
10	141.1, CH	6.69, t, (7.4)	141.0, CH	6.48, d, (7.8)	144.1, CH	6.72, t, (7.4)	144.6, CH	6.57, d, (9.28)
11	128.5, C	–	129.4, C	–	128.8, C	–	126.6, C	–
12	171.3, C	–	171.2, C	–	172.0, C	–	170.4, C	–
13	11.2, CH_3	1.80, s	11.8, CH_3	1.84, s	12.3, CH_3	1.78, s	11.4, CH_3	1.84, d, (1.5)
14	12.3, CH_3	0.85, s	22.4, CH_3	1.25, s	17.0, CH_3	0.86, s	18.5, CH_3	1.07, s
15a 15b	15.5, CH_3	0.93, s	63.4, CH_2	3.64, d, (11.0) 3.47, d, (11.0)	65.1, CH_2	3.59, d, (11.0) 3.51, d, (11.0)	16.1, CH_3	0.91, s

Craterodoratin B (**2**). Colorless crystals (MeOH), mp 138.2–140.3 °C; $[\alpha]_D^{21}$ − 11.4 (*c* 0.09, MeOH); UV (MeOH) λ_{max} (log ε) 215 (1.88) nm; IR (KBr) ν_{max} 3442, 2951, 2841, 1647, 1018 cm^{-1}; ^1H and ^{13}C NMR spectroscopic data, see Table 1; HRESIMS *m/z* 267.15903 $[M + H]^+$ (calcd for $C_{15}H_{23}O_4^+$, 267.15909).

Craterodoratin C (**3**). Colorless crystals (MeOH), mp 180.7–182.4 °C; $[\alpha]_D^{21}$ − 14.5 (*c* 0.11, MeOH); UV (MeOH) λ_{max} (log ε) 230 (3.12) nm; IR (KBr) ν_{max} 3464, 2924, 1647, 1280, 1265 cm^{-1}; ^{1}H and ^{13}C NMR spectroscopic data, see Table 1; HRESIMS *m*/*z* 291.15650 [M + Na]$^+$ (calcd for $C_{15}H_{24}O_4Na^+$, 291.15668).

Craterodoratin D (**4**). Colorless crystals (MeOH), mp 159.1–162.5 °C; $[\alpha]_D^{21}$ + 32 (*c* 0.10, MeOH); UV (MeOH) λ_{max} (log ε) 225 (3.02) nm; IR (KBr) ν_{max} 3468, 2967, 2541, 1647, 1016 cm^{-1}; ^{1}H and ^{13}C NMR spectroscopic data, see Table 1; HRESIMS *m*/*z* 267.14712 [M − H]$^-$ (calcd for $C_{15}H_{23}O_4{}^-$, 267.16018).

Craterodoratin E (**5**). Colorless crystals (MeOH), mp 140.9–142.4 °C; $[\alpha]_D^{21}$ + 6.9 (*c* 0.10, MeOH); UV (MeOH) λ_{max} (log ε) 210 (1.27) nm; IR (KBr) ν_{max} 3408, 2951, 2841, 1651, 1018 cm^{-1}; ^{1}H and ^{13}C NMR spectroscopic data, see Table 2; HRESIMS *m*/*z* 307.15146 [M + Na]$^+$ (calcd for $C_{15}H_{24}O_5Na^+$, 307.15159).

Table 2. ^{1}H (600 MHz) and ^{13}C (150 MHz) NMR Spectroscopic Data for **5**–**8** in Methanol-d_4.

Entry	5		6		7		8	
	δ_C, Type	δ_H (*J* in Hz)	δ_C, Type	δ_H (*J* in Hz)	δ_C, Type	δ_H (*J* in Hz)	δ_C, Type	δ_H (*J* in Hz)
1	49.9, C	–	49.8, C	–	52.0, CH	2.64, t, (5.3, 10.8)	49.1, CH	2.79, t, (5.4)
2	86.0, CH	3.34, d (1.3)	82.2, CH	3.47, d (1.3)	150.0, C	–	151.0, C	–
3a 3b	47.3, C	–	47.8, C	–	23.1, CH_2	2.61, m 2.33, m	24.3, CH_2	2.65, m 2.40, m
4a 4b	44.0, CH	2.06, d, (3.8)	44.7, CH	2.23, d, (3.5)	23.5, CH_2	2.11, m 1.82, m	24.7, CH_2	1.96, m
5a 5b	27.0, CH_2	1.71, m 1.47, m	27.6, CH_2	1.65, m 1.39, m	39.4, CH	2.32, m	40.6, CH	2.27, m
6a 6b	26.7, CH_2	1.72, m 0.99, m	26.7, CH_2	1.66, m 0.99, m	26.5, CH_2	2.20, m 1.47, d, (9.7)	29.1, CH_2	2.28, m 1.42, m
7a 7b	41.9, CH_2	1.38, d, (10.6) 1.13, d, (10.6)	41.5, CH_2	1.47, d, (9.7) 1.07, d, (9.7)	54.7, C	–	56.5, C	–
8a 8b	47.0, CH_2	1.70, m 1.53, dd, (2.6, 14.8)	80.5, CH	3.39, d, (6.0)	40.4, CH_2	2.50, dd, (12.4, 9.7) 2.02, dd, (12.4, 4.2)	39.4, CH_2	2.46, dd, (6.7, 11.9) 1.92, dd, (9.5, 11.9)
9	66.6, CH	4.60, m	70.6, CH	4.38, dd, (9.1, 6.0)	73.7, CH	4.91, m	75.7, CH	4.90, s
10	142.6, CH	6.49, d (8.2)	138.4, CH	6.53, d (8.6)	143.0, CH	6.74, dd, (1.5, 8.0)	144.2, CH	6.65, d, (8.2)
11	131.5, C	–	135.7, C	–	128.1, C	–	130.2, C	–
12	174.9, C	–	176.1, C	–	169.8, C	–	171.3, C	–
13	13.1, CH_3	1.80, s	14.3, CH_3	1.86, s	11.4, CH_3	1.85, d, (1.5)	12.8, CH_3	1.85, s
14a 14b	64.0, CH_2	3.64, d, (11.7) 3.53, d, (11.7)	19.9, CH_3	1.04, s	106.4, CH	4.59, s	106.8, CH	4.57, s
15a 15b	19.4, CH_3	1.03, s	12.2, CH_3	0.81, s	107.0, CH_2	4.74, q, (1.8) 4.68, dt, (3.0, 1.5)	107.8, CH_2	4.66, s 4.63, s
OCH_3	–	–	–	–	53.3, CH_3	3.29, s	55.4, CH_3	3.28, s

Craterodoratin F (**6**). Colorless oil; $[\alpha]_D^{21}$ + 18 (*c* 0.10, MeOH); UV (MeOH) λ_{max} (log ε) 210 (0.84) nm; IR (KBr) ν_{max} 3400, 2949, 2837, 1653, 1456, 1411, 1112, 1024 cm^{-1}; ^{1}H and ^{13}C NMR spectroscopic data, see Table 2; HRESIMS *m*/*z* 307.15161 [M + Na]$^+$ (calcd for $C_{15}H_{24}O_5Na^+$, 307.15159).

Craterodoratin G (**7**). Colorless oil; $[\alpha]_D^{21}$ + 1.10 (*c* 0.10, MeOH); UV (MeOH) λ_{max} (log ε) 210 (1.74) nm; IR (KBr) ν_{max} 3444, 2933, 1647, 1091, 1016 cm^{-1}; ^1H and ^{13}C NMR spectroscopic data, see Table 2; HRESIMS *m*/*z* 279.15903 $[M + H]^+$ (calcd for $C_{15}H_{23}O_4{}^+$, 279.15909).

Craterodoratin H (**8**). Colorless oil; $[\alpha]_D^{21}$ − 30.6 (*c* 0.16, MeOH); UV (MeOH) λ_{max} (log ε) 210 (0.68) nm; IR (KBr) ν_{max} 3398, 2949, 2835, 1653, 1452, 1112, 1031 cm^{-1}; ^1H and ^{13}C NMR spectroscopic data, see Table 2; HRESIMS *m*/*z* 279.15920 $[M + H]^+$ (calcd for $C_{15}H_{23}O_4{}^+$, 279.15909).

Craterodoratin I (**9**). Colorless oil; $[\alpha]_D^{21}$ + 31.6 (*c* 0.10, MeOH); UV (MeOH) λ_{max} (log ε) 230 (3.01) nm; IR (KBr) ν_{max} 3379, 2947, 2835, 1653, 1456, 1114, 1031 cm^{-1}; ^1H and ^{13}C NMR spectroscopic data, see Table 3; HRESIMS *m*/*z* 251.16409 $[M + H]^+$ (calcd for $C_{15}H_{23}O_3{}^+$, 251.16417).

Table 3. ^1H (600 MHz) and ^{13}C (150 MHz) NMR Spectroscopic Data for **9–12** in Methanol-d_4.

Entry	9		10		11		12	
	δ_C, Type	δ_H (*J* in Hz)	δ_C, Type	δ_H (*J* in Hz)	δ_C, Type	δ_H (*J* in Hz)	δ_C, Type	δ_H (*J* in Hz)
1	51.0, CH	2.10, t, (5.4)	48.0, CH	2.29, t, (5.6)	51.3, CH	2.09, t (5.7)	47.4, CH	2.70, t, (5.3)
2	87.4, C	–	88.2, C	–	87.3, C	–	89.2, C	–
3a 3b	32.0, CH_2	1.88, m 1.61, m	27.4, CH_2	1.92, m 1.56, td, (10.2, 4.2)	72.2, CH	3.63, d, (6.8)	29.5, CH_2	2.06, m 1.83, m
4a 4b	22.4, CH_2	1.88, m 1.76, m	22.0, CH_2	1.91, m 1.82, m	32.8, CH_2	2.25, m 1.85, m	22.4, CH_2	1.95, m 1.81, m
5	39.2, CH	2.24, q, (5.4)	39.2, CH	2.32, m	37.8, CH	2.27, m	40.2, CH	2.43, q, (5.3)
6a 6b	22.2, CH_2	2.13, m 1.50, d, (10.1)	22.3, CH_2	2.14, m 1.51, d, (10.2)	21.7, CH_2	2.11, m 2.05, m	22.1, CH_2	2.28, m 1.75, d, (10.9)
7	55.3, C	–	53.7, C	–	54.7, C	–	53.6, C	–
8a 8b	32.4, CH_2	1.91, m 1.72, m	40.2, CH_2	2.01, dd, (14.2, 4.2) 1.84, m	40.1, CH_2	2.00, dd, (14.2, 4.0) 1.81, dd, (14.2, 4.0)	43.4, CH_2	3.15, q, (3.2)
9a 9b	24.5, CH_2	2.19, m 2.10, m	65.7, CH	4.45, td, (8.7, 3.8)	65.8, CH	4.42, td, (9.0, 4.0)	199.4, C	–
10	141.7, CH	6.77, t, (7.5)	143.6, CH	6.67, d, (7.7)	141.7, CH	6.57, d, (9.0)	130.6, CH	7.11, s
11	127.9, C	–	127.3, C	–	129.3, C	–	142.4, C	–
12	170.4, C	–	170.5, C	–	172.2, C	–	169.9, C	–
13	11.0, CH_3	1.80, s	11.4, CH_3	1.83, s	11.8, CH_3	1.83, s	13.4, CH_3	2.15, s
14a 14b	70.2, CH_2	3.82, d, (8.8) 3.47, d, (8.8)	71.7, CH_2	3.94, d, (9.4) 3.63, d, (9.4)	71.5, CH_2	3.85, d, (9.4) 3.61, d, (9.4)	179.3, C	–
15a 15b	24.0, CH_3	1.24, s	65.9, CH_2	3.49, d, (11.4) 3.45, d, (11.4)	21.0, CH_3	1.31, s	23.6, CH_3	1.51, s

Craterodoratin J (**10**). Colorless oil; $[\alpha]_D^{21}$ + 52 (*c* 0.10, MeOH); UV (MeOH) λ_{max} (log ε) 210 (1.48) nm; IR (KBr) ν_{max} 3379, 2947, 2835, 1653, 1456, 1114, 1031 cm^{-1}; ^1H and ^{13}C NMR spectroscopic data, see Table 3; HRESIMS *m*/*z* 305.13602 $[M + Na]^+$ (calcd for $C_{15}H_{22}O_5Na^+$, 305.13594).

Craterodoratin K (**11**). Colorless oil; $[\alpha]_D^{21}$ + 37 (*c* 0.10, MeOH); UV (MeOH) λ_{max} (log ε) 210 (1.34) nm; IR (KBr) ν_{max} 3456, 2924, 2815, 1647, 1396, 1018 cm^{-1}; ^1H and ^{13}C NMR spectroscopic data, see Table 3; HRESIMS *m*/*z* 283.15136 $[M + H]^+$ (calcd for $C_{15}H_{23}O_5{}^+$, 283.15400).

Craterodoratin L (**12**). Colorless oil; $[\alpha]_D^{21}$ + 53 (*c* 0.11, MeOH); UV (MeOH) λ_{max} (log ε) 235 (1.12) nm; IR (KBr) ν_{max} 3390, 2947, 2833, 1653, 1456, 1112, 1031 cm^{-1}; ^1H and

^{13}C NMR spectroscopic data, see Table 3; HR-ESIMS *m/z* 279.12250 [M + H]$^+$ (calcd for $C_{15}H_{19}O_5^+$, 279.12270).

Craterodoratin M (**13**). Colorless oil; $[\alpha]_D^{21}$ − 23 (*c* 0.10, MeOH); UV (MeOH) λ_{max} (log ε) 215 (1.90) nm; IR (KBr) ν_{max} 3487, 1645 cm^{-1}; ^{1}H and ^{13}C NMR spectroscopic data, see Table 4; HRESIMS *m/z* 253.17992 [M + H]$^+$ (calcd for $C_{15}H_{25}O_3^+$, 253.17982).

Table 4. ^{1}H (600 MHz) and ^{13}C (150 MHz) NMR Spectroscopic Data for **13–16** in Methanol-d_4.

Entry	13		14		15		16	
	δ_C, Type	δ_H (*J* in Hz)	δ_C, Type	δ_H (*J* in Hz)	δ_C, Type	δ_H (*J* in Hz)	δ_C, Type	δ_H (*J* in Hz)
1	40.3, CH	2.04, m	73.5, C	–	43.2, CH	2.48, d, (6.7)	47.6, CH	1.87, m
2	39.2, CH	1.96, q, (5.1)	40.3, CH	1.38, m	149.4, C	–	150.2, C	–
3a 3b	37.3, CH_2	1.84, m 1.76, m	29.6, CH_2	1.86, m 0.93, m	24.7, CH_2	2.64, m 2.32, m	23.5, CH_2	2.63, m 2.33, m
4a 4b	23.9, CH_2	3.16, m	26.7, CH_2	1.79, m 1.08, m	30.2, CH_2	1.95, m 1.75, m	22.6, CH_2	1.95, m 1.88, m
5	37.3, CH	2.13, m	47.3, CH	1.33, m	74.9, C	–	37.6, CH	2.18, q, (5.4)
6a 6b	18.0, CH_2	1.63, m 1.27, m	25.9, CH_2	1.92, m 1.12, dd, (3.7, 12.5)	34.6, CH_2	2.33, m 1.79, d, (9.5)	25.8, CH_2	2.33, m 1.47, d, (10.1)
7	42.1, C	–	48.4, C	–	48.6, C	–	46.7, C	–
8a 8b	22.9, CH_2	2.01, d, (5.7) 1.37, m	38.1, CH_2	1.53, t, (1.5)	32.3, CH_2	1.82, m 1.68, m	29.9, CH_2	1.87, m
9a 9b	23.7, CH	1.84, m	22.3, CH_2	2.21, m	23.8, CH_2	2.28, m 2.21, m	23.5, CH_2	2.33, m 2.23, q, (8.0)
10	140.7, CH	6.71, dd, (7.4, 14.5)	138.9, CH	6.60, t, (6.6)	142.5, CH	6.81, t, (7.0)	142.8, CH	6.84, t, (8.0)
11	129.3, C	–	130.8, C	–	127.8, C	–	127.5, C	–
12	171.3, C	–	174.5, C	–	171.3, C	–	170.8, C	–
13	11.4, CH_3	1.81, s	11.8, CH_3	1.81, s	11.1, CH_3	1.82, s	11.0, CH_3	1.83, s
14a 14b	15.8, CH_3	0.89, s	22.0, CH_3	1.09, s	15.6, CH_3	0.81, s	61.1, CH_2	3.43, d, (11.7) 3.33, d, (11.7)
15a 15b	65.3, CH_2	3.34, dd, (2.6, 6.0)	67.3, CH_2	3.34, d, (3.3)	106.1, CH_2	4.63, d, (2.0)	105.9, CH_2	4.64, br s 4.60, br s

Craterodoratin N (**14**). Colorless oil; $[\alpha]_D^{21}$ − 11 (*c* 0.10, MeOH); UV (MeOH) λ_{max} (log ε) 210 (1.04) nm; IR (KBr) ν_{max} 3487, 1645 cm^{-1}; ^{1}H and ^{13}C NMR spectroscopic data, see Table 4; HRESIMS *m/z* 291.15658 [M + Na]$^+$ (calcd for $C_{15}H_{24}O_4Na^+$, 291.15668).

Craterodoratin O (**15**). Colorless oil; $[\alpha]_D^{21}$ − 28.6 (*c* 0.10, MeOH); UV (MeOH) λ_{max} (log ε) 210 (1.48) nm; IR (KBr) ν_{max} 3460, 2951, 2843, 1645, 1016 cm^{-1}; ^{1}H and ^{13}C NMR spectroscopic data, see Table 4; HRESIMS *m/z* 251.16431 [M + H]$^+$ (calcd for $C_{15}H_{23}O_3^+$, 251.16417).

Craterodoratin P (**16**). Colorless oil; $[\alpha]_D^{21}$ − 27.9 (*c* 0.10, MeOH); UV (MeOH) λ_{max} (log ε) 210 (1.77) nm; IR (KBr) ν_{max} 3367, 2943, 2833, 1654, 1456, 1112, 1031 cm^{-1}; ^{1}H and ^{13}C NMR spectroscopic data, see Table 4; HRESIMS *m/z* 251.16417 [M + H]$^+$ (calcd for $C_{15}H_{23}O_3H^+$, 251.16417).

Craterodoratin Q (**17**). Colorless oil; $[\alpha]_D^{21}$ + 22 (*c* 0.07, MeOH), UV (MeOH) λ_{max} (log ε) 210 (1.37) nm; IR (KBr) ν_{max} 3468, 1645, 1016 cm^{-1}; ^{1}H and ^{13}C NMR spectroscopic data, see Table 5; HRESIMS *m/z* 331.15140 [M + Na]$^+$ (calcd for $C_{17}H_{24}O_5Na^+$, 331.15159).

Table 5. ^{1}H (600 MHz) and ^{13}C (150 MHz) NMR Spectroscopic Data for **17–19** in Methanol-d_4.

Entry	17		18		19	
	δ_C, Type	δ_H (J in Hz)	δ_C, Type	δ_H (J in Hz)	δ_C, Type	δ_H (J in Hz)
1	48.2, CH	3.00, t, (5.3, 10.6)	45.0, CH	2.03, t, (6.2)	40.8, CH	2.31, br s
2	149.7, C	–	44.2, CH	1.94, q, (6.9)	42.2, CH	2.82, s
3a 3b	23.2, CH_2	2.64, m 2.34, m	96.5, C	–	159.1, C	–
4a 4b	22.6, CH_2	1.90, m	41.0, CH_2	2.24, m 1.87, d, (12.4)	58.7, CH_2	3.51, m 3.08, dd, (10.6, 2.1)
5	38.5, CH	2.21, m	37.7, CH_3	2.33, q, (6.2)	58.8, CH_2	3.51, m
6a 6b	44.1, C	–	35.6, CH_2	2.49, m 1.16, d, (9.7)	58.7, CH	1.99, br s
7a 7b	25.5, CH_2	2.43, m 1.51, d, (10.0)	39.2, C	–	48.4, C	–
8a 8b	37.9, CH_2	2.17, dd, (8.7, 15.0) 1.81, m	39.5, CH_2	1.85, m 1.71, dd, (14.8, 4.4)	40.8, CH_2	1.41, m 1.33, m
9a 9b	65.7, CH	4.60, m	65.7, CH	4.45, m	24.8, CH_2	1.63, m 1.08, m
10	143.1, CH	6.65, d, (8.4)	141.4, CH	6.52, d, (8.2)	44.7, CH	1.08, m
11	127.8, C	–	129.7, C	–	30.9, CH	1.35, m
12	171.6, C	–	172.6, C	–	20.2, CH_3	0.87, d, (6.5)
13	11.6, CH_3	1.83, s	12.0, CH_3	1.85, s	20.0, CH_3	0.81, d, (6.5)
14a 14b	64.8, CH_2	4.00, d, (11.9) 3.93, d, (11.9)	69.3, CH_2	4.01, d, (10.2) 3.94, d, (10.2)	103.4, CH_2	5.02, s 4.74, s
15a 15b	106.7, CH_2	4.67, br s 4.64, br s	13.8, CH_3	1.06, d, (6.9)	20.1, CH_3	0.98, s
-OOCCH_3	19.3, CH_3	2.00, s	45.0, CH	2.03, t, (6.2)	48.2, CH_2	3.26, t, (6.0)
-OOCCH_3	171.6, C	–	44.2, CH	1.94, q, (6.9)	174.0, C	–

Craterodoratin R (**18**). Colorless oil; $[\alpha]_D^{21}$ − 42 (*c* 0.10, MeOH); UV (MeOH) λ_{max} (log ε) 215 (1.82) nm; IR (KBr) ν_{max} 3412, 2924, 1637, 1574, 1435, 1065 cm^{-1}; ^{1}H and ^{13}C NMR spectroscopic data, see Table 5; HRESIMS *m/z* 305.13596 [M +Na]$^+$ (calcd for $C_{15}H_{22}O_5Na^+$, 305.13594).

Craterodoratin S (**19**). Colorless oil; $[\alpha]_D^{21}$ − 125 (*c* 0.10, MeOH); UV (MeOH) λ_{max} (log ε) 210 (1.04) nm; IR (KBr) ν_{max} 3412, 2924, 1637, 1574, 1435, 1065 cm^{-1}; ^{1}H and ^{13}C NMR spectroscopic data, see Table 5; HRESIMS *m/z* 278.21140 [M +H]$^+$ (calcd for $C_{17}H_{28}NO_2^+$, 278.21146).

X-ray Crystallographic Data for Craterodoratin B (**2**). $C_{15}H_{22}O_4{\cdot}H_2O$, M = 284.34, a = 8.6658(3) Å, b = 10.0982(4) Å, c = 17.1595(6) Å, α = 90°, β = 90°, γ = 90°, V = 1501.61(9) Å^3, T = 100(2) K, space group *P*212121, Z = 4, μ(Cu Kα) = 0.768 mm^{-1}, 12029 reflections measured, 2967 independent reflections (R_{int} = 0.0391). The final R_1 values were 0.0314 ($I > 2\sigma(I)$). The final $wR(F^2)$ values were 0.0802 ($I > 2\sigma(I)$). The final R_1 values were 0.0325 (all data). The final $wR(F^2)$ values were 0.0813 (all data). The goodness of fit on F^2 was 1.049. Flack parameter = 0.06(7). CCDC: 2059695 (https://www.ccdc.cam.ac.uk (accessed on 13 November 2021)).

X-ray Crystallographic Data for Craterodoratin C (**3**). $C_{15}H_{24}O_4$, M = 268.34, a = 9.9718(3) Å, b = 7.2863(2) Å, c = 10.4844(3) Å, α = 90°, β = 111.5850(10)°, γ = 90°, V = 708.35(4) Å^3, T = 100(2) K, space group *P*1211, Z = 2, μ(Cu Kα) = 0.729 mm^{-1}, 13370 reflections measured, 2751 independent reflections (R_{int} = 0.0334). The final R_1 values were 0.0290 ($I > 2\sigma(I)$). The final $wR(F^2)$ values were 0.0749 ($I > 2\sigma(I)$). The final R_1 values were 0.0291 (all data).

The final $wR(F^2)$ values were 0.0750 (all data). The goodness of fit on F^2 was 1.039. Flack parameter = 0.06(6). CCDC: 2059696 (https://www.ccdc.cam.ac.uk (accessed on 13 November 2021)).

X-ray Crystallographic Data for Craterodoratin D (**4**). $C_{15}H_{24}O_4$, $M = 268.34$, $a = 6.4802(4)$ Å, $b = 10.8539(7)$ Å, $c = 10.8186(7)$ Å, $\alpha = 90.00°$, $\beta = 101.555(2)°$, $\gamma = 90.00°$, $V = 745.51(8)$ Å^3, $T = 297(2)$ K, space group $P1211$, $Z = 2$, μ(Cu Kα) = 1.54178, 15119 reflections measured, 3109 independent reflections ($R_{int} = 0.0263$). The final R_1 values were 0.0293 (all data). The final $wR(F^2)$ values were 0.0798 (all data). The goodness of fit on F^2 was 1.051. Flack parameter = 0.09(4). CCDC: 2059697 (https://www.ccdc.cam.ac.uk (accessed on 13 November 2021)).

X-ray Crystallographic Data for Craterodoratin E (**5**). $C_{15}H_{24}O_5 \cdot H_2O$, $M = 302.36$, $a = 25.5045(11)$ Å, $b = 6.4589(3)$ Å, $c = 19.4274(8)$ Å, $\alpha = 90°$, $\beta = 97.923(2)°$, $\gamma = 90°$, $V = 3169.7(2)$ Å^3, $T = 100(2)$ K, space group C121, $Z = 8$, μ(Cu Kα) = 0.805 mm^{-1}, 50,328 reflections measured, 6168 independent reflections ($R^{int} = 0.0956$). The final R_1 values were 0.0680 ($I > 2\sigma(I)$). The final $wR(F^2)$ values were 0.1739 ($I > 2\sigma(I)$). The final R_1 values were 0.0741 (all data). The final $wR(F^2)$ values were 0.1821 (all data). The goodness of fit on F^2 was 1.085. Flack parameter = 0.26(11). CCDC: 2059698 (https://www.ccdc.cam.ac.uk (accessed on 13 November 2021)).

2.4. NMR and ECD Calculations

Details of NMR and ECD calculations for compounds **1**, **6–8**, **10**, **11** and **18** were given in the Supplementary Materials.

2.5. Cytotoxicity Assay

There are five human cancer cell lines using in this assay including human myeloid leukemia HL-60, human breast cancer MCF-7, human colon cancer SW480, human hepatocellular carcinoma SMMC-7721, and human lung cancer A-549 cells. All selected cells were stored in DMEM medium or RPMI-1640, supplemented with 10% fetal bovine serum (Hyclone, Logan, UT, USA) at 37 °C. The MTT (3-(4,5-dimethylthiazol-2-yl)-2,5-diphenyl tetrazolium bromide) method in 96-well microplates was used for cytotoxicity assay [22]. In brief, 100 µL adherent cells were seeded into each well for 12 h before drug addition, while suspended cells were seeded just before adding the test compound with initial density of 1×10^5 cells/mL. The tumor cell line was exposed to the test compound at concentrations of 0.0625, 0.32, 1.6, 8 and 40 µM in triplicates for 48 h. Taxol (Sigma, St. Louis, MO, USA) was used as a positive control. The cell viability was detected, and the cell growth curve was graphed after compound treatment. The IC_{50} values were calculated by the Reed and Muench's method [23].

2.6. Immunosuppressive Activities Assay

2.6.1. Preparation of Spleen Cells from Mice

Female BALB/c mice were sacrificed by cervical dislocation, and their spleens were aseptically removed. After cell debris, mononuclear cell suspensions were prepared, and clumps were removed. Erythrocytes were depleted with ammonium chloride buffer solution. Lymphocytes were washed and resuspended in RPMI 1640 medium supplemented with 10% FBS, penicillin (100 U/mL), and streptomycin (100 mg/mL).

2.6.2. Cytotoxicity Assay

We used the Cell Counting Kit-8 (CCK-8) assay to test cytotoxicity. In brief, fresh spleen cells were obtained from female BALB/c mice (18–20 g). Spleen cells (1×10^6 cells) were seeded in triplicate in 96-well flat plates and cultured at 37 °C for 48 h, with or without compounds of various concentrations, in a humidified and 5% CO_2-containing incubator. A certain amount of CCK-8 was added to each well during the last 8–10 h of culture. At the end of the culture, we use a microplate reader (Bio-Rad 650) to measure the OD value at 450 nm. Cyclosporin A (CsA) is an immunosuppressant agent which is used as a positive control with definite activity. Only the OD value of culture medium is used as background. The cytotoxicity of compounds was expressed as the concentration of the compounds that reduces cell viability to 50% (CC_{50}).

2.6.3. T and B Cell Function Assay

As mentioned above, fresh spleen cells were obtained from female BALB/c mice (18–20 g). Similarly, the 5×10^5 spleen cells were cultured at the same conditions. The cultures, with or without various concentrations of compounds, were stimulated with 5 μg/mL of concanavalin A (ConA) to induce the T cells' proliferative response or 10 μg/mL of lipopolysaccharide (LPS) to induce B cells' proliferative response. Proliferation was assessed in terms of uptake of [^{3}H]-thymidine during 8 h of pulsing with 25 μL/well of [^{3}H]-thymidine, then cells will be harvested onto glass fiber filters. A Beta scintillation counter was used to count the incorporated radioactivity. Cells treated without any stimuli were used as a negative control. The immunosuppressive activity was expressed as the concentration of the compound that inhibited T or B cell proliferation to 50% (IC_{50}) of the control value. Both the cytotoxicity and proliferation assessment were repeated twice.

3. Results and Discussion

Compound **1** was isolated as a colorless oil. Its molecular formula of $C_{15}H_{24}O_3$ was determined by positive high-resolution electrospray ionization mass spectrometry (HRESIMS) analysis, corresponding to four degrees of unsaturation. In the ^{1}H NMR data, three singlets for methyl groups were readily observed at δ_H 1.80 (3H, s, Me-13), δ_H 0.93 (3H, s, Me-15), δ_H 0.85 (3H, s, Me-14) (Table 1). The ^{13}C NMR and DEPT data revealed 15 carbon resonances including three CH_3, five CH_2, three CH, and four non-protonated carbons (Table 1). Of them, two olefinic carbons at δ_C 141.1 (C-10) and 128.5 (C-11), and one carboxy carbon at δ_C 171.3 (C-12) occupied two degrees of unsaturation, which suggested that **1** should be a bicycle sesquiterpenoid. Preliminary analysis of 1D and 2D NMR data, with respect to those in previous isolated from the same source, suggested that **1** might be a bergamotane-type sesquiterpenoid. The ^{1}H-^{1}H COSY spectrum afforded evidence to form a fragment as shown in Figure 2. Through the HMBC correlations from δ_H 0.85 (3H, s, Me-14) to δ_C 49.9 (C-1), 76.3 (C-2), 25.8 (C-6), and 50.9 (C-7), as well as correlations from δ_H 0.93 (3H, s, Me-15) to δ_C 49.9 (C-1), 41.9 (C-4), 50.9 (C-7), and 30.9 (C-8), the positions of two methyl groups including Me-14 at C-1 and Me-15 at C-7 were established (Figure 2). Furthermore, the carboxyl group was identified to be conjugated with the double bond by HMBC correlations from δ_H 1.80 (3H, s, Me-13) to δ_C 141.1 (C-10), 128.5 (C-11), and 171.3 (C-12). In the ROESY spectrum (Figure 3), the cross peaks of H_3-15/H-5a (δ_H 1.69), H_3-15/H-6b (δ_H 1.23), and H-3a (δ_H 2.16)/H-8a (δ_H 1.46) indicated that Me-15 and CH_2-5, CH_2-6 were on the same side. The cross peaks of H_2-9/H_3-13 and the absent correlation between H_3-13 and H-10 confirmed the *E*-configured double bond. However, the relative configuration of 2-OH could not be determined by the ROESY data. The theoretical NMR calculations and DP4+ probability analyses were employed on two possible structures of (1*S**,2*R**,4*S**,7*R**)-**1a** and (1*S**,2*S**,4*S**,7*R**)-**1b**, and the calculations messages suggested that (1*S**,2*R**,4*S**,7*R**)-**1a** was the correct relative configuration for **1** (see Table S3 in Section S1). Finally, the absolute configuration of **1** was established to be 1*S*,2*R*,4*S*,7*R* by ECD calculations (Figure 5). Therefore, compound **1** was identified and named as craterodoratin A.

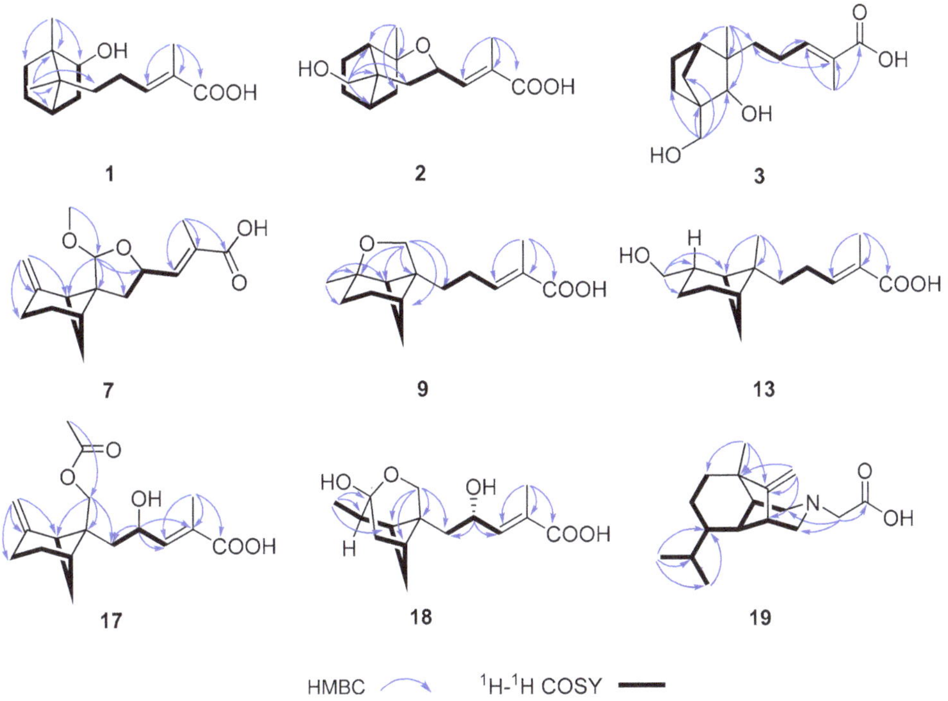

Figure 2. ^{1}H-^{1}H COSY and key HMBC correlations of **1–3**, **7**, **9**, **13**, and **17–19**.

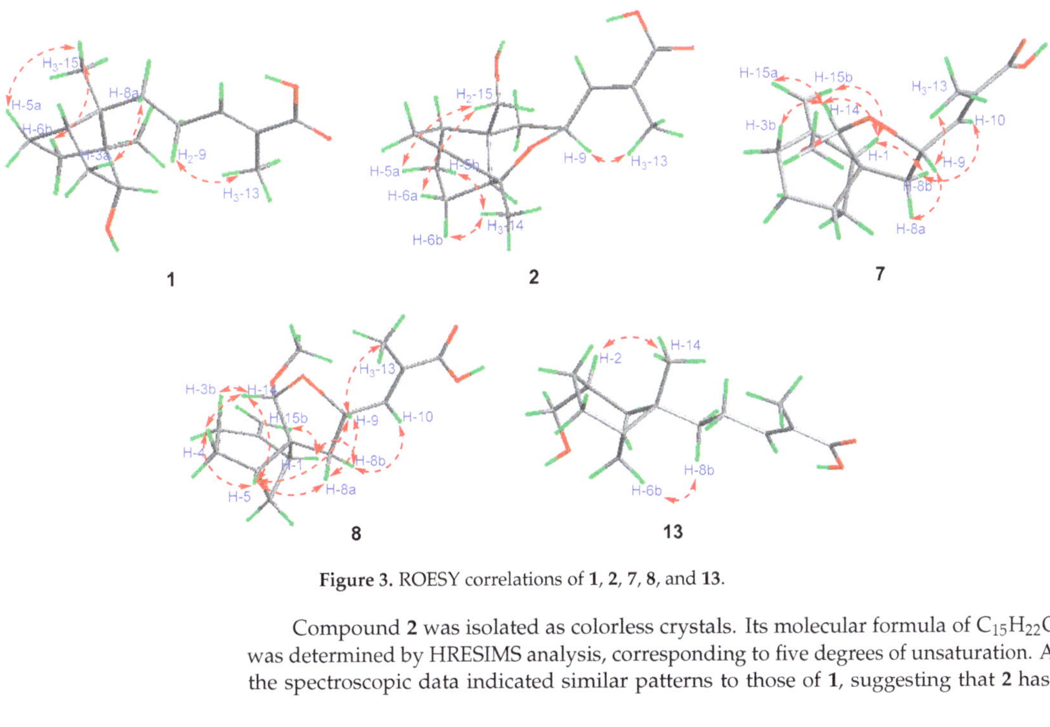

Figure 3. ROESY correlations of **1**, **2**, **7**, **8**, and **13**.

Compound **2** was isolated as colorless crystals. Its molecular formula of $C_{15}H_{22}O_4$ was determined by HRESIMS analysis, corresponding to five degrees of unsaturation. All the spectroscopic data indicated similar patterns to those of **1**, suggesting that **2** has a similar structure to **1**. Detailed analysis of 1D and 2D NMR data revealed the differences. Firstly, Me-15 was oxygenated into a hydroxymethylene in **2** as suggested by the HMBC correlations from δ_H 3.64 (1H, d, *J* = 11.0 Hz, H-15a) and δ_H 3.47 (1H, d, *J* = 11.0 Hz, H-15b) to δ_C 43.5 (C-1), 42.3 (C-4), 52.5 (C-7) and 33.7 (C-8) (Figure 2). Secondly, Me-14 shifted from C-1 to C-2 in **2** which supported by HMBC correlations from δ_H 1.25 (3H, s, Me-14) to δ_C 43.5 (C-1), 81.8 (C-2), and 48.2 (C-3). Thirdly, analysis of the MS data and carbon shifts at δ_C 81.8 (C-2) and δ_C 66.4 (C-9), as well as the HMBC from δ_H 4.64 (1H, m, H-9) to δ_C 81.8 (C-2) indicated an ether bond between C-2 and C-9. The ROESY (Figure 3) cross peaks of H_3-14/H-5b, and H_3-14/H-6b indicated Me-14, H-5b, and H-6b were on the same side, cross peaks of H_2-15/H-6a, and H_2-15/H-5a indicated CH_2-15, H-5a, and H-6a were on the same side, while correlations between H-9/H_3-13 and the absence of the ROESY correction of between H_3-13/H-10 confirmed the *E*-configured double bond. Finally, the single-crystal X-ray diffraction established the structure of **2** with the absolute configuration (Flack parameter = 0.06 (7), CCDC: 2059695. Figure 4). Therefore, compound **2** was identified and named as craterodoratin B.

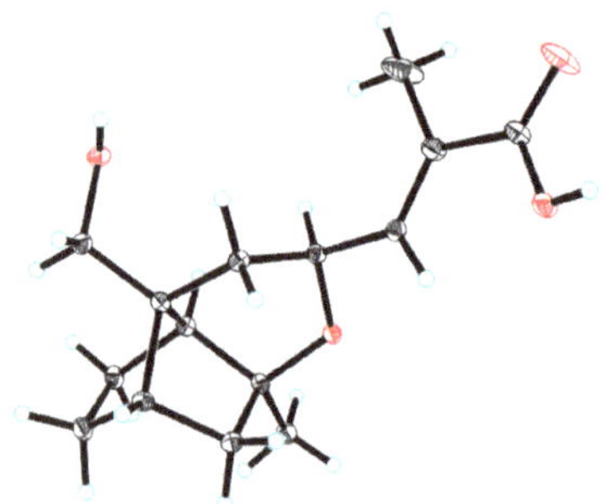

2

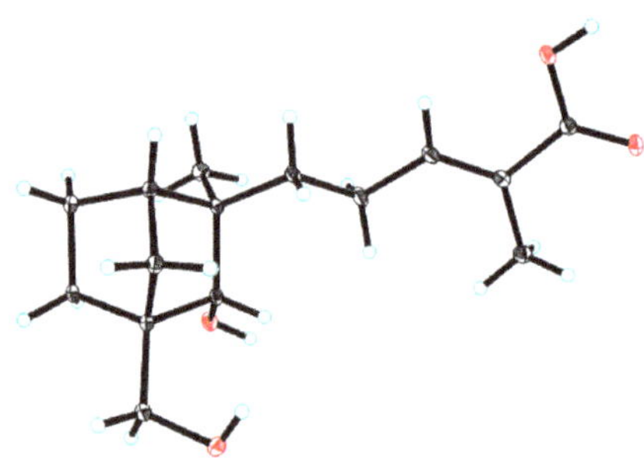

3

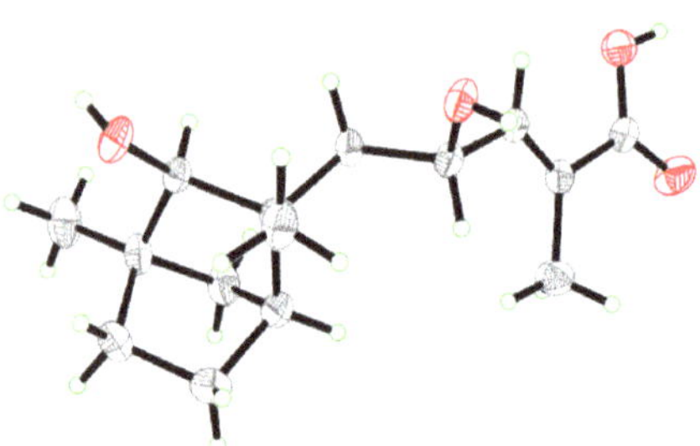

4

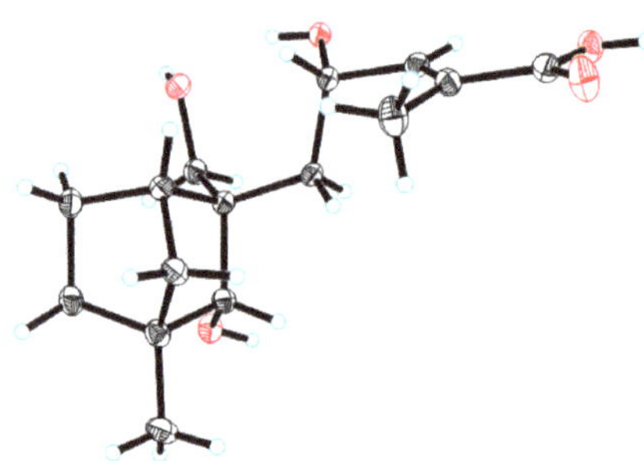

5

Figure 4. ORTEP diagrams of **2–5** showing absolute configurations.

Compound **3** was isolated as colorless crystals. Its molecular formula of $C_{15}H_{24}O_4$ was determined by HRESIMS data, corresponding to four degrees of unsaturation. The ^{13}C NMR and DEPT data (Table 1) displayed 15 carbon resonances including two methyl carbons (δ_C 12.3 and 17.0), one olefinic methine carbons (δ_C 144.1), one oxygenated methine carbon (δ_C 79.7), six methylenes, one methine, two sp^3 quaternary carbons (δ_C 55.8 and 43.2), and one carboxyl carbon (δ_C 172.0). Analysis of 2D NMR data including ^1H-^1H COSY and HMBC correlations as shown in Figure 2 suggested that **3** should be a bergamotane-type sesquiterpenoid related to (*Z*)-2α-hydroxyalbumol [24]. The differences were that C-12 and C-15 in **3** were oxidized into a carboxy group and a hydroxymethylene group, respectively. In the ROESY spectrum, cross peaks of H-7a and H-8 supported that C-7 and C-8 were on the same side. Besides, the double bond between C-10 and C-11 in **3** was assigned as *E* geometry by the ROESY correlation of H-9/H_3-13. The single-crystal X-ray diffraction established the absolute configuration of **3** (Flack parameter = 0.06(6), CCDC: 2059696, Figure 4). Therefore, compound **3** was identified and named as craterodoratin C.

Compound **4** was isolated as colorless crystals. Its molecular formula of $C_{15}H_{24}O_4$ was determined by HRESIMS data, the same as that of **3**. All the spectroscopic data showed similar patterns to those of **3**, except one oxygenated methylene group of C-15 was reduced to a methyl group in **4**. This was supported by the key HMBC correlations from δ_H 0.91 (3H, s, Me-15) to δ_C 41.5 (C-1), 83.5 (C-2), and 25.0 (C-6). In addition, C-9 was oxidized into a hydroxymethine group, as demonstrated by the HMBC correlations from δ_H 4.58 (1H, m, H-9) to δ_C 48.9 (C-8), 144.6 (C-10), and 126.6 (C-11). The single-crystal X-ray diffraction not only confirmed the planar structure as elucidated above but also established the absolute configuration (Flack parameter = 0.09(4), CCDC: 2059697, Figure 4). Therefore, compound **4** was identified and named as craterodoratin D.

Compound **5** was isolated as colorless crystals. Its molecular formula of $C_{15}H_{24}O_5$ was determined by HRESIMS data, corresponding to four degrees of unsaturation. The 1D NMR data (Table 2) indicated that **5** has a closely related structure to that of **4**, except that one methyl group of C-14 was oxygenated into a hydroxyl methylene group in **5**. It was supported by the loss of one methyl signal in the ^{1}H NMR spectrum and the HMBC correlations from δ_H 3.64 (1H, d, J = 11.7 Hz, H-14a) and δ_H 3.53 (1H, d, J = 11.7 Hz, H-14b) to δ_C 86.0 (C-2), 47.3 (C-3), 44.0 (C-4), and 47.0 (C-8). The single-crystal X-ray diffraction confirmed the planar structure and established the absolute configuration (Flack parameter = 0.26(11), CCDC: 2059698. Figure 4). Thus, compound **5** was identified and named as craterodoratin E.

Compound **6** was isolated as a colorless oil. Its molecular formula of $C_{15}H_{24}O_5$ was determined by HRESIMS data, corresponding to four degrees of unsaturation. All the spectroscopic data indicated that **6** had a similar structure to that of **4**. Detailed analysis of 1D and 2D NMR data revealed that C-8 was oxygenated into a hydroxymethine group in **6** as established by the shift at δ_C 80.5 (C-8) and MS data. The coupling constant ($^3J_{H\text{-}H}$ = 6.0 Hz) of H-8/H-9 of **6** observed in ^{1}H NMR spectrum and a strong ROESY correlation between H-9 and H-2 suggested *anti* configuration of H-8 and H-9 [25]. To determine the stereochemistry of C-8, the theoretical NMR calculations and DP4 + probability analyses were employed on two possible relative structures (1*R**,2*R**,3*R**,4*S**,8*S**,9*S**)-**6a** and (1*R**,2*R**,3*R**,4*S**,8*R**,9*S**)-**6b**. The results suggested that (1*R**,2*R**,3*R**,4*S**,8*R**,9*S**)-**6b** was the correct relative configuration for **6** (see Table S6 in Section S2). Based on this, the absolute configuration of **6** was suggested to be 1*R*,2*R*,3*R*,4*S*,8*R*,9*S* by the ECD calculations (Figure 5). Therefore, compound **6** was identified and named as craterodoratin F.

Compounds **7** and **8** were isolated as a pair of epimers. They possessed the same molecular formula of $C_{16}H_{22}O_4$, on the basis of HRESIMS data. Analysis of 1D (Table 1) and 2D NMR data suggested that the planar structures of **7** and **8** were similar to that of donacinoic acid A (**23**) [9], except the ether bond between C-2 and C-14 was cut off to give a methoxy group at C-14. In the ROESY spectra, the observed correlations of H-14/H-15a in **7** and H-5/H-14 in **8** suggested that the stereo-configurations of C-14 in **7** and **8** were different. The theoretical NMR calculations and DP4 + probability analyses were employed to elucidate the relative configurations of **7** and **8** (see Tables S9 and S10). Then, the ECD calculations established the absolute configurations to be 7*S*,9*S*,14*S* for **7** and 7*S*,9*S*,14*R* for **8** (Figure S9 in Section S3). Finally, compounds **7** and **8** were identified and named as craterodoratins G and H, respectively.

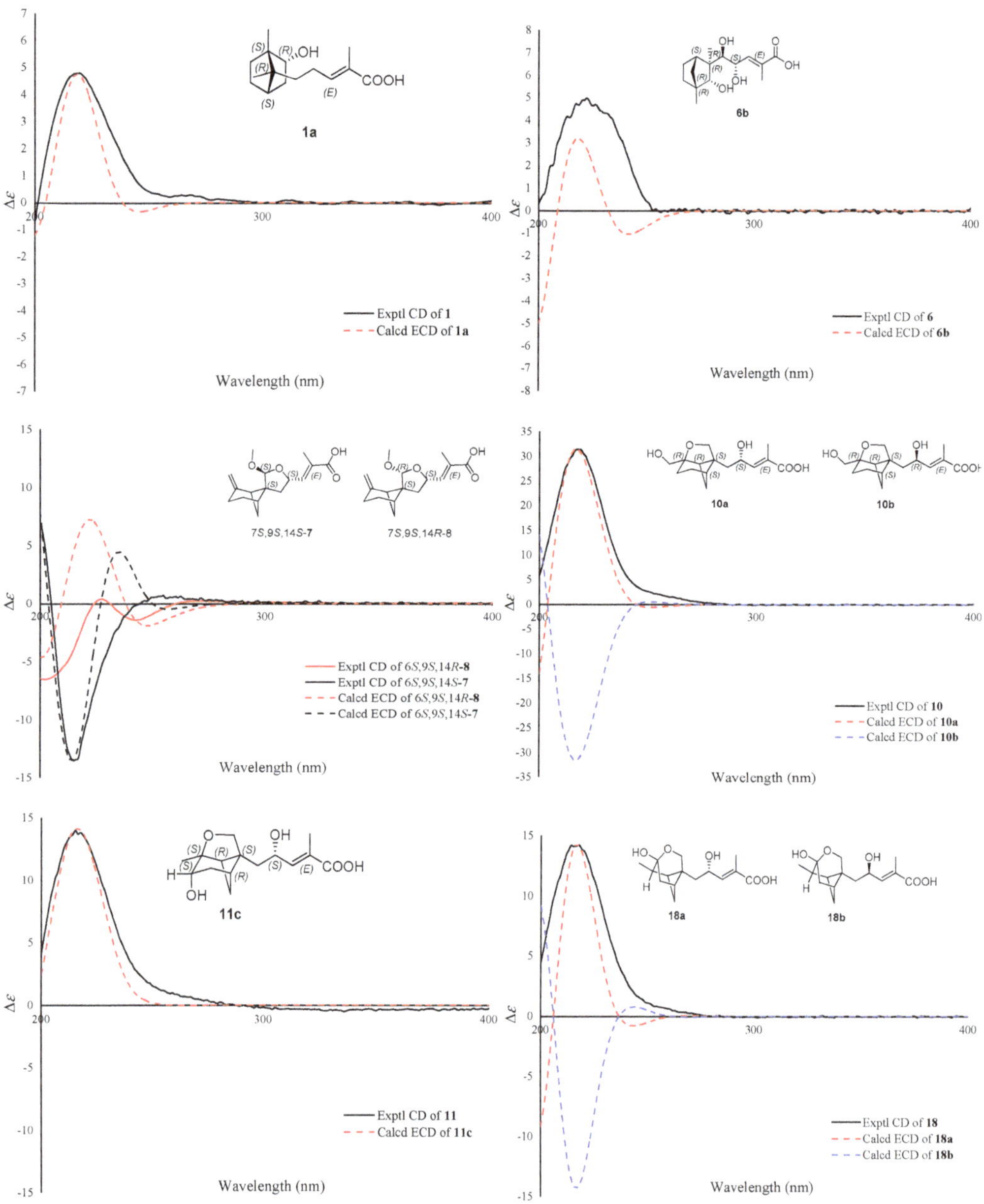

Figure 5. ECD calculations for compounds **1**, **6**, **7**, **8**, **10**, **11**, and **18**.

Compound **9** was isolated as a colorless oil. Its molecular formula of $C_{15}H_{22}O_3$ was determined on the basis of the HRESIMS data, corresponding to five degrees of unsaturation. Analysis of its NMR data (Table 3) indicated similar patterns to those of massarinolin B [15], except that C-9 in massarinolin B was reduced to a methylene in **9** as determined by the MS data and ^1H–^1H COSY correlations (Figure 2). Therefore, compound **9** was identified and named as craterodoratin I.

Compound **10** was isolated as a colorless oil. Its molecular formula of $C_{15}H_{22}O_5$ was determined by HRESIMS analysis, corresponding to five degrees of unsaturation. The ^1H and ^{13}C NMR (Table 3) spectroscopic characteristics were similar to those of **9**, except for two hydroxy groups placed at C-15 and C-9 in **10**. These were determined by the HMBC correlations from δ_H 4.45 (1H, td, *J* = 8.7, 3.8 Hz, H-9) to δ_C 40.2 (C-8), 143.6 (C-10), 127.3 (C-11), and from δ_H 3.49 (1H, d, *J* = 11.4 Hz, H-15a), 3.45 (1H, d, *J* = 11.4 Hz, H-15b) to δ_C 88.2 (C-2), 27.4 (C-3), 48.0 (C-1). According to the ROESY correlation between H-9 and H-13, the double bond was established as *E* form. The absolute configuration was established by ECD calculations as shown in Figure 5. Therefore, compound **10** was identified and named as craterodoratin J.

Compound **11** was isolated as a colorless oil. Its molecular formula of $C_{15}H_{22}O_5$ was determined by HRESIMS analysis, corresponding to five degrees of unsaturation. The ^1H and ^{13}C NMR and DEPT (Table 3) data of **11** displayed signals for structural features similar to **10**, except that one hydroxy was transfered from C-15 to C-3. The location was determined by the HMBC correlations from δ_H 1.31 (3H, s, Me-15) to δ_C 87.3 (C-2), 72.2 (C-3), 51.3 (C-1), and from δ_H 3.63 (1H, d, *J* = 6.8 Hz, H-3) to δ_C 87.3 (C-2), 32.8 (C-4). The stereo-configurations for C-3 and C-9 could not be established according to the ROESY data. Thus, the theoretical NMR calculations and DP4 + probability analyses were employed on four possible structures (see Table S17 in Section S5). Based on these data, the absolute configuration of **11** was suggested to be 1*R*,2*S*,3*S*,5*R*,7*S*,9*S* by the ECD calculations (Figure 5). Therefore, compound **11** was identified and named as craterodoratin K.

Compound **12** was isolated as a colorless oil. Its molecular formula of $C_{15}H_{18}O_5$ was determined by HRESIMS analysis, corresponding to seven degrees of unsaturation. The ^1H and ^{13}C NMR and DEPT (Table 3) data of **12** displayed signals for structural features similar to **9**, except that C-14 and C-9 were oxidized into two carbonyl carbons in **12**. The locations were determined by the HMBC correlations from δ_H 3.15 (2H, q, *J* = 3.2 Hz, H-8) to δ_C 40.2 (C-5), 53.6 (C-7), 179.3 (C-14), and 199.4 (C-9). Therefore, compound **12** was identified and named as craterodoratin L.

Compound **13** was isolated as a colorless oil. Its molecular formula of $C_{15}H_{24}O_3$ was determined by HRESIMS analysis, corresponding to four degrees of unsaturation. The ^{13}C NMR and DEPT spectrum (Table 4) displayed 15 carbon resonances including two methyl carbons, one olefinic methine carbon, one quaternary olefinic carbon, five methylenes, three methines, and one quaternary carbon. Analysis of 1D and 2D NMR data suggested that **13** had a similar structure to that of massarinolin C (**21**) [15]. The locations of 7-Me and 2-CH_2OH were determined by the HMBC correlations from δ_H 0.89 (3H, s, Me-14) to δ_C 40.3 (C-1), 39.2 (C-5), 42.1 (C-7), 37.3 (C-8), and from δ_H 3.34 (2H, dd, *J* = 6.0, 2.6 Hz, H-15) to δ_C 40.3 (C-1), 37.2 (C-2), 18.0 (C-3). In the ROESY spectrum (Figure 3), the cross peak of H_3-14/H-2 indicated that Me-14 and H-2 were on the same side, while the cross peak of H-9/H-13 indicated the *E* form of the double bond. Therefore, compound **13** was identified as craterodoratin M.

Compound **14** was isolated as a colorless oil. Its molecular formula of $C_{15}H_{24}O_4$ was determined by HRESIMS analysis, corresponding to four degrees of unsaturation. All 1D and 2D NMR data suggested that **14** had a structure closely related to that of **13** except that one more hydroxy group at C-1 in **14**, which was supported by the HMBC correlations from δ_H 1.09 (3H, s, Me-14) to δ_C 73.5 (C-1), 47.3 (C-5), and 48.4 (C-7). Detailed analysis of 2D NMR data suggested that the other parts of **14** were the same as those of **13**. Therefore, compound **14** was identified and named as craterodoratin N.

Compound **15** was isolated as a colorless oil. Its molecular formula of $C_{15}H_{22}O_3$ was determined by HRESIMS analysis, corresponding to five degrees of unsaturation. The ^1H and ^{13}C NMR and DEPT (Table 4) data of **15** displayed a close resemblance to those of **13**. One significant difference was that the hydroxy group was substituted at C-5, as evidenced by HMBCs from δ_H 2.33 (1H, m, H-6a) and 1.79 (1H, d, *J* = 9.5 Hz, H-6b) to δ_C 43.2 (C-1), 74.9 (C-5), and 48.6 (C-7). In addition, one terminal double bond was established between C-2 and C-15, as proved by HMBC correlations from δ_H 4.63 (2H, d, *J* = 2.0 Hz, H-15) to δ_C

43.2 (C-1), 149.4 (C-2), and 24.7 (C-3). Therefore, compound **15** was identified and named as craterodoratin O.

Compound **16** was isolated as a colorless oil. Its molecular formula of $C_{15}H_{22}O_3$ was determined to be the same to that of **15** by HRESIMS data. Detailed analysis of 1D and 2D NMR data suggested that **16** had a very similar structure to that of **15** except for the hydroxy group at C-14 in **16** replaced at C-5 in **15**. It was supported by the HMBC correlations from δ_H 3.43 (1H, d, J = 11.7 Hz, H-14a) and δ_H 3.33 (1H, d, J = 11.7 Hz, H-14b) to δ_C 47.6 (C-1), 37.6 (C-5), 29.9 (C-8). Therefore, compound **16** was identified and named as craterodoratin P.

Compound **17** was isolated as a colorless oil. Its molecular formula of $C_{17}H_{24}O_5$ was determined by HRESIMS analysis, corresponding to six degrees of unsaturation. The ^1H and ^{13}C NMR and DEPT (Table 5) data of **17** displayed a close resemblance to those of massarinolin C (**21**) [15]. Analysis of 2D NMR data suggested that one more *O*-acetyl group was substituted at C-14, as evidenced by HMBC correlations from δ_H 4.00 (1H, d, J = 11.9 Hz, H-14a) and 3.93 (1H, d, J = 11.9 Hz, H-14b) to δ_C 48.2 (C-1), 38.5 (C-5), 37.9 (C-8), 171.6 (-$OOCCH_3$). The absolute configuration of **17** was established by ECD calculations as shown in Figure S18 in Section S6. Therefore, compound **17** was identified and named craterodoratin Q.

Compound **18** was isolated as a colorless oil. Its molecular formula of $C_{15}H_{22}O_5$ was determined by HRESIMS data, corresponding to five degrees of unsaturation. The 1D NMR data (Table 5) indicated that **18** should have a closely related structure to that of brasilamide A [26]. One significant difference was that the amide group was replaced by a carboxy group at δ_C 172.6 (C-12), as evidenced by the MS data. In addition, C-9 and C-14 in **18** was reduced into a hydroxymethine group (δ_C 65.7) and a methylene (δ_C 69.3), respectively, as supported by the HMBC correlations from δ_H 4.45 (1H, m, H-9) to δ_C 39.5 (C-8) and 141.4 (C-10), and from δ_H 4.01 (1H, d, J = 10.2 Hz, H-14a) and 3.94 (1H, d, J = 10.2 Hz, H-14b) to δ_C 45.0 (C-1), 39.2 (C-7), 96.5 (C-3), and 37.7 (C-5). Based on the ROESY data, cross peaks of H-9/H-14 supported that H-9 and C-14 were on the same side; cross peaks of H-2/H-6b supported that H-2 and C-14 were on the same side. Finally, the absolute configuration of **18** was suggested to be 1*S*, 2*S*, 3*R*, 5*R*, 7*S*, 9*S* by the ECD calculations (Figure 5). Therefore, compound **18** was identified and named as craterodoratin R.

Compound **19** was isolated as a colorless oil. Its molecular formula of $C_{17}H_{27}NO_2$ was determined by HRESIMS data, which suggested an *N*-containing structure. The 1D and 2D NMR data of **19** were similar to those of victoxinine [19,27], except that one oxygenated methylene carbon at C-17 in victoxinine was replaced by a carboxyl group in **19** as proved by HMBC correlation from δ_H 3.26 (2H, t, J = 6.0 Hz, H-16) to δ_C 174.0 (C-17). Detailed analysis of 2D NMR data suggested that the other parts of **19** were the same as that of victoxinine. Therefore, compound **19** was identified and named as craterodoratin S.

In addition to the new compounds as described above, four known sesquiterpenoids obtained in this study were identified as massarinolin B (**20**) [15,28], massarinolin C (**21**) [15], massarinolin A (**22**) [15], and donacinoic acid A (**23**) [9] by comparison of their spectroscopic data with those reported in the literature. The biosynthesis of the isolated compounds **1**–**18** and **20**–**23** was proposed as shown in Scheme 1. Bergamotane sesquiterpenoids, bearing a borneol ring system or a β-pinene ring system, are naturally occurring in plants and fungi (Scheme 1) [9,14–17,29–34]. In this study, compounds **1** and **2** possess a rare carbon skeleton while the methyl group (C-14) in **2** had a further 1,2-migration. The sesquiterpenes **7–18** belong to β-pinene derivatives (Scheme 1), which produced compounds **3–6** by an alkyl migration.

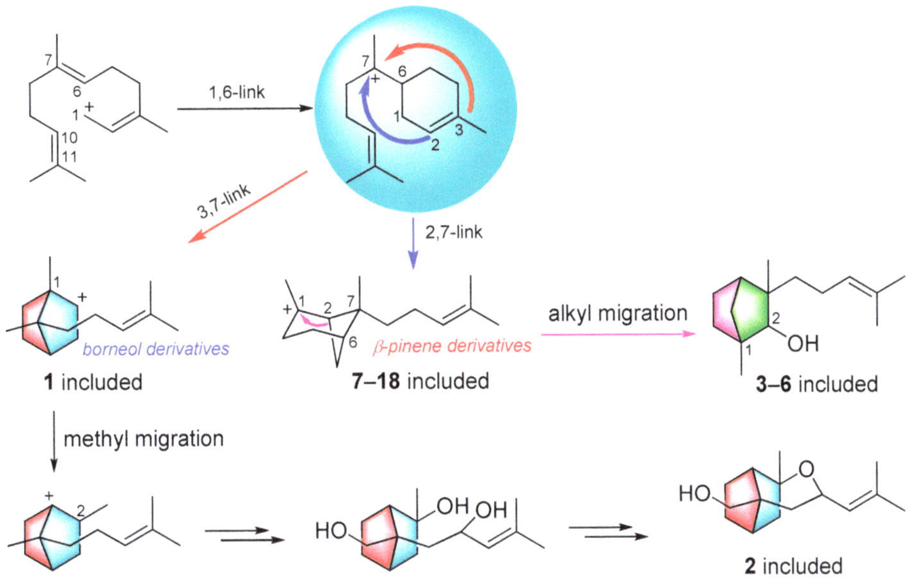

Scheme 1. Proposed Biosynthetic Pathways for Bergamotane-type Sesquiterpenes.

All compounds were evaluated for their cytotoxicity to five human cancer cell lines (HL-60, A-549, SMMC-7721, SW480, MCF-7). However, no compounds are active at the concentration of 40 μM. In addition, all the compounds were investigated for their in vitro immunomodulatory effect on BALB/c mice T and B lymphocyte proliferation. Compounds **3**, **10**, **12–15**, **19**, **20**, and **23** exhibited potent inhibitory activity against LPS-induced proliferation of B lymphocyte cell with IC_{50} values ranging from 0.67 to 22.68 μM. Compounds **17** and **20** exhibited inhibitions on ConA-induced proliferation of T lymphocyte cells with IC_{50} values of 31.50 and 0.98 μM, respectively (Table 6).

Table 6. Immunosuppressive Tests of the Isolates.

Entry	CC_{50} (μM)	Con A-Induced T-Cell Proliferation		LPS-Induced B-Cell Proliferation	
		IC_{50} (μM)	SI [a]	IC_{50} (μM)	SI [a]
3	>40	–	–	12.62 ± 1.14	>3.17
10	>40	–	–	19.40 ± 0.48	>2.06
12	>40	–	–	13.71 ± 0.65	>2.92
13	>40	–	–	15.43 ± 1.03	>2.59
14	>40	–	–	13.26 ± 1.29	>3.02
15	>40	–	–	17.12 ± 1.14	>2.34
17	>40	31.50 ± 1.79	>1.27	–	–
19	>40	–	–	22.68 ± 1.67	>1.76
20	>40	0.98 ± 0.01	>40.82	0.67 ± 0.004	>59.26
23	>40	–	–	13.23 ± 0.97	>3.02
CsA	>2.80	0.04	>70.00	0.47	>5.95

[a] SI (selectivity index) is determined as the ratio of the concentration of the compounds that reduced cell viability to 50% (CC_{50}) to the concentration of the compounds needed to inhibit the proliferation by 50% relative to the control value (IC_{50}).

Structurally, almost all bergamotane sesquiterpenoids in this study contain an α,β-unsaturated carboxylic acid moiety, which might be the key functional group for their immunosuppressive activity. However, it seems that the compounds (**10** and **12–15**) with a basic core of a β-pinene showed a wider range of biological activities. Immunosuppressants are a kind of drug that can inhibit human immunity. They are mainly used in organ transplantation to combat rejection and autoimmune diseases. At present, there are several immunosuppressive drugs that work by inhibiting T cell proliferation, but new, efficient, and safe immunosuppressive drugs inhibiting B cell proliferation are still unavailable [35,36]. In this study, many bergamotane sesquiterpenoids were found to have potential inhibition of B cell proliferation. To the best of our knowledge, bergamotane sesquiterpenoids were reported for their immunosuppressive activity for the first time.

4. Conclusions

In summary, a total of 23 sesquiterpenoids including 19 new ones were isolated from the edible fungus *C. ordoratus*. Of them, 22 compounds belong to bergamotane sesquiterpenoids in three carbon skeletons, while compounds **1** and **2** possess a rare ring-rearranged backbone. In addition, many bergamotane sesquiterpenoids exhibited selective inhibitions on LPS-induced B cell proliferation. This study suggests that *C. odoratus* is rich in bergamotane sesquiterpenoids with promising immunosuppressive activity and provides strong support for the further development and utilization of the edible mushroom *C. odoratus*.

Supplementary Materials: The following are available online at https://www.mdpi.com/article/10.3390/jof7121052/s1, Section S1: Calculational details for **1**, Section S2: Calculational details for **6**, Section S3: Calculational details for 6*S*,9*S*,14*S*-**7** and 6*S*,9*S*,14*R*-**8**, Section S4: Calculational details for **10**, Section S5: Calculational details for **11**, Section S6: Calculational details for **17**,Section S7: Calculational details for **18**, Section S8: NMR and MS spectra for **1**, Section S9: NMR and MS spectra for **2**, Section S10: NMR and MS spectra for **3**, Section S11: NMR and MS spectra for **4**, Section S12: NMR and MS spectra for **5**, Section S13: NMR and MS spectra for **6**, Section S14: NMR and MS spectra for 6*S*,9*S*,14*S*-**7**, Section S15: NMR and MS spectra for 6*S*,9*S*,14*R*-**8**, Section S16: NMR and MS spectra for **9**, Section S17: NMR and MS spectra for **10**, Section S18: NMR and MS spectra for **11**, Section S19: NMR and MS spectra for **12**, Section S20: NMR and MS spectra for **13**, Section S21: NMR and MS spectra for **14**, Section S22: NMR and MS spectra for **15**, Section S23: NMR and MS spectra for **16**, Section S24: NMR and MS spectra for **17**, Section S25: NMR and MS spectra for **18**, and Section S26: NMR and MS spectra for **19**.

Author Contributions: Conceptualization, Q.D. and T.F.; methodology, Q.D., F.-L.Z., Z.-H.L. and T.F.; resources, Z.-H.L., J.H. and T.F.; data curation, Q.D.; writing—original draft preparation, Q.D. and F.-L.Z.; writing—review and editing, J.H. and T.F.; project administration, T.F.; funding acquisition, T.F. and J.H.; All authors have read and agreed to the published version of the manuscript.

Funding: This research was funded by the National Natural Science Foundation of China (grant numbers 81872762 and 22177139) and the Research Fund for Central University, South-Central University for Nationalities (grant number CZP21001).

Institutional Review Board Statement: Not applicable.

Informed Consent Statement: Not applicable.

Data Availability Statement: X-ray crystallographic data of **2–5** (CIF) is available free of charge at https://www.ccdc.cam.ac.uk.

Acknowledgments: The authors thank the Analytical & Measuring Centre, South-Central University for Nationalities for the spectra measurements.

Conflicts of Interest: The authors declare no conflict of interest. The funders had no role in the design of the study; in the collection, analyses, or interpretation of data; in the writing of the manuscript, or in the decision to publish the results.

References

1. Zhang, Y.; Mo, M.Z.; Yang, L.; Mi, F.; Cao, Y.; Liu, C.L.; Tang, X.Z.; Wang, P.F.; Xu, J.P. Exploring the species diversity of edible mushrooms in Yunnan, Southwestern China, by DNA barcoding. *J. Fungi* **2021**, *7*, 310. [CrossRef] [PubMed]
2. Feng, T.; Cai, J.L.; Li, X.M.; Zhou, Z.Y.; Li, Z.H.; Liu, J.K. Chemical constituents and their bioactivities of mushroom *Phellinus rhabarbarinus*. *J. Agric. Food Chem.* **2016**, *64*, 1945–1949. [CrossRef] [PubMed]
3. Huang, Y.; Zhang, S.B.; Chen, H.P.; Zhao, Z.Z.; Zhou, Z.Y.; Li, Z.H.; Feng, T.; Liu, J.K. New acetylenic acids and derivatives from the edible mushroom *Craterellus lutescens* (Cantharellaceae). *J. Agric. Food Chem.* **2017**, *65*, 3835–3841. [CrossRef] [PubMed]
4. Chen, H.P.; Zhao, Z.Z.; Li, Z.H.; Huang, Y.; Zhang, S.B.; Tang, Y.; Yao, J.N.; Chen, L.; Isaka, M.; Feng, T.; et al. Anti-proliferative and anti-inflammatory lanostane triterpenoids from the Polish edible mushroom *Macrolepiota procera*. *J. Agric. Food Chem.* **2018**, *66*, 3146–3154. [CrossRef] [PubMed]
5. Zhong, X.R.; Li, T.H.; Jiang, Z.D.; Deng, W.Q.; Huang, H. A new yellow species of *Craterellus* (Cantharellales, Hydnaceae) from China. *Phytotaxa* **2018**, *360*, 35–44. [CrossRef]
6. Bijeesh, C.; Manoj Kumar, A.; Vrinda, K.B.; Pradeep, C.K. Two new species of *Craterellus* (Cantharellaceae) from tropical India. *Phytotaxa* **2018**, *346*, 157–168. [CrossRef]
7. Dahlman, M.; Danell, E.; Spatafora, J.W. Molecular systematics of *Craterellus*: Cladistic analysis of nuclear LSU rDNA sequence data. *Mycol. Res.* **2000**, *104*, 388–394. [CrossRef]
8. Tripathy, S.S.; Rajoriya, A.; Gupta, N. Nutritional and biochemical analysis of *Craterellus odoratus*: Rare chanterelle basidiomycetes from similipal biosphere reserve of odisha. *Ann. Appl. Bio-Sci.* **2015**, *2*, A1–A6.
9. Zhao, Z.Z.; Zhao, K.; Chen, H.P.; Bai, X.; Zhang, L.; Liu, J.K. Terpenoids from the mushroom-associated fungus *Montagnula donacina*. *Phytochemistry* **2018**, *147*, 21–29. [CrossRef] [PubMed]
10. Zhang, L.; Shen, Y.; Wang, F.; Leng, Y.; Liu, J.K. Rare merosesquiterpenoids from basidiomycete *Craterellus odoratus* and their inhibition of 11β-hydroxysteroid dehydrogenases. *Phytochemistry* **2010**, *71*, 100–103. [CrossRef] [PubMed]
11. Zhang, L.; Yao, J.N.; Bai, X.; Li, Z.H.; Dong, Z.J.; Liu, J.K. Two new 4,6-dimethyl-3,4-dihydrochromen-2-one derivatives from *Craterellus odoratus*. *J. Asian Nat. Prod. Res.* **2017**, *19*, 241–246. [CrossRef] [PubMed]
12. Guo, H.; Feng, T.; Li, Z.H.; Liu, J.K. Five new polyketides from the basidiomycete *Craterellus odoratus*. *Nat. Prod. Bioprospect.* **2012**, *2*, 170–173. [CrossRef]
13. Guo, H.; Feng, T.; Li, Z.H.; Liu, J.K. Four new compounds from the basidiomycete *Craterellus odoratus*. *J. Asian Nat. Prod. Res.* **2012**, *14*, 950–955. [CrossRef] [PubMed]
14. Ying, Y.M.; Fang, C.A.; Yao, F.Q.; Yu, Y.; Shen, Y.; Hou, Z.N.; Wang, Z.; Zhang, W.; Shan, W.G.; Zhan, Z.J. Bergamotane sesquiterpenes with alpha-glucosidase inhibitory activity from the plant pathogenic fungus *Penicillium expansum*. *Chem. Biodivers.* **2017**, *14*, e1600184. [CrossRef]
15. Oh, H.; Gloer, J.B.; Shearer, C.A. Massarinolins A−C: New bioactive sesquiterpenoids from the aquatic fungus *Massarina tunicata*. *J. Nat. Prod.* **1999**, *62*, 497–501. [CrossRef] [PubMed]
16. Che, Y.; Gloer, J.B.; Koster, B.; Malloch, D. Decipinin A and decipienolides A and B: New bioactive metabolites from the coprophilous fungus *Podospora decipiens*. *J. Nat. Prod.* **2002**, *65*, 916–919. [CrossRef] [PubMed]
17. Wang, Y.N.; Xia, G.Y.; Wang, L.Y.; Ge, G.B.; Zhang, H.W.; Zhang, J.F.; Wu, Y.Z.; Lin, S. Purpurolide A, 5/5/5 spirocyclic sesquiterpene lactone in nature from the endophytic fungus *Penicillium purpurogenum*. *Org. Lett.* **2018**, *20*, 7341–7344. [CrossRef] [PubMed]
18. Pringle, R.B.; Beaun, A.C. Isolation of victoxinine from cultures of *Helminthosporium victoriae*. *Phytopathology* **1960**, *50*, 324–325.
19. Dorn, F.; Arigoni, D. Structure of victoxinine. *J. Chem. Soc. Chem. Commun.* **1972**, *1*, 1342–1343. [CrossRef]
20. Pringle, R.B. Comparative biochemistry of the phytopathogenic fungus *Helminthosporium*. XVI. the production of victoxinine by *H. sativum* and *H. victoriae*. *Can. J. Biochem.* **1976**, *54*, 783–787. [CrossRef] [PubMed]
21. Li, Y.Y.; Tan, X.M.; Yang, J.; Guo, L.P.; Ding, G. Naturally occurring *seco*-sativene sesquiterpenoid: Chemistry and biology. *J. Agric. Food Chem.* **2020**, *68*, 9827–9838. [CrossRef]
22. Mosmann, T. Rapid colorimetric assay for cellular growth and survival: Application to proliferation and cytotoxicity assays. *J. Immunol. Methods* **1983**, *65*, 55–63. [CrossRef]
23. Reed, L.J.; Muench, H. A simple method of estimating fifty percent endpoints. *Am. J. Hygiene* **1938**, *27*, 493–497.
24. Ochi, T.; Shibata, H.; Higuti, T.; Kodama, K.H.; Kusumi, T.; Takaishi, Y. Anti-*Helicobacter pylori* compounds from *Santalum album*. *J. Nat. Prod.* **2005**, *68*, 819–824. [CrossRef]
25. Matsumori, N.; Kaneno, D.; Murata, M.; Nakamura, H.; Tachibana, K. Stereochemical determination of acyclic structures based on carbon−proton spin-coupling constants. A method of configuration analysis for natural products. *J. Org. Chem.* **1999**, *64*, 866–876. [CrossRef] [PubMed]
26. Liu, L.; Gao, H.; Chen, X.; Cai, X.; Yang, L.; Guo, L.; Yao, X.; Che, Y. Brasilamides A–D: Sesquiterpenoids from the plant endophytic fungus *Paraconiothyrium brasiliense*. *Eur. J. Org. Chem.* **2010**, *17*, 3302–3306. [CrossRef]
27. Kono, Y.; Takeuchi, S.; Daly, J.M. Isolation and structure of a new victoxinine derivative produced by *Helminthosporium victoriae*. *Agric. Biol. Chem.* **1983**, *47*, 2701–2702. [CrossRef]
28. Rodríguez López, M.; Bermejo, F.A. Total synthesis of (+)-massarinolin B and (+)-4-*epi*-massarinolin B, fungal metabolites from *Massarina tunicata*. *Tetrahedron* **2006**, *62*, 8095–8102. [CrossRef]
29. Cane, D.E. Enzymic formation of sesquiterpenes. *Chem. Rev.* **1990**, *90*, 1089–1103. [CrossRef]

30. Fraga, B.M. Natural sesquiterpenoids. *Nat. Prod. Rep.* **2013**, *30*, 1226–1264. [CrossRef]
31. Massias, M.; Rebuffat, S.; Molho, L.; Chiaroni, A.; Riche, C.; Bodo, B. Expansolides A and B: Tetracyclic sesquiterpene lactones from *Penicillium expansum*. *J. Am. Chem. Soc.* **1990**, *112*, 8112–8115. [CrossRef]
32. MacíAs, F.A.; Varela, R.M.; Simonet, A.M.; Cutler, H.G.; Cutler, S.J.; Hill, R.A. Absolute configuration of bioactive expansolides A and B from *Aspergillus fumigatus* Fresenius. *Tetrahedron Lett.* **2003**, *44*, 941–943. [CrossRef]
33. Zhang, L.H.; Feng, B.M.; Chen, G.; Li, S.G.; Sun, Y.; Wu, H.H.; Bai, J.; Hua, H.M.; Wang, H.F.; Pei, Y.H. Sporulaminals A and B: A pair of unusual epimeric spiroaminal derivatives from a marine-derived fungus *Paraconiothyrium sporulosum* YK-03. *RSC. Adv.* **2016**, *6*, 42361–42366. [CrossRef]
34. Davis, D.C.; Walker, K.L.; Hu, C.; Zare, R.N.; Waymouth, R.M.; Dai, M. Catalytic carbonylative spirolactonization of hydroxycyclopropanols. *J. Am. Chem. Soc.* **2016**, *138*, 10693–10699. [CrossRef]
35. Dangroo, N.A.; Singh, J.; Dar, A.A.; Gupta, N.; Chinthakindi, P.K.; Kaul, A.; Khuroo, M.A.; Sangwan, P.L. Synthesis of α-santonin derived acetyl santonous acid triazole derivatives and their bioevaluation for T and B-cell proliferation. *Eur. J. Med. Chem.* **2016**, *120*, 160–169. [CrossRef] [PubMed]
36. Feng, T.; Duan, K.T.; He, S.J.; Wu, B.; Zheng, Y.S.; Ai, H.L.; Li, Z.H.; He, J.; Zuo, J.P.; Liu, J.K. Ophiorrhines A and B, two immunosuppressive monoterpenoid indole alkaloids from *Ophiorrhiza japonica*. *Org. Lett.* **2018**, *20*, 7926–7928. [CrossRef] [PubMed]

Journal of
Fungi

MDPI

Article

Sesquiterpenoids and Xanthones from the Kiwifruit-Associated Fungus *Bipolaris* sp. and Their Anti-Pathogenic Microorganism Activity

Jun-Jie Yu, Ying-Xue Jin, Shan-Shan Huang and Juan He *

National Demonstration Center for Experimental Ethnopharmacology Education, School of Pharmaceutical Sciences, South-Central University for Nationalities, Wuhan 430074, China; junjieyu98@outlook.com (J.-J.Y.); Yancy6020@163.com (Y.-X.J.); HuangSS1998@126.com (S.-S.H.)

* Correspondence: 2015048@mail.scuec.edu.cn

Abstract: Nine previously undescribed sesquiterpenoids, bipolarisorokins A–I (**1–9**); two new xanthones, bipolarithones A and B (**10** and **11**); two novel sativene-xanthone adducts, bipolarithones C and D (**12** and **13**); as well as five known compounds (**14–18**) were characterized from the kiwifruit-associated fungus *Bipolaris* sp. Their structures were elucidated by extensive spectroscopic methods, electronic circular dichroism (ECD), ^{13}C NMR calculations, DP4+ probability analyses, and single crystal X-ray diffractions. Many compounds exhibited anti-pathogenic microorganism activity against the bacterium *Pseudomonas syringae* pv. *actinidiae* and four pathogenic microorganisms.

Keywords: *Bipolaris* sp.; kiwi-associated fungus; sesquiterpenoid; xanthone; anti-pathogenic microorganism activity

Citation: Yu, J.-J.; Jin, Y.-X.; Huang, S.-S.; He, J. Sesquiterpenoids and Xanthones from the Kiwifruit-Associated Fungus *Bipolaris* sp. and Their Anti-Pathogenic Microorganism Activity. *J. Fungi* **2022**, *8*, 9. https://doi.org/10.3390/jof8010009

Academic Editor: Frank Surup

Received: 13 December 2021
Accepted: 21 December 2021
Published: 23 December 2021

1. Introduction

Kiwifruit (*Actinidia chinensis* Planch., Actinidiaceae) is an emerging, healthy, and economical fruit which has become increasingly popular worldwide owing to its flavor and nutritional properties [1]. It is an excellent source of vitamin C and provides balanced nutritional components of minerals, dietary fiber, folate, and health-promoting metabolites [2,3]. China is the leading kiwifruit producer in the world, followed by Italy and New Zealand. The cultivation area and annual output reached 243,000 hm^2 and 2,500,000 tons at the end of 2020 [4]. Nevertheless, as the cultivation of kiwifruit expands rapidly, many serious diseases such as bacterial canker, soft rot, bacterial blossom blight, brown spot, and root rot are a serious and ongoing threat to kiwifruit production [5–12]. Particularly, the destructive bacterial canker disease, which is associated with an infection by *P. syringae* pv. *actinidiae* (Psa), has led to reduced kiwifruit production and huge economic losses worldwide [13,14]. Although the application of copper-based chemicals and streptomycin have played a positive role in the prevention and treatment of bacterial canker, these chemical residues are extremely threatening to human health and the ecological environment [15,16]. Additionally, chemical fungicides easily induce pathogen resistance [17,18]. Thus, it is urgent to develop safer and more effective biological pesticides.

Endophytic microorganisms reside within different tissues of the host plant without causing any disease symptoms and produce various metabolites with different activities [19,20]. Therefore, the endophytic fungi have been proved to be valuable sources of important natural products [21,22]. Some natural products from endophytic fungi play important roles in plant defense systems. Therefore, we carried out the excavation of anti-Psa active substances from metabolites of kiwifruit endophytes and harvested a number of bioactive molecules. For instance, 3-decalinoyltetramic acids and cytochalasins from the kiwifruit endophytic fungus *Zopfiella* sp showed anti-Psa activity [23,24], while imidazole alkaloids ether were characterized as anti-Psa agents from *Fusarium tricinctum* [25]. These discoveries prompted us to search for more novel and bioactive metabolites from kiwifruit-associated fungi. In the current study, a total of eighteen compounds have been isolated from the large-scale fermentation of the kiwifruit-associated fungus *Bipolaris* sp. (Figure 1), which included nine new sativene or longifolene sesquiterpenoids, bipolarisorokins A–I (**1–9**); two new xanthones, bipolarithones A and B (**10** and **11**); two novel sativene-xanthone adducts, bipolarithones C and D (**12** and **13**); as well as five known ones (**14–18**). Their structures were established by means of spectroscopic methods, namely, ECD and ^{13}C NMR calculations, DP4+ probability analyses, and single crystal X-ray diffractions. All compounds were evaluated for their inhibitory activities against Psa. Additionally, their inhibitory activity against four phytopathogens (*Phytophthora infestans*, *Alternaria solani*, *Rhizoctonia solani*, and *Fusarium oxysporum*) were assessed. Here, the details of isolation, structural elucidation, and bioactivity evaluations for **1–18** are reported.

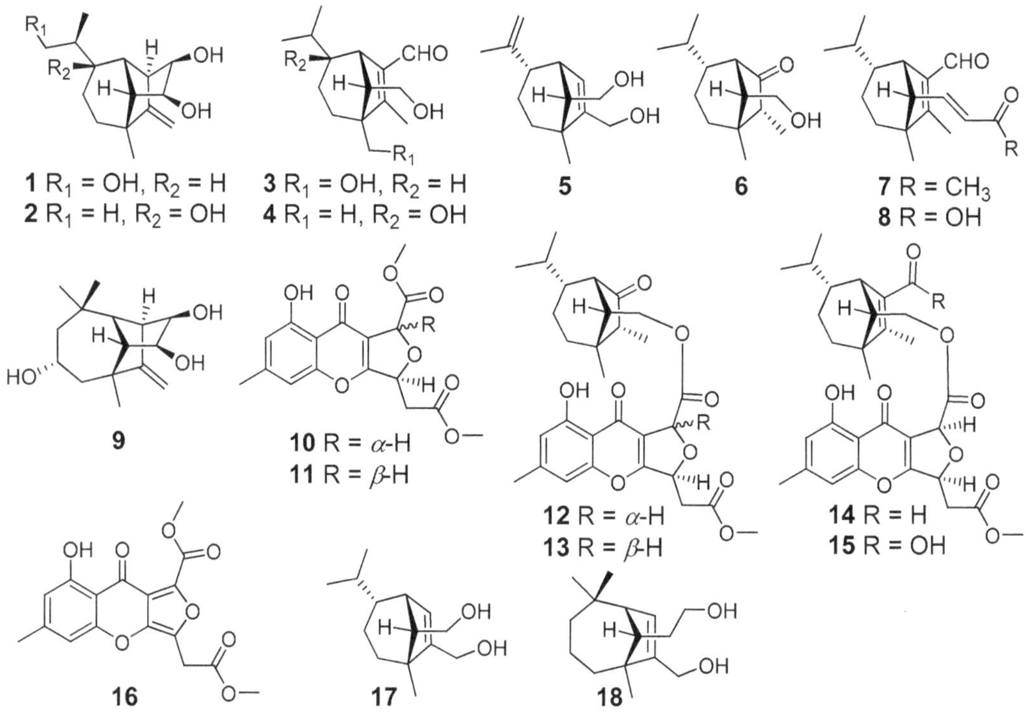

Figure 1. Structures of compounds **1–18**.

2. Materials and Methods

2.1. General Experimental Procedures

Melting points were obtained on an X-4 micro melting point apparatus. Optical rotations were measured with an Autopol IV polarimeter (Rudolph, Hackettstown, NJ, USA). UV spectra were obtained using a double beam spectrophotometer UH5300 (Hitachi High-Technologies, Tokyo, Japan). IR spectra were obtained by a Shimadzu IRTracer-100 spectrometer using KBr pellets. 1D and 2D NMR spectra were run on a Bruker Avance III 600 MHz spectrometer with TMS as an internal standard. Chemical shifts (δ) were expressed in ppm with references to the solvent signals. High resolution electrospray ionization mass spectra (HR-ESIMS) were recorded on a LC-MS system consisting of a Q Exactive™ Orbitrap mass spectrometer with an HRESI ion source (ThermoFisher Scientific, Bremen, Germany) used in ultra-high-resolution mode (140,000 at m/z 200) and a UPLC system

(Dionex UltiMate 3000 RSLC, ThermoFisher Scientific, Bremen, Germany). Column chromatography (CC) was performed on silica gel (200–300 mesh, Qingdao Marine Chemical Ltd., Qingdao, China), RP-18 gel (20–45 μm, Fuji Silysia Chemical Ltd., Kasugai, Japan), and Sephadex LH-20 (Pharmacia Fine Chemical Co. Ltd., Uppsala, Sweden). Medium-pressure liquid chromatography (MPLC) was performed on a Büchi Sepacore System equipped with a pump manager C-615, pump modules C-605, and a fraction collector C-660 (Büchi Labortechnik AG, Flawil, Switzerland). Preparative high-performance liquid chromatography (prep-HPLC) was performed on an Agilent 1260 liquid chromatography system equipped with Zorbax SB-C18 columns (5 μm, 9.4 mm × 150 mm, or 21.2 mm × 150 mm) and a DAD detector. Chiral resolution was achieved by HPLC equipped with a Daicel AD-H column. Fractions were monitored by TLC (GF 254, Qingdao Haiyang Chemical Co. Ltd. Qingdao, China), and spots were visualized by heating silica gel plates sprayed with 10% H_2SO_4 in EtOH.

2.2. Fermentation, Extraction, and Isolation

The fungus *Bipolaris* sp. was isolated from fresh and healthy stems of kiwifruit plants (*Actinidia chinensis* Planch., Actinidiaceae), which were collected from the Cangxi county of the Sichuan Province (GPS: N 31°12′, E 105°76′) in July 2018. Each fungus was obtained simultaneously from at least three different healthy tissues. The fungus was identified as one species of the genus *Bipolaris* by observing the morphological characteristics and analysis of the internal transcribed spacer (ITS) regions. A living culture (internal number HFG-20180727-HJ32) has been deposited at the School of Pharmaceutical Sciences, South-Central University for Nationalities, China.

This fungal strain was cultured on a potato dextrose agar (PDA) medium at 24 °C for 10 days. The agar plugs were inoculated in 500 mL Erlenmeyer flasks, each containing 100 mL potato dextrose media. Flask cultures were incubated at 28 °C on a rotary shaker at 160 rpm for two days as the seed culture. Four hundred 500 mL Erlenmeyer flasks, each containing 150 mL potato dextrose broth (PDB), were individually inoculated with 25 mL of seed culture and were incubated at 25 °C on a rotary shaker at 160 rpm for 25 days.

The cultures of *Bipolaris* sp. (60 L) were extracted four times by EtOAc to afford a crude extract (32.0 g) which was subjected to CC over silica gel eluted with a gradient of $CHCl_3$-MeOH (a gradient from 1:0 to 0:1) to give six fractions, A–F. Fraction B (13.0 g) was fractionated by MPLC CC over RP-18 eluted with MeOH–H_2O (from 10:90 to 100:0, v/v) to give twelve sub-fractions (B_1–B_{12}). Fraction B_3 (1.2 g) was applied to Sephadex LH-20 eluting with $CHCl_3$–MeOH (1:1, v/v) and was further purified by preparative HPLC with MeCN–H_2O (19:81, v/v, 4.0 mL/min) to obtain compounds **9** (18.6 mg, retention time (t_R) = 40 min), **18** (22.6 mg, t_R = 15.8 min), **2** (3.3 mg, t_R = 32 min), and **1** (5.4 mg, t_R = 36 min). Fraction B_5 (2.1 g) was separated by CC over silica gel with a gradient elution of the $CHCl_3$–MeOH system (50:1→0:1) and was prepared by HPLC with MeCN–H_2O (12:88, v/v, 4.0 mL/min) to obtain **3** (4.9 mg, t_R = 36 min), **4** (14.4 mg, t_R = 46 min), **17** (28.3 mg, t_R = 43 min), and **5** (2.1 mg, t_R = 40 min). Fraction B_6 (1.8 g) was purified over Sephadex LH-20 eluted with MeOH to give four subfractions ($B_{6.1}$–$B_{6.4}$). Fraction $B_{6.2}$ (210 mg) was purified using semipreparative HPLC with MeOH-H_2O (28:72, v/v, 3.0 mL/min) to afford **8** (8.8 mg, t_R = 17.8 min) and **7** (9.6 mg, t_R = 21.1 min). Fraction $B_{6.3}$ (170 mg) was purified by preparative HPLC with MeCN–H_2O (23:77, v/v, 4 mL/min) to yield **6** (4.3 mg, 26 min). Fraction C (4.3 g) was separated by CC over silica gel with a gradient elution of PE-acetone (50:1→0:1) to afford subfractions C_1–C_8. Fraction C_2 (340 mg) was purified by preparative HPLC with MeCN-H_2O (55:45, v/v, 4 mL/min) to give **12** (10.3 mg, t_R = 38 min), **13** (3.7 mg, t_R = 39 min), **14** (3.1 mg, t_R = 36 min) and **15** (3.4 mg, t_R = 34 min). Fraction C_5 (230 mg) was isolated by CC over Sephadex LH-20 (MeOH) and was prepared by HPLC (32:68, v/v, 4 mL/min) to give **10** (3.7 mg, t_R = 28 min), **11** (4.2 mg, t_R = 29 min), and **16** (5.1 mg, t_R = 24 min).

Bipolarisorokin A (**1**): colorless crystals; mp 145–148 °C; $[\alpha]^{20}_D$ + 67.8 (c 0.01, MeOH); UV (MeOH) λ_{max} (log ε) 205 (3.30); IR (KBr) ν_{max} 3360, 2947, 2833, 1651, 1454, 1114,

1031 cm^{-1}; ^{1}H NMR (600 MHz, $CDCl_3$) and ^{13}C NMR (150 MHz, $CDCl_3$) data, see Table 1; positive ion HRESIMS *m/z* 251.16624 [M–H]$^+$, (calculated for $C_{15}H_{23}O_3^-$ 251.16527).

Table 1. ^{1}H (600 MHz) and ^{13}C (150 MHz) NMR Spectroscopic Data for **1–3**.

No.	1 [a]		2 [a]		3 [b]	
	δ_C, Type	δ_H (*J* in Hz)	δ_C, Type	δ_H (*J* in Hz)	δ_C, Type	δ_H (*J* in Hz)
1	54.3, CH	2.71, s	55.4, CH	2.59, br s	140,0, C	
2	156.8, C		155.7, C		167.0, C	
3	43.0, C		42.8, C		58.0, C	
4a	39.9, CH_2	1.50, m	36.7, CH_2	1.23, m	29.6, CH_2	1.42, dd (13.2, 6.0)
4b		1.36, m		1.74, m		1.55, dd (12.8, 6.0)
5a	25.2, CH_2	1.58, m	32.3, CH_2	1.44, m	25.9, CH_2	0.90, m
5b		1.24, m		1.55, m		1.80, m
6	37.6, CH	1.65, m	73.7, C		46.4, CH	1.05, m
7	42.2, CH	2.46, s	47.8, CH	2.44, br s	42.7, CH	3.06, br s
8a	20.8, CH_3	1.05, s	20.8, CH_3	1.06, s	64.6, CH_2	3.63, d (11.6)
8b						3.71, d (11.6)
9	40.5, CH	1.46, m	36.9, CH	1.57, m	32.9, CH	1.02, m
10	15.4, CH_3	0.92, d (6.8)	16.2, CH_3	0.88, d (6.9)	21.1, CH_3	0.78, d (6.4)
11	66.9, CH_2	3.64, overlap	16.4, CH_3	0.94, d (6.9)	22.1, CH_3	1.06, d (6.4)
12a	103.5, CH_2	4.94, s	105, CH_2	4.69, s	11.0, CH_3	2.13, s
12b		4.62, s		4.97, s		
13	58.2, CH	1.70, br s	54.7, CH	1.97, br s	60.8, CH	1.82, m
14a	69.6, CH	4.02, d (5.9)	69.5, CH	4.07, d (6.1)	62.9, CH_2	3.34, dd (11.0, 6.8)
14b						3.61, dd (11.2, 6.8)
15	74.9, CH	3.65, overlap	74.8, CH	3.68, d (6.1)	190.0, CH	10.02, s

[a] Measured in $CDCl_3$; [b] Measured in methanol-d_4.

Bipolarisorokin B (**2**): colorless oil; $[\alpha]^{22}{}_D$ − 100.1 (*c* 0.05, MeOH); UV (MeOH) λ_{max} (log ε) 210 (3.23); ^{1}H NMR (600 MHz, $CDCl_3$) and ^{13}C NMR (150 MHz, $CDCl_3$) data, see Table 1; positive ion HRESIMS *m/z* 275.16166 [M+Na]$^+$, (calculated for $C_{15}H_{24}O_3Na^+$ 275.16177).

Bipolarisorokin C (**3**): colorless, needle-like crystals (MeOH); mp 135–138 °C; $[\alpha]^{22}{}_D$ − 21.8 (*c* 0.05, MeOH); UV (MeOH) λ_{max} (log ε) 265 (3.49) nm; ^{1}H NMR (600 MHz, methanol-d_4) and ^{13}C NMR (150 MHz, methanol-d_4) data, see Table 1; positive ion HRESIMS *m/z* 253.17971 [M+H]$^+$ (calculated for $C_{15}H_{25}O_3^+$ 253.17982).

Bipolarisorokin D (**4**): colorless oil; $[\alpha]^{25}{}_D$ + 32.0 (*c* 0.05, MeOH); UV (MeOH) λ_{max} (log ε) 255 (3.65); ^{1}H NMR (600 MHz, methanol-d_4) and ^{13}C NMR (150 MHz, methanol-d_4) data, see Table 2; positive ion HRESIMS *m/z* 275.16153 [M+Na]$^+$ (calculated for $C_{15}H_{24}NaO_3^+$ 275.16177). *Bipolarisorokin E* (**5**): colorless oil; $[\alpha]^{25}{}_D$ − 22.7 (*c* 0.05, MeOH); UV (MeOH) λ_{max} (log ε) 210 (3.24); ^{1}H NMR (600 MHz, methanol-d_4) and ^{13}C NMR (150 MHz, methanol-d_4) data, see Table 2; positive ion HRESIMS *m/z* 221.15529 [M–H]$^-$ (calculated for $C_{14}H_{21}O_2^-$ 221.15470).

Bipolarisorokin F (**6**): white powder; $[\alpha]^{20}{}_D$ − 3.3 (*c* 0.04, MeOH); UV (MeOH) λ_{max} (log ε) 215 (3.72); ^{1}H NMR (600 MHz, $CDCl_3$) and ^{13}C NMR (150 MHz, $CDCl_3$) data, see Table 2; positive ion HRESIMS *m/z* 225.18506 [M+H]$^+$ (calculated for $C_{14}H_{25}O_2^+$ 225.18491).

Bipolarisorokin G (**7**): colorless oil; $[\alpha]^{20}{}_D$ + 17.2 (*c* 0.02, MeOH); UV (MeOH) λ_{max} (log ε) 230 (3.21); ^{1}H NMR (600 MHz, $CDCl_3$) and ^{13}C NMR (150 MHz, $CDCl_3$) data, see Table 3; positive ion HRESIMS *m/z* 275.20059 [M+H]$^+$ (calculated for $C_{18}H_{27}O_2^+$ 275.20056).

Table 2. ^{1}H (600 MHz) and ^{13}C (150 MHz) NMR Spectroscopic Data for **4–6**.

No.	4 [b]		5 [b]		6 [a]	
	δ_C, Type	δ_H (J in Hz)	δ_C, Type	δ_H (J in Hz)	δ_C, Type	δ_H (J in Hz)
1	140.5, C		124.2, CH	5.56, br s	212.0, C	
2	170.4, C		147.2, C		50.7, CH	2.10, m
3	52.0, C		47.7, C		41.8, C	
4a	32.4, CH_2	1.38, m	35.2, CH_2	1.34, m	36.1, CH_2	1.44, m
4b		1.71, m		1.41, dd (12.5, 5.2)		1.66, dd (13.7,5.7)
5a	32.8, CH_2	1.25, m	26.0, CH_2	1.56, m	26.0, CH_2	1.80, m
5b		1.61, m				0.87, m
6	73.5, C		45.2, CH	2.03, m	50.1, CH	1.33, m
7	47.9, CH	3.16, br s	45.3, CH	2.74, br s	51.3, CH	2.70, brs
8	18.7, CH_3	1.07, s	18.9, CH_3	0.99, s	22.1, CH_3	1.09, s
9	37.2, CH	1.28, m	150.3, C		29.9, CH	1.55, m
10	17.1, CH_3	1.02, d (6.6)	109.2, CH_2	4.69, d (5.1)	20.3, CH_3	1.03, d (6.5)
11	16.4, CH_3	0.80, d (6.6)	22.7, CH_3	1.74, s	21.4, CH_3	0.86, d (6.5)
12	11.3, CH_2	2.06, s	59.8, CH_2	4.06, m	6.3, CH_3	0.96, d (7.2)
13	55.3, CH	2.43, dd (9.1, 5.4)	64.3, CH	1.64, dd (9.6, 4.9)	54.9, C	1.72, dd (7.9, 5.0)
14a	62.3, CH_2	3.19, dd (10.5, 9.1)	62.5, CH_2	3.38, m	62.0, CH_2	3.85, dd (10.7, 5.0)
14b		3.61, dd (10.5, 5.4)		3.65, dd (10.5, 5.0)		3.50, dd (10.7, 7.9)
15	189.7, CH	9.97, s				

[a] Measured in $CDCl_3$; [b] Measured in methanol-d_4.

Table 3. ^{1}H (600 MHz) and ^{13}C (150 MHz) NMR Spectroscopic Data for **7–9**.

No.	7 [a]		8 [b]		9 [b]	
	δ_C, Type	δ_H (J in Hz)	δ_C, Type	δ_H (J in Hz)	δ_C, Type	δ_H (J in Hz)
1	137.5, C		137.4, C		57.5, CH	2.54, br s
2	165.3, C		165.3, C		163.7, C	
3	52.6, C		52.5, C		41.8, C	
4a	33.7, CH_2	1.41, dd (13.4, 5.9)	33.6, CH_2	1.40, dd (13.3, 5.9)	53.2, CH_2	1.66, dd (13.2, 10.4)
4b		1.50, dd (13.4, 6.4)		1.48, dd (13.3, 6.5)		2.10, dd (13.2, 10.4)
5a	25.2, CH_2	0.91, m	25.2, CH_2	0.90, m	67.0, CH	3.84, m
5b		1.80, m		1.78, m		
6a	44.3, CH	1.06, m	44.2, CH	1.06, m	47.2, CH_2	1.21, m
6b						1.98, m
7	44.7, CH	3.06, br s	44.5, CH	3.04, br s	32.2, C	
8	19.7, CH_3	0.97, s	19.6, CH_3	0.96, s	28.7, CH_3	0.99, s
9	31.6, CH	1.03, m	31.6, CH	1.03, m	55.0, CH	2.02, br s
10	21.7, CH_3	1.06, d (5.9)	21.7, CH_3	1.04, d (5.8)	30.3, CH_3	1.09, s
11	20.8, CH_3	0.77, d (5.9)	20.8, CH_3	0.76, d (5.8)	31.7, CH_3	0.95, s
12a	11.0, CH_3	2.06, s	10.9, CH_3	2.04, s	103.9, CH_2	4.75, br s
12b						4.97, br s
13	63.6, CH	2.22, d (9.6)	63.4, CH	2.23, d (9.8)	53.2, CH	2.01, br s
14	147.9, CH	6.55, dd (15.9, 9.6)	151.5, CH	6.80, dd (15.4, 9.9)	70.5, CH	4.13, d (6.2)
15	188.1, CH	10.08, s	188.1, CH	10.05, s	74.9, CH	3.59, d (6.2)
16	132.2, CH	6.08, d (15.9)	122.1, CH	5.81, d (15.5)		
17	198.6, C		171.1, C			
18	27.5, CH_3	2.20, s				

[a] Measured in $CDCl_3$; [b] Measured in methanol-d_4.

Bipolarisorokin H (**8**): colorless oil; $[\alpha]^{25}_D$ − 136.9 (c 0.05, MeOH); UV (MeOH) λ_{max} (log ε) 225 (3.93); ^{1}H NMR (600 MHz, $CDCl_3$) and ^{13}C NMR (150 MHz, $CDCl_3$) data, see Table 3; positive ion HRESIMS m/z 277.17984 $[M+H]^+$ (calculated for $C_{17}H_{25}O_3{}^+$ 277.17982).

Bipolarisorokin I (**9**): colorless crystals; mp 191–194 °C; $[\alpha]^{22}_D$ + 8.8 (c 0.05, MeOH); UV (MeOH) λ_{max} (log ε) 210 (3.46); ^{1}H NMR (600 MHz, methanol-d_4) and ^{13}C NMR (150 MHz, methanol-d_4) data, see Table 3; positive ion HRESIMS m/z 251.16621 $[M–H]^-$, (calculated for $C_{21}H_{23}O_3{}^-$ 251.16527).

Bipolarithone A (**10**): colorless oil; $[\alpha]^{23}_{D}$ + 136.0 (*c* 0.05, MeOH); UV (MeOH) λ_{max} (log ε) 245 (3.30); ^{1}H NMR (600 MHz, $CDCl_3$) and ^{13}C NMR (150 MHz, $CDCl_3$) data, see Table 4; positive ion HRESIMS *m/z* 349.09143 $[M+H]^{+}$, (calculated for $C_{17}H_{17}O_8{}^{+}$ 349.09179).

Table 4. ^{1}H (600 MHz) and ^{13}C (150 MHz) NMR Spectroscopic Data for **10** and **11**.

No.	10 [a]		11 [a]	
	δ_C, Type	δ_H (*J* in Hz)	δ_C, Type	δ_H (*J* in Hz)
1	161.1, C		161.1, C	
2	113.7, CH	6.68, s	113.7, CH	6.69, s
3	147.7, C		147.8, C	
4	108.1, CH	6.75, s	108.1, CH	6.76, s
4a	157.4, C		157.4, C	
5	78.2, CH	5.73, ddd (6.6, 4.4, 3.9)	78.6, CH	5.62, ddd (8.4, 3.8, 1.7)
6a	37.7, CH_2	3.01, dd (16.2, 4.4)	39.3, CH_2	3.10, dd (16.3, 8.4)
6b		2.85, dd (16.2, 6.6)		2.99, dd (16.3, 3.8)
7	170.0, C		170.2, C	
8	79.4, CH	5.64, d (3.9)	79.8, CH	5.63, d (1.7)
8a	114.7, C		114.6, C	
9	178.3, C		178.2, C	
9a	109.0, C		109.0, C	
10a	167.7, C		167.4, C	
1′	22.5, CH_3	2.41, s	22.5, CH_3	2.42, s
2′	169.5, C		170.1, C	
3′	52.4, CH_3	3.73, s	52.5, CH_3	3.78, s
4′	53.0, CH_3	3.81, s	53.1, CH_3	3.83, s
1-OH		12.06, s		12.01, s

[a] Measured in $CDCl_3$; [b] Measured in methanol-d_4.

Bipolarithone B (**11**): colorless oil; $[\alpha]^{23}_{D}$ − 24.2 (*c* 0.05, MeOH); UV (MeOH) λ_{max} (log ε) 245 (3.30); ^{1}H NMR (600 MHz, $CDCl_3$) and ^{13}C NMR (150 MHz, $CDCl_3$) data, see Table 4; positive ion HRESIMS *m/z* 349.09157 $[M+H]^{+}$, (calculated for $C_{17}H_{17}O_8{}^{+}$ 349.09179).

Bipolarithone C (**12**): colorless oil; $[\alpha]^{25}_{D}$ + 52.9 (*c* 0.5, MeOH); UV (MeOH) λ_{max} (log ε) 245 (4.06); ^{1}H NMR (600 MHz, $CDCl_3$) and ^{13}C NMR (150 MHz, $CDCl_3$) data, see Table 5; positive ion HRESIMS *m/z* 541.24310 $[M+H]^{+}$, (calculated for $C_{30}H_{37}O_9{}^{+}$ 541.24321).

Bipolarithone D (**13**): colorless oil; $[\alpha]^{25}_{D}$ + 10.2 (*c* 0.5, MeOH); UV (MeOH) λ_{max} (log ε) 245 (3.88); ^{1}H NMR (600 MHz, $CDCl_3$) and ^{13}C NMR (150 MHz, $CDCl_3$) data, see Table 5; positive ion HRESIMS *m/z* 541.24316 $[M+H]^{+}$, (calculated for $C_{30}H_{37}O_9{}^{+}$ 541.24321).

Crystal data for Cu_**1**_0m: $C_{15}H_{24}O_3$, M = 252.34, a = 9.7038(6) Å, b = 13.7866(8) Å, c = 16.6333(10) Å, α = 95.329(3)°, β = 104.898(2)°, γ = 102.525(3)°, V = 2073.0(2) Å^3, T = 100(2) K, space group P 1, Z = 6, μ(Cu Kα) = 1.54178 mm^{-1}, $F(000)$ = 828, 82979 reflections measured, 16831 independent reflections (R_{int} = 0.0695). The final R_1 values were 0.0437 ($I > 2\sigma(I)$). The final $wR(F^2)$ values were 0.1047 ($I > 2\sigma(I)$). The final R_1 values were 0.0531 (all data). The final $wR(F^2)$ values were 0.1143 (all data). The goodness of fit on F^2 was 1.039. Flack parameter = −0.10(7). CCDC: 2124305. Available online: https://www.ccdc.cam.ac.uk (accessed on 11 December 2021).

Crystal data for Cu_**3**_0m: $C_{15}H_{24}O_3$, M = 252.34, a = 7.0044(5) Å, b = 10.1468(8) Å, c = 20.1433(14) Å, α = 90.00°, β = 90.00°, γ = 90.00°, V= 1431.63(18) Å^3, T = 295(2) K, space group P 21 21 21, with Z = 4, μ(Cu Kα) = 1.54178 mm^{-1}, $F(000)$ = 552, 6263 reflections measured, 2527 independent reflections (R_{int} = 0.0500). The final R_1 values were 0.0519 ($I > 2\sigma(I)$). The final $wR(F^2)$ values were 0.1538 ($I > 2\sigma(I)$). The final R_1 values were 0.0719 (all data). The final $wR(F^2)$ values were 0.2087 (all data). The goodness of fit on F^2 was 1.117. Flack parameter = −0.40(17). CCDC: 2124306. Available online: https://www.ccdc.cam.ac.uk (accessed on 11 December 2021).

Table 5. ^{1}H (600 MHz) and ^{13}C (150 MHz) NMR Spectroscopic Data for **12** and **13**.

No.	**12** [a]		**13** [a]	
	δ_C, Type	δ_H (J in Hz)	δ_C, Type	δ_H (J in Hz)
1	221.6, C		221.4, C	
2	50.6, CH	2.16, m	50.6, CH	2.13, m
3	42.1, C		42.0, C	
4a	36.0, CH2	1.45, m	36.1, CH_2	1.47, m
4b		1.66, m		1.67, m
5a	26.0, CH2	0.84, m	26.0, CH_2	0.83, m
5b		1.78, m		1.79, m
6	50.2, CH	1.29, m	50.1, CH	1.28, m
7	51.5, CH	2.56, br s	51.5, CH	2.62, br s
8	22.1, CH3	1.08, s	22.1, CH_3	1.09, s
9	29.9, CH	1.41, m	30.0, CH	1.43, m
10	20.4, CH3	0.77, d (6.6)	20.4, CH_3	0.78, d (6.7)
11	21.3, CH3	0.89, d (6.4)	21.3, CH_3	0.92, d (6.5)
12	6.5, CH3	0.95, d (7.2)	6.5, CH_3	0.96, d (7.2)
13	51.6, CH	1.90, m	51.6, CH	1.94, m
14a	65.3, CH2	4.05, dd (11.3, 5.1)	65.5, CH_2	4.04, dd (11.3, 5.2)
14b		4.35, dd (11.3, 5.1)		4.37, dd (11.3, 5.2)
1′	161.1, C		161.1, C	
2′	113.6, CH	6.67, s	113.7, CH	6.68, s
3′	147.6, C		147.8, C	
4′	108.1, CH	6.75, s	108.1, CH	6.75, s
4a′	157.3, C		157.4, C	
5′	78.2, CH	5.67, ddd (6.4, 4.3, 3.9)	78.5, CH	5.59, ddd (8.2, 3.9, 1.8)
6′a	37.7, CH2	2.99, dd (16.1, 4.3)	39.2, CH_2	3.07, dd (16.3, 8.2)
6′b		2.84, dd (16.1, 6.4)		2.99, dd (16.3, 3.9)
7′	169.5, C		170.0, C	
8′	79.5, CH	5.59, d (3.9)	79.9, CH	5.58, d (1.8)
8a′	114.5, C		114.5, C	
9′	178.2, C		178.2, C	
9a′	109.0, C		109.0, C	
10a′	167.7, C		167.3, C	
1″	22.5, CH3	2.40, s	22.5, CH_3	2.41, s
2″	169.4, C		169.7, C	
3″	52.4, CH3	3.72, s	52.5, CH_3	3.72, s
1′-OH		12.06, s		12.06, s

[a] Measured in $CDCl_3$; [b] Measured in methanol-d_4.

Crystal data for Cu_**9**_0m: $C_{15}H_{24}O_3$, M = 252.34, a = 6.8634(2) Å, b = 15.0872(4) Å, c = 13.5156(3) Å, α = 90.00°, β = 90.4010(10)°, γ = 90.00°,V = 1399.50(6) Å^3, T = 295(2) K, space group P 1 21 1, with Z = 4, μ(Cu Kα) = 1.54178 mm^{-1}, F(000) = 552, 32232 reflections measured, 5982 independent reflections (R_{int} = 0.0279). The final R_1 values were 0.0300 ($I > 2\sigma(I)$). The final $wR(F^2)$ values were 0.0808 ($I > 2\sigma(I)$). The final R_1 values were 0.0304 (all data). The final $wR(F^2)$ values were 0.0812 (all data). The goodness of fit on F^2 was 1.057. Flack parameter = −0.01(3). CCDC: 2124307. Available online: https://www.ccdc.cam.ac.uk (accessed on 11 December 2021).

Crystal data for Cu_**17**_0m: $C_{14}H_{24}O_2$, M = 224.33, a = 13.6388(2) Å, b = 13.6388(2) Å, c = 13.0174(2) Å, α = 90.00°, β = 90.00°, γ = 90.00°, V = 2097.04(7) Å^3, T = 296(2) K, space group P 31 2 1, with Z = 6, μ(Cu Kα) = 1.54178 mm^{-1}, F(000) = 744, 39026 reflections measured, 3033 independent reflections (R_{int} = 0.0459). The final R_1 values were 0.0353 ($I > 2\sigma(I)$). The final $wR(F^2)$ values were 0.0988 ($I > 2\sigma(I)$). The final R_1 values were 0.0366 (all data). The final $wR(F^2)$ values were 0.1003 (all data). The goodness of fit on F^2 was 1.047. Flack parameter =0.01(5). CCDC: 2126101. Available online: https://www.ccdc.cam.ac.uk (accessed on 11 December 2021).

Crystal data for Cu_**18**_0m: $C_{15}H_{26}O_2$, M = 238.36, a = 13.1977(2) Å, b = 13.1977(2) Å, c = 8.49040(10) Å, α = 90.00°, β = 90.00°, γ = 90.00°, V = 1478.85(5) Å^3, T = 297(2) K, space group P 43, with Z = 4, μ(Cu Kα) = 1.54178 mm^{-1}, F(000) = 528, 14568 reflections measured, 3063 independent reflections (R_{int} = 0.0269). The final R_1 values were 0.0534 ($I > 2\sigma(I)$). The final $wR(F^2)$ values were 0.1525 ($I > 2\sigma(I)$). The final R_1 values were 0.0541 (all data). The final $wR(F^2)$ values were 0.1539 (all data). The goodness of fit on F^2 was 1.051. Flack parameter =0.12(7). CCDC: 2126105. Available online: https://www.ccdc.cam.ac.uk (accessed on 11 December 2021).

2.3. ECD Calculations

The ECD calculations were carried out using the Gaussian 16 software package [26]. Systematic conformational analyses were performed via SYBYL-X 2.1 using the MMFF94 molecular mechanics force field calculation with 10 kcal/mol of cutoff energy [27,28]. The optimization and frequency of conformers were calculated on the B3LYP/6-31G(d) level in the Gaussian 09 program package. The ECD (TDDFT) calculations were performed on the B3LYP/6-311G(d) level of theory with an IEFPCM solvent model (MeOH). The ECD curves were simulated in SpecDis V1.71 using a Gaussian function [29]. The calculated ECD data of all conformers were Boltzmann averaged by Gibbs free energy.

2.4. NMR Calculations

All the optimized conformers in an energy window of 5 kcal/mol (with no imaginary frequency) were subjected to gauge-independent atomic orbital (GIAO) calculations of their ^{13}C NMR chemical shifts, using density functional theory (DFT) at the mPW1PW91/6-311+G (d,p) level with the PCM model. The calculated NMR data of these conformers were averaged according to the Boltzmann distribution theory and their relative Gibbs free energy. The ^{13}C NMR chemical shifts for TMS were also calculated by the same procedures and used as the reference. After the calculation, the experimental and calculated data were evaluated by the improved probability DP4^{+} method [30].

2.5. Antibacterial Activity Assay

The bacterium *P. syringae* pv. *actinidiae* was donated by Dr. He Yan of Northwest A&F University, China. A sample of each culture was then diluted 1000-fold in fresh Luria-Bertani (LB) (Beijing Solarbio Science & Technology. Co. Ltd., Beijing, China) and incubated with shaking (160 rpm) at 27 °C for 10 h. The resultant mid-log phase cultures were diluted to a concentration of 5 × 10^5 CFU/mL, then 160 μL was added to each well of the compound-containing plates. Subsequently, 1:1 serial dilutions with sterile PBS of each compound were performed, giving a final compound concentration range from 4 to256 μg/mL. The minimum inhibitory concentration (MIC, with an inhibition rate of ≥90%) was determined by using photometry at OD_{600} nm after 24 h. Streptomycin was used as the positive control.

2.6. Anti-Phytopathogens Assay

Four phytopathogens (*Phytophthora infestane, Alternaria solani, Rhizoctonia solani*, and *Fusarium oxysporum*) were cultured in PDA with micro glass beads at 27 °C for a week, as well as shaking (160 rpm). Ninety microliters of PDA, together with a 10 μL volume of an aqueous test sample solution, was added into each well of the 96-well plate. The test solutions contained different concentrations of the sample being tested. Then, agar plugs (1 mm^3) with fresh phytopathogens were inoculated into each well. Subsequently, a two-fold serial dilution in the microplate wells was performed over a concentration range of 4 to 256 μg/mL. Plates were covered and incubated at 27 °C for 24 h. Finally, the minimum inhibitory concentration was determined by observing the plates, with no growth in the well taken as that value. Hygromycin B was used as the positive control.

3. Results and Discussion

Bipolarisorokin A (**1**) was isolated as colorless crystals. Its molecular formula of $C_{15}H_{24}O_3$ was determined on the basis of the HR-ESIMS data (measured at *m/z* 251.16624

[M–H]$^-$, calculated for $C_{15}H_{23}O_3{}^-$ 251.16527), corresponding to four degrees of unsaturation. The ^{1}H and ^{13}C NMR spectra, in association with the HSQC spectrum, revealed two methyls, four methenes, seven methines, and two quaternary carbons (Table 1). Of them, signals at δ_C 66.9 (t, C-11), 69.6 (d, C-14), and 74.9 (d, C-15) were identified as the oxygenated methylene and methines. Two olefinic carbons at δ_C 156.8 (s, C-2) and 103.5 (t, C-12) corresponded to a double bond, which suggested that **1** possessed a tricyclic system. Considering the 15 carbons in **1**, as well as those isolates from the same source, compound **1** was suggested to be a tricyclic sesquiterpenoid. In the ^{1}H–^{1}H COSY spectrum, a fragment was revealed, as shown with bold lines in Figure 2. The HMBC correlations from to δ_H 4.94 (H, s, H-12a) and 4.62 (H, s, H-12b), to δ_C 156.8 (s, C-2), 54.3 (d, C-1) and 43.0 (s, C-3), established the connections between C-12, C-2, and C-1. Further analyses of ^{1}H–^{1}H COSY, as well as HMBC correlations from δ_H 0.92 (3H, d, J = 6.8 Hz, H-10) to δ_C 37.6 (d, C-6), 40.5 (d, C-9) and 66.9 (t, C-11), indicated a hydroxy group at C-11. In addition, the connections of C-8/C-3, C-3/C-4, C-3/C-2, and C-3/C-13 were deduced from HMBC correlations from δ_H 1.05 (3H, s, H-8) to δ_C 43.0 (s, C-3), 39.9 (t, C-4), 156.8 (s, C-2), and 58.2 (d, C-13). Moreover, the proton of an oxygenated methine at δ_H 4.02 (H, d, J = 5.9 Hz, H-14) showed key correlations to C-13, C-3, and δ_C 42.2 (d, C-7), which indicated that δ_C 69.6 (d, C-14) should be placed at C-13. The above 2D NMR data analysis suggested that compound **1** possessed a sativene type sesquiterpene backbone. A ROESY experiment was carried out to establish the relative configuration of **1** (Figure 3). The key correlations of H-13/H-8, H-13/H-6, H-8/H-14, and H-7/H-13 suggested that H-6, H-7, H-8, and H-13 were β oriented, while the correlation of H-1/H-9 indicated that H-1 and H-9 were α-oriented. Because of the rigid structure and the ROESY correlation of H-8/H-14, both H-14 and H-15 were assigned as an α orientation [31]. Finally, the single-crystal X-ray diffraction not only confirmed the planar structure, as elucidated above, but also established the absolute configuration of **1** (Flack parameter = −0.10(7), CCDC: 2124305; Figure 4).

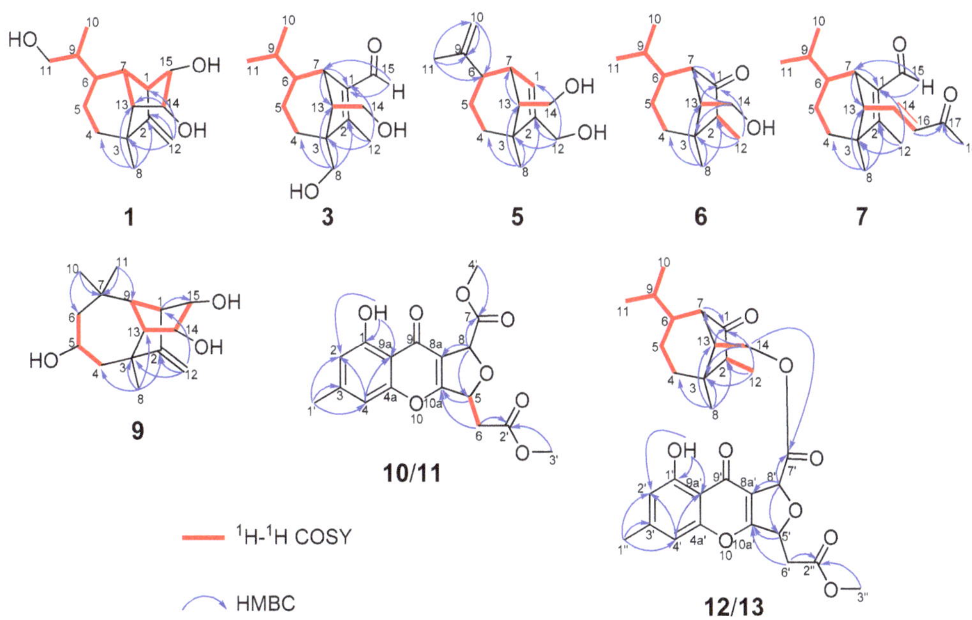

Figure 2. Key ^{1}H–^{1}H COSY and HMBC correlations for **1**, **3**, **5**, **6**, **7**, and **9–13**.

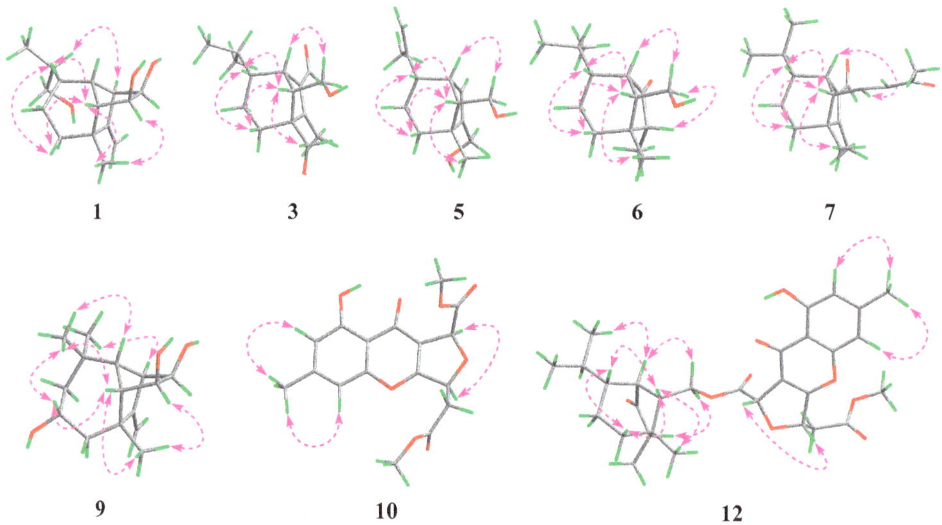

Figure 3. Key ROESY correlations for **1**, **3**, **5**, **6**, **7**, **9**, **10**, and **12**.

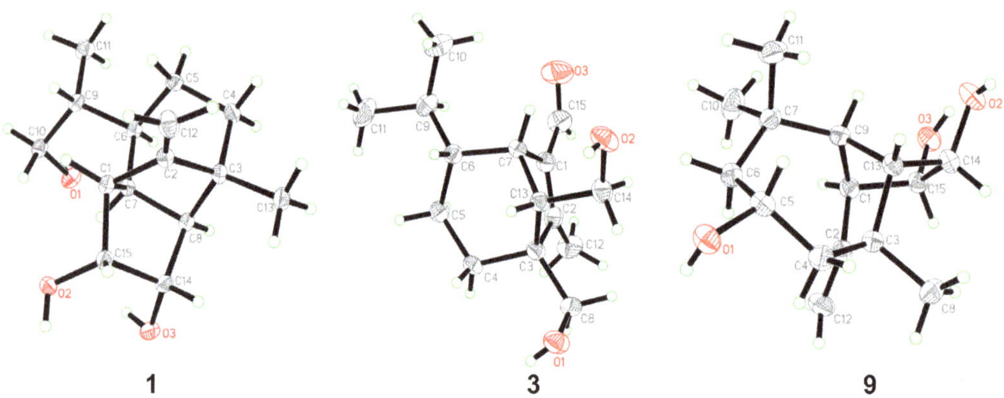

Figure 4. ORTEP diagrams of **1**, **3**, and **9**.

The molecular formula of bipolarisorokin B (**2**) was determined to be $C_{15}H_{24}O_3$ from the HRESIMS data (measured at *m/z* 275.16166 $[M+Na]^+$, calculated for $C_{15}H_{24}O_3Na^+$ 275.16177). Close similarities were observed in the 1D NMR data (Table 1) of compound **1**. However, signals for a methyl (δ_H 0.94, d, J = 6.9 Hz, H-11; δ_C 16.4, C-11) and an oxygenated quaternary carbon (δ_C 73.7, C-6) in **2** was suggested to replace the oxymethylene (δ_H 3.64, overlap, H-11; δ_C 66.9, C-11) and the methine (δ_H 1.65, m, H-6; δ_C 37.6, C-6) in **1**. These observations indicated that the hydroxy group at C-10 in **1** migrated to C-6 in **2**. The observed $^1H-^1H$ COSY cross-peak of H-10 (δ_H 0.88, 3H, d, J = 6.9 Hz) and H-9 (δ_H 1.57, 1H, m), and H-9/H-11, along with the HMBC correlations from H-10 to C-6, C-9, and C-11 confirmed the above deduction (Figure 2). Furthermore, ROESY correlations of H-13/H-8, H-8/H-14, H-7/H-13, and H-1/H-9 revealed that compounds **2** and **1** shared the same relative configuration. In consideration of its biosynthetic origin, the absolute configuration of compound **2** was identified the same as that of **1**.

Bipolarisorokin C (**3**) was obtained as colorless needles. Its molecular formula of $C_{15}H_{24}O_3$ was determined on the basis of the HR-ESIMS data (measured at *m/z* 253.17971 $[M+H]^+$, calculated for $C_{15}H_{25}O_3^+$ 253.17982), corresponding to four degrees of unsaturation. The ^1H NMR data (Table 1) showed characteristic signals, including three methyls at δ_H 0.78 (3H, d, *J* = 6.4 Hz, H-10), 1.06 (3H, d, *J* = 6.4 Hz, H-11), and 2.13 (3H, s, H-12), and the proton of an aldehyde group at δ_H 10.02 (H, s, H-15). The ^1H and ^{13}C NMR data, in association with the HSQC data, revealed three methyls, four methenes, five methines, and three nonprotonated carbons (Table 1). Preliminary analyses on the 1D NMR data revealed that **3** was likely to be a seco-sativene type sesquiterpenoid. Detailed analyses of the 2D NMR data indicated that the majority of the data of **3** was the same as those of helminthosporol [32], except for a hydroxy group at C-8 (t, δ_C 64.6) in **3**, which was confirmed by the HMBC correlations from δ_H 3.63 (H, d, *J* = 11.6 Hz, H-8a) and 3.71 (H, d, *J* = 11.6 Hz, H-8b) to δ_C 58.0 (s, C-3), 29.6 (t, C-4), 167.0 (s, C-2), and 60.8 (d, C-13) (Figure 2). A ROESY experiment was carried out to establish the relative configuration of **3** (Figure 3). The cross peaks of H-13/H-8a, H-13/H-4b, H-4b/H-6, and H-7/H-14b were observed, which indicated that H-6, H-7, H-8, and H-13 were β oriented. Furthermore, single crystal X-ray diffraction established the relative configuration (Flack parameter = −0.40(17), CCDC: 2124306; Figure 4), and the absolute configuration of **3** was determined by ECD calculations, as shown in Figure 5.

Bipolarisorokin D (**4**) was isolated as a colorless oil. The molecular formula was determined to be $C_{15}H_{24}O_3$ according to the HRESIMS spectra (measured at *m/z* 275.16153 $[M+Na]^+$, calculated for $C_{15}H_{24}NaO_3^+$ 275.16177). Compound **4** had the same molecular formula and NMR spectral patterns to that of **3** (Table 2). The key difference was an oxygenated quaternary carbon (δ_C 73.5, s) in **4** instead of the methine in **3** (δ_C 46.4, d). The HMBC correlations from H-4a (δ_H 1.38, m), H-5b (δ_H 1.61, m), H-7 (δ_H 3.16, br s), H-10 (δ_H 1.02, d, *J* = 6.6 Hz), and H-11 (δ_H 0.80, d, *J* = 6.6 Hz) to δ_C 73.5 established the quaternary carbon to be C-6. In addition, a methyl (s, δ_H 1.07, H-8; δ_C 18.7, C-8) in **4** replaced the oxygenated methylene (δ_C 64.6) of C-8 in **3**, which was verified by HMBC correlations from H-8 (δ_H 1.07, s) to C-2 (δ_C 170.4, s), C-3 (δ_C 52.0, s), C-4 (δ_C 32.4, t), and C-13 (δ_C 55.3, d). Detailed analyses of 2D NMR (HSQC, HMBC, ^1H-^1H COSY and ROESY) data confirmed that the other fragments of **4** were the same as those of **3**.

Bipolarisorokin E (**5**) was obtained as a colorless oil. Its molecular formula $C_{14}H_{22}O_2$ was characterized according to HRESIMS (measured at *m/z* 221.15529 $[M-H]^-$, calculated for $C_{14}H_{21}O_2^-$ 221.15470), implying four degrees of unsaturation. The general features of its NMR data closely resembled that of **3** (Table 2). Detailed analyses of 1D and 2D NMR data revealed the differences. At first, the loss of the aldehyde group at C-1 was revealed by the chemical shift of C-1 at δ_C 124.2, along with the data from ^1H–^1H COSY and HMBC spectra as shown in Figure 2. Second, the hydroxy migrated from C-8 to C-12 (δ_C 59.8, t) as identified by the HMBC correlation from δ_H 4.06 (2H, m, H-12) to δ_C 124.2 (d, C-1), 147.2 (s, C-2), and 47.7 (s, C-3). Third, one double bond between C-9 and C-10 was built by HMBC correlations from δ_H 4.69 (2H, d, *J* = 5.1 Hz, H-10) to δ_C 22.7 (q, C-11) and 45.2 (d, C-6). The other parts of **5** were elucidated as the same as those of **3** by a detailed analysis of 2D NMR data.

Bipolarisorokin F (**6**) was purified as white powder, and its molecular formula $C_{14}H_{24}O_2$ was determinded according to HRESIMS (measured at *m/z* 225.18506 $[M+H]^+$, calculated for $C_{14}H_{25}O_2^+$ 225.18491). Analyses of the 1D and 2D NMR data (Table 2) suggested that **6** showed structural similarities to **3**. The distinction between the two compounds was that the α,β-unsaturated aldehyde group (δ_C 140.0, C-1; δ_C 167.0, C-2; δ_C 190.0, C-15) in **3** was replaced by a carbonyl (δ_C 212.0, C-1) and a methylene group (δ_C 50.7, C-2) in **6**. It was supported by HMBC correlations from δ_H 2.70 (H, br s, H-7), 0.96 (3H, q, *J* = 7.2 Hz, H-12), and 1.72 (H, dd, *J* = 7.9, 5.0 Hz, H-13) to δ_C 212.0 (s, C-1), 50.7(d, C-2), and the COSY cross-peak of δ_H 2.10 (1H, m, H-2) and H-12. The hydroxymethyl group (C-8) in **3** was replaced by a methyl group at C-8 (δ_C 22.1, q) in **6**, as well as the HMBC correlations from δ_H 1.09 (3H, s, H-8) to C-2, δ_C 41.8 (s, C-3), δ_C 36.1 (t, C-4), and δ_C 54.9 (d, C-13). The key

ROESY cross-peak (Figure 3) of H-2/Ha-14 (H, dd, *J* = 10.7, 5.0 Hz, δ_H 3.85) suggested that H-2 was *β* oriented. Other ROESY data revealed the same patterns to **3**. Finally, regarding the same origin of **6** and **3**, the absolute configuration of **6** was identified to be the same as that of **3**, as depicted.

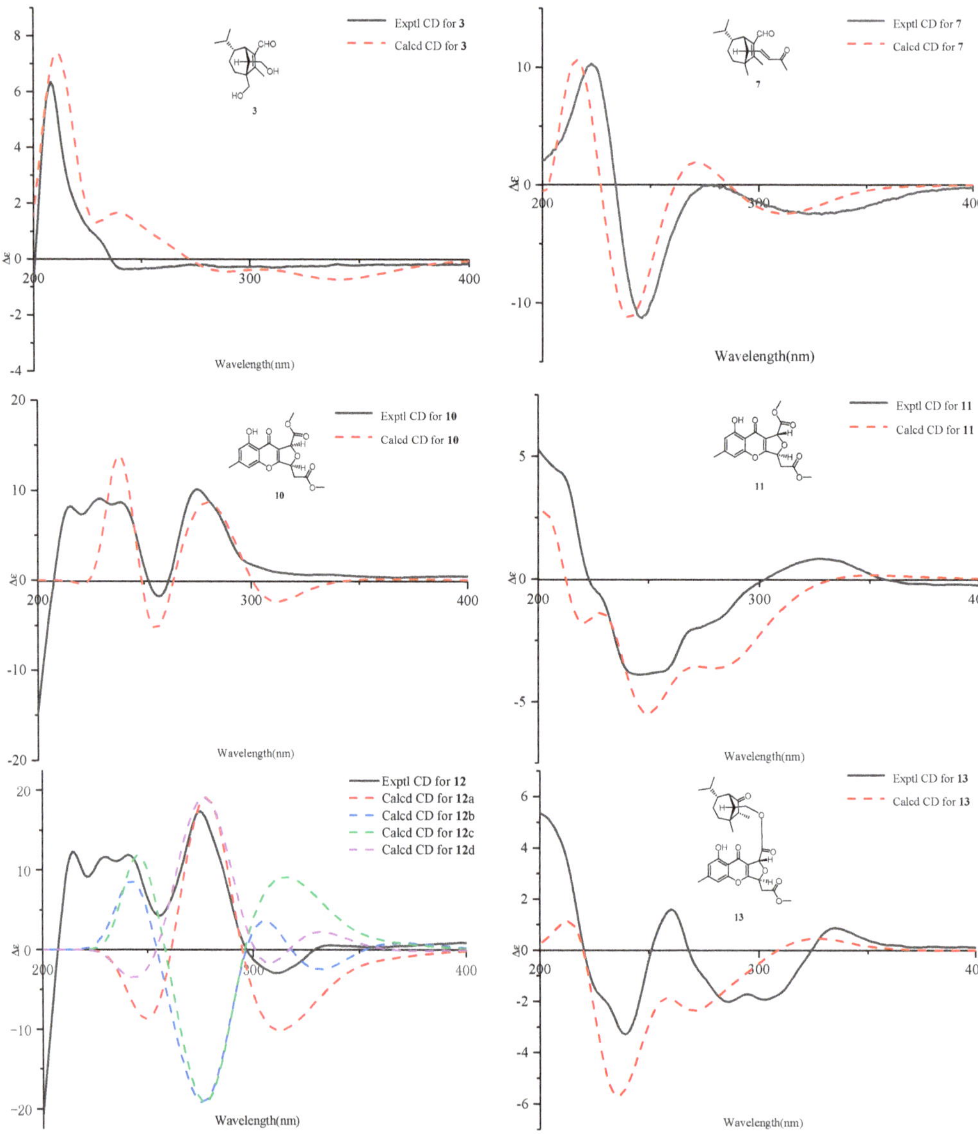

Figure 5. ECD calculations for **3**, **7**, and **10–13**.

The molecular formula of bipolarisorokin G (**7**) was assigned as $C_{18}H_{26}O_2$ based on its HRESIMS spectrum (measured at *m/z* 275.20059 $[M+H]^+$, calculated for $C_{18}H_{27}O_2^+$, 275.20056), which contained three more carbon atoms than **3**. The interpretation of the ^1H and ^{13}C NMR data of **7** (Table 3) indicated the same structure skeleton to that of **3**. Analyses of 2D NMR spectra revealed modifications in **7** (Figure 2). HMBC correlations from δ_H

0.97 (3H, s, H-8) to δ_C 165.3 (s, C-2), 52.6 (s, C-3), 33.7 (t, C-4), and 63.6 (d, C-13) suggested that a hydroxy group was missing in **7**. In addition, an α,β-unsaturated ketone group was identified by the HMBC correlations from δ_H 6.55 (H, dd, J = 15.9, 9.6 Hz, H-14), 6.08 (H, d, J = 15.9 Hz, H-16), and 2.20 (3H, s, H-18) to δ_C 198.6 (s, C-17). In the $^1H-^1H$ COSY spectrum, correlations from H-14 to δ_H 2.22 (H, d, J = 9.6 Hz, H-13) and H-16 indicated that the α,β-unsaturated carboxyl moiety was located at C-13. Finally, the absolute configuration of **7** can be fully resolved by the ECD calculation, as shown in Figure 5.

Bipolarisorokin H (**8**) was obtained as a colorless oil. Its molecular formula, $C_{17}H_{24}O_3$, was inferred from the pseudomolecular ion peak at m/z 277.17984 $[M+H]^+$ in the HRESIMS (calculated for $C_{17}H_{25}O_3^+$ 277.17982). The NMR data of **8** (Table 3) resembled that of **7**, except for the presence of a carboxyl (δ_C 171.1, C-17) in **8** instead of a carbonyl (δ_C 198.6, C-17) in **7**, as well as the loss of a methyl group. This was supported by HMBC correlations from δ_H 6.80 (H, dd, J = 15.4, 9.9 Hz, H-14) and 5.81 (H, d, J = 15.5 Hz, H-16) to δ_C 171.1 (s, C-17). Detailed analyses of 2D NMR data suggested that the other data were the same as those of **7**.

Bipolarisorokin I (**9**) was isolated as colorless crystals. Its molecular formula was identified as $C_{15}H_{24}O_3$ by HRESIMS (measured at *m/z* 251.16621 $[M–H]^-$, calculated for $C_{21}H_{23}O_3^-$ 251.16527). All the spectroscopic data indicated similar patterns to those of longifolene [33]. Detailed analyses of 1D and 2D NMR data revealed the differences. Signals at δ_C 67.0 (d, C-5), 70.5 (d, C-14), and 74.9 (d, C-15) were identified as the oxygenated methines. Therefore, three hydroxyls were suggested to be placed at C-5, C-14, and C-15, respectively, which were identified by the HMBC and $^1H–^1H$ COSY correlations, as shown in Figure 2. Comprehensive analyses of other data suggested that the other parts of **9** were the same as those of longifolene. The relative configuration of **9** was revealed by a ROESY experiment, as shown in Figure 3. The ROESY correlations of Me-10/H-13, H-13/H-5, Me-8/H-13, Me-10/H-9, and Me-10/H-5 indicated these groups were cofacial (assigned as β orientation). In addition, the Me-11/H-1 interaction suggested that H-1 should be α oriented. Moreover, the coupling constant between H-14 and H-15 ($J_{14,15}$ = 6.2 Hz), as well as the ROESY correlations of Me-8/H-14 and Me-8/H-15, suggested that H-14 and H-15 were α oriented. Finally, the single-crystal X-ray diffraction not only confirmed the planar structure but also established the absolute configuration of **9** (Flack parameter =0.01(3), CCDC: 2124307; Figure 4).

Bipolarithone A (**10**) was isolated as a yellow oil, and its molecular formula was determined to be $C_{17}H_{16}O_8$ by HRESIMS (measured at *m/z* 349.09143 $[M+H]^+$, calculated for $C_{17}H_{17}O_8^+$ 349.09179). The NMR data (Table 4) of **10** were similar to those of the dechlorinated methyl ester (**16**) isolated in this study [34]. The major difference was that **10** exhibited a dihydrofuran ring rather than a furan ring. HMBC correlations from H-8 (H, d, J = 3.9 Hz, δ_H 5.64) to C-8a (δ_C 114.7, s), C-7 (δ_C 170.0, s), C-9 (δ_C 178.3, s), and C-10a (δ_C 167.7, s), together with H-5 (H, ddd, J = 6.6, 4.4, 3.9 Hz, δ_H 5.73) to C-10a, C-8a, C-6 (δ_C 37.7, t), and C-2′ (δ_C 169.5, s), supported the above assignment. The relative configuration of **10** was identified by the analysis of its ROESY data. The ROESY correlation between H-8 and H-5 indicated that H-8 had the same orientation as H-5 (assigned as an α orientation). The calculated ECD of **10** established the configuration of **10,** as shown in Figure 5. Therefore, the structure of **10** was characterized as depicted.

Bipolarithone B (**11**) was isolated as a yellow oil. The HRESIMS spectrum of **11** suggested a molecular formula of $C_{17}H_{16}O_8$ (measured at *m/z* 349.09157 $[M+H]^+$, calculated for $C_{17}H_{17}O_8^+$ 349.09179), the same as that of **10**. The planar structure of **11** was elucidated to be the same as that of **10** by the analysis of its 1D and 2D NMR data. The main difference was suggested as its stereochemistry at C-8 (δ_C 79.8, d). Analyses of the ^1H NMR information showed that the coupling constants of H-8, H-5, and H-6 were significantly different from those of **11,** as shown in the Table 4. Furthermore, the ROESY correlation of H-8 (δ_H 5.63, 1H, d, J = 1.7 Hz)/H-5 (δ_H 5.62, 1H, ddd, J = 8.4, 3.8, 1.7 Hz) was not observed in **11**. These data suggested that **11** was an epimer of **10**. The ECD calculation for **11** was performed, and the results of **11** matched well with the experimental ECD curve (Figure 5). Hence, the absolute configuration of **11** can be fully assigned, as shown.

Bipolarithone C (**12**) was assigned a molecular formula of $C_{30}H_{36}O_9$ based on its HRESIMS data (measured at *m/z* 541.24310 $[M+H]^+$, calculated for $C_{30}H_{37}O_9{}^+$ 541.24321). The NMR data of **12** were very similar to those of bipolenin I (**14**) (Table 5), a novel sesquiterpenoid-xanthone adduct isolated from the fungus *Bipolaris eleusines* [35]. The significant differences were that there was an absence of an aldehyde group and two olefinic carbons, as well as the presence of an additional methine and carbonyl, in **12**. These data suggested that the α,β-unsaturated aldehyde moiety disappeared in **12**. This assignment was confirmed by the HMBC correlations of δ_H 2.16 (H, m, H-2), 1.29 (H, m, H-6), 2.56 (1H, br s, H-7), 0.95 (3H, d, J = 7.2 Hz, C-12), and 1.90 (H, m, H-13) to δ_C 50.6 (d, C-2) and 221.6 (s, C-1). The ROESY spectrum displayed similar patterns to those of **14**. Furthermore, a cross peak between H-2 and H-14a (δ_H 4.05, 1H, dd, J = 11.3, 5.1 Hz) confirmed the relative configuration of C-2, as shown. The absolute configuration of **12** was elucidated by the quantum chemistry calculations. At first, the ECD calculations were conducted on the four possible conformers (**12a–d**), using time-dependent density functional theory (TDDFT) at the B3LYP/6-311G (d) level in methanol with the PCM model. The overall calculated ECD spectrum of each configuration was then generated according to the Boltzmann weighting of the conformers. As a result, the calculated ECD curves of **12a** and **12d** matched well with the experimental data (Figure 5). To determine its final structure, the theoretical NMR calculations and DP4+ probabilities were employed. The ^{13}C NMR chemical shifts of **12a** and **12d** were calculated at the mPW1PW91/6-311+G (d,p) level in the gas phase. According to the DP4+ probability analyses, **12a** was assigned with 100% probability (see data in the Supporting Information). Structurally, compound **12** comprised of a seco-sativene sesquiterpenoid unit and a xanthone unit, whose absolute configurations were in accord with compound **6** and compound **10**, respectively. Therefore, the structure of **12** was established as depicted.

Bipolarithone D (**13**) had the same molecular formula ($C_{30}H_{36}O_9$) as that of **12**, according to their HRESIMS spectra (measured at *m/z* 541.24316 $[M + H]^+$, calculated for $C_{30}H_{37}O_9{}^+$ 541.24321). The NMR resonances for **13** (Table 5) resembled those of **12**, except that the resonances of C-6′ ($\Delta\delta_C$ + 1.5), H-6′a ($\Delta\delta_H$ + 0.08), and H-6′b ($\Delta\delta_H$ + 0.15) were shifted downfield, while the data H-5′ ($\Delta\delta_H$ − 0.08) were shifted upfield. A detailed comparison of the 1D and 2D NMR data of **13** with that of **12** indicated that the two compounds possessed the same planar structure. The main difference was the stereochemistry at C-8′. A key ROESY correlation of H-5′/H-8′ could be detected in **12** but not in **13.** In addition, the coupling constants of H-8′ in **13** (J = 1.8 Hz) were different from that in **12** (J = 3.9 Hz). All the data suggested that compound **13** was a C-8′ epimer of **12**. Finally, the absolute configuration of **13** was confirmed by ECD calculations (Figure 5).

Five known compounds were determined as bipolenins I and J (**14** and **15**), dechlorinated methyl ester (**16**), drechslerines A (**17**), and (+)-secolongifolene diol (**18**) by the comparison of their spectral data with that reported in the literature [32,34,35]. In this study, the absolute configurations of compounds **17** and **18** were confirmed by single crystal X-ray diffractions (Figure 6), which could support the absolute configurations of **1–9**, **12,** and **13** as depicted in the text, since they were obtained from the same source.

All compounds (**1–18**) were evaluated for their anti-Psa activity. As a result, compounds **10** and **15** showed significant inhibitory activity, with MICs of 64 and 16 μg/mL, respectively, while compounds **7**, **11**, **13**, and **16** showed moderate activity, with MICs of 128 μg/mL (Table 6).

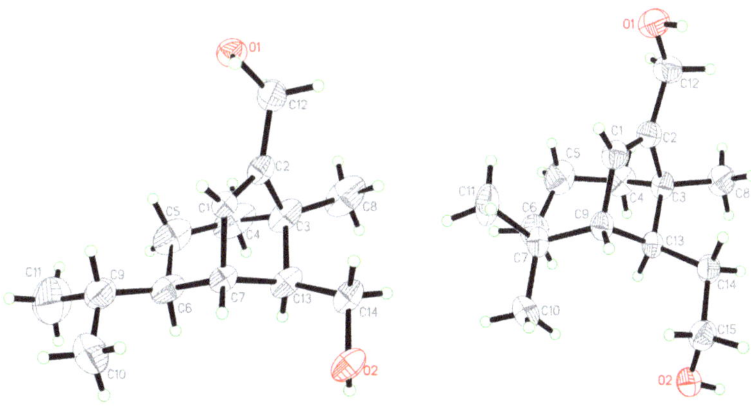

17 **18**

Figure 6. ORTEP diagrams of **17** and **18**.

Table 6. Inhibitory effects of the isolates against five plant pathogens (MIC, μg/mL) [a].

Compd	Psa	*P. infestans*	*A. solani*	*R. solani*	*F. oxysporum*
3	256	NA	128	256	NA
4	NA [c]	128	NA	NA	NA
7	128	NA	64	128	256
8	256	NA	256	NA	NA
9	NA	128	NA	NA	NA
10	64	128	NA	NA	NA
11	128	64	NA	NA	NA
12	256	64	NA	64	NA
13	128	32	NA	NA	NA
14	NA	NA	8	NA	128
15	16	NA	16	NA	NA
16	128	128	128	256	NA
Streptomycin [b]	8	–	–	–	–
Hygromycin B [b]	–	8	4	16	32

[a] Compounds without any bioactivity are not listed; [b] Positive controls; [c] NA = no activity at 256 μg/mL.

In addition, our previous study on chemicals from *B. eleusines* suggested that sativene-xanthone adducts have promising inhibitory activity against plant pathogenic microorganisms [35]. Therefore, all compounds were evaluated for their inhibitory activity against four plant pathogenic microorganisms, including *P. infestane*, *A. solani*, *R. solani*, and *F. oxysporum*. As a result, many compounds showed certain inhibitory activity, as given in Table 6.

A brief structure–activity relationship analysis suggested that the aldehyde-containing sativene sesquiterpenoids were more active than the others, while the xanthones or their derivatives showed better inhibitory activities than sativene sesquiterpenoids.

4. Conclusions

A total of 18 compounds, including 13 new ones, were characterized from the kiwifruit-associated fungus *Bipolaris* sp. Their structures, with absolute configurations, were established by means of spectroscopic methods. Many compounds possessed anti-Psa activity and inhibitory activity against plant pathogens. It is concluded that *Bipolaris* sp. is rich in sativene sesquiterpenoids and xanthones, and both sativene sesquiterpenoids and xanthones possess potential antimicrobial application prospects. This study also suggested that it is an effective way to find natural anti-Psa agents from kiwifruit-associated fungi.

Supplementary Materials: The following are available online at https://www.mdpi.com/article/10.3390/jof8010009/s1, Section S1: Supplementary of NMR, HRESIMS and CD spectra for **1–13**, Section S2: Calculational details for **3**, **7**, **10**, **11**, **12** and **13**.

Author Contributions: Conceptualization, J.H.; methodology, J.-J.Y., Y.-X.J., S.-S.H. and J.H.; resources, J.H.; data curation, J.-J.Y.; writing—original draft preparation, J.-J.Y.; writing—review and editing, J.H.; project administration, J.H.; funding acquisition, J.H. All authors have read and agreed to the published version of the manuscript.

Funding: This research was funded by the National Natural Science Foundation of China (grant number 22177139).

Institutional Review Board Statement: Not applicable.

Informed Consent Statement: Not applicable.

Data Availability Statement: X-ray crystallographic data of **1**, **3**, **9**, and **18** (CIF) is available free of charge at https://www.ccdc.cam.ac.uk (accessed on 1 December 2021).

Acknowledgments: The authors thank the Analytical & Measuring Centre, South-Central University for Nationalities for the spectra measurements.

Conflicts of Interest: The authors declare no conflict of interest.

References

1. Dolly, S.; Kaur, J.; Bhadariya, V.; Sharma, K. *Actinidia deliciosa* (Kiwi fruit): A comprehensive review on the nutritional composition, health benefits, traditional utilization and commercialization. *J. Food Process. Preserv.* **2021**, *45*, e15588. [CrossRef]
2. Wang, S.N.; Qiu, Y.; Zhu, F. Kiwifruit (*Actinidia* spp.): A review of chemical diversity and biological activities. *Food Chem.* **2021**, *350*, 128469. [CrossRef]
3. Richardson, D.P.; Ansell, J.; Drummond, L.N. The nutritional and health attributes of kiwifruit: A review. *Eur. J. Nutr.* **2018**, *57*, 2659–2676. [CrossRef] [PubMed]
4. Wang, Q.; Zhang, C.; Long, Y.; Wu, X.; Su, Y.; Lei, Y.; Ai, Q. Bioactivity and control efficacy of the novel antibiotic tetramycin against various kiwifruit diseases. *Antibiotics* **2021**, *10*, 289. [CrossRef] [PubMed]
5. Kim, G.H.; Jung, J.S.; Koh, Y.J. Occurrence and epidemics of bacterial canker of kiwifruit in Korea. *Plant Pathol. J.* **2017**, *33*, 351–361. [CrossRef] [PubMed]
6. Wang, Q.; Zhang, C.; Wu, X.; Long, Y.; Su, Y. Chitosan augments tetramycin against soft rot in kiwifruit and enhances its improvement for kiwifruit growth, quality and aroma. *Biomolecules* **2021**, *11*, 1257. [CrossRef]
7. Lee, Y.S.; Han, H.S.; Kim, G.H.; Koh, Y.J.; Hur, J.S.; Jung, J.S. Causal agents of blossom blight of kiwifruit in Korea. *Plant Pathol. J.* **2009**, *25*, 220–224. [CrossRef]
8. Balestra, G.M.; Mazzaglia, A.; Rossetti, A. Outbreak of bacterial blossom blight caused by *Pseudomonas viridiflava* on *Actinidia chinensis* kiwifruit plants in Italy. *Plant Dis.* **2008**, *92*, 1707. [CrossRef]
9. Jeong, I.H.; Lim, M.T.; Kim, G.H.; Han, T.W.; Kim, H.C.; Kim, M.J.; Park, H.S.; Shin, S.H.; Hur, J.S.; Shin, J.S.; et al. Incidences of leaf spots and blights on kiwifruit in Korea. *Plant Pathol. J.* **2008**, *24*, 125–130. [CrossRef]
10. Li, H.H.; Tang, W.; Liu, K.; Zhang, L.; Tang, X.F.; Miao, M.; Liu, Y.S. First report of *Fusarium fujikuroi* causing brown leaf spot on kiwifruit. *Plant Dis.* **2020**, *104*, 1560. [CrossRef]
11. Polat, Z.; Awan, Q.N.; Hussain, M.; Akgul, D.S. First report of *Phytopythium vexans* causing root and collar rot of kiwifruit in Turkey. *Plant Dis.* **2017**, *101*, 1058. [CrossRef]
12. Wang, K.X.; Xie, Y.L.; Yuan, G.Q.; Li, Q.Q.; Lin, W. First report of root and collar rot caused by *Phytopythium helicoides* on kiwifruit (*Actinidia chinensis*). *Plant Dis.* **2015**, *99*, 725. [CrossRef]
13. McCann, H.C.; Li, L.; Liu, Y.F.; Li, D.W.; Pan, H.; Zhong, C.H.; Rikkerink, E.H.A.; Templeton, M.D.; Straub, C.; Colombi, E.; et al. Origin and evolution of the kiwifruit canker pandemic. *Genome Biol. Evol.* **2017**, *9*, 932–944. [CrossRef]
14. Vanneste, J.L. The scientific, economic, and social impacts of the New Zealand outbreak of bacterial canker of kiwifruit (*Pseudomonas syringae* pv. *actinidiae*). *Annu. Rev. Phytopathol.* **2017**, *55*, 377–399. [CrossRef]
15. Wicaksono, W.A.; Jones, E.E.; Casonato, S.; Monk, J.; Ridgway, H.J. Biological control of *Pseudomonas syringae* pv. *actinidiae* (Psa), the causal agent of bacterial canker of kiwifruit, using endophytic bacteria recovered from a medicinal plant. *Biol. Control* **2018**, *116*, 103–112. [CrossRef]
16. Scortichini, M. Aspects still to solve for the management of kiwifruit bacterial canker caused by *Pseudomonas syringae* pv. *actinidiae* biovar 3. *Eur. J. Hortic. Sci.* **2018**, *83*, 205–211. [CrossRef]
17. Bardas, G.A.; Veloukas, T.; Koutita, O.; Karaoglanidis, G.S. Multiple resistance of botrytis cinerea from kiwifruit to SDHIs, QoIs and fungicides of other chemical groups. *Pest Manag. Sci.* **2010**, *66*, 967–973. [CrossRef] [PubMed]

18. Colombi, E.; Straub, C.; Kunzel, S.; Templeton, M.D.; McCann, H.C.; Rainey, P.B. Evolution of copper resistance in the kiwifruit pathogen *Pseudomonas syringae* pv. *actinidiae* through acquisition of integrative conjugative elements and plasmids. *Environ. Microbiol.* **2017**, *19*, 819–832. [CrossRef]
19. Gupta, S.; Chaturvedi, P.; Kulkarni, M.G.; van Staden, J. A critical review on exploiting the pharmaceutical potential of plant endophytic fungi. *Biotechnol. Adv.* **2020**, *39*, 107462. [CrossRef]
20. Kusari, S.; Hertweck, C.; Spitellert, M. Chemical ecology of endophytic fungi: Origins of secondary metabolites. *Chem. Biol.* **2012**, *19*, 792–798. [CrossRef]
21. Strobel, G.; Daisy, B.; Castillo, U.; Harper, J. Natural products from endophytic microorganisms. *J. Nat. Prod.* **2004**, *67*, 257–268. [CrossRef] [PubMed]
22. Helaly, S.E.; Thongbai, B.; Stadler, M. Diversity of biologically active secondary metabolites from endophytic and saprotrophic fungi of the ascomycete order *Xylariales*. *Nat. Prod. Rep.* **2018**, *35*, 992–1014. [CrossRef]
23. Zhang, J.Y.; He, J.; Li, Z.H.; Feng, T.; Liu, J.K. Zopfiellasins A–D, two pairs of epimeric cytochalasins from kiwi-associated fungus *Zopfiella* sp. and their antibacterial assessment. *Molecules* **2021**, *26*, 5611. [CrossRef]
24. Yi, X.W.; He, J.; Sun, L.T.; Liu, J.K.; Wang, G.K.; Feng, T. 3-Decalinoyltetramic acids from kiwi-associated fungus *Zopfiella* sp. and their antibacterial activity against *Pseudomonas syringae*. *RSC Adv.* **2021**, *11*, 18827–18831. [CrossRef]
25. Ma, J.T.; Du, J.X.; Zhang, Y.; Liu, J.K.; Feng, T.; He, J. Natural imidazole alkaloids as antibacterial agents against *Pseudomonas syringae* pv. *actinidiae* isolated from kiwi endophytic fungus *Fusarium tricinctum*. *Fitoterapia* **2022**, *156*, 105070. [CrossRef]
26. Frisch, M.J.T.; Trucks, G.W.; Schlegel, H.B.; Scuseria, G.E.; Robb, M.A.; Cheeseman, J.R.; Scalmani, G.; Barone, V.; Mennucci, B.; Petersson, G.A.; et al. *Gaussian 09*; Revision D. 01; Gaussian Inc.: Wallingford, CT, USA, 2010.
27. Shao, Y.; Molnar, L.F.; Jung, Y.; Kussmann, J.; Ochsenfeld, C.; Brown, S.T.; Gilbert, A.T.B.; Slipchenko, L.V.; Levchenko, S.V.; O'Neill, D.P.; et al. Advances in methods and algorithms in a modern quantum chemistry program package. *Phys. Chem. Chem. Phys.* **2006**, *8*, 3172–3191. [CrossRef] [PubMed]
28. Hehre, W.J. *A Guide to Molecular Mechanics and Quantum Chemical Calculations*; Wavefunction Inc.: Irvine, CA, USA, 2003; Volume 51, pp. 1–812. [CrossRef]
29. Bruhn, T.; Schaumlöffel, A.; Hemberger, Y.; Bringmann, G. SpecDis: Quantifying the comparison of calculated and experimental electronic circular dichroism spectra. *Chirality* **2013**, *25*, 243–249. [CrossRef] [PubMed]
30. Grimblat, N.; Zanardi, M.M.; Sarotti, A.M. Beyond DP4: An improved probability for the stereochemical assignment of isomeric compounds using quantum chemical calculations of NMR shifts. *J. Org. Chem.* **2015**, *80*, 12526–12534. [CrossRef] [PubMed]
31. Nukina, M.; Hattori, H.; Marumo, S. Cis-Sativenediol, a plant growth promotor, produced by fungi. *J. Am. Chem. Soc.* **1975**, *97*, 2542–2543. [CrossRef]
32. Osterhage, C.; König, G.M.; Höller, U.; Wright, A.D. Rare sesquiterpenes from the algicolous fungus *Drechslera dematioidea*. *J. Nat. Prod.* **2002**, *65*, 306–313. [CrossRef]
33. Dorn, F.; Arigoni, D. Ein bicyclischer Abkömmling von (−) longifolen aus *Helminthosporium sativum* und *H. victoriae*. *Experientia* **1974**, *30*, 851–852. [CrossRef]
34. Han, J.; Zhang, J.; Song, Z.; Liu, M.; Hu, J.; Hou, C.; Zhu, G.; Jiang, L.; Xia, X.; Quinn, R.; et al. Genome- and MS-based mining of antibacterial chlorinated chromones and xanthones from the phytopathogenic fungus *Bipolaris sorokiniana* strain 11134. *Appl. Microbiol. Biotechnol.* **2019**, *103*, 5167–5181. [CrossRef] [PubMed]
35. He, J.; Yang, M.S.; Wang, W.X.; Li, Z.H.; Elkhateeb, W.; Wen, T.C.; Ai, H.L.; Feng, T. Anti-phytopathogenic sesquiterpenoid-xanthone adducts from potato endophytic fungus *Bipolaris eleusines*. *RSC Adv.* **2019**, *9*, 128–131. [CrossRef]

Journal of *Fungi*

MDPI

Article

Anti-Adipogenic Lanostane-Type Triterpenoids from the Edible and Medicinal Mushroom *Ganoderma applanatum*

Xing-Rong Peng [1,2,†], Qian Wang [3,†], Hai-Guo Su [1,2], Lin Zhou [1,2], Wen-Yong Xiong [1,3,*] and Ming-Hua Qiu [1,2,*]

1 State Key Laboratory of Phytochemistry and Plant Resources in West China, Kunming Institute of Botany, Chinese Academy of Sciences, Kunming 650201, China; pengxingrong@mail.kib.ac.cn (X.-R.P.); suhg@mail.sysu.edu.cn (H.-G.S.); zhoulin@mail.kib.ac.cn (L.Z.)

2 University of the Chinese Academy of Sciences, Beijing 100049, China

3 Key Laboratory of Medicinal Chemistry for Natural Resource, Ministry of Education, Yunnan Provincial Center for Research & Development of Natural Products, School of Chemical Science and Technology, Yunnan University, Kunming 650091, China; 20210209@ynu.edu.cn

* Correspondence: xwy@ynu.edu.cn (W.-Y.X.); mhchiu@mail.kib.ac.cn (M.-H.Q.)

† These authors contributed equally to this work.

Abstract: Our previous research has shown that lanostane triterpenoids from *Ganoderma applanatum* exhibit significant anti-adipogenesis effects. In order to obtain more structurally diverse lanostane triterpenoids to establish a structure–activity relationship, we continued the study of lanostane triterpenoids from the fruiting bodies of *G. applanatum*, and forty highly oxygenated lanostane-type triterpenoinds (**1–40**), including sixteen new compounds (**1–16**), were isolated. Their structures were elucidated using NMR spectra, X-ray crystallographic analysis, and Mosher's method. In addition, some of their parts were evaluated to determine their anti-adipogenesis activities in the 3T3-L1 cell model. The results showed that compounds **16**, **22**, **28**, and **32** exhibited stronger anti-adipogenesis effects than the positive control (LiCl, 20 mM) at the concentration of 20 μM. Compounds **15** and **20** could significantly reduce the lipid accumulation during the differentiation process of 3T3-L1 cells, comparable to the untreated group. Their IC_{50} values were 6.42 and 5.39 μM, respectively. The combined results of our previous and present studies allow us to establish a structure-activity relationship of lanostane triterpenoids, indicating that the A-*seco*-23→26 lactone skeleton could play a key role in anti-adipogenesis activity.

Keywords: *Ganoderma applanatum*; lanostane triterpenoid; Mosher's method; anti-adipogenesis activity; structure–activity relationship

Citation: Peng, X.-R.; Wang, Q.; Su, H.-G.; Zhou, L.; Xiong, W.-Y.; Qiu, M.-H. Anti-Adipogenic Lanostane-Type Triterpenoids from the Edible and Medicinal Mushroom *Ganoderma applanatum*. *J. Fungi* **2022**, *8*, 331. https://doi.org/10.3390/jof8040331

Academic Editors: Tao Feng and Frank Surup

Received: 22 February 2022
Accepted: 18 March 2022
Published: 22 March 2022

1. Introduction

Macro-fungi provide crucial food and medicinal resources [1]. *Tricholoma matsutake*, [2,3] *Lentinula edodes*, [4,5], and *Collybia albuminosa* [6] are delicious mushrooms which contain plentiful amino acids, fatty acids, vitamins, crude fiber, and protein. In addition, *Fomitopsis pinicola* (SW.) [7] Karst, *Inonotus obliquus* [8,9], *Phellinus igniarius* [10,11], *Ganoderma lucidum* [12,13], and *Ganoderma sinense* [14,15] have been used as edible and medicinal mushrooms for preventing and treating various diseases. The *Ganoderma* genus plays an important role in the history of Chinese medicine [16]. *Shennong's Herbal Classics* recorded its traditional effects to include improving eyesight, strengthening muscles and bones, reinforcing kidney function, soothing the nerves, and prolonging the lifespan. *G. lucidum* and *G. sinense* have been registered in the *Chinese Pharmacopoeia* (2015 version). Meanwhile, *G. lucidum* was also included in the catalog of the latest edition of "Homology of medicine and Food" in 2020. Modern pharmacological research has further demonstrated that *Ganoderma* has a variety of pharmacological activities [17–21]. Thus, *Ganoderma* has great prospects in preventing and treating diseases.

Ganoderma applanatum, belonging to the genus *Ganoderma*, has traditionally been used to treat various chronic diseases, such as chronic hepatitis, immunological disorders, neurasthenia, arthritis, and nephritis [22]. Meanwhile, *G. applanatum* has been made into capsules and injections to cure chronic liver fibrosis and inflammation in a clinical setting [23–26]. *G. applanatum* is rich in chemical constituents, including polysaccharides, triterpenoids, meroterpenoids, alkaloids, and steroids. The majority of studies relating to it focus on the application and development of polysaccharides [23,27–31]. However, our previous research proved that highly oxygenated triterpenoids showed significant anti-adipogenesis activities [27,28]. In order to search for more active compounds to clarify the structure–activity relationship to lay the foundations for the discovery of lead compounds, we continued to investigate triterpenoids isolated from *G. applanatum* and 40 lanostane-type triterpenoids; of these (**1**–**40**), 16 were new compounds (**1**–**16**, Figure 1). Furthermore, their anti-adipogenesis effects were evaluated in the 3T3-L1 cell model, and their structure–activity relationship was established.

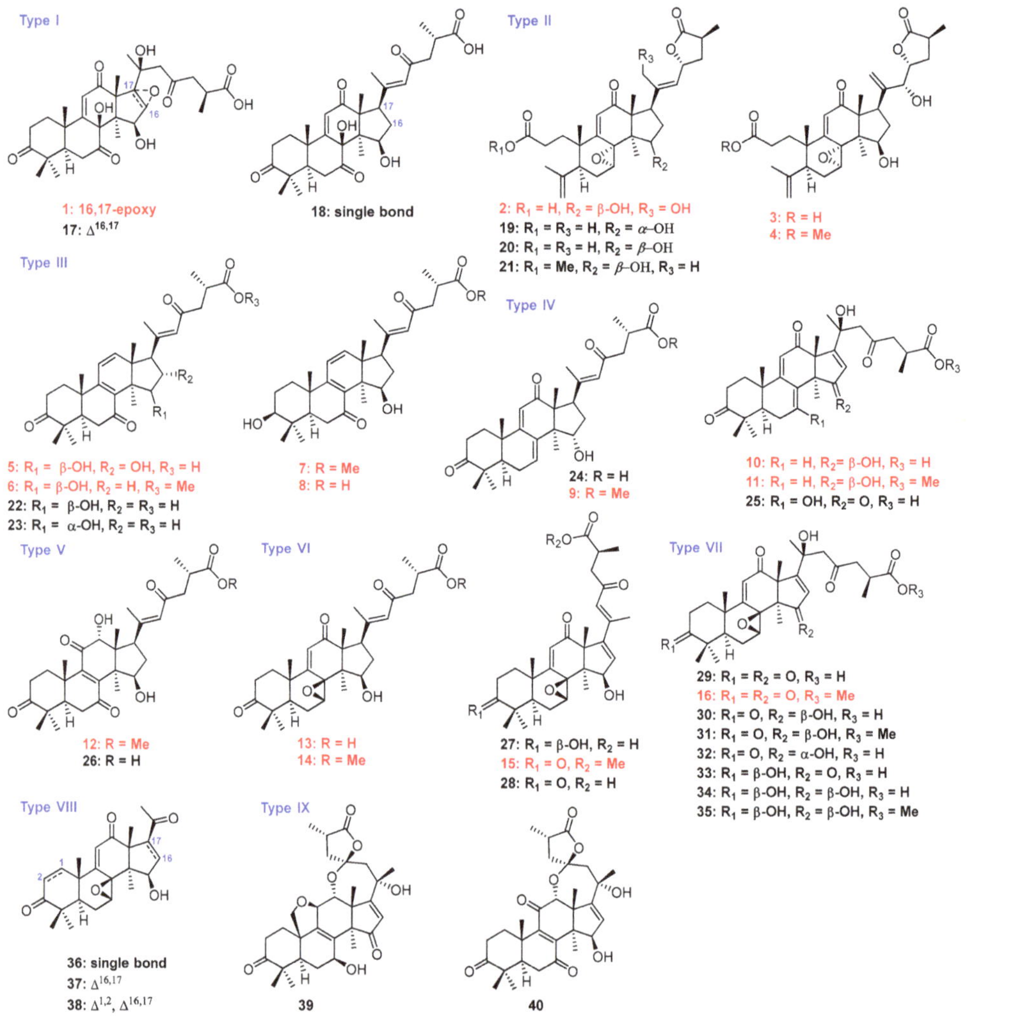

Figure 1. Structures and classifications of isolates from *G. applanatum* (red: new compounds).

2. Materials and Methods

2.1. General Experimental Procedures

NMR spectra were recorded on a Bruker AV-600 MHz (Bruker, Zurich, Switzerland) using TMS as an internal standard for chemical shifts with reference to the TMS resonance. ESIMS and HRTOF-ESIMS were measured on an API QSTAR Pulsar spectrometer. UV spectra were recorded on a Shimadzu UV-2401PC spectrometer. IR was recorded on the Bruker Tensor-27 instrument using KBr pellets. Optical rotations were recorded on a Horiba SEPA-300 polarimeter. CD spectra were measured on a Chirascan instrument. An Agilent 1100 series instrument equipped with an Agilent ZORBAX SB-C18 column (5 μm, 9.6 mm × 250 mm) was used for high-performance liquid chromatography (HPLC) separation.

TLC was performed on precoated TLC plates (200–250 μm thickness, F254 Si gel 60, Qingdao Marine Chemical, Inc., Qingdao, China), with compounds visualized by spraying the dried plates with 10% aqueous H_2SO_4 followed by heating until they were dry. Silica gel ((200–300) mesh, Qingdao Marine Chemical, Inc.), Lichroprep RP-18 (40–63 μm, Fuji), and Sephadex LH-20 (20–150 μm, Pharmacia) were used for column chromatography. Methanol, chloroform, ethyl acetate, acetone, petroleum ether, n-hexane, and 2-propanol were purchased from Tianjing Chemical Reagents Co. (Tianjing, China). All other materials were of the highest grade available.

2.2. Fungal Materials

Ganoderma applanatum (39 kg) was purchased in December 2019 from a traditional Chinese medicine market in Kunming, Yunnan, China, which was identified by Prof. Yang Zhuliang, Kunming Institute of Botany, Chinese Academy of Science (voucher No. 19122201).

2.3. Extraction and Isolation

G. applanatum (39 kg) was chipped and extracted with 95% EtOH under reflux three times (three hours each time). The combined EtOH extracts were evaporated under reduced pressure. The residue was suspended in H_2O and extracted with EtOAc. The volume of the combined EtOAc extracts was reduced to one-third under a reduced pressure. The residue was fractionated by macroporous resin (D-101; MeOH−H_2O, 50:50, 70:30, and 90:10, v/v): fractions I−III. Fraction III (245 g) was further fractioned by a silica gel column with petroleum ether (PE)/ethyl acetate (EA) as the mobile phase, which gave six subfractions (Fr. III-1→Fr. III-6).

Fr. III-2 (156 g) was treated by a silica gel column and $CHCl_3$/MeOH (80:1→20:1, v/v) was used as an eluent. Ten fractions (Fr. III-2-1→Fr. III-2-10) were obtained, of which Fr. III-2-4 (20 g) was separated using Sephadex LH-20 (MeOH) to obtain three subfractions (Fr. III-2-4a→Fr. III-2-4c). Compound **29** (235 mg) was purified through recrystallization from Fr. III-2-4b. The remaining solution was isolated using semi-preparative HPLC (CH_3CN/H_2O = 52%, v/v) to gain compound **6** (8 mg, t_R = 28.6 min). Fr. III-2-5 (10 g) was treated by a silica gel column, being eluted with PE/EA (20:1, v/v) to obtain five parts (Fr. III-2-5a→Fr. III-2-5e). Subsequently, 5b and 5d were purified using P-TLC ($CHCl_3$/MeOH = 40:1, v/v) to obtain compounds **18** (11 mg), **16** (26 mg), and **21** (9.2 mg). Fr. III-2-6 (12 g) was separated by Rp-C18 with the elution of MeOH/H_2O (50%→55%) to obtain five fractions. Compounds **36** (6.2 mg) and **13** (12 mg) were obtained from Fr. III-2-6c and Fr. III-2-6d through P-TLC ($CHCl_3$/MeOH = 40:1, v/v), respectively.

Fr. III-2-7 (25 g) was fractioned by an Rp-C18 column, being eluted with MeOH/H_2O (50%→65% containing 0.3% CF_3COOH, v/v); nine subfractions (7a→7i) were obtained. Furthermore, 7d, 7g, and 7h were purified by semi-preparative HPLC (CH_3CN/H_2O = 45%→60%, v/v) to obtain compounds **14** (5.3 mg, t_R = 19.1 min), **4** (8.3 mg, t_R = 19.1 min), **8** (3.4 mg, t_R = 14.8 min), **9** (3.2 mg, t_R = 17.4 min), **12** (4.2 mg, t_R = 21.3 min), and **17** (5.1 mg, t_R = 22.2 min). Similarly, Fr. III-2-8 (31 g) was also treated using an Rp-C18 column with MeOH/H_2O (50%→55%) to obtain nine subfractions (8a→8i), from which compounds **37** (3.2 mg, t_R = 26.6 min), **38** (3.6 mg, t_R = 27.6 min), **31** (6.1 mg, t_R = 22.1 min), **22** (3.1 mg, t_R = 21.5 min), **25** (14 mg, t_R = 25.9 min), **19** (12.5 mg, t_R = 27.5 min), and **20** (7.2 mg,

t_R = 20.7 min) were purified by semi-preparative HPLC (CH_3CN/H_2O = 43%→60% containing 0.3% CF_3COOH, *v*/*v*). The Rp-C18 column and semi-preparative HPLC were used to treat Fr. III-2-9 (20 g), and compounds **35** (2.5 mg, t_R = 20.8 min), **28** (2.9 mg, t_R = 20.3 min), **13** (2.1 mg, t_R = 18.3 min), **7** (2.2 mg, t_R = 19.5 min), and **23** (2.9 mg, t_R = 18.5 min) were isolated from 9d-1 and 9d-2. 9e (15 mg) was purified by P-TLC ($CHCl_3$/MeOH = 30:1, *v*/*v*) to obtain compounds **39** (4.2 mg) and **40** (2.8 mg).

The combination of Fr. III-4 and Fr. III-5 weighing 52 g was fractioned using Rp-C18 column elution with MeOH/H_2O (35%→100%, *v*/*v*) to obtain six subfractions (Fr. III-4-1→Fr. III-4-6). Among these, Fr. III-4-2→Fr. III-4-5 were treated using Sephadex LH-20 (MeOH). Subsequently, the triterpenoid parts were purified by P-TLC ($CHCl_3$/MeOH = 20:1 containing 0.3% CF_3COOH, *v*/*v*) and semi-preparative HPLC (CH_3CN/H_2O = 38%→53% containing 0.3% CF_3COOH, *v*/*v*) to obtain compounds **27** (4.2 mg), **24** (2.1 mg), **30** (6.2 mg), **5** (3.1 mg), **33** (3 mg), **26** (4.8 mg), **17** (5 mg, t_R = 23.5 min), **3** (7.2 mg, t_R = 22.5 min), **34** (6.1 mg, t_R = 12.1 min), **32** (3.3 mg, t_R = 17.8 min), **1** (3.0 mg, t_R = 17.5 min), **10** (2.9 mg), and **2** (3.7 mg).

Ganoapplic acid A (**1**): white powder (MeOH); $[\alpha]^{28}_{D}$ −1.2 (*c* 0.25, MeOH); UV (MeOH); λ_{max} (log ε): 230 (3.36), and 196 (3.33); IR (KBr) v_{max} 3428, 2953, 2943, 1653, 1636, 1423, 1364, 1212, and 1131 cm^{-1}; ^1H NMR and ^{13}C NMR data: see Tables 1 and 2; HRMS (ESI-TOF) *m*/*z*: 561.2696 $[M + H]^+$ (calcd for $C_{30}H_{40}O_{10}$, 561.2694).

Table 1. ^1H NMR spectra of compounds **1–8** (600 MHz, methanol-d_4, *J* in Hz, δ in *ppm*).

Position	1	2	3	4	5	6	7	8
1	1.87 dt (13.7 4.5) 2.21 dd (15.4 2.8)	1.86 m 2.20 m	1.82 m 2.16 m	1.85 m 2.17 m	1.91 m 2.29 m	1.87 m 2.16 m	2.02 m 1.51 m	1.50 m 2.02 m
2	2.40 m 2.95 m	2.19 m 2.36 m	2.17 m 2.34 m	2.25 m 2.40 m	2.55 m 2.74 m	2.51 m 2.71 m	1.78 m	1.77 m
3							3.24 t (8.4)	3.23 t (8.4)
5	1.74 dd (14.8 2.8)	2.97 t (8.3)	2.94 t (8.4)	2.95 dd (9.3 7.5)	2.28 m	2.19 m	1.66 dd (14.4 3.6)	1.66 dd (14.4 3.6)
6	2.30 dd (15.4 2.8) 3.30 m	2.03 m	1.99 m	2.03 m	2.44 dd (16.3 3.3)	2.52 m 2.66 m	2.62 m 2.48 m	2.48 m 2.59 m
7		4.93 s	4.95 s	4.98 s				
11	5.98 s	6.04 s	6.02 s	6.04 s	6.28 d (9.9)	6.14 d (10.2)	6.31 d (10.2)	6.31 d (10.2)
12					6.59 d (9.9)	6.59 d (10.2)	6.64 d (10.2)	6.65 d (10.2)
15	4.83 s	3.95 d (6.5)	3.91 d (6.5)	3.94 d (6.4)	4.20 s	4.44 d (7.8)	4.39 d (7.2)	4.39 d (7.2)
16	3.36 s	1.80 m 4.65 m	1.73 m 2.64 m	1.76 m 2.67 m	4.57 d (7.8)	2.20 m 2.42 m	2.16 m 2.51 m	2.16 m 2.51 m
17		3.33 m	3.21 dd (10.7 8.7)	3.23 m	2.73 m	2.73 m	2.87 m	2.83 m
18	1.97 s	1.30 s	1.37 s	1.39 s	0.99 s	1.01 s	1.00 s	0.99 s
19	1.65 s	1.01 s	1.04 s	1.05 s	1.14 s	1.27 s	1.16 s	1.16 s
21	1.34 s	4.37 s	5.33 s 5.39 s	5.36 s 5.43 s	1.21 s	2.23 s	2.19 s	2.19 s
22	2.66 d (14.9) 3.00 d (14.9)	5.78 d (8.6)	4.47 d (5.6)	4.50 d (5.6)	6.36 s	6.22 s	6.40 s	6.32 s
23		5.59 dd (8.6 6.6)	4.72 m	4.73 ddd (8.6 5.6 3.2)				
24	2.62 dd (18.3 5.6) 3.07 dd (18.3 7.7)	2.19 m	1.97 m 2.16 m	2.00 m 2.56 m	2.61 m 2.90 m	2.54 m 2.94 m	2.64 m 2.90 m	2.53 m 2.55 m
25	1.74 dd (14.8 2.8)	2.82 q (7.7)	2.84 m	2.86 m	2.74 m	2.97 m	2.85 m	2.84 m
26	1.16 d (6.6)	1.29 d (7.3)	1.22 d (6.5)	1.25 d (6.5)	1.18 d (7.0)	1.19 d (7.2)	1.17 d (7.2)	1.17 d (7.2)
28	1.14 s	4.80 s 4.98 s	4.77 s 4.95 s	4.77 s 4.95 s	1.12 s	1.13 s	1.00 s	1.00 s
29	1.07 s	1.80 s	1.78 s	1.80 s	1.10 s	1.14 s	0.90 s	0.90 s
30	0.74 s	1.06 s	1.01 s	1.04 s	1.10 s	0.94 s	0.95 s	0.95 s
OMe				3.65 s		3.68 s	3.65 s	

Table 2. ^{13}C NMR spectra of compounds **1–8** (150 MHz, methanol-d_4).

Position	1	2	3	4	5	6	7	8
1	37.6 CH_2	37.7 CH_2	37.7 CH_2	37.5 CH_2	36.5 CH_2	35.6 CH_2	36.3 CH_2	36.3 CH_2
2	35.1 CH_2	30.4 CH_2	30.5 CH_2	30.5 CH_2	35.0 CH_2	34.0 CH_2	28.1 CH_2	28.1 CH_2
3	215.8 C	177.4 C	177.4 C	175.6 C	217.1 C	214.3 C	78.3 CH	78.3 CH
4	48.7 C	146.5 C	146.5 C	146.3 C	51.9 C	46.9 C	39.7 C	39.8 C
5	48.5 CH	45.0 CH	45.3 CH	45.3 CH	50.7 CH	49.3 CH	50.8 CH	50.8 CH
6	35.8 CH_2	28.2 CH_2	28.1 CH_2	28.1 CH_2	37.5 CH_2	36.8 CH_2	37.0 CH_2	37.0 CH_2
7	205.9 C	64.7 CH	64.8 CH	65.0 CH	203.0 C	210.4 C	204.3 C	204.3 C
8	85.5 C	67.4 C	67.5 C	67.5 C	135.8 C	134.9 C	135.7 C	135.7 C
9	165.0 C	167.2 C	167.8 C	167.3 C	161.4 C	160.6 C	164.5 C	164.5 C
10	41.0 C	45.0 C	45.2 C	45.3 C	39.2 C	37.9 C	39.9 C	39.7 C
11	127.3 CH	130.7 CH	130.6 CH	130.6 CH	123.2 CH	122.0 CH	123.3 CH	123.3 CH
12	204.0 C	206.5 C	207.3 C	207.0 C	147.4 CH	147.5 CH	148.3 CH	148.3 C
13	62.3 C	60.8 C	60.4 C	60.4 C	52.2 C	50.2 C	51.4 C	51.4 C
14	47.0 C	54.4 C	54.4 C	54.4 C	48.0 C	52.5 C	53.9 C	53.8 C
15	80.0 CH	77.4 CH	77.2 CH	77.2 CH	84.9 CH	75.0 CH	76.5 CH	76.6 CH
16	61.2 CH	41.3 CH_2	43.7 CH_2	43.5 CH_2	83.6 CH	36.2 CH_2	37.2 CH_2	37.2 CH_2
17	71.4 C	43.3 CH	41.6 CH	41.6 CH	58.7 CH	48.5 CH	49.7 CH	49.7 C
18	24.6 CH_3	20.2 CH_3	20.2 CH_3	20.2 CH_3	19.2 CH_3	17.1 CH_3	20.2 CH_3	21.2 CH_3
19	19.9 CH_3	24.2 CH_3	23.4 CH_3	23.4 CH_3	21.1 CH_3	19.5 CH_3	18.0 CH_3	18.0 CH_3
20	73.4 C	144.1 C	148.7 C	148.7 C	156.5 C	157.3 C	158.8 C	158.4 C
21	27.8 CH_3	61.8 CH_2	117.6 CH_2	117.5 CH_2	19.7 CH_3	21.4 CH_3	21.4 CH_3	21.3 CH_3
22	53.9 CH_2	130.8 CH	76.5 CH	76.5 CH	126.9 CH	124.8 CH	126.0 CH	126.4 CH
23	209.9 C	76.8 CH	80.9 CH	80.9 CH	200.8 C	198.3 C	200.6 C	201.3 C
24	48.7 CH_2	38.3 CH_2	32.2 CH_2	32.2 CH_2	48.7 CH_2	47.7 CH_2	48.8 CH_2	49.0 CH_2
25	35.8 CH	35.7 CH	35.3 CH	35.3 CH	36.2 CH	34.8 CH	36.6 CH	36.6 CH
26	17.4 CH_3	15.9 CH_3	16.3 CH_3	16.0 CH_3	17.5 CH_3	17.1 CH_3	17.4 CH_3	17.7 CH_3
27	179.7 C	182.7 C	183.1 C	183.2 C	179.7 C	176.4 C	178.1 C	180.8 C
28	22.3 CH_3	115.9 CH_2	115.9 CH_2	115.9 CH_2	21.0 CH_3	25.6 CH_3	27.8 CH_3	27.8 CH_3
29	24.9 CH_3	23.5 CH_3	23.4 CH_3	23.4 CH_3	25.9 CH_3	20.4 CH_3	15.7 CH_3	15.7 CH_3
30	30.7 CH_3	21.3 CH_3	21.4 CH_3	21.4 CH_3	19.7 CH_3	20.6 CH_3	21.2 CH_3	20.2 CH_3
OMe				52.3 CH_3		51.8 CH_3	52.3 CH_3	

Ganoapplic acid B (**2**): white powder (MeOH); $[\alpha]^{28}_D$ +48.0 (c 0.13, MeOH); UV (MeOH); λ_{max} (log ε): 251 (3.75), and 196 (4.08); IR (KBr) ν_{max} 3430, 2953, 2928, 1653, 1636, 1473, 1344, 1211, and 1147 cm^{-1}; ^{1}H NMR and ^{13}C NMR data: see Tables 1 and 2; HRMS (ESI-TOF) m/z: 527.2653 $[M-H]^-$ (calcd for $C_{30}H_{40}O_8$, 527.2650).

Ganoapplic acid C (**3**): white powder (MeOH); $[\alpha]^{28}_D$ +42.0 (c 0.09, MeOH); UV (MeOH); λ_{max} (log ε): 251 (3.59), and 196 (3.93); IR (KBr) ν_{max} 3423, 2955, 2930, 1673, 1635, 1428, 1380, 1219, and 1132 cm^{-1}; ^{1}H NMR and ^{13}C NMR data: see Tables 1 and 2; HRMS (ESI-TOF) m/z: 527.2654 $[M-H]^-$ (calcd for $C_{30}H_{40}O_8$, 527.2650).

Methyl ganoapplate C (**4**): white powder (MeOH); $[\alpha]^{28}_D$ +39.5 (c 0.07, MeOH); UV (MeOH); λ_{max} (log ε): 251 (3.52), and 196 (3.85); IR (KBr) ν_{max} 3458, 2957, 2928, 1689, 1606, 1473, 1375, 1210, and 1132 cm^{-1}; ^{1}H NMR and ^{13}C NMR data: see Tables 1 and 2; HRMS (ESI-TOF) m/z: 565.2773 $[M+Na]^+$ (calcd for $C_{31}H_{42}O_8Na$, 565.2772).

Ganoapplic acid D (**5**): white powder (MeOH); $[\alpha]^{28}_D$ −20.1 (c 0.21, MeOH); UV (MeOH); λ_{max} (log ε): 242 (3.71), and 195 (3.79); IR (KBr) ν_{max} 3503, 2967, 2935, 1663, 1635, 1452, 1390, 1200, and 1125 cm^{-1}; ^{1}H NMR and ^{13}C NMR data: see Tables 1 and 2; HRMS (ESI-TOF) m/z: 511.2707 $[M-H]^-$ (calcd for $C_{30}H_{40}O_7$, 511.2701).

Methyl gibbosate M (**6**): white powder (MeOH); $[\alpha]^{28}_D$ −49.64 (c 0.11, MeOH); UV (MeOH); λ_{max} (log ε): 307 (3.21), 243 (3.80), and 196 (3.77); IR (KBr) ν_{max} 3438, 2966, 2912, 1688, 1634, 1453, 1364, 1212, and 1132 cm^{-1}; ^{1}H NMR and ^{13}C NMR data: see Tables 1 and 2; HRMS (ESI-TOF) m/z: 533.2873 $[M+Na]^+$ (calcd for $C_{31}H_{42}O_6Na$, 533.2874).

Methyl ganoapplate E (**7**): white powder (MeOH); $[\alpha]^{28}_D$ −69.3 (c 0.13, MeOH); UV (MeOH); λ_{max} (log ε): 316 (3.13), 243 (3.74), and 196 (3.72); IR (KBr) ν_{max} 3413, 2953, 2916,

1657, 1620, 1454, 1374, 1209, and 1142 cm^{-1}; ^{1}H NMR and ^{13}C NMR data: see Tables 1 and 2; HRMS (ESI-TOF) *m/z*: 535.3033 [M + Na]$^{+}$ (calcd for $C_{31}H_{44}O_6Na$, 535.3030).

Ganoapplic acid E (**8**): white powder (MeOH); $[\alpha]^{28}{}_{D}$ −143.5 (*c* 0.19, MeOH); UV (MeOH); λ_{max} (log ε): 317 (3.57), 244 (4.15), and 196 (4.05); IR (KBr) v_{max} 3445, 2980, 2906, 1712, 1690, 1458, 1380, 1214, and 1028 cm^{-1}; ^{1}H NMR and ^{13}C NMR data: see Tables 1 and 2; HRMS (ESI-TOF) *m/z*: 497.2911 [M − H]$^{-}$ (calcd for $C_{30}H_{41}O_6$, 497.2909).

Methyl gibbosate L (**9**): white powder (MeOH); $[\alpha]^{28}{}_{D}$ +20.27 (c 0.15, MeOH); UV (MeOH); λ_{max} (log ε): 291 (3.70), 244 (3.94), and 196 (3.85); IR (KBr) v_{max} 3444, 2976, 2911, 1673, 1658, 1423, 1374, 1219, and 1140 cm^{-1}; ^{1}H NMR and ^{13}C NMR data: see Tables 3 and 4; HRMS (ESI-TOF) *m/z*: 533.2876 [M + Na]$^{+}$ (calcd for $C_{31}H_{42}O_6Na$, 533.2874).

Ganoapplic acid F (**10**): white powder (MeOH); $[\alpha]^{28}{}_{D}$ +20.86 (*c* 0.14, MeOH); UV (MeOH); λ_{max} (log ε): 299 (3.77), and 196 (3.98); IR (KBr) v_{max} 3437, 2953, 2924, 1678, 1656, 1433, 1374, 1215, and 1132 cm^{-1}; ^{1}H NMR and ^{13}C NMR data: see Tables 3 and 4; HRMS (ESI-TOF) *m/z*: 511.2708 [M − H]$^{-}$ (calcd for $C_{30}H_{40}O_7$, 511.2701).

Methyl ganoapplate F (**11**): white powder (MeOH); $[\alpha]^{28}{}_{D}$ +5.80 (*c* 0.10, MeOH); UV (MeOH); λ_{max} (log ε): 301 (3.46), and 196 (3.87); IR (KBr) v_{max} 3439, 2953, 2929, 1723, 1638, 1445, 1334, 1219, and 1138 cm^{-1}; ^{1}H NMR and ^{13}C NMR data: see Tables 3 and 4; HRMS (ESI-TOF) *m/z*: 549.2827 [M + Na]$^{+}$ (calcd for $C_{31}H_{42}O_7Na$, 549.2823).

Table 3. ^{1}H NMR spectra of compounds **9–16** (600 MHz, *J* in Hz, *δ* in *ppm*).

Position	9 [a]	10 [b]	11 [b]	12 [b]	13 [b]	14 [b]	15 [b]	16 [a]
1	2.27 m 1.86 m	1.83 m 2.39 m	1.83 m 2.38 m	2.80 m 2.95 m	1.82 m 2.26 m	1.79 m 2.23 m	1.81 m 2.23 m	1.80 m 1.80 m
2	2.42 m 2.78 m	2.37 m 2.93 m	2.37 m 2.94 m	2.53 m 2.59 m	2.31 m 2.95 m	2.27 m 2.94 m	2.28 m 2.93 m	2.36 m 2.89 m
5	1.65 m	1.78 dd (12.8 3.1)	1.76 dd (11.4 3.60	2.36 dd (15.0 3.0)	1.29 dd (12.4 5.9)	1.67 dd (12.6 6.0)	1.66 m	1.56 dd (12.66.0)
6	1.11 m 2.28 m	2.33 m 2.49 m	2.34 m 2.53 m	2.54 m 2.68 m	2.25 m	2.25 m	2.30 m	2.16 m
7	6.50 br s	6.62 m	6.63 d (7.2)		3.96 d (5.7)	3.93 d (5.4)	3.90 d (3.6)	4.45 d (5.4)
11	5.65 s	5.74 s	5.74 s		6.04 s	6.01 s	5.98 s	6.03 s
12				3.70 s				
15	4.58 t (8.3)	4.47 d (3.0)	4.47 d (3.0)	4.40 d (7.8)	4.08 d (6.0)	4.05 d (6.0)	4.35 d (3.0)	
16	1.80 m 2.48 m	5.75 d (3.0)	5.74 overlapped	2.07 m 2.43 m	1.94 m 2.43 m	1.91 m 2.42 m	6.20 d (3.0)	5.66 s
17	3.26 m			3.28 t (9.6)	3.24 dd (10.7 7.3)	3.21 dd (10.8 7.8)		
18	0.80 s	1.54 s	1.54 s	−0.096 s	1.46 s	1.43 s	1.90 s	1.76 s
19	1.28 s	1.39 s	1.40 s	1.31 s	1.47 s	1.45 s	1.47 s	1.44 s
20								
21	2.22 s	1.42 s	1.40 s	2.24 s	2.28 s	2.25 s	2.30 s	1.49 s
22	6.37 s	2.82 d (14.2) 3.01 d (14.2)	2.79 d (14.4) 2.79 dd (14.4)	6.31 s	6.56 s	6.52 s	6.50 s	2.92 d (15.0) 2.98 d (15.0)
24	2.54 m 2.94 m	2.66 dd (18.4 5.2) 3.02 m	2.79 dd (18.6 5.4) 3.02 dd (18.6 8.4)	2.54 m 2.94 m	2.62 m 2.92 m	2.62 m 2.88 m	2.58 m 2.93 m	3.04 m 2.59 dd (18.0 5.4)
25	2.93 m	2.80 m	2.81 m	1.19 m	2.89 m	2.89 m	2.89 m	2.89 m
26	1.17 d (6.7)	1.15 d (7.2)	113 d (7.2)	1.19 d (7.2)	1.20 d (7.0)	1.16 d (7.2)	1.20 d (7.0)	1.15 d (7.2)
28	1.11 s	1.20 s	1.20 s	1.15 s	1.09 s	1.09 s	1.10 s	1.11 s
29	1.15 s	1.11 s	1.06 s	1.13 s	1.13 s	1.13 s	1.38 s	1.12 s
30	1.08 s	1.06 s	3.63 s	1.24 s	0.96 s	0.96 s	1.00 s	1.28 s
OMe	3.68 s		3.56 s	3.69 s		3.65 s	3.63 s	3.65 s

[a] Measured in $CDCl_3$; [b] measured in methanol-d_4.

Table 4. ^{13}C NMR spectra of compounds **9–16** (150 MHz).

Position	**9** [a]	**10** [b]	**11** [b]	**12** [b]	**13** [b]	**14** [b]	**15** [b]	**16** [a]
1	35.6 CH_2	36.7 CH_2	36.8 CH_2	35.1 CH_2	38.4 CH_2	38.4 CH_2	38.4 CH_2	37.2 CH_2
2	34.2 CH_2	35.4 CH_2	35.4 CH_2	34.3 CH_2	34.9 CH_2	34.9 CH_2	34.8 CH_2	33.8 CH_2
3	214.8 C	217.2 C	217.2 C	214.9 C	216.4 C	216.4 C	216.5 C	213.6 C
4	47.0 C	48.4 C	48.8 C	47.1 C	48.7 C	48.5 C	48.8 C	47.8 C
5	49.5 CH	51.1 CH	51.2 CH	49.8 CH	50.7 CH	50.7 CH	51.2 CH	49.9 CH
6	23.9 CH_2	25.2 CH_2	25.2 CH_2	37.6 CH_2	22.6 CH_2	22.6 CH_2	22.4 CH_2	21.6 CH_2
7	130.9 CH	135.5 CH	135.5 CH	203.4 C	59.7 CH	59.7 CH	58.8 CH	57.1 CH
8	139.8 C	137.3 C	137.3 C	150.2 C	65.7 C	65.8 C	64.4 C	59.0 C
9	161.0 C	165.5 C	165.6 C	151.9 C	162.2 C	162.3 C	163.0 C	165.0 C
10	38.0 C	39.6 C	39.7 C	39.3 C	39.0 C	39.0 C	38.6 C	38.2 C
11	118.2 CH	118.3 CH	118.3 CH	203.9 C	127.6 CH	127.6 CH	127.2 CH	125.0 CH
12	202.5 C	206.5 C	206.5 C	79.6 CH	204.6 C	204.6 C	201.9 C	200.6 C
13	52.3 C	64.0 C	64.0 C	52.9 C	59.4 C	59.4 C	63.1 C	61.8 C
14	57.8 C	54.3 C	54.3 C	50.0 C	51.3 C	51.3 C	50.0 C	54.4 C
15	72.9 CH	79.3 CH	79.4 CH	77.5 CH	79.0 CH	79.0 CH	79.9 CH	202.7 C
16	36.5 CH_2	127.5 CH	127.6 CH	34.2 CH_2	38.2 CH_2	38.2 CH_2	134.9 CH	124.4 CH
17	45.7 CH	158.9 C	159.0 C	46.9 CH	49.6 CH	49.6 CH	148.8 C	181.9 C
18	17.8 CH_3	29.0 CH_3	29.1 CH_3	18.6 CH_3	20.3 CH_3	20.3 CH_3	27.0 CH_3	30.9 CH_3
19	21.3 CH_3	21.5 CH_3	21.5 CH_3	19.0 CH_3	21.0 CH_3	21.0 CH_3	21.2 CH_3	20.8 CH_3
20	157.6 C	72.6 C	72.6 C	157.5 C	159.0 C	159.2 C	156.0 C	72.6 C
21	20.4 CH_3	29.0 CH_3	29.1 CH_3	20.3 CH_3	21.5 CH_3	21.5 CH_3	17.5 CH_3	29.1 CH_3
22	125.7 CH	54.6 CH_2	54.7 CH_2	125.1 CH	127.5 CH	127.5 CH	127.1 CH	52.7 CH_2
23	198.3 C	209.8 C	209.7 C	198.6 C	201.0 C	201.0 C	201.7 C	206.3 C
24	47.7 CH_2	48.8 CH_2	48.7 CH_2	47.9 CH_2	48.7 CH_2	48.7 CH_2	48.9 CH_2	47.8 CH_2
25	34.7 CH	35.8 CH	35.9 CH	35.0 CH	36.3 CH	36.3 CH	36.3 CH	34.5 CH
26	17.0 CH_3	17.4 CH_3	17.3 CH_3	16.9 CH_3	17.4 CH_3	17.4 CH_3	17.5 CH_3	17.0 CH_3
27	176.4 C	179.7 C	178.2 C	176.7 C	179.7 C	178.2 C	180.1 C	176.2 C
28	25.2 CH_3	22.9 CH_3	25.5 CH_3	26.9 CH_3	24.9 CH_3	24.9 CH_3	24.9 CH_3	24.5 CH_3
29	22.3 CH_3	25.2 CH_3	22.9 CH_3	20.4 CH_3	22.3 CH_3	22.3 CH_3	22.5 CH_3	22.1 CH_3
30	18.1 CH_3	29.0 CH_3	29.1 CH_3	27.4 CH_3	22.5 CH_3	22.5 CH_3	24.9 CH_3	25.9 CH_3
OMe	51.8 CH_3		52.2 CH_3	52.0 CH_3		52.3 CH_3	52.0 CH_3	51.9 CH_3

[a] Measured in $CDCl_3$; [b] measured in methanol-d_4.

Methyl gannosate I (**12**): white powder (MeOH); $[\alpha]^{28}_D$ +95.0(*c* 0.14, MeOH); UV (MeOH); λ_{max} (log ε): 249 (3.99), and 195 (3.73); IR (KBr) v_{max} 3452, 2985, 2921, 1673, 1618, 1425, 1376, 1221, and 1132 cm^{-1}; ^1H NMR and ^{13}C NMR data: see Tables 3 and 4; HRMS (ESI-TOF) *m/z*: 565.2769 $[M + Na]^+$ (calcd for $C_{31}H_{42}O_8Na$, 565.2772).

Ganoapplic acid G (**13**): white powder (MeOH); $[\alpha]^{28}_D$ −32.86 (*c* 0.14, MeOH); UV (MeOH); λ_{max} (log ε): 242 (3.99), and 196 (3.86); IR (KBr) v_{max} 3440, 2978, 2965, 1683, 1628, 1403, 1364, 1200, and 1151 cm^{-1}; ^1H NMR and ^{13}C NMR data: see Tables 3 and 4; HRMS (ESI-TOF) *m/z*: 535.0000 $[M + Na]^+$ (calcd for $C_{30}H_{40}O_7Na$, 535.0000).

Methyl ganoapplate G (**14**): white powder (MeOH); $[\alpha]^{28}_D$ +0.80 (*c* 0.07, MeOH); UV (MeOH); λ_{max} (log ε): 243 (3.79), and 196 (3.77); IR (KBr) v_{max} 3445, 2963, 2931, 1683, 1638, 1453, 1384, 1209, and 1142 cm^{-1}; ^1H NMR and ^{13}C NMR data: see Tables 3 and 4; HRMS (ESI-TOF) *m/z*: 549.2823 $[M + Na]^+$ (calcd for $C_{31}H_{42}O_7Na$, 549.2823).

Methyl applate C (**15**): white powder (MeOH); $[\alpha]^{28}_D$ +43.44 (*c* 0.18, MeOH); UV (MeOH); λ_{max} (log ε): 248 (3.99), and 196 (3.95); IR (KBr) v_{max} 3443, 2956, 2915, 1665, 1624, 1433, 1376, 1205, and 1132 cm^{-1}; ^1H NMR and ^{13}C NMR data: see Tables 3 and 4; HRMS (ESI-TOF) *m/z*: 547.2672 $[M + Na]^+$ (calcd for $C_{31}H_{40}O_7Na$, 547.2666).

Methyl gibbosate A (**16**): white powder (MeOH); $[\alpha]^{28}_D$ +18.94 (*c* 0.15, MeOH); UV (MeOH); λ_{max} (log ε): 237 (3.78), and 196 (3.82); IR (KBr) v_{max} 3430, 2953, 2929, 1703, 1628, 1433, 1354, 1229, and 1147 cm^{-1}; ^1H NMR and ^{13}C NMR data: see Tables 3 and 4; HRMS (ESI-TOF) *m/z*: 563.2612 $[M + Na]^+$ (calcd for $C_{31}H_{40}O_8Na$, 563.2615)

X-ray *Crystallographic Data for Ganoapplic acid A* (**1**): $C_{30}H_{40}O_{10}\cdot CH_4O\cdot H_2O$, M = 610.68, a = 13.9761(7) Å, b = 6.9089(3) Å, c = 15.8732(7) Å, α = 90°, β = 90.948(2)°, γ = 90°,

$V = 1532.50(12)$ Å^3, $T = 100.(2)$ K, space group $P1211$, $Z = 2$, μ(Cu Kα) = 0.844 mm^{-1}, 26,208 reflections measured, 5973 independent reflections (R_{int} = 0.0593). The final R_1 values were 0.0504 ($I > 2\sigma(I)$). The final $wR(F^2)$ values were 0.1381 ($I > 2\sigma(I)$). The final R_1 values were 0.0543 (all data). The final $wR(F^2)$ values were 0.1436 (all data). The goodness of fit for F^2 was 1.042. Flack parameter = 0.07(9).

2.4. Mosher's Method

The specific esterification of compound **4** was performed based on the previous method [32]. The ^{1}H NMR spectroscopic data of the (R)-MTPA ester derivative (**4r**) of **4** (600 MHz, pyridine-d_5; data were obtained from the reaction NMR tube directly and assigned on the basis of correlations of the ^{1}H-^{1}H COSY spectrum): δ 4.253 (1H, m, H-15), δ 2.158 (1H, m, H-16a), δ 2.756 (1H, m, H-16b), δ 3.566 (1H, m, H-17), δ 5.619 (1H, s, H-21a), δ 5.632 (1H, s, H-21b), δ 5.007 (1H, m, H-22), δ 5.067 (1H, m, H-23), δ 2.759 (1H, m, H-24a), mboxemphδ 1.905 (1H, m, H-24b), δ 3.080 (1H, m, H-25), δ 1.198 (3H, s, Me-26). Meanwhile, the ^{1}H NMR spectroscopic data of the (S)-MTPA ester derivative (**4s**) of **4** were: δ 4.255 (1H, m, H-15), δ 2.156 (1H, m, H-16a), δ 2.756 (1H, m, H-16b), δ 3.568 (1H, m, H-17), δ 5.639 (1H, s, H-21a), δ 5.633 (1H, s, H-21b), δ 5.003 (1H, m, H-22), δ 5.062 (1H, m, H-23), δ 2.756 (1H, m, H-24a), δ 1.903 (1H, m, H-24b), δ 3.078 (1H, m, H-25), δ 1.196 (3H, s, Me-26) (see Figures S29–S31 in Section S10).

2.5. Inhibition of Lipogenesis Assay

2.5.1. Cell Culture and Adipocyte Differentiation

3T3-L1 cells were purchased from the American Type Culture Collection (ATCC, Manassas, VA, U.S.A.). The culture and differentiation of 3T3-L1 cells were performed based on the description reported previously [28]. Firstly, Dulbecco's modified Eagle's medium (DMEM) containing 10% bovine calf serum (CS) was used to cultivated 3T3-L1 cells. The whole system was incubated at 37 °C in a humidified atmosphere of 5% CO_2 and 95% air. Secondly, different medium systems were used for the different differentiated phases. Confluent cells were grown in DMEM medium containing 1 µg/mL insulin, 1 µM dexamethasone (DEX), 0.5 mM 3-isobutyl-1-methylxanthine (IBMX), and 1 µM rosiglitazone (Rosi). Then, on the third day, post-differentiation medium—namely, DMEM with 10% fetal bovine serum (FBS) and 1 µg/mL insulin—was used to continually cultivate the cells. From the fourth day, DMEM + 10% FBS was used as a maintenance medium for cell differentiation. In this process, it commonly takes two days for the mature adipocytes to form. During the whole differentiation process, the tested compounds or 0.1% DMSO were added to the differentiated 3T3-L1 cells, for which the 0.1% DMSO group was used as the vehicle.

2.5.2. Cell Viability Assay

The viability of cell treated with compounds **15** and **20** was determined by the MTS method. The detailed experimental procedures were similar to those described in our previous study [29].

2.5.3. Lipid Content Analysis

The intracellular lipid contents of 3T3-L1 adipocytes were determined by Oil Red O staining [33]. Briefly, differentiated 3T3-L1 cells were washed twice with PBS and fixed with 10% formaldehyde for 1 h. After another washing with PBS, the fixed cells were stained with 0.5% Oil Red O in 3:2 of Oil Red O/H_2O for 15 min at room temperature and then washed with 60% isopropanol and distilled water. The lipid content was imaged with an inverted light microscope Nikon TS100 (Tokyo, Japan). Finally, 100% isopropanol was used to elute Oil Red O dye and it was quantified at 492 nm absorbance.

3. Results and Discussion

The molecular formula of ganoapplic acid A (**1**) was established to be $C_{30}H_{40}O_{10}$ by HRESIMS ion at m/z 561.2696 $[M + H]^+$ (calcd. 561.2694), suggesting 11 degrees of unsaturation. Its ^{13}C NMR spectra displayed 30 carbon resonances, of which seven methyls, three ketone carbonyls (δ_C 215.8, δ_C 205.9, and δ_C 209.9), an α,β-unsaturated carbonyl (δ_C 165.0, δ_C 127.3, and δ_C 204.0), one carboxyl (δ_C 179.7), two oxygenated methines (δ_C 80.0 and δ_C 61.2), and three quaternary carbons containing oxygen (δ_C 85.5, δ_C 71.4, and δ_C 73.4) were assigned based on the HSQC and ^{13}C-DEPT NMR spectra (see Figures S2 and S4 in Section S1). These data indicated that the structure of compound **1** was similar to that of gibbosic acid G with a 7,12,23-trioxo-8,20-dihydroxy-lanosta-9,11-en-26-oic acid skeleton [34], except for the presence of one oxygenated methine and one oxygenated quaternary carbon and the absence of a double bond at C-16 and C-17 in **1**. Furthermore, the HMBC spectrum (Figure 2A) of **1** revealed the correlations of H_3-21 with C-20 (δ_C 73.4), C-22, and the oxygenated quaternary carbon (δ_C 71.4); of H_3-30 with C-15 (δ_C 80.0); and of H-15 (δ_H 4.83, s) with the oxygenated methine (δ_C 61.2) and quaternary carbons (δ_C 71.4). Meanwhile, the proton of the methine containing oxygen (δ_H 3.36, s) showed the HMBC correlations with C-13, C-14, C-15, and C-20, as well as the ^1H-^1H COSY correlation with H-15 (Figure 2A), which certified that C-16 and C-17 was substituted by hydroxyls. According to the molecular formula of **1**, an epoxy was present in **1** at C-16 and C-17.

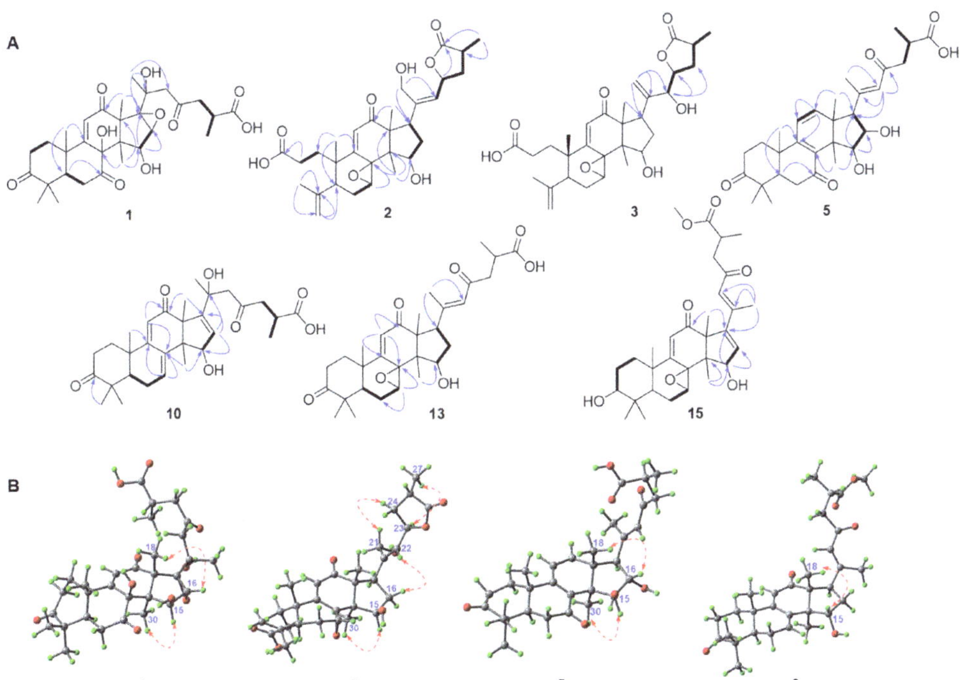

Figure 2. (**A**) Selected HMBC (H→C) and ^1H-^1H COSY (H▬H) correlations of compounds **1–3**, **5**, **10**, **13**, and **15**. (**B**) Selected ROESY correlations of compounds **1**, **2**, **5**, and **9**.

In the ROESY spectrum (Figure 2B) of **1**, H-15 showed an apparent cross peak with H_3-30, while H-16 correlated with H_3-18, suggesting that H-15 and H-16 were α- and β-oriented, respectively. The X-ray crystallographic analysis (Figure 3A) of **1** (Cu $\kappa\alpha$) further confirmed that the absolute configurations of C-15, C-16, C-20, and C-25 were *R*,

S, *S*, and *S* (see Table S1 in Section S1). Finally, the structure of **1** was determined to be (20*S*,25*S*)-7,12,23-trioxo-8β,20-dihydroxy-16α,17α-epoxy-lanosta-9,11-en-26-oic acid.

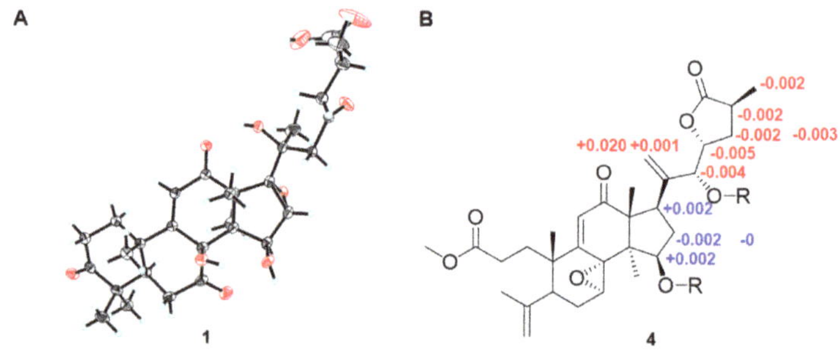

Figure 3. (**A**) X-ray crystallographic structure of **1**; (**B**) values of δ_S–δ_R of the MTPA esters of **4**.

Ganoapplic acid B (**2**) was isolated as a colorless solid. Its molecular formula was determined to be $C_{30}H_{40}O_8$ by the HRESIMS. The 1D NMR and HSQC data of **2** supported the presence of five methyls, an α,β-unsaturated ketone, two carboxyl or ester carbonyls, one oxygenated methylene (δ_H 4.37, s; δ_C 61.8), three oxygenated methines (δ_H 3.95, d, J = 6.5 Hz, δ_C 77.4; δ_H 4.93, s, δ_C 64.7; δ_H 5.59, dd, J = 8.6 and 6.6 Hz, δ_C 76.8), one quaternary carbon containing oxygen (δ_C 67.4), a terminal double bond (δ_C 146.5; δ_C 115.9), and a double bond (δ_C 144.1; δ_H 5.78, d, J = 8.6 Hz, δ_C 130.8), which indicated that compound **2** was an A-*seco*-lanostane triterpenoid and had the similar structure to that of gibbosicolid F. [35] However, five methyls and an oxygenated methylene were observed in **2**, rather than the six methyls found in gibbosicolid F. The detailed analysis of the 2D NMR spectra (see Figures S10–S13 in Section S2) exhibited the long-range HMBC correlations (Figure 2A) of the oxygenated methylene with C-17, C-20, and C-22; of H_3-18 with C-12, C-13, C-14, and C-17; and of H-22 with C-17, C-20, C-23, and C-24, together with the ^{1}H-^{1}H COSY correlations of H-22/H-23/H_2-24/H-25/H_3-26. These pieces of evidence confirmed that the hydroxyl was connected to C-21.

The ROESY cross peaks (Figure 2B) of H-7/H_3-18 and H-15/H_3-30 demonstrated that both H-15 and the epoxy between C-7 and C-8 were α. H_2-21 showed a ROESY correlation with H-23, suggesting that the double bond at C-20 and C-22 was *Z*. In addition, H_3-26 displayed the ROESY correlation with H-23, hinting that H_3-26 and H-23 were on the same face. Moreover, the *dd*-peak type of H-23 (J = 8.6 and 6.6 Hz) was consistent with that of gibbosicolid B (δ_H 5.31, dd, J = 13.2 and 8.0 Hz), which further indicated that the absolute configurations of C-23 and C-25 were *R* and *S*, respectively. [35] Therefore, the structure of **2** was assigned as (23*R*,25*S*)-15β,21-dihydroxyl-7α,8α-epoxy-12-oxo-3,4-*seco*-lanosta-4(28), 9-(11),20*E*(22)-trien-23,26-olide-3-oic acid and named ganoapplic acid B (**2**).

The molecular formula of ganoapplic acid C (**3**) was established to be $C_{30}H_{40}O_8$ by the HRESIMS. The 1D NMR spectra (see Figures S15 and S16 in Section S6) of **3** showed a high similarity with those of ganoapplic acid B (**2**), suggesting that compound **3** was an A-*seco* lanostane triterpenoid. However, the comparison of the 1D NMR spectroscopic data of **2** and **3** revealed that another terminal double bond (δ_C 117.6 and δ_C 148.7) and oxygenated methine (δ_C 76.5) were present in **3**, while a double bond (C-20-C-22) and an oxygenated methylene (C-21) were observed in **2**. Furthermore, the obvious ^{1}H-^{1}H COSY correlations of H_3-26/H-25/H_2-24/H-23/H-22, as well as the HMBC correlations of H_3-26 with C-24, C-25, and C-27 and of H_2-24 with C-22, C-23, and C-26, indicated that the hydroxyl was located at C-22 (Figure 2A). Meanwhile, H-22 (δ_H 4.47, d, J = 5.6 Hz) and H-17 displayed HMBC correlations of the terminal double bond, proving that the terminal double bond was at C-21 and C-20.

The 1D NMR spectra and molecular weight (see Figures S22 and S23 in Section S8, and Figure S28 in Section S9) of methyl ganoapplate C (**4**) showed that compound **4** was the ester derivative of **3**, which was confirmed by the HMBC correlation of OMe with C-27 (δ_C 175.6). The analysis of the ROESY spectra of **3** and **4** exhibited cross peaks of H-7/H_3-18, H-23/H_3-26, and H-22/H-25, indicating that the epoxy between C-7 and C-8 was α; meanwhile, H-23 and H_3-26 were cofacial (Figure 2B). Biogenetically, the absolute configuration of C-25 from *G. applanatum* was *S*. Thus, C-23 was determined to be *R*. The absolute configuration of C-22 was established as *R* based on the revised Mosher's method (Figure 3B) [33]. Thus, the structures of **3** and **4** were elucidated as (22*R*,23*R*,25*S*)-15β,22-dihydroxyl-7α,8α-epoxy-12-oxo-3,4-*seco*-lanosta-4(28),9(11),20(21)-trien-23,26-olide-3-oic acid and methyl (22*R*,23*R*,25*S*)-15β,22-dihydroxyl-7α,8α-epoxy-12-oxo-3,4-*seco*-lanosta-4(28),9(11),20(21)-trien-23,26-olide-3-oate, respectively.

Ganoapplic acid D (**5**) was isolated as a white powder and its molecular formula was determined to be $C_{30}H_{40}O_7$ based on the HRESIMS at *m*/*z* 511.2701 $[M - H]^-$ (calcd. 511.2707). Its ^{13}C-DEPT spectra (see Figures S32 and S33 in Section S11) showed thirty carbon resonances belonging to seven methyls, four methylenes, eight methines (including three sp^2 and two oxygenated), and eleven quaternary carbons (including three ketone carbonyls, one carboxyl, and three sp^2). These data indicated that compound **5** was a lanostane-type triterpenoid and had similar structure to that of gibbosic acid M (**22**) [35], except for the replacement of the methylene (C-16) in **22** with an oxygenated methine in **5**. The HMBC spectrum of **5** revealed the correlations of H_3-30 with C-15 (δ_C 84.9), C-13, and C-14; of H_3-18 with C-12, C-13, C-14, and C-17; and of H-17 with C-16, C-15, C-20, C-21, and C-22, suggesting that C-16 was an oxygenated methine in **5**, together with the ^{1}H-^{1}H COSY correlations of H-15/H-16/H-17 (Figure 2A). The ROESY correlations of H-15/H_3-30 and of H-16/H_3-18 illustrated that H-15 and H-16 were α- and β-oriented, respectively. The *E*-configuration of $\Delta^{20,22}$ was determined by the ROESY correlation of H-22/H-17/H-16. Therefore, the structure of **5** was assigned as 15β,16α-dihydroxy-3,7,23-trioxolanosta-8,11,20*E*(22)-trien-26-oic acid and named ganoapplic acid D (**5**).

Methyl gibbosate M (**6**) had the molecular formula of $C_{31}H_{42}O_6$ based on the positive HRESIMS at *m*/*z* 533.2874 $[M + Na]^+$ (calcd. 533.2873). The 1D NMR spectra (see Figures S39 and S40 in Section S13) of **6** were the same as those of gibbosic acid M (**22**) [35], except that the carboxyl at C-27 in **22** was replaced by the ester carbonyl in **6**. The key HMBC correlation of OMe with C-27 confirmed the above deduction. The characteristic *d*-coupling type of H-15 and the ROESY correlation of H-15/H_3-30 indicated that the 15-OH was β-oriented [35]. Finally, the structure of **6** was established to be methyl 15β-hydroxy-3,7,23-trioxolanosta-8,11,20*E*(22)-trien-26-oate and named methyl gibbosate M (**6**).

Methyl ganoapplate E (**7**) was isolated as a white powder and its molecular formula was determined to be $C_{31}H_{44}O_6$ by the HRESIMS. The analysis of the 1D NMR spectra (see Figures S46 and S47 in Section S15) of **7** showed that compound **7** had a similar structure to that of **6**, with the only difference being in the replacement of the ketone carbonyl at C-3 in **6** with the oxygenated methine in **7**, which was confirmed by the HMBC correlations of H_3-28, H_3-29, and H-5 with the oxygenated methine (δ_C 78.3). The ROESY correlations of H-3/H-5 and of H-15/H_3-30 indicated that 3-OH and 15-OH were β.

Ganoapplic acid E (**8**) was deduced to be the demethylated derivative of **7** on the basis of the HMBC correlation regarding the lack of the OMe at C-27 and the low-field shift of C-27. Thus, the structures of compounds **7** and **8** were elucidated as methyl (25*S*)-3β,15β-dihydroxy-7,23-dioxolanosta-8,11,20*E*(22)-trien-26-oate and (25*S*)-3β,15β-dihydroxy-7,23-dioxolanosta-8,11,20*E*(22)-trien-26-oic acid, respectively.

Methyl gibbosate O (**9**) was found to be similar to the known compound gibbosic acid O (**24**) [35] based on the 1D NMR spectroscopic data, except for distinct differences in the chemical shift of C-27 and the presence of an additional methoxyl. The HMBC spectrum of **9** showed the correlation of OMe with C-27, suggesting that **9** was a methyl ester derivative of gibbosic acid O (**24**). Therefore, the structure of **9** was determined to be methyl 15α-hydroxy-3,12,23-trioxolanosta-7,9(11),20*E*(22)-trien-26-oate.

The molecular formula of ganoapplic acid F (**10**) was deduced to be $C_{30}H_{40}O_7$ based on the HRESIMS and NMR data. Its 1D NMR spectra (see Figures S67 and S68 in Section S21) showed a similar tetracyclic skeleton to that of gibbosic acid O (**24**) [35] with a 15-hydroxy-

3,12-dioxolanosta-7(8),9(11)-diene skeleton, which was confirmed by the 2D NMR spectra. In addition, an oxygenated quaternary carbon signal (δ_C 72.6) and two sp^2 carbon signals (δ_C 127.5 and δ_C 158.9) were characteristic for the quaternary carbon containing oxygen at C-20 and the double bond at C-16 and C-17. Furthermore, the HMBC correlations (Figure 2A) of H_3-30 with C-15; of H-15 with C-16 and C-17; of H_3-18 with C-17; of H_3-21 with C-17, C-20, and C-22; and of H-22 with C-20, C-23, and C-24 confirmed the above deduction.

Methyl ganoapplate F (**11**) was an ester derivative at C-27 of ganoapplic acid H (**10**), according to the HMBC correlation of OMe with C-27. The ROESY spectra of **10** and **11** showed cross peaks of H-15/H_3-30, indicating the β-orientation of 15-OH. Therefore, the structures of **10** and **11** were determined to be 15β,20-dihydroxy-3,12, 23-trioxo-5α-lanosta-7,9(11),16-trien-26-oic acid and methyl 15β,20-dihydroxy-3,12,23-trioxo-5α-lanosta-7,9(11),16-trien-26-oate, respectively.

Based on the NMR data analysis, methyl gibbosate I (**12**) was found to be close to that of **26** [35], with a 12,15-dihydroxy-3,7,11,23-tetraoxolanosta-8,20(22)-dien structure. The 2D NMR spectra further confirmed its structure and **12** had an additional methoxyl at C-27, which was proven by the key HMBC correlation of OMe with C-27. Moreover, the ROESY correlations of H-12/H_3-18 and H-15/H_3-30 demonstrated that 12-OH was α while 15-OH was β. Thus, the structure of **12** was deduced to be methyl 12α,15β-dihydroxy-3,7,11,23-tetraoxolanosta-8,20(22)-dien-26-ate.

Ganoapplic acid G (**13**) was isolated as a white powder and its molecular formula was determined to be $C_{30}H_{40}O_7$ based on the HRMS (ESI-TOF) m/z 535.0000 $[M + Na]^+$ (calcd. 535.0000). The 1D NMR spectra of **13** showed the presence of the ketone carbonyl at C-3, 7,8-epoxyl, α,β-unsaturated ketones (C-9/C-11/C-12 and C-20/C-22/C-23), 15-OH, and 27-oic acid, which was further confirmed by the 2D NMR spectra (Figure 2A). The aforementioned information indicated that compound **13** had the same planar structure as gibbosic acid N. [35] The comparison of the ROESY spectra of **13** and gibbosic acid N revealed that they were 15-isomers due to the existence of the ROESY correlation of H-15/H_3-30 and the *d*-coupling of H-15 [35]. Therefore, the structure of **13** was established to be 15β-hydroxy-7β,8β-epoxy-3,12,23-trioxolanosta-9(11),20*E*(22)-dien-26-oic acid. In addition, methyl ganoapplate G (**14**) was deduced to be the methylation product of **13** on the basis of the HMBC correlation of OMe with C-27.

Methyl applanate C (**15**) was found to have a similar structure to methyl ganoapplate F (**14**), except for the presence of a double bond in **15**, rather than one methylene and one methine in **14**. Furthermore, in the HMBC spectrum of **15**, the correlations (Figure 2A) of H_3-30 with C-15, of H-15 with the sp^2 methine and quaternary carbon, and of H_3-18 and H_3-21 with the sp^2 quaternary carbon were observed, which proved that the double bond was located at C-16 and C-17. The ROESY correlation of H-16/H_3-21 and H_3-18/H-22 suggested that the geometry of the 16,20(22)-conjugated diene was 17,20-*Z*-(16*Z*, 20*E*). Additionally, the ROESY correlation of H_3-30/H-15 demonstrated that 15-OH was β. Finally, the structure of **15** was determined to be methyl 15β-hydroxy-7β,8β-epoxy-3,12,23-trioxolanosta-9(11),16Z,20*E*(22)-trien-26-oate and named methyl applanate C (**15**).

Methyl gibbosate A (**16**) was considered to be the methylation derivative of gibbosic acid A (**29**) [34] because of their similar 1D and 2D spectra (see Figures S104–S109 in Section S33) and the HMBC correlation of OMe with C-27. 7β,8β-epoxy was proven by the ROESY correlation of H-17/H_3-30. Thus, the structure of **16** was established to be methyl 20-hydroxy-7β,8β-epoxy-3,12,15,23-tetraoxo-lanosta-9,16-dien-26-oate.

In addition, 24 known compounds were identified by comparing their 1D NMR spectra with those reported in the literature, and they were assigned as gibbosic acid G (**17**) [34], applanoic acid B (**18**) [36], gibbosicolid E (**19**) [35], gibbosicolid F (**20**) [35], gibbosicolid G (**21**) [35], gibbosic acid M (**22**) [35], gibbosic acid L (**23**) [35], gibbosic acid O (**24**) [35], applanoic acid D (**25**) [36], gibbosic acid I (**26**) [35], ganodapplanoic acid D (**27**) [27], applanoic acid C (**28**) [36], ganoapplanic acid F (**29**) [37], elfvingic acid B (**30**) [37], applanoxidic acid G methyl ester (**31**) [37], gibbosic acid C (**32**) [34], gibbosic acid B (**33**) [34], elfvingic acid C (**34**) [38], methyl ganoapplaniate D (**35**) [37], applanone E (**36**) [36], ganoapplanoid K (**37**) [28], ganoapplanoid L (**38**) [28], ganoapplanilactone B (**39**) [37], and ganoapplanilactone A (**40**) [37].

Parts of the isolated compounds were evaluated to determine their anti-adipogenesis activities. At a concentration of 20 μM, compounds **16**, **22**, **28**, and **32** showed comparable inhibition for lipid accumulation compared to the positive control (LiCl, 20 mM). Meanwhile, compounds **15** and **20** displayed stronger inhibitory effects than the positive control, even resembling the untreated group (Figure 4). Furthermore, compounds **15** and **20** did not show any toxicity for the 3T3-L1 cells when the concentration was less than 100 μM. At the concentrations of 1.25, 2.5, 5, 10, 20, and 30 or 40, compounds **15** and **20** showed significantly inhibitory activities in a dose-dependent manner, with IC_{50} values of 6.42 and 5.39 μM, respectively (Figure 5).

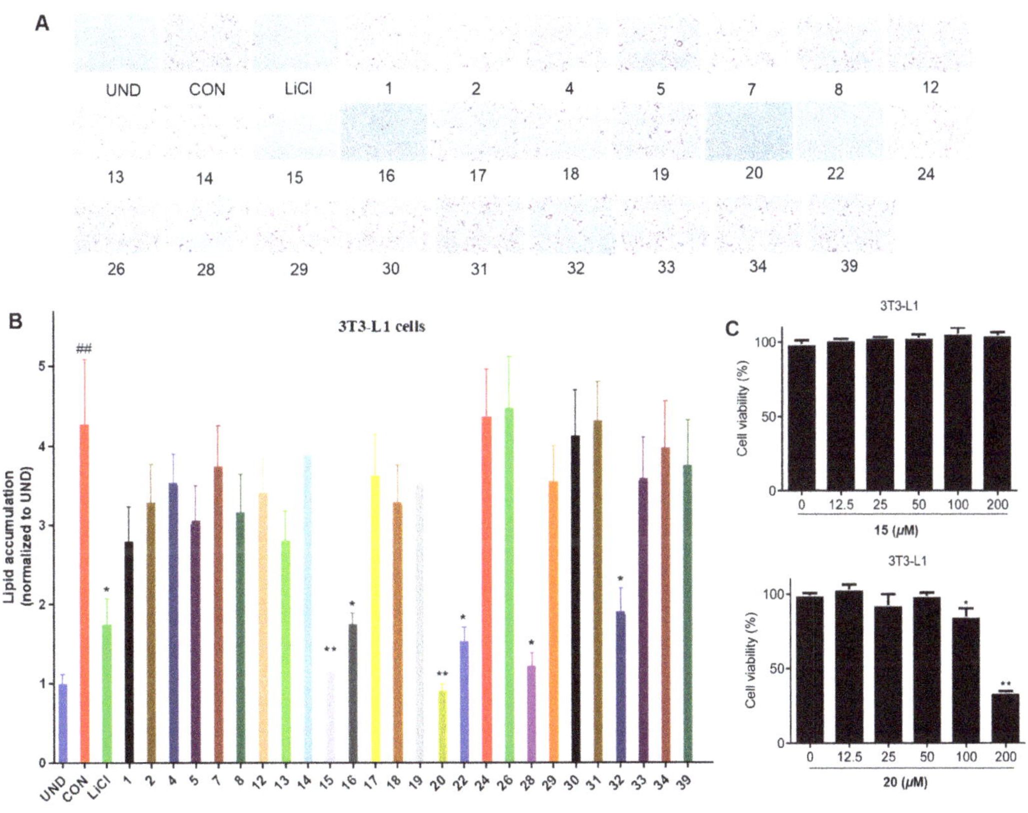

Figure 4. Effects of compounds (**1**, **2**, **4**, **5**, **7**, **8**, **12–20**, **22**, **24**, **26**, **28–34**, and **39**) at a level of 20 μM on lipid accumulation during 3T3-L1 adipocyte differentiation (**A**). LiCl (20 mM) was used as a positive control. Quantification of intracellular lipids in Oil Red O-stained adipocytes (**B**). Cell viability of compounds **15** and **20** on 3T3-L1 pre-adipocytes when treated for 24 h with an MTS assay (**C**). Data are representative results from three independent experiments. Data are shown as mean ± SD (n = 3) versus undifferentiated cells (UND). (##) $p < 0.01$ versus undifferentiated cells (UND). (*) $p < 0.05$ and (**) $p < 0.01$ versus fully differentiated cells (CON).

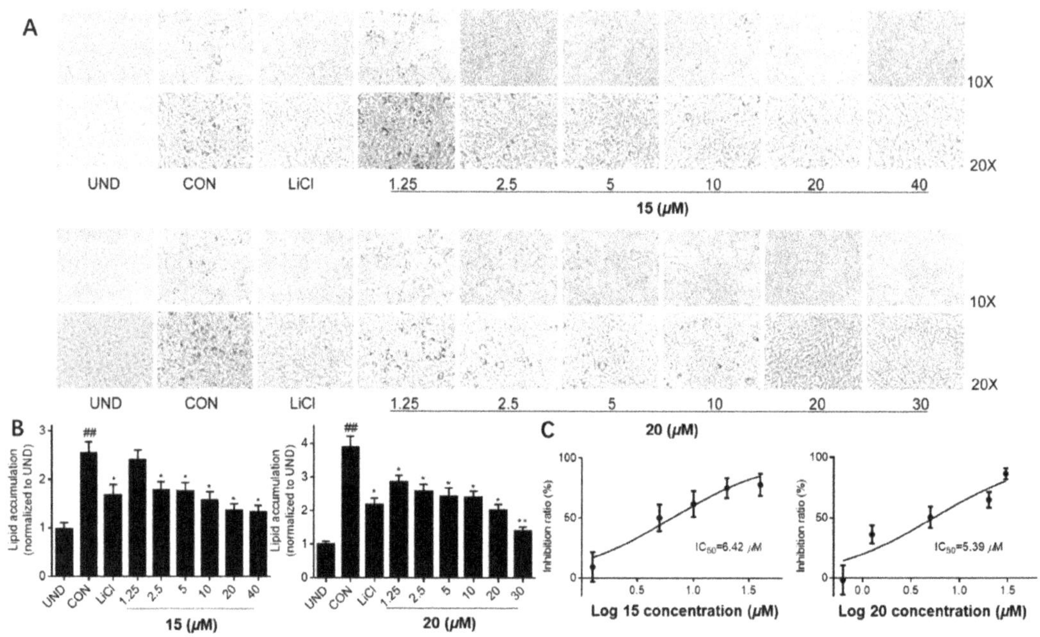

Figure 5. Effects of compounds **15** and **20** on lipid accumulation in 3T3-L1 adipocytes. (**A**) Oil Red O staining of cells administrated with serial doses of compounds **15** and **20**. (**B**) Quantification of intracellular lipid in Oil Red O-stained adipocytes. (C) The IC_{50} values of compounds **15** and **20**. LiCl (20 mM) was used as a positive control. Data are representative results from three independent experiments. Data are shown as mean ± SD (n = 3), versus undifferentiated cells (UND). (##) $p < 0.01$ versus undifferentiated cells (UND). (*) $p < 0.05$ and (**) $p < 0.01$ versus fully differentiated cells (CON).

The structures of the isolates were divided into nine types, including type I with a 7,12-dioxo-8-hydroxy-9,11-en fraction, type II with a A-seco-7,8-epoxy-9,11-en-12-oxo-23→27 lactone fraction, type III with a 7,23-dioxo-8(9),11(12),20(22)-trien fraction, type IV with a 12-oxo-7(8),9(11)-dien fraction, type V with a 7,11-dioxo-12-hydroxy-8(9)-en fraction, type VI with a 7,8-epoxy-12,23-dioxo-9(11),16(17),20(22)-trien fraction, type VII with a 20-hydroxy-7,8-epoxy-12,23-dioxo-9(11)-en fraction, type VIII with a 12-oxo-7,8-epoxy-9(11)-en-21,22,23,24,25,26,27-norlanostane, and type IX with a 12,23-epoxy-23→27 lactone fraction. The combined results of the previous and present studies showed that bioactive compounds were mainly present in type II (**20** and **21**), type III (**22**), type VI (**15** and **28**), type VII (**16** and **32**), and type VIII (**37**). For type II, when the relative configuration of 15-OH was α, the activity was decreased, similar to compound **19**, while any changes in the side chain decreased their activities, such as compounds **2–4**. For type III, no matter which reactions happened in type III, compounds **5–8** and **23** did not show inhibitory activity. For type VI, compound **27** was the 3-OH analogue of **15** and **28**, leading to a decrease in inhibition. In type VII, the carbonyl at C-3 and the carbonyl or hydroxyl at C-15 could be the crucial active functionalities. C24 lanostane triterpenoids possessing a double bond at C-16 and C-17 displayed anti-adipogenesis activity [27,28]. Compared to the other compounds, compound **20** belonging to type II showed the strongest inhibitory activity, suggesting that A-*seco*-15β-hydroxy-7,8-epoxy-12-oxolanosta-9,11-en-23→27 lactone could play a significant role in the anti-adipogenesis effect (Figure 6).

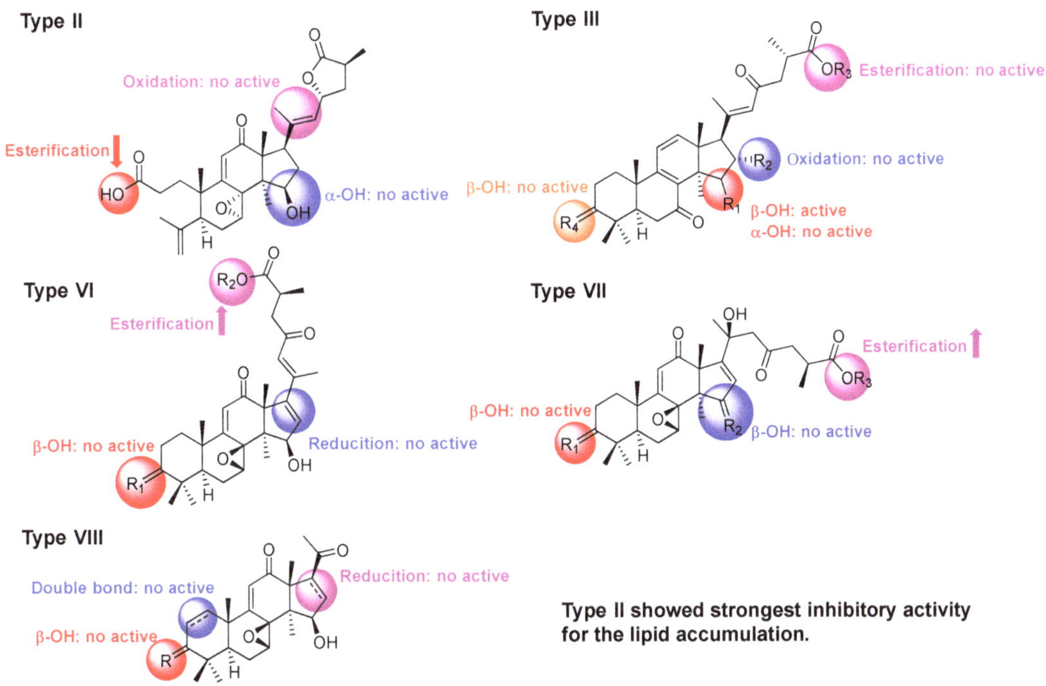

Figure 6. The proposed structure–activity relationship of triterpenoids from *G. applanatum*.

4. Conclusions

Overall, inspired by our previous studies, we investigate the lanostane-type triterpenoids of *G. applanatum*; 40 triterpenoids, including 16 new compounds, were isolated. Their anti-adipogenesis activities were evaluated and the results showed that compounds **15** and **20** can significantly inhibit lipid accumulation, with the IC_{50} values of 6.42 and 5.39 μM, respectively. Furthermore, we established a structure–activity relationship for the lanostane-type triterpenoids from *G. applanatum*, suggesting that the structure skeleton (A-*seco*-15β-hydroxy-7,8-epoxy-12-oxolanosta-9,11-en-23→27 lactone) could be of importance for the anti-adipogenic effect. In the next step, we can use type III as a template for further structural modification in order to find the lead compound.

Supplementary Materials: The following supporting information can be downloaded at: https://www.mdpi.com/article/10.3390/jof8040331/s1, Section S1: 1D and 2D NMR spectra of compound **1**; Section S2: HRESIMS spectrum of **1**; Section S3: X-ray crystallographic data of **1**; Section S4: 1D and 2D NMR spectra of **2**; Section S5: HRESIMS spectrum of **2**; Section S6: 1D and 2D NMR spectra of **3**; Section S7: HRESIMS spectrum of **3**; Section S8: 1D and 2D NMR spectra of **4**; Section S9: HRESIMS spectrum of **4**; Section S10: Comparison of ^{1}H NMR and ^{1}H-^{1}H COSY spectra between **4r** and **4s**; Section S11: 1D and 2D NMR spectra of **5**; Section S12: HRESIMS spectrum of **5**; Section S13: 1D and 2D NMR spectra of **6**; Section S14: HRESIMS spectrum of **6**; Section S15: 1D and 2D NMR spectra of **7**; Section S16: HRESIMS spectrum of **7**; Section S17: 1D and 2D NMR spectra of **8**; Section S18: HRESIMS spectrum of **8**; Section S19: 1D and 2D NMR spectra of **9**; Section S20: HRESIMS spectrum of **9**; Section S21: 1D and 2D NMR spectra of **10**; Section S22: HRESIMS spectrum of **10**; Section S23: 1D and 2D NMR spectra of **11**; Section S24: HRESIMS spectrum of **11**; Section S25: 1D and 2D NMR spectra of **12**; Section S26: HRESIMS spectrum of **12**; Section S27: 1D and 2D NMR spectra of **13**; Section S28: HRESIMS spectrum of **13**; Section S29: 1D and 2D NMR spectra of **14**; Section S30: HRESIMS spectrum of **14**; Section S31: 1D and 2D NMR spectra of **15**; Section S32: HRESIMS spectrum of **15**; Section S33: 1D and 2D NMR spectra of **16**; Section S34: HRESIMS spectrum of **16**.

Author Contributions: Conceptualization, X.-R.P. and M.-H.Q.; methodology, Q.W., X.-R.P., H.-G.S., W.-Y.X. and M.-H.Q.; resources, X.-R.P., L.Z., W.-Y.X. and M.-H.Q.; data curation, X.-R.P. and Q.W.; writing—original draft preparation, X.-R.P.; writing—review and editing, X.-R.P., W.-Y.X. and M.-H.Q.; project administration, X.-R.P.; funding acquisition, X.-R.P. All authors have read and agreed to the published version of the manuscript.

Funding: This research was funded by the Basic Research Project of Yunnan Province (201901T070239) and the Youth Innovation Promotion Association of CAS (2019383).

Institutional Review Board Statement: Not applicable.

Informed Consent Statement: Not applicable.

Data Availability Statement: X-ray crystallographic data of **1** (CIF) are available free of charge.

Acknowledgments: This work was supported by grants from the Basic Research Program of Yunnan Province (202001AT070070), and the Youth Innovation Promotion Association of CAS (2019383). The authors are grateful to the Analytical and Testing Center at Kunming Institute of Botany for the NMR and ECD data collection.

Conflicts of Interest: The authors declare no conflict of interest. The funders had no role in the design of the study; in the collection, analysis, or interpretation of data; in the writing of the manuscript; or in the decision to publish the results.

References

1. Sudheer, S.; Bai, R.G.; Muthoosamy, K.; Tuvikene, R.; Gupta, V.K.; Manickam, S. Biosustainable production of nanoparticles via mycogenesis for biotechnological applications: A critical review. *Environ. Res.* **2022**, *204*, 111963–111979. [CrossRef] [PubMed]
2. Li, M.; Lv, R.; Xu, X.; Ge, Q.; Lin, S. *Tricholoma matsutake*-Derived Peptides Show Gastroprotective Effects against Ethanol-Induced Acute Gastric Injury. *J. Agric. Food Chem.* **2021**, *69*, 14985–14994. [CrossRef] [PubMed]
3. Li, M.; Ge, Q.; Du, H.; Lin, S. *Tricholoma matsutake*-derived peptides ameliorate inflammation and mitochondrial dysfunction in RAW264.7 macrophages by modulating the NF-κB/COX-2 pathway. *Foods* **2021**, *10*, 2680. [CrossRef] [PubMed]
4. Yu, H.; Shen, X.; Chen, H.; Dong, H.; Zhang, L.; Yuan, T.; Zhang, D.; Shang, X.; Tan, Q.; Liu, J.; et al. Analysis of heavy metal content in *Lentinula edodes* and the main influencing factors. *Food Control* **2021**, *130*, 108198–108205. [CrossRef]
5. Shi, D.; Yin, C.; Fan, X.; Yao, F.; Qiao, Y.; Xue, S.; Lu, Q.; Feng, C.; Meng, J.; Gao, H. Effects of ultrasound and gamma irradiation on quality maintenance of fresh *Lentinula edodes* during cold storage. *Food Chem.* **2022**, *373*, 131478. [CrossRef]
6. Chen, Y.-h.; Chen, J. Optimization of extraction conditions for polysaccharides from *Collybia albuminosa* via response surface methodology. *Mod. Food Sci. Technol.* **2012**, *28*, 541.
7. Peng, X.R.; Su, H.G.; Liu, J.H.; Huang, Y.J.; Yang, X.Z.; Li, Z.R.; Zhou, L.; Qiu, M.H. C30 and C31 triterpenoids and triterpene sugar esters with cytotoxic activities from edible mushroom *Fomitopsis pinicola* (Sw. Ex Fr.) Krast. *J. Agric. Food. Chem.* **2019**, *67*, 10330–10341. [CrossRef]
8. Ying, Y.M.; Yu, H.F.; Tong, C.P.; Shan, W.G.; Zhan, Z.J. Spiroinonotsuoxotriols A and B, two highly rearranged triterpenoids from *Inonotus obliquus*. *Org. Lett.* **2020**, *22*, 3377–3380. [CrossRef]
9. Yu, J.; Xiang, J.Y.; Xiang, H.; Xie, Q. Cecal butyrate (Not Propionate) was connected with metabolism-related chemicals of mice, based on the different effects of the two *Inonotus obliquus* extracts on obesity and their mechanisms. *ACS Omega* **2020**, *5*, 16690–16700. [CrossRef]
10. Li, Y.-T.; Zhang, Z.; Feng, Y.; Cheng, Y.; Li, S.; Li, C.; Tian, L.-W. Cardioprotective 22-hydroxylanostane triterpenoids from the fruiting bodies of *Phellinus igniarius*. *Phytochemistry* **2021**, *191*, 112907–112914. [CrossRef]
11. Cao, Y.; Liu, Y.; Wang, G.; Wang, W.; Li, Y.; Xuan, L. Styryl pyranones with apoptosis activities from the sporocarps of *Phellinus igniarius*. *Phytochem. Lett.* **2021**, *44*, 154–159. [CrossRef]
12. Shao, W.; Xiao, C.; Yong, T.; Zhang, Y.; Hu, H.; Xie, T.; Liu, R.; Huang, L.; Li, X.; Xie, Y.; et al. A polysaccharide isolated from *Ganoderma lucidum* ameliorates hyperglycemia through modulating gut microbiota in type 2 diabetic mice. *Int. J. Biol. Macromol.* **2022**, *197*, 23–38. [CrossRef] [PubMed]
13. Wen, L.; Sheng, Z.; Wang, J.; Jiang, Y.; Yang, B. Structure of water-soluble polysaccharides in spore of *Ganoderma lucidum* and their anti-inflammatory activity. *Food Chem.* **2022**, *373*, 131374. [CrossRef] [PubMed]
14. Han, W.; Chen, H.; Zhou, L.; Zou, H.; Luo, X.; Sun, B.; Zhuang, X. Polysaccharides from *Ganoderma Sinense*—rice bran fermentation products and their anti-tumor activities on non-small-cell lung cancer. *BMC Complementary Med. Ther.* **2021**, *21*, 169–179. [CrossRef]
15. Teseo, S.; Houot, B.; Yang, K.; Monnier, V.; Liu, G.; Tricoire, H. *G. sinense* and *P. notoginseng* extracts improve healthspan of aging flies and provide protection in a huntington disease model. *Aging Dis.* **2021**, *12*, 425–440. [CrossRef]

16. Ahmad, R.; Riaz, M.; Khan, A.; Aljamea, A.; Algheryafi, M.; Sewaket, D.; Alqathama, A. *Ganoderma lucidum* (Reishi) an edible mushroom; a comprehensive and critical review of its nutritional, cosmeceutical, mycochemical, pharmacological, clinical, and toxicological properties. *Phytother. Res.* **2021**, *35*, 6030–6062. [CrossRef]
17. Zeng, M.; Qi, L.; Guo, Y.; Zhu, X.; Tang, X.; Yong, T.; Xie, Y.; Wu, Q.; Zhang, M.; Chen, D. Long-term administration of triterpenoids From *Ganoderma lucidum* mitigates age-associated brain physiological decline via regulating sphingolipid metabolism and enhancing autophagy in Mice. *Front. Aging Neurosci.* **2021**, *13*, 628860–628881. [CrossRef] [PubMed]
18. Xu, J.; Xiao, C.; Xu, H.; Yang, S.; Chen, Z.; Wang, H.; Zheng, B.; Mao, B.; Wu, X. Anti-inflammatory effects of *Ganoderma lucidum* sterols via attenuation of the p38 MAPK and NF-κB pathways in LPS-induced RAW 264.7 macrophages. *Food Chem. Toxicol.* **2021**, *150*, 112073–112082. [CrossRef]
19. Shao, C.S.; Feng, N.; Zhou, S.; Zheng, X.X.; Wang, P.; Zhang, J.S.; Huang, Q. Ganoderic acid T improves the radiosensitivity of HeLa cells via converting apoptosis to necroptosis. *Toxicol. Res.* **2021**, *10*, 531–541. [CrossRef]
20. Kou, R.W.; Gao, Y.Q.; Xia, B.; Wang, J.Y.; Liu, X.N.; Tang, J.J.; Yin, X.; Gao, J.M. Ganoderterpene A, a New Triterpenoid from *Ganoderma lucidum*, Attenuates LPS-Induced Inflammation and Apoptosis via Suppressing MAPK and TLR-4/NF-kappaB Pathways in BV-2 Cells. *J. Agric. Food. Chem.* **2021**, *69*, 12730–12740. [CrossRef]
21. Wang, Y.-Q.; Wang, N.-X.; Luo, Y.; Yu, C.-Y.; Xiao, J.-H. Ganoderal A effectively induces osteogenic differentiation of human amniotic mesenchymal stem cells via cross-talk between Wnt/β-catenin and BMP/SMAD signaling pathways. *Biomed. Pharmacother.* **2020**, *123*, 109807–109817. [CrossRef] [PubMed]
22. Chen, Y.; Gao, J.; Chen, Q.; Liu, W.; Qi, Y.; Aisa, H.A.; Yuan, T. Applanaic acids A-C, three new highly oxygenated lanostane triterpenoids from the fruiting bodies of *Ganoderma applanatum*. *Nat. Prod. Res.* **2020**, *35*, 3918–3924. [CrossRef] [PubMed]
23. Song, X.; Cui, W.; Gao, Z.; Zhang, J.; Jia, L. Structural characterization and amelioration of sulfated polysaccharides from *Ganoderma applanatum* residue against CCl_4-induced hepatotoxicity. *Int. Immunopharmacol.* **2021**, *96*, 107554–107563. [CrossRef]
24. Mfopa, A.; Mediesse, F.K.; Mvongo, C.; Nkoubatchoundjwen, S.; Lum, A.A.; Sobngwi, E.; Kamgang, R.; Boudjeko, T. Antidyslipidemic potential of water-soluble polysaccharides of *Ganoderma applanatum* in MACAPOS-2-induced obese rats. *Evid. Based Complement. Alternat. Med.* **2021**, *2021*, 2452057. [CrossRef] [PubMed]
25. Hossen, S.M.M.; Islam, M.J.; Hossain, M.R.; Barua, A.; Uddin, M.G.; Emon, N.U. CNS anti-depressant, anxiolytic and analgesic effects of *Ganoderma applanatum* (mushroom) along with ligand-receptor binding screening provide new insights: Multi-disciplinary approaches. *Biochem. Biophys. Rep.* **2021**, *27*, 101062. [CrossRef]
26. Hossain, M.S.; Barua, A.; Tanim, M.A.H.; Hasan, M.S.; Islam, M.J.; Hossain, R.M.; Emon, N.U.; Hossen, S.M.M. *Ganoderma applanatum* mushroom provides new insights into the management of diabetes mellitus, hyperlipidemia, and hepatic degeneration: A comprehensive analysis. *Food Sci. Nutr.* **2021**, *9*, 4364–4374. [CrossRef] [PubMed]
27. Su, H.G.; Wang, Q.; Zhou, L.; Peng, X.R.; Xiong, W.Y.; Qiu, M.H. Functional triterpenoids from medicinal fungi *Ganoderma applanatum*: A continuous search for antiadipogenic agents. *Bioorg. Chem.* **2021**, *112*, 104977–104986. [CrossRef]
28. Su, H.G.; Wang, Q.; Zhou, L.; Peng, X.R.; Xiong, W.Y.; Qiu, M.H. Highly oxygenated lanostane triterpenoids from *Ganoderma applanatum* as a class of agents for inhibiting lipid accumulation in adipocytes. *Bioorg. Chem.* **2020**, *104*, 104263–104276. [CrossRef]
29. Peng, X.R.; Wang, Q.; Wang, H.R.; Hu, K.; Xiong, W.Y.; Qiu, M.H. FPR2-based anti-inflammatory and anti-lipogenesis activities of novel meroterpenoid dimers from *Ganoderma*. *Bioorg. Chem.* **2021**, *116*, 105338–105348. [CrossRef]
30. Peng, X.; Su, H.; Wang, H.; Hu, G.; Hu, K.; Zhou, L.; Qiu, M. Applanmerotic acids A and B, two meroterpenoid dimers with an unprecedented polycyclic skeleton from *Ganoderma applanatum* that inhibit formyl peptide receptor 2. *Org. Chem. Front.* **2021**, *8*, 3381–3389. [CrossRef]
31. Lee, S.Y.; Kim, J.S.; Lee, S.; Kang, S.S. Polyoxygenated ergostane-type sterols from the liquid culture of *Ganoderma applanatum*. *Nat. Prod. Res.* **2011**, *25*, 1304–1311. [CrossRef]
32. Peng, X.; Liu, J.; Xia, J.; Wang, C.; Li, X.; Deng, Y.; Bao, N.; Zhang, Z.; Qiu, M. Lanostane triterpenoids from *Ganoderma hainanense* J. D. Zhao. *Phytochemistry* **2015**, *114*, 137–145. [CrossRef]
33. Wang, Q.; Mu, R.F.; Liu, X.; Zhou, H.M.; Xu, Y.H.; Qin, W.Y.; Yang, C.R.; Wang, L.B.; Li, H.Z.; Xiong, W.Y. Steaming changes the composition of saponins of *Panax notoginseng* (Burk.) F.H. Chen that function in treatment of hyperlipidemia and obesity. *J. Agric. Food. Chem.* **2020**, *68*, 4865–4875. [CrossRef] [PubMed]
34. Pu, D.; Li, X.; Lin, J.; Zhang, R.; Luo, T.; Wang, Y.; Gao, J.; Zeb, M.A.; Zhang, X.; Li, X.; et al. Triterpenoids from *Ganoderma gibbosum*: A class of sensitizers of FLC-Resistant candida albicans to fluconazole. *J. Nat. Prod.* **2019**, *82*, 2067–2077. [CrossRef] [PubMed]
35. Pu, D.-B.; Zheng, X.; Gao, J.-B.; Zhang, X.-J.; Qi, Y.; Li, X.-S.; Wang, Y.-M.; Li, X.-N.; Li, X.-L.; Wan, C.-P.; et al. Highly oxygenated lanostane-type triterpenoids and their bioactivity from the fruiting body of *Ganoderma gibbosum*. *Fitoterapia* **2017**, *119*, 1–7. [CrossRef] [PubMed]
36. Peng, X.; Li, L.; Dong, J.; Lu, S.; Lu, J.; Li, X.; Zhou, L.; Qiu, M. Lanostane-type triterpenoids from the fruiting bodies of *Ganoderma applanatum*. *Phytochemistry* **2019**, *157*, 103–110. [CrossRef] [PubMed]
37. Li, L.; Peng, X.-R.; Dong, J.-R.; Lu, S.-Y.; Li, X.-N.; Zhou, L.; Qiu, M.-H. Rearranged lanostane-type triterpenoids with anti-hepatic fibrosis activities from *Ganoderma applanatum*. *RSC Adv.* **2018**, *8*, 31287–31295. [CrossRef]
38. Yoshikawa, K.; Nishimura, N.; Bando, S.; Arihara, S.; Matsumura, E.; Katayama, S. New lanostanoids, elfvingic acids A-H, from the fruit body of *Elfvingia applanata*. *J. Nat. Prod.* **2002**, *65*, 548–552. [CrossRef]

Journal of *Fungi*

MDPI

Article

Cytochalasans from the Endophytic Fungus *Phomopsis* sp. shj2 and Their Antimigratory Activities

Bing-Chao Yan [1,2,3], Wei-Guang Wang [1,2], Ling-Mei Kong [1,2], Jian-Wei Tang [1,2,3], Xue Du [1,2,3], Yan Li [1,2] and Pema-Tenzin Puno [1,2,3,*]

1 State Key Laboratory of Phytochemistry and Plant Resources in West China, Kunming Institute of Botany, Chinese Academy of Sciences, Kunming 650201, China; yanbingchao@mail.kib.ac.cn (B.-C.Y.); wwg@live.cn (W.-G.W.); konglingmei@mail.kib.ac.cn (L.-M.K.); tangjianwei@mail.kib.ac.cn (J.-W.T.); duxue@mail.kib.ac.cn (X.D.); liyan@mail.kib.ac.cn (Y.L.)

2 Yunnan Key Laboratory of Natural Medicinal Chemistry, Kunming 650201, China

3 University of Chinese Academy of Sciences, Beijing 100049, China

* Correspondence: punopematenzin@mail.kib.ac.cn

Abstract: Cytochalasans from the endophytic fungi featured structure diversity. Our previous study has disclosed that cytochalasans from the endophytic fungus *Phomopsis* sp. shj2 exhibited an antimigratory effect. Further chemical investigation on *Phomopsis* sp. shj2 has led to the discovery of seven new cytochalasans (**1–7**), together with four known ones. Their structures were elucidated through extensive spectroscopic data interpretation and single-crystal X-ray diffraction analysis. Compounds **1–3** and **8–11** exhibited antimigratory effects against MDA-MB-231 in vitro with IC_{50} values in the range of 1.01−10.42 μM.

Keywords: endophytic fungus; *Phomopsis*; cytochalasan; antimigratory activity

Citation: Yan, B.-C.; Wang, W.-G.; Kong, L.-M.; Tang, J.-W.; Du, X.; Li, Y.; Puno, P.-T. Cytochalasans from the Endophytic Fungus *Phomopsis* sp. shj2 and Their Antimigratory Activities. *J. Fungi* **2022**, *8*, 543. https://doi.org/10.3390/jof8050543

Academic Editors: Tao Feng and Frank Surup

Received: 1 May 2022
Accepted: 20 May 2022
Published: 23 May 2022

1. Introduction

Endophytic fungi are emerging as rich resources for structurally unique and bioactive secondary metabolites, which arouse increasing research interest in the past decades [1–3]. Cytochalasans represent a large class of fungal polyketide synthase-nonribosomal peptide synthetase (PKS-NRPS) hybrid secondary metabolites. Recently, plenty of polycyclic cytochalasans have been identified [4–8] and synthesised [9,10]; moreover, they exhibited a broad spectrum of interesting biological activities, such as cytotoxic [4,5,8], immunoregulatory [11], and antimicrobial [6] activities. To date, more than 400 cytochalasans have been isolated from various fungal sources, such as *Phomopsis* [4], *Xylaria* [12], *Chaetomium* [13] and *Phoma* [14] genera.

Tumour spread is a major concern in cancer therapeutics as cancer metastasis is responsible for 90% of deaths from solid tumours [15]. Natural products with antimigratory activity represent a highly interesting field to explore for cancer chemoprevention and therapy. Fungi are emerging as a natural source, such as *Diaporthe* [16], *Isaria* [17], and *Phenicillium* [18,19] genera. Chemical investigations on endophytes of *Isodon* species have disclosed structurally diverse and bioactive natural products [19–22]. Phomopchalasins B and C were isolated from the endophytic fungus *Phomopsis* sp. shj2 from the stems of *Isodon eriocalyx* var. *laxiflora* and exhibited in vitro antimigratory effects against MDA-MB231 [19]. In our continuous efforts for more bioactive structures, the strain was further investigated by one strain-many compounds strategy (OSMAC), which led to the isolation of seven new cytochalasans (**1–7**), along with four known ones (Figure 1). Herein, we report the isolation, structure elucidation, and antimigratory activities of these cytochalasans.

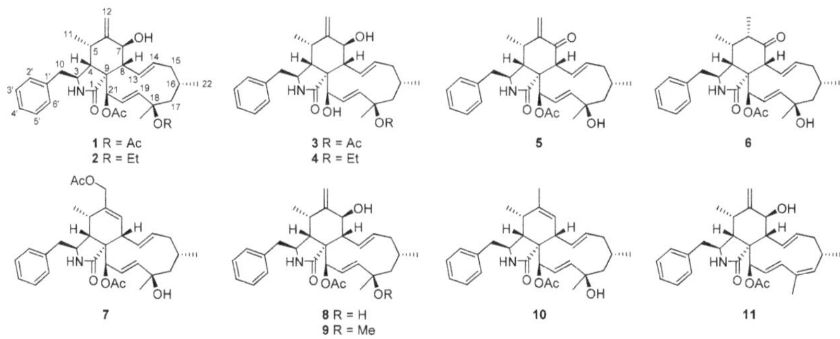

Figure 1. Structures of compounds **1–11**.

2. Materials and Methods

2.1. General Experimental Procedures

Column chromatography (CC) was performed with silica gel (100–200 mesh, Qingdao Marine Chemical, Inc., Qingdao, China), Lichroprep RP-18 gel (40–63 μm, Merck, Darmstadt, Germany). Preparative HPLC and semi-preparative HPLC were performed on an Agilent 1200 liquid chromatograph with a Zorbax SB-C18 (9.4 mm × 25 cm) column. Fractions were monitored by TLC, and spots were visualized by heating silica gel plates sprayed with 10% H_2SO_4 in EtOH. Petroleum ether (PE, 60–90 °C), EtOAc, $CHCl_3$, acetone, MeOH, and EtOH were of analytical grade and purchased from Sinopharm Chemical Reagent Co. Ltd., China. All solvents were distilled before use. NMR spectra were recorded on Bruker DRX-500, AV-600, and 800 spectrometers. ESIMS and HRESIMS experiments were performed on a Bruker HCT/Esquire spectrometer and a Waters AutoSpec Premier P776 spectrometer. CD spectra were measured on an Applied Photophysics Chirascan spectrophotometer. Optical rotations were measured with a JASCO P-1020 polarimeter. UV spectra were obtained using a Shimadzu UV-2401A spectrophotometer.

2.2. Fungal Material

The culture of *Phomopsis* sp. shj2 was isolated from the stems of *Isodon eriocalyx* var. *laxiflora* collected from Kunming Botanical Garden, Kunming, People's Republic of China, in December 2012. The isolate was identified based on sequence (GenBank Accession No. KU533636) analysis of the ITS region of the rDNA. The fungal strain was cultured on slants of potato dextrose agar at 25 °C for 7 days. Agar plugs were cut into small pieces (about 0.5 × 0.5 × 0.5 cm^3) under aseptic conditions, and 15 pieces were used to inoculate three Erlenmeyer flasks (250 mL), each containing 50 mL of media (0.4% glucose, 1% malt extract, and 0.4% yeast extract); the final pH of the media was adjusted to 6.5, and the flasks were sterilized by autoclave. Three flasks of the inoculated media were incubated at 28 °C on a rotary shaker at 180 rpm for 5 days to prepare the seed culture. Fermentation was carried out in 125 Fernbach flasks (500 mL), each containing 80 g of rice. Spore inoculum was prepared in sterile, distilled H_2O to give a final spore/cell suspension of 1 × 10^6/mL. Distilled H_2O (120 mL) was added to each flask, and the contents were soaked overnight before autoclaving at 15 psi for 30 min. After cooling to room temperature, each flask was inoculated with 5.0 mL of the spore inoculum and incubated at 28 °C for 42 days.

2.3. Extraction and Isolation

The fermented material was extracted with EtOAc (4 × 10.0 L) and the organic solvent was evaporated to dryness under vacuum to afford a crude extract (170 g). The crude extract was purified by CC (column chromatography on SiO_2 with $CHCl_3$/acetone gradient system 1:0, 9:1, 8:2, 7:3, 6:4 and 1:1) to yield six main fractions, Fr.s A–F. Fr. B was subjected to chromatography over silica gel CC (petroleum ether-EtOAc) to give subfractions B1–B9. Fr. B2 was further purified by silica gel CC (petroleum ether-acetone) to give **1** (10.7 mg).

Fr. B8 was purified by semi-preparative HPLC (3 mL/min, detector UV λ_{max} 210 nm, MeCN-H_2O) to afford **11** (3.2 mg), **8** (25.1 mg), and **10** (3.7 mg). Fr. C was purified by chromatography over silica gel CC (petroleum ether-acetone) to give subfractions Fr.s C1–C10. The subfraction C8 was recrystallized to give **7** (20.5 mg). Fr. C5 was separated by semi-preparative HPLC (3 mL/min, detector UV λ_{max} 210 nm, MeCN-H_2O) to afford **3** (4.7 mg) and **9** (10.2 mg). Fr. D was subjected to Sephadex LH-20 (CH_3Cl-MeOH) to yield subfractions D1–D6. The subfraction D5 was purified by recrystallization to afford **4** (1.2 mg). And Fr. D5 was further purified to afford **2** (20.3 mg). Fr. E was purified by semi-preparative HPLC (3 mL/min, detector UV λ_{max} 210 nm, MeCN-H_2O) to afford **5** (1.5 mg) and **6** (1.6 mg).

18-Acetoxycytochalasin H (**1**): white powder (MeOH); $[\alpha]_D^{20}$ = +44.2 (c 0.23, MeOH), UV (MeOH) λ_{max} (log ε): 203.2 (0.5151); ^1H and ^{13}C NMR data, see Tables 1 and 2; HRESIMS $[M + Na]^+$ m/z 558.2826 (calcd for $C_{32}H_{41}NO_6Na$, 558.2826).

Table 1. ^1H NMR data ($CDCl_3$, δ in ppm) of compounds **1–7**.

No.	1 [a,b]	2 [a,c]	3 [c,d]	4 [a,c]	5 [a,c]	6 [a,e]	7 [a,c]
3	3.25 (m)	3.26 (m)	3.33 (m)	3.27 (m)	3.23 (dt, 9.4, 4.3)	3.54 (dt, *J* = 9.4, 4.3)	3.28 (overlap)
4	2.15 (m)	2.14 (m)	2.65 (m)	2.58 (m)	2.35 (t, 4.3)	2.25 (t, 4.2)	2.19 (t, 4.3)
5	2.76 (m)	2.78 (m)	2.72 (m)	2.90 (m)	3.08 (m)	2.11 (m)	2.53 (m)
6						2.01 (m)	
7	3.84 (d, 10.5)	3.83 (d, 10.5)	3.79 (d, 10.5)	3.82 (d, 10.5)			5.72 (s)
8	2.93 (d, 10.5)	2.96 (d, 10.5)	2.94 (d, 10.5)	2.90 (d, 10.5)	3.94 (d, 9.3)	3.79 (d, 9.4)	3.26 (overlap)
10	2.85 (dd, 13.5, 4.5) 2.65 (dd, 13.5, 9.6)	2.86 (dd, 13.3, 3.8) 2.67 (m)	2.81 (m) 2.79 (m)	2.58 (m) 1.71 (m)	2.90 (dd, 13.5, 4.3) 2.65 (dd, 13.5, 9.4)	2.92 (dd, 13.5, 4.3) 2.64 (dd, 13.5, 9.4)	2.91 (dd, 13.5, 4.3) 2.60 (dd, 13.5, 10.2)
11	0.99 (d, 6.7)	1.01 (d, 6.7)	0.84 (d, 6.8)	1.10 (d, 6.7)	1.12 (d, 6.7)	0.98 (d, 6.7)	1.17 (d, 7.3)
12	5.33 (s) 5.10 (s)	5.35 (s) 5.11 (s)	5.18 (s) 4.95 (s)	5.32 (s) 5.11 (s)	6.25 (s) 5.29 (s)	1.12 (d, 7.0)	4.53 (d, 12.8) 4.48 (d, 12.8)
13	5.74 (dd, 15.5, 9.7)	5.73 (dd, 15.1, 10.0)	5.70 (dd, 15.0, 9.2)	5.71 (dd, 15.5, 9.8)	5.81 (dd, 15.6, 9.3)	5.69 (dd, 15.5, 9.4)	5.84 (dd, 15.3, 10.3)
14	5.38 (m)	5.43 (m)	5.22 (m)	5.35 (m)	5.19 (m)	5.16 (m)	5.24 (m)
15	2.01 (overlap) 1.79 (d, 12.4)	2.00 (overlap) 1.79 (d, 11.3)	1.90 (dd, 13.9, 3.1) 1.80 (m)	1.98 (dd, 10.4, 4.7) 1.78 (m)	2.04 (dd, 12.9, 4.4) 1.89 (m)	2.01 (m) 1.81 (m)	1.99 (m) 1.77 (overlap)
16	1.65 (m)	1.78 (m)	1.66 (m)	1.75 (m)	1.76 (m)	1.75 (m)	1.75 (m)
17	2.05 (dd, 14.3, 3.7), 1.75 (dd, 14.3, 3.0	1.69 (m)	1.90 (m) 1.75 (m)	1.78 (overlap)	1.85 (overlap) 1.54 (dd, 14.3, 3.2)	1.85 (m) 1.53 (dd, 14.3, 3.1)	1.88 (dd, 14.3, 2.7) 1.54 (d, 14.3)
19	5.56 (d, 16.6)	5.52 (d, 16.7)	5.84 (d, 16.7)	5.73 (d, 16.7)	5.52 (d, 16.6)	5.49 (d, 16.6)	5.52 (d, 16.6)
20	5.85 (dd, 16.6, 2.3)	5.79 (dd, 16.7, 2.4)	5.97 (dd, 16.7, 2.2)	5.99 (dd, 16.7, 2.6)	5.90 (dd, 16.6, 2.6)	5.85 (dd, 16.6, 2.5)	5.91 (dd, 16.6, 2.6)
21	5.63 (t, 2.3)	5.54 (t, 2.4)	4.02 (t, 2.2)	4.12 (t, 2.6)	5.65 (t, 2.6)	5.60 (t, 2.5)	5.68 (t, 2.6)
22	1.02 (d, 6.9)	1.01 (d, 6.5)	0.99 (d, 6.3)	1.01 (d, 6.3)	1.04 (d, 7.0)	1.03 (d, 6.9)	1.04 (d, 6.3)
23	1.58 (s)	1.26 (s)	1.53 (s)	1.28 (s)	1.34 (s)	1.32 (s)	1.34 (s)
2′, 6′	7.14 (d, 7.4)	7.15 (d, 7.4)	7.21 (d, 7.3)	7.15 (d, 7.4)	7.12 (d, 7.4)	7.15 (d, 7.3)	7.14 (d, 7.2)
3′, 5′	7.31 (t, 7.4)	7.32 (t, 7.4)	7.29 (t, 7.3)	7.31 (t, 7.4)	7.32 (t, 7.4)	7.33 (t, 7.3)	7.31 (t, 7.5)
4′	7.25 (t, 7.4)	7.25 (t, 7.4)	7.26 (d, 7.3)	7.24 (t, 7.4)	7.25 (t, 7.4)	7.25 (t, 7.3)	7.24 (t, 7.2)
12-OAc							2.04, s
18-OR	R = Ac 2.00 (s)	R = Et 3.38 (m), 2.65 (m) 1.14 (t, 6.9)	R = Ac 1.96 (s)	R = Et 3.41 (m), 3.37 (m) 1.17 (t, 7.0)	R = H	R = H	R = H
21-OAc	2.24 (s)	2.25 (s)			2.30 (s)	2.28 (s)	2.25 (s)

[a] Recorded in $CDCl_3$. [b] Recorded at 800 MHz. [c] Recorded at 600 MHz. [d] Recorded in acetone-d_6. [e] Recorded at 500 MHz.

Table 2. ^{13}C NMR data ($CDCl_3$, δ in ppm) of compounds **1–7**.

No.	1 [a,b]	2 [a,c]	3 [d,e]	4 [a,e]	5 [a,e]	6 [a,f]	7 [a,e]
1	174.3 s	174.3 s	176.7 s	175.8 s	172.8 s	173.5 s	174.9 s
3	53.9 d	53.9 d	54.2 d	50.6 d	54.0 d	53.3 d	56.1 d
4	50.5 d	50.9 d	50.2 d	53.9 d	50.6 d	51.1 d	53.8 d
5	33.0 d	33.0 d	33.8 d	33.1 d	34.2 d	35.7 d	34.7 d
6	148.0 s	148.0 s	152.1 s	148.6 s	143.9 s	45.8 d	135.8 s
7	70.0 d	69.9 d	71.6 d	70.1 d	198.7 s	214.0 s	134.6 d
8	47.3 d	47.4 d	46.8 d	46.0 d	53.1 d	52.0 d	43.4 d
9	52.1 s	51.9 s	54.5 s	53.0 s	52.9 s	53.6 s	56.2 s
10	45.7 t	45.8 t	45.5 t	45.8 t	46.0 t	46.3 t	46.1 t
11	14.1 q	14.2 q	14.2 q	14.1 q	14.4 q	15.9 q	13.2 q
12	114.3 t	114.3 t	112.0 t	113.9 t	121.0 t	16.0 q	64.9 t
13	127.6 d	127.2 d	130.2 d	128.0 d	123.0 d	123.2 d	128.3 d
14	138.2 d	138.6 d	135.7 d	137.8 d	138.4 d	137.9 d	136.3 d
15	42.7 t	43.1 t	43.7 t	42.8 t	43.1 t	42.9 t	42.8 t
16	28.6 d	28.0 d	29.2 d	27.8 d	28.6 d	28.5 d	28.7 d
17	51.5 t	51.8 t	52.9 t	51.1 t	53.7 t	53.5 t	53.5 t
18	84.4 s	78.5 s	85.0 s	78.5 s	74.6 s	74.5 s	74.6 s
19	136.6 d	138.9 d	135.1 d	137.3 d	137.7 d	137.7 d	137.3 d
20	124.9 d	125.8 d	131.8 d	130.8 d	125.9 d	125.9 d	126.5 d
21	77.4 d	78.1 d	76.7 d	77.0 d	77.9 d	77.8 d	77.0 d
22	25.5 q	26.2 q	25.9 q	26.0 q	26.6 q	26.6 q	26.6 q
23	26.3 q	25.2 q	27.0 q	25.2 q	31.5 q	31.4 q	31.6 q
1′	137.5 s	137.6 s	138.9 s	137.7 s	137.0 s	137.1 s	137.7 s
2′, 6′	129.1 d	129.2 d	130.8 d	129.1 d	129.2 d	129.2 d	129.1 d
3′, 5′	129.1 d	129.1 d	129.2 d	128.9 d	129.1 d	129.1 d	129.0 d
4′	127.2 d	127.3 d	127.4 d	127.1 d	127.4 d	127.4 d	127.3 d
12-OR							170.6 s, 21.1 q
18-OR	R = Ac 170.1 s, 21.0 q	R = Et 57.9 t, 16.3 q	R = Ac 170.4 s, 22.2 q	R = Et 57.5 t, 16.0 q	R = H	R = H	
21-OAc	170.3 s, 22.4 q	170.3 s, 21.1 q			170.1 s, 21.1 q	170.1 s, 21.1 q	170.2 s, 21.2 q

[a] Recorded in $CDCl_3$. [b] Recorded at 100 MHz. [c] Recorded at 150 MHz. [d] Recorded in acetone-d_6. [e] Recorded at 150 MHz. [f] Recorded at 125 MHz.

18-Ethoxycytochalasin H (**2**): white solid; $[\alpha]_D^{18}$ = +39.0 (*c* 0.15, MeOH), UV (MeOH) λ_{max} (log ε): 204.0 (0.4717); ^1H and ^{13}C NMR data, see Tables 1 and 2; HRESIMS $[M + Na]^+$ *m/z* 544.3030 (calcd for $C_{32}H_{43}NO_5Na$, 544.3039).

18-Acetoxycytochalasin J (**3**): $[\alpha]_D^{22}$ = +22.0 (*c* 0.24, MeOH), UV (MeOH) λ_{max} (log ε): 204.0 (0.5467); ^1H and ^{13}C NMR data, see Tables 1 and 2; HRESIMS $[M + Na]^+$ *m/z* 516.2726 (calcd for $C_{30}H_{39}NO_5Na$, 516.2720).

18-Ethoxycytochalasin J (**4**): $[\alpha]_D^{18}$ = +50.0 (*c* 0.19, MeOH), UV (MeOH) λ_{max} (log ε): 203.8 (0.5179); ^1H and ^{13}C NMR data, see Tables 1 and 2; HRESIMS $[M + H]^+$ *m/z* 480.3111 (calcd for $C_{30}H_{42}NO_4$, 480.3108).

7-Oxocytochalasin H (**5**): $[\alpha]_D^{24}$ = −12.3 (*c* 0.19, MeOH), UV (MeOH) λ_{max} (log ε): 205.0 (0.5284); ^1H and ^{13}C NMR data, see Tables 1 and 2; HRESIMS $[M + Na]^+$ *m/z* 514.2559 (calcd for $C_{30}H_{37}NO_5Na$, 514.2564).

Cytochalasin H_3 (**6**): $[\alpha]_D^{24}$ = −63.2 (*c* 0.19, MeOH), UV (MeOH) λ_{max} (log ε): 205.5 (0.5241); ^1H and ^{13}C NMR data, see Tables 1 and 2; HRESIMS $[M + Na]^+$ *m/z* 516.2719 (calcd for $C_{30}H_{39}NO_5Na$, 516.2720).

Cytochalasin H_4 (**7**): white solid; $[\alpha]_D^{20}$ = −28.4 (*c* 0.18, MeOH), UV (MeOH) λ_{max} (log ε): 203.2 (0.5250); IR (KBr) λ_{max} 3471, 2958, 2925, 1740, 1688, 1640, 1454, 1441, 1384, 1232 cm^{-1}; ^1H and ^{13}C NMR data, see Tables 1 and 2; HRESIMS $[M + H]^+$ *m/z* 536.3014 (calcd for $C_{32}H_{42}NO_6$, 536.3007).

2.4. X-ray Crystal Structure Analysis

The intensity data for **1** and **3** were collected on a Bruker APEX DUO diffractometer using graphite-monochromated Cu Kα radiation. The structures of these compounds were solved by direct methods (SHELXS97), expanded using difference Fourier techniques, and refined by the program and full-matrix least-squares calculations. The non-hydrogen atoms were refined anisotropically, and hydrogen atoms were fixed at calculated positions. Crystallographic data for the structures of **1** (deposition number CCDC 2169670) and **3** (deposition number CCDC 2169671) have been deposited in the Cambridge Crystallographic Data Centre database. Copies of the data can be obtained free of charge from the CCDC at www.ccdc.cam.ac.uk (accessed on 1 May 2022).

Crystal data for **1**: $C_{32}H_{41}NO_6$, M = 535.66, orthorhombic, a = 9.5019 (6) Å, b = 15.8046 (9) Å, c = 19.6446 (11) Å, α = 90.00°, β = 90.00°, γ = 90.00°, V = 2950.1 (3) Å^3, T = 100 (2) K, space group P212121, Z = 4, μ (CuKα) = 0.664 mm^{-1}, 11817 reflections measured, 4764 independent reflections (R_{int} = 0.0569). The final R_1 values were 0.0917 (I > 2σ (I)). The final wR (F^2) values were 0.2549 (I > 2σ (I)). The final R_1 values were 0.0937 (all data). The final wR (F^2) values were 0.2565 (all data). The goodness of fit on F^2 was 1.163. Flack parameter = 0.3 (5). The Hooft parameter is 0.15 (11) for 1883 Bijvoet pairs.

Crystal data for **3**: $C_{30}H_{39}NO_5{\cdot}H_2O$, M = 511.64, monoclinic, a = 9.7873 (3) Å, b = 9.4430 (3) Å, c = 15.5029 (4) Å, α = 90.00°, β = 103.6560 (10)°, γ = 90.00°, V = 1392.30 (7) Å^3, T = 100 (2) K, space group P21, Z = 2, μ (CuKα) = 0.678 mm^{-1}, 9344 reflections measured, 3802 independent reflections (R_{int} = 0.0470). The final R_1 values were 0.0600 (I > 2σ (I)). The final wR (F^2) values were 0.1736 (I > 2σ (I)). The final R_1 values were 0.0681 (all data). The final wR (F^2) values were 0.2047 (all data). The goodness of fit on F^2 was 1.093. Flack parameter = 0.1 (3). The Hooft parameter is 0.31 (9) for 1240 Bijvoet pairs.

2.5. Antimigration Assay

Cell migration was determined using the Oris™ Pro Cell Migration Assay (Platypus Technologies, Madison, WI, USA), according to the manufacturer's protocol. Briefly, MDA-MB-231 cells were seeded and incubated (37 °C, 5% CO_2) for 1 h, and then indicated concentrations of compounds were added and incubated with cells for an additional 18 h. At the end of incubation, the cell viability was evaluated with MTS assays and the migration area of each group was calculated and analysed, and the results of each subgroup were presented as a percentage of DMSO-treated cells.

3. Results and Discussion

3.1. Structure Elucidation

The molecular formula of 18-acetoxycytochalasin H (**1**) was determined to be $C_{32}H_{41}NO_6$ on the basis of HRESIMS ion at m/z 558.2826 [M + Na]$^+$ (calcd. 558.2826), indicating 13 degrees of unsaturation. Its ^{1}H NMR data (Table 1) showed typical signals of three tertiary methyl groups (δ_H 2.24, s; δ_H 2.00, s; δ_H 1.58, s), two secondary methyl groups (δ_H 0.99, d, J = 6.7 Hz; δ_H 1.02, d, J = 6.9 Hz), six olefinic protons (δ_H 5.85, dd, J = 16.6, 2.3 Hz; δ_H 5.74, dd, J = 15.5, 9.7 Hz; δ_H 5.56, d, J = 16.6 Hz; δ_H 5.38, m; δ_H 5.33, s; δ_H 5.10, s), two oxygenated methine groups (δ_H 5.63, d, J = 2.3 Hz; δ_H 3.84, d, J = 10.5 Hz), and one single-substituted phenyl (δ_H 7.31, t, J = 7.4 Hz, 2H; δ_H 7.25, t, J = 7.4 Hz, 1H; δ_H 7.14, d, J = 7.4 Hz, 2H). The ^{13}C NMR data (Table 2) displayed resonances for 32 carbons, ascribed to 5 methyls, 4 methylenes (including 1 olefinic), 11 methines (4 olefinic and 2 oxygenated), 61 quaternary carbons (1 olefinic, 1 amide and 2 ester carbonyls), and 6 other signals assignable to the single-substituted phenyl group. Thus, the above-mentioned results indicated that **1** should be a new tetracyclic cytochalasin including a benzene ring, with structural similarity with cytochalasin H [23]. The manifest difference of the structure of **1** from that of cytochalasin H was an additional acetoxy group linked at C-18 (δ_C 84.4) in **1**, which was further supported by the HMBC correlation from OAc (δ_H 2.24, s) to C-18. And the planar structure of **1** was established by extensive analysis of its 2D NMR spectra (Figure 2); its relative configuration was determined by the ROESY correlations

(Figure 3) and comparative analysis of those of cytochalasin H. Fortunately, suitable crystals of **1** were obtained and subjected to X-ray diffraction analysis using Cu Kα radiation (Figure 4), which confirmed the above deductions and unambiguously determined the absolute configuration of **1** as *3S,4R,5S,7S,8R,9R,16S,18R,21R* with the Hooft parameter 0.15 (11) for 1883 Bijvoet pairs (CCDC 2169670).

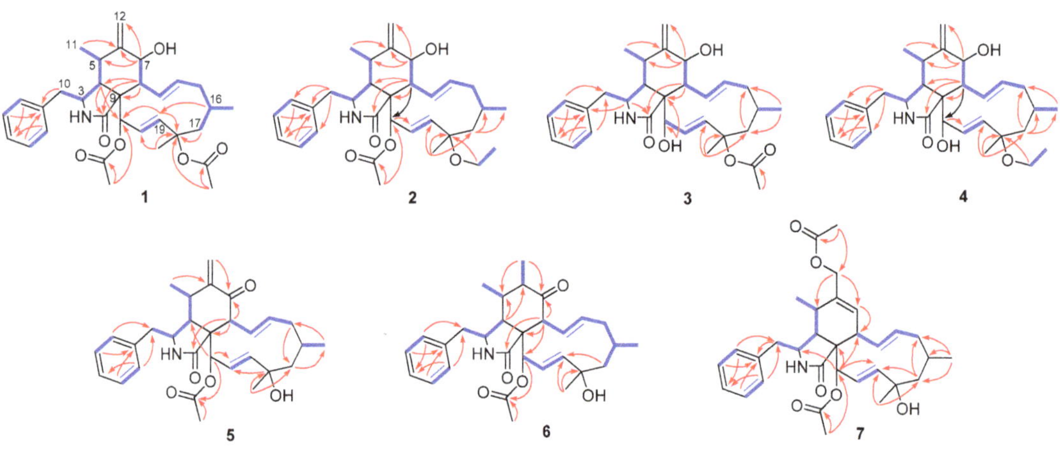

Figure 2. Key HMBC (red arrows) and ^{1}H-^{1}H COSY (blue bold) correlations of compounds **1**–**7**.

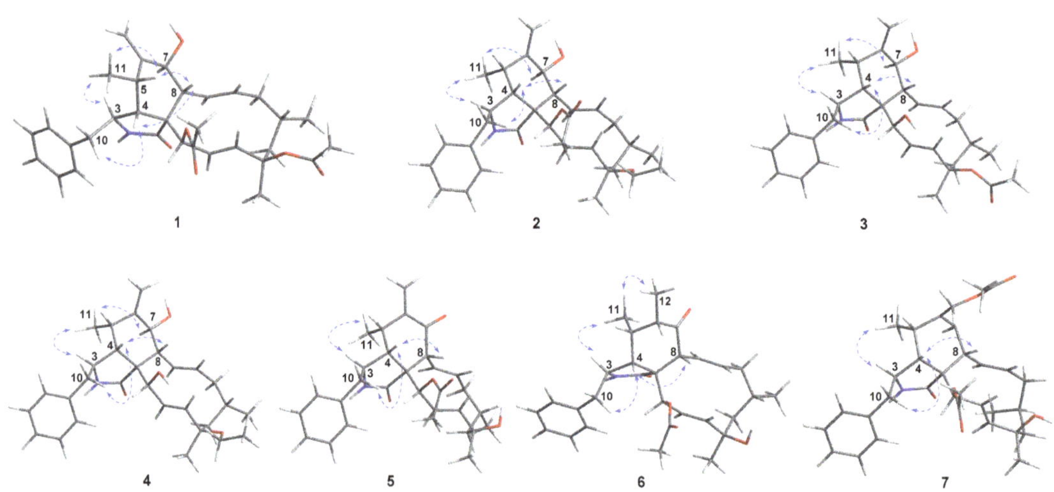

Figure 3. Key ROESY correlations of compounds **1**–**7**.

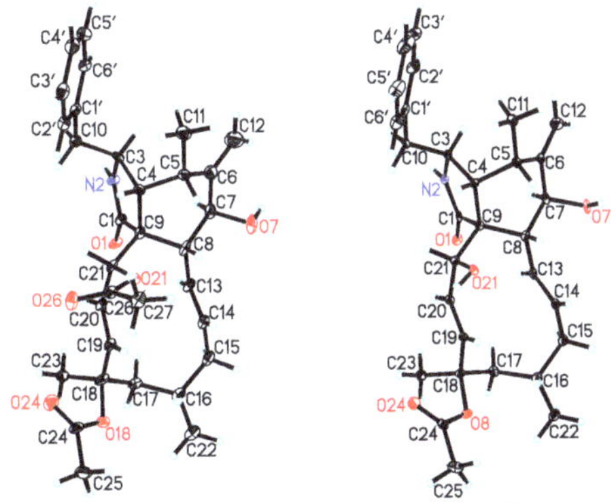

Figure 4. X-ray crystallographic structures of compounds **1** and **3**.

18-Ethoxycytochalasin H (**2**) was obtained as a white powder; its molecular formula was established as $C_{32}H_{43}NO_5$ on the basis of the HRESIMS ion peak at *m/z* 544.3030 [M + Na]$^+$ (calcd for $C_{32}H_{43}NO_5Na$, 544.3039), indicating 12 degrees of unsaturation. Analyses of the NMR data of **2** with those of **1** indicated their structural similarities, except for an ethoxy group located at C-18 in **2** rather than the 18-OAc group in **1**, which was confirmed by the ^{1}H-^{1}H COSY correlation of CH_2 (δ_H 3.38, m; 2.65, m)/CH_3 (1.14, t, J = 6.9 Hz) in the ethoxy group and the HMBC correlations from CH_2-18-OEt (δ_H 3.38, m; 2.65, m) to C-18 (δ_C 78.5) (Figure 2). The relative configurations of C-3, C-4, C-5, C-7, and C-8 in **2** were determined to be the same as those of **1** by analysis of the ROESY spectrum (Figure 3). Considering the almost complete consistent CD spectra of **1** and **2** (see Supplementary Materials), the absolute configuration of **2** was determined as shown.

18-Acetoxycytochalasin J (**3**) had the molecular formula of $C_{30}H_{39}NO_5$ based on the positive HRESIMS at *m/z* 516.2726 [M + Na]$^+$ (calcd 516.2720), corresponding to 12 degrees of unsaturation. The 1D NMR data (Tables 1 and 2) of **3** were similar to those of cytochalasin J [24], except for an additional acetoxy group located at C-18 in **3**. The above deduction was further confirmed by the changed chemical shift of C-18, compared with the ^{13}C NMR data of cytochalasin J [24], and the HMBC correlation from CH_3-18-OAc (δ_H 1.96, s) to 18-OAc carbonyl (δ_C 170.4) (Figure 2); its structure including the relative configuration was finally established as shown by X-ray diffraction analysis (Figure 4). Considering the similar CD spectra of **1** and **3** (SI), the absolute configuration of **3** was determined to be *3S,4R,5S,7S,8R,9R,16S,18R,21R*.

18-Ethoxycytochalasin J (**4**) had the molecular formula of $C_{30}H_{41}NO_4$ on the basis of the positive HRESIMS (*m/z* 480.3111 [M + H]$^+$, calcd 480.3108), corresponding to 11 degrees of unsaturation. Careful comparison of the ^{1}H and ^{13}C NMR spectra of **4** and **3** (Tables 1 and 2) suggested their similar structures, except for an ethoxy group located at C-18 in **4** rather than an acetoxy group in **3**. The above deduction was supported by the ^{1}H-^{1}H COSY correlations of CH_2/CH_3 (18-OEt) and the HMBC correlation from CH_2-18-OEt (δ_H 3.41, m; 3.37, m) to C-18 (δ_C 78.5) (Figure 2). The absolute configuration of **4** was determined to be the same as that of **1** by analysis of their CD spectra; thus, the structure of **4** was established as shown (Figure 1).

7-Oxocytochalasin H (**5**) possessed the molecular formula of $C_{30}H_{37}NO_5$ with 13 degrees of unsaturation, which was determined by the positive HRESIMS (*m/z* 514.2559 [M + Na]$^+$, calcd 514.2564). Analysis of the ^{1}H and ^{13} C NMR data (Tables 1 and 2) of **5** and cytochalasin H [23] indicated their structural similarity. The manifest differences were that the C-7 oxymethine group in cytochalasin H was replaced by the C-7 carbonyl group (δ_C 198.7) in **5**. The HMBC correlations from H-8 (δ_H 3.94, d, J = 9.3 Hz) and H_2-12 (δ_H 6.25, s) to C-7 and other correlations in the 2D spectra of **5** confirmed the above deduction (Figure 2).

The correlations of H-4/H-8, H_2-10/H-4, and H_3-11/H-3 in the ROESY spectrum indicated that H-3 was *a*-oriented and H-4, H-5, H-8 were *β*-oriented (Figure 3). Considering the same biogenetic pathway of **1** and **5**, the structure of **5** was determined as shown (Figure 1).

Cytochalasin H_3 (**6**) had the molecular formula of $C_{30}H_{39}NO_5$ with 12 degrees of unsaturation, which was determined by the positive HRESIMS (*m/z* 516.2719 $[M + Na]^+$, calcd 516.2720). Detailed analysis of ^1H and ^{13}C NMR data of **6** (Tables 1 and 2) indicated that **6** possessed a similar structure to that of **5**, except for the presence of a saturated C–C bond between C-6 (δ_C 45.8) and C-12 (δ_C 16.0) in **6** rather than a terminal double bond in **5**, which was further confirmed by the ^1H–^1H COSY correlation of H-6/H_3-12 and the HMBC correlations from H_3-12 (δ_H 1.12, d, J = 7.0 Hz) to C-5 (δ_C 35.7) and C-7 (δ_C 214.0) (Figure 2). The ROESY correlations of H-3/H_3-11, H_3-11/H_3-12, H_2-10/H-4, and H-4/H-8 implied the *α*-orientation of H-3 and *β*-orientations of H-4, H-5, H-6, and H-8 (Figure 3). By analysis of the similar CD spectra of **6** and **1** and biogenetic consideration, the structure of **6** was determined as shown.

The molecular formula of cytochalasin H_4 (**7**) was deduced to be $C_{32}H_{41}NO_6$ with 13 degrees of unsaturation based on the positive HRESIMS (*m/z* 536.3014 $[M + H]^+$, calcd 536.3007). The ^{13}C NMR data (Tables 1 and 2 of **7** displayed resonances for 32 carbons, ascribed to five methyls, four methylenes (including one oxygenated), 11 methines (5 olefinic and one oxygenated), six quaternary carbons (one olefinic, one amide and two ester carbonyls), and 6 other signals assignable to the single-substituted phenyl group. The above-mentioned results indicated the presence of an additional acetoxy group and an oxymethine group compared to those of the known RKS-1778 (**10**) [25]. The ROESY correlations of H-3/H_3-11, H_2-10/H-4, and H-4/H-8 implied the *α*-orientation of H-3 and *β*-orientations of H-4, H-5, and H-8 (Figure 3). The absolute configuration of **7** was determined as shown by analysis of their similar CD spectra of **7** and **10** and biogenetic consideration.

Compounds **8–11** were identified as cytochalasin H (**8**) [23], cytochalasin J_1 (**9**) [24], RKS-1778 (**10**) [25], 21-acetoxycytochalasin J_2 (**11**) [26] on the basis of their spectroscopic features and by comparison with the published data in the literature. Biogenetically, compounds **1–11** might be derived from a polyketide chain (octaketide) and an amino acid building block (phenylalanine) through a number of steps involving cycloaddition, oxidation, reduction, dehydration, acetylation, ethylation and methylation [27,28].

3.2. Antimigratory Activity

Our previous studies have revealed that phomopchalasins B and C displayed antimigratory effects [19]. In order to explore the potential of the cytochalasans on antimigration against tumours, eight compounds in sufficient natural amounts (Table 3) were evaluated for antimigratory activities against MDA-MB-231 in vitro. As a result, **1–3** and **8–11** exhibited in vitro antimigratory effects with IC_{50} values in the range of 1.01–10.42 μM (cytochalasin D as the positive control); it suggested the activity decreased when the C-18 hydroxy group was substituted with the acetoxy, ethoxy or methoxy group (**8** vs. **1**, **2**, and **9**). When a double bond was introduced between C-17 and C-18 rather than an ethoxy or methoxy group at C-18, the activity slightly improved (**11** vs. **2** and **9**). Compound **3** displayed antimigratory activity with an IC_{50} value of 6.38 μM. The introduction of an acetoxy group at C-21 may enhance the activity (**1** vs. **3**). When the unit of a terminal double bond (C6-C12) and a hydroxy group at C-7 was replaced by a trisubstituted alkene (C12-C6-C7), the activity slightly improved (**8** vs. **10**), but the further introduction of an acetoxy group at C-12 decreased the activity (**10** vs. **7**).

Table 3. Antimigratory activities of the compounds against MDA-MB-231 in vitro.

Compounds	IC_{50} (μM)	Compounds	IC_{50} (μM)
Cytochalasin D [a]	0.78	**8**	1.25
1	3.14	**9**	7.31
2	10.42	**10**	1.01
3	6.38	**11**	6.41
7	>25		

[a] Positive control.

4. Conclusions

Seven new cytochalasans (**1–7**), together with four known ones, cytochalasin H (**8**), cytochalasin J_1 (**9**), RKS-1778 (**10**), and 21-acetoxycytochalasin J_2 (**11**), were isolated from *Phomopsis* sp. shj2. Their structures were elucidated through extensive spectroscopic data interpretation and single-crystal X-ray diffraction analysis. In the present study, eight cytochalasans were evaluated for their antimigratory activity. Compounds **1–3** and **8–11** exhibited antimigratory activity against MDA-MB-231 in vitro with IC_{50} values in the range of 1.01 − 10.42 μM. The results will lay a foundation for further study of the structure–activity relationship for the discovery of antitumour lead compounds.

Supplementary Materials: The following supporting information can be downloaded at: https://www.mdpi.com/article/10.3390/jof8050543/s1, Section S1: NMR, HRESIMS, UV, ORD, and CD spectra of compound **1**; Section S2: NMR, HRESIMS, UV, ORD, and CD spectra of compound **2**; Section S3: NMR, HRESIMS, UV, ORD, and CD spectra of compound **3**; Section S4: NMR, HRESIMS, UV, ORD, and CD spectra of compound **4**; Section S5: NMR, HRESIMS, UV, ORD, and CD spectra of compound **5**; Section S6: NMR, HRESIMS, UV, ORD, and CD spectra of compound **6**; Section S7: NMR, HRESIMS, UV, ORD, and CD spectra of compound **7**.

Author Contributions: Conceptualization, B.-C.Y. and P.-T.P.; methodology, W.-G.W., B.-C.Y., L.-M.K., J.-W.T., X.D., Y.L. and P.-T.P.; resources, W.-G.W., B.-C.Y., Y.L. and P.-T.P.; data curation, B.-C.Y. and L.-M.K.; writing—original draft preparation, B.-C.Y.; writing—review and editing, B.-CY. and P.-T.P.; project administration, B.-C.Y. and P.-T.P.; funding acquisition, B.-C.Y. and P.-T.P. All authors have read and agreed to the published version of the manuscript.

Funding: This research was funded by the National Natural Science Foundation of China (No. 81874298), the NSFC-Joint Foundation of Yunnan Province (U2002221), and the Yunnan Science Fund for Distinguished Young Scholars (2019FJ002), and the Postdoctoral Directional Training Foundation of Yunnan Province (B.-C.Y.).

Institutional Review Board Statement: Not applicable.

Data Availability Statement: X-ray crystallographic data of **1** and **3** (CIF) are available free of charge from the CCDC at https://www.ccdc.cam.ac.uk (accessed on 1 May 2022).

Conflicts of Interest: The authors declare no conflict of interest.

References

1. Rai, N.; Keshri, P.K.; Verma, A.; Kamble, S.C.; Mishra, P.; Barik, S.; Singh, S.K.; Gautam, V. Plant associated fungal endophytes as a source of natural bioactive compounds. *Mycology* **2021**, *12*, 139–159. [CrossRef] [PubMed]
2. Deshmukh, S.K.; Dufossé, L.; Chhipa, H.; Saxena, S.; Mahajan, G.B.; Gupta, M.K. Fungal endophytes: A potential source of antibacterial compounds. *J. Fungi* **2022**, *8*, 164. [CrossRef] [PubMed]
3. Xiao, Y.; Liang, W.; Zhang, Z.; Wang, Y.; Zhang, S.; Liu, J.; Chang, J.; Ji, C.; Zhu, D. Polyketide derivatives from the endophytic fungus *Phaeosphaeria* sp. LF5 isolated from *Huperzia serrata* and their acetylcholinesterase inhibitory activities. *J. Fungi.* **2022**, *8*, 232. [CrossRef] [PubMed]
4. Chen, Y.; Yang, W.; Zou, G.; Wang, G.; Kang, W.; Yuan, J.; She, Z. Cytotoxic bromine- and iodine-containing cytochalasins produced by the mangrove endophytic fungus *Phomopsis* sp. QYM-13 using the OSMAC approach. *J. Nat. Prod.* **2022**. [CrossRef]
5. Miao, S.; Liu, M.; Qi, S.; Wu, Y.; Sun, K.; Zhang, Z.; Zhu, K.; Cai, G.; Gong, K. Cytochalasins from coastal saline soil-derived fungus *Aspergillus flavipes* RD-13 and their cytotoxicities. *J. Antibiot.* **2022**. [CrossRef]
6. Zhang, J.-Y.; He, J.; Li, Z.-H.; Feng, T.; Liu, J.-K. Zopfiellasins A–D, two pairs of epimeric cytochalasins from kiwi-associated fungus *Zopfiella* sp. and their antibacterial assessment. *Molecules* **2021**, *26*, 5611. [CrossRef]

7. Zhang, X.; Wu, Z.; Bao, A.; Zhao, Z.; Chen, Y.; Zhao, H.; Wang, J.; Chen, C.; Tong, Q.; Zhu, H.; et al. Asperflavipines C–E and aspermichalasine A: Three cytochalasan heterotetramers and an unusual cytochalasan monomer from *Aspergillus micronesiensis*. *Org. Chem. Front.* **2022**, *9*, 2585–2592. [CrossRef]
8. Yang, X.; Wu, P.; Xue, J.; Li, H.; Wei, X. Cytochalasans from endophytic fungus *Diaporthe* sp. SC-J0138. *Fitoterapia* **2020**, *145*, 104611. [CrossRef]
9. Long, X.; Wu, H.; Ding, Y.; Qu, C.; Deng, J. Biosynthetically inspired divergent syntheses of merocytochalasans. *Chem* **2021**, *7*, 212–223. [CrossRef]
10. Bao, R.; Tian, C.; Zhang, H.; Wang, Z.; Dong, Z.; Li, Y.; Gao, M.; Zhang, H.; Liu, G.; Tang, Y. Total syntheses of asperchalasines A–E. *Angew. Chem. Int. Ed.* **2018**, *57*, 14216–14220. [CrossRef]
11. Hua, C.; Yang, Y.; Sun, L.; Dou, H.; Tan, R.; Hou, Y. Chaetoglobosin F, a small molecule compound, possesses immunomodulatory properties on bone marrow-derived dendritic cells via TLR9 signaling pathway. *Immunobiology* **2013**, *218*, 292–302. [CrossRef] [PubMed]
12. Ye, K.; Ai, H.-L.; Liu, J.-K. Identification and bioactivities of secondary metabolites derived from endophytic fungi isolated from ethnomedicinal plants of Tujia in Hubei Province: A review. *Nat. Prod. Bioprospect.* **2021**, *11*, 185–205. [CrossRef] [PubMed]
13. Zheng, Q.-C.; Kong, M.-Z.; Zhao, Q.; Chen, G.-D.; Tian, H.-Y.; Li, X.-X.; Guo, L.-D.; Li, J.; Zheng, Y.-Z.; Gao, H. Chaetoglobosin Y, a new cytochalasan from *Chaetomium globosum*. *Fitoterapia* **2014**, *93*, 126–131. [CrossRef]
14. Evidente, A.; Andolfi, A.; Vurro, M.; Zonno, M.C.; Motta, A. Cytochalasins Z4, Z5, and Z6, three new 24-oxa[14]cytochalasans produced by *Phoma exigua* var. *heteromorpha*. *J. Nat. Prod.* **2003**, *66*, 1540–1544. [CrossRef] [PubMed]
15. Gupta, G.P.; Massagué, J. Cancer metastasis: Building a framework. *Cell* **2006**, *127*, 679–695. [CrossRef]
16. Nakashima, K.; Tomida, J.; Kamiya, T.; Hirai, T.; Morita, Y.; Hara, H.; Kawamura, Y.; Adachi, T.; Inoue, M. Diaporthols A and B: Bioactive diphenyl ether derivatives from an endophytic fungus *Diaporthe* sp. *Tetrahedron Lett.* **2018**, *59*, 1212–1215. [CrossRef]
17. Yahagi, H.; Yahagi, T.; Furukawa, M.; Matsuzaki, K. Antiproliferative and antimigration activities of beauvericin isolated from *Isaria* sp. on pancreatic cancer cells. *Molecules* **2020**, *25*, 4586. [CrossRef]
18. Monteillier, A.; Allard, P.-M.; Gindro, K.; Wolfender, J.-L.; Cuendet, M. Lung cancer chemopreventive activity of patulin isolated from *Penicillium vulpinum*. *Molecules* **2018**, *23*, 636. [CrossRef]
19. Yan, B.-C.; Wang, W.-G.; Hu, D.-B.; Sun, X.; Kong, L.-M.; Li, X.-N.; Du, X.; Luo, S.-H.; Liu, Y.; Li, Y.; et al. Phomopchalasins A and B, two cytochalasans with polycyclic-fused skeletons from the endophytic fungus *Phomopsis* sp. shj2. *Org. Lett.* **2016**, *18*, 1108–1111. [CrossRef]
20. Tang, J.-W.; Kong, L.-M.; Zu, W.-Y.; Hu, K.; Li, X.-N.; Yan, B.-C.; Wang, W.-G.; Sun, H.-D.; Puno, P.-T. Isopenicins A–C: Two types of antitumor meroterpenoids from the plant endophytic fungus *Penicillium* sp. sh18. *Org. Lett.* **2019**, *21*, 771–775. [CrossRef]
21. Xia, J.-N.; Hu, K.; Su, X.-Z.; Tang, J.-W.; Li, X.-N.; Sun, H.-D.; Puno, P.-T. Discovery of *ent*-kaurane diterpenoids, characteristic metabolites of *Isodon* species, from an endophytic fungal strain *Geopyxis* sp. XY93 inhabiting *Isodon parvifolia*. *Fitoterapia* **2022**, *158*, 105160. [CrossRef] [PubMed]
22. Su, X.-Z.; Zhu, Y.-Y.; Tang, J.-W.; Hu, K.; Li, X.-N.; Sun, H.-D.; Li, Y.; Puno, P.-T. Pestaloamides A and B, two spiro-heterocyclic alkaloid epimers from the plant endophytic fungus *Pestalotiopsis* sp. HS30. *Sci. China Chem.* **2020**, *63*, 1208–1213. [CrossRef]
23. Izawa, Y.; Hirose, T.; Shimizu, T.; Koyama, K.; Natori, S. Six new 10-pheynl-[11]cytochalasans, cytochalasins N–S from *Phomopsis* sp. *Tetrahedron* **1989**, *45*, 2323–2335. [CrossRef]
24. Shang, Z.; Raju, R.; Salim, A.A.; Khalil, Z.G.; Capon, R.J. Cytochalasins from an Australian marine sediment-derived *Phomopsis* sp. (CMB-M0042F): Acid-mediated intramolecular cycloadditions enhance chemical diversity. *J. Org. Chem.* **2017**, *82*, 9704–9709. [CrossRef] [PubMed]
25. Kakeya, H.; Morishita, M.; Onozawa, C.; Usami, R.; Horikoshi, K.; Kimura, K.; Yoshihama, M.; Osada, H. RKS-1778, a new mammalian cell-cycle inhibitor and a key intermediate of the [11] cytochalasin group. *J. Nat. Prod.* **1997**, *60*, 669–672. [CrossRef]
26. Huang, X.; Zhou, D.; Liang, Y.; Liu, X.; Cao, F.; Qin, Y.; Mo, T.; Xu, Z.; Li, J.; Yang, R. Cytochalasins from endophytic *Diaporthe* sp. GDG-118. *Nat. Prod. Res.* **2021**, *35*, 3396–3403. [CrossRef]
27. Scherlach, K.; Boettger, D.; Remme, N.; Hertweck, C. The chemistry and biology of cytochalasans. *Nat. Prod. Rep.* **2010**, *27*, 869–886. [CrossRef]
28. Qiao, K.; Chooi, Y.-H.; Tang, Y. Identification and engineering of the cytochalasin gene cluster from *Aspergillus clavatus* NRRL 1. *Metab. Eng.* **2011**, *13*, 723–732. [CrossRef]

Journal of *Fungi*

Article

[20(22)*E*]-Lanostane Triterpenes from the Fungus *Ganoderma australe*

Lin Zhou [1,†], Li-Li Guo [1,†], Masahiko Isaka [2], Zheng-Hui Li [1,*] and He-Ping Chen [1,*]

[1] School of Pharmaceutical Sciences, South-Central Minzu University, Wuhan 430074, China; 2019110443@mail.scuec.edu.cn (L.Z.); g15513998615@163.com (L.-L.G.)
[2] National Center for Genetic Engineering and Biotechnology (BIOTEC), 113 Thailand Science Park, Phaholyothin Road, Klong Luang, Pathumthani 12120, Thailand; isaka@biotec.or.th
* Correspondence: lizhenghui@mail.scuec.edu.cn (Z.-H.L.); chenhp@mail.scuec.edu.cn (H.-P.C.)
† These authors contributed equally to this work.

Abstract: Twelve new lanostane triterpenoids (**1–5**, **7–13**) were isolated from the fruiting bodies of the fungus *Ganoderma australe*. The structures of the new compounds were elucidated by extensive 1D and 2D NMR, and HRESIMS spectroscopic analysis. All the triterpenes are featured by 20(22)*E* configurations which are uncommon in the *Ganoderma* triterpene family. The absolute configuration of the C-25 of compounds **1**, **2**, and **6** were determined by the phenylglycine methyl ester (PGME) method. A postulated biosynthetic pathway for compound **1** was discussed. This study opens new insights into the secondary metabolites of the chemically underinvestigated fungus *G. australe*.

Keywords: *Ganoderma australe*; triterpene; 20(22)*E* configuration; PGME method

Citation: Zhou, L.; Guo, L.-L.; Isaka, M.; Li, Z.-H.; Chen, H.-P. [20(22)*E*]-Lanostane Triterpenes from the Fungus *Ganoderma australe*. *J. Fungi* **2022**, *8*, 503. https://doi.org/10.3390/jof8050503

Academic Editors: Tao Feng and Frank Surup

Received: 31 March 2022
Accepted: 11 May 2022
Published: 12 May 2022

Publisher's Note: MDPI stays neutral with regard to jurisdictional claims in published maps and institutional affiliations.

1. Introduction

Mushrooms are popular in the food market due to their delicious taste and nutrition values. Mushroom-derived secondary metabolites have contributed lots of lead compounds for medical and agricultural use. Psilocybin, a specialized compound from the genus *Psilocybe*, is a naturally occurring hallucinogenic prodrug for treating psychiatric disorders [1]. Strobilurins, firstly originated from the mushroom *Strobilurus tenacellus*, are a group of natural products and their synthetic analogs are used in agriculture as fungicides [2,3]. More and more attention has been paid to mining promising lead compounds from the mushroom natural product reservoir in recent years.

Ganoderma, called "*lingzhi*" in China, is a group of wood-decaying mushrooms with hard fruiting bodies which grow mostly in spare scatting sunshine, on the trees, and on open grounds [4]. It is a genus of notable medicinal fungi and traditional herbal medicine for the treatment of diseases such as hepatopathy, nephritis, neurasthenia, and asthma [5–10]. The *Shennong Ben Cao Jing*, an ancient Chinese medicinal book, documented that *Ganoderma* was effective for maintaining health, prolonging life, boosting memory, and relieving stress. *Ganoderma lucidum* and *G. sinense* are two registered species recorded in the *Chinese Pharmacopoeia* (2015). Many studies show that triterpenoids and polysaccharides are the main bioactive substances in *Ganoderma* [11–17]. *Ganoderma australe* is a species used in folk medicine as the alternative of *G. lucidum*. However, this fungus has rarely been chemically investigated compared to other *Ganoderma* species, such as *G. lucidum*, *G. cochlear*, and *G. sinense*. Previous studies on this fungus have led to the isolation of lanostane triterpenes [18–21], meroterpenoids [22,23], and alkaloids [22]. The lanostanoids from this species are over-oxygenated compared to the ones isolated from other species of *Ganoderma*, especially the position of C-20 [18,20]. The quaternary hydroxy group substituted at C-20 led to the introduction of an additional chiral carbon of which the stereochemistry was difficult to be assigned even by chemical derivatization. Moreover, this substituted pattern of the C-20 hydroxy group always triggered to dehydration between C-21 to produce the

20(22)-double bond, which always incorporated into an α,β-unsaturated ketone group with the C-23 carbonyl group. In this study, we have investigated the secondary metabolite profiles of *G. australe*, which led to the isolation of twelve new highly oxygenated lanostane triterpenes with uncommon 20(22)*E* configurations. We, herein, report the isolation and structural elucidation of the new compounds.

2. Experimental Section

2.1. General Experimental Procedures

Optical rotations were obtained on an Autopol IV-T digital polarimeter (Rudolph, Hackettstown, NJ, USA). UV spectra were recorded on a Hitachi UH5300 spectrophotometer (Hitachi, Tokyo, Japan). CD spectra were measured on a Chirascan Circular Dichroism spectrometer (Applied Photophysics Limited, Leatherhead, Surrey, UK). The 1D and 2D spectra were obtained on the Bruker Avance III 500 MHz and 600 MHz spectrometers (Bruker Corporation, Karlsruhe, Germany). HRESIMS spectra were measured on a Q Exactive Orbitrap mass spectrometer (Thermo Fisher Scientific, Waltham, MA, USA). Single-crystal X-ray diffraction data were recorded on the BRUKER D8 QUEST diffractometer (Bruker Corporation, Karlsruhe, Germany). Medium pressure liquid chromatography (MPLC) was performed on an Interchim system equipped with a column packed with RP-18 gel (40–75 μm, Fuji Silysia Chemical Co., Ltd., Kasugai, Japan). Preparative high performance liquid chromatography (prep-HPLC) was performed on an Agilent 1260 Infinity II liquid chromatography system equipped with a Zorbax SB-C_{18} column (particle size 5 μm, dimensions 150 mm × i.d. 9.4 mm, flow rate 5 $mL \cdot min^{-1}$) and a DAD detector (Agilent Technologies, Santa Clara, CA, US). Sephadex LH-20 (GE Healthcare, Stockholm, Sweden) and silica gel (200–300 mesh, Qingdao Haiyang Chemical Co., Ltd., Qingdao, China) were used for column chromatography (CC). (*S*)- and (*R*)-phenylglycine methyl ester were bought from Sigma-Aldrich (St. Louis, MO, USA).

2.2. Fungal Material

The fruiting bodies of *G. australe* were collected in Tongbiguan Natural Reserve, Dehong, Yunnan Province, China, in 2016, and identified by Prof. Yu-Cheng Dai (Institute of Microbiology, Beijing Forestry University). A voucher specimen of *G. australe* was deposited in the Mushroom Bioactive Natural Products Research Group of South-Central University for Nationalities.

2.3. Extraction and Isolation

The dry fruiting bodies of *G. australe* (3.26 kg) were extracted four times with $CHCl_3$:MeOH (1:1) at room temperature to obtain crude extract, which was further suspended in distilled water and partitioned against EtOAc to afford EtOAc layer extract (130 g). The EtOAc layer extract was eluted on MPLC with a stepwise gradient of MeOH−H_2O (20:80−100:0) to afford eight fractions (A−H).

Fraction C was subjected to Sephadex LH-20 (MeOH) and obtained 16 subfractions (C1-C16), and C2 was separated by prep-HPLC (MeCN−H_2O: 20:80−40:60, 25 min, 4 mL/min) to yield compound **2** (6.4 mg, t_R = 20.5 min).

Fraction D was separated by Sephadex LH-20 (MeOH) to give eight subfractions (D1–D8). Subfraction D4 was subjected to silica gel CC (petroleum ether−acetone from *v/v* 6:1 to 1:1) and yielded eleven subfractions (D4a–D4k). Subfraction D4d was purified by prep-HPLC (MeCN−H_2O: 20:80–40:60, 25 min, 4 mL/min) to yield compound **3** (2.0 mg, t_R = 19.0 min).

Fraction E was separated by Sephadex LH-20 ($CHCl_3$:MeOH = 1:1) to afford four subfractions (E1–E4). E2 was separated by CC on silica gel (petroleum ether−acetone from *v/v* 15:1 to 1:1) to obtain 10 subfractions (E2a–E2j). E2b was subjected to prep-HPLC (MeCN–H_2O: 70:30–90:10, 25.0 min, 4 mL/min) to obtain compound **7** (3.8 mg, t_R = 14.0 min) and **8** (3.7 mg, t_R = 15.0 min). Compound **10** (3.7 mg, t_R = 21.0 min) was purified from E2f by prep-HPLC (MeCN–H_2O: 70:30–90:10, 25 min, 4 mL/min). E4 was subjected to CC on silica gel (petroleum ether–acetone from *v/v* 10:1 to 1:1) to obtain 12 fractions. Compound **12** (6.0 mg, t_R = 18.0 min) was purified from E4c by prep-HPLC (MeCN–H_2O: 40:60–60:80, 25 min, 4 mL/min). Compound **6** (5.1 mg, t_R = 19.1 min) was purified from E4d by prep-HPLC (MeCN–H_2O: 40:60–60:80, 25.2 min, 4 mL/min). Compound **9** (21.4 mg, t_R = 20.0 min) was purified from E4f by prep-HPLC (MeCN–H_2O: 40:60–60:80, 25 min, 4 mL/min). Compound **5** (7.3 mg, t_R = 30.0 min) was purified from E4g by prep-HPLC (MeCN–H_2O: 45:55–73:27, 35.0 min, 4 mL/min). Compound **11** (2.6 mg, t_R = 27.0 min) was purified from E4h by prep-HPLC (MeCN–MeOH–H_2O: 40:20:40, isocratic, 30 min, 4 mL/min). E7 was subjected to column chromatography (CC) on silica gel (petroleum ether–acetone from *v/v* 15:1 to 1:1) to obtain 13 fractions. E7j was purified by prep-HPLC (MeCN–H_2O: 30:70–50:50, 25 min, 4 mL/min) to yield compound **1** (8.6 mg, t_R = 14.0 min).

Fraction G was separated by column chromatography (CC) on silica gel (petroleum ether–acetone from *v/v* 15:1 to 1:1) to obtain 10 subfractions (G1–G10). Compound **4** (13.3 mg, t_R = 12.7 min) was purified from G7 by prep-HPLC (MeCN–H_2O: 50:50–30:70, 25 min, 4 mL/min). Compound **13** (4.0 mg, t_R = 13.2 min) was purified from G7 by prep-HPLC (MeCN–H_2O: 50:50–30:70, 25 min, 4 mL/min).

2.4. Spectroscopic Data

Ganoaustralenone A (**1**): yellow oil. $[\alpha]^{24}_{D}$ +8.9 (*c* 0.09, MeOH); UV (MeOH) λ_{max} (log ε) 250.0 (4.22); ^{1}H NMR (600 MHz, CD_3OD) data, see Table 1, ^{13}C NMR (150 MHz, CD_3OD) data, see Table 2 HRESIMS *m/z* 535.26685 $[M + Na]^+$ (calcd for $C_{30}H_{40}O_7Na$, 535.26717).

Table 1. ^{1}H NMR Spectroscopic Data of Compounds **1**–**5**.

No.	**1** [ad]	**2** [ac]	**3** [bc]	**4** [ae]	**5** [ac]
1	2.87, ddd (14.3, 8.5, 5.8)	2.85, ddd (14.2, 9.8, 7.6)	2.93, overlapped	2.75, ddd (13.8, 9.7, 7.8)	2.95, ddd (14.2, 8.4, 6.2)
	1.82, ddd (14.3, 9.7, 6.2)	1.84, ddd (14.2, 11.9, 2.1)	1.70, overlapped	1.94, ddd (13.8, 12.0, 2.5)	1.73, ddd (14.2, 9.2, 6.7)
2	2.71, ddd (15.2, 9.7, 5.8)	2.74, ddd (14.4, 11.9, 7.6)	2.60, ddd (15.5, 9.8, 6.2)	2.86, ddd (14.5, 12.0, 7.8)	2.60, ddd (15.5, 9.2, 6.2)
	2.45, ddd (15.2, 8.5, 6.2)	2.33, ddd (14.4, 9.8, 2.1)	2.44, ddd (15.0, 8.3, 6.2)	2.16, ddd (14.5, 9.7, 2.5)	2.50, ddd (15.5, 8.4, 6.7)
5	2.38, dd (14.7, 2.4)	2.31, d (13.7)	2.11, dd (9.9, 5.5)	2.46, d (13.6)	2.25, dd (15.3, 2.6)
6	2.68, t (14.7)	4.44, d (13.7)	1.70, overlapped, 2H	4.57, d (13.6)	2.53, dd (15.3, 14.7)
	2.36, dd (14.7, 2.4)				2.40, dd (14.7, 2.6)
7			4.47, t (2.3)		
12	2.96, d (16.3)	2.83, d (17.0)	2.77, d (17.3)	3.02, d (16.9)	2.81, d (16.2)
	2.41, d (16.3)	2.51, d (17.0)	2.38. d (17.3)	2.45, d (16.9)	2.47, d (16.2)
15	2.24, m	2.29, overlapped	1.86, m	2.22, overlapped	2.30, ddd (12.4, 8.3, 2.7)
	1.87, overlapped	1.74, m	2.05, m	1.81, td (12.0, 7.0)	1.83, m
16	2.04, m	1.95, overlapped, 2H	1.98, m, 2H	2.07, overlapped	1.92, overlapped, 2H
	1.91, dt (11.3, 2.6)			1.93, overlapped	
17	2.99, overlapped	2.91, t (9.2)	2.95, overlapped	3.06, overlapped	2.87, t (9.1)
18	0.75, s	0.69, s	0.67, s	0.77, s	0.70, s
19	1.28, s	1.23, s	1.00, s	1.23, s	1.27, s
21	2.16, s	2.18, s	2.17, s	2.18, s	2.19, s
22	6.51, s	6.23, br.s	6.25, s	6.56, s	6.22, br.s
24	4.21, d (4.9)	4.29, d (2.9)	4.33, d (3.1)	4.22, d (4.0)	4.23, dd (5.7, 3.2)
25	2.93, dd (7.3, 4.9)	3.01, dd (7.2, 2.9)	3.00, dd (7.2, 3.1)	3.04, dd (7.1, 4.0)	2.99, dd (7.2, 3.2)
27	1.18, d (7.3)	1.28, d (7.2)	1.24, d (7.2)	1.18, d (7.1)	1.28, d (7.2)
28	1.13, s	1.34, s	1.13, s	1.30, s	1.13, s
29	1.10, s	1.42, s	1.06, s	1.38, s	1.11, s
30	1.35, s	1.38, s	1.35, s	1.42, s	1.31, s
OCH_3				3.57, s	3.63, s
6-OH				4.17, br.s	
24-OH					3.89, d (5.7)

[a] Measured at 600 MHz; [b] Measured at 500 MHz; [c] Measured in $CDCl_3$; [d] Measured in CD_3OD; [e] Measured in acetone-d_6.

Table 2. ^{13}C NMR Spectroscopic Data of Compounds **1–5**, and **7**.

No.	**1** ^{be}	**2** ^{bd}	**3** ^{ad}	**4** ^{bf}	**5** ^{bd}	**7** ^{cd}
1	36.1, CH_2	35.3, CH_2	34.9, CH_2	35.8, CH_2	35.0, CH_2	36.0, CH_2
2	35.0, CH_2	33.4, CH_2	34.2, CH_2	33.7, CH_2	34.1, CH_2	34.5, CH_2
3	218.6, C	216.8, C	218.6, C	216.3, C	215.6, C	214.7, C
4	47.8, C	47.7, C	46.5, C	48.1, C	46.8, C	47.9, C
5	50.8, CH	54.9, CH	45.0, CH	54.8, CH	49.8, CH	49.9, CH
6	37.7, CH_2	72.2, CH	29.3, CH_2	72.8, CH	37.1, CH_2	24.3, CH_2
7	202.7, C	202.6, C	67.1, CH	203.4, C	201.1, C	131.2, CH
8	151.3, C	148.3, C	160.3, C	149.2, C	150.9, C	138.3, C
9	152.4, C	150.0, C	140.1, C	150.1, C	150.2, C	160.4, C
10	40.1, C	39.9, C	37.9, C	40.5, C	39.0, C	38.2, C
11	202.8, C	200.1, C	200.1, C	200.7, C	200.8, C	119.1, CH
12	50.9, CH_2	49.8, CH_2	50.1, CH_2	50.5, CH_2	50.0, CH_2	203.3, C
13	49.7, C	48.2, C	48.9, C	48.9, C	48.5, C	57.8, C
14	49.7, C	48.1, C	50.8, C	48.9, C	48.4, C	56.4, C
15	33.5, CH_2	31.8, CH_2	30.4, CH_2	32.6, CH_2	32.4, CH_2	77.0, CH
16	24.2, CH_2	23.4, CH_2	23.3, CH_2	23.8, CH_2	23.4, CH_2	36.9, CH_2
17	54.6, CH	53.5, CH	54.4, CH	53.9, CH	53.6, CH	47.8, CH
18	19.1, CH_3	18.7, CH_3	18.8, CH_3	18.9, CH_3	18.6, CH_3	17.3, CH_3
19	18.4, CH_3	19.3, CH_3	17.7, CH_3	19.7, CH_3	18.0, CH_3	21.3, CH_3
20	161.3, C	161.7, C	162.6, C	160.1, C	161.5, C	157.9, C
21	22.2, CH_3	22.2, CH_3	21.9, CH_3	22.0, CH_3	22.3, CH_3	21.0, CH_3
22	122.0, CH	120.0, CH	119.6, CH	121.4, CH	120.1, CH	126.2, CH
23	202.0, C	198.7, C	199.0, C	200.3, C	198.8, C	199.0, C
24	80.0, CH	78.2, CH	78.2, CH	79.4, CH	78.5, CH	47.6, CH_2
25	44.6, CH	43.1, CH	43.3, CH	44.0, CH	43.3, CH	35.2, CH
26	176.9, C	177.1, C	177.1, C	173.6, C	173.5, C	176.3, C
27	13.8, CH_3	13.3, CH_3	13.0, CH_3	13.9, CH_3	13.6, CH_3	20.0, CH_3
28	27.7, CH_3	31.2, CH_3	27.7, CH_3	31.2, CH_3	27.6, CH_3	25.2, CH_3
29	20.7, CH_3	19.5, CH_3	20.6, CH_3	19.6, CH_3	20.5, CH_3	22.6, CH_3
30	26.4, CH_3	26.4, CH_3	27.6, CH_3	26.4, CH_3	26.3, CH_3	26.5, CH_3
-O$\underline{C}H_3$				51.8, CH_3	52.1, CH_3	
-O$\underline{C}H_2CH_3$						60.6, CH_2
-O$CH_2\underline{C}H_3$						14.3, CH_3

^{a} Measured at 500 MHz; ^{b} Measured at 600 MHz; ^{c} Measured at 800 MHz; ^{d} Measured in $CDCl_3$; ^{e} Measured in CD_3OD; ^{f} Measured in acetone-d_6.

Ganoaustralenone B (**2**): white powder. $[\alpha]_D^{24}$ +81.3 (*c* 0.06, MeOH); UV (MeOH) λ_{max} (log ε) 250.0 (4.00); ^{1}H NMR (600 MHz, $CDCl_3$) data, see Table 1, ^{13}C NMR (150 MHz, $CDCl_3$) data, see Table 2; HRESIMS *m/z* 551.26422 [M + Na]$^+$ (calcd for $C_{30}H_{40}O_8Na$, 551.26209).

Ganoaustralenone C (**3**): white powder. $[\alpha]_D^{24}$ +178.4 (*c* 0.05, MeOH); UV (MeOH) λ_{max} (log ε) 250.0 (4.20); ^{1}H NMR (500 MHz, $CDCl_3$) data, see Table 1, ^{13}C NMR (125 MHz, $CDCl_3$) data, see Table 2; HRESIMS *m/z* 537.28180 [M + Na]$^+$ (calcd for $C_{30}H_{42}O_7Na$, 537.28227).

Ganoaustralenone D (**4**): yellow oil. $[\alpha]_D^{24}$ +102.2 (*c* 0.13, MeOH); UV (MeOH) λ_{max} (log ε) 250.0 (4.24); ^{1}H NMR (600 MHz, CD_3COCD_3) data, see Table 1, ^{13}C NMR (150 MHz, CD_3COCD_3) data, see Table 2; HRESIMS *m/z* 565.27728 [M + Na]$^+$ (calcd for $C_{31}H_{42}O_8Na$, 565.27774).

Ganoaustralenone E (**5**): yellow oil. $[\alpha]_D^{24}$ +25.3 (*c* 0.07, MeOH); UV (MeOH) λ_{max} (log ε) 250.0 (4.27); ^{1}H NMR (600 MHz, $CDCl_3$) data, see Table 1, ^{13}C NMR (150 MHz, $CDCl_3$) data, see Table 2; HRESIMS *m/z* 549.28210 [M + Na]$^+$ (calcd for $C_{31}H_{42}O_7Na$, 549.28282).

Ganoaustralenone F (**7**): yellow oil. $[\alpha]_D^{24}$ +5.7 (*c* 0.04, MeOH); UV (MeOH) λ_{max} (log ε) 245.0 (3.53); ^{1}H NMR (800 MHz, $CDCl_3$) data, see Table 3, ^{13}C NMR (150 MHz, $CDCl_3$) data, see Table 2; HRESIMS *m/z* 547.30286 [M + Na]$^+$ (calcd for $C_{32}H_{44}O_6Na$, 547.30356).

Table 3. ^{1}H NMR Spectroscopic Data of Compounds **7**–**10**.

No.	**7** [ac]	**8** [ac]	**9** [bc]	**10** [bc]
1	2.28, overlapped	2.10, ddd (13.9, 10.9, 6.2)	2.61, ddd (15.5, 9.7, 6.0)	2.94, ddd (14.4, 8.3, 6.1)
	1.86, m	1.95, ddd (13.9, 9.2, 5.0)	2.46, ddd (15.5, 8.5, 6.5)	1.77, ddd (14.4, 9.1, 7.0)
2	2.81, td (14.6, 5.6)	2.69, overlapped	2.97, ddd (14.3, 8.5, 6.0)	2.62, ddd (15.3, 9.1, 6.1)
	2.43, overlapped	2.48, overlapped	1.70, ddd (14.3, 9.7, 6.5)	2.52, ddd (15.3, 8.3, 7.0)
5	1.70, dd (12.1, 3.6)	2.75, dd (13.0, 3.0)	2.09, dd (11.8, 3.9)	2.26, dd (15.3, 2.8)
6	2.44, overlapped	2.17, dt (15.0, 3.0)	1.70, overlapped	2.62, t (15.3)
	2.28, overlapped	1.75, dd (15.0, 13.0)		2.49, dd (15.3, 2.8)
7	6.27, d (6.7)	4.68, d (3.0)	4.47, m	
11	5.69, s	6.05, s		
12			2.76, d (17.3)	2.87, d (16.9)
			2.38, d (17.3)	2.44, d (16.9)
15	4.31, d (6.2)		2.04, overlapped	4.40, ddd (10.2, 5.6, 1.8)
			1.84, m	
16	2.52, ddd (15.4, 8.8, 6.2)	2.66, dd (12.7, 9.3)	2.03, overlapped	2.49, overlapped
	2.04, dd (15.4, 8.8)	2.46, dd (12.7, 9.3)	1.94, overlapped	1.82, ddd (15.0, 10.2, 5.6)
17	3.23, t (8.8)	3.51, t (9.3)	2.88, t (8.7)	2.92, overlapped
18	1.17, s	1.13, s	0.67, s	0.73, s
19	1.36, s	1.09, s	1.02, s	1.28, s
21	2.28, s	2.26, s	2.11, s	2.10, s
22	6.44, s	6.34, s	6.12, s	6.09, s
24	2.94, overlapped	2.95, overlapped	2.95, ddd (20.5, 14.2)	2.94, overlapped
	2.55, dd (20.4, 8.6)	2.54, dd (20.5, 8.5)	2.52, dd (20.5, 8.6)	2.51, overlapped
25	2.94, overlapped	2.94, overlapped	2.96, ddd (14.2, 8.6, 6.8)	2.94, overlapped
27	1.18, s	1.19, d (6.9)	1.19, d (6.8)	1.19, d (6.8)
28	1.12, s	1.12, s	1.07, s	1.15, s
29	1.17, s	1.08, s	1.14, s	1.12, s
30	1.00, s	1.29, s	1.35, s	1.26, s
-O<u>CH</u>$_3$			3.68, s	
-O<u>CH</u>$_2$CH$_3$	4.13, overlapped, 2H	4.13, overlapped, 2H		4.13, m, 2H
-OCH$_2$<u>CH</u>$_3$	1.25, overlapped	1.25, overlapped		1.25, t (7.1)
15-OH				4.48, d (1.8)

[a] Measured at 800 MHz; [b] Measured at 600 MHz; [c] Measured in $CDCl_3$.

Ganoaustralenone G (**8**): yellow oil. $[\alpha]_D^{24}$ +17.1 (*c* 0.04, MeOH); UV (MeOH) λ_{max} (log ε) 245.0 (3.62); ^{1}H NMR (800 MHz, $CDCl_3$) data, see Table 3, ^{13}C NMR (200 MHz, $CDCl_3$) data, see Table 4; HRESIMS *m/z* 561.28223 $[M + Na]^+$ (calcd for $C_{32}H_{42}O_7Na$, 561.28282).

Ganoaustralenone H (**9**): yellow oil. $[\alpha]_D^{24}$ +31.3 (*c* 0.11, MeOH); UV (MeOH) λ_{max} (log ε) 250.0 (3.39); ^{1}H NMR (600 MHz, $CDCl_3$) data, see Table 3, ^{13}C NMR (150 MHz, $CDCl_3$) data, see Table 4; HRESIMS *m/z* 513.32135 $[M + H]^+$ (calcd for $C_{31}H_{45}O_6$, 513.32161).

Ganoaustralenone I (**10**): pale yellow oil. $[\alpha]_D^{24}$ +67.7 (*c* 0.05, MeOH); UV (MeOH) λ_{max} (log ε) 245.0 (4.02); ^{1}H NMR (600 MHz, $CDCl_3$) data, see Table 3, ^{13}C NMR (150 MHz, $CDCl_3$) data, see Table 4; HRESIMS *m/z* 563.29749 $[M + Na]^+$ (calcd for $C_{32}H_{44}O_7Na$, 563.29847).

Ganoaustralenone J (**11**): yellow oil. $[\alpha]_D^{24}$ +119.56 (*c* 0.05, MeOH); UV (MeOH) λ_{max} (log ε) 245.0 (4.17); ^{1}H NMR (600 MHz, $CDCl_3$) data, see Table 5, ^{13}C NMR (150 MHz, $CDCl_3$) data, see Table 4; HRESIMS *m/z* 549.28210 $[M + Na]^+$ (calcd for $C_{31}H_{42}O_7Na$, 549.28282).

Ganoaustralenone K (**12**): pale yellow oil. $[\alpha]_D^{24}$ +149.2 (*c* 0.06, MeOH); UV (MeOH) λ_{max} (log ε) 250.0 (4.25); ^{1}H NMR (600 MHz, $CDCl_3$) data, see Table 5, ^{13}C NMR (150 MHz, $CDCl_3$) data, see Table 4; HRESIMS *m/z* 565.31287 $[M + Na]^+$ (calcd for $C_{32}H_{46}O_7Na$, 565.31412).

Table 4. ^{13}C NMR Spectroscopic Data of Compounds **8**–**13**.

No.	**8** ^{cd}	**9** ^{bd}	**10** ^{bd}	**11** ^{bd}	**12** ^{bd}	**13** ^{ad}
1	36.1, CH_2	34.2, CH_2	35.1, CH_2	35.2, CH_2	34.9, CH_2	35.1, CH_2
2	33.7, CH_2	34.9, CH_2	34.0, CH_2	37.0, CH2	34.2, CH_2	34.1, CH_2
3	216.4, C	218.0, C	214.9, C	215.1, C	218.0, C	215.6, C
4	45.9, C	46.5, C	46.6, C	46.7, C	46.5, C	46.8, C
5	40.7, CH	45.2, CH	49.2, CH	49.3, CH	45.1, CH	49.8, CH
6	22.8, CH_2	29.4, CH_2	36.9, CH_2	34.1, CH_2	29.4, CH_2	37.1, CH_2
7	58.9, CH	67.3, CH	204.4, C	204.5, C	67.3, CH	201.2, C
8	62.7, C	159.5, C	150.4, C	150.5, C	159.5, C	150.9, C
9	164.3, C	140.2, C	152.8, C	153, C	140.2, C	150.2, C
10	40.4, C	38.0, C	39.2, C	39.3, C	38.0, C	39.0, C
11	129.9, CH	199.5, C	200.3, C	200.4, C	199.3, C	200.7, C
12	200.7, C	50.3, CH_2	50.5, CH_2	50.6, CH_2	50.3, CH_2	50.0, CH_2
13	57.7, C	48.6, C	48.9, C	49.0, C	48.9, C	48.6, C
14	55.2, C	50.6, C	52.3, C	52.4, C	50.7, C	48.5, C
15	209.9, C	30.4, CH_2	72.4, CH	72.5, CH	30.4, CH_2	32.4, CH_2
16	38.5, CH_2	23.2, CH_2	32.0, CH_2	32.1, CH_2	23.3, CH_2	23.4, CH_2
17	42.9, CH	54.1, CH	51.9, CH	52.0, CH	54.5, CH	53.9, CH
18	18.3, CH_3	18.6, CH_3	18.9, CH_3	19.0, CH_3	18.8, CH_3	18.6, CH_3
19	25.0, CH_3	17.7, CH_3	17.7, CH_3	17.8, CH_3	17.7, CH_3	18.1, CH_3
20	154.0, C	158.0, C	155.8, C	156.1, C	162.0, C	162.9, C
21	20.6, CH_3	21.4, CH_3	21.1, CH_3	21.2, CH_3	22.2, CH_3	22.5, CH_3
22	127.0, CH	123.8, CH	124.5, CH	124.5, CH	119.6, CH	119.6, CH
23	198.8, C	198.4, C	198.2, C	198.3, C	198.8, C	198.7, C
24	47.9, CH_2	47.9, CH_2	47.7, CH_2	47.9, CH_2	78.6, CH	77.3, CH
25	35.1, CH	35.0, CH	35.0, CH	34.9, CH	43.2, CH	42.8, CH
26	176.1, C	176.7, C	176.0, C	176.6, C	173.0, C	173.4, C
27	17.3, CH_3	17.3, CH_3	17.2, CH_3	17.3, CH_3	13.8, CH_3	9.5, CH_3
28	28.8, CH_3	27.8, CH_3	27.4, CH_3	27.5, CH_3	27.7, CH_3	27.6, CH_3
29	21.7, CH_3	20.6, CH_3	20.4, CH_3	20.5, CH_3	20.6, CH_3	20.5, CH_3
30	17.8, CH_3	27.8, CH_3	20.7, CH_3	20.9, CH_3	27.7, CH_3	26.4, CH_3
-O$\underline{CH_3}$		52.0, CH_3		52.1, CH_3		
-O$\underline{CH_2}CH_3$	60.7, CH_2		60.6, CH_2		61.0, CH_2	61.3, CH_2
-O$CH_2\underline{CH_3}$	29.8, CH_3		14.2, CH_3		14.3, CH_3	14.3, CH_3

^{a} Measured at 500 MHz; ^{b} Measured at 600 MHz; ^{c} Measured at 800 MHz; ^{d} Measured in $CDCl_3$.

Table 5. ^{1}H NMR Spectroscopic Data of Compounds **11**–**13** ($CDCl_3$).

No.	**11** ^{bc}	**12** ^{bc}	**13** ^{ac}
1	2.96, overlapped	2.98, overlapped	2.87, overlapped
	1.76, ddd (17.3, 9.9, 3.2)	1.70, overlapped	1.68, overlapped
2	2.60, overlapped	2.61, ddd (14.9, 9.0, 5.4)	2.55, ddd (15.5, 9.6, 6.0)
	2.48, overlapped	2.45, ddd (14.9, 8.6, 6.5)	2.45, overlapped
5	2.26, dd (15.1, 2.6)	2.09, dd (9.9, 5.4)	2.19, dd (15.1, 2.7)
6	2.60, overlapped	1.70, overlapped	2.48, dd (15.1, 14.4)
	2.52, overlapped	1.24, m	2.34, dd (14.4, 2.7)
7		4.47, br.s	
12	2.87, d (16.8)	2.78, d (17.4)	2.76, d (16.3)
	2.44, d (16.8)	2.37, d (17.4)	2.40, d (16.3)
15	4.39, dd (9.7, 5.7)	2.05, m	2.27, overlapped
		1.87, m	1.77, overlapped
16	1.82, ddd (15.1, 10.2, 5.5)	2.00, overlapped	1.89, overlapped
	2.49, overlapped	1.96, overlapped	1.68, overlapped
17	2.92, overlapped	2.96, overlapped	2.80, overlapped
18	0.73, s	0.68, s	0.63, s

Table 5. *Cont.*

No.	11 [bc]	12 [bc]	13 [ac]
19	1.27, s	1.02, s	1.21, s
21	2.10, s	2.22, s	2.15, s
22	6.08, m	6.25, s	6.16, s
24	2.52, overlapped 2.95, overlapped	4.21, br.s	4.62, dd (5.2, 2.9)
25	2.94, overlapped	2.96, overlapped	2.78, overlapped
27	1.18, d (6.7)	1.30, d (7.3)	0.92, d (7.1)
28	1.14, s	1.14, s	1.07, s
29	1.12, s	1.07, s	1.04, s
30	1.26, s	1.36, s	1.25, s
-O$\underline{CH_3}$	3.68, s		
-O$\underline{CH_2}CH_3$		4.08, m, 2H	4.14, q (7.1), 2H
-$OCH_2\underline{CH_3}$		1.19, t (7.3)	1.24, t (7.1)
24-OH		3.93, br.s	

[a] Measured at 500 MHz; [b] Measured at 600 MHz; [c] Measured in $CDCl_3$.

Ganoaustralenone L (**13**): yellow oil. $[\alpha]^{24}_{D}$ +44.67 (*c* 0.05, MeOH); UV (MeOH) λ_{max} (log ε) 250.0 (3.91); ^{1}H NMR (500 MHz, $CDCl_3$) data, see Table 5, ^{13}C NMR (125 MHz, $CDCl_3$) data, see Table 4; HRESIMS *m/z* 563.29688 [M + Na]$^+$ (calcd for $C_{32}H_{44}O_7Na$, 563.29792).

2.5. Synthesis of the Phenylglycine Methyl Ester (PGME) Derivatives

To a solution of **1** (2.0 mg, 3.9 μmol) in DMF (0.5 mL) on ice add PyBOP (2.5 mg, 4.8 μmol), HBTU (1.9 mg, 5.0 μmol), *N*-methylmorpholine 100 μL, and (*S*)-PGME (1.0 mg, 4.9 μmol). The reaction mixture was stirred at room temperature for 3 h. The reaction was stopped by adding 1 mL of EtOAc and then washed with H_2O. The EtOAc layer was concentrated under reduced pressure to obtain a pale yellow oil sample, which was purified by HPLC to furnish (*S*)-PGME amide product **1a**. Similarly, (*R*)-PGME amide product **1b** was prepared from **1** (2.0 mg) and (*R*)-PGME (1.0 mg). NMR assignments of the protons for **1a** and **1b** were achieved by analysis of the ^{1}H-^{1}H COSY spectra.

Similarly, **2a** was prepared from **2** (0.5 mg) and (*S*)-PGME, **2b** was prepared from **2** (0.5 mg) and (*R*)-PGME. **3a** was prepared from **3** (1 mg) and (*S*)-PGME, **3b** was prepared from **3** (1 mg) and (*R*)-PGME. NMR assignments of the protons for **2a** and **2b** were achieved by analysis of the ^{1}H-^{1}H COSY spectra.

To a solution of **6** (2.5 mg) in THF (1 mL) was added 1 mL of LiOH (1 mol/L). The reaction mixture was stirred at room temperature overnight. The reaction was stopped by concentrating under reduced pressure to obtain a pale yellow oil sample, which was purified by HPLC to obtain **6H** (0.4 mg). Then **6Ha** was prepared from **6H** (0.2 mg) with (*S*)-PGME, **6Hb** was prepared from **6H** (0.2 mg) with (*R*)-PGME. NMR assignments of the protons for **6Ha** and **6Hb** were achieved by analysis of the ^{1}H-^{1}H COSY spectra.

1a: ^{1}H NMR (600 MHz, $CDCl_3$) 2.824 (1H, m, H-1a), 1.814 (1H, m, H-1b), 1.887 (1H, m H-2a), 2273 (1H, m, H-2b), 2.393 (1H, m, H-5), 2.247 (1H, m, H-6a), 2.530 (1H, m, H-6b), 2.480 (1H, m, H-12a), 2.770 (1H, m, H-12b), 2.500 (1H, m, H-15a), 2.923 (1H, m, H-15b), 2.608 (1H, m, H-16a), 1.737 (1H, m H-16b), 2.962 (1H, m, H-17), 0.667 (3H, s, H-18), 1.266 (3H, s, H-19), 2.141 (3H, s, H-21), 6.321 (1H, s, H-22), 4.123 (1H, dd, *J* = 6.0, 3.3 Hz, H-24), 2.924 (1H, m, H-25), 1.377 (3H, d, *J* = 7.2 Hz, H-27), 1.133 (3H, s, H-28), 1.108 (3H, s, H-29), 1.291 (1H, s, H-30), 6.892 (1H, d, *J* = 7.0 Hz, NH), 5.423 (1H, d, *J* = 7.0 Hz, H-2′ of PGME), 7.343–7.307 (5H, overlapped, phenyl group), 3.709 (3H, s, OCH_3). HRESIMS *m/z* 682.33380 [M + Na]$^+$ (calcd for $C_{39}H_{49}O_8NNa$, 682.33559).

1b: ^{1}H NMR (600 MHz, $CDCl_3$) 2.818 (1H, m, H-1a), 1.855 (1H, m, H-1b), 1.910 (1H, m H-2a), 2265 (1H, m, H-2b), 2.388 (1H, m, H-5), 2.267 (1H, m, H-6a), 2.522 (1H, m, H-6b), 2.500 (1H, m, H-12a), 2.743 (1H, d, *J* = 16.6 Hz, H-12b), 2.515 (1H, m, H-15a), 2.927 (1H, m, H-15b), 2.613 (1H, m, H-16a), 1.744 (1H, m H-16b), 2.972 (1H, m, H-17), 0.536 (3H, s, H-18),

1.266 (3H, s, H-19), 2.098 (3H, s, H-21), 6.308 (1H, s, H-22), 4.132 (1H, dd, J = 6.0, 3.3 Hz, H-24), 2.927 (1H, m, H-25), 1.387 (3H, d, J = 7.2 Hz, H-27), 1.133 (3H, s, H-28), 1.110 (3H, s, H-29), 1.290 (1H, s, H-30), 7.061 (1H, d, J = 6.7 Hz, NH), 5.428 (1H, d, J = 6.7 Hz, H-2′ of PGME), 7.352–7.309 (5H, overlapped, phenyl group), 3.699 (3H, s, OCH_3). HRESIMS m/z 682.33392 $[M + Na]^+$ (calcd for $C_{39}H_{49}O_8NNa$, 682.33559).

2a: ^{1}H NMR (600 MHz, $CDCl_3$) 2.782 (1H, m, H-1a), 1.838 (1H, m, H-1b), 2.324 (1H, m H-2a), 2720 (1H, m, H-2b), 2.297 (1H, d, J = 13.5, H-5), 4.408 (1H, dd, J = 13.5, 3.1 Hz, H-6), 2.505 (1H, d, J = 17.0 Hz, H-12a), 2.784 (1H, m, H-12b), 1.708 (1H, m, H-15a), 2.249 (1H, m, H-15b), 1.906 (1H, m, H-16a), 2.835 (1H, m H-16b), 1.932 (1H, m, H-17), 0.658 (3H, s, H-18), 1.219 (3H, s, H-19), 2.131 (3H, s, H-21), 6.328 (1H, s, H-22), 4.115 (1H, dd, J = 6.0, 3.4 Hz, H-24), 2.932 (1H, dd, J = 7.0, 3.4 Hz, H-25), 1.378 (3H, d, J = 7.0 Hz, H-27), 1.343 (3H, s, H-28), 1.438 (3H, s, H-29), 1.362 (1H, s, H-30), 6.875 (1H, d, J = 7.0 Hz, NH), 5.417 (1H, d, J = 6.8 Hz, H-2′ of PGME), 7.352–7.301 (5H, overlapped, phenyl group), 3.704 (3H, s, OCH_3). HRESIMS m/z 698.33087 $[M + Na]^+$ (calcd for $C_{39}H_{49}O_9NNa$, 698.33050).

2b: ^{1}H NMR (600 MHz, $CDCl_3$) 2.873 (1H, m, H-1a), 1.838 (1H, m, H-1b), 2.328 (1H, m H-2a), 2722 (1H, m, H-2b), 2.293 (1H, d, J = 13.5, H-5), 4.400 (1H, dd, J = 13.5, 3.1 Hz, H-6), 2.399 (1H, d, J = 17.0 Hz, H-12a), 2.754 (1H, m, H-12b), 1.700 (1H, m, H-15a), 2.243 (1H, m, H-15b), 1.906 (1H, m, H-16a), 2.839 (1H, m H-16b), 1.958 (1H, m, H-17), 0.519 (3H, s, H-18), 1.221 (3H, s, H-19), 2.093 (3H, s, H-21), 6.315 (1H, s, H-22), 4.128 (1H, dd, J = 5.0, 3.5 Hz, H-24), 2.935 (1H, dd, J = 7.3, 3.5 Hz, H-25), 1.386 (3H, d, J = 7.2 Hz, H-27), 1.343 (3H, s, H-28), 1.440 (3H, s, H-29), 1.348 (1H, s, H-30), 7.047 (1H, d, J = 6.6 Hz, NH), 5.419 (1H, d, J = 6.6 Hz, H-2′ of PGME), 7.347–7.277 (5H, overlapped, phenyl group), 3.696 (3H, s, OCH_3). HRESIMS m/z 698.32990 $[M + Na]^+$ (calcd for $C_{39}H_{49}O_9NNa$, 698.33050).

3a: ^{1}H NMR (600 MHz, $CDCl_3$) 2.976 (1H, m, H-1a), 1.709 (1H, m, H-1b), 2.474 (1H, m H-2a), 2595 (1H, m, H-2b), 2.084 (1H, overlapped H-5), 2.902 (1H, overlapped, H-6a), 1.687 (1H, overlapped, H-6b), 4.459 (1H, overlapped, H-7), 2.749 (1H, d, J = 17.3 Hz, H-12a), 2.375 (1H, d, J = 17.3 Hz, H-12b), 1.701 (1H, m, H-15a), 2.439 (1H, m, H-15b), 1.978 (1H, m, H-16a), 2.026 (1H, m H-16b), 2.981 (1H, m, H-17), 0.642 (3H, s, H-18), 1.022 (3H, s, H-19), 2.137 (3H, s, H-21), 6.343 (1H, s, H-22), 4.137 (1H, dd, J = 6.0, 3.4 Hz, H-24), 2.939 (1H, dd, J = 7.3, 3.4 Hz, H-25), 1.385 (3H, d, J = 7.3 Hz, H-27), 1.147 (3H, s, H-28), 1.073 (3H, s, H-29), 1.344 (1H, s, H-30), 6.903 (1H, d, J = 6.9 Hz, NH), 5.426 (1H, d, J = 6.9 Hz, H-2′ of PGME), 7.343–7.306 (5H, overlapped, phenyl group), 3.703 (3H, s, OCH_3). HRESIMS m/z 684.35034 $[M + Na]^+$ (calcd for $C_{39}H_{51}O_8NNa$, 684.35124).

3b: ^{1}H NMR (600 MHz, $CDCl_3$) 2.976 (1H, m, H-1a), 1.692 (1H, m, H-1b), 2.609 (1H, m H-2a), 2.458 (1H, m, H-2b), 2.073 (1H, overlapped H-5), 1.691 (1H, overlapped, H-6a), 1.638 (1H, overlapped, H-6b), 4.438 (1H, t, J = 4.2 Hz, H-7), 2.711 (1H, d, J = 16.9 Hz, H-12a), 2.692 (1H, d, J = 16.9 Hz, H-12b), 1.809 (1H, m, H-15a), 2.021 (1H, m, H-15b), 1.977 (1H, m, H-16a), 1.911 (1H, m H-16b), 2.904 (1H, m, H-17), 0.463 (3H, s, H-18), 1.025 (3H, s, H-19), 2.096 (3H, s, H-21), 6.316 (1H, s, H-22), 4.136 (1H, dd, J = 4.5, 3.7 Hz, H-24), 2.948 (1H, dd, J = 7.3, 3.7 Hz, H-25), 1.402 (3H, d, J = 7.3 Hz, H-27), 1.145 (3H, s, H-28), 1.076 (3H, s, H-29), 1.326 (1H, s, H-30), 7.116 (1H, d, J = 6.9 Hz, NH), 5.420 (1H, d, J = 6.9 Hz, H-2′ of PGME), 7.333–7.265 (5H, overlapped, phenyl group), 3.697 (3H, s, OCH_3). HRESIMS m/z 684.35022 $[M + Na]^+$ (calcd for $C_{39}H_{51}O_8NNa$, 684.35124).

6Ha: ^{1}H NMR (600 MHz, $CDCl_3$) 2.260 (1H, m, H-1a), 2.015 (1H, m, H-1b), 2.953 (1H, m H-2a), 2.347 (1H, m, H-2b), 1.861 (1H, overlapped H-5), 2.300 (1H, overlapped, H-6a), 2.295 (1H, overlapped, H-6b), 6.510 (1H, overlapped, H-7), 5.666 (1H, overlapped, H-11), 4.585 (1H, m, H-15), 2.500 (1H, m, H-16a), 1.812 (1H, m H-16b), 3.308 (1H, m, H-17), 0.907 (3H, s, H-18), 1.095 (3H, s, H-19), 2.273 (3H, s, H-21), 6.365 (1H, s, H-22), 2.540 (1H, m H-24a), 2.540 (1H, m H-24b), 2.937 (1H, m, H-25), 1.139 (3H, d, J = 6.8 Hz, H-27), 1.122 (3H, s, H-28), 1.161 (3H, s, H-29), 1.298 (1H, s, H-30), 6.914 (1H, d, J = 7.0 Hz, NH), 5.487 (1H, d, J = 7.0 Hz, H-2′ of PGME), 7.355–7.338 (5H, overlapped, phenyl group), 3.686 (3H, s, OCH_3). HRESIMS m/z 666.33997 $[M + Na]^+$ (calcd for $C_{39}H_{49}O_7NNa$, 666.34067).

6Hb: ^{1}H NMR (600 MHz, $CDCl_3$) 2.267 (1H, m, H-1a), 1.790 (1H, m, H-1b), 2.790 (1H, m H-2a), 2.875 (1H, m, H-2b), 1.875 (1H, overlapped H-5), 2.286 (1H, overlapped, H-6a),

2.301 (1H, overlapped, H-6b), 6.494 (1H, overlapped, H-7), 5.662 (1H, s, H-11), 4.553 (1H, m, H-15), 2.453 (1H, m, H-16a), 1.477 (1H, m H-16b), 3.267 (1H, m, H-17), 1.295 (3H, s, H-18), 1.080 (3H, s, H-19), 2.197 (3H, s, H-21), 6.289 (1H, s, H-22), 2.510 (1H, m H-24a), 2.510 (1H, m H-24b), 2.914 (1H, m, H-25), 1.200 (3H, d, J = 7.2 Hz, H-27), 1.120 (3H, s, H-28), 1.160 (3H, s, H-29), 0.821 (1H, s, H-30), 6.803 (1H, d, J = 7.2 Hz, NH), 5.516 (1H, d, J = 7.2 Hz, H-2′ of PGME), 7.360–7.293 (5H, overlapped, phenyl group), 3.722 (3H, s, OCH_3). HRESIMS *m/z* 666.33978 $[M + Na]^+$ (calcd for $C_{39}H_{49}O_7NNa$, 666.34067).

2.6. Biological Activity Assays

Biological activity assays, including the cytotoxicity against five human cancer cell lines [24], α-glucosidase inhibition [25], protein tyrosine phosphatase 1 β (PTP1B) [26], dipeptidyl peptidase 4 (DDP4) [27], and angiotensin converting enzyme 2 (ACE2) [28], were screened according to the protocols in the Supplementary Materials.

3. Results and Discussion

Compound **1** (Figure 1), obtained as a yellow oil, gave an $[M + Na]^+$ ion peak at *m/z* 535.26685 in the HRESIMS (calcd for $C_{30}H_{40}O_7Na$, 535.26717). The ^1H NMR spectroscopic data (Table 1) displayed six methyl singlets at δ_H 1.28 (Me-19), δ_H 0.75 (Me-18), δ_H 2.16 (Me-21), δ_H 1.13 (Me-28), δ_H 1.10 (Me-29), and δ_H 1.35 (Me-30), one methyl doublet at δ_H 1.18 (d, J = 7.3 Hz, Me-27), an olefinic proton at δ_H 6.51 (s, H-22), and an oxygenated methine proton at δ_H 4.21 (d, J = 4.9 Hz, H-24). The ^{13}C NMR and DEPT spectroscopic data (Table 2) of **1** showed 30 carbon resonances which were ascribed to seven methyl carbons at δ_C 19.1 (C-18), 18.4 (C-19), 22.2 (C-21), 13.8 (C-27), 27.7 (C-28), 20.7 (C-29), and 26.4 (C-30), six methylenes at δ_C 36.1 (C-1), 35.0 (C-2), 37.7 (C-6), 50.9 (C-12), 33.5 (C-15), and 24.2 (C-16), four methines at δ_C 50.8 (C-5), 54.6 (C-17), 88.0 (C-24), and 44.6 (C-25), two pairs of olefinic carbons at δ_C 151.3 (C-8), 152.4 (C-9), 161.3 (C-20), and 122.0 (C-22), four sp^3-quaternary carbons at δ_C 47.8 (C-4), 40.1 (C-10), 49.7 (C-13, C-14), and four carbonyls at δ_C 218.6 (C-3), 202.7 (C-7), 202.8 (C-11), 202.0 (C-23), and 176.9 (C-27). The chemical shifts of 1D NMR of **1** indicated that it was a lanostane triterpenoid similar to resinacein N, except for the substitutions at C-3, C-7, and C-15 [29]. In the HMBC spectrum of **1**, the correlations from Me-29 to the carbonyl C-3, from Me-30 to the methylene carbon C-15, and from H-5 (δ_H 2.38) and H-6 (δ_H 2.36, 2.68) to the carbonyl C-7, along with the ^1H-^1H COSY correlation of H-15 (δ_H 2.24, 1.87)/H-16 (δ_H 2.04, 1.91) (Figure 2), suggested that C-3 and C-7 were ketone carbons and C-15 was a methylene instead of being a hydroxylated methine in resinacein N. Therefore, the planar structure of **1** was elucidated as shown in Figure 1.

The key ROESY correlations between H-22 (δ_H 6.51) and H-16a/b (δ_H 2.04, 1.91) allowed the assignment of the *E* configuration of the C-20–C-22 double bond (Figure 3). The absolute configuration of the chiral center C-25 was determined by the PGME method (Figure 4). The (*R*)- and (*S*)-PGME amide derivatives were chemically synthesized, and the $\Delta\delta_H$ ($\delta_S - \delta_R$) values indicated that C-25 was the *S* configuration. The attempt to assign the absolute configuration of C-24 by Mosher's method failed, probably due to the bulky groups around the hydroxy group. Therefore, the configuration of C-24 remained unassigned. Compound **1** was elucidated as [20(22)*E*,24*R*,25*R*]-24-hydroxy-3,7,11,23-tetraoxolanosta-8,20-dien-26-oic acid, and was given the trivial name ganoaustralenone A.

Compound **2**, obtained as a white powder, displayed an $[M + Na]^+$ peak at *m/z* 551.26422 in the HRESIMS (calcd for $C_{30}H_{40}O_8Na$, 551.26209). The 1D NMR data of **2** (Tables 1 and 2) showed a resemblance to those of compound **1**, implying the analogous structures of the two compounds. Analysis of the 1D NMR data suggested that the only difference between **1** and **2** was C-6. The HMBC correlation from H-5 to a hydroxymethine at δ_C 72.2 (C-6), as well as the ^1H-^1H COSY correlation of H-5 and the proton at δ_H 4.44 (H-6) (Figure 2), revealed that the C-6 in **2** attached to a hydroxy group compared to that of **1**. These assignments are consistent with the HRESIMS result. The absolute configuration of C-25 was determined by the PGME method, as in the case of compound

1 (Figure 4). Therefore, compound 2 was determined as shown in Figure 1, and trivially named ganoaustralenone B.

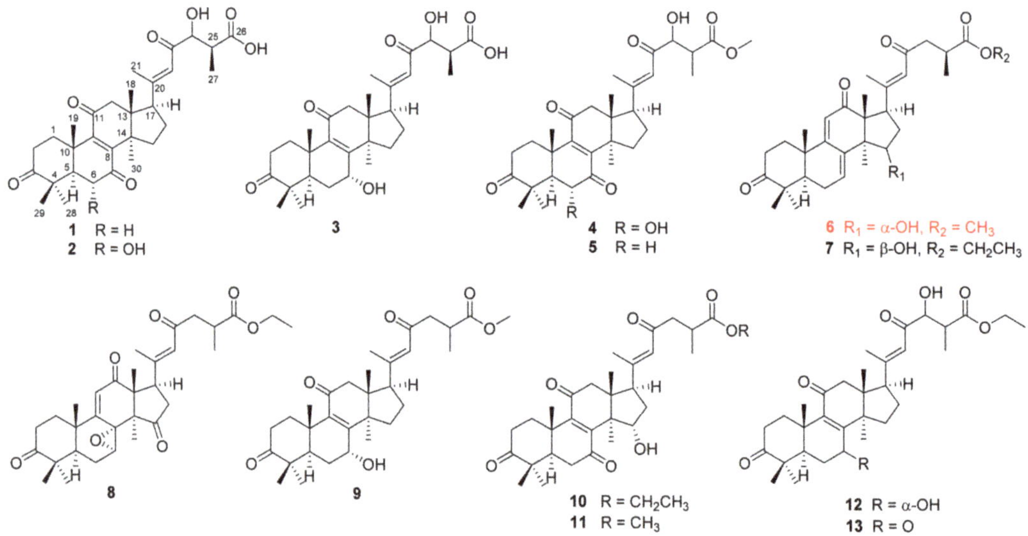

Figure 1. Chemical structures of compounds **1–13**. (Red: known compounds.)

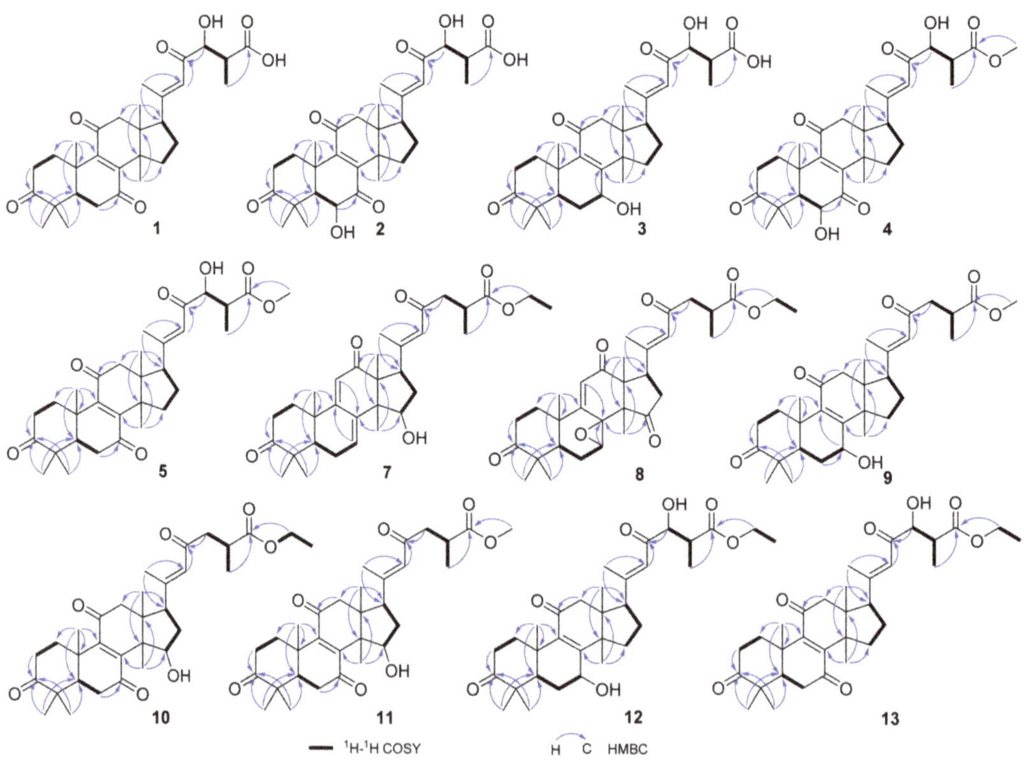

Figure 2. Key ^{1}H-^{1}H COSY and HMBC correlations of compounds **1−5**, **7−13**.

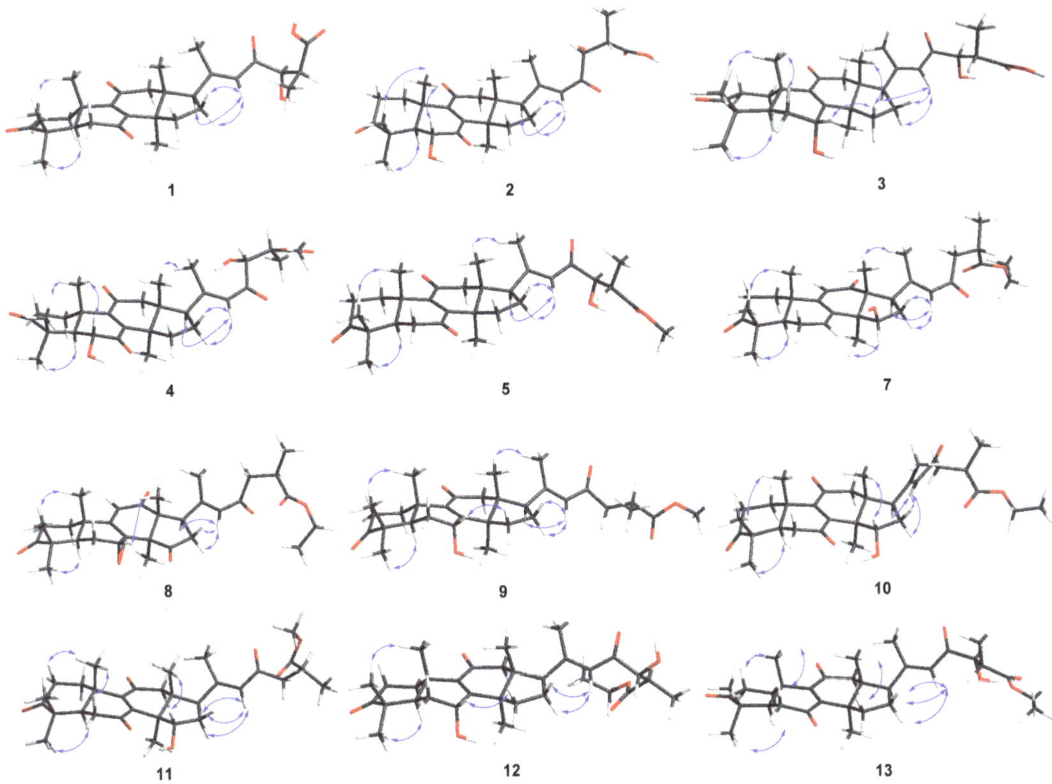

Figure 3. Key ROESY correlations of **1–5**, **7–13**.

Compound **3**, obtained as a yellow oil, displayed an $[M + Na]^+$ ion peak at *m/z* 537.28180 in the HRESIMS analysis (calcd for $C_{30}H_{42}O_7Na$, 537.28227). The ^1H NMR and ^{13}C NMR data of **3** (Tables 1 and 2) highly resemble those of **1**, except for the chemical shift of C-7. The key ^1H-^1H COSY correlations H-5 (δ_H 2.11)/H-6 (δ_H 1.70)/H-7 (δ_H 4.47), as well as the HMBC correlation from H-7 (δ_H 4.47) and C-8 (δ_C 160.3) (Figure 2), implied the presence of a hydroxyl group at C-7. The key ROESY correlations of H-7/H-15β/H_3-18 indicated the α orientation of 7-OH (Figure 3). The absolute configurations of C-25 were determined by the PGME method, as in the case of compound **1** (Figure 4). Therefore, compound **3** was determined as shown in Figure 1, and identified as ganoaustralenone C.

The yellowish oil compounds **4** and **5** gave the sodium adduct ion peaks of *m/z* 565.27728 and *m/z* 549.28210 in the HRESIMS analysis, corresponding to the molecular formulas of $C_{31}H_{42}O_8$ and $C_{31}H_{42}O_7$ (calcd for $C_{31}H_{42}O_8Na$ 565.27774; $C_{31}H_{42}O_7Na$, 549.28282), respectively. The 1D NMR spectra of the two compounds (Tables 1 and 2) showed characteristic signals of triterpene, indicating the same skeletons of **1–5**. Analysis of the 1D NMR spectra of **4** and **5** suggested that the two compounds were highly similar to those of **1** and **2**, respectively. The differences between these two pairs of compounds (**1** vs. **4**, **2** vs. **5**) were the status of C-26 carboxylic group. The correlations from the methoxy singlets to the carbonyl group (C-26) in the HMBC spectra of **4** and **5** (Figure 2) indicated that C-26 of **4** and **5** have been methyl esterified instead of being free carboxylic groups in **1** and **2**. Therefore, compounds **4** and **5** were elucidated as the C-26 methyl ester derivatives of **1** and **2**, respectively. However, these changes hampered the absolute configuration determination of C-25 of **4** and **5** by the PGME method. The relative configurations of C-24 and C-25 were assigned as *R** and *S**, respectively, by analysis of the Newman projections of

C-24–C-25 and the coupling constants of H-24 (4.0 Hz, and 3.1 Hz). Therefore, compounds **4** and **5** were named ganoaustralenones D and E, respectively.

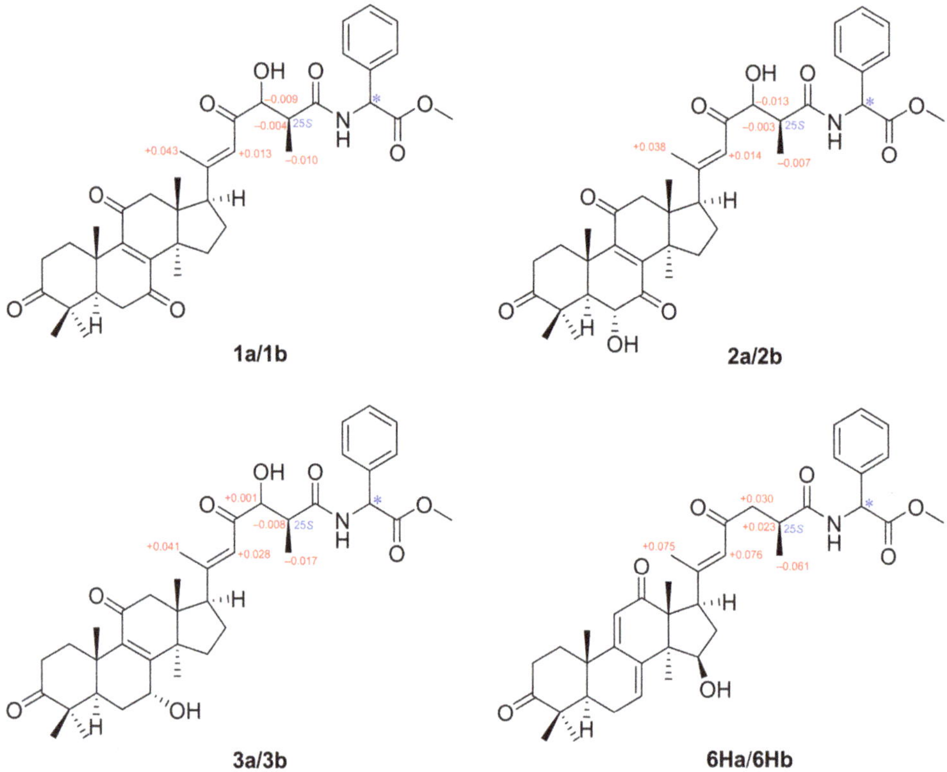

Figure 4. The structures and $\Delta\delta_H$ ($\delta_S - \delta_R$) of (*S*)/(*R*)-PGME derivative of **1**, **2**, **3**, and **6H**.

Compound **6** was determined to be methyl gibbosate O by comparison with the NMR spectroscopic data (Supplementary Materials) [30,31]. However, the chemical shifts of C-13 and C-14 of methyl gibbosate O have been erroneously assigned previously [30]. The key HMBC correlation of H-11 (δ_H 5.66, s) to an sp^3-quaternary carbon at δ_C 58.0, together with the HMBC correlation from H-7 (δ_H 6.50, m) to an sp^3-quaternary carbon at δ_C 52.5 enabled the correct assignment of the chemical shifts of C-13 (58.0 ppm) and C-14 (52.5 ppm). Moreover, the absolute configuration of C-25 of gibbosic acid O was assigned as *S* without any evidence [31], while for methyl gibbosate O, the C-25 configuration was assigned to be same with gibbosic acid O only by comparison with the chemical shifts [30]. However, C-25 is far away from any other chiral centers in the structure, so the chemical shift deviation is inadequate to discriminate the *S* and *R* configuration of C-25. Therefore, more solid evidence should be presented to corroborate the real configuration of C-25. In order to determine the absolute configuration of the chiral center C-25, compound **6** was firstly hydrolyzed by LiOH to obtain the previously reported compound gibbosic acid O (**6H**). Then, the (*R*)- and (*S*)-PGME amide derivatives of **6H** were chemically synthesized (Scheme 1), and the $\Delta\delta_H$ ($\delta_S - \delta_R$) values indicated that C-25 was the *S* configuration (Figure 4). Therefore, the absolute configuration of compound **6** has been fully assigned.

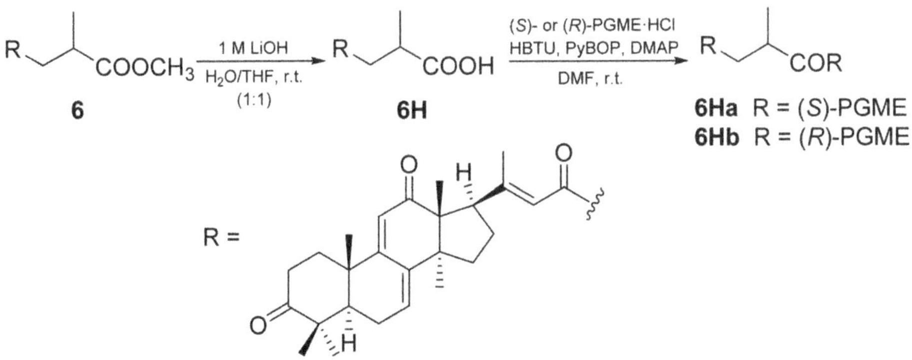

6 **6H** **6Ha** R = (*S*)-PGME **6Hb** R = (*R*)-PGME

Scheme 1. Alkaline hydrolysis and PGME derivatization of compound **6**.

The HRESIMS analysis of **7**, a yellow oily compound, gave a sodium adduct ion peak at *m/z* 547.30286, corresponding to the molecular formula of $C_{32}H_{44}O_6$ (calcd for $C_{32}H_{44}O_6Na$, 547.30356) with 11 double bond equivalences. Comparing the 1D NMR data of **7** (Tables 2 and 3) with those of **6** suggested that **7** differed from **6** with the presence of an oxygenated methylene and a triplet methyl group with the absence of the methoxy group. These signals were assigned to be ethyl ester moiety of the C-26 carbonyl group instead of the methyl ester moiety in **6**. The ^1H-^1H COSY correlation of $O\underline{CH_2}CH_3$ (δ_H 4.13)/$OCH_2\underline{CH_3}$ (δ_H 1.25), and the HMBC correlation from $O\underline{CH_2}CH_3$ (δ_H 4.13) to C-27 (δ_C 176.3) (Figure 2), confirmed the above assignments. Notably, 15-OH was assigned to be β orientation by the key ROESY correlation of H-15 (δ_H 4.31)/Me-30 (δ_H 1.00) (Figure 3). Therefore, compound **7** was named ganoaustralenone F.

Compound **8** had an $[M + Na]^+$ ion peak at *m/z* 561.28223 ($C_{32}H_{42}O_7Na$) in the HRESIMS analysis (calcd for $C_{32}H_{42}O_7Na$, 561.28282). The molecular formula of **8** is two oxygen atoms more than that of **7**, indicating the existence of more oxygenated carbons in **8** than those of **7**. The ^1H NMR spectra of **8** (Table 3) displayed six methyl singlets (δ_H 1.12, 1.08, 2.26, 1.12, 1.09, and 1.29). The ^{13}C NMR (Table 4) and DEPT spectra of **8** exhibited signals for eight methyls, six methylenes (one was oxygenated at δ_C 60.7), six methines including two sp^2-ones and four sp^3-ones (one was oxygenated, δ_C 58.9), and eleven quaternary carbons (four carbonyls, five sp^3-ones, and two sp^2-ones). Further analysis of the 1D NMR data (Tables 3 and 4) allowed the assignment of **8** to be an analog of **7**, except for the positions of C-7, C-8, and C-15. The ^{13}C NMR chemical shifts of these three positions (δ_C 58.9, C-7; δ_C 62.7, C-8; δ_C 209.9, C-15) of **8** implied that an epoxy ring was located at C-7 and C-8, while C-15 was a ketone compared to that of **7**. These assignments were corroborated by the ^1H-^1H COSY correlations of H-5/H-6/H-7 and the HMBC correlations from H-7 to C-8, and from Me-30 to C-15 (Figure 2). Thus, compound **8** was trivially named ganoaustralenone G.

Compound **9** showed an $[M + H]^+$ peak at *m/z* 513.32135 in the HRESIMS, indicating the molecular formula $C_{31}H_{44}O_6$ (calcd for $C_{31}H_{45}O_6$, 513.32161). The 1D NMR data of **9** (Tables 3 and 4) displayed thirty-one carbon resonances, which were categorized into seven methyl carbons at δ_C 17.7 (C-19), 18.6 (C-18), 21.4 (C-21), 17.3 (C-27), 27.8 (C-28), 20.6 (C-29), and 27.8 (C-30), one methoxy carbon at δ_C 52.0, seven methylenes at δ_C 34.2 (C-1), 34.9 (C-2), 29.4 (C-6), 50.3 (C-12), 30.4 (C-15), 23.2 (C-16), and 47.9 (C-24), five methines at δ_C 45.2 (C-5), 67.3 (C-7), 54.1 (C-17), 123.8 (C-22), and 35.0 (C-25), and eleven proton-free carbons at δ_C 218.0 (C-3), 46.5 (C-4), 159.5 (C-8), 140.2 (C-9), 38.0 (C-10), 195.5 (C-11), 48.6 (C-13), 50.6 (C-14), 158.0 (C-20), 198.4 (C-23), and 176.7 (C-26). The above data suggested that **9** was a similar structure to 7β-hydroxy-3,11,15,23-tetraoxolanosta-8,20*E*(22)-diene-26-oic acid methyl ester, except for the position at C-15 and the configuration of 7-OH [32]. The ^1H-^1H COSY correlations of H-15/H-16/H-17, as well as the HMBC correlation from Me-30 (δ_H

1.35, s) to the methylene carbon at δ_C 30.4 (C-15) (Figure 2), suggested that C-15 in **9** was a methylene instead of being a ketone carbon in 7β-hydroxy-3,11,15,23-tetraoxolanosta-8,20*E*(22)-diene-26-oic acid methyl ester. In addition, the key ROESY correlations of H-7 (δ_H 4.47)/H-15β (δ_H 2.04)/Me-18 (δ_H 0.67) (Figure 3) suggested that 7-OH was an α orientation. Therefore, compound **9** was named ganoaustralenone H.

Compound **10**, a pale yellow oil, gave an $[M + Na]^+$ ion peak at *m/z* 563.29749 ($C_{32}H_{44}O_7Na$) in the HRESIMS (calcd for $C_{32}H_{44}O_7Na$, 563.29847). The ^1H and ^{13}C NMR spectroscopic data of **10** (Tables 3 and 4) showed high similarity to those of the structure 15α-hydroxy-3,11,23-trioxolanosta-8,20*E*(22)-dien-26-oic acid methyl ester, a lanostane triterpenoid isolated from the *G. lucidum* [33]. Further analysis of the 2D NMR spectra revealed that the only difference between these two structures was C-7. The diagnostic HMBC correlations from the protons at δ_H 2.26 (H-5), 2.49 (H-6a), 2.62 (H-6b) to a carbonyl group at δ_C 204.4 (Figure 2) suggested that C-7 was a carbonyl group in **10** instead of being a methylene group in 15α-hydroxy-3,11,23-trioxolanosta-8,20*E*(22)-dien-26-oic acid methyl ester. In addition, the alcohol for forming the C-26 ester group was ethanol in **10** instead of methanol in the reported structure, as supported by the two chemical shifts at δ_C 60.6 (-O$\underline{CH_2}$CH$_3$) and 14.2 (-OCH$_2\underline{CH_3}$). Therefore, compound **10** was identified as ethyl 20(22)*E*-15α-hydroxy-3,7,11,23-tetraoxolanosta-8,20(22)-dien-26-oate, and was trivially named ganoaustralenone I.

Compound **11**, obtained as a yellow oil, displayed an $[M + Na]^+$ ion peak at *m/z* 549.28210 in the HRESIMS analysis (calcd for $C_{31}H_{42}O_7Na$, 549.28282), revealing the molecular formula of $C_{31}H_{42}O_7$. The 1D NMR data of **11** (Tables 4 and 5) showed 30 carbon resonances with high resemblance to those of compound **10**. Further analysis of the 2D NMR data (Figures 2 and 3) suggested that **11** differed from **10** by the presence of the methyl ester group. The significant HMBC correlation from the methoxy group (δ_H 3.68) to the carbonyl group C-26 (δ_C 176.6) (Figure 2) verified the terminal carboxylic group in **11** has been methyl esterified instead of being ethyl esterified in **10**. Therefore, compound **11** was identified as ganoaustralenone J.

The pale yellow oil compound **12** exhibited an $[M + Na]^+$ ion peak at *m/z* 565.31287 in the HRESIMS analysis (calcd for $C_{32}H_{46}O_7Na$, 565.31412). The NMR spectroscopic data of **12** (Tables 4 and 5) highly resemble those of **9**, except for the chemical shifts of C-24 and the alcoholic part of the C-26 ester. The important HMBC correlations from Me-27 (δ_H 1.30) to a hydroxymethine at δ_C 78.6 (C-24) (Figure 2), together with the chemical shifts of the alcoholic part [δ_C 61.0 (-O$\underline{CH_2}$CH$_3$) and 14.3 (-OCH$_2\underline{CH_3}$)], indicated that a hydroxy group situated at C-24 and the presence of ethyl ester of C-26 in **12** compared to those of **9**. Therefore, compound **12** was identified as ganoaustralenone K.

Compound **13**, obtained as a yellow oil, displayed an $[M + Na]^+$ ion peak at *m/z* 563.29688 in the HRESIMS analysis (calcd for $C_{32}H_{44}O_7Na$, 563.29792). Analysis of the ^1H and ^{13}C NMR data (Tables 4 and 5) revealed that this compound was a structural analog to **12**. The main difference between the NMR data of the two analogs was the position C-7 (δ_C 201.2), which indicated that C-7 was a carbonyl carbon. In the HMBC spectrum of compound **13**, significant correlations from H$_2$-6 (δ_H 2.48, 2.34) to C-7 (δ_C 201.2) (Figure 2) indicated that C-7 was a carbonyl carbon. Therefore compound **13** was identified as ganoaustralenone L.

The identification of a series of 20(22)*E*-lanostanes from this species of *Ganoderma* inspired a proposal of the possible biosynthetic pathways. Take compound **1** as an example, as shown in Scheme 2, the common precursor squalene, which was derived from two molecules of farnesyl pyrophosphate, which was oxygenated and followed by function migration to give the lanostane scaffold. The lanosterol was oxygenated at the positions of C-3, C-7, C-11, C-20, C-23, and C-26 to give the key intermediate **A**, which underwent an elimination reaction by the E1cb mechanism to yield **B**. Finally, a nucleophilic attack at C-24 by a hydroxy group produced compound **1**.

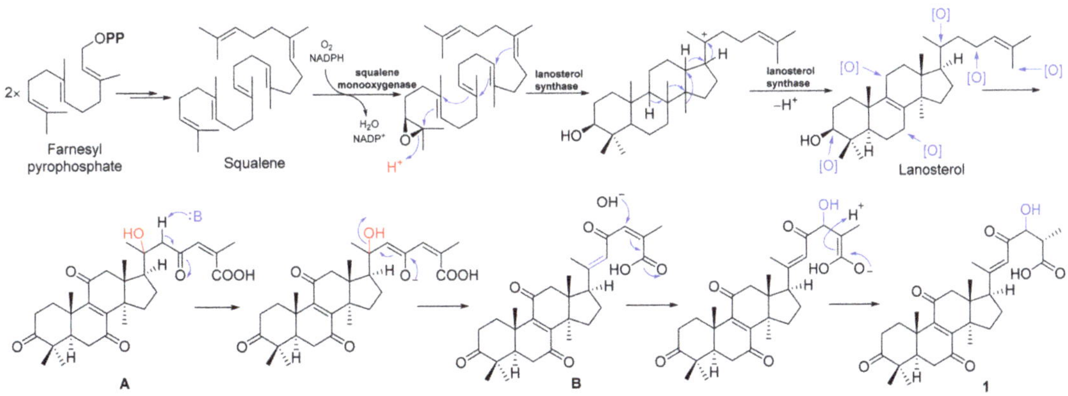

Scheme 2. Plausible biosynthetic pathways to **1**.

All the isolates were subjected to evaluate their cytotoxicity against the five human cancer cell lines (the HL-60 (ATCC CCL-240), the human myeloid leukemia cell line, the SMMC-7721 human hepatocellular carcinoma cell line, the A549 (ATCC CCL-185) lung cancer cell line, the MCF-7 (ATCC HTB-22) breast cancer cell line, and the SW-480 (ATCC CCL-228) human colon cancer cell line, as well as the inhibitory activity against α-glucosidase, protein tyrosine phosphate 1 β (PTP1B), dipeptidyl peptidase 4 (DDP4), and angiotensin-converting enzyme 2 (ACE2). However, no significant bioactivity was observed.

4. Conclusions

In conclusion, twelve previously undescribed lanostane-type triterpenes were obtained from the medicinal mushroom *Ganoderma australe*. By using the NMR and HRESIMS techniques for structural elucidation, the structures of twelve triterpenes were determined, and the absolute configurations of **1**, **2**, and **6** were assigned by the phenylglycine methyl ester (PGME) method. Ganoderma triterpenes have been reported to have more than 400 chemical entities to date [15]. Most of them were oxygenated at the positions of C-3, C-7, C-11, C-15, and C-26. Interestingly, more and more studies have revealed that there was an oxygenated position bias that differed from species to species. The triterpenes described here are featured by an unusual 20(22)-*trans* double bond, which has rarely been found in the *Ganoderma* lanostanoid family. Although no significant biological activities were found in this study, the results have also initiated the understanding of the structural diversity of *Ganoderma*-derived triterpenoids.

Supplementary Materials: The following supporting information can be downloaded at: https://www.mdpi.com/article/10.3390/jof8050503/s1, including the NMR, HRMS data of compounds **1–13**.

Author Contributions: H.-P.C. and Z.-H.L. designed and supervised the project; L.Z. and L.-L.G. performed the experiments; L.Z., L.-L.G., M.I., Z.-H.L. and H.-P.C. analyzed the data, discussed the results, and wrote the paper; L.Z. and L.-L.G. contributed equally to this work. All authors have read and agreed to the published version of the manuscript.

Funding: This research was funded by the National Natural Science Foundation of China (21961142008, 81903512) and the Thailand Research Fund (DBG6280008).

Institutional Review Board Statement: Not applicable.

Informed Consent Statement: Not applicable.

Data Availability Statement: The data presented in this study are available in this manuscript and can be requested from the corresponding author.

Acknowledgments: We thank the Analytical and Measuring Center, School of Pharmaceutical Sciences, South-Central University for Nationalities for the spectra measurement. We thank the Bioactivity Screening Center of Natural Products, Kunming Institute of Botany, Chinese Academy of Sciences for the biological activity tests of the isolated compounds.

Conflicts of Interest: The authors declare no conflict of interest.

References

1. Dinis-Oliveira, R.J. Metabolism of psilocybin and psilocin: Clinical and forensic toxicological relevance. *Drug Metab. Rev.* **2017**, *49*, 84–91. [CrossRef] [PubMed]
2. Anke, T.; Oberwinkler, F.; Steglich, W.; Schramm, G. The strobilurins-New antifungal antibiotics from the basidiomycete Strobilurus tenacellus. *J. Antibiot.* **1977**, *30*, 806–810. [CrossRef] [PubMed]
3. Chen, L.; Guo, X.F.; Fan, Z.J.; Zhang, N.L.; Zhu, Y.J.; Zhang, Z.M.; Khazhieva, I.; Yurievich, M.Y.; Belskaya, N.P.; Bakulev, V.A. Synthesis and fungicidal activity of 3,4-dichloroisothiazole based strobilurins as potent fungicide candidates. *RSC Adv.* **2017**, *7*, 3145–3151. [CrossRef]
4. Richter, C.; Wittstein, K.; Kirk, P.M.; Stadler, M. An assessment of the taxonomy and chemotaxonomy of Ganoderma. *Fungal Divers.* **2014**, *71*, 1–15. [CrossRef]
5. Fatmawati, S.; Shimizu, K.; Kondo, R. Ganoderic acid Df, a new triterpenoid with aldose reductase inhibitory activity from the fruiting body of *Ganoderma lucidum*. *Fitoterapia* **2010**, *81*, 1033–1036. [CrossRef]
6. Mizushinaa, Y.; Hanashimaa, L.; Yamaguchib, T.; Takemurac, M.; Sugawaraa, F.; Saneyoshib, M.; Matsukaged, A.; Yoshidac, S.; Sakaguchi, K. A Mushroom Fruiting Body-Inducing Substance Inhibits Activities of Replicative DNA Polymerases. *Biochem. Biophys. Res. Commun.* **1998**, *249*, 17–22. [CrossRef]
7. Nishitoba, T.; Oda, K.; Sato, H.; Sakamura, S. Novel triterpenoids from the fungus *Ganoderma lucidum*. *Agric. Biol. Chem.* **1988**, *52*, 367–372. [CrossRef]
8. Wasser, S.P. Reishi or ling zhi (*Ganoderma lucidum*). *Encycl. Diet. Suppl.* **2005**, *1*, 603–622. [CrossRef]
9. Wasser, S.P.; Weis, A.L. Therapeutic effects of substances occurring in higher *Basidiomycetes* mushrooms: A modern perspective. *Crit. Rev. Immunol.* **1999**, *19*, 65–96. [CrossRef]
10. Peng, X.-R.; Liu, J.-Q.; Wang, C.-F.; Li, X.-Y.; Shu, Y.; Zhou, L.; Qiu, M.-H. Hepatoprotective effects of triterpenoids from *Ganoderma cochlear*. *J. Nat. Prod.* **2014**, *77*, 737–743. [CrossRef]
11. Boh, B.; Berovic, M.; Zhang, J.; Zhi-Bin, L. *Ganoderma lucidum* and its pharmaceutically active compounds. *Biotechnol. Annu. Rev.* **2007**, *13*, 265–301. [CrossRef] [PubMed]
12. Amaral, A.E.; Carbonero, E.R.; Simão, R.D.C.G.; Kadowaki, M.K.; Sassaki, G.L.; Osaku, C.A.; Gorin, P.A.; Iacomini, M. An unusual water-soluble β-glucan from the basidiocarp of the fungus *Ganoderma resinaceum*. *Carbohydr. Polym.* **2008**, *72*, 473–478. [CrossRef]
13. Peng, X.-R.; Liu, J.-Q.; Han, Z.-H.; Yuan, X.-X.; Luo, H.-R.; Qiu, M.-H. Protective effects of triterpenoids from *Ganoderma resinaceum* on H2O2-induced toxicity in HepG2 cells. *Food Chem.* **2013**, *141*, 920–926. [CrossRef] [PubMed]
14. Zengin, G.; Sarikurkcu, C.; Gunes, E.; Uysal, A.; Ceylan, R.; Uysal, S.; Gungor, H.; Aktumsek, A. Two *Ganoderma* species: Profiling of phenolic compounds by HPLC–DAD, antioxidant, antimicrobial and inhibitory activities on key enzymes linked to diabetes mellitus, Alzheimer's disease and skin disorders. *Food Funct.* **2015**, *6*, 2794–2802. [CrossRef] [PubMed]
15. Baby, S.; Johnson, A.J.; Govindan, B. Secondary metabolites from *Ganoderma*. *Phytochemistry* **2015**, *114*, 66–101. [CrossRef] [PubMed]
16. Xia, Q.; Zhang, H.; Sun, X.; Zhao, H.; Wu, L.; Zhu, D.; Yang, G.; Shao, Y.; Zhang, X.; Mao, X.; et al. A Comprehensive Review of the Structure Elucidation and Biological Activity of Triterpenoids from *Ganoderma* spp. *Molecules* **2014**, *19*, 17478–17535. [CrossRef]
17. Paterson, R.R.M. *Ganoderma*–A therapeutic fungal biofactory. *Phytochemistry* **2006**, *67*, 1985–2001. [CrossRef]
18. Guo, J.-C.; Yang, L.; Ma, Q.-Y.; Ge, Y.-Z.; Kong, F.-D.; Zhou, L.-M.; Zhang, F.; Xie, Q.-Y.; Yu, Z.-F.; Dai, H.-F.; et al. Triterpenoids and meroterpenoids with α-glucosidase inhibitory activities from the fruiting bodies of *Ganoderma australe*. *Bioorganic Chem.* **2021**, *117*, 105448. [CrossRef]
19. León, F.; Valencia, M.; Rivera, A.; Nieto, I.; Quintana, J.; Estévez, F.; Bermejo, J. Novel Cytostatic Lanostanoid Triterpenes from *Ganoderma australe*. *Helvetica Chim. Acta* **2003**, *86*, 3088–3095. [CrossRef]
20. Isaka, M.; Chinthanom, P.; Mayteeworakoon, S.; Laoteng, K.; Choowong, W.; Choeyklin, R. Lanostane triterpenoids from cultivated fruiting bodies of the basidiomycete *Ganoderma australe*. *Nat. Prod. Res.* **2017**, *32*, 1044–1049. [CrossRef]
21. Jain, A.C.; Gupta, S.K. The isolation of lanosta-7,9(11),24-trien-3β,21-diol from the fungus *Ganoderma australe*. *Phytochemistry* **1984**, *23*, 686–687. [CrossRef]
22. Zhang, J.-J.; Dong, Y.; Qin, F.-Y.; Yan, Y.-M.; Cheng, Y.-X. Meroterpenoids and alkaloids from *Ganoderma australe*. *Nat. Prod. Res.* **2019**, *35*, 3226–3232. [CrossRef] [PubMed]
23. Zhang, J.-J.; Dong, Y.; Qin, F.-Y.; Cheng, Y.-X. Australeols A–F, neuroprotective meroterpenoids from *Ganoderma australe*. *Fitoterapia* **2019**, *134*, 250–255. [CrossRef] [PubMed]
24. Chen, H.-P.; Zhao, Z.-Z.; Li, Z.-H.; Dong, Z.-J.; Wei, K.; Bai, X.; Zhang, L.; Wen, C.-N.; Feng, T.; Liu, J.-K. Novel Natural Oximes and Oxime Esters with a Vibralactone Backbone from the Basidiomycete *Boreostereum vibrans*. *ChemistryOpen* **2016**, *5*, 142–149. [CrossRef] [PubMed]

25. Chen, L.; Li, Z.-H.; Yao, J.-N.; Peng, Y.-L.; Huang, R.; Feng, T.; Liu, J.-K. Isoindolinone-containing meroterpenoids with α-glucosidase inhibitory activity from mushroom *Hericium caput-medusae*. *Fitoterapia* **2017**, *122*, 107–114. [CrossRef]
26. Ren, Y.-M.; Zhang, R.; Feng, Z.; Ke, C.-Q.; Yao, S.; Tang, C.; Lin, L.; Ye, Y. Macrocephatriolides A and B: Two Guaianolide Trimers from *Ainsliaea macrocephala* as PTP1B Inhibitors and Insulin Sensitizers. *J. Org. Chem.* **2021**, *86*, 17782–17789. [CrossRef]
27. Zhu, Y.; Zhang, P.; Yu, H.; Li, J.; Wang, M.W.; Zhao, W. Anti-*Helicobacter pylori* and Thrombin Inhibitory Components from Chinese Dragon's Blood, *Dracaena cochinchinensis*. *J. Nat. Prod.* **2007**, *70*, 1570–1577. [CrossRef]
28. Liao, W.; Bhullar, K.S.; Chakrabarti, S.; Davidge, S.T.; Wu, J. Egg White-Derived Tripeptide IRW (Ile-Arg-Trp) Is an Activator of Angiotensin Converting Enzyme 2. *J. Agric. Food Chem.* **2018**, *66*, 11330–11336. [CrossRef]
29. Chen, X.-Q.; Zhao, J.; Chen, L.-X.; Wang, S.-F.; Wang, Y.; Li, S.-P. Lanostane triterpenes from the mushroom *Ganoderma resinaceum* and their inhibitory activities against α-glucosidase. *Phytochemistry* **2018**, *149*, 103–115. [CrossRef]
30. Peng, X.-R.; Wang, Q.; Su, H.-G.; Zhou, L.; Xiong, W.-Y.; Qiu, M.-H. Anti-Adipogenic Lanostane-Type Triterpenoids from the Edible and Medicinal Mushroom *Ganoderma applanatum*. *J. Fungi* **2022**, *8*, 331. [CrossRef]
31. Pu, D.; Li, X.; Lin, J.; Zhang, R.; Luo, T.; Wang, Y.; Gao, J.; Zeb, M.A.; Zhang, X.; Li, X.; et al. Triterpenoids from *Ganoderma gibbosum*: A Class of Sensitizers of FLC-Resistant Candida albicans to Fluconazole. *J. Nat. Prod.* **2019**, *82*, 2067–2077. [CrossRef] [PubMed]
32. Shim, S.H.; Ryu, J.; Kim, J.S.; Kang, S.S.; Xu, Y.; Jung, S.H.; Lee, Y.S.; Lee, S.; Shin, K.H. New Lanostane-Type Triterpenoids from *Ganoderma applanatum*. *J. Nat. Prod.* **2004**, *67*, 1110–1113. [CrossRef] [PubMed]
33. Su, H.-G.; Peng, X.-R.; Shi, Q.-Q.; Huang, Y.-J.; Zhou, L.; Qiu, M.-H. Lanostane triterpenoids with anti-inflammatory activities from *Ganoderma lucidum*. *Phytochemistry* **2020**, *173*, 112256. [CrossRef] [PubMed]

Journal of *Fungi*

Article

Four New Highly Oxygenated Eremophilane Sesquiterpenes from an Endophytic Fungus *Boeremia exigua* Isolated from *Fritillaria hupehensis*

Hong-Lian Ai †, Xiao Lv †, Ke Ye, Meng-Xi Wang, Rong Huang, Bao-Bao Shi * and Zheng-Hui Li *

School of Pharmaceutical Sciences, South-Central MinZu University, Wuhan 430074, China; aihonglian05@163.com (H.-L.A.); 2019110403@mail.scuec.edu.cn (X.L.); 2019110410@mail.scuec.edu.cn (K.Y.); 2021110499@mail.scuec.edu.cn (M.-X.W.); konglingniao1988@163.com (R.H.)
* Correspondence: 2021068@mail.scuec.edu.cn (B.-B.S.); lizhenghui@mail.scuec.edu.cn (Z.-H.L.)
† These authors contributed equally to this work.

Abstract: Four new eremophilane-type sesquiterpenes, boeremialanes A–D (**1–4**) were obtained from solid substrate cultures of *Boeremia exigua* (Didymellaceae), an endophytic fungus isolated from *Fritillaria hupehensis* (Liliaceae). Boeremialanes A–C (**1–3**) are highly oxygenated eremophilanes with a benzoate unit attached at the C-13 position and are rarely found in nature. Their structures and absolute configurations were determined by extensive spectroscopic methods, electronic circular dichroism (ECD), and nuclear magnetic resonance (NMR) calculations with DP4+ analysis. Boeremialane D (**4**) potently inhibited nitric oxide production in lipopolysaccharide-treated RAW264.7 macrophages with an IC_{50} of 8.62 μM and was more potent than the positive control, pyrrolidinedithiocarbamate (IC_{50} = 23.1 μM).

Keywords: *Boeremia exigua*; *Fritillaria hupehensis*; eremophilanes; boeremialanes; anti-inflammatory; NO production inhibition

Citation: Ai, H.-L.; Lv, X.; Ye, K.; Wang, M.-X.; Huang, R.; Shi, B.-B.; Li, Z.-H. Four New Highly Oxygenated Eremophilane Sesquiterpenes from an Endophytic Fungus *Boeremia exigua* Isolated from *Fritillaria hupehensis*. *J. Fungi* **2022**, *8*, 492. https://doi.org/10.3390/jof8050492

Academic Editors: Frank Surup

Received: 18 April 2022
Accepted: 5 May 2022
Published: 8 May 2022

1. Introduction

Eremophilane-type derivatives are structurally irregular and bicyclic natural products belonging to a small sesquiterpene family [1,2]. These eremophilane sesquiterpenes are biogenetically derived from farnesyl diphosphate in association with a methyl migration [3] and consist of three isoprene subunits [4]. The structural diversity of eremophilane analogs is due to oxidation occurring at different sites along the isopropyl side chain and bicyclic backbone to generate alcohol [5], acid [6], ester [7–9], furan [10,11], and lactone functionalities, with some of the alcohols further glycosylated [12]. Since the first eremophilane-type sesquiterpene was isolated from the wood oil of *Eremophila mitchellii* in 1932 [13], more than 650 biologically active eremophilane derivatives have been obtained [2,14]. In addition to the related analogs obtained from terrestrial plants [15,16] and marine fungi [17,18], plant endophytic fungi are recognized as a new source of derivatives eremophilane [19,20]. Due to their special structural features and various functional groups, eremophilane-type sesquiterpenes possess a lot of biological activities such as anti-inflammatory [21], antitumor [10], and antibacterial [22,23] activities, which have received increasing interest in the recent years. As part of our ongoing efforts to discover bioactive terpenoids derived from endophytic fungi [24–27], a chemical investigation on the cultural broth of *B. exigua* in rice medium was carried out. As a result, four new highly oxygenated eremophilane-type sesquiterpenes, boeremialanes A–D (**1–4**), were isolated from cultures of the fungus *B. exigua*. The new structures were established by extensive spectroscopic methods, ECD and NMR calculations, as well as DP4+ analysis. All compounds were tested for their anti-inflammatory activities on nitric oxide production in LPS-induced RAW264.7 macrophages. Herein, details of the isolation, structural elucidation and bioactivities of the compounds are reported.

2. Materials and Methods

2.1. General Experimental Procedures

Optical rotations were measured with an Autopol IV polarimeter (Rudolph, Hackettstown, NJ, USA). UV spectra were measured on a UV-2450 spectrometer (Hitachi High-Technologies, Tokyo, Japan). CD spectra were recorded with an Applied Photophysics spectrometer (Chirascan, New Haven, CT, USA). One-dimensional and 2D spectra were recorded on a Bruker AV-600 spectrometer (Bruker, Karlsruhe, Germany) with TMS as an internal standard. HRESIMS spectra were recorded on Q Exactive Obitrap mass spectrometer (ThermoFisher Scientific, Waltham, MA, USA). Medium pressure liquid chromatography (MPLC) was performed on a Biotage SP1 System and column packed with RP-18 gel (Biotage, Uppsala, Sweden). Silica gel (Qingdao Marine Chemical Factory, Qingdao, China), RP-18 gel (Fuji Silysia Chemical Factory, Kasugai, Japan), and Sephadex LH-20 (Pharmacia Fine Chemical Factory, Uppsala, Sweden) were used for column chromatography (CC). Semi-preparative HPLC experiments were carried on Agilent 1260 HPLC with Zorbax SB-C_{18} column (Agilent, Palo Alto, CA, USA, 5 μm, 9.4 mm × 150 mm). Fractions were monitored by TLC (GF 254, Qingdao Haiyang Chemical Factory, Qingdao, China), and spots were visualized by heating silica gel plates sprayed with vanillin and 10% H_2SO_4 in EtOH.

2.2. Culture and Fermentation of Fungal Material

The strain *B. exigua* was isolated from the healthy leaf tissue of *Fritillaria hupehensis* Hsiao. It was identified by Dr. Hong-Lian Ai (South-Central MinZu University). The ITS sequence of this strain is almost identical to the strain deposited in Genbank with accession number MT154621.1 (max identity: 100%, query cover: 100%). The fungal specimen is deposited at South-Central MinZu University, China. The strain was cultured on PDA medium for 8 days, and then was cut into small pieces to incubate solid rice medium to culture for further 30 days at 25 °C (50 g rice, 50 mL water, in each 500 mL Erlenmeyer flask, the total weight of rice was 17 kg).

2.3. Extraction and Isolation

The rice fermentation product of *B. exigua* (17 kg) was extracted five times with methanol to yield a crude extract after evaporation under vacuum. The crude extract was partitioned between water and EtOAc to give an EtOAc layer. The extract (800 g) of the organic layer was subjected to column chromatography over silica gel (200–300 mesh, CH_2Cl_2-MeOH, step gradient elution 1:0, 20:1, 10:1, 5:1, 2:1, 1:1, and 0:1) to obtain six fractions (A–F). Fr. C (35.8 g) was fractionated by MPLC over an RP-18 silica gel column and eluted with MeOH-H_2O (from 20:80 to 90:10, v/v) to yield five subfractions (C_1–C_5). Fraction C_1 (5.8 g) was separated on a silica gel column (200–300 mesh, CH_2Cl_2-MeOH, step gradient elution 10:1, 4:1, 2:1, and 1:1) to give four subfractions (Fr. $C_{1\text{-}1}$–Fr. $C_{1\text{-}4}$). Then, Fr. $C_{1\text{-}1}$ was purified by semi-preparative HPLC (CH_3CN/H_2O from 25:75 to 35:65 over 30 min) to obtain compound **4** (2.0 mg, retention time (t_R) = 12.0 min). Fr. $C_{1\text{-}2}$ was purified by semi-preparative HPLC (CH_3CN/H_2O from 28:72 to 36:64 over 28 min) to obtain compound **1** (5.1 mg, t_R = 15.6 min), compound **2** (7.2 mg, t_R = 17.8 min), and compound **3** (8.1 mg, t_R = 19.7 min).

Boeremialane A (**1**): Yellowish oil; $[\alpha]_D^{27}$ 102.5 (*c* 0.1, MeOH); UV (MeOH) λ_{max} (log ε) 205 (3.61), 230 (3.46) nm; ^{1}H NMR (600 MHz) and ^{13}C NMR (150 MHz, methanol-d_4), see Table 1; HRESIMS (positive) *m/z* 483.16220 [M + Na]$^+$ (calcd for $C_{24}H_{28}O_9Na^+$, 483.16255).

Boeremialane B (**2**): Yellowish oil; $[\alpha]_D^{27}$ 50.0 (*c* 0.1, MeOH); UV (MeOH) λ_{max} (log ε) 210 (3.44), 235 (3.16) nm; ^{1}H NMR (600 MHz) and ^{13}C NMR (150 MHz, methanol-d_4), see Table 1; HRESIMS (positive) *m/z* 441.15195 [M + Na]$^+$ (calcd for $C_{22}H_{26}O_8Na^+$, 441.15199).

Boeremialane C (**3**): Yellowish oil; $[\alpha]_D^{27}$ 216.0 (*c* 0.1, MeOH); UV (MeOH) λ_{max} (log ε) 205 (3.75), 250 (3.80) nm; ^{1}H NMR (600 MHz) and ^{13}C NMR (150 MHz, methanol-d_4), see Table 1; HRESIMS (positive) *m/z* 441.15182 [M + Na]$^+$ (calcd for $C_{22}H_{26}O_8Na^+$, 441.15199).

Boeremialane D (**4**): Yellow amorphous powder; $[\alpha]_D^{27}$ 136.0 (*c* 0.1, MeOH); UV (MeOH) λ_{max} (log ε) 240 (3.47) nm; ^{1}H NMR (600 MHz) and ^{13}C NMR (150 MHz, methanol-d_4), see Table 1; HRESIMS (positive) *m/z* 345.13064 [M + Na]$^+$ (calcd for $C_{17}H_{22}O_6Na^+$, 345.13086).

Table 1. ^{1}H and ^{13}C NMR Spectroscopic Data for **1** and **2** in Methanol-d_4 (δ in ppm, *J* in Hz).

No.	δ_H (1) [a]	δ_C (1) [b]	δ_H (2) [a]	δ_C (2) [b]
1	2.41 (tdd, 14.4, 5.0, 1.9) 2.24 (dt, 14.4, 4.1)	31.6, CH_2	2.51 (tdd, 14.4, 5.0, 1.9) 2.28 (dt, 14.4, 3.5)	31.6, CH_2
2	2.03 (dd, 12.3, 4.4) 1.27 (ddd, 12.3, 5.0, 4.1)	36.2, CH_2	2.07 (dd, 12.2, 4.4) 1.29 (ddd, 12.2, 5.0, 3.5)	36.3, CH_2
3	3.44 (td, 10.5, 4.4)	71.1, CH	3.53 (td, 10.5, 4.4)	71.1, CH
4	1.63 (dq, 10.5, 6.7)	45.9, CH	1.70 (dq, 10.5, 6.7)	46.0, CH
5		42.2, C		42.3, C
6	3.80 (s)	63.9, CH	3.91 (s)	64.1, CH
7		62.9, C		62.8, C
8		195.5, C		195.4, C
9	5.61 (d, 1.9)	121.8, CH	5.66 (d, 1.9)	121.8, CH
10		166.5, C		166.5, C
11		73.9, C		74.2, C
12	4.19 (d, 11.6) 3.76 (d, 11.6)	65.6, CH_2	4.17 (d, 11.6) 3.81 (d, 11.6)	65.5, CH_2
13	4.64 (d, 11.5) 4.59 (d, 11.5)	69.3, CH_2	4.83 (d, 11.7) 4.44 (d, 11.7)	67.5, CH_2
14	0.64 (s)	18.3, CH_3	1.03 (s)	19.0, CH_3
15	1.16 (d, 6.7)	11.6, CH_3	1.23 (d, 6.7)	11.6, CH_3
1′		169.6, C		168.1, C
2′		134.3, C		132.4, C
3′		132.8, C	7.37 (dd, 2.6, 1.3)	117.3, CH
4′	7.77 (dd, 6.4, 2.1)	130.3, CH		158.8, C
5′	7.62 (td, 6.4, 2.7)	129.7, CH	7.00 (dd, 7.9, 2.6)	121.3, CH
6′	7.61 (td, 6.4, 2.1)	132.9, CH	7.25 (t, 7.9)	130.5, CH
7′	7.59 (dd, 6.4, 2.7)	132.3, CH	7.44 (dd, 7.9, 1.3)	121.8, CH
COO<u>C</u>H$_3$	3.85 (s)	53.5, CH_3		
<u>C</u>OOCH$_3$		168.9, C		

[a] Recorded at 600 MHz, [b] Recorded at 150 MHz.

2.4. Quantum Chemical Calculations

The initial conformational analysis of compounds **1–4** was performed using the Monte Carlo search algorithm via the MMFF94 molecular mechanics force field [28], with the aid of the Spartan 16 program package that resulted in some relatively favorable conformations with an energy range of 3 kcal/mol above the global minimum. The minimum energy conformers of the resulting force field were optimized in vacuum with the M06-2X/def2-SVP level, and implemented in the Gaussian 09 software package by the Density functional theory [29]. At the same time, harmonic vibrational frequencies were also measured to confirm the lack of imaginary frequencies of the finally optimized conformers. These primary conformations were subjected to theoretical calculations of ECD utilizing time-dependent density functional theory (TDDFT) calculations at the M06-2X/def2-SVP level in MeOH using the polarizable continuum model (PCM) solvent model. The energies, oscillator strengths, and rotational strengths of each conformation were determined with the Gaussian 09 software package. Theoretical calculations of ECD spectra for each part were then approximated by the Gaussian distribution. The final ECD spectrum of the individual conformers was summed up on the basis of the Boltzmann-weighed population contribution by the SpecDisv1.71 [30]. DFT GIAO ^{13}C NMR calculations were performed on the mPW1PW91/6-31 + G(d,p)//M06-2X/def2-SVP level of theory [31]. The solvent effect was accounted for by using methanol in the calculations to mimic the experimental conditions. The ^{13}C NMR chemical shifts in compound **1** were considered the average values of the same atoms in the different conformers. We took the relative Gibbs free energy as the weighting factor and used the Boltzmann distribution to find the average values. The overall theoretical NMR data were analyzed using DP4+ probability [32].

2.5. Nitric Oxide Production Inhibitory Assay

The anti-inflammatory effect of Raw264.7 macrophages was studied and cultured in Dulbecco's modified eagle medium (DMEM, HyClone, Logan, UT, USA) with 10% fetal bovine serum (FBS, PAN, Aidenbach, Germany) in a humidified incubator (5% CO_2, 37 °C). RAW264.7 cells (5×10^4 cells/well) were seeded into a 96-well multiplate for 12 h. After 12 h of incubation, the cells were treated with LPS (1 μg/mL) and different concentrations of the tested compounds (**1–4**, 20 μM) for 18 h. A Griess reagent kit (Promega, Madison, WI, USA) was used to measure the amount of nitrite, a stable metabolite of Nitric Oxide (NO), in the supernatants. Briefly, 50 μL of each culture medium was added to a 96-well plate, and then the same volume of sulfanilamide solution was added. After incubation at room temperature for 5 min, 50 μL of N-1-naphthylethylenediamine dihydrochloride solution was added to all wells. The absorption at 540 nm was measured by a microplate reader after 10 min incubation at room temperature [33]. The IC_{50} values were calculated by GraphPad Prism 6 software. Cell viability was determined with the MTT (3-[4,5-dimethylthiazol-2-yl]-2,5 diphenyl tetrazolium bromide) assay. Pyrrolidine dithiocarbamate (PDTC, Sigma–Aldrich, St Louis, MO, USA) was used as a positive control.

3. Results and Discussion

Boeremialane A (**1**) was obtained as a yellowish oil, and the molecular formula of compound **1** was determined to be $C_{24}H_{28}O_9$ from the HRESI mass spectrum ($[M + Na]^+$ data, found 483.16220, calcd. 483.16255). The ^1H and ^{13}C NMR data of compound **1** indicated the presence of two methyl groups (δ_C 11.6 and 18.3), four methylene groups (δ_C 31.6, 36.2, 65.6, and 69.3), eight methine groups (δ_C 71.1, 45.9, 63.9, 121.8, 130.3, 129.7, 132.9, and 132.3), one carbonyl (δ_C 195.5), two ester carbonyls (δ_C 169.6 and 168.9), three sp3 quaternary carbons (δ_C 42.2, 62.9, and 73.9), and three sp2 quaternary carbons (δ_C 166.5, 134.3, and 132.8) (Table 1, Figures S5–S10). In the HMBC spectrum (Figures 1 and 2), a singlet for the Me-14 at δ_H 0.64 (3H, s, H-14) showed correlations to C-4 (δ_C 45.9), C-6 (δ_C 63.9), C-10 (δ_C 166.5), and a sp^3 quaternary carbon at δ_C 42.2 (C-5). This was very important for the establishment of the three C-C bonds of C-4, C-10, and C-6 with C-5. In addition, the HMBC spectrum showed correlations from H-1 (δ_H 2.41 and 2.24) to C-5 and C-10, and the ^1H-^1H COSY spectrum analysis (H-1/H-2/H-3/H-4/H-15) together with a characteristic oxygenated methine carbon (δ_C 71.1, C-3) determined a 1,2,3,3,4-pentasubstituted cyclohexane ring of compound **1**. A 2-cyclohexen-1-one ring was inferred by the HMBC correlations from H-6 (δ_H 3.80) to C-5, C-7, C-8, and C-10 and from H-9 (δ_H 5.61) to C-5 and C-7, with the connection to the cyclohexane ring by the C-5/C-10 position on the basis of the HMBC correlations of H-1/C-9 and H-4/C-6 (Figure 2). The HMBC correlations from H-12 (δ_H 4.19 and 3.76) to C-11 and C-13 and from H-13 (δ_H 4.64 and 4.59) to C-11 and C-12 together with the downfield shifts of C-11 (δ_C 73.9), C-12 (δ_C 65.6), and C-13 (δ_C 69.3) indicated the existence of a highly oxidized propane group, which was linked to the position of C-7, as evidenced by the HMBC correlations from H-12 to C-7 and from H-13 to C-7. These data, as well as other HMBC correlations, suggested that unit A was a tetrol phaseolinone [34], which had been previously isolated from *Macrophomina phaseolina*.

For unit B, the ^1H NMR spectrum of compound **1** revealed the signals for four aromatic protons (δ_H 7.77, 7.62, 7.61, and 7.59). In the ^1H-^1H COSY spectrum, a disubstituted benzene ring was identified by four continuous aromatic protons at δ_H 7.77 (1H, d, H-4′), 7.62 (1H, t, H-5′), 7.61 (1H, t, H-6′), and 7.59 (1H, d, H-7′), and two aromatic doublets and two aromatic triplets with the same coupling constant (J = 6.4 Hz) indicated an ortho-disubstituted benzene group. A carbomethoxy substituent in the benzene ring was identified by the HMBC correlations from H-4′ to the carbomethoxy substituent (δ_C 168.9). Similarly, an ester carbonyl carbon (δ_C 169.6) was positioned at C-2′ based on observed cross-peaks at H-7′/C-1′. The HMBC correlations from H-13 to the ester carbonyl carbon

(C-1′) confirmed the 13,1′-ester linkage of the two substructures. Thus, the planar structure of compound **1** was elucidated as shown in Figure 1.

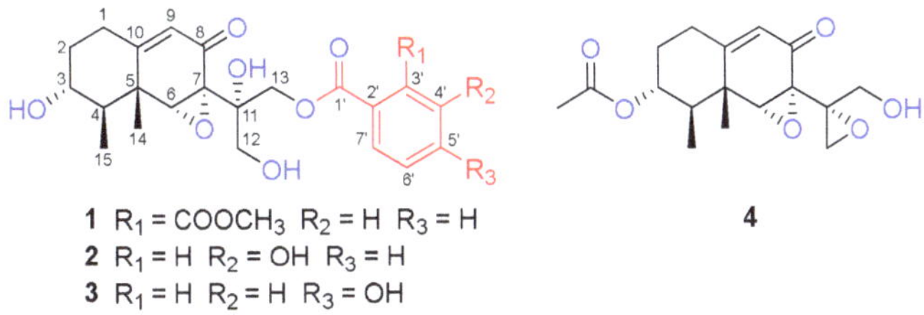

Figure 1. Chemical structures of compounds **1–4**.

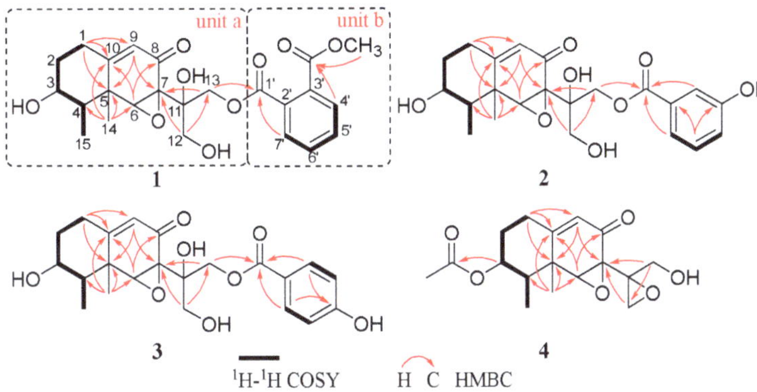

Figure 2. Key HMBC and ^{1}H-^{1}H COSY correlations of compounds **1–4**.

The configuration of boeremialane A (**1**) was established by ROESY experiments and quantum chemistry calculations. The ROESY correlations of H-3/H_3-15, H-3/H_3-14, H-6/H_3-14, and H-6/H_3-15 suggested that they were β-oriented (Figure 3). In addition, to determine the configuration of C-11 in the flexible bond, nuclear magnetic resonance (NMR) calculations of two epimers, 11*S*-**1** and 11*R*-**1**, were carried out. The two epimers were subjected to a strict conformational screening procedure; then, the NMR chemical shifts were calculated at the mPW1PW91/6-31 + G(d,p)//M06-2X/def2-SVP level of theory with the PCM solvent in methanol. The DP4+ analysis identified 11*S*-**1** as the most likely structure of compound **1** with 100.00% DP4+ probability (all data) (Figure 4 and Table S1). Finally, the absolute configuration of compound **1** was resolved by comparing the calculated and experimental ECD data using time-dependent density-functional theory (TDDFT). The theoretical spectrum of compound **1** showed an excellent fit with the experimental plot recorded in MeOH (Figures 5 and S1), which supported an absolute configuration of 3*R*, 4*R*, 5*R*, 6*R*, 7*S*, and 11*S*. Thus, the structure of compound **1** was determined, and it was named boeremialane A.

Boeremialane B (**2**) was obtained as a yellowish oil, and the molecular formula was determined to be $C_{22}H_{26}O_8$ from the HRESI mass spectrum data ($[M + Na]^+$, found 441.15195, calcd. 441.15199). The ^{1}H and ^{13}C NMR data of compound **2** indicated the presence of two methyl groups (δ_C 11.6 and 19.0), four methylene groups (δ_C 31.6, 36.3, 65.6, and 67.5), eight methine groups (δ_C 71.1, 46.0, 64.1, 121.8, 117.3, 121.3, 130.5, and 121.8), one carbonyl (δ_C 195.4), one ester carbonyl (δ_C 168.1), three sp3 quaternary carbons (δ_C 42.3, 62.8, and 74.2), and three sp2 quaternary carbons (δ_C 166.5, 132.4, and 158.8) (Table 1 and Figures S12–S17). The ^{1}H and ^{13}C NMR data of compound **2** were structurally similar to those of compound **1**, except for the absence of a carbomethoxy group at δ_C 168.9 and 53.5 in compound **1** and the presence of an additional hydroxy group in compound **2**. The hydroxyl group at C-4′ was evident from the downfield shift of C-4′ (δ_C 158.8) as well as the HMBC correlations from H-13 to the sp2 quaternary carbon (C-4′) (Figure 2). The relative configuration of compound **2** was the same as that found in compound **1** based on the ROESY correlations of H-3/H_3-15, H-3/H_3-14, H-6/H_3-14, and H-6/H_3-15 (Figure 3). Finally, the absolute configuration of **2** was determined by ECD calculations on the M06-2X/def2-SVP (IEFPCM, MeOH) level of theory. The experimental ECD spectrum of compound **2** fits well with the calculated spectrum of 3*R*, 4*R*, 5*R*, 6*R*, 7*S*, and 11*S*-**2** (Figures 5 and S2). Therefore, the structure of compound **2** was determined, and it was given the name boeremialane B.

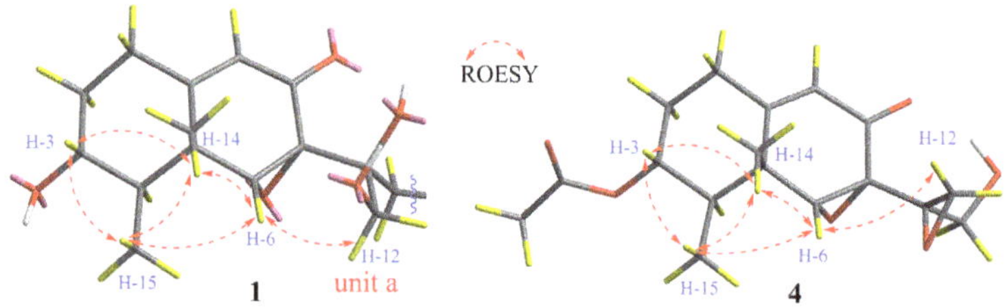

Figure 3. Key ROESY correlations of compounds **1** and **4**.

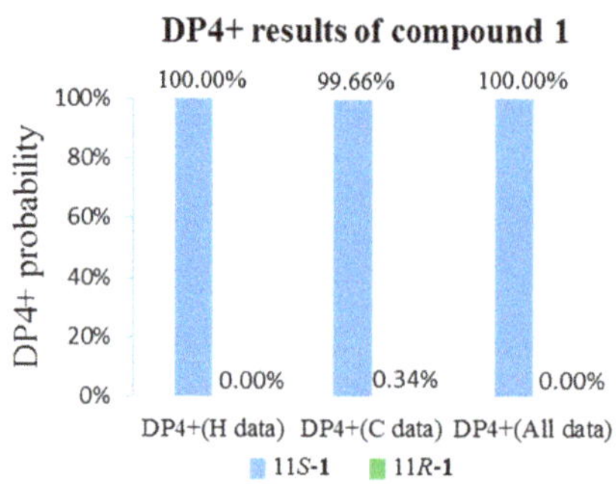

Figure 4. qccNMR coupled with DP4+ probability analysis of compound **1**.

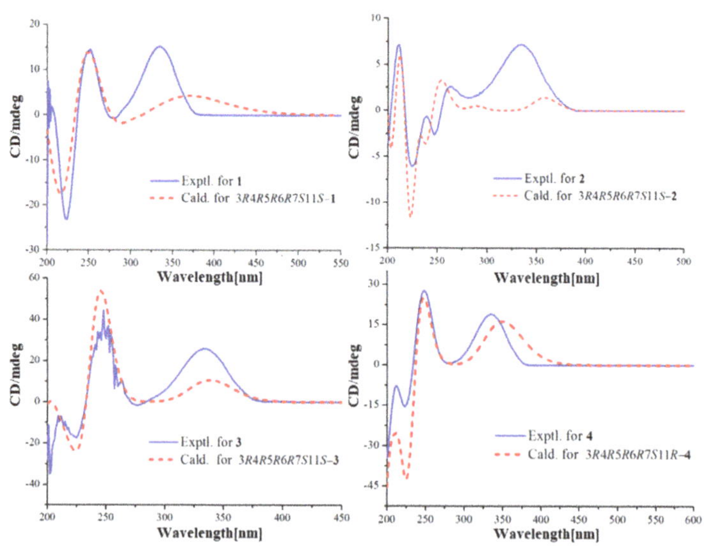

Figure 5. Experimental and calculated ECD spectra of compounds **1–4** at the M06-2X/def2-SVP level in methanol.

Boeremialane C (**3**) has a molecular formula of $C_{35}H_{40}O_8$ according to its HRESIMS ion at *m/z* 441.15182 $[M + Na]^+$ (calcd for $C_{22}H_{26}O_8Na$, 441.15199). The ^1H and ^{13}C NMR data of **3** (Table 2 and Figures S19–S24) were structurally similar to those of compound **2**, except for the presence of a para-substituted benzene ring of the benzoate unit. This difference was supported by the HMBC correlations from H-3′ (7′) (δ_H 7.81) to C-1′ (δ_C 168.4) and C-5′ (δ_C 165.9) along with the COSY correlations between H-3′ (7′)/H-4′ (6′) (δ_H 6.74) (Figure 2). The ECD spectrum of compound **3** was similar to that of compound **1** with negative exciton coupling at 211 nm and positive exciton coupling at 241 nm (Figure S33), which indicated that they share the identical absolute configuration. Therefore, the absolute configuration of **3** was defined as 3*R*, 4*R*, 5*R*, 6*R*, 7*S*, and 11*S*. This presumption was confirmed by comparative analysis of calculated and experimental ECD spectra. The experimental ECD spectrum of **3** fits well with the calculated spectrum of 3*R*, 4*R*, 5*R*, 6*R*, 7*S*, and 11*S*-**3** (Figures 5 and S3). Thus, the structure of **3** was determined and named boeremialane C.

Table 2. ^1H and ^{13}C NMR Spectroscopic Data for **3** and **4** in Methanol-d_4 (δ in ppm, *J* in Hz).

No.	δ_H (3) [a]	δ_C (3) [b]	δ_H (4) [a]	δ_C (4) [b]
1	2.50 (tdd, 14.4, 4.8, 1.8) 2.28 (dt, 14.4, 3.5)	31.6, CH_2	2.41 (tdd, 14.6, 5.0, 1.8) 2.39 (dt, 14.6, 4.0)	31.2, CH_2
2	2.07 (dd, 12.5, 4.4) 1.29 ddd, 12.5, 4.8,3.5	36.3, CH_2	2.15 (dd, 12.3, 4.4) 1.40 (ddd,12.3, 5.0, 4.0)	32.5, CH_2
3	3.53 (td, 10.6, 4.4)	71.1, CH	4.91 (td, 10.5, 4.4)	74.2, CH
4	1.69 (dq, 10.6, 6.8)	45.9, CH	1.95 (dq, 10.5, 6.8)	43.1, CH
5		42.2, C		42.4, C
6	3.89 (s)	64.1, CH	3.63 (s)	65.5, CH
7		62.8, C		62.1, C
8		195.3, C		194.2, C
9	5.65 (d, 1.8)	121.8, CH	5.75 (d, 1.8)	121.7, CH
10		166.4, C		165.4, C
11		74.3, C		59.0, C
12	4.17 (d, 11.6) 3.81 (d, 11.6)	65.5, CH_2	2.87 (d, 5.1) 2.66 (d, 5.1)	48.3, CH_2
13	4.82 (d, 11.7) 4.39 (d, 11.7)	67.2, CH_2	4.07 (d, 12.3) 3.72 (d, 12.3)	62.0, CH_2
14	1.02 (s)	19.1, CH_3	1.26 (s)	18.6, CH_3
15	1.22 (d, 6.8)	11.6, CH_3	1.14 (d, 6.8)	11.4, CH_3
1′		168.4, C		
2′		120.5, C		
3′	7.81 (d, 8.8)	133.1, CH		
4′	6.74 (d, 8.8)	116.8, CH		
5′		165.9, C		
6′	6.74 (d, 8.8)	116.8, CH		
7′	7.81 (d, 8.8)	133.1, CH		
$\underline{C}H_3CO$			2.06 (s)	21.0, CH_3
$CH_3\underline{C}O$				172.4, C

[a] Recorded at 600 MHz, [b] Recorded at 150 MHz.

Boeremialane D (**4**) was obtained as a yellow amorphous powder, and the molecular formula, $C_{17}H_{22}O_6$, was determined by (+)-HRESIMS, which showed an $[M + Na]^+$ ion at *m/z* 345.13064 (calcd for $C_{17}H_{22}O_6Na$: 345.13086). The ^1H and ^{13}C NMR data of compound **4** indicated the presence of three methyl groups (δ_C 18.6, 11.4, and 21.0), four methylene groups (δ_C 31.2, 32.5, 48.3, and 62.0), four methine groups (δ_C 74.2, 43.1, 65.5, and 121.7), one carbonyl (δ_C 194.2), one ester carbonyl (δ_C 172.4), three sp3 quaternary carbons (δ_C 42.4, 62.1, and 59.0), and one sp2 quaternary carbon (δ_C 165.4) (Table 2 and Figures S26–S31). The ^1H and ^{13}C NMR data of compound **4** were structurally similar to those of phaseolinone [35], except for the appearance of an additional acetyl group in compound **4**. The attachment of this acetyl group at C-3 was supported by the HMBC correlation from the H-3 to the ester carbonyl carbons (δ_C 172.4). The relative configuration of compound **4** was the same as that found in compound **1** based on the ROESY correlations of H-3/H_3-15, H-3/H_3-14, H-6/H_3-14, and H-6/H_3-15 (Figure 3). Similar to compound **3**, the tendencies of the ECD curves of compounds **4** and **1** with negative exciton coupling at 225 nm and positive exciton coupling at 250 and 337 nm were relatively consistent (Figure S33, Supporting Information), which indicated that they have an the identical absolute configuration. In addition, the identity of the measured ECD and calculated ECD spectrum of compound **4** further confirmed this conclusion (Figures 5 and S4). Therefore, the structure of compound **4** was determined, and it was given the name boeremialane D.

All compounds were evaluated for their inhibition of NO production in LPS-treated RAW264.7 macrophages. As a result, compound **4** showed certain inhibitory activity with IC_{50} values of 8.62 μM, which was more potent than the positive control, pyrrolidinedithiocarbamate (IC_{50} = 23.1 μM) (Figure 6).

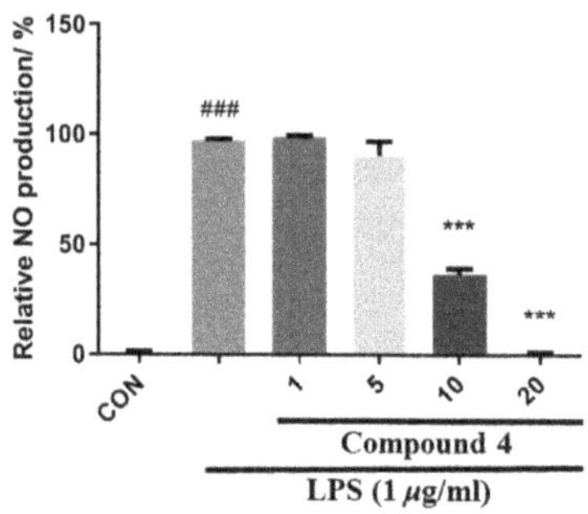

Figure 6. Effects of compound **4** isolated from *B. exigua* on NO production in LPS-stimulated RAW 264.7 macrophages. Cells were pretreated with the indicated concentrations of the isolates for 1 h and then stimulated with LPS (1 μg/mL) for 24 h. The NO levels in the culture medium were measured by the MTT assay. ### $p < 0.0001$ vs. control. *** $p < 0.0001$ vs. LPS-stimulated group.

4. Conclusions

In summary, the structures of four new eremophilane-type sesquiterpenes (**1**–**4**) were unambiguously determined by analyses of their HRESI and NMR spectroscopic data, with the absolute configuration being determined by quantum chemistry calculations. Boeremialanes A–C (**1**–**3**) are highly oxygenated eremophilanes with the benzoate unit attached at the C-13 position, and only one such natural compound has been discovered to date [35]. Compound **4** exhibited potent inhibition against NO production in LPS-activated RAW 264.7 macrophages, suggesting that it is a new chemical entity for anti-inflammatory effects. The present research provides new insights into understanding the structural diversity and interesting biological activities of eremophilane sesquiterpenes.

Supplementary Materials: The following are available online at https://www.mdpi.com/article/10.3390/jof8050492/s1. Table S1. DP4+ analysis results of **1a** (Isomer 1) and **1b** (Isomer 2). Table S2. Experimental and calculated ^{13}C NMR chemical shifts of **1a** and **1b**. Table S3. Important thermodynamic parameters of the M06-2X/def2-SVP optimized conformers of **1a** in the gas phase. Table S4. Conformational analysis of the M06-2X/def2-SVP optimized conformers of **1a** in the gas phase (T = 298.15 K). Table S5. Important thermodynamic parameters of the M06-2X/def2-SVP optimized conformers of **2** in the gas phase. Table S6. Conformational analysis of the M06-2X/def2-SVP optimized conformers of **2** in the gas phase (T = 298.15 K). Table S7. Important thermodynamic parameters of the M06-2X/def2-SVP optimized conformers of **3** in the gas phase. Table S8. Conformational analysis of the M06-2X/def2-SVP optimized conformers of **3** in the gas phase (T = 298.15 K). Table S9. Important thermodynamic parameters of the M06-2X/def2-SVP optimized conformers of **4** in the gas phase. Table S10. Conformational analysis of the M06-2X/def2-SVP optimized conformers of **4** in the gas phase (T = 298.15 K). Table S11. Cartesian coordinates for the low-energy optimized conformers of **1**–**4** at M06-2X/def2-SVP level. Figure S1. Experimental ECD spectra and calculated ECD spectra of **1**. Figure S2. Experimental ECD spectra and calculated ECD spectra of **2**. Figure S3. Experimental ECD spectra and calculated ECD spectra of **3**. Figure S4. Experimental ECD spectra and calculated

ECD spectra of **4**. Figure S5. ^{1}H NMR spectrum of **1** in CD_3OD. Figure S6. ^{13}C NMR spectrum of **1** in CD_3OD. Figure S7. HSQC spectrum of **1** in CD_3OD. Figure S8. HMBC spectrum of **1** in CD_3OD. Figure S9. COSY spectrum of **1** in CD_3OD. Figure S10. ROESY spectrum of **1** in CD_3OD. Figure S11. HRMS spectrum of **1.** Figure S12. ^{1}H NMR spectrum of **2** in CD_3OD. Figure S13. ^{13}C NMR spectrum of **2** in CD_3OD. Figure S14. HSQC spectrum of **2** in CD_3OD. Figure S15. HMBC spectrum of **2** in CD_3OD. Figure S16. COSY spectrum of **2** in CD_3OD. Figure S17. Roesy spectrum of **2** in CD_3OD. Figure S18. HRMS spectrum of **2.** Figure S19. ^{1}H NMR spectrum of **3** in CD_3OD. Figure S20. ^{13}C NMR spectrum of **3** in CD_3OD. Figure S21. HSQC spectrum of **3** in CD_3OD. Figure S22. HMBC spectrum of **3** in CD_3OD. Figure S23. COSY spectrum of **3** in CD_3OD. Figure S24. Roesy spectrum of **3** in CD_3OD. Figure S25. HRMS spectrum of **3**. Figure S26. ^{1}H NMR spectrum of **4** in CD_3OD. Figure S27. ^{13}C NMR spectrum of **4** in CD_3OD. Figure S28. HSQC spectrum of **4** in CD_3OD. Figure S29. HMBC spectrum of **4** in CD_3OD. Figure S30. COSY spectrum of **4** in CD_3OD. Figure S31. Roesy spectrum of **4** in CD_3OD. Figure S32. HRMS spectrum of **4**. Figure S33. CD spectrum of **1–4** in MeOH.

Author Contributions: H.-L.A. contributed to isolation and cultivation of fungi. X.L. contributed to the extraction, isolation, and identification of the samples. K.Y. contributed to the isolation of the samples. M.-X.W. and R.H. contributed to the bioactivity tests. B.-B.S. contributed to the quantum chemical calculation and the preparation of the manuscript. H.-L.A. and Z.-H.L. designed the experiments. All authors have read and agreed to the published version of the manuscript.

Funding: This work was financially supported by the National Natural Science Foundation of China (31870513) and the Fundamental Research Funds for the Central Universities, South-Central MinZu University (CZD21003).

Institutional Review Board Statement: Not applicable.

Informed Consent Statement: Not applicable.

Data Availability Statement: Not applicable.

Acknowledgments: The authors thank the Analytical & Measuring Centre, South-Central MinZu University for the spectral measurements.

Conflicts of Interest: The authors declare no conflict of interest. The funders had no role in the design of the study; in the collection, analyses, or interpretation of data; in the writing of the manuscript, or in the decision to publish the results.

References

1. Fraga, B.M. Natural sesquiterpenoids. *Nat. Prod. Rep.* **2009**, *26*, 1125–1155. [CrossRef] [PubMed]
2. Wu, L.; Liao, Z.X.; Liu, C.; Jia, H.Y.; Sun, J.Y. Eremophilane Sesquiterpenes from the Genus *Ligularia*. *Chem. Biodivers.* **2016**, *13*, 645–671. [CrossRef] [PubMed]
3. Schenk, D.J.; Starks, C.M.; Manna, K.R.; Chappell, J.; Noel, J.P.; Coates, R.M. Stereochemistry and deuterium isotope effects associated with the cyclization-rearrangements catalyzed by tobacco *epiaristolochene* and *hyoscyamus premnaspirodiene* synthases, and the chimeric CH_4 hybrid cyclase. *Arch. Biochem. Biophys.* **2006**, *448*, 31–44. [CrossRef] [PubMed]
4. Hou, C.J.; Kulka, M.; Zhang, J.Z.; Li, Y.M.; Guo, F.J. Occurrence and biological activities of eremophilane-type sesquiterpenes. *Mini Rev. Med. Chem.* **2014**, *14*, 664–677. [CrossRef] [PubMed]
5. Becker, K.; Wongkanoun, S.; Wessel, A.C.; Bills, G.F.; Stadler, M.; Luangsa-ard, J.J. Phylogenetic and chemotaxonomic studies confirm the affinities of *Stromatoneurospora phoenix* to the *coprophilous xylariaceae*. *J. Fungi* **2020**, *6*, 144. [CrossRef]
6. Li, C.S.; Ding, Y.Q.; Yang, B.J.; Hoffman, N.; Yin, H.Q.; Mahmud, T.; Turkson, J.; Cao, S.G. Eremophilane sesquiterpenes from Hawaiian endophytic fungus *Chaetoconis* sp. FT087. *Phytochemistry* **2016**, *126*, 41–46. [CrossRef]
7. Khan, B.; Zhao, S.S.; Wang, Z.Y.; Ye, Y.H.; Rajput, N.A.; Yan, W. Eremophilane sesquiterpenes and benzene derivatives from the endophyte *microdiplodia* sp. WGHS5. *Chem. Biodivers.* **2021**, *18*, e2000949. [CrossRef]
8. Cheng, Z.B.; Zhao, J.J.; Liu, D.; Proksch, P.; Zhao, Z.M.; Lin, W.H. Eremophilane-type sesquiterpenoids from an *Acremonium* sp. fungus isolated from deep-sea sediments. *J. Nat. Prod.* **2016**, *79*, 1035–1047. [CrossRef]
9. Wu, Q.X.; Shi, Y.P.; Yang, L. Unusual sesquiterpene lactones from *Ligularia virgaurea* spp. *oligocephala*. *Org. Lett.* **2004**, *6*, 2313–2316. [CrossRef]
10. Liu, J.M.; Zhang, D.W.; Zhang, M.; Zhao, J.L.; Chen, R.D.; Wang, N.; Zhang, D.; Dais, J.G. Eremophilane Sesquiterpenes from an Endophytic Fungus Periconia Species. *J. Nat. Prod.* **2016**, *79*, 2229–2235. [CrossRef]
11. Kato, T.; Hirota, H.; Kuroda, C.; Gong, X.; Ohsaki, A. New eremophilane-type sesquiterpenes from *Ligularia cymbulifera*. *Nat. Prod. Commun.* **2017**, *12*, 1165–1167. [CrossRef]

12. Silchenko, A.S.; Kalinovsky, A.I.; Ponomarenko, L.P.; Avilov, S.A.; Andryjaschenko, P.V.; Dmitrenok, P.S.; Gorovoy, P.G.; Kim, N.Y.; Stonik, V.A. Structures of eremophilane-type sesquiterpene glucosides, alticolosides A-G, from the Far Eastern endemic *Ligularia Alticola Worosch*. *Phytochemistry* **2015**, *111*, 169–176. [CrossRef] [PubMed]
13. Bradfield, A.E.; Penfold, A.R.; Simonsen, J.L. The constitution of eremophilone and of two related hydroxy-ketones from the wood oil of *Eremophila mitchelli*. *J. Chem. Soc.* **1932**, 2744–2759. [CrossRef]
14. Yuyama, K.T.; Fortkamp, D.; Abraham, W.R. Eremophilane-type sesquiterpenes from fungi and their medicinal potential. *Biol. Chem.* **2018**, *399*, 13–28. [CrossRef]
15. Zhou, M.; Duan, F.F.; Gao, Y.; Peng, X.G.; Meng, X.G.; Ruan, H.L. Eremophilane sesquiterpenoids from the whole plant of *Parasenecio albus* with immunosuppressive activity. *Bioorg. Chem.* **2021**, *115*, 105247. [CrossRef]
16. Chen, J.; Chen, C.; Yao, X.; Jin, X.; Gao, K. Eremophilane-type sesquiterpenoids with diverse skeletons from *Ligularia sagittal*. *J. Nat. Prod.* **2014**, *77*, 1329–1335. [CrossRef]
17. Liu, M.Y.; Li, P.L.; Tang, X.L.; Luo, X.C.; Liu, K.C.; Zhang, Y.; Wang, Q.; Li, G.Q. Lemnardosinanes A-I: New bioactive sesquiterpenoids from soft coral *Lemnalia* sp. *J. Org. Chem.* **2021**, *86*, 970–979. [CrossRef]
18. Wu, G.W.; Lin, A.Q.; Gu, Q.Q.; Zhu, T.J.; Li, D.H. Four new chloro-eremophilane sesquiterpenes from an antarctic deep-sea derived fungus, *penicillium* sp. PR19N-1. *Mar. Drugs.* **2013**, *4*, 1399–1408. [CrossRef]
19. Lin, L.B.; Jian, X.; Qiang, Z.; Rui, H.; Xu, B.; Yang, S.X.; Han, W.B.; Tang, J.J.; Gao, J.M. Eremophilane sesquiterpenoids with antibacterial and anti-inflammatory activities from the endophytic fungus *Septoria rudbeckiae*. *J. Agric. Food Chem.* **2021**, *69*, 11878–11889.
20. Wang, A.; Yin, R.Y.; Zhou, Z.Y.; Gu, G.; Dai, J.G.; Lai, D.W.; Zhou, L.G. Eremophilane-type sesquiterpenoids from the endophytic fungus *Rhizopycnis vagum* and their antibacterial, cytotoxic, and phytotoxic activities. *Front. Chem.* **2020**, *8*, 596889. [CrossRef]
21. Amaral, L.S.; Rodrigues, E. Two novel eremophilane sesquiterpenes from an endophytic xylariaceous fungus isolated from leaves of *Cupressus lusitanica*. *J. Braz. Chem. Soc.* **2010**, *21*, 1446–1450. [CrossRef]
22. Xu, Y.J.; Nan, Z.D.; Li, W.H.; Huang, H.L.; Yuan, C.S. New eremophilanolides from *Ligularia hodgsonii*. *Helv. Chim. Acta* **2009**, *92*, 209–216. [CrossRef]
23. Liu, Q.; Shen, L.; Wang, T.T.; Chen, C.J.; Qi, W.Y.; Gao, K. Novel modified furanoeremophilane-type sesquiterpenes and benzofuran derivatives from *Ligularia veitchiana*. *Food Chem.* **2010**, *122*, 55–59. [CrossRef]
24. Ye, K.; Lv, X.; Zhang, X.; Wei, P.P.; Li, Z.H.; Ai, H.L.; Zhao, D.K.; Liu, J.K. Immunosuppressive Isopimarane Diterpenes from Cultures of the Endophytic Fungus *Ilyonectria robusta*. *Front. Pharmacol.* **2022**, *12*, 766441. [PubMed]
25. Yang, H.X.; Wu, X.; Chi, M.J.; Li, Z.H.; Feng, T.; Ai, H.L.; Liu, J.K. Structure and cytotoxicity of trichothecenes produced by the potato-associated fungus *Trichothecium crotocinigenum*. *Bioorg. Chem.* **2021**, *111*, 104874. [CrossRef] [PubMed]
26. Zhang, X.; Yang, H.X.; Ye, K.; Wei, P.P.; Lv, X.; Fan, Y.Z.; Yang, Y.L.; Ai, H.L.; Liu, J.K. Oblongolides from endophytic fungus *Phoma bellidis* Neerg. harbored in *Tricyrtis maculata* (D. Don) J.F.Macbr. *Phytochemstry* **2022**, *198*, 113126. [CrossRef]
27. Chen, Y.; Sun, L.T.; Yang, H.X.; Li, Z.H.; Liu, J.K.; Ai, H.L.; Wang, G.K.; Feng, T. Depsidones and diaryl ethers from potato endophytic fungus *Boeremia exigua*. *Fitoterapia* **2020**, *141*, 104483.
28. Wavefunction Inc. *Spartan 14*; Wavefunction Inc.: Irvine, CA, USA, 2014.
29. Frisch, M.J.; Trucks, G.W.; Schlegel, H.B.; Scuseria, G.E.; Robb, M.A.; Cheeseman, J.R.; Scalmani, G.; Barone, V.; Mennucci, B.; Petersson, G.A.; et al. *Gaussian 09, Revision D.01*; Gaussian, Inc.: Wallingford, CT, USA, 2010.
30. Bruhn, T.; Schaumlöffel, A.; Hemberger, Y.; Bringmann, G. SpecDis: Quantifying the comparison of calculated and experimental electronic circular dichroism spectra. *Chirality* **2013**, *25*, 243–249. [CrossRef]
31. Jain, R.J.; Bally, T.; Rablen, P.R. Calculating accurate proton chemical shifts of organic molecules with density functional methods and modest basis sets. *J. Org. Chem.* **2009**, *74*, 4017–4023. [CrossRef]
32. Grimblat, N.; Zanardi, M.M.; Sarotti, A.M. Beyond DP4: An improved probability for the stereochemical assignment of isomeric compounds using quantum chemical calculations of NMR shifts. *J. Org. Chem.* **2015**, *80*, 12526–12534. [CrossRef]
33. Yu, W.W.; Ma, J.T.; He, J.; Li, Z.H.; Liu, J.K.; Feng, T. Cadinane sesquiterpenoids from the fungus *Antrodiella albocinnamomea* and their inhibitory activity against nitric oxide production. *Phytochemistry* **2022**, *196*, 113081. [CrossRef] [PubMed]
34. Bhattacharya, G.; Dhar, T.K.; Bhattacharyya, F.K.; Siddiqui, K.A. Mutagenic action of phaseolinone, a mycotoxin isolated from *Macrophomina phaseolina*. *Aust. J. Biol. Sci.* **1987**, *40*, 349–353. [CrossRef] [PubMed]
35. Dhar, T.K.; Siddiqui, K.A.I.; Ali, E. Structure of phaseolinone, a novel phytotoxin from *Macrophomina phaseolina*. *Tetrahedron Lett.* **1982**, *23*, 5459–5462.

Journal of *Fungi*

MDPI

Article

Exploring Verrucosidin Derivatives with Glucose-Uptake-Stimulatory Activity from *Penicillium cellarum* Using MS/MS-Based Molecular Networking

Junjie Han [1,†], Baosong Chen [1,†], Rui Zhang [2], Jinjin Zhang [1], Huanqin Dai [1], Tao Wang [1], Jingzu Sun [1], Guoliang Zhu [3], Wei Li [1], Erwei Li [4], Xueting Liu [3], Wenbing Yin [1] and Hongwei Liu [1,2,*]

[1] State Key Laboratory of Mycology, Institute of Microbiology, Chinese Academy of Sciences, Beijing 100101, China; hanjj@im.ac.cn (J.H.); chenbs@im.ac.cn (B.C.); zjjsmilence93@163.com (J.Z.); daihq@im.ac.cn (H.D.); wangtao@im.ac.cn (T.W.); sunjz@im.ac.cn (J.S.); liw@im.ac.cn (W.L.); yinwb@im.ac.cn (W.Y.)
[2] School of Medical Devices, Shenyang Pharmaceutical University, Shenyang 110016, China; Raynaymond@outlook.com
[3] State Key Laboratory of Bioreactor Engineering, East China University of Science and Technology, Shanghai 200237, China; zhuguoliang@ecust.edu.cn (G.Z.); liuxueting@ecust.edu.cn (X.L.)
[4] Institutional Center for Shared Technologies and Facilities, Institute of Microbiology, Chinese Academy of Sciences, Beijing 100101, China; liew@im.ac.cn
* Correspondence: liuhw@im.ac.cn; Tel.: +86-10-64806074
† These authors contributed equally to this work.

Abstract: Under the guidance of LC-MS/MS-based molecular networking, seven new verrucosidin derivatives, penicicellarusins A-G (**3–9**), were isolated together with three known analogues from the fungus *Penicillium cellarum*. The structures of the new compounds were determined by a combination of NMR, mass and electronic circular dichroism spectral data analysis. The absolute configuration of penicyrone A (**10**) was corrected based on X-ray diffraction analyses. Bioactivity screening indicated that compounds **1**, **2**, and **4** showed much stronger promising hypoglycemic activity than the positive drug (rosiglitazone) in the range of 25–100 μM, which represents a potential new class of hypoglycemic agents. Preliminary structure-activity relationship analysis indicates that the formation of epoxy ring on C_6-C_7 in the structures is important for the glucose uptake-stimulating activity. The gene cluster for the biosynthesis of **1–12** is identified by sequencing the genome of *P. cellarum* and similarity analysis with the gene cluster of verrucosidins in *P. polonicum*.

Keywords: verrucosidins; *Penicillium cellarum*; glucose uptake-stimulating activity; molecular networking

Citation: Han, J.; Chen, B.; Zhang, R.; Zhang, J.; Dai, H.; Wang, T.; Sun, J.; Zhu, G.; Li, W.; Li, E.; et al. Exploring Verrucosidin Derivatives with Glucose-Uptake-Stimulatory Activity from *Penicillium cellarum* Using MS/MS-Based Molecular Networking. *J. Fungi* **2022**, *8*, 143. https://doi.org/10.3390/jof8020143

Academic Editors: Tao Feng and Frank Surup

Received: 3 January 2022
Accepted: 29 January 2022
Published: 30 January 2022

1. Introduction

Fungi have attracted much attention of chemists and biologists due to their potential in producing bioactive secondary metabolites with diverse chemical skeletons [1,2]. Verrucosidins produced by *Penicillium* strains belong to a family of highly reducing fungal polyketides that are characterized with 2*H*-pyran-2-one and dicyclic fused 3,6-dioxabicyclo[3.1.0]hexane moieties interlinked by a polyene chain [3–6]. They have been reported to display important bioactivities, such as antitumor [7,8], antivirus [9], antibacterial [3,10], and neurological activities [11]. In order to explore in depth this kind of compounds with unique chemical structure and diverse biological activities, we explored *Penicillium* strains collected in our lab searching for verrucosidin analogues.

Molecular networking analyses include acquisition and similarity comparison of mass spectral fragment data, cluster grouping and visualization [12,13]. More recently, the MS/MS-based molecular networking has been demonstrated to be powerful in dereplicating known natural products from a targeted extract and searching for new analogues with the specific skeleton.Examples included thermoactinoamide A with moderate antiproliferative activity from *Thermoactinomyces vulgaris* DSM 43016 [14], suffranidines A-C with significant neuritogenic activity from *Flueggea suffruticosa* [15], and trilliumoside D with strong cytotoxicity against MOLT-4 cell lines from *Trillium tschonoskii* maxim [16]. To explore new reducing fungal polyketides from fungi, we applied the LC-MS/MS-based molecular networking for new verrucosidins from *Penicillium* strains using deoxyverrucosidin that was deposited in our compound library as the probing agent.

The EtOAc extracts of *Penicillium* strains fermented on solid culture were first analyzed by high performance liquid chromatography (HPLC) with UV diode array detection (DAD) to find fungi potentially producing verrucosidin derivatives (Figure S1). In this work, the target isolation was further conducted on the selected fungus *P. cellarum* YM1 under the guidance of LC-MS/MS-based molecular networking (Figure S2). As a result, seven new verrucosidins, penicicellarusins A-G (**3–9**), as well as five known verrucosidins (compounds **1**, **2** and **10–12**) were identified from the culture of *P. cellarum* YM1 (Figure 1). The isolated compounds were evaluated for anti-bacterial effect, cytotoxicity, and glucose uptake-stimulating activities. This work described the details of the isolation, structure elucidation, and biological activities of the isolated secondary metabolites from *P. cellarum* YM1.

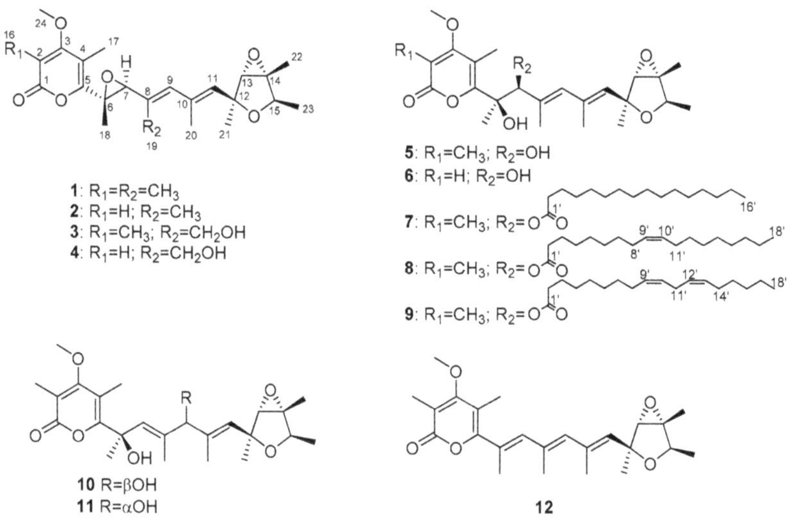

Figure 1. Structures of compounds **1–12**.

2. Materials and Methods

2.1. General

NMR spectral data were obtained with an AVANCE-500 spectrometer (Bruker, Bremen, Germany) ($CDCl_3$, δ_H 7.26/δ_C 77.16, and CD_3OD, δ_H 3.30/δ_C 49.9). High-resolution electrospray ionization mass spectrometry (HRESIMS) data and LC-MS/MS measurements were procured on a Q Exactive Orbitrap mass spectrometer (Thermo Fisher Scientific, Waltham, MA, USA) coupled with a LC-30AD series UPLC (Shimadzu, Kyoto, Japan) equipped with an ACQUITY BEH C18 column (Waters, MA, USA; 2.1 × 100 mm, 1.7 μm). UV data, optical rotation, and IR data, were recorded on Genesys-10S UV-Vis spectrophotometer (Thermo Fisher Scientific, Waltham, MA, USA), MCP 200 Automatic Polarimeter (Anton Paar, Graz, Austria) and IS5 FT-IR spectrophotometer (Thermo Fisher Scientific, Waltham, MA, USA) respectively. The CD spectra were measured by a J-815 spectropolarimeter (JASCO, Tsukuba, Japan). Silica gel (Qingdao Haiyang Chemical Co., Ltd., Qingdao, China, 200–300 mesh), Sephadex LH-20 (GE Healthcare, Uppsala, Sweden), and ODS (50 μm, YMC Co., Ltd., Kyoto, Japan) were used for column chromatography. Semi-preparative

HPLC was performed on an Agilent 1200 HPLC system equipped with a DAD UV−vis spectrometric detector (Agilent Technologies Inc., CA, USA) using a reversed-phase Eclipse XDB-C8 column (5 μm, 9.4 × 250 mm, Agilent) with a flow rate of 2.0 mL/min and a CHIRALPAK IC column (5 μm, 4.6 × 250 mm, Daicel, Osaka, Japan) with a flow rate of 0.8 mL/min. For gas chromatography-mass spectrometry (GC-MS) a Shimadzu GCMS-QP2010 Ultra system (Shimadzu, Kyoto, Japan) was used.

2.2. Fungal Material

The strain *Penicillium* sp. YM1 used in this work was isolated from mildewed corn, collected in China, in September 2017. The sequences of RPB2 (MT898427), Ben A (MT898428), and CaM (MT898429) of our fungus were deposited in GenBank and employed for phylogenetic analysis. The fungus is similar to *P. cellarum* in forming hyaline, roughened stipes with bearing terminal terverticillate penicillii; and producing typically two rami per stipe, which are usually hyaline, roughened, appressed or only narrowly divergent; and having four to five metulae typically per ramus, which are usually hyaline, roughened, appressed or only narrowly divergent as well; and producing typically six to eight per metula phialides, which are usually hyaline, smooth, ampulliform, slender; and with pale green conidia that were typically smooth, globose to sometimes subglobose [17,18]. The phylogenetic analyses based on a combined dataset of RPB2, Ben A, and CaM was conducted by using PhyML v.3.0, with 1000 bootstrap replicates presented that our taxon grouped with the other taxa of *P. cellarum* with strongly maximum likelihood bootstrap proportions value (Figure S3). In consideration of the morphological features and phylogeny, this fungus was identified as *P. cellarum* YM1.

2.3. Fermentation and Extraction

P. cellarum was cultured on slant of PDA at 28 °C for 10 days. To prepare inoculum, the spores of the strain on the plate were collected with 0.01% sterile solution of Tween 80 (BTL, Warsaw, Poland) and adjusted to 1×10^6 CFU/mL. A large-scale fermentation was done in 40 × 500 mL Fernbach culture flasks containing 80 g of rice in 110 mL of distilled water (each with 0.5 mL of spore suspension) and incubated at 28 °C for 3 weeks. The fermented rice substrates were extracted with EtOAc (3 × 4 L) with the aid of ultrasonication, and the organic solvent was filtered and evaporated to dryness under vacuum to afford the crude extract (33.7 g).

2.4. LC-MS/MS and Molecular Metworking Analysis

LC-MS/MS (MS/MS scan 100−1500 Da) was performed with a Waters ACQUITY BEH C_{18} column (2.1 × 100 mm, 1.7 μm particles) eluted by MeCN−H_2O (0.005% TFA) (0.01−8 min 5−80% 8−12 min 80−99%, 12−15 min 99%) in a gradient manner. All the MS/MS data files were converted to ".mzML" format files using MSConver software and uploaded on the GNPS Web platform (http://gnps.ucsd.edu (accessed on 6 October 2021)) for MN analysis using Classic mode. For the network creation, a parent mass tolerance of 0.02 Da and a fragment ion tolerance of 0.05 Da were applied. The generated molecular network was visualized in Cytoscape 3.8.2 (www.cytoscape.org (accessed on 6 October 2021)) and guided the isolation of **1–12**. The MS/MS molecular network can be browsed and downloaded on the GNPS Web site with the following link: https://gnps.ucsd.edu/ProteoSAFe/status.jsp?task=8716192add914a1fb3bd8f469f7d2d81 (accessed on 6 October 2021).

2.5. Isolation and Characterization Data

The EtOAc fraction was subjected to a silica gel column chromatography (CC) eluting witH-N-hexane/ether-ethyl acetate (*v*/*v*, 100:0, 100:1, 100:2, 100:4, 100:10) and dichloromethane/methanol (*v*/*v*, 100:0, 100:1, 100:2, 100:4, 100:8, 100:12, 100:20, 0:100) to give 13 fractions (PC.1–PC.13). Fractions PC.6, PC.8, and PC.12 containing secondary metabolites with similar UV spectra were selected for further purification.

Fraction PC.6 (1.5 g) was further separated on silica gel column by a gradient elution with methanol-dichloromethane to give 25 fractions (PC.6-1–PC.6-25). PC.6–8 (60 mg) was purified finally by RP-HPLC with acetonitrile-water (63:37) to give **1** (13.5 mg, t_R 42.3 min) and **2** (6.6 mg, t_R 31.5 min). Compound **12** (44.5 mg) was obtained from subfractions PC.6–11 (65 mg) by Sephadex LH-20 chromatography eluting with methanol. Compounds **7** (8.6 mg, t_R 31.1 min), **8** (9.5 mg, t_R 40.5 min), and **9** (1.6 mg, t_R 42.5 min) were obtained from PC.6–20 (75 mg) by RP-HPLC using 86% acetonitrile in acidic water (0.005% TFA).

Fraction PC.8 (3.5 g) was separated on an ODS column using a gradient elution with methanol (35%, 55%, 70%, and 100%) in acidic water (0.005%TFA) to afford 15 subfractions (PC.8-1–PC.8–15). Compounds **5** (14.2 mg) and **6** (12.8 mg) were obtained from subfractions PC.8–9 and PC.8–10 by Sephadex LH-20 chromatography eluting with methanol, respectively.

Fraction PC.12 (4.3 g) eluted with CH_2Cl_2-acetone (v/v 20:1) was first separated by ODS using a gradient of increasing methanol (35%, 55%, 70%, and 100%) in water to afford 21 subfractions (PC.12-1–PC.12-21). Compounds **3** (2.5 mg, t_R 62.5 min) and **4** (3.5 mg, t_R 30.5 min) were yielded from PC.12-15 (35 mg) by RP-HPLC using 23% acetonitrile in acidic water (0.005% TFA). Subfractions PC.12-6 (60 mg) was purified by RP-HPLC using 27% acetonitrile in water to afford a mixture of **10** and **11** (35.0 mg, t_R 28.3 min). Enantioseparation of the mixture was carried out on CHIRALPAK IC using isopropanol/*n*-hexane (15:85) as mobile phase to afford **10** (10.0 mg, t_R 21.5 min) and **11** (11.5 mg, t_R 23.5 min).

Penicicellarusin A (3). light yellow oil, $[\alpha]_D^{25}$ + 55.6 (c 0.1 MeOH); UV (MeOH) λ_{max} (log ε) 240 (3.11), 302 (1.24) nm; IR (neat) v_{max} 3429, 2972, 2931, 1703, 1574, 1450, 1378, 1210, 1042, 811 cm^{-1}; Positive HRESIMS: *m/z* 455.2041 $[M+Na]^+$ (calcd. for $C_{24}H_{32}O_7Na$, 455.2040). ^1H-NMR and ^{13}C-NMR, see Table 1.

Table 1. ^1H and ^{13}C-NMR Data for compounds **3–4** in CD_3OD.

Pos.	3		4		Pos.	3		4	
	δ_C	δ_H (*J* in Hz)	δ_C	δ_H (*J* in Hz)		δ_C	δ_H (*J* in Hz)	δ_C	δ_H (*J* in Hz)
1	167.5		166.5		14	68.7		68.7	
2	111.6		89.5	5.66 s	15	78.4	4.09 q (6.8)	78.4	4.09 q (6.8)
3	170.2		173.2		16	10.4	2.04 s		
4	112.2		109.9		17	9.7	2.12 s	8.9	2.07 s
5	157.3		159.3		18	15.7	1.44 s	15.6	1.44 s
6	62.2		62.2		19	59.6	4.36 d (12.2)	59.6	4.36 d (12.2)
7	64.6	3.81 s	64.5	3.81 s			4.41 d (12.2)		4.40 d (12.2)
8	133.5		133.5		20	18.7	1.98 s	18.7	1.98 s
9	134.9	5.98 brs	134.9	5.98 brs	21	22.1	1.40 s	22.1	1.40 s
10	135.6		135.6		22	13.8	1.48 s	13.8	1.48 s
11	133.9	5.60 brs	133.9	5.60 brs	23	19.2	1.23 d (6.8)	19.2	1.23 d (6.8)
12	81.4		81.4		24	61.3	3.91 s	57.4	3.92 s
13	68.7	3.60 s	68.7	3.59 s					

Penicicellarusin B (4). light yellow oil, $[\alpha]_D^{25}$ + 51.0 (c 0.1 MeOH); UV (MeOH) λ_{max} (log ε) 241 (3.35), 297 (1.50) nm; IR (neat) v_{max} 3420, 2971, 2931, 1700, 1572, 1450, 1377, 1209, 1054, 812 cm^{-1}; Positive HRESIMS: *m/z* 441.1890 $[M+Na]^+$ (calcd. for $C_{23}H_{30}O_7Na$, 441.1884). ^1H-NMR and ^{13}C-NMR, see Table 1.

Penicicellarusin C (5). light yellow oil, $[\alpha]_D^{25}$ + 45.0 (c 0.1 MeOH); UV (MeOH) λ_{max} (log ε) 236 (3.56), 303 (2.57) nm; IR (neat) v_{max} 3410, 2974, 2930, 1685, 1560, 1450, 1378, 1224, 1089, 1043, 812 cm^{-1}; Positive HRESIMS: *m/z* 443.2047 $[M+Na]^+$ (calcd. for $C_{23}H_{32}O_7Na$, 443.2040). ^1H-NMR and ^{13}C-NMR, see Table 2.

Table 2. ^{1}H and ^{13}C-NMR Data for compounds **5–6**.

Pos.	Verrucisidinol [a,c]		5 [a]		5 [b]		6 [b]	
	δ_C	δ_H (*J* in Hz)	δ_C	δ_H (*J* in Hz)	δ_C	δ_H (*J* in Hz)	δ_C	δ_H (*J* in Hz)
1	165.0		165.5		166.1		166.6	
2	110.4		110.2		109.4		88.5	5.60 s
3	169.2		170.4		170.3		174.3	
4	111.8		112.7		112.4		111.5	
5	159.7		160.9		160.6		163.8	
6	78.8		79.1		78.9		81.4	
7	79.8	4.61 s	79.3	4.73 s	79.8	4.34 s	80.4	4.35 s
8	133.9		134.1		132.5		133.9	
9	134.3	5.87 s	134.5	5.91 brs	135.1	5.69 brs	136.5	5.71 brs
10	134.4		134.6		135.2		136.6	
11	133.0	5.43 s	133.0	5.43 brs	131.4	5.42 brs	132.8	5.43 brs
12	80.1		80.3		80.7		82.1	
13	67.5	3.43 s	67.6	3.43 s	67.3	3.54 s	68.8	3.55 s
14	67.4		67.5		67.2		68.7	
15	76.7	4.12 q (7.0)	76.8	4.12 q (6.8)	76.8	4.05 q (6.8)	78.3	4.06 q (6.8)
16	10.2	2.01 s	10.3	2.04 s	8.8	2.00 s		
17	9.9	2.21 s	10.1	2.22 s	8.9	2.28 s	9.4	2.25 s
18	23.4	1.40 s	23.6	1.40 s	21.5	1.48 s	22.9	1.49 s
19	14.8	1.82 d	14.9	1.83 d	13.5	1.82 s	14.9	1.83 s
20	18.6	1.89 s	18.7	1.88 s	17.4	1.86 s	18.8	1.87 s
21	21.9	1.41 s	22.0	1.41 s	20.7	1.37 s	22.1	1.38 s
22	13.8	1.47 s	14.0	1.46 s	12.4	1.47 s	13.8	1.48 s
23	18.8	1.18 d (7.0)	19.0	1.18 d (6.8)	17.8	1.17 d (6.8)	19.2	1.19 d (6.8)
24	60.3	3.79 s	60.6	3.79 s	59.6	3.84 s	57.3	3.90 s

[a] NMR data were measured in CD_3Cl; [b] NMR data were measured in CD_3OD; [c] NMR data reported in literature.

Penicicellarusin D (6). light yellow oil, $[\alpha]_D^{25}$ + 32.9 (c 0.1 MeOH); UV (MeOH) λ_{max} (log ε) 235 (3.56), 299 (2.54) nm; IR (neat) v_{max} 3412, 2974, 2932, 1683, 1559, 1450, 1378, 1225, 1089, 1043, 812 cm^{-1}; Positive HRESIMS: *m/z* 457.2190 [M+Na]$^+$ (calcd. for $C_{24}H_{34}O_7Na$, 457.2197). ^{1}H-NMR and ^{13}C-NMR, see Table 2.

Penicicellarusin E (7). light yellow oil, $[\alpha]_D^{25}$ +7 8.5 (c 0.1 MeOH); UV (MeOH) λ_{max} (log ε) 235 (3.65), 299 (2.58) nm; IR (neat) v_{max} 2975, 2933, 1695, 1555, 1450, 1378, 1224, 1086, 1043, 812 cm^{-1}; Positive HRESIMS: *m/z* 695.4490 [M+Na]$^+$ (calcd. for $C_{40}H_{64}O_8Na$, 695.4493). ^{1}H-NMR and ^{13}C-NMR, see Table 3.

Table 3. ^{1}H and ^{13}C-NMR Data for compounds **7–9** in CD_3OD.

Pos.	7		8		9	
	δ_C	δ_H (*J* in Hz)	δ_C	δ_H (*J* in Hz)	δ_C	δ_H (*J* in Hz)
1	167.0		167.0		167.0	
2	111.2		111.2		111.2	
3	171.3		171.3		171.3	
4	114.2		114.2		114.2	
5	160.4		160.4		160.4	
6	79.3		79.3		79.3	
7	83.8	5.39 s	83.8	5.39 s	83.3	5.39 s
8	132.8		132.8		132.8	
9	135.8	5.79 brs	135.8	5.79 brs	135.8	5.79 brs
10	136.2		136.2		136.2	
11	133.5	5.43 brs	133.5	5.43 brs	133.5	5.43 brs
12	81.4		81.4		81.4	
13	68.7	3.54 s	68.7	3.54 s	68.7	3.54 s
14	68.7		68.7		68.7	

Table 3. *Cont.*

Pos.	7		8		9	
	δ_C	δ_H (*J* in Hz)	δ_C	δ_H (*J* in Hz)	δ_C	δ_H (*J* in Hz)
15	78.3	4.05 q (6.8)	78.3	4.05 q (6.8)	78.2	4.06 q (6.8)
16	10.3	2.01 s	10.3	2.01 s	10.3	2.01 s
17	10.4	2.28 s	10.4	2.28 s	10.4	2.29 s
18	23.7	1.56 s	23.8	1.56 s	23.6	1.56 s
19	15.9	1.83 s	15.9	1.83 s	15.9	1.83 s
20	18.7	1.87 s	18.7	1.87 s	18.8	1.87 s
21	22.1	1.36 s	22.1	1.36 s	22.1	1.36 s
22	13.8	1.47 s	13.8	1.47 s	13.8	1.47 s
23	19.2	1.18 d (6.8)	19.2	1.18 d (6.8)	19.2	1.18 d (6.8)
24	61.1	3.84 s	61.1	3.84 s	61.1	3.84 s
1′	174.2		174.2		174.1	
2′	35.2	2.35 dt (2.4, 7.3)	35.2	2.35 dt (2.3, 7.3)	35.2	2.35 dt (2.4, 7.3)
3′	26.1	1.59 m	26.1	1.59 m	26.1	1.59 m
9′			130.8	5.36 m	130.8	5.36 m
10′			130.9	5.36 m	130.9	5.36 m
12′					129.0	5.36 m
13′					129.1	5.36 m
16′	14.5	0.92 t (6.8)				
18′			14.5	0.91 t (6.8)	14.5	0.92 t (6.8)
Others	29.1–30.9	1.30 m	29.1–30.9	1.30 m	29.1–30.9	1.30 m

"m" means multiplet or overlapped with other signals.

Penicicellarusin F (8). light yellow oil, $[\alpha]_D^{25}$ + 81.0 (c 0.1 MeOH); UV (MeOH) λ_{max} (log ε) 236 (3.38), 290 (2.54) nm; IR (neat) v_{max} 2972, 2932, 1689, 1557, 1450, 1378, 1225, 1088, 1045, 811 cm^{-1}; Positive HRESIMS: *m/z* 721.4658 [M+Na]$^+$ (calcd. for $C_{42}H_{66}O_8Na$, 721.4650). ^{1}H-NMR and ^{13}C-NMR, see Table 3.

Penicicellarusin G (9). light yellow oil, $[\alpha]_D^{25}$ + 76.5 (c 0.1 MeOH); UV (MeOH) λ_{max} (log ε) 236 (3.38), 290 (2.54) nm; IR (neat) v_{max} 2974, 2932, 1720, 1570, 1455, 1378, 1209, 1054, 812 cm^{-1}; Positive HRESIMS: *m/z* 719.4500 [M+Na]$^+$ (calcd. for $C_{42}H_{64}O_8Na$, 719.4493). ^{1}H-NMR and ^{13}C-NMR, see Table 3.

2.6. X-ray Crystallographic Analysis of Compound **1** and **10**

2.6.1. Penicicellarusin A (**1**)

Colorless needles of compound **1** were obtained from ethyl ether. Data collection was performed on a Eos CCD (Bruker, Bremen, Germany) using graphite-monochromated Cu K_α radiation, λ = 1.54184 Å at 100.00(10) K. Crystal data: $C_{24}H_{32}O_6$, *M* = 416.49, space group orthorhombic, $P2_12_12_1$; unit cell dimensions were determined to be a = 5.70420(10) Å, b = 11.7760(2) Å, c = 33.3228(6) Å, $\alpha = \beta = \gamma$ = 90.00°, *V* = 2238.38(7) Å^3, Z = 4, ρ_{calc} =1.236 mg/mm^3, *F* (000) = 896.0, μ (Cu K_α) = 0.715 mm^{-1}. 16083 unique reflections were collected to $2\theta_{max}$ = 144.15°, in which 4334 reflections were observed [$F^2 > 4\sigma(F^2)$]. The structure refinements were conducted by a previously reported method [19]. The final refinement gave R_1 = 0.0357, wR_2 = 0.0852 ($w = 1/\sigma|F|^2$), and S = 1.047. CCDC 2039557 contains the supplementary crystallographic data for **1**. These data can be obtained from The Cambridge Crystallographic Data Centre via www.ccdc.cam.ac.uk/data_request/cif (accessed on 20 October 2020).

2.6.2. Penicyrone A (**10**)

Colorless needles of compound **10** from methanol were obtained. Data collection was performed on a Eos CCD using graphite-monochromated Cu K_α radiation, λ = 1.54184 Å at 100.00(10) K. Crystal data: $C_{24}H_{34}O_7$, *M* = 434.519, space group monoclinic, $P2_1$; unit cell dimensions were determined to be a = 5.96230(10) Å, b = 14.1879(3) Å, c = 14.0680(6) Å,

$\alpha = \gamma = 90.00°$, $\beta = 95.885°$ $V = 1183.78(4)$ Å^3, Z = 2, ρ_{calc} =1.219 mg/mm^3, F (000) = 468.0, μ (Cu K$_\alpha$) = 0.728 mm^{-1}. 12357 unique reflections were collected to $2\theta_{max}$ = 140.124°, in which 4418 reflections were observed [$F^2 > 4\sigma$ (F^2)]. The structure refinements were conducted by the same method as described for compound **1**. The final refinement gave R_1 = 0.0341, wR_2 = 0.0842 ($w = 1/\sigma|F|^2$), and S = 1.048. CCDC 2039558 contains the supplementary crystallographic data for **10**. These data can be obtained from The Cambridge Crystallographic Data Centre via www.ccdc.cam.ac.uk/data_request/cif (accessed on 20 October 2020).

2.7. *Alkaline Hydrolysis of Compound* **8** *and* **9**

Alkaline hydrolysis reaction was carried out following a previously described method [20]. Each compound (2.0 mg) was dissolved and hydrolyzed with 2 M NaOH/MeOH at 25 °C for 3 h. Then neutralized with 1 N HCl/MeOH and extracted with chloroform for two times (10 mL × 2). Methyl esters of the fatty acids were identified by GC-MS. The GC-MS was operated in EI mode (70 eV) scanning from 40 to 500 amu.

2.8. *Bioinformatic Analyses*

To identify Biosynthetic Gene Clusters (BGCs) in the genomes of *P. cellarum* YM1, antiSMASH 6.2 was used and only clusters containing a putative PKS similar to both VerA and CtvA protein were further considered [21,22]. The proteins in these clusters were additionally blasted against *P. polonicum* and *Aspergillus terreus* var. *aureus* to verify their presence. To find functional domains and predict a putative function, we resort to NCBI BLAST using Non-Redundant database and Interproscan.

2.9. *Computation Section*

Systematic conformational analyses for **5a**, **5b**, **5c** and **5d** were performed using the CONFLEX softwre (version 7 Rev. A; CONFLEX Corporation, Tokyo, Japan) via the MMFF94 molecular mechanics force field. Using TDDFT at B3LYP/6-31+G(d,p) basis set level, the MMFF94 conformers were further optimized in methanol with PCM model. The stationary points have been checked as the true minima of the potential energy surface by verifying that they do not exhibit vibrational imaginary frequencies. ECD spectra were calculated by TD-DFT using a Gaussian function at the PBE1PBE/6-311G* level. Using Boltzmann statistics, equilibrium populations of conformers at 298.15 K were calculated from their relative free energies (ΔG). According to Boltzmann weighting of main conformers, the overall ECD spectra were then generated [23].

2.10. *Evaluation of Biological Activities*

2.10.1. Antimicrobial Bioassay

Assay for antibacterial activities including *Staphylococcus aureus* (ATCC 6538 and CGMCC 1.2465), meticillin-resistant *S. aureus* (MRSA, clinical isolates, Beijing Chao-yang Hospital, Beijing, China), *Enterococcus faecalis* (clinical isolates, Beijing Chao-yang Hospital), *Bacillus subtilis* (ATCC 6633), and antifungal activities including *Candida albicans* (ATCC 18804), and *Aspergilus fumigatus* (CGMCC 3.5835) were carried out as previously described method [24]. The inhibition rate was calculated and plotted versus test concentrations to afford the MIC. MIC values were defined as the minimum concentration of compounds that inhibited visible microbial growth. All the experiments were performed in triplicate.

2.10.2. Cytotoxicity Assay

Cytotoxicity test against A549, HepG2, and K562 cell lines was carried out as previously described method [25]. Taxol, 5-flourouracil, and cisplatin were used as the positive controls.

2.10.3. 2-[N-(7-Nitrobenz-2-oxa-1,3-diazol-4-yl)amino]-2-deoxy-D-glucose (2-NBDG) Glucose Uptake Assay

This experiment was consistent with those reported in our previous work [26]. The HepG2 hepatoma cells were cultured in DMEM supplemented with 10% fetal bovine serum (FBS; Gibco, NY, USA), 100 U/mL penicillin/streptomycin. The cells reaching confluence were treated with 10^{-6} M insulin for 24 h to generate insulin resistance. Compounds or positive drug (rosiglitazone) were mixed and incubated for 24 h, with the final concentration of 100, 50, 25, and 12.5 μM; then, 100 nM of insulin was added and incubated for 30 min at 37 °C followed by addition of 50 μM (2-NBDG). After that, cells were washed with ice-cold PBS and 100 μL FBS-free DMEM was added to each well. The level of 2-NBDG uptake was determined on microplate reader (Bio-Tek Instruments, VT, USA) at 485 nm excitation and 528 nm emission. All data were handled with GraphPad Prism 5 and reported as mean ± SD of three independent experiments.

3. Results

In this study, the MS/MS-based molecular networking strategy was applied for target isolation of new verrucosidins. First, *Penicillium* strains were cultured on rice substrates and the resulting EtOAc extracts was screened by HPLC-UV-DAD analysis (Figure S1). Then, the ethyl acetate extract of *P. cellarum* YM1 that produced secondary metabolites with similar retention time and UV characteristics to those of deoxyverrucosidin was further investigated by UPLC-HRMS/MS. The LC-MS/MS data were used to generate a visualized molecular networking that was further annotated by Cytoscape 3.8.2 (Figure S2).

In details, the HPLC-HRMS/MS analysis in the positive ion mode was conducted on the ethyl acetate extract from *P. cellarum* YM1 with deoxyverrucosidin as the phishing probe. The obtained fragmentation data were organized by molecular networking, yielding a metabolite-level view of the data. Individual MS/MS spectrum was organized into 106 clusters consisting of 899 connected nodes (Figure S2). Using the MS/MS data of deoxyverrucosidin as "seed" spectra, an initial focal point (a blue hexagon with *m/z* 401.232) was generated in the global molecular networking. A close examination of the molecular network indicated some nodes connected to deoxyverrucosidin (Figure 2), which predicted the presence of potential natural analogs. Under the guidance of MS/MS-based molecular networkings, seven new verrucosidins, namely penicicellarusins A-I (**3–9**), in addition to five known polyketides verrucosidin (**1**) [4,27], normethylverrucosidin (**2**) [27], penicyrone A-B (**10–11**) [28], and deoxyverrucosidin (**12**) [29] were obtained by the isolation workflow. The structures of known compounds were determined by comparing their NMR and MS data with literature data.

Compound **1** was obtained as white needle-like crystals and identified as verrucosidin by comparison of the NMR data reported in the literature [4,27] and the single-crystal X-ray crystallographic analysis (Figure 3).

Penicicellarusin A (**3**) was isolated as yellow oil with a molecular formula $C_{24}H_{32}O_7$ (indicating nine degrees of unsaturation) as deduced by HRESIMS data ($[M+Na]^+$ *m/z* 455.2041; calcd. 455.2040). The ^1H-, ^{13}C-NMR and HSQC spectra of **3** revealed the presence of eight methyls, including one oxygenated one [δ_H/δ_C 1.23 (3H, d, 6.8 Hz) /19.2, 1.40 (3H, s)/22.1, 1.44 (3H, s)/15.7, 1.48 (3H, s)/13.8, 1.98 (3H, s)/18.7, 2.04 (3H, s)/10.4, 2.12 (3H, s)/9.7, and 3.91(3H, s)/61.3], one hydroxymethyl [δ_H/δ_C 4.36 (d, J = 12.2 Hz), 4.41 (d, J = 12.2 Hz)/59.6], five methines including three oxygenated [δ_H/δ_C 3.60 (s)/68.7, 3.81 (s)/64.6, 4.09 (d, 6.8 Hz)/78.4] and two sp^2 methines [δ_H/δ_C 5.60 (brs) /133.9, 5.98 (brs)/134.9]. The NMR data of compound **3** were similar with those of verrucosidin (**1**) [27], indicating the presence of a 3,6-dioxabicyclic[3.1.0]hexane moiety, a 2*H*-pyran-2-one moiety, and a polyene chain in **3** (Figure 1).

The HMBC correlations were detected from H_3-21 to C-11 and C-12, from H-13 to C-12 and C-14, from H_3-22 to C-14, from H_3-23 to C-14 and C-15, from H-15 to C-12, C-14 and C-23, which together with the ^1H-^1H COSY correlations of H-15-H_3-23 confirmed the presence of the 3,6-dioxabicyclic[3.1.0]hexane moiety. The HMBC correlations of H_3-16 to

C-1, C-2, and C-3, H_3-24 to C-3, H_3-17 to C-3, C-4, and C-5, and as well as the chemical shifts of C-1 (δ 167.5), C-2 (δ 111.6), C-3 (δ 170.2), C-4 (δ 112.2), and C-5 (δ 157.3) completed the assignment of α-pyranone moiety. Furthermore, the ^{1}H-^{1}H COSY correlations of H-7-H_3-19-H-9-H_3-20-H-11 together with the HMBC correlations from H_3-18 to C-5 and C-6, from H-7 to C-5, C-6, and C-8, from H_3-19 to C-7 and C-8, from H-9 to C-7, C-8, and C-10, from H_3-20 to C-9, C-10, and C-11, and from H-11 to C-10 and C-12 supported a heptadiene moiety which was connected with α-pyranone moiety through C-5 and linked with 3,6-dioxabicyclic[3.1.0]hexane moiety through C-12 (Figure 4).

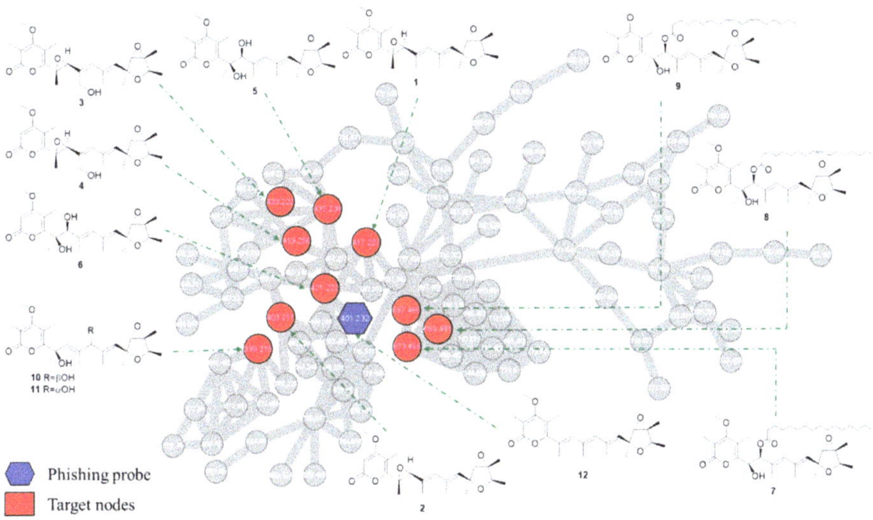

Figure 2. The subnetwork of tandem MS/MS molecular working for crude extracts of the fungus *P. cellarum*. The entire network and subnetwork are presented in Figure S2.

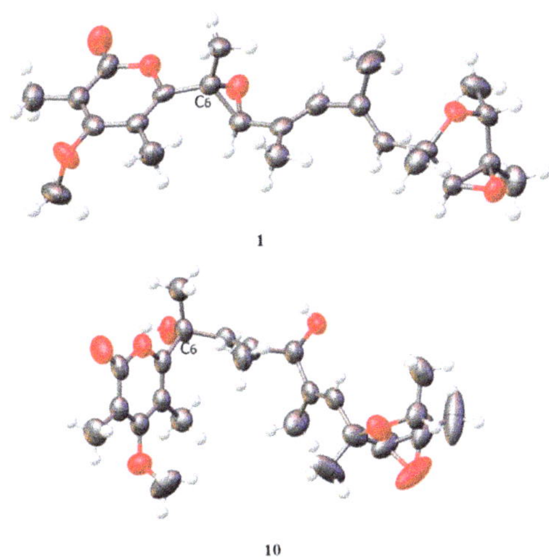

Figure 3. The X-ray crystallographic structure of **1** and **10**.

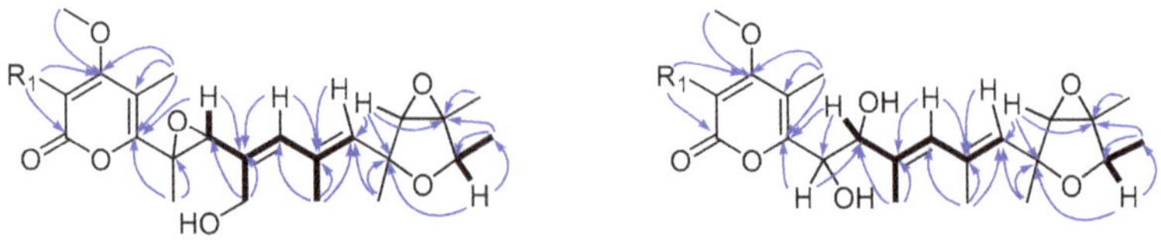

3: R_1=CH_3
4: R_1=H

5: R_1=CH_3
6: R_1=H

COSY(H ⟷ H) HMBC(H ⟶ C)

Figure 4. Selected key HMBC and ^{1}H-^{1}H COSY correlations of **3–6**.

The relative configuration of **3** was confirmed by NOESY experiment (Figure 5). The NOE correlations H-11 (δ 5.60) to H_3-23 (δ 1.23) and H-13 (δ 3.60), H_3-22 (δ 1.48) to H-13 and H_3-23, and H-15 (δ 4.09) to H_3-21 (δ 1.40), indicated that H_3-22, H_3-23, H-11, and H-13 were on the same face, while H_3-21 and H-15 were on the opposite face. The geometry of C_8=C_9 and C_{10}=C_{11} were confirmed to be *E* by the NOE correlations of H-7 (δ 3.81) with H-9 (δ 5.98) and H_3-20 with H_3-21.

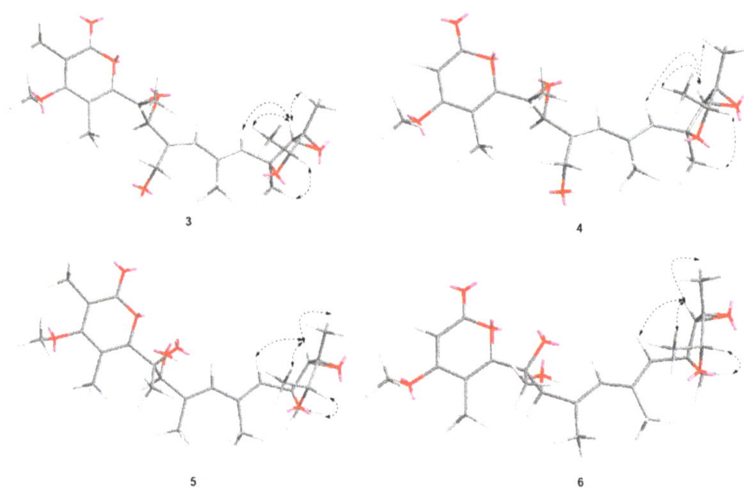

Figure 5. Selected key NOE correlations of **3–6**.

In the experimental ECD spectrum, compound **3** showed similar Cotton effects as verrucosidin (**1**) (Figure 6), supporting the same configuration at C-6 and C-7 between **1** and **3**. Thus, compound **3** was assigned a 6*R*, 7*S*, 12*S*, 13*S*, 14*R*, 15*R* configuration and named penicicellarusin A.

Penicicellarusin B (**4**) was obtained as yellow oil with the molecular formula $C_{23}H_{30}O_7$ and nine degrees of unsaturation, as deduced from HRESIMS data. The 1D NMR spectroscopic data of **4** (Table 1) were similar with those of **3**, except for the lack of a singlet methyl group and the presence of one additional olefinic proton at δ 5.66 in **4**. HMBC correlations of H-2 (δ 5.66) to C-1, C-3 and C-4, H_3-17 to C-3 and C-5, H_3-24 to C-3 confirmed the structural changes on the 2*H*-pyran-2-one moiety in **4**. A further comprehensive analysis of its ^{1}H-^{1}H COSY, HMQC, and HMBC spectra assigned the planar structure of **4** (Figure 4).

The NOESY correlations of H-11 with H-13 and H_3-22, H_3-23 with H-13 and H_3-22, H-15 with H_3-21, H-7 with H-9 and H_3-17, H_3-20 with H_3-21 supported the same relative configurations for double bonds and 3,6-dioxabicyclic[3.1.0]hexane moiety between 4 and compound **3** (Figure 5). Compound **4** showed similar Cotton effects in the experimental electronic circular dichroism (ECD) spectrum with those of **3** (Figure 6), indicating it has the absolute configuration of 6*R*, 7*S*, 12*S*, 13*S*, 14*R*, and 15*R*, as described in **3**.

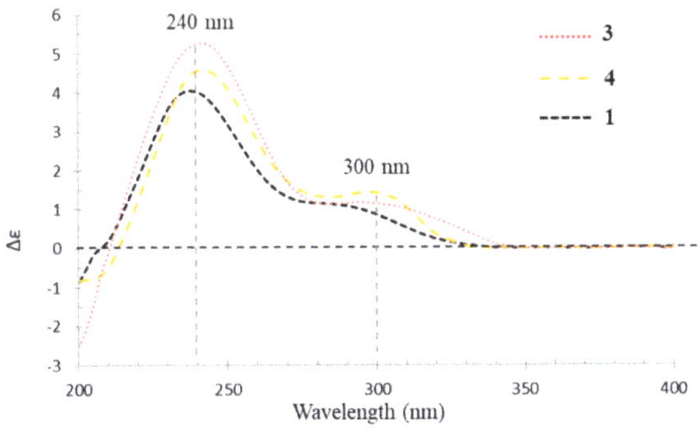

Figure 6. Experimental CD spectra of **1**, **3**, and **4** in MeOH.

Compound **5** was assigned the molecular formula of $C_{24}H_{34}O_7$ (eight degree of unsaturation) on the basis of its HRESIMS at *m/z* 457.2190 $[M+Na]^+$ and NMR data (Table 2). The ^{1}H-, ^{13}C-NMR, and UV spectra of **5** were similar with those of verrucisidinol [6], with the notable difference in the ^{1}H-NMR data of C-3, C-4, C-5, and H-7 (Table 2). A comprehensive analysis of its 2D NMR spectra including ^{1}H-^{1}H COSY, HMQC, and HMBC experiments confirmed the planar structure of **5** (Figure 4).

The partial relative configuration of **5** was confirmed by a NOESY experiment (Figure 5). The geometry of C_8=C_9 and C_{10}=C_{11} were confirmed to be *E* by analysis of the NOESY observations. The key NOESY correlations of H-11 with H-13 and H_3-22, H_3-23 with H-13 and H_3-22, and H-15 with H_3-21 supported the same relative configurations on furan ring as verrucisidinol [6]. Considering the same biosynthesis origin, compound **5** is deduced to share the same absolute configuration with those of **1–4** in the furan ring. In addtion, the optical rotation data of **5** ($[\alpha]_D^{25}$ = +45.0, *c* = 0.1, MeOH) were opposite to that of verrucisidinol ($[\alpha]_D^{25}$ = −10.0, *c* = 0.1, MeOH), implying the enantiomeric relationship between them. To determine the absolute configurations at C-6 and C-7, ECD calculation method was applied. The four configurations (**5a**, **5b**, **5c** and **5d**, Figure 7) were calculated using time-dependent density functional theory (TDDFT) at PBE1PBE/6-311 G* level with PCM model in methanol, and 60 exciting states were calculated. By comparison of the experimental and simulated ECD curves (Figure 7), the experimental ECD was match better with **5a** (6*R*, 7*S*, 12*S*, 13*S*, 14*R*, and 15*R*). Thus, the compound **5** was assigned as 6*R*, 7*S*, 12*S*, 13*S*, 14*R*, and 15*R*, and named as penicicellarusin C.

The molecular formula of penicicellarusin D (**6**) was determined to be $C_{23}H_{32}O_7$ with the unsaturation degrees of eight on the basis of the HRESIMS data at *m/z* 443.2047 $[M+Na]^+$ (calcd. for $C_{23}H_{32}O_7Na$ *m/z* 443.2040) and NMR data (Table 2). The NMR data of **6** were similar to those of **5** except for the absence of one singlet methyl group. The key HMBC correlations from H-2 (δ 5.60) to C-1, C-3 and C-4, as well as the upfield shift of C-2 (δ 88.5) confirmed the disappearance of the methyl group on the C-2 position in **6**. Furthermore, Compound **6** showed similar Cotton effects in the experimental CD spectrum with those of **5** (Figure S5), which assigned the absolute configurations of **6** as 6*R*, 7*S*, 12*S*, 13*S*, 14*R*, and 15*R*. It was designated as penicicellarusin D.

Penicicellarusins E-G (compounds **7–9**) were determined to be fatty acid esters of **5** by interpretation of the HRESIMS, 1D and 2D NMR data (Table 3, Figures S6 and S7), and ECD spectra (Figure S8). The MS/MS data of **7–9** confirmed the presence of the fatty acid moiety in their structures. The pseudo molecular ion peaks $[M+H]^+$ at *m/z* 673.4669 in **7**, *m/z* 699.4819 in **8**, and *m/z* 697.4668 in **9** together with the fragment ion peaks $[M+H-C_{16}H_{32}O_2]^+$ at *m/z* 435.2381 in **7**, $[M+H-C_{18}H_{32}O_2]^+$ at *m/z* 435.2375 in **8**, $[M+H-C_{18}H_{30}O_2]^+$ at *m/z* 435.2374 in **9** due to the loss of the corresponding fatty acid moiety. To assign the structure of fatty acid moieties, compounds **7–9** was hydrolyzed with alkaline solution followed by methyl esterification. The fatty acid chain in **7–9** was determined to be the palmitic acid, the oleic acid, and the linoleic acid, respectively, by comparison of the retention time and

MS spectrum with those of standards by GC-MS analysis (Figure S9). Compounds **7–9** showed similar Cotton effects in the experimental CD spectrum with those of **5** (Figure S8), which assigned their absolute configurations as 6*R*, 7*S*, 12*S*, 13*S*, 14*R*, and 15*R*.

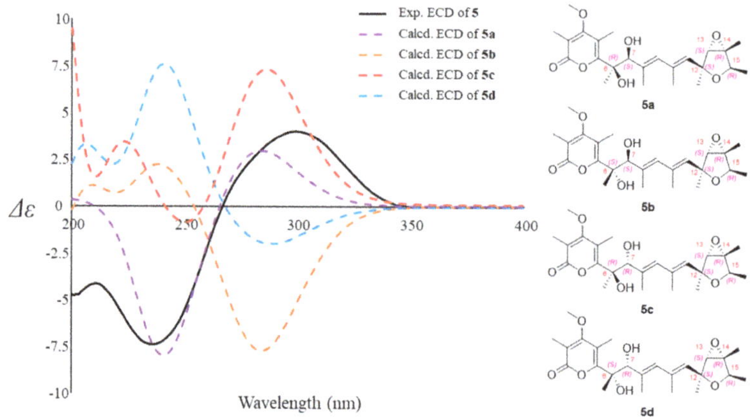

Figure 7. Experimental CD spectra of **5**, the calculated ECD spectra and the structure of **5a**, **5b**, **5c** and **5d** (bandwidth σ = 0.30 eV).

With the help of single-crystal X-ray crystallographic analysis, the 6*R*, 9*R*, 12*S*, 13*S*, 14*R*, and 15*R* absolute configuration of penicyrone A (**10**) was determined. The value of the Flack absolute structure parameter 0.03 (8) was obtained, and a perspective ORTEP plot was shown in Figure 3 (CCDC 2039558). According to X-ray diffraction analysis, the configuration at the C-6 positions in **10** was 6*R*, instead of 6*S* reported in the literature [28]. The structures of other known compounds were determined by comparing spectroscopic data with those in the literature.

To explore the bioactivities of verrucosidins, compounds **1–12** were evaluated for the antimicrobial effect, cytotoxic activity, and hypoglycemic activity. As a result, **1–12** showed no significant bioactivity in the antimicrobial assays and cytotoxicity assays at the dose of 100 μM. However, Compounds **1–4** were found to enhance the insulin-stimulated uptake of 2-NBDG in insulin-resistant HepG2 cells with the EC_{50} values at 47.2 ± 1.2, 9.9 ± 2.5, 93.2 ± 1.2 and 40.2 ± 1.3 μM, respectively, while the other compounds showed no significant activity (Figure 8). In particular, compounds **1**, **2**, and **4** showed much stronger activity than the positive drug (rosiglitazone) in the range of 25–100 μM.

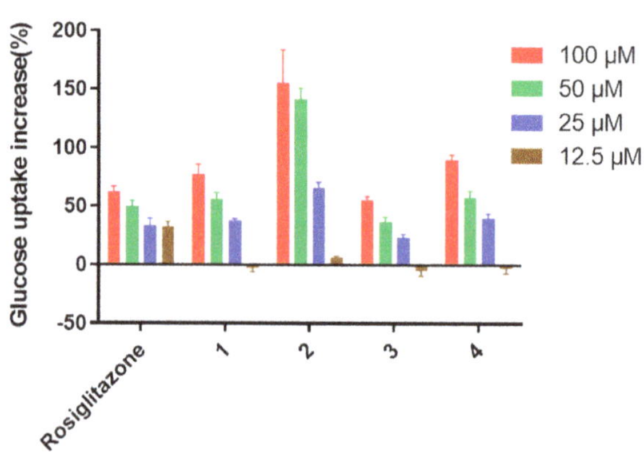

Figure 8. Stimulation on 2-NBDG glucose uptake in insulin-resistant HepG2 cells.

4. Discussion

Up to now, less than 20 verrucosidins and structurally-related compounds have been found in fungi. In this study, seven new verrucosidin derivatives (compounds **3–9**), were isolated together with five previously identified compounds from the fermentation products of fungus *P. cellarum*, suggesting that this fungus is an important producer of verrucosidins.

Verrucosidins share similar structural features with the citreoviridins, including a methylated α-pyrone, a polyene linker, and a tetrahydrofuran ring. The citreoviridin biosynthetic gene cluster containing a polyketide synthase (CtvA), a SAM-dependent methyltransferase (CtvB), a flavin-dependent monooxygenase (CtvC), and a hydrolase (CtvD) has been identified in *Aspergillus terreus var. aureus* [21]. CtvC is the only monooxygenase in the cluster, which can iteratively oxidize the terminal triene portion of the precursor into a bisepoxide moiety. As a regioselective hydrolase, CtvD can transform the bisepoxide moiety into a tetrahydrofuran ring moiety [21,30]. In addition, the verrucosidin bosynthetic gene cluster was confirmed by constructing deletion mutants for *verA* gene coding for HR-PKS known to be the key enzyme of the biosynthesis. Different from citreoviridin, the bosynthetic gene cluster for verrucosidin in the genome of *P. polonicum* contains two flavin-dependent monooxygenases, VerC1 and VerC2, which means that the cluster can synthesize compounds with higher oxidation degree [22]. However, the enzymes involved in the biosynthesis of verrucosidin are largely uncharacterized.

In our work, with the genome of *P. cellarum* YM1 sequenced in our group, we searched for the gene cluster for penicicellarusins. By similarity analysis with the polyketide synthase gene *ctvA* and *verA*, the putative gene cluster *celA* was found in the genome of *P. cellarum* (Table 4 and Figure S11). Further bioinformatic analysis revealed seven genes, the polyketide synthase gene (*celA*), the SAM-dependent methyltransferase gene (*celB*), the flavin-dependent monooxygenase (*celC1 and celC2*), the cytochrome P450 gene (*celD*), the acyl-acyltransferase gene (*celF*), and the lyase gene (*celH*) potentially involved in the biosynthesis of penicicellarusins in *P. cellarum*. Based on above evidence, we propose the biosynthetic pathway of **1–12** (Figures 9 and S10). Verrucosidin (**1**) could be formed from **12** by oxidation of the olefinic bond, and further oxidation produces **3–6**. Compounds **7–9** were biosynthesized by esterification of **5** with different fatty acids. Compounds **10** and **11** can be transformed from **5** or **6** by dehydration reaction.

Table 4. Penicicellarusins biosynthetic genes and gene function prediction in *P. cellarum* and their homologs in other fungal species.

P. cellarum Gene	*P. polonicum* Homologue aa (Identity/Similarity, %), Gene (Identity/Similarity, %)	*A. terreus* Homologue aa (Identity/Similarity, %)	Putative Function
celA	VerA (86/98) PENPOL_C002G03804 (99/98)	CtvA (42/77)	Polyketide synthase
celB	VerB (79/29) PENPOL_C002G01780 (98/41)	CtvB (30/41)	Methyltransferase
celC1	VerC1 (99/93) PENPOL_c002G07872 (99/45)	CtvC (50/56)	FAD monooxygenase
celC2	VerC2 (91/50) PENPOL_C002G07909 (99/57)	-	FAD monooxygenase
celD	VerH (99/95) PENPOL_c002G07307 (98/69)	-	Cytochrome P450
celE	VerF (78/96) PENPOL_c002G07024 (99/95)	-	Transcription factor domain
celF	VerG (87/100) PENPOL_c002G05667 (97/99)	-	Acyl-acyltransferase
celH	-	-	Lyase

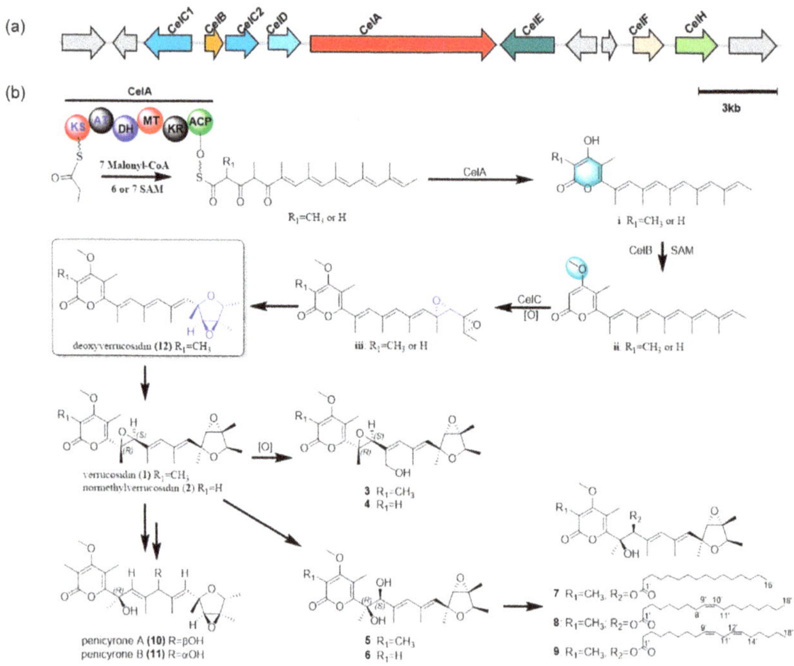

Figure 9. The biosynthetic gene clusters and postulated biogenetic pathway of **1–12**. (**a**) The penicicellarusin biosynthesis gene cluster in *P. cellarum*. *celA*: polyketide synthase gene, *celB*: the SAM-dependent methyltransferase gene, *celC1/celC2*: the flavin-dependent monooxygenase gene, *celD*: the cytochrome P450 gene, *celF*: the acyl-acyltransferase gene, *celE*: transcriptional factor gene, and *celH*: lyase gene. (**b**) postulated biogenetic pathway of **1–12**. PKS domain abbreviations: KS ketosynthase, AT acyltransferase, DH dehydratase, MT methyltransferase, KR keroreductase, ACP acyl carrier protein.

Early studies have demonstrated that verrucosidins and structurally-related compounds are endowed with several interesting bioactivities, such as antibacterial activities [3,10], antitumor [7,8], antiviral [9], and neurological activities [11]. In this work, it was found that compounds **1–4** show promising hypoglycemic activity, especially compounds **2** and **4**. Preliminary structure-activity relationship showed that the formation of epoxy three-membered ring on C_6-C_7 in the structures contributes greatly for the glucose uptake-enhancing activity in insulin-resistant HepG2 cells. The promising hypoglycemic activity is an interesting new bioactivity for this class of compounds.

5. Conclusions

In summary, a MS/MS-based molecular networking for the target discovery of verrucosidin-like polyketides was established in this study. The stereochemistry of the new compounds was determined by electronic circular dichroism (ECD) methods or comparison of experimental ECD spectra. The absolute configuration of penicyrone A (**10**) was corrected based on X-ray diffraction analyses. Bioactivity screening indicated that compounds **1**, **2**, and **4** showed much stronger promising hypoglycemic activity than the positive drug (rosiglitazone) in the range of 25–100 μM. The promising hypoglycemic activity is an interesting new bioactivity for this class of compounds. This work further proved the efficacy of the molecular networking in discovering natural products with unique structural features.

Supplementary Materials: The following are available online at https://www.mdpi.com/article/10.3390/jof8020143/s1, Figure S1: The HPLC profiles of metabolites extracted from the culture medium of *Penicillium* strains, Figure S2: The molecular network obtained by combining the LC-MS/MS analyses of extracts from *P. cellarum* YM1, Figure S3: Phylogenetic analysis and morphological characters of *P. cellarum* YM1, Figure S4: Most stable conformers of **5** in solvated model calculations at the B3LYP/6-31+G(d,p) level (d), Figure S5: Experimental CD spectra of **5** and **6** in MeOH, Figure S6: Selected key HMBC and ^{1}H-^{1}H COSY correlations of **7–9**, Figure S7: Selected key NOE correlations of **7–9**, Figure S8: Experimental CD spectra of **5** and **7–9** in MeOH, Figure S9: GC-MS analysis of methyl linoleate, methyl oleate and products of alkaline hydrolysis-methyl esterification of compounds **8** and **9**, Figure S10: Gene cluster schematic illustrating comparative organization of the penicicellarusin, verrucosidin, and citreoviridin, Figures S11–S40: NMR spectra for compounds **3–9**.

Author Contributions: Conceptualization, J.H.; validation, B.C.; software, R.Z., G.Z., data curation, J.H., B.C., R.Z., J.Z. and H.D.; formal analysis, W.L., X.L. and W.Y.; investigation, J.H. and H.L.; resources, T.W. and J.S.; writing—original draft preparation, J.H. and B.C.; writing—review and editing, H.L.; project administration, H.L.; funding acquisition, J.H., E.L., H.L. and H.D. All authors have read and agreed to the published version of the manuscript.

Funding: This project was supported by the National Special Project for Key Science and Technology of Food Safety (grant No. 2017YFC1601302), the National Key R&D Program of China (grant No. 2017YFE0108200), and the National Natural Science Foundation (Grant Nos. 22177131, and 82073723).

Institutional Review Board Statement: Not applicable.

Informed Consent Statement: Not applicable.

Data Availability Statement: Not applicable.

Acknowledgments: The authors thank the National Special Project and the National Natural Science Foundation for funding.

Conflicts of Interest: The authors declare no conflict of interest.

References

1. Ortega, H.; Torres-Mendoza, D.; Caballero, E.Z.; Cubilla-Rios, L. Structurally uncommon secondary metabolites derived from endophytic fungi. *J. Fungi* **2021**, *7*, 570. [CrossRef] [PubMed]
2. Al-Fakih, A.A.; Almaqtri, W.Q.A. Overview on antibacterial metabolites from terrestrial *Aspergillus* spp. *Mycol.* **2019**, *10*, 1–19.
3. Pan, C.Q.; Shi, Y.T.; Auckloo, B.; Chen, X.G.; Chen, C.T.; Tao, X.Y.; Wu, B. An Unusual conformational isomer of verrucosidin backbone from a hydrothermal vent fungus, *Penicillium* sp. Y-50-10. *Mar. Drugs* **2016**, *14*, 156. [CrossRef] [PubMed]
4. Burka, L.T.; Ganguli, M.; Wilson, B.J. Verrucosidin, a tremorgen from *Penicillium-verrucosum* var cyclopium. *J. Chem. Soc. Chem. Comm.* **1983**, *9*, 544–545. [CrossRef]
5. El-Banna, A.A.; Pitt, J.I.; Leistner, L. Production of mycotoxins by *Penicillium* species. *System. Appl. Microbiol.* **1987**, *10*, 42–46. [CrossRef]
6. Yu, K.; Ren, B.; Wei, J.L.; Chen, C.X.; Sun, J.S.; Song, F.H.; Dai, H.Q.; Zhang, L.X. Verrucosidinol and verrucosidinol acetate, two pyrone-type polyketides isolated from a marine derived fungus *Penicillum aurantiogriseum*. *Mar. Drugs* **2010**, *8*, 2744–2754. [CrossRef] [PubMed]
7. Park, H.R.; Ryoo, I.J.; Choo, S.J.; Hwang, J.H.; Kim, J.Y.; Cha, M.R.; Shin-Ya, K.; Yoo, I.D. Glucose-deprived HT-29 human colon carcinoma cells are sensitive to verrucosidin as a GRP78 down-regulator. *Toxicology* **2007**, *229*, 253–261. [CrossRef]
8. Thomas, S.; Sharma, N.; Gonzalez, R.; Pao, P.W.; Hofman, F.M.; Chen, T.C.; Louie, S.G.; Pirrung, M.C.; Schönthal, A.H. Repositioning of verrucosidin, a purported inhibitor of chaperone protein GRP78, as an inhibitor of mitochondrial electron transport chain complex I. *PLoS ONE* **2013**, *8*, 1–14. [CrossRef]
9. Vieta, I.; Savarino, A.; Papa, G.; Vidotto, V.; Cantamessa, C.; Pugliese, A. In vitro inhibitory activity of citreoviridin against HIV-1 and an HIV-associated opportunist: *Candida albicans*. *J. Chemother.* **1996**, *8*, 351–357. [CrossRef]
10. Zhang, Y.; Feng, Y.; Kramer, M.; Essmann, F.; Grond, S. A new acetylenic compound and other bioactive metabolites from a shark gillderived *Penicillium* strain. *Rec. Nat. Prod.* **2017**, *11*, 31–36.
11. Fink-Gremmels, J.; Henning, J.; Leistner, A. Quantitative determination of verrucosidin producing *Penicillium aurantiogriseum*. *Microbiol. Aliments. Nutr. Microbiol. Foods Feed. Nutr.* **1991**, *5*, 155–160.
12. Gerwick, W.H. The face of a molecule. *J. Nat. Prod.* **2017**, *80*, 2583–2588. [CrossRef] [PubMed]
13. Yang, J.Y.; Sanchez, L.M.; Rath, C.M.; Liu, X.T.; Boudreau, P.D.; Bruns, N.; Glukhov, E.; Wodtke, A.; de Felicio, R.; Fenner, A.; et al. Molecular networking as a dereplication strategy. *J. Nat. Prod.* **2013**, *76*, 1686–1699. [CrossRef] [PubMed]

14. Sala, G.D.; Mangoni, A.; Costantino, V.; Teta, R. Identification of the biosynthetic gene cluster of thermoactinoamides and discovery of new congeners by integrated genome mining and MS-Based molecular networking. *Front. Chem.* **2020**, *8*, 397. [CrossRef] [PubMed]
15. He, Q.F.; Wu, Z.L.; Li, L.R.; Sun, W.Y.; Wang, G.Y.; Jiang, R.W.; Hu, L.J.; Shi, L.; He, R.R.; Wang, Y.; et al. Discovery of neuritogenic securinega alkaloids from *Flueggea suffruticosa* by a building blocks-based molecular network strategy. *Angew. Chem. Int. Ed.* **2021**, *60*, 1–6. [CrossRef]
16. Li, D.; Liu, H.; Ni, W.; Xiao, W.L.; He, L.; Guo, Z.Y.; Qin, X.J.; Liu, H.Y. Molecular networking-based strategy for the discovery of polyacetylated 18-norspirostanol saponins from *Trillium tschonoskii* maxim. *Phytochemistry* **2019**, *168*, 112125. [CrossRef]
17. Frisvad, J.C.; Samson, R.A. *Penicillium subgenus* Penicillium–A guide to identification of food and air-borne terverticillate Penicillia and their mycotoxins. *Stud. Mycol.* **2004**, *49*, 1–173.
18. Strausbaugh, C.A.; Dugan, F. A novel *Penicillium* sp. causes rot in stored sugar beet roots in Idaho. *Plant. Dis.* **2017**, *101*, 1781–1787. [CrossRef]
19. Sheldrick, G.M. *SHELXL-97, Program for X-ray Crystal Structure Solution and Refinement*; University of Göttingen: Göttingen, Germany, 1997.
20. Rajan, R.; Venkataraman, R.; Baby, S. A new lupane-type triterpenoid fatty acid ester and other isolates from *Ophiorrhiza shendurunii. Nat. Prod. Res.* **2016**, *30*, 2197–2203. [CrossRef]
21. Lin, T.S.; Chiang, Y.M.; Wang, C.C.C. Biosynthetic pathway of the reduced polyketide product citreoviridin in *Aspergillus terreus* var. aureus revealed by heterologous expression in *Aspergillus nidulans. Org. Lett.* **2016**, *18*, 1366–1369. [CrossRef]
22. Valente, S.; Piombo, E.; Schroeckh, V.; Meloni, G.R.; Heinekamp, T.; Brakhage, A.A.; Spadaro, D. CRISPR-Cas9-Based discovery of the verrucosidin biosynthesis gene cluster in Penicillium polonicum. *Front. Microbiol.* **2021**, *12*, 660871. [CrossRef] [PubMed]
23. Han, J.J.; Liu, C.C.; Li, L.; Zhou, H.; Liu, L.; Bao, L.; Chen, Q.; Song, F.H.; Zhang, L.X.; Li, E.W.; et al. Decalin-containing tetramic acids and 4-hydroxy-2-pyridones with antimicrobial and cytotoxic activity from the fungus *Coniochaeta cephalothecoides* collected in Tibetan Plateau (Medog). *J. Org. Chem.* **2017**, *82*, 11474–11486. [CrossRef] [PubMed]
24. Ibraheim, Z.Z.; Abdel-Mageed, W.M.; Dai, H.Q.; Guo, H.; Zhang, L.X.; Jaspars, M. Antimicrobial antioxidant daucane sesquiterpenes from *Ferula hermonis* Boiss. *Phytother. Res.* **2012**, *26*, 579–586. [CrossRef]
25. Han, J.J.; Bao, L.; Tao, Q.Q.; Yao, Y.J.; Liu, X.Z.; Yin, W.B.; Liu, H.W. Gloeophyllins A–J, cytotoxic ergosteroids with various skeletons from a chinese tibet fungus *Gloeophyllum abietinum. Org. Lett.* **2015**, *17*, 2538–2541. [CrossRef] [PubMed]
26. Chen, B.S.; Han, J.J.; Wang, M.M.; Dai, H.Q.; Zhang, J.J.; Cai, L.; Wei, S.L.; Zhang, X.; Liu, H.W. Amplisins A–E, chromone methide polymers with hypoglycemic activity from a new fungicolous fungus *Amplistroma fungicola. Org. Chem. Front.* **2020**, *7*, 2761–2769. [CrossRef]
27. Hodge, R.P.; Harris, C.M.; Harris, T.M. Verrucofortine, a major metabolite of *Penicillium verrucosum* var. cyclopium, the fungus that produces the mycotoxin verrucosidin. *J. Nat. Prod.* **1988**, *51*, 66–73. [CrossRef]
28. Bu, Y.Y.; Yamazaki, H.; Takahashi, O.; Kirikoshi, R.; Ukai, K.; Namikoshi, M. Penicyrones A and B, an epimeric pair of α-pyrone-type polyketides produced by the marine-derived *Penicillium* sp. *J. Antibiot.* **2016**, *69*, 57–61. [CrossRef]
29. Choo, S.J.; Park, H.R.; Ryoo, I.J.; Kim, J.P.; Yun, B.S.; Kim, C.J.; Shin-Ya, K.; Yoo, I.D. Deoxyverrucosidin, a novel GRP78/BiP down-regulator produced by *Penicillium* sp. *J. Antibiot.* **2005**, *58*, 210–213. [CrossRef]
30. Asai, T.; Luo, D.; Yamashita, K.; Oshima, Y. Structures and biomimetic synthesis of novel α-pyrone polyketides of an endophytic *Penicillium* sp. in *Catharanthus roseus. Org. Lett.* **2013**, *15*, 1020–1023. [CrossRef]

Journal of *Fungi*

MDPI

Article

Sorbicillinoid Derivatives with the Radical Scavenging Activities from the Marine-Derived Fungus *Acremonium chrysogenum* C10

Chengbao Duan [1,2,†], Shiyuan Wang [1,2,†], Ruiyun Huo [1,2,†], Erwei Li [3], Min Wang [4], Jinwei Ren [1], Yuanyuan Pan [1,*], Ling Liu [1,2,*] and Gang Liu [1,2,*]

1 State Key Laboratory of Mycology, Institute of Microbiology, Chinese Academy of Sciences, Beijing 100101, China; 18813111804@163.com (C.D.); wangshiyuan20@mails.ucas.ac.cn (S.W.); huory@im.ac.cn (R.H.); renjw@im.ac.cn (J.R.)
2 College of Life Sciences, University of Chinese Academy of Sciences, Beijing 100049, China
3 China Institutional Center for Shared Technologies and Facilities, Institute of Microbiology, Chinese Academy of Sciences, Beijing 100101, China; liew@im.ac.cn
4 State Key Laboratory of Microbial Resources, Institute of Microbiology, Chinese Academy of Sciences, Beijing 100101, China; hualuozhisheng@126.com
* Correspondence: panyy@im.ac.cn (Y.P.); liul@im.ac.cn (L.L.); liug@im.ac.cn (G.L.); Tel.: +86-10-64806113 (Y.P.); +86-10-64807043 (L.L.); +86-10-64806017 (G.L.)
† These authors contributed equally to this work.

Citation: Duan, C.; Wang, S.; Huo, R.; Li, E.; Wang, M.; Ren, J.; Pan, Y.; Liu, L.; Liu, G. Sorbicillinoid Derivatives with the Radical Scavenging Activities from the Marine-Derived Fungus *Acremonium chrysogenum* C10. *J. Fungi* **2022**, *8*, 530. https://doi.org/10.3390/jof8050530

Academic Editors: Tao Feng and Frank Surup

Received: 26 April 2022
Accepted: 16 May 2022
Published: 20 May 2022

Abstract: Sorbicillinoids are a class of structurally diverse hexaketide metabolites with good biological activities. To explore new structural sorbicillinoids and their bioactivities, the marine-derived fungus *Acremonium chrysogenum* C10 was studied. Three new sorbicillinoid derivatives, acresorbicillinols A–C (**1–3**), along with five known ones, trichotetronine (**4**), trichodimerol (**5**), demethyltrichodimerol (**6**), trichopyrone (**7**) and oxosorbicillinol (**8**), were isolated. The structures of new sorbicillinoids were elucidated by analysis of nuclear magnetic resonance (NMR) and high-resolution electrospray ionization mass spectroscopy (HRESIMS). The absolute configurations of compounds **1–3** were determined by comparison of the experimental and calculated electronic circular dichroism (ECD) spectra. Compound **3** exhibited a strong 2,2-diphenyl-1-picrylhydrazyl (DPPH) radical scavenging activity, with the IC_{50} value ranging from 11.53 ± 1.53 to 60.29 ± 6.28 μM in 24 h. Additionally, compounds **2** and **3** showed moderate activities against *Staphylococcus aureus* and *Cryptococcus neoformans*, with IC_{50} values of 86.93 ± 1.72 and 69.06 ± 10.50 μM, respectively. The boundary of sorbicillinoid biosynthetic gene cluster in *A. chrysogenum* was confirmed by transcriptional analysis, and the biosynthetic pathway of compounds **1–8** was also proposed. In summary, our results indicated that *A. chrysogenum* is an important reservoir of sorbicillinoid derivatives, and compound **3** has the potential for new natural agents in DPPH radical scavenging.

Keywords: sorbicillinoids; *Acremonium chrysogenum*; structure elucidation; DPPH radical scavenging activity; antimicrobial compounds; marine natural products

1. Introduction

Marine-derived fungi can thrive in the extreme environments including salinity, high pressure, low temperature and oligotrophic conditions compared to their terrestrial counterparts, which makes them able to produce structurally diverse bioactive compounds more easily [1–3]. Meanwhile, these compounds usually have unique structures that also provide the possibility for structural design and modification of the leading compounds [4]. As one special marine-derived fungus, *Acremonium chrysogenum* has made irreplaceable contributions to controlling the bacterial infections and saving countless patients for production of the β-lactam antibiotic cephalosporin C (CPC) and its derivatives [5]. The genomic sequences and annotation of *A. chrysogenum* was first completed in 2014, and a total of

42 secondary metabolite biosynthetic gene clusters, including 14 polyketide synthetase (PKS) clusters, 10 terpene synthase clusters, 8 hybrid clusters, 7 nonribosomal peptide synthetase clusters and 3 non-identified secondary metabolite clusters, were predicted [6].

Sorbicillinoids are a class of structurally diverse hexaketide metabolites with a characteristic sorbyl side chain residue [7–9]. They were first isolated from *Penicillium notatum* in 1948 and structurally elucidated by Cram and Tishler [10,11]. Up until now, more than 159 naturally occurring sorbicillinoids have been isolated and have displayed good biological activities in cytotoxic, antimicrobial and phytotoxic activities [7–9]. Because free radicals play an important role in the development of aging and many diseases, including cancer, arthritis and atherosclerosis, exploring the novel radical scavengers is crucial for developing new drugs to slow down the aging process and treat these diseases. Some sorbicillinoid derivatives have shown great antioxidative application prospects, such as bisorbicillinol (ED_{50} = 31.4 μM) and bisorbibetanone (ED_{50} = 62.5 μM), etc. [8]. Additionally, there is an urgent need to find more novel compounds for the emergence of microbial resistance. Some sorbicillinoids showed significant antimicrobial activity, indicating their potential as candidates [7]. Meanwhile, the sorbicillinoid biosynthetic gene clusters from *Penicillium chrysogenum* and *Trichoderma reesei* have been identified and their biosynthetic pathway has been partially revealed [12–14]. Generally, two PKSs SorA and SorB are responsible for the formation of sorbicillin and dihydrosorbicillin, which are then oxidative dearomatized to give sorbicillinol and dihydrosorbicillinol by the FAD-dependent monooxygenase SorC, respectively [15]. Sorbicillinol is regarded as the precursor of most sorbicillinoids since it is condensed with its derivatives or other compounds to form the dimeric and hybrid sorbicillinoids by Diels-Alder or Michael-addition-like reactions [16,17]. The sorbicillinoid biosynthetic gene cluster in *A. chrysogenum* has been regarded as the most ancient, based on evolutionary origin, and carries more modifier than other species [13], and disruption of these two PKS encoding genes results in the abolishment of sorbicillinoids [18]. However, there is lack of a systematic investigation about sorbicillinoids produced by *A. chrysogenum*.

Based on the chemical investigations in this study, the resulting crude extracts of *A. chrysogenum* C10 from the rice solid fermentation, which has a higher accumulation of compounds and reproducibility than submerged fermentation [19], had afforded three structurally unique compounds: acresorbicillinols A–C (**1**–**3**) and five known sorbicillinoids including trichotetronine (**4**) [20], trichodimerol (**5**) [21], demethyltrichodimerol (**6**) [21], trichopyrone (**7**) [22] and oxosorbicillinol (**8**) [20] (Figure 1). Compounds **1**–**8** were evaluated for their DPPH radical scavenging abilities and antimicrobial activities. In addition, the boundary of sorbicillinoid biosynthetic gene cluster (*Acsor*) was confirmed and its biosynthetic pathway was proposed. This study reported the isolation, structural elucidation and bioactivities of the isolated compounds from *A. chrysogenum* C10.

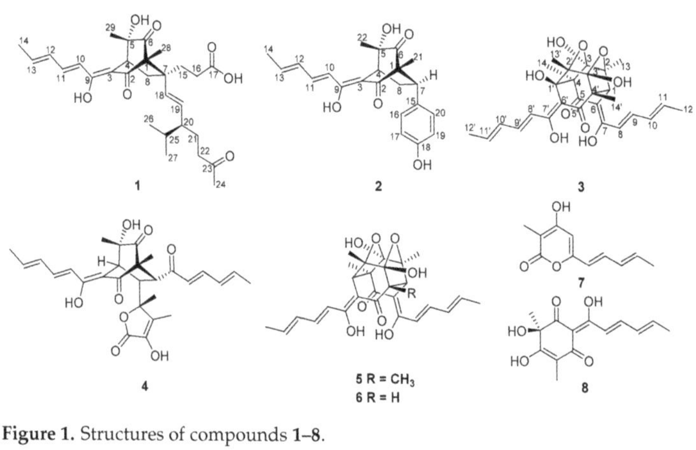

Figure 1. Structures of compounds **1**–**8**.

2. Materials and Methods

2.1. General Experimental Procedure

Optical rotations, ECD spectra, UV and IR data were measured on the Austria Anton Paar MCP 200 Automatic Polarimeter, the Applied Photophysics Chirascan circular dichroism spectrometer, the Thermo Scientific GENESYS 10S UV-Vis and the Thermo Scientific Nicolet IS5 spectrophotometers, respectively. HRESIMS data and MS were obtained using an Agilent 6520B Q-TOF Mass instrument equipped with an ESI source. All MS experiments were performed in positive ion mode. NMR data were acquired with the AVANCE-500 spectrometer (Bruker, Bremen, Germany) using solvent signals (CD_3OD, δ_H 3.30/δ_C 49.9, DMSO, δ_H 2.50, 3.30/δ_C 39.5, and $CDCl_3$, δ_H 7.26/δ_C 77.16) as references. Octadecylsilyl (ODS, 50 μm, YMC Co., Ltd. Japan) and Sephadex^{TM} LH-20 (Cytiva, Uppsala, Sweden) were used for column chromatography. High performance liquid chromatography (HPLC) was performed on the SHIMADZU LC20AT system equipped with UV diode array detector using the Thermo Hypersil Gold-C18 columns (5 μm, 250 mm × 4.6 mm) at a flow rate of 1 mL/min. For semi-preparative HPLC, Waters 1525 system equipped with the UV/Visible detector and the Thermo Hypersil Gold-C18 columns (5 μm, 250 mm × 10 mm) was used and performed at a flow rate of 2 mL/min. Solvents including methanol and ethyl acetate (EtOAc) for extraction and chromatographic separation were analytical grade. HPLC-grade solvents (acetonitrile and formic acid) were used for the HPLC and semi-preparative HPLC analysis.

2.2. Fungal Materials and Fermentation

One high CPC-producing strain of *A. chrysogenum* C10 (ATCC 48272) was released by PanLab. This fungus was inoculated on the rice solid medium in 500 mL Erlenmeyer flasks containing 80 g of rice and 120 mL of H_2O, and cultivated at 28 °C for 7 days for the production of sorbicillinoids. A total of 10 kg fermentation sample was harvested.

2.3. Extraction and Isolation

The rice solid fermentation of *A. chrysogenum* was extracted with EtOAc (3 × 5 L) under the ultrasonication processing. The organic solvents were filtered and evaporated by the vaccum to get the crude extracts (25 g). Extracts were fractionated by ODS reverse silica gel using the gradient MeOH/H_2O (v/v, 10%, 20%, 30%, 40%, 50%, 55%, 60%, 65%, 70%, 75%, 80%, 85%, 90%, 95%, 100%) to afford 15 fractions (Fr.1–Fr.15). Fr.8 (MeOH/H_2O (v/v, 65%)) (150 mg) was further subjected to the Sephadex^{TM} LH-20 and eluted with MeOH to give 30 subfractions. Fr.8–24 (30 mg) was purified by semi-preparative RP-HPLC using 50% acetonitrile in acidic water (0.1% formic acid) to give compounds **2** (5.0 mg, t_R = 23 min), **4** (4.0 mg, t_R = 24.1 min) and **8** (2.0 mg, t_R = 22.2 min). Fr.10 (MeOH/H_2O (v/v, 75%)) (133 mg) was subjected to the Sephadex^{TM} LH-20 and eluted with MeOH to give 30 subfractions. Fr.10–12 (56 mg) was purified by semi-preparative RP-HPLC using 58% acetonitrile in acidic water (0.1% formic acid) to yield compounds **1** (3.5 mg, t_R = 27.2 min) and **6** (2.5 mg, t_R = 26.4 min). Fr.11 (MeOH/H_2O (v/v, 80%)) (220 mg) was subjected to the Sephadex^{TM} LH-20 and eluted with MeOH to give 30 subfractions. Fr.11–24 (60 mg) was purified by semi-preparative RP-HPLC using 60% acetonitrile in acidic water (0.1% formic acid) to yield compounds **3** (10.0 mg, t_R = 28.5 min) and **5** (2.9 mg, t_R = 28.3 min). Fr.5 (MeOH/H_2O (v/v, 50%)) (96 mg) was subjected to the Sephadex^{TM} LH-20 and eluted with MeOH to give 25 subfractions. Fr.5-16 (10 mg) was purified by semi-preparative RP-HPLC using 37% acetonitrile in acidic water (0.1% formic acid) to yield compound **7** (2.0 mg, t_R = 17.5 min).

Acresorbicillinol A (**1**): pale yellow solid; $[\alpha]_D^{25}$ +81 (c 0.1, MeOH); UV (MeOH) λ_{max} (log ε) 223 (3.65), 367 (1.72) nm; ECD (c 3.0 × 10^{-3} M, MeOH) λ_{max} ($\Delta\varepsilon$) 200 (+4.51), 228 (–8.29), 270 (–12.89), 315 (–34.60), 352 (+24.06) nm; IR (neat) ν_{max} 3399, 2956, 2872, 1722, 1601, 1446, 1381, 1258 cm^{-1}; ^1H and ^{13}C NMR, Table 1; HRESIMS at m/z 501.2850 $[M + H]^+$ (calcd for $C_{29}H_{41}O_7$, 501.2847).

Table 1. ^{1}H NMR (500 MHz) and ^{13}C NMR data (125 MHz) for **1** and **2**.

Position	**1** [a]		**2** [a]	
	δ_H (*J* in Hz)	δ_C	δ_H (*J* in Hz)	δ_C
1		70.3, qC		66.7, qC
2		200.3, qC		199.7, qC
3		112.3, qC		113.8, qC
4	3.18, t (2.8)	41.5, CH	3.30, t (2.7)	42.3, CH
5		75.4, qC		75.2, qC
6		212.3, qC		211.4, qC
7		47.8, qC	3.09, dd (10.6, 6.1)	47.5, CH
8a	2.38, m	30.6, CH_2	3.00, ddd (13.6, 10.6, 2.7)	32.7, CH_2
8b	1.97, dd (13.3, 2.8)		1.80, ddd (13.6, 6.1, 2.7)	
9		167.6, qC		167.7, qC
10	6.42, d (14.6)	119.5, CH	6.48, d (14.6)	119.6, CH
11	7.26, dd (14.6, 10.9)	142.9, CH	7.37, dd (14.6, 11.0)	143.3, CH
12	6.39, dd (14.6, 10.9)	132.3, CH	6.41, dd (14.6, 11.0)	132.3, CH
13	6.20, dq (14.6, 7.0)	140.0, CH	6.23, dq (14.6, 7.0)	140.1, CH
14	1.89, d (7.0)	18.9, CH_3	1.90, d (7.0)	18.9, CH_3
15a 15b	1.81, td (13.2, 4.8) 1.50, m	34.1, CH_2		133.9, qC
16	2.16, m	31.4, CH_2	6.80, d (8.4)	130.5, CH
17		178.3, qC	6.67, d (8.4)	116.2, CH
18	5.18, d (15.6)	135.4, CH		157.7, qC
19	5.13, dd (15.6, 9.0)	135.9, CH	6.67, d (8.4)	116.2, CH
20	1.68, m	50.6, CH	6.80, d (8.4)	130.5, CH
21a 21b	1.64, m 1.23, m	27.2, CH_2	0.80, s	11.4, CH_3
22a 22b	2.42, m 2.30, m	42.4, CH_2	1.21, s	24.0, CH_3
23		212.4, qC		
24	2.16, s	30.0, CH_3		
25	1.54, m	33.4, CH		
26	0.86, d (7.0)	21.2, CH_3		
27	0.81, d (7.0)	19.7, CH_3		
28	1.16, s	7.4, CH_3		
29	1.12, s	24.5, CH_3		

[a] Recorded in CD_3OD.

Acresorbicillinol B (**2**): pale yellow solid; $[\alpha]_D^{25}$ +5 (*c* 0.1, MeOH); UV (MeOH) λ_{max} (log ε) 221 (2.72), 322 (0.68), 351 (0.54) nm; ECD (*c* 3.0 × 10^{-3} M, MeOH) λ_{max} ($\Delta\varepsilon$) 215 (–14.07), 245 (+36.00), 315 (–75.16) nm, 360 (+13.47) nm; IR (neat) ν_{max} 3413, 1724, 1624, 1440, 1378, 1243 cm^{-1}; ^{1}H and ^{13}C NMR, Table 1; HRESIMS at m/z 369.1696 $[M + H]^+$ (calcd for $C_{22}H_{25}O_5$, 369.1697).

Acresorbicillinol C (**3**): bright yellow solid; $[\alpha]_D^{25}$ −1048 (*c* 0.1, MeOH); UV (MeOH) λ_{max} (log ε) 207 (1.82), 37 (2.12), 278 (2.55), 375 (3.23) nm; ECD (*c* 3.0 × 10^{-3} M, MeOH) λ_{max} ($\Delta\varepsilon$) 221 (–22.24), 275 (+38.06), 345 (+51.25) nm, 405 (–88.67) nm; IR (neat) ν_{max} 3420, 1664,

1606, 1556, 1412, 1347, 1209 cm^{-1}; ^1H and ^{13}C NMR, Table 2; HRESIMS at m/z 513.2116 $[M + H]^+$ (calcd for $C_{28}H_{33}O_9$, 513.2119).

Table 2. ^1H NMR (500 MHz) and ^{13}C NMR data (125 MHz) for **3**.

Position	3 [b]	
	δ_H (J in Hz)	δ_C
1	3.71, s	53.9, CH
2		78.2, qC
3		107.8, qC
4		59.2, qC
5		190.9, qC
6		100.6, qC
7		167.9, qC
8	6.49, d (14.6)	120.6, CH
9	7.12, dd (14.6, 10.9)	137.8, CH
10	6.38, overlap	131.1, CH
11	6.10, (14.6, 6.8)	136.2, CH
12	1.83, d (6.8)	18.4, CH_3
13	1.30, s	25.2, CH_3
14	1.29, s	18.8, CH_3
1′		78.3, qC
2′		78.7, qC
3′		103.5, qC
4′		59.2, qC
5′		199.3, qC
6′		108.0, qC
7′		185.2, qC
8′	7.38, d (14.6)	122.4, CH
9′	7.48, dd (14.6, 10.9)	146.5, CH
10′	6.38, overlap	131.1, CH
11′	6.42, overlap	143.4, CH
12′	1.89, d (6.8)	18.3, CH_3
13′	1.17, s	22.2, CH_3
14′	1.31, s	18.8, CH_3
OH-7	16.38, s	
OH-7′	18.02, s	

[b] Recorded in DMSO:$CDCl_3$ = 3:1.

2.4. ECD Calculations

Conformational analyses were performed using Maestro 10.2 in the OPLS3 molecular mechanics force-field within an energy window of 5.0 or 3.0 kcal/mol. The conformers were then further optimized with the software package Gaussian 09 at the B3LYP/6-31G(d) level for compounds **1**–**3**, respectively, and the harmonic vibrational frequencies were also calculated to confirm their stability. The TDDFT methods at the CAM-B3LYP/6-31G(d) and B3LYP/6-31G(d) level were applied to calculate the 60 lowest electronic transitions to obtain conformers in a vacuum, respectively. The Gaussian function was applied to simulate the ECD spectrum of the conformers. The calculated ECD spectra were obtained according to the Boltzmann weighting of each conformer's ECD spectrum [23].

2.5. Antimicrobial Activity Assay

The bacterial strains (*Staphylococcus aureus* CGMCC 1.89, *Pseudomonas aeruginosa* ATCC 15692) and the fungal strains (*Cryptococcus neoformans* W1585, *Candida albicans* SC5314) were used in this study. The concentration of 50 mM compounds was prepared using dimethyl sulfoxide (DMSO). The bacterial and fungal strains were streaked onto Mueller–Hinton Agar (MHA) and Potato Dextrose Agar (PDA) for growth at 37 °C and 28 °C, respectively. Single colony was picked and adjusted to 2×10^5 CFU/mL by Mueller–Hinton Broth (MHB) or Potato Dextrose Broth (PDB). The stock solutions of compounds were diluted into 500, 250, 125, 62.5 and 31.25 μM by MHB or PDB, successively. Fifty microliters of serial dilutions of each compound and 50 μL of microbial suspension were added to the 96-well plates and incubated at 37 °C or 28 °C for 24 h until the results were recorded. IC_{50} was defined as the half maximal inhibitory concentrations of the compounds that inhibited the visible microbial growth after 24 h of incubation. Ampicillin and amphotericin B were used as the positive control for detecting the activities of these compounds against bacteria and fungi, respectively.

2.6. DPPH Radical Scavenging Assay

The DPPH radical scavenging activity of the compounds was carried out as previously described [24,25]. The modified parameter was the reaction time from 0.5 h to multiple time-points including 0.5, 1, 4, 6, 8 and 24 h. Ascorbic acid and ethanol were used as the positive and negative control, respectively. All experiments were replicated at least three times.

2.7. RNA Isolation and Real-Time RT-PCR Analysis

The mycelia of *A. chrysogenum* C10 grown on the modified MDFA medium were collected at different time-points [26]. RNA isolation and real-time RT-PCR were performed as described previously [27,28]. All primers used in this study were listed in Table S1.

3. Results and Discussion

3.1. Isolation and Structure Elucidation

Acresorbicillinol A (**1**) was obtained as a pale yellow solid, and its molecular formula was established as $C_{29}H_{40}O_7$ based on HRESIMS data at 501.2850 $[M + H]^+$ (calcd for $C_{29}H_{41}O_7$, 501.2847), indicating 10 degrees of unsaturation. The IR spectrum indicated the presence of hydroxy (3399 cm^{-1}) and ketone (1722 cm^{-1}) groups. The ^1H NMR data (Table 1 and Figure S1) of **1** showed signals for six methyl signals [δ_H 2.16 (s, H_3-24), 1.89 (d, J = 7.0 Hz, H_3-14), 1.16 (s, H_3-28), 1.12 (s, H_3-29), 0.86 (d, J = 7.0 Hz, H_3-26), and 0.81 (d, J = 7.0 Hz, H_3-27)], five methylene protons [δ_H 2.42 (m, H-22a), 2.38 (m, H-8a), 2.30 (m, H-22b), 2.16 (m, H_2-16), 1.97 (dd, J = 13.3, 2.8 Hz, H-8b), 1.81 (td, J = 13.2, 4.8 Hz, H-15a), 1.64 (m, H-21a), 1.50 (m, H-15b), and 1.23 (m, H-21b)], three methine protons [δ_H 3.18 (t, J = 2.8 Hz, H-4), 1.68 (m, H-20), and 1.54 (m, H-25)], six olefinic protons [δ_H 7.26 (dd, J = 14.6, 10.9 Hz, H-11), 6.42 (d, J = 14.6 Hz, H-10), 6.39 (dd, J = 14.6, 10.9 Hz, H-12), 6.20 (dq, J = 14.6, 7.0 Hz, H-13), 5.18 (d, J = 15.6 Hz, H-18), and 5.13 (dd, J = 15.6, 9.0 Hz, H-19)]. Detailed interpretation of the ^{13}C NMR and HSQC data (Table 1, Figures S2 and S4) of **1** revealed the presence of 29 carbon resonances corresponding to six methyls, five sp^3 methylenes, three sp^3 methines, six sp^2 methines, three sp^3 quarternary carbons with one oxygenated, two sp^2 non-protonated carbons and four carbonyl carbons (δ_C 212.4, 212.3, 200.3 and 178.3, respectively). These data accounted for all ^1H and ^{13}C NMR resonances of **1** except for three unobserved exchangeable protons, suggesting that **1** was a bicyclic compound. The planar structure of **1** was assigned through detailed analysis of the ^1H-^1H COSY and HMBC correlations (Figures 2, S3 and S5). The ^1H-^1H COSY (Figure 2) correlations of H-10/H-11/H-12/H-13/H_3-14, combined with the HMBC correlations from H-10 to the olefinic carbons C-3 (δ_C 112.3) and C-9 (δ_C 167.6) and from H-11 to C-9, suggested the presence of the enolic sorbyl side chain. The HMBC correlations (Figure 2) from H-4 to C-3, the sp^3 quaternary carbon C-5 (δ_C 75.4) and two ketone carbons C-2 (δ_C 200.3) and C-6 (δ_C 212.3), from H_3-28 to the sp^3 quaternary carbon C-1 (δ_C 70.3), C-2 and C-6, and from H_3-29 to C-4, C-5, and C-6 permitted the completion of the cyclohexandione ring, with the enolic sorbyl unit positioned at C-3 and two methyl groups located at C-1 and C-5, respectively. Meanwhile, the ^{1}H-^{1}H COSY (Figure 2) correlations of H-18/H-19/H-20/H_2-21/H_2-22 and of H-20/H-25/H_3-26/H_3-27, as well as the HMBC correlations from H_2-22 to the ketone carbon C-23 (δ_C 212.4) and C-24 (δ_C 30.0), and from H_3-24 to C-22 and C-23, established the 3-isopropyl-6-oxohept-1-en-1-yl (C-18–C-27) subunit. Moreover, the ^{1}H-^{1}H COSY (Figure 2) correlations of H_2-15/H_2-16, and the HMBC correlations from H_2-15 and H_2-16 to the carbonyl carbon C-17 (δ_C 178.3), indicated that carbonyl carbon C-17 was attached to C-16 directly. Additional HMBC correlations from H_2-15 to the sp^3 quaternary carbon C-7 (δ_C 47.8) and the olefinic carbon C-18 (δ_C 135.4), from H_2-16 and H-19 to C-7, and from H-18 to C-7 and C-15, indicated that C-7 was located between C-15 and C-18. Key HMBC correlations from H-4, H_2-8 and H_3-28 to C-7, and from H_2-15 and H-18 to C-1 and C-8, along with the ^{1}H-^{1}H COSY correlations of H-4/H_2-8 implied that C-1 and C-8 were all connected to C-7, permitting the completion of the bridged bicyclo [2.2.2]octane-2,6-dione core structure. By consideration of the molecular formula and the chemical shifts of C-5 (δ_C 75.4) and C-17 (δ_C 178.3), these two carbons should be hydroxylated. Thus, the planar structure of **1** was established as shown (Figure 1).

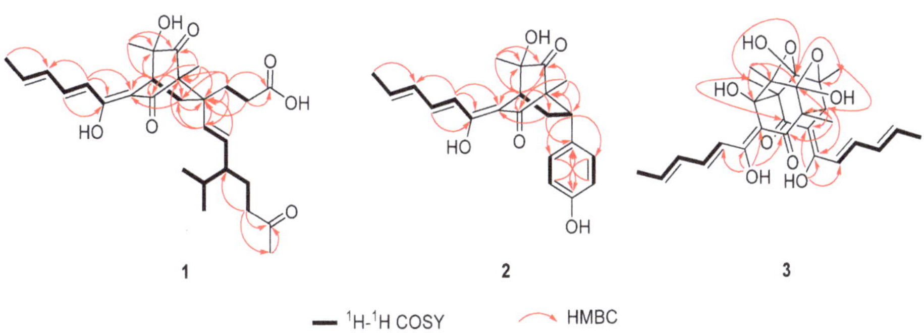

Figure 2. Key COSY and HMBC correlations of compounds **1–3**.

The relative configuration of **1** was determined by NOESY correlations, coupling constants and HMBC correlations. The NOESY correlation (Figures 3 and S6) of H-10 with H-4 assigned the olefin C-3/C-9 as *Z* geometry. The geometry of the conjugated diene was assigned as 10*E*, 12*E* by the large coupling constants ($J_{H\text{-}10/H\text{-}11}$ = 14.6 Hz and $J_{H\text{-}12/H\text{-}13}$ = 14.6 Hz) along with the NOESY correlations of H_3-14 with H-12 and of H-13 with H-11. The *E* geometry of the C-18/C-19 double bond was also deduced by the large coupling constant between H-18 and H-19 (15.6 Hz). The NOESY correlations of H-10 with H-4 and H_3-29 suggested that these protons were close in space. Moreover, the strong HMBC correlations from H-8a to C-3 and C-15, and from H-8b to C-5, and the weak correlation from H-8a to C-5, as well as the lack of HMBC correlation from H-8b to C-3 and C-15, indicated that H-8a and C-15 were eclipsed and that H-8b and C-3 were gauche [20,29]. Meanwhile, the NOESY correlations of H_3-28 with H-15a, and of H-8a with H-15b, assigned the relative configurations of C-1 and C-7. However, the relative configuration for C-20 could not be established by the NOESY data. The absolute configuration for **1** was assigned by a comparison of the experimental and calculated ECD spectra of two pairs of enantiomers, (1*R*,4*S*,5*S*,7*R*,20*S*)-**1** (**1a**), (1*S*,4*R*,5*R*,7*S*,20*R*)-**1** (**1b**), (1*R*,4*S*,5*S*,7*R*,20*R*)-**1** (**1c**), and (1*S*,4*R*,5*R*,7*S*,20*S*)-**1** (**1d**). The ECD calculations were conducted using time-dependent density functional theory (TDDFT) at the CAM-B3LYP/6-

31G(d) level. The overall calculated ECD spectrum of **1a–1d** was then generated according to Boltzmann weighting of the conformers (Figure S19). For compound **1** the experimental first positive (200 nm), second negative (228 nm), third negative (270 nm), fourth negative (315 nm) and fifth positive (352 nm) Cotton effects compared well with the calculated ECD curve for (1*R*,4*S*,5*S*,7*R*,20*S*)-**1** (**1a**), which showed five corresponding Cotton effects around 200, 222, 270, 315 and 350 nm (Figure 4). Therefore, qualitative analysis of the result allowed the assignment of the absolute configuration of **1** as 1*R*,4*S*,5*S*,7*R*,20*S*.

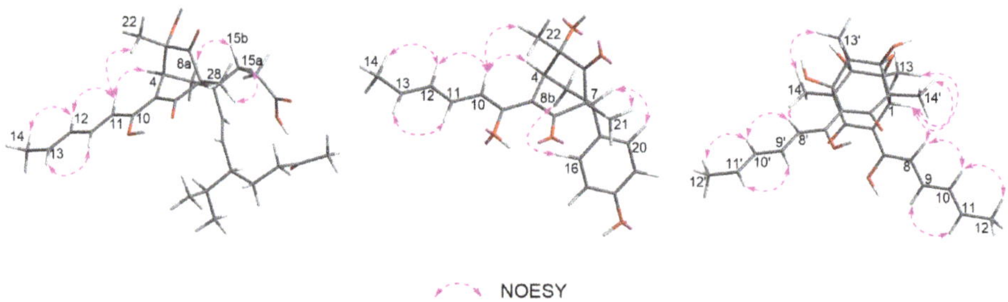

Figure 3. Key NOESY correlations of compounds **1–3**.

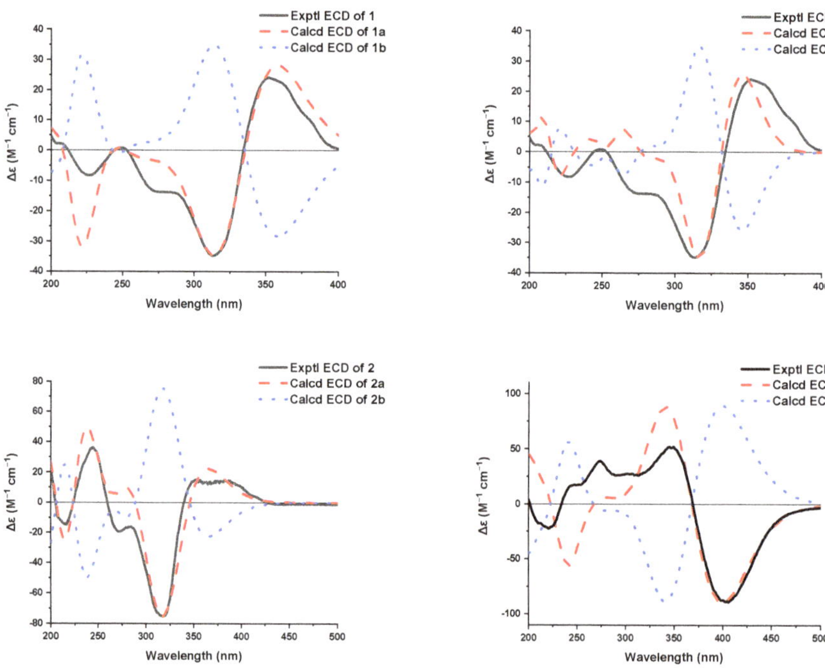

Figure 4. Calculated and experimental ECD spectra of compounds **1–3**.

Acresorbicillinol B (**2**) was obtained as a pale yellow solid. The molecular formula of **2** was assigned as $C_{22}H_{24}O_5$ (11 degrees of unsaturation) based on its HRESIMS data at *m/z* 369.1696 $[M + H]^+$ (calcd for $C_{22}H_{25}O_5$, 369.1697). The ^1H and ^{13}C NMR spectroscopic data (Table 1, Figures S7 and S8), in association with the HSQC spectrum (Figure S10), indicated 22 carbon resonances including 3 methyl groups, 1 sp^3 methylenes, 2 sp^3 methines, 2 sp^3 non-protonated carbons with 1 oxygenated, 12 olefinic or aromatic carbons (8 protonated), and 2 carbonyl carbons (δ_C 211.4, and 199.7, respectively), which were similar to those of **1**.

Analysis of the ^1H-^1H COSY and HMBC data (Figure 2, Figures S9 and S11) of **2** determined the same bicyclo [2.2.2]octane-2,6-dione moiety with the enolic sorbyl substituted at C-3. However, the substitutes at C-7 of **2** were different from those of **1**. The HMBC correlations from H_2-8 to C-15 (δ_C 133.9), from H-7 to C-15, C-16 (δ_C 130.5) and C-20 (δ_C 130.5), from H-16/H-20 to C-7 (δ_C 47.5) and C-18 (δ_C 157.7) and from H-17/H-19 to C-15 completed the *para*-hydroxyphenyl group located at C-7. On the basis of these data, the planar structure of **2** was established as shown (Figure 1).

The relative stereochemistry of **2** was determined by NOESY correlations and coupling constants as well as by comparison with those of **1** and the known compound sorbicatechol C [30]. The large coupling constants ($J_{H-10/H-11}$ = 14.6 Hz and $J_{H-12/H-13}$ = 14.6 Hz), along with NOESY correlations (Figures 3 and S12) of H_3-14 with H-12 and of H-13 with H-11 indicated that the geometry of the conjugated diene was 10*E*, 12*E*. Furthermore, the NOESY correlation (Figure 3) of H-4 with H-10 implied a *Z* geometry of the C3/C9 double bond. Other NOESY correlations of H-10 with H-4 and H_3-22, and of H-4 with H_3-22, placed these protons on the same side. While NOESY correlations of H-8b (δ_H 1.80, ddd, *J* = 13.6, 6.1, 2.7 Hz) with H-16 (H-20), and of H-7 with H_3-21, combined with the strong HMBC correlations from H-8b to C-5 and C-15, the weak correlation from H-8b to C-3 and lack of HMBC correlation from H-8a to C-5 and C-15 determined the relative stereochemistry of C-7 and C-1 as shown. The absolute configuration of **2** was also determined by a comparison of the experimental and calculated ECD spectra for enantiomers (1*R*,4*S*,5*S*,7*R*)-**2** (**2a**) and (1*S*,4*R*,5*R*,7*S*)-**2** (**2b**). As shown in Figure 4, the experimental ECD spectrum of **2** showed good agreement with the calculated ECD spectrum of (1*R*,4*S*,5*S*,7*R*)-**2** (**2a**), suggesting the absolute configuration of 1*R*,4*S*,5*S*,7*R* for **2**. Thus, the structure of **2** was defined as shown.

Acresorbicillinol C (**3**) was obtained as a bright yellow solid, and its molecular formula was deduced to be $C_{28}H_{32}O_9$ (13 degrees of unsaturation) on the basis of the HRESIMS data at *m/z* 513.2116 $[M + H]^+$ (calcd for $C_{28}H_{33}O_9$, 513.2119). The IR absorptions suggested the presence of hydroxy (3420 cm^{-1}) and ketone (1664 cm^{-1}) groups. Its ^1H NMR data (Table 2 and Figure S13) revealed signals of eight olefinic protons [δ_H 6.10–7.49], one methine proton [δ_H 3.71 (s, H-1)] and six methyls [δ_H 1.89 (d, *J* = 6.8 Hz, H_3-12'), 1.83 (d, *J* = 6.8 Hz, H_3-12), 1.31 (s, H_3-14'), 1.30 (s, H_3-13), 1.29 (s, H_3-14), 1.17 (s, H_3-13')]. The ^{13}C NMR spectrum (Table 2 and Figure S14) and the HSQC data (Figure S16) displayed a total of 28 carbon resonances, which were assignable to 6 methyl groups, 8 sp^2 methines, 1 sp^3 methines, 13 non-protonated carbons containing 2 carbonyls (δ_C 199.3, and 190.9), 4 sp^2 non-protonated with two oxygenated, 7 sp^3 non-protonated carbon with 5 oxygenated. These signals (Table 2 and Figures S13–S16) were very similar to those of trichodimerol (**5**) [31,32], except that the proton at the C-1′ position in **5** was changed to a hydroxy moiety in **3**. This was evidenced by the HRESIMS data and HMBC correlations (Figures 2 and S17) from H_3-13′ and H_3-14 to C-1′ (δ_C 78.3). Therefore, **3** was 1′-hydroxylated analogue of **5**.

The relative configuration of **3** was confirmed by NOESY correlations and coupling constants. The NOESY correlations (Figures 3 and S18) of H-9/H-11, of H-8/H-10/H-12, of H-9′/H-11′ and of H-8′/H-10′/H-12′, along with the large coupling constants ($J_{H-8/H-9}$ = $J_{H-10/H-11}$ = $J_{H-8'/H-9'}$ = $J_{H-10'/H-11'}$ = 14.6 Hz) suggested the 8*E*, 10*E*, 8′*E* and 10′*E* configurations of the conjugated dienes in the sorbyl side chains. Meanwhile, the NOESY correlations of H-1/H-8 and H_3-14/H-8′ suggested the *Z* geometry of C-6/C-7 and C-6′/C-7′ double bonds. Furthermore, the NOESY correlations of H-1/H_3-13, of H_3-13′/H_3-14 and of H_3-14′/H-1 inferred that these protons were in close proximity to their related functional groups, respectively. The similar Cotton effects in the ECD spectra of **3** and **5** deduced the absolute configuration of **3** to be the same as that of **5**, which was further verified by ECD calculations (Figure 4). The calculated ECD curve of (1*S*,2*S*,3*R*,4*R*,1′*R*,2′*S*,3′*R*,4′*R*)-**3** (**3a**) matched well with the experimental data, suggesting the absolute configuration to be 1*S*,2*S*,3*R*,4*R*,1′*R*,2′*S*,3′*R*,4′*R*. Thus, the structure of **3** was defined as depicted.

Except for the new compounds **1–3**, the structure of five known sorbicillinoids isolated in this study were confirmed by comparison of the spectroscopic data with those in the literature [20–22]. The resulting EtOAc extracts of *A. chrysogenum* cultivated on the rice were screened by HPLC analysis (Figure S20).

3.2. Biological Activities Evaluation

To explore the bioactivities of compounds **1–8**, their abilities of anti-microorganisms and DPPH radical scavenging were evaluated. The results showed that compounds **2** and **3** exhibited the moderate activities against *S. aureus* and *C. neoformans* with the IC_{50} values of 86.93 ± 1.72 and 69.06 ± 10.50 μM, respectively. However, other compounds did not give IC_{50} value at a concentration below 100 μM (Table 3). No candidate compounds could significantly inhibit the growth of *C. albicans* and *P. aeruginosa*. Compound **3** might function as the β-1,6-glucan inhibitor to inhibit the fungal growth as its structural analogue bisvertinolone [33]. Bisvertinolone also exhibited significant inhibitory activity against *S. aureus* with the minimal inhibitory concentration (MIC) value of 30 μg/mL [34]. However, only several monomeric sorbicillinoids from *Scytalidium album* exhibited the weak activity against *C. neoformans* with the MIC value of over 38 μg/mL [35].

Table 3. Anti-microbial inhibitory activities of compounds **1–8**.

Compounds	*S. aureus*	*C. neoformans*
	IC_{50} (μM)	
1	>100	>100
2	86.93 ± 1.72	>100
3	>100	69.06 ± 10.50
4	>100	>100
5	>100	>100
6	>100	>100
7	>100	>100
8	>100	>100
Ampicillin	0.016 ± 0.004	–
Amphotericin B	–	0.018 ± 0.003

Through the DPPH radical scavenging assay, compound **3** exhibited strong activity with the IC_{50} value of 60.29 ± 6.28 μM after standing for 0.5 h, and then we continued to record its radical scavenging activity for 24 h (at 1, 4, 6, 8 and 24 h). Compound **3** gave the significant activity with the IC_{50} values of 43.52 ± 5.93, 22.57 ± 7.34, 15.85 ± 5.94, 12.30 ± 5.74 and 11.53 ± 1.53 μM, respectively, indicating that **3** displays the time-dependent manner for DPPH radical scavenging. Compared with the IC_{50} value of ascorbic acid as the positive control, which was 25.36 ± 3.82 to 28.45 ± 3.04 μM, compound **3** represents one novel DPPH radical scavenging agent (Figure 5 and Table 4). Compound **8** exhibited the radical scavenging activity with the IC_{50} values of 155.40 ± 12.42 and 55.36 ± 14.92 μM for 0.5 and 24 h, respectively. Although the IC_{50} values of **4**, **5** and **6** were over 200 μM for 0.5 h, their radical scavenging activity significantly enhanced at 24 h, and the IC_{50} values were 151.87 ± 15.63, 116.83 ± 3.93 and 102.48 ± 5.04 μM, respectively (Table 4). Compounds **4**, **5**, **6** and **8** also displayed the time-dependent manner as compound **3**. The time-dependent manner of sorbicillinoids for radical scavenging was previously reported, including for oxosorbicillinol, trichotetronine, bisorbicillinolide and methylbisorbibutenolide [22,36,37]. There was a different scavenging values of **4** and **8** between this study and the reports in Hirota's Lab, and the reaction buffer might be the key determination factor. Additionally, the IC_{50} values of compounds **1**, **2** and **7** exceeded 200 μM, even standing for 24 h, indicating that they did not have DPPH radical scavenging ability (Table 4). DPPH radical scavenging activity of other representative sorbicillinoids has been reported, including for bisorbicillinol, bisvertinolone and bisorbibetanone, which showed ED_{50} values of 31.4, 44.3 and 62.5 μM, respectively [21,37]. To date, compound **3** displayed the best DPPH radical scavenging activity for 24 h among all reported sorbicillinoids.

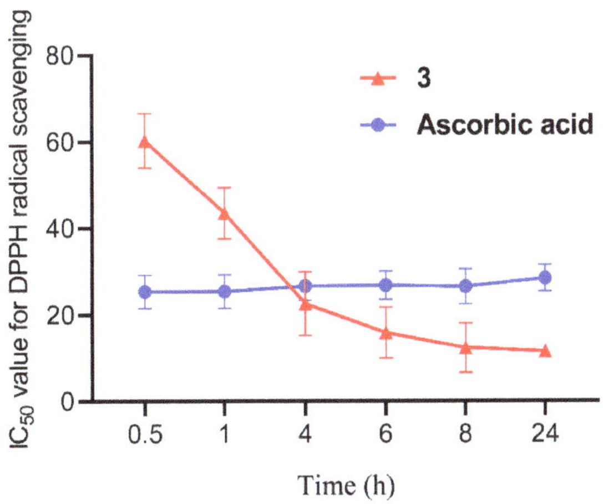

Figure 5. DPPH radical scavenging activity of compound **3** and ascorbic acid as the positive control at 0.5, 1, 4, 6, 8 and 24 h.

Table 4. DPPH radical scavenging activities of compounds **1–8**.

Compounds	**IC_{50} Value (μM)**					
	0.5 h	**1 h**	**4 h**	**6 h**	**8 h**	**24 h**
1	>200	>200	>200	>200	>200	>200
2	>200	>200	>200	>200	>200	>200
3	60.29 ± 6.28	43.52 ± 5.93	22.57 ± 7.34	15.85 ± 5.94	12.30 ± 5.74	11.53 ± 1.53
4	>200	>200	>200	>200	>200	151.87 ± 15.63
5	>200	>200	>200	>200	>200	116.83 ± 3.93
6	>200	>200	>200	>200	197.73 ± 27.70	102.48 ± 5.04
7	>200	>200	>200	>200	>200	>200
8	155.40 ± 12.42	129.87 ± 12.09	88.38 ± 16.29	77.20 ± 15.38	71.00 ± 14.56	55.36 ± 14.92
Ascorbic acid	25.36 ± 3.82	25.42 ± 3.85	26.65 ± 3.29	26.77 ± 3.24	26.48 ± 4.03	28.45 ± 3.04

3.3. Determination of Acsor Cluster Boundary and Its Proposed Biosynthetic Pathway of Sorbicillinoid

To confirm the boundary of the sorbicillinoid biosynthetic gene cluster, the total RNA was isolated from *A. chrysogenum* C10 after incubation in the modified MDFA medium (also producing sorbicillinoids as in the rice solid medium) for 1, 3 and 5 days, and used as a template for real-time RT-PCR, the transcriptions of all 10 genes, including *orf2* (ACRE_048080), *AcsorD* (ACRE_048110), *AcsorR2* (ACRE_048120), *AcsorT* (ACRE_048130), *AcsorE* (ACRE_048140), *AcsorR1* (ACRE_048150), *AcsorC* (ACRE_048160), *AcsorB* (ACRE_048170), *AcsorA* (ACRE_048180) and *orf1* (ACRE_048200), were analysed (Figure 6A). Transcriptional results showed that *AcsorA*, *AcsorB*, *AcsorC*, *AcsorD*, *AcsorE*, *AcsorT*, *AcsorR1* and *AcsorR2* displayed a similar transcriptional pattern. In other words, the transcriptional level gradually increases during the fermentation. However, *orf2* was silent during fermentation. Although *orf1* was transcribed, the transcriptional trend was significantly different from other genes in the *Acsor* cluster. Thus, *orf1* and *orf2* are considered to be situated outside the *Acsor* cluster (Figure 6B). Combining with the results from bioinformatic analysis, a 35.5 kb *Acsor* cluster was identified that contains eight genes encoding one high-reducing polyketide synthase AcsorA, one non-reducing PKS AcsorB, two FAD-dependent monooxygenases AcsorC and AcsorD, one major facilitator superfamily transporter AcsorT, two putative regulators AcsorR1 and AcsorR2 and one putative serine hydrolase AcsorE.

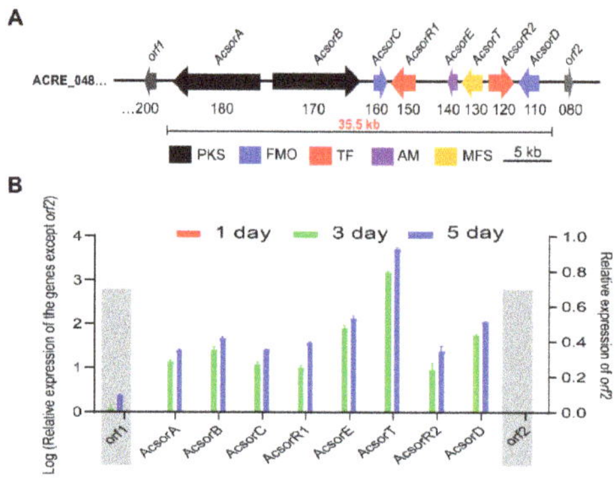

Figure 6. (**A**) Organization of the sorbicillinoid biosynthetic gene cluster. FMO, FAD-dependent monooxygenase; PKS, polyketide synthase; TF, transcriptional factor; AM, auxiliary modifier; MFS, major facilitator superfamily transporter. (**B**) Transcriptional profiles of the *Acsor* genes during fermentation.

Based on the confirmation of *Acsor* cluster, the biosynthetic pathway of compounds **1**–**8** was proposed. Sorbicillinoid biosynthesis starts from the formation of the polyketide backbone via condensation of acetate units catalyzed by AcsorA and AcsorB to generate sorbicillin and dihydrosorbicillin, and then they are oxidative dearomatized by AcsorC to form the common precursor-sorbicillinol and dihydrosorbicillinol. Sorbicillinol and its derivatives can be converted to **1**, **2** and **4** by a Diels–Alder reaction. Compounds **3**, **5** and **6** were biosynthesized by a Michael addition of sorbicillinol. Compounds **7** and **8** could be formed from sorbicillinol by an oxidation reaction (Figure 7). The structure diversification of sorbicillinoid derivatives was likely due to the multi-functions of *AcsorD* in *A. chrysogenum*.

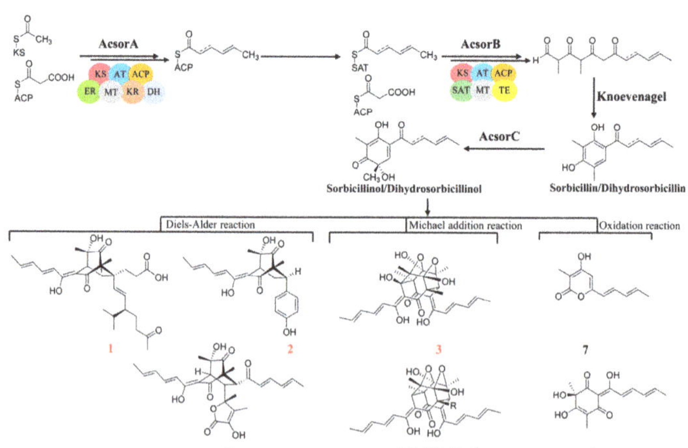

Figure 7. Proposed biosynthetic pathway of compounds **1**–**8**.

4. Conclusions

In summary, eight sorbicillinoid derivatives including three new ones, acresorbicillinols A–C (**1–3**), were isolated from the marine-derived fungus *A. chrysogenum*. The absolute configurations of compounds **1–3** were determined by ECD calculations. Compound **3** exhibited strong DPPH radical scavenging, indicating that it can be regarded as one novel DPPH radical scavenging agent. Compounds **2** and **3** exhibited the moderate activities against *S. aureus* and *C. neoformans*, respectively. Meanwhile, the boundary of the *Acsor* cluster was confirmed and the biosynthetic pathway of compounds **1–8** was also proposed. This study suggests that *A. chrysogenum* is a potential pool for novel sorbicillinoids and radical scavenging agents.

Supplementary Materials: The following supporting information can be downloaded at: https://www.mdpi.com/article/10.3390/jof8050530/s1, Figure S1: ^{1}H NMR spectrum of acresorbicillinol A (**1**; 500 MHz, CD_3OD), Figure S2: ^{13}C NMR spectrum of acresorbicillinol A (**1**; 125 MHz, CD_3OD), Figure S3: ^{1}H-^{1}H COSY spectrum of acresorbicillinol A (**1**, CD_3OD), Figure S4: HSQC spectrum of acresorbicillinol A (**1**, CD_3OD), Figure S5: HMBC spectrum of acresorbicillinol A (**1**, CD_3OD), Figure S6: NOESY spectrum of acresorbicillinol A (**1**, CD_3OD), Figure S7: ^{1}H NMR spectrum of acresorbicillinol B (**2**; 500 MHz, CD_3OD), Figure S8: ^{13}C NMR spectrum of acresorbicillinol B (**2**; 125 MHz, CD_3OD), Figure S9: ^{1}H-^{1}H COSY spectrum of acresorbicillinol B (**2**, CD_3OD), Figure S10: HSQC spectrum of acresorbicillinol B (**2**, CD_3OD), Figure S11: HMBC spectrum of acresorbicillinol B (**2**, CD_3OD), Figure S12. NOESY spectrum of acresorbicillinol B (**2**, CD_3OD) Figure S13: ^{1}H NMR spectrum of acresorbicillinol C (**3**; 500 MHz, DMSO:$CDCl_3$ = 3:1), Figure S14: ^{13}C NMR spectrum of acresorbicillinol C (**3**; 125 MHz, DMSO:$CDCl_3$ = 3:1), Figure S15: ^{1}H-^{1}H COSY spectrum of acresorbicillinol C (**3**, DMSO:$CDCl_3$ = 3:1), Figure S16: HSQC spectrum of acresorbicillinol C (**3**, DMSO:$CDCl_3$ = 3:1), Figure S17: HMBC spectrum of acresorbicillinol C (**3**, DMSO:$CDCl_3$ = 3:1), Figure S18: NOESY spectrum of acresorbicillinol C (**3**, DMSO:$CDCl_3$ = 3:1), Figure S19: ECD conformers of acresorbicillinols A–C (**1–3**). Figure S20: HPLC profiles of the extracts from the rice solid medium of *A. chrysogenum* after 7 days fermentation. Tale S1: Primers used in this study.

Author Contributions: Conceptualization, C.D. and S.W.; methodology, C.D., S.W. and J.R.; software, R.H.; validation, S.W.; formal analysis, E.L., M.W. and R.H.; investigation, C.D. and S.W.; resources, G.L.; data curation, C.D.; writing—original draft preparation, C.D., S.W. and R.H.; writing—review and editing, Y.P., L.L. and G.L.; visualization, R.H.; supervision, Y.P., L.L. and G.L.; project administration, Y.P., L.L. and G.L.; funding acquisition, Y.P. and L.L. All authors have read and agreed to the published version of the manuscript.

Funding: This research was supported by the National Key Research and Development Program of China (2021YFC2300400) and the National Natural Science Foundation of China (31770056, 32022002, 21977113 and 32170067).

Institutional Review Board Statement: Not applicable.

Informed Consent Statement: Not applicable.

Data Availability Statement: All data generated or analyzed in this study are available within the manuscript and are available from the corresponding authors upon request.

Acknowledgments: We thank Francisco Fierro (Universidad Autónoma Metropolitana-Unidad Iztapalapa, Mexico) for providing the *A. chrysogenum* C10 and Wenzhao Wang (Institute of Microbiology, CAS) for HRESIMS analysis. We also appreciate Guanghua Huang (Fudan University, Shanghai, China) and Linqi Wang (Institute of Microbiology, CAS) providing the strains of *Candida albicans* SC5314 and *Cryptococcus neoformans* W1585, respectively.

Conflicts of Interest: The authors declare no conflict of interest.

References

1. Wang, H.N.; Sun, S.S.; Liu, M.Z.; Yan, M.C.; Liu, Y.F.; Zhu, Z.; Zhang, Z. Natural bioactive compounds from marine fungi (2017–2020). *J. Asian Nat. Prod. Res.* **2022**, *24*, 203–230. [CrossRef] [PubMed]
2. Karthikeyan, A.; Joseph, A.; Nair, B.G. Promising bioactive compounds from the marine environment and their potential effects on various diseases. *J. Genet. Eng. Biotechnol.* **2022**, *20*, 14. [CrossRef] [PubMed]
3. Julianti, E.; Abrian, I.A.; Wibowo, M.S.; Azhari, M.; Tsurayya, N.; Izzati, F.; Juanssilfero, A.B.; Bayu, A.; Rahmawati, S.I.; Putra, M.Y. Secondary metabolites from marine-derived fungi and actinobacteria as potential sources of novel colorectal cancer drugs. *Mar. Drugs* **2022**, *20*, 67. [CrossRef] [PubMed]
4. Zhao, J.C.; Li, X.M.; Gloer, J.B.; Wang, B.G. First total syntheses and antimicrobial evaluation of penicimonoterpene, a marine-derived monoterpenoid, and its various derivatives. *Mar. Drugs* **2014**, *12*, 3352–3370. [CrossRef]
5. Terfehr, D.; Dahlmann, T.A.; Kück, U. Transcriptome analysis of the two unrelated fungal β-lactam producers *Acremonium chrysogenum* and *Penicillium chrysogenum*: Velvet-regulated genes are major targets during conventional strain improvement programs. *BMC Genom.* **2017**, *18*, 272. [CrossRef]
6. Terfehr, D.; Dahlmann, T.A.; Specht, T.; Zadra, I.; Kürnsteiner, H.; Kück, U. Genome sequence and annotation of *Acremonium chrysogenum*, producer of the β-lactam antibiotic cephalosporin C. *Genome Announc.* **2014**, *2*, 5. [CrossRef]
7. Hou, X.; Zhang, X.; Xue, M.; Zhao, Z.; Zhang, H.; Xu, D.; Lai, D.; Zhou, L. Recent advances in sorbicillinoids from fungi and their bioactivities (Covering 2016–2021). *J. Fungi* **2022**, *8*, 62. [CrossRef]
8. Harned, A.M.; Volp, K.A. The sorbicillinoid family of natural products: Isolation, biosynthesis, and synthetic studies. *Nat. Prod. Rep.* **2011**, *28*, 1790–1810. [CrossRef]
9. Meng, J.; Wang, X.; Xu, D.; Fu, X.; Zhang, X.; Lai, D.; Zhou, L.; Zhang, G. Sorbicillinoids from fungi and their bioactivities. *Molecules* **2016**, *21*, 715. [CrossRef]
10. Cram, D.J.; Tishler, M. Mold metabolites; isolation of several compounds from clinical penicillin. *J. Am. Chem. Soc.* **1948**, *70*, 4238. [CrossRef]
11. Cram, D.J. Mold metabolites; the structure of sorbicillin, a pigment produced by the mold *Penicillium notatum*. *J. Am. Chem. Soc.* **1948**, *70*, 4240–4243. [CrossRef] [PubMed]
12. Guzmán-Chávez, F.; Salo, O.; Nygård, Y.; Lankhorst, P.P.; Bovenberg, R.A.L.; Driessen, A.J.M. Mechanism and regulation of sorbicillin biosynthesis by *Penicillium chrysogenum*. *Microb. Biotechnol.* **2017**, *10*, 958–968. [CrossRef] [PubMed]
13. Derntl, C.; Rassinger, A.; Srebotnik, E.; Mach, R.L.; Mach-Aigner, A.R. Identification of the main regulator responsible for synthesis of the typical yellow pigment produced by *Trichoderma reesei*. *Appl. Environ. Microbiol.* **2016**, *82*, 6247–6257. [CrossRef] [PubMed]
14. Salo, O.; Guzmán-Chávez, F.; Ries, M.I.; Lankhorst, P.P.; Bovenberg, R.A.L.; Vreeken, R.J.; Driessen, A.J.M. Identification of a polyketide synthase involved in sorbicillin biosynthesis by *Penicillium chrysogenum*. *Appl. Environ. Microbiol.* **2016**, *82*, 3971–3978. [CrossRef] [PubMed]
15. Alfahad, A.; Abood, A.; Fisch, K.M.; Osipow, A.; Davison, J.; Avramović, M.; Butts, C.P.; Piel, J.; Simpson, T.J.; Cox, R.J. Oxidative dearomatisation: The key step of sorbicillinoid biosynthesis. *Chem. Sci.* **2014**, *5*, 523–527.
16. Kahlert, L.; Bassiony, E.F.; Cox, R.J.; Skellam, E.J. Diels-Alder reactions during the biosynthesis of sorbicillinoids. *Angew. Chem. Int. Ed. Engl.* **2020**, *59*, 5816–5822. [CrossRef] [PubMed]
17. Kahlert, L.; Cox, R.J.; Skellam, E. The same but different: Multiple functions of the fungal flavin dependent monooxygenase SorD from *Penicillium chrysogenum*. *Chem. Commun.* **2020**, *56*, 10934–10937. [CrossRef]
18. Chen, G.; Chu, J. Characterization of two polyketide synthases involved in sorbicillinoid biosynthesis by *Acremonium chrysogenum* using the CRISPR/Cas9 system. *Appl. Biochem. Biotechnol.* **2019**, *188*, 1134–1144. [CrossRef]
19. Krishna, C. Solid-state fermentation systems—An overview. *Crit. Rev. Biotechnol.* **2005**, *25*, 1–30. [CrossRef]
20. Ngo, M.T.; Nguyen, M.V.; Han, J.W.; Park, M.S.; Kim, H.; Choi, G.J. In vitro and in vivo antifungal activity of sorbicillinoids produced by *Trichoderma longibrachiatum*. *J. Fungi* **2021**, *7*, 428. [CrossRef]
21. Abe, N.; Murata, T.; Hirota, A. Novel DPPH radical scavengers, bisorbicillinol and demethyltrichodimerol, from a fungus. *Biosci. Biotechnol. Biochem.* **1998**, *62*, 661–666. [CrossRef] [PubMed]
22. Washida, K.; Abe, N.; Sugiyama, Y.; Hirota, A. Novel DPPH radical scavengers, demethylbisorbibutenolide and trichopyrone, from a fungus. *Biosci. Biotechnol. Biochem.* **2007**, *71*, 1052–1057. [CrossRef] [PubMed]
23. Frisch, M.J.; Trucks, G.W.; Schlegel, H.B.; Scuseria, G.E.; Robb, M.A.; Cheeseman, J.R.; Scalmani, G.; Barone, V.; Mennucci, B.; Petersson, G.A.; et al. *Gaussian 09*; Gaussian, Inc.: Wallingford, CT, USA, 2009.
24. Fan, W.; Li, E.; Ren, J.; Wang, W.; Liu, X.; Zhang, Y. Cordycepamides A-E and cordyglycoside A, new alkaloidal and glycoside metabolites from the entomopathogenic fungus *Cordyceps* sp. *Fitoterapia* **2020**, *142*, 104525. [CrossRef] [PubMed]
25. Guo, L.; Lin, J.; Niu, S.; Liu, S.; Liu, L. Pestalotiones A-D: Four new secondary metabolites from the plant endophytic fungus *Pestalotiopsis Theae*. *Molecules* **2020**, *25*, 470. [CrossRef] [PubMed]
26. Long, L.K.; Yang, J.; An, Y.; Liu, G. Disruption of a glutathione reductase encoding gene in *Acremonium chrysogenum* leads to reduction of its growth, cephalosporin production and antioxidative ability which is recovered by exogenous methionine. *Fungal Genet. Biol.* **2012**, *49*, 114–122. [CrossRef]
27. Li, J.; Pan, Y.; Liu, G. Disruption of the nitrogen regulatory gene *AcareA* in *Acremonium chrysogenum* leads to reduction of cephalosporin production and repression of nitrogen metabolism. *Fungal. Genet. Biol.* **2013**, *61*, 69–79. [CrossRef]

28. Li, H.; Gao, W.; Cui, Y.; Pan, Y.; Liu, G. Remarkable enhancement of bleomycin production through precise amplification of its biosynthetic gene cluster in *Streptomyces verticillus*. *Sci. China Life Sci.* **2021**, *64*, 1–9. [CrossRef]
29. Washida, K.; Abe, N.; Sugiyama, Y.; Hirota, A. Novel secondary metabolites, spirosorbicillinols A, B, and C, from a fungus. *Biosci. Biotechnol. Biochem.* **2009**, *73*, 1355–1361. [CrossRef]
30. Xie, C.L.; Zhang, D.; Lin, T.; He, Z.H.; Yan, Q.X.; Cai, Q.; Zhang, X.K.; Yang, X.W.; Chen, H.F. Antiproliferative sorbicillinoids from the deep-sea-derived *Penicillium allii-sativi*. *Front. Microbiol.* **2020**, *11*, 636948. [CrossRef]
31. Liu, W.; Gu, Q.; Zhu, W.; Cui, C.; Fan, G. Dihydrotrichodimerol and tetrahydrotrichodimerol, two new bisorbicillinoids, from a marine-derived *Penicillium terrestre*. *J. Antibiot.* **2005**, *58*, 621–624. [CrossRef]
32. Pang, X.; Zhou, X.; Lin, X.; Yang, B.; Tian, X.; Wang, J.; Xu, S.; Liu, Y. Structurally various sorbicillinoids from the deep-sea sediment derived fungus *Penicillium* sp. SCSIO06871. *Bioorg. Chem.* **2021**, *107*, 104600. [CrossRef] [PubMed]
33. Kontani, M.; Sakagami, Y.; Marumo, S. First β-1,6-glucan biosynthesis inhibitor, bisvertinolone isolated from fungus, *Acremonium strictum* and its absolute stereochemistry. *Tetrahedron Lett.* **1994**, *35*, 2577–2580. [CrossRef]
34. Corral, P.; Esposito, F.P.; Tedesco, P.; Falco, A.; Tortorella, E.; Tartaglione, L.; Festa, C.; D'Auria, M.V.; Gnavi, G.; Varese, G.C.; et al. Identification of a sorbicillinoid-producing *Aspergillus* strain with antimicrobial activity against *Staphylococcus aureus*: A new polyextremophilic marine fungus from Barents Sea. *Mar. Biotechnol.* **2018**, *20*, 502–511. [CrossRef] [PubMed]
35. El-Elimat, T.; Raja, H.A.; Figueroa, M.; Swanson, S.M.; Falkinham, J.O.; Lucas, D.M.; Grever, M.R.; Wani, M.C.; Pearce, C.J.; Oberlies, N.H. Sorbicillinoid analogs with cytotoxic and selective anti-*Aspergillus* activities from *Scytalidium album*. *J. Antibiot.* **2015**, *68*, 191–196. [CrossRef] [PubMed]
36. Abe, N.; Yamamoto, K.; Hirota, A. Novel fungal metabolites, demethylsorbicillin and oxosorbicillinol, isolated from *Trichoderma* sp. USF-2690. *Biosci. Biotechnol. Biochem.* **2000**, *64*, 620–622. [CrossRef]
37. Abe, N.; Murata, T.; Hirota, A. Novel oxidized sorbicillin dimers with 1,1-diphenyl-2-picrylhydrazyl-radical scavenging activity from a fungus. *Biosci. Biotechnol. Biochem.* **1998**, *62*, 2120–2126. [CrossRef]

Journal of *Fungi*

MDPI

Article

New Antibacterial Chloro-Containing Polyketides from the Alga-Derived Fungus *Asteromyces cruciatus* KMM 4696

Olesya I. Zhuravleva [1,2,*], Galina K. Oleinikova [1], Alexandr S. Antonov [1], Natalia N. Kirichuk [1], Dmitry N. Pelageev [1], Anton B. Rasin [1], Alexander S. Menshov [1], Roman S. Popov [1], Natalya Yu. Kim [1], Ekaterina A. Chingizova [1], Artur R. Chingizov [1], Olga O. Volchkova [2], Gunhild von Amsberg [3,4], Sergey A. Dyshlovoy [2,3,4], Ekaterina A. Yurchenko [1], Irina V. Guzhova [5] and Anton N. Yurchenko [1]

[1] G.B. Elyakov Pacific Institute of Bioorganic Chemistry, Far Eastern Branch of the Russian Academy of Sciences, Prospect 100-Letiya Vladivostoka, 159, 690022 Vladivostok, Russia; oleingk@mail.ru (G.K.O.); antonov_as@piboc.dvo.ru (A.S.A.); sheflera@bk.ru (N.N.K.); pelageev@mail.ru (D.N.P.); abrus__54@mail.ru (A.B.R.); menshov90@piboc.dvo.ru (A.S.M.); popov_rs@piboc.dvo.ru (R.S.P.); kim_ny@piboc.dvo.ru (N.Y.K.); martyyas@mail.ru (E.A.C.); chingizov_ar@piboc.dvo.ru (A.R.C.); eyurch@piboc.dvo.ru (E.A.Y.); yurchenkoan@piboc.dvo.ru (A.N.Y.)

[2] Institute of High Technologies and Advanced Materials, Far Eastern Federal University, 10 Ajax Bay, Russky Island, 690922 Vladivostok, Russia; olavol1309@gmail.com (O.O.V.); s.dyshlovoy@uke.de (S.A.D.)

[3] Laboratory of Experimental Oncology, Department of Oncology, Hematology and Bone Marrow Transplantation with Section Pneumology, Hubertus Wald-Tumorzentrum, University Medical Center Hamburg-Eppendorf, 20246 Hamburg, Germany; g.von-amsberg@uke.de

[4] Martini-Klinik Prostate Cancer Center, University Hospital Hamburg-Eppendorf, 20246 Hamburg, Germany

[5] Institute of Cytology Russian Academy of Sciences, Tikhoretskiy Ave. 4, 194064 St. Petersburg, Russia; ina.guzhova@incras.ru

* Correspondence: zhuravleva.oi@dvfu.ru; Tel.: +7-423-231-1168

Citation: Zhuravleva, O.I.; Oleinikova, G.K.; Antonov, A.S.; Kirichuk, N.N.; Pelageev, D.N.; Rasin, A.B.; Menshov, A.S.; Popov, R.S.; Kim, N.Y.; Chingizova, E.A.; et al. New Antibacterial Chloro-Containing Polyketides from the Alga-Derived Fungus *Asteromyces cruciatus* KMM 4696. *J. Fungi* **2022**, *8*, 454. https://doi.org/10.3390/jof8050454

Academic Editors: Tao Feng and Gary A. Strobel

Received: 14 February 2022
Accepted: 26 April 2022
Published: 27 April 2022

Publisher's Note: MDPI stays neutral with regard to jurisdictional claims in published maps and institutional affiliations.

Abstract: Six new polyketides acrucipentyns A–F (**1–6**) were isolated from the alga-derived fungus *Asteromyces cruciatus* KMM 4696. Their structures were established based on spectroscopic methods. The absolute configurations of acrucipentyn A was assigned by the modified Mosher's method and ROESY data analysis. Acrucipentyns A–E were identified to be the very first examples of chlorine-containing asperpentyn-like compounds. The cytotoxic and antimicrobial activities of the isolated compounds were examined. Acrucipentyns A–F were found as antimicrobial agents, which inhibited sortase A enzyme activity, bacterial growth and biofilm formation of *Staphylococcus aureus* and decreased LDH release from human keratinocytes HaCaT in *S. aureus* skin infection in an in vitro model.

Keywords: *Asteromyces cruciatus*; marine fungi; secondary metabolites; polyketides; sortase A; chlore-containing metabolites; *Staphylococcus aureus*; antibacterial activity; biofilm formation

1. Introduction

Microfilamentous fungi isolated from a variety of marine environments can be divided into facultative and obligate groups [1]. Facultative marine fungi are those found in both marine and terrestrial sources. The metabolism of such fungi species adapts to marine environment conditions and produces secondary metabolites, which are unusual for these species. For example, the well-known and widespread fungus *Penicillium chrysogenum*, isolated from marine sediments, was reported to be a producer of the unique dimeric nitrophenyl *trans*-epoxyamides, chrysamides A–C [2]. Obligate marine fungi are exclusively found in marine sources and have never been isolated from terrestrial samples. The metabolism of obligate marine fungi is more dramatically altered by the saline stress and other factors. The biosynthesis of previously undescribed chemical structures is the consequence [3].

Obligate marine fungus *Asteromyces cruciatus* C. Moreau et Moreau ex Hennebert is a widespread species, which can be found in the tropical and temperate zones of the Pacific, Atlantic and Indian oceans. This species was initially discovered in the sand sample of the coastal area and was validly described in 1961. *A. cruciatus* can be found drifting or washed up on the sore wood, algae, in sediment bottom samples. Currently, this species is still poorly studied, and its position in the fungal classification tree remains uncertain [4].

Nevertheless, a few chemical studies have shown that *A. cruciatus* is a promising source of new secondary metabolites. Thus, diketopiperazine gliovictin was the first compound isolated from *A. cruciatus* [5]. The fungus *A. cruciatus* 763 yielded the new pentapeptide lajollamide A, which exhibited a weak antibacterial activity, along with several known sulfur-containing diketopiperazines [6]. Two new polyketides, primarolides A and B, were isolated from an A. cruciatus culture treated with suberoylanilide hydroxamic acid and high concentrations of NaCl [7].

Being a part of the microbial community, both facultative and obligate marine fungi produce various bioactive secondary metabolites, which help them to interact and fight with other species. By 2021, about 300 small molecules possessing a potent antimicrobial activity and isolated from various marine fungi had been reported [8,9]. Marine fungal secondary metabolites exhibit their antibacterial effects by directly inhibiting bacterial growth, as well as by decreasing virulence or biofilms formation [10]. A membrane-associated enzyme, sortase A, is responsible for the covalent attachment of many virulent Gram-positive bacteria, including *Staphylococcus aureus*, to the mammalian cell wall [11,12]. Thus, the sortase A enzyme is an attractive target for new drugs against virulent and antibiotic-resistant Gram-positive bacteria, which are known to be one of the main causes of infectious disease worldwide [13]. The compounds capable of sortase A inhibition and lacking, at the same time, cytotoxicity to mammalian cells are of particular interest, because they can inhibite bacterial biofilm formation and decrease the virulence and toxicity of bacteria.

In the current research, as a part of our continuing efforts to search for new antibacterial metabolites in marine fungi [14,15], we isolated new isoprenylated cyclohexanols acrucipentyns A–F (**1–6**) from a culture of *Asteromyces cruciatus* KMM 4696 associated with brown alga *Sargassum pallidum* (Vostok Bay, the Sea of Japan). The effect of acrucipentyns on enzymatic activity of sortase A from *Staphylococcus aureus*, growth and biofilm formation of *S. aureus*, as well as toxicity of acrucipentyns to various mammalian (human) cells, were tested. Finally, the effect of isolated compounds on human keratinocytes HaCaT co-cultured with *S. aureus* was investigated.

2. Materials and Methods

2.1. General Experimental Procedures

Optical rotations were measured on a Perkin-Elmer 343 polarimeter (Perkin Elmer, Waltham, MA, USA). UV spectra were recorded on a Shimadzu UV-1601PC spectrometer (Shimadzu Corporation, Kyoto, Japan) in methanol. CD spectra were measured with a Chirascan-Plus CD spectrometer (Leatherhead, UK) in methanol. NMR spectra were recorded in $CDCl_3$, acetone-d_6 and DMSO-d_6, on a Bruker DPX-300 (Bruker BioSpin GmbH, Rheinstetten, Germany), a Bruker Avance III-500 (Bruker BioSpin GmbH, Rheinstetten, Germany) and a Bruker Avance III-700 (Bruker BioSpin GmbH, Rheinstetten, Germany) spectrometer, using TMS as an internal standard. HRESIMS spectra were measured on a Maxis impact mass spectrometer (Bruker Daltonics GmbH, Rheinstetten, Germany). Microscopic examination and photography of fungal cultures were performed with Olympus CX41 microscope equipped with an Olympus SC30 digital camera. Detailed examination of ornamentation of the fungal conidia was performed using scanning electron microscopy (SEM) EVO 40.

Low-pressure liquid column chromatography was performed using silica gel (60/100 µm, Imid Ltd., Krasnodar, Russia) and Gel ODS-A (12 nm, S—75 um, YMC Co., Ishikawa, Japan). Plates precoated with silica gel (5–17 µm, 4.5 cm × 6.0 cm, Imid Ltd., Russia) and silica gel 60

RP-18 $F_{254}S$ (20 cm × 20 cm, Merck KGaA, Darmstadt, Germany) were used for thin-layer chromatography. Preparative HPLC was carried out on an Agilent 1100 chromatograph (Agilent Technologies, Santa Clara, CA, USA) with an Agilent 1100 refractometer (Agilent Technologies, Santa Clara, CA, USA) and a Shimadzu LC-20 chromatograph (Shimadzu USA Manufacturing, Canby, OR, USA) with a Shimadzu RID-20A refractometer (Shimadzu Corporation, Kyoto, Japan) using YMC ODS-AM (YMC Co., Ishikawa, Japan) (5 μm, 10 mm × 250 mm), YMC ODS-AM (YMC Co., Ishikawa, Japan) (5 μm, 4.6 mm × 250 mm) and Hydro-RP (Phenomenex, Torrance, CA, USA) (4 μm, 250 mm × 10 mm) columns.

2.2. Fungal Strain

The brown algae samples were collected in Vostok Bay (Sea of Japan) in sterile plastic bags. Before use, they were stored in a freezer at −18 °C. Isolation of fungi from algae samples was carried out by the plate method using Tubaki agar medium. The fungus was isolated into a pure culture by transferring the inoculum from a Petri dish onto a slant wort agar, where it was further stored. Microscopic examination of the strain was performed using Olympus CX41.

For DNA isolation, a fungus culture grown at 25 °C for 7 days was used. Isolation of genomic DNA was carried out using a commercial DNA kit (DNA-Technology Ltd., Moscow, Russia) in accordance with the protocol. Amplification and sequencing of the ITS genes were performed using ITS1 and ITS4 gene-specific primers [16]. The obtained sequence was compared with the GenBank sequence dataset and registered under accession number OL477331.

2.3. Cultivation of Fungus

The fungus was cultured at 22 °C for three weeks in 60 × 500 mL Erlenmeyer flasks, each containing rice (20.0 g), yeast extract (20.0 mg), KH_2PO_4 (10 mg) and natural seawater from the Marine Experimental Station of PIBOC, Troitsa (Trinity) Bay, Sea of Japan (40 mL).

2.4. Extraction and Isolation

At the end of the incubation period, the mycelia and medium were homogenized and extracted with EtOAc (1 L). The obtained extract was concentrated to dryness. The residue (17.5 g) was dissolved in H_2O–EtOH (4:1) (100 mL) and extracted successively with *n*-hexane (0.2 L × 3), EtOAc (0.2 L × 3) and BuOH (0.2 L × 3). After evaporation of the ethyl acetate layer, the residual material (5.5 g) was subjected to column chromatography on silica gel, which was eluted with a gradient of *n*-hexane in ethyl acetate (100:1, 95:5, 90:10, 80:20, 70:30, 50:50). Fractions of 250 mL were collected and combined on the basis of TLC (silica gel, toluene–isopropanol 6:1 and 3:1, *v/v*).

The fractions of *n*-hexane-EtOAc (95:5, 80 mg) and *n*-hexane-EtOAc (90:10, 200 mg) were separated on a Gel ODS-A column (1.5 cm × 8 cm), which was eluted by a step gradient from 40% to 80% CH_3OH in H_2O (total volume 1 L), to afford subfractions I and II. Subfraction I (40% CH_3OH, 146 mg) was purified by RP HPLC on a YMC ODS-AM column eluted with CH_3OH-H_2O (90:10) and then with CH_3OH-H_2O (60:40) to yield **2** (1.8 mg) and **4** (6.0 mg). Subfraction II (60% CH_3OH, 110 mg) was purified by RP HPLC on a YMC ODS-AM column eluted with CH_3OH-H_2O (80:20) and then with CH_3OH-H_2O (55:45) to yield **1** (7 mg).

The *n*-hexane-EtOAc (80:20, 470 mg) fraction was separated on a Gel ODS-A column (1.5 cm × 8 cm), which was eluted with a step gradient from 40% to 80% CH_3OH in H_2O (total volume 1 L) to give subfraction III. Subfraction III (40% CH_3OH, 250 mg) was separated by RP HPLC on a YMC ODS-AM column eluting with CH_3OH-H_2O (90:10) and then with CH_3OH-H_2O (60:40) to yield **6** (58 mg).

After evaporation of the BuOH layer, the residual material (0.98 g) was passed through a silica column (3 cm × 14 cm), which was separated in the same way as the ethyl acetate extract.

The *n*-hexane-EtOAc (50:50, 252 mg) fraction was purified by RP HPLC on a YMC ODS-A column eluted with CH_3OH-H_2O (45:1055) and then with CH_3CN-H_2O (25:75) to yield **3** (3.9 mg) and **5** (5.2 mg).

2.5. Spectral Data

Acrucipentyn A (**1**): amorphous solids; $[\alpha]_D^{20}$ − 36.0 (*c* 0.09 CH_3OH); CD (*c* 9.6 × 10^{-4}, CH_3OH), λ_{max} ($\Delta\varepsilon$) 210 (−0.97), 266 (−0.14) nm, see Supplementary Figure S1; UV (CH_3OH) λ_{max} (log ε) 271 (2.54), 256 (2.50) and 224 (3.77) nm, see Supplementary Figure S7; ^1H and ^{13}C NMR data, see Tables 1 and 2, Supplementary Figures S13–S22; HRESIMS *m*/*z* 229.0625 $[M - H]^-$ (calcd. for $C_{11}H_{14}ClO_3$, 229.0637, Δ5.0 ppm), 253.0599 $[M + Na]^+$ (calcd. for $C_{11}H_{15}ClO_3Na$, 253.0602, Δ1.2 ppm).

Table 1. ^1H NMR data (δ in ppm, *J* in Hz) for compounds (**1–6**).

Position	1 [a]	2 [a]	3 [b]	4 [c]	5 [c]	6 [d]
1	3.81, brs	3.70, m	3.68, dt (10.4, 5.4)	4.06, brt (4.2)	4.05, t (8.2)	4.50, brs
2	3.57, ddd (10.2, 6.8, 2.6)	3.83, brs	3.91, t (9.9)	3.87, ddd (8.5, 5.2, 4.0)	3.69, dd (11.2, 8.6)	3.60, t (3.3)
3	4.00, dd (10.3, 2.3)	4.18, t (4.1)	3.32, td (9.8, 4.6)	4.25, dd (8.6, 3.9)	3.31, ddd (11.3, 7.7, 6.2)	3.50, m
4	3.92, brs	4.03, dq (12.0, 3.9)	3.81, m	4.37, brq (4.4)	3.97, m	4.53, d (4.8)
5	α: 1.92, td (12.9, 2.1) β: 1.65, dt (13.0, 3.2)	1.74, dt (12.4, 3.9) 1.65, q (12.1)	1.63, ddd (13.3, 11.8, 4.1) 2.01, ddd (13.3, 4.5, 3.2)	5.92, d (4.2)	5.82, brt (2.0)	6.04, dt (5.0, 2.0)
6	2.96, dt (12.6, 2.5)	2.73, td (12.0, 3.9)	3.21, q (4.0)			
4′	a: 5.14, s b: 5.18, s	a: 5.16, s b: 5.19, s	5.21, m	a: 5.27, s b: 5.32, s	a: 5.27, s b: 5.31, s	a: 5.30, s b: 5.36, s
5′	1.80, s	1.81, s	1.86, t (1.2)	1.86, s	1.86, s	1.92, s
1-OH	4.89, d (4.3)	4.77, d (6.7)	4.38, d (5.6)	5.04, d (6.7)	5.73, d (7.9)	
2-OH	4.92, d (7.0)	5.40, brs		5.17, d (5.5)		
3-OH			4.50, d (4.4)		5.48, d (6.0)	
4-OH	5.06, d (4.2)	4.90, d (4.9)	4.04, d (3.4)	5.25, brd (5.8)	5.31, d (5.4)	

[a] Chemical shifts were measured at 700.13 MHz in DMSO-d_6. [b] Chemical shifts were measured at 700.13 MHz in acetone-d_6. [c] Chemical shifts were measured at 500.13 MHz in DMSO-d_6. [d] Chemical shifts were measured at 500.13 MHz in $CDCl_3$.

Acrucipentyn B (**2**): amorphous solids; $[\alpha]_D^{20}$ +46.2 (*c* 0.09 CH_3OH); CD (*c* 8.7 × 10^{-4}, CH_3OH), λ_{max} ($\Delta\varepsilon$) 223 (+0.87) nm, see Supplementary Figure S2; UV (CH_3OH) λ_{max} (log ε) 224 (3.93), 200 (3.41) and 194 (3.51) nm, see Supplementary Figure S8; ^1H and ^{13}C NMR data, see Tables 1 and 2, Supplementary Figures S23–S29; HRESIMS *m*/*z* 229.0634 $[M - H]^-$ (calcd. for $C_{11}H_{14}ClO_3$, 229.0637, Δ1.4 ppm), 253.0601 $[M + Na]^+$ (calcd. for $C_{11}H_{15}ClO_3Na$, 253.0602, Δ0.5 ppm).

Table 2. ^{13}C NMR data (δ in ppm) for compounds **1–6**.

Position	1 [a]	2 [a]	3 [b]	4 [c]	5 [c]	6 [d]
1	71.9, CH	68.7, CH	74.2, CH	67.9, CH	72.9, CH	65.2, CH
2	70.2, CH	72.5, CH	68.9, CH	68.6, CH	68.3, CH	53.5, CH
3	65.7, CH	66.1, CH	79.5, CH	63.3, CH	74.8, CH	55.3, CH
4	68.2, CH	64.0, CH	71.1, CH	65.1, CH	71.2, CH	63.0, CH
5	33.1, CH_2	33.4, CH_2	34.2, CH_2	136.0, CH	137.2, CH	131.4, CH
6	28.5, CH	30.3, CH	35.0, CH	123.5, C	123.8, C	123.5, C
1′	91.1, C	91.7, C	88.4, C	88.2, C	86.8, C	85.1, C
2′	81.7, C	81.7, C	86.3, C	89.9, C	90.7, C	93.5, C
3′	126.8, C	126.7, C	128.6, C	126.2, C	126.2, C	126.2, C
4′	120.7, CH_2	120.7, CH_2	121.4, CH_2	122.2, CH_2	122.2, CH_2	123.5, CH_2
5′	23.5, CH_3	23.5, CH_3	23.8, CH_3	23.0, CH_3	23.0, CH_3	23.2, CH_3

[a] Chemical shifts were measured at 176.04 MHz in DMSO-d_6. [b] Chemical shifts were measured at 75.47 MHz in acetone-d_6. [c] Chemical shifts were measured at 125.77 MHz in DMSO-d_6. [d] Chemical shifts were measured at 125.77 MHz in $CDCl_3$.

Acrucipentyn C (**3**): amorphous solids; $[\alpha]_D^{20}$ −127.9 (*c* 0.07 CH_3OH); CD (*c* 1.2 × 10^{-3}, CH_3OH), λ_{max} (Δε) 223 (−0.89), 297 (−0.11), 305 (−0.12) nm, see Supplementary Figure S3; UV (CH_3OH) λ_{max} (log ε) 270 (3.06), 251 (2.63) and 223 (3.70) nm, see Supplementary Figure S9; ^1H and ^{13}C NMR data, see Tables 1 and 2, Supplementary Figures S30–S35; HRESIMS *m/z* 229.0631 $[M-H]^-$ (calcd. for $C_{11}H_{14}ClO_3$, 229.0637, Δ2.5 ppm), 253.0595 $[M+Na]^+$ (calcd. for $C_{11}H_{15}ClO_3Na$, 253.0602, Δ2.7 ppm).

Acrucipentyn D (**4**): amorphous solids; $[\alpha]_D^{20}$ − 94.3 (*c* 0.06 CH_3OH); CD (*c* 8.8 × 10^{-4}, CH_3OH), λ_{max} (Δε) 210 (−3.73), 227 (−1.44), 238 (−1.17) nm, see Supplementary Figure S4; UV (CH_3OH) λ_{max} (log ε) 259 (3.90), 226 (3.56) and 198 (3.80) nm, see Supplementary Figure S10; ^1H and ^{13}C NMR data, see Tables 1 and 2, Supplementary Figures S36–S42; HRESIMS *m/z* 227.0471 $[M-H]^-$ (calcd. for $C_{11}H_{12}ClO_3$, 227.0480, Δ4.0 ppm), 251.0441 $[M+Na]^+$ (calcd. for $C_{11}H_{13}ClO_3Na$, 251.0445, Δ1.6 ppm).

Acrucipentyn E (**5**): amorphous solids; $[\alpha]_D^{20}$ −60.5 (*c* 0.09 CH_3OH); CD (*c* 8.8 × 10^{-4}, CH_3OH), λ_{max} (Δε) 199 (−5.11), 226 (−1.19), 237 (−0.83) nm, see Supplementary Figure S5; UV (CH_3OH) λ_{max} (log ε) 260 (4.22), 226 (3.77) and 198 (4.10) nm, see Supplementary Figure S11; ^1H and ^{13}C NMR data, see Tables 1 and 2, Supplementary Figures S43–S49; HRESIMS *m/z* 227.0470 $[M-H]^-$ (calcd. for $C_{11}H_{12}ClO_3$, 227.0480, Δ4.5 ppm), 251.0441 $[M+Na]^+$ (calcd. for $C_{11}H_{13}ClO_3Na$, 251.0445, Δ1.6 ppm).

Acrucipentyn F (**6**): amorphous solids; $[\alpha]_D^{20}$ +40.3 (*c* 0.11 CH_3OH); CD (*c* 1.0 × 10^{-3}, CH_3OH), λ_{max} (Δε) 199 (−4.56), 226 (−0.72), 259 (+1.13) nm, see Supplementary Figure S6; UV (CH_3OH) λ_{max} (log ε) 259 (4.05), 226 (3.59) and 201 (3.88) nm, see Supplementary Figure S12; ^1H and ^{13}C NMR data, see Tables 1 and 2, Supplementary Figures S50–S56; HRESIMS *m/z* 191.0704 $[M-H]^-$ (calcd. for $C_{11}H_{11}O_3$, 191.0714, Δ5.2 ppm), 215.0677 $[M+Na]^+$ (calcd. for $C_{11}H_{12}O_3Na$, 215.0679, Δ0.8 ppm).

2.6. Preparation of Acetonides of **1a** and **4a**

To a DMFA solution of **1** (4.0 mg) 2,2-dimethoxypropane (0.5 mL) and catalyst *p*-toluenesulfonic acid (0.8 mg) at room temperature were added and the solution was stirred for 24 h. After the evaporation of the solvent, the product was dissolved in MeOH and purified by reversed-phase HPLC (YMC ODS-A column) eluted with CH_3CN-H_2O (50:50) to yield the acetonide product **1a** (1.5 mg). Compound **4** (4.0 mg) was treated similarly and yielded the acetonide product **4a** (1.4 mg).

Acetonide of acrucipentyn A (**1a**): amorphous solids; ^1H NMR (Acetone-d_6, 500.13 MHz) δ: 5.20 (1H, s, H-4a′), 5.19 (1H, s, H-4b′), 4.37 (1H, t, *J* = 4.3 Hz, H-1), 4.20 (1H, dd, *J* = 8.3;

4.8 Hz, H-2), 4.11 (1H, m, H-4), 4.00 (1H, dd, J = 8.3; 2.6 Hz, H-3), 3.50 (1H, dt, J = 12.2; 4.4 Hz, H-6), 2.03 (1H, m, H-5b), 1.97 (1H, dt, J = 13.7; 4.7 Hz, H-5a), 1.85 (3H, s, H_3-5′), 1.47 (3H, s, H_3-3″), 1.34 (3H, s, H_3-2″); ^{13}C NMR (Acetone-d_6, 125.77 MHz) δ: 128.9 (C-3′), 122.0 (C-4′), 110.3 (C-1″), 90.3 (C-1′), 84.1 (C-2′), 80.5 (C-2), 77.4 (C-1), 70.7 (C-4), 67.1 (C-3), 35.0 (C-6), 29.3 (C-3″), 27.2 (C-2″), 26.5 (C-5), 24.6 (C-5′), see Supplementary Figures S57–S58; HRESIMS m/z 293.0912 [M + Na]$^+$ (calcd. for $C_{14}H_{19}ClO_3Na$, 293.0915, Δ0.9 ppm).

Acetonide of acrucipentyn D (**4a**): amorphous solids; ^{1}H NMR (Acetone-d_6, 700.13 MHz) δ: 5.33 (1H, t, J = 0.9 Hz, H-4a′), 5.31 (1H, t, J = 1.7 Hz, H-4b′), 6.11 (1H, d, J = 3.0 Hz, H-5), 4.63 (1H, dd, J = 5.7; 0.9 Hz, H-1), 4.58 (1H, m, H-4), 4.57 (1H, t, J = 5.7 Hz, H-2), 4.38 (1H, dd, J = 5.6; 3.5 Hz, H-3), 1.91 (3H, t, J = 1.4 Hz, H_3-5′), 1.37 (3H, s, H_3-2″/H_3-3″), 1.33 (3H, s, H_3-2″/H_3-3″); ^{13}C NMR (Acetone-d_6, 125.77 MHz) δ: 137.7 (C-5), 130.0 (C-6), 129.3 (C-3′), 122.8 (C-4′), 110.6 (C-1″), 91.3 (C-2′), 87.9 (C-1′), 77.1 (C-2), 74.6 (C-1), 66.0 (C-4), 62.8 (C-3), 27.8 (C-2″/C-3″), 26.0 (C-2″/C-3″), 23.4 (C-5′), see Supplementary Figures S63–S64; HRESIMS m/z 291.0758 [M + Na]$^+$ (calcd. for $C_{14}H_{17}ClO_3Na$, 291.0758, Δ0.9 ppm).

2.7. Preparation of (S)-MTPA and (R)-MTPA Esters of **1a**

To a pyridine solution of **1a** (0.7 mg) 4-dimethylaminopyridine (a few crystals) and (*S*)-MTPA-Cl (10 μL) at room temperature were added, and the solution was stirred for 24 h. After the evaporation of the solvent, the residue was purified by RP HPLC on a YMC ODS-AM column eluted with CH_3CN-H_2O (70:30) to afford the (*R*)-MTPA ester (**1a-1**). The (*S*)-MTPA ester (**1a-2**) was prepared in a similar manner using (*R*)-MTPACl. ^{1}H NMR and COSY data, see Supplementary Figure S59–S62.

(*R*)-MTPA ester of **1a**: ^{1}H NMR (Acetone-d_6, 500.13 MHz) δ: 5.56 (1H, brs, H-4), 5.22 (1H, s, H-4a′), 5.20 (1H, s, H-4b′), 4.32 (1H, brs, H-3), 4.30 (1H, brs, H-1), 4.09 (1H, dd, J = 8.1; 4.9 Hz, H-2), 2.80 (1H, m, H-6), 2.24 (1H, t, J = 13.7 Hz, H-5a), 2.09 (1H, m, H-5b), 1.84 (3H, s, H_3-5′), 1.50 (3H, s, H_3-3″), 1.33 (3H, s, H_3-2″). HRESIMS m/z 509.1310 [M + Na]$^+$ (calcd for $C_{24}H_{26}ClF_3O_5$, 509.1313)

(*S*)-MTPA ester of **1a**: ^{1}H NMR (Acetone-d_6, 500.13 MHz) δ: 5.65 (1H, brs, H-4), 5.24 (1H, s, H-4a′), 5.21 (1H, s, H-4b′), 4.29 (1H, d, J = 7.9 Hz, H-3), 4.44 (1H, t, J = 4.1 Hz, H-1), 4.06 (1H, dd, J = 8.0; 5.2 Hz, H-2), 3.27 (1H, dt, J = 12.8; 4.1 Hz, H-6), 2.33 (1H, t, J = 13.4 Hz, H-5a), 2.22 (1H, dt, J = 14.6; 4.7 Hz, H-5b), 1.85 (3H, s, H_3-5′), 1.50 (3H, s, H_3-3″), 1.34 (3H, s, H_3-2″). HRESIMS m/z 509.1307 [M + Na]$^+$ (calcd for $C_{24}H_{26}ClF_3O_5$, 509.1313).

2.8. An Epoxy Ring-Opening Reaction

Next, 2,2-dimethylpropanoyl chloride (0.5 mL) was added to an aqueous solution of **6** (12.0 mg) at room temperature, and the solution was stirred for 12 h. After the evaporation of the solvent, the product was dissolved in MeOH and purified by reversed-phase HPLC (Phenomenex Hydro-RP column) eluted with CH_3OH-H_2O (50:50) to yield compound **4** (3.7 mg) and compound **5** (6.4 mg).

2.9. Cell Lines and Culture Conditions

The human normal cell lines HEK 293T and MRC-9 cell lines were purchased from ECACC (Salisbury, UK). Human normal cell lines PNT2 and RWPE-1 as well as human cancer cell lines PC-3, DU145, 22Rv1, VCaP and LNCaP, as well as a human normal prostate line were purchased from ATCC (Manassas, VA, USA). Human normal cell line HUVEC (passage 11) was kindly donated by Prof. Sonja Loges (University Medical Center Hamburg-Eppendorf, Hamburg, Germany). The human keratinocytes cell line HaCaT was kindly provided by Prof. N. Fusenig (Cancer Research Centre, Heidelberg, Germany). All the cells had a passage number $\leq$ 30.

Cells were incubated in humidified 5% CO_2 at 37 °C. The cells were continuously kept in a culture for 3 months maximum and regularly checked for mycoplasma infection using MycoAlert™ PLUS Mycoplasma Detection Kit (Lonza, Karlsruhe, Germany) and stable phenotype using light microscopy [17].

The following culture media were used: RPMI medium supplemented with Glutamax^{TM}-I (Invitrogen, Paisley, UK) with 10% fetal bovine serum (FBS, Invitrogen) and 1% penicillin/streptomycin (Invitrogen) for PC-3, DU145, LNCaP, 22Rv1 and PNT2 cells. DMEM medium supplemented with Glutamax^{TM}-I (Invitrogen) containing 10% FBS and 1% penicillin/streptomycin (Invitrogen) for MRC-9, HEK 293 and VCaP cells. Clonetics® EGM^{TM}-2 SingleQuots® medium (Lonza, Walkersville, MD, USA) containing 10% FBS for RWPE-1 cells. Clonetics® EGM^{TM}-2 SingleQuots® medium (Lonza, Walkersville, MD, USA) containing 10% FBS for HUVEC cells. DMEM medium (BioloT, St. Petersburg, Russia) containing 10% FBS and 1% penicillin/streptomycin (Invitrogen) for HaCaT cells.

2.10. MTT Assay

Cytotoxicity of the isolated compounds to mammalian cells was evaluated using an MTT assay as previously reported with minor modification [18]. In brief, 6000 cells/well were seeded in 96-well planes in 100 μL/well and were incubated overnight. Then the media were exchanged with fresh media containing tested compounds in different concentrations. Following 72 h of incubation, 10 μL of MTT solution (5 mg/mL, Sigma-Aldrich, Munich, Germany) was added to each well, the cells were incubated for 2–4 h. Then the media was carefully aspirated and the plates were dried for 2 h. Then 50 μL/well of DMSO was added to each well to dissolve formazan crystals and the absorbance was measured using plate reader according to the manufacture's protocol. The data were analyzed and the IC_{50}s values were calculated using GraphPad Prism software v.9.1.1 (GraphPad Software, San Diego, CA, USA).

2.11. Sortase Activity Inhibition Assay

The enzymatic activity of sortase A from *Staphylococcus aureus* was determined using SensoLyte 520 Sortase A Activity Assay Kit * Fluorimetric * (AnaSpec AS-72229, AnaSpec, San Jose, CA, USA) in accordance with the manufacturer's instructions. Substances **1–6** were dissolved in DMSO and diluted with reaction buffer to obtain a final concentration of 0.8% DMSO, which did not affect enzyme activity. DMSO at a concentration of 0.8% was used as a control. PCMB (4-(hydroxymercuri)benzoic acid) was used as sortase A enzyme activity inhibitor. Fluorescence was measured with a plate reader PHERAStar FS (BMG Labtech, Offenburg, Germany) for 60 min, with a time interval of 5 min. The data were processed with MARS Data Analysis v. 3.01R2 (BMG Labtech, Offenburg, Germany). The results were presented as relative fluorescent units (RFUs) and percentage of the control data, calculated using STATISTICA 10.0 software [14].

2.12. Antimicrobial Activity

The antibacterial activity of compounds **1–6** was evaluated as described previously [19].

The bacterial culture of *Staphylococcus aureus* ATCC 21027 (Collection of Marine Microorganisms PIBOC FEBRAS) was cultured in a Petri dish at 37 °C for 24 h on solid medium Mueller Hinton broth with agar—16.0 g/L.

The assays were performed in 96-well microplates in appropriate Mueller Hinton broth. Each well contained 90 μL of bacterial suspension (10^9 CFU/mL). Then, 10 μL diluted at concentrations from 1.5 μM to 100.0 μM using two-fold dilution was added to compounds **1–6** (DMSO concentration < 1%). Culture plates were incubated overnight at 37 °C, and the OD_{620} was measured using a Multiskan Spectrum spectrophotometer (Thermo Labsystems Inc., Beverly, MA, USA). Gentamicin was used as a positive control in concentration 1 mg/mL; 1% DMSO solution in PBS as a negative.

2.13. Biofilm Formation

The inhibition of the reducing biofilm formation and growth was assessed using the crystal violet biofilm assay as described [20]. Mueller Hinton broth was inoculated with 10^9 CFU/mL of *S. aureus* overnight cultures. A total of 90 μL of this cell suspension was then dispensed into 96-well microtiter plates containing 10 μL of different concentrations of compounds **1–6**. After 24 h growth at 37 °C the plates were washed with PBS to remove unbound cells. Next, the wells were stained with 0.1% crystal violet solution for 10 min at 37 °C. At the completion of the incubation, plates were washed 3 times with PBS and dried. Then, the crystal violet dye from the biofilm was solubilized with 100 μL of ethanol. A total of 100 μL of this solution was then moved to a new microtiter plate for absorbance measurement at λ = 570 nm. The results were reported as percent inhibition normalized to the wild-type control.

2.14. *Co-Cultivation of HaCaT Cells with S. aureus*

Co-cultivation of HaCaT cells with *S. aureus* was carried out as described [21]. HaCaT cells at a concentration of 1.5×10^4 cells per well were seeded in 96-well plates for 24 h. Then, culture medium in each well was changed with *S. aureus* suspension (10^2 CFU/mL) in full DMEM medium. Fresh DMEM medium without *S. aureus* suspension was added in other wells as needed. Compounds **1–6** at a concentration of 10 μM were added in wells after 1 h. HaCaT cells and *S. aureus* were cultured at 37 °C in a humidified atmosphere with 5% (*v*/*v*) CO_2 for 48 h.

After incubation, the plate was centrifuged at 250× *g* for 10 min and 50 μL of supernatant from each well was transferred into the corresponding wells of an optically clear 96-well plate. An equal volume of the reaction mixture (50 μL) from LDH Cytotoxicity Assay Kit (Abcam, Cambridge, UK) was added to each well and incubated for up to 30 min at room temperature. The absorbance of all samples was measured at λ = 450 nm using a Multiskan FC microplate photometer (Thermo Scientific, Waltham, MA, USA) and expressed in optical units (o.u.).

2.15. *Statistical Analysis*

All the experiments were performed in biological triplicates. Statistical analyses were performed using GraphPad Prism v.9.1.1 (GraphPad Software, San Diego, CA, USA) or STATISTICA 10.0 software. The data are reported as mean ± SD (standard deviation). For the analysis of statistical differences between the control and drug-exposed group, a one-way ANOVA test followed by Dunnett's post-hoc test was used. Asterisk (*) indicates statistically significant difference between the treated group and control group if $p < 0.05$.

3. Results and Discussion

3.1. *Isolated Compounds from Asteromyces cruciatus*

The fungus *Asteromyces cruciatus* was cultivated on a solid rice medium for 21 days. The ethyl acetate extract of the mycelium was fractionated on silica gel, followed by C_{18}-SiO_2-column chromatography, and reversed-phase HPLC to produce compounds **1–6** (Figure 1).

3.2. *Structural Characterization of New Compounds*

The molecular formula of **1** was determined as $C_{11}H_{15}ClO_3$, based on the analysis of HRESIMS (*m/z* 229.0625 $[M - H]^-$ calcd for $C_{11}H_{14}ClO_3$, 229.0637), showing the characteristic isotope pattern with one chlorine atom, and confirmed by NMR data. A close inspection of the ^{1}H and ^{13}C NMR data (Tables 1 and 2; Figures S13–S19) of **1** by DEPT and HSQC revealed the presence of one methyl (δ_H 1.80, δ_C 23.5), two methylenes (δ_H 1.65, 1.92, δ_C 33.1; δ_H 5.14, 5.18, δ_C 120.7), and five methines (δ_H 2.96, δ_C 28.5; δ_H 4.00, δ_C 65.7; δ_H 3.92, δ_C 68.2; δ_H 3.57, δ_C 70.2 and δ_H 3.81, δ_C 71.9), including three oxygen-bearing methines, one sp^2 quaternary carbon and one triple bond (δ_C 81.7 and 91.1). The ^{1}H–^{1}H COSY correlations of H-1(OH)/H-2(OH)/H-3/H-4(OH)/H_2-5/H-6/H-1 together with the ^{1}H-^{13}C HMBC correlations (Figure 2) OH-1 (δ_H 4.89)/C-6 (δ_C 28.5); OH-2 (δ_H 4.92)/C-1 (δ_C 71.9), C-2 (δ_C 70.2), C-3 (δ_C 65.7) and OH-4 (δ_H 5.06)/C-4 (δ_C 68.2), C-5 (δ_C 33.1) indicated the presence of a penta-substituted cyclohexane ring and the location of the hydroxy groups at C-1, C-2 and C-4 in **1**. These data, as well as the chemical shifts of CH-3 (δ_H 4.00, δ_C 65.7), indicated the location of the chlorine atom at C-3.

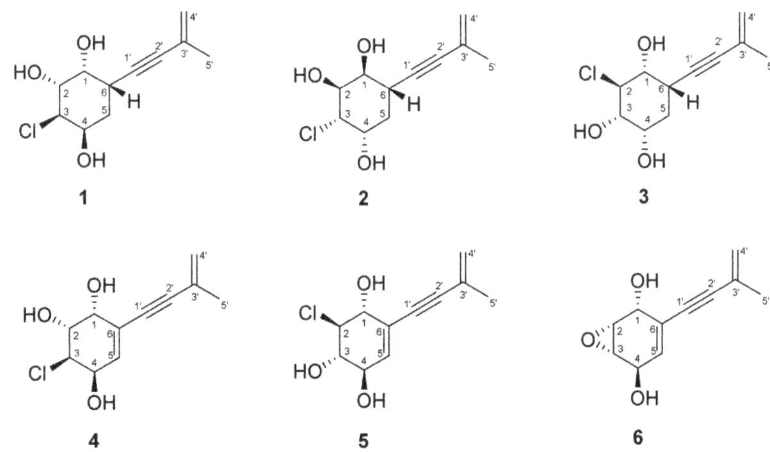

Figure 1. Chemical structures of **1–6**.

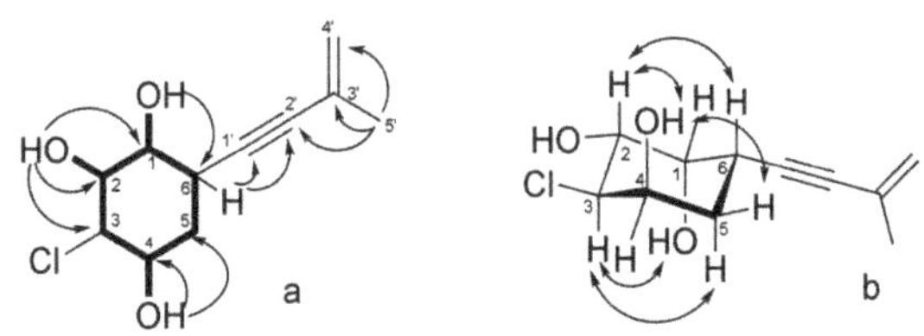

Figure 2. Key ^{1}H–^{1}H COSY, ^{1}H–^{13}C HMBC (**a**) and ROESY (**b**) correlations of **1**.

The HMBC correlations from H-4′a (δ_H 5.14) to C-2′ (δ_C 81.7) and C-3′ (δ_C 126.8), from H_3-5′ (δ_H 1.80) to C-2′, C-3′ and C-4′ (δ_C 120.7) and cross ^{1}H–^{1}H COSY correlations between H_2-4′ and H_3-5′ revealed the presence of a 3-methyl-3-buten-1-ynyl side chain in **1**. The correlations H-6 (δ_H 3.19)/C-1′ (δ_C 87.5) and C-2′ (δ_C 84.1) observed in the HMBC spectrum, recorded in $CDCl_3$ (Figures S20–S22), established the position of the side chain at C-6.

The relative configurations of **1** were assigned based on ROESY correlations (Figure 2) H-6 (δ_H 2.96)/H-2 (δ_H 3.57); OH-4 (δ_H 5.06)/H-5β (δ_H 1.65) and H-2; H-3 (δ_H 4.00)/H-2β (δ_H 1.92), OH-1 (δ_H 4.89) and ^{1}H-^{1}H coupling constants (Table 1). For further investigation, we analyzed the stereoconfigurations of diol at C-1 and C-2 and for protection of these groups before MTPA-esters obtaining the acetonide derivative (**1a**) of compound **1** (Figure 3) was prepared. The small coupling constant ($J_{1,2}$ = 4.3 Hz) and dissimilar magnetic environment of acetonide methyls (Δ = 0.13 ppm) (Figure S57) indicate an *erythro* configuration of the diol group at C-1 and C-2 [22]. The absolute configuration of **1** was established by the modified Mosher's method [23]. Esterification of **1a** with (*S*)- and (*R*)-MTPA chloride occurred at the C-4 hydroxy group to yield the (*R*)-and (*S*) MTPA esters **1a-1** and **1a-2**, respectively. The observed chemical shift differences $\Delta\delta(\delta_S$-$\delta_R)$ (Figure 3) indicated the 4*R* configuration and, therefore, the absolute configurations of **1** were established as 1*R*,2*R*,3*R*,4*R*,6*S*. Compound **1** was named acrucipentyn A.

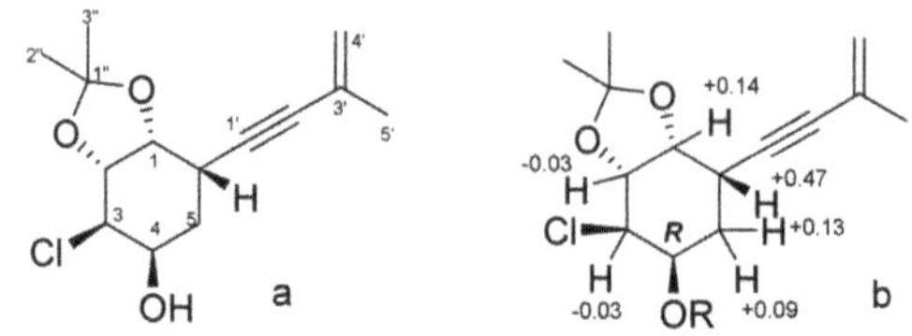

Figure 3. Chemical structure of **1a** (**a**) and $\Delta\delta(\delta_S\text{-}\delta_R)$ values (in ppm) for MTPA esters of **1a** (**b**).

The HRESIMS of **2** and **3** showed the peaks of $[M–H]^-$ at *m/z* 229.0634 and *m/z* 229.0631, respectively. These data, coupled with ^{13}C NMR spectral data (DEPT), established the molecular formulas of **2** and **3** as $C_{11}H_{15}ClO_3$ for both. A close inspection of the ^1H and ^{13}C NMR data (Tables 1 and 2 and Figures S23–S35) of **2** and **3** by DEPT and HSQC revealed the presence of a penta-substituted cyclohexane ring with three hydroxy groups and a 3-methyl-3-buten-1-ynyl side chain.

The main ^1H–^1H COSY and HMBC correlations (Figures S26 and S28) indicated that compound **2** has the same planar structure as **1**. The relative configuration of **2** was assigned based on ^1H-^1H vicinal coupling constants (Table 1) and ROESY (Figure S29) correlations H-6 (δ_H 2.87)/H-4 (δ_H 4.22) and H-5β (δ_H 1.91)/H-1 (δ_H 3.92). Due to the small amount of compound **2,** the absolute configuration establishing by Mosher's method was impossible. Compound **2** was named acrucipentyn B.

The ^1H–^1H COSY correlations of H-1(OH)/H-2/H-3(OH)/H-4(OH)/H_2-5/H-6 together with the ^1H-^{13}C HMBC correlations (Figure S34) OH-1 (δ_H 4.38)/C-1 (δ_C 74.2), C-2 (δ_C 68.9) and C-6 (δ_C 35.0); OH-3 (δ_H 4.50)/C-2, C-3 (δ_C 79.5) and C-4 (δ_C 71.1), and OH-4 (δ_H 4.04)/C-3, C-4 and C-5 (δ_C 34.2) indicated the location of the hydroxy groups at C-1, C-3, C-4 and a chlorine atom at C-2 in a penta-substituted cyclohexane ring of **3**. The structure of the 3-methyl-3-buten-1-ynyl side chain and its position at C-6 in **3** were determined by HMBC correlations (Figure S34), as for acrucipentyn A (**1**).

The relative configurations of the chiral centers in **3** were determined based on ^1H-^1H coupling constants (Table 1). Using the Mosher's method to determine absolute configurations of compound **3** was unsuccessful, due to lability in this compound. Compound **3** was named acrucipentyn C.

The HRESIMS of **4** showed the peak of $[M - H]^-$ at *m/z* 227.0471. These data, coupled with ^{13}C NMR spectral data (DEPT), established the molecular formula of **4** as $C_{11}H_{13}ClO_3$. The ^1H and ^{13}C NMR (Tables 1 and 2 and Figures S36–S42), DEPT and HSQC spectra showed the presence of three hydroxy protons (δ_H 5.25, 5.17, 5.04), one methyl group (δ_H 1.86, δ_C 23.0), one olefinic methylene (δ_H 5.32, 5.27, δ_C 122.2) and five methines (δ_H 4.25, δ_C 63.3; δ_H 4.37, δ_C 65.1; δ_H 4.06, δ_C 67.9; δ_H 3.87, δ_C 68.6 and δ_H 5.92, δ_C 136.0), including three oxygen-bearing methines and one olefinic methine, two sp^2 quaternary carbons and one triple bond (δ_C 88.2 and 89.9).

The HMBC correlations (Figure 4) from H-3 (δ_H 4.25) to C-1 (δ_C 67.9), C-2 (δ_C 68.6), C-4 (δ_C 65.1) and C-5 (δ_C 136.0); from H-5 (δ_H 5.92) to C-1, C-4, C-6 (δ_C 123.5) and C-1′ (δ_C 88.2), and from OH-4 (δ_H 5.25) to C-4, C-5, together with ^1H–^1H COSY correlations of H-1(OH)/H-2(OH)/H-3/H-4(OH)/H-5, indicated the presence of a penta-substituted cyclohexene ring with $\Delta^{5,6}$ double bond, the location of the hydroxy groups at C-1, C-2, C-4 and a chlorine atom at C-3 in **4**. The structure of the 3-methyl-3-buten-1-ynyl side chain and its position at C-6 in **4** were determined by HMBC correlations (Figure S41), as for acrucipentyn A (**1**).

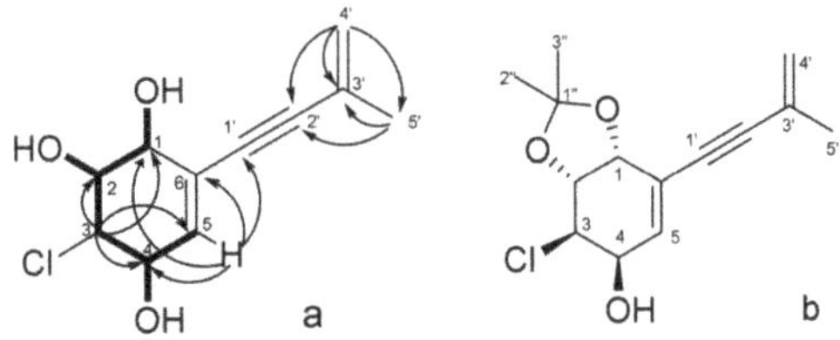

Figure 4. Key ^1H–^1H COSY, ^1H–^{13}C HMBC correlations of **4** (**a**) and chemical structure of acetonide derivatives **4a** (**b**).

The relative configuration of **4** was assigned based on ^1H-^1H coupling constants (Table 1) and ROESY correlation (Figure S42) H-2 (δ_H 3.87)/OH-4 (δ_H 5.25). The acetonide derivative (**4a**) of compound **4** (Figure 4) was prepared for further investigation of the stereochemistry at the diol position. The small coupling constant ($J_{1,2}$ = 5.7 Hz) and dissimilar magnetic environment of acetonide methyls (Δ = 0.01 ppm) (Figure S63) indicate an *erythro* configuration of the diol group at C-1 and C-2 [21]. Esterification of **4a** with (*S*)- and (*R*)-MTPA-Cl led to destruction of the compound. Etherification of compound **4** with (*S*)- and (*R*)-MTPA chloride resulted in the formation of esters at three hydroxyl groups, which made it impossible to establish the absolute configuration using the modified Mosher's method. Compound **4** was named acrucipentyn D.

The HRESIMS of **5** showed the peak of $[M - H]^-$ at *m/z* 227.0468. These data, coupled with ^{13}C NMR spectral data (DEPT), suggested the molecular formula of **5** as $C_{11}H_{13}ClO_3$. The ^1H and ^{13}C NMR data (Tables 1 and 2 and Figures S43–S49) of **5** revealed the presence of a penta-substituted cyclohexene ring with three hydroxy groups and a 3-methyl-3-buten-1-ynyl side chain, the same as in **4**. The location of the hydroxy groups at C-1, C-3, C-4, a chlorine atom at C-2 and a 3-methyl-3-buten-1-ynyl side chain at C-6 in **5** were determined by ^1H–^1H COSY and HMBC correlations (Figures S46 and S48), as for acrucipentyn C (**3**).

The relative configuration of **5** was assigned based on ROESY correlations (Figure S49) H-2 (δ_H 3.69)/H-4 (δ_H 3.97), OH-1 (δ_H 5.73), OH-3 (δ_H 5.48); H-3 (δ_H 3.31)/H-1 (δ_H 4.05), OH-4 (δ_H 5.31) and ^1H-^1H coupling constants (Table 1). The attempts to obtain an acetonide of compound **5** were unsuccessful. Etherification of compound **5** with (*S*)- and (*R*)-MTPA-Cl resulted in the formation of esters at three hydroxyl groups, which made it impossible to establish the absolute configuration by the modified Mosher's method. Compound **5** was named acrucipentyn E.

The molecular formula of compound **6** was determined as $C_{11}H_{12}O_3$, based on the analysis of HRESIMS (*m/z* 191.0705 $[M - H]^-$, calcd for $C_{11}H_{11}O_3$ 191.0714) and NMR data (Tables 1 and 2; Figures S50–S56). The ^1H and ^{13}C NMR data for this compound were similar to those obtained for (+)-asperpentyn [24,25], with the exception of the CH-2 and CH-3 proton and carbon signals. These data, as well as the biogenetic relationship of compound **6** with acrucipentyns D (**4**) and E (**5**), led us to suggest a configuration of asymmetric centers for it, different from the known asperpentyns.

The ROESY spectrum data and ^1H-^1H coupling constants were useless to establish the relative stereochemistry of **6** unambiguously. Therefore, an epoxy ring-opening reaction was carried out in **6**. The reaction of compound **6** with 2,2-dimethylpropanoyl chloride in an aqueous medium (Figure 5) yielded two products, the spectra of which (^1H, ^{13}C NMR, HRESIMS and CD) were identical to acrucipentyns D (**4**) and E (**5**). The presence of only two reaction products confirmed the S_{N2} mechanism of the epoxy ring opening, which corresponded to the literature data [26]. The orientation of the hydroxyl groups in the reaction products corresponds with the configuration of the epoxy ring in the initial compound. These data made it possible to establish the relative configuration of **6**. Compound **6** was named acrucipentyn F.

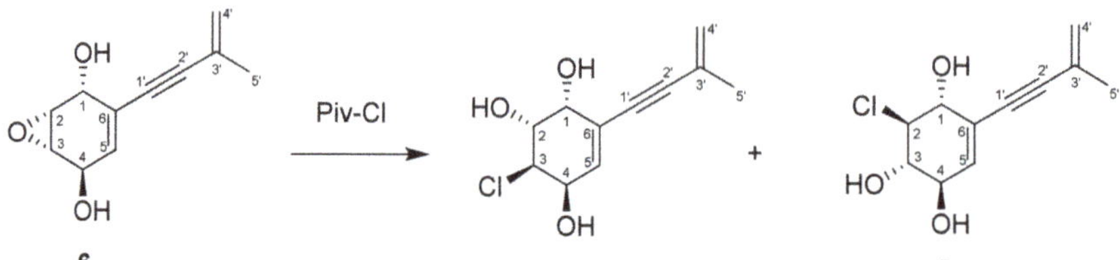

Figure 5. Scheme of an epoxy ring reaction in compound **6**.

It should be noted that acrucipentyn F is a stereoisomer of well-known fungal metabolites (−)-asperpentyn [27,28] and (+)-asperpentyn [24]. To the best of our knowledge, acrucipentyns are the first chlorine-containing asperpentyn-like compounds. However, it should be noted that several other related groups of 3-methylbutenynyl cyclohexanols, e.g., truncateols from the marine-derived fungi *Truncatella angustata* [29,30] and oxirapentyns from the marine-derived fungi *Beauveria felina* KMM 4639 [31], also have chloro-containing members.

3.3. Biological Activity

We evaluated the safety and toxicity of compounds **1–6** in various human cells. As such, we examined cytotoxicity in ten different human cell lines, including human prostate cells PNT2 and RWPE-1, human embryonic kidney cells HEK 293T, human fibroblast cells MRC-9 and human umbilical vascular endothelial cell line HUVEC, human keratinocytes HaCaT, as well as human prostate cancer cells PC-3, DU145, 22Rv1, VCaP and LNCaP using MTT assay. Indeed, none of the investigated compounds exhibited any significant cytotoxicity at concentrations up to 100 µM, following 72 h of treatment (IC_{50} > 100 µM, Figure S65). Additionally, no morphological changes of the cells exposed to the isolated compounds (100 µM for 72 h) could be detected (Figure S66). Therefore, the isolated acrucipentyns A–F were assumed to be nontoxic to mammalian (human) cells.

The inhibitory effect of compounds **1–6** on sortase A enzyme from *Staphylococcus aureus* activity was investigated to detect their antibacterial potential (Figure 6).

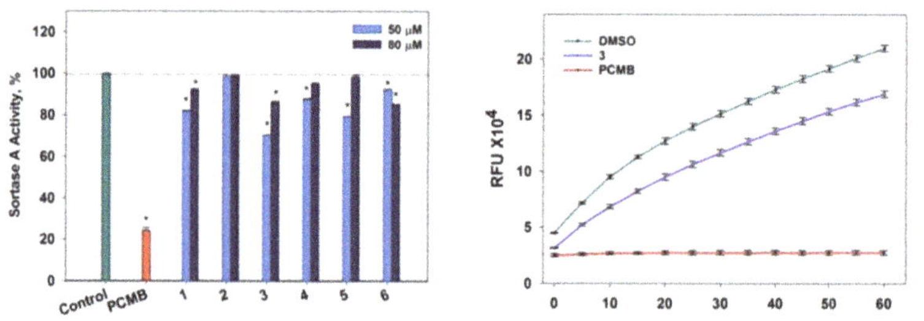

Figure 6. The effect of compounds **1–6** on sortase A enzymatic activity. (**a**) The effect of compounds **1–6** on sortase A enzymatic activity measured after 10 min of incubation with the substrate. (**b**) The time-dependent effect of compound **3** (50 µM) on sortase A enzymatic activity. Data presented as relative fluorescent units (RFU). DMSO (0.8%) did not show any inhibition activity in comparison with sortase A assay buffer and was used as a control. The sortase inhibitor—4-(hydroxymercuri)benzoic acid (PCMB) in DMSO 0.8% was used as a positive control. All experiments were performed in three independent replicates and the data presented as a mean ± standard error mean (SEM). * indicates the significant differences with $p \leq 0.05$.

Compounds **1**, **3**, and **5**, at a concentration of 50 µM, significantly decreased sortase A activity by 18%, 30%, and 21%, respectively (Figure 6a). Compounds **4** and **6** showed less significant effects on sortase A enzymatic activity and compound **2** was inactive in this test. It was observed that an increase in concentrations of **1–6** up to 80 µM resulted in some decrease in their sortase A inhibitory activity, which was sometimes detected [32]. The inhibitory effect of compound **3** on sortase A activity was detected throughout the entire period of data acquisition (Figure 6b) and the effects of other studied compounds were similar.

To detect the antibacterial activity of isolated acrucipentyns A–E (**1–6**), their inhibitory effects on bacterial growth and biofilm formation of *Staphylococcus aureus* were investigated (Figure 7).

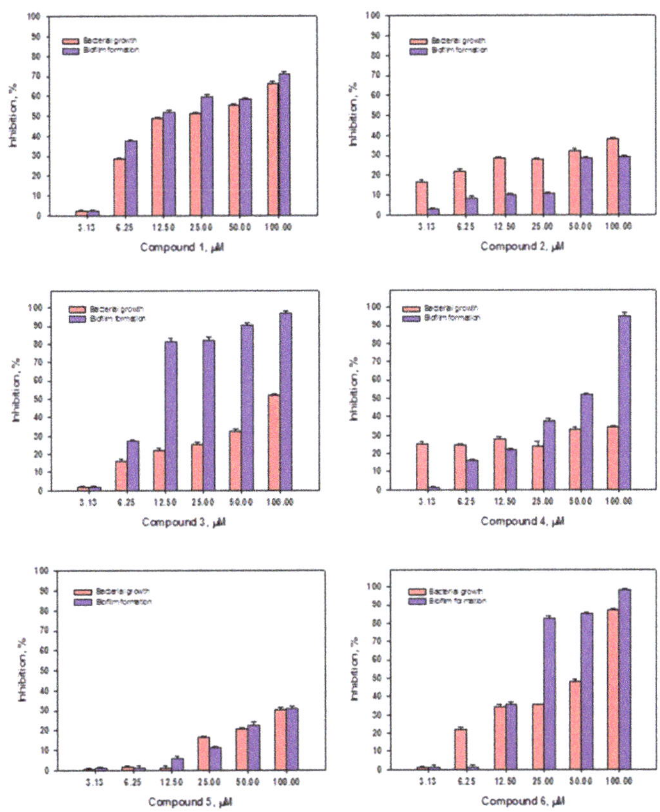

Figure 7. Effect of acrucipentyns A–F (**1–6**) on growth and biofilm formation of *Staphylococcus aureus*. All experiments were performed in three independent replicates and data were presented as a mean ± SEM.

Compounds **1** and **6** have shown the most pronounced antimicrobial effects against Gram-positive bacterium *S. aureus*. Compound **6** almost completely inhibited the growth of *S. aureus* at a concentration of 100 µM; a two-fold decrease in the concentration of the substance also halved the antimicrobial activity. Compound **1** at a concentration of 100 µM reduced the bacterial growth by 60%. A decrease in concentration to 12.5 µM reduced antimicrobial activity up to 50%. Compound **3** at 100 µM inhibited *S. aureus* growth by 50%. A decrease in the concentration of compound **3** led to a two-fold decrease in activity.

The antimicrobial effects of **2**, **4**, and **5**, even at the highest used concentration of 100 μM, do not exceed 50% inhibition of bacterial growth.

When studying the effect of compounds **1–6** on the ability to inhibit the biofilm formation by Gram-positive bacteria *S. aureus*, it was noted that compounds **3**, **4**, and **6** have the most pronounced inhibitory activity at a concentration of 100 μM, in the case of which the formation of biofilms is almost absent. A high level of inhibition of biofilm formation in the case of substances **3** and **6** is kept up to concentrations of 12.5 and 25 μM, respectively. When substance **4** is diluted twice, its effect on biofilm formation is halved. Compound **1** inhibits the biofilm formation at concentrations of 12.5–100 μM by 50–70%, respectively. Compounds **2** and **5** at a concentration of 100 μM inhibited biofilm formation by 30%.

Sortase A is an essential component of *S. aureus* virulence because it is responsible for the covalent anchoring of many virulent factors of Gram-positive bacteria onto the cell wall (Appendix A) and, as a result, sortase A plays a key role in the pathogenic processes of *S. aureus* infection [33]. The decrease in sortase activity leads to the abolition of bacterial adhesion to mammalian cells and, thus, is one of the mechanisms preventing the formation of biofilms, which are the predominant form of bacterial existence [34].

The inhibitory effect of investigated compounds **1**, **3**, **4**, and **6** on the biofilm formation correlated with an ability to affect the activity of sortase A. Compound **2**, which did not show a significant effect on biofilm formation, also did not have any effect on the sortase A activity. Opposite to this, compound **5**, in the same experiments, had a significant effect on the activity of sortase A, but had a weaker inhibitory effect on the biofilm formation, in comparison with compounds **1**, **3**, **4**, and **6**. Thus, substances **1–6** can be assumed as anti-Staphylococcal agents.

To further confirm their antibacterial properties in a model of infectious damage to human cells, we investigated their effects on lactate dehydrogenase (LDH) release from human keratinocytes HaCaT, co-cultivated with *S. aureus* (Figure 8).

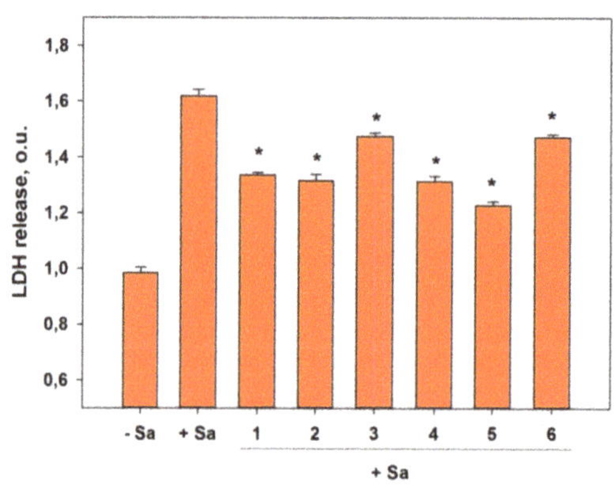

Figure 8. Effect of acrucipentyns A–F (**1–6**) on LDH release from human keratinocytes HaCaT co-cultivated with *Staphylococcus aureus* (Sa) for 48 h. All compounds were tested at a concentration of 10 μM. All experiments were performed in triplicates and data are presented as a mean ± SEM. The difference between control (without Sa) and HaCaT/Sa co-cultivation was statistically significant with $p < 0.05$ (one-way ANOVA test). Asterisk (*) indicates significant differences ($p < 0.05$) between HaCaT/Sa without compounds and HaCaT/Sa with compounds variants.

In normal conditions, LDH weakly releases from cells to culture media. *S. aureus* caused a significant increase in the LDH release from keratinocytes during co-cultivation. The addition of compounds **1–6** at a concentration of 10 μM reduced the LDH release by 30–50%. The greatest effect was registered for compounds **1**, **2**, **4**, and **5**.

Interestingly, compounds **2** and **4**, which showed significant activity in co-cultivation HaCaT cells with *S. aureus*, did not show high antibacterial and anti-biofilm-forming activity, in contrast to substances **1** and **5**. We assume that the cytoprotective effect of substances **1–6** in the in vitro infectious skin lesions could be due not only to their anti-*S. aureus* effects, but also to their anti-inflammatory and other cytoprotective effects. Recently, we observed similar dual effects during an investigation on flavuside B, an inhibitor of sortase A enzymatic activity derived from fungi. Flavuside B was able to inhibit *S. aureus* growth and biofilm formation, as well as protect HaCaT keratinocytes against *S. aureus* infection in a co-culture model via an anti-inflammatory pathway [14].

Our work is the very first detailed investigation of anti-Staphylococcal activity of isoprenylated cyclohexanols. To the best of our knowledge, earlier, only oxirapentyns A and D were found as antimicrobial agents among related compounds [35]. Thus, this group of secondary metabolites of marine fungi is interesting for future study, including their structure–activity relationships.

4. Conclusions

Six new isoprenylated cyclohexanols acrucipentyns A–F (**1–6**) were isolated from the alga-derived fungus *Asteromyces cruciatus* KMM 4696. The absolute configuration of acrucipentyn A was assigned by the modified Mosher's method and ROESY data. Acrucipentyns A–E are the very first members of chlorine-contained monocyclic cyclohexanols containing a 3-methylbutenynyl unit. The compounds have shown inhibitory activity against sortase A from *Staphylococcus aureus*, as well as inhibition of *S. aureus* growth and biofilm formation, while no cytotoxicty to mammalian cells was observed. Moreover, acrucipentyns A–F (**1–6**) protected human keratinocytes HaCaT from *S. aureus* toxicity in skin infection in an in vitro model. Thus, the isolated compounds hold a good potential as antimicrobal agents and should be further investigated.

Supplementary Materials: The following are available online at https://www.mdpi.com/article/10.3390/jof8050454/s1, Figures S1–S12: CD and UV spectra of compound **1–6**, Figures S13–S22, S23–S28, S29–S35, S36–S43, S44–S49 and S50–S56: NMR spectra of compounds **1–6**, Figures S57–S58, S59–S60, S61–S62 and S63–S64: NMR spectra of compounds **1a**, **1a-1**, **1a-2**, **4a**, Figure S65: Viability of the various cells exposed to the tested compounds, Figure S66: Microphotographs of human cells exposed to the tested compounds.

Author Contributions: Conceptualization, O.I.Z.; Investigation, G.K.O., A.S.A., N.N.K., D.N.P., A.B.R., A.S.M., R.S.P., N.Y.K., A.R.C., E.A.C., O.O.V. and S.A.D.; Methodology, E.A.Y., D.N.P. and A.N.Y.; Project administration, O.I.Z.; Resources, O.I.Z., I.V.G., G.v.A. and A.N.Y.; Supervision, G.v.A. and A.N.Y.; Validation, G.v.A. and A.N.Y.; Visualization, E.A.C.; Writing—Original draft, O.I.Z.; Writing—Review and editing, G.v.A., S.A.D. and A.N.Y. All authors have read and agreed to the published version of the manuscript.

Funding: This research was funded by the Russian Science Foundation, grant number 19-74-10014.

Institutional Review Board Statement: Not applicable.

Informed Consent Statement: Not applicable.

Data Availability Statement: Data are contained within the article or Supplementary Material.

Acknowledgments: The study was carried out on equipment at the Collective Facilities Center "The Far Eastern Center for Structural Molecular Research (NMR/MS) PIBOC FEB RAS". The study was carried out using the Collective Facilities Center "Collection of Marine Microorganisms PIBOC FEB RAS".

Conflicts of Interest: The authors declare no conflict of interest.

Appendix A

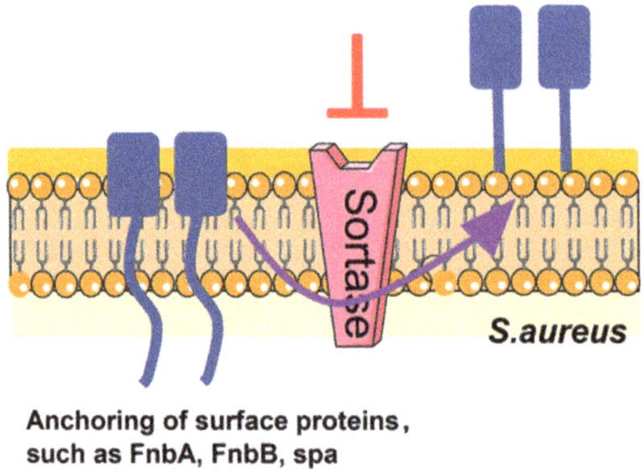

Figure A1. Representation of sortase A role [36].

References

1. Pang, K.L.; Overy, D.P.; Jones, E.B.G.; Calado, M.D.L.; Burgaud, G.; Walker, A.K.; Johnson, J.A.; Kerr, R.G.; Cha, H.J.; Bills, G.F. 'Marine fungi' and 'marine-derived fungi' in natural product chemistry research: Toward a new consensual definition. *Fungal Biol. Rev.* **2016**, *30*, 163–175. [CrossRef]
2. Chen, S.; Wang, J.; Lin, X.; Zhao, B.; Wei, X.; Li, G.; Kaliaperumal, K.; Liao, S.; Yang, B.; Zhou, X.; et al. Chrysamides A-C, three dimeric nitrophenyl trans-epoxyamides produced by the deep-sea-derived fungus *Penicillium chrysogenum* SCSIO41001. *Org. Lett.* **2016**, *18*, 3650–3653. [CrossRef] [PubMed]
3. Jones, E.B.G.; Stanley, S.J.; Pinruan, U. Marine endophyte sources of new chemical natural products: A review. *Bot. Mar.* **2008**, *51*, 163–170. [CrossRef]
4. Jones, E.B.G.; Pang, K.L.; Abdel-Wahab, M.A.; Scholz, B.; Hyde, K.D.; Boekhout, T.; Ebel, R.; Rateb, M.E.; Henderson, L.; Sakayaroj, J.; et al. An online resource for marine fungi. *Fungal Divers.* **2019**, *96*, 347–433. [CrossRef]
5. Shin, J.H.; Fenical, W. Isolation of gliovictin from the marine deuteromycete *Asteromyces cruciatus*. *Phytochemistry* **1987**, *26*, 3347. [CrossRef]
6. Gulder, T.A.M.; Hong, H.; Correa, J.; Egereva, E.; Wiese, J.; Imhoff, J.F.; Gross, H. Isolation, structure elucidation and total synthesis of lajollamide A from the marine fungus *Asteromyces cruciatus*. *Mar. Drugs* **2012**, *10*, 2912–2935. [CrossRef]
7. Igboeli, H.A.; Marchbank, D.H.; Correa, H.; Overy, D.; Kerr, R.G. Discovery of Primarolides A and B from Marine Fungus Asteromyces cruciatus Using Osmotic Stress and Treatment with Suberoylanilide Hydroxamic Acid. *Mar. Drugs* **2019**, *17*, 435. [CrossRef]
8. Wang, C.; Tang, S.; Cao, S. Antimicrobial compounds from marine fungi. *Phytochem. Rev.* **2020**, *20*, 85–117. [CrossRef]
9. Gomes, N.G.M.; Madureira-Carvalho, Á.; Dias-da-Silva, D.; Valentão, P.; Andrade, P.B. Biosynthetic versatility of marine-derived fungi on the delivery of novel antibacterial agents against priority pathogens. *Biomed. Pharmacother.* **2021**, *140*, 111756. [CrossRef]
10. Nweze, J.A.; Mbaoji, F.N.; Huang, G.; Li, Y.; Yang, L.; Zhang, Y.; Huang, S.; Pan, L.; Yang, D. Antibiotics development and the potentials of marine-derived compounds to stem the tide of multidrug-resistant pathogenic bacteria, fungi, and protozoa. *Mar. Drugs* **2020**, *18*, 145. [CrossRef]
11. Mazmanian, S.K.; Ton-That, H.; Su, K.; Schneewind, O. An iron-regulated sortase anchors a class of surface protein during *Staphylococcus aureus* pathogenesis. *Proc. Natl. Acad. Sci. USA* **2002**, *99*, 2293–2298. [CrossRef] [PubMed]
12. Hendrickx, A.P.; Budzik, J.M.; Oh, S.Y.; Schneewind, O. Architects at the bacterial surface-sortases and the assembly of pili with isopeptide bonds. *Nat. Rev. Microbiol.* **2011**, *9*, 166–176. [CrossRef] [PubMed]
13. Nitulescu, G.; Margina, D.; Zanfirescu, A.; Olaru, O.T.; Nitulescu, G.M. Targeting bacterial sortases in search of anti-virulence therapies with low risk of resistance development. *Pharmaceuticals* **2021**, *14*, 415. [CrossRef] [PubMed]
14. Chingizova, E.A.; Menchinskaya, E.S.; Chingizov, A.R.; Pislyagin, E.A.; Girich, E.V.; Yurchenko, A.N.; Guzhova, I.V.; Mikhailov, V.V.; Aminin, D.L.; Yurchenko, E.A. Marine fungal cerebroside flavuside B protects HaCaT keratinocytes against *Staphylococcus aureus* induced damage. *Mar. Drugs* **2021**, *19*, 553. [CrossRef]
15. Girich, E.V.; Rasin, A.B.; Popov, R.S.; Yurchenko, E.A.; Chingizova, E.A.; Trinh, P.T.H.; Ngoc, N.T.D.; Pivkin, M.V.; Zhuravleva, O.I.; Yurchenko, A.N. New tripeptide derivatives asterripeptides A-C from vietnamese mangrove-derived fungus *Aspergillus terreus* LM.5.2. *Mar. Drugs* **2022**, *20*, 77. [CrossRef]
16. White, T.J.; Bruns, T.D.; Lee, S.B.; Taylor, J.W. Amplification and direct sequencing of fungal ribosomal RNA genes for phylogenetics. In *PCR Protocols: A Guide to Methods and Applications*; Innis, M.A., Gelfand, D.H., Sninsky, J.J., White, T.J., Eds.; Academic Press: London, UK, 1990; pp. 315–322.

17. Dyshlovoy, S.A.; Pelageev, D.N.; Hauschild, J.; Borisova, K.L.; Kaune, M.; Krisp, C.; Venz, S.; Sabutskii, Y.E.; Khmelevskaya, E.A.; Busenbender, T.; et al. Successful targeting of the warburg effect in prostate cancer by glucose-conjugated 1,4-naphthoquinones. *Cancers* **2019**, *11*, 21. [CrossRef]
18. Dyshlovoy, S.A.; Pelageev, D.N.; Hauschild, J.; Sabutskii, Y.E.; Khmelevskaya, E.A.; Krisp, C.; Kaune, M.; Venz, S.; Borisova, K.L.; Busenbender, T.; et al. Inspired by sea urchins: Warburg effect mediated selectivity of novel synthetic non-glycoside 1,4-naphthoquinone-6S-glucoseconjugates in prostate cancer. *Mar. Drugs* **2020**, *18*, 31. [CrossRef]
19. Macià, M.D.; Rojo-Molinero, E.; Oliver, A. Antimicrobial susceptibility testing in biofilm-growing bacteria. *Clin. Microbiol. Infect.* **2014**, *20*, 981–990. [CrossRef]
20. Thappeta, K.R.; Zhao, L.N.; Nge, C.E.; Crasta, S.; Leong, C.Y.; Ng, V.; Kanagasundaram, Y.; Fan, H.; Ng, S.B. In-Silico identified new natural sortase A inhibitors disrupt *S. aureus* biofilm formation. *Int. J. Mol. Sci.* **2020**, *21*, 8601. [CrossRef]
21. Wiegand, C.; Abel, M.; Ruth, P.; Hipler, U.C. HaCaT keratinocytes in co-culture with *Staphylococcus aureus* can be protected from bacterial damage by polihexanide. *Wound Repair Regen.* **2009**, *17*, 730–738. [CrossRef]
22. Nagle, D.G.; Gerwick, W.H. Structure and stereochemistry of constanolactones A-G, lactonized cyclopropyl oxylipins from the red marine alga *Constantinea simplex*. *J. Org. Chem.* **1994**, *59*, 7227–7237. [CrossRef]
23. Kusumi, T.; Ooi, T.; Ohkubo, Y.; Yabuuchi, T. The modified Mosher's method and the sulfoximine method. *Bull. Chem. Soc. Jpn.* **2006**, *79*, 965–980. [CrossRef]
24. Rukachaisirikul, V.; Rungsaiwattana, N.; Klaiklay, S.; Phongpaichit, S.; Borwomwiriyapan, K.; Sakayaroji, J. gamma-Butyrolactone, cytochalasin, cyclic carbonate, eutypinic acid, and phenalenone derivatives from the soil fungus *Aspergillus* sp. PSU-RSPG185. *J. Nat. Prod.* **2014**, *77*, 2375–2382. [CrossRef]
25. Li, J.; Park, S.; Miller, R.L.; Lee, D. Tandem enyne metathesis-metallotropic [1,3]-shift for a concise total syntheses of (+)-asperpentyn, (−)-harveynone, and (−)-tricholomenyn A. *Org. Lett.* **2009**, *11*, 571–574. [CrossRef] [PubMed]
26. Rao, C.B.; Rao, D.C.; Venkateswara, M.; Venkateswarlu, Y. Protective opening of epoxide using pivaloyl halides under catalyst-free conditions. *Green Chem.* **2011**, *13*, 2704–2707. [CrossRef]
27. Muhlenfeld, A.; Achenbach, H. Asperpentyn, a Novel Acetylenic Cyclohexene Epoxide from *Aspergillus duricaulis*. *Phytochemistry* **1988**, *27*, 3853–3855. [CrossRef]
28. Yurchenko, A.N.; Smetanina, O.F.; Khudyakova, Y.V.; Kirichuk, N.N.; Yurchenko, E.A.; Afiyatullov, S.S. Metabolites of the marine isolate of the fungus *Curvularia inaequalis*. *Chem. Nat. Compd.* **2013**, *49*, 163–164. [CrossRef]
29. Zhao, Y.; Si, L.; Liu, D.; Proksch, P.; Zhou, D.; Lin, W. Truncateols A–N, new isoprenylated cyclohexanols from the sponge-associated fungus *Truncatella angustata* with anti-H1N1 virus activities. *Tetrahedron* **2015**, *71*, 2708–2718. [CrossRef]
30. Zhao, Y.; Liu, D.; Proksch, P.; Zhou, D.; Lin, W. Truncateols O-V, further isoprenylated cyclohexanols from the sponge-associated fungus *Truncatella angustata* with antiviral activities. *Phytochemistry* **2018**, *155*, 61–68. [CrossRef]
31. Yurchenko, A.N.; Smetanina, O.F.; Kalinovsky, A.I.; Pushilin, M.A.; Glazunov, V.P.; Khudyakova, Y.V.; Kirichuk, N.N.; Ermakova, S.P.; Dyshlovoy, S.A.; Yurchenko, E.A.; et al. Oxirapentyns F-K from the marine-sediment-derived fungus *Isaria felina* KMM 4639. *J. Nat. Prod.* **2014**, *77*, 1321–1328. [CrossRef]
32. Volynets, G.; Vyshniakova, H.; Nitulescu, G.; Nitulescu, G.M.; Ungurianu, A.; Margina, D.; Moshynets, O.; Bdzhola, V.; Koleiev, I.; Iungin, O.; et al. Identification of novel antistaphylococcal hit compounds targeting sortase a. *Molecules* **2021**, *26*, 7095. [CrossRef] [PubMed]
33. Guo, Y.; Cai, S.; Gu, G.; Guo, Z.; Long, Z. Recent progress in the development of sortase A inhibitors as novel anti-bacterial virulence agents. *RSC Adv.* **2015**, *5*, 49880–49889. [CrossRef]
34. Mazmanian, S.K.; Liu, G.; Jensen, E.R.; Lenoy, E.; Schneewind, O. *Staphylococcus aureus* sortase mutants defective in the display of surface proteins and in the pathogenesis of animal infections. *Proc. Natl. Acad. Sci. USA* **2000**, *97*, 5510. [CrossRef]
35. Smetanina, O.F.; Yurchenko, A.N.; Afiyatullov, S.S.; Kalinovsky, A.I.; Pushilin, M.A.; Khudyakova, Y.V.; Slinkina, N.N.; Ermakova, S.P.; Yurchenko, E.A. Oxirapentyns B-D produced by a marine sediment-derived fungus *Isaria felina* (DC.) Fr. *Phytochem. Lett.* **2012**, *5*, 165–169. [CrossRef]
36. Wang, L.; Wang, G.; Qu, H.; Wang, K.; Jing, S.; Guan, S.; Su, L.; Li, Q.; Wang, D. Taxifolin, an Inhibitor of Sortase A, Interferes With the Adhesion of Methicillin-Resistant *Staphylococcal aureus*. *Front. Microbiol.* **2021**, *12*, 686864. [CrossRef] [PubMed]

Journal of *Fungi*

MDPI

Article

Gram-Level Production of Balanol through Regulatory Pathway and Medium Optimization in Herb Fungus *Tolypocladium ophioglossoides*

Rui-Qi Li [1,2], Xiang Liu [1,2], Min Zhang [1,2], Wei-Qun Xu [1], Yong-Quan Li [2] and Xin-Ai Chen [1,2,*]

1 School of Medicine and the Children's Hospital, Zhejiang University, Hangzhou 310058, China; 21918020@zju.edu.cn (R.-Q.L.); 22018007@zju.edu.cn (X.L.); zjuzm@outlook.com (M.Z.); vicky__xu@zju.edu.cn (W.-Q.X.)

2 Institute of Pharmaceutical Biotechnology, Zhejiang University, Hangzhou 310058, China; lyq@zju.edu.cn

* Correspondence: biolab@zju.edu.cn; Tel.: +86-571-8820-8569

Abstract: As a potential protein kinase C inhibitor, the fungus metabolite balanol has become more attractive in recent decades. In our previous work, we revealed its biosynthetic pathway through overexpression of the cluster-situated regulator gene *blnR* in Chinese herb fungus *Tolypocladium ophioglossoides*. However, information on the regulation of *blnR* is still largely unknown. In this study, we further investigated the regulation of balanol biosynthesis by BlnR through the analysis of affinity binding using EMSA and RNA-seq analysis. The results showed that BlnR positively regulates balanol biosynthesis through binding to all promoters of *bln* gene members, including its own promoter. Microscopic observation revealed *blnR* overexpression also affected spore development and hypha growth. Furthermore, RNA-seq analysis suggested that BlnR can regulate other genes outside of the balanol biosynthetic gene cluster, including those involved in conidiospore development. Finally, balanol production was further improved to 2187.39 mg/L using the optimized medium through statistical optimization based on response surface methodology.

Keywords: balanol biosynthesis; protein kinase C inhibitor; Zn_2Cys_6; regulator BlnR; medium optimization; fermentation; *Tolypocladium ophioglossoides*

Citation: Li, R.-Q.; Liu, X.; Zhang, M.; Xu, W.-Q.; Li, Y.-Q.; Chen, X.-A. Gram-Level Production of Balanol through Regulatory Pathway and Medium Optimization in Herb Fungus *Tolypocladium ophioglossoides*. *J. Fungi* **2022**, *8*, 510. https://doi.org/10.3390/jof8050510

Academic Editor: Ulrich Kück

Received: 19 April 2022
Accepted: 12 May 2022
Published: 16 May 2022

1. Introduction

The fungal metabolite balanol was isolated as a potent ATP-competitive inhibitor of protein kinase C (PKC) from *Verticillium balanoides*, which was the same compound as that previously reported as azepinostatin from *Fusarium merisomides* and ophiocordin from *Tolypocladium ophioglossoides* (its synonymic name is *Cordyceps ophioglossoides)* [1–4]. It was shown to have inhibitory activity toward PKC isozymes in the nanomolar range, with better potency than the reported product staurosporine, as well as showing some activity against PKA. Protein kinase C (PKC) group can regulate the conformation and activity of target proteins by phosphorylating their serine or threonine residuals. PKC is the receptor for phorbol esters that promote tumor formation, playing crucial roles in cell proliferation and differentiation [5]. The upregulated activation of PKC has been related to a range of disease states, including central nervous system (CNS) diseases, cardiovascular disorders, diabetes, asthma and HIV infections [6,7]. The role of these enzymes in the development of cancer makes them an ideal target for screening their interesting inhibitory compounds. The specific molecular structure of balanol is similar to that of ATP, making it the ATP competitor binding to the PKC enzyme, thus inhibiting PKC activity.

T. ophioglossoides is a kind of parasitic mushroom of certain Elaphomyces, and it has been used to relieve postmenopausal syndrome in women in Chinese traditional medicine throughout history. Several active compounds have been identified from *T. ophioglossoides*, such as balanol, cordycepol and tyrosol [8–10]. Compared with the annotated gene clusters based on the genome sequence of *T. ophioglossoides*, many other gene clusters are still cryptic during laboratory culture. Although balanol has been produced through liquid fermentation, its yield was very low, even in large-scale fermentation. In recent decades, our group focused on the chemical synthesis of balanol, as well as its analogs, before elucidating f its biosynthetic pathway through overexpression of its cluster-situated regulator gene, *blnR* [10].

Several efficient strategies have been developed to activate cryptic gene clusters for the production of new compounds, such as microbial cocultivation [11,12], promoter engineering [13], ribosome engineering [14], epigenetic regulation [15,16] and transcription regulation by regulatory proteins, including global and pathway-specific regulators [17,18]. Transcription factors are usually involved in various important processes during microorganism growth and development via regulation of a series of target genes. Microbial genome sequence analysis reveals that transcription regulator-encoding genes are present within many individual gene clusters. Switching on the regulator gene in its active state is considered an important strategy to activate cryptic gene clusters to produce new secondary metabolites (SMs) via genetic engineering or improve the production of valuable products [19]. Compared with the global regulator, the pathway-specific activator gene is usually cluster-located, and its overexpression using a strong promoter or knockout is a simple and efficient strategy to improve the product of interest or activate the cryptic gene cluster. For example, the overexpression of the StrR family regulator in *Streptomyces* significantly increased ristomycin A content [20]. Moreover, SM production by microbes has been observed to vary with the composition of culture media and culture conditions [21]. Based on this concept, the approach of 'one strain, many compounds' (OSMAC) was developed by changing culture medium and has been widely used in many bacteria and fungi to efficiently mine novel SMs, including polyketides, non-ribosomal peptides and terpenes, in recent years [22–24]. Hence, changing culture medium is also a classic approach to activating cryptic gene clusters. In addition, cultivation parameters, including temperature, salinity and dissolved oxygen, are considered effective ways to trigger cryptic biosynthetic pathways in *Aspergillus ochraceus*.

With respect to the biosynthesis pathway of balanol, we previously characterized several key biosynthetic enzymes, including BlnJ, BlnF, BlnO and BlnP [10]. However, there is still a lack of regulatory information about the balanol biosynthesis. According to the annotation, there is only one regulatory gene, *blnR* was found within the gene cluster *bln*. BlnR is a Zn_2Cys_6 family regulator, but the role of BlnR in balanol production has not been investigated. Therefore, it is important to understand its regulatory function in balanol biosynthesis in order to improve its production. In this study, the regulation of balanol biosynthesis was first investigated by analyzing the affinity-binding ability between regulator BlnR and target genes using an EMSA experiment. Then, RNA-seq analysis was performed to investigate the effect of *blnR* overexpression on *T. ophioglossoides*. Additionally, the culture conditions for balanol production by strain *blnR*OE were further optimized based on statistical experimentation through a one-factor experiment and response surface methodology (RSM). As a consequence, the cluster-situated regulator, BlnR, positively regulated balanol biosynthesis by binding to all the promoters of gene cluster members, as well as its own promoter. The transcriptomic data showed that BlnR is broadly involved in both the primary and secondary metabolites. The concentration of balanol in a 5 L fermenter tank was improved to 2187.39 mg/L from 700 mg/L through statistical optimization using the optimized medium.

2. Materials and Methods

2.1. Chemicals

Potato dextrose agar (PDA) and chlorimuron-ethyl were purchased from Sigma (St. Louis, MO, USA). Polypeptone, imidazole, Tris-HCl solution, kanamycin PMSF and IPTG were purchased from Sangon Biotechnology Incorporation (Shanghai, China). Yeast extract was purchased from Oxoid (England, UK). Sucrose (99.5%), $MgSO_4{\cdot}7H_2O$ (99%),

KH_2PO_4 (99%), NaOH (96%), NaCl (99.5%) and trifluoroacetic acid (TFA, 99.5%) were purchased from China National Pharmaceutical Group Corporation (Shanghai, China). Acetonitrile (99.9%) was obtained from Astoon Chemical Technology Incorporation (Wilmington, DE, USA).

2.2. Strain, Media and Culture Conditions

Escherichia coli DH5α was used for routine DNA manipulation, and BL21 (DE3)/Origami B was used as a host for protein expression. *blnR* overexpression strain *T. ophioglossoides blnR*OE was used as the balanol-producing strain, which was routinely supplemented with 4 mg/L of chlorimuron-ethyl (Sigma, USA) if necessary, as previously described [10]. *E. coli* cells were cultured in LB broth at 37 °C and 200 rpm. *Agrobacterium tumefaciens* EHA105 was used to transfer DNA into *T. ophioglossoides* via T-DNA transformation, as described previously [25]. COB liquid medium (sucrose 30 g/L, polypeptone 5 g/L, yeast extract 5 g/L, $MgSO_4.7H_2O$ 1 g/L, KH_2PO_4 0.5 g/L, pH 5.5) was used as the seed culture and starter fermentation medium.

For balanol production, seed cultures were prepared by inoculating 2×10^5 spores/mL of *blnR*OE strain in COB medium in 250 mL shake flasks with 80 mL medium and incubated at 26 °C and 160 rpm for 3–4 days. Then, seed cultures were transferred to the fermentation medium using 2.5% (*v*/*v*) inoculum. For each strain, a shake flask assay was carried out in triplicate parallel bottles. Batch fermentation was carried out in a 15 L jar fermenter at 26 °C (BLBIO-3GJ, China) containing 8 L of culture medium. The seed culture was prepared and inoculated into the fermenter jar at 2% (*v*/*v*). The pH of the medium was maintained at 4.9 via automatic addition of 2 M NaOH. The dissolved oxygen concentration was maintained at 20% air saturation by automatically increasing the agitation speed. Culture samples were periodically taken to analyze balanol concentration and CDWs. Both the batch and fed-batch fermentation experiments were carried out in triplicate.

2.3. Heterologous Expression of blnR in E. coli and Its Purification

Total RNA extracted from mycelia of the *blnR*OE strain was used as a template for first-strand cDNA by SuperScript™ IV reverse transcriptase (Invitrogen). A cDNA fragment encoding the whole length of *blnR* and its DNA binding domain (*blnR*^{DBD}) of *T. ophioglossoides* was subcloned into the pET-32a vector (Novagen, Darmstadt, Germany) by infusion cloning technology (Vazyme, Nanjing, China). The BlnR^{DBD} and BlnR protein were produced in *E. coli* BL21 (DE3) cells by the addition of IPTG and grown overnight at 16 °C in a 1 L flask in the presence of 50 mg/L kanamycin. The pellet cells were collected by centrifugation, resuspended in 200 mL lysis buffer (20 mM Tris-HCl, 500 mM NaCl, 10 mM imidazole, 1 mM PMSF, pH 8.0) and disrupted by sonication. After centrifugation at 12,000× rpm at 4 °C for 5 min, the cleared cellular extract was adjusted to pH 8.0 and loaded on a Ni-agarose column (GE, Munich, Germany), which was previously equilibrated with 20 mM Tris-HCl, 500 mM NaCl and 20 mM imidazole (pH 8.0). The impurity protein was washed with 20 mM Tris-HCl, 500 mM NaCl and 100 mM imidazole (pH 8.0), and the BlnR^{DBD}-containing fraction was eluted with elution buffer containing 500 mM imidazole (pH 8.0). The elution fraction was loaded on a desalting column with 3 kDa to be concentrated and redissolved in 20 mM Tris-HCl, 500 mM NaCl and 1 mM PMSF (pH 8.0).

2.4. Affinity Analysis by Electrophoretic Mobility Shift Assay (EMSA)

An electrophoretic mobility shift assay was performed to investigate the affinity of regulator BlnR with DNA fragments as described previously [26]. An FM-labeled DNA fragment was prepared by PCR amplification, directly using FM-labeled primers as EMSA probes. The purified PCR product was then employed as a template to generate the FM-labeled DNA probe using the corresponding FM-labeled primer, as shown in Table S1. A volume of 1 μg salmon sperm DNA was used as a non-specific competitor in the binding mixture. The FM-labeled probes were detected with an LAS4000 machine (GE, Boston, MA, USA).

2.5. *Phylogenetic Analysis*

All the Zn_2Cys_6 transcription factors used for alignment were collected from the NCBI protein database by blast alignment. All the protein sequences were aligned using ClustalW in MEGA 7, and a phylogenetic tree was constructed with the maximum likelihood method based on the JTT matrix-based model. Protein domain architecture analysis was performed by conducting a search of the Conserved Domains Database (https://www.ncbi.nlm.nih.gov/cdd), as well as online analysis with Pfam (www.pfam.org).

2.6. *mRNA-Seq Analysis and Differential Gene Expression Analysis*

Total RNA was isolated from *T. ophioglossoides* mycelia grown in COB medium for 4 days at 26 °C by RNA extraction kits according to the manufacturer's instruction (Takara, Japan). The genomic DNA in the RNA samples was digested by RNase-free DNase I (Takara, Japan). The first-strand cDNA was reverse-transcribed from total RNA with SuperScript™ IV reverse transcriptase (18090010, Invitrogen). qRT-PCR was performed using SYBR Premix Ex Taq II (Takara, Japan), and PCR procedures were performed at 95 °C for 5 min, 40 cycles of 95 °C for 15 s, 56 °C for 40 s and 72 °C for 20 s. The *Totef1* gene encoding housekeeping translational elongation factor was used as the internal control. The changes in target gene expression were quantified as $2^{-\Delta\Delta Ct}$ according to the manufacturer's instructions (Takara, Japan). The primers used are listed in Table S1.

For mRNA-seq analysis, 1 μg of total RNA was used as the input template for the RNA sample preparations. The sequencing libraries were generated using the NEBNext UltraTM RNA Library Prep Kit for Illumina (NEB, Ipswich, MA, USA) following the manufacturer's recommendations, and index codes were added to attribute sequences to each sample. First-strand cDNA was synthesized using a random hexamer primer and M-MuLV reverse transcriptase. Second-strand cDNA synthesis was subsequently performed using DNA polymerase I and RNase H. Remaining overhangs were converted into blunt ends via exonuclease/polymerase activities. After adenylation of the 3′ ends of DNA fragments, NEBNext Adaptors with hairpin loop structure were ligated to prepare for hybridization. The library fragments were purified with an AMPure XP system (Beckman Coulter, Beverly, MA, USA) to choose cDNA fragments with a preferential length of 240 bp. Then, USER enzyme (NEB, Ipswich, MA, USA) was used with adaptor-ligated cDNA at 37 °C for 15 min, followed by 5 min at 95 °C before PCR. PCR was performed using Phusion high-fidelity DNA polymerase, universal PCR primers and index (X) Primer. Finally, PCR products were purified (AMPure XP system), and their quality was evaluated by the Agilent Bioanalyzer 2100 system. Clustering of the index-coded samples was performed on a cBot cluster-generation system using a TruSeq PE v4-cBot-HS cluster kit (Illumina) according to the manufacturer's instructions. After cluster generation, the library preparations were sequenced on an Illumina NovaSeq 6000 platform, and paired-end reads were generated. The reference genome of *T. ophioglossoides* was predefined for analysis and mapping of RNA-seq reads with an HISAT2 system.

Differential expression analysis of genes between two samples was performed using the EdgeR bioconductor package and a dispersion parameter of 0.1. EdgeR provided statistical routines to determine differential expression in digital gene expression data using a model based on the negative binomial distribution. The resulting P values were adjusted using the Benjamini–Hochberg approach to control the false discovery rate. Genes with an adjusted p-value < 0.05, as determined by EdgeR, were classified as differentially expressed.

2.7. *Optimization of Medium Components for Balanol Production by Response Surface Methodology (RSM)*

The Box−Behnken statistical method was used for the optimization of the medium components. Critical parameters were observed, namely sucrose, polypeptone and initial pH. Design-Expert® 10.0.0 software was employed for experimental design and analysis. A total of 12 run experiments were tested based on the design matrix with three center points to minimize the experimental error. A model was generated based on the response values of balanol production, and statistical significance was tested by analysis of variance (ANOVA). The predicted combination of medium components for maximum balanol production was further validated experimentally.

The optimal medium was statistically optimized through response surface methodology in the *blnR*OE strain. The effect of one factor on balanol production was first examined, and then PB design was applied to determine the significant components according to the balanol production, including 12 run experiments and 6 variables, including sucrose, polypeptone, yeast extract, KH_2PO_4, $MgSO_4$ $7H_2O$ and pH (Table 1). Based on the results of the PB, the significant factors were further optimized based on the RSM coupled with BB design using Design-Expert software to determine the final optimal fermentation medium. The quality of fit of the second-order polynomial model equation was determined via a coefficient of determination (R^2) and the adjusted R^2. ANOVA (analysis of variance) was used for graphical analyses to estimate the interaction between the component variables and balanol production. The components in the culture media that showed confidence levels >95% were considered to exhibit significant responses to balanol production.

Table 1. PB experimental design and observed balanol production.

Run	A	B	C	D	E	F	G	Balanol (mg/L)
1	1	−1	−1	−1	1	−1	1	1218.48
2	1	−1	1	1	−1	1	1	586.54
3	−1	−1	−1	−1	−1	−1	−1	832.88
4	−1	−1	−1	1	−1	1	1	633.36
5	−1	1	−1	1	1	−1	1	400.22
6	−1	1	1	1	−1	−1	−1	318.68
7	1	1	−1	−1	−1	1	−1	638.92
8	1	1	1	−1	−1	−1	1	1197.94
9	1	1	−1	1	1	1	−1	579.02
10	−1	−1	1	−1	1	1	−1	405.68
11	1	−1	1	1	1	−1	−1	1633.4
12	−1	1	1	−1	1	1	1	143.46

A: sucrose (g/L); B: polypeptone; C: yeast extract; D: KH_2PO_4; E: $MgSO_4$ $7H_2O$; F: pH, G: inoculum volume (*v*/*v*).

2.8. Analysis of Balanol Production by HPLC

The concentration of balanol was determined according to the method described by He et al. (2018) [10]. Culture broth was sampled for analysis of balanol by HPLC using a reverse-phase C18 column (Agilent Eclipse Plus C18, 4.6 × 250 mm, 5 μm) (1260 Infinity, Agilent Technologies, Santa Clara, CA, USA). Chromatographic conditions were composed of solvent A and solvent B. Solvent A comprised water with 0.001 M trifluoroacetic acid (TFA), and solvent B comprised acetonitrile-0.001 M TFA; the solvent gradient was 5% B in the first 5 min and increased to 58% at 35 min and to 95% B at 36 min, followed by 4 min with 95% B, with a flow rate of 1 mL/min and UV detection at 254 nm. The structure-identified balanol was used as the standard control. Through analysis, the peak area of balanol with different concentrations was determined by HPLC, and the standard curve of balanol concentration was established (Figure 1). The concentration of balanol production in culture broth was determined with the regression equation from the standard curve: $Y = 31.764X - 203.51$ ($R^2 = 0.9993$), where Y indicates the concentration of balanol (mg/L), and X is the peak area of balanol.

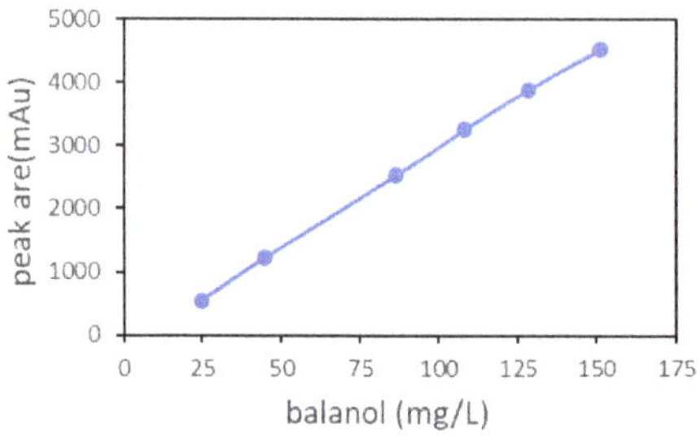

Figure 1. Standard curve of the balanol. Balanol was prepared in a solution from 25 mg/L to 180 mg/L for HPLC analysis. The peak area of balanol was used to establish the standard curve of its concentration. In the regression equation, Y = 31.764x − 203.51 (R^2 = 0.9993), Y indicates the concentration of balanol, and X indicates the peak area of balanol.

3. Results

3.1. Characterization of Regulator BlnR within the Gene Cluster bln in T. ophioglossoides

The gene *blnR* was found as an orphan regulatory gene situated within the *bln* cluster in *T. ophioglossoides*, which connects the PKS and NRPS part of gene cluster *bln* containing 18 genes, with a length of 79 kb. The *blnR* gene is 1443 bp long without intron and encodes a protein BlnR containing 380 amino acids. Through bioinformatic analysis, BlnR was determined to be a putative transcription regulator featuring a typical N-terminal GAL4-type Zn_2Cys_6 DNA-binding domain and a C-terminal AflR domain (Figure 2A). Like other GAL4 domains among the aligned proteins, the six cysteine residues were conserved in the putative GAL4 domain of BlnR (Figure 2B). The transcription factor AflR contains a GAL4-type binuclear zinc finger cluster domain, CX2CX6CX6CX2CX6CX2, which plays a key role in aflatoxin biosynthesis, especially in *Aspergillus* taxa [27,28]. The six-cysteine (Zn_2Cys_6) binuclear cluster DNA binding domain was first characterized in the GAL4 protein of *Saccharomyces cerevisiae*. To date, this domain-containing protein has been identified exclusively in the fungus kingdom [29,30]. Phylogenic analysis showed that BlnR belongs to a separate clade from *Aspergillus* AflR regulator protein, showing it to be different from the AflR of Aspergillus with a different function (Figure 2C).

3.2. BlnR Positively Regulates Balanol Biosynthesis by Binding All the Promoters of bln Gene Members

Usually, a typical transcriptional activator consists of a DNA-binding domain (DBD), which is responsible for promoter recognition in order to regulate gene transcription. Nothing is known yet about the regulation of balanol biosynthesis, although we have elucidated its biosynthetic pathway in *T. ophioglossoides*. Therefore, we first attempted to heterologously express the whole length of BlnR in the *E. coli* system. Unfortunately, both $BlnR^{360aa}$ and $BlnR^{180aa}$ were expressed as inclusion bodies in *E. coli*, even with the help of solubilizing tag protein GST. Therefore, we attempted to only express its N-terminal GAL4 domain with 90 aa in *E. coli*. As shown in Figure 3B, the fusion GST-$BlnR^{90aa}$ was expressed successfully as a soluble protein. After purification by Ni-affinity agarose, GST-$BlnR^{90aa}$ was used for the DNA-binding experiment. Using FM-labelled primer pairs, 12 promoter fragments of the *bln* gene cluster were amplified for EMSA assay to examine the affinity of BlnR protein to their promoters of the *bln* gene members. As the results revealed, $BlnR^{90aa}$ showed a strong affinity with all the tested promoters, as well as its own promoter (Figure 3C). These results are consistent with the upregulation of their transcription level in the *blnR*OE strain as compared with the wild type [10]. It is reasonable to speculate that there a conserved binding site exists in the promoter region of all *bln* gene members. Therefore, we carried out motif investigation to mine DNA-binding motifs by multiple expectation maximizations for motif elicitation (MEME) (https://meme-suite.org). It was demonstrated that there is a conserved motif in all promoter regions with GAGCCAAT (Figure 3D).

3.3. BlnR Is a Positive Regulator toward Balanol Biosynthesis in T. ophioglossoides

In *T. ophioglossoides*, there are another 34 gene clusters aside from *bln* according to analysis by AntiSMASH (https://fungismash.secondarymetabolites.org/, accessed on 18 April 2022). We showed that the overexpression of the *blnR* gene significantly upregulated the transcription level of all the *bln* member genes and activated the biosynthesis of balanol. The results suggest that BlnR has a positive regulatory effect on balanol biosynthesis. The *blnR*OE strain remained stable for balanol production after 10 generations grown on a PDA plate without selective pressure. Furthermore, we examined the effect of *blnR* overexpression on the transcriptional level of all other gene clusters by comparing their core gene in the wild-type and *blnR*OE strain. The results demonstrate that *blnR* overexpression exclusively improved the core gene expression of the balanol gene cluster and insignificantly ffected the expression of other gene clusters (Figure 4A). The metabolite profiles by HPLC analysis also exhibite that no other compounds were producedexcept balanol and its intermediates (Figure 4B).

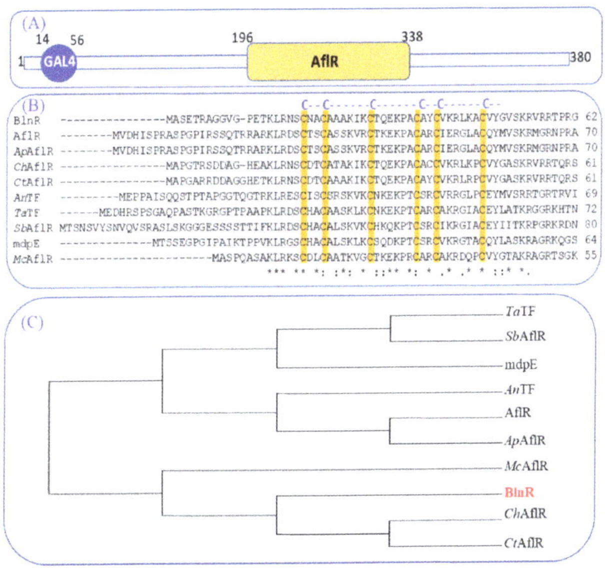

Figure 2. Characterization of regulator BlnR and its phylogenic analysis. (**A**) Domain characterization of regulator BlnR. (**B**) Alignment analysis of the conserved cysteine amino acids. (**C**) Evolutionary phylogenetic analysis by maximum likelihood method using MEGA 7.0 software. The evolutionary history was inferred using the maximum likelihood method based on the JTT matrix-based model: *Ta*TF (accession no. XP_040731924.1), *Sb*AflR (accession no. ESZ98975.1), mdpE (accession no. AN0148), *An*TF(accession no. AAC49195), AflR (accession no. P43651.3), *Ap*AflR (accession no. AAM02999.1), *Mc*AflR (accession no. XP_002844737.1), *Ch*AlfR (accession no. XP_018155792.1), *Ct*AflR (Accession no. TKW57950.1).

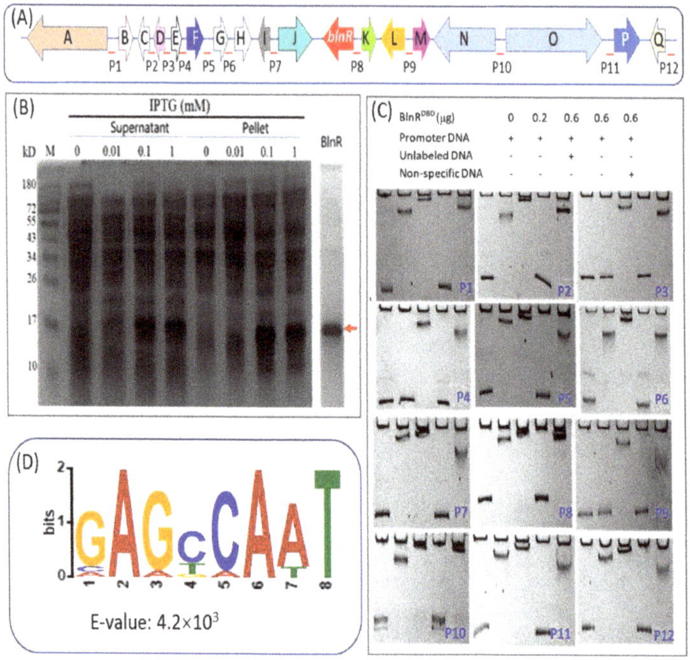

Figure 3. Affinity binding analysis of BlnR with promoter DNA through EMSA experiment. (**A**) Schematic diagram of promoter design in the *bln* gene cluster. (**B**) Heterologous expression of BlnR in *E. coli*. M: protein marker. (**C**) Affinity binding analysis of BlnR to gene promoters within *bln* by EMSA. (**D**) The conserved binding motif of BlnR was predicted by MEME analysis.

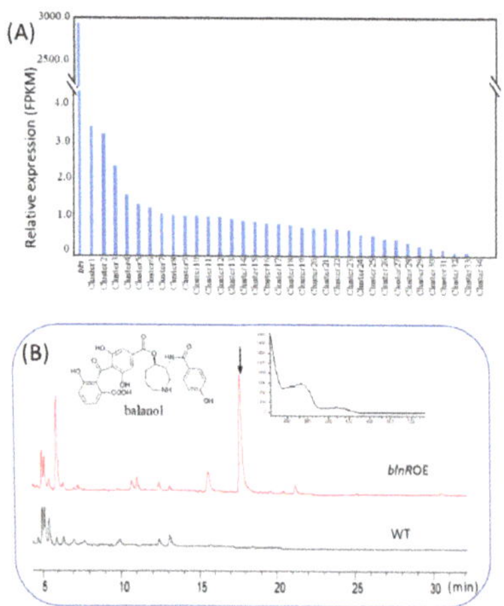

Figure 4. BlnR is a positive regulator of balanol biosynthesis. (**A**) Effect of *blnR* overexpression on the transcriptional level of other gene clusters. (**B**) Metabolite profile of wild-type and *blnR*OE strains after cultivation in COB medium for 10 days.

3.4. BlnR Is Involved in the Crosstalk between Primary and Secondary Metabolism

When grown on a PDA plate, we found *blnR* overexpression changed its morphologic phenotype. In addition to the production of light-yellow pigment of stable balanol, the *blnR*OE strain also showed slowed growth as compared with the wild-type strain

(Figure 5A,B). Through scanning electron microscopy, the number of conidiospores was found to be reduced significantly, and its slimly hyphae elongated and became curly in the *bln*ROE strain, whereas the grown hyphae were linearly elongated with numerous conidia visible in the wild-type strain with typical morphology (Figure 5C). Therefore, we further analyzed the key central regulators involved in conidiation development, such as *brlA* and *abaA*, as well as their upstream genes, *fluG*, *flbC* and *flbD*. *abaA*, *fluG* and *flbD* were upregulated, whereas *brlA* and *flbC* were slightly downregulated in the *bln*ROE strain (Figure 5D).

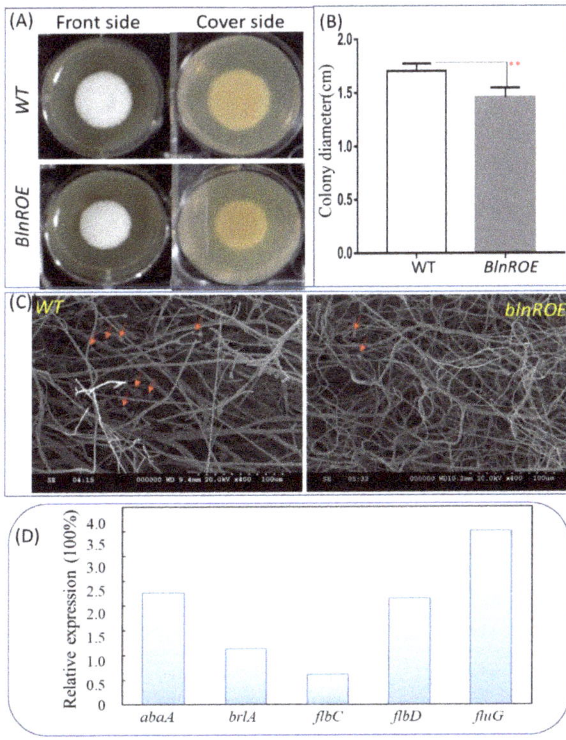

Figure 5. Effects of BlnR on the development of asexual conidiospore and filamentous growth. (**A**) A total of 10^3 spores were spotted on a PDA plate for growth at 26 °C for 7 days. (**B**) Comparison of colony size between the wild-type and *bln*ROVE (n = 10) strains. "**" indicates the significant difference with the *p*-value < 0.01. (**C**) Microscopic observation of hyphae and spores. The red arrows indicate spores. (**D**) qRT-PCR analysis of spore-development-related regulators (n = 3).

We further investigated the differential expression pattern between *bln*ROE and the wild-type strain through mRNA-seq analysis. As shown in Figure 6, *blnR* overexpression upregulated the transcription level of 498 genes and downregulated the transcription level of 503 genes, whereas 8316 genes maintained their expression at a regular level. RNA-seq data analysis revealed the differential expression of numerous genes belonging to various pathways of primary or secondary metabolism. Pathways of primary metabolism include starch and sucrose metabolism, fatty acid metabolite, TCA cycle, lysine biosynthesis, ether lipid metabolism and aromatic amino acid (phe, tyr, trp) biosynthesis (Table 2). Among them, the expression of most genes in starch and sucrose metabolism, lysine biosynthesis and aromatic amino acid biosynthesis were significantly upregulated in the *bln*ROE strain, whereas genes in fatty acid metabolite and TCA cycle were up- or downregulated. The genes expression in the pathway of spore development was significantly downregulated, except for the *bln* gene cluster, which was activated strongly in *bln*ROE as compared with the wild-type strain. Our results show that as a cluster-situated regulator, the overexpression of the *blnR* gene is involved in the control of secondary metabolism, as well as primary metabolism, possibly by manipulating the distant genes.

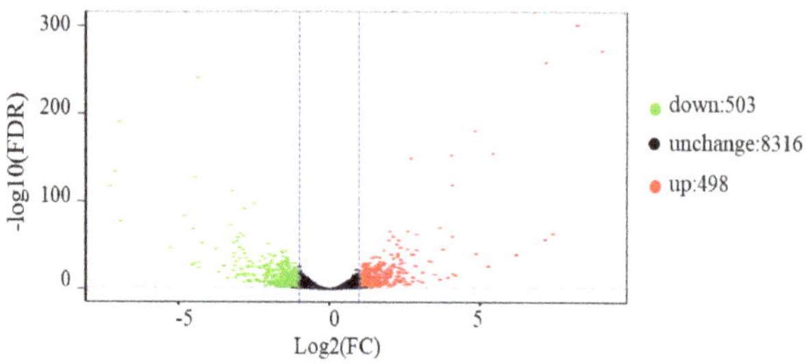

Figure 6. Differential expression analysis between *blnR*OE and the wild-type strain.

Table 2. List of genes up- or down regulated in *blnR*OE as compared with the wild-type strain based on RNA-seq analysis.

Gene ID	Protein Name	log2FC
Starch and sucrose metabolism		
g1571	putative betaglucosidase I	1.797
g2561	putative sucrose utilization protein SUC1	1.574
g2754	Hexokinase-1	1.715
g6339	Alphaamylase A type-3	1.338
g6369	Alphaglucosidase	1.387
g6411	Endoglucanase EG-II	−1.405
g6822	Probable betaglucosidase A	1.301
g7696	alphatrehalose-phosphate synthase	1.388
g857	endo-1,3-betaglucosidase eglC	1.360
g9295	Glucose-6-phosphate isomerase	−1.422
Fatty acid metabolism		
g507	Cytochrome P450	−2.278
g672	putative aldehyde dehydrogenase	−1.201
g2909	3-ketoacyl-CoA thiolase	−1.931
g3641	Acetyl-CoA acetyltransferase	−1.198
g3915	Enoyl-CoA isomerase/hydratase	−1.153
g5070	Enoyl-(Acyl carrier protein) reductase	2.751
g5094	Short-chain-type dehydrogenase/reductase	11.281
g5215	Acyl-CoA dehydrogenase family member 10	1.212
g6087	Phosphotransferase	−1.056
g6743	Aldehyde dehydrogenase	1.447
g7131	Acetoacetyl-CoA reductase	1.059
g7814	Isotrichodermin C-15 hydroxylase	−1.498
g9557	Short/branched-chain-specific acyl-CoA dehydrogenase	−1.433
Citrate cycle (TCA cycle)		
g3392	2-methylcitrate synthase	−1.435
g8077	Succinyl-CoA ligase	1.935
g9576	putative succinate dehydrogenase	−1.168
Lysine biosynthesis		
g3241	Homoaconitase, mitochondrial	1.077
g6260	Homocitrate synthase, mitochondrial	1.129
Phenylalanine, tyrosine and tryptophan biosynthesis		
g6745	Fungal-specific transcription factor	2.344
MAPK signaling pathway		
g7635	Catalase	−1.13221
g9623	Catalase	−3.62842
Spore development		
g6286	Spore development regulator vosA	−1.22895
g655	Outer spore wall protein RRT8	−0.94555

3.5. High Production of Balanol at the Gram Level through Medium Optimization via Response Surface Methodology (RSM)

In order to further enhance balanol production, the effect of condition parameters, including inoculum dosage, pH, carbon source and nitrogen source on balanol production was investigated through PB experiments. As shown in Figure 7, sucrose, which varied from 30 g/L to 120 g/L, showed a significant effect on balanol production and cell growth, whereas 105 g/L of sucrose was determined to be the optimal concentration (Figure 7A), and 15 g/L polypeptone as the nitrogen source (Figure 7B) or 10 g/L of the yeast extract (Figure 7C) was the best concentration with the maximum titer of balanol. The effect of the combined nitrogen source further demonstrates that balanol production reached the maximum at 5 g/L yeast extract and 10 g/L of polypeptone (Figure 7D).

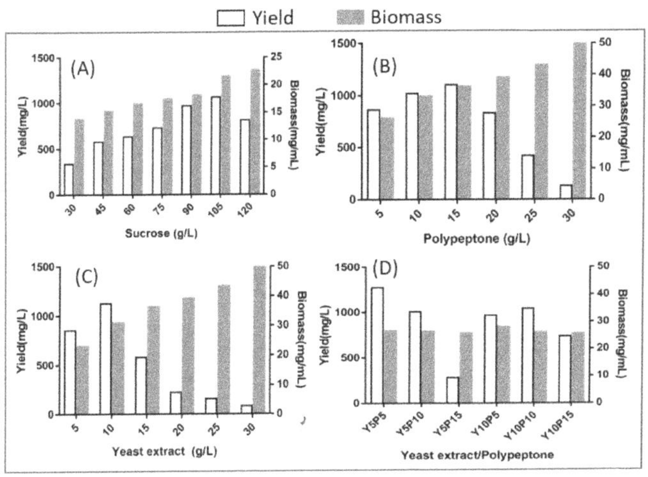

Figure 7. Effect of nutritional factor on balanol production including sucrose (**A**), polypeptone (**B**), yeast extract (**C**) and combined yeast extract and polypeptone (**D**).

The significant factors affecting production as observed in OFAT, namely (A) sucrose, (B) polypeptone and (C) yeast extract, were studied for their optimum combination using RSM and adhering to the PB and BB design matrix. A total of 12 run experiments were performed according to the PB matrix to investigate the significant components in balanol production (Table 1). Regression statistics were performed to examine the model feasibility. As shown in Table 3, a model with a p-value lower than 0.05 was considered significant, and sucrose, polypeptone and initial pH were determined to be the main significant components affecting balanol production. Therefore, a 17-full factorial BB experimental design was further implemented to determine their optimal values based on RSM. The experimental design matrices that included all the variables and balanol titers are shown in Table S2. The resulting fermentation titers of balanol were used to fit a quadratic model using regression analysis, yielding the following response equation to predict balanol production in terms of coded variables:

$$Y = 1814.86 - 21.55A - 103.12B + 332.95C + 107.44AB - 33.82AC + 54.17BC - 585.06A^2 - 365.13B^2 - 434.71C^2$$

where Y indicates the balanol production (mg/L); and A, B and C are sucrose, polypeptone and pH, respectively. Moreover, we used this statistical model to evaluate the relationship between different variables and their interactive effects on balanol production, as summarized in Table 4. Regression analysis of BB design showed that the model's F value was 32.35, and the model's p-value was lower than 0.0001 with statistical significance, suggesting that this model is fit well to describe the response of balanol production of these variable components.

Table 3. Signification analysis of the PB experiments on balanol production.

Source	SS	DF	MS	*F*-Value	*p*-Value
Model	4.336×10^5	7	6.821×10^5	6.82	0.0413 *
A	2.028×10^5	1	2.028×10^5	20.27	0.0108 *
B	8.603×10^4	1	8.603×10^5	8.60	0.04727
C	4.005×10^5	1	6.15	6.145×10^{-4}	0.9814
D	1.706×10^3	1	1.706×10^3	0.12	0.7009
E	5.495×10^4	1	6.159×10^2	0.044	0.8163
F	6.16×10^2	1	1.424×10^5	10.18	0.0188 *
G	1.424×10^5	1	4.390×10^5	1.55	0.2677
Residual	4.003×10^5	4	1.001×10^5		
Cor Total	5.175×10^5	11			

* indicates that the effect of the variable is significant.

Table 4. Variance analysis of the binary regression equation.

Source	SS	DF	MS	*F*-Value	Pr > F
Model	4.148×10^6	9	4.609×10^5	32.35	<0.0001 **
A	3.715×10^4	1	3.715×10^4	0.26	0.6253
B	8.506×10^5	1	8.506×10^5	5.97	0.0445 *
C	8.869×10^5	1	8.869×10^5	62.25	<0.0001 **
AB	4.618×10^5	1	4.618×10^5	3.24	0.1148
AC	4.575×10^3	1	4.575×10^3	0.32	0.5886
BC	1.174×10^5	1	1.174×10^5	0.82	0.3942
A2	1.441×10^6	1	1.441×10^6	101.15	<0.0001 **
B2	5.614×10^5	1	5.614×10^5	39.40	0.0004 **
C2	7.957×10^5	1	7.957×10^5	55.85	0.0001 **
Lack of Fit	2.485×10^5	3	8.283×10^3	0.44	0.7355
Pure Error	7.488×10^5	4	1.872×10^5		
Cor Total	4.248×10^6	16			

* and ** indicate that the effect of the variable is significant and more significant, respectively.

According to the BB model, the individual and interactive effects of the independent component on the response of balanol production are further shown by contour plots to predict the response surface; the balanol titer was a function of two tested variables, whereas the other independent variable was kept constant at the zero level. Among these components, the interactive effect between two components on balanol production is given in Figure 8. Balanol titers initially increased with concentrations of both polypeptone and pH, then declined after reaching the maximum point in each variable, showing that there was a curve relationship between the independent variables and balanol production (Figure 8A). A similar interaction was independently observed between sucrose and polypeptones, sucrose and pH (Figure 8B,C). However, the *p*-values for the interactive term of these pairs of variables were higher than 0.05, suggesting statistically insignificant interactions between these three components. It can be inferred that the central values of these three variants subjected to BB design were close to their optimal values. According to the RSM model, the optimal values for the maximum production of balanol were 100 g/L sucrose, 13.6 g/L polypeptone and an initial pH of 4.9.

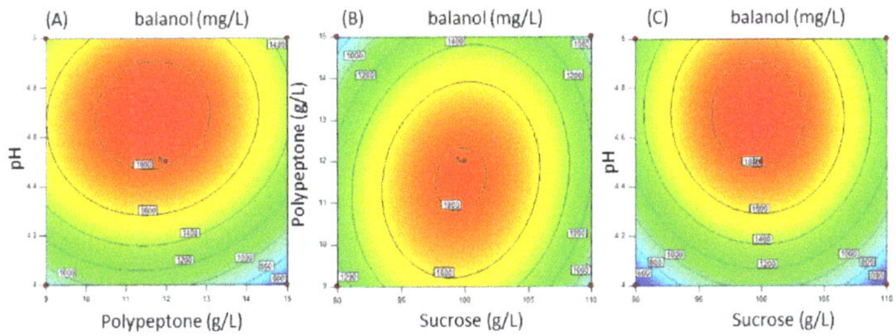

Figure 8. Contour plots for response interaction between two variables. (**A**) Interaction effect of pH and polypeptone on balanol production; (**B**) Interaction effect of sucrose and polypeptone on balanol production; (**C**) Interaction effect of pH and sucrose on balanol production.

3.6. Batch Fermentation for Balanol Production in a Scaled-Up 15 L Tank

Using the optimal medium compositions, including 100 g/L sucrose, 13.6 g/L polypeptone, 5 g/L yeast extract, 0.6 g/L KH_2PO_4, 1.0 g/L $MgSO_4{\cdot}7H_2O$, initial pH and 2% inoculum volume, we carried out a scaled-up batch fermentation with the *bln*ROE strain in a 15 L fermenter. As can be seen from the time course of the fermentative profile shown in Figure 9A, the maximum concentration of balanol reached 2187.39 mg/L after culture for 10 days, and the cell biomass was found to be the highest after 8 d growth. The results matched well with the predicted value of the developed statistical model, suggesting that this model truly reflects the effect of medium components on balanol production. It provides a feasible practical attempt for large-scale industry production. We also analyzed the effect of optimized medium on the expression of gene members of the *bln* gene cluster (Figure 9B). The results show that their expression did not differ significantly when cultured in the optimized medium.

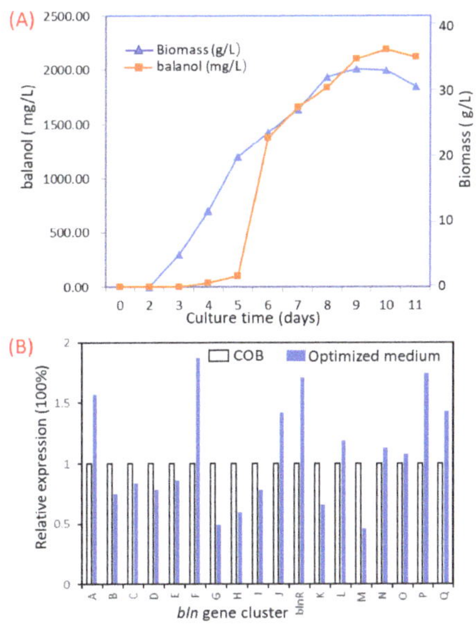

Figure 9. Time profile of balanol production by *bln*ROE strain using the optimized medium (**A**) and analysis of the expression level of the *bln* gene cluster (**B**). The *bln*ROE strain was cultured in a 15 L tank at 26 °C for 11 days with 8 L of medium. The culture broth was sampled for analysis of cell growth by determination of dried weight and balanol production every two days. The dissolved oxygen was maintained at 30% through cultivation. For RNA extraction, 4-day mycelium was used to analyze the expression level.

4. Discussion

Fungi micro-organisms are a rich source for producing novel compounds with potential bioactivity. With the development of sequencing technology, bioinformatic analysis has revealed that a far larger number of gene clusters is cryptic in the genomic sequence. Mining the genomic source to discover novel compounds is attractive. The secondary metabolism is a complex network regulated by global regulators, such as LaeA and velvet [31,32]. In the developed strategies, transcription regulation has proven to be a simple and feasible method to activate cryptic biosynthetic gene clusters or improve the production of compounds of interest. It is known that transcription factors can directly regulate the transcription of their target genes by binding to promoters, leading to an improvement or reduction in the production of target compounds. Recently, a targeted and high-throughput activation of silent gene clusters using transcription factor decoys was applied in Streptomycetes [33]. In this study, we identified a cluster-situated Zn_2Cys_6 family regulator, BlnR, which plays a positive role in regulation of balanol biosynthesis by binding to all promoters of gene cluster members, as well as its own promoter, within the *bln* (Figure 3C). Overexpression of the *blnR* gene significantly activated the transcription levels of all gene members within the *bln* gene cluster.

Additionally, mRNA-seq analysis exhibited that the overexpression of *blnR* led to 1001 differentially expressedgenes, which are involved in different pathways, including primary and secondary metabolism, in the *blnR*OE strain (Table 2). For primary metabolism, such as starch and sucrose metabolism, fatty acid metabolism and TCA cycle, there are up-regulated genes and downregulated genes. Aromatic amino acid biosynthesis and lysin-biosynthesis-related genes were upregulated significantly, which was reasonable because lysine and phenylalanine are the substrate of NRPS for balanol biosynthesis within the gene *bln* cluster. All these data indicate that BlnR is involved in the coordination of secondary metabolism and primary metabolism to promote balanol biosynthesis.

We also found that *blnR* overexpression resulted in changes in its physiological phenotype in the *blnR*OE strain. Conidiospore development was blocked, with a reducing number, and the hyphae grew curly with retarded growth (Figure 5). Conidiation is considered the most common asexual reproductive mode for many filamentous fungi, and its developmental mechanisms have been characterized in *A. nidulans* and *Neurospora crassa* [34,35]. Transcription analysis showed that regulators involved in spore development, such as AbaA, FluG and FlbD, were upregulated. RNA-seq data also revealed that outer spore wall protein RRT8 and spore development regulator vosA were downregulated significantly in the *blnR*OE strain (Figure 5 and Table 2). These data suggest that the *blnR* gene is also involved in spore development with the slowed growth of the hyphae.

Previous studies have shown that AflR can regulate the expression of genes outside of the aflatoxin biosynthetic cluster under conditions conducive to aflatoxin production in *A. parasiticus* and *A. flavus* [36,37], suggesting that AflR may have a broad function and regulate other distant genes. Consistent with these results, our study also showed that BlnR can regulate the genes within the balanol biosynthetic cluster, as well as other distant genes involved in many other metabolic pathways (Figure 6 and Table 2), which are directly regulated by BlnR and will be crucial for further studies.

Genetic manipulation was proven to be an efficient technique to activate and prove the biosynthesis of secondary metabolites. Meanwhile, the OSMAS cultivation-based technique can also powerfully activate or increase SM production by changing the culture conditions, including the proper ratio of carbon to nitrogen, metal ions, pH and temperature. Many novel compounds were found through this simple strategy. In large-scale industrial production, the optimization of culture conditions is used as a feasible way to increase the production of target products in order to reduce economic costs. In this study, we carried out medium optimization using RSM-based statistics and further improved balanol production by 3.12-fold to 2187.39 mg/L in a 15 L fermenter by increasing the transcription level of gene members of the *bln* gene cluster (Figure 9A). The optimized medium did not further enhance the expression of all *bln* gene members, revealing that changing culture conditions possibly led to the alteration of enzyme activity or other metabolite pathways.

5. Conclusions

In conclusion, the cluster-situated Zn_2Cys_6-family regulator, BlnR, has a positive and specific regulatory effect on balanol biosynthesis in *T. ophioglossoides*. BlnR was found to regulate the other genes outside of the balanol biosynthetic gene cluster, including the primary and secondary metabolite pathways. In addition, BlnR was also found to be involved in the development of asexual conidiospores and mycelium growth. Furthermore, statistical methods based on RSM were used to determine optimal medium compositions with a maximum titer of balanol in the *bln*ROE strain. Using these optimized components, the highest concentration of balanol was determined to be 2187.39 mg/L after 10 d cultivation in a 15 L batch tank.

Supplementary Materials: The following supporting information can be downloaded at: https://www.mdpi.com/article/10.3390/jof8050510/s1. Table S1: Primers used in this study. Table S2: Experimental scheme of response surface design and results.

Author Contributions: Conceptualization, X.-A.C., R.-Q.L. and M.Z.; methodology, R.-Q.L. and M.Z.; software, M.Z.; validation, X.L., W.-Q.X. and Y.-Q.L.; formal analysis, R.-Q.L. and X.L.; investigation, R.-Q.L. and X.L.; resources, Y.-Q.L.; data curation, M.Z.; writing—original draft preparation, X.-A.C.; writing—review and editing, X.-A.C., R.-Q.L., X.L. and W.-Q.X.; supervision, X.-A.C.; project administration, X.-A.C.; funding acquisition, X.-A.C. All authors have read and agreed to the published version of the manuscript.

Funding: This research was financially supported by the National Key R&D Program of China (grant numbers "2018YFA0903200" and "2021YFC2100600").

Institutional Review Board Statement: Not applicable.

Informed Consent Statement: Not applicable.

Data Availability Statement: The data presented in this study are available upon request from the corresponding author. The RNA-seq data are not publicly available because other data from these whole-genome transcriptomes are being used for other analyses to be published independently of this one.

Acknowledgments: We thank Han-min Chen from the College of Life Sciences, Zhejiang University, for his technical support in relation to scanning microscopy.

Conflicts of Interest: The authors declare no conflict of interest. The funders had no role in the design of the study; in the collection, analyses, or interpretation of data; in the writing of the manuscript, or in the decision to publish the results.

References

1. Kulanthaivel, P.; Hallock, Y.F.; Boros, C.; Hamilton, S.M.; Janzen, W.P.; Ballas, L.M.; Loomis, C.R.; Jiang, J.B. Balanol: A novel and potent inbibitor of protein kinase C from the fungus *Verticum balanoides*. *J. Am. Chem. Soc.* **1993**, *115*, 6452–6453. [CrossRef]
2. Boros, C.; Hamilton, S.M.; Katz, B.; Kulanthaivel, P. Comparison of balanol from *Verticillium balanoides* and ophiocordin from *Cordyceps ophioglossoides*. *J. Antibiot.* **1994**, *9*, 1010–1016. [CrossRef] [PubMed]
3. Ohshima, S.; Yanagisawa, M.; Katoh, A.; Fujii, T.; Sano, T.; Matsukuma, S.; Furumai, T.; Fujiu, M.; Watanabe, K.; Yokose, K. Fusarium merismoides Corda NR 6356, the source of the protein kinase C inhibitor, azepinostatin. Taxonomy, yield improvement, fermentation and biological activity. *J. Antibiot.* **1994**, *47*, 639–647. [CrossRef] [PubMed]
4. Kneifel, H.; König, W.A.; Loeffler, W.; Müller, R. Ophiocordin, an antifungal antibiotic of *Cordyceps ophioglossoides*. *Arch. Microbiol.* **1977**, *113*, 121–130. [CrossRef]
5. Nishizuka, Y. The molecular heterogeneity of protein kinase C and its implications for cellular regulation. *Nature* **1988**, *334*, 661–665. [CrossRef]
6. Castagna, M.; Takai, Y.; Kaibuchi, K.; Sano, K.; Kikkawa, U.; Nishizuka, Y. Direct activation of calcium-activated, phospholipid-dependent protein kinase by tumor-promoting phorbol esters. *J. Biol. Chem.* **1982**, *257*, 7847–7851. [CrossRef]
7. Bradshaw, D.; Hill, C.H.; Nixon, J.S.; Wilkinson, S.E. Therapeutic potential of protein kinase C inhibitors. *Agents Actions* **1993**, *38*, 135–147. [CrossRef]

8. Hirokazu, K.; Kentaro, O.; Fumio, K.; Noriko, K. Estrogenic substances from the mycelia of medicinal fungus *Cordyceps ophioglossoides* (Ehrh.) Fr. (Ascomycetes). *Int. J. Med. Mushrooms* **2004**, *6*, 249–251.
9. Sun, Y.; Zhao, Z.; Feng, Q.; Xu, Q.; Lv, L.; Liu, J.K.; Zhang, L.; Wu, B.; Li, Y.Q. Unusual spirodecane sesquiterpenes and a sumagillol analogue from *Cordyceps ophioglossoides*. *Helv. Chim. Acta* **2013**, *96*, 76–84. [CrossRef]
10. He, X.; Zhang, M.; Guo, Y.Y.; Mao, X.M.; Chen, X.A.; Li, Y.Q. Revelation of the balanol biosynthetic pathway in *Tolypocladium ophioglossoides*. *Org. Lett.* **2018**, *20*, 6323–6326. [CrossRef]
11. Schroeckh, V.; Scherlach, K.; Nützmann, H.W.; Shelest, E.; Schmidt-Heck, W.; Schuemann, J.; Martin, K.; Hertweck, C.; Brakhage, A.A. Intimate bacterial-fungal interaction triggers biosynthesis of archetypal polyketides in *Aspergillus nidulans*. *Proc. Natl. Acad. Sci. USA* **2009**, *106*, 14558–14563. [CrossRef] [PubMed]
12. Moody, S.C. Microbial co-culture: Harnessing intermicrobial signaling for the production of novel antimicrobials. *Future Microbiol.* **2014**, *9*, 575–578. [CrossRef] [PubMed]
13. Jin, L.Q.; Jin, W.R.; Ma, Z.C.; Shen, Q.; Cai, X.; Liu, Z.Q.; Zheng, Y.G. Promoter engineering strategies for the overproduction of valuable metabolites in microbes. *Appl. Microbiol. Biotechnol.* **2019**, *103*, 8725–8736. [CrossRef] [PubMed]
14. Lopatniuk, M.; Myronovskyi, M.; Nottebrock, A.; Busche, T.; Kalinowski, J.; Ostash, B.; Fedorenko, V.; Luzhetskyy, A. Effect of "ribosome engineering" on the transcription level and production of *S. albus* indigenous secondary metabolites. *Appl. Microbiol. Biotechnol.* **2019**, *103*, 7097–7110. [CrossRef] [PubMed]
15. Shwab, E.K.; Bok, J.W.; Tribus, M.; Galehr, J.; Graessle, S.; Keller, N.P. Histone deacetylase activity regulates chemical diversity in *Aspergillus*. *Eukaryot. Cell* **2007**, *6*, 1656–1664. [CrossRef]
16. Mao, X.M.; Xu, W.; Li, D.; Yin, W.B.; Chooi, Y.H.; Li, Y.Q.; Tang, Y.; Hu, Y. Epigenetic genome mining of an endophytic fungus leads to the pleiotropic biosynthesis of natural products. *Angew. Chem. Int. Ed. Engl.* **2015**, *54*, 7592–7596. [CrossRef]
17. Chiang, Y.M.; Szewczyk, E.; Davidson, A.D.; Entwistle, R.; Keller, N.P.; Wang, C.C.; Oakley, B.R. Characterization of the *Aspergillus nidulans* monodictyphenone gene cluster. *Appl. Environ. Microbiol.* **2010**, *76*, 2067–2074. [CrossRef]
18. Chung, C.H.; Lin, D.W.; Eames, A.; Chandrasekaran, S. Next-generation genome-scale metabolic modeling through integration of regulatory mechanisms. *Metabolites* **2021**, *11*, 606. [CrossRef]
19. Martín, J.F.; Liras, P. Engineering of regulatory cascades and networks controlling antibiotic biosynthesis in Streptomyces. *Curr. Opin. Microbiol.* **2010**, *13*, 263–273. [CrossRef]
20. Liu, K.; Hu, X.R.; Zhao, L.X.; Wang, Y.; Deng, Z.; Tao, M. Enhancing ristomycin a production by overexpression of ParB-like StrR family regulators controlling the biosynthesis genes. *Appl. Environ. Microbiol.* **2021**, *87*, e0106621. [CrossRef]
21. Bode, H.B.; Bethe, B.; Höfs, R.; Zeeck, A. Big effects from small changes: Possible ways to explore nature's chemical diversity. *ChemBioChem* **2002**, *3*, 619–627. [CrossRef]
22. Hewage, R.T.; Aree, T.; Mahidol, C.; Ruchirawat, S.; Kittakoop, P. One strain-many compounds (OSMAC) method for production of polyketides, azaphilones, and an isochromanone using the endophytic fungus *Dothideomycete* sp. *Phytochemistry* **2014**, *108*, 87–94. [CrossRef] [PubMed]
23. Romano, S.; Jackson, S.A.; Patry, S.; Dobson, A.D.W. Extending the "One Strain Many Compounds" (OSMAC) Principle to Marine Microorganisms. *Mar. Drugs* **2018**, *16*, 244. [CrossRef] [PubMed]
24. Supratman, U.; Suzuki, T.; Nakamura, T.; Yokoyama, Y.; Harneti, D.; Maharani, R.; Salam, S.; Abdullah, F.F.; Koseki, T.; Shiono, Y. New metabolites produced by endophyte *Clonostachys rosea* B5-2. *Nat. Prod. Res.* **2021**, *35*, 1525–1531. [CrossRef] [PubMed]
25. Chen, X.A.; He, X.; Zhang, M.; Mao, X.M.; Li, Y.Q. An efficient genetic transformation system for Chinese medicine fungus *Tolypocladium ophioglossoides*. *J. Microbiol. Methods* **2020**, *176*, 106032. [CrossRef] [PubMed]
26. Hellman, L.M.; Fried, M.G. Electrophoretic mobility shift assay (EMSA) for detecting protein-nucleic acid interactions. *Nat. Protoc.* **2007**, *2*, 1849–1861. [CrossRef] [PubMed]
27. Chang, P.K.; Cary, J.W.; Bhatnagar, D.; Cleveland, T.E.; Bennett, J.W.; Linz, J.E.; Woloshuk, C.P.; Payne, G.A. Cloning of the *Aspergillus parasiticus* apa-2 gene associated with the regulation of aflatoxin biosynthesis. *Appl. Environ. Microbiol.* **1993**, *59*, 3273–3279. [CrossRef]
28. Payne, G.A.; Nystrom, G.J.; Bhatnagar, D.; Cleveland, T.E.; Woloshuk, C.P. Cloning of the afl-2 gene involved in aflatoxin biosynthesis from *Aspergillus flavus*. *Appl. Environ. Microbiol.* **1993**, *59*, 156–162. [CrossRef]
29. Harrison, S.C. A structural taxonomy of DNA-binding domains. *Nature* **1991**, *353*, 715–719. [CrossRef]
30. Anderson, S.F.; Steber, C.M.; Esposito, R.E.; Coleman, J.E. UME6, a negative regulator of meiosis in *Saccharomyces cerevisiae*, contains a C-terminal Zn_2Cys_6 binuclear cluster that binds the URS1 DNA sequence in a zinc-dependent manner. *Protein Sci.* **1995**, *4*, 1832–1843. [CrossRef]
31. Keller, N.P.; Turner, G.; Bennett, J.W. Fungal secondary metabolism from biochemistry to genomics. *Nat. Rev. Microbiol.* **2005**, *3*, 937–947. [CrossRef] [PubMed]
32. Zhang, C.; Zhang, H.; Zhu, Q.Q.; Hao, S.; Chai, S.Y.; Li, Y.H.; Jiao, Z.; Shi, J.C.; Sun, B.G.; Wang, C.T. Overexpression of global regulator LaeA increases secondary metabolite production in *Monascus purpureus*. *Appl. Microbiol. Biotechnol.* **2000**, *104*, 3049–3060. [CrossRef] [PubMed]
33. Wang, B.; Guo, F.; Dong, S.H.; Zhao, H. Activation of silent biosynthetic gene clusters using transcription factor decoys. *Nat. Chem. Biol.* **2019**, *15*, 111–114. [CrossRef] [PubMed]
34. Rerngsamran, P.; Murphy, M.B.; Doyle, S.A.; Ebbole, D.J. Fluffy, the major regulator of conidiation in Neurospora crassa, directly activates a developmentally regulated hydrophobin gene. *Mol. Microbiol.* **2005**, *56*, 282–297. [CrossRef] [PubMed]

35. Park, H.S.; Yu, J.H. Genetic control of asexual sporulation in filamentous fungi. *Curr. Opin. Microbiol.* **2012**, *15*, 669–677. [CrossRef]
36. Price, M.S.; Yu, J.; Nierman, W.C.; Kim, H.S.; Pritchard, B.; Jacobus, C.A.; Bhatnagar, D.; Cleveland, T.E.; Payne, G.A. The aflatoxin pathway regulator AflR induces gene transcription inside and outside of the aflatoxin biosynthetic cluster. *FEMS Microbiol. Lett.* **2006**, *255*, 275–279. [CrossRef]
37. Kong, Q.; Chang, P.K.; Li, C.; Hu, Z.; Zheng, M.; Sun, Q.; Shan, S. Identification of AflR binding sites in the genome of *Aspergillus flavus* by ChIP-Seq. *J. Fungi* **2020**, *6*, 52. [CrossRef]

Journal of *Fungi*

Article

Trichoderma reesei Contains a Biosynthetic Gene Cluster That Encodes the Antifungal Agent Ilicicolin H

Mary L. Shenouda [1,2], Maria Ambilika [1] and Russell J. Cox [1,*]

1 Institute for Organic Chemistry and Biomolekulares Wirkstoffzentrum (BMWZ), Schneiderberg 38, 30167 Hannover, Germany; mary.shenouda@oci.uni-hannover.de (M.L.S.); m.ambilika@gmail.com (M.A.)
2 Pharmacognosy Department, Faculty of Pharmacy, Alexandria University, Alexandria 21521, Egypt
* Correspondence: russell.cox@oci.uni-hannover.de

Abstract: The *trili* biosynthetic gene cluster (BGC) from the well-studied organism *Trichoderma reesei* was studied by heterologous expression in the fungal host *Aspergillus oryzae*. Coexpression of *triliA* and *triliB* produces two new acyl tetramic acids. Addition of the ring-expanding cytochrome P450 encoded by *triliC* then yields a known pyridone intermediate to ilicicolin H and a new chain-truncated shunt metabolite. Finally, addition of the intramolecular Diels-Alderase encoded by *triliD* affords a mixture of 8-epi ilicicolin H and ilicicolin H itself, showing that the *T. reesei trili* BGC encodes biosynthesis of this potent antifungal agent. Unexpected *A. oryzae* shunt pathways are responsible for the production of the new compounds, emphasising the role of fungal hosts in catalysing diversification reactions.

Keywords: heterologous expression; PKS-NRPS; ilicicolin H; shunt metabolite; *Trichoderma reesei*

Citation: Shenouda, M.L.; Ambilika, M.; Cox, R.J. *Trichoderma reesei* Contains a Biosynthetic Gene Cluster That Encodes the Antifungal Agent Ilicicolin H. *J. Fungi* **2021**, *7*, 1034. https://doi.org/10.3390/jof7121034

Academic Editors: Tao Feng and Frank Surup

Received: 8 November 2021
Accepted: 29 November 2021
Published: 1 December 2021

1. Introduction

Trichoderma reesei (*Hypocrea jecorina*) is a filamentous fungus widely used for the industrial production of cellulase [1]. The organism can grow to high cell density and can grow robustly on a variety of low-quality substrates. It has been subject to a wide variety of biotechnological improvements since the discovery of the wild-type QM6A strain in the 1940s, and many molecular tools are available for its genetic manipulation. Many *Trichoderma* species have been used in the remediation of fungal diseases. For example, Foster and coworkers have shown that some *Trichoderma* spp. can effectively treat *Armillaria mellea* (honey fungus) that causes root rot in many tree species [2]. *T. reesei* is also well-known for the production of a number of secondary metabolites such as sorbicillinol **1** [3,4], and numerous multimeric derivatives such as bisvertinolone **2** [5].

Other members of the *Trichoderma* genus are well-known producers of many secondary metabolites [6]. In particular, compounds such as acyl tetramic acids (acyl 2-pyrollones) and acyl 2-pyridones, synthesised by hybrid polyketide synthase-nonribosomal peptide synthetases (PKS-NRPS) are commonly produced by various *Trichoderma* species. These include metabolites such as harzianic acid **3** (*T. harzianum*) [7,8], and deacetyl-18-deoxy cytochalasin H **4** [9] among the pyrollones; and trichodin A **5** [10] and harzianopyridone **6** [11,12] among the 2-pyridones (Figure 1). The biosynthesis of related pyrollones such as fusarin C **7** in *Fusarium* spp [13] and acurin A **8** in *Aspergillus aculeatus*, [14] and 2-pyridones such as tenellin **19** in *Beauveria bassiana* [15] and ilicicolin H **10** in *Talaromyces variabilis* and *Nectria* spp [16,17] are well-understood.

However, despite decades of study, *T. reesei* itself has not been reported to produce any PKS-NRPS compounds. Since we have a broad interest in fungal PKS-NRPS compounds and we recently made detailed studies on the biosynthesis of sorbicillinoids in *T. reesei* [3,4], we set out to determine if this organism can also produce PKS-NRPS compounds.

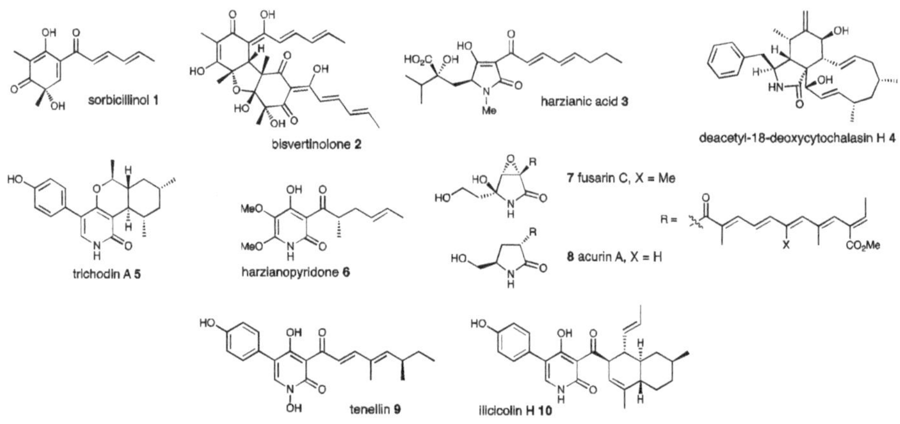

Figure 1. Typical compounds found in various *Trichoderma* species (**1–6**) and related fungal metabolites (**7–10**).

2. Materials and Methods

All chemicals and media ingredients used in this work were purchased from Duchefa Biochemie, Roth, VWR, Fisher scientific, Sigma Aldrich, abcr and Formedium. All media, buffers, solutions and antibiotics were prepared with double-distilled water (ddH_2O), where deionized water was further purified by GenPure Pro UV/UF Milipore device from Thermo scientific. All growth media and sterile solutions were autoclaved at 121 °C for 15 min using Systec VX150 or a classic Prestige Medical 2100 autoclave. Antibiotic solutions were sterilized using a sterile syringe filter (0.45 µm pore size, Roth). See ESI for details of all growth media, buffers and solutions, enzymes and antibiotics.

2.1. Microbiological Methods

Bacterial and fungal strains and all microbiological methods used in this work are summarized in the Electronic Supplementary Information (ESI).

2.2. Construction of Heterologous Expression Vectors

The genomic DNA (gDNA) of *T. reesei* QM6a Δ*tmus53* Δ*pyr4* [18] was used as a template to amplify genes for yeast homologous recombination (without removal of introns). The vector pEYA was used to assemble the PKS-NRPS (*TRIREDRAFT_58285, triliA*) gene (~12 kb) in four different DNA fragments of around 3 kb each using yeast homologous recombination. This was followed by LR recombination between the entry plasmid pEYA·*triliA* (MSIII139) and the expression vector pTYGS·*arg* to yield vector pTYGS·*argB*·*triliA* (MSIII144). The enoyl reductase gene (*TRIREDRAFT_58289, triliB*) was also amplified from the gDNA of *T. reesei* using primers with overhang to P_{gpdA} and T_{gpdA}. Using yeast homologous recombination, the final vector pTYGS·*argB*·*triliA*·*triliB* (MSIII152) was constructed which contains the PKS-NRPS (*triliA*)- and ER (*triliB*)-encoding genes from the specified cluster. A vector for the expression of the first three genes of the cluster; PKS/NRPS (TRIREDRAFT_58285, *triliA*), ER (TRIREDRAFT_58289, *triliB*), $P450_{RE}$ (*TRIREDRAFT_58953, triliC*) was constructed using yeast homologous recombination using gDNA of *T. reesei* QM6a·Δ*tmus53*·Δ*Pyr4* as a template for the amplification of the genes. A vector for the expression of the first four genes of the cluster; PKS/NRPS (TRIREDRAFT_58285, *triliA*), ER (TRIREDRAFT_58289, *triliB*), $P450_{RE}$ (*TRIREDRAFT_58953, triliC*) and *Diels-Alderase* (*triliD*) was constructed via yeast homologous recombination using gDNA of *T. reesei* QM6a·Δ*tmus53*·Δ*Pyr4* as template for the amplification of the genes.

2.3. Chemical Analysis

All the chemicals and materials were purchased from one of the following companies: Bio-Rad (München, Germany), New England Biolabs (Beverly, MA, USA), Roth (Karlsruhe, Germany), Sigma Aldrich (Steinheim, Germany), and Thermo Fisher Scientific (Waltham, MA, USA).

2.3.1. Liquid Chromatography–Mass Spectrometry (LC–MS)

Analytical LC–MS data of the organic extracts were obtained by either a Waters mass-directed autopurification system (Waters, Wilmslow, UK) including a Waters 2767 autosampler, a Waters 2545 pump and Phenomenex Kinetex column (2.6 µm, C_{18}, 100 Å, 4.6 mm × 100 mm; Phenomenex) equipped with a Phenomenex Security Guard precolumn (Luna, C_5, 300 Å). Detection was performed by a diode array detector from 210–600 nm (DAD; Waters 2998 or Waters 996), an electron light-scattering detector (ELSD; Waters 2424) and an electrospray ionisation mass detector in the range of 100–1000 *m/z* (Waters SQD-2). Gradient was run over 15 min starting at 10% acetonitrile/ 90% HPLC grade water (+ 0.05% formic acid) and ramping to 90% acetonitrile/ 10% HPLC grade water (+ 0.05% formic acid). Flowrate was 1 mL/min and 20 µL of the sample was injected. Data were displayed using the software MassLynx. The LC–MS program is summarized in the ESI.

Single compounds were purified from the raw organic extracts by a Waters mass-directed autopurification system. It comprises a Waters 2767 autosampler, a Waters 2545 binary gradient module system and a Phenomenex Kinetex Axia column (5 µm, C_{18}, 100 Å, 21.2 mm × 250 mm) equipped with a Phenomenex Security Guard precolumn (Luna, C_5, 300 Å). A water/acetonitrile gradient was run over 15 min with a flowrate of 20 mL/min and a post-column flow split of 100:1. The minority flow was applied for simultaneous analysis by a diode array detector (Waters 2998) in the range 210 to 1600 nm, an evaporative light-scattering detector (ELSD; Waters 2424) and a Waters SQD-2 mass detector, operating in ES+ and ES− modes between 100 and 1000 *m/z*. Selected peaks were collected into test tubes and solvent was evaporated under reduced pressure. Compounds were dissolved in methanol (1 mg/mL). High-Resolution Mass Spectrometry was performed on a Q-Tof Premier mass spectrometer (Waters) coupled to an Acquity UPLC–domain (Waters). Electron spray Ionisation (EsI) mass spectroscopy was measured in positive or negative mode depending on the compound.

2.3.2. Nuclear Magnetic Resonance (NMR) Analysis

Bruker Ascend 400 MHz, Bruker DRX 500 MHz or a Bruker Ascend 600 MHz Spectrometer (Bruker) were used for NMR measurements of the samples. Raw data were then analyzed using the software Bruker TopSpin 3.5. Chemical shifts are expressed in parts per million (ppm) in comparison to the Tetramethylsilane (TMS) standard and are referenced to the deuterated solvent.

2.3.3. Extraction of Fungal Cultures

For *A. oryzae* transformants, a small cell sample was dried by Büchner filtration or gravity filtration and used for gDNA analysis. Mycelia were ground using a hand blender and removed from the culture supernatant by Büchner filtration. Supernatant was acidified with 2 M HCL to pH 3–4 and extracted twice with ethyl acetate. Combined organic layers were dried over $MgSO_4$ and solvents were removed under vacuum. Organic residue was dissolved in methanol or acetonitrile to a concentration of 5–10 mg/mL (analytical) or 50 mg/mL (preparative) and filtered over glass wool.

3. Results

The genome of *T. reesei* QM6A, the wild-type isolate, was sequenced in 2008 [19]. We examined this genome sequence using a combination of automated AntiSMASH [20] and manual screening, with a particular focus on PKS and PKS-NRPS encoding genes. Bioinformatic analysis of all PKS and PKS-NRPS genes in *T. reesei* QM6a revealed the presence of 11 PKS and two hybrid PKS-NRPS encoding genes (see ESI for details). Of these, only the two PKS genes responsible for the sorbicillins and related compounds have been identified and confirmed (TRIREDRAFT_73618 and 73621) [4,21,22]. The HR-PKS TRIREDRAFT_65116 shows a very high similarity to the recently investigated PKS from *Trichoderma virens*, which is encoded in the biosynthetic gene cluster (BGC) that directs the production of virensols and trichoxide [23]. However, the other PKS and PKS-NRPS BGCs could not be linked to the production of any specific compound.

Although the *T. reesei* genome contains two hybrid PKS-NRPS encoding genes, no PKS-NRPS secondary metabolites have been reported from this fungus and these BGCs are likely to be silent under laboratory conditions. The two PKS-NRPS themselves show similarity (>60%) to proteins involved in fusarin C **8** biosynthesis in *Fusarium moniliforme* [13] and illicicolin H **10** biosynthesis in *Penicillium variabile* (= *Talaromyces variabilis*) [16]. Ilicicolin H **10** is an interesting antifungal agent that targets the cytochrome bc1 respiration complex in the 3–5 nM range [24], and it is conceivable that ilicicolin H **10** production may be responsible for the observed antifungal effects of *Trichoderma* species in general [2] and *T. reesei* in particular [25]. We therefore selected the putative *T. reesei* ilicicolin H BGC (*trili*) for further investigation.

Based on the NCBI prediction, the *trili* BGC (Figure 2) consists of a hybrid PKS/NRPS (*triliA*), an ER (*triliB*), and a putative cytochrome P450 (*triliC*). Reannotation of the cluster by FungiSMASH and Clinker [26] and Artemis comparison to a closely related BGC in *T. citrinoviride* resulted in the identification of a putative *S*-adenosylmethionine (SAM)-dependent Diels-Alderase in the cluster (*triliD*), that shows high similarity to *iccD* from the illicicolin-H BGC in *Talaromyces variabilis* [16]. Finally, *triliE* encodes a putative TIM-like epimerase homologous to IccE, known to epimerise epi-ilicicolin H in *T. variabilis*. Multigene analysis using cblaster [27] identified highly similar *ili* BGCs in other strains of *T. reesei* (RUT C-30) and other *Trichoderma* species, including *T. parareesei*, *T. virens* Gv29-8, *T. citrinoviride* TUCIM6016 and *T. longibrachiatum* ATCC 18648 (see ESI).

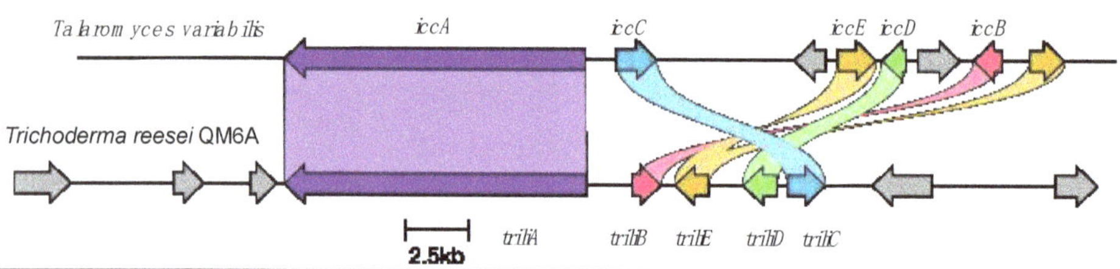

	Gene TRIREDRAFT	Protein	Predicted Function	Closest Blastp	Identity /%	Ref
triliA	58285	XP_006963788.1	PKS-NRPS	IliA (P0DO30.1)	66.4	[17]
triliB	58289	XP_006963601.1	Trans-ER	IccB (QBQ83710.1)	63.6	[16]
triliE	76204	XP_006963789.1	TIM-like epimerase	IccE (QBQ83707.1)	54.9	[16]
triliD	-	-	SAM-dependent Diels Alderase	IccD (QBQ83708.1)	55.9	[16]
triliC	58953	XP_006963602.1	Cytochrome P450	IliC (P0DO32.1)	70.4	[17]

Figure 2. Annotation of *T. reesei* ilicicolin H BGC (*trili*) and comparison to the *Talaromyces variabilis icc* BGC. Ili proteins are from the *Neonectria* sp. DH2 ilicicolin H biosynthetic pathway [17]; Icc proteins are from the *T. variabilis* ilicicolin H biosynthetic pathway [16].

In an initial heterologous expression experiment, *triliA* and *triliB* were cloned into the vector pTYGS·*argB* [28] using rapid recombination of genome-derived PCR fragments in yeast and *in vitro* (Gateway) recombination. The PKS-NRPS-encoding gene *triliA* was placed downstream of the inducible *A. oryzae amyB* promoter (P_{amyB}), while *triliB* was placed under the control of the *Aspergillus gpdA* promoter (P_{gpdA}). The resulting vector pTYGS·*argB*·*triliA*·*triliB* was transformed into the fungal host *Aspergillus oryzae* NSAR1 [29]. Comparison of the organic extracts of transformants with those of untransformed *A. oryzae*

NSAR1 revealed the production of new compounds by LC–MS analysis eluting at 6.7 min, 8.0 min and 11.0 min (Figure 3).

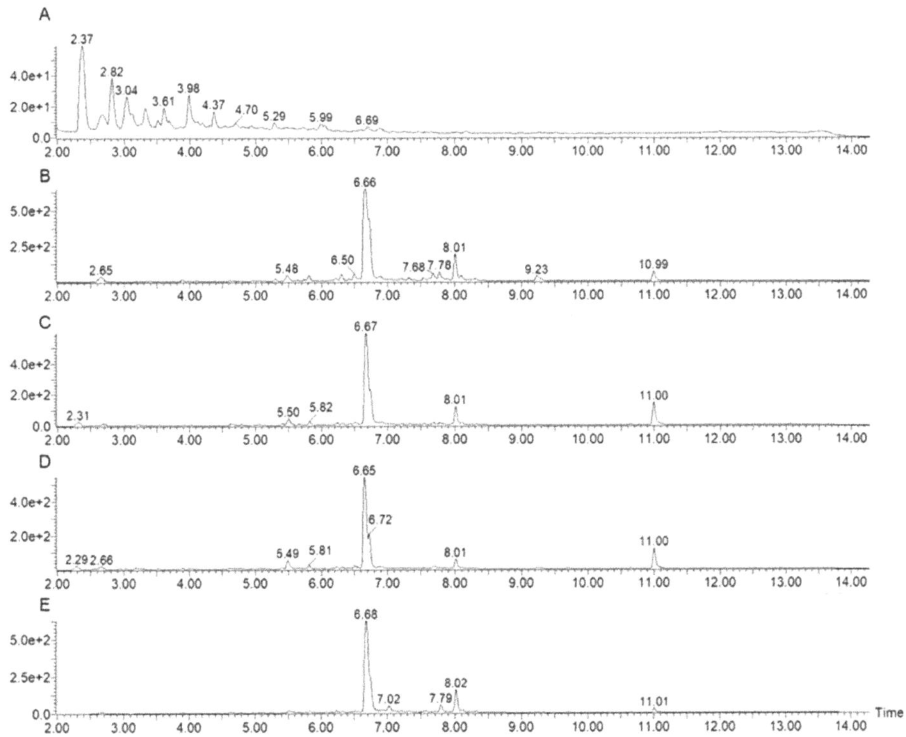

Figure 3. LC–MS traces (DAD 210–600 nm) of transformed *A. oryzae* with *illiA* and *illiB.* (**A**), the extract of the wild-type *A. oryzae* NSAR1. (**B–E**), extracts of four transformants of *A. oryzae* NSAR1 with *triliA* and *triliB* showing the production of different PKS-NRPS-related compounds at *RT* 6.7, 8.0 and 11.0 min.

Compound **11** (*RT* 6.7 min, 30 mg) was isolated by preparative LC–MS, and the structure was elucidated using 1D and 2D NMR. HRMS analysis of **11** confirmed a molecular formula of $C_{23}H_{28}NO_6$ ([M+H]+ calculated 414.1917, found 414.1917). Compound **12** (*RT* 8.0 min, 4 mg) was isolated by preparative LC–MS and the structure was elucidated using NMR. HRMS analysis of **12** confirmed a molecular formula of $C_{27}H_{32}NO_6$ ([M+H]+ calculated 466.2230, found 466.2237).

NMR analysis of these compounds (see ESI for full details) showed that **11** is a linear acyl tetramic acid with a C_{12} sidechain terminating in an unusual (at this position) carboxylic acid. Minor compound **12** was shown to be a related acyl tetramic acid featuring a decalin, with an unusual 3-propenoic acid substituent. Although the minor compound **13** at 11.0 min could not be fully identified by NMR, mass spectroscopic and UV analysis suggested that it is likely to be a third related tetramic acid with a C_{16} sidechain, first reported by Tang and coworkers during biosynthetic studies of ilicicolin H (Scheme 1) [16].

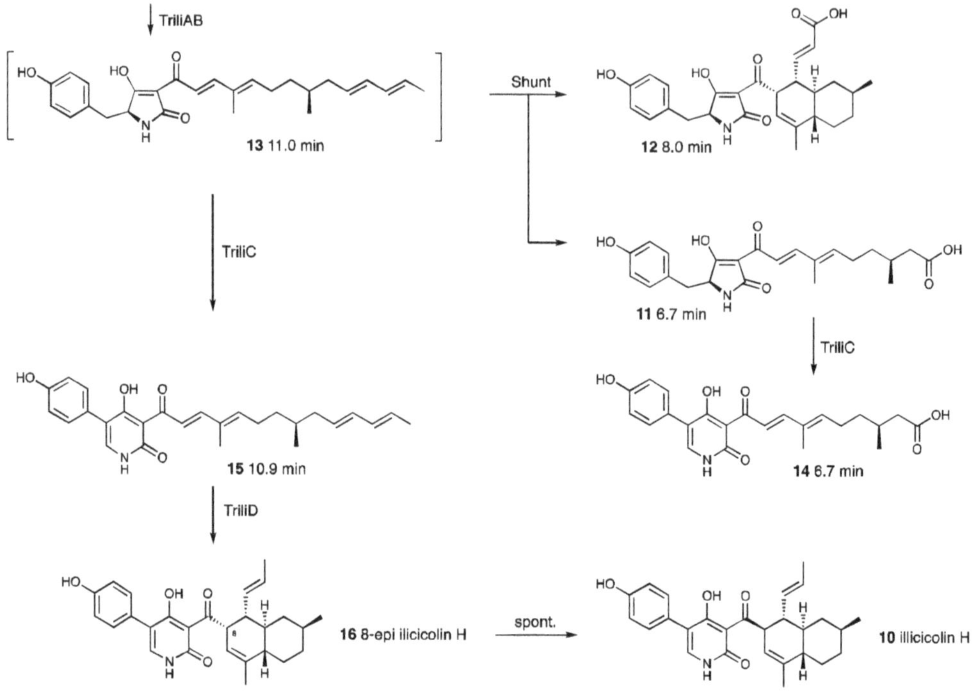

Scheme 1. Overall predicted pathway showing production of new compounds **11**, **12** and **14** in *A. oryzae* NSAR1.

In the next round of experiments, *triliC*, encoding a cytochrome P450 oxidase, was added to the *A. oryzae* strains, driven by P_{adh}. TriliC is related to ring-expanding cytochrome P450 oxidases first characterised from the tenellin pathway [15]. Coexpression of *triliA*, *triliB* and *triliC* resulted in production of two new compounds in comparison to the previous transformants and the untransformed *A. oryzae* (Figure 4), eluting at 6.7 (**14**), and 10.8 (**15**) min. Full NMR analysis of the *RT* 6.7 min compound showed it to be the pyridone homolog of **11**. Similarly, the 10.8 min compound was shown to be the pyridone homolog of **13**.

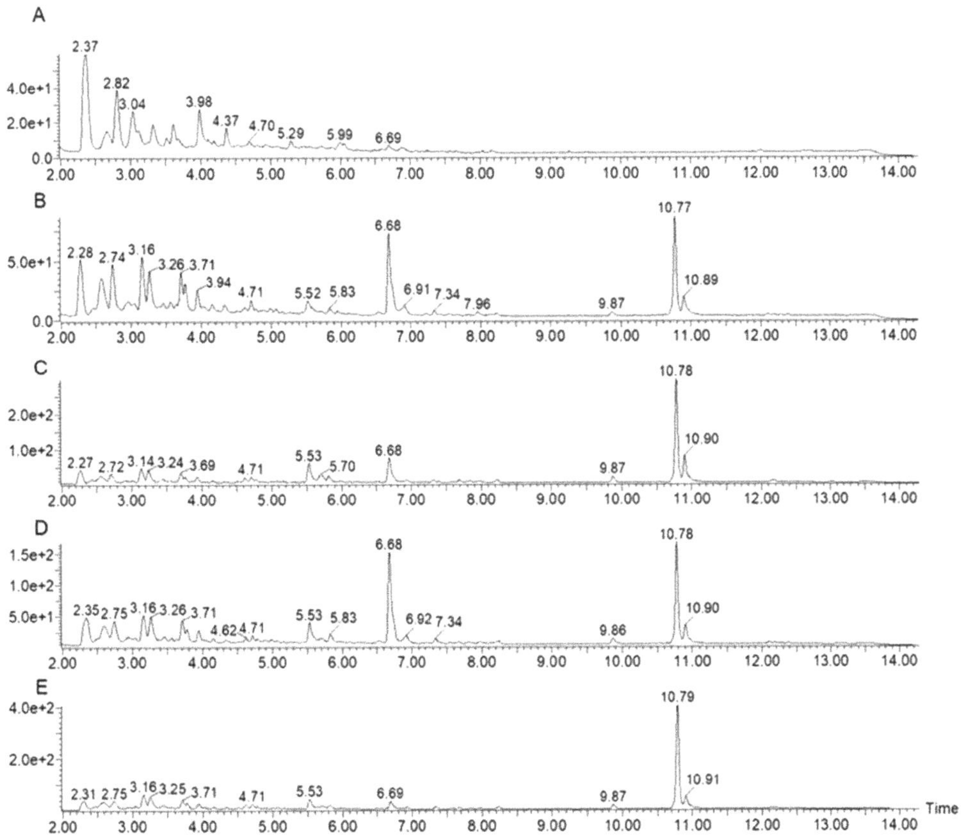

Figure 4. LC–MS traces (DAD 210–600 nm) of transformed *A. oryzae* NSAR1 with *triliA*, *triliB* and *triliC*. (**A**), the extract of untransformed *A. oryzae* NSAR1; (**B–E**), extracts of four transformants of *A. oryzae* with *triliA*, *triliB* and *triliC*. New compounds were observed at RT 6.7 and 10.8 min.

Finally, a gene homologous to Diels-Alderase *iccD* (*triliD*) was added to the expression system downstream of P_{eno} and transformed into *A. oryzae*. This experiment again produced **11** (*RT* 6.7 min). Additionally, two new peaks in comparison to the previous transformants and the untransformed *A. oryzae* were also observed by LC–MS (Figure 5) at *RT* 10.76 and 11.0 min. The minor compound at 11.0 min was again assigned as **13** after analysis of partial data. NMR and HRMS analysis (see ESI for full details) showed the peak at 10.76 min consists of an almost coeleuting mixture of the known 8-epi ilicicolin H **16** and ilicicolin H **10** itself.

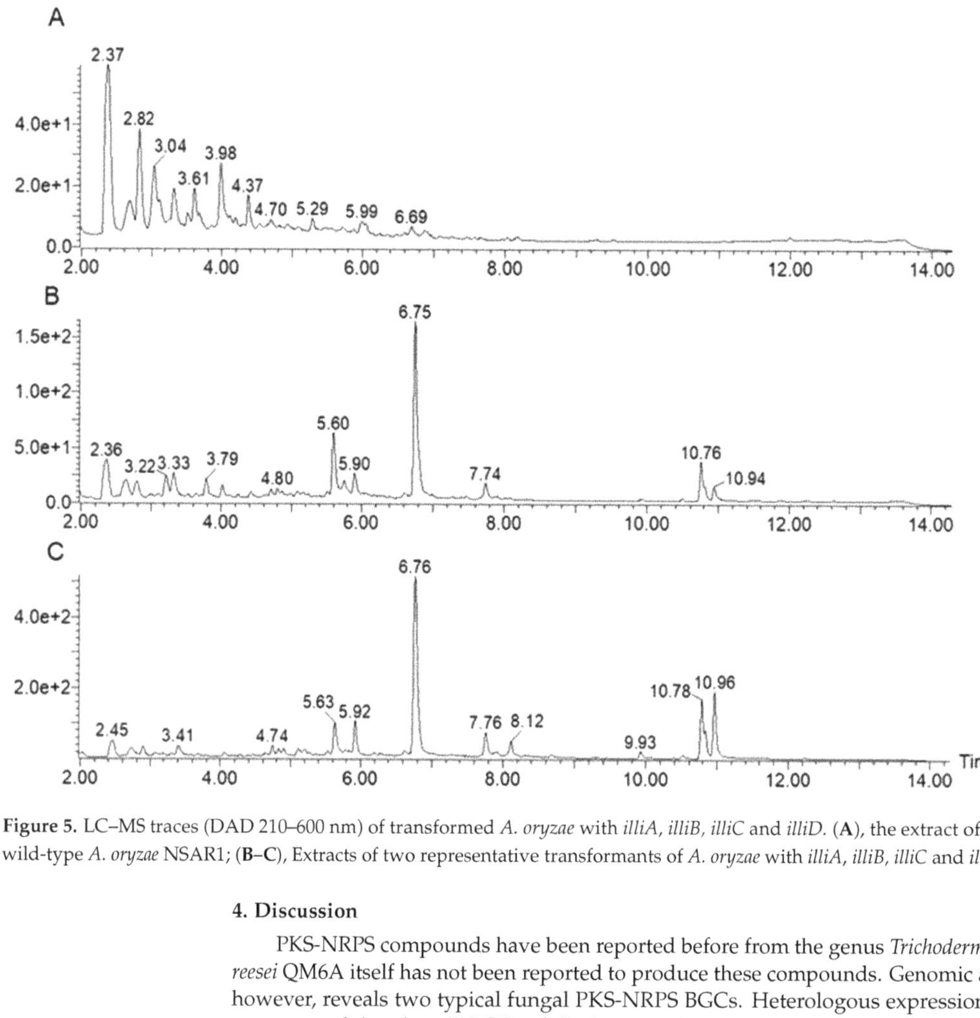

Figure 5. LC–MS traces (DAD 210–600 nm) of transformed *A. oryzae* with *illiA, illiB, illiC* and *illiD.* (**A**), the extract of the wild-type *A. oryzae* NSAR1; (**B–C**), Extracts of two representative transformants of *A. oryzae* with *illiA, illiB, illiC* and *illiD.*

4. Discussion

PKS-NRPS compounds have been reported before from the genus *Trichoderma*, but *T. reesei* QM6A itself has not been reported to produce these compounds. Genomic analysis, however, reveals two typical fungal PKS-NRPS BGCs. Heterologous expression experiments reveal that the *trili* BGC is fully functional, but apparently silent in *T. reesei* QM6A under laboratory conditions. Initial expression of the PKS-NRPS (*triliA*) and *trans*-ER (*triliB*) genes in *A. oryzae* led to production of two new acyl tetramic acids (**11** and **12**) and a third compound **13** that is likely to be a precursor of illicicolin H, recently described by Tang and coworkers [16] from *Nectria sp. B13* and *T. variabilis* and Gao and coworkers [17] in *Neonectria sp. DH2*. Decalin **12** appears to have been oxidised to a carboxylic acid at its terminal polyketide methyl group. Addition of the ring-expanding P450 oxidase then produces the corresponding 2-pyridones **14** and **15**. Compound **15** is identical to an ilicicolin H precursor previously identified by Tang and coworkers [16], but **14** is new. Finally, addition of the Diels-Alderase encoded by *triliD* produces a mixture of 8-epi ilicicolin H **16** and ilicicolin H **10** itself.

The *T. variabilis icc* [16] and *Neonectra DH2 ili* [17] BGCs were previously investigated by expression in *A. nidulans*. Our results are highly similar to those obtained from these studies, except that *A. oryzae* as a host appears to catalyse some unexpected steps. For example, in *A. oryzae* the initial tetramic acid precursor **13** appears to be oxidatively degraded to remove the terminal four polyketide carbons and leaves a carboxylic acid in the structure of **11**. Previous heterologous expression studies on *A. oryzae* also indicate that this host has a propensity for the oxygenation of polyenes [30,31] and oxidative cleavage of olefins [32]. Likewise, oxygenation of the terminal methyl of **12** to a carboxylic acid is newly observed in *A. oryzae*. This is similar to the BueE-catalysed oxidation recently observed by Piggott and coworkers in the case of the fungal polyketide decalin burnettiene A in *Aspergillus burnettii* [33]. It also appears that a spontaneous (or *A. oryzae*-catalysed) Diels–Alder reaction can occur to produce a decalin that is then oxidised to a carboxylic acid at the terminal methyl to generate new compound **12**. In previous studies on ilicicolin H biosynthesis, using *A. nidulans* as the host, a specific Diels–Alder catalyst was required for this step. We also observed the chain-truncated pyridone **14**—again a new natural product. In *A. nidulans* it appears that 8-epi ilicicolin H **16** is a stable compound, requiring expression of *iccE* for the conversion to ilicicolin H. However, in *A. oryzae*, observable amounts of ilicicolin H **10** are produced in the absence of this epimerase catalyst.

Our results therefore illustrate that *T. reesei* QM6A has a fully functional PKS-NRPS BGC that encodes the biosynthesis of ilicicolin H, a potent antifungal agent. The ability of *T. reesei* QM6A to make this compound has never before been observed [34], despite nearly 70 years of laboratory-based investigations. It is likely that this BGC can be activated under unknown environmental conditions in its native habitat and may contribute to its utility as an agricultural antifungal agent, but it appears to be silent under laboratory conditions. Finally, our results also show that different fungal heterologous hosts can introduce unexpected shunt pathways which can diversify known secondary metabolites and produce new-to-nature compounds.

Supplementary Materials: The following are available online at https://www.mdpi.com/article/10.3390/jof7121034/s1. All experimental details including growth media, buffers, enzyme and antibiotic solutions, bacterial and fungal strains, vectors and oligonucleotides, bioinformatic procedures and results, structure elucidation details, tabulated NMR data, NMR spectra, and HRMS data.

Author Contributions: M.L.S. and R.J.C. designed the study. Experimental work was performed by M.L.S., assisted by M.A. The paper was written by R.J.C. and M.L.S., R.J.C. was responsible for providing physical resources. Funding was gained by M.L.S. (DAAD GERLS) and R.J.C. All authors have read and agreed to the published version of the manuscript.

Funding: M.L.S. thanks the German Academic Exchange Service (DAAD) and the Egyptian Ministry of Higher Education and Scientific Research (MHESR) for funding (GERLS programme, 2017, 57311832). This work was supported by DFG [CO 1328/2-1, INST 187/621-1, INST 187/686-1)]. The APC was funded by the Open Access Fund of the Leibniz Universität Hannover.

Acknowledgments: The authors would like to thank Eric Kuhnert and Kevin Becker for the helpful discussions and Astrid Mach-Aigner for donating *T. reesei* strain.

Conflicts of Interest: The authors declare no conflict of interest. The funders had no role in the design of the study, in the collection, analyses, or interpretation of data, in the writing of the manuscript, or in the decision to publish the results.

References

1. Beier, S.; Hinterdobler, W.; Monroy, A.A.; Bazafkan, H.; Schmoll, M. The Kinase USK1 Regulates Cellulase Gene Expression and Secondary Metabolite Biosynthesis in Trichoderma reesei. *Front. Microbiol.* **2020**, *11*, 974. [CrossRef] [PubMed]
2. Rees, H.J.; Bashir, N.; Drakulic, J.; Cromey, M.G.; Bailey, A.M.; Foster, G.D. Identification of native endophytic Trichoderma spp. for investigation of in vitro antagonism towards Armillaria mellea using synthetic- and plant-based substrates. *J. Appl. Microbiol.* **2021**, *131*, 392–403. [CrossRef] [PubMed]
3. Al Fahad, A.; Abood, A.; Fisch, K.M.; Osipow, A.; Davison, J.; Avramović, M.; Butts, C.P.; Piel, J.; Simpson, T.J.; Cox, R.J. Oxidative dearomatisation: The key step of sorbicillinoid biosynthesis. *Chem. Sci.* **2014**, *5*, 523–527. [CrossRef] [PubMed]
4. Kahlert, L.; Bassiony, E.F.; Cox, R.; Skellam, E.J. Diels–Alder Reactions during the Biosynthesis of Sorbicillinoids. *Angew. Chem. Int. Ed.* **2020**, *59*, 5816–5822. [CrossRef]
5. Harned, A.M.; Volp, K.A. The sorbicillinoid family of natural products: Isolation, biosynthesis, and synthetic studies. *Nat. Prod. Rep.* **2011**, *28*, 1790–1810. [CrossRef]
6. Shenouda, M.L.; Cox, R.J. Molecular methods unravel the biosynthetic potential of Trichoderma species. *RSC Adv.* **2021**, *11*, 3622–3635. [CrossRef]

7. Sawa, R.; Mori, Y.; Iinuma, H.; Naganawa, H.; Hamada, M.; Yoshida, S.; Furutani, H.; Kajimura, Y.; Fuwa, T.; Takeuchi, T. Harzianic acid, a new antimicrobial antibiotic from a fungus. *J. Antibiot.* **1994**, *47*, 731–732. [CrossRef]
8. Xie, L.; Zang, X.; Cheng, W.; Zhang, Z.; Zhou, J.; Chen, M.; Tang, Y. Harzianic Acid from Trichoderma afroharzianum Is a Natural Product Inhibitor of Acetohydroxyacid Synthase. *J. Am. Chem. Soc.* **2021**, *143*, 9575–9584. [CrossRef]
9. Chen, H.; Daletos, G.; Okoye, F.; Lai, D.; Dai, H.; Proksch, P. A New Cytotoxic Cytochalasin from the Endophytic Fungus Trichoderma harzianum. *Nat. Prod. Commun.* **2015**, *10*, 585–587. [CrossRef]
10. Wu, B.; Oesker, V.; Wiese, J.; Schmaljohann, R.; Imhoff, J.F. Two New Antibiotic Pyridones Produced by a Marine Fungus, Trichoderma sp. Strain MF106. *Mar. Drugs* **2014**, *12*, 1208–1219. [CrossRef]
11. Dickinson, J.M.; Hanson, J.R.; Hitchcock, P.B.; Claydon, N. Structure and biosynthesis of harzianopyridone, an antifungal metabolite of Trichoderma harzianum. *J. Chem. Soc. Perkin Trans. 1* **1989**, *1*, 1885–1887. [CrossRef]
12. Bat-Erdene, U.; Kanayama, D.; Tan, D.; Turner, W.C.; Houk, K.N.; Ohashi, M.; Tang, Y. Iterative Catalysis in the Biosynthesis of Mitochondrial Complex II Inhibitors Harzianopyridone and Atpenin B. *J. Am. Chem. Soc.* **2020**, *142*, 8550–8554. [CrossRef]
13. Song, Z.; Cox, R.J.; Lazarus, C.M.; Simpson, T.J. Fusarin C Biosynthesis in Fusarium moniliforme and Fusarium venenatum. *ChemBioChem* **2004**, *5*, 1196–1203. [CrossRef]
14. Wolff, P.B.; Nielsen, M.L.; Slot, J.C.; Andersen, L.N.; Petersen, L.M.; Isbrandt, T.; Holm, D.K.; Mortensen, U.H.; Nødvig, C.S.; Larsen, T.O.; et al. Acurin A, a novel hybrid compound, biosynthesized by individually translated PKS- and NRPS-encoding genes in Aspergillus aculeatus. *Fungal Genet. Biol.* **2020**, *139*, 103378. [CrossRef]
15. Halo, L.M.; Heneghan, M.N.; Yakasai, A.A.; Song, Z.; Williams, K.; Bailey, A.; Cox, R.; Lazarus, C.M.; Simpson, T.J. Late Stage Oxidations during the Biosynthesis of the 2-Pyridone Tenellin in the Entomopathogenic FungusBeauveria bassiana. *J. Am. Chem. Soc.* **2008**, *130*, 17988–17996. [CrossRef]
16. Zhang, Z.; Jamieson, C.S.; Zhao, Y.-L.; Li, D.; Ohashi, M.; Houk, K.N.; Tang, Y. Enzyme-Catalyzed Inverse-Electron Demand Diels–Alder Reaction in the Biosynthesis of Antifungal Ilicicolin H. *J. Am. Chem. Soc.* **2019**, *141*, 5659–5663. [CrossRef]
17. Lin, X.; Yuan, S.; Chen, S.; Chen, B.; Xu, H.; Liu, L.; Li, H.; Gao, Z. Heterologous Expression of Ilicicolin H Biosynthetic Gene Cluster and Production of a New Potent Antifungal Reagent, Ilicicolin J. *Molecules* **2019**, *24*, 2267. [CrossRef]
18. Derntl, C.; Kiesenhofer, D.P.; Mach, R.L.; Mach-Aigner, A.R. Novel Strategies for Genomic Manipulation of Trichoderma reesei with the Purpose of Strain Engineering. *Appl. Environ. Microbiol.* **2015**, *81*, 6314–6323. [CrossRef]
19. Chia-Ling, C.; Huang, C.-H.; Chen, C.-L.; Tung, S.-Y.; Wang, T.-F. Trichoderma reesei complete genome sequence, repeat-induced point mutation, and partitioning of CAZyme gene clusters. *Biotechnol. Biofuels* **2017**, *10*, 170. [CrossRef]
20. Blin, K.; Shaw, S.; Kloosterman, A.M.; Charlop-Powers, Z.; van Wezel, G.P.; Medema, M.H.; Weber, T. antiSMASH 6.0: Improving cluster detection and comparison capabilities. *Nucleic Acids Res.* **2021**, *49*, W29–W35. [CrossRef]
21. Derntl, C.; Rassinger, A.; Srebotnik, E.; Mach, R.L.; Mach-Aigner, A.R. Identification of the Main Regulator Responsible for Synthesis of the Typical Yellow Pigment Produced by Trichoderma reesei. *Appl. Environ. Microbiol.* **2016**, *82*, 6247–6257. [CrossRef]
22. Derntl, C.; Guzman-Chavez, F.; Mello-De-Sousa, T.M.; Busse, H.-J.; Driessen, A.J.M.; Mach, R.; Mach-Aigner, A.R. In Vivo Study of the Sorbicillinoid Gene Cluster in Trichoderma reesei. *Front. Microbiol.* **2017**, *8*, 2037. [CrossRef]
23. Liu, L.; Tang, M.-C.; Tang, Y. Fungal Highly Reducing Polyketide Synthases Biosynthesize Salicylaldehydes That Are Precursors to Epoxycyclohexenol Natural Products. *J. Am. Chem. Soc.* **2019**, *141*, 19538–19541. [CrossRef]
24. Singh, S.B.; Liu, W.; Li, X.; Chen, T.; Shafiee, A.; Card, D.; Abruzzo, G.; Flattery, A.; Gill, C.; Thompson, J.R.; et al. Antifungal Spectrum, In Vivo Efficacy, and Structure–Activity Relationship of Ilicicolin H. *ACS Med. Chem. Lett.* **2012**, *3*, 814–817. [CrossRef]
25. Damodaran, T.; Rajan, S.; Muthukumar, M.; Gopal, R.; Yadav, K.; Kumar, S.; Ahmad, I.; Kumari, N.; Mishra, V.K.; Jha, S.K. Biological Management of Banana Fusarium Wilt Caused by Fusarium Oxysporum f. sp. cubense Tropical Race 4 Using Antagonistic Fungal Isolate CSR-T-3 (Trichoderma Reesei). *Front. Microbiol.* **2020**, *11*, 595845. [CrossRef]
26. Gilchrist, C.L.M.; Chooi, Y.-H. Clinker & clustermap.js: Automatic generation of gene cluster comparison figures. *Bioinformatics* **2021**, *37*, 2473–2475. [CrossRef]
27. Gilchrist, C.L.M.; Booth, T.J.; van Wersch, B.; van Grieken, L.; Medema, M.H.; Chooi, Y.-H. cblaster: A remote search tool for rapid identification and visualization of homologous gene clusters. *Bioinform. Adv.* **2021**, *1*, vbab016. [CrossRef]
28. Pahirulzaman, K.A.K.; Williams, K.; Lazarus, C.M. A Toolkit for Heterologous Expression of Metabolic Pathways in Aspergillus oryzae. *Methods Enzymol.* **2012**, *517*, 241–260. [CrossRef]
29. Jin, F.J.; Maruyama, J.-I.; Juvvadi, P.; Arioka, M.; Kitamoto, K. Development of a novel quadruple auxotrophic host transformation system byargBgene disruption usingadeAgene and exploiting adenine auxotrophy inAspergillus oryzae. *FEMS Microbiol. Lett.* **2004**, *239*, 79–85. [CrossRef]
30. Song, Z.; Bakeer, W.; Marshall, J.W.; Yakasai, A.A.; Khalid, R.M.; Collemare, J.; Skellam, E.; Tharreau, D.; Lebrun, M.-H.; Lazarus, C.M.; et al. Heterologous expression of the avirulence gene ACE1 from the fungal rice pathogen Magnaporthe oryzae. *Chem. Sci.* **2015**, *6*, 4837–4845. [CrossRef]
31. Heneghan, M.N.; Yakasai, A.A.; Williams, K.; Kadir, K.A.; Wasil, Z.; Bakeer, W.; Fisch, K.M.; Bailey, A.M.; Simpson, T.J.; Cox, R.J.; et al. The programming role of trans-acting enoyl reductases during the biosynthesis of highly reduced fungal polyketides. *Chem. Sci.* **2011**, *2*, 972–979. [CrossRef]
32. Fujii, R.; Ugai, T.; Ichinose, H.; Hatakeyama, M.; Kosaki, T.; Gomi, K.; Fujii, I.; Minami, A.; Oikawa, H. Reconstitution of biosynthetic machinery of fungal polyketides: Unexpected oxidations of biosynthetic intermediates by expression host. *Biosci. Biotechnol. Biochem.* **2016**, *80*, 426–431. [CrossRef]

33. Roux, I.; Bowles, S.; Kalaitzis, J.A.; Vuong, D.; Lacey, E.; Chooi, Y.-H.; Piggott, A.M. Characterisation and Heterologous Biosynthesis of Burnettiene A, a New Polyene-Decalin Polyketide from Aspergillus Burnettii. *Org. Biomol. Chem.* **2021**, *19*, 9506–9513. [CrossRef] [PubMed]
34. Rush, T.A.; Shrestha, H.K.; Meena, M.G.; Spangler, M.K.; Ellis, J.C.; Labbé, J.L.; Abraham, P.E. Bioprospecting Trichoderma: A Systematic Roadmap to Screen Genomes and Natural Products for Biocontrol Applications. *Front. Fungal Biol.* **2021**, *2*, 716511. [CrossRef]

Journal of *Fungi*

MDPI

Article

Heterologous Expression of Secondary Metabolite Genes in *Trichoderma reesei* for Waste Valorization

Mary L. Shenouda [1,2], Maria Ambilika [1], Elizabeth Skellam [1,3] and Russell J. Cox [1,*]

1 Institute for Organic Chemistry and Biomolekulares Wirkstoffzentrum (BMWZ), Schneiderberg 38, 30167 Hannover, Germany; mary.shenouda@oci.uni-hannover.de (M.L.S.); m.ambilika@gmail.com (M.A.); elizabeth.skellam@unt.edu (E.S.)
2 Pharmacognosy Department, Faculty of Pharmacy, Alexandria University, Alexandria 21521, Egypt
3 Department of Chemistry and BioDiscovery Institute, University of North Texas, 1155 Union Circle, Denton, TX 76201, USA
* Correspondence: russell.cox@oci.uni-hannover.de

Abstract: *Trichoderma reesei* (*Hypocrea jecorina*) was developed as a microbial cell factory for the heterologous expression of fungal secondary metabolites. This was achieved by inactivation of sorbicillinoid biosynthesis and construction of vectors for the rapid cloning and expression of heterologous fungal biosynthetic genes. Two types of megasynth(et)ases were used to test the strain and vectors, namely a non-reducing polyketide synthase (nr-PKS, *aspks1*) from *Acremonium strictum* and a hybrid highly-reducing PKS non-ribosomal peptide synthetase (hr-PKS-NRPS, *tenS* + *tenC*) from *Beauveria bassiana*. The resulting engineered *T. reesei* strains were able to produce the expected natural products 3-methylorcinaldehyde and pretenellin A on waste materials including potato, orange, banana and kiwi peels and barley straw. Developing *T. reesei* as a heterologous host for secondary metabolite production represents a new method for waste valorization by the direct conversion of waste biomass into secondary metabolites.

Keywords: heterologous expression; PKS-NRPS; PKS; waste valorization; microbial cell factory; *Trichoderma reesei*

Citation: Shenouda, M.L.; Ambilika, M.; Skellam, E.; Cox, R.J. Heterologous Expression of Secondary Metabolite Genes in *Trichoderma reesei* for Waste Valorization. *J. Fungi* **2022**, *8*, 355. https://doi.org/10.3390/jof8040355

Academic Editors: Tao Feng and Frank Surup

Received: 9 March 2022
Accepted: 28 March 2022
Published: 30 March 2022

1. Introduction

Heterologous expression of fungal biosynthetic gene clusters (BGC) is an effective method for the synthesis of known, and engineering of new, natural products [1–3]. Heterologous hosts include *Escherichia coli* [4], *Saccharomyces cerevisiae* [5], *Aspergillus oryzae* [1] and *Aspergillus nidulans* [6]. Many common host organisms such as *S. cerevisiae* and *E. coli* are conveniently manipulated, but require tight control of fermentation conditions and media components [1,5,7–11]. However, filamentous fungi offer high potential as hosts for secondary metabolite production as they can grow rapidly on a wide variety of substrates. In particular, *Trichoderma reesei* (*Hypocrea jecorina*) is a filamentous fungus well known for its industrial production of cellulases and other cell wall-degrading enzymes [12]. *T. reesei* has long been used as a heterologous host for enzyme production due to its excellent protein production capabilities and its Generally Recognized as Safe (GRAS) status [13]. However, the use of *T. reesei* as a heterologous host for secondary metabolite production has not been reported so far.

The ability of *T. reesei* to produce high levels of cellulases has exploited the organism's ability to grow to high cell mass on low-value materials such as waste streams from agriculture, food production and paper and cardboard production [14–16]. This contrasts with other fungi used in biotechnology such a *A. oryzae* or *A. nidulans* that usually require high-quality or food-grade substrates. The use of *T. reesei* in waste valorization by converting cellulosic biomass to fuels and chemicals was pioneered by Reese and Mandels after the discovery of *T. reesei* cellulases that degraded the tent canvas of the US army during

World War II [17]. Since then, enhancing the production levels of cellulases from *T. reesei* has been a significant focus of research [18]. We reasoned that *T. reesei* should also make an effective host for heterologous secondary metabolite production and the aim of this work is to develop *T. reesei* as a microbial cell factory for secondary metabolite production by direct fermentation of different types of waste materials.

2. Materials and Methods

All chemicals and media ingredients used in this work were purchased from Duchefa Biochemie (Haarlem), Roth (Karlsruhe), VWR (Darmstadt), Fisher scientific (Schwerte), Sigma Aldrich (Taurkirchen), abcr (Kahlsruhe) and Formedium (Hunstanton). Double-distilled water (dd H_2O) was used for the preparation of all media, buffers, solutions, and antibiotics. Sterilization of all growth media and solutions was achieved by autoclaving at 121 °C for 15 min using Systec VX150 or Prestige Medical 2100 autoclaves. Sterilization of antibiotic solutions was achieved using a sterile syringe filter (0.45 μm pore size, Roth). For details of all growth media, buffers and solutions, enzymes, and antibiotics, see electronic Supplementary Information (ESI Tables S1.1–S1.5).

2.1. Microbiological Methods

Bacterial and fungal strains used in this work are summarized in the ESI (Table S2.1). All the microbiological methods used were the same as those published previously [10].

2.2. Molecular Biology Methods

All enzymes were purchased from NEB and Thermo Fisher Scientific and were used according to the manufacturer's protocols using the supplied buffers. All the vectors and oligonucleotides used in this work are summarized in the ESI (Tables S3.1–S3.3). All molecular biology methods used were the same as published previously [10].

2.3. Fermentation

For potato, orange, banana and kiwi peels, 50 g of the peels were cut into small pieces and autoclaved with 100 mL pure water at 121 °C for 15 min. For coffee press, 10 g coffee press left-over from coffee machine was autoclaved (121 °C, 15 min) with 100 mL pure water. For barley straw, 5 g barley straw (donated from a horse stable near Hannover, Germany) was cut into small pieces and autoclaved with 100 mL pure water at 121 °C for 15 min. *Trichoderma reesei* transformants were grown on minimal media agar plates for 3–5 days. Mycelia and spores were scraped off using a sterile spatula. Finally, 250 μL of this suspension were inoculated on the waste media. The flasks were incubated for 14 days at 28 °C without shaking.

2.4. Extraction of Transformants Grown on Waste Materials

For potato, orange, banana, and kiwi peels as well as coffee press, the fungal mycelia together with the media components were homogenized at the end of the fermentation period using a hand blender. This was followed by filtration and the filtrate was acidified with 2 M HCl before extraction twice with ethyl acetate. The ethyl acetate fraction was then dried over anhydrous $MgSO_4$ and evaporated under reduced pressure. The residue was dissolved in acetonitrile to reach a final concentration of 5 mg/mL.

For barley straw, 200 mL of ethyl acetate was added to the flasks containing the fungal mycelia and the straw and stirred using a magnetic stirrer for 4 h before filtration. The ethyl acetate fraction was then separated, dried over anhydrous $MgSO_4$, and evaporated under reduced pressure. The residue was dissolved in acetonitrile to reach a final concentration of 5 mg/mL.

2.5. Chemical Analysis

All the chemicals and materials were purchased from one of the following companies: Bio-Rad (München, Germany), New England Biolabs (Beverly, MA, USA), Roth (Karlsruhe,

Germany), Sigma Aldrich (Steinheim, Germany), and Thermo Fisher Scientific (Waltham, MA, USA). All chemical analysis methods used were the same as previously described [10].

3. Results

3.1. Construction of a Host with a Cleaner Secondary Metabolic Background

T. reesei is a prolific producer of sorbicillinoids, derived from sorbicillin **1** and sorbicillinol **2** (Figure 1). The sorbicillin pathway is well understood and the encoding sorbicillin (*sor*) BGC has been investigated in detail (Figure 1A) [19–21]. Sorbicillinoid compounds make up the majority of the natural products produced in *T. reesei* fermentations and include previously identified compounds such as epoxysorbicillinol **3**, bisorbicillinol **4** and bisvertinolone **5**. We decided to prevent the biosynthesis of sorbicillinoids to give a cleaner secondary metabolic background and potentially enable a higher proportion of metabolic precursors to be used for the production of desired metabolites. Therefore, we aimed to knock out key *sor* biosynthetic genes using the classical bipartite method of Nielsen that dramatically decreases the chances of false-positive transformants [22].

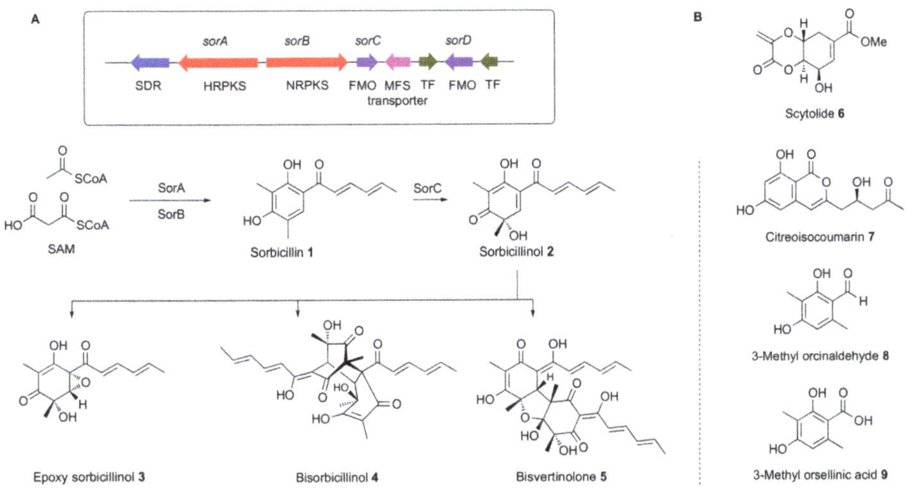

Figure 1. (**A**), the sorbicillin (*sor*) biosynthetic gene cluster in *T. reesei*, and known pathway to monomeric (**1–3**) and dimeric (**4–5**) sorbicillinoids; (**B**), compounds isolated from *T. reesei* QM6a·Δ*tmus53*·Δ*pyr4*·Δ*sorBC* and *T. reesei* QM6a·Δ*tmus53*·Δ*sorBC*·P_{pdc}·*aspks1*.

A vector was constructed using yeast homologous recombination to knock out (KO) the adjacent *sorB* and *sorC* genes simultaneously. The vector was constructed using the pEYA backbone [23] and a hygromycin resistance gene (*hph* [24]) inserted between P_{gpdA} and T_{trpC} with ≈1 kb regions homologous to the 5′ end of *sorB* and 3′ end of *sorC* (see ESI for details, Figure S7.1). This vector was used as a template for the construction of overlapping PCR fragments for the bipartite KO of the *sorB/sorC* region. *T. reesei* QM6a·Δ*tmus53*·Δ*pyr4* (gift of Prof. Dr. Astrid Mach-Aigner, TU Wien [25]) was transformed with these PCR fragments. After three rounds of selection on PDA containing hygromycin, eighteen hygromycin-resistant transformants were selected. Chemical analysis of six of them showed that one transformant, *T. reesei* QM6a·Δ*tmus53*·Δ*pyr4*·Δ*sorBC*, showed no production of any sorbicillin-related compounds after cultivation for seven days in DPY + 1% glucose (LC–MS analysis, Figure 2). Cultivation of this transformant on different media revealed its inability to produce sorbicillinoids under all tested conditions. PCR analysis and partial sequencing confirmed the expected loss of the *sorB* and *sorC* genes.

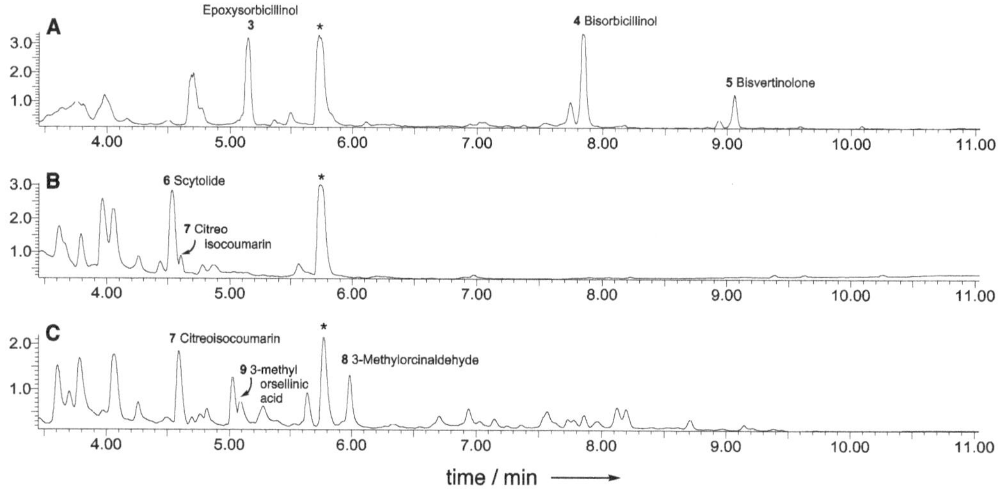

Figure 2. LC–MS traces of *T. reesei* strains grown in on DPY + 1% glucose media. (**A**), *T. reesei* QM6a·Δ*tmus53*·Δ*pyr4*; (**B**), *T. reesei* QM6a·Δ*tmus53*·Δ*pyr4*·Δ*sorBC*; (**C**), *T. reesei* QM6a·Δ*tmus53*·Δ*sorBC*·P_{pdc}·*aspks1*. * = unrelated compound.

Although the *T. reesei* QM6a·Δ*tmus53*·Δ*pyr4*·Δ*sorBC* strain did not produce any sorbicillin-related compounds, it was able to produce scytolide **6** as previously observed [20]. In addition, new compounds were produced that were not previously observed in *T. reesei* QM6a·Δ*tmus53*·Δ*pyr4*. One of these compounds was present in a concentration high enough to allow its isolation and structure elucidation using NMR and comparison to published data (See ESI Figures S11.1–S11.5). The compound was identified as the known citreoisocoumarin **7** (Figure 1B) [26].

3.2. Construction of a *T. reesei* Heterologous Expression Vector

In our previous work, we have made extensive use of the modular vector system developed by Lazarus and coworkers for use in *Aspergillus oryzae* [27,28]. Due to the low availability of vectors for gene integration in *T. reesei*, an *A. oryzae* vector was adapted as the backbone to construct a new *T. reesei* expression system. The vector was constructed based on the pTYGS·*argB* vector [29], where the inducible *Aspergillus oryzae amyB* promoter (P_{amyB}) was replaced with the pyruvate decarboxylase promoter (P_{pdc}) from *T. reesei* itself and the *Aspergillus argB* selection marker was replaced with the *pyr4* marker using homologous recombination in yeast (Figure 3). These changes were designed to allow easy and high-throughput heterologous expression of megasynth(et)ases in *T. reesei*.

The constructed vector, pTYGS·*pyr4*·P_{pdc}, has *att*R sites that allow in vitro recombination of a target gene present on a Gateway entry vector, downstream of P_{pdc}. In an initial experiment, we wished to use a reliable and well-understood synthase to test the system. We therefore selected *aspks1* that encodes the non-reducing polyketide synthase (nr-PKS) methylorcinaldehyde synthase (MOS) from the xenovulene biosynthetic pathway of *Acremonium strictum* [30]. The entry vector pEYA·*aspks1* [27] was recombined in vitro with pTYGS·*pyr4*·P_{pdc} to insert *aspks1* into the cloning site downstream of P_{pdc}. The resulting vector, pTYGS·*pyr4*·P_{pdc}·*aspks1*, was confirmed by PCR and sequencing.

The vector pTYGS·*pyr4*·P_{pdc}·*aspks1* was transformed into *T. reesei* QM6a·Δ*tmus53*·Δ*pyr4*·Δ*sorBC* using standard PEG-mediated transformation (See ESI) [31]. After three rounds of selection on minimal media lacking uridine, eight transformants were obtained and cultivated on PDB media for 48 h at 28 °C and 110 rpm. Extraction of all of the transformants showed that seven out of the eight transformants produce the expected natural product 3-methylorcinaldehyde **8** in addition to minor amounts of the corresponding

3-methylorsellinic acid **9** (Figure 2C). Longer cultivation periods and fermentation on different media showed an increase in the production levels of the acid **9** and a decrease in the aldehyde **8** production (see ESI Figure S9.3). Analysis of the gDNA of all eight transformants showed the correct insertion of *aspks1* in seven out of the eight transformants (see ESI Figure S9.5), in agreement with the LC–MS analysis.

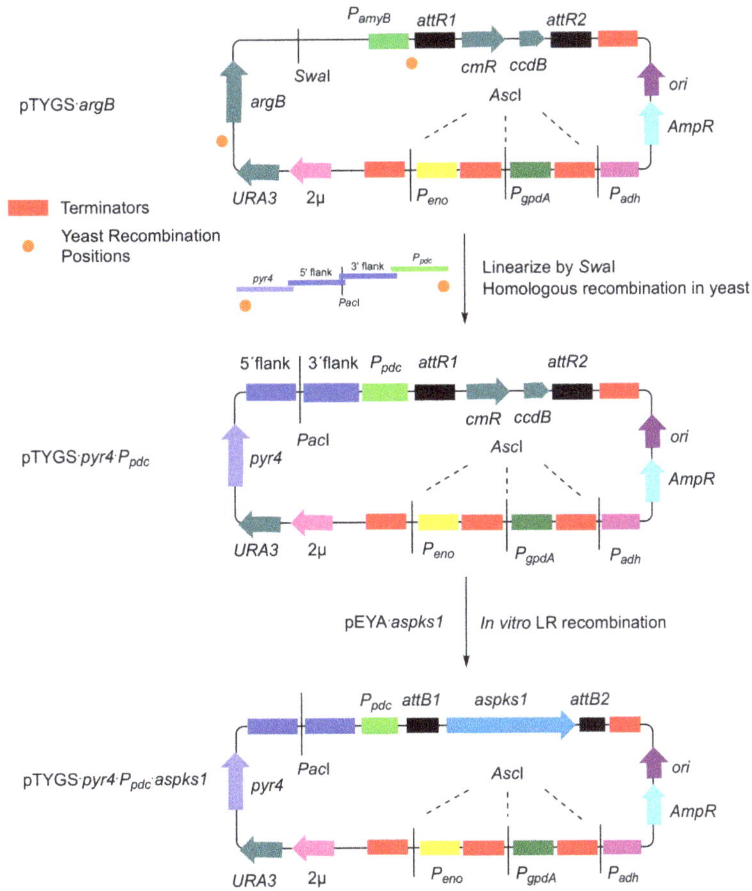

Figure 3. Construction of general expression vector pTYGS·*pyr4*·P_{pdc} and nr-PKS expression system pTYGS·*pyr4*·P_{pdc}·*aspks1*.

3.3. Growing Recombinant T. reesei Strains on Different Waste Materials

The ability of the new transformant, *T. reesei* QM6a·Δ*tmus53*·Δ*sorBC*·P_{pdc}·*aspks1*, to grow on different waste materials and produce **8** and **9** was tested. Different substrates were used such as potato peel, orange peel, banana peel, kiwi peel, coffee grinds and barley straw. The fungus grew quickly on all substrates, except orange peels. Fermentations were extracted after 14 days of cultivation with ethyl acetate.

LC–MS analysis of the extracts showed the ability of the transformed strain to produce methylorcinaldehyde **8** and methyl orsellinic acid **9** on four out of the five tested media but in different ratios (Figure 4). Fermentation on coffee grinds did not appear to produce either of **8** or **9**. Quantification of the production of these compounds on potato peels showed that the strain was able to produce up to 371 mg·kg^{-1} dry weight of combined **8** and **9** (See ESI Figures S9.7 and S9.8) without further optimization. The orange peel fermentation grew very slowly and was incubated at 28 °C for four months. Extraction of this culture showed the production of 3-methylorcinaldehyde **8**, albeit at low titer (see ESI Figure S9.1).

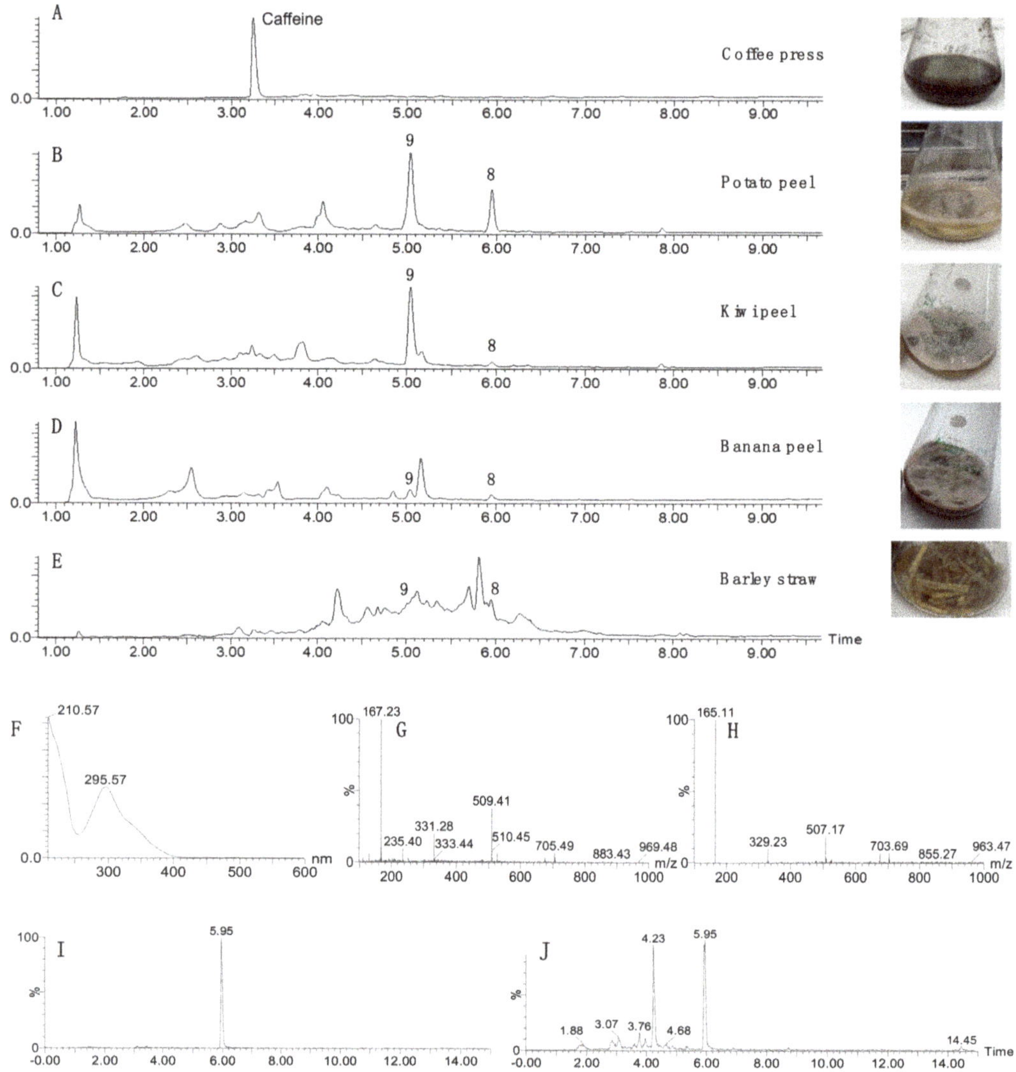

Figure 4. LC–MS traces of *T. reesei* QM6a·Δ*tmus53*·Δ*sorBC*·P_{pdc}·*aspks1* cultivated on different media after 14 days of growth. (**A**), grown on coffee press; (**B**), grown on potato peel; (**C**), grown on kiwi peel; (**D**), grown on banana peel; (**E**), grown on barley straw; (**F**), uv diode array detector (DAD) trace of 5.95 min peak; (**G**), ES+ spectrum of 5.95 min peak; (**H**), ES- spectrum of 5.95 min peak; (**I**), extracted ion chromatogram (EIC, ES- 165) for barley straw fermentation; (**J**), EIC chromatogram (ES+ 167) for barley straw fermentation.

3.4. Construction of a Multiple-Gene Expression System

We were next interested to test the ability of *T. reesei* to express more than one biosynthetic gene in parallel. Additional native promoters were therefore chosen to be inserted in the pTYGS vector. The strongest known constitutive promoters in *T. reesei* are those driving expression of cDNA1 (P_{cDNA1}) and enolase (P_{TReno}) [32]. Therefore, the two promoters, P_{cDNA1} and P_{TReno} were chosen to expand the vector pTYGS·*pyr4*·P_{pdc}.

P_{cDNA1} and P_{TReno} coding regions were amplified from *T. reesei* gDNA using primers with overhangs homologous to the pTYGS·*pyr4*·P_{pdc} vector backbone and to the terminators T_{adh} and T_{eno}, respectively. Inclusion of a *Swa*I restriction site downstream of P_{cDNA1} and P_{TReno} facilitated later gene insertion by yeast homologous recombination. The vector was then linearized using *Asc*I and the new vector was constructed by yeast homologous recombination between the linearized vector pTYGS·*pyr4*·P_{pdc}, the coding sequence of P_{cDNA1} and P_{TReno} and PCR-derived patches to repair unused *Asc*I restriction sites. The resultant vector was confirmed by PCR and sequencing and named pTYGS·*pyr4*·P_{pdc}·P_{cDNA1}·P_{TReno} (Figure 5).

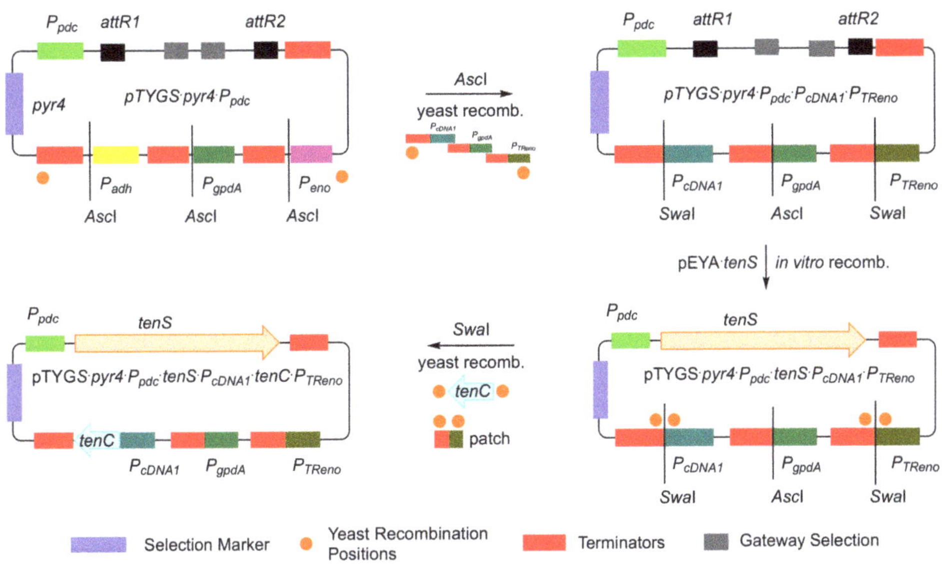

Figure 5. Construction of pTYGS·*pyr4*·P_{pdc}·P_{cDNA1}·P_{Treno} and pTYGS·*pyr4*·P_{pdc2}·*tenS*·P_{cDNA1}·*tenC*·P_{Treno}.

To test the ability of the pTYGS·*pyr4*·P_{pdc}·P_{cDNA1}·P_{TReno} system, a two-gene biosynthetic pathway, was selected. The well-studied tenellin BGC was chosen as it requires the cooperation of a PKS-NRPS encoded by *tenS* with a *trans*-acting enoyl reductase encoded by *tenC* [33]. Therefore, a new vector (pTYGS·*pyr4*·P_{pdc}·*tenS*·P_{cDNA1}·*tenC*·P_{TReno}) was constructed with *tenS* under the control of P_{pdc} and *tenC* driven by P_{cDNA1} (Figure 5). Transformation of *T. reesei* QM6a Δ*tmus53*·Δ*pyr4*·Δ*sorBC* with pTYGS·*pyr4*·P_{pdc2}·*tenS*·P_{cDNA1}·*tenC*·P_{TReno} resulted in the production of nine transformants. The transformants were selected three times on minimal media and finally transferred onto PDA plates. Five out of the nine transformants were cultivated in DPY + 1% glucose for three days followed by extraction with EtOAc and LC–MS analysis. This showed the production of the expected natural product pretenellin A **10** in all the tested transformants (Figure 6A) [34].

PCR analysis of the gDNA of four different *T. reesei* QM6a·Δ*tmus53*·Δ*sorBC*·P_{pdc}·*tenS*·P_{cDNA1}·*tenC* transformants showed the correct insertion of *tenS* and *tenC* in all the tested transformants (ESI Figure S10.3). The best-producing transformant was then cultivated on autoclaved banana peels to test the ability of the transformed strain to grow on waste materials and produce the expected compound. The LC–MS chromatogram of the transformant after 12 days of growth showed the production of the expected compound **10** (Figure 6B).

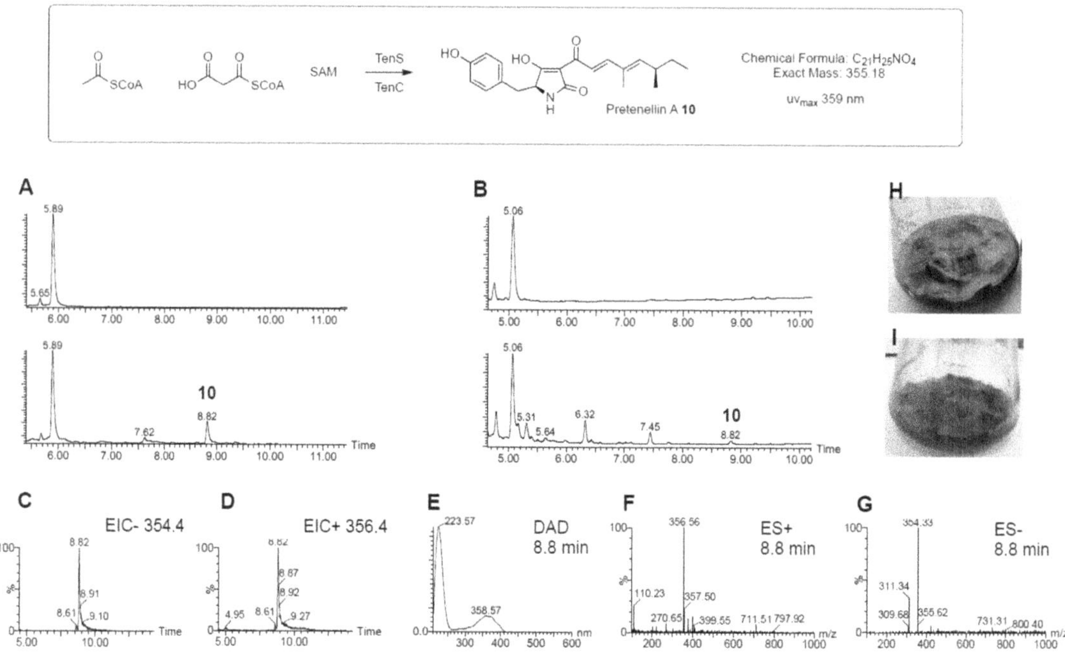

Figure 6. Production of pretenellin A **10**. (**A**), LC–MS traces of *T. reesei* QM6a·Δ*tmus53*·Δ*sorBC*·P_{pdc}·*tenS*·P_{cDNA1}·*tenC* transformant (lower) compared to *T. reesei* QM6a·Δ*tmus53*·Δ*pyr4* Δ*sorBC* strain (upper) on DPY+ 1% glucose after 3 days of cultivation; (**B**), DAD chromatogram of transformant *T. reesei* QM6a·Δ*tmus53*·Δ*sorBC*·P_{pdc}·*tenS*·P_{cDNA1}·*tenC* on banana peels (lower) in comparison to *T. reesei* QM6a·Δ*tmus53*·Δ*pyr4* Δ*sorBC* strain (upper); (**C**,**D**), extracted ion chromatograms (EIC) for the expected masses of pretenellin A **10** in the extract of *T. reesei* QM6a·Δ*tmus53*·Δ*sorBC*·P_{pdc}·*tenS*·P_{cDNA1}·*tenC* transformant on DPY + 1% glucose; (**E**), diode array detector (DAD) data for 8.8 min peak; (**F**), ES+ spectrum for 8.8 min peak; (**G**), ES- spectrum for 8.8 min peak; (**H**), growth of *T. reesei* QM6a·Δ*tmus53*·Δ*sorBC* on banana peel; (**I**), growth of *T. reesei* QM6a·Δ*tmus53*·Δ*sorBC*·P_{pdc}·*tenS*·P_{cDNA1}·*tenC* on banana peel.

4. Discussion

Due to its impressive ability to produce high amounts of cellulases, *T. reesei* has been developed as an effective heterologous host for protein production. Its safety and high production capacity are very appealing and have led to increased attention on *T. reesei* in recent decades, especially after the publication of its full genome sequence in 2008 [35]. Therefore, many toolkits for transforming *T. reesei* and for the expression of proteins and reporter genes have been developed [24,36,37]. These include the development of different auxotrophic strains to facilitate the transformation, identification of many native promoters to increase and control protein expression and different methods for transformation [19].

Despite this great interest in *T. reesei* as a heterologous host for protein production, almost no research has been performed on *T. reesei* as a heterologous host for secondary metabolite production. We aimed to test the suitability of *T. reesei* as a heterologous host for the expression of fungal genes to produce two model natural products. *T. reesei* is evidently able to effectively produce secondary metabolites, being a well-known producer of the sorbicillinoids [19,20,38–40]. Several other secondary metabolites are known including the polyketide trichodermatides A-D [41], and the cyclic tetrapeptide trichoderide A [42], and we have recently demonstrated that the organism harbours a fully functional ilicicolin H BCG that is silent under laboratory conditions [10].

The production of sorbicillinoids is known to hamper the production of secondary metabolites in industrial filamentous fungi. Sorbicillin production was previously eliminated in the beta-lactam producer *Penicillium chrysogenum* [43] as part of classical strain-improvement strategies. Knock out of the *T. reesei* sorbicillin BGC would result in several advantages. First, sorbicillin production likely consumes available acetyl and malonyl-CoA and cofactors that could be better directed towards production of desired metabolites. Secondly, the high levels of sorbicillinoids in the extract complicate the isolation of desired metabolites. Finally, recent studies have shown negative effects of sorbicillinoid production on growth, conidiation, cell wall integrity and cellulase production in *T. reesei* [2,44]. Therefore, a *T. reesei* strain lacking sorbicillin synthesis was desired. Previous attempts to knock out this gene cluster from *T. reesei* have been successful but they were achieved in the wild-type strain [19,20]. However, we required a strain possessing auxotrophic selection markers to allow the later selection of expressed biosynthetic genes.

Bipartite knock out of the *sorB*/*sorC* region was easily achieved and the resulting strain *T. reesei* QM6a·Δ*tmus53*·Δ*pyr4*·Δ*sorBC* strain could not produce sorbicillinoids under any tested conditions. However, some new compounds, including citreoisocoumarin **7**, were produced in observable amounts in this strain, possibly due to the increased availability of acetyl-CoA and other building blocks. Although this compound was not previously reported from *T. reesei*, it has been isolated from the sponge derived *Trichoderma* HPQJ-34 and is a common natural product isolated from other fungi and is likely to be a shunt from the alternariol biosynthetic pathway [45]. Bioinformatic analysis of all PKS genes from *T. reesei* [10] showed the presence of a gene encoding a PKS with high similarity (67%) to the nr-PKS (PkgA) from *A. nidulans*, which was reported to produce citreoisocoumarin and related compounds [46] and the PKS is closely related to snPKS19 from *Parastagonospora nodorum* that is known to synthesize alternariol [47].

The vector pTYGS·*pyr4*·P_{pdc} was then constructed from the well-known *Aspergillus* expression vector pTYGS·*argB*. It has all the advantages of the pTYGS vectors including the Gateway® in vitro recombination system and the ability to shuttle between yeast and *E. coli* to facilitate gene insertion by yeast homologous recombination. The *aspks1* gene was then used to test this system. *T. reesei* QM6a·Δ*tmus53*·Δ*pyr4*·Δ*sorBC* strains that were transformed with pTYGS·*pyr4*·P_{pdc}·*aspks1* were able to produce the expected compounds 3-methylorcinaldehyde **8** and 3-methylorsellinic acid **9** that is presumably derived by facile oxidation of the aldehyde. The same compounds were observed when *aspks1* was expressed in *A. oryzae* [29]. As expected, the new *T. reesei* transformant expressing *aspks1* was able to produce 3-methylorcinaldehyde **8** on different waste materials such as potato, banana, kiwi and orange peels and barley straw. However, the strain showed delayed growth on autoclaved fresh orange peels that might be attributed to the reported antifungal activity of essential oils of orange peels [48,49].

To expand the new system, additional native promoters of *T. reesei* were added to the vector using yeast homologous recombination. The new vector (pTYGS·*pyr4*·P_{pdc}·P_{cDNA1}·P_{TReno}) contains three native constitutive promoters (P_{TReno}, P_{cDNA1} and P_{pdc}) and one *A. nidulans* constitutive promoter (P_{gpdA}). Constitutive promoters were used instead of potentially stronger inducible promoters involved in cellulase expression to allow for constitutive production of the produced compounds under different cultivation conditions. The constructed vector was used to express the megasynthetase PKS-NRPS gene (*tenS*) together with its *trans*-acting ER (*tenC*) in *T. reesei* by adding the *tenS* gene under control of P_{pdc} and *tenC* gene under the control of P_{cDNA1}. *T. reesei* transformants containing this construct were able to produce the expected pretenellin A **10** on different media and on banana peels.

5. Conclusions

We have demonstrated that *T. reesei* can be engineered to produce fungal natural products using an adaptation of the highly successful *A. oryzae* pTY expression system developed by Lazarus and coworkers [1,28]. Simple exchange of selection markers and promoters led to expression vectors functional in *T. reesei*. Entry vectors previously constructed for use in the *A. oryzae* system, or rapid recombination in yeast, can easily supply genes that are functionally expressed in *T. reesei*. Since *T. reesei* is easily grown on a range of waste materials, this research paves the way for use of *T. reesei* as a microbial cell factory for waste valorization. The ability of *T. reesei* as a host to grow on low-value and waste substrates offers an important advantage over host organisms such as *E. coli* and *S. cerevisiae*. Production of high-value secondary metabolites in *T. reesei* represents a promising, low-cost, fast, sustainable, and green alternative for synthetic chemistry in the production of secondary metabolites. Removal of sorbicillin metabolites from the expression host makes both analysis by chromatography and chromatographic purification of desired products simpler.

In this research, waste materials such as potato peel, kiwi and banana peel were used as a cultivation media for the producing strain without any pre-treatment. However, further experiments are required to enhance the system such as testing different pre-treatment methods to allow better and faster production of the expected natural products on waste. The promoters used in this work were constitutive promoters to allow the production of natural products under all cultivation conditions, and it may be that inducible promoters previously successfully used for cellulase production could be effectively exploited [35]. Therefore, future experiments would be further enhancement of the system by the construction of new vectors with stronger promoters such as the tunable cellulase promoters *cbh1* and *cbh2* [35]. These promoters could also allow the system to produce a higher concentration of the expected compounds on lignocellulosic waste and other biomass [50].

So far, we have shown that one-gene (e.g., *aspks1*) and two-gene (e.g., *tenS* + *tenC*) biosynthetic pathways can be expressed successfully in *T. reesei*. In the future, expansion of the numbers of genes expressed in parallel should enable more complex pathways to be investigated. Impressive work by Abe and coworkers in *A. oryzae* [51] has shown that up to 12 biosynthetic genes can be co-expressed in parallel to produce highly complex meroterpenoid toxins. It seems feasible that *T. reesei* could also be used to produce such complex metabolites directly from waste materials. Future work will inevitably explore an expansion of the preliminary results demonstrated here.

Supplementary Materials: The following are available online at https://www.mdpi.com/article/10.3390/jof8040355/s1: all experimental details including growth media, buffers, enzyme, and antibiotic solutions; bacterial and fungal strains; vectors and oligonucleotides; tabulated NMR data; NMR spectra; and HRMS data. References [25,28–30,32,33,45,52–62] are cited in Supplementary Materials.

Author Contributions: M.L.S., E.S. and R.J.C. designed this study. Experimental work was performed by M.L.S., assisted by M.A. This paper was written by R.J.C. and M.L.S., R.J.C. was responsible for providing physical resources. Funding was gained by M.L.S. (DAAD GERLS) and R.J.C. All authors have read and agreed to the published version of the manuscript.

Funding: M.L.S. thanks the German Academic Exchange Service (DAAD) and the Egyptian Ministry of Higher Education and Scientific Research (MHESR) for funding (GERLS programme, 2017, 57311832). This work was supported by DFG (CO 1328/2-1, INST 187/621-1, INST 187/686-1). The APC was funded by the Open Access Fund of the Leibniz Universität Hannover.

Institutional Review Board Statement: Not applicable.

Informed Consent Statement: Not applicable.

Data Availability Statement: Not applicable.

Acknowledgments: The authors thank Astrid Mach-Aigner for *T. reesei* QM6a·Δ*tmus53*·Δ*pyr4*, and Katharina Schmidt (LUH) for a sample of pretenellin A **10**.

Conflicts of Interest: The authors declare no conflict of interest. The funders had no role in the design of the study; in the collection, analyses, or interpretation of data; in the writing of the manuscript, or in the decision to publish the results.

References

1. Lazarus, C.M.; Williams, K.; Bailey, A.M. Reconstructing fungal natural product biosynthetic pathways. *Nat. Prod. Rep.* **2014**, *31*, 1339–1347. [CrossRef] [PubMed]
2. Shenouda, M.L.; Cox, R.J. Molecular methods unravel the biosynthetic potential of *Trichoderma* species. *RSC Adv.* **2021**, *11*, 3622–3635. [CrossRef]
3. Kahlert, L.; Schotte, C.; Cox, R.J. Total Mycosynthesis: Rational Bioconstruction and Bioengineering of Fungal Natural Products. *Synthesis* **2021**, *53*, 2381–2394. [CrossRef]
4. Yuet, K.P.; Khosla, C. Challenges and opportunities for engineering assembly-line polyketide biosynthesis in *Escherichia coli*. *Metab. Eng. Commun.* **2020**, *10*, e00106. [CrossRef] [PubMed]
5. Harvey, C.J.B.; Tang, M.; Schlecht, U.; Horecka, J.; Fischer, C.R.; Lin, H.-C.; Li, J.; Naughton, B.; Cherry, J.; Miranda, M.; et al. HEx: A Heterologous Expression Platform for the Discovery of Fungal Natural Products. *Sci. Adv.* **2018**, *4*, eaar5459. [CrossRef]
6. Chiang, Y.M.; Oakley, C.E.; Ahuja, M.; Entwistle, R.; Schultz, A.; Chang, S.L.; Sung, C.T.; Wang, C.C.C.; Oakley, B.R. An efficient system for heterologous expression of secondary metabolite genes in *Aspergillus nidulans*. *J. Am. Chem. Soc.* **2013**, *135*, 7720–7731. [CrossRef]
7. Vandova, G.A.; O'Brien, R.V.; Lowry, B.; Robbins, T.F.; Fischer, C.R.; Davis, R.W.; Khosla, C.; Harvey, C.J.; Hillenmeyer, M.E. Heterologous Expression of Diverse Propionyl-CoA Carboxylases Affects Polyketide Production in *Escherichia coli*. *J. Antibiot.* **2017**, *70*, 859–863. [CrossRef]
8. Feng, J.; Hauser, M.; Cox, R.J.; Skellam, E. Engineering *Aspergillus oryzae* for the Heterologous Expression of a Bacterial Modular Polyketide Synthase. *J. Fungi* **2021**, *7*, 1085. [CrossRef]
9. Song, Z.; Bakeer, W.; Marshall, J.W.; Yakasai, A.A.; Khalid, R.M.; Collemare, J.; Skellam, E.; Tharreau, D.; Lebrun, M.H.; Lazarus, C.M.; et al. Heterologous expression of the avirulence gene ACE1 from the fungal rice pathogen *Magnaporthe oryzae*. *Chem. Sci.* **2015**, *6*, 4837–4845. [CrossRef]
10. Xiang, P.; Ludwig-Radtke, L.; Yin, W.B.; Li, S.M. Isocoumarin formation by heterologous gene expression and modification by host enzymes. *Org. Biomol. Chem.* **2020**, *18*, 4946–4948. [CrossRef]
11. Shenouda, M.L.; Ambilika, M.; Cox, R.J. *Trichoderma reesei* Contains a Biosynthetic Gene Cluster That Encodes the Antifungal Agent Ilicicolin H. *J. Fungi* **2021**, *7*, 1034. [CrossRef] [PubMed]
12. Schuster, A.; Schmoll, M. Biology and biotechnology of *Trichoderma*. *Appl. Microbiol. Biotechnol.* **2010**, *87*, 787–799. [CrossRef] [PubMed]
13. Nevalainen, H.; Suominen, P.; Taimisto, K. On the safety of *Trichoderma reesei*. *J. Biotechnol.* **1994**, *37*, 193–200. [CrossRef]
14. Weiss, R.; Eischer, A.; Tadic, T.; Gritsch, S.M.; Ortner, M.; Prall, K.; Neunteufel, E.; Putz, R.F.; Guebitz, G.M.; Nyanhongo, G.S. Valorisation of slaughterhouse and deinking paper waste streams for the production of enzyme by *Trichoderma reesei*. *J. Clean. Prod.* **2020**, *275*, 122882. [CrossRef]
15. Wang, H.; Kaur, G.; Pensupa, N.; Uisan, K.; Du, C.; Yang, X.; Lin, C.S.K. Textile waste valorization using submerged filamentous fungal fermentation. *Process. Saf. Environ. Prot.* **2018**, *118*, 143–151. [CrossRef]
16. Lai, T.T.; Pham, T.T.H.; Adjallé, K.; Montplaisir, D.; Brouillette, F.; Barnabé, S. Production of *Trichoderma reesei* RUT C-30 Lignocellulolytic Enzymes Using Paper Sludge as Fermentation Substrate: An Approach for On-Site Manufacturing of Enzymes for Biorefineries. *Waste Biomass Valorization* **2016**, *8*, 1081–1088. [CrossRef]
17. Mandels, M.; Hontz, L.; Nystrom, J.; Lee, R.; Lynd, I. Enzymatic hydrolysis of waste cellulose. *Biotechnol. Bioeng.* **2010**, *105*, 3–25. [CrossRef]
18. Liu, P.; Lin, A.; Zhang, G.; Zhang, J.; Chen, Y.; Shen, T.; Zhao, J.; Wei, D.; Wang, W. Enhancement of Cellulase Production in Trichoderma Reesei RUT-C30 by Comparative Genomic Screening. *Microb. Cell Factories* **2019**, *18*, 81. [CrossRef]
19. Derntl, C.; Rassinger, A.; Srebotnik, E.; Mach, R.L.; Mach-Aigner, A.R. Identification of the main regulator responsible for synthesis of the typical yellow pigment produced by *Trichoderma reesei*. *Appl. Environ. Microbiol.* **2016**, *82*, 6247–6257. [CrossRef]
20. Derntl, C.; Guzmán-Chávez, F.; Mello-de-Sousa, T.M.; Busse, H.J.; Driessen, A.J.M.; Mach, R.L.; Mach-Aigner, A.R. In vivo study of the sorbicillinoid gene cluster in *Trichoderma reesei*. *Front. Microbiol.* **2017**, *8*, 2037. [CrossRef]
21. Kahlert, L.; Bassiony, E.F.; Cox, R.J.; Skellam, E.J. Diels–Alder Reactions During the Biosynthesis of Sorbicillinoids. *Angew. Chem.* **2020**, *132*, 5865–5871. [CrossRef]
22. Nielsen, M.L.; Albertsen, L.; Lettier, G.; Nielsen, J.B.; Mortensen, U.H. Efficient PCR-Based Gene Targeting with a Recyclable Marker for *Aspergillus nidulans*. *Fung. Genet. Biol.* **2006**, *43*, 54–64. [CrossRef] [PubMed]
23. Williams, K.; Szwalbe, A.J.; Mulholland, N.P.; Vincent, J.L.; Bailey, A.M.; Willis, C.L.; Simpson, T.J.; Cox, R.J. Heterologous Production of Fungal Maleidrides Reveals the Cryptic Cyclization Involved in Their Biosynthesis. *Angew. Chem.* **2016**, *55*, 6784–6788. [CrossRef] [PubMed]
24. Gritz, L.; Davies, J. Plasmid-encoded hygromycin B resistance: The sequence of hygromycin B phosphotransferase gene and its expression in *Escherichia coli* and *Saccharomyces cerevisiae*. *Gene* **1983**, *25*, 179–188. [CrossRef]
25. Derntl, C.; Kiesenhofer, D.P.; Mach, R.L.; Mach-Aigner, A.R. Novel Strategies for Genomic Manipulation of *Trichoderma reesei* with the Purpose of Strain Engineering. *Appl. Environ. Microbiol.* **2015**, *81*, 6314–6323. [CrossRef]
26. Watanabe, A.; Ono, Y.; Fujii, I.; Sankawa, U.; Mayorga, M.E.; Timberlake, W.E.; Ebizuka, Y. Product Identification of Polyketide Synthase Coded by *Aspergillus nidulans* WA Gene. *Tetrahedron Lett.* **1998**, *39*, 7733–7736. [CrossRef]

27. Nofiani, R.; de Mattos-Shipley, K.; Lebe, K.E.; Han, L.-C.; Iqbal, Z.; Bailey, A.M.; Willis, C.L.; Simpson, T.J.; Cox, R.J. Strobilurin biosynthesis in Basidiomycete fungi. *Nat. Commun.* **2018**, *9*, 3940. [CrossRef]
28. Schor, R.; Schotte, C.; Wibberg, D.; Kalinowski, J.; Cox, R.J. Three previously unrecognised classes of biosynthetic enzymes revealed during the production of xenovulene A. *Nat. Commun.* **2018**, *9*, 1963. [CrossRef]
29. Pahirulzaman, K.A.K.; Williams, K.; Lazarus, C.M. A Toolkit for Heterologous Expression of Metabolic Pathways in *Aspergillus oryzae*. *Methods Enzymol.* **2012**, *517*, 241–260. [CrossRef]
30. Bailey, A.M.; Cox, R.J.; Harley, K.; Lazarus, C.M.; Simpson, T.J.; Skellam, E. Characterisation of 3-methylorcinaldehyde synthase (MOS) in *Acremonium strictum*: First observation of a reductive release mechanism during polyketide biosynthesis. *Chem. Commun.* **2007**, 4053–4055. [CrossRef]
31. Penttilä, M.; Nevalainen, H.; Rättö, M.; Salminen, E.; Knowles, J. A versatile transformation system for the cellulolytic filamentous fungus *Trichoderma reesei*. *Gene* **1987**, *61*, 155–164. [CrossRef]
32. Li, J.; Wang, J.; Wang, S.; Xing, M.; Yu, S.; Liu, G. Achieving efficient protein expression in *Trichoderma reesei* by using strong constitutive promoters. *Microb. Cell Factories* **2012**, *11*, 84. [CrossRef] [PubMed]
33. Halo, L.M.; Marshall, J.W.; Yakasai, A.A.; Song, Z.; Butts, C.P.; Crump, M.P.; Heneghan, M.; Bailey, A.M.; Simpson, T.J.; Lazarus, C.M.; et al. Authentic Heterologous Expression of the Tenellin Iterative Polyketide Synthase Nonribosomal Peptide Synthetase Requires Coexpression with an Enoyl Reductase. *ChemBioChem* **2008**, *9*, 585–594. [CrossRef] [PubMed]
34. Heneghan, M.N.; Yakasai, A.A.; Halo, L.M.; Song, Z.; Bailey, A.M.; Simpson, T.J.; Cox, R.J.; Lazarus, C.M. First heterologous reconstruction of a complete functional fungal biosynthetic multigene cluster. *ChemBioChem* **2010**, *11*, 1508–1512. [CrossRef]
35. Martinez, D.; Berka, R.M.; Henrissat, B.; Saloheimo, M.; Arvas, M.; Baker, S.E.; Chapman, J.; Chertkov, O.; Coutinho, P.M.; Cullen, D.; et al. Genome sequencing and analysis of the biomass-degrading fungus *Trichoderma reesei* (syn. *Hypocrea jecorina*). *Nat. Biotechnol.* **2008**, *26*, 553–560. [CrossRef]
36. Fitz, E.; Wanka, F.; Seiboth, B. The promoter toolbox for recombinant gene expression in *Trichoderma reesei*. *Front. Bioeng. Biotechnol.* **2018**, *6*, 135. [CrossRef]
37. Jørgensen, M.S.; Skovlund, D.A.; Johannesen, P.F.; Mortensen, U.H. A novel platform for heterologous gene expression in *Trichoderma reesei* (Teleomorph *Hypocrea jecorina*). *Microb. Cell Factories* **2014**, *13*, 33. [CrossRef]
38. Marra, R.; Nicoletti, R.; Pagano, E.; DellaGreca, M.; Salvatore, M.M.; Borrelli, F.; Lombardi, N.; Vinale, F.; Woo, S.L.; Andolfi, A. Inhibitory effect of trichodermanone C, a sorbicillinoid produced by *Trichoderma citrinoviride* associated to the green alga *Cladophora sp.*, on nitrite production in LPS-stimulated macrophages. *Nat. Prod. Res.* **2018**, *33*, 3389–3397. [CrossRef]
39. Sugaya, K.; Koshino, H.; Hongo, Y.; Yasunaga, K.; Onose, J.; Yoshikawa, K.; Abe, N. The biosynthesis of sorbicillinoids in *Trichoderma* sp. USF-2690: Prospect for the existence of a common precursor to sorbicillinol and 5-epihydroxyvertinolide, a new sorbicillinoid member. *Tetrahedron Lett.* **2008**, *49*, 654–657. [CrossRef]
40. Druzhinina, I.S.; Kubicek, E.M.; Kubicek, C.P. Several steps of lateral gene transfer followed by events of "birth-and-death" evolution shaped a fungal sorbicillinoid biosynthetic gene cluster. *BMC Evol. Biol.* **2016**, *16*, 269. [CrossRef]
41. Sun, Y.; Tian, L.; Huang, J.; Ma, H.Y.; Zheng, Z.; Lv, A.L.; Yasukawa, K.; Pei, Y.H. Trichodermatides A-D, novel polyketides from the marine-derived fungus *Trichoderma reesei*. *Org. Lett.* **2008**, *10*, 393–396. [CrossRef] [PubMed]
42. Sun, Y.; Tian, L.; Huang, Y.F.; Sha, Y.; Pei, Y.H. A new cyclotetrapeptide from marine fungus *Trichoderma reesei*. *Pharmazie* **2006**, *61*, 809–810. [CrossRef] [PubMed]
43. Salo, O.; Guzmán-Chávez, F.; Ries, M.I.; Lankhorst, P.P.; Bovenberg, R.A.L.; Vreeken, R.J.; Driessen, A.J.M. Identification of a Polyketide Synthase Involved in Sorbicillin Biosynthesis by *Penicillium chrysogenum*. *Appl. Environ. Microbiol.* **2016**, *82*, 3971–3978. [CrossRef]
44. Zhang, W.; An, N.; Guo, J.; Wang, Z.; Meng, X.; Liu, W. Influences of genetically perturbing synthesis of the typical yellow pigment on conidiation, cell wall integrity, stress tolerance, and cellulase production in *Trichoderma reesei*. *J. Microbiol.* **2021**, *59*, 426–434. [CrossRef] [PubMed]
45. Fang, F.; Zhao, J.; Ding, L.; Huang, C.; Naman, C.B.; He, S.; Wu, B.; Zhu, P.; Luo, Q.; Gerwick, W.H.; et al. 5-Hydroxycyclopenicillone, a New β-Amyloid Fibrillization Inhibitor from a Sponge-Derived Fungus *Trichoderma* sp. HPQJ-34. *Mar. Drugs* **2017**, *15*, 260. [CrossRef] [PubMed]
46. Ahuja, M.; Chiang, Y.-M.; Chang, S.-L.; Praseuth, M.B.; Entwistle, R.; Sanchez, J.F.; Lo, H.-C.; Yeh, H.-H.; Oakley, B.R.; Wang, C.C.C. Illuminating the Diversity of Aromatic Polyketide Synthases in *Aspergillus nidulans*. *J. Am. Chem. Soc.* **2012**, *134*, 8212–8221. [CrossRef]
47. Chooi, Y.-H.; Muria-Gonzalez, M.; Mead, O.L.; Solomon, P.S. SnPKS19 Encodes the Polyketide Synthase for Alternariol Mycotoxin Biosynthesis in the Wheat Pathogen *Parastagonospora nodorum*. *Appl. Environ. Microbiol.* **2015**, *81*, 5309–5317. [CrossRef]
48. Velázquez-Nuñez, M.J.; Avila-Sosa, R.; Palou, E.; López-Malo, A. Antifungal activity of orange (*Citrus sinensis* var. Valencia) peel essential oil applied by direct addition or vapor contact. *Food Control* **2013**, *31*, 1–4. [CrossRef]
49. Caccioni, D.R.L.; Guizzardi, M.; Biondi, D.M.; Renda, A.; Ruberto, G. Relationship between volatile components of citrus fruit essential oils and antimicrobial action on *Penicillium digitatum* and *Penicillium italicum*. *Int. J. Food Microbiol.* **1998**, *43*, 73–79. [CrossRef]
50. Dashtban, M.; Schraft, H.; Syed, T.A.; Qin, W. Fungal Biodegradation and Enzymatic Modification of Lignin. *Int. J. Biochem. Mol. Biol.* **2010**, *1*, 36–50.

51. Matsuda, Y.; Wakimoto, T.; Mori, T.; Awakawa, T.; Abe, I. Complete Biosynthetic Pathway of Anditomin: Nature's Sophisticated Synthetic Route to a Complex Fungal Meroterpenoid. *J. Am. Chem. Soc.* **2014**, *136*, 15326–15336. [CrossRef] [PubMed]
52. Van Dijken, J.P.; Bauer, J.; Brambilla, L.; Duboc, P.; Francois, J.M.; Gancedo, C.; Giuseppin, M.L.F.; Heijnen, J.J.; Hoare, M.; Lange, H.C.; et al. An interlaboratory comparison of physiological and genetic properties of four Saccharomyces cerevisiae strains. *Enzyme Microb. Technol.* **2000**, *26*, 706–714. [CrossRef]
53. Yang, X.-L.; Friedrich, S.; Yin, S.; Piech, O.; Williams, K.; Simpson, T.J.; Cox, R.J. Molecular basis of methylation and chain-length programming in a fungal iterative highly reducing polyketide synthase. *Chem. Sci.* **2019**, *10*, 8478–8489. [CrossRef] [PubMed]
54. Gietz, R.D.; Schiestl, R.H. High-efficiency yeast transformation using the LiAc/SS carrier DNA/PEG method. *Nat. Protoc.* **2007**, *2*, 31–34. [CrossRef] [PubMed]
55. Gruber, F.; Visser, J.; Kubicek, C.P.; de Graaff, L.H. The development of a heterologous transformation system for the cellulolytic fungus *Trichoderma reesei* based on a pyrG-negative mutant strain. *Curr. Genet.* **1990**, *18*, 71–76. [CrossRef]
56. Dockrey, S.A.B.; Narayan, A.R.H. Photocatalytic Oxidative Dearomatization of Orcinaldehyde Derivatives. *Org. Lett.* **2020**, *22*, 3712–3716. [CrossRef]
57. Shang, Z.; Khalil, Z.; Li, L.; Salim, A.A.; Quezada, M.; Kalansuriya, P.; Capon, R.J. Roseopurpurins: Chemical Diversity Enhanced by Convergent Biosynthesis and Forward and Reverse Michael Additions. *Org. Lett.* **2016**, *18*, 4340–4343. [CrossRef]
58. Hantke, V.; Wang, C.; Skellam, E.J.; Cox, R.J. Function of pathway specific regulators in the ACE1 and pyrichalasin H biosynthetic gene clusters. *RSC Adv.* **2019**, *9*, 35797–35802. [CrossRef]
59. Uzbas, F.; Sezerman, U.; Hartl, L.; Kubicek, C.P.; Seiboth, B. A homologous production system for *Trichoderma reesei* secreted proteins in a cellulase-free background. *Appl. Microbiol. Biotechnol.* **2012**, *93*, 1601–1608. [CrossRef]
60. Liu, M.; Ohashi, M.; Hung, Y.-S.; Scherlach, K.; Watanabe, K.; Hertweck, C.; Tang, Y. AoiQ Catalyzes Geminal Dichlorination of 1,3-Diketone Natural Products. *J. Am. Chem. Soc.* **2021**, *143*, 7267–7271. [CrossRef]
61. Nielsen, K.F.; Månsson, M.; Rank, C.; Frisvad, J.C.; Larsen, T.O. Dereplication of Microbial Natural Products by LC-DAD-TOFMS. *J. Nat. Prod.* **2011**, *74*, 2338–2348. [CrossRef] [PubMed]
62. Nakari-Setala, T.; Penttila, M. Production of *Trichoderma reesei* cellulases on glucose-containing media. *Appl. Environ. Microbiol.* **1995**, *61*, 3650–3655. [CrossRef] [PubMed]

Journal of *Fungi*

MDPI

Article

Engineering *Aspergillus oryzae* for the Heterologous Expression of a Bacterial Modular Polyketide Synthase

Jin Feng [1], Maurice Hauser [1], Russell J. Cox [1,*] and Elizabeth Skellam [1,2,*]

1 Institute for Organic Chemistry and Biomolekular Wirkstoff Zentrum, Leibniz University Hannover, Schneiderberg 38, 30167 Hannover, Germany; jin.feng@oci.uni-hannover.de (J.F.); maurice.hauser@oci.uni-hannover.de (M.H.)
2 Department of Chemistry, BioDiscovery Institute, University of North Texas, 1155 Union Circle, Denton, TX 76201, USA
* Correspondence: russell.cox@oci.uni-hannover.de (R.J.C.); elizabeth.skellam@unt.edu (E.S.)

Abstract: Microbial natural products have had phenomenal success in drug discovery and development yet form distinct classes based on the origin of their native producer. Methods that enable metabolic engineers to combine the most useful features of the different classes of natural products may lead to molecules with enhanced biological activities. In this study, we modified the metabolism of the fungus *Aspergillus oryzae* to enable the synthesis of triketide lactone (TKL), the product of the modular polyketide synthase DEBS1-TE engineered from bacteria. We established (2S)-methylmalonyl-CoA biosynthesis via introducing a propionyl-CoA carboxylase complex (PCC); reassembled the 11.2 kb DEBS1-TE coding region from synthetic codon-optimized gene fragments using yeast recombination; introduced bacterial phosphopantetheinyltransferase SePptII; investigated propionyl-CoA synthesis and degradation pathways; and developed improved delivery of exogenous propionate. Depending on the conditions used titers of TKL ranged from <0.01–7.4 mg/L. In conclusion, we have demonstrated that *A. oryzae* can be used as an alternative host for the synthesis of polyketides from bacteria, even those that require toxic or non-native substrates. Our metabolically engineered *A. oryzae* may offer advantages over current heterologous platforms for producing valuable and complex natural products.

Keywords: propionyl-CoA metabolism; modular polyketide synthase; propionyl-CoA carboxylase; heterologous expression

Citation: Feng, J.; Hauser, M.; Cox, R.J.; Skellam, E. Engineering *Aspergillus oryzae* for the Heterologous Expression of a Bacterial Modular Polyketide Synthase. *J. Fungi* **2021**, *7*, 1085. https://doi.org/10.3390/jof7121085

Academic Editors: Tao Feng and Frank Surup

Received: 25 November 2021
Accepted: 14 December 2021
Published: 17 December 2021

1. Introduction

Recent estimates suggest that over 25% of drugs approved between 1981 and 2014 are natural products or directly derived from natural products [1]. The bacterial polyketide class of compounds constitutes a significant percentage of natural products used in clinical or agricultural settings. This class of compounds includes the antibiotic erythromycin B and the antifungal compound amphotericin B, among many others (Figure 1). Fungi are also prolific producers of valuable antibiotics such as the polyketides griseofulvin and strobilurin A. Industrial production of these compounds often relies on large-scale fermentation and semi-synthesis as the most efficient routes [2]. Thus, the development of new biosynthetic engineering platforms that can rationally manipulate and improve the production of these compounds is an important current goal.

Polyketides are produced in two phases: a polyketide synthase (PKS) first assembles a carbon skeleton that is later modified by tailoring enzymes that can introduce a high diversity of chemical functionalization. Type I PKS consist of large multifunctional proteins with individual functional domains that are covalently linked. Type I PKS are further sub-divided into iterative PKS (iPKS) and modular PKS (mPKS, Scheme 1). iPKS consists of a single module (Scheme 1A), where each catalytic domain typically accepts, extends, and processes different substrates during each cycle of chain extension. iPKS use the same set of functional domains repeatedly until chain extension is complete.

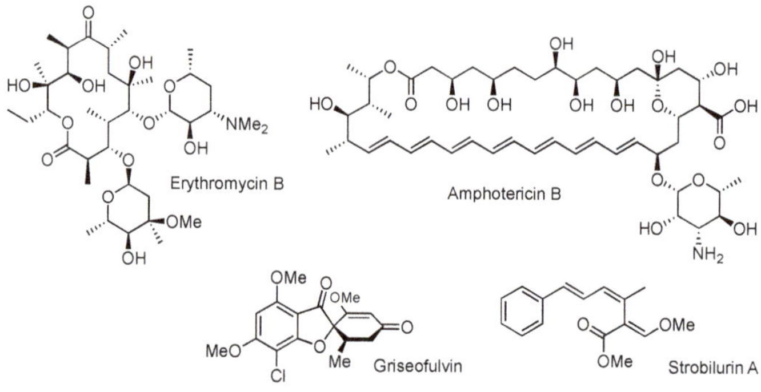

Figure 1. Examples of bioactive polyketide natural products from bacteria and fungi.

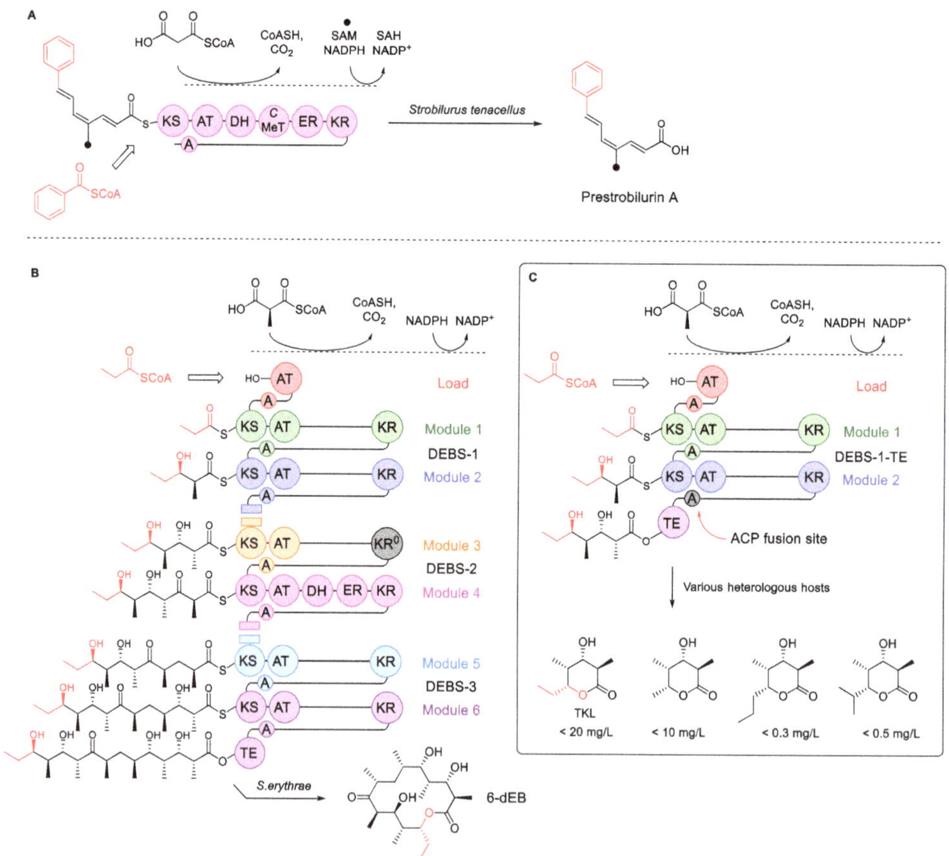

Scheme 1. Type I PKS. (**A**) The iterative PKS involved in the biosynthesis of prestrobilurin A; (**B**) 6-module DEBS system that synthesizes 6-dEB; (**C**) engineered DEBS1-TE that synthesizes triketide lactones (TKL). Abbreviations: KS, ketosynthase; AT, acyl transferase; A, acyl carrier protein (ACP); DH, dehydratase; C-MeT, C-methyl transferase; ER, enoyl reductase; KR, keto-reductase; KR^0-inactive KR domain; TE, thiolesterase. Diagrams inspired by Keatinge-Clay [3].

In contrast, mPKS consists of multiple modules. Each module typically accepts, extends, and processes a single substrate, and passes it to the next module for further processing. Type I iPKS are typical of fungi but are also known in bacteria, while type I mPKS are almost exclusively limited to bacteria.

Heterologous expression has emerged as a highly successful strategy for the production and engineering of microbial natural products in both bacteria and fungi [4]. Heterologous expression systems usually rely on host strains that are phylogenetically close to the producing organism to increase the likelihood that native transcriptional elements function properly, promoters remain functional, translation is efficient, the resulting proteins fold correctly, and codon usage is relatively conserved. In addition, post-translational modification processes, such as phosphopantetheinylation of acyl carrier proteins (ACP), must remain effective. Phylogenetically diverse organisms are also used as heterologous expression hosts for complex polyketides. Such organisms include *Escherichia coli* [5] and *Saccharomyces cerevisiae* [6,7], however, extensive host-engineering is often required, and titers can be low [8].

Aspergillus oryzae is a filamentous fungus commonly used in the fermentation industry for the production of sake, miso, and soy sauce from rice [9]. It has a generally regarded as safe (GRAS) status and is easily cultured in the lab, amenable to genetic modification using a variety of techniques, and is known for its high protein production. Recently, *A. oryzae* has proven itself to be an extremely capable host for heterologous expression of complete and partial fungal biosynthetic gene clusters (BGC), and for their systematic engineering [10–17]. *A. oryzae* does not produce significant amounts of its own secondary metabolites, meaning it is a clean host allowing facile detection and purification of heterologously produced compounds. Moreover, there are a number of examples where heterologous expression of a specific pathway in *A. oryzae* has exceeded the titer of the natural product reported in the native producer [18,19]. Advantages of fungi over other heterologous hosts are: the use of monocistronic operons, meaning separate genes can be individually controlled; numerous cellular compartments, where selective reactions may be contained to overcome toxicity; and tailoring enzymes that catalyze diverse and unique chemical modifications not accessible to other organisms [20]. However, *A. oryzae* has not yet been explored as a host for the production of bacterial polyketides. In particular, it is notable that reports of the presence of modular PKS in fungi are extremely rare [21]. However, successful expression of modular PKS in fungi could open up significant opportunities for the production of hybrid bacterial-fungal compounds unknown in nature, with potentially unique properties. In addition, fungi are capable of numerous oxidative tailoring modifications unknown in bacteria, offering further possibilities for the generation of novel compounds [22].

The biosynthesis of 6-deoxyerythronolide B (6-dEB) in *Saccharopolyspora erythraea* has long been used as a model mPKS system. Here, three large multimodular proteins (DEBS1, DEBS2, and DEBS3, Scheme 1B) work together to make the hexaketide macrolide 6-deoxyerythronolide B (6-dEB). DEBS1-TE is a well-studied simplified mPKS derived by fusing the C-terminal thiolesterase (TE) release domain of DEBS3 to the C-terminus of DEBS1 [23]. DEBS1-TE thus consists of a loading module and two extending modules that synthesize triketide lactone (TKL) from propionyl CoA and 2*S*-methylmalonyl CoA (Scheme 1C). The observed titers vary (0.5–20 $mg{\cdot}L^{-1}$) depending on the host used (Scheme 1) [24–27]. The acyltransferase (AT) domain within the DEBS1-TE loading module usually selects and activates propionyl-CoA, although other small acyl CoAs can be used if propionate is lacking.

A number of factors are likely to impede the use of fungal hosts for the successful expression of mPKS. For example, high concentrations of propionyl CoA are toxic to fungi through inhibition of crucial metabolic pathways [28,29] and 2*S*-methylmalonyl-CoA, required by mPKS extender modules, is not known to be synthesized by fungi. Additionally, actinobacterial genes are typically high GC% and it is unknown whether transcription and translation would be effective in *A. oryzae*. Furthermore, DEBS1-TE requires post-translational modification of the ACP domains by a phosphopantetheinyltransferase enzyme (PPTase), and it is unknown whether the native *A. oryzae* PPTase is capable of post-translationally modifying the ACP domains of mPKS.

In this study, we modified the metabolism of *A. oryzae* to enable the synthesis of TKL by the DEBS1-TE modular PKS. We established 2*S*-methylmalonyl-CoA biosynthesis *via* the introduction of a bacterial propionyl-CoA carboxylase complex (PCC); reassembled the 11.2 kb DEBS1-TE-coding region from synthetic codon-optimized gene fragments using rapid yeast recombination; introduced the bacterial phosphopantetheinyltransferase (PPTase) SePptII; investigated the propionyl-CoA synthesis and degradation pathways; and developed improved delivery of exogenous propionate. Overall, we demonstrated that *A. oryzae* can be used as an effective alternative host for the synthesis of bacterial polyketides that require toxic or non-native substrates, requiring minimal metabolic modification. *A. oryzae* may thus offer future advantages over other heterologous platforms for producing valuable and complex bacterial natural products.

2. Materials and Methods

See Supplementary Material for tables of media, strains, plasmids, and oligonucleotides.

2.1. Strains and Culture Conditions

One Shot^{TM} Top10 chemically competent *Escherichia coli*, One Shot^{TM} *ccdB* Survival^{TM} 2 T1^{R}-competent *E. coli*, and One Shot™ OmniMAX™ 2 T1^{R} *E. coli* were used as hosts for the construction of general plasmids and grown in LB medium with 50 mg·mL^{-1} of the antibiotic carbenicillin at an incubation temperature of 37 °C. *Saccharomyces cerevisiae* strain CEN.PK was used to prepare competent yeast cells in YPAD medium and then utilized as the host for the expression plasmid assembly by homologous recombination in SM-URA medium at 28 °C. *Aspergillus oryzae* NSAR1 was used as the heterologous host for fungal transformation and metabolite production in DPY medium at 28 °C.

2.2. Propionyl-CoA Toxicity Assessment to *A. oryzae NSAR1*

Methionine, isoleucine, arginine, and sodium propionate were individually dissolved in deionized water and sterilized. The prepared solutions were then mixed into DPY agar media at increasing concentrations of 0, 10 mM, 25 mM, 50 mM, and 100 mM, respectively. The same amount of *A. oryzae* spore suspension was inoculated on the face of each DPY plate bearing the supplementary compound. The culture plates were grown in the incubator at 28 °C. After 2, 4, and 7 days, the growth state of each sample was recorded.

2.3. PCR-Based Gene Identification

Fungal genomic DNA for PCR was extracted from mycelia of *A. oryzae* NSAR1 grown in DPY medium by using the GenElute plant genomic DNA kit (Sigma-Aldrich, Darmstadt, Germany). Using genomic DNA as the PCR template, a pair of specific 5′ and 3′ primers for gene identification were designed. PCR product purification was carried out by using the Wizard SV gel and PCR clean-up system (Promega, Madison, WI, USA). The PCR product was further identified by gene sequencing.

Total RNA was isolated from fresh 2-day-old cultured mycelia of *A. oryzae* NSAR1 grown in DPY medium by using the Quick-RNA™ Fungal/Bacterial Miniprep Kit according to the manufacturer's protocol (Zymo Research, Irvine, CA, USA). Single-stranded cDNA was prepared from total RNA by using the High-Capacity RNA-to-cDNA™ Kit (ThermoFisher, Waltham, MA, USA). Using the cDNA as a template, the PCR reaction with specific 5′ and 3′ primers for gene identification was carried out. For the gene identification under the condition of adding propionate, 50 mM of sodium propionate was supplemented into the *A. oryzae* culture one day before total RNA isolation.

2.4. Construction of *pTYGS·arg·pccABE*

The vector pTYGS·*arg* was fully digested using *Asc*I. All codon-optimized *pcc* gene fragments (http://genomes.urv.es/OPTIMIZER/, accessed on: 1 January 2020) flanked with overlaps (by *ca* 30 bp) were synthesized commercially. The plasmids bearing *pccABE* gene fragments were prepared by using double restriction enzymatic digestion (*pccA-EcoRI/BamHI*, *pccB-XbaI/BamHI*, *pccE-BamHI/XbaI*) for the plasmid reassembly in yeast. The competent *Saccharomyces cerevisiae* CEN.PK cells were removed from −80 °C storage and placed on ice to thaw. The following components were added to the yeast pellet in order: 240 μL PEG solution (50% (*w*/*v*) polyethylene glycol 3350); 36 μL LiAc (1 M); 50 μL denatured salmon testis DNA (2 mg·mL^{-1} in TE buffer) and 34 μL DNA fragments containing the linearized pTYGS·*arg* and three desired inserts in equimolar concentration (the uncut plasmid was used as the positive control and the linearized plasmid was used as the negative control). The mixture was resuspended and incubated for 50 min at 42 °C. Cells were pelleted by centrifugation at 11,000× *g* for 15 s and the supernatant was removed. The pellet was resuspended in 500 μL deionized H_2O, and 100 μL suspension was spread on selective SM-URA plates, which were incubated for 3 days at 28 °C.

Yeast colonies were gathered. Plasmids were extracted from yeast colonies by using the Zymoprep™ Yeast Plasmid Miniprep II kit (Zymo Research, Freiburg Germany). Subsequently, using standard heat-shock protocols, the extracted plasmid mixture was introduced into *E. coli ccdB* survival 2 $T1^R$ with the antibiotic carbenicillin (50 mg·mL^{-1}). Large amounts of *E. coli* colonies were generated, and then five colonies were picked up and identified by the colony PCR using specific 5′ and 3′ test primers. Plasmids were then extracted from positive colonies by using the Nucleospin® Extract Kit (Machery-Nagel, Düren, Germany) and identified using the double restriction enzymatic digestion (*Nhe*I+*Swa*I). The construct was checked by gene sequencing for correct construction.

2.5. Construction of pTYGS·*arg*·*debs1te*·*pccABE*

DEBS1-TE, as a non-native gene, was designed in silico using data from its original construction by Leadlay and coworkers [23]. The whole DEBS1-TE sequence (11.2 kb) was codon-optimized (http://genomes.urv.es/OPTIMIZER/, accessed on: 1 January 2020). Meanwhile, the original start codon GTG was replaced with ATG. For synthetic convenience, the entire DEBS1-TE sequence was designed as four fragments (~2.8 kb each). Each fragment was designed to overlap with adjacent fragments or the expression vector at both ends (by *ca* 30 bp) for the subsequent yeast recombination. Two different unique restriction enzymes were adhered to ends of each fragment (*debs1te-f1-Sca*I/*BamHI*; *debs1te-f2-Sma*I/*Hand*III; *debs1te-f3-EcoRV*/*Xba*I; *debs1te-f4-Dra*I/*EcoRI*).

The four DEBS1-TE fragments were synthesized commercially in four separate pUC57 vectors. All desired DEBS1-TE fragments and the linearized pEYA vector (fully digested by *Not*I) were prepared by restriction-enzyme digestion and then introduced into competent yeast for homologous recombination as described above. After two days of incubation, yeast colonies were collected. Plasmids were extracted from yeast colonies and immediately introduced into *E. coli* Top10 competent cells with the antibiotic kanamycin (50 mg·mL^{-1}). *E. coli* colonies were screened by PCR identification. In the resulting positive *E. coli* colonies, plasmids were extracted and then identified by restriction enzyme digestion. The empty vector pEYA was used as the control. The extracted plasmids were verified by the restriction enzyme digestion using *Sca*I. The construct pEYA-*debs1te* was finally confirmed by gene sequencing for no mutation.

The pEYA·*debs1te* construct was then used as the entry vector and pTYGS·*arg*·*pccABE* was used as the destination vector in a Gateway in vitro LR recombination reaction according to the manufacturer's instructions of the Gateway™ LR Clonase™ II Enzyme Mix Kit (Invitrogen). Transformation of the recombination mix into *E. coli* One Shot™ OmniMAX™ 2 $T1^R$ with the carbenicillin selection afforded the plasmid construction, which was then confirmed as pTYGS·*arg*·*debs1te*·*pccABE* by colony PCR and restriction enzyme digestion as described above. Lastly, the construct was checked by gene sequencing for correct construction.

2.6. PEG-Mediated Transformation of A. oryzae NSAR1

A. oryzae NSAR1 conidia from sporulating plates were inoculated into 50 mL GN medium and incubated overnight at 28 °C with 110 rpm shaking. Germinated conidia were collected and incubated in 10 mL of filter-sterilized lysing solution at room temperature with gentle mixing for 3 h. The protoplasts were released and filtered through a sterile Miracloth. The filtrate was centrifuged at 3000× *g* for 5 min to pellet the protoplasts which were resuspended in solution I. The plasmid pTYGS·*arg*·*debs1te*·*pccABE* (~1 µg) was added to 100 µL of the protoplast suspension and incubated on ice for 2 min. One milliliter of solution II was added and the transformation mixture was incubated at room temperature for 20 min. Five milliliters of molten CZD/S top medium with 0.8% agar was added to the transformation mixture and overlaid onto prepared CZD/S bottom medium plates with 1.50% agar. Plates were incubated at 28 °C for 3 days. For the introduction of co-expression plasmids pTYGS·*arg*·*debs1te-egfp*·*pccABE* and pTYGS·*met*·*sepptII*, the selection medium CZD/S without methionine was used in fungal transformation. After 3 days of incubation, the single transformant was transferred onto the new selection medium plate for 2 more rounds of screening. The pure transformants were inoculated onto the DPY plates for incubation at 28 °C for 5 days. Generated conidia were collected and transferred to the DPY liquid media for fermentation.

2.7. TKL Fermentation, Extraction and Analysis

A. oryzae transformants were fermented in the liquid DPY media (100 mL in 500 mL Erlenmeyer flask) at 28 °C for 5 days. Sodium propionate was added to a final concentration of 50 mM, either in one batch, in five equal batches over 5 days, or continuously over 5 days by a syringe pump. Cultures were grown and harvested. Cultures (100 mL) were homogenized by an electric blender. Solids were removed by vacuum filtration. The filtrate was extracted two times with EtOAc (100 mL). The combined organic extract was dried (anhydrous $MgSO_4$), filtered, and evaporated to dryness in vacuo and weighed. Methanol (1 mL) was added to dissolve the extract. The impurity was removed through spinning at 14,000× *g* for 5 min, and the sample was subjected to LCMS analysis.

Analytical LCMS data for 20 µL samples were obtained using a Waters LCMS system consisting of a Waters 2767 autosampler, Waters 2545 pump system, a Phenomenex Kinetex column (2.6 µm, C_{18}, 100 Å, 4.6 × 100 mm) equipped with a Phenomenex Security Guard precolumn (Luna, C_5, 300 Å) at a flow rate of 1 mL·min^{-1}. Detection was carried out by a diode array detector (Waters 2998) in the range 210 to 600 nm and an ELSD detector (Waters 2424) connected to a mass spectrometry, Waters SQD-2 mass detector, operating simultaneously in ES+ and ES- modes between 100 and 1000 *m/z*. The mobile phase was composed of HPLC-grade water mixed with 0.05% formic acid (solvent A) and HPLC-grade acetonitrile mixed with 0.045% formic acid (solvent B). A solvent gradient was run over 15 min starting at 10% B and ramping up to 90% B. In the case of the competition assay, a shallower gradient was applied ramping from 10 to 30% B in 15 min.

2.8. TKL Quantification

The synthetic TKL was dissolved in methanol and diluted to different concentrations from 2.5 µg·mL^{-1} to 200 µg·mL^{-1}. Each concentration was analyzed and the corresponding product peak on SIR (single ion recording) LCMS chromatogram was integrated. A calibration curve was made based on the concentrations and peak areas of the synthetic TKL. Then, a standard extraction workflow was determined. The transformant was inoculated in 100 mL DPY media. After 4 days of cultivation, 50 mM sodium propionate was added into the culture. The culture with propionate was incubated overnight. Next, the five-day culture was extracted with the same volume of ethyl acetate twice. The organic phase was gathered and evaporated to dryness. The concentrated extract was re-dissolved in 1.5 mL methanol. Subsequently, the extract suspension was diluted 10 times with methanol. The sample was centrifuged at 14,000× *g* for 5 min. Lastly, the supernatant was collected and analyzed by LCMS using the SIR detection method. The extract from

100 mL of the transformant culture was applied to the quantitative analysis based on the calibration curve. By calculation, the titer of the triketide lactone product TKL in *A. oryzae* was determined.

2.9. Gene Knockout of Degradation Pathways

Specific primers for *degradation gene* knockout cassette construct were designed. Primers were used to individually amplify the 5′ and 3′ termini (750 bp each) of *degradation gene* from *A. oryzae* genomic DNA. The promoter P_{trpC} and selection marker *argB* were also flanked by overlaps (30 bp) with *degradation gene* fragments by PCR. Then, all fragments were reassembled to the *Not*I-digested pEYA to make the *degradation gene* knockout cassette by yeast recombination. Subsequently, the bipartite gene knockout-based fungal transformation was carried out as previously published protocols. After transformant screening, single transformants were obtained followed by fluorescence imaging. Positive transformants were selected and subject to genomic DNA extraction. All genomic samples were identified by PCR using multiple specific primers.

3. Results

3.1. Investigating Propionyl-CoA Metabolism in A. oryzae

Propionyl-CoA is synthesized in fungi *via* the degradation of odd-chain fatty acids, amino acids such as methionine, isoleucine, and valine, or by thiolesterification of exogenous propionic acid (Scheme 2). However, sodium propionate is also a useful antifungal agent because high levels of propionyl-CoA are toxic to many fungi due to the inhibition of pyruvate dehydrogenase and succinyl-CoA synthetase [28,29]. This inhibition affects glucose and acetyl-CoA metabolism, respectively. In order to test the toxic effects of elevated levels of propionyl-CoA, we grew *A. oryzae* on dextrin-peptone-yeast Extract (DPY) agar supplemented with varying concentrations (0–100 mM) of methionine, isoleucine, and sodium propionate.

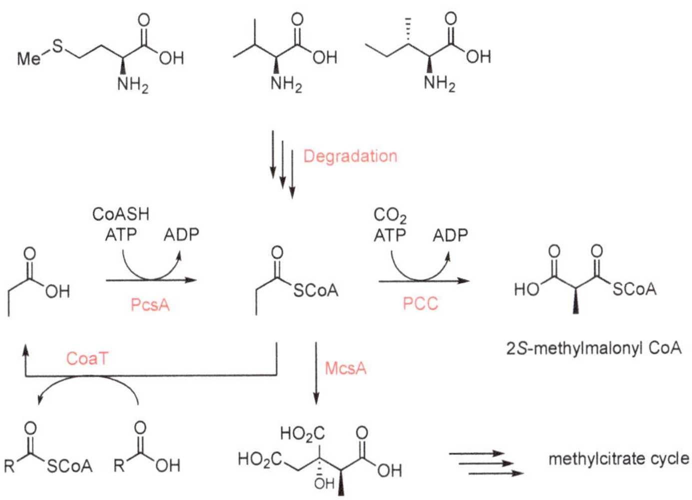

Scheme 2. Propionate metabolism relevant to this work. Abbreviations: PcsA, propionyl-CoA synthetase; PCC, propionyl-CoA carboxylase, McsA, methylcitrate synthase; CoaT, acyl-CoA transferase.

As a control, we also supplemented the medium with arginine, an amino acid not converted to propionyl-CoA. DPY is a rich medium often used with the pTYGS·*arg* heterologous expression system, developed specifically for *A. oryzae* NSAR1 [30].

All supplements inhibited the growth of *A. oryzae* NSAR1 at 100 mM, evident from as early as two days (Figure 2A–D). The control, arginine, had the least effect on the growth of

A. oryzae at any concentration, whereas methionine exhibited toxic effects at concentrations above 50 mM, isoleucine exhibited toxic effects at concentrations above 25 mM, and sodium propionate inhibited growth at 10 mM. At 25 mM, sodium propionate had a drastic effect on fungal growth and at concentrations above 50 mM sodium propionate completely abolished the ability of *A. oryzae* NSAR1 to grow on DPY. Since sodium propionate is the most direct precursor to propionyl-CoA and induces inhibitory effects at relatively low concentrations, we selected this to study propionyl-CoA metabolism in *A. oryzae* NSAR1.

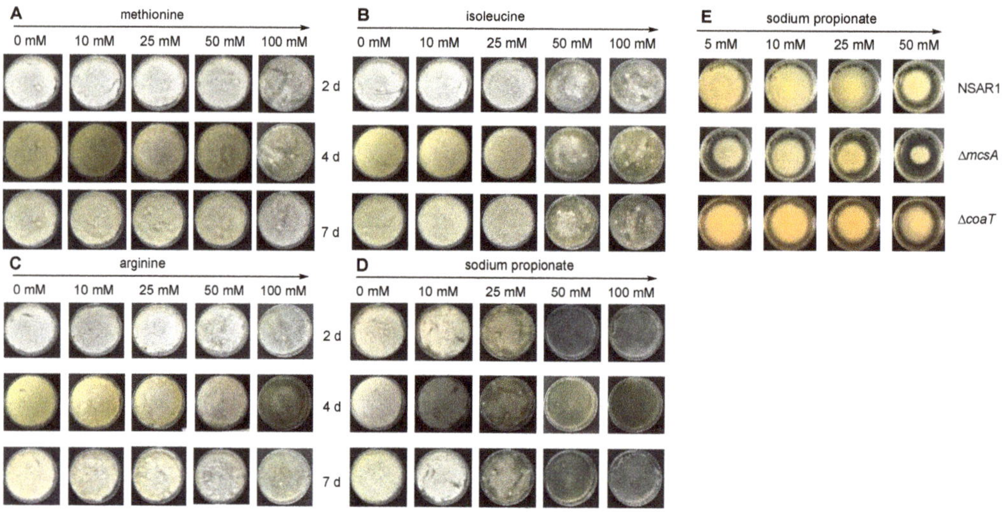

Figure 2. Susceptibility of *A. oryzae* NSAR1 to different amino acids and propionate at the indicated concentrations and times. (**A**) Methionine; (**B**), isoleucine; (**C**) arginine; (**D**) sodium propionate; (**E**) susceptibility of indicated mutants to indicated concentrations of sodium propionate after 5 days.

In *Aspergillus nidulans*, propionyl-CoA is synthesized from propionate, coenzyme A, and ATP by propionyl-CoA synthetase (PcsA), encoded by *pcsA* (Scheme 2) [31]. *A. oryzae* NSAR1 is a quadruple auxotroph (*argB−*, *niaD−*, *sC−*, and *adeA−*) derived from parental strain *A. oryzae* RIB40, the genome of which is fully sequenced. *A. oryzae* ORF XP_001826479 was identified as the closest homolog of *pcsA* by direct BLAST query of the *A. oryzae* RIB40 genome. In *Aspergillus nidulans*, two routes are known for the degradation of propionyl CoA. First, methylcitrate synthase (McsA) condenses propionyl-CoA with oxaloacetate to yield 2*S*,3*S*-methylcitrate (Scheme 2) [32]. Second, a broadly selective CoA-transferase (CoaT) transfers CoASH from various acyl thiolesters to abundant carboxylic acids [33]. Both McsA and CoaT are compartmentalized in mitochondria [32,33]. *A. oryzae* RIB40 ORFs XP_001817006 and XP_001817633 were identified as *mcsA* and *coaT* homologs, respectively.

To determine if *A. oryzae pcsA* is specifically induced by propionate, we conducted RT-PCR from cDNA. Transcription of *pcsA* was detected when sodium propionate was added at 50 mM but not when omitted from solid growth medium (see Electronic Supplementary Information [ESI]). Likewise, *mcsA* could only be amplified from cDNA when propionate was added to the culture. In contrast, *coaT* could be amplified from cDNA without the addition of exogenous propionate (ESI). These results confirm that PcsA and McsA are induced by propionate, but CoaT is active regardless of the presence of propionate.

3.2. *Introducing a Methylmalonyl-CoA Pathway into A. oryzae NSAR1*

Confident that propionyl-CoA metabolism is active in *A. oryzae* NSAR1, and that the propionyl-CoA starter unit would be available for DEBS1-TE, we turned our attention to the production of 2*S*-methylmalonyl-CoA. We opted to add the well-understood bacterial propionyl-CoA carboxylase (PCC) pathway to *A. oryzae*. PCC consists of α, β, and ε subunits: the ε-subunit interacts with the β-subunit to dramatically increase the specificity of the enzyme complex [34]. A recent study of various PCC systems in *E. coli* indicated that PCC from *Streptomyces coelicolor* is the most effective for increasing product titers by mPKS [35]. Typical of genes from actinobacteria, *S. coelicolor pccA*, *pccB*, and *pccE* exhibit higher GC content than typical *A. oryzae* genes (e.g., *pccA* is 74.6% GC; *A. oryzae* ORFs have a median of 47.4% GC) and demonstrate significantly different codon bias (ESI). Therefore, *pccA*, *pccB*, and *pccE* from *S. coelicolor* were synthesized using codon-optimized DNA and the start codon was adjusted for each from GTG to ATG preferred by fungi. However, despite codon preference being optimized to *A. oryzae*, the GC% remained high at around 70%.

The genes encoding the PCC α, β, and ε subunits were individually cloned into fungal expression vector pTYGS·*arg* [36], under the control of the amylose-inducible *A. oryzae amyB* promoter (P_{amyB}). In each case, these were fused in-frame at their 3′-termini with *egfp*. The resulting vectors were individually transformed into *A. oryzae* NSAR1 and after several rounds of screening on minimal media lacking arginine, selected transformants were induced in DPY media. Individual transformants were analyzed by fluorescence microscopy. Green fluorescence was detected in all cases (Figure 3), indicating that the high GC % genes are successfully transcribed and translated in *A. oryzae*. The diffuse fluorescence observed indicates that all three PCC components are expressed in the cytoplasm. However, RT-PCR for determining the expression of *pccA*, *B* and *E* was unsuccessful and we attribute this to the high GC content of the corresponding mRNA that may form complex secondary structures preventing successful amplification.

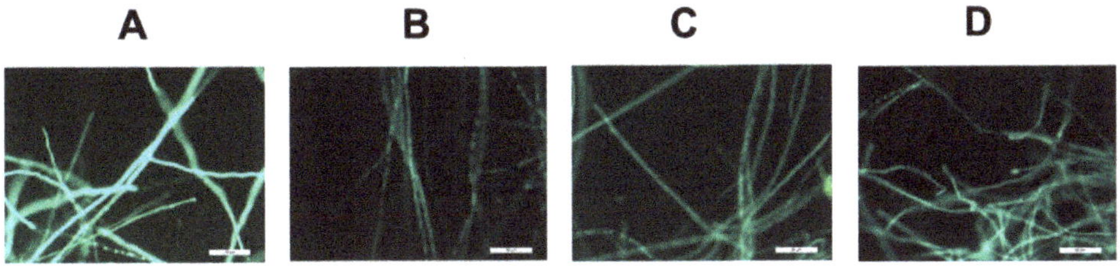

Figure 3. Green fluorescent imaging of: (**A–C**) transformants with PCC subunit genes fused to *eGFP*; (**D**) DEBS1-TE fused to eGFP. Scale bar 50 μm.

3.3. Heterologous Expression of DEBS1-TE in A. oryzae to form TKL

We replicated the original DEBS1-TE construction strategy[23] *in silico* and, based on this sequence, we designed four overlapping DNA fragments spanning the 11.2 kb coding region. The four fragments were codon-optimized for *A. oryzae* and the start codon adjusted to ATG. However, like the codon-optimized *pcc* genes, the overall GC content remained high at 73%. The four fragments were re-assembled using rapid yeast recombination downstream of P_{amyB} in pTYGS·*arg*, fused in-frame with 3′ *egfp*, and then transformed into *A. oryzae* NSAR1. The resulting transformants were inspected microscopically after induction in DPY medium and diffuse green fluorescence was detected (Figure 3D) indicative of successful cytoplasmic expression of the DEBS1-TE protein.

We then built a single expression vector to co-express DEBS1-TE with the PCC components (Figure 4A). This time the eGFP fusions were omitted for the *pcc* components since we had established that they could be transcribed. The *debs1te-egfp* gene was placed under the control of P_{amyB}, while the PCC components were driven by strong constitutive promoters known to be active in *A. oryzae* [36]. The transformants were cultured for 5 days and propionate (50 mM) was added the day before extraction. The cultures were extracted with ethylacetate and screened by LCMS. No new metabolites were detected in the transformant extracts (ESI). We performed single-ion monitoring for the expected mass of TKL (*m/z* 172) but no new metabolites were detected. Since it is known that DEBS1-TE is also able to accept acetate starter units we also scanned for metabolites with m/z = 158 but could not detect any.

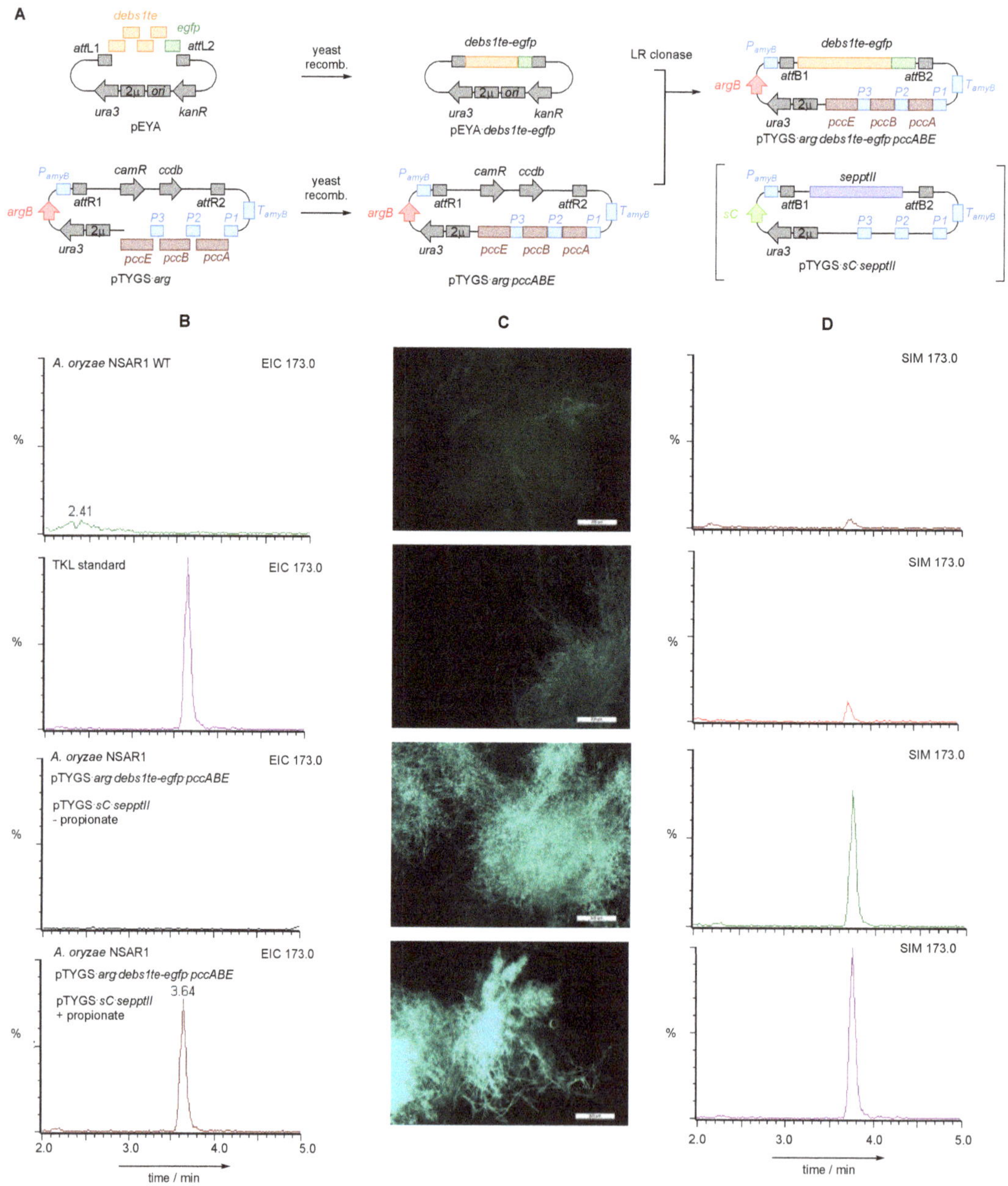

Figure 4. Construction of functional modular PKS in *A. oryzae*. (**A**) Construction of vectors; (**B**) LCMS analysis of *A. oryzae* NSAR1, standard TKL, and transformants in the absence and presence of propionate as indicated ES+ at m/z 173.0; (**C**) microscopic images of four *A. oryzae* NSAR1 transformants showing variability of eGFP intensity; (**D**) quantitative selective ion monitoring (SIM, ES+ at m/z 173.0) of extracts of the corresponding *A. oryzae* strains shown in panel C. Scale bars 200 µm.

We considered the most likely reason for the inactivity of the DEBS1-TE system to be the lack of phosphopantetheinylation of its three ACP domains. The *A. oryzae* PPTase NpgA homolog shares very low (7–14%) sequence identity with bacterial PPTases such as Sfp from *Bacillus subtilis* and SePptII from the erythromycin-producing strain *S. erythraea* [23,37], despite being in the same structural class (Figure S6.3) [38]. Both Sfp and SePptII are known to activate DEBS1-TE but their activities in *A. oryzae* are unknown [24,39,40].

The gene encoding SePptII from *S. erythraea* was codon optimized, synthesized, and cloned into a pTYGS vector containing an *sC* selection marker to give pTYGS·*sC*·*sepptII* [36]. The *sC* gene encodes sulfate adenyl transferase, part of the sulfur assimilation pathway of *A. oryzae*, and is normally complemented by the addition of methionine. pTYGS·*sC*·*sepptII* was then co-transformed into *A. oryzae* with pTYGS·*arg*·*pccABE*·*debs1te-egfp*. Transformants that grew on minimal media lacking arginine and methionine and that also exhibited fluorescence in inducing DPY media were selected for fermentation and chemical analysis (Figure 4). Propionate (50 mM) was added to the cultures 24 h before harvest. Cultures were extracted with ethylacetate and analyzed by LCMS. The EIC chromatogram of $[M + H]^+$ 173, corresponding to TKL, showed a single new distinct peak not present in control strains. The *m/z* 173 peak was only detected when exogenous propionate was added to the culture (Figure 4B). The peak corresponded to standard synthetic TKL in terms of retention time, UV absorption, and mass fragmentation. For further clarification synthetic TKL was mixed 1:1 with the crude extract of a transformant and analyzed by LCMS, confirming precise co-elution (see ESI). Examination of randomly selected cotransformants showed a variation in the intensity of the detected eGFP fluorescence (Figure 4C). Quantitative analysis of the 173 peak for these transformants using selected ion monitoring (SIM) showed that increased fluorescence is correlated with increased production of TKL (Figure 4C,D). Synthetic TKL was then used to generate a calibration curve (see ESI) and TKL production was quantified at 0.6 $mg \cdot L^{-1}$ under these conditions from the most productive transformant.

3.4. Improving Titers of TKL through Metabolic Engineering of *A. oryzae* Primary Metabolism and Sodium Propionate Feeding Methods

A series of experiments were then devised in an attempt to optimize TKL production. The effects of different exogenous propionate feeding regimes were investigated for their effect on TKL production. Propionate was added at different concentrations (5–50 mM), at different time points, and in single batch-fed, pulse-fed, or continuously fed systems (Tables 1 and 2, Figure 5). Titers of TKL were increased to 3.6 $mg \cdot L^{-1}$ by pulse-feeding 10 mM of sodium propionate to shake flasks at 24 h intervals over five days before extraction. Batch feeding a single high concentration of propionate or continual infusion of lower concentrations over longer periods was ineffective (Figure 5).

Table 1. Description of sodium propionate feeding regimes.

	Time after Inoculation (Days)/Sodium Propionate Inoculation Concentration (mM)										
Condition	1	2	3	4	5	6	7	8	9	10	11
A	-	-	-	50	Extract						
B	12.5	12.5	12.5	12.5	Extract						
C	50 mM infusion with pump				Extract						
D	10	10	10	10	10	Extract					
E	50 mM infusion with pump					Extract					
F	6.25	6.25	6.25	6.25	6.25	6.25	6.25	6.25	Extract		
G	5	5	5	5	5	5	5	5	5	5	Extract
H	2	2	2	2	2	Extract					
I	10 mM infusion with pump					Extract					

Table 2. Description of different feeding experiments perfomed on various *A. oryzae* host strains.

Experiment ID	*A. oryzae* Host	Expression Vectors	Condition
1	NSAR1	pTYGS-*arg-pccABE-debs1te-egfp* + pTYGS-*sC-sepptII*	A
2			B
3			C
4			D
5			E
6			F
7			G
8			H
9			I
10	NSAR1	pTYGS-*arg-pccABE-debs1te* + pTYGS-*sC-sepptII*	A
11			D
12	NSAR1	pTYGS-*arg-pccABE-debs1te-egfp* + pTYGS-*sC-sepptII-pcsA*	A
13			D
14	NSAR1 Δ*mcsA*	pTYGS-*arg-pccABE-debs1te-egfp* + pTYGS-*sC-sepptII*	D

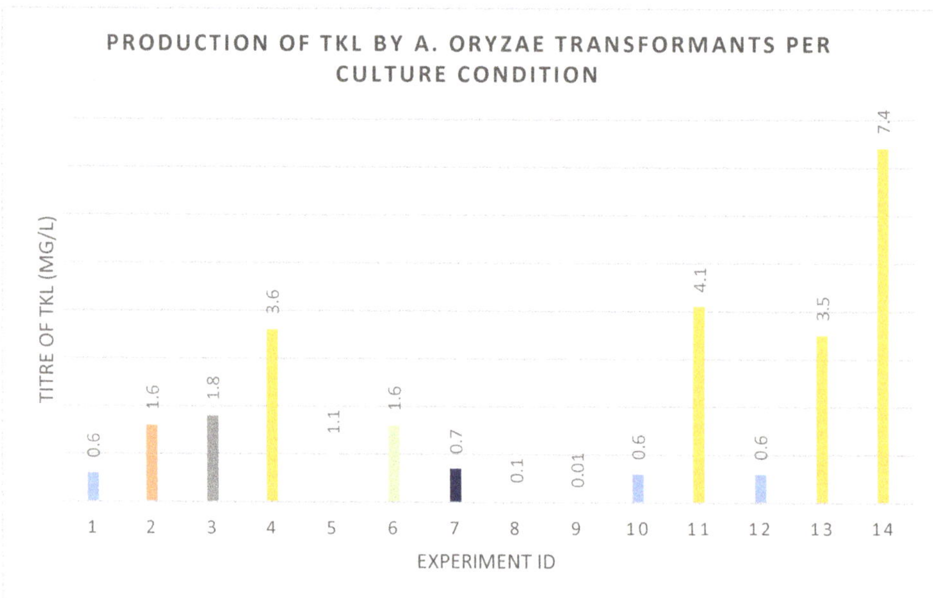

Figure 5. Optimization of TKL production and resulting titers of TKL. Culture condition and experiment ID described in Tables 1 and 2.

To determine whether the eGFP tag was interfering with the activity of DEBS1-TE, a new vector omitting eGFP was built (pTYGS·*arg*·*pccABE*·*debs1te*) and co-expressed with pTYGS·*sC*·*sepptII*. TKL titers remained at 0.6 mg·L^{-1} when strains were batch fed 50 mM propionate (Figure 5) but could be increased to 4.1 mg·L^{-1} when pulse fed with 10 mM propionate over five days.

We next wanted to determine whether propionyl-CoA synthesis was a rate-limiting factor. In an effort to ensure adequate supplies of propionyl-CoA, we over-expressed *pcsA*, encoding propionyl-CoA synthetase, while co-expressing DEBS1-TE, the PCC components, and SePptII. This was achieved by inserting a genomic copy of *A. oryzae pcsA* into the vector pTYGS·*sC*·*sepptII* under the control of a constitutive *Aspergillus* promoter. However, TKL titers remained at 0.6 mg·L^{-1} when strains were batch fed 50 mM propionate (Figure 5). Titers of TKL could be increased to 3.5 mg·L^{-1} through pulse feeding propionate at 10 mM over five days (Figure 5).

The MCS pathway is proposed as being the major pathway for the degradation of propionyl-CoA in fungi [41]. We inactivated *mcsA* in *A. oryzae* NSAR1 using the bipartite gene knockout strategy [42] by inserting *argB* as the selection marker into the target gene. Transformants were selected on minimal media lacking arginine and disruption was confirmed by PCR. The *mcsA* mutants were investigated for sensitivity to propionate. Compared to *A. oryzae* NSAR1, *mscA* mutants exhibited increased sensitivity to propionate, with observable growth inhibition at concentrations as low as 5 mM (Figure 2E).

We then co-transformed pTYGS·*argB*·*pccABE*·*debs1te-egfp* and pTYGS·*sC*·*sepptII* into the *mcsA* mutant using *sC* selection and screened for eGFP fluorescence. TKL production remained low under all feeding regimes, except when propionate was pulse-fed (5 × 10 mM aliquots, Figure 5). Under these conditions, the titer was raised to 7.4 mg·L^{-1}. In a separate series of experiments, we also knocked out *coaT* in *A. oryzae* NSAR1 using the same methods. In this case, the KO strains did not exhibit a significantly different response to exogenous propionate as the *A. oryzae* NSAR control strain. Attempts were made to transform *A. oryzae* Δ*coaT* with pTYGS·*argB*·*pccABE*·*debs1te-egfp* and pTYGS·*sC*·*sepptII*, but transformants could not be prepared. We also attempted a double knockout to disrupt both *coaT* and *mscA* using either mutant strain as the parent strain. However, we were unable to produce the desired double mutants.

4. Discussion

A. oryzae is a highly useful host organism for the heterologous expression and engineering of fungal secondary metabolites. In particular, it has been extensively used for the production of fungal metabolites stemming from iterative type I PKS that use predominantly acetyl-CoA and malonyl-CoA for their construction [17]. Bacterial modular PKS, on the other hand, often use propionyl-CoA and methylmalonyl-CoA for the construction of polyketides [23–27]. The fact that propionyl-CoA is toxic to fungi, and that methylmalonyl-CoA pathways do not exist in fungi, may explain why modular PKS systems are almost unknown in these organisms.

Our results show that propionate induces PcsA and McsA in *A. oryzae*, consistent with the creation and degradation of propionyl-CoA *via* the methyl citrate cycle. The broadly selective CoaT system also seems to be constitutively active, but this system does not actually degrade the carbon skeleton of propionate and is therefore a less effective resistance mechanism. However, we assume that propionyl-CoA can be produced as a potential feedstock for modular PKS activity.

Experiments in which we fused the *S. coelicolor* PCC genes and the *S. erythraea debs1te* to *egfp* showed that all systems produced apparently full-length proteins in *A. oryzae*, showing that bacterial genes with GC% as high as 70% can be successfully transcribed and translated. Although the genes were codon optimized to *A. oryzae*, they were, for the most part, unevenly skewed towards the highest GC content codons. This adversely affected our ability to confirm transcription by RT-PCR and therefore may indicate that translation of DEBS1-TE is still inefficient. Heterologous production of non-fungal genes in *A. oryzae* has determined that codon optimization increases steady-state mRNA levels [43]. Even so, translation was possible as evidenced by GFP detection and TKL production. It will be interesting to observe whether native actinobacterial sequences will also be successfully transcribed and translated in *A. oryzae*.

Removal of the *egfp* tags and expression of *debs1te* with *pccA*, *pccB* and *pccE*, however, was not sufficient for the production of TKL in *A. oryzae*. It is common to observe that the post-translational modification of PKS ACP domains is inefficient in heterologous hosts [44]. We, therefore, co-expressed the bacterial PPTase SePptII in parallel. Under these conditions, we observed effective TKL production. Compared to other hosts used to express mPKS that lack the key acyl-CoA metabolic pathways, *A. oryzae* performs better in synthesizing TKL [39,45,46]. (Scheme 2 and Figure 5). Considering that DEBS1-TE is a chimeric mPKS, possibly lacking correct protein-protein interactions at the fusion site, we may expect to see improved titers for native mPKS systems, as observed when all three DEBS proteins were expressed in engineered *E. coli* [24].

Experiments in which all genes were fused with *egfp* indicate that DEBS1-TE and the PCC components are present in the cytoplasm of *A. oryzae*, as is PcsA [31], suggesting that propionyl-CoA should be directly available to DEBS1-TE and PCC. In contrast, studies by Brock et al. have determined that the propionyl-CoA degradation pathways are confined to the mitochondria. The MCS degradation pathway has only been functionally investigated in a few species of fungi including *Aspergillus* sp. [47], *Fusarium* sp. [48], *Cochliobolus* sp., *Trichoderma atroviride* [49], *Gibberella zeae* [50], and *Pyricularia oryzae* [51]. Here, we report the first attempts to engineer the MCS pathway with the specific aim of enabling high concentrations of propionyl-CoA for use by DEBS1-TE and PCC. Knockout of *mcsA* led to strains that were more sensitive to propionate but could produce higher titers of TKL (up to 7.4 $mg{\cdot}L^{-1}$) under pulse-feeding conditions. These results suggest that the MCS pathway is probably the major route for the degradation of propionyl CoA in *A. oryzae*.

A recent proteomic study of propionyl-CoA metabolism in the pathogenic fungus *Paracoccidioides lutzii* demonstrated that a carnitine *O*-acetyltransferase (PAAG_06224) is upregulated in response to propionate [52]. This study also identified an acetate kinase (PAAG_07180), that was only detected when *P. lutzii* was grown on propionate, which is homologous to propionate kinases in bacteria. In the bacterium *Neisseria meningitidis*, propionate kinases generate propionyl phosphate that is proposed to be converted to propionyl-CoA via phosphotransacetylase [53]. *A. oryzae* NSAR1 has homologs of both the carnitine *O*-acetyltransferase (XP_001821862.3) and putative propionate kinase (XP_02309003) with 71% and 59% similarity, respectively. Additionally, *Candida albicans* has been shown to use a modified β-oxidation pathway to degrade propionyl-CoA to acetyl-CoA [54]. Future work, therefore, could focus on characterizing these hypothetical propionyl-CoA metabolic pathways in *A. oryzae* to provide adequate substrates for PCC and DEBS1-TE.

Continuous infusion of propionate does not result in increased titers of TKL, presumably because low propionyl CoA concentrations do not provide adequate methylmalonyl-CoA for TKL construction because propionate is cleared from the system quickly by McsA and CoaT. Similarly, feeding propionate in a single batch is not effective, presumably because propionyl-CoA concentrations do not rise high enough for a long enough period. However, pulse feeding of several high concentrations of propionate consistently gave the highest titers of TKL, particularly in Δ*mcsA* strains. Our interpretation is that high concentrations of propionate temporarily overwhelm the ability of the cells to degrade propionyl-CoA, allowing a brief temporal window when PCC can create sufficient 2*S*-methylmalonyl CoA for TKL synthesis. This suggests that the system could be more productive if elevated 2*S*-methylmalonyl CoA concentrations could be maintained over longer periods. Future investigations will examine the production of this metabolite *via* isomerization of succinyl-CoA by methylmalonyl-CoA mutase. This enzyme creates 2*R*-methylmalonyl-CoA, but rapid epimerization can be catalyzed by methylmalonyl-CoA epimerase and has previously been used in *E. coli* for this purpose [55]. This system could therefore bypass the requirement for high propionate concentrations and provide 2*S*-methylmalonyl-CoA more consistently.

Recombination of synthetic gene fragments in yeast was used to rapidly build the 11.2 kb DEBS1-TE. We previously demonstrated that yeast recombination offers an excellent method for the reconstruction of large modular synthetases (e.g., a five module non-ribosomal peptide synthetase, NRPS) [56] as well as being ideally suited to the precise engineering of PKS. For example, we have achieved precise swaps of individual 12-residue protein helices in the tenellin fungal PKS-NRPS to reprogram its function [57]. Thus, there is considerable scope for the future precise engineering of mPKS using a combination of synthetic DNA and yeast recombination to achieve their rational reprogramming.

There are very few reports on PPTases in fungi and, so far, include those involved in lysine biosynthesis in yeasts, e.g., Lys5p/Lys7p, or those involved in both lysine biosynthesis and polyketide synthesis in filamentous fungi, i.e., CfwA/NpgA [58], and PPT1 from various species [59–63]. These PPTases have been investigated in terms of their roles in primary and secondary metabolism and the effects on fungal virulence. Here, we determined that the Sfp-type PPTase in *A. oryzae* is unable to post-translationally modify the ACP domains of DEBS1-TE. This is somewhat surprising considering that iPKS and mPKS are believed to have evolved from the same ancestor [64]. *A. oryzae* has been used as a host for heterologous expression of PKS from the lichen-forming fungus *Cladonia uncialis* but no polyketide products were detected despite correctly spliced mRNA being detected, even when codon-optimized PKS genes were utilized [65]. It was proposed that *A. oryzae*'s native PPTase is unable to post-translationally modify PKS from lichens and our investigation with DEBS1-TE would support that. Therefore, for developing any heterologous host, understanding the limitations of the native PPTase and identifying functional alternatives should not be underestimated.

In conclusion, we have metabolically engineered the fungus *A. oryzae* NSAR1 and, for the first time, show that it is capable of expressing a bacterial mPKS for polyketide production using minimal modifications of its metabolism. This approach required the introduction of the non-native propionyl-CoA carboxylase enzyme complex from *S. coelicolor* and the *S. erythraea* PPTase SePptII. Increased titers of TKL were observed in *mcsA* KO strains, but this modification is not essential. *A. oryzae* is easily able to express actinobacterial components for polyketide production and there seems to be no fundamental reason why native filamentous fungi cannot exploit modular PKS systems—making their absence from fungi all the more mysterious. Our work shows that the expression and engineering of bacterial components in a fungal host is simple, rapid, and effective and should offer the opportunity to create new hybrid fungal–bacterial biosynthetic pathways in the future.

Supplementary Materials: The following are available online at https://www.mdpi.com/article/10.3390/jof7121085/s1: RT-PCR study of expression of *pcsA*, *mcsA* and *coaT* in *A. oryzae* NSAR1; codon optimization details for DEBS1-TE and PCC; sequences of DEBS1-TE and PCC; DEBS1-TE expression cassette construction; gene expression and identification of DEBS1-TE and PCC; fungal and bacterial PPTase alignment; synthesis and characterisation of of TKL; quantification of TKL; deletion of propionyl-CoA degradation pathways; deyaols of plasmids and primers used in the study.

Author Contributions: Conceptualization, E.S. and R.J.C.; methodology, E.S., R.J.C. and J.F.; formal analysis, E.S., R.J.C., J.F. and M.H.; investigation, E.S., R.J.C., J.F. and M.H.; resources, R.J.C.; data curation, E.S., R.J.C., J.F. and M.H.; writing—original draft preparation, E.S., R.J.C. and J.F.; writing—review and editing, E.S., R.J.C., J.F. and M.H.; supervision, E.S. and R.J.C.; project administration, E.S. and R.J.C.; funding acquisition, R.J.C. and J.F. All authors have read and agreed to the published version of the manuscript.

Funding: J.F. thanks the China Scholarship council for funding (201709110130). We thank DFG for the funding of LCMS (INST 187/626-1) and 600 MHz NMR instruments (INST 187/686-1). The publication of this article was funded by the Open Access Fund of the Leibniz Universität Hannover.

Conflicts of Interest: The authors declare no conflict of interest.

References

1. Newman, D.J.; Cragg, G.M. Natural products as sources of new drugs from 1981 to 2014. *J. Nat. Prod.* **2016**, *79*, 629–661. [CrossRef]
2. Wu, J.; Zhang, Q.; Deng, W.; Qian, J.; Zhang, S.; Liu, W. Toward Improvement of Erythromycin A Production in an IndustrialSaccharopolyspora erythraea Strain via Facilitation of Genetic Manipulation with an Artificial attB Site for Specific Recombination. *Appl. Environ. Microbiol.* **2011**, *77*, 7508–7516. [CrossRef]
3. Miyazawa, T.; Fitzgerald, B.J.; Keatinge-Clay, A.T. Preparative production of an enantiomeric pair by engineered polyketide synthases. *Chem. Commun.* **2021**, *57*, 8762–8765. [CrossRef]
4. Cook, T.B.; Pfleger, B.F. Leveraging synthetic biology for producing bioactive polyketides and non-ribosomal peptides in bacterial heterologous hosts. *MedChemComm* **2019**, *10*, 668–681. [CrossRef]

5. Yang, D.; Park, S.Y.; Park, Y.S.; Eun, H.; Lee, S.Y. Metabolic Engineering of Escherichia coli for Natural Product Biosynthesis. *Trends Biotechnol.* **2020**, *38*, 745–765. [CrossRef]
6. Bond, C.; Tang, Y.; Li, L. Saccharomyces cerevisiae as a tool for mining, studying and engineering fungal polyketide synthases. *Fungal Genet. Biol.* **2016**, *89*, 52–61. [CrossRef]
7. Harvey, C.J.B.; Tang, M.; Schlecht, U.; Horecka, J.; Fischer, C.R.; Lin, H.-C.; Li, J.; Naughton, B.; Cherry, J.; Miranda, M.; et al. HEx: A heterologous expression platform for the discovery of fungal natural products. *Sci. Adv.* **2018**, *4*, eaar5459. [CrossRef] [PubMed]
8. Xu, X.; Liu, Y.; Du, G.; Ledesma-Amaro, R.; Liu, L. Microbial Chassis Development for Natural Product Biosynthesis. *Trends Biotechnol.* **2020**, *38*, 779–796. [CrossRef] [PubMed]
9. Daba, G.M.; Mostafa, F.A.; Elkhateeb, W.A. The ancient koji mold (*Aspergillus oryzae*) as a modern biotechnological tool. *Bioresour. Bioprocess.* **2021**, *8*, 1–17. [CrossRef]
10. He, Y.; Cox, R.J. The molecular steps of citrinin biosynthesis in fungi. *Chem. Sci.* **2016**, *7*, 2119–2127. [CrossRef] [PubMed]
11. Williams, K.; Szwalbe, A.J.; Mulholland, N.P.; Vincent, J.L.; Bailey, A.M.; Willis, C.L.; Simpson, T.J.; Cox, R.J. Heterologous Production of Fungal Maleidrides Reveals the Cryptic Cyclization Involved in their Biosynthesis. *Angew. Chem. Int. Ed.* **2016**, *55*, 6784–6788. [CrossRef] [PubMed]
12. Kahlert, L.; Bassiony, E.F.; Cox, R.; Skellam, E.J. Diels–Alder Reactions during the Biosynthesis of Sorbicillinoids. *Angew. Chem. Int. Ed.* **2020**, *59*, 5816–5822. [CrossRef]
13. Kahlert, L.; Villanueva, M.; Cox, R.J.; Skellam, E.J. Biosynthesis of 6-Hydroxymellein Requires a Collaborating Polyketide Synthase-like Enzyme. *Angew. Chem. Int. Ed.* **2021**, *60*, 11423–11429. [CrossRef]
14. Feng, J.; Surup, F.; Hauser, M.; Miller, A.; Wennrich, J.-P.; Stadler, M.; Cox, R.J.; Kuhnert, E. Biosynthesis of oxygenated brasilane terpene glycosides involves a promiscuous *N*-acetylglucosamine transferase. *Chem. Commun.* **2020**, *56*, 12419–12422. [CrossRef]
15. Schor, R.; Schotte, C.; Wibberg, D.; Kalinowski, J.; Cox, R.J. Three previously unrecognised classes of biosynthetic enzymes revealed during the production of xenovulene A. *Nat. Commun.* **2018**, *9*, 1–9. [CrossRef] [PubMed]
16. Nofiani, R.; De Mattos-Shipley, K.; Lebe, K.E.; Han, L.-C.; Iqbal, Z.; Bailey, A.; Willis, C.L.; Simpson, T.J.; Cox, R.J. Strobilurin biosynthesis in Basidiomycete fungi. *Nat. Commun.* **2018**, *9*, 1–11. [CrossRef]
17. Kahlert, L.; Schotte, C.; Cox, R.J. Total Mycosynthesis: Rational Bioconstruction and Bioengineering of Fungal Natural Products. *Synthesis* **2021**, *53*, 2381–2394. [CrossRef]
18. Alberti, F.; Khairudin, K.; Venegas, E.R.; Davies, J.; Hayes, P.M.; Willis, C.L.; Bailey, A.M.; Foster, G.D. Heterologous expression reveals the biosynthesis of the antibiotic pleuromutilin and generates bioactive semi-synthetic derivatives. *Nat. Commun.* **2017**, *8*, 1–9. [CrossRef]
19. Bailey, A.; Alberti, F.; Kilaru, S.; Collins, C.; De Mattos-Shipley, K.; Hartley, A.J.; Hayes, P.; Griffin, A.; Lazarus, C.M.; Cox, R.; et al. Identification and manipulation of the pleuromutilin gene cluster from Clitopilus passeckerianus for increased rapid antibiotic production. *Sci. Rep.* **2016**, *6*, 1–11. [CrossRef] [PubMed]
20. Rendsvig, J.K.H.; Futyma, M.E.; Jarczynska, Z.D.; Mortensen, U.H. Filamentous Fungi as Hosts for Heterologous Production of Proteins and Secondary Metabolites in the Post-Genomic Era. In *Genetics and Biotechnology*; The Mycota (A Comprehensive Treatise on Fungi as Experimental Systems for Basic and Applied Research); Benz, J.P., Schipper, K., Eds.; Springer: Cham, Switzerland, 2020; Volume 2, pp. 227–265.
21. Thynne, E.; Mead, O.; Chooi, Y.-H.; McDonald, M.C.; Solomon, P.S. Acquisition and Loss of Secondary Metabolites Shaped the Evolutionary Path of Three Emerging Phytopathogens of Wheat. *Genome Biol. Evol.* **2019**, *11*, 890–905. [CrossRef]
22. Cox, R. Oxidative rearrangements during fungal biosynthesis. *Nat. Prod. Rep.* **2014**, *31*, 1405–1424. [CrossRef]
23. Cortes, J.; Wiesmann, K.E.H.; Roberts, G.A.; Brown, M.J.B.; Staunton, J.; Leadlay, P.F. Repositioning of a Domain in a Modular Polyketide Synthase to Promote Specific Chain Cleavage. *Science* **1995**, *268*, 1487–1489. [CrossRef]
24. Pfeifer, B.A.; Admiraal, S.J.; Gramajo, H.; Cane, D.E.; Khosla, C. Biosynthesis of Complex Polyketides in a Metabolically Engineered Strain of *E. coli*. *Science* **2001**, *291*, 1790–1792. [CrossRef]
25. Kao, C.M.; Luo, G.; Katz, L.; Cane, D.E.; Khosla, C. Manipulation of macrolide ring size by directed mutagenesis of a modular polyketide synthase. *J. Am. Chem. Soc.* **1995**, *117*, 9105–9106. [CrossRef]
26. Kim, B.S.; Cropp, T.A.; Florova, G.; Lindsay, Y.; Sherman, D.H.; Reynolds, K.A. An Unexpected Interaction between the Modular Polyketide Synthases, Erythromycin DEBS1 and Pikromycin PikAIV, Leads to Efficient Triketide Lactone Synthesis. *Biochemistry* **2002**, *41*, 10827–10833. [CrossRef] [PubMed]
27. Pieper, R.; Luo, G.; Cane, D.E.; Khosla, C. Cell-free synthesis of polyketides by recombinant erythromycin polyketide synthases. *Nat. Cell Biol.* **1995**, *378*, 263–266. [CrossRef] [PubMed]
28. Brock, M.; Buckel, W. On the mechanism of action of the antifungal agent propionate. *JBIC J. Biol. Inorg. Chem.* **2004**, *271*, 3227–3241. [CrossRef] [PubMed]
29. Zhang, Y.-Q.; Keller, N.P. Blockage of methylcitrate cycle inhibits polyketide production in Aspergillus nidulans. *Mol. Microbiol.* **2004**, *52*, 541–550. [CrossRef]
30. Lazarus, C.M.; Williams, K.; Bailey, A.M. Reconstructing fungal natural product biosynthetic pathways. *Nat. Prod. Rep.* **2014**, *31*, 1339–1347. [CrossRef]
31. Zhang, Y.-Q.; Brock, M.; Keller, N.P. Connection of Propionyl-CoA Metabolism to Polyketide Biosynthesis in Aspergillus nidulans. *Genetics* **2004**, *168*, 785–794. [CrossRef]

32. Brock, M.; Fischer, R.; Linder, D.; Buckel, W. Methylcitrate synthase from Aspergillus nidulans: Implications for propionate as an antifungal agent. *Mol. Microbiol.* **2000**, *35*, 961–973. [CrossRef] [PubMed]
33. Fleck, C.B.; Brock, M. Characterization of an acyl-CoA: Carboxylate CoA-transferase from Aspergillus nidulans involved in propionyl-CoA detoxification. *Mol. Microbiol.* **2008**, *68*, 642–656. [CrossRef]
34. Diacovich, L.; Peirú, S.; Kurth, D.; Rodríguez, E.; Podestá, F.; Khosla, C.; Gramajo, H. Kinetic and Structural Analysis of a New Group of Acyl-CoA Carboxylases Found in Streptomyces coelicolor A3 (2). *J. Biol. Chem.* **2002**, *277*, 31228–31236. [CrossRef]
35. Vandova, G.A.; O'Brien, R.V.; Lowry, B.; Robbins, T.F.; Fischer, C.R.; Davis, R.W.; Khosla, C.; Harvey, C.J.B.; Hillenmeyer, M.E. Heterologous expression of diverse propionyl-CoA carboxylases affects polyketide production in *Escherichia coli*. *J. Antibiot.* **2017**, *70*, 859–863. [CrossRef]
36. Pahirulzaman, K.A.K.; Williams, K.; Lazarus, C.M. A Toolkit for Heterologous Expression of Metabolic Pathways in *Aspergillus oryzae*. *Methods Enzymol.* **2012**, *517*, 241–260. [CrossRef] [PubMed]
37. Lambalot, R.H.; Gehring, A.M.; Flugel, R.S.; Zuber, P.; LaCelle, M.; Marahiel, M.A.; Reid, R.; Khosla, C.; Walsh, C.T. A new enzyme superfamily—the phosphopantetheinyl transferases. *Chem. Biol.* **1996**, *3*, 923–936. [CrossRef]
38. Kim, J.H.; Komatsu, M.; Shin-ya, K.; Omura, S.; Ikeda, H. Distribution and functional analysis of the phosphopantetheinyl transferase superfamily in Actinomycetales microorganisms. *Proc. Nat. Acad. Sci. USA* **2018**, *115*, 6828–6833. [CrossRef] [PubMed]
39. Mutka, S.C.; Bondi, S.M.; Carney, J.R.; Da Silva, N.A.; Kealey, J.T. Metabolic pathway engineering for complex polyketide biosynthesis inSaccharomyces cerevisiae. *FEMS Yeast Res.* **2006**, *6*, 40–47. [CrossRef] [PubMed]
40. Weissman, K.J.; Hong, H.; Oliynyk, M.; Siskos, A.P.; Leadlay, P.F. Identification of a Phosphopantetheinyl Transferase for Erythromycin Biosynthesis in *Saccharopolyspora erythraea*. *ChemBioChem* **2004**, *5*, 116–125. [CrossRef] [PubMed]
41. Brock, M. Role of Cellular Control of Propionyl-CoA Levels for Microbial Pathogenesis. In *Handbook of Hydrocarbon and Lipid Microbiology*; Timmis, K.N., Ed.; Springer: Berlin/Heidelberg, Germany, 2010; pp. 3279–3291.
42. Nielsen, M.L.; Albertsen, L.; Lettier, G.; Nielsen, J.B.; Mortensen, U.H. Efficient PCR-based gene targeting with a recyclable marker for *Aspergillus nidulans*. *Fungal Genet. Biol.* **2006**, *43*, 54–64. [CrossRef]
43. Tokuoka, M.; Tanaka, M.; Ono, K.; Takagi, S.; Shintani, T.; Gomi, K. Codon Optimization Increases Steady-State mRNA Levels in Aspergillus oryzae Heterologous Gene Expression. *Appl. Environ. Microbiol.* **2008**, *74*, 6538–6546. [CrossRef] [PubMed]
44. Cox, R.J.; Hitchman, T.S.; Byrom, K.J.; Findlow, I.C.; A Tanner, J.; Crosby, J.; Simpson, T.J. Post-translational modification of heterologously expressed Streptomyces type II polyketide synthase acyl carrier proteins. *FEBS Lett.* **1997**, *405*, 267–272. [CrossRef]
45. Gonzalez-Garcia, R.A.; Nielsen, L.K.; Marcellin, E. Heterologous Production of 6-Deoxyerythronolide B in *Escherichia coli* through the Wood Werkman Cycle. *Metabolites* **2020**, *10*, 228. [CrossRef]
46. Zhang, H.; Boghigian, B.A.; Pfeifer, B.A. Investigating the role of native propionyl-CoA and methylmalonyl-CoA metabolism on heterologous polyketide production in *Escherichia coli*. *Biotech. Bioeng.* **2010**, *105*, 567–573. [CrossRef]
47. Maerker, C.; Rohde, M.; Brakhage, A.A.; Brock, M. Methylcitrate synthase from *Aspergillus fumigatus*. Propionyl-CoA affects polyketide synthesis, growth and morphology of conidia. *FEBS J.* **2005**, *272*, 3615–3630. [CrossRef]
48. Domin, N.; Wilson, D.; Brock, M. Methylcitrate cycle activation during adaptation of Fusarium solani and Fusarium verticillioides to propionyl-CoA-generating carbon sources. *Microbiology* **2009**, *155*, 3903–3912. [CrossRef] [PubMed]
49. Dubey, M.K.; Broberg, A.; Jensen, D.; Karlsson, M. Role of the methylcitrate cycle in growth, antagonism and induction of systemic defence responses in the fungal biocontrol agent *Trichoderma atroviride*. *Microbiology* **2013**, *159*, 2492–2500. [CrossRef]
50. Lee, S.-H.; Han, Y.-K.; Yun, S.-H.; Lee, Y.-W. Roles of the Glyoxylate and Methylcitrate Cycles in Sexual Development and Virulence in the Cereal Pathogen *Gibberella zeae*. *Eukaryot. Cell* **2009**, *8*, 1155–1164. [CrossRef]
51. Yan, Y.; Wang, H.; Zhu, S.; Wang, J.; Liu, X.; Lin, F.; Lu, J. The Methylcitrate Cycle is Required for Development and Virulence in the Rice Blast Fungus *Pyricularia oryzae*. *Mol. Plant-Microbe Interact.* **2019**, *32*, 1148–1161. [CrossRef]
52. Santos, L.P.A.; Assunção, L.D.P.; Lima, P.D.S.; Tristão, G.B.; Brock, M.; Borges, C.L.; Silva-Bailao, M.; Soares, C.M.D.A.; Bailao, A. Propionate metabolism in a human pathogenic fungus: Proteomic and biochemical analyses. *IMA Fungus* **2020**, *11*, 1–16. [CrossRef]
53. Catenazzi, M.C.E.; Jones, H.; Wallace, I.; Clifton, J.; Chong, J.P.J.; Jackson, M.A.; Macdonald, S.; Edwards, J.; Moir, J.W.B. A large genomic island allows Neisseria meningitidis to utilize propionic acid, with implications for colonization of the human nasopharynx. *Mol. Microbiol.* **2014**, *93*, 346–355. [CrossRef] [PubMed]
54. Otzen, C.; Bardl, B.; Jacobsen, I.D.; Nett, M.; Brock, M. Candida albicans utilizes a modified β-oxidation pathway for the degradation of toxic propionyl-CoA. *J. Biol. Chem.* **2014**, *289*, 8151–8169. [CrossRef] [PubMed]
55. Dayem, L.C.; Carney, J.R.; Santi, D.V.; Pfeifer, B.A.; Khosla, C.; Kealey, J.T. Metabolic engineering of a methylmalonyl-CoA mutase-epimerase pathway for complex polyketide biosynthesis in *Escherichia coli*. *Biochemistry* **2002**, *41*, 5193–5201. [CrossRef]
56. De Mattos-Shipley, K.M.J.; Greco, C.; Heard, D.M.; Hough, G.; Mulholland, N.P.; Vincent, J.L.; Micklefield, J.; Simpson, T.J.; Willis, C.L.; Cox, R.J.; et al. The cycloaspeptides: Uncovering a new model for methylated nonribosomal peptide biosynthesis. *Chem. Sci.* **2018**, *9*, 4109–4117. [CrossRef]
57. Yang, X.-L.; Friedrich, S.; Yin, S.; Piech, O.; Williams, K.; Simpson, T.J.; Cox, R.J. Molecular basis of methylation and chain-length programming in a fungal iterative highly reducing polyketide synthase. *Chem. Sci.* **2019**, *10*, 8478–8489. [CrossRef] [PubMed]
58. Márquez-Fernández, O.; Trigos, A.; Ramos-Balderas, J.L.; Viniegra-González, G.; Deising, H.B.; Aguirre, J. Phosphopantetheinyl Transferase CfwA/NpgA Is Required for Aspergillus nidulans Secondary Metabolism and Asexual Development. *Eukaryot. Cell* **2007**, *6*, 710–720. [CrossRef]

59. Zainudin, N.A.I.M.; Condon, B.; De Bruyne, L.; Van Poucke, C.; Bi, Q.; Li, W.; Höfte, M.; Turgeon, B.G. Virulence, Host-Selective Toxin Production, and Development of Three Cochliobolus Phytopathogens Lacking the Sfp-Type 4′-Phosphopantetheinyl Transferase Ppt1. *Mol. Plant-Microbe Interact.* **2015**, *28*, 1130–1141. [CrossRef]
60. Albermann, S.; Elter, T.; Teubner, A.; Krischke, W.; Hirth, T.; Tudzynski, B. Characterization of novel mutants with an altered gibberellin spectrum in comparison to different wild-type strains of *Fusarium fujikuroi*. *Appl. Microbiol. Biotechnol.* **2013**, *97*, 7779–7790. [CrossRef]
61. Leng, Y.; Zhong, S. Sfp-type 4′-phosphopantetheinyl transferase is required for lysine synthesis, tolerance to oxidative stress and virulence in the plant pathogenic fungus *Cochliobolus sativus*. *Mol. Plant Pathol.* **2012**, *13*, 375–387. [CrossRef]
62. Velazquez-Robledo, R.; Contreras-Cornejo, H.A.; Macias-Rodriguez, L.; Hernandez-Morales, A.; Aguirre, J.; Casa-Flores, S.; Lopez-Bucio, J.; Herrera-Estrella, A. Role of the 4-Phosphopantetheinyl Transferase of *Trichoderma virens* in Secondary Metabolism and Induction of Plant Defense Responses. *MPMI* **2011**, *12*, 1459–1471. [CrossRef]
63. Horbach, R.; Graf, A.; Weihmann, F.; Antelo, L.; Mathea, S.; Liermann, J.C.; Opatz, T.; Thines, E.; Aguirre, J.; Deising, H.B. Sfp-Type 4′-Phosphopantetheinyl Transferase Is Indispensable for Fungal Pathogenicity. *Plant Cell* **2009**, *21*, 3379–3396. [CrossRef] [PubMed]
64. Jenke-Kodama, H.; Sandmann, A.; Müller, R.; Dittmann, E. Evolutionary Implications of Bacterial Polyketide Synthases. *Mol. Biol. Evol.* **2005**, *22*, 2027–2039. [CrossRef] [PubMed]
65. Bertrand, R.L.; Sorensen, J.L. Lost in Translation: Challenges with Heterologous Expression of Lichen Polyketide Synthases. *Chem. Sel.* **2019**, *4*, 6473–6483. [CrossRef]

Journal of *Fungi*

MDPI

Article

Antiproliferative and Cytotoxic Cytochalasins from *Sparticola triseptata* Inhibit Actin Polymerization and Aggregation

Katherine Yasmin M. Garcia [1,2], Mark Tristan J. Quimque [1,2,3], Christopher Lambert [4,5], Katharina Schmidt [5], Gian Primahana [4,6], Theresia E. B. Stradal [5], Andreas Ratzenböck [7], Hans-Martin Dahse [8], Chayanard Phukhamsakda [9,10], Marc Stadler [4,11], Frank Surup [4,11,*] and Allan Patrick G. Macabeo [2,*]

1 The Graduate School, University of Santo Tomas, España Blvd., Manila 1015, Philippines; katherineyasmin.garcia.gs@ust.edu.ph (K.Y.M.G.); mtjquimque@gmail.com (M.T.J.Q.)
2 Laboratory for Organic Reactivity, Discovery and Synthesis (LORDS), Research Center for the Natural and Applied Sciences, University of Santo Tomas, España Blvd., Manila 1015, Philippines
3 Chemistry Department, College of Science and Mathematics, Mindanao State University–Iligan Institute of Technology, Tibanga, Iligan City 9200, Philippines
4 Department of Microbial Drugs, Helmholtz Centre for Infection Research and German Centre for Infection Research (DZIF), Partner Site Hannover/Braunschweig, Inhoffenstraße 7, 38124 Braunschweig, Germany; christopher.lambert@helmholtz-hzi.de (C.L.); gian.primahana@lipi.go.id (G.P.); marc.stadler@helmholtz-hzi.de (M.S.)
5 Department of Cell Biology, Helmholtz Centre for Infection Research (HZI), Inhoffenstraße 7, 38124 Braunschweig, Germany; katharina.schmidt@helmholtz-hzi.de (K.S.); theresia.stradal@helmholtz-hzi.de (T.E.B.S.)
6 Research Center for Chemistry, National Research and Innovation Agency (BRIN), Kawasan Puspitek, Serpong, Tangerang Selatan 15314, Indonesia
7 Institut für Organische Chemie, Universität Regensburg, Universitätstrasse 31, 93053 Regensburg, Germany; andreas.ratzenboeck@chemie.uni-regensburg.de
8 Leibniz-Institute for Natural Product Research and Infection Biology, Hans Knöll Institute (HKI), 07745 Jena, Germany; hans-martin.dahse@hki-jena.de
9 Center of Excellence in Fungal Research, Mae Fah Luang University, Chiang Rai 57100, Thailand; chayanard91@gmail.com
10 Institute of Plant Protection, College of Agriculture, Engineering Research Center of Chinese Ministry of Education for Edible and Medicinal Fungi, Jilin Agricultural University, Changchun 130118, China
11 Institute of Microbiology, Technische Universität Braunschweig, Spielmannstraße 7, 38106 Braunschweig, Germany
* Correspondence: frank.surup@helmholtz-hzi.de (F.S.); agmacabeo@ust.edu.ph (A.P.G.M.)

Citation: Garcia, K.Y.M.; Quimque, M.T.J.; Lambert, C.; Schmidt, K.; Primahana, G.; Stradal, T.E.B.; Ratzenböck, A.; Dahse, H.-M.; Phukhamsakda, C.; Stadler, M.; et al. Antiproliferative and Cytotoxic Cytochalasins from *Sparticola triseptata* Inhibit Actin Polymerization and Aggregation. *J. Fungi* **2022**, *8*, 560. https://doi.org/10.3390/jof8060560

Academic Editor: Gary A. Strobel

Received: 27 April 2022
Accepted: 20 May 2022
Published: 25 May 2022

Abstract: Laying the groundwork on preliminary structure–activity relationship study relating to the disruptive activity of cytochalasan derivatives on mammalian cell actin cytoskeleton, we furthered our study on the cytochalasans of the Dothideomycetes fungus, *Sparticola triseptata*. A new cytochalasan analog triseptatin (**1**), along with the previously described cytochalasans deoxaphomin B (**2**) and cytochalasin B (**3**), and polyketide derivatives *cis*-4-hydroxy-6-deoxyscytalone (**4**) and 6-hydroxymellein (**5**) were isolated from the rice culture of *S. triseptata*. The structure of **1** was elucidated through NMR spectroscopic analysis and high-resolution mass spectrometry (HR-ESI-MS). The relative and absolute configurations were established through analysis of NOESY spectroscopic data and later correlated with experimental electronic circular dichroism and time-dependent density functional theory (ECD–TDDFT) computational analysis. Compounds **1** and **2** showed cytotoxic activities against seven mammalian cell lines (L929, KB3.1, MCF-7, A549, PC-3, SKOV-3, and A431) and antiproliferative effects against the myeloid leukemia K-562 cancer cell line. Both **1** and **2** were shown to possess properties inhibiting the F-actin network, prompting further hypotheses that should to be tested in the future to enable a well-resolved concept of the structural implications determining the bioactivity of the cytochalasin backbone against F-actin.

Keywords: *Sparticola triseptata*; structure elucidation; ECD–TDDFT; antiproliferative; cytotoxic; actin inhibitors

1. Introduction

Cytochalasans are a diverse group of biologically active fungal polyketide–amino acid hybrids featuring a tricyclic core structure composed of a 7- to 15-membered macrocycle fused to a highly substituted perhydroisoindolone moiety. They are biosynthesized via a PKS-NRPS hybrid pathway [1]. Since the first report of cytochalasins A and B from *Phoma* strain S 298 and *Helminthosporium dematioideum*, over 300 derivatives have been reported in several genera of Dothideomycetes, such as *Ascochyta*, *Preussia*, and *Phoma* [2–4], although producers mainly belong to Sordariomycetes, namely *Aspergillus*, *Daldinia*, and *Diaporthe* (formerly known as *Phomopsis* and *Xylaria*) [5–7]. Variations in the amino acid side chain, intramolecular rearrangements, and observation of unique substitution patterns in the macrocyclic ring have led to a diversity of cytochalasin structures. Cytochalasans exhibit antimicrobial, antiviral, and antiparasitic properties [8–12], regulate hormonal functions [13,14], inhibit cholesterol synthesis [15] and bacterial biofilms [16], and interfere with glucose transport proteins [17] and Ca^{2+} influx regulation [18]. By far, the most prominent and frequently reported biological effects of cytochalasans have been related to their interference with the actin cytoskeleton [5]. Interestingly, different derivatives of this compound class can either have strong and irreversible or weak and reversible activities. Studies on the structure–activity relationship of such mechanism are limited [6].

We have previously reported the actin depolymerization activity of 25 cytochalasans isolated from ascomata and mycelial cultures of different Ascomycota to establish a preliminary structure–activity relationship study [5]. We noted that the presence of hydroxyl group in the C7 and C18 of the cytochalasan backbone and their stereochemical configurations are important factors for actin cytoskeleton polymerization inhibition. When the reversibility of the actin-disrupting effects was evaluated, no direct correlations between potency and reversibility in the tested compound group were observed. As part of our ongoing efforts to explore biologically active Dothideomycetes secondary metabolites, the ascomycete *Sparticola triseptata* (Leuchtm.) Phukhamsakda & K. D. Hyde, *comb. nov.* obtained from the decomposing branches of *Tofieldia calyculata* (L.) Wahlenb. was investigated for its cytotoxic chemical constituents. In this study, we carried out the isolation and structure elucidation of one cytotoxic cytochalasin derivative from the solid rice culture of *S. triseptata*, hitherto referred to as triseptatin (**1**), along with deoxaphomin B (**2**), cytochalasin B (**3**), *cis*-4-hydroxy-6-deoxyscytalone (**4**), and 6-hydroxymellein (**5**). To complement and further investigate the cytotoxic effect on the F-actin network of mammalian cell lines of cytochalasin **1**–**3**, 1 h endpoint assays using fluorescence microscopy were also carried out.

2. Materials and Methods

2.1. General Experimental Procedures

Specific optical rotations ($[\alpha]^D$) were measured on a Perkin Elmer 241 polarimeter in a 100 mm × 2 mm cell at 20 °C. Nuclear magnetic resonance (NMR) spectra were obtained either on a Bruker Ascend 600 MHz spectrometer equipped with a 5 mm TXI cryoprobe (^{1}H 600 MHz, ^{13}C 150 MHz) or a Varian VNMRS-500 MHz (^{1}H 500 MHz, ^{13}C 125 MHz). Spectra were acquired at 25 °C (unless otherwise specified) in MeOH-d_4 with reference to residual ^{1}H or ^{13}C signals in the deuterated solvent. HR-ESI mass spectra were measured using Agilent 6200 series TOF and 6500 series Q-TOF LC/MS systems. The HPLC-DAD purification was performed on a Shimadzu Prominence liquid chromatograph LC-20AT coupled with a SPD-M20A photodiode array detector (Shimadzu Corp., Tokyo, Japan) and the semipreparative reversed-phase C18 column Inertsil ODS-3 (10 mm I.D. × 250 mm, 5 µm, G.L. Sciences, Tokyo, Japan). The mobile phase was composed of ultrapure water (Milli-Q, Millipore, Schwalbach, Germany) as solvent A and acetonitrile (HPLC grade) as solvent B.

2.2. Fungal Material

The ascomycete *Sparticola triseptata* (Leuchtm.) Phukhamsakda & K.D. Hyde, which represents the ex-type strain of the species, was isolated from a decayed branch of

Tofieldia calyculata (L.) Wahlenb. [19]. Phukhamsakda et.al [20] recently transferred this species to the genus *Sparticola* based on a polyphasic taxonomic study that included molecular phylogenetic and morphological methods. The type strain is deposited at the KNAW Westerdijk Fungal Biodiversity Centre, Utrecht, the Netherlands (CBS 614.86).

2.3. Production, Extraction, Isolation, and Structural Characterization

The fungal strain was cultured on malt extract agar plates for 8–10 weeks until the culture developed its characteristic brownish pigmentation. The fungus was cultivated on a solid rice media (70 g brown rice, 0.3 g peptone, 0.1 g corn syrup, and 100 mL ultrapure water) in 15 × 1000 mL sterilized Fernbach culture flasks, followed by autoclaving (121 °C, 20 min). Five agar blocks of a well-grown fungal culture were inoculated in the culture flasks and incubated under static condition in a dark room at 25–30 °C for 12 weeks until the fungal hyphae proliferated and the rice medium turned black in color. The rice cultures were homogenized using a sterile metal spatula. Fermentation was terminated by the addition of ethyl acetate (EtOAc; 3 × 300 mL). The combined extracts were concentrated in a rotary evaporator to afford the crude extract (30 g). The crude EtOAc extract was reconstituted with 300 mL 10% aqueous methanol and partitioned with *n*-heptane (3 × 100 mL). The combined organic layer was concentrated *in vacuo* to afford a dark brown methanolic crude extract (8.8 g).

The methanolic crude extract was fractionated using silica gel column chromatography, and elution was carried out using the following solvent systems: petroleum ether-EtOAc (1:1, 2:3, 3:7, 1:4, 1:9), EtOAc, dichloromethane (DCM), DCM–MeOH (9:1, 1:4, 7:3, 3:2, 1:1), and methanol to afford five combined main fractions. Fraction 3 (3.40 g) was chromatographed with DCM–MeOH (5:1) to yield three subfractions. The second subfraction (3.29 g) was subsequently eluted with DCM–MeOH (40:1) to afford five subfractions. The third subfraction, fraction 3.2.3 (259 mg), was further purified using semipreparative RP-HPLC. The mobile phase was composed of ultrapure water (Milli-Q, Millipore, Schwalbach, Germany; solvent A) and acetonitrile (RCI Labscan Ltd., HPLC grade; solvent B). Purification was carried out using the following gradient: 40% solvent B for 5 min, 40–100% solvent B for 20 min, 100% solvent B for 5 min, and 100–40% solvent B for 5 min. This resulted in 13 fractions affording **3** (2.12 mg, flow rate = 4.0 mL min^{-1}, UV detection 200−600 nm, t_R = 10.03 min), **4** (2.30 mg, flow rate = 4.0 mL min^{-1}, UV–vis detection 200−600 nm, t_R = 15.69 min), and **2** (6.87 mg, flow rate = 4.0 mL min^{-1}, UV–vis detection 200−600 nm, t_R = 22.85 min). Fraction 3.2.1 (1.54 g) was further chromatographed using petroleum ether–EtOAc (2:1, 1:1) and DCM–MeOH (100:1, 80:1) to yield three homogenous subfractions. The last subfraction, fraction 3.2.1.3 (889 mg), was subjected to semipreparative reversed-phase HPLC using similar gradient solvent system composition as described above to yield 10 fractions. Fraction 3.2.1.3.7 (53.8 mg) was purified twice by employing a linear gradient of solvent B from 40% solvent B for 5 min, 40–100% solvent B for 25 min, 100% solvent B for 5 min, and 100–40% solvent B for 5 min to yield compound **2** (7.64 mg, flow rate = 4.0 mL min^{-1}, UV detection 200−600 nm, t_R = 14.24 min). Fraction 3.2.1.3.10 (7.58 mg) was also subjected to semipreparative reversed-phase HPLC using similar gradient conditions to afford **1** (3.65 mg, 4.0 mL min^{-1}, UV detection 200−600 nm, t_R = 18.65 min).

Triseptatin (**1**): white amorphous solid; $[\alpha]_D^{25}$ +30 (*c* 0.1, MeOH); UV (*c* 0.1, MeOH) λ_{max} (log ε) 258 (3.83), 289 (3.89) nm; HPLC–ECD data in acetonitrile as λ_{max} (Δε) 205 (0.0074), 216 (–0.0025), and 239 (–0.0011). ^1H and ^{13}C NMR data, Table 1; HR-ESI-MS *m/z*: $[M + H]^+$ calcd for $C_{31}H_{40}NO_5$, 506.2906; found, 506.2901 (Figure S7). IUPAC nomenclature: (3*S*,5*S*,7*S*,8a*R*, 9a*R*,13*E*,16*R*,20*R*,21*E*)-3-benzyl-6-hydroxy-4,10-dimethyl-5-methylene-1,17-dioxo-2,3,3a,4,5,6,6a,9,10,11,12,13,14,17-tetradecahydro-1H-cyclotrideca[d]isoindol-14-yl acetate.

Table 1. NMR spectroscopic data of **1** in MeOH-d_4.

Position	**1**	
	δ_H (mult., J in Hz)	δ_C, Type
1	-	176.4, C
3	3.38 (m)	54.0, CH
4	2.91 (m)	46.3, CH
5	2.72 (m)	33.2, CH
6	-	151.0, C
7	3.90 (d, 9.9)	72.8, CH
8	2.50 (t, 9.9)	51.9, CH
9	-	63.9, C
10a	2.46 (dd, 13.0, 7.3)	43.5, CH_2
10b	2.64 (dd, 13.0, 6.4)	
11	0.79 (d, 6.8)	13.3, CH_3
12a	5.03 (s)	114.1, CH
12b	5.17 (s)	
13	6.01 (ddd, 15.2, 9.6, 1.9)	128.3, CH
14	5.22 (m)	137.3, CH
15a	2.06 (m)	40.5, CH_2
15b	1.79 (dd, 11.2, 2.8)	
16	1.51 (m)	34.0, CH
17a	1.09 (m)	34.9, CH_2
17b	1.26 (m)	
18a	1.15 (m)	19.7, CH_2
18b	1.39 (m)	
19a	1.52 (m)	32.6, CH_2
19b	1.92 (m)	
20	5.20 (m)	76.4, CH
21	6.48 (dd, 15.4, 8.5)	144.0, CH
22	6.86 (d, 15.4)	128.0, CH
23	-	198.6, C
24	0.89 (d, 6.9)	20.8, CH_3
25	-	171.6, C
26	2.04 (s)	21.1, CH_3
1′	-	137.9, C
2′/6′	7.08 (d, 7.6)	130.7/131.0, CH
3′/5′	7.32 (t, 7.6)	129.7/129.6, CH
4′	7.23 (t, 7.6)	127.8, CH

Recorded at 600 MHz; carbon multiplicities were deduced from HSQC–DEPT-135 spectra.

2.4. Computational Calculations

A conformational analysis on triseptatin (**1**) was performed using the Avogadro (version 1.1.1) platform [21], which included a search for low-energy conformations using the MMFF94 molecular mechanics force field and conformer optimization following the steepest descent algorithm. All stable conformations were subjected to geometry reoptimizations via density functional theory calculations with B3LYP/6-31G(d) basis set on a polarizable continuum model (PCM) with methanol as the solvent model. The calculated energies, taken as the sum of electronic and zero-point energies, were used to estimate the Boltzmann population for each conformer. The optimized geometries were then subjected to time-dependent DFT (TDDFT) following the same functional, basis set, and PCM solvent model. A Gaussian distribution function was used to generate the ECD curve from the calculated rotatory strength values with 3000 cm^{-1} half-height width. All DFT calculations were carried out using Gaussian 16W [22], while the visualization of results was conducted on GaussView 6.0.

2.5. Antiproliferation and Cytotoxicity Assays

Compounds **1–3** were assayed against human umbilical vein endothelial cells (HUVEC) and K-562 human chronic myeloid leukemia cells (DSM ACC 10) for their antiproliferative effects (GI_{50}) [23]. Cytotoxicity properties of the compounds were also assessed against several mammalian cancer cell lines, including mouse fibroblast L929, HeLa (KB3.1), human breast adenocarcinoma (MCF-7), adenocarcinomic human alveolar basal epithelial cells (A549), human prostate cancer cells (PC-3), ovarian carcinoma (SKOV-3), and squamous cell carcinoma (A431), and expressed as IC_{50} in the MTT assay [24]. Inhibitory concentrations were calculated as 50% half-maximal inhibitory concentration (IC_{50}, concentration of the substance where a specific biological process is reduced by half), 50% inhibition of cell growth (GI_{50}, the concentration needed to reduce the growth of treated cells to half that of untreated cells), or 50% cytotoxic concentration (CC_{50}, the concentration that kills 50% of treated cells).

2.6. Cell Culture of U2-OS Cells and Actin Disruption Assay

The impact of **1** and **2** on the organization of filamentous actin in tissue culture cells was investigated in an actin disruption assay following the procedure presented by [6]. Cytochalasin B (**3**, previously isolated by [6]) and D (MP Biomedicals, Solon, OH, USA) were chosen as widely used controls for cytochalasan impact on F-actin network organization in DMSO (Carl Roth GmbH, Karlsruhe, Germany). DMSO was also used as a vehicle control. Concentrations to study the effects on the F-actin organization were estimated based on previously determined IC_{50} against L929 mouse fibroblasts ($1 \times IC_{50}$ and $5 \times IC_{50}$, low-dose and high dose (cf. Wang et al. [25]). Concurrently, a second testing strategy consisted of a titration experiment employing concentrations ranging from 30 to 0.03 μM of compounds **1**, **2**, and **3**. Cells of the osteosarcoma cell line U2-0S [ATCC HTB-96] were cultured in Dulbecco's modified minimum essential medium (DMEM, Life Technologies, Carlsbad, CA, USA) containing 10% fetal bovine serum (Sigma-Aldrich, St. Louis, MO, USA), 1% L-glutamine (Life Technologies, Carlsbad, CA, USA), 1% minimum essential medium nonessential amino acids (MEM NEAA, Life Technologies, Carlsbad, CA, USA), and 1% sodium pyruvate (Life Technologies, Carlsbad, CA, USA) at 37 °C and 7.5% CO_2 for the following experiments. Fibronectin (Roche, Mannheim, Germany)-coated cover slips were seeded with 20,000 cells and allowed to spread overnight and treated with overnight equilibrated culture medium spiked with cytochalasins or DMSO in the abovementioned concentrations. Cells were treated with the prepared medium and incubated for 1 h before being washed once with PBS (pH 7.4) and fixed with 4% prewarmed paraformaldehyde (AppliChem, Darmstadt, Germany) supplied in PBS for 20 min. Reversibility of the impact on the F-actin network organization was tested by washout with 3× prewarmed PBS prior to the addition of fresh medium and a recovery time of 1 h before fixation. Fixed cells were washed with PBS thrice and permeabilized using 0.1% Triton X-100 (Bio-Rad Laboratories, Hercules, CA, USA) in PBS for 1 min at room temperature. After three additional washing steps with PBS, fluorescently labelled phalloidin (ATTO-594, ATTO-Tec, Siegen, Germany) diluted in PBS (1:100) was used and incubated for an additional hour to probe for the F-actin cytoskeleton and mounted in ProLong Diamond Antifade Mountant (Invitrogen, Carlsbad, CA, USA) containing DAPI to probe for nuclear DNA. Samples were examined for epifluorescence using an inverted microscope (Nikon eclipse Ti2, Tokio, Japan) with a 60× Nikon oil immersion objective (Plan Apofluar, 1.4 NA) with pE-4000 (CoolLED, Andover, UK) as a light source. Pictures were recorded with a pco.edge back-illuminated sCMOS camera (Excelitas Technologies, Mississauga, ON, Canada) operated by NIS elements (Nikon, Tokio, Japan) and processed by Image J (NIH, Bethesda, MD, USA).

3. Results and Discussion

The EtOAc extract of *Sparticola triseptata* obtained from the rice culture was partitioned between *n*-heptane and 10% aqueous MeOH. The resulting aqueous methanolic crude extract was purified using silica gel column chromatography and semipreparative HPLC to afford metabolites **1–5** (Figure 1). The known compounds, deoxaphomin B (**2**), cytochalasin B (**3**), *cis*-4-hydroxy-6-deoxyscytalone (**4**), and 6-hydroxymellein (**5**), were identified by comparing their physicochemical and NMR spectroscopic data with those reported in the literature [4,26–29].

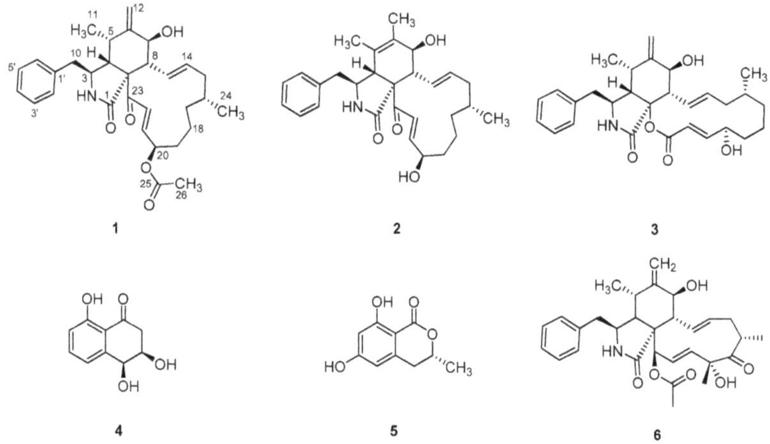

Figure 1. Secondary metabolites **1–5** from *Sparticola triseptata* and cytochalasin D (**6**).

Triseptatin (**1**) was obtained as an optically active whitish-yellow amorphous solid. The molecular formula $C_{31}H_{40}NO_5$, indicating 13 degrees of unsaturation, was established based on the protonated molecular ion peak at *m/z* $C_{31}H_{39}NO_5$ $[M + H]^+$ of its positive-ion HR-ESI-MS. This was consistent with the number of proton and carbon peaks detected in its ^1H and ^{13}C NMR spectroscopic data (Table 1). Preliminary spectroscopic analysis of its 1D NMR data showed identical resonances to that of deoxaphomin, which was first isolated from the organic culture extracts of *Ascochyta heteromorpha* and *Phoma multirostrata*.

The ^1H NMR spectrum of **1** (Figure S1) displayed three methyl proton signals at δ_H 0.79 (3H, *d*, *J* = 6.8 Hz, H-11), 0.89 (3H, *d*, *J* = 6.9 Hz, H-24), and 2.04 (3H, s, H-26); two exocyclic methylene proton signals at δ_H 5.03 (1H, *s*, H-12a) and 5.17 (1H, *s*, H-12b); four olefinic proton signals at δ_H 5.22 (1H, *m*, H-14), 6.01 (1H, *ddd*, *J* = 15.2, 9.6, 1.9 Hz, H-13), 6.48 (1H, *dd*, *J* = 15.4, 8.5 Hz, H-21), and 6.86 (1H, *d*, *J* = 15.4 Hz, H-22); two pairs of *ortho*-coupled aromatic proton signals at δ_H 7.08 (2H, *d*, *J* = 7.6 Hz, H-2′/H-6′) and 7.32 (2H, *t*, *J* = 7.6 Hz, H-3′/H-5′); and several aliphatic proton signals between δ_H 1.09 and 3.38 (Table 1). The ^{13}C and heteronuclear single quantum coherence–distortionless enhancement by polarization transfer (HSQC–DEPT) NMR spectroscopic data revealed the presence of three carbonyl carbons, one quaternary carbon, two sp^2 nonprotonated carbons, five aromatic methines, four olefinic methines, seven sp^3 methines (two oxygenated and five nonoxygenated), one sp^2 methylene, five sp^3 methylene carbons, and three methyl carbons.

Analysis of 1D NMR spectroscopic data revealed that **1** possessed a cytochalasin core skeleton, which was confirmed by its homonuclear and heteronuclear 2D NMR data (Figures S2–S5). The ^1H–^1H correlation spectroscopy (COSY) cross peaks of δ_H 3.38 (1H, *m*, H-3) with δ_H 2.91 (1H, *m*, H-4), δ_H 2.46 (1H, *dd*, *J* = 13.0, 7.3 Hz, H-10a), and δ_H 2.64 (1H, *dd*, *J* = 13.0, 6.4 Hz, H-10b), as well as the long-range HMBC coupling from H-3 to δ_C 176.4 (C-1) and δ_C 46.3 (C-4), from H-4 to C-1 and δ_C 63.9 (C-9), and from H_2-10 to δ_C 54.0 (C-3), revealed the presence of a 2-pyrrolidinone fragment. Key HMBC correlations from H-4 to δ_C 33.2 (C-5) and δ_C 151.0 (C-6); from H_3-11 to C-4, C-5, and C-6; from the exocyclic methylene H_2-12 to C-5 and δ_C 72.8 (C-7); and from δ_H 2.50 (1H, *t*, *J* = 9.9 Hz, H-8) to δ_C 63.9 (C-9) established the linkage between the methylidenecyclohexanol ring and 2-pyrrolidinone moiety forming the perhydroisoindolone subunit (Figure 2). Further analyses of the COSY spectrum led to the construction of an *ortho*-coupled (J = 7.6 Hz) three distinct proton spin systems illustrating homonuclear coupling correlations between H-2′/H-6′, and H-3′/H-5′, and of δ_H 7.23 (1H, *t*, J = 7.6 Hz, H-4′) corresponding to δ_C 130.7 (C-2′/C-6′), δ_C 129.7 (C-3′/C-5′), and δ_C 127.8 (C-4′) of the monosubstituted phenyl group, respectively. In addition, the HMBC correlations of H-2′/H-6′ to δ_C 43.5 (C-10) and from H_2-10 to δ_C 137.9 (C-1′), C-2′/C-6′, and C-3 resulted in the attachment of the monosubstituted phenyl group to the perhydroisoindolone residue forming the 10-phenylperhydroisoindolone moiety. The remaining portion of **1** was identified as a 13-membered carbocycle ring, which was elucidated by a combination of homonuclear COSY cross peaks establishing a 13-proton spin system resonating from H-7 to H-22 and based on the key HMBC correlations from δ_H 6.01 (1H, *ddd*, J = 15.2, 9.6, 1.9 Hz, H-13) to δ_C 40.5 (C-15), from δ_H 1.09 (1H, *m*, H-17a) and δ_H 1.26 (1H, *m*, H-17b) to δ_C 32.6 (C-19), from δ_H 1.39 (1H, *m*, H-18b) to δ_C 76.4 (C-20), and from δ_H 6.48 (1H, *dd*, J = 15.4, 8.5 Hz, H-21) and δ_H 6.86 (1H, *d*, J = 15.4 Hz, H-22) to δ_C 198.6 (C-23). Finally, HMBC correlations from H-13 to δ_C 51.9 (C-8) as well as the long-range coupling from H-4 and H-22 to C-23 suggested the connectivity of the 13-membered carbocyclic ring residue to the 10-phenylperhydroisoindolone moiety.

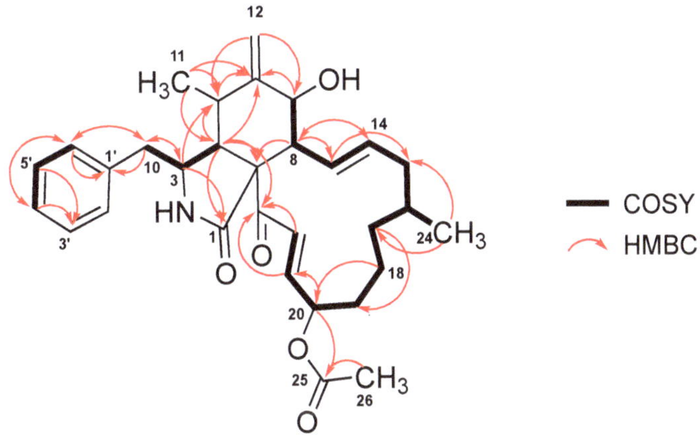

1

Figure 2. COSY and HMBC correlations in **1**.

Based on the COSY and HMBC correlations discussed above, the proton and carbon signals originating from the cytochalasin core structure and the macrocyclic ring fragment were found to have a close resemblance to deoxaphomin [4,29]. The main difference is an additional carboxylic carbon resonating at δ_C 171.6 (C-25) and a methyl carbon at δ_C 21.1 (C-26) and δ_H 2.04 (3H, *s*, H-26). Furthermore, the NMR signals of the acetyl group were evidenced by HMBC correlations from δ_H 5.20 (1H, *m*, H-20) to C-25 and from H-26 to C-25, indicating the attachment of the acetyl group at C-20.

The relative configuration of **1** was deduced through a combination of proton coupling constants and nuclear Overhauser effect spectroscopy (NOESY) data (Figure S6). In the ^{1}H NMR spectrum, the large coupling constant (J = 10.0 Hz) between H-7 and H-8 suggested coaxial positioning of this proton pair. The *trans*-geometric configurations of $\Delta^{13(14)}$ and $\Delta^{21(22)}$ were established based on the observed large coupling constants (H-13, H-14, J =15.1 Hz; H-21, H-22, J = 15.3 Hz). Additional NOESY interactions of H-3 with H_3-11 and H-7 revealed similar spatial orientation for C-3 benzyl and C-7 hydroxy moieties, whereas correlations of H-4 with H-5 and H-8 were β-oriented. Based on biosynthetic considerations and previous studies [30,31] naturally occurring cytochalasins possess similar α-orientation of the methyl groups attached to C-16, whereas the hydroxyl group at C-20 is β-oriented.

Thus, the relative configuration of 16-Me and 20-OH in **1** may follow that of known and related cytochalasin derivatives.

Triseptatin (**1**) was also subjected to ECD– TDDFT calculations to confirm its absolute configuration. MMFF94 conformational analysis and B3LYP/6-31G(d) geometry reoptimization of **1** afforded two energetically low-lying conformers with a preassigned (3*S*,4*S*,6*S*,6a*R*,7*E*,10*R*,14*R*,15*E*,17a*R*) configuration (Figure 3). All conformers were confirmed stable as per harmonic vibrational frequency calculations. The differences in spatial structural arrangement of compound **1** conformers were a result of the C3–C10 bond rotation, with the ωC-1′,C-10,C-3,C-4 torsion angles for conformers 1A (78.79%) and 1B (21.21%) being 69.38° and −177.72°, respectively. The theoretically obtained Boltzmann-averaged ECD spectrum of **1** at wB97XD/DGDZVP (PCM/MeCN) after wavelength correction showed a good correlation with the experimental data (Figure 4), confirming that the absolute configuration of triseptatin is (3*S*,5*S*,7*S*,8a*R*, 9a*R*, 13*E*,16*R*,20*R*,21*E*)-**1**.

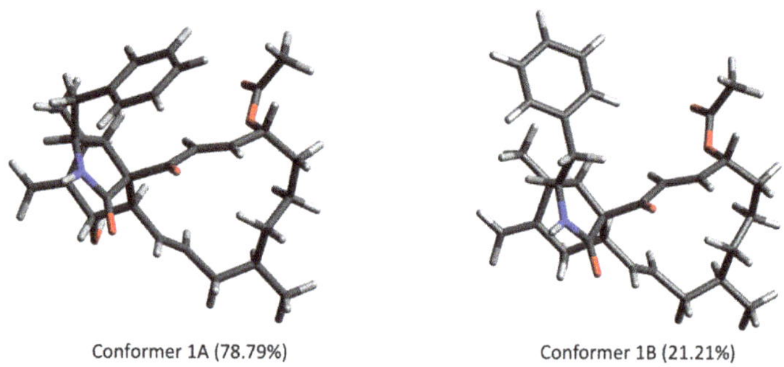

Figure 3. Low-energy conformers (>1%) of (3*S*,5*S*,7*S*,8a*R*, 9a*R*,13*E*,16*R*, 20*R*,21*E*)-**1** optimized at B3LYP/6-31G(d) (PCM/MeOH).

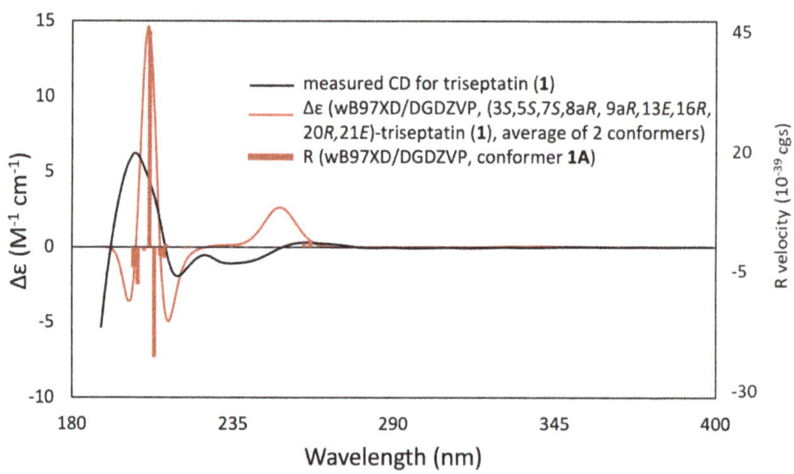

Figure 4. Experimental ECD spectrum of triseptatin (**1**, black solid curve) compared with wB97XD/DGDZVP-calculated ECD spectra (red solid curve) for the B3LYP/6-31G(d)-optimized conformers of (3*S*,5*S*,7*S*,8a*R*, 9a*R*, 13*E*,16*R*,20*R*,21*E*)-**1**.

Triseptatin (**1**) and deoxaphomin B (**2**) were biologically tested for their cytotoxicity in seven mammalian cell lines, including the mouse fibroblast line (L929), HeLa carcinoma line (KB3.1), human breast adenocarcinoma cell line (MCF-7), human lung carcinoma line (A549), human prostate cancer (PC-3), ovarian carcinoma cell line (SKOV-3), and squamous cell carcinoma (A431), using MTT assay with epothilone B as a positive control. Cytochalasans **1** and **2** exhibited cytotoxic activities against all cell lines with half-maximal inhibitory concentrations of 1.80 to 11.28 μM and 1.55 to 6.91 μM, respectively (Table 2). While comparable IC_{50} values were noted against most cancer cell lines, compound **2** exhibited five times potency versus the prostate cancer cells. However, the cytotoxic activity of the isolated cytochalasans were relatively weak in comparison to the half-maximal inhibitory concentration values obtained from epothilone B. Compounds **1–3** were also assessed for antiproliferative activity against human umbilical vein epithelial cell (HUVEC) and myelogenous leukemia cell (K-562) and cytotoxicity against HeLa cervical cancer cell line through CellTiter Blue assay (Table 2). All compounds exhibited antiproliferative effects and cytotoxic activities. Compounds **2** and **3** were cytotoxic against HeLa cells with IC_{50} values of 4.96 and 7.30 μM, respectively. Cytochalasins **1–3** showed moderate antileukemic activity, while strong antiproliferative effects were observed against HUVEC cells.

Table 2. Antiproliferative effect and cytotoxicity of **1–2** against mammalian cell ines.

Cell Line	Compound			Positive Control	
	1	2	3	Epothilone B	Imatinib
Cytotoxicity [a] IC_{50} (μM)					
Mouse fibroblast L929	4.16	5.18	N.D.	1.4×10^{-3}	N.D.
HeLa cells KB3.1	1.80	1.83	N.D.	8.9×10^{-5}	N.D.
Human breast adenocarcinoma MCF-7	1.86	1.79	N.D.	2.4×10^{-4}	N.D.
Human lung carcinoma A549	7.32	6.91	N.D.	6.9×10^{-5}	N.D.
Human prostate cancer PC-3	11.28	2.81	N.D.	1.6×10^{-3}	N.D.
Ovarian carcinoma SKOV-3	1.84	1.55	N.D.	2.8×10^{-4}	N.D.
Squamous cell carcinoma A431	2.17	1.60	N.D.	7.9×10^{-5}	N.D.
Antiproliferative Effect [b] GI_{50} (μM)					
HUVEC	4.55	1.08	8.31	N.D.	18.5
Myelogenous leukemia K-562	8.31	3.67	3.34	N.D.	0.17

[a] MTT assay, [b] CellTiter blue assay. The starting concentration for the cytotoxicity assay was 300 μg/mL, and substances were dissolved in MeOH (1 mg/mL). MeOH was used as the negative control and showed no activity against the tested mammalian cell lines. Results were expressed as IC_{50}: half-maximal inhibitory concentration, CC_{50}: half-maximal cytotoxicity concentration, GI_{50}: half-maximal cell proliferation (μM), N.D: not determined.

Compounds **1**, **2**, **3**, and cytochalasin D (**6**) were evaluated in 1 h endpoint assays with selected concentrations estimated based on previously determined half-maximal inhibitory concentrations, referred to as low dose (LD, $1 \times IC_{50}$) and high dose (HD, $5 \times IC_{50}$) for their impact on the F-actin network of U2-OS cells. F-actin-containing structures were monitored by staining with fluorescently coupled phalloidin (seen in red, Figures 5 and 6). LD of cytochalasin B and D (Figure 5c,d) led to F-actin networks devoid of lamellipodial structures with F-actin-rich protrusions at the cell periphery clearly visible in the DMSO vehicle control (Figure 5i) and F-actin-rich knot-like structures in the main cell body. Cytochalasin D (Figure 5h) in HD further increased the number of these aggregates, coinciding with a reduction in other visible F-actin-containing structures in the main cell body, while HD of cytochalasin B (Figure 5g)-treated cells showed a severe impediment but no total collapse of the F-actin network. The impact of **1** (Figure 5a,e) in this set of experiment most closely resembled the effect induced by cytochalasin B intoxication and was characterized by a lack of lamellipodial structures and the presence of F-actin-rich aggregates in the main cell body. This observation was much more pronounced in the high-dose experiment (Figure 5e), coinciding with a lower number of visible stress fibers. Deoxaphomin B (**2**) applied in LD (Figure 5b), however, led to strongly contrasted cable-like remnants in comparison to the dim fluorescence signal of phalloidin detected in the main cell body. A high dose (Figure 5f) led to a well-described endpoint of cytochalasin intoxication, with F-actin-containing aggregates dominating the otherwise collapsed F-actin network with no visible stress fibers. The impact of the tested compounds on the F-actin organization was fully reversible in all cases (compare Figure 5e,j, Figure 5f,k, Figure 5g,l, and Figure 5h,m with DMSO vehicle controls Figure 5i,n). Concurrently, the concentration-dependent effect of the cytochalasans on the F-actin cytoskeletal organization was examined incrementally (note that compound **1** was only assessed in a high-dose/low-dose assay due to scarcity of the compound). First, changes to the F-actin network presented themselves by a reduction in lamellipodia-like structures and formation of previously described phalloidin-stainable F-actin-containing aggregates (**2**: Figure 6d, 1 μM; **3**: Figure 6l, 1 μM; and **6**: Figure 6r, 0.1 μM). The effects did not grow more severe once a concentration was reached, which led to a full collapse of F-actin-containing macrostructures, with the F-actin aggregates remaining in the cell (**2**: Figure 6g, 30 μM; **3**: Figure 6n, 10 μM; and **6**: Figure 6l, 1 μM). Remarkably, the progression of F-actin network dispersal was different among cells treated with deoxaphomin B and cytochalasin B. Here, the periphery of the cells remained largely intact with strongly contrasted stress-fiber-like structures appearing at higher concentrations as had been observed before, which did not appear at any tested concentration after cytochalasin B and D treatment (compare Figure 6d–f with Figure 6m–o,k–w).

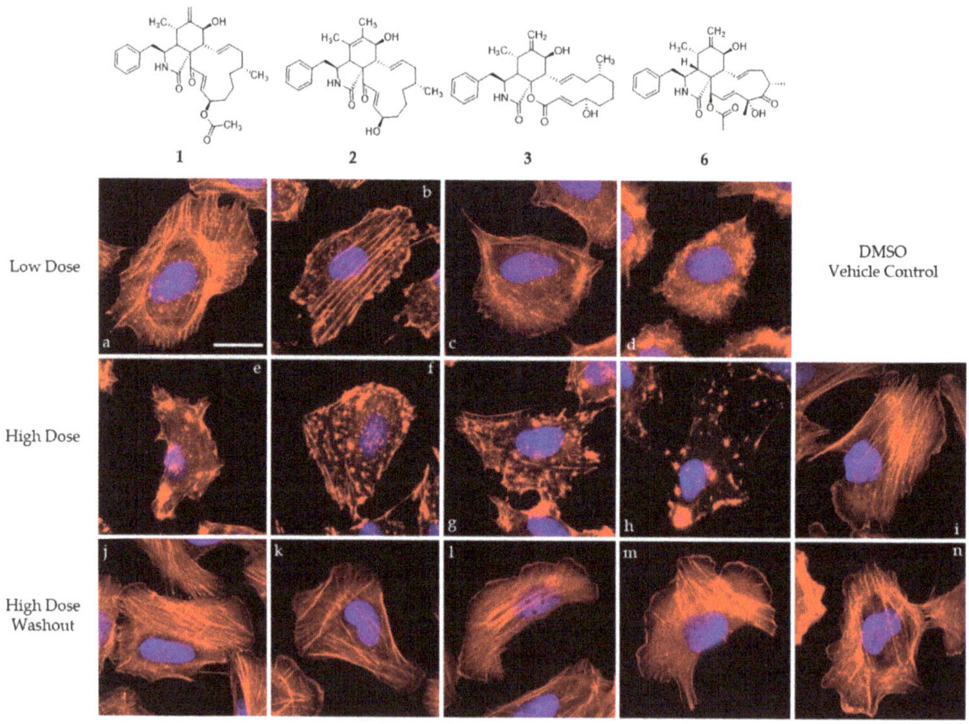

Figure 5. Overlay images of immobilized U2-OS cells treated with different concentrations of compounds **1–3** and **6** (**1**: (**a**,**e**,**j**), **2**: (**b**,**f**,**k**), **3**: (**c**,**g**,**l**), and **6**: (**d**,**h**,**m**)) normalized against their cytotoxicity against L929 cells for one hour (IC_{50}; low dose, $1 \times IC_{50}$ (**a**–**d**); high dose, $5 \times IC_{50}$ (**e**–**h**)) and the corresponding high-dose washout experiment after one hour recovery time (**j**–**m**). Volume of the DMSO vehicle control was adjusted to the highest volume of DMSO used in the corresponding experiment (**i**,**n**). Cells were stained for their F-actin cytoskeleton using fluorescently coupled phalloidin (pseudocoloured in red) and nuclear DNA using DAPI (pseudocoloured in blue). Representative scale bar in image (**a**) correspond to 25 μm.

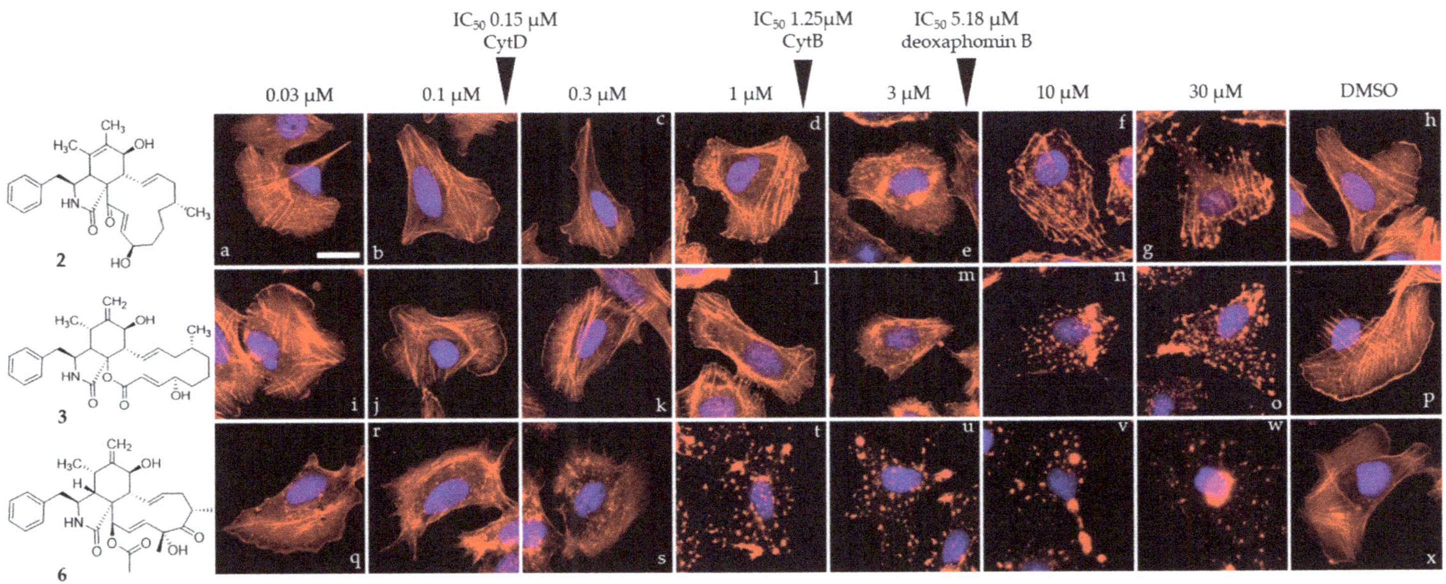

Figure 6. Overlay images of immobilized U2-OS cells treated with a gradually increasing concentration (0.03–30 μM) of compounds **2**, **3**, and **6** (**2**: (**a**–**g**), **3**: (**i**–**o**), and **3**: (**q**–**w**)) for one hour. IC_{50} values determined against L929 cells are denoted with arrows. Volume of the DMSO vehicle control was adjusted to the highest volume of DMSO used in the corresponding experiment (**h**,**p**,**x**). Cells were stained for their F-actin cytoskeleton using fluorescently coupled phalloidin (pseudocoloured in red) and nuclear DNA using DAPI (pseudocoloured in blue). Representative scale bar shown in image (**a**) represent 25 μm.

The observations fit well with the well-described phenotype of cytochalasin-induced F-actin disorganization [32,33]. Notably, deoxaphomin B seemed to have a stronger effect on stress fiber organization even in low doses compared to the other cytochalasans tested here. The reason for this behavior has to be scrutinized in the future once more material of the compound becomes available.

The deoxaphomin B reported here differed from deoxaphomin by the presence of a double bond inside the six-ring between C-5 and C-6 instead of deoxaphomin with a methylidene group of C-12, while triseptatin featured an additional O-acetyl group at the C-20 position. Interestingly, desoxaphomin was reported to exhibit irreversible effects on the F-actin organization of U2-OS cells, while the triseptatin tested here and the closely related deoxaphomin B were both shown to exhibit fully reversible effects even when applied in high concentrations [6]. Notably, Lambert et al. [7] reported that the newly described pseudofuscochalasin A exhibited irreversible effects, while cells treated with cytochalasin C, which differ from the former by an additional acetyl group at the ketone at C-23, were able to fully recover after one hour. The results presented here indicate that the α–β unsaturated bond next to the ketone should not be considered the solely decisive factor steering an unrecoverable impact on F-actin organization. To further understand the phenomenon of reversible and irreversible changes towards the F-actin network induced by cytochalasins, two emerging hypotheses have to be examined in the future: (1) the configuration of the six-ring impacting reversibility and (2) the stereochemistry of the hydroxy group at C-20 also exhibiting influences on reversibility. Recent findings comparing cytochalasin E derivatives indicated that a change from a methylidene group to two methyl groups influencing the stereochemistry at the six-ring found in cytochalasin K had a profound impact on its cytotoxicity, being 80 times less toxic than the former. Moreover, the effect was found to be partially reversible without observation of a full recovery after one hour (C. L. et al., unpublished data). Deoxaphomin B features the previously mentioned methylidene group but exhibits acetylation at the hydroxyl group with unknown consequences for the bioactivity in question. In the light of the current findings, hypothesis 1 seems more tempting, but further research focusing on finding a producer of a triseptatin featuring a C-20-deacetyl deoxaphomin B backbone is necessary to formally exclude hypothesis 2.

4. Conclusions

This study described the biologically active chemical constituents of the rice fermentation culture of *Sparticola triseptata*, where one new cytochalasin derivative, triseptatin (**1**), along with deoxaphomin B (**2**), cytochalasin B (**3**), *cis*-4-hydroxy-6-deoxyscytalone (**4**), and 6-hydroxymellein (**5**), were isolated and identified. Compounds **1** and **2** displayed cytotoxic activities against seven mammalian cell lines (L929, KB3.1, MCF-7, A549, PC-3, SKOV-3, and A431) and antiproliferative effects against myeloid leukemia K-562 cancer cell line. Further investigation on the mechanism of their cytotoxicity showed the reversible effects of cytochalasans **1–3** on the disruption of the actin cytoskeleton in vivo. Our findings indicate that the presence of α–β unsaturated bond adjacent to the ketone group at C-23 does not directly correlate with the reversibility of this effect. Rather, two previously unrecognized structural features of the cytochalasin framework could have significant impact on the reversible actin disruption, namely the configuration of the cyclohexene ring and the chirality of the hydroxy group attached at C-20. In general, we have established the potential of cytochalasin derivatives as a possible drug inspiration for anticancer agents.

Supplementary Materials: The following are available online at https://www.mdpi.com/article/10.3390/jof8060560/s1, Figure S1: 1H NMR spectrum (MeOH-d_4, 600 MHz) of triseptatin (**1**); Figure S2: ^{13}C NMR spectrum (MeOH-d_4, 600 MHz) of triseptatin (**1**); Figure S3: HSQC–DEPT spectrum of triseptatin A (**1**); Figure S4: COSY spectrum of triseptatin (**1**); Figure S5: HMBC spectrum of triseptatin (**1**); Figure S6: ROESY spectrum of triseptatin (**1**); Figure S7: HR-ESI-MS spectrum of triseptatin (**1**); Figure S8: Low-energy conformers (>1%) of (*S*)-**1** optimized at B3LYP/6-31G(d) (PCM/MeOH); Figure S9: 1H NMR spectrum (MeOH-d_4, 600 MHz) of deoxaphomin B (**2**); Figure S10: 13C NMR spectrum (MeOH-d_4, 600 MHz) of deoxaphomin B (**2**); Figure S11: ^{1}H NMR spectrum (MeOH-d_4,

600 MHz) of cytochalasin B (**3**); Figure S12: 13C NMR spectrum (MeOH-d_4, 600 MHz) of cytochalasin B (**3**); Figure S13: ^{1}H NMR spectrum (MeOH-d_4, 600 MHz) of cis-4-hydroxy-6-deoxyscytalone (**4**); Figure S14: ^{13}C NMR spectrum (MeOH-d_4, 600 MHz) of cis-4-hydroxy-6-deoxyscytalone (**4**); Figure S15: 1H NMR spectrum (MeOH-d_4, 500 MHz) of 6-hydroxymellein (**5**); Figure S16: ^{13}C NMR spectrum (MeOH-d_4, 125 MHz) of 6-hydroxymellein (**5**). References [26–33] are cited in the supplementary materials.

Author Contributions: Conceptualization, K.Y.M.G., M.T.J.Q., M.S., F.S. and A.P.G.M.; methodology, K.Y.M.G., M.T.J.Q., C.L., K.S., G.P., T.E.B.S., A.R., H.-M.D. and C.P.; software, M.T.J.Q.; validation, K.Y.M.G., M.T.J.Q., M.S., H.-M.D., F.S. and A.P.G.M.; formal analysis, K.Y.M.G., M.T.J.Q., C.L., K.S., G.P., T.E.B.S., A.R., H.-M.D., C.P., F.S. and A.P.G.M.; investigation, K.Y.M.G., M.T.J.Q., C.L., G.P., K.S., T.E.B.S., A.R., H.-M.D., C.P., F.S. and A.P.G.M.; resources, K.Y.M.G., M.T.J.Q., M.S., A.R., H.-M.D., F.S. and A.P.G.M.; data curation, K.Y.M.G., M.T.J.Q., C.L., K.S. and T.E.B.S.; writing—original draft preparation, K.Y.M.G., M.T.J.Q., C.L., T.E.B.S., C.P., K.S., F.S. and A.P.G.M.; writing—review and editing, M.S., H.-M.D., F.S. and A.P.G.M.; visualization, K.Y.M.G. and M.T.J.Q.; supervision, M.S., F.S. and A.P.G.M.; project administration, A.P.G.M.; funding acquisition, K.Y.M.G. and A.P.G.M. All authors have read and agreed to the published version of the manuscript.

Funding: K.Y.M.G. acknowledges a scholarship grant from the Department of Science and Technology (DOST) through its Accelerated Science and Technology Human Resource Development Program (ASTHRDP). C.L. is grateful for a stipend from the Life Science Foundation, Braunschweig, Germany. G.P. is grateful for a PhD stipend from the German Academic Exchange Service. T.E.B.S., K.S., and M.S. are grateful for funding from the Deutsche Forschungsgemeinschaft (Research Unit CytoLabs; FOR 5170). A.P.G.M and M.S. are also indebted to the Alexander von Humboldt Foundation for a Digital Cooperation Fellowship.

Institutional Review Board Statement: Not applicable.

Informed Consent Statement: Not applicable.

Data Availability Statement: Not applicable.

Acknowledgments: We thank Wera Collisi and Christel Kakoschke for expert technical assistance.

Conflicts of Interest: The authors declare no conflict of interest.

References

1. Skellam, E. The biosynthesis of cytochalasans. *Nat. Prod. Rep.* **2017**, *34*, 1252–1263. [CrossRef] [PubMed]
2. Rothweiler, W.; Tamm, C. Isolation and structure of phomin. *Experentia* **1966**, *22*, 750–752. [CrossRef]
3. Aldridge, D.C.; Armstrong, J.J.; Speake, R.N.; Turner, W.B. The cytochalasins, a new class of biologically active mould metabolites. *Chem. Comm.* **1967**, *26*, 1667–1676. [CrossRef]
4. Capasso, R.; Evidente, A.; Ritieni, A.; Randazzo, G.; Vurro, M.; Bottalico, A. Ascochalasin, a new cytochalasin from *Ascochyta heteromorpha*. *J. Nat. Prod.* **1988**, *51*, 567–571. [CrossRef] [PubMed]
5. Scherlach, K.; Boettger, D.; Remme, N.; Hertweck, C. The chemistry and biology of cytochalasans. *Nat. Prod. Rep.* **2010**, *27*, 869–886. [CrossRef] [PubMed]
6. Kretz, R.; Wendt, L.; Wongkanoun, S.; Luangsa-ard, J.J.; Surup, F.; Helaly, S.E.; Noumeur, S.R.; Stadler, M.; Stradal, T.E.B. The effect of cytochalasans on the actin cytoskeleton of eukaryotic cells and preliminary structure–activity relationships. *Biomolecules* **2019**, *9*, 73. [CrossRef]
7. Lambert, C.; Pourmoghaddam, M.J.; Cedeño-Sanchez, M.; Surup, F.; Khodaparast, S.A.; Krisai-Greilhuber, I.; Voglmayr, H.; Stradal, T.E.B.; Stadler, M. Resolution of the *Hypoxylon fuscum* complex (Hypoxylaceae, Xylariales) and discovery and biological characterization of two of its prominent secondary metabolites. *J. Fungi* **2021**, *7*, 131. [CrossRef]
8. Betina, V.; Micekova, D.; Nemec, P. Antimicrobial properties of cytochalasins and their alteration of fungal morphology. *J. Gen. Microbiol.* **1972**, *71*, 343–349. [CrossRef]
9. Horn, W.S.; Simmonds, M.S.J.; Blaney, W.M. Phomopsichalasin, a novel antimicrobial agent from an endophytic *Phomopsis* sp. *Tetrahedron* **1995**, *51*, 3969–3978. [CrossRef]
10. Makioka, A.; Kumagai, M.; Kobayashi, S.; Takeuchi, T. Effect of proteasome inhibitors on the growth, encystation, and excystation of *Entamoeba histolytica* and *Entamoeba invadens*. *Parasitol. Res.* **2004**, *93*, 68–71. [CrossRef]
11. Jayasuriya, H.; Herath, K.B.; Ondeyka, J.G.; Polishook, J.D.; Bills, G.F.; Dombrowski, A.W.; Springer, M.S.; Siciliano, S.; Malkowitz, L.; Sanchez, M.; et al. Isolation structure of antagonists of chemokine receptor (CCR5). *J. Nat. Prod.* **2004**, *67*, 1036–1038. [CrossRef] [PubMed]

12. Rochfort, S.; Ford, J.; Ovenden, S.; Wan, S.S.; George, S.; Wildman, H.; Tait, R.M.; Meurer-Grimes, B.; Cox, S.; Coates, J.; et al. A novel aspochalasin with HIV-1 integrase inhibitory activity from *Aspergillus flavipes*. *J. Antibiot.* **2005**, *58*, 279–283. [CrossRef] [PubMed]
13. Schofield, J.G. Cytochalasin B and release of growth hormone. *Nat. New Biol.* **1971**, *234*, 215–216. [CrossRef] [PubMed]
14. Williams, J.A.; Wolff, J. Cytochalasin B inhibits thyroid secretion. *Biochem. Biophys. Res. Commun.* **1971**, *44*, 422–425. [CrossRef]
15. Crivello, J.F.; Jefcoate, C.R. Intracellular movement of cholesterol in rat adrenal cells. Kinetics and effects of inhibitors. *J. Biol. Chem.* **1980**, *255*, 8144–8151. [CrossRef]
16. Yuyama, K.T.; Wendt, L.; Surup, F.; Chepkirui, C.; Wittstein, K.; Boonlarppradab, C.; Wongkanoun, S.; Luangsa-ard, J.; Stadler, M.; Abraham, W.-R. Cytochalasans act as inhibitors of biofilm formation of *Staphylococcus aureus*. *Biomolecules* **2018**, *8*, 129. [CrossRef]
17. Rampal, A.L.; Pinkofsky, H.B.; Jung, C.Y. Structure of cytochalasins and cytochalasin B binding sites in human erythrocyte membranes. *Biochemistry* **1980**, *19*, 679–683. [CrossRef]
18. George, T.P.; Cook, H.W.; Byers, D.M.; Palmer, F.B.S.C.; Spence, M.W. Inhibition of phosphatidylcholine and phosphatidylethanolamine biosynthesis by cytochalasin B in cultured glioma cells: Potential regulation of biosynthesis by Ca^{2+}-dependent mechanisms. *Biochim. Biophys. Acta.* **1991**, *1084*, 185–193. [CrossRef]
19. Leuchtmann, A. *Phaeosphaera padellana* und *Massariosphaeria triseptata*, zwei neue bitunicate Ascomyceten aus den Alpen. *Mycol. Helv.* **1987**, *2*, 183–191.
20. Phukhamsakda, C.; Ariyansawa, H.A.; Phillips, A.J.L.; Wanasinghe, D.N.; Bhat, D.J.; McKenzie, E.H.C.; Hyde, K.D. Additions to sporormiaceae: Introducing two novel genera, *Sparticola* and *Forliomyces*, from *Spartium*. *Cryptogam. Mycol.* **2016**, *37*, 75–97. [CrossRef]
21. Hanwell, M.D.; Curtis, D.E.; Lonie, D.C.; Vandermeerschd, T.; Zurek, E.; Hutchison, G.R. Avogadro: An advanced semantic chemical editor, visualization, and analysis platform. *J. Cheminf.* **2012**, *4*, 17–34. [CrossRef] [PubMed]
22. Frisch, M.J.; Trucks, G.W.; Schlegel, H.B.; Scuseria, G.E.; Robb, M.A.; Cheeseman, J.R.; Scalmani, G.; Barone, V.M.; Petersson, G.A.; Nakatsuji, H.; et al. *Gaussian16, Revision A.03*; Gaussian, Inc.: Wallingford, CT, USA, 2016.
23. Karuth, F.; Dahse, H.-M.; Rüttinger, H.-H.; Frohberg, P. Synthesis and characterization of novel 1,2,4-triazine derivatives with antiproliferative activity. *Bioorg. Med. Chem.* **2010**, *18*, 1816–1821. [CrossRef] [PubMed]
24. Becker, K.; Wessel, A.C.; Luangsa-ard, J.J.; Stadler, M. Viridistratins A–C, antimicrobial and cytotoxic benzo[j]fluoranthenes from stromata of *Annulohypoxylon viridistratum* (Hypoxylaceae, Ascomycota). *Biomolecules* **2020**, *10*, 805. [CrossRef] [PubMed]
25. Wang, C.; Lambert, C.; Hauser, M.; Deuschmann, A.; Zeilinger, C.; Rottner, K.; Stradal, T.E.B.; Stadler, M.; Skellam, E.J.; Cox, R.J. Diversely Functionalised Cytochalasins via Mutasynthesis and Semi-Synthesis. *Chem. Eur. Chem.* **2020**, *26*, 13578–13583. [CrossRef]
26. Kim, E.L.; Li, J.L.; Dang, H.T.; Hong, J.; Lee, C.O.; Kim, D.K.; Yoon, W.D.; Kim, E.; Liu, Y.; Jung, J.H. Cytotoxic cytochalasins from the endozoic fungus *Phoma* sp. of the giant jellyfish *Nemopilema nomurai*. *Bioorg. Med. Chem. Lett.* **2012**, *22*, 3126–3129. [CrossRef]
27. Couché, E.; Fkyerat, A.; Tabacchi, R. Stereoselective Synthesis of *cis*- and *trans*-3,4-dihydro-3,4,8-trihydroxynaphthalen-1(2H)-one. *Helv. Chim. Acta.* **2009**, *92*, 903–917. [CrossRef]
28. Islam, M.S.; Ishigami, K.; Watanabe, H. Synthesis of (-)-Mellein (I), (+)-Ramulosin (II), and related natural products. *Tetrahedron* **2007**, *63*, 1074–1079. [CrossRef]
29. Chen, Z.; Chen, H.-P.; Li, Y.; Feng, T.; Liu, J.-K. Cytochalasins from cultures of endophytic fungus *Phoma multirostrata* EA-12. *J. Antibiot.* **2014**, *68*, 23–26. [CrossRef]
30. Chen, C.; Wang, J.; Liu, J.; Zhu, H.; Sun, B.; Wang, J.; Zhang, J.; Luo, Z.; Yao, G.; Xue, Y.; et al. Armochaetoglobins A–J: Cytochalasan alkaloids from *Chaetomium globosum* TW1-1, a fungus derived from the terrestrial arthropod *Armadillidium vulgare*. *J. Nat. Prod.* **2015**, *78*, 1193–1201. [CrossRef]
31. Evidente, A.; Andolfi, A.; Vurro, M.; Zonno, M.C.; Motta, A. Ascochalasin, a new cytochalasin from *Ascochyta heteromorpha*. *J. Nat. Prod.* **2003**, *66*, 1540–1544. [CrossRef]
32. Yahara, I.; Harada, F.; Sekita, S.; Yoshihira, K.; Natori, S. Correlation between effects of 24 different cytochalasins on cellular structures and cellular events and those on actin in vitro. *J. Cell Biol.* **1982**, *92*, 69–78. [CrossRef] [PubMed]
33. Cooper, J.A. Effects of cytochalasin and phalloidin on actin. *J. Cell Biol.* **1987**, *105*, 1473–1478. [CrossRef] [PubMed]

Journal of
Fungi

MDPI

Article

Polyketide Derivatives from the Endophytic Fungus *Phaeosphaeria* sp. LF5 Isolated from *Huperzia serrata* and Their Acetylcholinesterase Inhibitory Activities

Yiwen Xiao [1,2,†], Weizhong Liang [2,†], Zhibin Zhang [1], Ya Wang [2], Shanshan Zhang [2], Jiantao Liu [2], Jun Chang [2], Changjiu Ji [3] and Du Zhu [1,2,*]

1 Key Laboratory of Protection and Utilization of Subtropic Plant Resources of Jiangxi Province, College of Life Sciences, Jiangxi Normal University, Nanchang 330022, China; xyw1152858687@163.com (Y.X.); zzbbio@jxnu.edu.cn (Z.Z.)
2 Key Laboratory of Bioprocess Engineering of Jiangxi Province, College of Life Sciences, Jiangxi Science and Technology Normal University, Nanchang 330013, China; lwz271508312@163.com (W.L.); jxwangya@126.com (Y.W.); zss18135437140@163.com (S.Z.); liujt@jxstnu.edu.cn (J.L.); changjun@jxstnu.edu.cn (J.C.)
3 College of Chemistry and Biological Engineering, Yichun University, Yichun 336000, China; jichjiu999@jxycu.edu.cn
* Correspondence: zhudu@jxstnu.edu.cn; Tel.: +86-79-188121934
† These authors contributed equally to this work.

Abstract: The secondary metabolites of *Phaeosphaeria* sp. LF5, an endophytic fungus with acetylcholinesterase (AChE) inhibitory activity isolated from *Huperzia serrata*, were investigated. Their structures and absolute configurations were elucidated by means of extensive spectroscopic data, including one- and two-dimensional nuclear magnetic resonance (NMR), high-resolution electrospray ionization mass spectrometry (HR-ESI-MS) analyses, and calculations of electronic circular dichroism (ECD). A chemical study on the solid-cultured fungus LF5 resulted in 11 polyketide derivatives, which included three previously undescribed derivatives: aspilactonol I (**4**), 2-(1-hydroxyethyl)-6-methylisonicotinic acid (**7**), and 6,8-dihydroxy-3-(1′*R*, 2′*R*-dihydroxypropyl)-isocoumarin (**9**), and two new natural-source-derived aspilactonols (G, H) (**2**, **3**). Moreover, the absolute configuration of de-*O*-methyldiaporthin (**11**) was identified for the first time. Compounds **4** and **11** exhibited inhibitory activity against AChE with half maximal inhibitory concentration (IC_{50}) values of 6.26 and 21.18 μM, respectively. Aspilactonol I (**4**) is the first reported furanone AChE inhibitor (AChEI). The results indicated that *Phaeosphaeria* is a good source of polyketide derivatives. This study identified intriguing lead compounds for further research and development of new AChEIs.

Keywords: *Phaeosphaeria* sp.; secondary metabolite; polyketide; AChE inhibitor; biosynthetic pathways

Citation: Xiao, Y.; Liang, W.; Zhang, Z.; Wang, Y.; Zhang, S.; Liu, J.; Chang, J.; Ji, C.; Zhu, D. Polyketide Derivatives from the Endophytic Fungus *Phaeosphaeria* sp. LF5 Isolated from *Huperzia serrata* and Their Acetylcholinesterase Inhibitory Activities. *J. Fungi* **2022**, *8*, 232. https://doi.org/10.3390/jof8030232

Academic Editors: Tao Feng and Frank Surup

Received: 3 February 2022
Accepted: 24 February 2022
Published: 26 February 2022

1. Introduction

Natural products are important sources of active compounds and play important roles in modern drug research and development. Fungi are considered an important group of microorganisms in the production of antitumor, immunosuppressant, antibiotic, antifungal, antiparasitic, anti-inflammatory, enzyme-inhibiting, and other active secondary metabolites [1,2]. Endophytic fungi reside in the internal tissues of living plants without causing apparent disease. Due to their unique ecological niche, endophytic fungi have become important sources of natural products to be screened for with unique chemical structures and biological activity [3,4]. Therefore, the natural product screening of endophytic fungi is currently a hot research topic [5–9]. In this sense, it is worth undertaking a constant search for novel compounds from endophytic fungal sources and paying attention to discovering potential drug candidates.

Huperzia serrata is a member of the *Lycophyllaceae* family called shezucao in China [10]. Huperzine A (HupA) was first isolated from *H. serrata* in the 1980s and was approved in the 1990s in China as an acetylcholinesterase inhibitor (AChEI) to treat Alzheimer's disease (AD). These promising studies showed that endophytic fungi of *H. serrata* can synthesize HupA and similar compounds in host plants and also contains many novel compounds [10]. Thus far, many studies have been conducted on the diversity of endophytic fungi of *H. serrata*, but there are few studies on the isolation and screening of AChEIs in these endophytic fungi, and therefore further studies are needed. In our previous study, a total of 22 endophytic fungal strains showed strong AChE inhibitory activity (≥50%) [11]. As part of our ongoing research, we are currently characterizing the bioactive secondary metabolites of these endophytic fungi.

Phaeosphaeria sp. LF5 is an endophytic fungus isolated from the leaves of *H. serrata* [12]. Members of *Phaeosphaeria* have afforded a variety of natural products, such as polyketides, peptides, and terpenes [13,14]. Herein, *Phaeosphaeria* sp. LF5 was selected for screening for new AChEI natural products. We refermented the strain in solid substrate fermentation medium and then isolated 11 polyketide derivatives, which included three new compounds, aspilactonol I (**4**), 2-(1-hydroxyethyl)-6-methylisonicotinic acid (**7**), and 6,8-dihydroxy-3-(1′*R*, 2′*R*-dihydroxypropyl)-isocoumarin (**9**), and two new natural source-derived aspilactonols (G, H) (**2**, **3**). We also identified the absolute configuration of de-*O*-methyldiaporthin (**11**) for the first time (Figure 1). In addition, we detected their AChE inhibitory activity. Herein, the isolation, structural elucidation, and bioactivities of these isolated compounds are described.

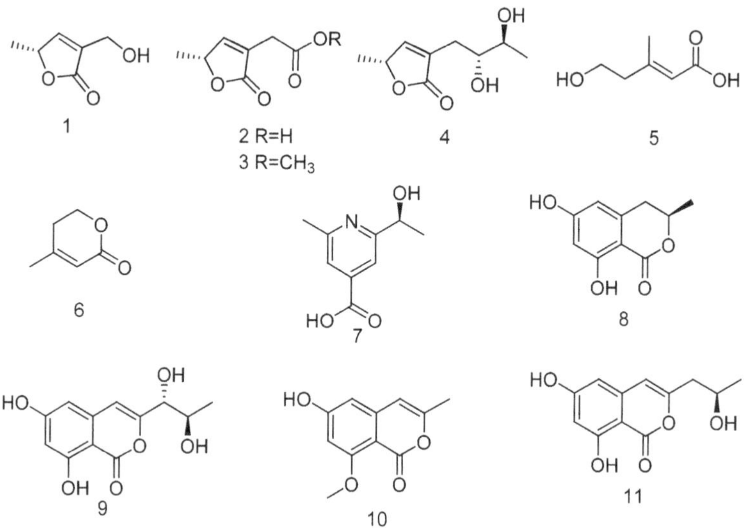

Figure 1. Chemical structure of compounds **1–11**.

Generally, furanone derivatives are polyketide metabolites found in *Aspergillus* [15,16]. They are classified into three structural types: linear (aspinonene), δ-lactones (aspinrone), and γ-lactones (isoaspinonene and aspilactonols) [17]. To date, it has been a challenging task to assign the absolute configurations of furanone derivatives due to the flexibility of their aliphatic sidechain in the partial polyketide unit [18,19]. Isocoumarins comprise a six-membered oxygen heterocycle (α-pyranone) with one benzene ring. Isocoumarins represent a group of natural compounds rich in lactones, which are mainly derived from the fungal polyketone pathway. These compounds have exhibited a wide range of biological functions, including antifungal, anti-inflammatory, insecticidal, and hepatoprotective activities [20]. However, the determination of their absolute configuration becomes quite challenging due to the high degree of free rotation of the steric centers on the chain, with the side chains connected to the nuclei of isocoumarin derivatives [21]. In the present study, the structures and absolute configurations of polyketide derivatives isolated from *Phaeosphaeria* sp. LF5 were elucidated by means of extensive spectroscopic data, including one- and two-dimensional nuclear magnetic resonance (NMR) spectrometry, high-resolution electrospray ionization mass spectrometry (HR-ESI-MS) analyses, and calculations of electronic circular dichroism (ECD).

2. Materials and Methods

2.1. General Experimental Procedures

Optical rotation values were determined on a JASCO P-1010 polarimeter (Jasco, Tokyo, Japan). UV spectra were recorded on a PerkinElmer Lambda 365 UV-Vis spectrophotometer (PerkinElmer, Hopkinton, MA, USA). High-resolution electrospray ionization mass spectrometry (HR-ESI-MS) data were measured on a Waters ACQUITY UPLC H-Class Q-TOF LC-MS spectrometer (Waters, Milford, MA, USA). High-performance liquid chromatography (HPLC) analysis was carried out on an ACQUITY UPLC H-Class System (quaternary solvent manager, sample manager, PDA detector, and ELS detector) using a YMC ODS (4.6 × 250 mm, 5 μm, 1 mL/min) column. MPLC was performed on a PuriFlash450 (Interchim, Los Angeles, CA, USA) with a Flash C18 cartridge (50 μm, 40 g, YMC, Kyoto, Japan). Semipreparative HPLC was performed on a Waters 2535 Quaternary gradient module with a FlexInject, 2489 UV–VIS detector and Fraction Collector III (Waters, Milford, MA, USA). The NMR spectra were recorded on a Bruker Avance 400 MHz spectrometer using tetramethylsilane as the internal standard (Bruker, Ettlingen, Germany). Thin-layer chromatography (TLC) analyses were performed on glass precoated with silica gel GF254 glass plates. All reagents for the analysis were purchased from Xilong Scientific Co., Ltd. (Guangdong, China).

2.2. Fungal Material

The endophytic fungus *Phaeosphaeria* sp. LF5 was isolated from the leaves of *H. serrata* at the Chinese Academy of Sciences' Lushan Botanical Garden in Jiangxi Province, China [12]. This strain was deposited in the culture collection of the Key Laboratory of Protection and Utilization of Subtropical Plant Resources of Jiangxi Province, Jiangxi Normal University.

2.3. Fermentation and Extraction

The endophytic fungus LF5 was cultivated in 100 Erlenmeyer flasks (1000 mL); each flask contained 80 g of rice and 120 g of H_2O to create solid rice medium. The flasks were then static incubated at 28 °C for 40 days.

After the mycelia entered the static growth state, the rice solid fermentation was taken out of the Erlenmeyer flask, dried at 45 °C to remove the water, crushed before adding 80% ethanol, and ultrasonically agitated for 1 h. The static precipitation was filtered, and the above steps were repeated four times to obtain an ethanol extract. The filtrate was removed with a rotary evaporator (35 °C, 160 rpm). The ethanol crude extract was placed in 2000 mL of water and transferred to a separatory funnel. In turn, petroleum ether (PE), ethyl acetate (EA), and water-saturated butanol were used for extraction four times and were concentrated in vacuo to yield the combined crude extracts, PE extract (16.2 g), EA extract (76 g), n-butanol extract (105 g), and water extract (450 g).

2.4. Isolation and Purification

The EA extract (76.0 g) was dried and subjected to column chromatography on 200–300 mesh silica gel with different solvents of increasing polarity from PE to EA to MeOH to obtain eight fractions (Frs. 1–8) on the basis of TLC analysis.

Fraction 6 was purified by Sephadex LH-20 (GE Healthcare, Pittsburgh, PA, USA) (MeOH) to obtain three subfractions: Fr. 6.1–Fr. 6.3. Fraction 6.1 was further purified by semipreparative HPLC (CH_3OH/H_2O, 30:70, *v/v*) to yield **1** (12 mg, t_R = 4.0 min),

2 (9.5 mg, t_R = 2.9 min), and **3** (7 mg, t_R = 7.0 min). Fraction 8 was loaded onto a Sephadex LH-20 column and eluted with EA/CH_3OH (8:2) to yield two subfractions: Fr. 8.1 and Fr. 8.2. Subfraction Fr. 8.2 was separated by semi-preparative RP-HPLC (CH_3OH/H_2O, 10:90) to generate **4** (6.8 mg, t_R = 11.0 min) and **5** (4.0 mg, t_R = 8.0 min). Fraction 3 was separated by Sephadex LH-20 using CH_3OH as the eluting solvent and was then further purified via semipreparative HPLC CH_3OH/H_2O (30:70, *v/v*) to obtain compound **6** (2.9 mg, t_R = 6.0 min). Fraction 7 was further purified using Sephadex LH-20 (EA/CH_3OH, 90:10, *v/v*) to yield two subfractions: Frs. 7.1 and 7.2. Subfraction Fr. 7.1 was separated by semipreparative reversed-phase HPLC (RP-HPLC) (CH_3OH-H_2O, 5:95) to produce **7** (4.5 mg, t_R = 4.5 min). Fraction 2 (PE/EA 7:3) was separated into three subfractions (Fr. 2.1–2.3) with Sephadex LH-20 using MeOH as a mobile phase. Subfraction Fr. 2.1 was further purified via semipreparative HPLC using CH_3OH:H_2O (30:70) as a mobile phase at a flow rate of 5 mL/min to yield **8** (10 mg, t_R = 36.0 min). Subfraction Fr. 6.3 was separated by preparative HPLC (CH_3OH/H_2O, 30:70, *v/v*) to yield compounds **9** (8 mg, t_R = 17 min), **10** (5 mg, t_R = 30.0 min), and **11** (4 mg, t_R = 45.0 min).

2.5. *Acetylcholinesterase Inhibitory Activity In Vitro Assay*

The determination of the in vitro AChE inhibitory activity of the endophytic fungal extracts and compounds **1–11** was performed according to the spectrophotometry method developed by Ellman et al. [22] and modified by Ortiz et al. [23]. Rivastigmine and HupA, two known AChEIs, were used as positive controls. The assay was carried out in a 96-well microtiter plate reader. In brief, a preincubation solution of 250 μL of phosphate buffer (200 mM, pH 7.7) that contained 15 μL of purified compounds/HupA, 80 μL of DTNB (3.96 mg of DTNB and 1.5 mg of sodium bicarbonate dissolved in 10 mL of phosphate buffer, pH 7.7), and 10 μL of AChE was prepared. The mixture was incubated for 5 min at 25 °C. After preincubation, 15 μL of the substrate AChI (10.85 mg in 5 mL of phosphate buffer) was added and incubated again for 5 min. The color developed was measured in a microwell plate reader at 412 nm (Molecular Devices, SpectraMax M2, San Jose, CA, USA). Percent inhibition was calculated through the following formula: (control absorbance–sample absorbance)/control absorbance × 100. The IC_{50} values were the means ± SD of three determinations.

2.6. *ECD Calculations*

In general, conformational analyses were performed by random searching in Sybyl-X 2.0 using the MMFF94S force field with an energy cut-off of 5 kcal/mol (Sybyl Software, version X 2.0, 2013) [24]. The results showed the five lowest-energy conformers for compounds **4**, **7**, **9**, and **11**. Subsequently, geometric optimizations and frequency analyses were implemented at the B3LYP-D3(BJ)/6-31G* level in PCM MeOH using ORCA4.2.1 [25,26]. All conformers used for property calculations in this study were characterized as stable points on a potential energy surface with no imaginary frequencies. The excitation energies, oscillator strengths, and rotational strengths (velocity) of the first 60 excited states were calculated by the time-dependent density-functional theory (TD-DFT) at the PBE0/def2-TZVP level in MeOH. The ECD spectra were simulated by the overlapping Gaussian function (half the bandwidth at 1/e peak height, sigma = 0.30 for all) [27]. The Gibbs free energies for the conformers were determined by using thermal correction at the B3LYP-D3(BJ)/6-31G** level, and electronic energies were evaluated at the wB97M-V/def2-TZVP level in PCM MeOH using ORCA4.2.1 [25,26]. To obtain the final spectra, we used the Boltzmann distribution theory and the conformers' relative Gibbs free energy (ΔG) to average the simulated spectra. The absolute configuration of the only chiral center was determined by comparing the experimental spectra to the calculated molecular models.

3. Results and Discussion

Structure Elucidation

Compound **4**, a white powder soluble in methanol (MeOH), exhibited a pseudo-molecular ion peak at *m/z* 187.0965 $[M + H]^+$ (calculated for $C_9H_{15}O_4^+$: 187.0926) in the HR-ESI-MS spectrum, indicating a molecular formula of $C_9H_{14}O_4$, and two degrees of unsaturation. The ^1H NMR data indicated two methyls at δ_H 1.20 (3H, d, J = 6.1 Hz, H-10) and 1.39 (3H, d, J = 6.8 Hz, H-6); one olefinic methane at δ_H 7.36 (1H, br s, H-4); and three oxymethines at δ_H 3.58 (1H, m, H-9), 3.60 (1H, m, H-8), and 5.09 (1H, br q, J = 6.8 Hz, H-5). The ^{13}C-NMR spectra revealed one ester carbonyl (δ_C 176.5), one olefinic methine (δ_C 154.2), one nonprotonated sp^2 carbon (δ_C 131.7), three oxygenated methines (δ_C 71.6, 74.8, and 79.8), one methylene (δ_C 29.6), and two methyls (δ_C 18.9 and 19.1).

The HMBC spectrum showed correlations between H-10 (δ_H 1.20)/C-8 (δ_C 74.8) and H-10 (δ_H 1.20)/C-9 (δ_C 71.6), as well as H-9 (δ_H 3.58)/C-7 (δ_C 29.6), H-9 (δ_H 3.58)/C-8, and H-9 (δ_H 3.58)/C-10. C-8 was correlated with C-3, C-7, and C-9 (Figure 2). When combined with the peak shape analysis of H-10 (δ_H 1.20, d, J = 6.1), C-9 was found to be connected to C-10, and C-8 was connected to C-9. Since C-8 and C-9 are methylene carbons and there is no nitrogen in the molecular formula, when combined with the chemical shift value, C-8 and C-9 were found to be connected to hydroxyl groups. H-7 is related to C-8, C-9, C-3, and C-4, and H-7 is methylene, but there were two groups of different H signals. Therefore, it can be inferred that one side of C-7 was connected to C-8. C-3 had the same characteristic signals as H-4, H-7, and H-8. The carbon shift signals of C-3 and C-4 were δ_C 131.7 and δ_C 154.2, respectively. HSQC indicated that C-4 was connected by protons, and it can be concluded that C-3 and C-4 were connected by a double bond and that C-4 was a quaternary carbon with two substitutions, one of which was connected with C-7. According to the HMBC cross-peak correlation of H-4/C-2 (δ_C 176.5), H-7/C-2, and C-3, we were able to infer that the side-chain fragment was attached to α,β-unsaturated-γ-lactone. HSQC indicated that C-2 was not connected to a proton. When combined with the molecular formula, C-4 was found to be a carbonyl group, H-4 was related to C-5, H-5 was related to C-6, and H-4 was related to HMBC. It was concluded that C-5 was connected to C-4, C-6 was connected to C-5, and the chemical shift value of C-5 was δ_C 79.8. When combined with the molecular formula, C-5 was also found to be connected to oxygen. Since the unsaturation degree of the compound was 2, the double bond between C-4 and C-3 occupied an unsaturation. An unsaturation remained, and there was no other double-bond carbon signal in the compound: hence, it is inferred that there was a cyclization system in which the chemical shift of C-2 was lower-field than that of the conventional carbonyl group. It was inferred that the other side of C-2 was connected to the oxygen and that C-5 and C-2 passed through the oxygen to form a lactone ring. The ^1H (CD_3OD, 400 MHz) and ^{13}C-NMR (CD_3OD, 100 MHz) data are listed in Table 1.

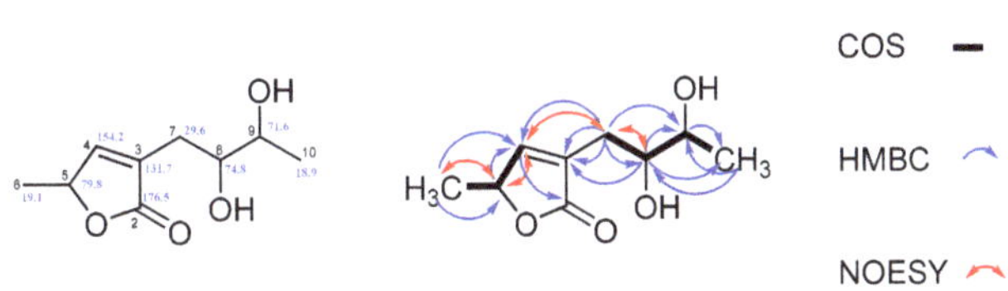

Figure 2. Signal assignments of compound **4** via ^{13}C and 2D NMR.

The ^{13}C chemical shift calculation was carried out at the B3LYP-D3(BJ)/6-31G** level to obtain the accurate relative configuration of **4**. In addition, the absolute configurations of **4** (5*R*, 8*R*, and 9*S*) were established by comparing electronic circular dichroism (ECD) calculations at the PBE0/def2-TZVP level with the experimental one (Figure 3). In addition, high correlation coefficients (R^2) between experimental and calculated chemical shifts were shown, with 0.9985 for **4** (Figure 4), indicating that the δ_C of **4** matched the calculated δ_C very well, which confirmed the framework of **4**. The structure of compound **4** was determined to be (*R*)-5-((8*R*, 9*S*)-8, 9-dihydroxybutyl)-5-methylfuran-2(5H)-one, so compound **4** was named aspilactonol I, as shown in Figure 1 (See Supplementary material).

Table 1. The ^{1}H, ^{13}C, HMBC, and HSQC NMR data for compound **4** in CD_3OD.

No.	^{1}H NMR	^{13}C NMR	HMBC	HSQC
2	—	176.5 (s)	H-4, H-7	—
3	—	131.7 (s)	H-4, H-7, H-8	—
4	7.36 (1H, br s)	154.2 (d)	H-5, H-6, H-7	7.36
5	5.09 (1H, br q, 6.8)	79.8 (d)	H-4, H-6	5.09
6	1.39 (3H, d, 6.8)	19.1 (q)	H-5	1.39
7	2.30 (1H, br dd, 15.2, 9.1); 2.59 (1H, br d, 15.2)	29.6 (t)	H-8, H-9	2.30, 2.59
8	3.60 (1H, m)	74.8 (d)	H-7, H-9, H-10	3.60
9	3.58 (1H, m)	71.6 (d)	H-8, H-10	3.58
10	1.20 (3H, d, 6.1)	18.9 (q)	H-8, H-9	1.20

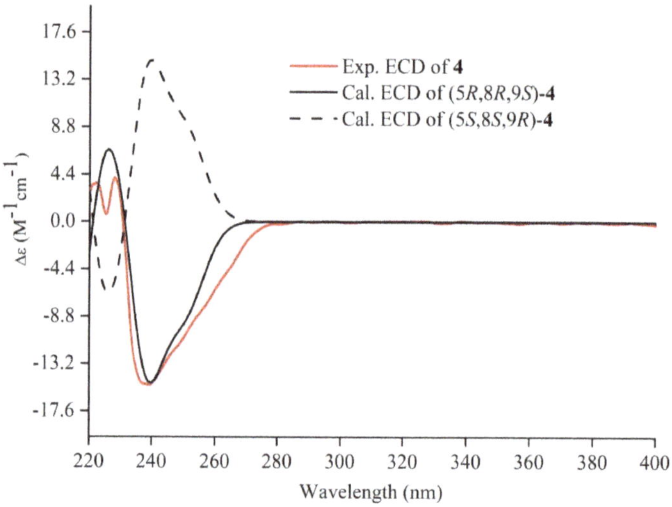

Figure 3. Experimental ECD spectra (200–400 nm) of **4** in MeOH and the calculated ECD spectra of the model molecules of **4** at the PBE0/def2-TZVP level.

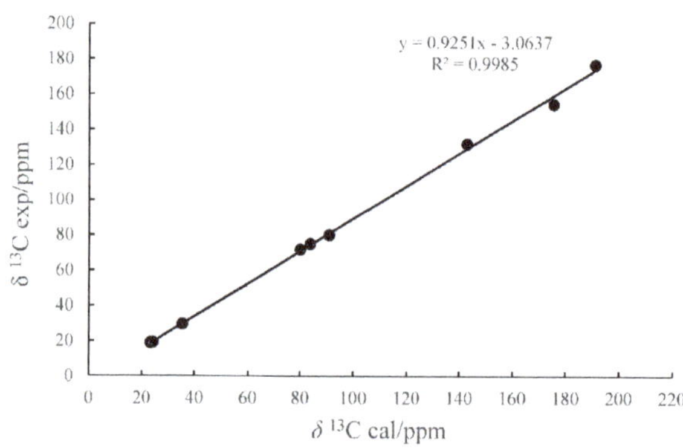

Figure 4. Calculated ^{13}C chemical shifts against the experimental data of **4**.

Compound **7**, a white powder soluble in MeOH, has the molecular formula of $C_9H_{11}NO_3$ with four degrees of unsaturation from the protonated molecular ion at *m/z* 182.0810 $[M + H]^+$ (calculated for $C_9H_{12}NO_3^+$, 182.0812), as evidenced by HR-ESI-MS; the combination of ^{13}C-NMR and ^{1}H-NMR spectra showed that the compound contained six low-field carbon signals, including four substituted low-field carbons, two unsubstituted aromatic carbons, one submethyl carbon, and two methyl carbons, including one methyl carbon directly connected to the aromatic ring. The compound is an alkaloid, as verified by the bismuth potassium iodide reaction. It is inferred that the compound contained a nitrogen-containing heterocyclic ring. According to the HMBC signal (Table 2), H-5 (δ_H 7.67) was correlated with C-6, C-3, and C-9, and H-3 was strongly correlated with C-2, C-5, and C-9. Furthermore, the HMBC-related signal intensity was weakly correlated. Combined with hydrogen spectrum signal splitting, C-3 and C-5 were interpositionally substituted, and H-3 and H-5 were weakly correlated with the C-9 signal intensity. H-5 was correlated with C-6 and C-10, and H-10 was strongly correlated with C-6, indicating that C-10 was linked to C-6, and the chemical shift of C-6 was lower than that of conventional aromatic carbonization. It is inferred that C-6 was linked to heteroatom N, resulting in a low chemical shift field, and H-3 and C-2 (δ_C 166.9) were detected. The correlation signal of H-7 with C-3 and C-8 and the correlation hydrogen spectrum signal of H-8 split (d) indicated that C-2 was connected with a hydroxyethyl, and the abnormal chemical shift of C-2 indicates that it was connected to N. After assignment of the related signals, it was found that C-4 (δ_C 142.2) did not generate any related signals, and there were two oxygens in the molecular formula of the compound that had not been attributed. Thus, it was inferred that the compound contained carboxylic acid groups. Since the unsaturation degree of the compound was 4 and it was a nitrogen-containing alkaloid, the carboxylic acid groups should be connected with the pyridine ring (Figure 5). The position of C-9 was determined according to the signal correlation between H-3, H-5, and C-9, and the paraposition substitution of the carbonyl group and paraposition N led to the chemical shift of C-4 moving to the lower field: ^{1}H-NMR (400 MHz, CD_3OD) δ_H 7.89 (br s, 1H, H-3), 7.67 (br s, 1H, H-5), 4.89 (1H, overlapped, H-7), 2.59 (s, 3H, H-10) and 1.46 (d, J = 6.6, H-8); ^{13}C-NMR (100 MHz; CD_3OD) δ_C 168.4 (C-9), 166.9 (C-1), 159.5 (C-6), 142.2 (C-4), 122.6 (C-5), 117.5 (C-3), 71.1 (C-3), 24.4 (C-8), and 23.7 (C-10).

Table 2. The ^{1}H, ^{13}C, HMBC, and HSQC NMR data for compound **7** in CD_3OD.

No.	^{1}H NMR	^{13}C NMR	HMBC	HSQC(δ)
2	—	166.9 (s)	H-3, H-8	—
3	7.89 (1H, br s)	117.5 (d)	H-5	7.89
4	—	142.2 (s)	—	—
5	7.67 (1H, br s)	122.6 (d)	H-3, H-10	7.67
6	—	159.5 (s)	H-10, H-5	—
7	4.89 (1H, overlapped)	71.1 (d)	H-8	4.89
8	1.46 (3H, d, 6.6)	24.4 (q)	H-7	1.36
9	—	168.4 (s)	H-3, H-5	—
10	2.59 (3H, s)	23.7 (q)	H-5	2.59

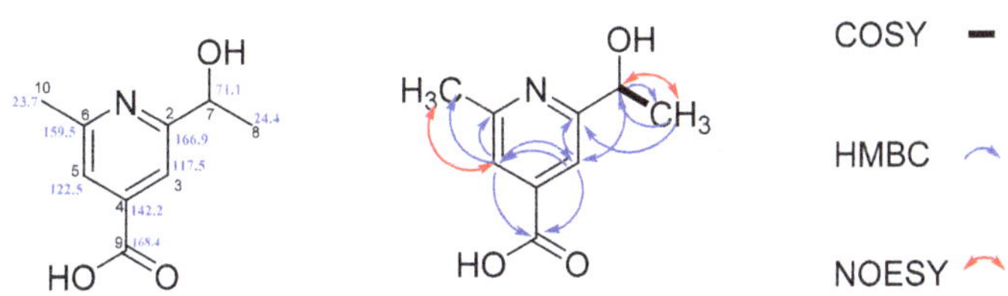

Figure 5. Signal assignment of compound **7** via ^{13}C and 2D NMR.

To confirm the stereochemical assignments of **7**, we carried out the ECD calculation at the PBE0/def2-TZVP level. The experimental ECD spectrum of **7** exhibited a negative Cotton effect at 258 nm ($\Delta\varepsilon$ −10.54) and a positive Cotton effect at 286 nm ($\Delta\varepsilon$ +7.76), which displayed strong agreement with the calculated ECD curve of *S*-**7** (Figure 6). Thus, the absolute configuration at the stereogenic center in **7** was (7*S*). The structure of compound **7** was determined to be 2-(1-hydroxyethyl)-6-methylisonicotinic acid.

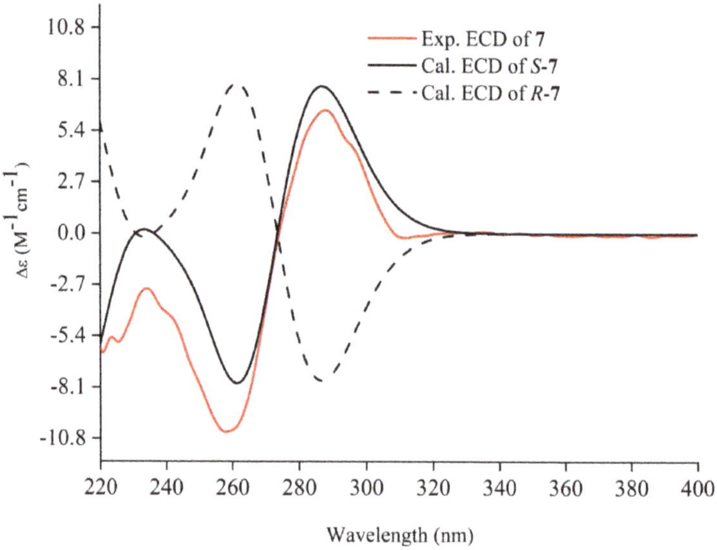

Figure 6. Experimental ECD spectra (200–400 nm) of **7** in methanol and the calculated ECD spectra of the model molecules of **7** at the PBE0/def2-TZVP level.

Compound **9** was isolated as a light yellow powder, and the HR-ESI-MS data showed a molecule peak at *m/z* 253.0655 $[M + H]^+$ (calculated for $C_{12}H_{13}O_6{}^+$, 253.0707), which indicated the molecular formula as $C_{12}H_{12}O_6$ and 7 as the index of hydrogen deficiency; ^{1}H-NMR (400 MHz; DMSO-d_6) δ_H 1.12–1.14 (d, *J* = 6 Hz, 3H, H-3′), 3.77–3.85 (m, *J* = 6.2 Hz, 1H, H-2′), 3.97–4.00 (t, *J* = 5.6 Hz, 1H, H-3′), 3.75–3.77 (d, 1H, OH), 5.64–5.55 (d, 1H, OH), 6.34 (d, *J* = 1.2 Hz, 1H, H-5), 6.44 (d, *J* = 1.2 Hz, 1H, H-7), 6.61 (s, 1H, H-4), 10.85 (s, 1H, H-6), 11.00 (s, 1H, H-8); and ^{13}C-NMR (400 MHz; DMSO-d_6) δ_C 165.5(C-1), 165.3 (C-6), 162.5 (C-8), 157.6 (C-3), 139.2 (C-4a), 104.4 (C-4), 102.9 (C-5), 101.5 (C-7), 97.8 (C-8a), 74.4 (C-1′), 67.4 (C-2′), and 19.7 (C-3′) (Figure 7). The ^{1}H-NMR showed that the compound contained one methyl group, two methoxymethyl groups, two alcohol hydroxyl groups, one phenolic hydroxyl group, one phenolic hydroxyl group forming an intramolecular hydrogen bond, and two double-bond protons. The ^{13}C-NMR spectrum (DMSO-d_6) of the compound showed 12 carbon signals, including five oxygen-linked aromatic carbons, four oxygen-free aromatic carbons, two oxygen-linked methylene carbons, and one methyl carbon. The unsaturation of the compound was calculated as 6 according to the molecular formula of the compound, and it was inferred that the compound contained two amphioxic heterocyclic systems. H-7 (δ 6.33) was correlated with C-5 (δ 102.9), C-8a (δ 97.8), C-8 (δ 162.5), and C-6 (δ 165.3), and H-5 was correlated with C-4 (δ 104.4), C-6 (δ 165.3), and C-8a (δ 97.8). Among them, oxygen substitution existed in C-6 (δ 165.3), C-8 (δ 162.5), and C-1 (δ 165.54). According to a comparison of the hydrogen spectra of CD_3OD, C-6 and C-8 are hydroxyl substitutions, and C-1 is the carbonyl group. The abnormal chemical shift of the C-8 hydroxyl group (δ 11.00) indicated that it was greatly shifted to the lower field under the influence of the neighboring carbonyl group. The low-field shift for C-3 (δ 157.6) and the HSQC signal suggested the existence of substitution. On the basis of a combination with the HMBC and chemical shift characteristics of C-3 and C-1, we inferred that the two carbons were connected by oxygen, resulting in a large chemical shift to the low field (Table 3). On the basis of the above information, we inferred that the compound was a derivative of an isocoumarin skeleton. There was a correlation between C-4, C-3, and C-1, and it was inferred that C-3 had a branched chain substitution: H-1′, C-2′, C-3′, and H-3′ were correlated with C-1′ and C-2′. A combination of the H spectrum splitting characteristics and hydrogen integral values of H-1, H-2, and H-3 (dd, dq, d peaks; 1H, 1H, 3H, respectively) and the chemical shifts of C-1′, C-2′, and C-3′ (74.4, 67.4, and 19.7, respectively) determined that C-1′ and C-2′ were hydroxyl substituted and C-3′ was a methyl, and the compound signal was assigned. The absolute configurations of compound **9** were established to be 1′*R* and 2′*R* by the ECD calculations (Figure 8). Finally, compound **9** was named 6,8-dihydroxy-3-(1′*R*, 2′*R*-dihydroxypropyl)-isocoumarin (Figure 1).

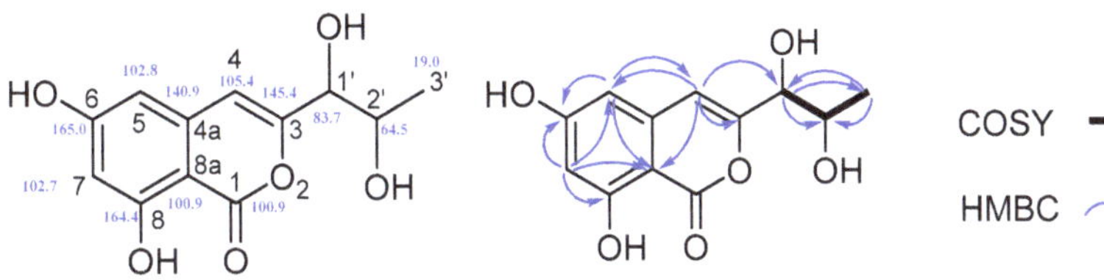

Figure 7. Signal assignment of compound **9** via ^{13}C and HMBC.

Table 3. The ^{1}H, ^{13}C, HMBC, and HSQC NMR data for compound **9** in DMSO-*d6*.

No.	^{1}HNMR	^{13}C NMR	HMBC	HSQC
1	—	165.54	—	—
3	—	157.6	H-4	—
4	6.61 (1H, s)	104.4	—	6.61
4a	—	139.2	—	—
5	6.43 (1H, s)	102.9	H-4, H-7	6.43
6	10.88 (OH, s)	165.3	H-7	—
7	6.33 (1H, s)	101.5	—	6.33
8	11.00 (OH, s)	162.5	H-7	—
8a	—	97.8	H-4, H-5, H-7	—
1′	3.97 (1H, dd)	74.7	H-4, H-3′	3.97
2′	3.80 (1H, dq)	67.4	H-1′, H-3′	3.80
3′	1.12 (3H, d)	19.7	H-1′	1.12

In addition to the new compounds described above, eight known compounds obtained in this study were identified as compounds **1–3**, **5–6**, **8**, and **10–11** by comparing their spectroscopic data to those reported in the literature. Details of NMR and MS data for compounds **1–11** were given in the Supplementary Materials.

Compound **1** was obtained as a colorless powder, and the HR-ESI-MS data showed a molecule peak at *m/z* 129.0552 $[M + H]^+$ (calculated for $C_6H_9O_3^+$, 129.0546), which indicated a molecular formula of $C_6H_8O_3$ with three degrees of unsaturation. In examining the proton nuclear magnetic resonance (^{1}H-NMR) data, we found signals for methyl protons δ_H 1.41 (3H d, J = 6.8 Hz, H-CH_3), one methylene proton δ_H 4.28 (2H, d, J = 1.7 Hz), and two methines (two oxygenated sp^3 and one sp^2): δ_H 5.11–5.16 (1H, m, H-5) and δ_H 7.42–7.47 (1H, m, H-4). The ^{13}C-NMR spectra revealed six carbon signals: one ester carbonyl (δ_C 174.53), one olefinic methine (δ_C 152.93), one nonprotonated sp^2 carbon (δ_C 134.75), two oxygenated methines (δ_C 80.13), one oxygenated methylene (δ_C 56.95), and methyls (δ_C 19.02). The structure of **1** was determined as 3-(hydroxymethyl)-5-methylfuran-2(5H)-one [28], as shown in Figure 1.

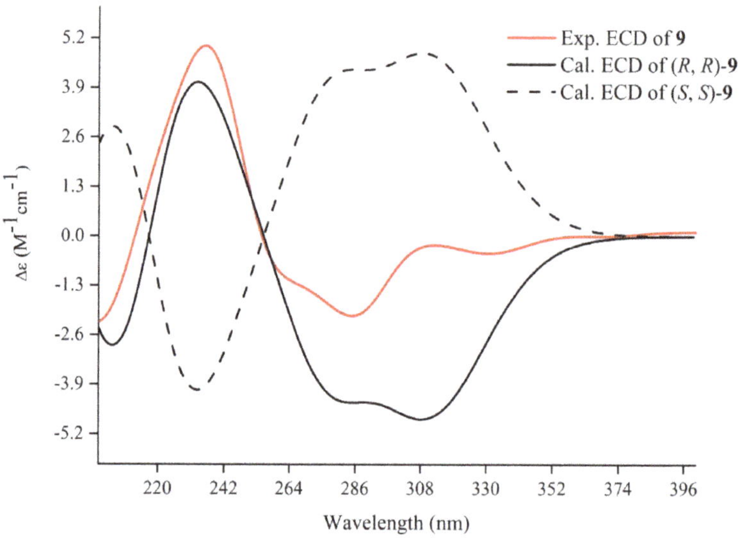

Figure 8. Experimental ECD spectra (200–400 nm) of **9** in methanol and the calculated ECD spectra of the model molecules of **9** at the PBE0/def2-TZVP level.

Compound **2** was a white powder with the molecular formula ($C_7H_8O_4$): HR-ESI-MS *m/z* 157.0493 $[M + H]^+$ (calculated for $C_7H_9O_4^+$, 157.0495); ^{1}H-NMR (400 MHz, CD_3OD) δ_H 1.33 (d, *J* = 6.8 Hz, 3H, H-CH_3), 5.05–5.09 (m, 1H, H-5), and 7.41–7.45 (m, 1H, H-4); and ^{13}C-NMR (100 MHz, CD_3OD) δ_C 18.93 (C-CH_3), 31.08 (C-2′), 80.00 (C-5), 127.99 (C-3), 155.38 (C-4), 173.10 (C-2), and 175.47 (C-1′). Thus, compound **2** was identified as [2-(5-methyl-2-oxo-2,5-dihydrofuran-3-yl)-acetic acid] by comparing NMR reference data [29]. Previously, compound **2** had only been obtained through chemical synthesis and was isolated as a natural product for the first time [29]. Thus, we named it aspilactonol G.

Compound **3** was a colorless oil with the molecular formula ($C_8H_{10}O_4$): HR-ESI-MS (*m/z* 171.0654 $[M + H]^+$ (calculated for $C_8H_{10}O_4^+$, 171.0652)). ^{1}H-NMR data (400 MHz, CD_3OD) δ_H 1.33 (d, *J* = 6.8 Hz, 3H, H-5), 5.05–5.09 (m, 2H, H-2), and 7.41–7.45 (m, 1H, H-4); and ^{13}C-NMR (100 MHz, CD_3OD) δ_C 18.92, 30.09, 52.68, 80.05, 127.55, 155.58, 171.64, and 175.25. Thus, compound **3** was identified as [methyl-2-(5-methyl-2-oxo-2,5-dihydrofuran -3-yl)-acetate] by comparing NMR reference data [29]. Compound **3** was also obtained as a natural product for the first time [29]; thus, we named it aspilactonol H.

Compound **5** yielded the following data: HR-ESI-MS *m/z* 131.0663 $[M + H]^+$ (calculated for $C_6H_{11}O_3^+$, 131.0703); ^{1}H-NMR (400 MHz, CD_3OD) δ_H 2.11 (d, 3H, C-6), 2.30–2.34 (t, 2H, H-4), 3.67–3.70 (t, 2H, H-5), and 5.76 (s, 1H, H-2); and ^{13}C-NMR (400 MHz, CD_3OD) δ_C 18.5, 44.6, 60.9, 120.4, 152.7, and 172.1. Compound **5** is E-Δ^2-anhydromevalonic acid [30].

Compound **6** yielded the following data: HR-ESI-MS *m/z* 113.0604 $[M + H]^+$ (calculated for $C_6H_9O_2^+$, 113.0597); 1H-NMR (400 MHz, CD_3OD), δ_H 5.79 (1H, q, *J* = 1.5 Hz, H-3), 4.38 (2H, t, *J* = 6.0 Hz, H-6), 2.41 (2H, br.t, *J* = 6.0 Hz, H-5), and 2.02 (3H, s, H-7); and ^{13}C-NMR (100 MHz, CD_3OD) δ_C 22.6, 28.8, 65.6, 116.2, 158.0, and 164.4. Compound **6** is 4-methyl-5,6-dihydropyren-2-one [31].

Compound **8** was a white amorphous powder. The molecular formula, $C_{10}H_{10}O_4$, was determined by HR-ESI-MS 195.0651 *m/z* $[M + H]^+$ (calculated for $C_{10}H_{11}O_4^+$, 195.0652) and ^{13}C-NMR data, corresponding to six degrees of unsaturation; ^{1}H-NMR (400 MHz, CD_3OD) δ_H 6. 21 (1H, s, H-5), 6.20 (1H, s, H-7), 4.63–4.67 (1H, m, H-3), 2.92 (1H, dd, *J* = 16.4, 3.6 Hz, H-4a), 2.82 (H, dd, *J* = 16.4, 11.2 Hz, H-4b), and 1.45 (3H, d, *J* = 6.3 Hz, H-9); and ^{13}C-NMR (CD_3OD, 100 MHz) δ_C 170.30 (C-1), 164.99 (C-8), 164.23 (C-6), 142.08 (C-4a), 106.53

(C-5), 100.82 (C-8a), 100.0 (C-7), 75.77 (C-3), 34.13 (C-4), and 19.44 (C-9). Compound **8** was identified as (*R*)-6-hydroxymellein [32].

Compound **10**, $C_{11}H_{10}O_4$, was a yellow powder: HR-ESI-MS *m/z* 207.0651 $[M + H]^+$ (calculated for $C_{11}H_{11}O_4^+$, 207.0652); ^{1}H-NMR (400 MHz, CD_3OD, δ, ppm) δ_H 6.39 (^{1}H, d, J = 1.5 Hz, H-7), 6.29 (1H, d, J = 1.5 Hz, H-5), 6.24 (1H, s, H-4), 3.78 (3H, s, H-10), and 2.10 (3H, s, H-9); and ^{13}C-NMR (100 MHz, CD_3OD) δ_C 165.1 (C, C-1), 163.6 123 (C, C-8), 158.3 (C, C-6), 155.0 (C, C-3), 142.2 (C, C-4a), 103.5 (CH, C-4), 102.9 (CH, C-5), 100.4 (C, C-8a), 99.2 (CH, C-7), 56.8 (CH_3, C-10), and 19.4 (CH_3, C-9). Compound **10** is 6-hydroxy-8-methoxy-3-methylisocoumarin [33].

Compound **11**, $C_{12}H_{12}O_5$, was a yellow powder: ESI-MS *m/z* 237.0765 $[M + H]^+$ (calculated for $C_{12}H_{13}O_5^+$, 237.0757); ^{1}H-NMR (400 MHz, CD_3OD, δ, ppm), δ_H 6.40 (2H, d, J = 2.1, H-5, 7), 6.37 (1H, d, J = 2.1, H-4), 4.69 (1H, m, H-2′), 2.59 (2H, m, H-1′), and 1.26 (3H, d, J = 6.2, H-3′); and ^{13}C-NMR (100 MHz, CD_3OD, δ, ppm), δ_C 167.8 (C, C-1), 167.3 (C, C-8), 164.8 (C, C-6), 156.2 (C, C-3), 141.3 (CH, C-4), 107.0 (C, C-7), 103.7 (C, C-10), 102.6 (CH, C-9), 99.8 (CH, C-5), 66.2 (CH, C-2′), 43.8 (CH_2, C-1′), and 23.3 (CH_3, C-3′). The absolute configuration of compound **11** was established to be 2′*R* by the ECD calculations (Figure 9). Thus, compound **11** was identified as de-*O*-methyldiaporthin by comparing NMR reference data [34,35].

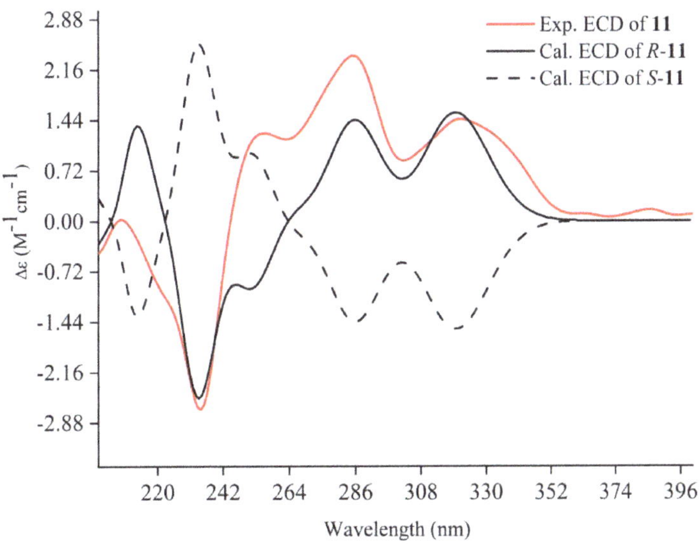

Figure 9. Experimental ECD spectra (200–400 nm) of **11** in methanol and the calculated ECD spectra of the model molecules of 11 at the PBE0/def2-TZVP level.

The biosynthesis of the isolated compounds **1–4** and **8–11** was proposed as shown in Schemes 1 and 2, respectively. Furan ring groups are abundant in natural products and play important roles in the pharmacophore of bioactive substances. Furanone and its derivatives have been shown to inhibit the formation of bacterial biofilms; interfere with bacterial population effects; and have analgesic, anti-inflammatory, anticancer, anticonvulsive, antibacterial, antifungal, antioxidation, and other activities. Most of the furanone compounds are synthesized by a single polyketone pathway, although chain fusion of furanone has also been reported in recent years [19,36]. In this study, compounds **4** and **9** possessed o-diol side chains. Different carbon skeletons with the same o-diol side chains suggested the presence of specific hydroxylating enzymes. A plausible biosynthetic pathway for compounds **1–4** is proposed (Scheme 1). Furanone **1–4** are derivatives of α,β-unsaturated γ-lactone. Their synthesis begins with the condensation of five molecules of acetyl-CoA to form the intermediate **a**, which is reduced to generate the critical intermediate **b**, and forms the intermediate **c** under the action of cyclase. Then, **c** undergoes ring-opening and oxidation to generate the intermediate **d**, which is dehydrated to produce compound **4**. Compound **2** is synthesized from **d** through an undefined pathway, then methylated to compound **3** and decarboxylated to compound **1** [19,36]. Compounds **8–11** are isocoumarin derivatives (Scheme 2), whose biosynthesis by the polyketone synthesis pathway begins with acetyl-CoA. Isocoumarin derivatives have been detected in both plants and microorganisms. The C-8 of isocoumarins does not have the oxidation found in plants, while the isocoumarins commonly found in microorganisms have oxidation at the C-8 position, which is considered to be the biological source of the two types of isocoumarin [37,38].

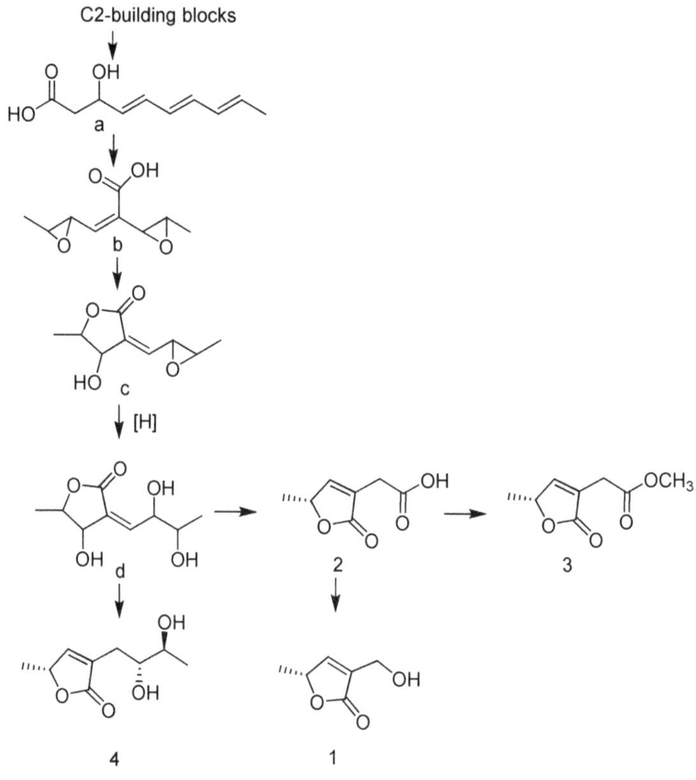

Scheme 1. Proposed biosynthetic pathways for furanone compounds (**1–4**).

The AChE inhibitory activities of the crude extracts were evaluated using Ellman's method, with Rivastigmine and Hup A as the control groups [22,39]. Ethyl acetate, with an inhibition effect value of 82.68%, exhibited better inhibition against AChE than did either petroleum ether extract (47.23%) or buthanol extract (15.82%) (Table S1). In addition, all of the compounds were investigated for their anti-AChE activities. Compounds **1–3** and **5–10** exhibited no inhibitory activity against AChE. Compounds **4** and **11** displayed moderate inhibitory effect on AChE activities with IC_{50} values of 6.26 and 21.18 μM, respectively (Table 4).

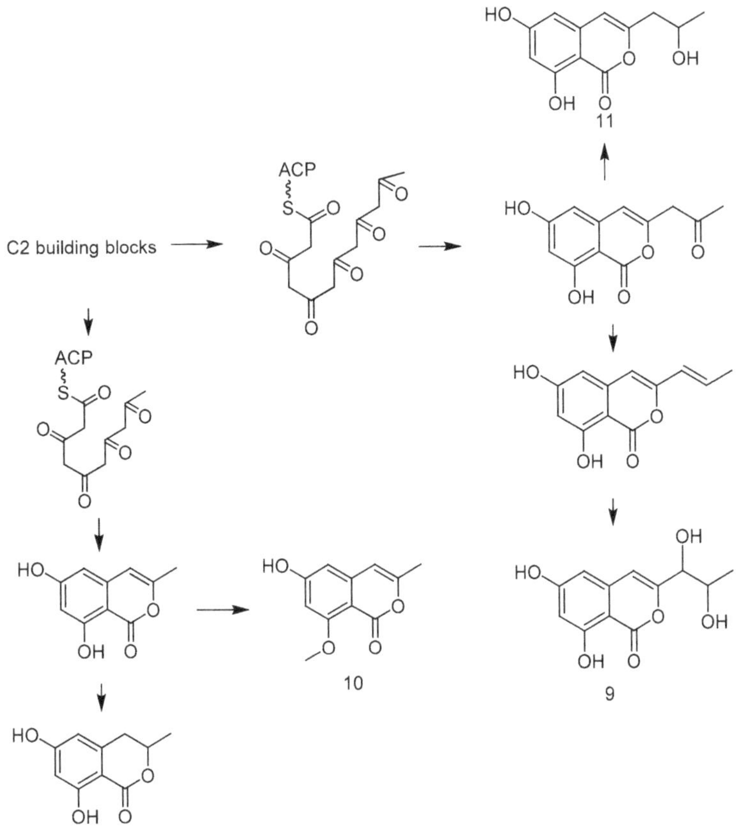

Scheme 2. Proposed biosynthetic pathways for isocoumarin compounds (**8–11**).

Table 4. Acetylcholinesterase inhibitory activities of the secondary metabolites of *Phaeosphaeria* sp. LF5, expressed as IC_{50} values.

Compound	IC_{50} (μM) [a]	Compounds	IC_{50} (μM) [a]
1	>100	**6**	>100
2	>100	**7**	>100
3	>100	**8**	>100
4	6.26 ± 0.15	**9**	>100
5	>100	**10**	>100
Rivastigmine	1.82 ± 0.13	**11**	21.18 ± 1.53
Hup-A	0.045 ± 0.01		

[a] Expressed as the mean ± SD of three parallel measurements ($p < 0.05$).

Structurally, almost all of the furanone compounds in this study contained an α,β-unsaturated carboxylic acid lactone moiety, which might be the key functional group for their biological activity. De-*O*-methyldiaporthin (**11**) was first reported in 1988 [34,35]. It can be used as a microbial herbicide due to its very strong phytotoxic activity [40]. AChEIs are drugs that can be used clinically to treat or alleviate symptoms of AD. They are primarily associated with the direction and efficacy of AD drug development based on the cholinergic injury hypothesis. Thus far, two generations of five AChEI drugs (Tacrine, Donepezil, Rivastigmine, Galantamine, and HupA) have been successfully developed and have become the first choice for clinical treatment or mitigation of AD [10]. Existing clinical AChEI drugs have limitations such as limited efficacy, significant toxicity, and drug resistance. In this study, furanone compound **4** and isocoumarin compound **11** were found to have the potential to inhibit AChE. To the best of our knowledge, furanone compounds were reported here for the first time for their AChE inhibitory activity. Their mechanisms of action and structure–activity relationships in inhibiting AChE require further study by inhibition kinetics analysis and molecular docking methods.

4. Conclusions

To summarize, 11 polyketide derivatives, which included three new compounds, aspilactonol I (**4**), 2-(1-hydroxyethyl)-6-methylisonicotinic acid (**7**), and 6, 8-dihydroxy-3-(1′R, 2′R-dihydroxypropyl)-isocoumarin (**9**), and two new natural-sources-derived aspilactonols (G, H) (**2**, **3**) were isolated from an endophytic fungus *Phaeosphaeria* sp. LF5 of *H. serrata*. Their absolute configurations of three new compounds (**4**, **7**, and **9**) and known compound **11** were determined by ECD calculations. Furanone compound **4** and isocoumarin compound **11** exhibited potent AChE inhibitory activities. This study indicates that *Phaeosphaeria* sp. LF5 from *H. serrata* may contain various AChEI compounds, which is a potential resource pool for bioprospecting and isolating AChEIs. Furthermore, this research also provided a material basis for the development of new and efficient AChEI drugs.

Supplementary Materials: The following supporting information can be downloaded at https://www.mdpi.com/article/10.3390/jof8030232/s1. Table S1: Acetylcholinesterase inhibitory activities of the crude extracts, Figure S1: ^{1}H NMR spectrum of compound **1** in CD_3OD, Figure S2: ^{13}C NMR spectrum of compound **1** in CD_3OD, Figure S3: ^{1}H NMR spectrum of compound **2** in CD_3OD, Figure S4: ^{13}C NMR spectrum of compound **2** in CD_3OD, Figure S5: ^{1}H NMR spectrum of compound **3** in CD_3OD, Figure S6: ^{13}C NMR spectrum of compound **3** in CD_3OD, Figure S7: ^{1}H NMR spectrum of compound **4** in CD_3OD, Figure S8: ^{13}C NMR spectrum of compound **4** in CD_3OD, Figure S9: ^{1}H-^{1}H COSY spectrum of compound **4** in CD_3OD, Figure S10: DEPT 90 spectrum of compound **4** in CD_3OD, Figure S11: DEPT 135 spectrum of compound **4** in CD_3OD, Figure S12: HSQC spectrum of compound **4** in CD_3OD, Figure S13: HMBC spectrum of compound **4** in CD_3OD, Figure S14: NOESY spectrum of compound **4** in CD_3OD, Figure S15: HRESIMS spectrum of compound **4**, Figure S16: ^{1}H NMR spectrum of compound **5** in CD_3OD, Figure S17: ^{13}C NMR spectrum of compound **5** in CD_3OD, Figure S18: ^{1}H NMR spectrum of compound **6** in CD_3OD, Figure S19: ^{13}C NMR spectrum of compound **6** in CD_3OD, Figure S20: ^{1}H NMR spectrum of compound **7** in CD_3OD, Figure S21: ^{13}C NMR spectrum of compound **7** in CD_3OD, Figure S22: ^{1}H-^{1}H COSY spectrum of compound **7** in CD_3OD, Figure S23: HSQC spectrum of compound **7** in CD_3OD, Figure S24: HMBC spectrum of compound **7** in CD_3OD, Figure S25: NOESY spectrum of compound **7** in CD_3OD, Figure S26: HRESIMS spectrum of compound **7**, Figure S27: ^{1}H NMR spectrum of compound **8** in CD_3OD, Figure S28: ^{13}C NMR spectrum of compound **8** in CD_3OD, Figure S29: ^{1}H NMR spectrum of compound **9** in DMSO-d_6, Figure S30: ^{13}C NMR spectrum of compound **9** in DMSO-d_6, Figure S31: DEPT 90 spectrum of compound 9 in DMSO-d_6, Figure S32: DEPT 135 spectrum of compound **9** in DMSO-d_6, Figure S33: ^{1}H-^{1}H COSY spectrum of compound **9** in DMSO-d_6, Figure S34: HSQC spectrum of compound **9** in DMSO-d_6, Figure S35: HMBC spectrum of compound **9** in DMSO-d_6, Figure S36: NOESY spectrum of compound **9** in DMSO-d_6, Figure S37: HRESIMS spectrum of compound **9**, Figure S38: ^{1}H NMR spectrum of compound **10** in CD_3OD, Figure S39: ^{13}C NMR spectrum of compound **10** in CD_3OD, Figure S40: ^{1}H NMR spectrum of compound **11** in CD_3OD, Figure S41: ^{13}C NMR spectrum of compound **11** in CD_3OD.

Author Contributions: Y.X.: conceptualization, methodology, validation, investigation, and writing—original draft; W.L.: investigation, methodology, and validation; S.Z.: data curation and software; Z.Z.: investigation; Y.W.: investigation and methodology; D.Z.: conceptualization, methodology, writing—review and editing, project administration, and funding acquisition; J.L.: resources and methodology; J.C.: resources; C.J.: chemical structure analysis. All authors have read and agreed to the published version of the manuscript.

Funding: This research was supported by the National Natural Science Foundation of China (81760649) and the Natural Science Foundation of Jiangxi Province of China (20181BAB215044).

Institutional Review Board Statement: Not applicable.

Informed Consent Statement: Not applicable.

Data Availability Statement: All data generated or analyzed in this study are available within the manuscript and are available from the corresponding authors upon request.

Conflicts of Interest: The authors declare no conflict of interest.

References

1. Newman, D.J.; Cragg, G.M. Plant endophytes and epiphytes: Burgeoning sources of known and "unknown" cytotoxic and antibiotic agents. *Planta Med.* **2020**, *86*, 13–14. [CrossRef] [PubMed]
2. Zou, Z.B.; Zhang, G.; Li, S.M.; He, Z.H.; Yan, Q.X.; Lin, Y.K.; Xie, C.L.; Xia, J.M.; Luo, Z.H.; Luo, L.Z.; et al. Asperochratides A–J, Ten new polyketides from the deep-sea-derived *Aspergillus ochraceus*. *Bioorg. Chem.* **2020**, *105*, 104349. [CrossRef] [PubMed]
3. Venugopalan, A.; Srivastava, S. Endophytes as in vitro production platforms of high value plant secondary metabolites. *Biotechnol. Adv.* **2015**, *33*, 873–887. [CrossRef]
4. Gakuubi, M.M.; Munusamy, M.; Liang, Z.X.; Ng, S.B. Fungal endophytes: A promising frontier for discovery of novel bioactive compounds. *J. Fungi* **2021**, *7*, 786. [CrossRef]
5. Deshmukh, S.K.; Gupta, M.K.; Prakash, V.; Saxena, S. Endophytic fungi: A source of potential antifungal compounds. *J. Fungi* **2018**, *4*, 77. [CrossRef]
6. Pal, P.P.; Shaik, A.B.; Begum, A.S. Prospective leads from endophytic fungi for anti-inflammatory drug discovery. *Planta Med.* **2020**, *86*, 941–959. [CrossRef] [PubMed]
7. Keshri, P.K.; Rai, N.; Verma, A.; Kamble, S.C.; Barik, S.; Mishra, P.; Singh, S.K.; Salvi, P.; Gautam, V. Biological potential of bioactive metabolites derived from fungal endophytes associated with medicinal plants. *Mycol. Prog.* **2021**, *20*, 577–594. [CrossRef]
8. Zheng, R.; Li, S.; Zhang, X.; Zhao, C. Biological activities of some new secondary metabolites isolated from endophytic fungi: A review study. *Int. J. Mol. Sci.* **2021**, *22*, 959. [CrossRef] [PubMed]
9. Bang, S.; Baek, J.Y.; Kim, G.J.; Kim, J.; Kim, S.J.; Deyrup, S.T.; Choi, H.; Kang, K.S.; Shim, S.H. Azaphilones from an endophytic *Penicillium* sp. prevent neuronal cell death via inhibition of MAPKs and reduction of Bax/Bcl-2 ratio. *J. Nat. Prod.* **2021**, *84*, 2226–2237. [CrossRef] [PubMed]
10. Cao, D.; Sun, P.; Bhowmick, S.; Wei, Y.H.; Guo, B.; Wei, Y.H.; Mur, L.A.J.; Sun, Z.L. Secondary metabolites of endophytic fungi isolated from *Huperzia serrata*. *Fitoterapia* **2021**, *155*, 104970. [CrossRef] [PubMed]
11. Wang, Y.; Lai, Z.; Li, X.X.; Yan, R.M.; Zhang, Z.B.; Yang, H.L.; Zhu, D. Isolation, diversity and acetylcholinesterase inhibitory activity of the culturable endophytic fungi harboured in *Huperzia serrata* from Jinggang Mountain, China. *World J. Microbiol. Biotechnol.* **2016**, *32*, 20. [CrossRef] [PubMed]
12. Wang, Y.; Zeng, Q.G.; Zhang, Z.B.; Yan, R.M.; Zhu, D. Isolation and characterization of endophytic huperzine a-producing fungi from *Huperzia serrata*. *J. Ind. Microbiol. Biotechnol.* **2010**, *38*, 1267–1278. [CrossRef] [PubMed]
13. Singh, S.B.; Ondeyka, J.; Harris, G.; Herath, K.; Zink, D.; Vicente, F.; Bills, G.; Collado, J.; Platas, G.; González del Val, A.; et al. Isolation, structure, and biological activity of phaeofungin, a cyclic lipodepsipeptide from a *Phaeosphaeria* sp. using the genome-wide *Candida albicans* fitness test. *J. Nat. Prod.* **2013**, *76*, 334–345. [CrossRef] [PubMed]
14. Karakoyun, Ç.; Küçüksolak, M.; Bilgi, E.; Doğan, G.; Çömlekçi, Y.E.; Bedir, E. Five new cardenolides transformed from oleandrin and nerigoside by *Alternaria eureka* 1e1bl1 and *Phaeosphaeria* sp. 1e4cs-1 and their cytotoxic activities. *Phytochem. Lett.* **2021**, *41*, 152–157. [CrossRef]
15. Zhan, Z.J.; Jin, J.P.; Ying, Y.M.; Shan, W.G. Furanone derivatives from *Aspergillus* sp. xw-12, an endophytic fungus in *Huperzia serrata*. *Helv. Chim. Acta* **2011**, *94*, 1454–1458. [CrossRef]
16. Husain, A.; Khan, S.A.; Iram, F.; Iqbal, M.A.; Asif, M. Insights into the chemistry and therapeutic potential of furanones: A versatile pharmacophore. *Eur. J. Med. Chem.* **2019**, *191*, 66–92. [CrossRef] [PubMed]
17. Yurchenko, A.N.; Trinh, P.T.H.; Ivanets, E.V.G.; Smetanina, O.F.; Afiyatullov, S.S. Biologically active metabolites from the marine sediment-derived fungus *Aspergillus flocculosus*. *Mar. Drugs* **2019**, *17*, 579. [CrossRef] [PubMed]
18. Liu, Y.; Li, X.M.; Meng, L.H.; Wang, B.G. Polyketides from the marine mangrove-derived fungus *Aspergillus ochraceus* MA-15 and their activity against aquatic pathogenic bacteria. *Phytochem. Lett.* **2015**, *12*, 232–236. [CrossRef]
19. Hu, H.C.; Li, C.Y.; Tsai, Y.H.; Yang, D.Y.; Chang, F.R. Secondary metabolites and bioactivities of *Aspergillus ochraceopetaliformis* isolated from *Anthurium brownii*. *ACS Omega* **2020**, *5*, 20991–20999. [CrossRef] [PubMed]
20. Tianpanich, K.; Prachya, S.; Wiyakrutta, S.; Mahidol, C.; Ruchirawat, S.; Kittakoop, P. Radical scavenging and antioxidant activities of isocoumarins and a phthalide from the endophytic fungus *Colletotrichum* sp. *J. Nat. Prod.* **2011**, *74*, 79–81. [CrossRef] [PubMed]
21. Gao, W.; Wang, X.; Chen, F.; Li, C.; Cao, F.; Luo, D. Setosphlides A-D, New Isocoumarin Derivatives from the Entomogenous Fungus *Setosphaeria rostrate* LGWB-10. *Natur. Prod. Bioprosp.* **2021**, *11*, 137–142. [CrossRef]
22. Ellman, G.L.; Courtney, K.D.; Andres, V.; Featherstone, R.M. A new and rapid colorimetric determination of acetylcholinesterase activity. *Biochem. Pharmacol.* **1961**, *7*, 88–95. [CrossRef]

23. Ortiz, J.E.; Berkov, S.; Pigni, N.B.; Theoduloz, C.; Roitman, G.; Tapia, A.; Bastida, J.; Feresin, G. Wild Argentinian Amaryllidaceae, a new renewable source of the acetylcholinesterase inhibitor galanthamine and other alkaloids. *Molecules* **2012**, *17*, 13473–13482. [CrossRef] [PubMed]
24. *Sybyl Software*; Version X 2.0; Tripos Associates Inc.: St. Louis, MO, USA, 2013.
25. Neese, F. The ORCA program system. *WIREs Comput. Mol. Sci.* **2012**, *2*, 73–78. [CrossRef]
26. Neese, F. Software update: The ORCA program system, version 4.0. *WIREs Comput. Mol. Sci.* **2017**, *8*, e1327. [CrossRef]
27. Stephens, P.J.; Harada, N. ECD cotton effect approximated by the Gaussian curve and other methods. *Chirality* **2010**, *22*, 229–233. [CrossRef] [PubMed]
28. National Center for Biotechnology Information. PubChem Compound Summary for CID 234614, 3-(Hydroxymethyl)-5-methylfuran-2(5h)-one. 2022. Available online: https://pubchem.ncbi.nlm.nih.gov/compound/3-_Hydroxymethyl_-5-methylfuran-2_5h_-one (accessed on 24 January 2022).
29. Patel, R.M.; Puranik, V.G.; Argade, N.P. Regio- and stereoselective selenium dioxide allylic oxidation of (E)-dialkyl alkylidenesuccinates to (Z)-allylic alcohols: Synthesis of natural and unnatural butenolides. *Org. Biomol. Chem.* **2011**, *9*, 6312–6322. [CrossRef] [PubMed]
30. Widmer, J.; Keller-Schierlein, W. Stoffwechselprodukte von Mikroorganismen. 139. Mitteilung. Synthesen in der Sideramin-Reihe: Rhodotorulasäure und Dimerumsäure. *Helv. Chim. Acta* **1974**, *57*, 1904–1912. [CrossRef]
31. Li, L.J.; Tao, M.H.; Chen, Y.C.; Zheng, C.X.; Huo, G.H.; Zhang, W.M. Secondary metabolites from the solid culture of marine fungal strain *Penicillium sclerotiorum* FS50. *Mycosystema* **2015**, *34*, 117–123. Available online: http://en.cnki.com.cn/Article_en/CJFDTOTAL-JWXT201501016.htm (accessed on 12 January 2022).
32. de Sá, J.D.M.; Pereira, J.A.; Dethoup, T.; Cidade, H.; Sousa, M.E.; Rodrigues, I.C.; Costa, P.M.; Mistry, S.; Silva, A.M.S.; Kijjoa, A. Anthraquinones, diphenyl ethers, and their derivatives from the culture of the marine sponge-associated fungus *Neosartorya spinosa* KUFA 1047. *Mar. Drugs* **2021**, *19*, 457. [CrossRef] [PubMed]
33. Feng, L.X.; Zhang, B.Y.; Zhu, H.J.; Pan, L.; Cao, F. Bioactive metabolites from *Talaromyces purpureogenus*, an endophytic fungus from *Panax notoginseng*. *Chem. Nat. Comp.* **2020**, *56*, 974–976. [CrossRef]
34. Hallock, Y.F.; Clardy, J.; Kenfield, D.S.; Strobel, G. De-O-methyldiaporthin, a phytotoxin from *Drechslera siccans*. *Phytochemistry* **1988**, *27*, 3123–3125. [CrossRef]
35. Zhang, X.Q.; Qu, H.R.; Bao, S.S.; Deng, Z.S.; Guo, Z.Y. Secondary metabolites from the endophytic fungus *Xylariales* sp. and their antimicrobial activity. *Chem. Nat. Comp.* **2020**, *56*, 530–532. [CrossRef]
36. Chen, X.W.; Li, C.W.; Cui, C.B.; Hua, W.; Zhu, T.J.; Gu, Q.Q. Nine new and five known polyketides derived from a deep sea-sourced *Aspergillus* sp. 16-02-1. *Mar. Drugs* **2014**, *12*, 3116–3137. [CrossRef]
37. Hussain, H.; Green, I.R. A patent review of two fruitful decades (1997–2016) of isocoumarin research. *Expert Opin. Ther. Pat.* **2017**, *27*, 1267–1275. [CrossRef]
38. Xiang, P.; Ludwig-Radtke, L.; Yin, W.B.; Li, S.M. Isocoumarin formation by heterologous gene expression and modification by host enzymes. *Org. Biomol. Chem.* **2020**, *18*, 4946–4948. [CrossRef]
39. Sonmez, F.; Zengin Kurt, B.; Gazioglu, I.; Basile, L.; Dag, A.; Cappello, V.; Ginex, T.; Kucukislamoglu, M.; Guccione, S. Design, synthesis and docking study of novel coumarin ligands as potential selective acetylcholinesterase inhibitors. *J. Enzyme Inhib. Med. Chem.* **2017**, *32*, 285–297. [CrossRef]
40. Yuan, C.; Guo, Y.H.; Li, G.; Wu, C.S.; Li, M. Metabolites of endophytic fungus *Elaphocordyceps* sp. *Microbiol. China* **2014**, *41*, 857–861.

Journal of *Fungi*

MDPI

Article

Exploring the Activity of Fungal Phenalenone Derivatives as Potential CK2 Inhibitors Using Computational Methods

Sabrin R. M. Ibrahim [1,2,*], Alaa A. Bagalagel [3], Reem M. Diri [3], Ahmad O. Noor [3], Hussain T. Bakhsh [3], Yosra A. Muhammad [4,5], Gamal A. Mohamed [6] and Abdelsattar M. Omar [4,5,7]

1 Department of Chemistry, Preparatory Year Program, Batterjee Medical College, Jeddah 21442, Saudi Arabia
2 Department of Pharmacognosy, Faculty of Pharmacy, Assiut University, Assiut 71526, Egypt
3 Department of Pharmacy Practice, Faculty of Pharmacy, King Abdulaziz University, Jeddah 21589, Saudi Arabia; abagalagel@kau.edu.sa (A.A.B.); rdiri@kau.edu.sa (R.M.D.); aonoor@kau.edu.sa (A.O.N.); htbakhsh@kau.edu.sa (H.T.B.)
4 Department of Pharmaceutical Chemistry, Faculty of Pharmacy, King Abdulaziz University, Jeddah 21589, Saudi Arabia; yosra.muhammad2017@gmail.com (Y.A.M.); asmansour@kau.edu.sa (A.M.O.)
5 Center for Artificial Intelligence in Precision Medicines, King Abdulaziz University, Jeddah 21589, Saudi Arabia
6 Department of Natural Products and Alternative Medicine, Faculty of Pharmacy, King Abdulaziz University, Jeddah 21589, Saudi Arabia; gahussein@kau.edu.sa
7 Department of Pharmaceutical Chemistry, Faculty of Pharmacy, Al-Azhar University, Cairo 11884, Egypt
* Correspondence: sabrin.ibrahim@bmc.edu.sa; Tel.: +966-581183034

Abstract: Cancer represents one of the most prevalent causes of global death. CK2 (casein kinase 2) activation boosted cancer proliferation and progression. Therefore, CK2 inhibition can have a crucial role in prohibiting cancer progression and enhancing apoptosis. Fungi have gained vast interest as a wealthy pool of anticancer metabolites that could particularly target various cancer progression-linked signaling pathways. Phenalenones are a unique class of secondary metabolites that possess diverse bioactivities. In the current work, the CK2 inhibitory capacity of 33 fungal phenalenones was explored using computational studies. After evaluating the usefulness of the compounds as enzyme inhibitors by ADMET prediction, the compounds were prepared for molecular docking in the CK2-α1 crystal structure (PDB: 7BU4). Molecular dynamic simulation was performed on the top two scoring compounds to evaluate their binding affinity and protein stability through a simulated physiological environment. Compound **19** had a superior binding affinity to the co-crystallized ligand (**Y49**). The improved affinity can be attributed to the fact that the aliphatic chain makes additional contact with Asp120 in a pocket distant from the active site.

Keywords: phenalenones; fungi; cancer; casein kinase; CK2 inhibitor; molecular docking

Citation: Ibrahim, S.R.M.; Bagalagel, A.A.; Diri, R.M.; Noor, A.O.; Bakhsh, H.T.; Muhammad, Y.A.; Mohamed, G.A.; Omar, A.M. Exploring the Activity of Fungal Phenalenone Derivatives as Potential CK2 Inhibitors Using Computational Methods. *J. Fungi* **2022**, *8*, 443. https://doi.org/10.3390/jof8050443

Academic Editors: Tao Feng and Frank Surup

Received: 18 March 2022
Accepted: 21 April 2022
Published: 24 April 2022

1. Introduction

Cancer is a complicated illness that is featured by uncontrolled cell proliferation and the development of tumors that may remarkably extend to the entire organ or other organs systemically [1]. It represents one of the most global deadly diseases, particularly in western countries. It was estimated that in 2020, ~ 9.9 million people have died due to cancer [2]. Its current treatment strategies include γ-radiation, chemotherapy, suicide gene therapy, or immunotherapy, which are fundamentally mediated by promoting apoptosis [3]. Chemotherapy is the most efficacious method for metastatic tumors treatment. However, the cancer cell's multidrug resistance, high cost, and untoward effects of these drugs represent the main holdbacks to the success of chemotherapeutic treatment [4]. Therefore, searching for sources of safe and effective treatment has become a crucial research area.

Casein kinase 2 (CK2) is one of the first discovered Ser/Thr kinases [5–7] that is involved in many cell processes from gene expression and protein synthesis to cell growth, proliferation, and differentiation [8]. The well-studied tetrameric form of this kinase is composed of two catalytic alpha proteins, CK2α and/or CK2α′, that differ only in their C-termini [9–11], and two regulatory CK2β proteins that are responsible for substrate specificity [11–13]. It has minimal substrate specificity; therefore, it is able to phosphorylate a large number of proteins involved in multiple signal transduction pathways [14]. This enzyme is ubiquitously expressed [7] and constitutively active in cells; hence, its activity is not relied on extracellular stimuli [5,15,16]. Many reports have found that the overexpression of CK2 in many cancer types leads to inhibiting apoptosis and promoting cancer cell survival [5,8]. It is clear that its downregulation revered cancer progression and enhanced apoptosis [17], suggesting that CK2 can be considered a valuable target for anticancer agents [8,18].

Natural metabolites and their derivatives have long been established as a wealthy source for the discovery of novel anticancer drugs [19]. It was reported that ~5% of the anticancer drugs approved by the FDA are unmodified natural metabolites, and ~50% of the approved drugs are either chemically-modified natural metabolites or synthesized relied on natural metabolites structures [20].

It is noteworthy that different metabolites belonging to various classes, such as flavonoids, coumarins, anthraquinones, pyrazolotriazines, polyhalogenated benzimidazoles, and benzotriazoles, have been known as CK2 inhibitors [21]. Additionally, CIGB-300 and CX-4945 are CK2 inhibitors that are already in human trials as anticancer drugs. CX-4945 (Silmitasertib) has been designated by the FDA for treating cholangiocarcinoma, and many clinical investigations are ongoing with it versus different types of cancers [22]. Furthermore, CIGB-300 is under investigation for cervical cancers [23].

Fungi are eukaryotic micro-organisms that inhabit nearly all kinds of environments in nature, where they play a fundamental role in maintaining eco-balance [24–28]. Fungi have drawn immense research attention since their cultivation, isolation, characterization, and purification demonstrated the existence of a vast number of metabolites with unique and diversified chemical classes and bioactivities [29–34]. Many reports revealed the characterization of diverse classes of fungal metabolites with CK2 inhibitory potential [35–37].

Phenalenones are a unique class of secondary metabolites, having a fused three-ring system [38]. They are produced by higher plants and fungi [39]. They are known as phytoalexins, which are biosynthesized by the plant in response to an external threat, such as mechanical injury or pathogenic infections [40]. Fungal phenalenones have immense structural diversity, such as hetero- and homo-dimerization and high degrees of nitrogenation and oxygenation, as well as being complexed with metals, incorporating with additional carbon frameworks, or isoprene unit by the formation of either a linear ether (e.g., **8** and **12**) or a trimethyl-hydrofuran (e.g., **3–7**, **9–11**, **13**, and **14**) moiety. It has been reported to possess various bioactivities, including antimicrobial, anti-plasmodial, anticancer, antidiabetic, antioxidant, and cytotoxic effects [40]. These metabolites exhibit cytotoxic capacities towards various cancer cell lines; however, there are limited or lacking studies that explore the mechanism of their anticancer properties. It is noteworthy that there is no available report on their CK2 inhibitory potential. The molecular docking-based virtual screening process, along with increasing the data about the structure of CK2 alone or in complex with various inhibitors, could become of particular relevance in the discovery of new lead compounds as CK2 inhibitors [41,42]. In continuation of our interest in exploring the bioactivities of fungal metabolites, the present work focuses on the in silico assessment of CK2 inhibitory capacity of 33 phenalenone derivatives reported from various fungal sources (Table 1, Figures 1 and 2).

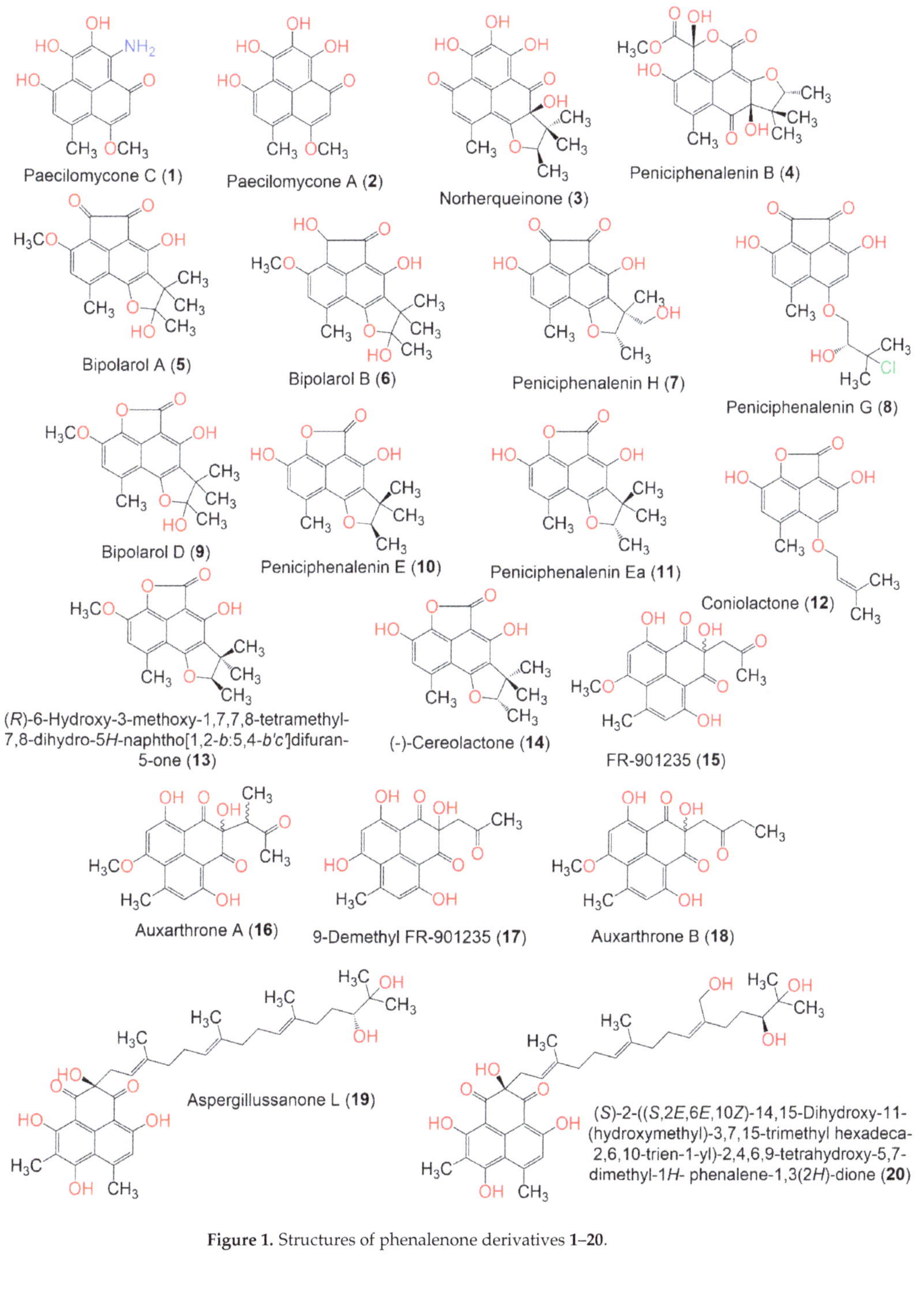

Figure 1. Structures of phenalenone derivatives **1–20**.

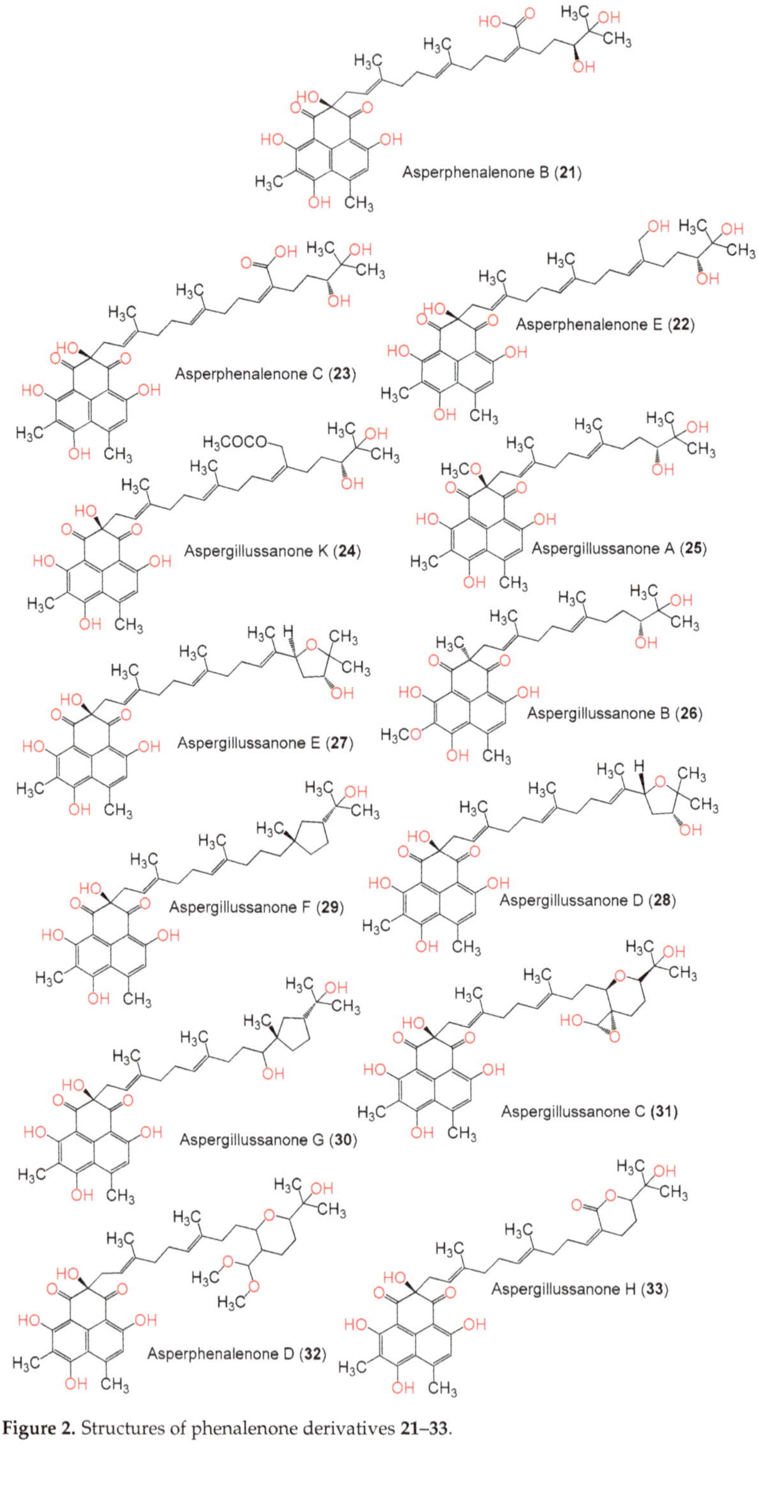

Figure 2. Structures of phenalenone derivatives **21–33**.

Computer-aided drug design/discovery (CADD) are useful tools for screening and identifying drug-like molecules in silico and thereby reducing the number of compounds to be tested experimentally. Several software and programs are used to filter and generate a group of compounds based on specified criteria, predict their physicochemical properties, predict suitable targets, and evaluate the binding affinity of the compounds to the predicted targets. One of the drug design and discovery approaches is structure-based drug design (SBDD). This approach relies on the knowledge of the 3D structure of the targets of interest, and it includes two common methods: molecular docking and molecular dynamic simulation. Molecular docking evaluates how tight the compound binds the target, as determined by the predicted binding affinity, while molecular dynamic (MD) simulations assess the behavior of the ligand-protein complex in terms of binding interactions and 3D conformation in aqueous conditions to mimic the physiological environment [43]. Several CADD methods and tools are used for investigating the phenalenone derivatives.

Table 1. List of phenalenone derivatives and their fungal source.

Compound Name	Fungal Source	Ref.
Paecilomycone C (**1**)	*Paecilomyces gunnii*	[44]
Paecilomycone A (**2**)	*Paecilomyces gunnii*	[44]
Norherqueinone (**3**)	*Penicillium* sp. G1071	[45]
Peniciphenalenin B (**4**)	*Penicillium* sp. ZZ901	[46]
Bipolarol A (**5**)	*Lophiostoma bipolare* BCC25910	[47]
Bipolarol B (**6**)	*Lophiostoma bipolare* BCC25910	[47]
Peniciphenalenin H (**7**)	*Pleosporales* sp. HDN1811400	[48]
Peniciphenalenin G (**8**)	*Pleosporales* sp. HDN1811400	[48]
Bipolarol D (**9**)	*Lophiostoma bipolare* BCC25910	[47]
Peniciphenalenin E (**10**)	*Penicillium* sp. ZZ901	[46]
Peniciphenalenin Ea (**11**)	*Penicillium* sp. ZZ901	[46]
Coniolactone (**12**)	*Chrysosporium lobatum* TM-237-S5	[49]
(R)-6-Hydroxy-3-methoxy-1,7,7,8-tetramethyl-7,8-dihydro-5H-naphtho[1,2-b:5,4-b′c′]difuran-5-one (**13**)	*Trypethelium eluteriae*	[50]
(-)-Cereolactone (**14**)	*Penicillium herquei* PSU-RSPG93	[51]
FR-901235 (**15**)	*Auxarthron pseudauxarthron* TTI-0363	[52]
Auxarthrone A (**16**)	*Auxarthron pseudauxarthron* TTI-0363	[52]
9-Demethyl FR-901235 (**17**)	*Talaromyces stipitatus*	[53]
Auxarthrone B (**18**)	*Auxarthron pseudauxarthron* TTI-0363	[52]
Aspergillussanone L (**19**)	*Aspergillus* sp.	[54]
(S)-2-((*S*,2*E*,6*E*,10*Z*)-14,15-Dihydroxy-11-(hydroxymethyl)-3,7,15-trimethylhexadeca-2,6,10-trien-1-yl)-2,4,6,9-tetrahydroxy-5,7-dimethyl-1*H*-phenalene-1,3(2*H*)-dione (**20**)	*Aspergillus* sp.	[54]
Asperphenalenone B (**21**)	*Aspergillus* sp. CPCC 400735	[55]
Asperphenalenone E (**22**)	*Aspergillus* sp. CPCC 400735	[55]
Asperphenalenone C (**23**)	*Aspergillus* sp. CPCC 400735	[55]
Aspergillussanone K (**24**)	*Aspergillus* sp.	[54]
Aspergillussanone A (**25**)	*Aspergillus* sp. PSU-RSPG185	[56]
Aspergillussanone B (**26**)	*Aspergillus* sp. PSU-RSPG185	[56]
Aspergillussanone E (**27**)	*Aspergillus* sp.	[54]
Aspergillussanone D (**28**)	*Aspergillus* sp.	[54]
Aspergillussanone F (**29**)	*Aspergillus* sp.	[54]
Aspergillussanone G (**30**)	*Aspergillus* sp.	[54]
Aspergillussanone C (**31**)	*Aspergillus* sp.	[54]
Asperphenalenone D (**32**)	*Aspergillus* sp. CPCC 400735	[55]
Aspergillussanone H (**33**)	*Aspergillus* sp.	[54]

2. Results and Discussion

2.1. AI (Artificial Intelligence)-Based Target Prediction for Phenalenone Derivatives

Choosing the appropriate target to investigate the inhibitory potential of these phenalenone derivatives relied on performing ligand-based-in silico target prediction [38]. The prediction webserver, SuperPred was the tool of choice to perform ATC (anatomical-therapeutic chemical) code and predict the potential targets for the investigated phenalenones [57]. After analyzing the results of the predicted target proteins (for example, Cathepsin D, mineralocorticoid receptor, and thyroid hormone receptor-α), the kinase CK2α was selected for the study due to having a high percent of probability and model accuracy (Table 2). After selecting the target and proper crystal structure, the listed phenalenones were docked in the protein crystal structure, after which the docking method was validated by redocking the co-crystallized ligand. Prediction of ADMET properties of the listed metabolites in silico and MD (molecular-dynamic) simulation for the two metabolites with the highest docking scores were followed.

Table 2. Prediction of target probability and model accuracy for phenalenone derivatives against CK2 using SuperPred target prediction webserver.

Compounds	Probability *	Model Accuracy **
1	61%	99%
2	65%	99%
3	89%	99%
4	88%	99%
5	83%	99%
6	88%	99%
7	83%	99%
8	50%	99%
9	85%	99%
10	94%	99%
11	94%	99%
12	91%	99%
13	92.9%	99%
14	94%	99%
15	75.7%	99%
16	75%	99%
17	76%	99%
18	76%	99%
19	75.7%	99%
20	75%	99%
21	79%	99%
22	75%	99%
23	79%	99%
24	80%	99%
25	76%	99%
26	76%	99%
27	82%	99%
28	82%	99%
29	81%	99%
30	86%	99%
31	83.39%	99.23%
32	83.6%	99%
33	88%	99%

* The probability of the test compound binding to a specific target, as determined by the respective target machine learning model. ** The accuracy of the performance of the prediction model displaying the 10-fold cross-validation score of the respective logistic regression model, as the model performance varied between different targets.

2.2. Ligands and Protein Preparation and Molecular Docking

Target identification filtered out > 100 phenalenone derivatives. These derivatives were prepared for docking, where their energy-minimized 3D structures were generated, and all possible ionization and tautomeric states were created.

For docking, the human CK2α1 crystal structure (PDB ID: 7BU4) was chosen for the study due to the structural similarity of the co-crystallized ligand (**Y49**) to the selected phenalenones. The **Y49** (4-(6-aminocarbonyl-8-oxidanylidene-9-phenyl-7*H*-purin-2-yl)benzoic acid) is made of three aromatic rings: a purine ring with two phenyl moieties attached to it, and polar groups present at the rings, with polar groups at different positions. The selected phenalenones have a nucleus made of three-fused rings substituted with multiple OH and carbonyl groups. Based on their structures, both the phenalenones and **Y49** were expected to have a similar 3D conformation in the binding pocket. The PDB file of the 7BU4 crystal structure was downloaded from the protein data bank (PDB) [58], which was then prepared and minimized using Schrodinger's Protein Preparation wizard [59–61].

The docking process started by generating a grid box around the binding site of the co-crystallized ligand to locate the pocket where the docking of the compounds occurs. A receptor-Grid-Generation tool in Maestro [62] was utilized for that purpose.

Re-docking of the co-crystallized ligand, 4-(6-aminocarbonyl-8-oxidanylidene-9-phenyl-7*H*-purin-2-yl)benzoic acid (PDB: **Y49**), was performed to evaluate the docking method. The re-docked reference had an identical pose (Figure 3C) and binding interactions to the co-crystallized structure (Figure 3A,B). For both, the backbone of Val116 is H-bonded with the oxygen of the amide moiety, and the purinone nitrogen interacts through the water bridge with Val116, Asn118, as well as with the amide oxygen. On the opposite side of the molecule, the carboxylate makes an ionic interaction with the side chain of the adjacent Lys68 (Figure 3A,B). The RMSD (root-mean-square deviation) of the re-docked ligand was minimal, with a value of 0.0744 Å, indicating the docking method is valid (Figure 3C). The molecular surface display in Figure 3D shows the re-docked reference **Y49** occupying the binding pocket of the crystal structure.

After docking validation, docking the 3D structures of the > 100 phenalenones that proceeded from the target prediction using the extra-precision (XP) mode was followed. The docking produced derivatives that are ranked based on their score and approximated the free energy of binding; the more negative the value, the stronger the binding. Different docking scores were generated, including the gscore (best for ranking different compounds), emodel (best for ranking conformers), and XP gscore. Glide uses emodel scoring to select the best poses of the docked compounds; then, it ranks the best poses based on the given gscores. XP gscore ranks the poses generated by the Glide XP mode. In general, Glide uses gscore to sort and rank the docked compounds. The 33 derivatives listed in Table 3 are the ones with gscores close to or better than the gscore of the reference **Y49** (−9.049 kcal/mol), with the top two compounds, **19** and **31**, scoring −12.878 and −12.521 kcal/mol, respectively.

Compound **19**, in addition to interacting directly with Val116 and Lys68 in the protein's binding pocket like the reference ligand, had a long chain that extended along the surface of the protein allowing the terminal (R,6*E*,10*E*,14*E*)-2,6,10,14-tetramethylhexadeca-6,10,14-triene-2,3-diol group to bind a distant binding pocket (Figure 4). Besides, compound **31** seems to have similar interactions with the protein; however, the tetrahydropyran at the end of the aliphatic chain did not occupy the distant pocket like **19** and remained exposed to the solvent (Figure 5). Figure 6 showed compounds **19** and **31** simultaneously superimposed on the reference **Y49** inside the binding pocket in the molecular surface display.

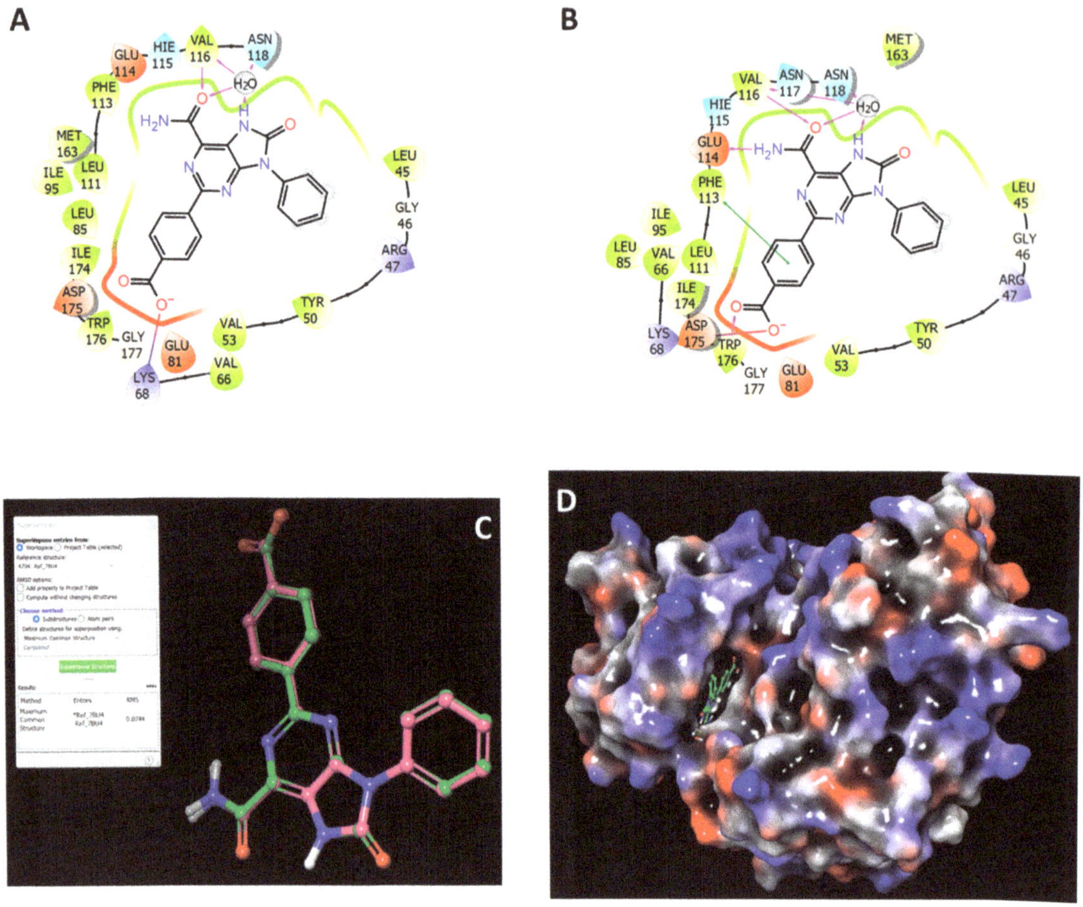

Figure 3. Re-docking of the co-crystallized ligand to validate the docking method. The figure showed the 2D view of the binding interactions of the reference ligand **Y49** complexed with CK2α1; (**A**) after minimization of the crystal structure 7BU4, and (**B**) after re-docking of ligand **Y49** into the CK2α1 crystal structure. (**C**) 3D structure of the re-docked **Y49** (pink) superimposed on the co-crystallized Y49 (green). (**D**) Molecular surface display with electrostatic potential color scheme for CK2α1 complexed with ligand **Y49** after re-docking.

3. In silico docking results of phenalenone derivatives with CK2α1 (PDB: 7BU4).

ıpounds	Docking Score	Glide GScore	Glide Emodel	XP GScore
19	−12.181	−12.878	−84.318	−12.878
31	−10.976	−12.521	−100.255	−12.521
20	−10.654	−12.303	−102.849	−12.303
23	−10.798	−11.495	−92.818	−11.495
26	−8.802	−10.538	−88.527	−10.538
33	−9.828	−10.526	−80.067	−10.526
22	−9.712	−10.409	−85.63	−10.409
27	−9.706	−10.403	−81.743	−10.403
24	−8.859	−10.403	−96.868	−10.403
25	−9.541	−10.366	−72.182	−10.366
3	−9.191	−10.232	−62.622	−10.232
17	−8.52	−10.194	−55.199	−10.194
28	−9.435	−10.132	−72.289	−10.132
21	−8.479	−10.128	−91.905	−10.128
12	−9.559	−9.935	−51.879	−9.935
29	−9.129	−9.825	−76.553	−9.825
1	−7.669	−9.8	−56.073	−9.8
2	−9.343	−9.753	−55.206	−9.753
32	−9.008	−9.704	−56.471	−9.704
30	−8.983	−9.679	−75.988	−9.679
15	−9.103	−9.624	−50.939	−9.624
14	−9.331	−9.492	−49.907	−9.492
11	−9.331	−9.492	−49.907	−9.492
18	−8.956	−9.485	−60.341	−9.485
4	−9.265	−9.341	−61.825	−9.341
10	−9.167	−9.328	−50.165	−9.328
9	−9.23	−9.31	−51.968	−9.31
13	−9.136	−9.216	−52.397	−9.216
7	−9.162	−9.162	−50.334	−9.162
6	−9.082	−9.154	−63.817	−9.154
5	−9.147	−9.147	−52.055	−9.147
16	−8.656	−9.106	−56.773	−9.106
8	−9.067	−9.068	−51.613	−9.068
Y49_7BU4	−9.049	−9.049	−93.921	−9.049

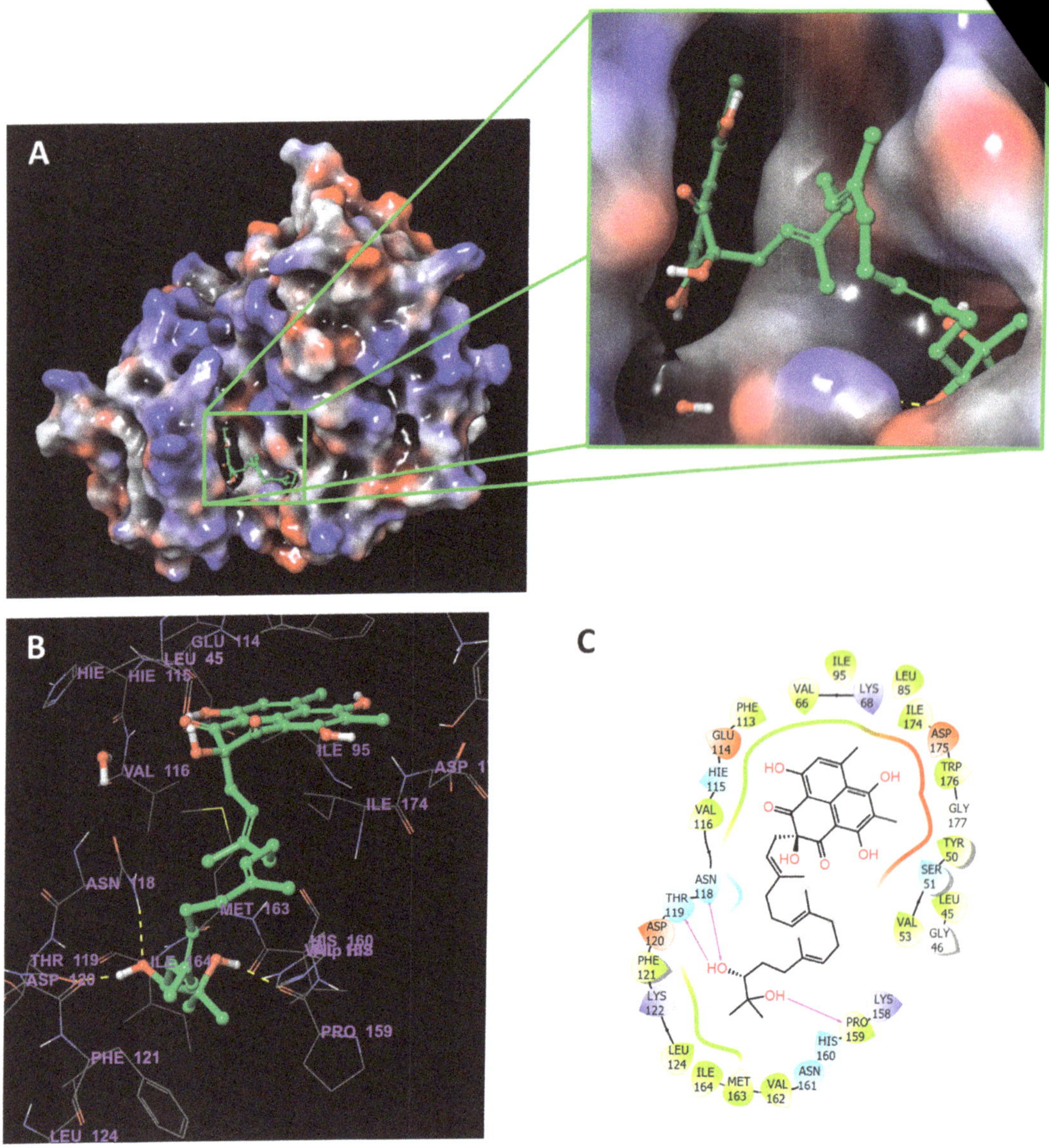

Figure 4. CK2α1 in complex with **19**. (**A**) Molecular surface display with electrostatic potential color scheme for CK2α1-**19** complex and the close-up view presented. (**B**) 3D presumed binding mode of **19** in the binding site of CK2 (PDB:7BU4). Compound **19** is displayed as green ball-and-sticks. The amino acids in the binding site are shown as grey sticks, and hydrogen bonds are represented in yellow dotted lines. (**C**) 2D depiction of the ligand-protein interactions.

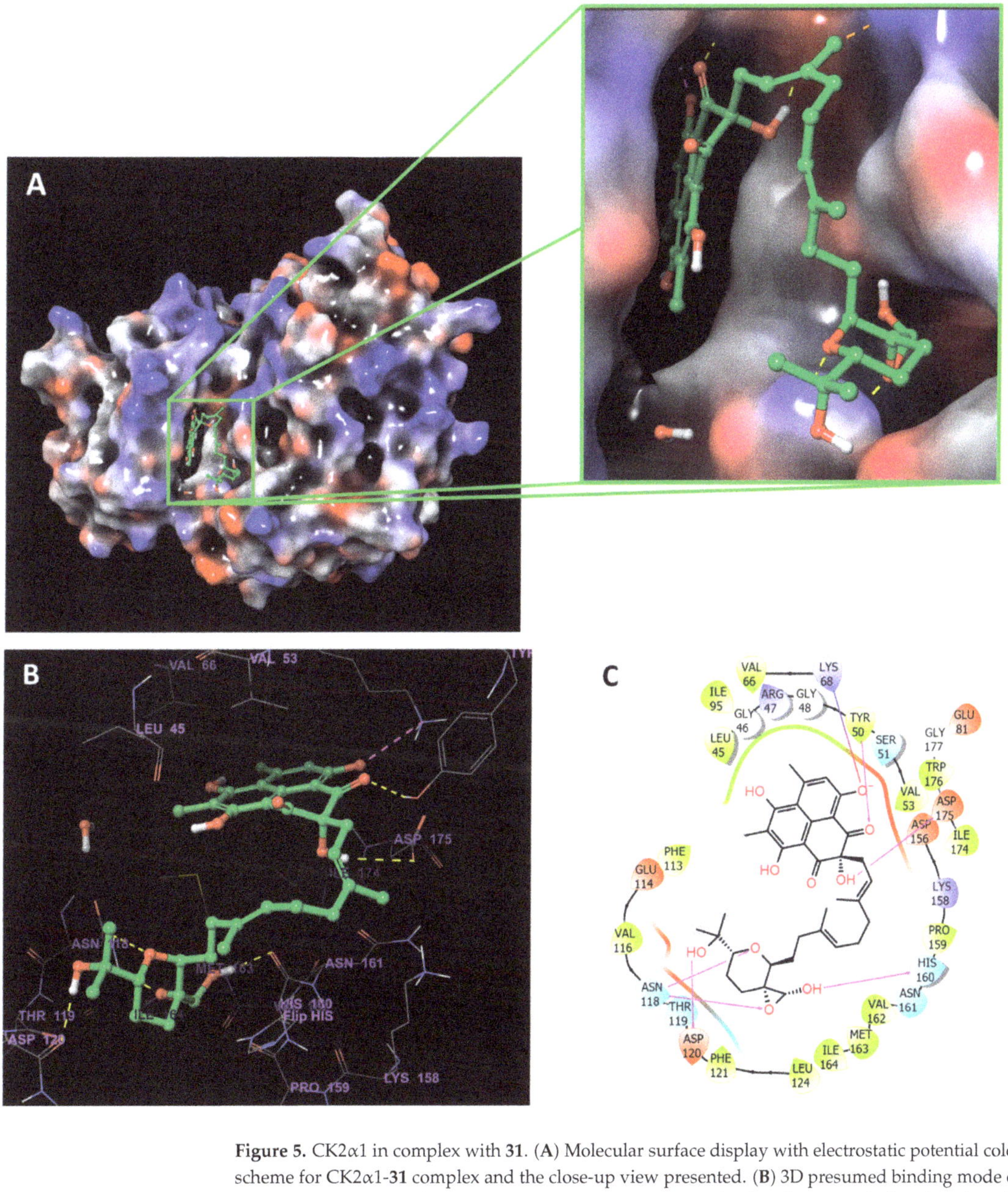

Figure 5. CK2α1 in complex with **31**. (**A**) Molecular surface display with electrostatic potential color scheme for CK2α1-**31** complex and the close-up view presented. (**B**) 3D presumed binding mode of **31** in the binding site of CK2 (PDB:7BU4). Compound **31** is displayed as green ball-and-sticks. The amino acids in the binding site are shown as grey sticks, and hydrogen bonds are represented in yellow dotted lines. (**C**) 2D depiction of the ligand-protein interactions.

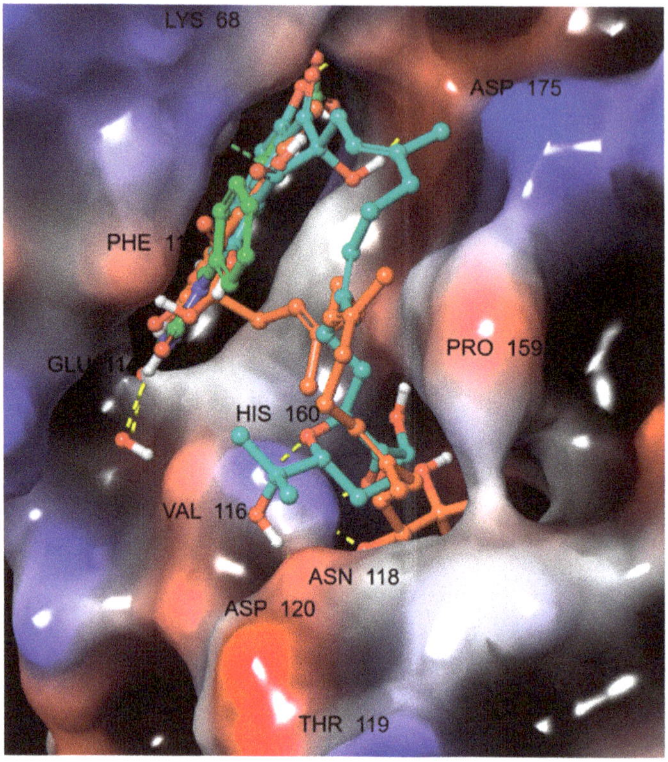

Figure 6. Molecular display with electrostatic potential color scheme for CK2α1 (PDB: 7B4U) showing **19** (orange) and **31** (cyan) superimposed on **Y49** (green) inside the binding pocket.

2.3. In Silico ADMET Properties of Selected Ligands

The drug-likeness and ADMET properties of the processed compounds were predicted using Maestro's QikProp Schrodinger module in terms of absorption, distribution, metabolism, excretion, and toxicity, among others [63]. The module can quickly and reliably predict many physicochemical properties and other descriptors, such as the number of possible metabolites and number of reactive functional groups, in order to detect and filter compounds that can be problematic during the late stages of drug discovery and development. Therefore, it can eliminate unnecessary tests and experiments that will ultimately fail in clinical trials [64]. The ADMET prediction evaluates the usefulness of the examined compounds by describing and determining their drug-likeness, physicochemical properties, and expected toxicity profiles. Several descriptors were predicted for these derivatives, and most of the predicted values of ADMET descriptors fell within the recommended range. The predicted ADMET properties and descriptors are presented in Table 4.

Table 4. Predicted in silico ADMET properties of the phenalenone derivatives.

Compounds *	Mol_MW	# Stars	Dipole	SASA	DonorHB	AccptHB	QPlogPo/w	QPlogS	QPlogKhsa	# Metab	QPlogBB	%Human Oral Absorption	QPlogHERG	CNS	# RtvFG
Recommended range	(130–725)	(0.0–5.0)	(1–12.50)	(300–1000)	(0–6)	(2.0–20.0)	(−2–6.5)	(−6.5–0.5)	(−1.5–1.5)	(1–8)	(−3–1.2)	(<25% poor; >80% high)	concern below −5	(−2 inactive) (+2 active)	(0–2)
1	287.271	0	2.269	482.741	4	5	0.637	−2.456	−0.357	6	−1.577	65.408	−3.892	−2	0
2	288.256	0	7.03	481.857	3	4.75	0.9	−2.549	−0.273	6	−1.552	67.526	−3.919	−2	0
3	358.347	0	4.081	543.911	2	5.75	1.679	−3.554	0.106	5	−1.509	71.209	−3.602	−2	0
4	404.373	0	9.43	616.737	3	10	0.76	−3.62	−0.273	3	−1.618	67.011	−4.155	−2	2
5	342.348	0	9.298	561.827	1	6	2.037	−3.9	0.109	3	−1.134	81.224	−3.898	−2	3
6	344.363	0	3.309	569.109	2	5.7	2.266	−3.979	0.158	4	−1.041	85.666	−3.946	−2	1
7	328.321	0	10.77	534.495	1	5.95	1.41	−3.29	−0.062	4	−1.567	69.494	−3.808	−2	2
8	364.782	1	12.714	596.584	1	5.95	2.161	−4.084	0.041	5	−1.884	72.456	−4.595	−2	2
9	330.337	0	7.925	528.447	1	4.5	2.59	−3.836	0.255	3	−0.728	91.237	−3.585	−1	1
10	300.31	0	5.524	498.414	1	3.75	2.366	−3.724	0.26	3	−0.796	86.825	−3.516	−1	0
11	300.31	0	5.616	500.081	1	3.75	2.398	−3.753	0.273	3	−0.795	87.024	−3.517	−1	0
12	300.31	0	9.368	548.255	1	3.75	2.634	−4.111	0.269	7	−1.29	85.236	−4.506	−2	0
13	314.337	0	7.36	523.476	0	3.75	3.022	3.848	0.268	3	−0.374	100	−3.613	0	0
14	300.31	0	5.617	499.932	1	3.75	2.399	−3.75	0.274	3	−0.794	87.031	−3.511	−1	0
15	344.32	1	13.714	566.499	0	6	1.573	−2.834	−0.308	6	−1.74	71.138	−4.202	−2	1
16	358.347	0	9.462	553.682	0	6	1.845	−2.591	−0.246	6	−1.417	76.489	−3.596	−2	1
17	330.293	1	15.038	518.961	1	6	0.984	−2.687	−0.224	6	−1.977	60.477	−3.804	−2	1
18	358.347	0	10.248	546.847	0	6	1.673	−2.299	−0.312	7	−1.588	72.711	−3.46	−2	1
19	594.744	2	9.484	789.214	4	7.45	5.322	−4.622	0.981	17	−2.174	57.778	−3.532	−2	0
20	610.743	7	9.975	1043.134	5	9.15	5.066	−7.89	0.907	17	−4.262	41.015	−6.435	−2	0
21	624.727	8	8.459	1026.611	5	9.45	5.31	−7.68	0.735	16	−4.35	28.338	−4.482	−2	1
22	610.743	7	9.879	1030.596	5	9.15	4.911	−7.677	0.918	17	−4.46	49.169	−6.255	−2	0
23	624.727	8	8.459	1026.616	5	9.45	5.31	−7.68	0.735	16	−4.351	28.338	−4.482	−2	1
24	652.78	8	12.335	1096.494	4	9.45	6.114	−8.908	1.288	17	−4.061	50.421	−6.561	−2	1
25	540.652	3	14.173	889.502	3	7.45	5.255	−6.976	0.997	13	−2.485	70.591	−5.556	−2	0

Table 4. *Cont.*

Compounds *	Mol_MW	# Stars	Dipole	SASA	DonorHB	AccptHB	QPlogPo/w	QPlogS	QPlogKhsa	# Metab	QPlogBB	%Human Oral Absorption	QPlogHERG	CNS	# RtvFG
26	540.652	2	7.928	919.785	3	7.45	5.282	−7.502	1.064	13	−2.834	67.029	−5.832	−2	0
27	592.728	6	12.392	1005.377	3	7.45	6.358	−9.167	1.54	17	−2.88	73.816	−6.238	−2	0
28	592.728	6	11.434	1011.989	3	7.45	6.379	−9.284	1.561	17	−2.954	73.171	−6.288	−2	0
29	592.771	5	11.447	968.98	3	5.75	7.264	−8.949	1.821	13	−2.355	85.225	−5.777	−2	0
30	608.77	5	12.352	986.08	4	7.45	5.988	−8.302	1.398	14	−3.232	53.649	−5.838	−2	0
31	624.727	4	11.95	974.752	4	11.15	4.249	−7.089	0.71	14	−3.346	53.716	−5.828	−2	3
32	654.796	7	10.567	1030.237	3	10.85	5.564	−7.91	1.034	14	−3.042	68.666	−6.017	−2	1
33	606.711	5	9.825	992.894	3	8.75	5.513	−8.33	1.235	15	−3.349	62.369	−6.083	−2	1

* Recommended range: for 95% of known drugs; #Stars: # of descriptors that fall outside the 95% range of same values for known drugs. Large star number indicates less drug-likeness, and vice versa; Dipole: computed dipole moment; SASA: Total solvent accessible surface area; DonorHB: estimated number H^+ to be donated in HB; AcceptHB: estimated number H^+ to be accepted in HB; QLogPo/w: predicted octanol/water partition coefficient; QPlogS: Predicted aqueous solubility; QPlogKhsa: Prediction of binding to human serum albumin; #Metab: number of possible metabolic reactions; QPlogBB: Predicted brain/blood partition coefficient; % Human Oral Absorption: Predicted human oral absorption on 0 to 100% scale; QPlogHERG: Predicted IC_{50} value for blockage of HERG K^+ channels; CNS: Predicted central nervous system activity; #RtvFG: Number of reactive functional groups.

2.4. Molecular Dynamic (MD) Simulation

The MD simulations are performed using Desmond software [65,66] to simulate the aqueous physiological environment and assess the changes in protein conformation and binding affinity during the simulation time compared to the original affinity and confirmation of the crystal structure [67]. Only the two top-scoring compounds from the docking study, i.e., compounds **19** and **31** along with reference **Y49**, were analyzed by MD. The root mean square deviation (RMSD) is a calculated value that compares the poses of investigated compounds to that of the co-crystalized ligand [43]. RMSD plots of the selected compounds complexed with the CK2α1 measure the average change in the positions of the atoms of the protein and ligand inside the binding pocket at the end of the simulation period (100 ns) compared to their starting positions before the simulation at 0 ns. The RMSD plot of **Y49** showed that both the protein and the reference **Y49** were stable, and the observed fluctuations were insignificant since they were within the acceptable range of 1–3 Å (Figure 7A). For compound **19**, the RMSD of the protein and **19** laid over each other, indicating increased binding affinity of **19** to the protein and stability of the CK2α1-**19** complex. Additionally, the fluctuation seen for both over the 100 ns was within the range as well (Figure 8A). A similar RMSD pattern was observed for **31** and CK2α1 complex, despite the sudden, non-significant fluctuation of **31** at around 90 ns, which is potentially a result of the compound adjusting its pose in the pocket (Figure 9A). When calculating the RMSD for the compounds, it is not uncommon to observe fluctuation in the plot for some time at the beginning of the simulation, as observed in Figure 7A, Figure 8A, and Figure 9A within the first 20 ns of the run. This expected fluctuation happens as the compound keeps adjusting its conformation inside the pocket to assume a pose that has the least free energy.

The secondary structure of the CK2α1 protein (PDB ID: 7BU4) was also evaluated throughout the simulation while it was complexed with each ligand. Figure 7B represented the protein evaluation while it was complexed with ligand **Y49**. The top plot showed the distribution of the SSE (α-helices and β-sheets) with the protein represented by the residue index. The middle plot summarized the SSE composition for each trajectory frame throughout the simulation, while the bottom plot monitored each residue and its SSE assignment over the simulation time. Both plots indicated that the overall %SSE of the protein was maintained, and each SSE was stable over the course of the simulation. Comparable results were obtained when **19** (Figure 8B) and **31** (Figure 9B) were complexed with the protein.

The MD study also evaluated the binding interactions of a protein-ligand complex. For ligand **Y49**, the interactions between the **Y49** and protein are presented in Figure 10A; the interaction types are color-coded. The stacked bar chart is normalized over the course of the trajectory: for example, a value of 0.8 suggested that the specific interaction was maintained during 80% of the simulation time. Values over 1.0 are possible and indicate that some protein residue may make multiple interactions of the same subtype with the ligand. As indicated in Figure 10A, Val116 made direct H-bonding as well as through water bridges with **Y49** and had a normalized value of ~1.9. The value >1 represented the combined value of >1 type of interaction, and it indicated that these interactions were maintained for ~190% of the simulation time. The other key interactions were with Glu114, Asn118, Lys68, and Asp175, having values of ~0.9, ~0.7, ~1.1, and ~1.5, respectively. Figure 10B showed only the interactions between **Y49** and the protein that occurred $\geq$ 30% of the simulation time. Figure 10C displayed the total specific interactions between ligand **Y49** and the protein (top plot), whilst the bottom panel demonstrated the protein residues that interacted with the ligand at each time point. As mentioned earlier, Val116 made > 1 interaction with the ligand, which was represented by the dark orange color in the plot throughout the trajectory. Other residues: Lys86, Glu114, and Asp175, also made specific interactions with the ligand.

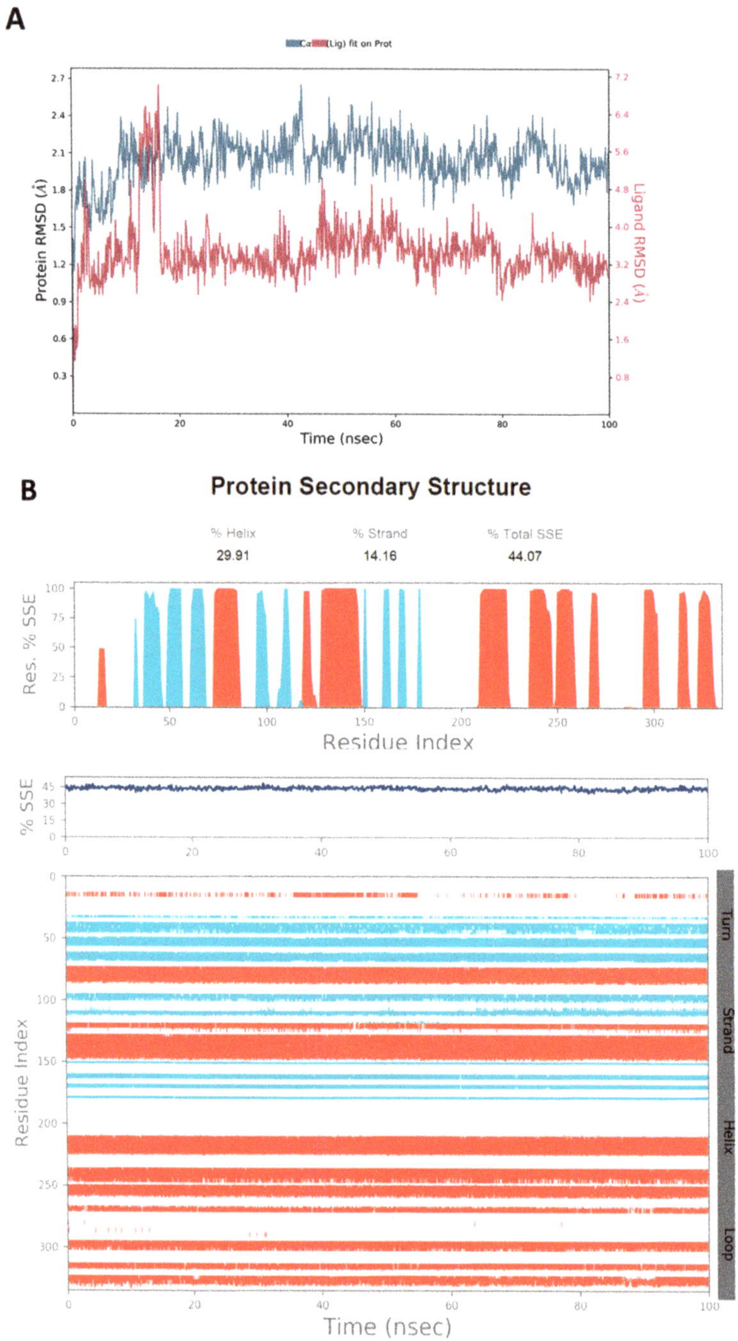

Figure 7. (**A**) The RMSD plot was obtained for the reference **Y49** complexed with CK2α1 protein (PDB-ID:7BU4). The simulation time (100 ns) reaffirmed the stability of the complex with no significant changes in the protein structure. (**B**) Stability of the secondary structure of CK2α1 protein (PDB ID: 7BU4) was evaluated by monitoring its SSE distribution (top plot), SSE composition (middle plot), and SEE assignment (bottom plot) over the 100 ns of MD simulation when complexed with **Y49**.

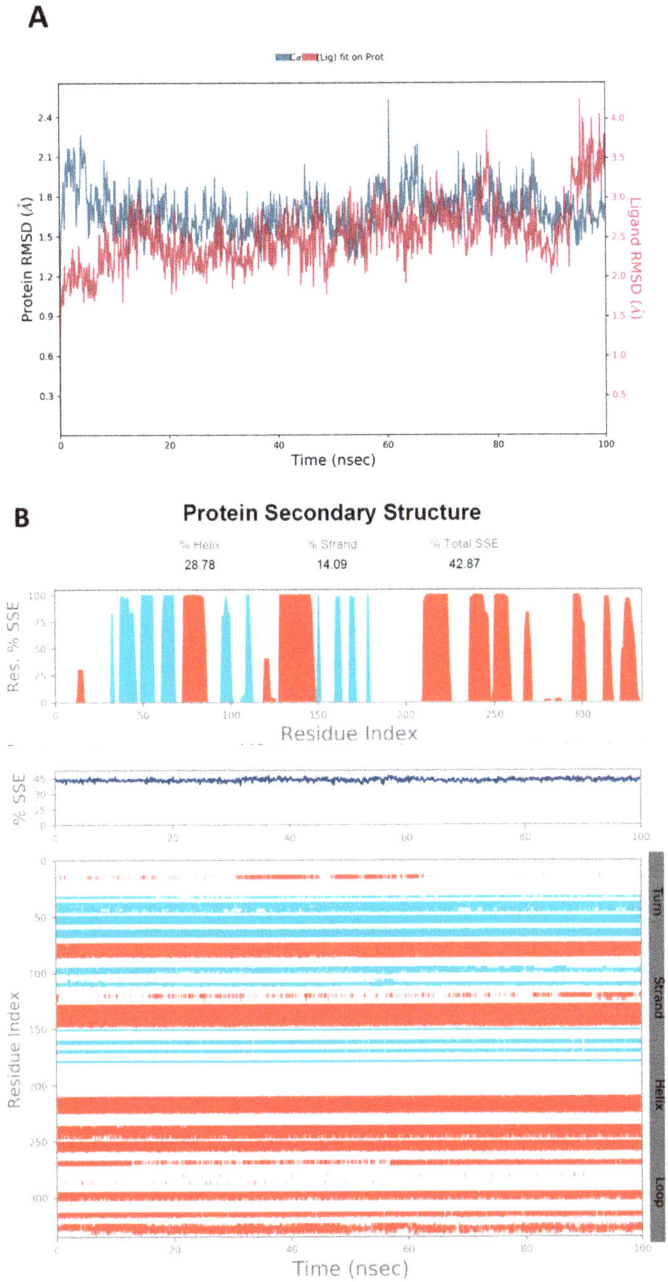

Figure 8. (**A**) The RMSD plot was obtained for compound **19** complexed with CK2α1 protein (PDB-ID:7BU4). The simulation time (100 ns) reaffirmed the stability of the complex with no significant changes in the protein structure. (**B**) Stability of the secondary structure of CK2α1 protein (PDB ID: 7BU4) was evaluated by monitoring its SSE distribution (top plot), SSE composition (middle plot), and SEE assignment (bottom plot) over the 100 ns of MD simulation when complexed with **19**.

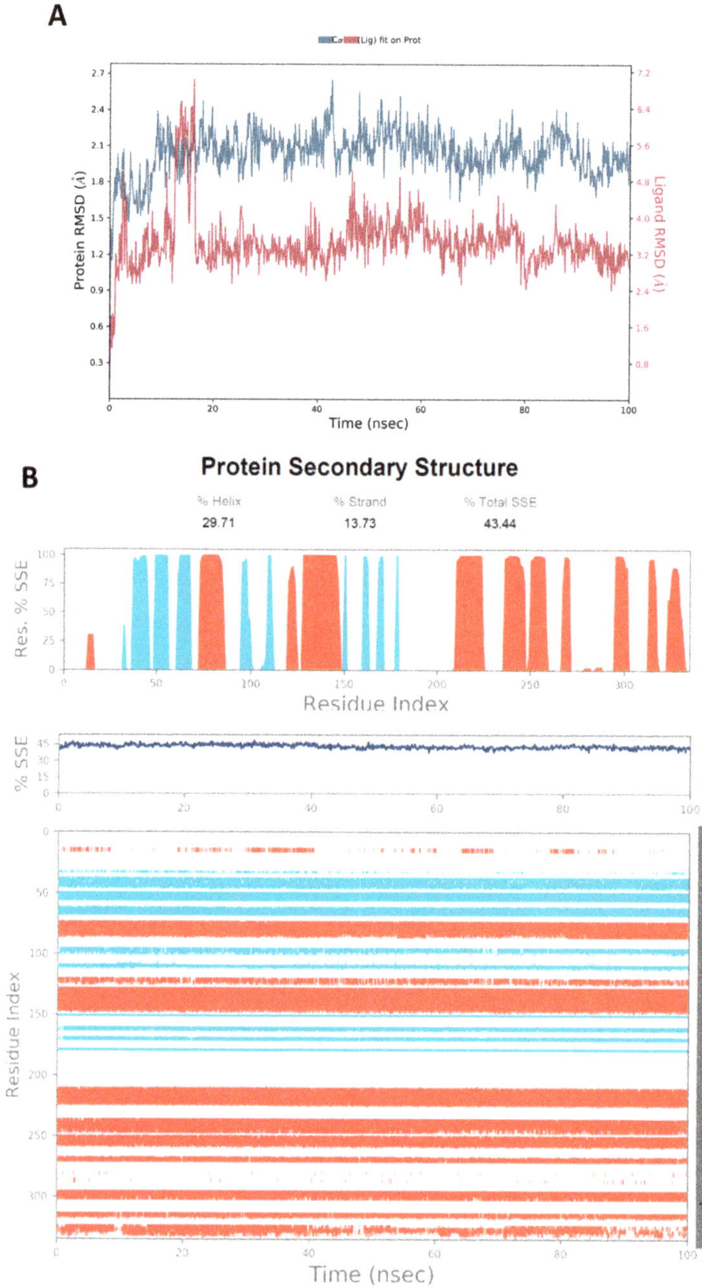

Figure 9. (**A**) Obtained RMSD plot for compound **31** complexed with CK2α1 protein (PDB-ID:7BU4). The simulation time (100 ns) reaffirmed the stability of the complex with no significant changes in the protein structure. (**B**) Stability of the secondary structure of CK2α1 protein (PDB ID: 7BU4) was evaluated by monitoring its SSE distribution (top plot), SSE composition (middle plot), and SEE assignment (bottom plot) over the 100 ns of MD simulation when complexed with **31**.

Figure 11 shows the amino acid residues of the protein binding pocket that interacted with compound **19**. The fused ring system of **19** made similar interactions with the pocket residues as **Y49**, where Val116 and Lys86 interacted through the H-bond with the ring system (Figure 11B). As previously seen in the molecular surface display (Figure 4A), the extended aliphatic chain occupied a distant pocket and created new interaction points between the two OH groups at the end of the chain and Asp120, where they occurred > 85% during the simulation (Figure 11B). This interaction was not present with **Y49** (Figure 3B,D). It might be safe to assume that the enhanced binding affinity and stability of the complex were due to this new interaction with Asp120, which can also be inferred from the RMSD plot (Figure 8A).

Compound **31** also created new interacting points with amino acids in the main binding pocket: the chiral OH of the fused ring system made a strong H-bonding with His160 and Asn161 through bridging water molecules with a value of ~1.18 and 0.6, respectively. An enhanced interaction with Arg47 was observed as well (~0.98) (Figure 12B). The tetrahydropyran ring at the end of the aliphatic chain of **31** also extended along the protein surface but did not occupy the distant pocket, as did compound **19** (Figure 5A). The reason for that might be the fact that the chain in compound **31** is 6-carbon shorter than that of **19**, so the group could not reach the second pocket. Another explanation could be the large size of the substituted tetrahydropyran hindered the group from occupying the pocket. Additionally, there is a high probability that the fluctuation observed in the RMSD of **31** towards the end of the simulation time (Figure 9A) might be a result of the inability of this group to bind to the second pocket. The L-RMSF (ligand-root-mean-square fluctuation) represents how the atoms of the ligand interact with the protein and the changes in the positions of the ligand atoms. As seen in the L-RMSF plot for compound **31** (Figure 13), the positions of atoms 29–45 were dramatically changed because of the free rotation around the aliphatic chain, which in turn decreased the interaction between this part of the molecule with the protein and was reflected by > 3 Å fluctuation in the RMSF plot. The time-depended representation of the CK2α1-**31** interactions showed that residues Arg47, His160, and Asp175 were the ones making specific interactions with the ligand, as indicated by the darker color in the plot (Figure 12C).

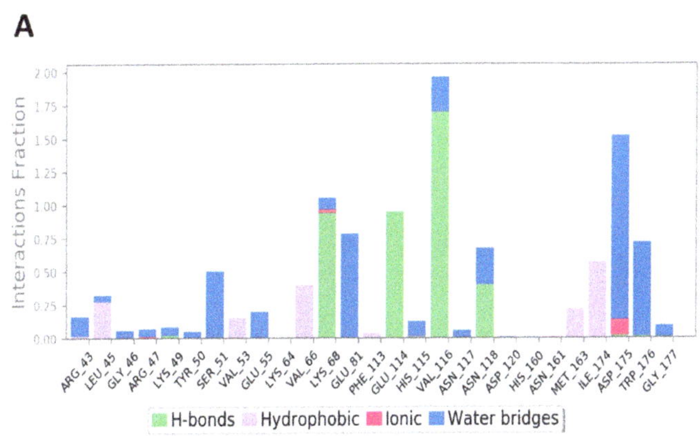

Figure 10. *Cont.*

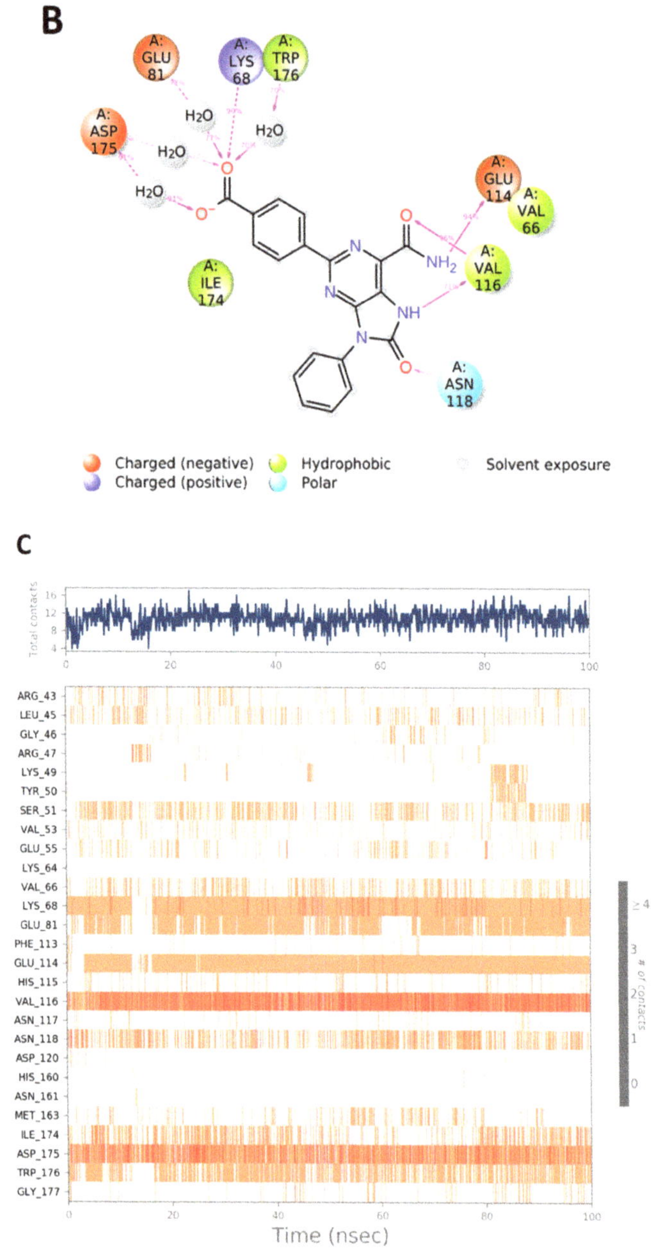

Figure 10. (**A**) Stacked bar graph for CK2 interactions with reference **Y49** throughout the simulation. (**B**) Schematic diagram shows the detailed 2D atomic interactions of **Y49** with CK2 that occurred > 30% of the simulation time in the selected trajectory. (**C**) A timeline representation of CK2-**Y49** interactions presented in (**A**). The top panel presents the total number of specific interactions the protein made with the ligand over the course of the trajectory. The bottom panel presents the residues interacting with the ligand in each trajectory frame. The dark orange color indicates more than one specific interaction is made between some residues and the ligand.

It was also noticed that the reference ligand, **Y49,** as well as the compounds **19** and **31,** all interacted with Asn118 through H-bonding; however, the contact points are different. While the residue interacted with **Y49** at the purine carbonyl oxygen (Figure 10B), it acted as an H-bond donor to the OH at the end of the aliphatic chain of **19** as well as to the OH at the fused ring system of **19** and **31** (Figures 11B and 12B). The reason for the different contact points of Asn118 with the compounds was probably attributed to the 3D conformation of the compounds inside the binding pocket, which is affected by the nature of the substitutions on the nucleus. Each compound assumed a pose that had the lowest possible free energy when interacting with the pocket's residues. Therefore, that pose with the different substitutions from one compound to the other created the unique binding interactions with the pocket amino acids.

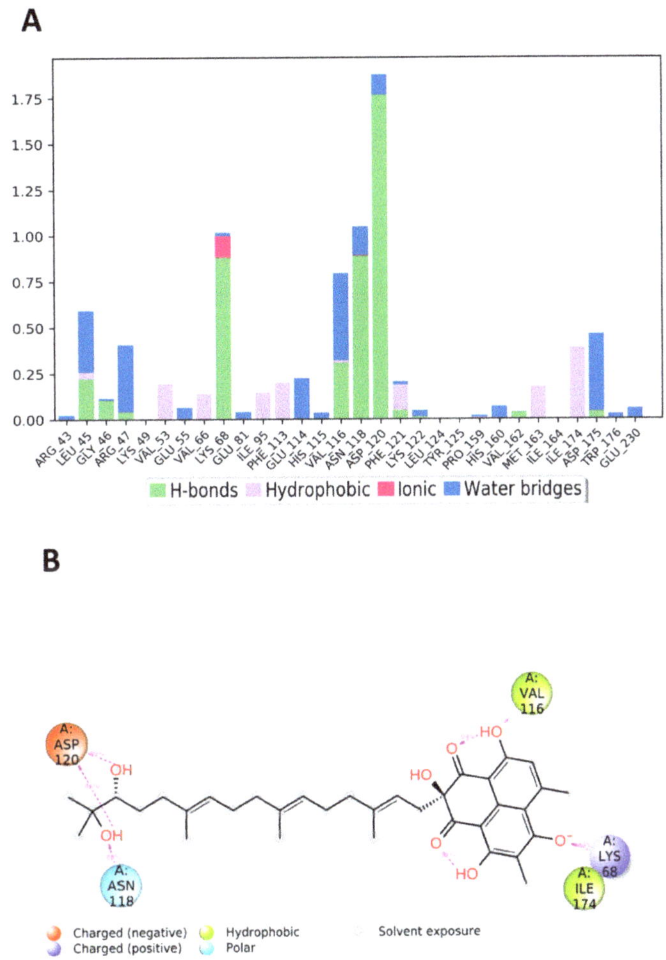

Figure 11. *Cont.*

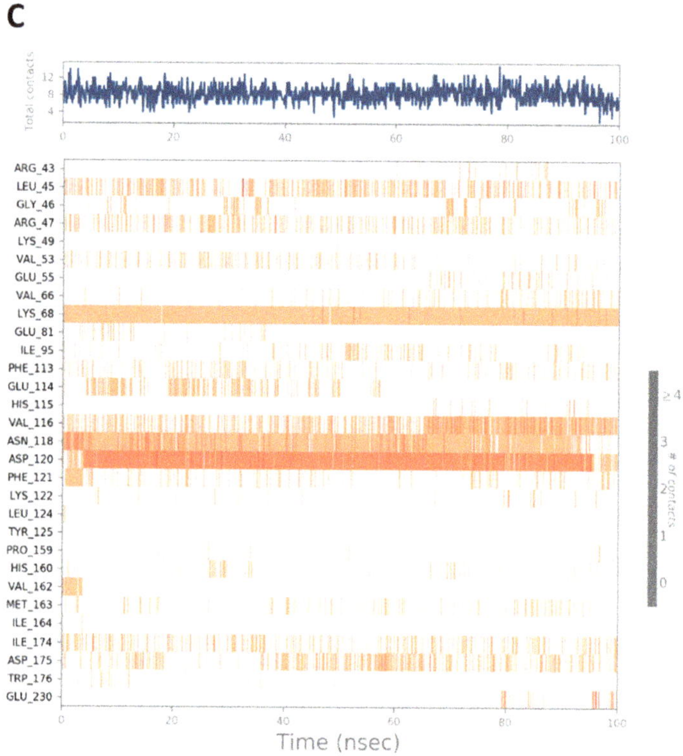

Figure 11. (**A**) CK2 interactions with compound **19** throughout the simulation. (**B**) Schematic diagram shows the detailed 2D atomic interactions of **19** with CK2 that occurred > 30% of the simulation time in the selected trajectory. (**C**) A timeline representation of CK2-**19** interactions presented in (**A**). The top panel presents the total number of specific interactions the protein made with the ligand over the course of the trajectory. The bottom panel presents residues interacting with the ligand in each trajectory frame. The dark orange color indicates that more than one specific interaction is made between some residues and the ligand.

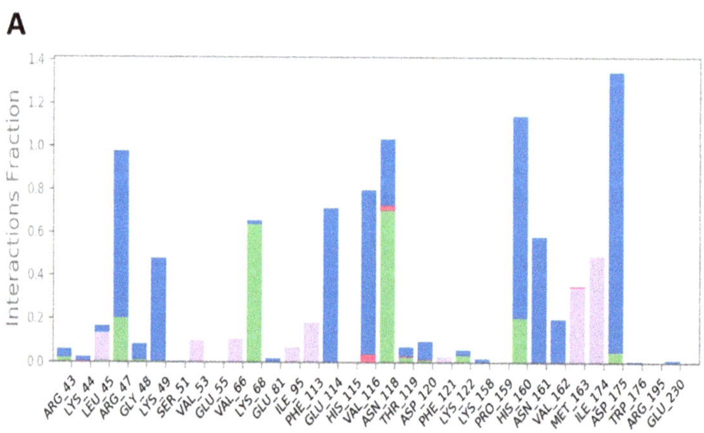

Figure 12. *Cont.*

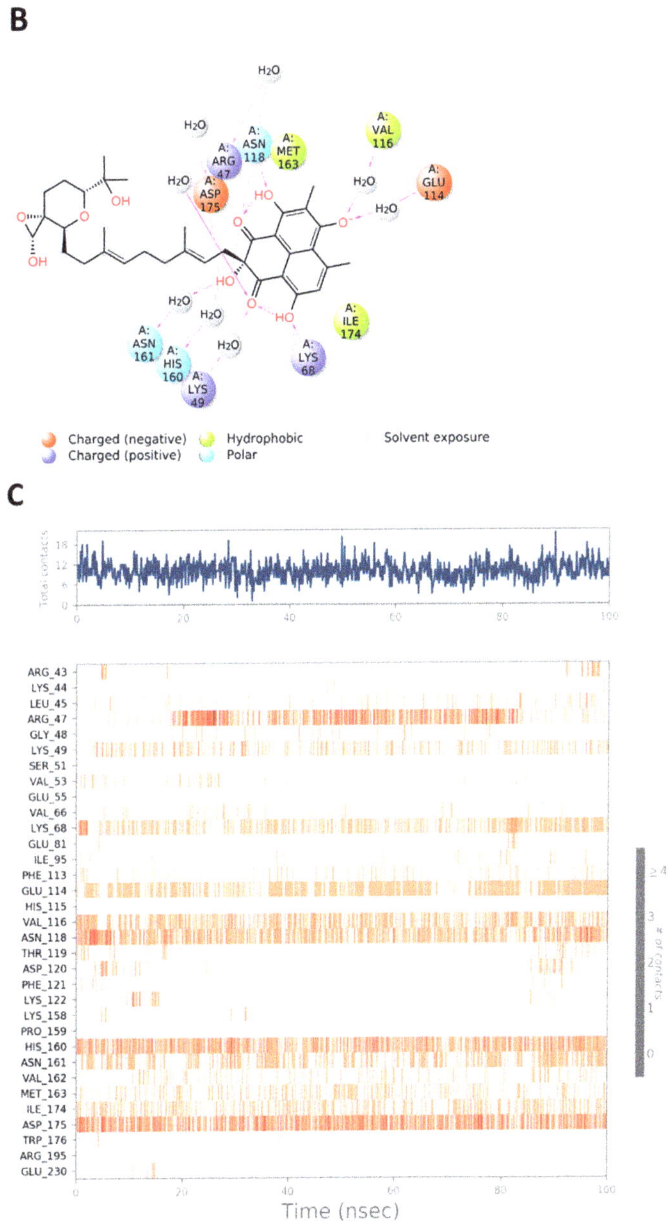

Figure 12. (**A**) CK2 interactions with compound **31** throughout the simulation. (**B**) Schematic diagram shows the detailed 2D atomic interactions of **31** with CK2 that occurred > 30% of the simulation time in the selected trajectory. (**C**) A timeline representation of CK2-**31** interactions presented in (**A**). The top panel presents the total number of specific interactions the protein made with the ligand over the course of the trajectory. The bottom panel presents residues interacting with the ligand in each trajectory frame. The dark orange color indicates more than one specific interaction is made between some residues and the ligand.

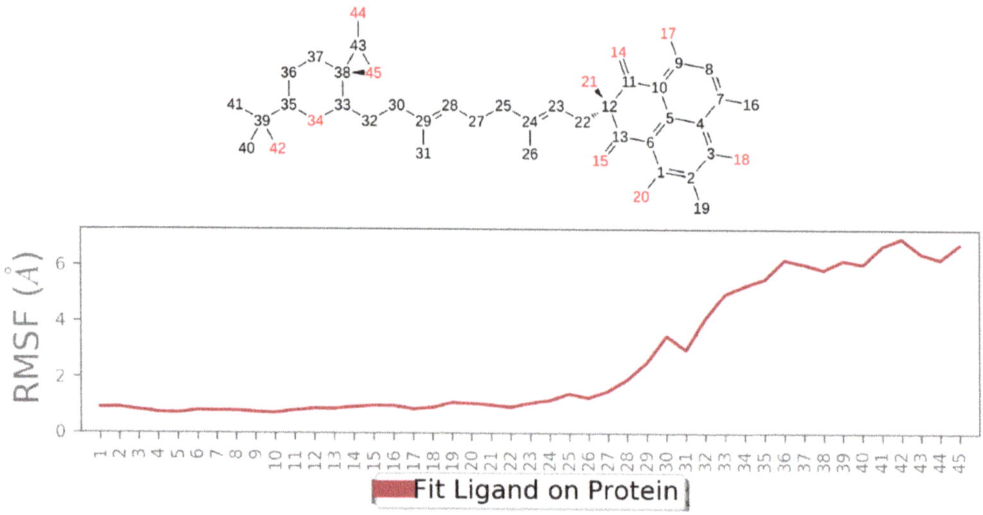

Figure 13. Ligand RMSF shows the fluctuations of **31** broken down by atom as represented by the compound's 2D structure. It provides ideas on how ligand atoms interact with the protein and their entropic role in the binding event. The 'Fit Ligand on Protein' line presents the ligand fluctuations, with respect to the protein. The CK2α1-**31** complex was first aligned on the protein backbone, and then the ligand RMSF was measured on the atoms of the ligand.

3. Conclusions

CK2 was related to many human illnesses, not only cancer, but also multiple sclerosis, cardiac hypertrophy, neurodegenerative and inflammatory disorders, cystic fibrosis, and virus infections [68]. It is noteworthy that the CK2 role is best recognized and investigated in cancer, where CK2 is almost positively upregulated, resulting in tumor progression because of its role in regulating nearly all the essential processes for developing apoptosis suppression [5,69]. It was reported that cancer cells rely on CK2 high levels compared to normal cells, which supports that the CK2 inhibitors can have a crucial contribution to cancer therapy development [5]. Recently, several compounds have been discovered and optimized via rational drug design approaches. Various structure-based drug design (SBDD) tools have been utilized for CK2 drug discovery for predicting possible compound and target interactions and their affinity. Various classes of natural metabolites, such as anthraquinones, benzoimidazoles, coumarins, pyrazolotriazines, and flavonoids, are recognized as CK2 inhibitors [70,71]. Fungal phenalenones are a fascinating class of fungal metabolites with diverse bioactivities that could be lead metabolites for drug discovery. With the aid of computational methods, i.e., ADMET, docking, and MD simulation, compounds **19** and **31** were identified as promising drug-like phenalenone derivatives that have better binding interactions and protein stability in a simulated aqueous physiological environment. The current work highlights the usefulness of these metabolites as lead for anticancer discovery. One of the important issues that require attentiveness is that several mechanistic studies are directed to the in silico methods because they provide information that cannot be obtained by other models and are less time-consuming. However, in vivo and in vitro investigations are warranted to strengthen the findings of in silico studies and provide opportunities for observing other mechanisms of the anticancer potential of these metabolites.

4. Materials and Methods

4.1. Target Prediction

The webserver SuperPred is a knowledge-based method that uses machine learning models for ATC code and target prediction of investigated compounds [57]. The machine learning model uses logistic regression and Morgan fingerprints of length 2048. The drugs approved by the WHO are classified by a drug classification system that connects the drugs' chemical properties and therapeutic properties and indications, where each classification is given a code called an Anatomical Therapeutic Chemical (ATC) code. Therefore, if a drug has more than one therapeutic indication, it is given an ATC code for each indication. The WHO has 6300 approved drugs that are linked to over 600,000 targets. Based on the hypothesis that compounds that have similar physiochemical properties exhibit similar biological effects, the webserver translates a user-defined compound into a structural fingerprint and compares this fingerprint to that of the WHO-approved drugs. When similarity is found, the webserver predicts the ATC code, the possible therapeutic target(s), and the putative therapeutic indication(s) for that compound. In other words, if an investigated compound is structurally similar to an approved drug, the compound is predicted to have biological activity on all possible targets of that drug. After targets are predicted, a probability score and a model accuracy score are reported. The probability represents the chance that the investigated compound will bind to a specific predicted target. The model accuracy reflects the performance accuracy of the used machine-learning model when predicting that specific target for the compound since the model performance differs between targets [57,72]. The targets and ATC codes for a library of investigated compounds were predicted using the SuperPred tool. The compounds that did not have the common target(s) of interest were excluded from further analysis. The ones sharing the common target(s) were advanced for the docking and further studies.

4.2. Preparation of PDB Structures

4.2.1. Ligand Preparation

Phenalenone derivatives were processed and prepared for docking using Schrodinger's LigPrep tool [40]. The 2D structures were converted to 3D and energy-minimized using OPLS3 force-field. After adding hydrogens, all possible ionization states and tautomeric forms were created at pH of 7.0 ± 0.2 by Epik; desalt option was also chosen. H-bonds were optimized by predicting the pKa of ionizable groups suing PROPKA [73].

4.2.2. Protein Preparation

CK2 crystal structure (PDB: 7BU4) was prepared using the Protein Preparation Wizard, added hydrogens to residues, changed covalent bonds to metal ions to zero-order, and created disulfide bonds. Water molecules > 5 Å from protein residues were deleted. Using Epik, the protonation state of residues was generated, and the formal charge on metal ions was adjusted. After removing the extra protein subunits of multi-subunit proteins and additional ligands, processing of the protein was refined by predicting the pKa of ionizable residues using PROPKA [73], and water molecules > 3 Å (not involved in water bridge) were removed. Finally, restrained minimization of the protein was applied using the OPLS4 force field.

4.3. Grid Generation and Docking

A grid box was generated around the co-crystallized ligand **Y49** in the protein crystal structure (PDB: 7BU4) binding site using Glide's Receptor-Grid-Generation tool [62]. Inside this box is where the docking of the phenalenone compounds was performed. The non-polar atoms were set for the VdW radii scaling factor by 1.0, and the partial charge cut-off was 0.25. Docking was then performed by the Schrodinger suite "Ligand Docking" tool [62,74]. The selected docking protocol was standard precision (SP), and the ligand sampling method was flexible. All other settings were default. Re-docking of the co-crystallized ligand (PDB: **Y49**) was performed to evaluate the docking method and docking of the investigated phenalenones followed.

4.4. ADMET Properties Prediction

The processed compounds were subjected to ADMET prediction using the QikProp-module of the Schrodinger suite [63]. The descriptors: molecular weight (mol_MW), drug-likeness (#Stars), dipole moment (dipole), total solvent accessible surface area (SASA), number of hydrogen bond donors and acceptors (donorHB and acceptHB), predicted octanol-water partitioning (QPlogPo/w), predicted aqueous solubility (QPlogS), estimated binding to human serum albumin (QPlogKhsa), number of the possible metabolites (# metab), predicted blood-brain partitioning (QPlogBB), percentage of human oral absorption, predicted IC_{50} for inhibiting HERG-K^+ channels (QPogHERG), central nervous system activity (CNS), and number of reactive functional groups present (#rtvFG), were predicted for these derivatives. The predicted values are compared to the recommended range derived from values determined/observed for 95% of known drugs.

4.5. MD Simulation

MD simulations were performed using Desmond software in the Schrodinger suite [65,66]. The protein-ligand complexes of interest were retrieved from the docking results where the force field was OPLS4. The complexes were tuned through the "System-Builder" tool to generate the solvated system for simulation. The solvent model was set as TIP3P, the selected box shape was orthorhombic, and the box dimensions were 10 Å. Na ions were added to neutralize the system. The simulation parameters were set up in the Molecular Dynamic tool, where the protein-ligand complexes were evaluated at pH 7.0 ± 0.2 over the 100 ns simulation time. The ensemble class was set as NPT in order to maintain the temperature and pressure constant during the run at 300 K and 1.01325 bar, respectively. After running the MD simulation, the generated results were analyzed.

Author Contributions: Conceptualization, G.A.M., A.M.O., and S.R.M.I.; methodology, G.A.M., A.M.O., S.R.M.I., and Y.A.M.; validation, A.A.B., R.M.D., A.O.N., and H.T.B.; resources, A.A.B., R.M.D., A.O.N., and H.T.B.; data curation, G.A.M., A.M.O., S.R.M.I., and Y.A.M.; software, A.M.O. and Y.A.M.; writing—original draft preparation, A.M.O., S.R.M.I., and Y.A.M.; writing—review and editing, A.A.B., R.M.D., A.O.N., and H.T.B. All authors have read and agreed to the published version of the manuscript.

Funding: This research received no external funding.

Institutional Review Board Statement: Not applicable.

Informed Consent Statement: Not applicable.

Data Availability Statement: Not applicable.

Conflicts of Interest: The authors declare no conflict of interest.

References

1. Martin, T.A.; Ye, L.; Sanders, A.J.; Lane, J.; Jiang, W.G. Cancer Invasion and Metastasis: Molecular and Cellular Perspective—Madame Curie Bioscience Database—NCBI Bookshelf. In *Cancer Invasion and Metastasis: Molecular and Cellular Perspective*; Jandial, R., Ed.; Landes Bioscience: Austin, TX, USA, 2013; pp. 135–168.
2. World Health Organization. Cancer. Available online: https://www.who.int/news-room/fact-sheets/detail/cancer (accessed on 5 March 2022).
3. Ghobrial, I.M.; Witzig, T.E.; Adjei, A.A. Targeting Apoptosis Pathways in Cancer Therapy. *CA Cancer J. Clin.* **2005**, *55*, 178–194. [CrossRef] [PubMed]
4. Li, S.-J.; Zhang, X.; Wang, X.-H.; Zhao, C.-Q. Novel natural compounds from endophytic fungi with anticancer activity. *Eur. J. Med. Chem.* **2018**, *156*, 316–343. [CrossRef] [PubMed]
5. Ruzzene, M.; Pinna, L.A. Addiction to protein kinase CK2: A common denominator of diverse cancer cells? *Biochim. Biophys. Acta* **2010**, *1804*, 499–504. [CrossRef] [PubMed]
6. Pinna, L.A. Protein kinase CK2: A challenge to canons. *J. Cell Sci.* **2002**, *115*, 3873–3878. [CrossRef] [PubMed]

7. Litchfield, D.W. Protein kinase CK2: Structure, regulation and role in cellular decisions of life and death. *Biochem. J.* **2003**, *369*, 1–15. [CrossRef] [PubMed]
8. Lolli, G.; Cozza, G.; Mazzorana, M.; Tibaldi, E.; Cesaro, L.; Donella-Deana, A.; Meggio, F.; Venerando, A.; Franchin, C.; Sarno, S.; et al. Inhibition of Protein Kinase CK2 by Flavonoids and Tyrphostins. A Structural Insight. *Biochemistry* **2012**, *51*, 6097–6107. [CrossRef]
9. Chua, M.M.J.; Lee, M.; Dominguez, I. Cancer-type dependent expression of CK2 transcripts. *PLoS ONE* **2017**, *12*, e0188854. [CrossRef]
10. Litchfield, D.W.; Bosc, D.G.; Canton, D.A.; Saulnier, R.B.; Vilk, G.; Zhang, C. Functional specialization of CK2 isoforms and characterization of isoform-specific binding partners. *Mol. Cell. Biochem.* **2001**, *227*, 21–29. [CrossRef]
11. Bibby, A.C.; Litchfield, D.W. The Multiple Personalities of the Regulatory Subunit of Protein Kinase CK2: CK2 Dependent and CK2 Independent Roles Reveal a Secret Identity for CK2β. *Int. J. Biol. Sci.* **2005**, *1*, 67–79. [CrossRef]
12. Venerando, A.; Ruzzene, M.; Pinna, L.A. Casein kinase: The triple meaning of a misnomer. *Biochem. J.* **2014**, *460*, 141–156. [CrossRef]
13. Trembley, J.H.; Wang, G.; Unger, G.; Slaton, J.; Ahmed, K. Protein Kinase CK2 in Health and Disease. *Cell. Mol. Life Sci.* **2009**, *66*, 1858–1867. [CrossRef] [PubMed]
14. Allende, J.E.; Allende, C.C. Protein kinase CK2: An enzyme with multiple substrates and a puzzling regulation. *FASEB J.* **1995**, *9*, 313–323. [CrossRef] [PubMed]
15. Pinna, L.A. Casein kinase 2: An "eminence grise" in cellular regulation? *Biochim. Biophys. Acta* **1990**, *1054*, 267–284. [CrossRef]
16. St-Denis, N.A.; Litchfield, D.W. Protein kinase CK2 in health and disease: From birth to death: The role of protein kinase CK2 in the regulation of cell proliferation and survival. *Cell. Mol. Life Sci.* **2009**, *66*, 1817–1829. [CrossRef]
17. Trembley, J.H.; Unger, G.M.; Tobolt, D.K.; Korman, V.L.; Wang, G.; Ahmad, K.A.; Slaton, J.W.; Kren, B.T.; Ahmed, K. Systemic administration of antisense oligonucleotides simultaneously targeting CK2α and α' subunits reduces orthotopic xenograft prostate tumors in mice. *Mol. Cell. Biochem.* **2011**, *356*, 21–35. [CrossRef]
18. Sarno, S.; Papinutto, E.; Franchin, C.; Bain, J.; Elliott, M.; Meggio, F.; Kazimierczuk, Z.; Orzeszko, A.; Zanotti, G.; Battistutta, R.; et al. ATP site-directed inhibitors of protein kinase CK2: An update. *Curr. Top. Med. Chem.* **2011**, *11*, 1340–1351. [CrossRef]
19. Evidente, A.; Kornienko, A.; Cimmino, A.; Andolfi, A.; Lefranc, F.; Mathieu, V.; Kiss, R. Fungal metabolites with anticancer activity. *Nat. Prod. Rep.* **2014**, *31*, 617–627. [CrossRef]
20. Kundu, M.; Das, S.; Dhara, D.; Mandal, M. Prospect of natural products in glioma: A novel avenue in glioma management. *Phytother. Res.* **2019**, *33*, 2571–2584. [CrossRef]
21. Borgo, C.; D'amore, C.; Sarno, S.; Salvi, M.; Ruzzene, M. Protein kinase CK2: A potential therapeutic target for diverse human diseases. *Signal Transduct. Target. Ther.* **2021**, *6*, 183. [CrossRef]
22. Search of: CX-4945—List Results—ClinicalTrials.gov, (n.d.). Available online: https://clinicaltrials.gov/ct2/results?cond=&term=CX-4945&cntry=&state=&city=&dist= (accessed on 12 April 2022).
23. Search of: CIGB-300—List Results—ClinicalTrials.gov, (n.d.). Available online: https://clinicaltrials.gov/ct2/results?cond=CIGB-300&term=&cntry=&state=&city=&dist= (accessed on 12 April 2022).
24. Ibrahim, S.R.M.; Mohamed, S.G.A.; Altyar, A.E.; Mohamed, G.A. Natural Products of the Fungal Genus *Humicola*: Diversity, Biological Activity, and Industrial Importance. *Curr. Microbiol.* **2021**, *78*, 2488–2509. [CrossRef]
25. Ibrahim, S.R.M.; Mohamed, S.G.A.; Sindi, I.A.; Gamal, A.M. Biologically active secondary metabolites and biotechnological applications of species of the family *Chaetomiaceae* (Sordariales): An updated review from 2016 to 2021. *Mycol. Prog.* **2021**, *20*, 595–639. [CrossRef]
26. Mohamed, G.A.; Ibrahim, S.R.M. Untapped Potential of Marine-Associated *Cladosporium* Species: An Overview on Secondary Metabolites, Biotechnological Relevance, and Biological Activities. *Mar. Drugs* **2021**, *19*, 645. [CrossRef] [PubMed]
27. Ibrahim, S.R.M.; Sirwi, A.; Eid, B.G.; Mohamed, S.G.A.; Mohamed, G.A. Bright Side of *Fusarium oxysporum*: Secondary Metabolites Bioactivities and Industrial Relevance in Biotechnology and Nanotechnology. *J. Fungi* **2021**, *7*, 943. [CrossRef] [PubMed]
28. Ibrahim, S.R.M.; Altyar, A.E.; Mohamed, S.G.A.; Mohamed, G.A. Genus *Thielavia*: Phytochemicals, industrial importance and biological relevance. *Nat. Prod. Res.* **2021**, 1–16. [CrossRef] [PubMed]
29. Ibrahim, S.R.M.; Sirwi, A.; Eid, B.G.; Mohamed, S.G.A.; Mohamed, G.A. Fungal Depsides—Naturally Inspiring Molecules: Biosynthesis, Structural Characterization, and Biological Activities. *Metabolites* **2021**, *11*, 683. [CrossRef]
30. Ibrahim, S.R.M.; Fadil, S.A.; Fadil, H.A.; Eshmawi, B.A.; Mohamed, S.G.A.; Mohamed, G.A. Fungal Naphthalenones; Promising Metabolites for Drug Discovery: Structures, Biosynthesis, Sources, and Pharmacological Potential. *Toxins* **2022**, *14*, 154. [CrossRef]
31. Omar, A.M.; Mohamed, G.A.; Ibrahim, S.R.M. Chaetomugilins and Chaetoviridins—Promising Natural Metabolites: Structures, Separation, Characterization, Biosynthesis, Bioactivities, Molecular Docking, and Molecular Dynamics. *J. Fungi* **2022**, *8*, 127. [CrossRef]
32. Noor, A.O.; Almasri, D.M.; Bagalagel, A.A.; Abdallah, H.M.; Mohamed, S.G.A.; Mohamed, G.A.; Ibrahim, S.R.M. Naturally Occurring Isocoumarins Derivatives from Endophytic Fungi: Sources, Isolation, Structural Characterization, Biosynthesis, and Biological Activities. *Molecules* **2020**, *25*, 395. [CrossRef]
33. Ibrahim, S.R.M.; Omar, A.M.; Bagalagel, A.A.; Diri, R.M.; Noor, A.O.; Almasri, D.M.; Mohamed, S.G.A.; Mohamed, G.A. Thiophenes—Naturally Occurring Plant Metabolites: Biological Activities and In Silico Evaluation of Their Potential as Cathepsin D Inhibitors. *Plants* **2022**, *11*, 539. [CrossRef]

34. Ibrahim, S.R.M.; Abdallah, H.M.; Elkhayat, E.S.; Al Musayeib, N.M.; Asfour, H.Z.; Zayed, M.F.; Mohamed, G.A. Fusaripeptide A: New antifungal and anti-malarial cyclodepsipeptide from the endophytic fungus *Fusarium* sp. *J. Asian Nat. Prod. Res.* **2018**, *20*, 75–85. [CrossRef]
35. Haidar, S.; Jürgens, F.; Aichele, D.; Jagels, A.; Humpf, H.-U.; Jose, J. Natural Compounds Isolated from *Stachybotrys chartarum* Are Potent Inhibitors of Human Protein Kinase CK. *Molecules* **2021**, *26*, 4453. [CrossRef] [PubMed]
36. Aichinger, G.; Dellafiora, L.; Pantazi, F.; del Favero, G.; Galaverna, G.; Dall'Asta, C.; Marko, D. Alternaria toxins as casein kinase 2 inhibitors and possible consequences for estrogenicity: A hybrid in silico/in vitro study. *Arch. Toxicol.* **2020**, *94*, 2225–2237. [CrossRef] [PubMed]
37. Baier, A.; Szyszka, R. Compounds from Natural Sources as Protein Kinase Inhibitors. *Biomolecules* **2020**, *10*, 1546. [CrossRef] [PubMed]
38. Elsebai, M.F.; Saleem, M.; Tejesvi, M.V.; Kajula, M.; Mattila, S.; Mehiri, M.; Turpeinen, A.; Pirttilä, A.M. Fungal phenalenones: Chemistry, biology, biosynthesis and phylogeny. *Nat. Prod. Rep.* **2014**, *31*, 628–645. [CrossRef]
39. Cooke, R.G.; Edwards, J.M. Naturally Occurring Phenalenones and Related Compounds. In *Fortschritte der Chemie Organischer Naturstoffe/Progress in the Chemistry of Organic Natural Products*; Herz, W., Grisebach, H., Kirby, G.W., Eds.; Springer: Vienna, Austria, 1981; pp. 153–190.
40. Song, R.; Feng, Y.; Wang, D.; Xu, Z.; Li, Z.; Shao, X. Phytoalexin Phenalenone Derivatives Inactivate Mosquito Larvae and Root-knot Nematode as Type-II Photosensitizer. *Sci. Rep.* **2017**, *7*, 42058. [CrossRef]
41. Otyepka, M.; Kryštof, V.; Havlíček, L.; Siglerová, V.; Strnad, M.; Koča, J. Docking-Based Development of Purine-like Inhibitors of Cyclin-Dependent Kinase. *J. Med. Chem.* **2000**, *43*, 2506–2513. [CrossRef]
42. Tripathi, S.K.; Muttineni, R.; Singh, S.K. Extra precision docking, free energy calculation and molecular dynamics simulation studies of CDK2 inhibitors. *J. Theor. Biol.* **2013**, *334*, 87–100. [CrossRef]
43. Leelananda, S.P.; Lindert, S. Computational methods in drug discovery. *Beilstein J. Org. Chem.* **2016**, *12*, 2694–2718. [CrossRef]
44. Lu, R.; Liu, X.; Gao, S.; Zhang, W.; Peng, F.; Hu, F.; Huang, B.; Chen, L.; Bao, G.; Li, C.; et al. New tyrosinase inhibitors from *paecilomyces gunnii*. *J. Agric. Food Chem.* **2014**, *62*, 11917–11923. [CrossRef]
45. Al Subeh, Z.Y.; Raja, H.A.; Burdette, J.E.; Falkinham, J.O.; Hemby, S.E.; Oberlies, N.H. Three diketomorpholines from a *Penicillium* sp. (strain G1071). *Phytochemistry* **2021**, *189*, 112830. [CrossRef]
46. Li, Q.; Zhu, R.; Yi, W.; Chai, W.; Zhang, Z.; Lian, X.-Y. Peniciphenalenins A–F from the culture of a marine-associated fungus *Penicillium* sp. ZZ901. *Phytochemistry* **2018**, *152*, 53–60. [CrossRef] [PubMed]
47. Intaraudom, C.; Nitthithanasilp, S.; Rachtawee, P.; Boonruangprapa, T.; Prabpai, S.; Kongsaeree, P.; Pittayakhajonwut, P. Phenalenone derivatives and the unusual tricyclic sesterterpene acid from the marine fungus *Lophiostoma bipolare* BCC. *Phytochemistry* **2015**, *120*, 19–27. [CrossRef] [PubMed]
48. Han, Y.; Sun, C.; Li, C.; Zhang, G.; Zhu, T.; Li, D.; Che, Q. Antibacterial phenalenone derivatives from marine-derived fungus *Pleosporales* sp. HDN1811400. *Tetrahedron Lett.* **2021**, *68*, 152938. [CrossRef]
49. Le Goff, G.; Lopes, P.; Arcile, G.; Vlachou, P.; Van Elslande, E.; Retailleau, P.; Gallard, J.-F.; Weis, M.; Benayahu, Y.; Fokialakis, N.; et al. Impact of the Cultivation Technique on the Production of Secondary Metabolites by *Chrysosporium lobatum* TM-237-S5, Isolated from the Sponge Acanthella cavernosa. *Mar. Drugs* **2019**, *17*, 678. [CrossRef] [PubMed]
50. Srinivasan, M.; Shanmugam, K.; Kedike, B.; Narayanan, S.; Shanmugam, S.; Neelakantan, H.G. Trypethelone and phenalenone derivatives isolated from the mycobiont culture of *Trypethelium eluteriae* Spreng. and their anti-mycobacterial properties. *Nat. Prod. Res.* **2020**, *34*, 3320–3327. [CrossRef] [PubMed]
51. Tansakul, C.; Rukachaisirikul, V.; Maha, A.; Kongprapan, T.; Phongpaichit, S.; Hutadilok-Towatana, N.; Borwornwiriyapan, K.; Sakayaroj, J. A new phenalenone derivative from the soil fungus *Penicillium herquei* PSU-RSPG. *Nat. Prod. Res.* **2014**, *28*, 1718–1724. [CrossRef]
52. Li, Y.; Yue, Q.; Jayanetti, D.R.; Swenson, D.C.; Bartholomeusz, G.A.; An, Z.; Gloer, J.B.; Bills, G.F. Anti-Cryptococcus Phenalenones and Cyclic Tetrapeptides from *Auxarthron pseudauxarthron*. *J. Nat. Prod.* **2017**, *80*, 2101–2109. [CrossRef]
53. Chaudhary, N.K.; Crombie, A.; Vuong, D.; Lacey, E.; Piggott, A.M.; Karuso, P. Talauxins: Hybrid Phenalenone Dimers from Talaromyces stipitatus. *J. Nat. Prod.* **2020**, *83*, 1051–1060. [CrossRef]
54. Gombodorj, S.; Yang, M.-H.; Shang, Z.-C.; Liu, R.-H.; Li, T.-X.; Yin, G.-P.; Kong, L.-Y. New phenalenone derivatives from Pinellia ternata tubers derived *Aspergillus* sp. *Fitoterapia* **2017**, *120*, 72–78. [CrossRef]
55. Pang, X.; Zhao, J.-Y.; Fang, X.-M.; Zhang, T.; Zhang, D.-W.; Liu, H.-Y.; Su, J.; Cen, S.; Yu, L.-Y. Metabolites from the Plant Endophytic Fungus *Aspergillus* sp. CPCC 400735 and Their Anti-HIV Activities. *J. Nat. Prod.* **2017**, *80*, 2595–2601. [CrossRef]
56. Rukachaisirikul, V.; Rungsaiwattana, N.; Klaiklay, S.; Phongpaichit, S.; Borwornwiriyapan, K.; Sakayaroj, J. γ-Butyrolactone, Cytochalasin, Cyclic Carbonate, Eutypinic Acid, and Phenalenone Derivatives from the Soil Fungus *Aspergillus* sp. PSU-RSPG185. *J. Nat. Prod.* **2014**, *77*, 2375–2382. [CrossRef] [PubMed]
57. Nickel, J.; Gohlke, B.; Erehman, J.; Banerjee, P.; Rong, W.W.; Goede, A.; Dunkel, M.; Preissner, R. SuperPred: Update on drug classification and target prediction. *Nucleic Acids Res.* **2014**, *42*, W26–W31. [CrossRef] [PubMed]
58. RCSB PDB: Homepage, (n.d.). Available online: http://www.rcsb.org/ (accessed on 15 May 2020).
59. *Schrödinger Release 2021-4: LigPrep*; Schrödinger, LLC: New York, NY, USA, 2021.

60. Houštecká, R.; Hadzima, M.; Fanfrlík, J.; Brynda, J.; Pallová, L.; Hánová, I.; Mertlíková-Kaiserová, H.; Lepšík, M.; Horn, M.; Smrčina, M.; et al. Biomimetic Macrocyclic Inhibitors of Human Cathepsin D: Structure-Activity Relationship and Binding Mode Analysis. *J. Med. Chem.* **2020**, *63*, 1576–1596. [CrossRef] [PubMed]
61. Sastry, G.M.; Adzhigirey, M.; Day, T.; Annabhimoju, R.; Sherman, W. Protein and ligand preparation: Parameters, protocols, and influence on virtual screening enrichments. *J. Comput. -Aided Mol. Des.* **2013**, *27*, 221–234. [CrossRef]
62. *Schrödinger Release 2021-4: Glide*; Schrödinger, LLC: New York, NY, USA, 2021.
63. *Schrödinger Release 2021-4: QikProp*; Schrödinger, LLC: New York, NY, USA, 2021.
64. Schrödinger Press. *QikProp 4.4 User Manual*; Schrödinger Press: New York, NY, USA, 2015.
65. *Schrödinger Release 2021-4: Desmond Molecular Dynamics System*; D.E. Shaw Research: New York, NY, USA, 2021.
66. *Maestro-Desmond Interoperability Tools*; Schrödinger: New York, NY, USA, 2021.
67. Hollingsworth, S.A.; Dror, R.O. Molecular Dynamics Simulation for All. *Neuron* **2018**, *99*, 1129–1143. [CrossRef] [PubMed]
68. Cozza, G. The Development of CK2 Inhibitors: From Traditional Pharmacology to in Silico Rational Drug Design. *Pharmaceuticals* **2017**, *10*, 26. [CrossRef] [PubMed]
69. Ortega, C.E.; Seidner, Y.; Dominguez, I. Mining CK2 in Cancer. *PLoS ONE* **2014**, *9*, e115609. [CrossRef]
70. Cozza, G.; Pinna, L.; Moro, S. Kinase CK2 inhibition: An update. *Curr. Med. Chem.* **2013**, *20*, 671–693. [CrossRef]
71. Cozza, G.; Pinna, L.A. Casein kinases as potential therapeutic targets. *Expert Opin. Ther. Targets* **2016**, *20*, 319–340. [CrossRef]
72. Dunkel, M.; Günther, S.; Ahmed, J.; Wittig, B.; Preissner, R. SuperPred: Drug classification and target prediction. *Nucleic Acids Res.* **2008**, *36*, W55–W59. [CrossRef]
73. Olsson, M.H.M.; Søndergaard, C.R.; Rostkowski, M.; Jensen, J.H. PROPKA3: Consistent treatment of internal and surface residues in empirical p K a predictions. *J. Chem. Theory Comput.* **2011**, *7*, 525–537. [CrossRef] [PubMed]
74. Friesner, R.A.; Banks, J.L.; Murphy, R.B.; Halgren, T.A.; Klicic, J.J.; Mainz, D.T.; Repasky, M.P.; Knoll, E.H.; Shelley, M.; Perry, J.K.; et al. Glide: A New Approach for Rapid, Accurate Docking and Scoring. Method and Assessment of Docking Accuracy. *J. Med. Chem.* **2004**, *47*, 1739–1749. [CrossRef] [PubMed]

MDPI
St. Alban-Anlage 66
4052 Basel
Switzerland
Tel. +41 61 683 77 34
Fax +41 61 302 89 18
www.mdpi.com

Journal of Fungi Editorial Office
E-mail: jof@mdpi.com
www.mdpi.com/journal/jof

www.ingramcontent.com/pod-product-compliance
Lightning Source LLC
Chambersburg PA
CBHW061506220926
51850CB00004B/183
9783036561448